XXIII International Baldin Seminar on High Energy Physics Problems (Baldin ISHEPP XXIII)

Relativistic Nuclear Physics and Quantum Chromodynamics

EPJ Web of Conferences Volume 138 (2017)

Dubna, Russia
19 - 24 September 2016

Editors:

Serge Bondarenko
Valery Burov
Alexander Malakhov

ISBN: 978-1-5108-3833-8

Printed from e-media with permission by:

Curran Associates, Inc.
57 Morehouse Lane
Red Hook, NY 12571

Some format issues inherent in the e-media version may also appear in this print version.

Printed by Curran Associates, Inc. (2017)

For additional information, please contact EDP Sciences – Web of Conferences at the address below.

EDP Sciences – Web of Conferences
17, Avenue du Hoggar
Parc d'Activité de Courtabœuf
BP 112
F-91944 Les Ulis Cedex A
France

Phone: +33 (0) 1 69 18 75 75
Fax: +33 (0) 1 69 28 84 91

contact-edps@webofconferences.org

Additional copies of this publication are available from:

Curran Associates, Inc.
57 Morehouse Lane
Red Hook, NY 12571 USA
Phone: 845-758-0400
Fax: 845-758-2633
Email: curran@proceedings.com
Web: www.proceedings.com

TABLE OF CONTENTS

edp sciences

EPJ Web of Conferences

Statement of Peer review

In submitting conference proceedings to EPJ Web of Conferences, I certify to the Publisher that I adhere to the **Policy on Publishing Integrity** of the journal in order to safeguard good scientific practice in publishing.

1. All articles have been subjected to peer review administered by the proceedings editors.
2. Reviews have been conducted by expert referees, who have been requested to provide unbiased and constructive comments aimed, whenever possible, at improving the work.
3. Proceedings editors have taken all reasonable steps to ensure the quality of the materials they publish and their decision to accept or reject a paper for publication has been based only on the merits of the work and the relevance to the journal.

Title, date and place of the conference

The XXIII International Baldin Seminar on High Energy Physics Problems
"Relativistic Nuclear Physics and Quantum Chromodynamics"
September 19-24, 2016, Dubna, Russia

Proceedings editor(s):

Serge Bondarenko, Valery Burov, Alexander Malakhov

Date and editor's signature

06/03/2017

Preface

The XXIII International Baldin Seminar on High Energy Physics Problems "Relativistic Nuclear Physics and Quantum Chromodynamics", organized by the Joint Institute for Nuclear Research was held in September 19-24, 2016 at Dubna, Russia (http://relnp.jinr.ru/ishepp-xxiii/).

This Seminar was devoted to the 90th anniversary of an outstanding scientist Academician A.M. Baldin (1926-2001).

The Seminar was supported by the Russian Foundation for Basic Research (RFBR) and the International Union of Pure and Applied Physics (IUPAP).

The Seminar continues the series of traditional conferences that was established by Alexander Mikhailovich Baldin in 1969 with support of M.A. Markov (1908-1994). They were given an informal, somewhat witty, name "Baldin autumn".

The first Seminar of this series was devoted to vector mesons and included a considerable number of talks aimed to create gauge field theory, in particular, to solve the problems of quantization of the Yang-Mills fields, their geometric interpretation and phenomenological applications of these theories to experiment. This was the first large-scale conference devoted to the main trend in theory of fundamental interactions. Among the participants were professors N.N. Bogoliubov, J. Bjorken, B. Zumino, M.A. Markov, S. Ting, L.D. Faddeev, V.A. Matveev and others, who made the basic contribution to this field of physics. The subsequent seminars of this series were devoted to the physics of strong interactions, multiparticle production and relativistic nuclear physics.

Since the second half of the seventies the gauge field theory greatly affected the physics of fundamental interactions and this was naturally reflected in the program of the Seminars. The quantum chromodynamics as a theory of strong interactions based on the first principles and the possibility of its experimental check were the subject at the previous Seminars. The problems of relativistic nuclear physics, multiple production processes in nucleus-nucleus collisions, traditionally discussed at the Seminar gained a large significance. Cumulative meson production, the laws governing the limiting fragmentation of nuclei and the nuclear reactions with large momentum transfer which were discovered in the early seventies became the main trends in the program of investigations of relativistic nuclear collisions at the JINR Synchrophasotron and were successfully interpreted by the language of quantum chromodynamics. The concept of the quark-parton structure functions of nuclei was considerably developed. At the VI-th Seminar, the dependence of the structure functions normalized to the atomic weight that was later named the EMC-effect, was discussed. The results of the experiments on deep inelastic muon scattering on nuclei in the cumulative region were first reported there, that gave convincing grounds for the conclusion about the properties of the quark-parton structure functions drawn on the basis of the study of the limiting fragmentation of nuclei.

At the VII-th Seminar attention was focused on the properties of the quark-parton structure functions of nuclei, and special sessions were devoted to this problem.

An essential part of the program of the previous Seminars was the phenomenological description of chromodynamics at large distances (bags, strings, multiquark systems, hidden colour, quark-gluon plasma, etc.).

In the past years the role of the colour degrees of freedom in nuclei was extensively studied and discussed at special conferences devoted to it. The effect of the quark degrees of freedom on the properties of nuclei and nuclear reactions is undoubtedly the main perspective of the fundamental studies in nuclear physics. Multiquark interactions and the systems with a nonstandard number of quarks, especially dibaryons, multiquark configurations in nuclei, systems with hidden colour, etc. are of special interest. The understanding of the spin content of the proton is the pressing problem of the day both for theory and experiment. All these problems are reflected in the suggested program.

 DOI: 10.1051/epjconf/201713800001

After the VIII-th Seminar physics of relativistic nuclei was essentially developed: beams of nuclei at energy of 200 GeV/nucleon and new large detectors began to work. Scores of big collaborations joined the research of processes of multiple particle production in relativistic nuclear collisions and the problem of chromoplasma. The natural phenomena in relativistic nuclear physics having an asymptotic character, played the decisive role in a detailed design and estimation of cost of the accelerator complex and construction of installations on nuclear beams. The principles of symmetry and self-similarity can be used to predict inclusive cross sections of particle production in the central rapidity region. Recent analysis makes it possible to conclude that the hopes for obtaining dense and hot matter in heavy ultrarelativistic nuclear collisions were not yet realized, but we can say that it was claimed at XX Baldin ISHEPP that hopes of physicists were really justified! An overall idea about the investigation in the field of relativistic nuclear physics can be got from the Proceedings of the Dubna International Seminar of the present series. The interplay of hadronic structure and QCD interactions is one of the important features of the transition regime.

The Proceedings of the XXIII Baldin ISHEPP are devoted to the progress of the experimental research on the accelerators of high energies. Talks on hot problems of the relativistic nuclear physics and reports on status and perspectives of the high-energy physics leading experiments (CMS/LHC, ALICE/LHC, ATLAS/LHC, STAR/RHIC, CLAS/JLAB, CBM/FAIR, PANDA/FAIR, NA61/CERN, BESIII/BEPCII, FLUKTON/IHEP, NICA/MPD/JINR, FAIR, CW-LINAC/GSI, PRAE, Nab, GlueX/JLAB, etc.)., which were devoted to the relativistic nuclear physics, are included in the Proceedings. During the Seminar about 157 reports of the 250 participants from 22 countries and international organizations were presented. We expect that the Proceedings will be interesting for scientists working in the relativistic nuclear physics.

Next XXIV seminar will be organized on September 17-22, 2018 and we hope that the scientific program will be even more intriguing and interesting because of progress of the relativistic nuclear physics and functioning of the new accelerators.

S.G. Bondarenko
V.V. Burov
A.I. Malakhov
September 2016, Dubna

DOI: 10.1051/epjconf/201713800001

JOINT INSTITUTE FOR NUCLEAR RESEARCH

XXIII INTERNATIONAL BALDIN SEMINAR ON HIGH ENERGY PHYSICS PROBLEMS

RELATIVISTIC NUCLEAR PHYSICS & QUANTUM CHROMODYNAMICS

dedicated to the 90th anniversary of Academician A.M. Baldin

Dubna, Russia, September 19-24, 2016

International Advisory Committee

A. Andrianov (Russia)
A. Antonov (Bulgaria)
Ts. Baatar (Mongolia)
J. Carbonell (France)
C. Ciofi degli Atti (Italy)
J. Cleymans (South Africa)
T. Donnelly (USA)
S. Dubnička (Slovak Republic)
M. Gazdzicki (Poland)
P. Giubellino (CERN)
H. Gutbrod (Germany)
P. W.-Y. Hwang (Taiwan)
V. Kekelidze (JINR)
A. Kurepin (Russia)
V. Kuvshinov (Belarus)
H. Machner (Germany)
L. Majling* (Czech Republic)
V. Matveev (JINR)
Y. Musakhanov (Uzbekistan)
Sh. Nagamiya (Japan)
V. Novozhilov (Russia)
Ch. Perdrisat (USA)
W. Plessas (Austria)
L. Pondrom (USA)
G. Roepke (Germany)
G. Salme (Italy)
V. Savrin (Russia)
L. Schroeder (USA)
P. Senger (Germany)
A. Shebeko (Ukraine)
A. Slavnov (Russia)
B. Slowinski (Poland)
H. Stoecker (Germany)
A. Thomas (USA)
H. Toki (Japan)
E. Tomasi-Gustafsson (France)
I. Tserruya (Israel)
V. Voronov (JINR)
Nu Xu (USA)
G. Zinovjev (Ukraine)

*deceased

Organizing Committee

V.V. Burov (Co-Chairman)
A.I. Malakhov (Co-Chairman)
S.G. Bondarenko (Vice-Chairman)
E.P. Rogochaya (Scientific Secretary)
E.N. Russakovich (Secretary)
Yu.S. Anisimov
A.A. Baldin
I.V. Boguslavsky
S.B. Gerasimov
A.D. Kovalenko
V.K. Lukyanov
I.I. Migulina
S.Z. Pakulyak
Yu.A. Panebratsev
N.M. Piskunov
E.B. Plekhanov
B.M. Starchenko
O.V. Teryaev
V.D. Toneev
A.S. Vodopyanov
P.I. Zarubin
I.G. Zarubina

E-mail: <ishepp@theor.jinr.ru>

Web-site: http://relnp.jinr.ru/ishepp/

Organized by

JOINT INSTITUTE FOR NUCLEAR RESEARCH

http://www.jinr.ru/

Supported by

INTERNATIONAL UNION OF PURE AND APPLIED PHYSICS

http://iupap.org/sponsored-conferences/approved-conferences/

Sponsored by

JOINT INSTITUTE FOR NUCLEAR RESEARCH

(Directorate Grant)
http://www.jinr.ru/

RUSSIAN FOUNDATION FOR BASIC RESEARCH

(Grant *N* 16-02-20311-г)
http://www.rfbr.ru/

JOINT INSTITUTE FOR NUCLEAR RESEARCH

XXIII INTERNATIONAL BALDIN SEMINAR ON HIGH ENERGY PHYSICS PROBLEMS

RELATIVISTIC NUCLEAR PHYSICS & QUANTUM CHROMODYNAMICS

September 19–24, 2016, Dubna, Russia

TO THE 90th ANNIVERSARY OF ACADEMICIAN **A.M. BALDIN**

Seminar Topics

- Quantum chromodynamics at large distances
- Relativistic heavy ion collisions
- Hadron spectroscopy, multiquarks
- Cumulative and subthreshold processes
- Structure functions of hadrons and nuclei
- Dynamics of multiparticle production
- Polarization phenomena, spin physics
- Studies of exotic nuclei in relativistic beams
- Applied use of relativistic beams
- Accelerator facilities: status and perspectives
- New project NICA/MPD (Nuclotron-based Ion Collider fAcility/ Multipurposed Detector) at JINR
- Progress in experimental studies in high energy centers — JINR, CERN, BNL, JLAB, GSI, etc.

Organizing Committee

Burov V. V. (co-Chairman)
Malakhov A. I. (co-Chairman)
Bondarenko S. G. (vice-Chairman)
Rogochaya E. P. (Scientific Secretary)
Russakovich E. N. (Secretary)

Anisimov Yu. S.
Baldin A. A.
Boguslavsky I. V.
Gerasimov S. B.
Kovalenko A. D.
Lukyanov V. K.
Migulina I. I.
Pakulyak S. Z.
Panebratsev Yu. A.
Piskunov N. M.
Plekhanov E. B.
Starchenko B. M.
Teryaev O. V.
Toneev V. D.
Vodopyanov A. S.
Zarubina I. G.
Zarubin P. I.

International Advisory Committee

A. Andrianov (Russia)
A. Antonov (Bulgaria)
Ts. Baatar (Mongolia)
J. Carbonell (France)
C. Ciofi degli Atti (Italy)
J. Cleymans (South Africa)
T. Donnelly (USA)
S. Dubnichka (Slovak Republic)
M. Gazdzicki (Poland/Germany)
P. Giubellino (CERN)
H. Gutbrod (Germany)
P. W-Y. Hwang (Taiwan)
V. Kekelidze (JINR)
A. Kurepin (Russia)
V. Kuvshinov (Belarus)
H. Machner (Germany)
L. Majling (Czech Republic)
V. Matveev (JINR)
Y. Musakhanov (Uzbekistan)
Sh. Nagamiya (Japan)
V. Novozhilov (Russia)
Ch. Perdrisat (USA)
W. Plessas (Austria)
L. Pondrom (USA)
G. Roepke (Germany)
G. Salme (Italy)
V. Savrin (Russia)
L. Schroeder (USA)
P. Senger (Germany)
A. Shebeko (Ukraine)
A. Slavnov (Russia)
B. Slowinski (Poland)
H. Stoecker (Germany)
A. Thomas (USA)
H. Toki (Japan)
E. Tomasi-Gustafsson (France)
I. Tserruya (Israel)
V. Voronov (JINR)
Nu Xu (USA)
G. Zinovjev (Ukraine)

Addresses for Correspondence:
Dr. Elena Rogochaya
Veksler and Baldin Laboratory of High Energy Physics
Joint Institute for Nuclear Research
141980 Dubna, Russia
ishepp@theor.jinr.ru

Visas, accomodation, travel details:
Mrs. Elena Russakovich
International Department
Joint Institute for Nuclear Research
141980 Dubna, Russia
rusakovich@jinr.ru
tel. +7 49621 63890, fax: +7 49621 65891

http://relnp.jinr.ru/ishepp

 DOI: 10.1051/epjconf/201713800002

Academician Alexander Mikhailovich Baldin (1926-2001)

Academician Alexander Mikhailovich Baldin, an outstanding scientist in the field of elementary particle physics and atomic nucleus, was born on February 26, 1926 in Moscow.

He graduated from locomotive technical secondary school and was enrolled in the Moscow Institute of Transport Engineers as a student. At that time, he had already realized that science is his true vocation. In 1946, he was invited, among other excellent students, to continue his education in a newly created Moscow Mechanics Institute of Ammunition, named later Moscow Engineering Physics Institute (MIFI).

After graduating from MIFI in 1949, he was assigned to the P.N. Lebedev Physical Institute of the USSR Academy of Sciences (FIAN), where he made his way from a junior research worker to a head of the theoretical division. He received the doctorate degree and became a full professor. He has matured as a scientist under the influence of a brilliant Pleiad of FIAN's physicists gathered by S.I. Vavilov. He considered D.V. Skobeltsyn and M.A. Markov to be his teachers. A.M. Baldin's first research work on the theory of particle motion in a cyclic accelerator had already then attracted the attention of specialists and first of all of V.I. Veksler. These investigations, carried out jointly with V.V. Mikhailov, under M.S. Rabinovich were aimed at the solution of a wide range of problems related to the theory of cyclic accelerators and had underlain the physical substantiation of the Synchrophasotron which was the biggest accelerator of this type in the word at that time. They were completed with the creation of a method which had become classic and is being widely used till now in the calculations of accelerators.

A.M. Baldin (left) and P.A. Cherenkov (right)

It was just at that time that he was keen on mountaineering, where he had achieved brilliant results and had become a recognized authority. Mountains favored him to develop such distinguishing traits of his character as inflexibility, manliness and the capability of making rational decisions which were used in full measure in his scientific leadership and like so much in his life trials.

The start up of electron and proton accelerators in the late forties and during the fifties in the USSR required highly qualified physicists engaged in the theoretical elaboration and planning of experimental research programs who would profoundly understand the status and prospects of the development of both modern theory and experimental and accelerator engineering. A.M. Baldin had become one of the most brilliant and fruitfully working personality who fell under this category of Soviet physicists.

G.N. Flerov, Yu.Ts. Oganessian
and A.M. Baldin (from left to right)

As long ago as in early fifties, due to the development of work at the FIAN electron synchrotron and on Markov's initiative, A.M. Baldin (partially in cooperation with V.V. Mikhailov) carried out pioneering calculations of the cross sections of meson production on nucleons and nuclei when exposed to high-energy photons. A pole model taking into account anomalous magnetic moments of nucleons introduced into these calculations was subsequently substantiated and became an integral part of the dispersion re-

 DOI: 10.1051/epjconf/201713800002

lation method. The research work along these lines, awarded the State Prize in 1973, played an appreciable role in the development of strong interaction theory and the formation of the ideas about the existence of an intrinsic structure and excited states of nucleons, as well as in the development of the methods of an adequate description of processes involving strongly interacting particles. In just the same way as classic Hofstadter's experiments on electron scattering on nucleons and nuclei led to the idea about spatial extension of the hadron electric charge distribution, the study of the Compton scattering performed at FIAN and interpreted by Baldin in terms of the coefficients of nucleon electromagnetic polarizability showed that the elementary particles, which the nucleons are usually attributed to, are not only spatially extended systems, but also dynamically deformed ones. The discovery of this effect by authors' group was registered in 1957. Baldin's dispersion sum rule for the sum of the coefficients of nucleon electric and magnetic polarizability suggested in 1960 served as a basis for the first realistic estimate of the proton electric polarizability and is presently the major and most universal ingredient of all modern analyses of the data on the low energy Compton effect on nucleons. A mutual transformability of the known elementary particles, the possibility of their production and annihilation in the intermediate state of a reaction in question, is the fundamental feature of relativistic quantum dynamics that leads to a mutual dependence of the properties of different elementary particles.

An experiment on lepton decay of the phi meson accompanied by an intermediate transition of the massive particle to a photon was performed at JINR, on Baldin's initiative and with his participation. This discovery, registered in 1971, confirmed, in particular, the presence of a "hadronic" photon component which stipulates a hadron-like behavior of the cross sections for high-energy photon interactions with nucleons and nuclei, determining, in a figurative sense, "nuclear" properties of light.

I.S. Baldina and A.M. Baldin

By generalizing the notions of molecular optics, A.M. Baldin introduced into nuclear physics the notions of tensor and vector polarizability of nuclei which describe "optical anisotropy" of atomic nuclei with a non-zero spin. One of the most striking manifestations of this property was found to be a splitting of a giant dipole resonance of strongly deformed nuclei into two maxima corresponding to two types of collective dipole oscillations of nucleons: along and across the nucleus symmetry axis. A.M. Baldin obtained the first numerical estimates of the parameters of the optical anisotropy of atomic nuclei and predicted in such a way a number of effects to be observed. Afterwards this result was proved experimentally by American physicists who recognized Baldin's priority.

On M.A. Markov's initiative and with his active support, Alexander Mikhailovich was elected a Director of the Laboratory of High Energies of the Joint Institute for Nuclear Research in 1968. A young man, as he was at that time, became responsible for the formulation of a topical scientific program to be realized by the Laboratory's staff formed by V.I. Veksler, for the conservation and increase of the research facilities available and, first of all, for the "Veksler's heritage" - the Synchrophasotron. This unique machine, the creation of which cost infinite labor in severe post-war years, gave pioneering results on elementary particle physics. A whole generation of experimenters thus got a start to their creative life.

As a scientific leader, A.M. Baldin had to determine priorities of the development of the accelerator and experimental complexes of the Laboratory which required all his scientific talent, life experience, health and soul. Nuclear interactions at relativistic energies were chosen as a major research direction. To this end, under the guidance of A.M. Baldin, the Synchrophasotron was modified to a specialized accelerator of relativistic and polarized nuclei.

In the early 70s, A.M. Baldin determined long-term research goals in relativistic nuclear physics - a scientific trend of paramount importance for the national science - which arose at the meeting point of atomic nuclear physics and elementary particle physics. It was just aimed at setting up the application limits for the proton-neutron model of the atomic nucleus and the creation of the physical picture of nuclear matter at the level of sub-nucleonic constituents, i.e. quarks and gluons. For the first time in the international practice of accelerator engineering, beams of relativistic nuclei moving with velocities near the speed of light and of an energy of order of several GeV per nucleon were obtained at the Synchrophasotron in LHE.

A.M. Baldin (left) and V.A. Matveev (right)

The study of nucleus-nucleus collisions was extended to the energy range where the principles of the theory of relativity became predominant. A new research trend, relativistic nuclear physics, arose on the foundation of the achievements obtained in quantum field theory, elementary particle physics, nuclear physics and acceleration physics. The whole intellectual potential of Alexander Michailovich was essentially required to formulate and promote this new trend. Followed by Dubna examples, relativistic nuclear physics has become a prominent part of programs at the biggest accelerator centers of the USA, Europe, Russia and the JINR Member-States. A wide experimental program on particle physics is progressing in collaboration with the Institute of High Energy Physics (Protvino), the E. Fermi Laboratory (USA) and the European Organization for Nuclear Research (CERN).

In the period of the formation of relativistic nuclear physics, A. Baldin displayed his creative energy, scientific foresight, capacity for work. He was able to communicate with large groups of people, to persuade and unite them. The importance of these qualities could scarcely be exaggerated. Being a broad-minded person of rare erudition in many facets of modern physics, deeply understanding modern physics experiments A.M. Baldin was highly respected by the scientific community all over the world. He did not only suggest ideas of new experiments, but he was also a direct organizer and a tireless participant in them.

A.M. Baldin

Baldin's prediction of the nuclear cumulative effect was the first success in this area of physics. The first experiments in Dubna showed that at an energy of several GeV the particle production in nuclear collisions is setting up in the asymptotic regime. It means the achievement of invariability of the physical picture of the secondary particle production in nuclear fragmentation with increasing collision energy or the achievement of the so-called limiting fragmentation of nuclei - the idea introduced by A.M. Baldin. The quark and gluon degrees of freedom begin to play an essential role in this area. A remarkable peculiarity of the nuclear fragmentation process is the extension of these properties also to particles produced behind the kinematic limit of free nucleon collisions, i.e. to cumulative particles. In the language of the parton model, this fact points out that nuclei contain multiquark states.

The picture of the relativistic energy nuclear fragmentation is found to be in a perfect analogy with the hitherto known particular features of high energy proton-proton interactions and deep inelastic scattering of electrons on protons. The discovered empirical regularities enabled A.M. Baldin to introduce a universal distribution of partons in nuclei - the quark-parton structure function of the nucleus. This aroused considerable interest in the development of a theory for the description of a nucleus at distances shorter than the nucleon size (models of fluctons, short-range nucleon correlations, multiquark states in nuclei and so on).

A.M. Baldin suggested universal approaches to the description of not only spectra of single particles, but also the whole picture of an act of multiple particle production in nuclear collisions. He proposed to describe the nuclear interaction process in the four-velocity space starting from the principles of symmetry - self-similarity symmetry. The picture of multiple particle production is found to be analogous to a microscopic process of "point-like explosion". There was discovered and is being studied a universal law which makes it possible to describe cumulative processes, sub-threshold processes, anti-nucleus production processes, as well as the ones that occur in the region of transition from nucleon to quark-gluon variables.

A.M. Baldin (left) and N.N. Bogoliubov (right)

The results of this new trend - relativistic nuclear physics - together with the fundamental works of theorists of the school of N.N. Bogoliubov were united to clear up a dynamic role of the new quantum number "color" and the appropriate symmetry in pursuing the observed scale-invariant behavior of hadron and nuclear interactions with large momentum transfer. They were awarded the Lenin Prize in 1988.

Under the guidance of A.M. Baldin, the Laboratory accelerator complex developed successfully: there were constructed new ion sources and experimental sites of accelerated beams. Experiments with polarized deuteron beams formed a special trend of research; unique polarized neutron beams were obtained. Unique information about the spin structure of the deuteron - this "hydrogen atom of nuclear physics" - at inter-nucleon distances less than the nucleon radius was extracted from these experiments.

The results of the first period of investigations with relativistic nuclei enabled A.M. Baldin to suggest and justify the idea about constructing of a specialized accelerator - the Nuclotron - the magnetic system of which is based on the phenomenon of superconductivity. A.M. Baldin succeeded in uniting people around this task due to his talent for organization and his purposeful activity. Under his guidance, exceptionally complicated engineering problems were solved and first of all fast-cycling superconducting magnets and a helium liquefaction complex were constructed. The Nuclotron launching and development in the 90s opened up the way to qualitatively new possibilities of studying the properties of atomic nuclei. The science of the JINR Member-States, including Russian physics, acquired a powerful basis for first-class investigations on strong interaction physics.

A.M. Baldin paid the most serious attention to applications of the achievements of relativistic nuclear physics, Nuclotron's technologies, for the design of accelerator controlled nuclear-energetic apparatus, for transmutation of nuclear wastes as well as for cosmic flight radiation studies.

Academician A.M. Baldin's scientific and organizational activities were extremely versatile. He was the President of the Council on Electromagnetic Interactions and a member of the Bureau of the Nuclear Physics Department of the Russian Academy of Sciences, Editor-in-Chief of the journals "Physics of Elementary Particles and Atomic Nuclei" and "Physics of Particles and Nuclei Letters" as well as a member of the Editorial Boards of many scientific publications. Among conferences that

 DOI: 10.1051/epjconf/201713800002

were organized by Alexander Michailovich of special importance are the International Seminars on High-Energy Physics Problems started in 1969 with M.A. Markov's support. They were given an unofficial, somewhat witty, name "Baldin autumn".

A.M. Baldin devoted much time and effort to the training of scientific staff. Among his disciples are dozens of holders of the doctoral degree and candidate of sciences. He gave lectures at Moscow State University, Moscow Engineering Physics Institute, the JINR University Center, as well as at many international schools of physics. Baldin did a lot for preparing highly skilled specialists in JINR Member-States. In Dubna he founded his scientific school which was intended for theorists, experimenters, specialists of accelerator engineering.

A.M. Baldin

In his talks and papers devoted to the general problems of the scientific research strategy, A.M. Baldin tirelessly emphasized the influence of the centers of fundamental science on scientific and technological progress and the potential of the country. His sense of high civic responsibility and deep understanding of the role of science in today's community impelled him, as a statesman, to recur to these problems.

The achievements of the outstanding scientist and a talented organizer of science were awarded with the Lenin Prize, the State Prize, the V.I. Veksler Prize of the Russian Academy of Sciences, as well as with governmental awards - the Order of the Red Banner of Labor, the Order of Merit, the Order of Service to the Fatherland, forth class. He was awarded orders and medals of the JINR Member States. A.M. Baldin was given the title of Honorary Citizen of the town of Dubna.

Alexander Michailovich gained fine reputation and sincere respect for his selfless service to the Science, high civic duty, true intelligence, as well as for his constant concern for people and social justice. He worked hard and effectively and he was very good for people.

Alexander Michailovich died on April 29, 2001 after serious illness.

V.V. Burov
A.I. Malakhov
September 2016, Dubna

Physics architecture

Nelly Konopleva[1,★]

[1] *"VNIIEM" Corporation", Moscow, Russia*

Abstract. Fundamental physical theory axiomatics is closely connected with methods of experimental measurements. The difference between the theories using global and local symmetries is explained. It is shown that symmetry group localization leads not only to the change of the relativity principle, but to the fundamental modification of experimental programs testing physical theory predictions. It is noticed that any fundamental physical theory must be consistent with the measurement procedures employed for its testing. These ideas are illustrated by events of my biography connected with Yang-Mills theory transformation from an ordinary phenomenological model to a fundamental physical theory based on local symmetry principles like the Einsteinian General Relativity. Baldin position in this situation is demonstrated.

1 Introduction

50 years ago physics theorists were in the permanent expectation of new discoveries in the elementary particle physics. That time there were two main scientific seminars in the P.N. Lebedev Physical Institute in Moscow where such problems were under regular discussion. They were the academician I.E. Tamm's seminar of the theoretical department of LPI and the seminar under the leadership of the academician M.A. Markov and professor A.M. Baldin. I frequented both these seminars.

I graduated the Moscow State M.V. Lomonosov University in 1964, and theoretical physics became my profession. The subject of my diploma thesis (1963) was the classical theory of gauge fields. That time they were called "the compensating fields". Without knowing Utiyama's paper of 1959 I obtained independently the same results that he did for Lagrangian form of the gauge field theory. My work was based on the book "Introduction to the quantum fields theory" (N.N. Bogoliubov and D.V. Shirkov, 1957, Moscow, GITTL). Moreover, I used the second Noether's theorem. This theorem was absent in the mentioned textbook. Geometrical interpretation of the gauge field vector-potentials as connection coefficients in some space of special construction was proposed by me, too. Geometrical part of my diploma thesis included the Einsteinian General Relativity and Weyl's theory of gravity, which is described by quadratic with respect to the Riemannian curvature tensor Lagrangian. So, the theories of gravity of Einstein and Weyl were connected by the local gauge invariance principle. Therefore, the well-known opposition of these great scientists could be finished.

Formally, my diploma thesis supervisor was professor K.P. Staniukowich. But he left the choice of the subject to me.

Attracted by the quantum physics, the theorists were a little interested in the classical field theory. Utiyama wrote about it in his book "Where did physics come to?" (in Japanese, 1983; in Russian,

★e-mail: nelly@theor.jinr.ru

DOI: 10.1051/epjconf/201713801001

1986). I saw it also. But I was inspired by the Einstein's, Weyl's and Cartan's works. If can geometrical equivalent of quantization exist? That was the question.

In 1964 the collection of papers on the classical gauge field theory was published in Russian translation edited by professor D.D. Ivanenko. This collection was named "Elementary particles and compensating fields". It helped Soviet scientists to go into these problems. The main problem is nuclear forces theory construction. Utiyama's paper was present in this book. The paper of Yang and Mills was also there. At that time the main works of professor D.D. Ivanenko were devoted to the elementary particle classification and gravitation theories.

At the same time at the mechanical-mathematics department of the M.V. Lomonosov Moscow State University the lectures on infinite Lie groups and fibre bundle space geometry were given. When I attended them I found that the local gauge groups are the special case of the infinite Lie groups, and the gauge field vector-potentials are the special case of the fibre bundle space connection coefficients. These generalizations of the classical gauge field theory and the statement that the gauge fields must be classified by representations of the infinite Lie groups became the base of my thesis for a candidate degree. So, the compensating procedure ceased to be actual. But the local gauge symmetry became the same fundamental as the global one.

The gauge fields were classified by infinite Lie group representations just as elementary particles are classified by finite Lie group representations. The finite and infinite Lie groups differ from each other in the sense that under the same local algebraic structure their transformation parameters behave differently. The transformation parameters of the finite Lie groups are independent on a space point (the are constants), whereas the transformation parameters of the infinite Lie groups depend on it. Such construction reflects transition from the global symmetry to the local one.

It was found that the system of the Lagrangian equations of the gauge field theory after symmetry localization contains less equations than before localization. The equations of motion became corollary of the conservation law of energy-momentum tensor. So, the Einstein's idea that particles can be considered as singularities of the gravitation field can be extended to the gauge field sources.

This approach helped to construct the theory of topological generators - the sources of electric current of new type. In that generators the electric current is determined by the motion of the magnetic flux instead of the electron motion in the usual generators. This theory was tested and confirmed experimentally with one percent uncertainty at IHEP (Protvino) in middle 1980s. The topological generators were made and used for the pump magnetic field by the real solenoid of the UNC superconducting circle (on the IHEP stand).

Now is 100th anniversary of the General Retivity, and we have unified the geometrical theory of all interactions, including gravity, thanks to Einsteinian stubborn struggle for the right of the local symmetry to be the basis of the fundamental physical theory.

2 A.M. Baldin influence on my life

Alexander Michailovich Baldin belonged to a circle of eminent scientists, which enclosed me since I was a student (when I became one of the authors of the proton-neutron nucleus model of professor D.D. Ivanenko, who was my father in law). But the interaction between theorists can be named a "long-range action". If they are not co-workers, they meet each other at seminars, lectures, conferences and so on, that is mainly in an official way.

Our first talk with Alexander Michailovich was in 1967 at the P.N. Lebedev Physical Institute. He invited me to give a talk about my geometrical interpretation of the gauge (at that time - compensating) fields at his seminar which he organized together with the academician M.A. Markov. I was very exited by this invitation. At the academician I.E. Tamm's seminar I could only listen and participate in discussions, but not to speak myself. After my talk the academician M.A. Markov said: "When a child arises, it is difficult to say what will with him/her later, but here, I think, something very interesting will come out". Since then they both followed my fortune.

Later M.A. Markov collaborators began to study the black holes, Einsteinian and Kaluza-Klein 5D unified classical theory of gravity and electromagnetism, and supergravity. Moses Alexandrovich proposed a model of elementary particle as a closed universe (simultaneously with professor K.P. Staniukowich). He assumed that the radius of such particle could be defined by the Plank length. He was also interested in description of a particle motion in the very small space-time domains (less than Plank length).

Alexander Michailovich looked for an experimental confirmation of the vector dominance idea. It was very interesting to me because it meant that the universal forces of new kind could exist. Not only gravity, but nuclear forces could also be described by some universal interaction, if it is found that the vector dominance idea is true. So, the quest for construction of the nuclear interaction theory by analogy with the General Relativity was opened.

A.M. Baldin was not only an excellent theorist, but he also had brilliant ideas about how to organize experiments. His activity inspired the hope for experimental progress in the theory of elementary particle interactions. It was very necessary because many different models of such interactions were proposed, but not a one selection principle was proposed. Sometimes even non-scientific arguments were used to support a concrete model. Yang-Mills model was not popular among Soviet theorists. Some scientists, as V.I. Ogievetskiy (JINR), E.S. Fradkin (LPI), B.L. Ioffe (ITEP) and others assumed that this model can never become a good physical theory. The time showed the opposite. The main selection principle for A.M. Baldin were symmetry principles. And for me, too.

So, in 1967 I submitted for consideration to the theoretical department of LPI my PhD (candidate) thesis on the classical gauge field theory "Geometrical description of interactions". I worked in another institute and was not a post-graduate student. The consideration was lasted two years but in the spring of 1969 it was successfully completed. My thesis was recommended for publication. On its basis I wrote my three-chapters part of the well-known book N.P. Konopleva and V.N. Popov "Gauge fields" (1972, Atomizdat, Moscow, in Russian). My chapters were devoted to the classical gauge field theory in Lagrangian and geometrical forms. The fourth chapter of this book was written by a mathematician from LOMI Victor Nicolaevich Popov. He wrote about the quantum theory of the gauge fields. This text became the basis of his Habilitation thesis. My Habilitation (doctor) thesis contained an extended version of these three chapters in the second edition of our book and it was devoted to the classical gauge fields plus the geometry with torsions. It was defended in 1983 at the Moscow State University.

My and V.N. Popov book arose due to the support of the academician M.A. Markov and professor A.M. Baldin. Moses Alexandrovich wrote to Atomizdat the recommendation to publish it. Alexander Michailovich became a reviewer. This book ran into three Russian (1972, 1980, 2000) and one English (1981) editions. Hence, something interesting, as M.A.Markov earlier said, really was presented! This book was the first to contain both classical and quantum theory of the gauge fields. It was the only textbook on the gauge field theory existing in Russia at that time.

After the defense in LPI A.M. Baldin invited me to participate in the International Seminar on electromagnetic interactions and vector dominance, which took place in 1969 in Dubna. My talk was entitled "Geometrical description of the gauge fields" (in Russian). Soon after this presentation I was forced to leave the K.P. Staniukowich's group as the financial support of theme "the gauge fields" was stopped. This theme was recognized to be not perspective. But geometrical methods in the field theory became active. When I left the K.P. Staniukowich's group I wrote my book to show them their mistake.

The International Seminar at JINR organized by M.A. Markov and A.M. Baldin in 1969 generated a set of regular Seminars "Relativistic nuclear physics and quantum chromodynamics", which was named "Baldin autumn". Now we have XXIII Seminar of this series. Gradually I became a permanent participant of these seminars. They became to me the necessary and sufficient condition of scientific existence.

In 1991 I began to collaborate with BLTP at JINR. Once when A.M. Baldin constructed his Nuclotron, he told me about the problem of magnetic flux pump to superconducting solenoids. I recommended him to make use our topological generators to solve this problem. Alexander Michailovich acquainted me with A.D. Kovalenko. Sometimes we discussed this problem together. Unfortunately, financial problems made impossible the application of this beautiful technical solution.

3 Conclusion

Alexander Michailovich Baldin and me were neither collaborators, nor friends, we did not have any relations. Nothing except science was connecting us with each other. But in critical moments we always proved to be at the same barricade side. Someone said: “To love is not every time to look at each other. It means to look in the same direction”. The direction, which interested all of us, was always science. It was the sense of our life.

A.M. Baldin was a well educated and gifted person. His aid was always disinterested. He supported young scientists and new ideas not only of his group but in others too, if he thought that it would be useful for science. He was a devoted son of his motherland. All his efforts were always directed to its prosperity.

In memory of Alexander Michailovich I proposed in 2006 to Anton Alexandrovich Baldin to organize the Baldin Physical Institute. From this idea the Advanced Studies Institute was born.

Acknowledgments. It is a pleasure to thank the Organizers of the International Seminar ISHEPP XXIII for their support.

References

[1] R. Utiyama, Phys. Rev. **101**, 1597 (1956)
[2] N.N. Bogoliubov, D.V. Shirkov, *Introduction to the quantum fields theory, 1th ed.* (GITTL, Moscow, 1957, in Russian)
[3] R. Utiyama, *Where did physics come to?* (Japan, 1983, in Japanese)
[4] Collection of papers, ed. by D.D. Ivanenko, *Elementary particles and compensating fields* (Mir, Moscow, 1964, in Russian)
[5] C.N. Yang, R.L. Mills, Phys. Rev. **96**, 191 (1954)
[6] N.P. Konopleva, V.N. Popov, *Gauge fields, 1th ed.* (Atonizdat, Moscow, 1972, in Russian). N.P. Konopleva, V.N. Popov, *Gauge fields, 2th ed.* (Energoatomizdat, Moscow, 1980, in Russian). N.P. Konopleva, V.N. Popov, *Gauge fields, 3th ed.* (URSS, Moscow, 2000, in Russian)
[7] N.P. Konopleva, V.N. Popov, *Gauge fields* (Harwood academic publisher, Chur-London-N.-Y., 1981)
[8] Proc. of Int. Sem., ed. by A.M. Baldin, *Vector mesons and electromagnetic interactions* (JINR, Dubna, 1969)
[9] Collection of papers, ed. by A.M. Baldin, *Electromagnetic interactions and elementary particles structure* (Mir, Moscow, 1969, in Russian)

Recent heavy ion results from the ATLAS experiment

Mariusz Przybycien[1,*] (on behalf of the ATLAS Collaboration)

[1]*AGH University of Science and Technology, Faculty of Physics and Applied Computer Science, Al. Mickiewicza 30, 30-059 Cracow, Poland*

Abstract. The ATLAS experiment at the LHC has undertaken a broad physics program to probe and characterize the hot nuclear matter created in relativistic lead-lead collisions. This overview presents recent results on production of electroweak bosons, charmonia, charged-particles, jets, and bulk particle collectivity from Pb+Pb and p+Pb collisions at 2.76 TeV and 5.02 TeV, respectively, as well as on electromagnetic processes in ultra-peripheral Pb+Pb collisions at 5.02 TeV.

1 Introduction

The Quark-Gluon Plasma (QGP) was first observed in experiments at RHIC [1–4] and since then it has been intensively studied in relativistic heavy-ion (HI) collisions. Significantly higher collision energies available at the LHC compared to RHIC, result in increased volume, lifetime and temperature of QGP. We are particularly interested in understanding of the collision geometry and centrality as well as in how the free nucleon parton distribution functions are modified in the nuclear environment. One can study properties of QGP using hard probes of different scales (electroweak bosons, jets, heavy quarks) or measuring parameters describing collective behaviour of the medium. Hard probes are produced early in the HI collision, in a process which cross section is not changed by presence of strongly interacting medium, i.e. it can by calculated using perturbative QCD. Passing through the medium hard probes interact weakly or strongly with it providing information on its properties. Due to their large electric charge heavy-ions are also strong sources of high energy photons. Thus, in ultra-peripheral collisions, in which strong interactions are suppressed, one can study two-photon interactions. A review of HI results from LHC experiments based on Run 1 data can be found in [5]. In these Proceedings an overview of the recent HI results from the ATLAS experiment [6] is presented based on Pb+Pb collision data at $\sqrt{s_{\rm NN}}$ = 2.76 TeV and p+Pb collision data at $\sqrt{s_{\rm NN}}$ = 5.02 TeV.

Yields of hard processes in HI collisions are expected to scale with the number of binary nucleon-nucleon collisions, $N_{\rm coll}$, which depends on the centrality of the collision. In the central collisions, when the overlap of nuclei is large, one expects a high number of $N_{\rm coll}$ and a high number of nucleons participating in the collision, $N_{\rm part}$. On the other hand, in more peripheral collisions, when nuclei overlap only partly, both $N_{\rm coll}$ and $N_{\rm part}$ are expected to be smaller.

A proxy to the QGP volume produced in a collision is the collision centrality. In ATLAS centrality of events is determined by the sum of transverse energy deposited in the forward calorimeter (FCal) covering the pseudorapidity range $3.2 < |\eta| < 4.8$. In case of Pb+Pb collisions the sum on both sides

*e-mail: Mariusz.Przybycien@agh.edu.pl

and in case of p+Pb collisions the sum on the Pb-going side only are taken into account [7, 8]. The centrality intervals are expressed in percentiles of the total inelastic non-Coulomb Pb+Pb or p+Pb cross sections, respectively. For the correlation between a centrality interval and the average number of participants, $\langle N_{\text{part}} \rangle$, the Glauber model is used for Pb+Pb collisions and both the Glauber and Glauber-Gribov colour fluctuation [9] models for p+Pb collisions.

A variable often used to quantify the modification of rates of hard processes in a nuclear environment with respect to pp collisions is the nuclear modification factor:

$$R_{\text{AA}} = \frac{1}{\langle T_{\text{AA}} \rangle} \frac{1/N_{\text{evt}}\, \mathrm{d}^2 N_{\text{Pb+Pb}}/\mathrm{d}\eta \mathrm{d}p_{\text{T}}}{\mathrm{d}^2 \sigma_{pp}/\mathrm{d}\eta \mathrm{d}p_{\text{T}}}$$

where in the numerator is the yield of a given process per minimum-bias (MB) event in HI collisions, possibly in a given centrality interval, and in the denominator the pp cross section for the same process. The T_{AA} is the mean nuclear thickness function and is a measure of the nuclear overlap i.e. of the number of nucleons which can participate in the hard scattering processes.

2 Electroweak bosons and charmonia

Electroweak (EW) bosons being colourless do not interact strongly with QGP, and their production rates are therefore expected to be directly sensitive to the overall thickness of the colliding nuclear matter. The rates are also expected to be sensitive to modifications of the partonic structure of nucleons bound in a nucleus, which are implemented as nuclear modifications to the parton distribution functions (PDFs) in the proton. These effects include nuclear shadowing (depletion of the parton densities at low Bjorken x), antishadowing (an enhancement at moderate x), the EMC effect and Fermi motion [10].

ATLAS has studied production of W and Z bosons in both Pb+Pb [11, 12] and in p+Pb collisions [13, 14]. In Pb+Pb collisions W bosons were measured in both electron and muon ($\ell = e, \mu$) decay channels in the fiducial phase space: $p_{\text{T}}^{\ell} > 25$ GeV, $p_{\text{T}}^{\text{miss}} > 25$ GeV, $m_{\text{T}} > 40$ GeV and extrapolated to $|\eta_\ell| < 2.5$. W boson production yields per binary collision for each charge separately as well as inclusively for the combined data are shown in figure 1(left) as a function of $\langle N_{\text{part}} \rangle$. Data are independent on centrality, what means they scale with $\langle N_{\text{coll}} \rangle$, and consistent with QCD NLO predictions obtained

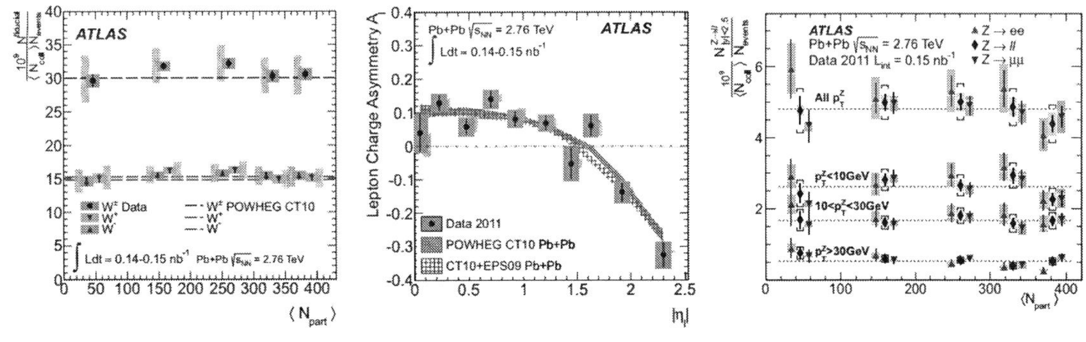

Figure 1. (left) W boson production yields per binary collision as a function of $\langle N_{\text{part}} \rangle$ for W^+, W^- and $W^\pm$ bosons for combined muon and electron decay channels in the fiducial region defined in the text [11]. (middle) The lepton charge asymmetry A_ℓ from $W^\pm$ bosons as a function of absolute pseudorapidity measured in the fiducial region defined in the text [11]. (right) Centrality dependence of Z boson yields divided by $\langle N_{\text{coll}} \rangle$. Results for ee and $\mu\mu$ decay channels are shifted left and right, respectively, from their weighted average [12].

from POWHEG [15] MC with CT10 PDF [16]. In figure 1(middle) the lepton charge asymmetry defined in terms of differential production yields as:

$$A_\ell(\eta_\ell) = \frac{\mathrm{d}N_{W^+\to\ell^+\nu_\ell}/\mathrm{d}\eta_\ell - \mathrm{d}N_{W^-\to\ell^-\bar{\nu}_\ell}/\mathrm{d}\eta_\ell}{\mathrm{d}N_{W^+\to\ell^+\nu_\ell}/\mathrm{d}\eta_\ell + \mathrm{d}N_{W^-\to\ell^-\bar{\nu}_\ell}/\mathrm{d}\eta_\ell}$$

is shown as a function of $|\eta_\ell|$ and compared to NLO predictions with the CT10 PDF set, both with and without EPS09 [10] nuclear modifications. Both predictions are in agreement with data, so at the current precision this measurement is insensitive to nuclear modifications to the PDF.

Per event yields of Z boson production in both electron and muon decay channels in Pb+Pb collisions are shown in figure 1(right) as a function of $\langle N_{\mathrm{part}}\rangle$ in bins of p_{T}^{Z}. The results in both channels are consistent with each other and the fits using a constant (shown as dashed lines) confirm that the Z boson per event yield obeys binary collision scaling.

ATLAS has also measured W boson production in p+Pb collisions in the muon channel in a similar fiducial region as in the case of Pb+Pb collisions [13]. The yields per MB events scaled by $\langle N_{\mathrm{coll}}\rangle$ are shown separately for W^+ and W^- in figure 2 for central (left) and peripheral (middle) events as a function of muon pseudorapidity $\eta^{\mu}_{\mathrm{lab}}$. Data are compared to POWHEG calculations with CT10 PDF. In peripheral collisions the model calculations are lower than data for all values of $\eta^{\mu}_{\mathrm{lab}}$. This may be due to the difficulty in definitively determining $\langle N_{\mathrm{coll}}\rangle$ for a given centrality bin; in addition to the $\langle N_{\mathrm{coll}}\rangle$ uncertainties shown in the figure, the normalization of each distribution depends on the choice of centrality model (Glauber or Glauber-Gribov Colour Fluctuation). There appears to be a dependence of the shape of the pseudorapidity distributions of both positively and negatively charged muons from W bosons on centrality. The middle panels, displaying the ratios of the measured muon pseudorapidity distribution coming from $W^{\pm}$ bosons to that calculated with the model, suggest the presence of a slope in the most central collisions, which does not appear in peripheral collisions. The

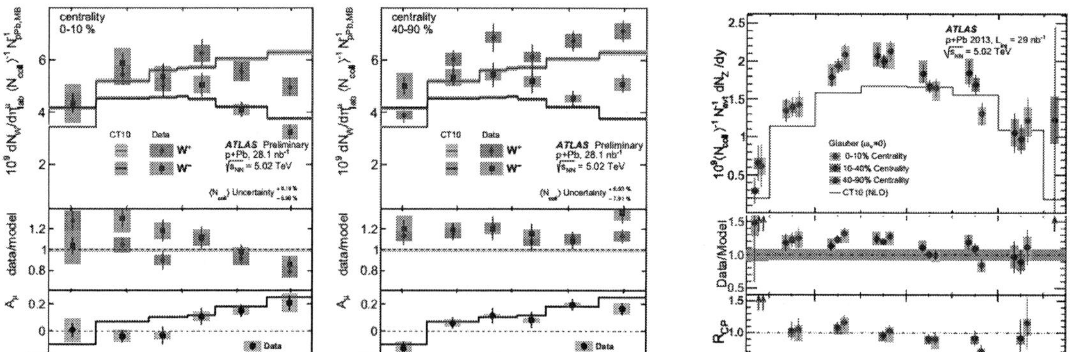

Figure 2. (left and middle) Production yields of $W \to \mu\nu$ in p+Pb collisions, scaled by $\langle N_{\mathrm{coll}}\rangle$ and per MB event are shown as a function of the muon pseudorapidity in the laboratory frame for the centralities indicated in the plots, separately for W^+ and W^-. The lower panels show the asymmetry defined in the text compared to the same model [13]. (right) Production yields of $Z \to \ell\ell$, ($\ell = e,\mu$), in p+Pb collisions, scaled by $\langle N_{\mathrm{coll}}\rangle$ and per MB event are shown as a function of the Z boson rapidity in the centre-of-mass frame for the centralities indicated in the plot. The lower panel shows the ratio of yields in a given centrality to the yield in most peripheral collisions [14]. In all plots: Results are compared to the POWHEG model using CT10 PDFs and the ratio of neutrons and protons in Pb nuclei. The data-to-model ratios are shown in the middle panels. The error bars correspond to the statistical uncertainties, and error boxes correspond to the total systematic uncertainties.

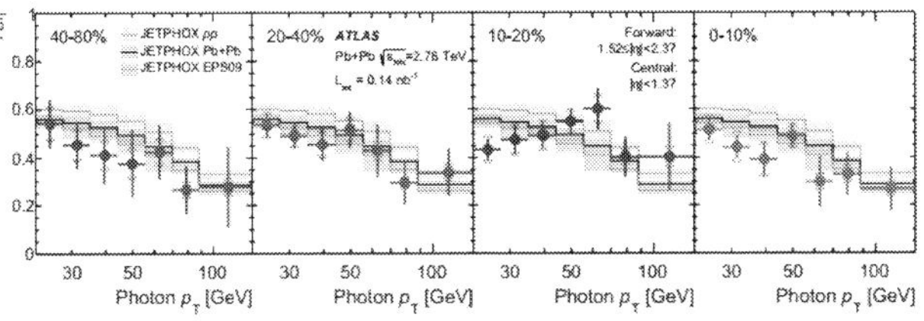

Figure 3. Ratios of yields of prompt photons as a function of p_T in forward ($1.52 < |\eta| < 2.37$) to central ($|\eta| < 1.37$) pseudorapidity regions measured in four centrality intervals [17]. Data are compared to MC predictions explained in text. Statistical and systematic uncertainties are shown as the bars and braces.

asymmetries shown in the lower panel of figure 2(left, middle) agree between the data and model, except in the Pb-going direction ($\eta^{\mu}_{\text{lab}} < 0$) in the most central collisions.

For completeness ATLAS has investigated production of Z bosons in p+Pb collisions [14]. It is worth noticing here that the asymmetric energy of the beams in p+Pb collisions results in a shift of the center-of-mass by 0.465 units of rapidity relative to the laboratory frame in the direction of the proton beam. Z bosons were reconstructed in both electron and muon channels. In figure 2(right) dependence of the Z boson yields in different centrality bins is shown as a function of the Z boson rapidity in the centre-of-mass system $y^{\star}_{Z}$. The ratio of central to peripheral data, R_{CP}, shown in the figure, is used to observe changes in rapidity distribution for different centrality bins in a model-independent way. Events with $40-90\%$ centrality define the peripheral event selection, and two central selections, $0-10\%$ and $10-40\%$, are compared with it. It is observed that the asymmetry in $y^{\star}_{Z}$ is slightly larger in more central events. Integrated over $y^{\star}_{Z}$, the centrality dependence appears to be consistent with binary scaling.

ATLAS has also studied production of prompt photons produced in Pb+Pb collisions [17]. Prompt photons have two primary sources: direct emission via quark-gluon Compton scattering or quark-antiquark annihilation, or in the fragmentation process through quark radiation of hard photons. To suppress the large background of non-prompt photons originating from the decays of neutral mesons in jets as well as photons from the fragmentation process, an isolation criterion has been applied. ATLAS has measured yields of prompt photons as a function of p_T in several centrality intervals in forward (F: $1.52 < |\eta| < 2.37$) and central (C: $|\eta| < 1.37$) pseudorapidity regions. In figure 3 the ratios of these yields are shown and compared to JETPHOX [18] predictions without and with nuclear modifications [10]. The precision of the measurement is not sufficient to indicate presence of nuclear modifications, however a slight preference of calculations incorporating isospin effects is observed.

ATLAS has performed the measurement of prompt and non-prompt J/Ψ production in p+Pb collisions [19]. The J/Ψ mesons were reconstructed using the dimuon decay mode in the fiducial phase space: $8 < p_T < 30$ GeV and $|y^{\star}| < 1.94$. A non-prompt J/Ψ contribution, i.e. J/Ψ from decay chains of b-quarks, was separated from prompt J/Ψ by measuring displaced decay vertices. The measured forward ($y^{\star} > 0$) to backward ($y^{\star} < 0$) production ratio R_{FB} is shown in figure 4 as a function of J/Ψ transverse momentum and the centre-of-mass rapidity. The measured distributions

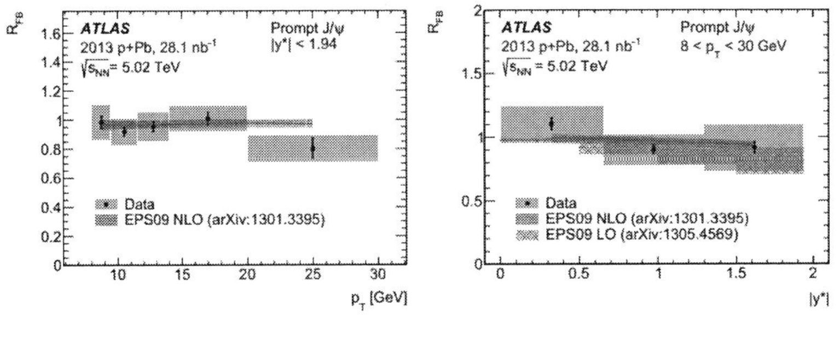

Figure 4. Forward-backward production ratio, R_{FB}, for prompt J/Ψ as a function of p_T (left) and $|y^{\star}|$ (right) [19]. Error bars show the statistical uncertainties and the shaded boxes show total uncertainties. EPS09 predictions are shown as indicated on the plots.

are consistent with unity and no significant dependence on p_T or $y^\star$ is found. Data are also in good agreement with calculations including nuclear modifications (EPS09) [10] in LO and NLO.

3 Charged-particle spectra and jets

Experimental studies of charged-particles and jet production in HI collisions can directly reveal the properties of the QGP created in the collisions. One of the predicted consequences of QGP formation is 'jet quenching' that refers to the modification of parton showers initiated by hard-scattering processes which take place in QGP. It is also expected that the resulting spectrum of hadrons, which originate from parton fragmentation, is also modified.

ATLAS has measured the nuclear modification factor in Pb+Pb, p+Pb and pp collisions both for charged particle spectra [8, 20, 21] and for jet yields [24, 25].

The nuclear modification factor for yields of charged-particles produced in Pb+Pb collisions has been studied by ATLAS in [21]. In the left panel of figure 5(upper) distributions of R_{AA} are shown for pseudorapidity range $|\eta| < 2$ and five centrality intervals in the p_T range $0.5 - 150$ GeV. The R_{AA} measured as a function of p_T shows a characteristic non-flat p_T shape which becomes more pronounced for the more central collisions. They first increase with increasing p_T reaching a maximum at $p_T \approx 2$ GeV, a feature commonly associated with the Cronin effect [22]. Then, R_{AA} decreases with higher p_T reaching a minimum at $p_T \approx 7$ GeV, where the charged-particle suppression is strongest. The suppression is strongest in the most central $0 - 5\%$ collisions, although it is still noticeable even in the $60 - 80\%$ centrality interval. Above this p_T, R_{AA} generally increases with increasing p_T up to approximately 60 GeV and then reaches a plateau. In the left panel of figure 5(lower), R_{AA} is shown as a function of the mean number of participating nucleons, $\langle N_{part}\rangle$, in the four characteristic transverse momentum intervals from the previous plot. In all these p_T intervals R_{AA} decreases with N_{part},

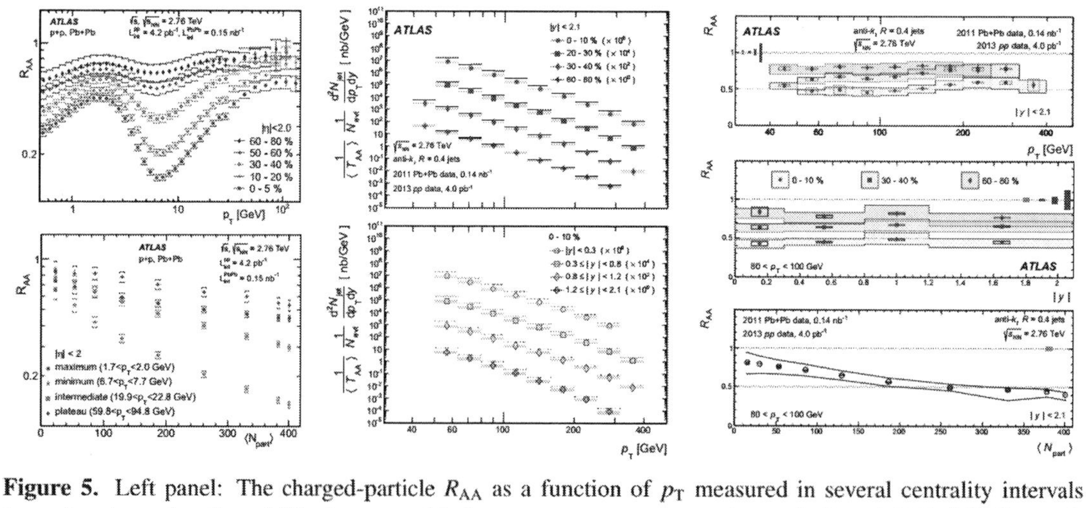

Figure 5. Left panel: The charged-particle R_{AA} as a function of p_T measured in several centrality intervals (upper) and as a function of $\langle N_{part}\rangle$ measured in four transverse momentum intervals (lower) [21]. Middle panel: The per-event jet yield in Pb+Pb collisions, multiplied by $1/\langle T_{AA}\rangle$, as a function of p_T. The upper plot shows the $|y| < 2.1$ rapidity range in different centrality intervals. The lower plot shows the $0 - 10\%$ centrality interval in different rapidity ranges. The solid and dashed lines represent the pp jet cross section for the same rapidity interval scaled by the same factor. Right panel: Jet R_{AA} as a function of p_T for $|y| < 2.1$ (upper), as a function of $|y|$ for $80 < p_T < 100$ GeV (middle) in different centrality bins and as a function of $\langle N_{part}\rangle$ [24].

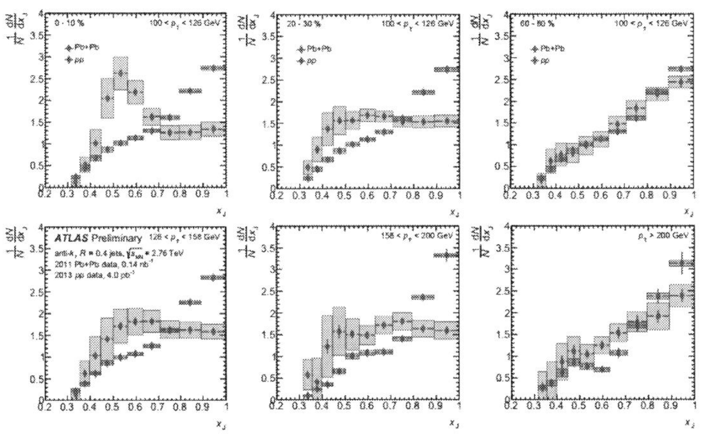

Figure 6. (upper) The x_J distributions for pairs with $100 < p_{T,1} < 126$ GeV for different collision centralities. (lower) The x_J distributions for $0-10\%$ centrality bin and for different ranges of the transverse momenta (indicated in the plots) of the leading jet. The Pb+Pb data are shown in red, while the pp distribution is shown for comparison in blue. Statistical uncertainties are indicated by the error bars while systematic uncertainties are shown with shaded boxes [26].

however the decrease is strongest for the p_T interval corresponding to the minimum and weakest in the plateau region. In the p_T ranges where the maximum and intermediate values of R_{AA} are measured the R_{AA} values are consistent for $N_{part} < 100$, but then deviate at larger N_{part}. In the most central collisions, R_{AA} in the maximum interval is comparable to that in the plateau region.

Nuclear modification factors for jets production in Pb+Pb collisions have been studied by the ATLAS experiment in [24]. In this analysis and also in the following ones discussed below, to reconstruct jets ATLAS uses anti-k_t algorithm [23] with the radius parameter $R = 0.4$.

The differential per-event jet yield in Pb+Pb collisions, multiplied by $1/\langle T_{AA}\rangle$ is shown in figure 5(middle panel), in selected rapidity and centrality bins. The dashed lines represent the pp jet cross sections for that same rapidity bin; the jet suppression is evidenced by the fact that the jet yields fall below these lines. In the right panel of figure 5, the jet R_{AA} is shown as a function of p_T, $|y|$ for different ranges in collision centrality and as a function of $\langle N_{part}\rangle$ for $80 < p_T < 100$ GeV. The R_{AA} is observed to increase weakly with p_T. The R_{AA} shows no significant rapidity dependence. The R_{AA} decreases smoothly from the most peripheral collisions (smallest $\langle N_{part}\rangle$) to central collisions, where it reaches a minimal value of approximately 0.4 in the most central 1% of collisions.

Quarks and gluons in the shower passing through QGP may be scattered elastically or inelastically resulting in both deflection and energy loss of the constituents of the shower. Therefore a complete characterization of the effects of jet quenching requires measurements of both the single jet suppression and of jet fragmentation distributions. ATLAS has recently studied the transverse momentum balance in dijet pairs [26]. The momentum balance of the dijet system is expressed by the variable $x_J \equiv p_{T,2}/p_{T,1}$, where $p_{T,1}$ and $p_{T,2}$ are the transverse momenta of the jets with the highest and second highest p_T in the event with $p_T > 25$ GeV and $|\eta| < 2.1$. The pair is also required to have $\Delta\phi > 7\pi/8$. The unfolded x_J distributions are shown in figure 6(upper), for pairs with $100 < p_{T,1} < 126$ GeV for different centrality intervals. The distribution in pp collisions is shown on each panel for comparison. In the $60-80\%$ centrality bin, where the effects of quenching are expected to be the smallest, the Pb+Pb data are generally consistent with the pp data. In more central Pb+Pb collisions, the distributions become significantly broader than that in pp collisions and the peak at $x_J \sim 1$, corresponding to nearly-symmetric dijet events, is reduced. The distribution becomes almost flat over the range $0.5 \lesssim x_J \lesssim 1$, before developing a peak at $x_J \sim 0.5$ in the most central collisions. At larger values of $p_{T,1}$ the x_J distributions, shown in figure 6(lower), are observed to narrow, indicating that higher p_T dijets tend to be better balanced in momentum. This is consistent with a picture in which the fractional energy loss decreases with jet p_T. The modifications observed in the Pb+Pb data lessen with increasing $p_{T,1}$ and for pairs with $p_{T,1} > 200$ GeV the maximum at $x_J \sim 1$ is restored.

 DOI: 10.1051/epjconf/201713801002

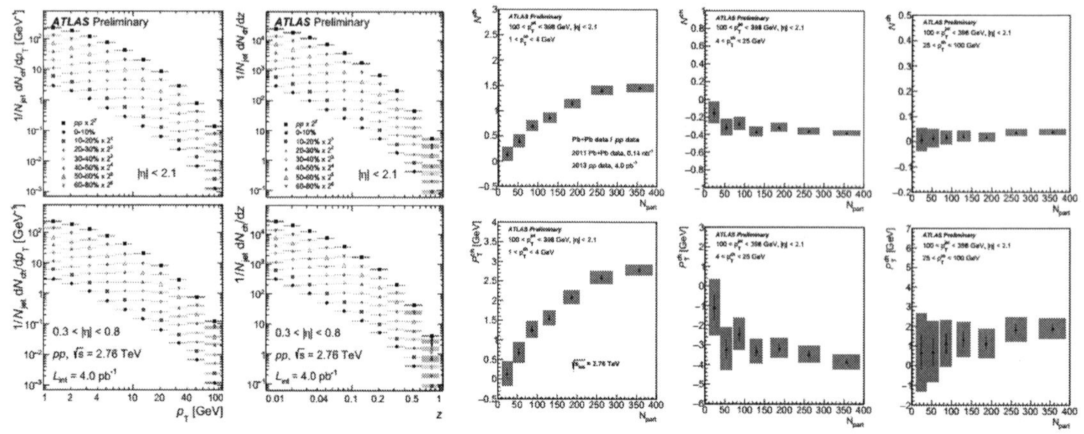

Figure 7. Left panel: Unfolded charged particle transverse momentum (left) and longitudinal momentum fraction (right) distributions, measured in pp collisions and for seven centrality bins measured in Pb+Pb collisions. Right panel: (upper) difference in the total yield of particles in a given p_{T} interval (indicated in the legend) measured in Pb+Pb and the total yield of particles in the same p_{T} interval measured in pp collisions, N^{ch}; (lower): difference in the total transverse momentum of particles in a given p_{T} interval measured in Pb+Pb and the total transverse momentum of particles measured in pp, $P_{\mathrm{T}}^{\mathrm{ch}}$. The differences are evaluated as a function of number of participating nucleons. In all plots: the error bars on the data points indicate statistical uncertainties while the shaded bands indicate systematic uncertainties [27].

Internal structure of jets in Pb+Pb and pp collisions ATLAS has studied in terms of longitudinal momentum fraction and transverse momenta of charged particles inside the jet [27]. Jet fragments having minimum p_{T} of 1 GeV were measured within an angular range $\Delta R = 0.4$ from the jet axis. The distributions of longitudinal momentum fraction and transverse momenta of charged particles inside jets with p_{T} in the interval 100 − 398 GeV and in two overlapping pseudorapidity ranges are shown in figure 7(left panel) for pp collisions and several centrality intervals for Pb+Pb collisions. The distributions exhibit a difference in shape between central HI collisions and peripheral HI collisions or pp reference. To better quantify the observed trends, the differences between integrals of distributions of charged particles transverse momenta measured in HI collisions and in pp collisions have been calculated and are shown in the right panel of figure 7(upper). The differences are shown for three ranges of charged particles transverse momenta (indicated in the plots) as a function of centrality expressed in terms of the number of participants. There is clear, almost logarithmic, increase of yields with centrality observed for particles with $1 < p_{\mathrm{T}} < 4$ GeV, a depletion is seen for particles with $4 < p_{\mathrm{T}} < 25$ GeV, and a slight enhancement is seen for particles with $25 < p_{\mathrm{T}} < 100$ GeV. Similarly, differences of the total transverse momentum of charged particles in HI and pp collisions have been calculated and are shown in figure 7(lower). The changes in the total transverse momentum follow the trends seen in the yields. The significance of the increase of the total transverse momentum in the high-$p_{\mathrm{T}}^{\mathrm{ch}}$ region is higher than the significance of the increase of yields of charged particles seen in the same region.

To help to disentangle the contributions to the observed dijet asymmetries from the unequal path lengths of the showers in the medium from fluctuations in the energy loss process itself, ATLAS measured the correlations between jets that are produced at small relative angles [28]. Neighbouring jet pairs include jets originating from the same hard interaction, but also jets from different hard interactions. The latter are not of interest in this analysis, and are subtracted statistically. The remaining neighbouring jet pairs result primarily from hard radiation by the parton that occurs early

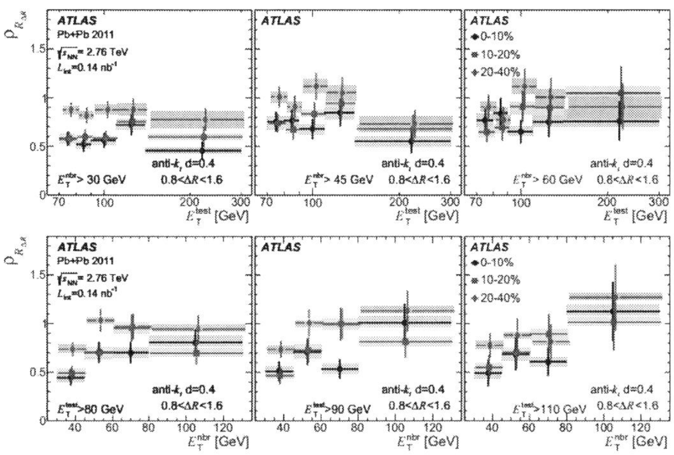

Figure 8. The ratio of $R_{\Delta R}$ for three bins of collision centrality to those in 40 − 80% collisions [28]. The ratio $\rho_{R_{\Delta R}}$ is evaluated as a function of E_T^{test} for three different choices of lower bound on E_T^{nbr} (upper) and as a function of E_T^{nbr} for three different choices of lower bound on E_T^{test} (lower). The error bars represent statistical uncertainties and the shaded bands the systematic uncertainties. The data points are shifted along the horizontal axis for clarity.

in the process of the shower formation. Generally, two neighbouring jets originating from the same hard scattering should have more similar path lengths in the medium compared to the two jets in the dijet measurement. Therefore measuring neighbouring jets could probe differences in their quenching that do not result primarily from difference in path length. More generally, measurements of the correlated production of jets in the same parton shower may provide more detailed insight into the modification of the parton shower in the QGP beyond the subsequent quenching of the resulting jets. The rate of neighbouring jets, $R_{\Delta R}$, that accompany 'test' jet within a given range of angular distance $\Delta R = \sqrt{(\Delta\phi)^2 + (\Delta\eta)^2}$ is measured in bins of transverse energies of test jets, E_T^{test}, and neighbouring jets, E_T^{nbr}, in several centrality intervals. In figure 8 the ratios of $R_{\Delta R}$ for three bins of collision centrality to those in 40 − 80% collisions is shown as a function of both E_T^{test} and E_T^{nbr}, in each case for different lower bounds of the other variable. The $\rho_{R_{\Delta R}}$ distributions do not exhibit any strong dependence on E_T^{test}. The suppression factor $\rho_{R_{\Delta R}}$ of the most central collisions is at the level of 0.5 − 0.7 for all three lower bounds on E_T^{nbr}. The suppression seems to become less pronounced with decreasing centrality. This is qualitatively consistent with the observation of the centrality-dependent suppression of inclusive jet yields. Contrary to a modest dependence of $\rho_{R_{\Delta R}}$ on the test-jet E_T, the $\rho_{R_{\Delta R}}$ evaluated as a function of E_T^{nbr} suggests a decrease of suppression with increasing E_T^{nbr}. Such a decrease in suppression with increasing E_T^{nbr} may in fact be expected. The jet quenching is generally expected to depend on the initial parton energy, but if the splitting happens such that the two partons have similar energy, their quenching would likely be comparable due to similar in-medium path-length traveled by the two partons forming neighbouring jets. Thus, in the configuration of $E_T^{test} \approx E_T^{nbr}$ the per-test-jet normalization effectively removes the impact of the suppression.

4 Bulk collectivity

The two-particle correlation function in pseudorapidity $C_N(\eta_1, \eta_2)$ has been measured by the ATLAS experiment in Pb+Pb, p+Pb and pp collisions [30]. In order that it contains only dynamical fluctuations, the correlation function has been corrected for any residual multiplicity dependence in the average shape $\langle N(\eta)\rangle$. In order to quantify the features of the correlation function, it is essential to estimate and separate the contributions from the short-range correlations (SRC) and the long-range correlations (LRC). The SRC reflect correlations within the same source, while the LRC reflects FB-asymmetry of the number of sources. A method based on different behaviour of the correlation function for SRC and LRC for the same-charge and opposite-charge particles [30], was developed and used to subtract the SRC from the total correlation function. The correlation function is measured

 DOI: 10.1051/epjconf/201713801002

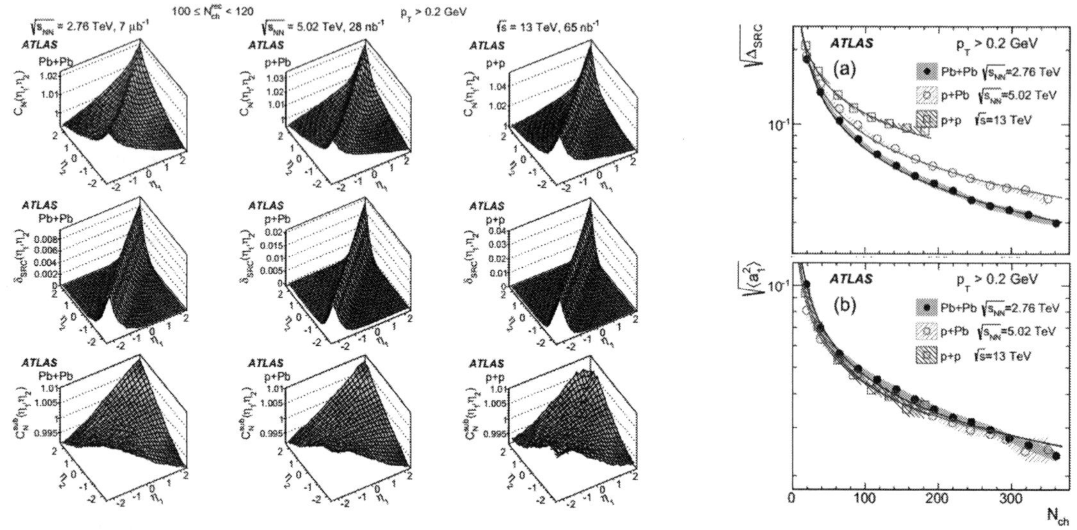

Figure 9. Left panel: The distributions of correlation functions $C_N(\eta_1, \eta_2)$ (top row), the estimated short-range component $\delta_{SRC}(\eta_1, \eta_2)$ (middle row) and long-range component C_N^{sub} (bottom row). They are shown for events with $100 \leq N_{ch}^{rec} < 120$ in Pb+Pb, p+Pb and pp collisions. Right panel: (a) The estimated magnitude of the short-range component $\sqrt{\Delta_{SRC}}$ and (b) the coefficient $\sqrt{\langle a_1^2 \rangle}$ as a function of N_{ch} for all-charge pairs in Pb+Pb, p+Pb and pp collisions. The shaded bands represent systematic uncertainties, and the statistical uncertainties are smaller than the symbols [30].

using charged particles in the pseudorapidity range $|\eta| < 2.4$ with transverse momentum $p_T > 0.2$ GeV, and it is measured as a function of event multiplicity, defined by a number of charged particles with $|\eta| < 2.5$ and $p_T > 0.4$ GeV. The measured $C_N(\eta_1, \eta_2)$ is shown in the upper row of the left panel of figure 9 for different collision systems and for the multiplicity range $100 < N_{ch}^{rec} < 120$. The resulting SRC denoted as $\delta_{SRC}(\eta_1, \eta_2)$ and LRC estimate denoted as $C_N^{sub}(\eta_1, \eta_2)$ are shown in the middle and bottom rows, respectively. The magnitude of the SRC in p+Pb is observed to be larger in the proton-going direction than in the lead-going direction, reflecting the fact that the particle multiplicity is smaller in the proton-going direction. However, this forward-backward asymmetry in p+Pb collisions is mainly associated with the SRC component, and the $C_N^{sub}(\eta_1, \eta_2)$ distribution shows very little asymmetry. $C_N(\eta_1, \eta_2)$ distributions show significant differences between the three systems, which is mainly due to their differences in $\delta_{SRC}(\eta_1, \eta_2)$. In fact the estimated long-range component $C_N^{sub}(\eta_1, \eta_2)$ shows similar shape and similar overall magnitude for the three systems.

The magnitude of the SRC is quantified by $\delta_{SRC}(\eta_1, \eta_2)$ averaged over two-particle pseudorapidity phase space $\Delta_{SRC} = (1/4Y^2) \int \delta_{SRC}(\eta_1, \eta_2) d\eta_1 d\eta_2$, where $Y = 2.4$. The corresponding contribution of the SRC at the single-particle level is $\sqrt{\Delta_{SRC}}$ and is shown in the right panel of figure 9(upper) as a function of N_{ch} for different charge combinations in the three collision systems. The strength of the SRC always decreases with N_{ch} and is larger for smaller collision systems.

To quantify the LRC the correlation function $C_N^{sub}(\eta_1, \eta_2)$ has been decomposed into orthogonal Legendre polynomials. From the fit to the measured distribution only the coefficient $\langle a_1^2 \rangle$ is significantly different from zero. Distributions of $\sqrt{\langle a_1^2 \rangle}$ as a function of N_{ch} for different collision systems are shown in the right panel of figure 9(bottom). The LRC signal is found to be similar between the three collision systems, which is a non-trivial result as the distribution of N_{ch} as well as the sizes of the overlap region are very different between the three collision systems.

 DOI: 10.1051/epjconf/201713801002

5 Ultra-peripheral collisions

In the case of nuclei colliding at large impact parameters, the strong interaction is not active. However, due to their strong electromagnetic fields, the nuclei can coherently produce high energy nearly real photons. That means that in these so-called ultra-peripheral collisions (UPC) one can study two-photon processes, because relatively small couplings are compensated with very high photon-photon luminosity which scales with the nuclear electric charge as Z^4. The maximum energy for coherent photons emitted from relativistic nucleus is $\gamma\hbar c/R$, where γ is the Lorentz factor of one nucleus and R is the nuclear radius. This is about 75 GeV for the Pb+Pb collisions at $\sqrt{s_{\rm NN}} = 5.02$ TeV.

ATLAS has measured cross sections for exclusive dimuon production in UPC of lead nuclei for large dimuon invariant mass $M_{\mu\mu} > 10$ GeV [31]. The cross sections were measured by selecting events using a single-muon trigger, in association with an otherwise low-multiplicity event. The events are required to have a primary vertex formed solely from two oppositely-charged muons. Events with QED final-state radiation (FSR) are subtracted by removing pairs with large acoplanarity (defined as $1 - |\Delta\phi|/\pi$). The events are then corrected for muon trigger and reconstruction efficiency, for vertex efficiency, and for the expected losses due to FSR. The results are compared directly with calculations from STARLIGHT 1.1 [32], as a function of the dimuon pair mass ($M_{\mu\mu}$) and pair rapidity ($Y_{\mu\mu}$) both defined in terms of the total four-momentum. The kinematic coverage is substantially larger than in any previous HI measurement, with pair rapidities out to $|Y_{\mu\mu}| < 2.4$ and pair masses ranging from $10 < M_{\mu\mu} < 100$ GeV.

The measured cross sections as a function of dimuon kinematic variables are shown in figure 10. The mass distribution is shown for two (overlapping) rapidity regions ($|Y_{\mu\mu}| < 2.4$ and $|Y_{\mu\mu}| > 1.6$) to show both the full distribution, and to focus on the forward region. The rapidity distribution is shown for three ranges of dimuon invariant mass ($10 < M_{\mu\mu} < 20$ GeV, $20 < M_{\mu\mu} < 40$ GeV, $40 < M_{\mu\mu} < 100$ GeV). Error bars show statistical uncertainties while the grey bands show the combined systematic uncertainties, including the 7% global luminosity uncertainty. The MC statistical uncertainties, which are negligible in most cases, are not shown. STARLIGHT is in good agreement with the data over the full kinematic range, within the quoted uncertainties. This suggests that the nuclear electromagnetic fields are reasonably described by the equivalent photon approximation and that the nuclear form factor is adequately implemented in STARLIGHT.

Integrating the distributions in figure 10 gives the total cross section in the fiducial acceptance $\sigma(\text{Pb}+\text{Pb} \rightarrow \mu^+{+}\mu^-{+}\text{Pb}^{(\star)}{+}\text{Pb}^{(\star)}) = 32.2\pm0.3\,(\text{stat})^{+4.0}_{-3.4}\,(\text{sys})$ μb to be compared with 31.64 ± 0.04 μb from the STARLIGHT prediction.

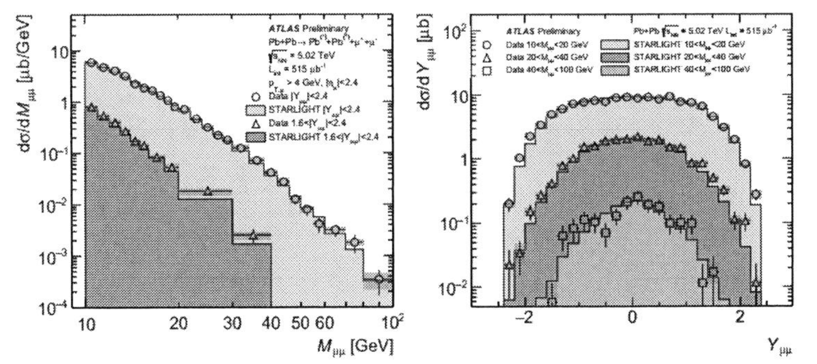

Figure 10. Cross section for dimuon production in UPC, as a function of pair mass (left) and pair rapidity (right) [31]. Data indicated by symbols are compared to STARLIGHT predictions shown by solid histograms. Error bars indicate statistical uncertainties while the gray band indicates the combined systematic uncertainties.

6 Summary

Recent results from the ATLAS experiment related to the study of Quark-Gluon Plasma via both hard and soft probes have been briefly presented. For more details and in depth discussion of the measurements the reader is referred to the original publications. All ATLAS heavy-ion results are available from https://twiki.cern.ch/twiki/bin/view/AtlasPublic/HeavyIonsPublicResults
This work was partly supported by the National Science Centre of Poland under grant number UMO-2012/05/B/ST2/02480.

References

[1] STAR Collaboration, J. Adams, et al., Nucl. Phys. A **757**, 102 (2005). [nucl-ex/0501009]
[2] BRAHMS Collaboration, I. Arsene, et al., Nucl. Phys. A **757**, 1 (2005). [nucl-ex/0410020]
[3] PHOBOS Collaboration, B.B. Back, et al., Nucl. Phys. A **757**, 28 (2005). [nucl-ex/0410022]
[4] PHENIX Collaboration, K. Adcox et al., Nucl. Phys. A **757**, 184 (2005). [nucl-ex/0410003]
[5] N. Armesto and E. Scomparin, Eur. Phys. J. Plus **131**, 52 (2016). [nucl-ex/1511.02151]
[6] ATLAS Collaboration, JINST **3**, S08003 (2008). https://cds.cern.ch/record/1129811
[7] ATLAS Collaboration, Phys. Lett. B **707**, 330–348 (2012). [nucl-ex/1108.6018]
[8] ATLAS Collaboration, Eur. Phys. J. C **76**, 199 (2016). [hep-ex/1508.00848]
[9] V. Guzey, M. Strikman, Phys. Lett. B **633** 245 (2006). [hep-ph/0505088]
[10] K. Eskola et al., J. High Energy Phys. **04**, 065 (2009). [hep-ph/0902.4154]
[11] ATLAS Collaboration, Eur. Phys. J. C **75** (2015) 23. [hep-ex/14084674]
[12] ATLAS Collaboration, Phys. Rev. Lett. **110**, 022301 (2013). [hep-ex/1210.6486]
[13] ATLAS Collaboration, ATLAS-CONF-2015-056, http://cds.cern.ch/record/2055677
[14] ATLAS Collaboration, Phys. Rev. C **92**, 044915 (2015). [hep-ex/1507.06232]
[15] S. Alioli, P. Nason, C. Oleari, E. Re, J. High Energy Phys. **07**, 60 (2008). [hep-ph/0805.4802]
[16] H.-L. Lai et al., Phys. Rev. D **82**, 074024 (2010). [hep-ph/1007.2241]
[17] ATLAS Collaboration, Phys. Rev. C **93**, 034914 (2016). [hep-ex/1506.08552]
[18] P. Aurenche et al., Phys. Rev. D **73**, 094007 (2006). [hep-ph/0602133]
[19] ATLAS Collaboration, Phys. Rev. C **92**, 034904 (2015). [hep-ex/1505.08141]
[20] ATLAS Collaboration, submitted to Phys. Lett. B. [hep-ex/1605.06436]
[21] ATLAS Collaboration, J. High Energy Phys. **09**, 50 (2015). [hep-ex/1504.04337]
[22] J.W. Cronin et al., Phys. Rev. D **11**, 3105 (1975)
[23] M. Cacciari, G.P. Salam, G. Soyez, J. High Energy Phys. **04**, 063 (2008). [hep-ph/0802.1189]
[24] ATLAS Collaboration, Phys. Rev. Lett. **114**, 072302 (2015). [hep-ex/1411.2357]
[25] ATLAS Collaboration, Phys. Lett. B **748** (2015) 392. [hep-ex/1412.4092]
[26] ATLAS Collaboration, ATLAS-CONF-2015-052, http://cds.cern.ch/record/2055673
[27] ATLAS Collaboration, ATLAS-CONF-2015-055, http://cds.cern.ch/record/2055676
[28] ATLAS Collaboration, Phys. Lett. B **751**, 376 (2015). [hep-ex/1506.08656]
[29] ATLAS Collaboration, submitted to Phys. Lett. B. [hep-ex/1605.06436]
[30] ATLAS Collaboration, submitted to Phys. Rev. C. [hep-ex/1606.08170]
[31] ATLAS Collaboration, ATLAS-CONF-2016-025, http://cds.cern.ch/record/2157689
[32] S. Klein and J. Nystrand, The STARLIGHT website, http://starlight.hepforge.org

Highlights in the hadron electromagnetic structure

Egle Tomasi-Gustafsson[1,⋆], *Simone* Pacetti[2], and *Andrea* Bianconi[3]

[1] *IRFU, CEA, Université Paris-Saclay, 91191 Gif-sur-Yvette Cedex, France*
[2] *Dipartimento di Fisica e Geologia, and INFN Sezione di Perugia, 06123 Perugia, Italy*
[3] *Dipartimento di Ingegneria dell'Informazione, Università di Brescia and Istituto Nazionale di Fisica Nucleare, via A. Bassi 6, 27100 Pavia, Italy*

Abstract. In frame of a general view of proton electromagnetic form factors, two recent findings related to reanalyses of data are presented. Recent experiments in the scattering and in the annihilation region provided us with more precise data and/or extending the kinematical region, allowing a deeper analysis and a common view of these fundamental quantities. We will discuss two issues: the discrepancy between the form factors extracted from unpolarized and polarized ep elastic scattering experiments, in connection with the commonly used dipole parametrization; peculiar oscillations in $e^+e^- \to \bar{p}p(\gamma)$ annihilation cross section, that become periodical when plotted as a function of the 3-momentum of the relative motion of the final proton and antiproton, after subtraction of a smooth function.

1 Introduction

Electromagnetic hadron form factors (FFs) are fundamental quantities that describe the internal structure of non-pointlike particles, their charge and magnetic currents. Assuming that the interaction occurs through one photon exchange, FFs can be measured (through electron proton elastic scattering and $e^+e^- \leftrightarrow \bar{p}p$ annihilations): in the first case, the transferred photon of virtual mass $Q^2 = -q^2 > 0$ is space-like, in the second case, the time-like photon has 4-momentum squared $q^2 > 0$. FFs are also directly calculated, as they enter in the expression of the electromagnetic current: hadron models, after reproducing the static properties of hadrons, as masses and magnetic moments, should reproduce their dynamical content, parametrized by FFs. Schematically, FFs at large momentum transfer probe the quark structure of the hadrons whereas at low momentum transfer, they probe the size of the hadron.

The analytical properties of FFs justify a common interpretation of scattering and annihilation experiments. It is tempting to have a global definition of FFs, in the space and time-like regions. In [1] the definition of FFs was generalized as follows:

$$F(q^2) = \int_{\mathcal{D}} d^4x e^{iq_\mu x^\mu} \rho(x), \; q_\mu x^\mu = q_0 t - \vec{q} \cdot \vec{x}, \tag{1}$$

where $\rho(x) = \rho(\vec{r}, t)$ is the space-time distribution of the electric charge in a space-time volume $\mathcal{D}$.

⋆e-mail: egle.tomasi@cea.fr

In the scattering channel, $e + p \to e + p$ and in the Breit frame, we recover the definition of FFs as Fourier transform of a charge density:

$$F(q^2) = \delta(q_0)F(Q^2),\ Q^2 = -(q_0^2 - |\vec{q}|^2) > 0, \tag{2}$$

where the zero energy transfer is implied.

In the annihilation channel:

$$F(q^2) = \int_{\mathcal{D}} dt e^{i\sqrt{q^2}t} \int d^3\vec{r}\rho(\vec{r},t) = \int_{\mathcal{D}} dt e^{i\sqrt{q^2}t} Q(t), \tag{3}$$

where $Q(t)$ can be interpreted as the time evolution of the charge distribution in the domain $\mathcal{D}$. In [1] a qualitative phenomenological picture of scattering and annihilation processes was presented, and the possibility of an observable dynamical polarization of the formed hadron pair was suggested.

Hadron FFs are investigated since decades, but recent findings and experimental possibilities have renewed interest in this field. High intensity electron beams, with large and stable polarization, together with large acceptance spectrometers and detectors, and the development of hadron polarimetry allowed in particular to apply to the Akhiezer-Rekalo recoil proton polarization method [2, 3], that requires to measure the recoil proton(neutron) polarization in the scattering of longitudinally polarized electrons. These experiments, performed at MIT Bates (USA), MAMI (Germany) and JLab (USA) at large momentum transfer, brought very precise information on the ratio of the electric to magnetic FF for protons and neutrons and showed that this ratio is not constant, but decreases for increasing momentum transfer approaching zero at 9 GeV^2. It was previously assumed that the electric and magnetic FFs follow a dipole dependence $\simeq (Q^2)^{-2}$. Still unpolarized ep scattering measurements seem to confirm such behavior. This discrepancy gave rise to a large number of theoretical papers and new experiments. We reanalyze the unpolarized cross section data in terms of FF ratio and show that the discrepancy between sets of data vanishes if one takes into account, besides effects of radiative corrections and parameter correlations, the effect of relative normalization.

The time-like region is accessible at e^+e^- colliders, as BABAR (USA), BESIII (China), and VEPPII (Russia), through the reaction $e^+e^- \to \bar{p}p(\gamma)$ and at $\bar{p}p$ colliders as LEAR (CERN) and FermiLab (USA). In next future a high energy and high intensity antiprotons beam will be available at PANDA (FAIR, Germany). The presence of oscillating structures in the time-like FF data has been recently highlighted in [4, 5]. In particular the data present a regularity when plotted as a function of the relative momentum of the outgoing hadrons. This pattern is tentatively associated with the interference between two sources, respectively of the quark and of the hadron size, bringing the signature of the first instants of the hadron formation from the excited vacuum.

2 Reanalysis of Rosenbluth data

It has been assumed for long time that the proton electric FF, as well as the magnetic FFs of the proton and neutron, normalized to their magnetic moment, have a Q^2-dipole dependence (Q is the four-momentum of the photon):

$$G_D(Q^2) = \left(1 + \frac{Q^2[\mathrm{GeV}]^2}{0.71}\right)^{-2},$$

whereas the neutron electric FF is essentially zero. The dipole parametrization was commonly accepted, because it was consistent with the non-relativistic approximation, where FFs are Fourier transforms of the spatial densities of electric charge and magnetization of the nucleon; in such frame, the

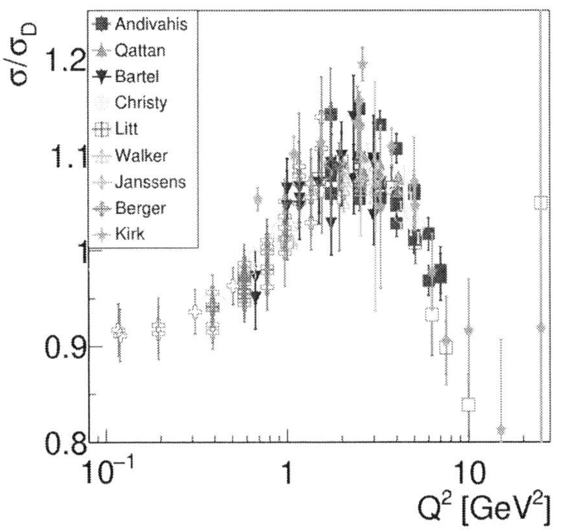

Figure 1. Cross section normalized to the dipole cross section, σ/σ_D, as a function of Q^2 for different experiments.

dipole approximation corresponds to an exponential charge distribution. Moreover, from the QCD point of view, as elastic FFs represent the probability that a proton remains in its ground state after that each of its valence quarks has received a momentum squared, Q^2, transferred by the virtual photon. Scaling laws predict a $(1/Q^2)^2$ dependence of the amplitude of the process [6, 7] (corresponding to two gluon exchange, the minimum number of exchanges needed for sharing the momentum among the three valence quarks).

However, several early experiments pointed out a deviation of the elastic *ep* cross section from the $(1/Q^2)^{-5}$ behavior. Quoting a presentation of the data at the highest available transferred momenta, from Nobel prize R. Taylor: "There appears to be definite evidence in the data for a significant deviation from the dipole fit" [8]. The dipole normalized cross section

$$\frac{\sigma}{\sigma_D} = \frac{\sigma_{\rm red}^{\rm exp}}{G_D^2(\epsilon/\mu^2 + \tau)},$$

being $\sigma_{\rm red}^{\rm exp}$ the measured reduced cross section, is reported in figure 1 as a function of Q^2, regardless of the value of ϵ. Here $\tau = Q^2/(4M^2)$, M is the proton mass, μ is the proton magnetic moment. The Q^2 coordinates for the data at different ϵ are seen as vertically quasi-aligned symbols. Note that if these points form a cluster with overlapping error bars, it means that they are compatible with the relation $G_E \simeq G_M/\mu \simeq G_D$. If points are not overlapping, then FFs do not follow a dipole behavior. In general, and particularly at large Q^2, one can see that the dipole fit is not a good representation of the data. The deviation at large Q^2 reaches 20-30% on the cross section and has to be attributed mainly to the magnetic term. The magnetic contribution to the elastic cross section becomes dominant at large Q^2. The *reduced* cross section of electron-proton elastic scattering $\sigma_{\rm red}$, in the one-photon exchange approximation, is linear in the variable $\epsilon = [1 + 2(1+\tau)\tan^2(\theta_e/2)]^{-1}$, being θ_e the electron scattering angle in the proton rest frame:

$$\sigma_{\rm red}(\theta_e, Q^2) = \left[1 + 2\frac{E}{M}\sin^2(\theta_e/2)\right]\frac{4E^2\sin^4(\theta_e/2)}{\alpha^2\cos^2(\theta_e/2)} \times \epsilon(1+\tau)\frac{d\sigma}{d\Omega} = \epsilon G_E^2 + \tau G_M^2, \qquad (4)$$

where E and $d\sigma/d\Omega$ are the electron initial energy and the differential cross section in the proton rest frame, and G_E and G_M are the proton electric and magnetic Sachs FFs. The measurement of the differential (reduced) cross section at fixed Q^2, for different angles, allows to extract the squared values of the FFs, G_E^2 and G_M^2, as the slope and the intercept (multiplied by τ), respectively, of this linear distribution (Rosenbluth separation [9]).

The doubts on the deviation from the dipole approximation for the electric FF, became evident with the advent of polarization experiments. In the 70's Akhiezer and Rekalo [2, 3] showed that the polarization of the scattered proton in the scattering of longitudinally polarized electrons on an unpolarized target (or the asymmetry with a transversely polarized target) contains an interference term proportional to the product $G_E G_M$. This observable would therefore be more sensitive to a small electric contribution, and even to its sign (particularly important for the neutron case). A measurement of the ratio of the transverse to the longitudinal polarization of the recoil proton gives a direct measurement of the ratio of electric and magnetic FFs, $R = G_E/G_M$:

$$\frac{P_t}{P_\ell} = -2\cot(\theta_e/2)\frac{M}{E+E'}\frac{G_E}{G_M}, \tag{5}$$

(E' is the scattered electron energy) and it is free from systematic errors coming from the beam polarization and the analyzing powers of the polarimeter. The data based on the Akhiezer-Rekalo method, mostly taken by the JLab GEp collaboration ([10] and references therein), showed with unprecedented precision that the ratio of electric to magnetic FFs decreases as Q^2 increases. Below it is shown that recent reanalyses of Rosenbluth data in electron proton elastic scattering in terms of the FF ratio, show agreement between the value of this ratio whenever extracted from polarized or unpolarized experiments [11]. In particular the large sensitivity of the ratio to relative normalization of the cross section data solves the discrepancy, within the assumption of the one photon exchange mechanism.

2.1 Reanalysis of existing data

Problems of parameter correlations and limits inherent to the Rosenbluth have been discussed in [12]. Previous global analyses were done, discussing in particular the problem of normalization among different sets of data, the omission of some of the data points [13] or reconsidering radiative corrections [14].

The following procedure to extract the FFs information from the unpolarized cross section has been suggested in [11]. Instead of extracting separately G_E and G_M, we write the reduced cross section given in (4) as

$$\sigma_{\rm red} = G_M^2(R^2\epsilon + \tau), \tag{6}$$

where G_M^2 and $R^2 = (G_E/G_M)^2$ are considered as independent parameters. The unpolarized data are fitted at fixed Q^2. The procedure has the advantage to extract directly the ratio, by automatically accounting for the effect of the correlations between G_E and G_M. The parameter R^2 represents directly the deviation of the linear dependence of the cross section from a constant ϵ value, whereas eventual general normalization and systematic errors are absorbed by G_M^2.

If for some of the data values and errors consistent with the original publication are recovered, the data from [15] deserve a specific discussion. This work is especially representative, as it extends the individual FF extraction by the Rosenbluth method to the largest values of Q^2.

2.1.1 Data from [15]

The data of [15], with eight points and two settings, span the region $1.75 \le Q^2 \le 8.83$ GeV^2. The two settings will be indicated as high energy (HE) and low energy (LE) experiments.

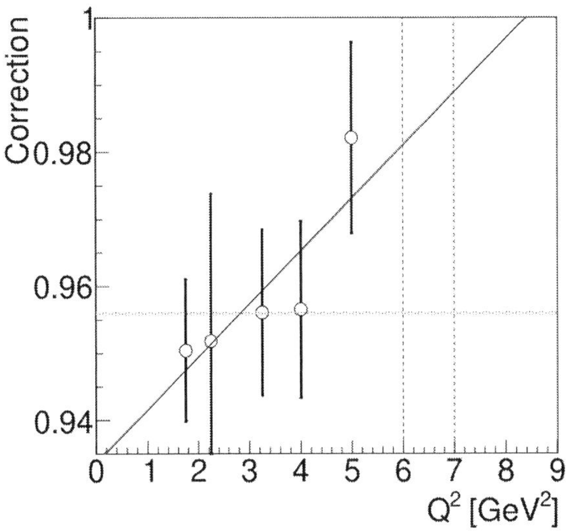

Figure 2. Correction factor as a function of Q^2. A linear fit (red line) shows an increasing of the factor. The dashed (blue) lines indicate that the extrapolated correction for the two HE points would be close to 1% instead than $\simeq 5\%$, as applied in the original paper.

In the original paper the measured cross sections were published, warning that an uncertainty of $\pm 5\%$ affected the second setting, due to a poor knowledge of the acceptance of the spectrometer. This error, however, was not added to the tabulated error. Instead, it was taken into account as a constant relative correction, fixed on the two lowest values $Q^2 = 1.75$ and 2.50 GeV^2 (the cross section was measured at each setting for the lowest ϵ value). The cross section from the LE setting was larger by 4-5%. Assuming a linear ϵ dependence of the reduced cross section (that implies the dominance of the one-photon exchange mechanism), a fit of the HE data was done and the LE energy point was renormalized to lie on the straight line. Then, the same constant normalization $C = 0.956$, fixed on the low Q^2 point, was applied to the cross section at all Q^2 (Analysis I). This procedure has the effect to enhance the slope, increasing the FF ratio. Note that for $Q^2 = 6$ and 7 GeV^2 only two points are present. The renormalization (lowering) of the first point changes essentially the slope of the linear fit.

The data were reanalyzed following several procedures. We recalculated the ratio using the data as published, without renormalizing the two settings and considering the LE points as additional, independent measurements. In this case the data points at $Q^2 = 1.75$ and 2.5 GeV^2 are both included in the fit, constraining the fit to an average value (Analysis II).

We fit only the HE points (excluding therefore the two points at $Q^2 = 6$ and 7 GeV^2). We find a slope consistent with analysis II, although affected by larger errors, as the number of points is smaller (Analysis III).

We repeat the normalization procedure, by aligning the LE point on the straight line fitting the HE points (Analysis IV). We note a systematic increase of the normalization factor (figure 2). Applying a normalization coefficient that is not constant with Q^2 but derived in order to align the LE point to the straight line defined by the HE points turns out to be equivalent to Analysis III in terms of slope and intercept.

The results are reported in figure 3 and compared to the ratio from polarization data. We may conclude that the results from Analysis II, III, and IV are consistent with the decreasing of the ratio indicated by the polarization data. Therefore a revision of the normalization factor brings the data into agreement. Moreover, at the light of all above, it is nonsense to use the FFs data from [15] to probe the two-photon effect, as they were extracted under the hypothesis of linearity of the reduced cross

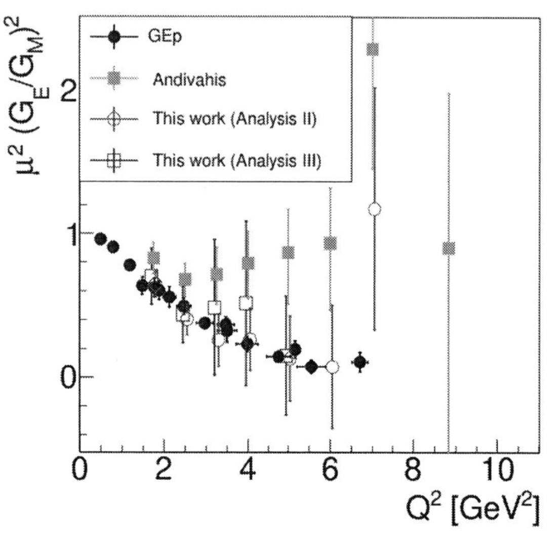

Figure 3. Color online. $\mu^2 R^2 = \mu^2 (G_E/G_M)^2$ as a function of Q^2 from (Andivahis: [15], green solid squares) as originally published; from Analysis II: without renormalization (red open circles); from Analysis III: omitting the lowest ϵ point (blue open squares) compared to the values from polarization experiments (GEp: [10], black solid circles).

section, by correcting the first point to be aligned. The results showed consistency with the hypothesis $\mu^2 R^2 \simeq 1$ at large Q^2, as expected at that time. The tendency of the first two points to deviate from unity was operatively corrected by the renormalization procedure.

Among the available data, three sets [16–18] show a particular behavior, that is not consistent with the previous finding, giving a value of the ratio exceeding unity and growing with Q^2.

For these experiments it was noted in [12] that radiative corrections and/or correlations are especially large. The data from [18] were extracted detecting the proton instead of the electron. Besides the above mentioned corrections, at large Q^2 the contamination of the elastic peak by the inelastic $e + p \to e + p + \pi^0$ reaction has to be carefully subtracted [10].

For [16, 17], G_M^2 extracted from the present analysis is systematically lower, showing that these measurements may be affected by a global systematic error probably due to normalization issues, whereas the results of [18] agree with the standard parametrization of the magnetic contribution.

Note that in [17] a somehow arbitrary renormalization was done by "changing the normalization of the small angle data from SLAC or DESY by ±1.5% with respect to the large angle data (Bonn)". This normalization increased the FFs ratio towards unity.

3 Periodic oscillations in the time-like region

In the time-like region, FFs become complex, due to unitarity, and the unpolarized cross section depends on their squared moduli, assuming one photon exchange approximation. The most recent data on electron-positron annihilation into a proton-antiproton pair have been collected by the BABAR collaboration, through initial state radiation (ISR): $e^+ + e^- \to \bar{p} + p(\gamma)$. When the photon is hard, the cross section can be factorized in a radiator factor, that depends on the hard photon kinematics only, multiplied by the cross section of the process of interest $e^+ + e^- \to \bar{p} + p$. The recent data by the BABAR collaboration [19, 20] cover with reasonable continuity a region ranging from slightly over the $\bar{p}p$ threshold to $q^2 \approx 36$ GeV^2. In particular about 30 data points have been extracted in the region $q^2 < 10$ GeV^2, with a relative error lower than 10 %. Due to the precision and the continuity of these data, oscillations of the cross section as a function of the q^2 become evident.

From the cross section, one extracts a generalized FF:

$$|F_p|^2 = \frac{3\beta q^2 \sigma}{2\pi\alpha^2 \left(2 + \frac{1}{\tau}\right)}, \tag{7}$$

(where $\alpha = e^2/(4\pi)$, $\beta = \sqrt{1 - 1/\tau}$, $\tau = \frac{q^2}{4M^2}$), M is the proton mass.

Plotting the generalized FF as a function of the 3-momentum of the relative motion of the final proton and antiproton, a systematic sinusoidal modulation appears in the near-threshold region, typical of an interference pattern, figure 4.

Through a Fourier analysis, the period is shown to correspond to a relative distance of 0.7-1.5 fm between the centers of the forming hadrons. It has been suggested that a rescattering process at the level of the newly formed hadrons interferes with the processes at a much smaller scale driven by the quark dynamics [4]. These oscillations may be interpreted as a signature of early mechanism of hadron creation.

We introduce a function of the form

$$F(p) = F_0(p) + F_{osc}(p), \tag{8}$$

where $p(q^2) = \sqrt{E^2 - M^2}$, $E = q^2/(2M) - M$, and p is the momentum of one of the two hadrons in the frame where the other one is at rest. The function $F_0(p)$ is a smooth background, expressing the regular behavior of the FF over a long q^2 range, whereas $F_{osc}(p)$ describes the overlapping oscillations.

The recent data are best reproduced by the function F_0 proposed in [21]:

$$|F_0(q^2)| = \frac{\mathcal{A}}{(1 + q^2/m_a^2)\left(1 - q^2[\mathrm{GeV}]^2/0.71\right)^2}, \quad \mathcal{A} = 7.7, \ m_a^2 = 14.8\ \mathrm{GeV}^2. \tag{9}$$

The best fit function is then subtracted from the data, leaving a regular oscillatory behavior, figure 4. It has magnitude $\sim 10\ \%$ of the regular term, and is well visible over the data uncertainties for $p > 3$ GeV. We have fitted the difference between the BABAR data and the regular background term $F_0(p)$ with the four-parameter function

$$F_{osc}(p) = A\, e^{-Bp} \cos(Cp + D). \tag{10}$$

This difference is plotted in the lower panel of figure 4. The values of the parameters are, respectively, $A = 0.05 \pm 0.01$, $B = (0.7 \pm 0.2)\ \mathrm{GeV}^{-1}$, $C = (5.5 \pm 0.2)\ \mathrm{GeV}^{-1}$, and $D = 0.0 \pm 0.3$. The relative errors in the parameters C and B show that the oscillation period is better defined than the damping coefficient e^{-Bp}. Two and a half oscillations are clearly visible over the reaction threshold, while for $p > 2.8$ GeV the vertical error bars overcome the oscillation amplitude $A\ e^{-Bp}$.

The parameter D defines the position of the first oscillation maximum that occurs at $p = 0$ within the error $\Delta D\ P/(2\pi)$, where P is the oscillation period. Estimating the oscillation period $P = 1.13$ GeV, the first oscillation maximum occurs at $p = 0$ within an error of 0.05 GeV.

Since p is a variable that is uniquely associated with the relative motion of the hadron, periodicity is associated to interactions (rescattering) between the forming hadrons after the virtual photon has been converted into quarks and antiquarks.

Let $\vec{r}$ be the space variable that is Fourier-conjugated to $\vec{p}$: $r = |\vec{r}|$ is the distance between the centers of mass of the two forming hadrons, in the frame where one is at rest. The observed behavior has been modeled via a two-stage process [22] where: 1) a $\bar{p}p$ pair is formed at a distance r with space distribution amplitude $M_0(r)$; 2) rescattering takes place between the newly formed hadrons (p and $\bar{p}$) according to an optical potential that is function of their distance r. Rescattering was modeled *via* the

 DOI: 10.1051/epjconf/201713801003

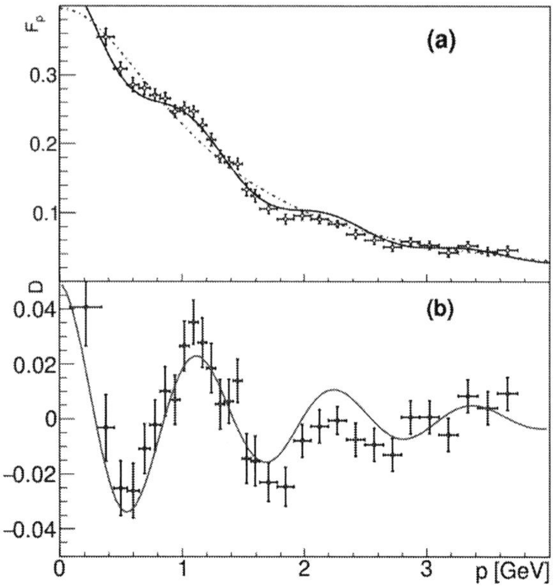

Figure 4. a) FF data from BABAR as a function of p_{Lab}, global fit: black solid line. b) after subtraction of a background function (blue dotted line), damped regular fit function (red solid line).

Distorted Wave Impulse Approximation (DWIA) formalism, following the scheme employed in [23]. The starting point is the Fourier transform

$$F_0(p) = \int d^3\vec{r} e^{i\vec{p}\cdot\vec{r}} M_0(r), \tag{11}$$

where we interpret $e^{i\vec{p}\cdot\vec{r}}$ as the plane wave final state of the $\bar{p}p$ pair in their center of mass, and $M_0(r)$ as a matrix element describing the earlier stage of the process. Plane Wave Impulse Approximation (PWIA) corresponds to the absence of rescattering: $\psi_f(\vec{r}) = e^{i\vec{p}\cdot\vec{r}}$. In the distorted DWIA formalism a simple factorized distortion $D(\vec{r})$ is introduced:

$$\psi_f(\vec{r}) = D(\vec{r}) e^{i\vec{p}\cdot\vec{r}}, \tag{12}$$

where $D(\vec{r})$ is calculated as a Glauber-like eikonal factor:

$$D(x, y, z) = e^{-ib\int_z^\infty \rho(x,y,z')dz'}, \tag{13}$$

where $\hat{z}$ // $\vec{p}$ and b is a complex number. When b is real, the rescattering potential is elastic, and it is attractive or repulsive (depending on the relative sign of $Re(b)$ and p_z), when b is pure imaginary, the potential generates flux absorption or flux creation. The origin of an optical potential with an imaginary part is related, in general, to the fact that a multi-channel process is inclusively projected onto one channel alone.

After testing several configurations, our best results have been obtained with double-layer imaginary potentials, presenting two r-ranges where the potential phase is opposite, *i.e.*, they present an inner region where the $\bar{p}p$ flux is produced and an outer region where the $\bar{p}p$ flux is absorbed, as illustrated in figure 5. All of them are purely imaginary ($Re(b) = 0$). The former two potentials have been calibrated to reproduce, as well as possible, BABAR data.

These potentials have been able to produce periodic oscillations with a period of 1 GeV or shorter, and of arbitrarily large amplitudes, depending on the parameters [24]. To reproduce LEAR elastic and annihilation data, phenomena taking place at small r have no relevance, since the surface interaction

 DOI: 10.1051/epjconf/201713801003

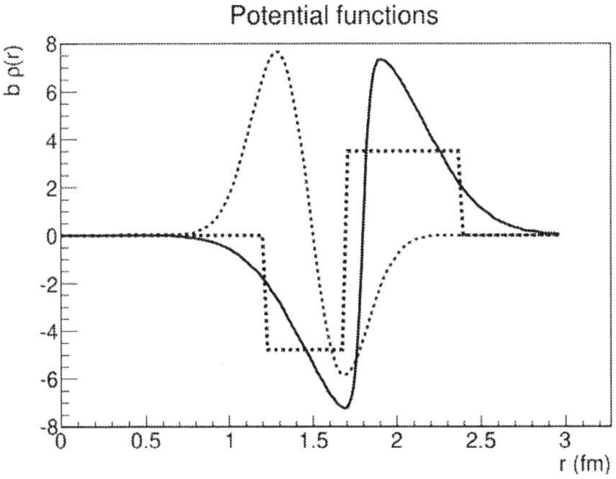

Figure 5. Three double-layer potentials used in (13) to calculate the final state distortion factor $D(x, y, z)$. They may reproduce the FFs data (background and oscillations), with an opportune choice of the parameters.

at $r \approx$ 1-2 fm prevents most of the initial $\bar{p}p$ channel wave function from entering the $r < 1$ fm region. The "regeneration" effect due to the inner potential introduced above would have little effect on the total elastic and inelastic cross sections, since it affects only a very small component of the wave function.

On the other side, this small component which is nonzero at small r is essential for the coupling of the proton-antiproton pair with a virtual photon with virtuality $q^2 > 4M^2$. The coupled regeneration/absorption mechanism produces, for $r < 2$ fm, an alternance of regions where this component of the $\bar{p}p$ wave function is enhanced or suppressed.

Specific conditions are necessary to reproduce the observed effect. An imaginary optical potential that is mainly flux-generating in a region of small distances between the centers of the forming (and still overlapping) proton and antiproton, and mainly flux-absorbing at larger distances, produces systematic oscillations of the effective proton TL FF, consistent with the observed ones. At distances $\approx$ 1-2 fm this potential behaves as the optical potentials ordinarily used to reproduce $\bar{p}p$ annihilation data: it damps the $\bar{p}p$ flux by annihilating $\bar{p}$ and p into multi-meson states. A possible explanation for the regeneration features of the potential at smaller distances could be in terms of coupling between the $\bar{p}p$ final channel and large-mass virtual states (like baryon-antibaryon) temporarily produced by the virtual photon. In order to reproduce the data, the transition from the flux-generating to the flux-absorbing region must be sudden.

4 Conclusion

Form factors data in the space and time-like regions have been critically revised and reanalyzed. In the space-like region, the conclusion is that no real discrepancy exists between the FF extraction from polarized and unpolarized experiments, once all corrections (radiative, normalizations, correlations) are taken carefully into account. There is no need to claim for a large contribution of two-photon exchange. The presence of such contribution (beyond the expected $\alpha - counting$ size) would make the formalism of ep scattering experiments very involved, and invalidate most of the information of the nucleon structure extracted so far.

In the time-like region, precise data open the possibility to access detailed features of the inner structure of the proton, and to shed light on the non-perturbative and pertubative aspects of the creation of matter from the quantum vacuum created by the annihilation process.

The perspectives of a global phenomenological frame and of new, precise data to be collected in next future open the way to access the details of the hadron formation and of the inner composition of the hadron.

References

[1] E. Kuraev, E. Tomasi-Gustafsson, A. Dbeyssi, Phys. Lett. B **712**, 240 (2012)
[2] A. Akhiezer, M. Rekalo, Sov.Phys.Dokl. **13**, 572 (1968)
[3] A. Akhiezer, M. Rekalo, Sov. J. Part. Nucl. **4**, 277 (1974)
[4] A. Bianconi, E. Tomasi-Gustafsson, Phys. Rev. Lett. **114**, 232301 (2015)
[5] A. Bianconi, E. Tomasi-Gustafsson, Phys. Rev. C **93**, 035201 (2016)
[6] V. Matveev, R. Muradian, A. Tavkhelidze, Lett. Nuovo Cim. **7**, 719 (1973)
[7] S.J. Brodsky, G.R. Farrar, Phys. Rev. Lett. **31**, 1153 (1973)
[8] R.E. Taylor, *Nucleon Form Factors above 6-GeV*, in Proc. of the Int. Symp. on Electron and Photon Interactions at High Energies, SLAC, 1967 (Stanford Univ., SLAC, 1967) 78-101
[9] M. Rosenbluth, Phys. Rev. **79**, 615 (1950)
[10] A. Puckett, E. Brash, O. Gayou, M. Jones, L. Pentchev et al., Phys. Rev. C **85**, 045203 (2012)
[11] S. Pacetti, E. Tomasi-Gustafsson, to appear in Phys. Rev. (2016). 1604.02421
[12] E. Tomasi-Gustafsson, Phys. Part. Nucl. Lett. **4**, 281 (2007)
[13] J. Arrington, Phys. Rev. C **68**, 034325 (2003)
[14] A.V. Gramolin, D.M. Nikolenko, Phys. Rev. C **93**, 055201 (2016)
[15] L. Andivahis et al., Phys. Rev. D **50**, 5491 (1994)
[16] R. Walker et al., Phys. Rev. D **49**, 5671 (1994)
[17] J. Litt et al., Physics Letters B **31**, 40 (1970)
[18] I. Qattan et al., Phys. Rev. Lett. **94**, 142301 (2005)
[19] J. Lees et al. (BaBar Collaboration), Phys. Rev. D **87**, 092005 (2013)
[20] J. Lees et al. (BaBar), Phys. Rev. D **88**, 072009 (2013)
[21] E. Tomasi-Gustafsson, M. Rekalo, Phys. Lett. B **504**, 291 (2001)
[22] A. Bianconi, E. Tomasi-Gustafsson, Phys. Rev. C **93**, 035201 (2016)
[23] A. Bianconi, M. Radici, Phys. Rev. C **54**, 3117 (1996)
[24] A. Bianconi et al., Phys. Lett. B **483**, 353 (2000)

Recent selected theory developments for NICA

David Blaschke[1,2,3,⋆]

[1] *Joint Institute for Nuclear Research, Dubna, Russia*
[2] *University of Wroclaw, Wroclaw, Poland*
[3] *National Research Nuclear University (MEPhI), Moscow, Russia*

Abstract. In this contribution I present a few selected topics of recent theoretical developments of relevance for the NICA facility under construction. In a first part, I discuss new aspects of the QCD phase diagram like the possible existence of a critical endpoint of first-order phase transitions from the perspective of generalizations of the 3-flavor PNJL model including the conjecture of a universal pressure for the onset of deconfinement in heavy-ion collisions and astrophysics. A second part is devoted to first results of the newly constructed event simulator (THESEUS) which is based on the particlization of the former three-fluid hydrodynamics code by Ivanov, Russkikh and Toneev that allows also to study the role of hadronic final state interactions. Possible signals of the mixed phase accessible at NICA are considered. In particular, the robustness of the baryon-stopping signal of deconfinement and the occurrence of antiflow for protons have been investigated for Au+Au collisions in the range of the NICA-MPD energy scan for $\sqrt{s} \sim 6 \ldots 8$ GeV. This signal is reflected also in the flow pattern of light nuclear clusters, in particular deuterons. The sharp peak for the K^+/π^+ ratio at $\sqrt{s} \sim 8$ GeV (the "horn" effect) is not obtained in the present version of THESEUS and calls for improvement of the equation of state input. I report the recent progress in developing a generalized Beth-Uhlenbeck approach to a unified description of quark-hadron matter which includes now strangeness and reveals a new mechanism for explaining the K^+/π^+ ratio due to the pronounced occurrence of an anomalous mode in the K^+ at finite baryochemical potentials.

1 Introduction

The White Paper on "Exploring Strongly Interacting Matter at High Densities" at the Nuclotron-based Ion Collider fAcility (NICA) has recently appeared as a Topical Issue in the European Physical Journal A [1]. It comprises 56 articles related to the most prospective physics aspects to be explored at this upcoming facility. I shall present a few selected topics of recent theoretical developments for NICA in which I have been involved.

First I consider key aspects of the QCD phase diagram like the possible existence of a critical endpoint (CEP) of first-order phase transitions that has been elaborated in [2] for nonlocal Polyakov-loop extended Nambu–Jona-Lasinio (PNJL) models constrained by lattice QCD with special emphasis on the role of a repulsive vector meanfield. The CEP is also discussed from the perspective of generalizations of the color superconducting 3-flavor PNJL model [3] including the conjecture of a universal

⋆e-mail: david.blaschke@ift.uni.wroc.pl

DOI: 10.1051/epjconf/201713801004

pressure for the onset of deconfinement in heavy-ion collisions and astrophysics [4]. In this context I find it worthwhile to quote results of [5] where consequences for the QCD phase diagram have been demonstrated of enforcing a coincidence between chiral symmetry restoration and deconfinement with a correspondingly adjusted bag pressure on top of a model with dynamical chiral symmetry restoration. These considerations of the QCD phase diagram in different dynamical models are important for the preparation of the NICA experiments as they can give theoretical estimates on the accessibility of certain QCD phases and what effects one may hope to detect in the given energy range.

Sect. 3 is devoted to the new event simulation program THESEUS. Within a collaboration involving a team of 9 scientists from Germany, France, Italy, Poland, Russia, Ukraine and the United States we have succeeded to create a startup version for a new **T**hree-fluid **H**ydrodynamics-based **E**vent **S**imulator **E**xtended by **U**rQMD final **S**tate interactions (THESEUS) and reported about its performance in [7]. In particular, the robustness of the baryon-stopping signal of deconfinement [8, 9] and the occurrence of antiflow for protons for Au+Au collisions in the range of the NICA-MPD energy scan for $\sqrt{s} \sim 6 \ldots 8$ GeV [7] have been reconsidered in THESEUS. For the first time it was possible to investigate the effect of hadronic rescattering on the results of the three-fluid hydrodynamics simulations. To this end the ultrarelativistic quantum molecular dynamics (UrQMD) program has been implemented. The antiflow signal for a first order phase transition is reflected also in the flow pattern of light nuclear clusters, in particular for deuterons [10]. THESEUS and its underlying three-fluid hydrodynamics model shall be developed further, in particular to be able to improve the description of the hadronization process and capture subtle effects like the "horn" effect for the K^+/π^+ ratio. A main goal of ongoing studies is to improve the equation of state (EoS) input. In this respect recent progress in developing a generalized Beth-Uhlenbeck (GBU) approach [11–13] to a unified description of quark-hadron matter [14] which includes now strangeness [15, 16] (see also [17]) has shown remarkable results. It has revealed a new mechanism for explaining the K^+/π^+ ratio due to the pronounced occurrence of an anomalous mode in the K^+ channel at finite baryochemical potentials [16]. The GBU approach on the basis of a PNJL model utilizes a generalization of the $\Phi-$ derivable approach that accounts for a spectrum of hadron resonances as bound states of quarks which get dissolved into the quark-gluon plasma continuum by the Mott effect encoded in medium-dependent hadronic phase shifts. It can describe the lattice QCD thermodynamics results on the temperature axis of the QCD phase diagram [14] and should therefore next be extended to finite baryochemical potentials.

2 The phase diagram as probed at NICA

This section selects a few aspects out of the rich thematics of the QCD phase diagram that have recently been discussed in the context of the upcoming NICA experiments. For a general introduction to the QCD phase diagram one may consult the review by Fukushima and Hatsuda [18] and the more recent one by Fukushima and Sasaki [19].

Interesting aspects that have been spared out here concern inhomogeneous condensates [20] for which instead of a CEP a Lifshitz point occurs in the phase diagram. Another point of interest included in the NICA White Paper is the behavior of scalar mesons in the phase diagram [21]. Their properties may trace the chiral restoration transition and provide a potentially observable signal in the two-photon channel [22] if the background can be sufficiently suppressed. Promising progress has also been made by the Functional Renormalization Group approach that provides spectral functions for the quark-meson model phase diagram [23]. In near future one can expect interesting predictions also for the part of the phase diagram probed at NICA providing the approach can be generalized to include baryons.

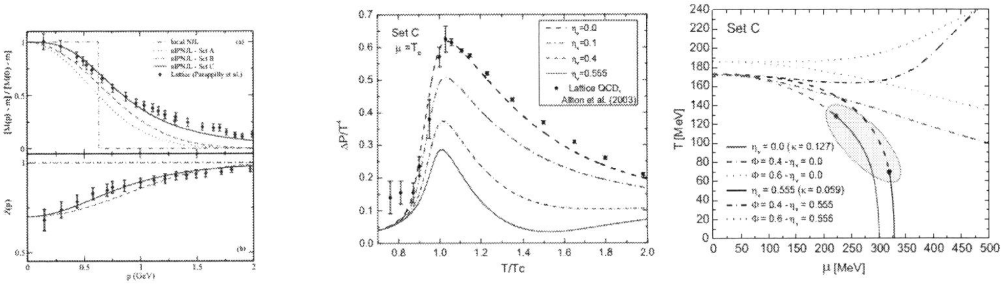

Figure 1. Left panel: Normalized dynamical masses for the different form factors under study (NJL, Set A - C) and wave function renormalization for Set B and Set C, fitted to lattice QCD data [28]. Middle panel: Comparison between Set C results for different values of the vector coupling parameter η_V and lattice QCD data [32]. Right panel: Phase diagrams with (pseudo)critical temperatures $T_c(\mu)$ and critical points for nonlocal rank-2 PNJL model (Set C). Dashed (full) lines correspond to crossover (first order) transitions. The corresponding dash-dotted and dotted lines represent the deconfinement transition range, i.e. $\Phi = 0.4$ and $\Phi = 0.6$, respectively. The highlighted region denotes the CEP position favored by the present study. The figures are from [2].

2.1 CEP in a nonlocal PNJL model constrained by lattice QCD

Dynamical models for quark hadron matter that have the potential to provide more insights to the microphysics of the hadron-to-quark matter phase transition are based on the hierarchy of the QCD Dyson Schwinger equations where hadrons appear as bound states of quarks due to their interactions via confining forces mediated by nonperturbative gluon propagators [24–26]. The key quantities of this approach are the dynamical mass function and the wave function renormalization of the quark propagator that are described in excellent agreement with lattice QCD data [27, 28].

A similar quality of the description of the quark propagator is achieved within the nonlocal chiral quark models that are defined by a covariant form factor which is calibrated with the same lattice QCD data, see the left panel of figure 1. For their finite temperature extension, the coupling to the Polyakov loop results in predictions of the pseudocritical temperature that are in excellent agreement with the lattice QCD data [29, 30]. The extension of these models to the whole $T - \mu$ plane has been accomplished in [31] with particular emphasis on the role that interactions in the Dirac vector channel play for the position and the very existence of the CEP. This discussion has been extended in [2] where the possibility of a very small or vanishing vector coupling at low chemical potentials as obtained on the basis of lattice QCD results for the imaginary chemical potential technique has been suggested as a possible solution of the problem to describe the lattice data for the temperature dependence of the pressure difference $\Delta P = P(T,\mu) - P(T,0)$ at finite $\mu = T_c$ [32], see the middle panel of figure 1. The favorable model is Set C which in this case predicts a critical endpoint at $T_{\rm CEP} \sim 130$ MeV and $\mu_{\rm CEP} = 225$ MeV. This corresponds to chemical freeze-out parameters for $\sqrt{s_{NN}} \sim 6$ GeV, i.e. below the range of the RHIC beam energy scan of the STAR experiment, but well in reach for the planned NICA energy scan.

2.2 Universal transition pressure in a 3-flavor color superconducting PNJL model

For the first time, the phase diagram of a 3-flavor color superconducting PNJL model has been obtained in [3], the results are shown in figure 2. The entropy per baryon isolines indicate paths for the dynamical evolution of a fireball at NICA/FAIR energies where upmost line ($s/n = 11$) corresponds to $\sqrt{s_{NN}} \sim 4.5$ GeV ($E_{\rm lab} \sim 9$ A GeV) [33]. The yellow region ($n_B < 0$) denotes an instability of

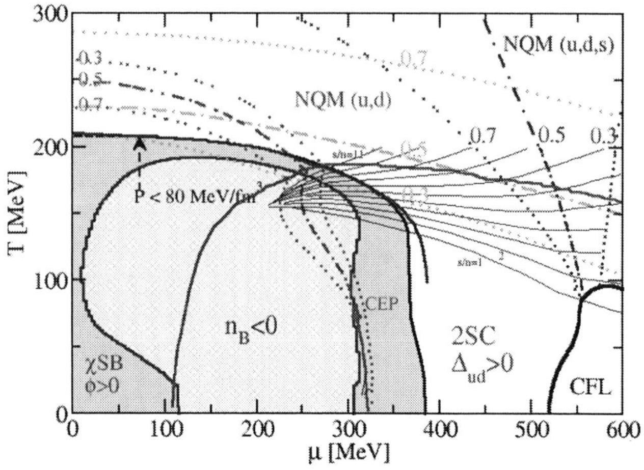

Figure 2. Phase diagram of the three-flavor color superconducting PNJL model for symmetric quark matter from [3]. For details see text in Subsect. 2.2.

the homogeneous meanfield approximation to quark matter which is cured when a phase transition to hadronic matter is invoked with a universal transition pressure of 80 MeV/fm^3 [34] delimiting the grey region. In the range of temperatures accessible in the energy scan programs of heavy-ion collisions for $\sqrt{s_{NN}} = 4 - 200$ GeV this coincides with the phase equilibrium between the PNJL model and the hadron resonance gas (solid blue line). The solid red line corresponds to the critical temperature for 2SC superconductivity in quark matter. The green (blue) lines correspond to equal light (strange) quark masses reduced relative to the vacuum value by a factor labelling these lines. Along the orange lines the traced Polyakov loop has the value that labels them.

2.3 Enforcing coincident chiral and deconfinement transitions in the phase diagram

A challenging question is whether chiral restoration and quark deconfinement transitions coincide in the QCD phase diagram or not. In lattice QCD simulations at low chemical potentials the peak positions of chiral and Polyakov-loop susceptibilities are to good accuracy coincident, but this may not be the case for high baryon densities, allowing for three alternatives: a massive quark matter phase [35, 36], a quarkyonic phase [37, 38] with chiral quarks confined in baryons (and mesons) and the case that chiral restoration and deconfinement coincide in the whole phase diagram. The latter case is based on the QCD Dyson-Schwinger equations where the Kugo-Ojima confinement mechanism is realized by the dynamical quark mass function which serves for chiral symmetry breaking and at the same time conspires with the wave function renormalization so as to remove the quasiparticle pole of the quark propagator [39, 40]. Dexheimer and Schramm [41] have constructed a model that realizes coincident chiral and deconfinement phase transition via the coupling of quark and hadron masses to the Polyakov loop. The position of the critical endpoint and the critical chemical potential at $T = 0$ is fixed by the behavior of the Polyakov-loop potential which they extend by adopting a chemical potential dependence.

In their contribution to the NICA White Paper, Fischer, Klähn and Hempel [5] realize the coincidence of chiral restoration and deconfinement in the QCD phase diagram by adopting a bag pressure B_{dc}, see also [6]. The hadronic phase is described by a generalized relativistic density functional model with light clusters [42] and for the phase transition a Maxwell construction is applied. In figure 3 the

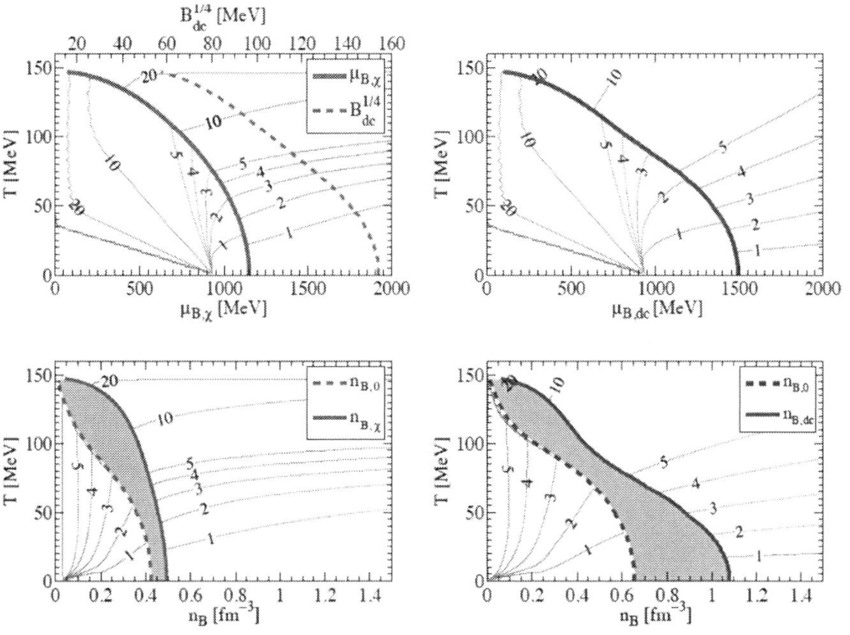

(a) vBag (b) 'standard' NJL approach

Figure 3. Phase diagram for a selected chiral bag constant $B_\chi^{1/4}$ = 152.7 MeV, comparing the new approach [6, 43] (left panels) with the "standard" NJL one (B_{dc} = 0, right panels). Top panels: Dependence on baryochemical potential. Bottom panels: Dependence on the baryon density (see text for definitions). Solid gray lines mark lines of constant entropy per baryon in units of the Boltzmann constant k_B. Typical trajectories of heavy-ion collisions in the NICA MPD experiment will lie between the lines labelled by "10" and "20"; for explanations see the text. Figure from [5].

phase diagram for this model (left panels) is contrasted to the case where $B_{dc} = 0$ (right panels). It is remarkable for the upcoming NICA experiments that the upper limiting density for the mixed phase is only about twice the nuclear saturation density for $\sqrt{s_{NN}} \sim 6$ GeV where $T \sim 130$ MeV when entering the mixed phase and after completion of the transition the matter is heated up to about 150 MeV. For an orientation in that figure, consider the range of entropy per baryon $s/n = 10 - 25$ [33] that will be covered by the NICA MPD collider experiment, where $\sqrt{s_{NN}} = 4 \ldots 11$ GeV [44]. This EoS fulfills also the constraint from recent precise mass measurements of pulsars which require the maximum mass of a neutron star or hybrid star predicted for the EoS to reach at least 2 $M_\odot$ [43].

2.4 High-mass twin stars support the CEP

Starting from the possible existence of a third family of compact hybrid stars at high mass as indicators of a strong first order phase transition in compact star interiors [45] (high mass twin stars [46]) from which the existence of a critical point in the QCD phase diagram follows, in the contribution [4] to the NICA White Paper the relation between the pressure at the onset of the deconfinement transition for stars with $M \sim 2\ M_\odot$ and their radius has been shown for the first time, see figure 4. If the maximum star radius from observations turns out to be less than 14 km, e.g. in the range 13.6 - 14.0 km this would indicate a limiting pressure of the hadronic phase of matter of 74 - 90 MeV/fm^3, in excellent agreement with the analysis of chemical freeze-out in heavy-ion collisions by Rafelski & Petran [34]

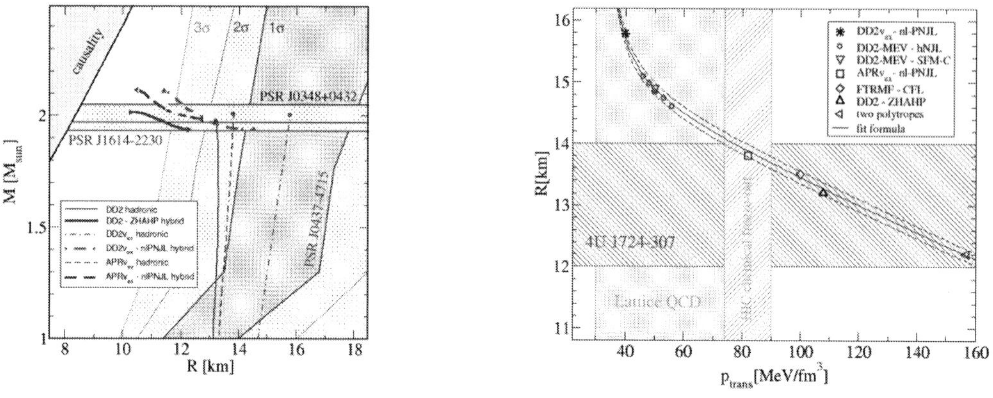

Figure 4. Left panel: Mass-radius relationships for three examples of a new class of hybrid neutron star equations of state with a strong first order phase transition exhibiting an endpoint of purely hadronic configurations at about 2 $M_\odot$ and a stable branch of compact hybrid stars, disconnected from the hadronic ones ("third family"). Right panel: The radius R at the endpoint of the hadronic configurations ($\sim$ 2 $M_\odot$) for different classes of hybrid equations of state versus the corresponding pressure at the phase transition $p_{\rm trans}$. All these equations of state allow for a third family of stars at high mass ("high-mass twin" stars). Figures from [4].

which gave 82 ± 8 MeV/fm^3. Such a universality of the limiting pressure of the hadronic phase in the QCD phase diagram, if it exists, would be of enormous heuristic value for interdisciplinary studies of the QCD phase diagram in heavy-ion collisions and in Astrophysics.

3 Three-fluid hydrodynamics based simulations and mixed phase signals

3.1 Baryon stopping signal

In figure 5 we show the NICA energy scan for the curvature Cy of the net proton rapidity distribution at midrapidity for central Au+Au collisions with impact parameter b = 2 fm (left panels), b = 6 fm (middle panels) and b = 11 fm (right panels). We compare the 3FH model result (black solid lines) with THESEUS (blue short-dashed lines) and THESEUS without UrQMD (red long-dashed lines). The results for the two-phase EoS (upper row of panels) are compared to those for the crossover EoS (lower row of panels). For noncentral collisions the curvature pattern is shifted towards positive values while the "wiggle" as a characteristic feature for the EoS with a first order phase transition [47, 48] remains rather robust [8, 9]. For details, see [7].

3.2 Antiflow of light clusters

Among the suggested signatures of a first order phase transition in heavy-ion collision experiments is the antiflow of nucleons, i.e., a negative slope for the rapidity dependence of the directed flow v_1 at midrapidity [49]. It has been shown in hydrodynamic calculations already in [50] that this feature is related to the softest point of the equation of state. Naturally, this sensitivity to the EoS could be nicely demonstrated within the three-fluid hydrodynamics approach [51, 52]. The proton antiflow feature has been obtained also in different transport theory based simulations that implement a first order phase transition, for instance the hybrid approach [53] which uses a hydrodynamic description

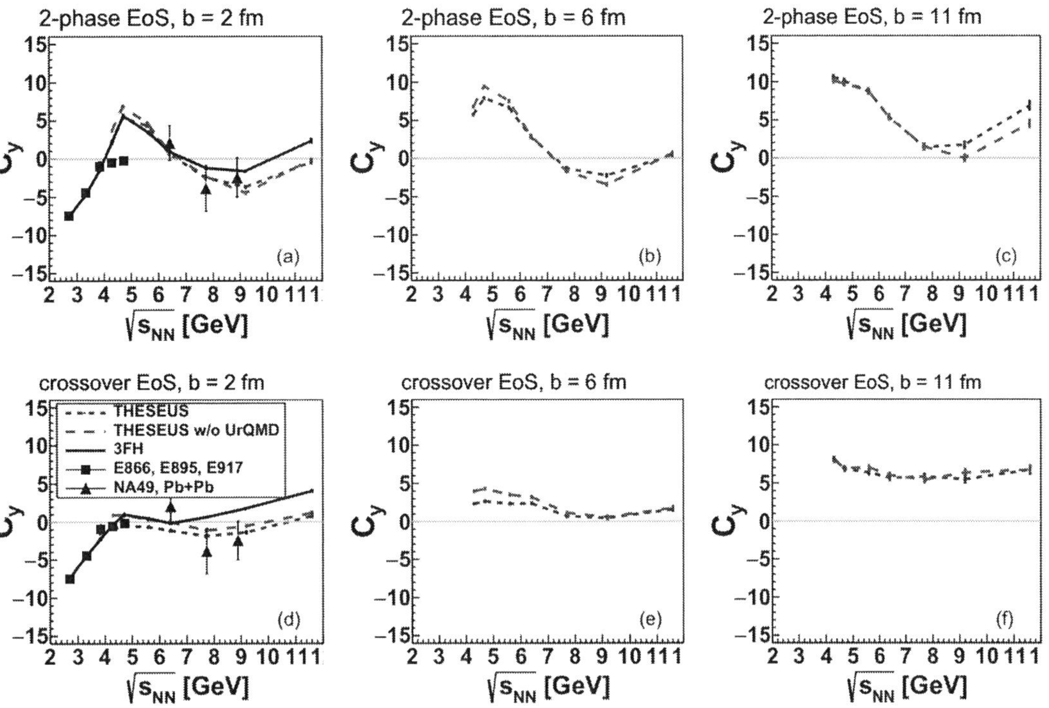

Figure 5. Energy scan for the curvature Cy of the net proton rapidity distribution at midrapidity for central Au+Au collisions with impact parameter b = 2 fm (panels (a) and (d)), b = 6 fm (panels (b) and (e)) and b = 11 fm (panels (c) and (f)). We compare the 3FH model result (black solid lines) with THESEUS (blue short-dashed lines) and THESEUS without UrQMD (red long-dashed lines). The results for the two-phase EoS (panels (a)-(c)) are compared to those for the crossover EoS (panels (d)-(f)). For noncentral collisions the curvature pattern is shifted towards positive values while the "wiggle" as a characteristic feature for the EoS with a first order phase transition remains rather robust. Figure from [7].

of the phase transition sandwiched by UrQMD modeling of initial and final stages, and the transport theoretical model JAM [54, 55].

In [10] it has been conjectured that the flow of light clusters such as deuterons may trace the early flow at the hadronization transition since by their inertia the heavier particles are less affected by hadronic rescattering effects. This has been checked with the THESEUS code and indeed, the antiflow of protons obtained in this program for the 2-phase EoS as a signature of the first order phase transition in this EoS in the energy range $\sqrt{s_{NN}} \sim 6 - 7$ GeV is reflected also in the deuteron flow for the same EoS while the crossover and pure hadronic EoS do not show the antiflow signature, see figure 6.

As for the proton antiflow, it would be premature to predict quantitatively the deuteron antiflow due to present ambiguities in the deuteron formation mechanism by coalescence and in the used EoS. In particular, the 2-phase EoS is too soft at high densities and has an unrealistically high onset density of the phase transition. With an improved EoS, the antiflow signature of the phase transition could occur in the beam energy range 4.7 GeV $< \sqrt{s_{NN}} <$ 7.7 GeV, where no experimental data exist yet but where the NICA MPD experiment will be operating [44].

 DOI: 10.1051/epjconf/201713801004

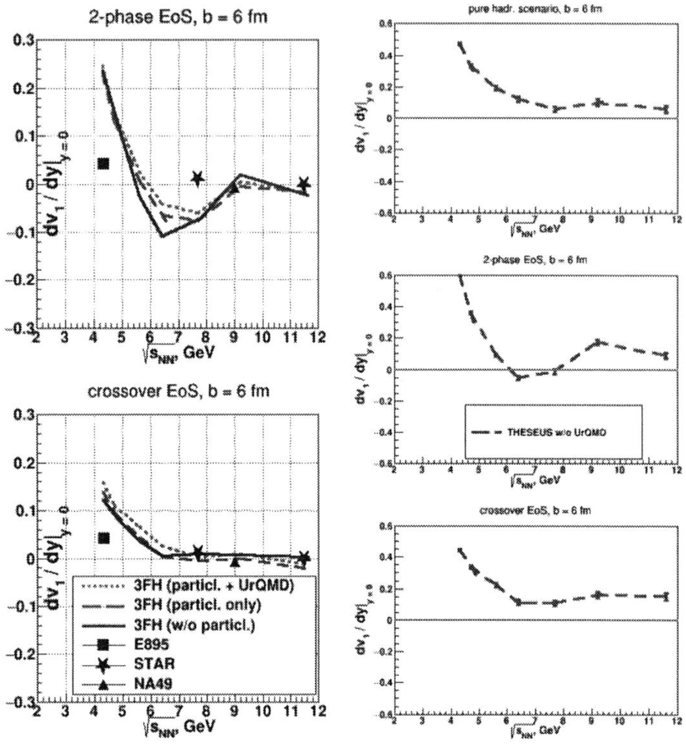

Figure 6. Energy scan of the slope of the directed flow for protons (left panel) and for deuterons (right panel). A comparison of results from THESEUS simulations with a 2-phase EoS that possesses a strong first-order phase transition to results for a crossover (or purely hadronic) EoS demonstrates that a dip at energies of $\sqrt{s_{NN}} = 6 - 8$ GeV could be regarded as a phase transition signal. Such "antiflow" signal is expected for protons as well as light clusters. For further details, see [10]; the figures are taken from that paper.

3.3 Marek's horn

In previous chapters we have praised the advantages of the new THESEUS simulation program for heavy-ion collisions in the NICA/FAIR energy range, where the transition from the baryon stopping to the transparency regime is expected. Promising signatures of a first order phase transition have been discussed. However, the "horn" effect which is observed in the energy scan of the K^+/π^+ ratio (for the data see, e.g., figure 11 of [56], partly displayed here in the left and middle panels of figure 7) is not reproduced in the present version of THESEUS and with the present EoS. This effect has been suggested by Marek Gazdzicki and Mark Gorenstein as a signature of the mixed phase in hadron-to-quark matter phase transition [57]. Its quantitative description in simulations of heavy-ion collisions is a notoriously difficult problem. Recently, in an upgrade of the PHSD program with chiral symmetry restoration effects, a satisfactory description of the "horn" effect could be given [58]. We advocate another possibility to solve the "horn" puzzle. In a recent work on the Mott dissociation of pions and kaons in hot, dense quark matter [16] which extends the GBU approach [11–13] to the strangeness sector, we have found that a stable mode of the K^+ meson occurs in the phase diagram at finite T and μ due to the compositeness and unequal masses of the quark constituents. This in-medium bound state is a good candidate for causing the enhancement of the K^+/π^+ ratio in a certain domain of the energy scan while the K^-/π^- ratio remains unaffected, see the right panel of figure 7. We plan to implement the corresponding EoS into an upgrade of the THESEUS program.

4 Conclusions

In this contribution I have discussed a few selected theoretical developments of relevance for the NICA facility under construction. First, I have picked new aspects of the QCD phase diagram like

DOI: 10.1051/epjconf/201713801004

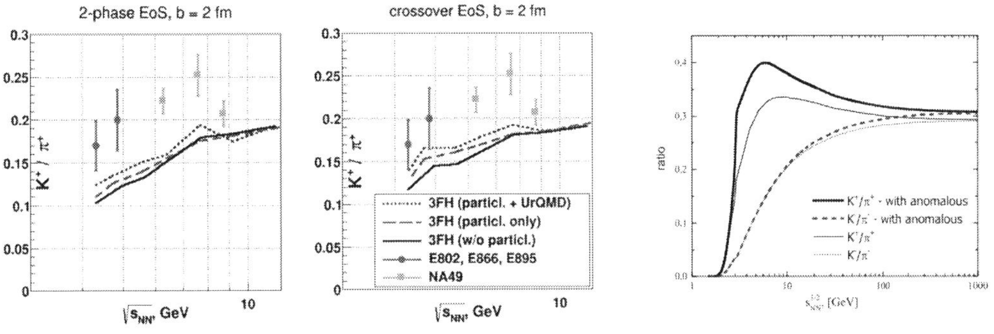

Figure 7. Energy scan for the particle ratio K^+/π^+ in the NICA energy range for central Au+Au collisions (impact parameter b = 2 fm) with (blue dotted lines) and without (red dashed lines) the UrQMD hadronic rescattering. The calculation with a first order phase transition in the EoS (left panel) is compared to that with the crossover EoS (middle panel). For comparison we show in both these panels the results without particlization and UrQMD rescattering (black solid lines) and experimental data, taken from figure 11 of [56]. Data from AGS experiments (E802, E866, E895) are shown by filled circles, data from NA49 by filled squares. In the right panel we demonstrate the effect of the anomalous modes on the K^+/π^+ and K^-/π^- ratios (thick lines), compared to the "normal" case without these modes (thin lines), from [16]. Upon implementation of this effect into the EoS used in the THESEUS simulation a decent improvement of the description of the "horn" effect will be achieved.

the possible existence of a critical endpoint of first-order phase transitions from the perspective of a nonlocal chiral quark model calibrated with lattice QCD simulations that strongly suggests a low or vanishing vector coupling at low chemical potentials; generalizations of the 3-flavor PNJL model including the conjecture of a universal pressure for the onset of deconfinement in heavy-ion collisions and astrophysics; and a new model that enforces the coincidence of chiral and deconfinement transitions predicting a low critical density, accessible already at NICA fixed target experiments. Next, I have discussed first results of the newly constructed event simulator THESEUS which is based on the particlization of the three-fluid hydrodynamics model by Ivanov, Russkikh and Toneev that allows also to study the role of hadronic final state interactions. Possible signals of the mixed phase accessible at NICA are considered. In particular, the robustness of the baryon-stopping signal of deconfinement and the occurrence of antiflow for protons and deuterons have been investigated for Au+Au collisions in the range of the NICA-MPD energy scan. The sharp peak for the K^+/π^+ ratio at $\sqrt{s} \sim 8$ GeV (the "horn" effect) is not obtained in the present version of THESEUS and calls for improvement of the equation of state input. In this context, I have pointed out recent progress in developing a generalized Beth-Uhlenbeck approach to a unified description of quark-hadron matter that reveals a new mechanism for explaining the K^+/π^+ ratio. One may look forward to further interesting developments of theory and simulations of heavy-ion collisions in the NICA energy range directed towards the discovery of a mixed phase in the QCD phase diagram.

Acknowledgements

I would like to thank the members of the board of guest editors for the EPJA Topical Issue on the "NICA White Paper", Jörg Aichelin, Elena Bratkovskaya, Volker Friese, Marek Gazdzicki, Jørgen Randrup, Oleg Rogachevsky, Oleg Teryaev and Vyacheslav Toneev, for their excellent work and indispensable help in making this volume appear. I am grateful to all my collaborators in the research reported here, in particular, David Alvarez-Castillo, Alexander Ayriyan, Niels-Uwe Bastian, Pavel

Batyuk, Sanjin Benic, Jens Berdermann, Marcus Bleicher, Gustavo Contrera, Pedro Costa, Pawel Danielewicz, Aleksandr Dubinin, Tobias Fischer, Hovik Grigorian, Gabriela Grunfeld, Sophia Han, Yuri Ivanov, Iuriy Karpenko,Thomas Klähn, Sergey Merts, Marlene Nahrgang, Hannah Petersen, Andrey Radzhabov, Gerd Röpke, Oleg Rogachevsky, Ludwik Turko, Dmitri Voskresensky, Agnieszka Wergieluk and Hermann Wolter. Finally, I would like to acknowledge numerous fruitful discussions with my colleagues at JINR Dubna, especially: Valery Burov, Alexandra Friesen, Yuri Kalinovsky, Sergey Nedelko, Alexander Sorin, Oleg Teryaev, Slava Toneev and at the University of Wroclaw: Pasi Huovinen, Jakub Jankowski, Alaksiej Kachanovich, Pok Man Lo, Krzysztof Redlich, Chihiro Sasaki and Jan Sobczyk. This work has been supported by the MEPhI Academic Excellence Project under contract No. 02.a03.21.0005 and by the Polish NCN under grant No. UMO-2011/02/A/ST2/00306.

References

[1] D. Blaschke, J. Aichelin, E. Bratkovskaya, V. Friese, M. Gazdzicki, J. Randrup, O. Rogachevsky, O. Teryaev and V. Toneev, Eur. Phys. J. A **52(8)**, 267 (2016)
[2] G. A. Contrera, A. G. Grunfeld and D. Blaschke, Eur. Phys. J. A **52(8)**, 231 (2016)
[3] A. Ayriyan, J. Berdermann, D. Blaschke and R. Lastowiecki, arXiv:1608.07875 [hep-ph]
[4] D. Alvarez-Castillo, S. Benic, D. Blaschke, S. Han and S. Typel, Eur. Phys. J. A **52(8)**, 232 (2016)
[5] T. Fischer, T. Klähn and M. Hempel, Eur. Phys. J. A **52(8)**, 225 (2016)
[6] T. Klähn, T. Fischer and M. Hempel, arXiv:1603.03679 [nucl-th]
[7] P. Batyuk et al., Phys. Rev. C **94**, 044917 (2016)
[8] Y. B. Ivanov and D. Blaschke, Phys. Rev. C **92(2)**, 024916 (2015)
[9] Y. B. Ivanov and D. Blaschke, Eur. Phys. J. A **52(8)**, 237 (2016)
[10] N.-U. Bastian et al., Eur. Phys. J. A **52(8)**, 244 (2016)
[11] M. Schmidt, G. Röpke and H. Schulz, Annals Phys. **202**, 57 (1994)
[12] J. Hüfner, S. P. Klevansky, P. Zhuang and H. Voss, Annals Phys. **234**, 225 (1994)
[13] D. Blaschke, M. Buballa, A. Dubinin, G. Röpke and D. Zablocki, Annals Phys. **348**, 228 (2014)
[14] D. Blaschke, A. Dubinin and L. Turko, arXiv:1611.09845 [hep-ph]
[15] K. Yamazaki and T. Matsui, Nucl. Phys. A **922**, 237 (2014)
[16] A. Dubinin, A. Radzhabov, D. Blaschke and A. Wergieluk, arXiv:1608.05383 [hep-ph]
[17] J. M. Torres-Rincon and J. Aichelin, arXiv:1601.01706 [nucl-th]
[18] K. Fukushima and T. Hatsuda, Rept. Prog. Phys. **74**, 014001 (2011)
[19] K. Fukushima and C. Sasaki, Prog. Part. Nucl. Phys. **72**, 99 (2013)
[20] M. Buballa and S. Carignano, Prog. Part. Nucl. Phys. **81**, 39 (2015)
[21] P. Costa, Eur. Phys. J. A **52(8)**, 228 (2016)
[22] M. K. Volkov, E. A. Kuraev, D. Blaschke, G. Röpke and S. M. Schmidt, Phys. Lett. B **424**, 235 (1998)
[23] R. A. Tripolt, N. Strodthoff, L. von Smekal and J. Wambach, Phys. Rev. D **89(3)**, 034010 (2014)
[24] F. Gao, J. Chen, Y. X. Liu, S. X. Qin, C. D. Roberts and S. M. Schmidt, Phys. Rev. D **93(9)**, 094019 (2016)
[25] C. D. Roberts, J. Phys. Conf. Ser. **706(2)**, 022003 (2016)
[26] I. C. Cloet and C. D. Roberts, Prog. Part. Nucl. Phys. **77**, 1 (2014)
[27] P. O. Bowman, U. M. Heller, D. B. Leinweber, M. B. Parappilly, A. G. Williams and J. b. Zhang, Phys. Rev. D **71**, 054507 (2005)
[28] M. B. Parappilly, P. O. Bowman, U. M. Heller, D. B. Leinweber, A. G. Williams and J. B. Zhang, Phys. Rev. D **73**, 054504 (2006)

[29] D. Horvatic, D. Blaschke, D. Klabucar and O. Kaczmarek, Phys. Rev. D **84**, 016005 (2011)
[30] A. E. Radzhabov, D. Blaschke, M. Buballa and M. K. Volkov, Phys. Rev. D **83**, 116004 (2011)
[31] G. A. Contrera, A. G. Grunfeld and D. B. Blaschke, Phys. Part. Nucl. Lett. **11**, 342 (2014)
[32] C. R. Allton, S. Ejiri, S. J. Hands, O. Kaczmarek, F. Karsch, E. Laermann and C. Schmidt, Phys. Rev. D **68**, 014507 (2003)
[33] Y. B. Ivanov and A. A. Soldatov, Eur. Phys. J. A **52(12)**, 367 (2016)
[34] M. Petran and J. Rafelski, Phys. Rev. C **88**, 021901 (2013)
[35] P. Castorina, K. Redlich and H. Satz, Eur. Phys. J. C **59**, 67 (2009)
[36] H. Schulz and G. Röpke, Z. Phys. C **35**, 379 (1987)
[37] L. McLerran and R. D. Pisarski, Nucl. Phys. A **796**, 83 (2007)
[38] A. Andronic et al., Nucl. Phys. A **837**, 65 (2010)
[39] D. Blaschke, C. D. Roberts and S. M. Schmidt, Phys. Lett. B **425**, 232 (1998)
[40] C. D. Roberts and S. M. Schmidt, Prog. Part. Nucl. Phys. **45**, S1 (2000)
[41] V. A. Dexheimer and S. Schramm, Nucl. Phys. Proc. Suppl. **199**, 319 (2010)
[42] S. Typel, G. Röpke, T. Klähn, D. Blaschke and H. H. Wolter, Phys. Rev. C **81**, 015803 (2010)
[43] T. Klähn and T. Fischer, Astrophys. J. **810(2)**, 134 (2015)
[44] V. D. Kekelidze, R. Lednicky, V. A. Matveev, I. N. Meshkov, A. S. Sorin and G. V. Trubnikov, Eur. Phys. J. A **52(8)**, 211 (2016)
[45] D. Blaschke, D. E. Alvarez-Castillo and S. Benic, PoS CPOD **2013**, 063 (2013) [arXiv:1310.3803 [nucl-th]]
[46] S. Benic, D. Blaschke, D. E. Alvarez-Castillo, T. Fischer and S. Typel, Astron. Astrophys. **577**, A40 (2015)
[47] Y. B. Ivanov, Phys. Lett. B **690**, 358 (2010)
[48] Y. B. Ivanov, Phys. Rev. C **87(6)**, 064904 (2013)
[49] H. Stoecker, Nucl. Phys. A **750**, 121 (2005)
[50] J. Brachmann, S. Soff, A. Dumitru, H. Stoecker, J. A. Maruhn, W. Greiner, L. V. Bravina and D. H. Rischke, Phys. Rev. C **61**, 024909 (2000)
[51] Y. B. Ivanov and A. A. Soldatov, Phys. Rev. C **91(2)**, 024915 (2015)
[52] Y. B. Ivanov and A. A. Soldatov, Eur. Phys. J. A **52(1)**, 10 (2016)
[53] J. Steinheimer, J. Auvinen, H. Petersen, M. Bleicher and H. Stöcker, Phys. Rev. C **89(5)**, 054913 (2014)
[54] Y. Nara, H. Niemi, A. Ohnishi and H. Stöcker, Phys. Rev. C **94(3)**, 034906 (2016)
[55] Y. Nara, H. Niemi, J. Steinheimer and H. Stöcker, arXiv:1611.08023 [nucl-th]
[56] Y. B. Ivanov, Phys. Rev. C **87(6)**, 064905 (2013)
[57] M. Gazdzicki and M. I. Gorenstein, Acta Phys. Polon. B **30**, 2705 (1999)
[58] A. Palmese, W. Cassing, E. Seifert, T. Steinert, P. Moreau and E. L. Bratkovskaya, Phys. Rev. C **94(4)**, 044912 (2016)

Emission of d, t, ^{3}He and α-particles from ^{12}C nuclei excited by intermediate energy photons

A.A. Turinge[1,⋆], *A.M.* Lapik[1], *A.N.* Mushkarenkov[1], *V.G.* Nedorezov[1], and *N.V.* Rudnev[1]
for the GRAAL collaboration

[1] *Institute for Nuclear Research, 117312, Moscow, Russia*

Abstract. New GRAAL data on the fragmentation of carbon nuclei by photons at 800 - 1500 MeV are discussed. In addition to previously published results for the proton and neutron emission probabilities at different multiplicities, preliminary results on the yields of heavier fragments as d, t, ^{3}He and α-particles are presented. The obtained results are compared with theoretical predictions and other fragmentation data measured in different reactions induced by photons, electrons and relativistic ions.

1 Introduction

Motivation to study the nuclear fragmentation by photons at intermediate energies is caused by several reasons. At first, that is a way to investigate the nuclear dynamics leading to the creation of nuclear clusters in the final state of interaction. Such projectiles as photons allow to describe the process qualitatively and to test the theoretical predictions [1] with high accuracy. The positive feature of modern photonuclear experiments like the GRAAL is the possibility to use a monochromatic photon beam and large aperture detector with high enough spatial and energy resolution. To create a beam of monochromatic photons within 800 - 1500 MeV energy range, the backward Compton scattering method was used at the ESRF storage ring [2].

In this work, we continue an analysis of the GRAAL data [3] devoted to the nucleon fragmentation of ^{12}C nuclei to study nuclear fragments heavier than nucleons. New information is analyzed in terms of the dynamics of nuclear excitations and correlations of nucleons in nuclei. A comparison of present results with literature data available for different projectiles is done.

In frame of the RELDIS model the time scale for creation of nuclear products is shown in figure 1, taking from [1].

The time scale is presented in fm/c, where fm is equal to 10^{-13} cm, c is the light velocity. Time of flight for the photon through the carbon nucleus is about 10^{-23} sec, 100 fm/s correspond to $3{\cdot}10^{-22}$ s. One can see that all the fragments are created in the final state of interaction when a nucleus undergoes the transfer from the liquid to gas state.

1.1 The experimental procedure and results

The experiment was performed at the GRAIL facility [2] using a beam of tagged photons, produced by backward Compton scattering of laser photons on the electron storage ring ESRF. The detector

⋆e-mail: turinge56@mail.ru

 DOI: 10.1051/epjconf/201713801005

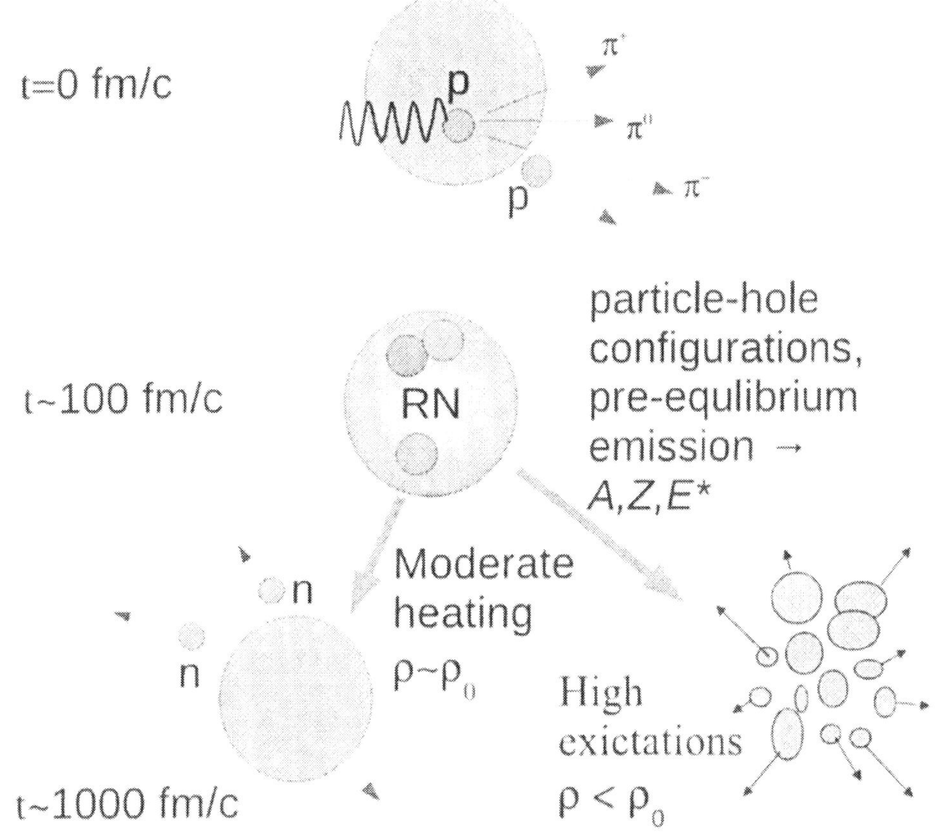

Figure 1. Dynamics of the multifragmentation process at intermediate energies [1].

LAGRANγE is described in detail in [4]. It consists of two main parts (see figure 2). The central part is organized to measure neutral and charged particles in a large solid angle close to 4π. Second part, situated at forward direction, allows measurement of particles using time of flight method with resolution about 0.6 ns, which corresponds to the energy resolution of $\approx$ 10% on the flight path of 3 m.

As seen in figure 1, the forward detector consists of the proportional chambers, two scintillation walls of plastic strips 10 cm wide and 3 cm thick, and showers calorimeter of five layers of plastic 5 cm thick and 5 mm thick lead. In previous work [3] to select the nucleons by shower calorimeter because the heavier fragments are absorbed by a plastic walls. In this paper, scintillation walls are included in the data analysis. Each scintillation strip, as well as shower calorimeter module was viewed from both ends by the photomultiplier tubes, what allowed determining the coordinates of the particles on the detector. So, the correction on the flight length for different emission angles of particles was done. The trigger for the data set served as a signal from the central part of LAGRANγE (BGO ball) in coincidence with the tagging system signal.

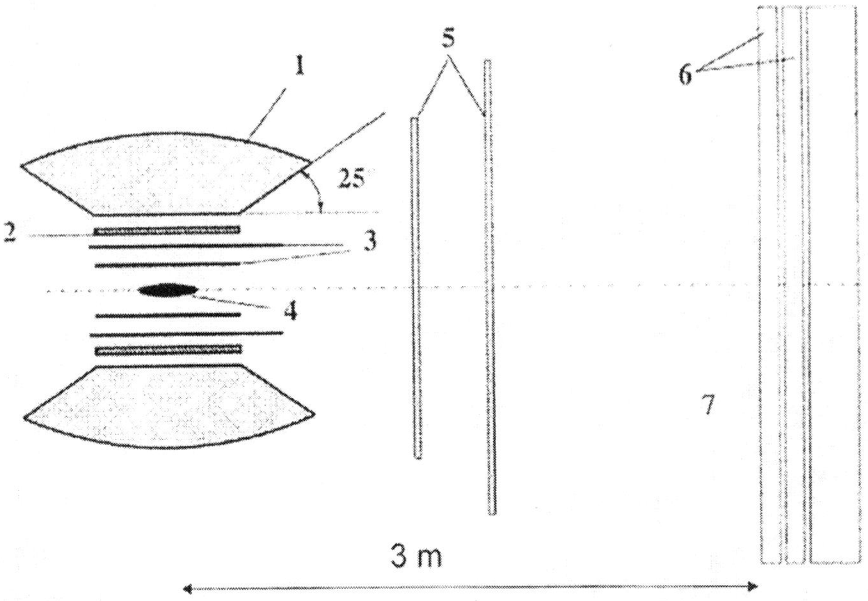

Figure 2. The detector LAGRANγE [6] of the experiment GRAAL [4]: BGO calorimeter (1), a plastic scintillation detector ring (2) is proportional to the cylindrical chamber (3), a target (4), flat proportional chamber (5), hodoscope of plastic scintillation detectors (6), shower detector (7).

The experimental results for the carbon target obtained with scintillation walls and shower calorimeter are shown in figure 3 and table 1 in comparison with simulations.

The simulation results for protons and fragments of different types are indicated by different colors. Such simulations take into account the intranuclear effects described by the RELDIS code [1], and the response function of the detector using the program LAGDIG [2].

Simulation and experiments show that fragments heavier than nucleons (^{2}D, ^{3}H, ^{4}He) are seen in the first scintillation wall clearly. The protons are separated in the calorimeter; heavier particles are absorbed in the walls.

Let us note some features of figure 3. Most of the fragments are concentrated within a band of (ΔE-TOF). The break in the slope of this band at the time of flight of about 18 ns means that the particle range exceeds the thickness of the detector, excepting the α - particles which have a high flight time (up to 100 ns) and consequently the small energy losses. The sharp peak at 12 ns, in the calorimeter corresponds to fast mesons and photons coming from the target to the wall with the speed of light. Scintillation walls do not indicate such peak because of low measurement efficiency for neutral particles. The significant background in all figures (black dots) is seen. This background corresponds to the scattered particles, probably neutrons.

As it is seen from figure 3, stripes of fragments of different types are overlapping. The main reason for that is the limited energy and ΔE resolution of the detectors. To extract the experimental values the same cuts as used in simulation were applied.

As simulations show, the energy resolution about 1% is required for a more reliable separation of the different observed fragments. Such resolution can be achieved using a magnetic spectrometer. This available facility is functioning at present in the collaboration BGO-OD in Bonn [5].

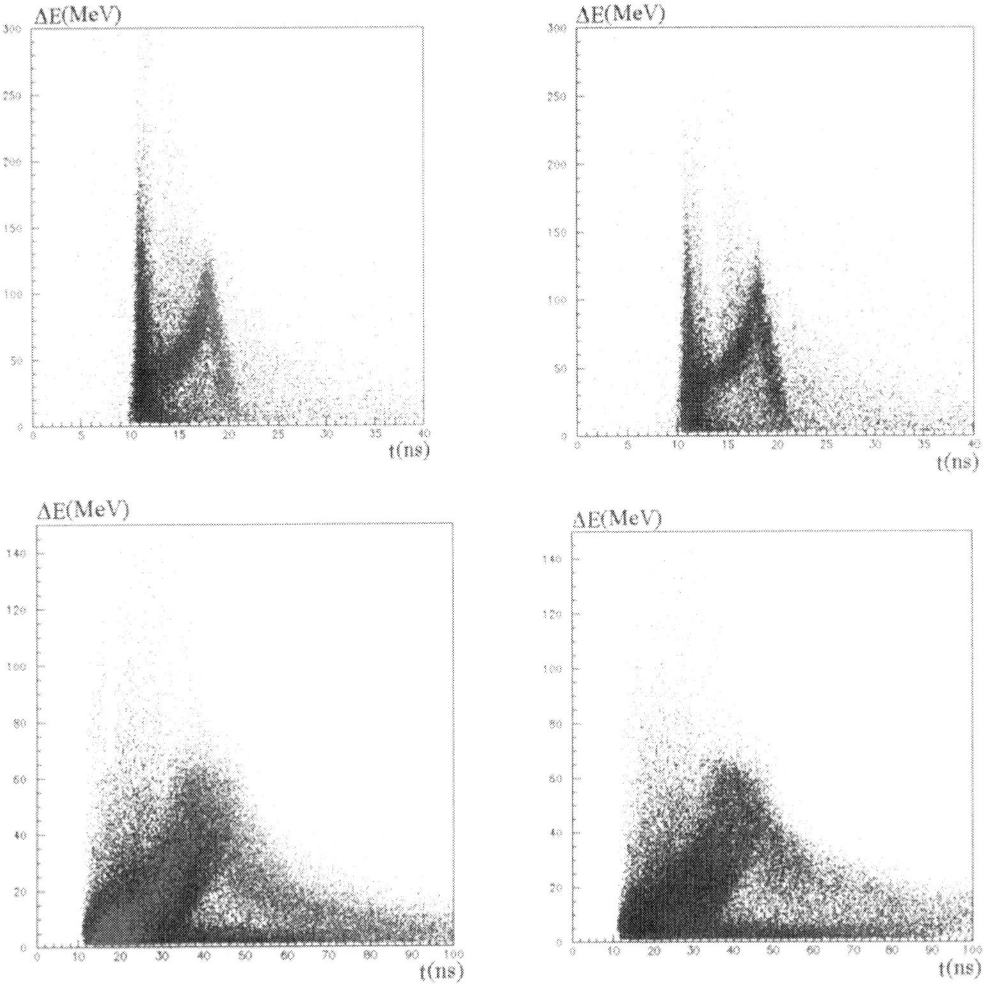

Figure 3. The dependence of the energy loss on the time of flight of the particles in the shower calorimeter (top) and scintillation wall (bottom). On the left: the results of simulations using the equal program RELDIS [5], on the right: the experimental data.

On the basis of the obtained data we can conclude that the experimental results for emission probabilities of deuterons, α- particles and tritium nuclei do not contradict the theoretical predictions [1]. The main feature is that the probability of fragmentation decreases with increasing mass of emitted fragments. This feature can be compared with another experimental results obtained with different projectiles.

1.2 Comparison of the nuclear fragmentation processes induced by real and virtual photons.

In accordance with the theoretical model [1], the mechanism of nuclear fragmentation depends primarily on their excitation energy. Below the meson production threshold, when the wavelength of the

DOI: 10.1051/epjconf/201713801005

Table 1. Relative probabilities of nuclear fragmentation evaluated from figure 3. Experimental values correspond to the yields from different detectors (two scintillation walls and shower calorimeter).

	First scintillation wall	First and second scintillation walls in coincidence	Shower calorimeter	Simulation
Protons	(53 ± 21)%	(73 ± 24)%	(91 ± 27)%	52%
Deuterons	(20 ± 11)%	(18 ± 12)%	(9 ± 6)%	18%
α-particles	(18 ± 11)%	(4 ± 3)%	-	18%
^{3}He	(5 ± 3)%	(2 ± 3)%	-	7%
^{3}H	(4 ± 3)%	(3 ± 3)%	-	5%

photon is comparable to the size of the nucleus, the most important role play the collective excitations, namely, giant resonances. Considering from this point of view, the data were obtained with beams of real and virtual photons (electrons and relativistic ions), when the excitation energy does not exceed 100 MeV.

So far, most of the experimental results on nuclear fragmentation by electrons and bremsstrahlung photons were obtained about 30 years ago (see for example the review [6]). The measured integrated yields were analyzed in term of real and virtual spectrum as seen in formula (1) :

$$\sigma_{e(q)}(E_e) = \frac{E_e \int_0^{E_e} N^{\lambda L}(E_e, E_\gamma)\sigma_\gamma(E_\gamma)dE_\gamma/E_\gamma}{\int_0^{E_e} N_V^{\lambda L}(E_e, E_\gamma)dE_\gamma} \quad . \tag{1}$$

In this formula, the symbols q and e correspond to real bremsstrahlung and virtual photons, respectively. So, the only difference is related to the photon spectra which are quite different for different projectiles, especially taking into account that λ means electrical or magnetic transition, L - multipolarity of interaction. In contrast to real photons having only transverse polarization, the virtual photons can have an admixture of the longitudinal components.

Qualitatively, in first approximation, the spectra of real and virtual photons are proportional to $1/E_\gamma$. Therefore, we would expect the similar behavior of fragmentation probabilities for different projectiles. In fact, it is not so. Figure 4 shows the output of light fragments from ^{27}Al, irradiated by electrons with an energy of 800 MeV [7]. In the same figure the spectrum of virtual photons calculated by the model Weizscker - Williams [8] is shown. One can see that this spectrum strongly depends on the multipole interaction, where E1 transition dominates. Experimentally, the role of high multipole interactions in nuclei photo - fragmentation is not studied yet.

As already noted, for light nuclei including the beryllium and carbon, the main mechanism of absorption of photons in this energy range is associated with the excitation of the giant dipole resonance. In such case the main decay channel is the neutron emission one.

We now consider the fragmentation of relativistic nuclei by virtual photons of relativistic nuclei. The yield also can be described by formula (1) but the spectrum of photons does not depend on the multipolarity of interaction.

Data for splitting fragments heavier than nucleons were obtained primarily in emulsions experiments. Table 1 shows the results of the collaboration BECQUEREL [9] in comparison with our data.

Table 2 shows that the probability of proton emission at Coulomb dissociation close to 50%, as well as for a bundle of real photons. The main difference is that in the real beam of photons, a signifi-

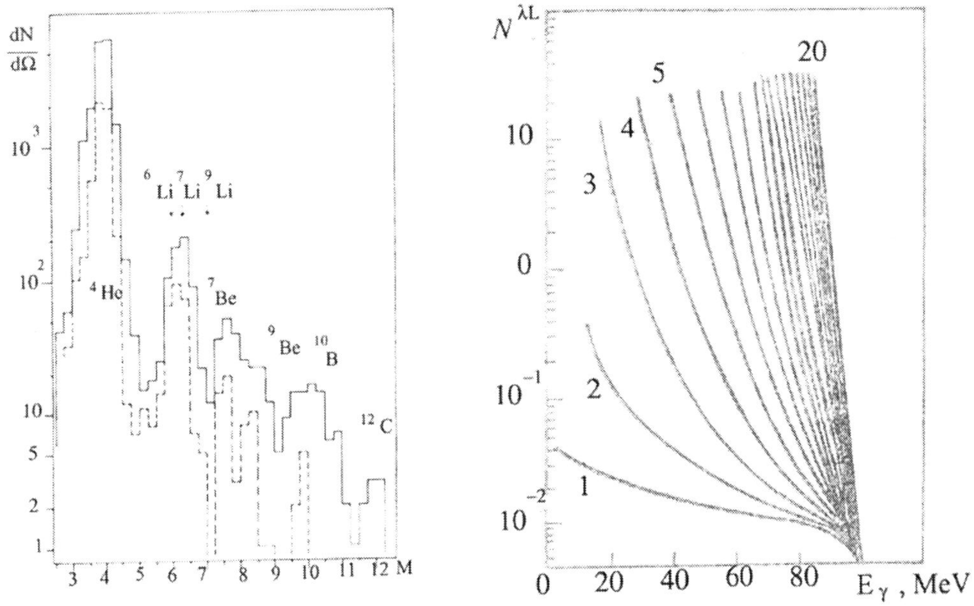

Figure 4. The yield of light fragments from the nucleus ^{27}Al by 800 MeV electrons [7]. Solid lines and dotted lines correspond to different methods of etching track detectors (experimental details see [7]). On the right, the spectrum of virtual photons for the uranium nucleus (Z = 92), is shown as the result of calculation in the long-wave Born approximation [8]. The numbers on the curves correspond to the multipole order of interactions.

Table 2. Distinguishing the probability of formation of light fragments in the Coulomb dissociation of relativistic nuclei ^{12}N and ^{11}C [9] at an energy of 1.2 GeV / nucleon in comparison with the data of the present work.

	^{12}C [simulation]	^{12}C Experiment (present work)	^{12}N Experiment[9]	^{11}C Experiment[9]
Proton	52%	53%	184 (68%)	204 (48%)
Deutron	18%	20%	0	0
^{3}H	5%	18%	0	0
^{3}He	7%	5%	0	0
^{4}He	18%	4%	75 (32%)	221 (52%)

cant contribution to output fragments comprise deuterons and tritium nuclei. In the case of relativistic nuclei are absent, while the departure α - particles of residual large and reaches approximately 20% relative to the total output. Obviously, this points to the fact that the dynamics of the processes of fragmentation under the influence of real and virtual photons in the Coulomb dissociation is different.

As it was already noted, the cause of the observed differences are likely to be found in the difference between the spectra of virtual and real photons, which have been shown in figure 4. However, this conclusion should be considered preliminary. For final output new experimental data and new theoretical assessment are required, primarily from the spectra of virtual photons.

The spectrum of virtual photons in the case of scattering of relativistic nuclei is usually expressed not dependent on the photon energy and the impact parameter b [10]:

$$N_{Z_1}(E_1,b) = \frac{\alpha Z_1^2}{\pi^2}\frac{x^2}{\beta^2 E_1 b^2}(K_1^2(x) + \frac{1}{\gamma^2}K_0^2(x)). \tag{2}$$

Here E_1 is the photon energy, Z - nuclear charge, γ- relativistic factor for the incident ion, K_0 and K_1 are the Bessel functions of the zero and first order, respectively. It can be seen that in all the above formulas, the photon energy is in the denominator. But the differences between the spectra of virtual photons to electrons and relativistic ions, as it is evident from the above formulas, very significant. It should be noted that they are all prepared using different model approximations and their accuracy is limited.

However, as it is already noted, even in light nuclei, such as beryllium or carbon basic mechanism of the Coulomb dissociation is associated with the excitation of the giant dipole resonance dominated photoneutron reaction. However, neutrons in the emulsion are not registered. The return target nuclei at these energies and momentum transfers of very small and also can not be registered. This casts doubt on the claims of orthodoxy coherence of the interaction of the incident photon with a nucleus. This raises also the question of why in the real photon coherent dissociation of sight.

Thus, since the conditions of experiments with real and virtual photons, which highlighted the "white star" essentially different, to compare them and make conclusions about the physical features of the mechanisms and dynamics of these reactions is difficult. This requires new experimental data carried out in the similar experimental conditions. This is even more needed more accurate theoretical evaluation of the spectra of virtual photons, or the results of their measurements, as proposed, for example, in [6].

2 Conclusion

In recent years, the level of photonuclear experiments at intermediate energies increased significantly. Firstly, this is due to the use of monochromatic photon beams, and secondly, with the creation of the detectors with a large solid angle and high energy and angular resolution. It is possible now to study photonuclear reactions with the formation of a large number of nucleons in the final state until the complete disintegration of the nucleus into separate nucleons. So, the new level to study the nucleon correlations in nuclei and cluster is available now. New developments are expected also with the help of a magnetic spectrometer in combination with a method of flight time.

An important motivation for these experiments is the presence of predictive framework based on modern theoretical models. In this study the dynamics of nuclear processes can be studied in photonuclear reactions for each of the partial meson photoproduction channel.

An additional incentive for the new experiments is possible to check the model calculations in the framework of quantum electrodynamics (the spectrum of virtual photons) to study the mechanisms of interaction of virtual photons with nuclei. Energy range of studies includes both the area of collective excitations of nuclei (giant resonances) and the region of nucleon resonances with photoproduction of mesons.

The authors thank I.Pshenichnov for providing programs to model calculations for the code RELDIS P.Zarubin and for useful discussions and to B.A.Tulupov for many valuable remarks and for the help in preparing the paper.

DOI: 10.1051/epjconf/201713801005

References

[1] I. Pshenichnov et al., Phys. Rev. C **57**, 1920 (1998)
[2] J. P.Bocquet et al., Nucl.Phys. A **622**, 124 (1997)
[3] V. Nedorezov et al. (GRAAL collaboration), Nucl.Phys. A **940**, 264 (2015)
[4] M. Castoldi et al., NIM A **403**, 22 (2002)
[5] T. Zimmerman et al., JPS Conf.Proc. **10**, 032003 (2016)
[6] V. Nedorezov, Yu.N. Ranyuk, *Photofission above the giant resonance* (Naukova Dumka, Kiev, 1984)
[7] S. di Lierto, F. Merdi, G. Romqanio et al., Nucl.Phys. A **296**, 519 (1978)
[8] P. Durgapara, D.S. Onley, Phys.Rev. C **27(2)**, 523 (1983)
[9] P.I. Zarubin, Lect.Notes Phys. **875**, 51-93 (2014)
[10] I. Pshenichnov, Phys.Atom.Nucl. **78(6)**, 794-799 (2015)

Nuclear parton distributions

S. A. Kulagin[1,*]

[1] *Institute for Nuclear Research of the Russian Academy of Sciences, Moscow 117312, Russia*

Abstract. We review a microscopic model of the nuclear parton distribution functions, which accounts for a number of nuclear effects including Fermi motion and nuclear binding, nuclear meson-exchange currents, off-shell corrections to bound nucleon distributions and nuclear shadowing. We also discuss applications of this model to a number of processes including lepton-nucleus deep inelastic scattering, proton-nucleus Drell-Yan lepton pair production at Fermilab, as well as $W^{\pm}$ and Z^0 boson production in proton-lead collisions at the LHC.

1 Introduction

The QCD factorization theorem [1] suggests that the parton distribution functions (PDFs) are universal process-independent characteristics of the target at high invariant momentum transfer Q. PDFs determine the leading contributions to the cross sections of various hard processes involving leptons and hadrons. PDFs cannot be reliably calculated in modern QCD, as they are driven by non-perturbative strong interactions, and QCD-based phenomenology remains to be the primary source of information on PDFs.

The deep-inelastic scattering (DIS) experiments with nuclear targets show significant nuclear effects with a rate that is more than one order of magnitude larger than the ratio of the nuclear binding energy to the nucleon mass (for a review see [2, 3]). These observations rule out the naive picture of the nucleus as a system of quasi-free nucleons and indicate that the nuclear environment plays an important role even at energies and momenta much higher than those involved in typical nuclear ground state processes.

The studies of the mechanisms responsible for observed effects provide a link between particle and nuclear physics and help to better understand both the intrinsic properties of the proton as well as the properties of hadrons in nuclear environment. A typical example in this context is the extraction of the d-quark PDF from the global fits involving the proton and the deuteron data. This procedure requires, in turn, a detailed knowledge of nuclear effects in order to control the corresponding systematic uncertainties.

A number of studies are available [4–6], in which the nuclear parton distributions (NPDFs) are extracted from global fits to nuclear high-energy data using the empirical parameterizations of nuclear correction factors for each PDF. Typically these studies involve a large number of free parameters and can be biased by the chosen parameterizations of the Bjorken x as well as the nuclear mass number A dependencies of nuclear correction factors. In this paper we follow a different approach and address

* e-mail: kulagin@ms2.inr.ac.ru

the NPDFs using the microscopic model developed in [7]. The model incorporates a number of nuclear corrections including the smearing with the energy-momentum distribution of bound nucleons (Fermi motion and binding), the off-shell correction to bound nucleon structure functions, corrections from nuclear meson exchange currents (MEC) and the propagation of the hadronic component of the virtual intermediate boson in the nuclear environment. In [8] this approach was applied to compute individual NPDFs. In Sec.2 we briefly review the approach in context of nuclear DIS. The applications to the nuclear Drell-Yan (DY) reaction in the context of the measurement of nuclear antiquark distributions in Fermilab E772 experiment is discussed in Sec.3. In Sec.4 we review recent studies of W and Z boson production in p+Pb collisions at LHC.

2 Nuclear DIS and PDFs

In a reference frame in which the hadron has a high (infinite) momentum, a PDF is the momentum distribution of the corresponding parton in a hadron in the units of hadron's momentum. However, the interpretation of PDFs is somewhat more complicated in the nucleus rest frame. We recall that in the target rest frame the characteristic DIS time (or longitudinal distance) can be estimated as $L \sim (Mx)^{-1}$, where M is the nucleon mass and x the Bjorken scaling variable [9]. Using L we identify two different kinematical regions for nuclear effects. In the region $L < d$, where d is the average distance between bound nucleons in a nucleus, i.e. for large values of $x > 0.1$, the nuclear DIS cross sections can be approximated by incoherent scattering from bound protons and neutrons. The dominant nuclear corrections in this region are due to energy-momentum distribution of bound nucleons (nuclear Fermi motion and binding [10–13]) and also off-shell correction to bound nucleon structure functions [14]. In the region of small $x < 0.1$ one has to account for a scattering corrections from nuclear meson fields [15, 16]. Also in the region $L \gg d$ (or $x \ll 0.1$) the propagation effects of virtual quark-gluon states in the nuclear environment results in essential corrections to the impulse approximation. For the leading contributions to the DIS structure functions the interference of multiple scattering terms results in a negative correction (nuclear shadowing, for a review see, e.g., [17]).

We use the notation $q_{a/h}(x, Q^2)$ for the distribution of (anti)quarks of the type $a = u, \bar{u}, d, \bar{d}, \ldots$ in a hadron h. Following [7, 8] we write the nuclear PDF $q_{a/A}$ as (for brevity, we suppress explicit dependencies on x and Q^2)

$$q_{a/A} = q_{a/A}^{\mathrm{IA}} + q_{a/A}^{\mathrm{coh}} + q_{a/A}^{\mathrm{MEC}}, \tag{1}$$

where the first term on the right side is the contribution from bound protons and neutrons in the impulse approximation, and the other terms are the corrections to the impulse approximation due to coherent nuclear interactions of the hadronic component of the virtual photon and to nuclear meson exchange currents, respectively. These contributions are reviewed below.

2.1 Impulse approximation

In the impulse approximation the nuclear PDFs can be written as a convolution of the proton (neutron) distribution of a nucleus with the corresponding parton distribution of a bound proton (neutron) [7, 14]:

$$q_{a/A}^{\mathrm{IA}} = \sum_{\tau=p,n} f_{\tau/A} \otimes q_{a/\tau} = \sum_{\tau=p,n} \int_{x<y} \frac{\mathrm{d}y\mathrm{d}p^2}{y} f_{\tau/A}(y, p^2) q_{a/\tau}(x/y, Q^2, p^2). \tag{2}$$

The nuclear convolution is an integration over both the nucleon light-cone momentum y and the nucleon invariant mass (virtuality) p^2, since PDFs of an off-shell nucleon generally depend on its virtuality [14]. The proton (neutron) distribution function in a nucleus is given in terms of the nuclear spectral function $\mathcal{P}$ (for brevity, we drop subscripts identifying the proton and the neutron distributions)

$$f(y, p^2) = \int [\mathrm{d}k] \left(1 + \frac{k_z}{M}\right) \mathcal{P}(\boldsymbol{k}, \varepsilon) \delta\left(y - \frac{k_0 + k_z}{M}\right) \delta\left(p^2 - k^2\right), \tag{3}$$

where the integration runs over the nucleon four-momentum $k = (M + \varepsilon, \boldsymbol{k})$ and $[\mathrm{d}k] = \mathrm{d}k_0 \mathrm{d}\boldsymbol{k}/(2\pi)^4$ and $k^2 = k_0^2 - \boldsymbol{k}^2$ is the invariant mass of the off-shell nucleon. The coordinate system is such that the momentum transfer is antiparallel to the z axis. The Bjorken variable of the nucleus is $x = Q^2/2Mq_0$ with q_0 the energy transfer in the target rest frame.

We refer (2) as a generalized impulse approximation because it accounts for the dependence of the nucleon PDF on the nucleon invariant mass p^2, which is usually not addressed in various versions of impulse approximation. A similar nuclear convolution equation but without off-shell effect was earlier discussed in [12, 13, 18]. Note that (2) and (3) were obtained starting from a Lorentz-covariant approach and using a systematic expansion of matrix elements in series of the small parameters $\boldsymbol{p}/M$ and ε/M, keeping terms of the order $\boldsymbol{p}^2/M^2$ and ε/M [7, 14, 18]. To this order a generalized nucleon distribution function (3) is given in terms of nonrelativistic spectral function

$$\mathcal{P}(\boldsymbol{k}, \varepsilon) = \int \mathrm{d}t\, e^{i\varepsilon t} \langle \psi^\dagger(\boldsymbol{k}, t) \psi(\boldsymbol{k}, 0) \rangle, \tag{4}$$

where $\psi(\boldsymbol{k}, t)$ is the nonrelativistic nucleon operator in the momentum-time representation (for more details see [7]). The spectral function describes the energy-momentum distribution of bound nucleons. Note that ε in (4) includes the recoil kinetic energy of the residual system of $A - 1$ nucleons, as it can be seen after inserting a complete set of states and integrating over the time. [1]

The proton (neutron) spectral function is normalized to the proton Z (neutron N) number, and using (3) we explicitly verify that the proton (neutron) distribution function is normalized as

$$\int \mathrm{d}y \mathrm{d}p^2 f_{p(n)/A}(y, p^2) = Z(N), \tag{5}$$

where the integral is taken over all possible light-cone momenta y and the nucleon virtuality p^2. Note that the term k_z/M in (3) gives vanishing contribution to the normalization (5) due to symmetry reason. The distribution function (3) is independent of Q^2 in the Bjorken limit and the Q^2 evolution of the NPDFs in the impulse approximation is governed by the evolution of the PDFs of the corresponding nuclear constituents. For the discussion of power corrections to the nuclear convolution (2) we refer to [7, 19, 20] (see also [21] for spin-dependent DIS).

The first moment of the nucleon distribution in y gives the fraction of the nuclear light-cone momentum carried by bound nucleons. The first moment in p^2 gives average nucleon virtuality. It will be convenient to discuss dimensionless virtuality $v = (p^2 - M^2)/M^2$. We have

$$\langle y \rangle_N = \frac{1}{A} \int \mathrm{d}y \mathrm{d}p^2 y f(y, p^2) = 1 + \frac{\langle \varepsilon \rangle + \frac{2}{3}\langle T \rangle}{M}, \tag{6}$$

$$\langle v \rangle_N = \frac{1}{A} \int \mathrm{d}y \mathrm{d}p^2 v f(y, p^2) = 2\frac{\langle \varepsilon \rangle - \langle T \rangle}{M}, \tag{7}$$

[1] Note that for the deuteron the spectral function is given in terms of the deuteron wave function as $\mathcal{P}_d = 2\pi |\Psi_d(\boldsymbol{k})|^2 \delta(\varepsilon - \varepsilon_d - \boldsymbol{k}^2/2M)$ with $\varepsilon_d = -2.22$ MeV the deuteron binding energy [7].

where we sum over protons and neutrons and $\langle\varepsilon\rangle$ and $\langle T\rangle = \langle k^2\rangle/2M$ are the nucleon energy and kinetic energy integrated with the nuclear spectral function (4) per one nucleon. Note that $\langle\varepsilon\rangle - \langle T\rangle = \langle V\rangle$ is the average nuclear potential energy per one nucleon. For the deuteron we can evaluate the moments in (6) and (7) using the deuteron wave function. For the Paris wave function we have $\langle y\rangle_N = 0.994$ and $\langle v\rangle_N = -0.044$. For ^{208}Pb, using the nuclear spectral function from [7], we have $\langle y\rangle_N = 0.963$ and $\langle v\rangle_N = -0.197$. Note that $\langle y\rangle_N < 1$ and $\langle v\rangle_N < 0$ because of nuclear binding. Also the inequality $\langle y\rangle_N < 1$ indicates that the bound nucleons carry only a part of the full nuclear light-cone momentum. In this approach the missing nuclear light-cone momentum is carried by the nuclear meson degrees of freedom which are responsible for nuclear binding, as discussed in Sec.2.2.

In terms of the variable y, the distribution function $f(y, p^2)$ is peaked about $y = \langle y\rangle_N$ with a width $\sim p_F/M$ with p_F the nuclear Fermi momentum. In terms of dimensionless virtuality v, the distribution $f(y, p^2)$ is located in a narrow region about $v = \langle v\rangle$.

Note that (2) predicts a finite nuclear PDF in the region $x > 1$, because of high momenta $p_z > p_F$. The rate of NPDF in this region is driven by the high-momentum part of the nuclear momentum distribution.[2] In applications we use a model spectral function, which includes both a mean field contribution dominant at low energy and momentum, and a high-momentum and high-energy component related to short range nucleon-nucleon correlations (SRC) [7]. In order to address the effect of nonisoscalarity (nuclear neutron excess) we use the isoscalar $\mathcal{P}_0 = \mathcal{P}_p + \mathcal{P}_n$ and the isovector $\mathcal{P}_1 = \mathcal{P}_p - \mathcal{P}_n$ combinations of the nuclear spectral function. We assume the SRC contribution to be similar for the proton and the neutron nuclear distribution such that the SRC term only contributes to the isoscalar spectral function $\mathcal{P}_0$ and that it cancels out in the isovector combination $\mathcal{P}_1$. This behavior is supported by the observation of the dominance of pn SRC pairs in nucleon knock-out experiments [25]. The isovector spectral function $\mathcal{P}_1$ is therefore calculated as the difference between the mean-field contributions to the proton and the neutron spectral functions and is proportional to the proton-neutron asymmetry $\beta = (Z - N)/A$ [7].

In off-mass-shell region the PDFs explicitly depend on the nucleon virtuality as indicated in (2). In order to address this effect we note that on average the nucleon virtuality v is small. We then expand the off-shell PDF $q(x, Q^2, p^2)$ in (2) about the on-shell value $p^2 = M^2$ in series in v [14]. Keeping the leading terms we have [7, 8]:

$$q(x, Q^2, p^2) \approx q(x, Q^2)(1 + \delta f(x, Q^2)v), \tag{8}$$

$$\delta f(x, Q^2) = \partial \ln q(x, Q^2, p^2)/\partial \ln p^2, \tag{9}$$

where the derivative is evaluated at $p^2 = M^2$ and $q(x, Q^2)$ is the PDF of the on-shell nucleon. The function δf describes the relative off-shell modification of the nucleon PDF. This special nucleon structure function does not contribute to the cross section of the physical nucleon, but it is relevant only for the bound nucleon and describes its response to the interaction in a nucleus. The function δf was studied phenomenologically[3] in an analysis of data on the nuclear DIS [7] (see Sec.2.4). This analysis suggests a common off-shell function for the quark and antiquark distributions, independent of Q^2 and the parton type. This observation is supported by a recent QCD global analysis of proton and deuteron data [36], as well as by the studies of the nuclear DY production of muon pair in Sec.3 and the W/Z boson production in p+Pb collisions in Sec.4.

2.2 Nuclear MEC correction

As it was noted in Sec.2.1 the nuclear binding effect causes $\langle y\rangle_N < 1$. Apparently the missing nuclear light-cone momentum should be carried by the fields responsible for nuclear binding [15]. The cor-

[2] Allowed region of nuclear Bjorken variable is $0 < x < M_A/M$, where M_A is the mass of the nucleus.
[3] Nuclear DIS with off-shell effect was discussed in a number of different approaches [14, 22–24].

rection from the nuclear virtual meson cloud was discussed in the context of the nuclear EMC effect in a number of papers [15, 16, 26–30]. These studies indicate an enhancement of the nuclear structure functions in the region $0.05 < x < 0.15$ due to nuclear meson cloud. However, the results of specific calculation are somewhat uncertain due to sensitivity to the details of the pion-nucleon form-factor, the treatment of the particle-hole nuclear excitations in intermediate state as well as the Δ-resonance degrees of freedom.

Here we briefly review the approach followed in [7, 8, 18] and consider constraints on the common nuclear meson cloud coming from the nuclear light-cone momentum sum rule together with the meson contribution to the nuclear potential energy. We write the MEC correction in a convolution form similar to (2):

$$q_{a/A}^{\mathrm{MEC}} = \sum_{m=\pi,\rho,\ldots} f_{m/A} \otimes q_{a/m} \tag{10}$$

where $f_{m/A}(y, k^2)$ is the nuclear distribution function of a meson type m over light-cone momentum y and invariant mass k^2 and $q_{a/m}$ is corresponding PDF in a virtual meson m with invariant mass k^2. The meson distribution function f_m is driven by the imaginary part of the full meson propagator in a nucleus.

For simplicity we will discuss the isoscalar part of $q_{a/A}^{\mathrm{MEC}}$ assuming the meson PDFs $q_{a/m}$ to be independent from the meson type and equal to the corresponding pion PDF. We also neglect possible off-shell dependence of $q_{a/m}$. This will allow us to discuss the light-cone distribution function corresponding to the sum over all mesons $f_M(y) = \sum_m \int \mathrm{d}k^2 f_m(y, k^2)$.

The nuclear light-cone momentum is shared between the nucleons and mesons and the corresponding balance equation can be written as [18]

$$\langle y \rangle_M + \langle y \rangle_N = \frac{M_A}{AM}, \tag{11}$$

Note the ratio of the nucleus mass $M_A = A(M + \varepsilon_B)$, where ε_B is the nuclear binding energy per nucleon, and the nucleon mass M in (11). This ratio appears because of the definition of the light-cone momentum y and the nuclear Bjorken variable x involves the nucleon mass M.

To further constrain the meson distribution, we consider a model nuclear Hamiltonian in effective meson-nucleon theory. Using equations of motion for interacting mesons and nucleons it was possible to obtain a relation [7, 8, 18]

$$3\langle y \rangle_M + \frac{m_M^2}{M^2}\langle y^{-1} \rangle_M = -\frac{\langle V \rangle}{M}, \tag{12}$$

where $\langle y \rangle_M$ and $\langle y^{-1} \rangle_M$ are the moments of the nuclear meson distribution function, m_M^2 is an average meson mass squared, and $\langle V \rangle$ is the average nuclear potential energy (see (7) and the discussion thereafter). Note that (12) was obtained in static approximation neglecting the contribution from the terms with time derivatives of meson fields (this is justified by the fact that a typical energy transfer in the nucleon–nucleon interaction in the nuclear ground state is much smaller than corresponding momentum transfer). If only the pion contribution is considered, then $m_M^2 = m_\pi^2$ and $\langle V \rangle$ corresponds to the pion contribution to the nuclear potential energy.

It is important to note that (11) and (12) allow us to overall constrain the nuclear meson distribution function in terms of average nucleon separation and kinetic energy $\langle \varepsilon \rangle$ and $\langle T \rangle$. In particular, for the deuteron with the Paris wave function we have $\langle y \rangle_M = 0.0045$ and $\langle y^{-1} \rangle_M = 0.390$, while for the ^{208}Pb nucleus we have 0.029 and 0.543, respectively.

In applications discussed below we consider a model meson distribution function which obeys the constraints discussed above. Following [7, 18] we assume $f_{M/A}(y)$ to scale as y as $y \to 0$. We also recall that large values of y are driven by configurations with k_0 and k_z comparable to or exceeding the nucleon mass M. We assume that such configurations are suppressed in nonrelativistic nuclei and consider the meson distribution in the region $0 < y < 1$ and use the following model

$$f_{M/A}(y) = c\,y(1-y)^n. \tag{13}$$

The parameters c and n depend on the specific nucleus and fixed from $\langle y\rangle_M$ and $\langle y^{-1}\rangle_M$ calculated by (11) and (12).

2.3 Correction from nuclear coherent processes

In the region of small x the DIS correlation length L exceeds typical distances between bound nucleons and intermediate quark-gluon states may develop multiple scattering interactions while propagating in nuclear environment. To the leading order this effect results in a negative correction known as the nuclear shadowing effect (for a review, see [17]).

In order to address this effect, in [7] we assume that the set of intermediate hadronic states can be approximated by a single effective state and describe its interaction with the nucleon by an effective scattering amplitude a. The rate of the contribution to DIS cross section is driven by $\operatorname{Im} a$. This amplitude depends on the type of the PDF and differ for the proton and the neutron.

It is convenient to discuss nuclear effects in terms of the ratio $R_{i/A} = q_{i/A}/(Zq_{i/p} + Nq_{i/n})$ for the given PDF type i. Using the optical theorem this ratio can be written in terms of effective cross sections, or the imaginary part of the effective amplitudes in the forward direction

$$\delta_{\rm coh} R_{i/A} = \operatorname{Im} \mathcal{A}_i/(Z \operatorname{Im} a_{ip} + N \operatorname{Im} a_{in}), \tag{14}$$

where $\mathcal{A}_i$ is the nuclear amplitude which describes propagation of intermediate states corresponding to PDF of type i in a nucleus, and a_{ip} and a_{in} are corresponding proton and neutron amplitudes, respectively. We compute the amplitude $\mathcal{A}$ in terms of the Glauber-Gribov multiple-scattering series [31, 32]. Note that the multiple-scattering series start from the double-scattering term, as the single-scattering term is already accounted in the impulse approximation of (2). For more detail discussion of $\mathcal{A}$ see [7, 8].

To have a closer link to the DIS structure functions, it is convenient to discuss the combinations of PDFs with definite C-parity, $q_i^\pm = q_i \pm \bar{q}_i$. Let $a_{ip}^\pm$ and $a_{in}^\pm$ be the corresponding proton and neutron effective amplitudes. For the u and d (anti)quark distributions we also consider the combinations with definite isospin $I = 0, 1$ and C parity q_I^C, i.e. $q_0^\pm = u^\pm + d^\pm$ and $q_1^\pm = u^\pm - d^\pm$.

We separate the isoscalar and the isovector contributions in $\mathcal{A}$ assuming the isospin symmetry for the scattering off protons and neutrons, i.e. $a_{up} = a_{dn}$ and $a_{dp} = a_{un}$, and write the amplitudes as $a_{up} = a_0 + \frac{1}{2}a_1$ and $a_{dp} = a_0 - \frac{1}{2}a_1$, where a_0 and a_1 are the isoscalar and isovector amplitudes, respectively. To the first order in $\beta = (Z-N)/A$ we have [34]

$$\mathcal{A}_{u,d} = \mathcal{A}(a_0) \pm \frac{\beta}{2} a_1 \mathcal{A}'(a_0), \tag{15}$$

where the sign $+-$ should be taken for the u and d quark, respectively, and $\mathcal{A}' = \partial\mathcal{A}/\partial a$. The first and the second terms in (15) drive the corrections to the isoscalar q_0 and the isovector q_1 quark distribution, respectively.

Let us consider first the isoscalar $I = 0$ case. For the nuclear corrections for the C-even and C-odd quark distributions we have

$$\delta R_0^+ = \operatorname{Im} \mathcal{A}(a_0^+)/(A \operatorname{Im} a_0^+), \tag{16a}$$

$$\delta R_0^- = \operatorname{Im}[a_0^- \mathcal{A}'(a_0^+)]/(A \operatorname{Im} a_0^-), \tag{16b}$$

where $a_0^\pm$ are the $I^C = 0^\pm$ amplitudes. We note that (16) are obtained by treating the C-odd amplitude as a small parameter and expanding the difference between the quark and antiquark nuclear amplitudes in series of a_0^- to the order $(a_0^-)^2$ [34]. The effective expansion parameter in (16) is the ratio of the amplitudes a_0^-/a_0^+. The smallness of this parameter can be justified within the Regge pole model of high-energy scattering amplitudes.

The nuclear corrections in the isovector term $I = 1$ can be calculated similarly to the isoscalar case discussed above. We consider $1^\pm$ channels and expand the corresponding nuclear amplitude in series of a_0^-. To the leading order we have

$$\delta R_1^\pm = \beta \,\mathrm{Im}\left[a_1^\pm \mathcal{A}'(a_0^+)\right] / (A\, \mathrm{Im}\, a_1^\pm), \tag{17}$$

where the superscript + and − corresponds to the channel 1^+ and 1^-, respectively. In the derivation of (17) we drop the terms of order $a_1^\pm a_0^-$. Note in this context that the effective amplitudes $1^\pm$ or 0^- are generally significantly smaller than the leading amplitude a_0^+, which drives multiple scattering corrections for all PDFs, as it can be seen from (16) and (17). Furthermore, in the considered approximation the ratio δR_0^- as well as the ratios $\delta R_1^\pm$ are independent of the effective cross sections in the corresponding channels and depend only on $\alpha = \mathrm{Re}\, a/\,\mathrm{Im}\, a$ of corresponding amplitude.

The individual corrections for u and d quarks and antiquarks can be derived from $\delta R_{0,1}^\pm$ ratios. For more detail see [8].

2.4 Discussion and comparison with data

A detailed analysis of data on the ratios of DIS structure functions $R(A/A') = F_2^A/F_2^{A'}$ for different nuclei was carried out in [7] in the context of the described model. The analysis included data with $Q^2 \geq 1\,\mathrm{GeV}^2$ for the full region of Bjorken x from CERN, FNAL and SLAC available before 1997 (see table 1 in [7]). The ratio $R(A/A')$ was computed including nuclear corrections discussed above. Also the target mass [37] and the higher-twist corrections [38] were applied to the structure functions. The predictions were then compared with data by evaluating χ^2 as discussed in [7].

The function δf, which determines off-shell correction (see (9)), and also the effective amplitude a_0^+, which drives coherent nuclear corrections for the structure function F_2 at small x, were determined phenomenologically from this analysis. The function δf was assumed universal, i.e. same for quark and antiquark PDFs independent of flavor, and also independent of scale Q^2, $\delta f = \delta f(x)$. We model $\delta f(x)$ as a third-order polynomial and extract parameters from analysis of $R(A/A')$. The effective amplitude a_0^+ is the subject of several constraints. In the region of low $Q^2 < 1\,\mathrm{GeV}^2$ the amplitude a_0^+ is constrained by vector meson dominance model [33].[4] Also in the region $Q^2 > 10\,\mathrm{GeV}^2$ the amplitude a_0^+ and the function δf are linked by the normalization condition on the nuclear valence quark distribution. The latter is used to determine the leading twist part of the amplitude a_0^+ [7, 8]. The available nuclear DIS data at small x constrains the effective amplitude a_0^+ in a region which is transitional between fully nonperturbative region of $Q^2 < 1\,\mathrm{GeV}^2$ and the region $Q^2 > 10\,\mathrm{GeV}^2$, which is driven by the leading twist contribution [7].

The results reported in [7] show an accurate description of the measured dependencies on x, Q^2 and the nuclear mass number A in the full kinematical region of data. The predictions of [7] were further verified [35] with the recent nuclear DIS data from HERA [47] and JLab [46]. Figure 1 summarizes the data on $R(A/D)$ from ^{4}He to ^{208}Pb together with the model calculations.

We would like to remark that in some cases the data points from different experiments are not fully consistent. In particular, the central points of ^{12}C/D and ^{40}Ca/D ratios from E665 experiment

[4] Although this region was excluded from our fit, it was used to verify the predictions of [7].

 DOI: 10.1051/epjconf/201713801006

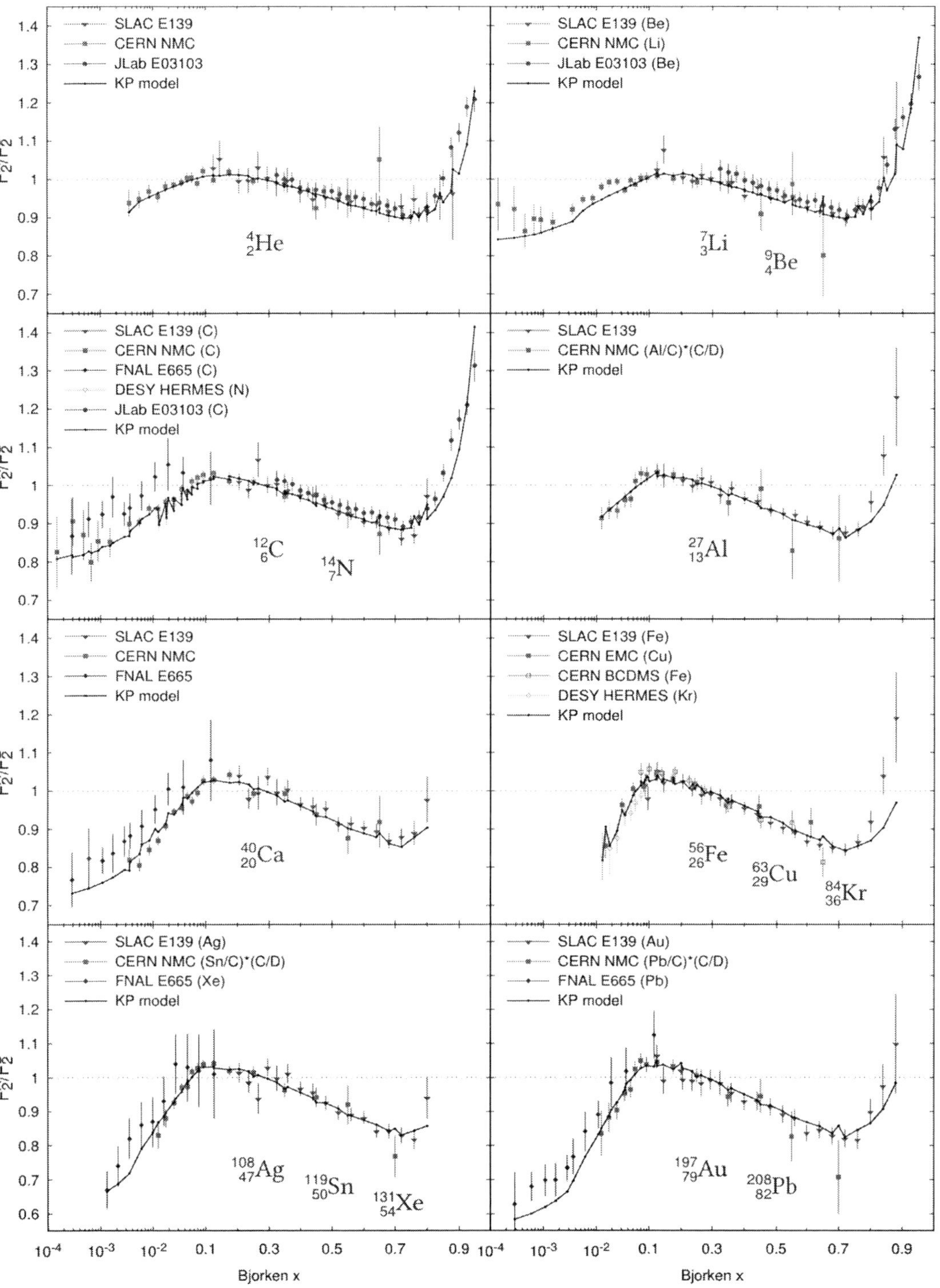

Figure 1. Summary of data on the ratio F_2^A/F_2^D for various nuclei from the measurements at CERN [39–43], SLAC [44], FNAL [45], JLab [46] and DESY [47]. The error bars represent the statistical and systematic uncertainties added in quadrature. The dots connected by the solid line are the predictions of [7] computed for the published values of (x, Q^2) of each data point (the wiggles are caused by Q^2 dependence and different values of Q^2 from different experiments). We use a logarithmic scale for $x < 0.1$ and a linear scale for $x > 0.1$ for a better display of both the small x and the large x regions.

[45] at low x are systematically above the corresponding NMC measurements, which have smaller uncertainties. Similarly, a normalization problem could be present for the E665 ^{208}Pb/D data. However, the double ratios (^{40}Ca/D)/(^{12}C/D) and (^{208}Pb/D)/(^{12}C/D) of the E665 measurement are in good agreement with the NMC data [41] as well as with our predictions. Also the central points of ^{4}He/D, ^{9}Be/D and ^{12}C/D ratios reported in [46] are systematically above the corresponding E139 [44] and NMC [39, 40] results for $x > 0.3$.[5] However, the slopes of the measured ratios for $x > 0.2$ seems to be in a good agreement for all experiments. We also comment that the analysis of [35] indicates that a common renormalization factor of 0.98 applied to the data points of [46] leads to a perfect statistical agreement of the discussed data sets.

In figure 2 we show nuclear effects on different PDFs computed for ^{208}Pb at $Q^2 = 16\,\mathrm{GeV}^2$ following [8]. The labels on the curves show different nuclear corrections included in turn: Fermi motion and binding (FMB), off-shell correction (OS), nuclear shadowing (NS), meson cloud (MEC).

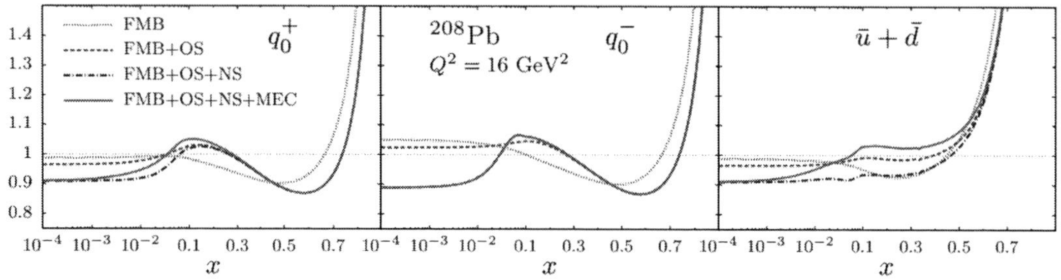

Figure 2. The ratio $R_i = q_{i/A}/(Zq_{i/p} + Nq_{i/n})$ computed for various combinations of parton distributions for the lead nucleus at fixed $Q^2 = 16\,\mathrm{GeV}^2$. The left panel shows different nuclear corrections for $q_0^+ = u + d + \bar{u} + \bar{d}$, the middle panel is for $q_0^- = u + d - \bar{u} - \bar{d}$, while the right panel is for antiquark distribution $\bar{u} + \bar{d}$.

Note that different nuclear effects outlined above are relevant in different regions of Bjorken x, as it can be seen from figure 2. In the region of large $x > 0.2$ the leading correction is due to nuclear Fermi motion and binding which is described in terms of the nuclear spectral function. The off-shell correction is also relevant in this region. The nuclear MEC correction is relevant for $x < 0.2$ while at small $x \ll 0.1$ the nuclear effects are dominated by multiple scattering of virtually produced quark-gluon intermediate states. Note also an interplay between different nuclear corrections on the antiquark distribution leading to a strong cancellation of nuclear effects in an intermediate region of x. The implication of this observation will be discussed in more detail in Sec.3 in the context of the measurement of nuclear sea-quark distribution in the Drell-Yan experiments.

We also remark that different nuclear effects are related through a number of sum rules which reflect different global symmetries. In particular, the conservation of the valence quark number links together the off-shell and the shadowing corrections to the valence quark distributions. The normalization conditions for the nuclear isoscalar q_0^- and the isovector q_1^- valence quark distributions were used in this analysis to constrain the unknown amplitudes $a_{0,1}^-$ controlling the nuclear correction in the small x region [8]. The conservation of the nuclear light-cone momentum causes the corresponding sum rule at two different levels. At the hadronic level, the nuclear light-cone momentum is shared between nucleons and mesons (see (11)), which allows us to constrain the MEC correction to NPDFs. At the partonic level, the light-cone momentum is balanced between quarks, antiquarks and gluons.

[5]Similarly about 3% mismatch in the normalization of data points from HERMES and JLab measurements is present for ^{3}He/D ratio.

The study of different contributions to the light-cone momentum sum rule can provide insights on modification of gluon distribution in nuclei.

3 Nuclear Drell-Yan process

The reaction of muon pair production in hadron-hadron collisions (Drell-Yan process) is an important source of information on the proton, pion and nuclear PDFs [48]. In the context of NPDFs, the use of DY data in combination with DIS data allows a separation of the nuclear valence and sea quark distributions. In the DY reaction with the proton beam by tuning the kinematics of the muon pair one can select a region in which the DY cross sections are driven by annihilation of valence quarks in the beam and antiquarks in the target. Then the ratio of the p+A DY cross sections off different nuclear targets provide a tool to measure the nuclear dependence of antiquark PDFs

$$\frac{\sigma_A^{DY}}{\sigma_B^{DY}} \approx \frac{\bar{u}_A(x_T, Q^2) + \bar{d}_A(x_T, Q^2)}{\bar{u}_B(x_T, Q^2) + \bar{d}_B(x_T, Q^2)}, \tag{18}$$

where x_T is the Bjorken variable of a nuclear target and Q is the mass of the lepton pair.

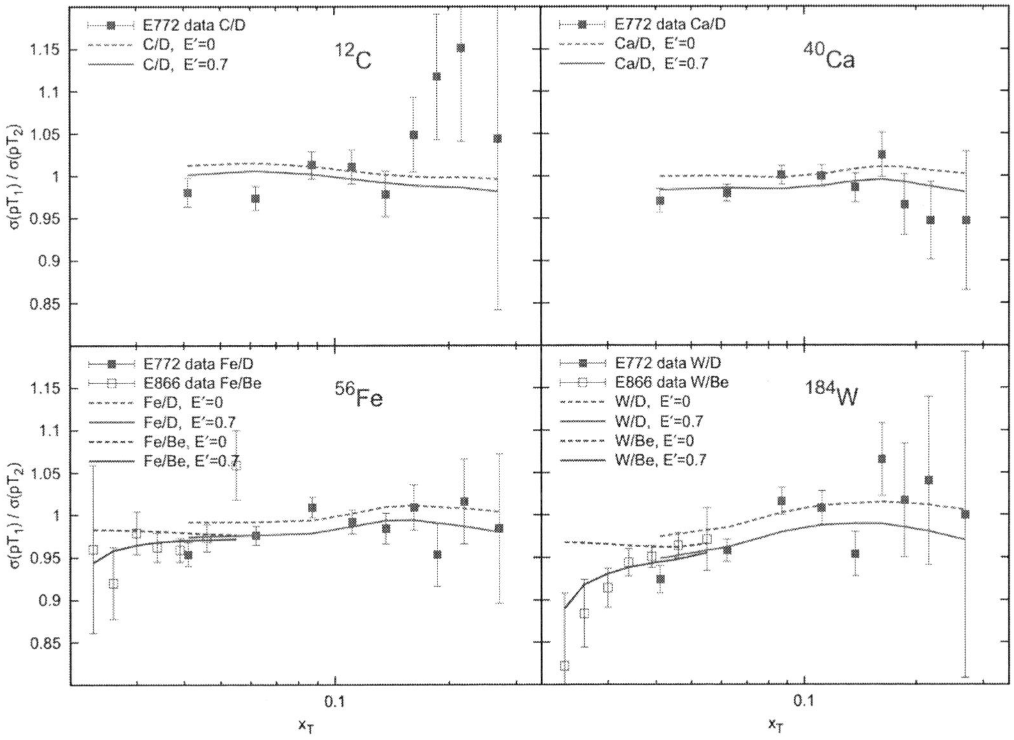

Figure 3. Ratio of the DY reaction cross sections for different nuclei vs. x_T. Data points are from the E772 experiment [49] (full squares) and E866 experiment [50] (open squares). Note that the ratio is normalized to one bound nucleon and taken relative to the deuteron for E772 and ^{9}Be for E866. The curves are the predictions of [8] for the deuteron (green) and beryllium (red) ratios with (full line) and without (dashed line) the projectile energy loss effect (see text and legend).

This ratio was measured in the experiments E772 [49] and E866 [50] with the proton beam momentum 800 GeV/c at Fermilab in the region $x_T < 0.15$ for a number of nuclear targets. Note that the scale Q^2 is determined by the dimuon mass and $Q^2 > 16\,\text{GeV}^2$ in [49], which is significantly higher than the corresponding scale from the fixed-target nuclear DIS in the sea region. In contrast to DIS, the data on DY nuclear ratios show no antishadowing (i.e. enhancement of nuclear antiquark distributions) at $x_T \sim 0.1$ that was a long standing puzzle since the nuclear binding should result in an excess of nuclear mesons, which is expected to produce a marked enhancement in the nuclear anti-quark distributions [51].

In figure 3 we show the data along with our predictions on the ratios of the DY cross sections [8]. Note that, as it can be seen from figure 2, various nuclear corrections to the antiquark distribution tend to cancel each other in the region $0.1 < x_T < 0.4$, in agreement with data on the nuclear DY process. We also remark that nuclear dependence of the DY process comes from two different sources: (i) the nuclear effects on the bound nucleon PDFs, and (ii) the initial state interaction of the projectile parton in nuclear environment that causes the parton energy loss [52] before annihilation into a dimuon pair. The rate of this effect is characterized by the parton energy loss in a nucleus per unit length E' and effectively results in a change of the projectile parton Bjorken x [53] which in turn affects the ratio in (18).

The data from the E866 experiment is shifted towards lower values of target's x_T and higher values of projectile's x_B with respect to the E772 data thus falling into a region where both the shadowing and the energy loss effects become more prominent. The analysis of [8] indicates that the model is in a good agreement with data at a moderate energy loss effect of order of 1 GeV/fm. The solid curves in figure 3 show our predictions with $E' = 0.7$ GeV/fm. Note also that the cross section ratios are taken relative to the deuterium for E772 and beryllium for E866. For this reason the corresponding curves in figure 3 are not identical in the overlap region.

4 Production of $W^{\pm}$ and Z bosons in p+Pb collisions at LHC

A study of the $W^{\pm}$ and Z^0 boson production cross sections in p+Pb collisions with $\sqrt{s} = 5.02\,\text{TeV}$ at the LHC was performed in [54] in terms of the NPDF model of [8]. Figure 4 shows the model predictions on the production cross sections as a function of the vector boson rapidity y in comparison with recent CMS data [55, 56].[6] Similar comparison with preliminary ATLAS data can be found in [54]. We found an excellent agreement between the theoretical predictions based on discussed NPDF model and the measured observables in the entire kinematic range accessible by the experiments. In particular, the model correctly describes the magnitude and the shape of rapidity distributions as well as the difference between W^+ and W^- boson distributions (flavor dependence).

This study clearly indicates the presence of nuclear modifications on the W/Z boson production cross sections with respect to the case of p+p collisions. The corresponding nuclear effect is illustrated in the lower panels of figure 4, in which we show the ratio of the result of the full calculation to that without nuclear corrections. In this study we use the ABMP15NNLO proton PDFs of [57]. For comparison we also show the results obtained with CT10NLO proton PDFs [58] with phenomenological nuclear corrections of [4].

[6] Experimentally, it is easier to measure the pseudorapidity of the charged lepton originated from the W boson decay, η^l, rather than the W boson rapidity y. The two variables are correlated and provide similar insights on the parton distributions.

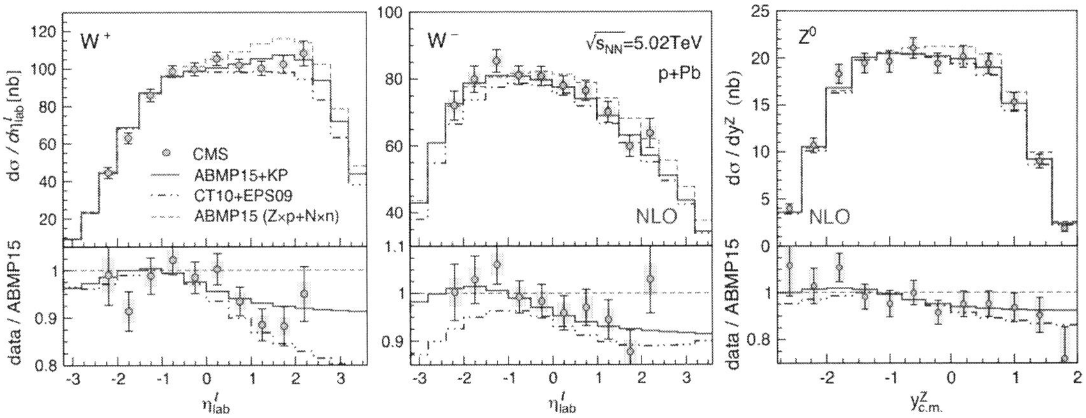

Figure 4. Top panels: differential cross sections for W^+ (left), W^- (middle), and Z^0 (right) production in p+Pb collisions at $\sqrt{s_{NN}}$ = 5.02 TeV, as a function of (pseudo)rapidity. The data points indicate the CMS measurements [55, 56], while the curves show the predictions based on different models: ABMP15+KP (solid), CT10+EPS09 (dashed-dotted), and ABMP15 without nuclear modifications (dashed). Bottom panels: ratios of the data points and the model predictions shown in the top panels with respect to the result obtained without nuclear modifications (ABMP15). For more details see [54].

In order to illuminate the link between the nuclear correction on the rapidity distributions and those on the PDFs, we recall that in the leading order we have[7]

$$d\sigma^{W^+}/dy \propto \bar{d}_p(x_p)u_A(x_A) + u_p(x_p)\bar{d}_A(x_A), \tag{19}$$

$$d\sigma^{W^-}/dy \propto \bar{u}_p(x_p)d_A(x_A) + d_p(x_p)\bar{u}_A(x_A), \tag{20}$$

$$x_p = x_0 \exp(y), \quad x_A = x_0 \exp(-y), \tag{21}$$

where x_p and x_A are the Bjorken variables for the proton and the nucleus, respectively, and $x_0 = M_W/\sqrt{s}$ corresponds to the central rapidity $y = 0$. At $\sqrt{s} = 5.02$ TeV we have $x_0 = 0.016$. Thus at $y > 0$ the rapidity distributions are determined by small-x_A region in a nucleus and subject to nuclear shadowing correction. Indeed, the lower panels in figure 4 show a clear suppression of p+A cross sections in this region. Note also that in this region the production of the W^+ and W^- is driven by, respectively, the u and d quarks in the proton. This explains a higher magnitude of W^+ boson production rate and also the different shape of W^- and W^+ boson rapidity distribution. The full nuclear correction on the vector boson production in p+Pb collisions is the result of an interplay of the various mechanisms discussed in Sec.2. It is interesting to note that in the backward rapidity region ($y < 0$) the off-shell correction plays an important role.

Finally, it is worth noting that the precision currently achieved by the LHC experiments with p+Pb and Pb+Pb collisions starts to be sensitive to the predicted nuclear corrections. A further improvement of the accuracy of future data would be extremely valuable in this context since it could provide a tool to disentangle the effect of different underlying mechanisms responsible for the nuclear modifications of PDFs and to study their flavor dependence.

[7]This discussion aims to clarify a link between W boson rapidity distributions and PDFs. The calculation of [54] was performed with the full account of NLO corrections.

Acknowledgements: I am grateful to R. Petti for fruitful collaboration on reviewed topics. I would like to thank the Organizing Committee of Baldin ISHEPP-23 Conference for warm hospitality. The work was supported by the Russian Science Foundation grant No. 14-22-00161.

References

[1] J. C. Collins, D. E. Soper and G. F. Sterman, Adv. Ser. Direct. High Energy Phys. **5**, 1 (1988)
[2] M. Arneodo, Phys. Rept. **240**, 301 (1994)
[3] P. R. Norton, Rept. Prog. Phys. **66**, 1253 (2003)
[4] K. J. Eskola, H. Paukkunen and C. A. Salgado, JHEP **0904**, 065 (2009)
[5] M. Hirai, S. Kumano and T. -H. Nagai, Phys. Rev. C **76**, 065207 (2007)
[6] D. de Florian, R. Sassot, P. Zurita and M. Stratmann, Phys. Rev. D **85**, 074028 (2012)
[7] S.A. Kulagin and R. Petti, Nucl. Phys. A **765**, 126 (2006)
[8] S. A. Kulagin and R. Petti, Phys. Rev. C **90**, 045204 (2014)
[9] B.L. Ioffe, V.A. Khoze and L.N. Lipatov, *Hard Processes. Vol. 1: Phenomenology, Quark Parton Model* (North-Holland, Amsterdam, 1984)
[10] G. B. West, Ann. Phys. **74**, 464 (1972)
[11] A. Bodek and J. L. Ritchie, Phys. Rev. D **23**, 1070 (1981)
[12] S. V. Akulinichev, G. M. Vagradov and S. A. Kulagin, JETP Lett. **42**, 127 (1985). [Pisma Zh. Eksp. Teor. Fiz. **42**, 105 (1985)]
[13] S. V. Akulinichev, S. A. Kulagin and G. M. Vagradov, Phys. Lett. B **158**, 485 (1985)
[14] S.A. Kulagin, G. Piller and W. Weise, Phys. Rev. C **50**, 1154 (1994)
[15] C. H. Llewellyn Smith, Phys. Lett. B **128**, 107 (1983)
[16] M. Ericson and A. W. Thomas, Phys. Lett. B **128**, 112 (1983)
[17] G. Piller and W. Weise, Phys. Rept. **330**, 1 (2000)
[18] S. A. Kulagin, Nucl. Phys. A **500**, 653 (1989)
[19] S. A. Kulagin, Nucl. Phys. A **640**, 435 (1998)
[20] S. A. Kulagin and A. V. Sidorov, Eur. Phys. J. A **9**, 261 (2000)
[21] S. A. Kulagin and W. Melnitchouk, Phys. Rev. C **78**, 065203 (2008)
[22] F. Gross and S. Liuti, Phys. Rev. C **45**, 1374 (1992)
[23] S. G. Bondarenko, V. V. Burov, A. V. Molochkov, G. I. Smirnov and H. Toki, Prog. Part. Nucl. Phys. **48**, 449 (2002)
[24] S. I. Alekhin, S. A. Kulagin and S. Liuti, Phys. Rev. D **69**, 114009 (2004)
[25] R. Subedi et al., Science **320**, 1476 (2008)
[26] B. L. Friman, V. R. Pandharipande and R. B. Wiringa, Phys. Rev. Lett. **51**, 763 (1983)
[27] E. E. Sapershtein and M. Z. Shmatikov, JETP Lett. **41**, 53 (1985). [Pisma Zh. Eksp. Teor. Fiz. **41**, 44 (1985)]
[28] H. Jung and G. A. Miller, Phys. Rev. C **41**, 659 (1990)
[29] L. P. Kaptari, A. I. Titov, E. L. Bratkovskaya and A. Y. Umnikov, Nucl. Phys. A **512**, 684 (1990)
[30] E. Marco, E. Oset and P. Fernandez de Cordoba, Nucl. Phys. A **611**, 484 (1996)
[31] R. J. Glauber and G. Matthiae, Nucl. Phys. B **21**, 135 (1970)
[32] V. N. Gribov, Sov. Phys. JETP **30**, 709 (1970). [Zh. Eksp. Teor. Fiz. **57**, 1306 (1969)]
[33] T. H. Bauer, R. D. Spital, D. R. Yennie and F. M. Pipkin, Rev. Mod. Phys. **50**, 261 (1978). **51**(E), 407 (1979)
[34] S. A. Kulagin and R. Petti, Phys. Rev. D **76**, 094023 (2007)

[35] S. A. Kulagin and R. Petti, Phys. Rev. C **82**, 054614 (2010)
[36] S.I. Alekhin, S.A. Kulagin and R. Petti, arXiv:1609.08463 [nucl-th]
[37] H. Georgi and H. D. Politzer, Phys. Rev. D **14**, 1829 (1976)
[38] S. Alekhin, S. A. Kulagin and R. Petti, AIP Conf. Proc. **967**, 215 (2007)
[39] M. Arneodo et al. [New Muon Collaboration.], Nucl. Phys. B **441**, 12 (1995)
[40] P. Amaudruz et al. [New Muon Collaboration], Nucl. Phys. B **441**, 3 (1995)
[41] M. Arneodo et al. [New Muon Collaboration], Nucl. Phys. B **481**, 3 (1996)
[42] J. Ashman et al. [European Muon Collaboration], Z. Phys. C **57**, 211 (1993)
[43] G. Bari et al. [BCDMS Collaboration], Phys. Lett. B **163**, 282 (1985)
[44] J. Gomez et al., Phys. Rev. D **49**, 4348 (1994)
[45] M. R. Adams et al. [E665 Collaboration], Z. Phys. C **67**, 403 (1995)
[46] J. Seely et al., Phys. Rev. Lett. **103**, 202301 (2009)
[47] K. Ackerstaff et al. [HERMES Collaboration], Phys. Lett. B **475**, 386 (2000). B **567** (E), 339 (2003)
[48] J.-C. Peng and J.-W. Qiu, Prog. Part. Nucl. Phys. **76**, 43 (2014)
[49] D. M. Alde et al., Phys. Rev. Lett. **64**, 2479 (1990)
[50] M. A. Vasilev et al., Phys. Rev. Lett. **83**, 2304 (1999)
[51] R. P. Bickerstaff, M. C. Birse and G. A. Miller, Phys. Rev. Lett. **53**, 2532 (1984)
[52] J. D. Bjorken, FERMILAB-PUB-82-059-THY, FERMILAB-PUB-82-059-T
[53] G. T. Garvey and J. C. Peng, Phys. Rev. Lett. **90**, 092302 (2003)
[54] P. Ru, S. A. Kulagin, R. Petti and B. W. Zhang, arXiv:1608.06835 [nucl-th]
[55] V. Khachatryan et al. [CMS Collaboration], Phys. Lett. B **750**, 565 (2015)
[56] V. Khachatryan et al. [CMS Collaboration], Phys. Lett. B **759**, 36 (2016)
[57] S. Alekhin, J. Bluemlein, S. Moch and R. Placakyte, arXiv:1508.07923 [hep-ph]
[58] J. Gao et al., Phys. Rev. D **89**, 033009 (2014)

Chiral imbalance in QCD

Alexander Andrianov[1,2,⋆], *Vladimir* Andrianov[1,⋆⋆], and *Domenec* Espriu[2,⋆⋆⋆]

[1] *Department of Physics, Saint Petersburg State University, Universitetskaya nab. 7/9, Saint Petersburg 199034, Russia,*
[2] *Institute of Cosmos Sciences (ICCUB), University of Barcelona, Martí i Franquès, 1, 08028 Barcelona, Spain*

Abstract. The chiral imbalance (ChI) is given by a difference between the numbers of RH and LH quarks which may occur in the fireball after heavy ion collision. To characterize it adiabatically a quark chiral (axial) chemical potential must be introduced taking into account emergence of a ChI in such a phase. In this report the phenomenology of formation of Local spatial Parity Breaking (LPB) in the hot and dense baryon matter is discussed and its simulation within a number of QCD-inspired models is outlined. The appearance of new states in the spectra of scalar, pseudoscalar and vector particles in such a matter is elucidated. In particular, from the effective vector meson theory in the presence of Chern-Simons interaction it is demonstrated that the spectrum of massive vector mesons splits into three polarization components with different effective masses. The asymmetry in production of longitudinally and transversely polarized states of ρ and ω mesons for various values of the dilepton invariant mass can serve as a characteristic indication of the LPB in PHENIX, STAR and ALICE experiments.

1 Introduction: Chiral Imbalance in Heavy Ions Collisions

The behaviour of baryonic matter under extreme conditions in HIC has received a lot of attention [1, 2]. New properties of QCD in the environment were tested in current accelerator experiments on RHIC and LHC [3]. A medium generated in these collisions (a fireball) may serve for experimental and theoretical studies of various phases of a matter.

The dedicated experimental study of hadron correlations in non-central heavy-ion collisions at RHIC [4] and LHC [5] revealed a signal of the separation of electric charges predicted in [6] as a signature of local P- and CP-odd fluctuations in QCD matter. The subsequent studies [7, 8] improved the theoretical understanding of the underlying phenomenon, the "chiral magnetic effect" (CME) in the reactions for peripheral ion collisions (see [9] for a review) .

On the contrary, a gradient density of isosinglet pseudoscalar condensate can be formed as a result of large, "long-lived" topological fluctuations of gluon fields in the fireball in central collisions (see [10] for details). To describe various effects of hadron matter in a fireball with parity breaking, we must introduce the axial/ chiral chemical potential [10]. At finite temperature, the transitions between

⋆e-mail: a.andrianov@spbu.ru
⋆⋆e-mail: v.andrianov@spbu.ru
⋆⋆⋆e-mail: espriu@icc.ub.edu

the vacuum states with different topological Chern-Simons numbers can be induced by a classical thermal activation process, so-called "sphaleron" [11]. In QCD matter, sphalerons are abundant [12] and induce the quark chirality non-conservation. There are some experimental indications of an abnormal dilepton excess in the range of low invariant masses and rapidities and moderate values of the transverse momenta (see the review in [13]), which can be thought of as a result of LPB in the medium (the details can be found in [14]). In particular, in heavy-ion collisions at high energies, with raising temperatures and baryon densities, metastable state can appear in the fireball with a nontrivial topological/axial charge T_5, which is related to the gluon gauge field G_i:

$$T_5(t) = \frac{1}{8\pi^2}\int_{\text{vol.}} d^3x\,\varepsilon_{jkl}\,\mathrm{Tr}\Big(G^j\partial^k G^l - i\frac{2}{3}G^jG^kG^l\Big), \quad j,k,l = 1,2,3, \tag{1}$$

where the integration is over the finite fireball volume. Its jump ΔT_5 can be associated with the space-time integral of the gauge-invariant Chern-Pontryagin density:

$$\Delta T_5 \;=\; T_5(t_f) - T_5(0) = \frac{1}{16\pi^2}\int_0^{t_f} dt \int_{\text{vol.}} d^3x\,\mathrm{Tr}(G^{\mu\nu}\widetilde{G}_{\mu\nu}) \tag{2}$$

It is known that the divergence of isosinglet axial quark current $J_{5,\mu} = \bar{q}\gamma_\mu\gamma_5 q$ is locally constrained via the relation of partial conservation of axial current (PCAC) affected by the gluon anomaly:

$$\partial^\mu J_{5,\mu} - 2i\widehat{m}_q\bar{q}\gamma_5 q = \frac{N_f}{2\pi^2}\mathrm{Tr}(G^{\mu\nu}\widetilde{G}_{\mu\nu}). \tag{3}$$

This relation allows to find the connection of a nonzero topological charge with a non-trivial quark axial charge Q_5^q. Namely, integrating (3) over a finite volume of fireball we come to the equality

$$\frac{d}{dt}(Q_5^q - 2N_f T_5) \simeq 2i\int_{\text{vol.}} d^3x\,\widehat{m}_q\bar{q}\gamma_5 q, \tag{4}$$

$$Q_5^q = \int_{\text{vol.}} d^3x\,q^\dagger\gamma_5 q = \langle N_L - N_R\rangle, \tag{5}$$

where $\langle N_L - N_R\rangle$ stands for the vacuum averaged difference between left and right chiral densities of baryon number (chiral imbalance). Therefrom it follows that in the chiral limit and for a finite fireball volume the axial quark current is conserved in the presence of non-zero topological charge. If for the lifetime of fireball and the size of hadron fireball of order $L = 5 - 10$ fm , the created topological charge is non-zero, $\langle\Delta T_5\rangle \neq 0$, then it may be associated with a topological chemical potential μ_θ or an axial chemical potential μ_5 [10]. Thus we have

$$\langle\Delta T_5\rangle \simeq \frac{1}{2N_f}\langle Q_5^q\rangle \iff \mu_5 \simeq \frac{1}{2N_f}\mu_\theta. \tag{6}$$

Thus adding to the QCD Lagrangian the term $\Delta\mathcal{L}_{\text{top}} = \mu_\theta\Delta T_5$ or $\Delta\mathcal{L}_q = \mu_5 Q_5^q$, we get the possibility of accounting for non-trivial topological fluctuations ("fluctons") in the nuclear (quark) fireball.

2 Effective meson theories with LPB

For the detection of LPB in the hadron fireball let us consider an effective theory describing the electromagnetic interactions in a fireball. In this case, we keep in mind mechanisms of hadron interactions which rely on vector dominance model [15].

The quark-meson interaction is described by the Lagrangian,

$$\mathcal{L}_{\rm int} = \bar{q}\gamma_\mu V^\mu q; \quad V_\mu \equiv -eA_\mu Q + \frac{1}{2}g_\omega \omega_\mu \mathbf{I}_q + \frac{1}{2}g_\rho \rho^0_\mu \lambda_3 + \frac{1}{\sqrt{2}}g_\phi \phi_\mu \mathbf{I}_s, \tag{7}$$

while $Q = \frac{\lambda_3}{2} + \frac{1}{6}\mathbf{I}_q - \frac{1}{3}\mathbf{I}_s$, $g_\omega \simeq g_\rho \equiv g \simeq 6 < g_\phi \simeq 7.8$ and the values of the constants are extracted from the decays of the vector mesons. Here, $\mathbf{I}_q$ and $\mathbf{I}_s$ are the unit matrices in the non-strange and strange quark sectors, and λ_3 is a corresponding Gell-Mann matrix. The parity-odd contribution is given by the Chern-Simons term,

$$\mathcal{L}_{CS}(k) = -\frac{1}{4}\varepsilon^{\mu\nu\rho\sigma}\,\mathrm{Tr}\left[\hat{\zeta}_\mu V_\nu(x)\, V_{\rho\sigma}(x)\right] = \frac{1}{2}\mathrm{Tr}\left[\hat{\zeta}\,\epsilon_{jkl}\, V_j\, \partial_k V_l\right], \tag{8}$$

which describes the mixing of photons and vector mesons under the local spatial parity breaking. We can obtain the relation $\zeta = N_c g^2 \mu_5 / 8\pi^2$ where N_c is a number of colors, and numerically $\zeta \simeq 1.5\mu_5$.

The analysis of the massive Chern-Simons electrodynamics [10] has shown that in the case of an isosinglet pseudoscalar background field the spectrum of massive vector mesons splits into three polarizations with the masses $m^2_{V,+} < m^2_{V,L} < m^2_{V,-}$. The position of resonance poles for transverse polarizations of ρ^0, ω - mesons is shifted with the wave vector $|\vec{k}|$ and also a resonance broadening occurs that leads to an increased contribution of the dilepton production compared with the case of resonances in vacuum (for details, see [16]). Then the question arises. Could the splitting be measured in experiments with heavy-ion collisions?

For this purpose there should be analyzed an effect of anomalous dilepton pair production in the range of low invariant masses and rapidities and moderate transverse momenta which was established in a series of experiments with heavy-ion collisions in recent years [13]. It is well known that the angular distribution of leptons carries the information on polarizations. However, the current angular distribution studies based on full angular average do not seem to detect possible parity-odd effects.

In order to isolate the transverse polarizations, we perform different cuts choosing the angle θ_A for the analysis (see figure 1) and study the variations of the ρ (and ω) spectral functions. A quite visible

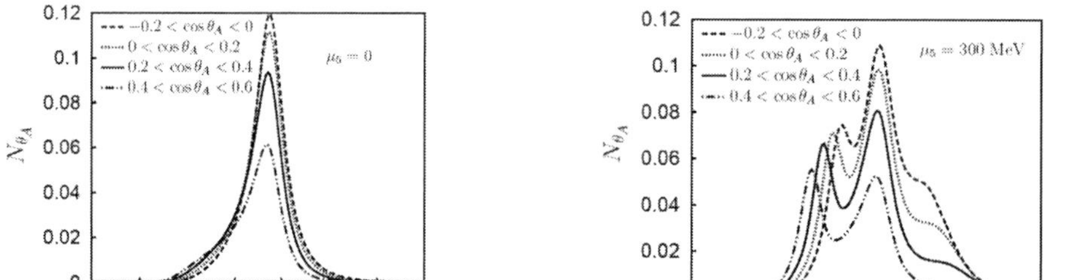

Figure 1. ρ spectral function depending on the dielectron invariant mass M in vacuum ($\mu_5 = 0$) and in a parity-breaking medium with $\mu_5 = 300$ MeV for different ranges of the angle θ_A between the two outgoing leptons in the laboratory frame [16].

secondary peak appears in a P-odd medium! To isolate the transverse polarizations in the spectrum, we selected different angle sectors and studied the changes in the ρ-meson spectral function. Various experimental possibilities for its identification were discussed in [16].

Thus a signal (a phase) with spatial parity breaking in heavy-ion collisions (in a fireball) can be sought in experiments "event by event" using an excess yield of dilepton pairs and predominantly with different circular polarizations outside the resonance region of the ρ- and ω-meson invariant masses.

3 Effective scalar meson Lagrangian with axial chemical potential

In the environment with chiral chemical potential the scalar sector can be described by using the spurion technique in the chiral Lagrangian with an isosinglet μ_5 [17],

$$D_\nu \Longrightarrow D_\nu - i\{\mathbf{I}_q\mu_5\delta_{0\nu},\cdot\} = D_\nu - 2i\mathbf{I}_q\mu_5\delta_{0\nu}, \quad (9)$$

where for three light flavors $\mathbf{I}_q \equiv \mathrm{diag}[1,1,0]$ in accordance with the preceding discussion. An effective Lagrangian is required to include the lightest isoscalar degrees of freedom such as σ and $a_0(980)$, which will be mixed with their pseudoscalar partners η, η' and π, respectively.

In the lowest order of chiral expansion such an effective Lagrangian reads,

$$\begin{aligned}\mathcal{L} &= \frac{1}{4}\mathrm{Tr}\left(D_\mu H D^\mu H^\dagger\right) + \frac{B}{2}\mathrm{Tr}\left[\hat{m}(H+H^\dagger)\right] + \frac{M^2}{2}\mathrm{Tr}\left(HH^\dagger\right) \\ &-\frac{\lambda_1}{2}\mathrm{Tr}\left[(HH^\dagger)^2\right] - \frac{\lambda_2}{4}\left[\mathrm{Tr}\left(HH^\dagger\right)\right]^2 + \frac{c}{2}(\det H + \det H^\dagger) \\ &+\frac{d_1}{2}\mathrm{Tr}\left[\hat{m}(HH^\dagger H + H^\dagger HH^\dagger)\right] + \frac{d_2}{2}\mathrm{Tr}\left[\hat{m}(H+H^\dagger)\right]\mathrm{Tr}\left(HH^\dagger\right) \end{aligned} \quad (10)$$

where the $SU(3)_L \times SU(3)_R$ chirally invariant part is parameterized by constants $M, \lambda_{1,2}, c$ and soft chiral symmetry breaking is realized with the help of vertices containing $\hat{m} = \mathrm{diag}[m_u, m_d, m_s]$ and constants $B, d_{1,2}$. The convenient splitting of scalar fields is given by

$$H = \xi\Sigma\xi, \quad \xi = \exp\left(i\frac{\Phi}{2f}\right), \quad \Phi = \lambda^a\phi^a, \quad \Sigma = \lambda^b\sigma^b$$

The v.e.v. of the neutral scalars are defined as $v_i = \langle\Sigma_{ii}\rangle$ where $i = u, d, s$, and, in the chiral limit $\hat{m} = 0$, satisfy the following gap equations:

$$(2\mu_5^2 + M^2) - 2(\lambda_1 + \lambda_2)v_q^2 - \lambda_2 v_s^2 + c v_s = 0;$$

$$(-2\lambda_2 + \frac{c}{v_s})v_q^2 + M^2 - (2\lambda_1 + \lambda_2)v_s^2 = 0,$$

where for non-vanishing isosinglet μ_5 we find our solutions to be $v_u = v_d \equiv v_q \neq v_s$.

Evidently for large μ_5 the correct solution has the following asymptotics,

$$v_q^2 = \frac{2\mu_5^2 + M^2 + \frac{c^2}{4\lambda^2}}{2(\lambda_1+\lambda_2)} + O(1/\mu_5^2), \qquad v_s = \frac{c}{2\lambda_2} + O(1/\mu_5^2).$$

These solutions are depicted in the figure 2.

We stress that the consistent solution with a finite limit for strange quark constituent mass $v_s \ll v_q$ exists if the term breaking $U(1)_A$ symmetry is in action. Meantime the constituent masses for light u, d flavours are growing with chiral chemical potential, first as $\sim \mu_5^2$ and next as $\sim \mu_5$. Moreover even if the spontaneous chiral symmetry breaking is absent for $M^2 < 0$ in parity-even environment this CSB will be triggered by chiral imbalance.

The lattice computations perfectly confirm [18] this behavior against chiral chemical potential.

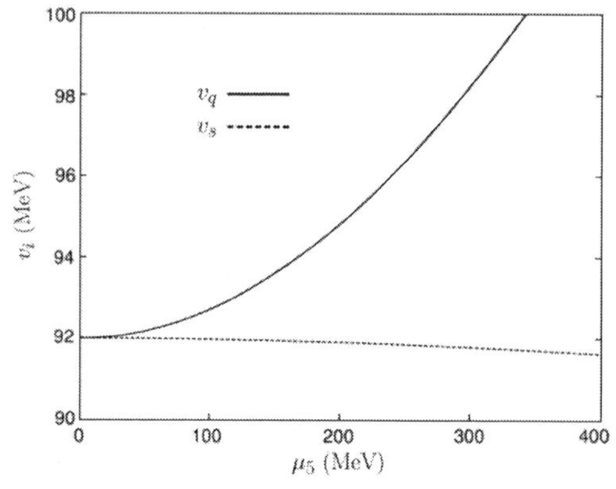

Figure 2. Vacuum condensates in the presence of chiral imbalance [17].

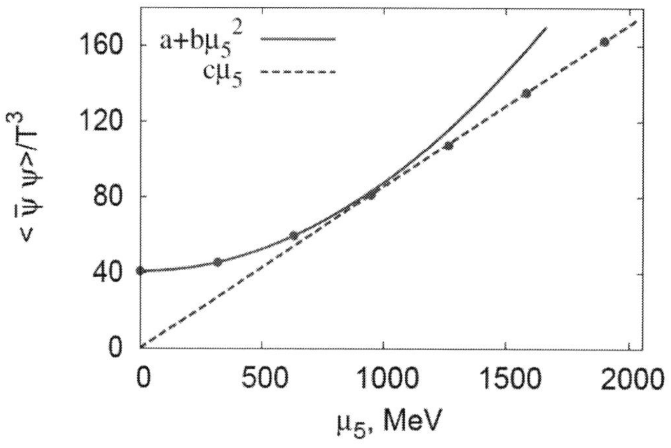

Figure 3. Chiral condensate in two-color QCD lattice (red dots) as a function of μ_5 in the confinement phase (T = 158 MeV) [18].

For further purposes we need the non-strange meson sector and η_s

$$\Phi = \begin{pmatrix} \eta_q + \pi^0 & \sqrt{2}\pi^+ & 0 \\ \sqrt{2}\pi^- & \eta_q - \pi^0 & 0 \\ 0 & 0 & \sqrt{2}\eta_s \end{pmatrix}, \Sigma = \begin{pmatrix} v_u + \sigma + a_0^0 & \sqrt{2}a_0^+ & 0 \\ \sqrt{2}a_0^- & v_d + \sigma - a_0^0 & 0 \\ 0 & 0 & v_s \end{pmatrix}.$$

$$\begin{pmatrix} \eta_q \\ \eta_s \end{pmatrix} = \begin{pmatrix} \cos\psi & \sin\psi \\ -\sin\psi & \cos\psi \end{pmatrix} \begin{pmatrix} \eta \\ \eta' \end{pmatrix}.$$

For $\mu_5 = 0$, in the chiral limit $\hat{m} = 0$ we adopt $v_u = v_d = v_s = v_0 \equiv f_\pi \approx 92$ MeV and the notations $m \equiv m_q = (m_u + m_d)/2$ and $m/m_s \simeq 1/25$.. The coupling constants (in units of MeV) were fitted to phenomenology assuming isospin symmetry with the help of χ^2 minimization (`MINUIT`).

One, of course, should not expect the range of applicability of dim-4 meson Lagrangian for chemical potentials larger than few hundreds of MeV. Nevertheless the predictions for lightest states, pions can be well controlled within the chiral perturbation theory.

Let us now consider the QCD-inspired scalar meson model with axial chemical potential [17] for a description of some properties of low-energy mesons in medium and how it manifests itself in LPB. First of all the axial chemical potential is introduced and treated as a constant time component of an isosinglet axial-vector field in the non-strange sector. For lightest isotriplet pseudoscalar π and scalar a_0 states the piece of the effective Lagrangian that is bilinear in fields looks as follows

$$\mathcal{L} = \frac{1}{2}(\partial a_0)^2 + \frac{1}{2}(\partial \pi)^2 - \frac{1}{2}m_1^2 a_0^2 - \frac{1}{2}m_2^2 \pi^2 - 4\mu_5 a_0 \dot{\pi}, \tag{11}$$

Evidently new eigenstates $\tilde{\pi}$ arise from the mixture of scalars and pseudoscalars then for big 3-momentum $\tilde{\pi}$ become massless and further on tachyonic [17]. For the same 3-momentum the quasi-scalar state σ looks stable.

4 Chemical potentials in the NJL model

In [19] we incorporated both a vector and an axial chemical potentials in the Nambu-Jona-Lasinio (NJL) model with the purpose of unraveling the landscape of different stable phases of the theory. It turns out that the inclusion of μ_5 changes radically the phase structure of the model and shows that μ is not a key player in ushering a thermodynamically stable phase where parity is violated in the NJL model, but μ_5 is. It leads to a non-trivial dependence of the scalar condensate in the chirally broken phase (see figure 4).

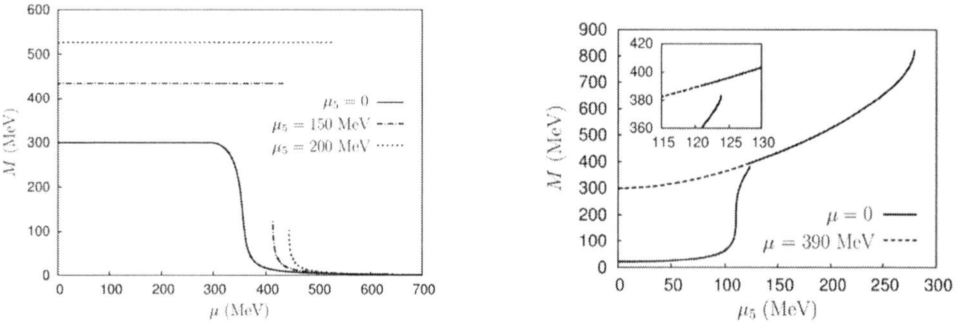

Figure 4. Evolution of the constituent quark mass M [19]: left panel- depending on μ for different values of the axial chemical potential μ_5; right panel- depending on μ_5 for different values of the chemical potential μ.

Critical temperature with axial potential switched on was computed on lattice [18] and a fairly well correspondence with the results of [19] was established. The main result is that with increasing chiral chemical potential the dynamical mass and critical temperature raise up.

5 Conclusions and outlook

In this talk we described a possibility of local spatial parity breaking emerging in a dense hot baryon medium (fireball) in heavy-ion collisions at high energies. We stress that LPB is not forbidden by any physical principle in QCD at finite temperature/density. We suggested a generalized Lagrangian of the vector meson dominance model in the presence of the Chern-Simons interaction. It turns out that the spectrum of massive vector mesons splits into three components with different polarizations and with different effective masses $m_{V,+}^2 < m_{V,L}^2 < m_{V,-}^2$ and a resonance broadening occurs that leads to an increase of the spectral contribution to the dilepton production as compared with the vacuum state. The proposed mechanism for generating local spatial parity breaking helps to explain qualitatively and quantitatively the anomalous yield of dilepton pairs in the CERES, PHENIX, STAR, NA60, and ALICE experiments.

Besides, LPB enhances dynamical chiral symmetry breaking in QCD: chiral condensates, critical temperature of chiral symmetry restoration are increasing with chiral chemical potential. Accordingly, LPB modifies dispersion laws for scalar and vector mesons: lightest "pseudoscalar" mesons tend to massless states in flight, vector meson polarizations split with different in-flight masses.

Acknowledgements

A.A. expresses his appreciation to the organizing Committee of the XXIII International Baldin Seminar "Relativistic Nuclear Physics and Quantum Chromodynamics", for the hospitality and organization of this conference. This work has been partially supported through grants FPA2013-46570, 2014-SGR-104 and Consolider CPAN. Funding was also provided by the Spanish MINECO under project MDM-2014-0369 of ICCUB (Unidad de Excelencia 'Maria de Maeztu'). A.A. and V.A. were supported by Grant RFBR project 16-02-00348 and also got a financial support of SPbSU, in particular, by projects 11.42.1049.2016, 11.42.1051.2016.

References

[1] P. Jacobs, D. Kharzeev, B. Muller, J. Nagle, K. Rajagopal and S. Vigdor, [arXiv:0705.1930 [nucl-ex]]
[2] J.-P. Blaizot, F. Gelis, J. -F. Liao, L. McLerran and R. Venugopalan, Nucl. Phys. A, **873**, 68 (2012)
[3] A. Andronic et al., Nucl. Phys. A, **837**, 65 (2010)
[4] B. Abelev, et al., Phys. Rev. C, **81**, 054908 (2010)
[5] B. Abelev et al. (ALICE Collaboration), Phys. Rev. Lett. **110**, 012301 (2013)
[6] D. Kharzeev, Phys. Lett. B **633**, 260 (2006)
[7] D. E. Kharzeev, L. D. McLerran, and H. J. Warringa, Nucl. Phys. A **803**, 227 (2008)
[8] K. Fukushima, D. E. Kharzeev, and H. J. Warringa, Phys. Rev. D **78**, 074033 (2008)
[9] D. E. Kharzeev, Prog. Part. Nucl. Phys. **75**, 133 (2014)
[10] A. A. Andrianov, V. A. Andrianov, D. Espriu and X. Planells, Phys. Lett. B **710**, 230 (2012)
[11] F.R. Klinkhamer, N.S. Manton, Phys. Rev. D **30**, 2212 (1984). V.A. Kuzmin, V.A. Rubakov, M.E. Shaposhnikov, Phys. Lett. B **155**, 36 (1985)
[12] L. D. McLerran, E. Mottola and M. E. Shaposhnikov, Phys. Rev. D **43**, 2027 (1991)
[13] I. Tserruya, Landolt-Börnstein **23**, 176 (2010)
[14] A. A. Andrianov, V. A. Andrianov, D. Espriu, X. Planells, Theor. Math. Phys., **170**, 17 (2012). A. A. Andrianov, V. A. Andrianov, Theor. Math. Phys. **185**, 1370 (2015)
[15] J. J. Sakurai, Ann. Phys. **11**, 1 (1960)
[16] A. A. Andrianov, V. A. Andrianov, D. Espriu, and X. Planells, Phys. Rev. D **90**, 034024 (2014)
[17] A. A. Andrianov, D. Espriu & X. Planells, Eur. Phys. J. C **73**, 2294 (2013)
[18] V. V. Braguta, V. A. Goy, E. M. Ilgenfritz, A. Yu. Kotov, A. V. Molochkov, M. Muller-Preussker, and B. Petersson. JHEP, **06**, 094 (2015). V.V. Braguta et al. Phys. Rev. D **93**, 034509 (2016). V.V. Braguta et al. AIP Conf. Proc., **1701**, 060002 (2016). V.V. Braguta and A.Y. Kotov, Phys. Rev. D **93**, 105025 (2016)
[19] A. A. Andrianov, D. Espriu, and X. Planells, Eur. Phys. J. C **74**, 2776 (2014)

Hyperons polarization in heavy-ion collisions

Mircea Baznat[1,2,⋆], *Konstantin* Gudima[1,⋆⋆], *Alexander* Sorin[2,3,4,⋆⋆⋆], and *Oleg* Teryaev[2,3,4,⋆⋆⋆⋆]

[1] *Institute of Applied Physics, Academy of Sciences of Moldova, MD-2028 Kishinev, Moldova*
[2] *Joint Institute for Nuclear Research, 141980 Dubna, Russia*
[3] *National Research Nuclear University MEPhI (Moscow Engineering Physics Institute), Kashirskoe Shosse 31, 115409 Moscow, Russia*
[4] *Dubna International University, 141980, Dubna, Russia*

Abstract. We study the structure of vorticity and hydrodynamic helicity fields in peripheral heavy-ion collisions using the kinetic Quark-Gluon Strings Model. The angular momentum which is a source of P-odd observables is preserved within this model with a good accuracy. We observe the formation of specific toroidal structures of vorticity field. Their existence is mirrored in the polarization of hyperons of the percent order. The observed qualitative energy dependence of polarization was predicted earlier and is quantified now.

1 Introduction

The local violation [1] of discrete symmetries in strongly interacting QCD matter is now under intensive theoretical and experimental investigations. The renowned Chiral Magnetic Effect (CME) uses the (C)P-violating (electro)magnetic field emerging in heavy ion collisions in order to probe the (C)P-odd effects in QCD matter.

There is an interesting counterpart of this effect, Chiral Vortical Effect (CVE)[2] due to coupling to P-odd medium vorticity leading to the induced electromagnetic and all conserved-charge currents [3], in particular the baryonic one.

Another important P-odd observable is the baryon polarizatiion. The mechanism analogous to CVE (known as axial vortical effect, see [4] and references therein) leads to induced axial current of strange quarks which may be converted to polarization of Λ-hyperons [3, 5, 6]. Another mechanism of this polarization is provided by so-called thermal vorticity in the hydrodynamical approach [7].

The zeroth component of axial current and correspondent axial charge are related to hydrodynamical helicity

$$H \equiv \int dV(\vec{v} \cdot \vec{w}),$$

⋆e-mail: baznat@theor.jinr.ru
⋆⋆e-mail: gudima@cc.acad.md
⋆⋆⋆e-mail: sorin@theor.jinr.ru
⋆⋆⋆⋆e-mail: teryaev@theor.jinr.ru

being the projection of velocity $\vec{v}$ to vorticity $\vec{w} = curl\vec{v}$. This quantity is manifesting the recently discovered [5] and confirmed [8] phenomenon of the separation, i.e. its mirror behavior with the same magnitudes but different signs in the half-spaces separated by the reaction plane.

The noncentral heavy ion collisions could naturally generate a rotation (global or local, both related to vorticity) with an angular velocity normal to the reaction plane, which is their generic qualitative feature. It is natural to expect that angular momentum conservation plays an essential role in the defining the quantitative properties of vortical effects. At the same time, it remains to be studied to which extent the particles carrying the main part of angular momentum participate in the collisions.

In the current paper we address this problem by performing the extensive numerical simulations. We explore the distribution of angular momentum and find that the role of participant nucleons is relatively small, albeit noticeable. We study in some detail the structure of vorticity field and compare different approaches to polarization calculation. We observed the peculiar toroidal "tire-like" structure manifesting themselves in the polarization of hyperons. We will explore different approaches to polarization calculation which will lead to qualitatively similar results.

2 Angular momentum conservation in the kinetic model

The natural source of the P-odd observables in heavy-ion collisions is the pseudovector of angular momentum. The question immediately emerges whether it is conserved in the course of evolution governed by Quark-Gluon String Model (QGSM)[9, 10]. To check this we calculated the angular momentum at various time moments of collision taking into account both the contributions of participants and spectators. We consider the $Au + Au$ collisions with $b = 8\ fm$ at $\sqrt{s} = 5\,GeV/u$ typical for future NICA collider. We observed (see figure 1) that the participants carry about 20% of angular momentum and that the total angular momentum of participants and spectators is conserved with a rather good accuracy.

One may conclude that the angular momentum is under the good control in QGSM model.

We also studied correlations of angular momentum and the hydrodynamic helicity and observed that these quantities vary in accordance with each other. In order to perform this comparison and further calculations, velocity, vorticity and helicity were determined following the earlier suggested [5] procedure when the respective quantities were properly averaged over events and particles within the three-dimensional cells providing the transition from the kinetic to hydrodynamic description.

Let us pass to the corresponding results.

3 Large-scale structures of vorticity fields

We start our studies with the qualitative structure of velocity and vorticity fields.

The general structure of velocity field follows the "little bang" pattern which may be quantified by the velocity dependence allowing to extract the "little Hubble" constant. We calculated the dependence of average cell velocity on the transverse distance $\rho = \sqrt{x^2 + y^2}$ and found (see figure 2) that it is consistent with the linear "Hubble" law

$$< v/c >= v_0/c + H\rho. \tag{1}$$

The Hubble constant H is measured in the units $10^{-22}\,sec = 30\,fm/c$ and changes in the range

$$H = 0.024 \div 0.028\,(fm/c)^{-1}$$

It corresponds to the "little Universe" lifetime of about $40\,fm/c$ which is only twice larger than the collision time.

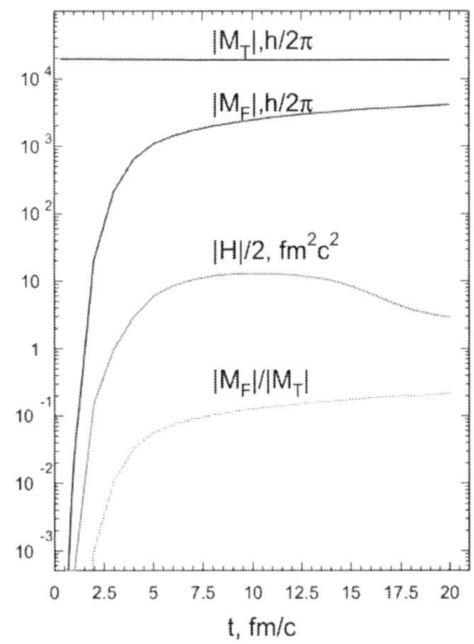

Figure 1. The time dependence of the total (M_T and fireball (M_F) angular momenta in Planck constant units and that of hydrodynamical helicity.

Our key observation is that while velocity field represent the "little bang" picture, vorticity field form the relatively thin toroidal "tire-like" structures (figure 3), which emerge in the layer (where velocity field changes rapidly) separating the "core" and "corona" regions [11, 12] and form the sort of vortex sheet [13].

An interesting property of these structures is that, while emerging due to angular momentum pseudovector $\vec{L}$ in the non-central collisions they do not "remember" the production plane and possess the cylindrical symmetry w.r.t. collisions axis z. This may be observed (figure 4) by considering the vortex sheet in the case of the particular direction of $\vec{L}$ along the y axis.

Such behaviour may resemble cyclones appearing at femtoscopic scale.

Let us now discuss the observable signatures of these nice structures.

4 Hyperon polarization

We consider hyperon polarization as the observable related to vorticity and helicity [3]. We shall concentrate mostly on Λ hyperon production, which has some advantages: they are produced in large numbers, their polarization may be easily determined in their weak decays, and their spin is carried by strange quark.

We compare the two rather distinct methods of determining the hyperon polarization. The first corresponds to its earlier suggested [3] and explored [5] relation to the induced axial current while the second one follows the procedure based on the thermal vorticity [7].

DOI: 10.1051/epjconf/201713801008

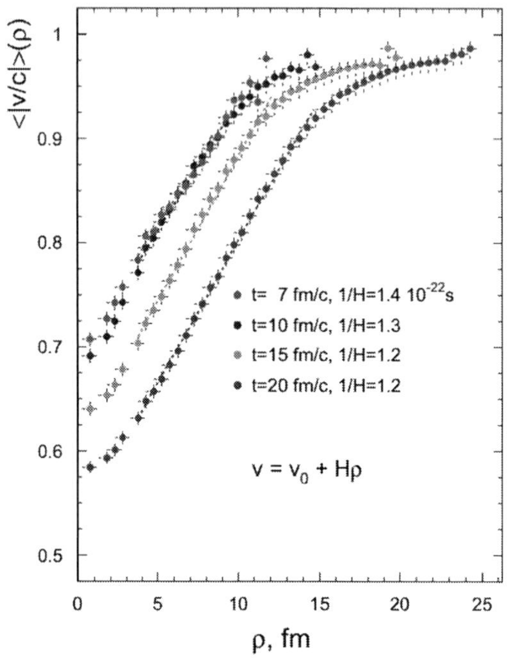

Figure 2. The cell velocity dependence on the transverse distance.

The first method is based on the calculation of strange axial charge

$$Q_5^s = \frac{N_c}{2\pi^2}\int d^3x\mu^2(x)\gamma^2\epsilon^{ijk}u_i\partial_j u_k$$
$$= \frac{<\mu^2\gamma^2> N_c H}{2\pi^2}. \quad (2)$$

In [5] we used the latter equality exploring the mean-value theorem, while here the spatial variation of strange chemical potential μ is taken into account. To do so, the description of kinetic distribution functions by the correspondent equilibrium equation was performed, providing the matching of kinetic and thermodynamical descriptions. The average polarization can be estimated by dividing Q_5^s (2) by the number of Λ's, assuming that the pseudovector of axial current is proportional to the pseudovector of polarization, $Q_5^s \sim <\Pi_0^{\Lambda,lab}>$. Selecting the axial charge related to the particles in the definite rapidity or transverse momentum interval, the respective dependence of polarization may be also obtained.

As the axial charge should be related to the zeroth component of hyperon polarization in laboratory frame Π_0^{lab}, the transformation to hyperon rest frame should be performed. Taking into account that

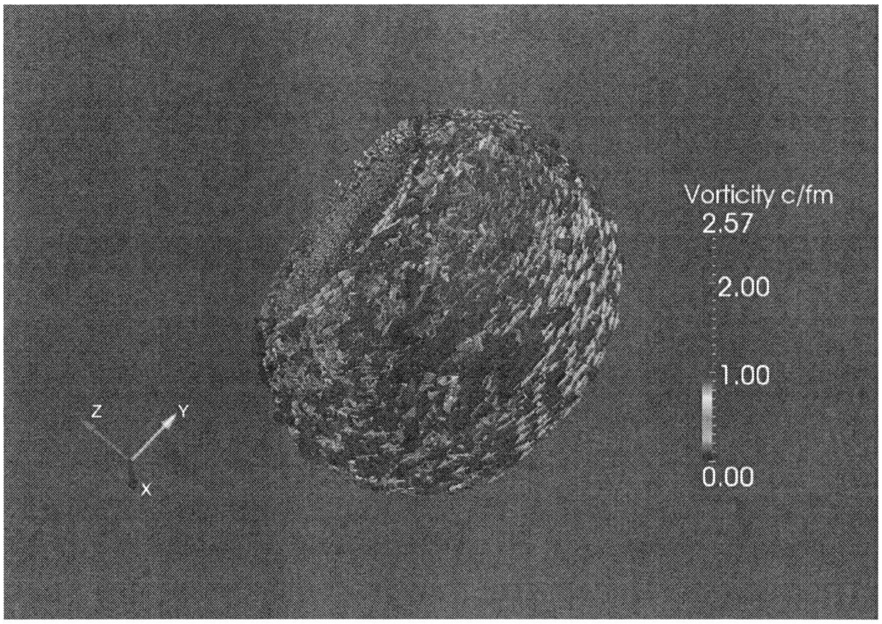

Figure 3. The vortex sheet.

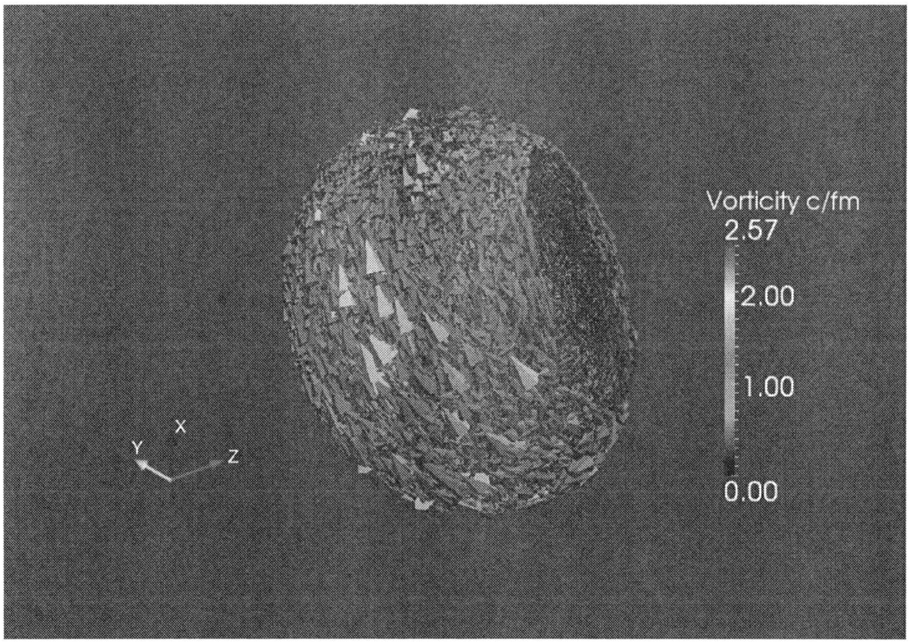

Figure 4. The vortex sheet for the particular direction of angular momentum.

 DOI: 10.1051/epjconf/201713801008

polarization pseudovector should be directed along y axis (as it has to be collinear to $\vec{L}$ pseudovector), one get

$$\Pi_0^{\Lambda,lab} = \frac{\Pi_0^{\Lambda} p_y}{M_{\Lambda}}, \tag{3}$$

so that the rest frame polarization can be obtained as

$$< \Pi_0^{\Lambda} > = Q_5^s < \frac{M_{\Lambda}}{N_{\Lambda} p_y} > . \tag{4}$$

The possible violation of positivity ($|\Pi_{\Lambda}| \geq 1$) or even the divergence is due to the fact, that hyperons with zero y component of the momentum should not have the zeroth component of polarization and therefore should not contribute to Q_5^s. To avoid this problem one may instead attribute the factor p_y/M to each hyperon in the denominator of (4). Nevertheless the comparison (see figure 5) of various approaches shows the similar scale and rapidity dependence of polarization.

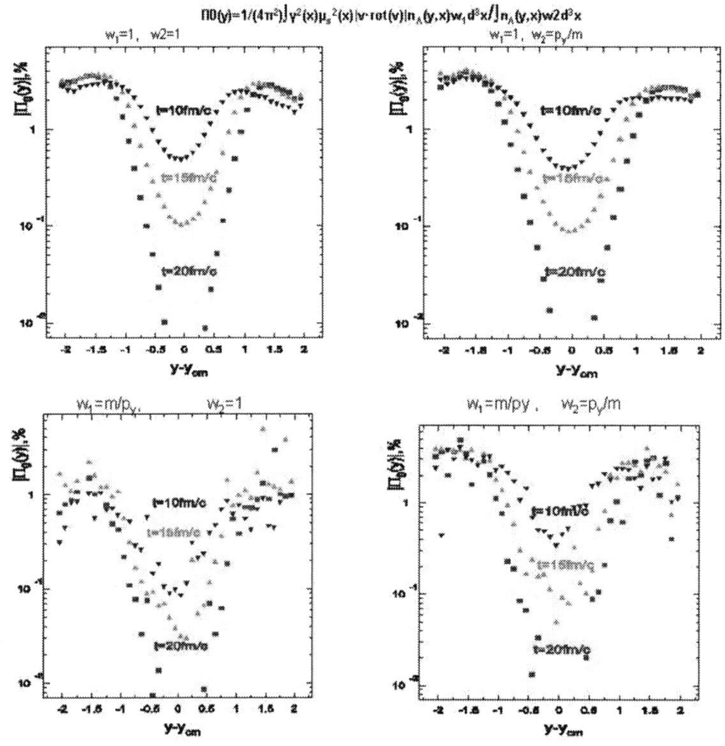

Figure 5. The rapidity dependence of polarization in helicity-based approach.

Another approach to polarization is based on the so-called thermal vorticity [7]. To provide the comparison we calculated (see figure 6) the thermal vorticity field and respective polarization. While a scale of the polarization in the thermal vorticity-based approach is several times larger, its rapidity dependence, surprisingly enough, appear to be similar in these rather distinct approaches.

The current STAR@RHIC measurements of polarization [14] manifested the decrease with energy which is naturally explained in the anomalous mechanism but also happens in the thermal vorticity mechanism. To quantify the predictions, we calculated the energy dependence in the anomalous mechanism, covering the experimental points (see figure 7).

DOI: 10.1051/epjconf/201713801008

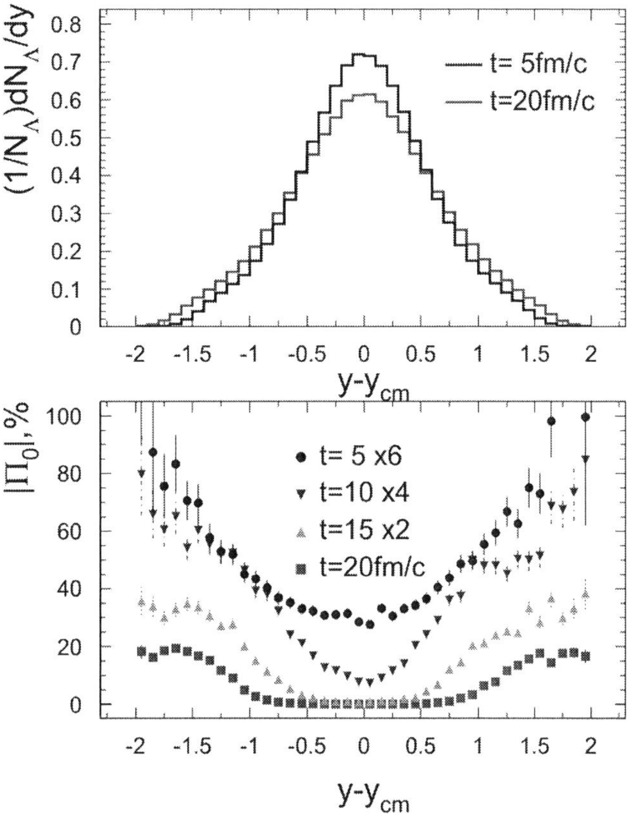

Figure 6. The rapidity dependence of polarization in thermal vorticity-based approach. The magnitudes of polarization at different times are rescaled by the identified factors.

5 Conclusions and Outlook

We investigated vorticity and hydrodynamical helicity in noncentral heavy-ion collisions in the framework of the kinetic Quark-Gluon String Model. We have confirmed our earlier observation that the vorticity field is predominantly localized in a relatively thin layer ($2 \div 3\ fm$) on the boundary between the participant and spectator nucleons and observed that it is forming the specific toroidal structures, which might be considered as vortex sheets with the unexpected cylindrical symmetry. They look as cyclones appearing at femtoscopic scale.

The vorticity and helicity fields are manifested in the Λ hyperon polarization. We performed its detailed calculations including the simulations of the strange chemical potential. We found that the polarization magnitude might reach a percent level. The comparison with the very different approach

DOI: 10.1051/epjconf/201713801008

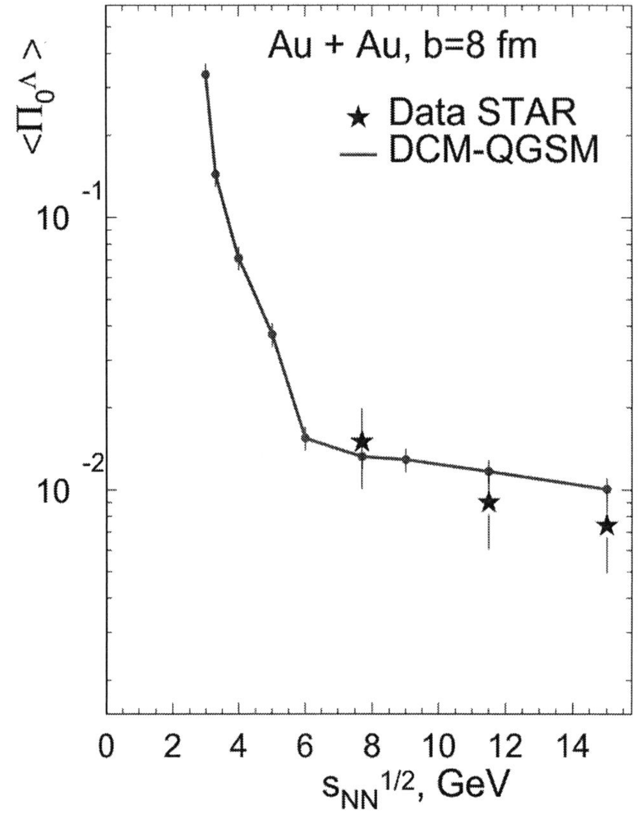

Figure 7. The energy dependence of polarization in anomaly-based approach compared with experimental STAR points.

exploring the thermal vorticity leads to qualitatively similar results, although the polarization scale is several times larger.

Acknowledgements

This work was supported in part by the Russian Foundation for Basic Research, Grant No. 14-01-00647.

References

[1] K. Fukushima, D.E. Kharzeev and H.J. Warringa, Phys. Rev. D **78**, 074033 (2008). [arXiv:0808.3382 [hep-ph]]
[2] D. Kharzeev and A. Zhitnitsky, Nucl. Phys. A **797**, 67 (2007). [arXiv:0706.1026 [hep-ph]]
[3] O. Rogachevsky, A. Sorin and O. Teryaev, Phys. Rev. C **82**, 054910 (2010). [arXiv:1006.1331 [hep-ph]]
[4] T. Kalaydzhyan, Phys. Rev. D **89 (10)**, 105012 (2014). [arXiv:1403.1256 [hep-th]]
[5] M. Baznat, K. Gudima, A. Sorin and O. Teryaev, Phys. Rev. C **88**, 061901 (2013). [arXiv:1301.7003 [nucl-th]]
[6] J. -H. Gao, Z. -T. Liang, S. Pu, Q. Wang and X. -N. Wang, Phys. Rev. Lett. **109**, 232301 (2012). [arXiv:1203.0725 [hep-ph]]
[7] F. Becattini, L. Csernai and D. J. Wang, Phys. Rev. C **88 (3)**, 034905 (2013). [arXiv:1304.4427 [nucl-th]]
[8] O. Teryaev and R. Usubov, Phys. Rev. C **92 (1)**, 014906 (2015)
[9] V. D. Toneev, N. S. Amelin, K. K. Gudima and S. Y. Sivoklokov, Nucl. Phys. A **519**, 463C (1990)
[10] N. S. Amelin, E. F. Staubo, L. P. Csernai, V. D. Toneev and K. K. Gudima, Phys. Rev. C **44**, 1541 (1991)
[11] J. Aichelin and K. Werner, Phys. Rev. C **79**, 064907 (2009). [Phys. Rev. C **81**, 029902 (2010)]. [arXiv:0810.4465 [nucl-th]]
[12] J. Steinheimer and M. Bleicher, Phys. Rev. C **84**, 024905 (2011). [arXiv:1104.3981 [hep-ph]]
[13] M. I. Baznat, K. K. Gudima, A. S. Sorin and O. V. Teryaev, Phys. Rev. C **93 (3)**, 031902 (2016). doi:10.1103/PhysRevC.93.031902. [arXiv:1507.04652 [nucl-th]]
[14] M. Lisa [for STAR Collaboration]. Talk at QCD Chirality Workshop, UCLA, February 2016

Probing composite models at the LHC with exotic quarks production

Romain Kukla[1,2,★]

[1]*Service de Physique des Particules, Institut de Recherche sur les lois Fondamentales de l'Univers, CEA Saclay, Gif sur Yvette, France*
[2]*Centre de Physique des Particules de Marseille, UMR CNRS/Université d'Aix-Marseille, Marseille, France*

Abstract. After the Higgs boson hunt, the LHC could be a powerful tool to unravel the mystery of which physics lies beyond the realm of the Standard Model. Different new sectors have been postulated to address naturalness: SUSY, extra dimensions and strong dynamics theories. Composite models extend EWSB to a global symmetry breaking whose pseudo-Goldstone boson is the SM Higgs boson. The resulting mass spectrum originates from a partial mixing between fundamental fermions and composite fields which creates massive states including new heavy quarks coupled preferentially to the top quark. Searches for these *top partners* have been carried out by the ATLAS and CMS collaborations, constraining the models. Other composite contributions are expected to enhance the 4-top production, which should be observable in the next years at the LHC.

1 Introduction

Even if the Standard Model (SM) seems to have reached its final form with the latest discoveries (1995: top quark, 2000: tau neutrino, 2012: Higgs boson), many open questions remain unsolved and physics Beyond the Standard Model (BSM) is necessary to answer them. For instance, the Higgs mass is stabilized in the SM thanks to very large quantum corrections from the loop diagrams. The hierarchy problem, or equivalently naturalness, is triggered by the existence of different scales in the SM especially between the fermion masses which should be at the same order of magnitude. The inclusion of dark matter candidates and strong CP violation to explain matter-antimatter asymmetry remain unsolved as for now. What is sure is that the SM needs to be extended and currently the two most explored ways to do so in the Higgs sector are: either considering that the Higgs boson is coupled with SUSY particles in the same way it is with the SM, with the help of superpartners which suppress the quadratic divergences of the Higgs mass and make the gauge couplings converging towards unification at high energy; or adding a strongly coupled theory which does not make the Higgs boson a fundamental object anymore and explains the fermion masses hierarchy with proto-Yukawa couplings depending on the linear mixing between the fundamental SM states and composite fields. In both cases, the top quark being the heaviest particle known in the SM and being coupled to the Higgs boson with a Yukawa coupling ~ 1, it is considered to be a logical probe to test BSM physics. Compositeness takes some inspiration from QCD pseudo-scalars which, as massive pseudo-Goldstone

★e-mail: romain.kukla@cern.ch

bosons of the spontaneous chiral symmetry breaking $SU(2)_L \times SU(2)_R \to SU(2)_V$, have a lower mass ($\sim$ 130 MeV) compared to the typical QCD mass range around 1 GeV (ρ mesons, neutron). This would be translated in terms of composite states around the TeV scale compared to a Higgs mass at $\sim$ 125 GeV. Many realizations of this paradigm exist and are eagerly studied by the theoretical and experimental communities as they are a compelling alternative to supersymmetry. It has been pointed out in the very first developments of holographic Higgs theories that the AdS/CFT duality would map compositeness to strongly coupled physics, through Randall-Sundrum warped extra dimensions [1]. This paper will present the partial compositeness framework which is constrained in many experimental searches, of which we will discuss some of them recently published by the ATLAS collaboration. We will focus on the pair production of heavy quarks. A prospective study will conclude our overview of compositeness at the LHC with the proposal to probe intermediate production of heavy states in four top final states at $\sqrt{s} = 14$ TeV.

2 The compositeness paradigm

The extension of the SM is done by adding a composite sector, which is assumed to have a strongly-coupled dynamics, characterized by an exact global symmetry $\mathcal{G}$ spontaneously broken into $\mathcal{H}$, where the electroweak group $G_{EW} = SU(2)_L \times U(1)_Y \in \mathcal{H}$. The $\mathcal{G}$ group is supposed to be large enough so that there is at least 4 Goldstone bosons generated in the coset $\mathcal{G}/\mathcal{H}$, which is usually done by choosing a $SO(5) \to SO(4)$ breaking. These four Goldstones bosons are defined as the Higgs doublet. In the elementary sector, which contains all other SM fields, $\mathcal{G}$ is explicitly broken and the coupling between the two sectors transmits this to the composite sector where the Higgs doublet acquires a mass and a potential, thus triggering EWSB. This big picture is summarized in figure 1.

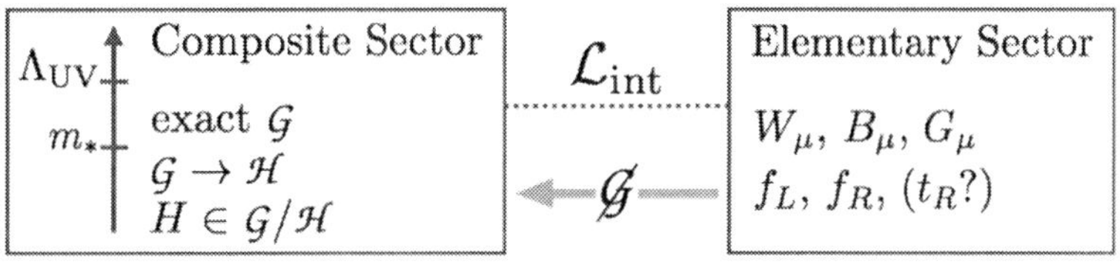

Figure 1. The composite Higgs paradigm realized with two sectors [2].

More precisely, the couplings between the two sectors are chosen to be linear, so that the mass eigenstates are a mixing of elementary and composite fields. In addition to the common SM particles, heavy resonances are expected, as described in Equations 1-2, where ϕ_n denotes the mixing angle for each particle. This is the *partial compositeness* scenario in which the mass hierarchy is transposed in a mixing hierarchy: indeed, the Higgs boson being part of the composite sector, the Yukawa coupling to the SM particles is achieved through the proto-Yukawa coupling to composite states (which acquired mass by another mechanism) appearing in the mixing, see figure 2. Thus, the higher the mixing angle is, the heavier the particle is.

$$|\text{SM}_n\rangle = \cos\phi_n |\text{elementary}_n\rangle + \sin\phi_n |\text{composite}_n\rangle \tag{1}$$

$$|\text{Heavy}_n\rangle = -\sin\phi_n |\text{elementary}_n\rangle + \cos\phi_n |\text{composite}_n\rangle \tag{2}$$

More details on the theoretical side can be found in the literature, such as [2]. What is of interest for us is the phenomenology arising from this new sector which is composed, as explained, of the Higgs

DOI: 10.1051/epjconf/201713801009

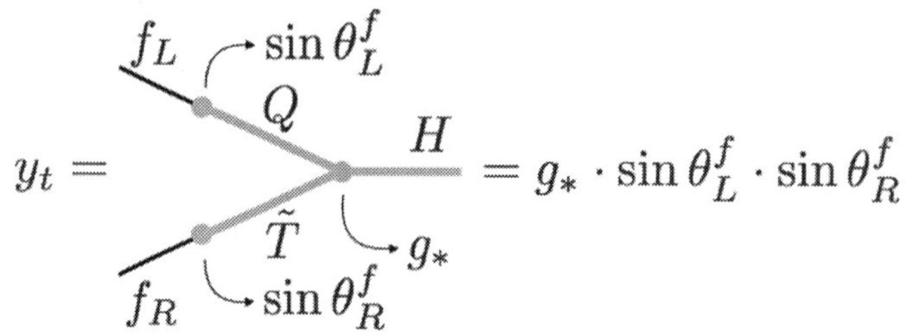

Figure 2. The effective Yukawa coupling between SM fermions f and the Higgs boson H via their composite fraction Q and $\tilde{T}$. The parameters g_* and $\sin\theta$ characterize the proto-Yukawa coupling in the composite sector and the mixing between the two sectors [2].

doublet plus heavy quarks coupled to the top (the light quarks are neglected due to tiny couplings and the composite bosons are not taken into account), written following the $5 = (2,2)\oplus(1,1)$ representation as a bi-doublet and a bi-singlet of the global symmetry: $Q = (2,2)$ et $\tilde{T} = (1,1)$, see Equation 3. The T, B and $\tilde{t}_1$ have the same quantum numbers as t_L, b_L and t_R. The heavy quarks doublet $\begin{pmatrix} T_{5/3} \\ T_{2/3} \end{pmatrix}$ (with respective charges of +5/3 and +2/3) and the singlet $\tilde{t}_2$ are consequences of the custodial symmetry introduced in $\mathcal{G}$ to stabilize the ρ parameter (or T operator) linked to $Zb_L\bar{b}_L$ vertex.

$$Q = \begin{bmatrix} T & T_{5/3} \\ B & T_{2/3} \end{bmatrix}, \quad \tilde{T} = [\tilde{t}_1, \tilde{t}_2] \quad , \quad H = \begin{bmatrix} \phi_0^* & \phi^+ \\ -\phi^- & \phi_0 \end{bmatrix} \tag{3}$$

These quarks couple to t and b via scalar and vector bosons:

- $T \to W^+b$, $T \to Zt$ and $T \to Ht$.
- $B \to W^-t$, $B \to Zb$ and $B \to Hb$.
- $T_{5/3} \to W^+t$.
- $T_{2/3} \to Zt$ and $T_{2/3} \to Ht$.

The main production diagram is the creation of a quark-antiquark pair initiated by QCD gluon fusion. An additional process leads to single production of these heavy quarks, highly dependent on the coupling, which can contribute at the same level of the pair production in the case of a very strong mixing parameter. These diagrams are shown in figure 3 for the $T_{5/3}$ case. For all the 4 heavy quarks, the pair production can give a peculiar signature with 2 same-charge leptons which is very rare in the SM, especially with associated b quarks (we will see later that this is characteristic of 4-top events). We will now describe in more details experimental searches for heavy quarks coming from the composite sector and which couple to the top quark.

3 Searches for top partners at the LHC in the 2 same-sign leptons channel

The very first analysis based on these models was performed at the Tevatron by the CDF collaboration, using 2.7 fb^{-1} of data collected at $\sqrt{s} = 0.98$ TeV [4]. The considered signal is the combined pair productions of B and $T_{5/3}$, assumed to have the same mass, which give the same signature. After an event selection requiring at least 2 jets with $p_T \geq 15$ GeV, with one b-tagged, 2 same-sign leptons with $p_T \geq 20$ GeV and at least 20 GeV of transverse missing p_T (nicknamed MET), they excluded

 DOI: 10.1051/epjconf/201713801009

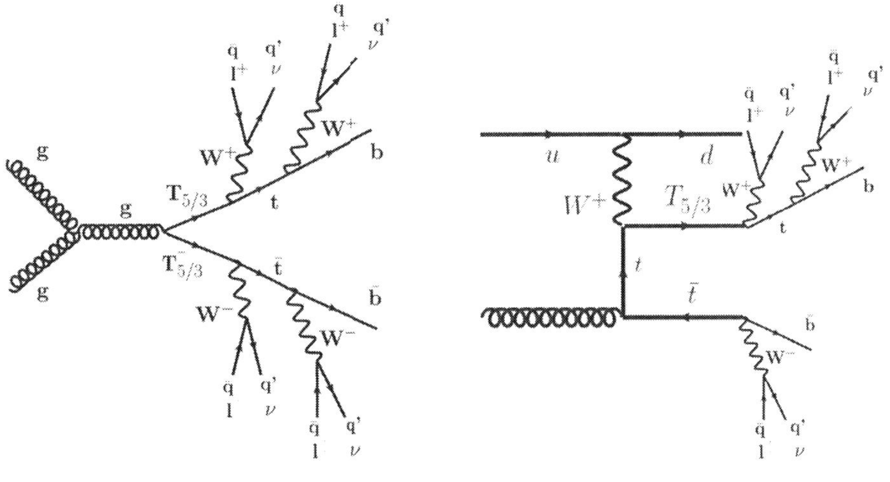

Figure 3. Inclusive final states for both pair and single productions of the $T_{5/3}$ exotic quark.

mass regions under 365 GeV. This very first limit was superseded with the first LHC data [3], namely 4.7 fb^{-1} at $\sqrt{s}$ = 7 TeV, for similar cuts except a new one on the scalar sum of the p_T of visible objects (jets, charged leptons): $H_T \geq 550$ GeV, justified by the intermediate production of two $T_{5/3}$ quarks whose mass is expected to be at least 350 GeV. This analysis used for the first time the single production for different regimes: very low and high couplings, so that its contribution ranged from marginal to dominant. The resulting exclusion plots from CDF and ATLAS are shown in figure 4.

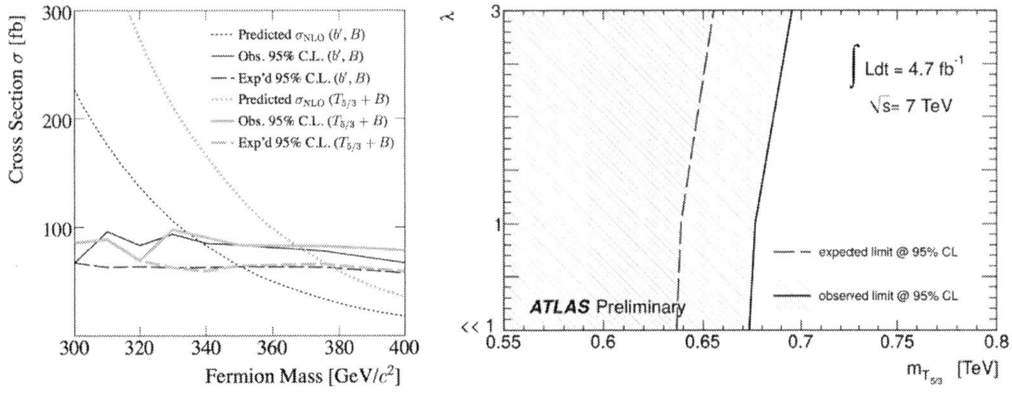

Figure 4. Exclusion limit on the $T_{5/3}$ mass done by the CDF collaboration on B and $T_{5/3}$ pair productions [4] (left) and by the ATLAS collaboration on $T_{5/3}$ pair and single productions depending on the λ coupling [3] (right).

Two successive analyses were done by the ATLAS collaboration at $\sqrt{s}$ = 8 and 13 TeV, respectively with 20.3 and 3.2 fb^{-1} [5, 6]. For them, the main SM backgrounds leading to a real same-sign pair of leptons come from two different processes: the dibosons (WW, WZ, ZZ) and the vector boson production associated to a top quark pair ($t\bar{t}W$, $t\bar{t}Z$). Other processes can contribute marginally (WWW, WH, ZH, $t\bar{t}WW$, $t\bar{t}H$). The corresponding diagrams are shown in figure 5. They are described using Monte-Carlo samples including simulation of the ATLAS detector by Geant 4. The scarcity of the final state makes instrumental effects much more important as a tiny probability of mis-reconstructing the charge of an electron, for instance, can lead to huge backgrounds due to large cross sections. This effect, called "*mis-id*", can turn a $t\bar{t}$ event into a signal-like final state. Another instrumental source of background comes from semi-leptonic events with additional non-prompt leptons and fake leptons which originate either from secondary decays (jet, kaon, $\pi^{\pm}$) or from π^0/γ: we will call it shortly "*fakes*". The Monte-Carlo simulation cannot reproduce correctly these effects, they have to be estimated directly in the data using data-driven methods.

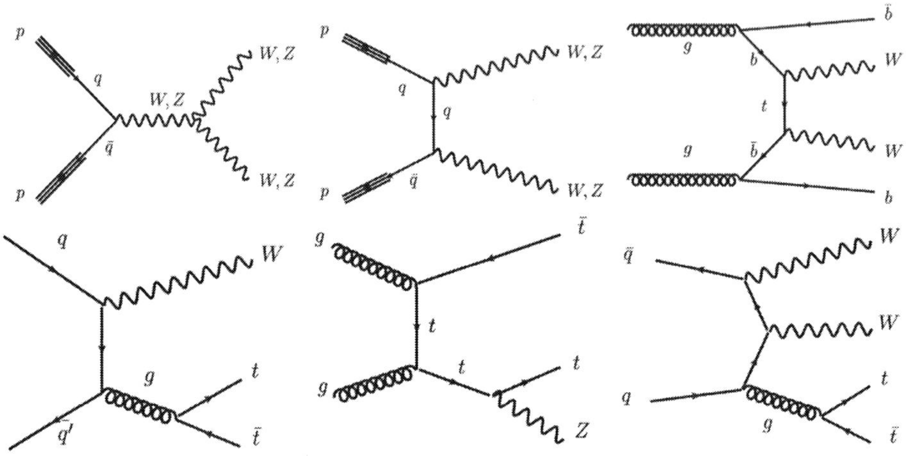

Figure 5. SM background processes giving 2 leptons of the same charge: dibosons (top) and $t\bar{t}V$ (bottom).

Several methods are used by experimentalists, depending on the quality of the objects, the available statistics, the selection used. Concerning the latest result from ATLAS:

- *mis-id*: due to a large lever arm, μ mis-id is neglected. For electron, the charge flip rate is estimated on $Z \to e^{\pm}e^{\pm}$ in a control region under the Z peak using a likelihood extraction method after residual background subtraction. This rate is then applied on opposite-sign leptons events.
- *fakes*: After defining two selection qualities (loose/tight), the probability that a real/fake loose lepton passes the tight criteria is estimated with a likelihood matrix method in control regions enriched in fakes backgrounds. These real/fake rates are then used on same-sign data events.

Table 1. Signal regions categorization with respect to H_T, MET and b-jet multiplicity cuts.

$e^{\pm}e^{\pm} + e^{\pm}\mu^{\pm} + \mu^{\pm}\mu^{\pm} + eee + ee\mu + e\mu\mu + \mu\mu\mu$, $N_j \geq 2$			
$400 < H_T < 700$ GeV	$N_b = 1$	$E_T^{miss} > 40$ GeV	SRVLQ0
	$N_b = 2$		SRVLQ1
	$N_b \geq 3$		SRVLQ2
$H_T \geq 700$ GeV	$N_b = 1$	$40 < E_T^{miss} < 100$ GeV	SRVLQ3
	$N_b = 1$	$E_T^{miss} \geq 100\ GeV$	SRVLQ4
	$N_b = 2$	$40 < E_T^{miss} < 100$ GeV	SRVLQ5
	$N_b = 2$	$E_T^{miss} \geq 100$ GeV	SRVLQ6
	$N_b \geq 3$	$E_T^{miss} > 40$ GeV	SRVLQ7

The discriminant variable used to separate control and signal regions is H_T, as it is shown in figure 6. Background events with a $t\bar{t}$ pair peak around 300 GeV while signal samples, with a heavy quark pair, will peak around 1 TeV. On top of this variable, the b-jet multiplicity can be used to reject dibosons events; even $t\bar{t}$ events (from fakes and mis-id) can be rejected with a $n_b \geq 3$ cut. As a result, a set of orthogonal signal regions is chosen, summarized in table 1. Systematic errors were put on different sources of uncertainties: theoretical cross sections, selection efficiencies, luminosity. Fakes and mis-id data-driven methods have a 70% and 40% systematic error, respectively.

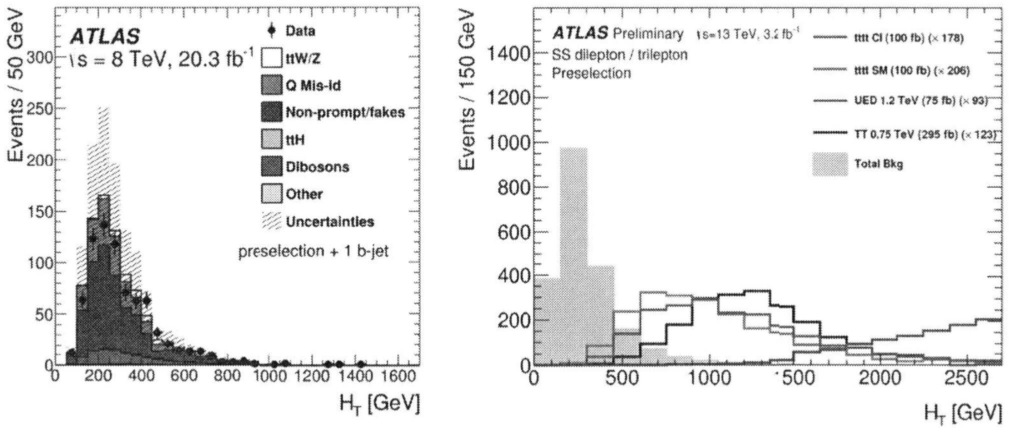

Figure 6. H_T comparison between background processes and signals respectively at 8 and 13 TeV [5, 6].

After comparison between data and background estimates in control regions showing good agreement, the signal regions were unblinded, giving the results plotted in figure 7. The 8 TeV analysis showed a 2 σ excess in high b-jet multiplicity regions: in SRVLQ6, 4.2 ± 0.44 ± 1.16 events were expected, 12 data events were seen; for SRVLQ7, data gave 6 events against 1.14 ± 0.26 ± 0.34 for the expectation. The latest data at 13 TeV seem to argue for a statistical fluctuation as there is no excess anymore.

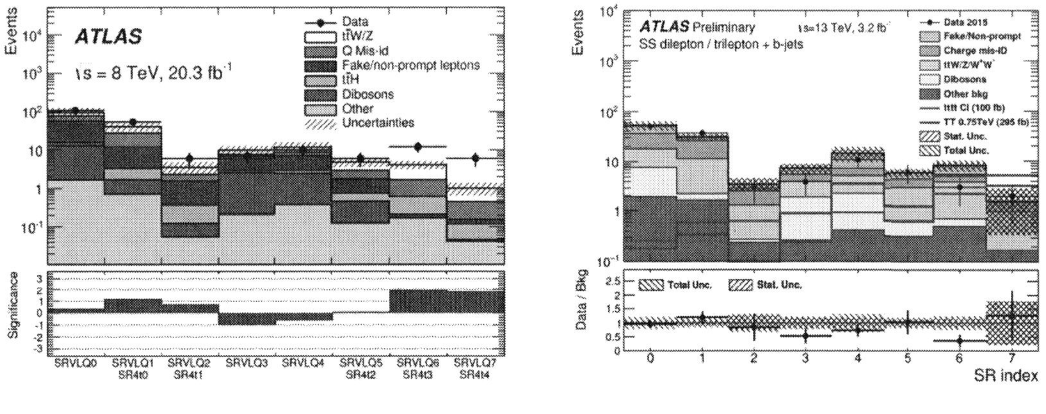

Figure 7. Background yields and observed data for each signal region in 8 TeV and 13 TeV analyses [5, 6].

DOI: 10.1051/epjconf/201713801009

These results were interpreted as experimental constraints on the $T_{5/3}$ mass using the CLs method. The exclusion plots in figure 8 show the limits obtained for two data sets (20.3 fb^{-1} at 8 TeV and 3.2 fb^{-1} at 13 TeV), the latest one being the most stringent one, excluding $m_{T_{5/3}} \leq 990$ GeV. With such mass limits reaching the TeV scale, composite models predicting $T_{5/3}$ are starting to be challenged. Indeed, if we continue to push higher the mass of these top partners, a *little hierarchy* problem will appear: the exotic quarks proposed to solve the hierarchy discrepancies between u and t should be around the TeV, because of the Higgs mass. With 100 fb^{-1} of data expected in the coming 2 years, one can expect to put very tight limits on these models and maybe jeopardize them.

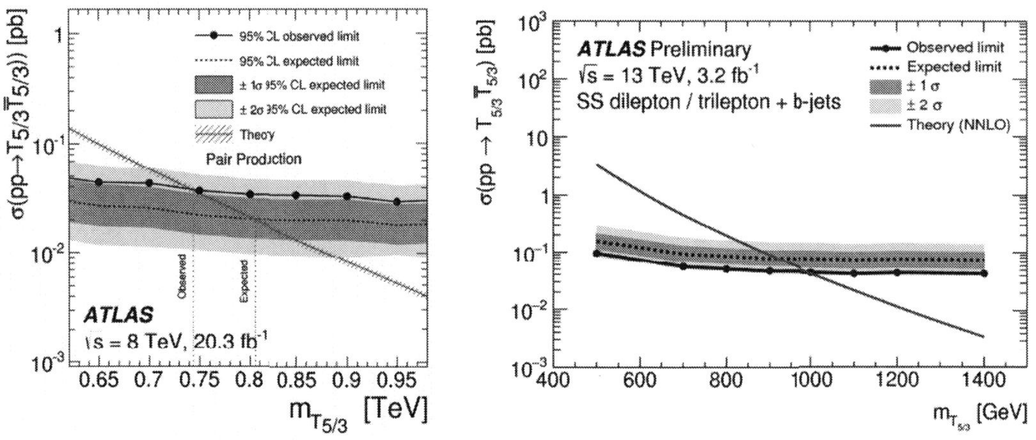

Figure 8. Exclusion limits at 95 % C.L. on the $T_{5/3}$ mass using 8 TeV (left) and 13 TeV (right) data [5, 6].

4 Four-top events at the LHC

The production of events with four tops in the final state occurs in the SM at a very low cross section (15 fb at $\sqrt{s} = 14$ TeV). This process has not yet been observed at the LHC, the most precise searches having put upper limit on its cross section to 16 times the SM. In the context of compositeness, one can expect couplings between composite bosons and the top quark. This would add BSM contributions to the 4-top production and thus enhance the cross section [7]. We can model this by extending the SM as $SU(3)_c \times SU(2)_L \times U(1)_Y \times U(1)_X$, where $U(1)_X$ is linked to a symmetry breaking giving rise to a massive particle Z' [8]. The mass of this generic resonance can be mapped to the scale of the new physics and its width is computed with Formula 4, with respect to the coupling g_{t_R} with the right-handed top [9, 10]. If this scale is too high, the process becomes a contact interaction between four top quarks. The corresponding diagrams are illustrated in figure 9. The final state $W^{\pm}W^{\pm}W^{\mp}W^{\mp}bbbb$ is very similar to the one studied for the pair production of $T_{5/3}$, which was $W^{\pm}W^{\pm}W^{\mp}W^{\mp}bb$.

$$\Gamma = \frac{g_{t_R}^2}{24\pi}\left(m_{Z'} - 2\frac{m_t^2}{m_{Z'}}\right)\sqrt{1 - 4\frac{m_t^2}{m_{Z'}^2}} \tag{4}$$

We try to estimate at 14 TeV the discovery potential in different leptonic channels for both SM and BSM signals, for 100 fb^{-1}, using Monte-Carlo simulations from MadGraph [11]: we scanned effective scales of $\frac{\Lambda}{\sqrt{|C_{4t}|}}$ =0.5, 1 and 1.5 TeV for $C_{4t} = 1$; and Z' mass points of 1 and 2 TeV for a coupling $g_{t_R} = 3$. The background taken into account in this analysis are: $t\bar{t}$, $t\bar{t}W$, $t\bar{t}Z$, $t\bar{t}H$. Dibosons samples were produced but as we require at least 3 b-jets, they are totally suppressed. BSM samples are simulated with interferences from the SM processes. The instrumental backgrounds are estimated on $t\bar{t}$ events using public results from ATLAS, for mis-id, and with the help of the DELPHES package for MadGraph which parametrizes the response of the detectors, for the fakes. The sensitivity is computed with $\frac{S}{\sqrt{B}}$ and the discovery (observation) luminosities correspond to $\sigma = 5$ (2).

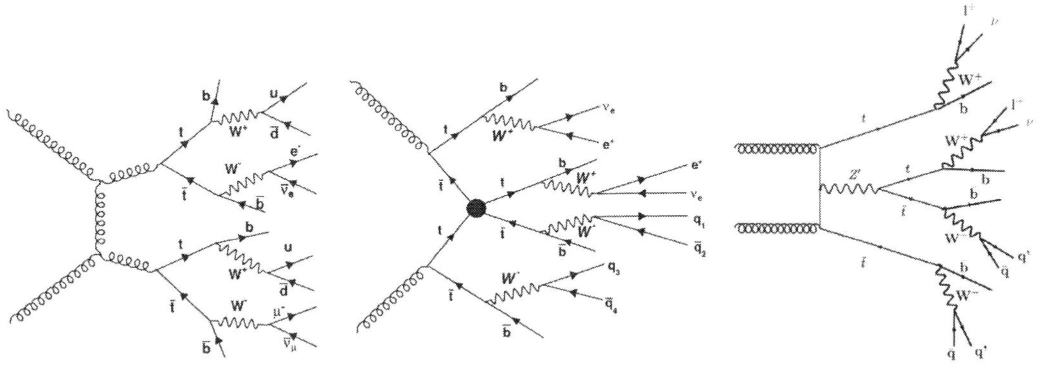

Figure 9. SM (left), contact interaction (middle) and resonant BSM (right) production diagrams for 4-top events.

The selection cuts, based on the variables n_j, H_T, MET and n_b, are optimized to maximize the significance. The figure 10 shows their distributions in the 2 same-sign leptons channel.

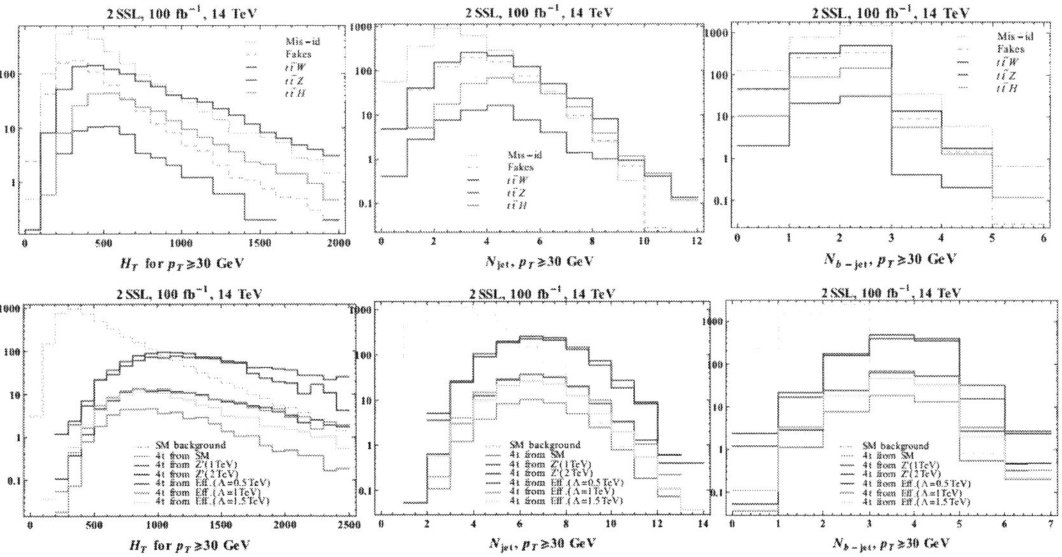

Figure 10. H_T, n_{jet} and n_{b-jet} distributions for the background (top) and for signals (bottom) at $\mathcal{L}$ = 100 fb^{-1}.

Ongoing work shows that the most sensitive channel in terms of significance is the 2 same-sign leptons one, though the trilepton channel gives also very promising results. Even if the semi-leptonic channel is very bad for discovering 4-top events, it is very useful in order to study the top quark properties. Indeed, the leptonic W boson can be reconstructed from the lepton and the MET which comes mainly from the neutrino. By choosing the b-jet which minimizes the difference between the reconstructed top invariant mass and the pole mass, one can compute the 4-momenta of the leptonic top quark. The top quark polarization is then estimated using the angle θ_{lep} between the lepton in the top rest frame and the top in the laboratory frame [12]. As a result, the differential cross-section with respect to the polarization angle is given in Equation 5. In the SM, the top quark is produced in the same proportions in both chiralities, so A~0.5; in composite models, where the right-handed top has a privileged coupling, A is around 0.75.

$$\frac{1}{\sigma}\frac{d\sigma}{d\cos\theta_{lep}} = \frac{1}{2} + (A - \frac{1}{2})\cos\theta_{lep} \qquad (5)$$

We reconstructed the leptonic top quark for SM processes and BSM 4-top production via a Z' resonance. The polarization angle distribution in figure 11 shows different slopes between standard tops and composite productions, as long as the Z' mass stays sufficient low so the SM contribution does not counterbalance the BSM slope. Fitting the slope on positive range improves the discrimination.

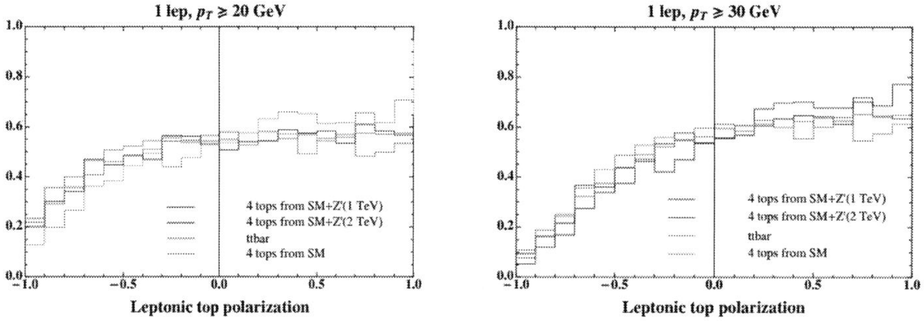

Figure 11. Top polarization for $t\bar{t}$, 4-top SM and BSM. A leptonic p_T cut is done at 20 and 30 GeV.

5 Conclusion

Composite models address the naturalness problem of the SM and solves the fine-tuning by considering the Higgs boson not as a fundamental object anymore. They predict the existence of heavy quarks which can be produced by pair at the LHC. Experimental searches have excluded mass ranges for the different top partners, especially in the 2 same-sign dilepton channel which is the most sensitive one, as the SM yields very few processes with this signature. Four-top events give a similar final state and can thus be searched in the same way. A prospective study showed that with optimized cuts, the SM process could be observed at 5 σ using few hundreds of fb^{-1} of data at $\sqrt{s}$ = 14 TeV. BSM production involving composite models can already be probed with 13 TeV data. Top polarization measurement using the reconstruction of the leptonic top in the semileptonic channel showed a nice sensitivity to composite physics.

Acknowledgments

The author is grateful to F. Deliot, J.-F. Laporte, A. Ouraou, L. Chevalier, C. Guyot, O. Mattelaer and J. Serra for fruitful discussions and crucial debugging help, as well to G. Servant for her guidance and financial support through the ERC Starting Grant Cosmo@LHC, without forgetting Professor E. Tomasi-Gustafsson, the Baldin seminar organization and JINR for their kind welcome in Dubna.

References

[1] M. Son, R. Sundrum, R. Contino et T. Kramer, JHEP **0705**, 074 (2007)
[2] G. Panico and A. Wulzer, *The Composite Nambu-Goldstone Higgs* (Springer Lecture Notes in Physics, 2016) 1-316
[3] The ATLAS Collaboration, ATLAS-CONF-2012-130 (2012)
[4] The CDF Collaboration, Phys. Rev. Lett. **104**, 091801 (2010). FERMILAB-PUB-09-628-E
[5] The ATLAS Collaboration, JHEP **10**, 150 (2015)
[6] The ATLAS Collaboration, ATLAS-CONF-2016-032 (2016)
[7] G. Servant, Physics at the LHC 2010 Proceedings. DESY-PROC-2010-01/251
[8] G. Brooijmans et al., CERN-PH-TH-2010-096 (2009)
[9] L. Gauthier, CERN-THESIS-2012-157, PhD: U. Paris-Sud 11, Dept. Phys., Orsay: 2012-09-15 (2012)
[10] R. Kukla, CERN-THESIS-2016-136, PhD Orsay: 2016-06-17, NNT: 2016SACLS146 (2016)
[11] G. Servant et R. Kukla, to be published (2016)
[12] A. Pomarol and J. Serra, Phys. Rev. D **78**, 074026 (2008)

EPJ Web of Conferences **138**, 01010 (2017) DOI: 10.1051/epjconf/201713801010
Baldin ISHEPP XXIII

Graviton mass bounds from an analysis of bright star trajectories at the Galactic Center

Alexander Zakharov[1,2,3,4,⋆], *Predrag* Jovanović[5,⋆⋆], *Dusko* Borka[6,⋆⋆⋆], and *Vesna* Borka Jovanović[6,⋆⋆⋆⋆]

[1] *Institute of Theoretical and Experimental Physics, 117259 Moscow, Russia*
[2] *Bogoliubov Laboratory for Theoretical Physics, JINR, 141980 Dubna, Russia*
[3] *National Research Nuclear University MEPhI (Moscow Engineering Physics Institute), 115409, Moscow, Russia*
[4] *North Carolina Central University, Durham, NC 27707, USA*
[5] *Astronomical Observatory of Belgrade, Volgina 7, 11060 Belgrade, Serbia*
[6] *Atomic Physics Laboratory (040), Vinča Institute of Nuclear Sciences, University of Belgrade, P.O. Box 522, 11001 Belgrade, Serbia*

Abstract. In February 2016 the LIGO & VIRGO collaboration reported the discovery of gravitational waves in merging black holes, therefore, the team confirmed GR predictions about an existence of black holes and gravitational waves in the strong gravitational field limit. Moreover, in their papers the joint LIGO & VIRGO team presented an upper limit on graviton mass such as $m_g < 1.2 \times 10^{-22}$ eV (Abbott et al. 2016). So, the authors concluded that their observational data do not show any violation of classical general relativity. We show that an analysis of bright star trajectories could constrain graviton mass with a comparable accuracy with accuracies reached with gravitational wave interferometers and the estimate is consistent with the one obtained by the LIGO & VIRGO collaboration. This analysis gives an opportunity to treat observations of bright stars near the Galactic Center as a useful tool to obtain constraints on the fundamental gravity law such as modifications of the Newton gravity law in a weak field approximation. In that way, based on a potential reconstruction at the Galactic Center we obtain bounds on a graviton mass.

1 Introduction. Centurial triumph of General Relativity

General relativity (GR) has been discovered in November 1915 in [1, 2] as it is well-known in the literature (see, [3–7] for reference). Gravitational field is described by the following expression

$$R_{\mu\nu} - \frac{1}{2}Rg_{\mu\nu} = \frac{8\pi G}{c^4}T_{\mu\nu}, \tag{1}$$

where $R_{\mu\nu}$ is the Ricci tensor, $g_{\mu\nu}$ is the metric tensor, R is the scalar curvature, $T_{\mu\nu}$ is the stress-energy tensor, G and c are the Newton constant and the speed of light, respectively.

⋆e-mail: zakharov@itep.ru
⋆⋆e-mail: pjovanovic@aob.rs
⋆⋆⋆e-mail: dusborka@vin.bg.ac.rs
⋆⋆⋆⋆e-mail: vborka@vinca.rs

1.1 Deflection of light and Mercury anomaly

In November 1915 (before the discovery of GR) in paper [8] A. Einstein calculated deflection of light which is double in respect to the Newtonian one (therefore, measurements of light deflection provide an effective test of a new theory of gravity) and explained the Mercury orbit anomaly found by U.J.J. Le Verrier in 1859 [9] (Le Verrier suggested also a way to resolve any anomaly, see [10] for a detailed discussion). At this period (before November 25, 1915) A. Einstein knew correct gravitational equations in vacuum

$$R_{\mu\nu} = 0, \tag{2}$$

and used them in his derivations of light deflection and the Mercury anomaly. So, Einstein derived an expression for light deflection

$$\Theta = \frac{4GM}{c^2 p}, \tag{3}$$

where M is a mass of gravitating body and p is an impact parameter. In the case, if we observe light deflection near a Solar disk during solar eclipse $M = M_\odot$ and $p = R_\odot$ and $\Theta = 1.75''$. The Einstein's prediction about the light deflection has been confirmed by British astronomers on 29 May, 1919 during solar eclipse in Sobral (Brazil) and island Principe (near Africa) [11] (see, also [12, 13]). Observations at Principe confirmed GR predictions while Sobral's observations showed $\Theta = 0.93''$, but astronomers decided that these observations are not reliable. After those observations mass media were very excited because a new (and more correct) theory of gravity has been found [13]. However, during solar eclipse on May 29, 1929 in Sumatra, E. Freundlich[1] obtained the light deflection angle $\Theta = 2.24 \pm 0.1''$ [18]. On December 11, 1931 Freundlich reported his results to the Royal Astronomical Society, he claimed that 1) a deflection exists; 2) it is not Newton's; 3) it seems to be greater than Einstein's, after that A. Eddington claimed that "I find it difficult to believe 1.75" can be wrong" [19]. Now the Einstein's expression (3) has been checked with a very high precision [20–22] and the post-Newtonian parameter γ is very close to 1 [23].

As it was mentioned earlier A. Einstein explained perihelion precession for Mercury [8]

$$\phi = \frac{24\pi^3 L^2}{T^2 c^2 (1 - e^2)}, \tag{4}$$

where L and T are semi-major axis and orbital period, respectively. The phenomenon is called the relativistic advance and it is extremely useful tool to evaluate parameters of the black hole, a stellar cluster and dark matter bulk distribution at the Galactic Center as we will discuss below.

1.2 Gravitational lensing

Gravitational lensing is based on phenomenon of light deflection in gravitational field and the theory started since papers by O. Chwolson [24], Einstein [25] and F. Zwicky [26] and now there are a number of papers describing different aspects of the phenomenon see [15, 16, 27–29] and references therein. Gravitational lens mapping is a specific case of Lagrange mapping and therefore, singularities

[1] In 1911 A. Einstein calculated a deflection of light in gravitational field $\Theta = \frac{2GM}{c^2 p}$ [14], which coincides with Newtonian one and in 1914 Einstein suggested his assistant E. Freundlich to check the result during Solar eclipse in Russian Crimea (near Feodosia) in August 1914. However, the observations were not done, Freundlich and his colleagues were taken as prisoner of war since they were German army reservists [15, 16], while American team led by W. W. Campbell (Lick Observatory) got rights to observe stars during the eclipse, however, they failed because there were clouds during the eclipse [17].

of the mapping play a key role because the highest amplification is in a region near singularities (caustics). All singularities of the Lagrange mappings up to four dimensions are classified by V. I. Arnold [30]. Singularities with the lowest co-dimensions are folds and caustics. Asymptotic behavior of magnification near folds is given in [15] and near cusps in [31]. If masses of lenses are around solar masses and sources are located in our Galaxy or in nearby galaxies, then angular distance between images are around a few angular milli-arcseconds and it is very hard to re-solve multiple images but it will be possible to observe a amplification of background stars as it was noted in [32, 33]. One can find a discussion of results of observations and their theoretical interpretation in reviews [34–38]. As it was noted in [39], microlensing is a very efficient tool to find low mass exoplanets near the habitable zone [40]. In observations of stars in Andromeda galaxy from the Northern hemisphere it is very hard to re-solve separate stars and to find manifestations of microlensing for them but it is possible to detect signature of so-called pixel lensing (or an amplification of a pixel). Such observations are described in a number of papers, see for instance [41]. Analyzing data observed by POINT-Agape collaboration, An et al. found anomaly [42] which was explained later as an exoplanet in the Andromeda galaxy [43]. If it is so, it will be the first exoplanet in another galaxy [44–47]. As it was noted [48–50] polarization observations for pixel lensing event play an extremely important role to clarify parameters of microlenses and sources which are giant stars for pixel lensing events (see also [51] for a review). Microlensing in X-ray band (including studies of microlensing for macrolensed objects) has been considered in [52–54].

1.3 Gravitational redshift and time dilation

Pound and Rebka found that a photon frequency is higher at the bottom of a tower than at the top of it [55, 56]. Vessot et al. measured a maser frequency shift aboard of a spacecraft and the shift was in agreement with predictions of GR [57]. In spite of the fact that the phenomenon can be easily calculated in the framework of the GR, there are discussions of physical interpretation of these experiments [58–64].

1.4 Inspiralling pulsars. An evidence of gravitational radiation

Soon after the discovery of GR, A. Einstein concluded that gravitational waves have to be emitted in the case if quadrupole moment for is changed with time [65, 66]. In his paper A. Eddington corrected a mistake in the expression describing gravitational waves, however, he believed that "they are not objective and (like absolute velocity) are not detectable by any conceivable experiment. There are merely sinuosities in the coordinate system and the only speed of propagation relevant to them is "the speed of thought" [67]. In 1936 A. Einstein and his young assistant N. Rosen submitted the paper under the "Do gravitational waves exist?" in *The Physical Review*. Based on the Einstein's letter to M. Born one can say that Einstein's answer was "No" [68]. However, a referee and on 23 July John Tate (who was the chief editor of PR) were not convinced and Tate asked Einstein to respond at remarks and criticism of a referee. On 27 July Einstein wrote to Tate [68]: "Dear Sir, We (Mr. Rosen and I) had sent you our manuscript for publication and had not authorized you to show it to specialists before it is printed. I see no reason to address the – in any case erroneous – comments of your anonymous expert. On the basis of this incident I prefer to publish the paper elsewhere. Respectfully, [signature]. P.S. Mr. Rosen, who has left for the Soviet Union, has authorized me to represent him in this matter." Tate answered that Einstein's unwilling contradicts to journal rules and he cannot accept such a paper. After that Einstein submitted the same paper in *the Journal of Franklin Institute* where the article has been published [69], but now conclusions in the paper were completely different, namely, authors claimed that gravitational waves do exist. As it was noted in [68], at the period between these two submissions in *The Physical Review* and *the Journal of Franklin Institute* Einstein tried to convince H. P. Robertson and L. Infeld (who substituted N. Rosen as Einstein's assistant) that the conclusions in the first version of the paper about an absence of gravitational waves are correct. At the time Robertson and Infeld were in Princeton and Einstein had very fruitful conversations with these two scientists and in the result of these conversations he arrived at the opposite conclusion that gravitational waves exist. As it was noted in [68], the important issue was that Robertson was the referee of the Einstein and Rosen paper submitted in *The Physical Review*. In spite of the claim about an existence of gravitational waves in their joint paper [69], N. Rosen was rather sceptical and he published a couple of papers (see, for instance [70, 71]) where he argued that gravitational waves do not exist. Theoretical and experimental aspects of the quadrupole expression are discussed in a book [72], where one could find an interesting historical development of a current understanding the issue. The discussion about validity of the quadrupole expression reached a consensus when it was found that an evolution of Hulse – Taylor pulsar PSR 1913+16 [73–75] is nicely described by a simple Peters – Mathews expression which is a version of quadrupole formula for two inspiraling points [76, 77].

1.5 Observational signatures of black holes

If we speak about observable manifestations of black hole features we need models with a strong gravitational field to describe 1) a final stage of inspiraling (merging and ring down) binary black holes; and 2) shapes of shadows around black holes. Perhaps, very soon observers will need GR corrections and later a full GR approach to fit observational data for bright stars near the Galactic Center. Assuming that a radiation in a spectral line is emitted from a disk near a black hole horizon, it was found (and after that it was observed the X-ray K_α-line) that an observed shape of the spectral line can be an important indicator of a strong gravitational field near a black hole, moreover, one can evaluate a black hole spin analyzing a spectral line structure [78–80][2] (see also more recent reviews [88, 89] on the subject). Another phenomenon, where one really needs a strong gravitational field approach, is simulations of a shadow formation started since [90–93] (see also calculations of shadows for different cases [94–99] and recent reviews on the subject [100, 101]).

The problem is connected with attempts to resolve the smallest spot at the Galactic Center with VLBI interferometry in mm-band [102]. As it was noted earlier, observations of bright star trajectories near the Galactic Center could provide an efficient tool to evaluate a gravitational potential, in particular, analyzing these trajectories one can obtain constraints on parameters of black hole and stellar cluster [103] and on parameters of dark matter distribution [104–107].

Two groups of astronomers with VLT and Keck telescopes observe stars near the Galactic Center, see [108–110] and references therein. An analysis of S2 like star trajectories gives an opportunity to obtain stringent constraints on alternative theories of gravity, including R^n theory which is a generalization of the classical GR and $n = 1$ corresponds to GR [111, 112] (there are also stringent constraints from Solar system data [113]), and Yukawa gravity [114]. In the paper we will obtain a graviton mass constraint from analysis of trajectories of bright stars at the Galactic Center.

2 Theories of gravity with massive graviton

A massive gravity theory was suggested in M. Fierz and W. Pauli [115]. So-called, the van Dam–Veltman–Zakharov (vDVZ) discontinuity has been found [116–118], or the theory for $m_g \to 0$ is different from the theory with vanishing graviton mass. However, Vainshtein found a nonlinear solution for the Schwarzschild problem [119] providing a continuity at $m \to 0$ with the massless graviton

[2]Results of iron K_α-line simulations in the framework of a simple model are given in [81–87].

theory (the classical Einstein's theory of relativity). Boulware and Deser found another unpleasant feature of a massive gravity such a presence of ghosts (and related instabilities) and other pathologies from quantum field theory point of view [120]. In the last years, there was a significant progress to overcome these problems and build a consistent theory without such defects like Boulware – Deser ghosts and discontinuities [121–124] (a great step has been done in the paper [125] where the authors developed a ghost free massive gravity). Here, we will not discuss theoretical aspects of massive gravity theory and we will consider only observational features of such an approach.

In spite of the problems of current theoretical models of massive gravity in seventies Goldhaber and Nieto found a graviton mass constraint based on the assumption that a Compton wave length of graviton is around $\lambda_g = 580$ kpc (it is around a typical distance between galaxies), and $m_g < 2 \times 10^{-62}$ g=1.1×10^{-29} eV [126]. In the relativistic theory of gravitation (RTG), developed by Logunov and his group, a non-vanishing mass of graviton substitutes Λ-term in the conventional ΛCDM cosmological model and one could find that a Compton wave length for graviton has a cosmological value, so that $m_g < (1.3 - 3.2) \times 10^{-66}$ g (depending on a way for an evaluation of the quantity) [127–129], see also estimates obtained recently $m_g < 5.2 \times 10^{-66} - 1.2 \times 10^{-65}$ g taking into account constraints on quintessence parameters [130]. Constraints on λ in Yukawa potential from Solar system data are given in [131] and analyzing these data, C. Will obtained a graviton mass constraint $m_g < 7.2 \times 10^{-23}$ eV at the 2σ level [132]. Analyzing weak gravitational lensing data (gravitational potential reconstruction for galactic clusters based on image deformations of background galaxies), Choudhury et al. found that a Compton wavelength of massive graviton has to be $\lambda > 100$ Mpc=3×10^{21} km [133], therefore, $m_g < 6 \times 10^{-32}$ eV. Finn and Sutton suggested to use binary pulsars PSR B1913+116 and PSR B1534+112 to evaluate a graviton mass and they found $m_g < 7.6 \times 10^{-20}$ eV with 90% confidence level [134]. Larson and Hiscock proposed to use future LISA data for observations of gravitational radiation from interacting white dwarf binary star systems, including helium cataclysmic variable (HeCV) systems and in this case one can expect to reach the following graviton mass bound $m_g < 1 \times 10^{-24}$ eV [135]. For a subsample of 400 close white dwarf binaries with high signal-to-noise ratio gravitational wave and optical data with magnitudes brighter than 25, the combined upper limit on the graviton mass is at the level of $m_g = 6 \times 10^{-24}$ eV [136]. One can expect even a better estimates for ASTROD-I mission [137] because it will be possibly a next generation of space borne gravitational interferometers in space for gravitational wave detections. Many years ago, Sazhin proposed to use pulsar timing for gravitational wave detection [138] (see, also a more detailed discussion in [139]). Graviton mass constraints were obtained in [140], but later in erratum the authors noted that their approach was not correct [141], however, in paper [142] it was concluded, that one can obtain a graviton mass bound at a level $m_g = 3 \times 10^{-22}$ eV, if bi-weekly observation of 60 pulsars is performed for 5 years with a pulsar rms timing accuracy of 100 ns and the estimate can be improved in the case if more observations for more pulsars a longer observation period would be done, moreover, such an estimate will be improved with a pulsar array technique [143]. There are other suggestions to evaluate a graviton mass, some of them are rather exotic and based on hardly verified assumptions [124]. Systematics of proposed experiments and observations is not well investigated, moreover, some weaknesses of the proposals for a graviton mass evaluation are pointed out in the review [124].

3 Graviton mass constraint from gravitational wave signal

If a graviton has a mass m_g, then a speed of gravitational wave propagation could differ from c and we have a dispersion relation [23, 132]

$$\frac{v_g^2}{c^2} = 1 - \frac{m_g^2 c^4}{E^2}, \tag{5}$$

where E is a graviton energy. Gravitons with different energies propagate with different velocities. Assume that we have gravitational waves and electromagnetic waves from the same source (from supernova explosion, for instance). In this case we have [23, 132]

$$1 - \frac{v_g}{c} = 5 \times 10^{-17} \left(\frac{200 \text{ Mpc}}{D} \right) \left(\frac{\Delta t}{1 \text{ s}} \right), \tag{6}$$

where $\Delta t = \Delta t_a - (1 + z)\Delta t_e$ is the time difference, where Δt_a and Δt_e are the differences in arrival time and emission time of the two signals, respectively, and z is the redshift of the source. Usually Δt_e is unknown, however, one could find an upper limit for Δt_e (for instance from a theoretical model), therefore, one could evaluate $1 - \frac{v_g}{c}$, therefore, m_g. Following [23, 132] and assuming that the frequency of gravitational wave is ν and $h\nu \gg m_g c^2$ (h is Planck's constant), therefore, we have $\frac{v_g}{c} \approx 1 - \frac{1}{2}\frac{h}{\lambda_g \nu}$, where $\lambda_g = \frac{h}{m_g c}$ or $\lambda_g \approx \frac{1}{2}\frac{1}{\sqrt{1 - v_g/c}}$. If one has an upper limit for $1 - v_g/c$, it can be re-written as a lower limit for λ_g, as the following expression [23, 132]

$$\lambda_g = 3 \times 10^{12} \text{km} \left(\frac{200 \text{ Mpc}}{D} \frac{\nu}{100 \text{Hz}} \right) \left(\frac{1}{\nu \Delta t} \right). \tag{7}$$

It is a lucky case if one observes electromagnetic and gravitational radiation from the same source. But even in the case if only gravitational radiation has been detected as it was noted [132] because gravitational wave signal with a massive graviton will be different from signal for a graviton with a vanishing mass and in this case for $D \approx 200\text{Mpc}, \nu \approx 100\text{Hz}, \nu\Delta t \sim \rho^{-1} \approx 0.1$ The result is $\lambda_g > 10^{13}$ km. Based on ideas expressed in [23, 132], the LIGO/VIRGO collaboration obtained the same estimate for the Compton wavelength of a massive graviton [144, 145] and detections of one additional event and one suspected case did not change the estimate [146].

4 Graviton mass constraint from an analysis of bright star trajectories at the Galactic Center

We use a modification of the Newtonian potential corresponding to a massive graviton case [23, 132, 147]:

$$V(r) = -\frac{GM}{(1 + \delta)r} \left[1 + \delta e^{-\left(\frac{r}{\lambda}\right)} \right], \tag{8}$$

where δ is a universal constant (we put $\delta = 1$). In our previous studies [114] we found constraints on parameters of Yukawa gravity. As it was described in [148, 149] we used observational data from NTT/VLT [108]. If we wish to find a limiting value for λ_x, so that $\lambda > \lambda_x$ with a probability $P = 1 - \alpha$ (where we select $\alpha = 0.1$) normalized depending on λ_x has to be equal to the threshold depending on degree of freedom ν and parameter α or in other words, $\chi^2(\lambda_x) = \chi^2_{\nu,\alpha}$. Computing these quantities we obtain $\lambda_x = 2900$ AU $\approx 4.3 \times 10^{11}$ km. Now we obtain the upper limit on a graviton mass and we could claim that with a probability $P = 0.9$, a graviton mass should be less than $m_g = 2.9 \times 10^{-21}$ eV (since $m_g = hc/\lambda_x$) in the case of $\delta = 1$ [148].

5 Conclusions

As it was noted earlier, our graviton mass estimate is slightly greater than estimate with LIGO interferometer, however, a) our approach has a minimal number of assumptions and our estimate was obtained in independent way; b) our estimate is consistent with LIGO's one; c) our estimate will be definitely improved with forthcoming facilities such as GRAVITY, E-ELT and TMT [150–152] because more precise observations of bright star orbits will give an opportunity to reconstruct a gravitational potential at the Galactic Center in a more accurate way, therefore, one can expect a better estimates for λ parameter and a graviton mass. However, such a progress will be not very rapid because of an exponential dependence of a potential on λ.

A.F.Z. thanks the organizers of the XXIII International Baldin Seminar for their kind invitation to present a plenary talk at the seminar. A.F.Z. appreciates NSF (HRD-1345219) and NASA (NNX09AV07A) at NASA CCFASE and NSF CREST Centers (NCCU, Durham, NC, USA) for a partial support. P.J., D.B.and V.B.J. wish to acknowledge the support by the Ministry of Education, Science and Technological Development of the Republic of Serbia through the project 176003 "Gravitation and the large scale structure of the Universe". P. J. also thanks ICTP - SEENET-MTP project PRJ-09 Cosmology and Strings for a partial support.

References

[1] A. Einstein, Sitzungsber. K. Preuss. Akad. Wiss. **48**, 844 (1915)
[2] D. Hilbert, Göttingen Nachr. **3**, 395 (1915)
[3] J. Mehra, *Einstein, Hilbert and the Theory of Gravitation* (Dordrecht, Boston, 1974)
[4] J. Earman and C. Glymour, Arch. Hist. and Exact. Sci. **19**, 291 (1978)
[5] V. P. Vizgin and Ya.A. Smorodinskii, Physics – Uspekhi **22**, 489 (1979)
[6] V. P. Vizgin, Physics – Uspekhi **44**, 1283 (2001)
[7] A. A. Logunov, M. A. Mestvirishvili and V. A. Petrov, Physics – Uspekhi **47**, 607 (2004)
[8] A. Einstein, Sitzungsber. K. Preuss. Akad. Wiss. **47**, 831 (1915)
[9] U. J. J. Le Verrier, R. Acad. Sci. Paris **59**, 379 (1859)
[10] A. F. Zakharov et al. Space Sci. Rev. **148**, 301 (2009)
[11] F. Dyson, A. S. Eddington and C. R. Davidson, Phil. Trans. Roy. Soc. A. **220**, 571 (1920)
[12] S. Chandrasekhar, Notes Rec. R. Soc. Lond. **30**, 249 (1976)
[13] J. Eisenstaedt, *The Curious History of Relativity* (Princeton University Press, Princeton, 2006) 167–195
[14] A. Einstein, Ann. Phys. **35**, 898 (1911)
[15] P. Schneider, J. Ehlers and E. E. Falco, *Gravitational Lenses* (Springer, Berlin, 1992)
[16] A. F. Zakharov, *Gravitational Lenses and Microlenses* (Yanus, Moscow, 1997) (in Russian)
[17] J. Crelinsten, Hist. Stud. Phys. Scien. **14**, 1 (1983)
[18] E. Freundlich, H. V. Klüber and A. V. Brunn, Zeitschrift für Astrophysik **3**, 171 (1931)
[19] H. Knox-Shaw, Meet. Roy. Astron. Soc. (December 11, 1931) The Observatory **55**, 1 (1932)
[20] S. S. Shapiro et al., Phys. Rev. Lett. **92**, 121101 (2004)
[21] S. B. Lambert and C. Le Poncin-Lafitte, Astron. Astrophys. **499**, 331 (2009)
[22] S. B. Lambert and C. Le Poncin-Lafitte, Astron. Astrophys. **529**, A70 (2011)
[23] C. Will, Living Rev. Relat. **17**, 4 (2014)
[24] O. Chwolson, Astron. Nachr. **221**, 329 (1924)
[25] A. Einstein, Science **84**, 506 (1936)
[26] F. Zwicky, Phys. Rev. **51**, 290 (1937)
[27] J. Wambsganss, Living Reviews in Relativity **1**, 12 (1998)
[28] A. Petters, H. Levine and J. Wambsganss, *Singularity theory and gravitational lensing* (Birkhäuser, Boston, 2001)

[29] S. Mollerach and E. Roulet, *Gravitational Lensing and Microlensing* (World Scientific, Singapore, 2002)
[30] V. I. Arnold, Soviet Math. Surveys **38**, 87 (1983)
[31] A. F. Zakharov, Astron. Astrophys. **293**, 1 (1995)
[32] A. V. Byalko, Sov. Astron. **13**, 784 (1970)
[33] B. Paczynski, Astrophys. J. **304**, 1 (1996)
[34] B. Paczynski, Ann. Rev. Astron. Astrophys. **34**, 419 (1996)
[35] E. Roulet and S. Mollerach, Phys. Rep. **279**, 67 (1997)
[36] A. F. Zakharov and M. V. Sazhin, Phys. - Usp. **41**, 945, (1998)
[37] A. F. Zakharov, Publ. Astron. Observ. Belgrade **75**, 27 (2003)
[38] A. F. Zakharov, Phys. Part. Nucl. **39**, 1176 (2008)
[39] S. Mao and B. Paczynski, Astrophys. J. **374**, L37 (1991)
[40] S. Mao, Res. Astron. Astrophys. **8**, 947 (2012)
[41] S. Calchi Novati et al., Astron. Astrophys. **469**, 115 (2007)
[42] J. H. An et al., Astrophys. J. **601**, 845 (2004)
[43] G. Ingrosso et al., Mon. Not. R. Astron. Soc. **399**, 219 (2009)
[44] A. F. Zakharov, New Astron. Rev. **53**, 202 (2009)
[45] G. Ingrosso et al., Gen. Relat. Grav. **43**, 1047 (2010). arXiv:1001.4342v1 [astro-ph.SR]
[46] A. F. Zakharov et al., Mem. della Soc. Astron. Ital. Suppl. **15**, 114 (2010)
[47] A. F. Zakharov et al. Publ. Astron. Observ. Belgrade **92**, 65 (2013)
[48] G. Ingrosso et al., Month. Not. R. Astron. Soc. **426**, 1496 (2012)
[49] A. F. Zakharov et al., Adv. Space Res. **54**, 1319 (2014). arXiv:1312.3468v1[astro-ph.SR]
[50] G. Ingrosso et al., Physica Scripta **89**, 084001 (2014). arXiv: 1310.5866v1[astro-ph.SR]
[51] A. F. Zakharov, J. Phys.: Conf. Ser. **678**, 012010 (2016)
[52] A. F. Zakharov, L. C. Popović and P. Jovanović, Astron. Astrophys. **420**, 881 (2004)
[53] L. C. Popovic et al., Astrophys. J. **637**, 630 (2006)
[54] P. Jovanović et al. Mon. Not. R. Astron. Soc. **386**, 397 (2008)
[55] R. V. Pound and G. A. Rebka, Phys. Rev. Lett. **3**, 439 (1959)
[56] R. V. Pound and G. A. Rebka, Phys. Rev. Lett. **4**, 337 (1960)
[57] R. F. C. Vessot et al., Phys. Rev. Lett. **45**, 2081 (1980)
[58] V. V. Okorokov, arXiv:gr-qc/9902033 (1999)
[59] V. V. Okorokov, Dokl. Phys. **46**, 400 (2001)
[60] L. B. Okun, K. G. Selivanov and V. L. Telegdi, Phys. – Usp. **42**, 1045 (1999)
[61] L. B. Okun, K. G. Selivanov and V. L. Telegdi, Am. J. Phys. **68**, 115 (2000)
[62] L. B. Okun, Mod. Phys. Lett. A **15**, 1941 (2000)
[63] L. B. Okun, Mod. Phys. Lett. A **15**, 2007 (2000)
[64] L. B. Okun and K. G. Selivanov, Dokl. Phys. **47**, 461 (2002)
[65] A. Einstein, Sitzungsber. K. Preuss. Akad. Wiss. **688** (1916)
[66] A. Einstein, Sitzungsber. K. Preuss. Akad. Wiss. **154** (1918)
[67] A. Eddington, Proc. R. Soc. London, Ser. A **154** (1922)
[68] D. Kennefick, Phys. Today **9**, 43 (2005)
[69] A. Einstein and N. Rosen, J. Franklin Inst. **223**, 43 (1937)
[70] N. Rosen, in *Jubilee of Relativity Theory* (Helv. Phys. Acta Suppl. IV), edited by A. Mercier and M. Kervaire (Birkhauser Verlag, Basel, 1956), 171

[71] N. Rosen, Nuovo Cim. **19**, 249 (1977)
[72] D. Kennefick, *Traveling at the Speed of Thought: Einstein and the Quest for Gravitational Waves* (Princeton University Press, Princeton, 2007)
[73] R. A. Hulse and J. H. Taylor, Astrophys. J. **195**, L51 (1975)
[74] J. H. Taylor and J. M. Weisberg, Astrophys. J. **253**, 908 (1982)
[75] J. M. Weisberg, D. J. Nice and J. H. Taylor, Astrophys. J. **722**, 1030 (2010)
[76] P. C. Peters and J. Mathews, Phys. Rev. **131**, 435 (1963)
[77] P. C. Peters, Phys. Rev. **136**, 1224 (1964)
[78] A. C. Fabian et al., Mon. Not. R. Astron. Soc. **238**, 729 (1989)
[79] L. Stella, Nature **344**, 747 (1990)
[80] Y. Tanaka et al., Nature **375**, 659 (1995)
[81] A. F. Zakharov and S. V. Repin, Astron. & Astrophys. **406**, 7 (2003)
[82] A. F. Zakharov and S. V. Repin, Astron. Rep. **47**, 733 (2003)
[83] A. F. Zakharov and S. V. Repin, Nuovo Cim. **118B**, 1193 (2003)
[84] A. F. Zakharov and S. V. Repin, Adv. Space Res. **34**, 2544 (2004)
[85] A. F. Zakharov and S. V. Repin, Mem. S. A. It. della Suppl. **7**, 60 (2005)
[86] A. F. Zakharov and S. V. Repin, New Astron. **11**, 405 (2006)
[87] A. F. Zakharov, Phys. Atom. Nucl. **70**, 159 (2007)
[88] A. C. Fabian and R. R. Ross, Space Sci. Rev. **157**, 167 (2010)
[89] P. Jovanović, New Astron. Rev. **56**, 37 (2012)
[90] P. Young, Phys. Rev. D **14**, 3281 (1976)
[91] S. Chandrasekhar, *Mathematical Theory of Black Holes* (Oxford: Clarendon Press, 1983)
[92] H. Falcke, F. Melia and E. Agol, Astrophys. J. **528**, L13 (2000)
[93] F. Melia and H. Falcke, Ann. Rev. Astron. Astrophys. **39**, 309 (2001)
[94] A. F. Zakharov et al., Astron. & Astrophys. **442**, 795 (2005)
[95] A. F. Zakharov et al., New Astron. **10**, 479 (2005). astro-ph/0411511
[96] F. De Paolis et al., Gen. Rel. Grav. **43**, 977 (2010)
[97] A. F. Zakharov et al., New Astron. Rev. **56**, 64 (2012)
[98] A. F. Zakharov, Phys. Rev. D **90**, 062007 (2014)
[99] A. F. Zakharov, J. Astrophys. Astr. **36**, 539 (2015)
[100] H. Falcke and S. Markoff, Class. Quan. Grav. **30**, 244003 (2013)
[101] T. Johannsen, Class. Quan. Grav. **33**, 113001 (2016)
[102] S. S. Doeleman et al., Nature **455**, 78 (2008)
[103] A. A. Nucita et al., Publ. Astron. Soc. Pac. **119**, 349 (2007)
[104] A. F. Zakharov et al., Phys. Rev. D **76**, 062001 (2007)
[105] A. F. Zakharov et al., Space Sci. Rev. **48**, 301 (2009)
[106] A. F. Zakharov et al. Phys. At. Nucl. **73** , 1870 (2010)
[107] V. I. Dokuchaev and Yu. N. Eroshenko Phys.-Usp. **58**, 772 (2015)
[108] S. Gillessen et al., Astrophys. J. **707**, L114 (2009)
[109] S. Gillessen et al., Nature **481**, 51 (2012)
[110] L. Meyer et al., Science **338**, 84 (2012)
[111] D. Borka et al., Phys. Rev. D **85**, 124004 (2012)
[112] A. F. Zakharov et al. Adv. Space Res. **54**, 1108 (2014)
[113] A. F. Zakharov et al., Phys. Rev. D **74**, 107101 (2006)

[114] D. Borka et al., J. Cosmol. Astropart. Phys. JCAP11 (2013) 050
[115] M. Fierz and W. Pauli, Proc. R. Soc. London, Ser. A **173**, 211 (1939)
[116] V. I. Zakharov, JETP Lett. **12**, 447 (1970)
[117] H. van Dam and M. Veltman, Nucl. Phys. B **22**, 397 (1970)
[118] Y. Iwasaki, Phys. Rev. D **2**, 2255 (1970)
[119] A. I. Vainshtein, Phys. Lett. B **39**, 393 (1972)
[120] D. G. Boulware and S. Deser, Phys. Rev. D **6**, 3368 (1972)
[121] A. I. Vainshtein, Surv. High Energy Phys. B **20**, 5 (2006)
[122] V. A. Rubakov and P. G. Tinyakov, Phys. – Usp. **51**, 759 (2008)
[123] C. de Rham, Liv. Rev. Relat. **17**, 7 (2014)
[124] C. de Rham et al., arXiv:1606.08462v1 [astro-ph.CO]
[125] C. de Rham, G. Gabadadze and A.G. Tolley, Phys. Rev. Lett. **106**, (2011)
[126] A. S. Goldhaber and M. M. Nieto, Phys. Rev. D **9**, 1119 (1974)
[127] S. S. Gershtein, A. A. Logunov and M. A. Mestvirishvili, Doklady Physics **48**, 282 (2003)
[128] S. S. Gershtein et al., Phys. At. Nucl. **67**, 1596 (2004)
[129] S. S. Gershtein et al., Phys. Usp. **49**, 1179 (2006)
[130] Yu. V. Chugreev, Phys. Part. Nucl. Lett. **13**, 38 (2016)
[131] C. Talmadge et al., Phys. Rev. Lett. **61**, 1159 (1998)
[132] C. Will, Phys. Rev. D **57**, 2061 (1998)
[133] S. R. Choudhury et al., Astropart. Phys. **21**, 559 (2004)
[134] L. S. Finn and P. J. Sutton, Phys. Rev. D **65**, 044022 (2002)
[135] S. L. Larson and W. A. Hiscock, Phys. Rev. D **61**, 104008 (2000)
[136] A. Cooray and N. Seto, Phys. Rev. D **69**, 103502 (2004)
[137] C. Braxmaier et al. Exper. Astron. **34**, 181 (2012)
[138] M. V. Sazhin, Sov. Astron. **22**, 36 (1978)
[139] S. Detweiler, Astrophys. J. **234**, 1100 (1979)
[140] D. Baskaran et al., Phys. Rev. D **78**, 044018 (2008)
[141] D. Baskaran et al., Phys. Rev. D **78**, 089901(E) (2008)
[142] K. Lee et al., Astrophys. J. **722**, 1589 (2010)
[143] K. Lee, Class. Quant. Grav. **30**, 224016 (2013)
[144] B. P. Abbott et al. (LIGO Scientific Collaboration and Virgo Collaboration), Phys. Rev. Lett. **116**, 061102 (2016)
[145] B. P. Abbott et al. (LIGO Scientific Collaboration and Virgo Collaboration) LIGO Document P1500213-v27. arXiv:1602.03841
[146] B. P. Abbott et al. (LIGO Scientific Collaboration and Virgo Collaboration) 2016 Phys. Rev. X **6**, 041015 (2016). LIGO-P1600088. arXiv:1606.04856
[147] M. Visser, Gen. Rel. Grav. **30**, 1717 (1998)
[148] A. F. Zakharov et al., J. Cosmol. Astropart. Phys. JCAP05 (2016) 045
[149] A. F. Zakharov et al., EPJ Web of Conferences **125**, 01011 (2016)
[150] N. Blind et al., arxiv:1503.07303 [astro-ph.IM]
[151] A. Ardeberg et al. 2014 *An Expanded View of the Universe Science with the European Extremely Large Telescope*, ed. M. Lyubenova and M. Kissler-Patig (ESO, Munich, 2014)
[152] W. Skidmore et al. *Thirty Meter Telescope Detailed Science Case: 2015* (TMT Observatory Corporation, 2014)

Baryon electromagnetic form factors at BESIII

Alaa Dbeyssi[1,*] (on behalf of BESIII collaboration)

[1] *Helmholtz-Institut Mainz, Germany*

Abstract. Electromagnetic form factors are fundamental quantities which parameterize the electric and magnetic structure of hadrons. This contribution reports on the measurements of baryon electromagnetic form factors at the BESIII experiment in Beijing. The Beijing e^+e^- collider BEPCII is a double-ring symmetric collider running at $\sqrt{s}$ between 2.0 and 4.6 GeV. Baryon electromagnetic form factors can be measured at BESIII in direct e^+e^--annihilation and in initial state radiation processes. Based on the data collected by the BESIII detector at 12 center of mass energies between 2.23 and 3.67 GeV, the $e^+e^- \to \bar{p}p$ cross section and the time-like proton form factor is measured. Preliminary results from the analysis of the initial state radiation process $e^+e^- \to \bar{p}p\gamma$ using a data set of 7.408 fb^{-1} collected at center-of-mass energies between 3.773 and 4.6 GeV, are also presented. The cross section for $e^+e^- \to \bar{\Lambda}\Lambda$ is measured based on 40.5 pb^{-1} data collected at 4 energy points from the threshold up to 3.08 GeV. Preliminary results on the total cross section and the Λ effective form factor are shown. Ongoing analysis based on the high luminosity energy scan from 2015 and from radiative return at different $\sqrt{s}$ are also described.

1 Introduction

Electromagnetic form factors (FFs) describe the modifications of the point-like photon-hadron vertex due to the structure of the hadrons. They provide information on the intrinsic electric and magnetic distributions of hadrons. Electromagnetic FFs have been associated in the Time-Like (TL) region with the time evolution of these distributions [1]. The number of electromagnetic FFs is determined by the spin of the hadron taking into account the symmetry properties of the electromagnetic interaction under the charge, parity, and time transformations. The structure of a non point-like particle of spin S is parameterized in terms of $(2S + 1)$ FFs.

The electromagnetic FFs of baryons, spin 1/2 particles, can be measured in TL region through the annihilation reactions $e^+e^- \leftrightarrow \bar{B}B$ using the energy scan technique, in which the center of mass energy of the collider ($\sqrt{s}$) is varied systematically, and at each energy point a measurement of the associated cross section can be carried out. In the Born approximation, the differential cross section for the annihilation reaction $e^+e^- \to \bar{B}B$ in the centre of mass system, is expressed as a function of

[*]e-mail: adbeyssi@uni-mainz.de

the baryon electromagnetic FFs $|G_E|$ and $|G_M|$ as [2]:

$$\frac{d\sigma}{d\cos\theta} = \frac{\pi\alpha^2\beta C}{2q^2}\left[(1+\cos^2\theta)|G_M|^2 + \frac{1}{\tau}\sin^2\theta|G_E|^2\right],$$
$$\beta = \sqrt{1-\frac{1}{\tau}},\ \tau = \frac{q^2}{4M^2}, C = \frac{y}{(1-exp(-y))},\ y = \frac{\alpha\pi}{\beta}, \tag{1}$$

where α is the electromagnetic coupling constant, and θ (M) is the scattering angle (mass) of the baryon. The Coulomb factor C makes the cross section non zero at threshold [3]. It can be expressed as a product between an enhancement factor $\mathcal{E} = \frac{\alpha\pi}{\beta}$ and a resummation term $\mathcal{R}$, i. e., the so called Sommerfeld-Schwinger-Sakharaov rescattering formula. In the limit $\beta \to 0$, the Coulomb factor behaves like the enhancement factor $\mathcal{E}$ so that, the factor β is cancelled and the cross section at threshold becomes finite. Equation 1 assumes that the interaction between baryons and leptons occurs through the exchange of one virtual photon, which carries a momentum transfer squared q^2.

The total cross section, integrated over the full range of the polar angle θ is:

$$\sigma(q^2) = \frac{2\pi\alpha^2\beta C}{3q^2\tau}\left(2\tau|G_M|^2 + |G_E|^2\right). \tag{2}$$

The measurement of the differential cross section (1) at a fixed energy allows the determination of the FF ratio $R = |G_E|/|G_M|$. With a precise knowledge of the normalization and the luminosity, a separate determination of $|G_E|$ and $|G_M|$ is possible. The quantity usually measured when the angular distributions cannot be precisely studied (due to the low statistics) is the effective FF, introduced as a linear combination of $|G_E|^2$ and $|G_M|^2$:

$$|G_{eff}| = \sqrt{\frac{2\tau|G_M|^2 + |G_E|^2}{2\tau+1}}. \tag{3}$$

The initial state radiation (ISR) technique is an effective tool to measure hadronic cross section at high luminosity e^+e^--storage rings, such as the Beijing Electron-Positron Collider II (BEPCII). Due to the fact that annihilation of e^+e^- is mostly accompanied by emission of one or several photons from the initial state, the high luminosity collected at the e^+e^- collider experiments, allows for a complementary approach to the energy scan technique in hadron cross section measurements: events with ISR photons lead to a reduction of the invariant mass (q^2) of the proton-antiproton system, and hence allow for a measurement of the baryonic FFs over a wide range of q^2 below $\sqrt{s}$.

2 BESIII detector and data sets

The BESIII detector [4] is a general purpose spectrometer located at BEPCII. BEPCII is a double ring e^+e^- collider running at center of mass energies ($\sqrt{s}$) between 2.0 and 4.6 GeV and reached a peak luminosity of 1.0×10^{33} cm^{-2}s^{-1} at $\sqrt{s}$ =3770 MeV. The cylindrical BESIII detector has an effective geometrical acceptance of 93% of 4π. It contains a small cell, helium-based (60% He, 40% C_3H_8) main drift chamber (MDC) which provides momentum measurements of charged particles with a resolution of 0.5% at 1 GeV/c in a 1 Tesla magnetic field. The energy loss (dE/dx) measurement provided by the MDC has a resolution better than 6%. A time-of-flight system (TOF) consisting of 5-cm-thick plastic scintillators with a time resolution of 80 pb in the barrel and 110 ps in the end caps. An electromagnetic calorimeter (EMC) consisting of 6240 CsI (Tl) in a cylindrical structure and two end caps is used to measure the energies of photons and electrons. The energy resolution of EMC (for 1 GeV electrons and photons) is 2.5% in the barrel and 5.0% in the end caps. The position resolution of EMC is 6 mm in the barrel and 9 mm in the end caps for 1 GeV electrons and photons. A muon system (MUC) consisting of 1000 m^2 of Resistive Plate Chambers (RPC) is used to identify muons and provides a spatial resolution better than 2 cm.

BESIII provided the world's largest samples of e^+e^- collisions in the τ-charm region. Data samples have been accumulated at different charmonium and X, Y, Z states, and high luminosity scan experiments have been performed in the region between 2.0 and 4.6 GeV. The data collected at BESIII allows the measurement of baryon FFs using the direct e^+e^- annihilations and the initial state radiation processes.

3 Proton form factors at BESIII

3.1 Measurement of the $e^+e^- \to \bar{p}p$ using the scan technique

Using the 157 pb^{-1} data samples collected in 2011 and 2012, BESIII measured the cross section of $e^+e^- \to \bar{p}p$ at 12 center of mass energies between 2.2 and 3.7 GeV [5]. The event selection criteria, the estimation of the background contamination and the evaluation of the signal efficiencies were obtained based on the Monte Carlo simulations for the signal and the different background processes. The Born cross section of $e^+e^- \to \bar{p}p$ is calculated by:

$$\sigma_{Born} = \frac{N_{obs} - N_{bkg}}{L\epsilon(1+\delta)}, \tag{4}$$

where N_{obs} is the number of selected events, N_{bkg} is the estimated number of background events, L is the integrated luminosity determined at each center of mass energy, ϵ is the detection efficiency of the signal and $(1+\delta)$ is the radiative correction factor calculated using the CONEXC event generator [6]. The corresponding effective electromagnetic FF of the proton is deduced from the measured cross section using 2 and 3 . The results on the Born cross section and the proton effective FF are shown in figure 1 together with the previous experimental measurements. Below $\sqrt{s}$ =3.08 GeV, BESIII improves the precision on the Born cross section by 30% compared to the most precise data measured by BaBar [7, 8]. The ratio of the proton FFs is extracted from the fit to the distribution of the proton polar angle using the large data samples collected at $\sqrt{s}$ = 2232.4, 2400.0 and a combined sample at $\sqrt{s}$=3050.0, 3060.0 and 3080.0 MeV. The measured ratios of the proton FFs are close to unity and they are consistent with the BABAR results in the same q^2 region.

3.2 Measurement of the $e^+e^- \to \bar{p}p$ using the ISR technique

Based on 7.408 fb^{-1} data samples collected with the BESIII detector at center of mass energies between 3.773 and 4.600 GeV, the proton FFs have been measured using the initial state radiation technique. In this work, the two charged tracks (proton and antiproton) are reconstructed with the MDC, and the ISR photon is assigned to the highest energy photon detected by the EMC. Only the events where the ISR photons are emitted within the acceptance of the EMC are selected (tagged ISR analysis). The ratio of the proton FFs has been extracted in six intervals of $\bar{p}p$-invariant mass from the threshold up to 3.0 GeV/c^2. The Born cross section of $e^+e^- \to \bar{p}p$ and the proton effective FF have been also measured in 31 intervals of $\bar{p}p$-invariant mass below 3.0 GeV/c^2.

Figures 2 and 3 show the preliminary results for the Born cross section, the proton effective FF and the proton FF ratio, respectively. The results are consistent with the previous experiments. The total uncertainty, dominated by the statistical one, is between 20% and 35% for the ratio and between 10% and 64% for the Born cross section.

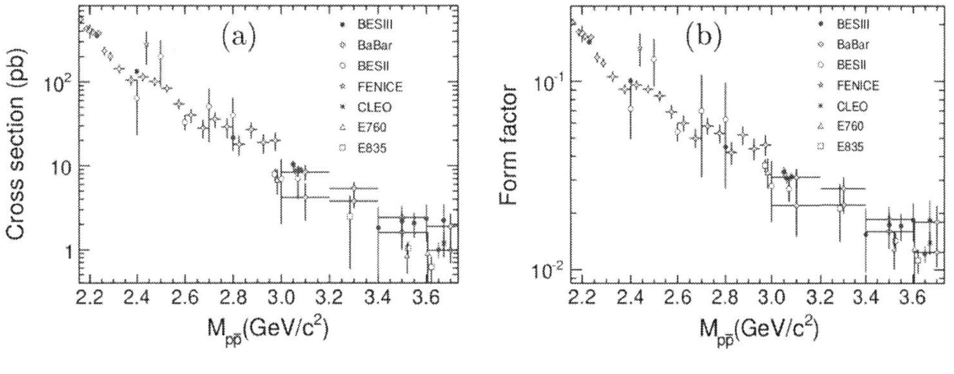

Figure 1. Comparison of the Born cross section of $e^+e^- \to \bar{p}p$ (a) and the proton effective FF (b) between BESIII (red points) and previous measurements [5].

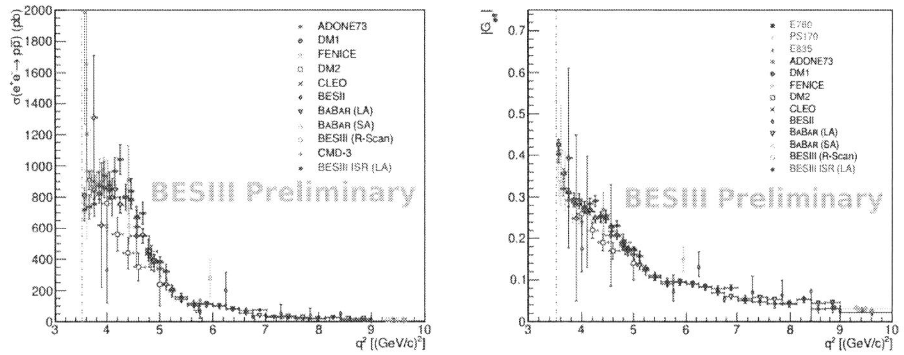

Figure 2. BESIII preliminary results on the Born cross section of $e^+e^- \to \bar{p}p$ (a) and the effective proton (b) from the tagged ISR analysis (BESIII ISR (LA), red points).

The angular distribution of ISR photon in $e^+e^- \to p\bar{p}\gamma$ is shown in figure 4. The differential cross section of $e^+e^- \to p\bar{p}\gamma$ reaches the maximum values at $\cos\theta_\gamma \to \pm 1$ ($\theta_\gamma \sim 0°, 180°$). These regions are not covered by the EMC. The $e^+e^- \to p\bar{p}\gamma$ events in which the ISR photons cannot be detected have been also analysed (untagged ISR analysis). The measurement of these events benefits from the fact that the number of statistics is increased by a factor ~ 3 compared to the tagged analysis. The difficulty of this measurement relays on the signal/background separation using only the information on two detected particles, proton and anti-proton. Signal and background processes have been simulated with the BESIII Offline Software System (BOSS) [9] which includes full detector simulation and event reconstruction. The background channels are almost suppressed by the event selection criteria keeping the signal efficiency around 16%. The same data samples (7.408 fb^{-1}) used for the tagged analysis have been analysed and the signal events in which the ISR photon is not detected have been selected. The results are not yet released by the BESIII collaboration. Based on the Monte Carlo simulations, using PHOKHARA 9.1 event generator [10], the Born cross section of $e^+e^- \to p\bar{p}$ can

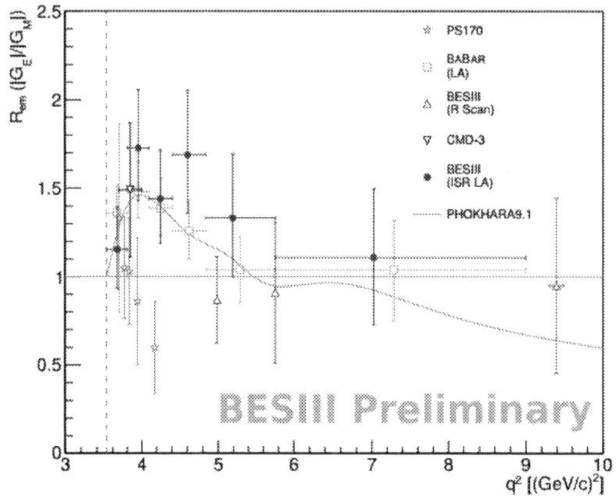

Figure 3. BESIII preliminary results on the ratio of the proton FFs from the tagged ISR measurement of $e^+e^- \to \bar{p}p\gamma$.

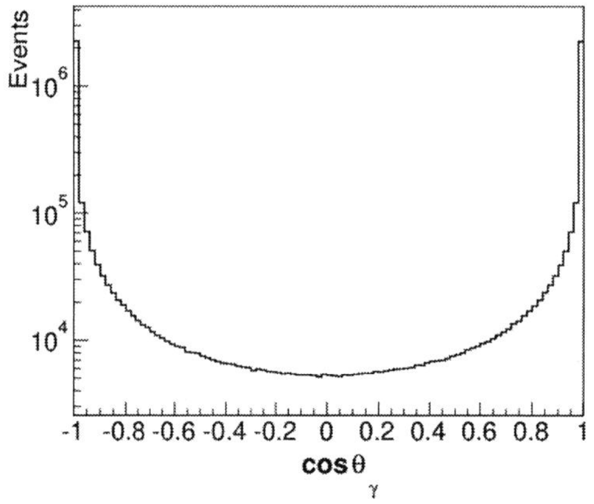

Figure 4. Angular distribution of the ISR photons ($\cos\theta_\gamma$) in $e^+e^- \to p\bar{p}\gamma$ at $\sqrt{s}$ = 4.23 GeV (simulated with PHOKHARA 9.1 event generator).

be measured in the region between 2 and ~ 4 GeV/c^2, with a statistical accuracy between 5% and 35%.

4 Hyperon form factors at BESIII

BESIII released recently preliminary results on the measurement of the $e^+e^- \to \bar{\Lambda}\Lambda$ cross section. Data samples of 40.5 pb^{-1} collected at $\sqrt{s}$ = 2.2324, 2.400, 2.800 and 3.080 GeV have been analysed. The lowest energy point, 2.2324 GeV, is only 1.0 MeV above the threshold of the $\bar{\Lambda}\Lambda$ production. At this energy, the $e^+e^- \to \bar{\Lambda}\Lambda$ process is analysed from two decay modes, $\bar{p}p\pi^+\pi^-$ and the inclusive $\bar{n}\pi^0+X$. The measurement of the cross section of $e^+e^- \to \bar{\Lambda}\Lambda$ process from the two decay modes gives

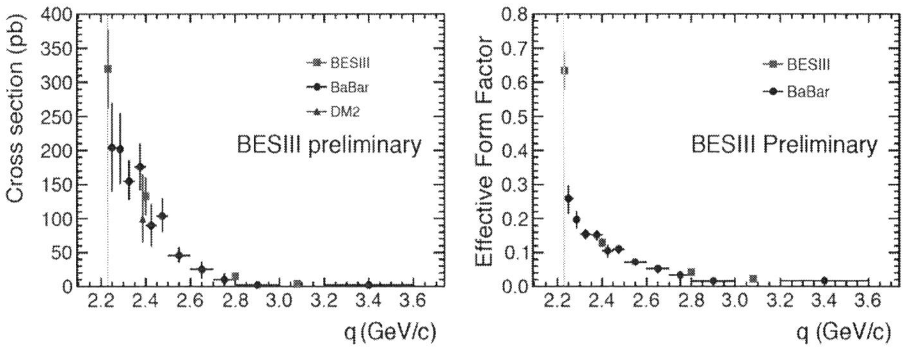

Figure 5. BESIII preliminary results (red points) on the Born cross section of $e^+e^- \to \bar{\Lambda}\Lambda$ (a) and the Λ effective FF (b).

consistent results. At $\sqrt{s}$ = 2.400, 2.800 and 3.080 GeV, the cross section is measured using only the charged decay mode ($\bar{p}p\pi^+\pi^-$). The preliminary results on the Born cross section and the hyperon effective FF are shown in figure 5. At $\sqrt{s}$ = 2.2324 GeV, the average cross section is 318 ± 60 pb, which is significantly larger than the expected value assuming no Coulomb interaction between the produced neutral hyperons. At the higher energy points, the results are consistent with the previous data.

5 Summary and prospects

The experimental measurements of the TL baryon FFs close to the threshold region shows some unexpected features [11, 12]. In this context, a precise and complete measurement of the baryon FFs in this region is of great importance. BESIII is an excellent laboratory for the measurement of baryon FFs, since both ISR and scan methods can be performed, and the kinematical threshold for the different baryon pair production is covered by the available data samples.

In 2015, BESIII performed high luminosity scan in 22 energy points between 2.0 and 3.08 GeV. Based on these data samples, the proton electromagnetic FFs $|G_E|$ and $|G_M|$ can be extracted separately. The expected precision on the ratio of the proton FFs is at the level of 10%. In addition to the proton, BESIII can provide the most precise measurements of the neutron FFs with the $e^+e^- \to \bar{n}n$ and $e^+e^- \to \bar{n}n\gamma$ processes. The determination of Λ FFs, $|G_E|$, $|G_M|$ and the relative phase between them is also possible. The polarization of Λ in $e^+e^- \to \bar{\Lambda}\Lambda$, which contains information on the relative phase between $|G_E|$ and $|G_M|$, can be experimentally accessed thanks to the self-analyzing weak decay of Λ, i.e $\Lambda \to p\pi^-$. The measurement of the hyperons FFs at BESIII can be extended to other channels, i.e. $e^+e^- \to \bar{\Lambda}_{\bar{c}}\Lambda_{c^+}$, $\Lambda\bar{\Sigma}^0$, ...

References

[1] E. A. Kuraev, E. Tomasi-Gustafsson, and A. Dbeyssi. Phys. Lett. B **712**, 240-244 (2012)
[2] A. Zichichi, S. M. Berman, N. Cabibbo, R Gatto, Nuovo Cim. **24**, 170 (1962)
[3] C. Tzara, Nucl. Phys. B **18**, 216-252 (1970)
[4] M. Abilikim et al. (BESIII Collaboration), Nucl. Instru. Meth. **A 614**, 345 (2010)
[5] M. Abilikim et al. (BESIII Collaboration), Phy. Rev. D **91**, 112004 (2015)

[6] R. G. Ping, Chin. Phys. C **38**, 083001 (2014)
[7] J. P. Lees et al. (BABAR Collaboration), Phys. Rev. D **87**, 092005 (2013)
[8] J. P. Lees et al. (BABAR Collaboration), Phys. Rev. D **88**, 072009 (2013)
[9] Z. Y. Deng et al., HEP & NP **30**, 371 (2006)
[10] H. Czyt, M. Gunia and J. H. Kuehn, JHEP **1308**, 110 (2013)
[11] A. Bianconi and E. Tomasi-Gustafsson. Phys. Rev. C **93**, 035201 (2016)
[12] J. Haidenbauer and U. G. Meissner, Phys. Lett. B **761**, 456 (2016)

EPJ Web of Conferences 138, 01012 (2017) DOI: 10.1051/epjconf/201713801012
Baldin ISHEPP XXIII

A new platform for research and applications with electrons: the PRAE project

Dominique Marchand[1]* on behalf of the PRAE collaboration

[1] *Institut de Physique Nucléaire d'Orsay, CNRS/IN2P3 and Université Paris-Sud, UMR8608, Université Paris-Saclay, 91406 Orsay, France*

Abstract. The future PRAE multi-disciplinary platform to be built in the campus of the Paris-Sud University (Orsay, France) is based on a high-quality pulsed electron beam in the energy range between 50 to 140 MeV. The beam will be delivered to three experimental beam lines, each of them being dedicated to specific research and applications projects related to instrumentation, radiobiology and nuclear physics.

1 Introduction

The PRAE[1] project aims at creating in the valley of the University Paris-Sud Campus a multi-disciplinary platform dedicated to research and applied physics. The idea of such a platform arose a year ago from the gathering of researchers and engineers from three nearby laboratories, all located in Orsay, namely the laboratory of Imaging and Modeling in Neurobiology and Cancer (IMNC), the Institute of Nuclear Physics of Orsay (IPNO) and the Laboratory of the Linear Accelerator (LAL). The PRAE facility is based on a high performance 50 Hz pulsed accelerator delivering electrons in the energy range from 50 to 140 MeV to three beamlines, one at a time. The direct beam line, which will benefit from the highest available energy electron beam, will be used to perform radiobiological effect studies to evaluate the potential of an innovative method in radiotherapy based on high energy electrons. A second beam line will be deviated to a fully equipped and instrumented test bench dedicated to detector R&D and quality checks. The third beam line and its related experimental setup will be devoted to nuclear physics research. The first goal in this field is to exploit the elastic electron-proton scattering to determine the proton electric form factor in a very low Q^2 range, from a few 10^{-5} to $10^{-4}(\mathrm{GeV}/c^2)^2$, providing unique data in an unexplored kinematical domain and contributing to the effort of the communities trying to solve the proton charge radius puzzle.

2 The electron accelerator

The PRAE accelerator will consist of a photo-injector, an acceleration section, the beam lines and the related instrumentation. Since a low emittance beam is required, a photo-injector has been chosen as the electron source. Risk minimization guides towards the construction of a gun similar to the one

*e-mail: marchand@ipno.in2p3.fr

[1] Platform for Research and Applications with Electrons

constructed for the CLIC "Test Facility 3" at CERN [1] and running successfully for more than four years. The same type of gun is operated at the PHoto-Injector at LAL (PHIL, [2]) platform and will be also used as the injector for the ThomX project now under construction [3] in Orsay. To obtain high charge per bunch, a metallic magnesium photocathode is employed, which can deliver more than 1 nC with a laser pulse energy of a few tens of a μJ at a wavelength of 260 nm. The magnesium photocathode could produce the specified charge without requiring an in-situ photo-cathode preparation chamber. The gun is made of 2.5 copper cells, magnetically coupled to a waveguide. To get 1 nC with an emittance lower than 5 mm mrad, an accelerating field of 80 MV/m is required, meaning a Radio Frequency (RF) power of 5 MW in a 3 μs pulse. The electron beam energy at the exit of the gun will be of the order of 5 MeV. An industrial laser, synchronized with the RF frequency and triggered at 50 Hz, will be used to extract electrons from the cathode by photoelectric effect. This laser will be installed in an optical room located outside the accelerator concrete shielding and the laser beam will be transported with dielectric mirrors and lenses. The PRAE acceleration section consists of a 3.4 m long compact accelerator structure based on high-gradient (HG) S-band (3 GHz) cavities. The HG is obtained by means of maximizing the RF efficiency and minimizing the surface fields and the modified Poynting vector at very high accelerating gradients. The HG accelerating structure will be a traveling wave, quasi-constant gradient section and will operate at 3 GHz (30°C in vacuum) in the $2\pi/3$ mode. The use of the S-band technology allows for a fast implementation enabling the rapid deployment of the first experiments and applications [4]. The final HG structure will provide an energy gain of 65 MeV for an input peak power of 22 MW with a RF flat top pulse length of 3 μs and a repetition rate of 50 Hz. The section will be located after the RF gun as shown in figure 1. After the first project phase

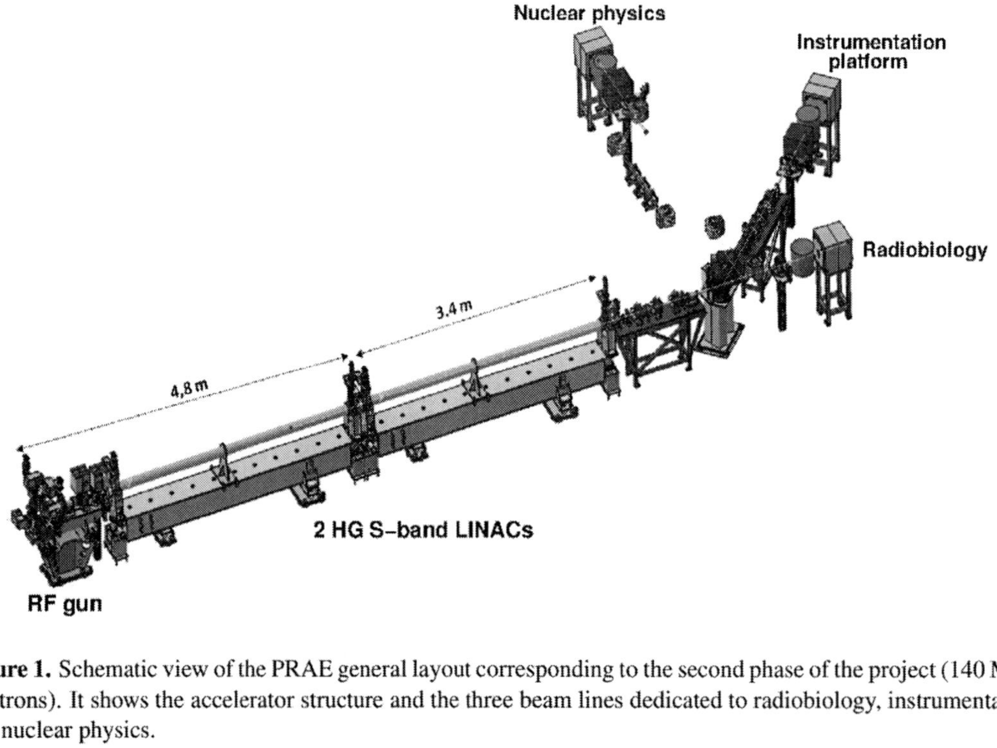

Figure 1. Schematic view of the PRAE general layout corresponding to the second phase of the project (140 MeV electrons). It shows the accelerator structure and the three beam lines dedicated to radiobiology, instrumentation and nuclear physics.

(2017-2019), our goal is to have the accelerator commissioned up to a maximum electron energy of 70 MeV and the three beam lines ready to test their own specific experimental set-up. In a second phase, after 2020, a second high-gradient cavity will be added to reach a 140 MeV electron beam

DOI: 10.1051/epjconf/201713801012

energy. The accelerator will be installed in the former site of the LAL linear accelerator benefiting from an existing dedicated infrastructure. The main design parameters of the PRAE accelerator are summarized in table 1.

Table 1. Principle design parameters of the PRAE electron accelerator related to the first project phase (2018-2019) and the second one in brackets (2020 →).

Beam parameters	Project phase 1 (2)
Energy range [MeV]	50 - 70 (100 - 140)
Charge (variable) [nC/bunch]	5×10^{-5} - 2.
Normalized emittance [mm.mrad]	3. - 10.
Radio Frequency [GHz]	3.
Repetition rate [Hz]	50.
Diameter [mm]	0.5
Bunch length [ps]	< 10.
Energy spread [%]	< 0.2

3 The instrumentation platform

A fully equipped and instrumented platform will be made available to a user community from the academic and the industrial media willing to test and optimize detectors.

The PRAE facility will provide an adjustable number of particles per pulse in a wide range of multiplicities (current/pulse: 1-10^{10} particles) and energies (50-140 MeV). Together with good spatial resolution (beam spot < 1 mm) and excellent timing properties (pulse width < 10 ps), this opens a wide range of applications for detector R&D. The test bench will be instrumented with a precise beam profile measurement device, a particle counter and a calorimeter. In order to operate remotely the beam diagnostic and detector positioning, we will design and produce a movable chamber at the beam exit. It will comprise a moveable screen for beam position estimate: the operator will observe with the embedded camera the spot produced by the beam on the screen (this technique has been proven at the PHIL/LEETECH facility [2][5]), and adjust the position of the chamber by remotely controlled motors. Then, the screen is removed and the beam profiler is inserted in the beam axis to perform the diagnostic. Once it is performed, the profiler and the screen are removed by a remotely controlled mechanical system. A laser beam for the alignment, fixed on the output of the chamber, will serve to drive the moving support of the detector under test and align it with the beam. In order to perform a fast transverse beam image, a TIMEPIX silicon pixel detector [6] will be made routinely available to the users. It will be integrated in the PRAE acquisition system. The sensors have square pixels with a side of 55 μm in a 256×256 pixels layout covering a square sensitive area of about 2 cm^2 and 300 μm thickness. TIMEPIX will be placed in front of the detector under test whenever asked by the user. The number of incident electrons impinging on the detector under tests will be measured by a counter placed just in front of it. This counter is composed of a radiator made of fused silica, where the light emitted by the interacting electrons will be measured by a fast photodetector. A timing resolution of 70 ps for a single electron is expected and the number of electrons will be determined with a precision of 5%. The setup calorimeter will be realized using BGO scintillator crystals read-out with PMTs or SiPMs. The BGO scintillators provide good stopping power at a relatively low production cost; non-hygroscopicity of the material will allow crystal arrangement in a compact matrix geometry, thus providing an optimal angular coverage and active detection area. The detector will be located along the beamline just behind the detector under test. The latter will be installed at the beam line exit on

DOI: 10.1051/epjconf/201713801012

a remotely controlled two-axis trolley with a displacement accuracy of 500 μm. High voltage power supply channels will be available for customers as well as coaxial 50 Ω cables to get the detector signals, and a multi-purpose data acquisition system (WaveCatcher [7] + NARVAL [8]) serving test detector, sensors, radiation monitors... This common acquisition will ease the event building and the analysis of the detector response under test signals. It has been implemented at LAL for the CORTO [9] test platform, and is currently being implemented at IPNO for the R&D of the GASPARD detector [10] (in the 64-channel version). The advantage of NARVAL is the definition of a user interface through reusable modules. The work achieved at IPNO, at CSNSM[2] and at LAL provides templates that will facilitate the installation. In order to test detectors with low intensities, the beamline will be instrumented with a spectrometer that will deliver electrons with adjustable energy from 10 to 140 MeV at intensities from 1 to 1000 particle(s)/sample. The design of this setup will follow the one of the LEETECH experimental platform that was built at the PHIL accelerator at LAL [2][5].

4 The radiobiological program

The motivation for the radiobiological program arose from the fact that whereas radiotherapy is the most frequently used method in cancer treatments [11] it is not fully satisfactory as the dose delivered to a tumor is conditioned by the dose that can be tolerated by the surrounding tissues. In this context, the main challenge in radiotherapy is to find novel approaches allowing the escalation of the dose delivered to the tumor, resulting in an improvement in the cure rate, while lowering the normal tissue complication probability.

In standard radiotherapy (RT), photon beams either X-rays or gamma rays are commonly used. RT with hadrons (protons and light ions) offers several advantages over photons, in particular because they deposit maximum energy at the end of their range, they can be shaped as narrow pencil beams of variable penetration depth and have been shown to be highly effective in treating certain radio-resistant tumors thanks to the radiobiological properties of light ions (see *e.g.* the reviews [12], [13]). However, the cost-effectiveness of Hadron-Therapy (HT) is discussed in literature for some tumor localizations (*e.g.* prostate cancer), as the construction and running costs of HT units are considerably greater than those of conventional facilities [14].

Electrons are also used in conventional RT, with energies ranging from 5 to 20 MeV. However, they are not suitable for the treatment of deep-seated tumors due to their short range and substantial lateral scattering. Since electron beams have a finite effectiveness range after which the dose falls off rapidly, electron beam therapy is typically used in the treatment of superficial tumors. This limitation can be overcome if the electron energy is increased above 70 MeV, whereby the penetration depth becomes longer and the transverse penumbra sharper, though the longitudinal penumbra is also increased. Radiation therapy delivered with very high-energy electrons (VHEE) presents an intriguing and possibly cost-effective option to treat cancer. Small diameter VHEE beams can be scanned, therefore, irradiation with VHEE beams requires, in general, less modulation of beam intensities than photon beams [15]. This is very advantageous for intensity-modulated radiation therapy. VHEE beams may be delivered with greater precision and accelerators may be constructed at significantly lower cost than proton beams.

Another approach to increase the therapeutic window for cancer treatment, is to better protect the healthy tissues by using the technique of Spatially Fractionated Radiotherapy (SFR) with submillimetric beam sizes. Such beam sizes allow exploiting the dose-volume effects [16]: the smaller the beam size, the higher the tolerances of the healthy tissue. In SFR, the resulting dose profiles consist of a pattern of peaks and valleys, *i.e.*, with high doses along the beam paths and low doses in

[2]Centre de Sciences Nucléaires et de Sciences de la Matière, Orsay, France.

the spaces between them. Such techniques have been proven to be safe and effective in delivering large cumulative doses of radiation in the treatment of cancer. This is due to the fact that areas receiving low (valley) doses serve as re-growth centers for normal tissue, which reduces healthy tissue complications [17]. Remarkably high healthy tissue dose tolerances ($\geqslant$ 100 Gy in one fraction) have been observed in numerous biological experiments using SFR techniques that employ micrometric beams in one direction (see references [18], [19] and [20]), like micro-beam radiation therapy (MRT) or mini-beam radiation therapy (MBRT). MRT and MBRT rendered significant tumor growth delay and, in some cases, complete tumor ablation, see references [18] to [20]. Nevertheless, the biological basis of normal tissue preservation after MBRT irradiations is not well understood yet. In fact, they appear to challenge many of the current paradigms in conventional radiation therapy, since they seem to implicate biological mechanisms different from those involved when direct damage by ionizing radiation takes place. It appears that high Peak-to-Valley Dose Ratio (PVDR) values and low valley doses are requested for healthy tissue sparing [20] while low PVDR values favor tumor control [18]. Considering that point, a possible way of further improving the therapy would be to explore the use of small grid sizes with charged particles, in order to reach very high spatial fractionation of dose in healthy tissue to maximize the sparing, and benefit from the lateral scattering of charged particles to reach a quasi-homogeneous dose in the tumor to increase the tumor control. The objective of the

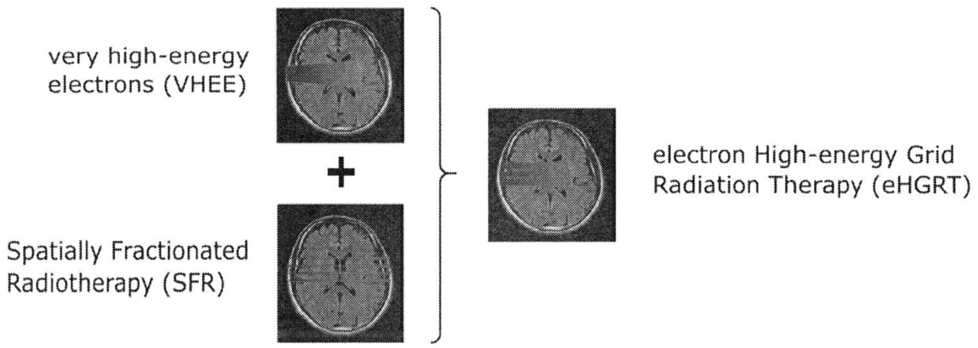

Figure 2. Schematic of the principle of the electron High-energy Grid Radiation Therapy (eHGRT) technique.

radiobiology axis of the PRAE project is to explore such a dose delivery method in electron therapy, combining the use of VHEE beams with the SFR strategy, as illustrated on figure 2. This novel approach will be called: electron High-energy Grid Radiation Therapy (eHGRT) and it is based on a sample or beam scanning used to perform a relative translation of the beam, with respect to the irradiated sample, following a virtual grid as illustrated on figure 3 (left). With this concept, the dose delivered to an hypothetic 8 cm-depth tumor would be quasi-homogeneous while a remarkable tissue sparing is observed at beam entrance, as shown on the figure 3 (right) obtained from simulation with a 150 MeV electron beam. The theoretical proof of concept of this innovative technique has been recently published in reference [21]. In this study, eHGRT has been compared at a dosimetric level with protons grid radiation therapy (pGRT), since the proton-minibeam technique has already been dosimetrically approved in previous study [22]. Both techniques were found suitable (with some configurations of beam size and grid spacing) to treat a 8 cm-depth tumor. The eHGRT shows the highest PVDR in the first centimeters of healthy tissues, improving the tissue sparing before the tumor while pGRT allows more homogeneous dose distribution in the tumor and better tissue sparing behind the tumor. As the dose homogeneity in tumor can be improved by interlacing several beams, these results

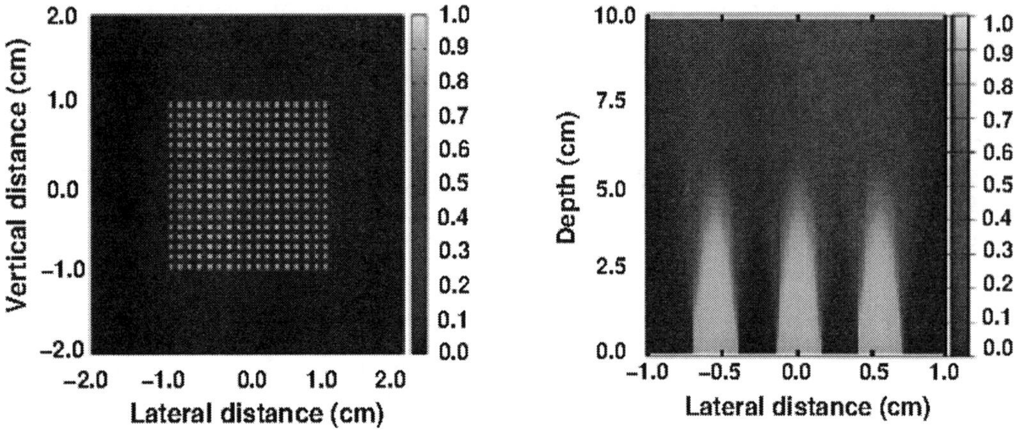

Figure 3. Illustration of the spatial fractionation of the dose with a grid pattern (left side) and the 2D depth-dose distribution in eHGRT (depth versus lateral distance) obtained from [21].

support the interest of performing biological experiments with eHGRT to evaluate this cost-effective grid radiation therapy.

At PRAE, feasibility studies will be performed to accurately evaluate radiobiological effects. A first series of in vivo experiments will also be performed with the main goal to confirm the gain in normal tissue sparing thanks to this novel approach. This program also includes the development of an innovative experimental methodology for a reliable dosimetry with the very small field sizes used, inspired by the dosimetry protocol that was developed for X-rays MBRT [23]. In addition a Monte-Carlo based dose calculation engine will be implemented to guide the pre-clinical trials.

5 The nuclear physics program

Our goal is to contribute to the current effort trying to solve the proton charge radius puzzle which arises from the 7σ disagreement between the proton charge radius value obtained from the most recent electron scattering experiments ($r_p = 0.879 \pm 0.008$ fm [24]) on the one hand, and from Lamb shift measurements with muonic atoms ($r_p = 0.84184 \pm 0.00067$ fm [25]) on the other hand as illustrated on figure 4 extracted from reference [26].

Several explanations have been proposed [26] to explain this striking discrepancy ranging from inaccurate radius extraction to lepton non-universality, genuine effects and physics beyond Standard Model. As yet no consensus has been reached and as the extrapolation down to $Q^2 = 0$ is a matter of controversy, the goal of PRAE is to provide high accuracy determination of the proton electric form factor at very low momentum transfer squared from a few 10^{-5} up to $10^{-4}(\text{GeV}/c^2)^2$ by measuring the elastic scattering of 50 to 70 MeV electrons off protons with with a 1% accuracy. In this unexplored very low Q^2 range, any deviation from unity of the electric form factor of the proton would indicate the existence of unexpected physics. Besides, the addition of accurate experimental data to existing ones, the Q^2 domain considered by PRAE should help to constrain the zero-extrapolation (see references [27], [28] and [29]) required to extract the proton charge radius.

The so-called ProRad (Proton Radius) experiment at PRAE consists of measuring the elastic electron proton scattering cross section over an angular range of the scattered electrons from 5° to 15°.

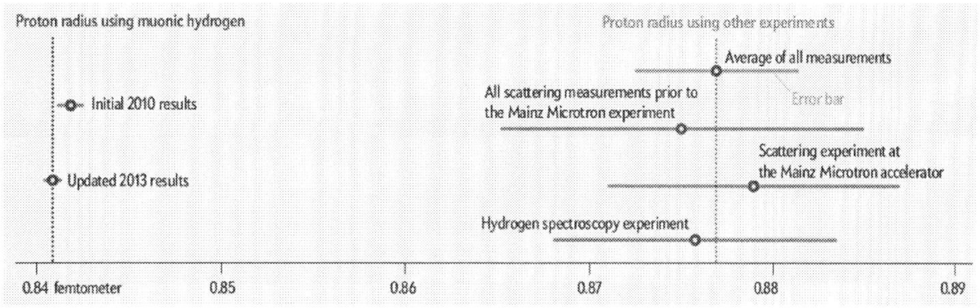

Figure 4. Summary from reference [26] of the proton charge radius values obtained from electron experiments (right) and muon experiments (left).

To select electrons scattered at specific angles we are currently planning to rely on a purely geometrical method avoiding the use of a magnetic field and tracking system. The energy of the electrons is measured in an electromagnetic calorimeter. To define the direction and to get rid of the background a thick shielding with small holes is placed before the detection. Simulations are on-going to define the optimized detection system. Concerning the hydrogen target, different options are considered all based on thin (10 to 20 μm) windowless solid hydrogen targets. Møller scattering electrons, although they represent background to elastic electrons, will be very useful to determine the absolute normalization. As a matter of fact, it is kinematically possible to disentangle Møller from elastic scattered electrons emitted in the same direction by measuring their energy deposition in the same crystal. From figure 5, one can see how much the Møller and the elastic peaks are separated for an electron scattered at 5° on the left and 10° on the right for a 50 MeV electron beam energy.

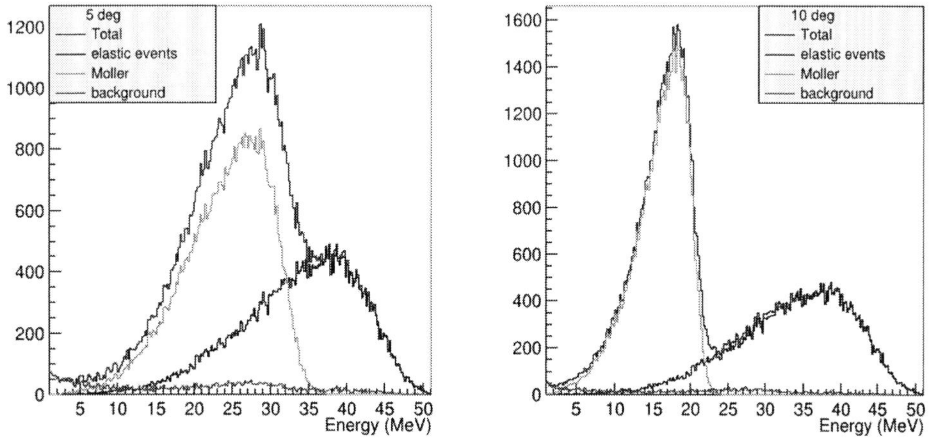

Figure 5. Projection of the total event distribution (blue) in a CSI crystal versus the energy deposit corresponding to 5° (left) and 10° (right) for a 50 MeV electron beam energy. The different contributions are: scattered electrons from Møller scattering (green), elastically scattered electrons (black) and background (red).

Besides the requirements on the position of the beam ($\sigma_{x,y}$ < 0.5mm) and the angular resolution of scattered electrons ($\Delta\theta$ < 1mrad), an accuracy better than 10^{-3} is mandatory on the incident beam energy and the radiative effects need to be well evaluated to meet the foreseen sub-percent level on the measurement of the proton electric form factor.

6 Conclusion

In a near future, the PRAE multi-disciplinary facility will greatly contribute to the fields of fundamental, as well as applied physics, and will also provide valuable tools for education. Based on a high performance electron accelerator, feasibility studies will be carried out to evaluate the potential of a new innovative method in radiotherapy. A high quality fully instrumented test bench will serve to investigate the performance of detectors and accurate measurements of the proton electric form factor will be performed in an unexplored range of momentum transfer squared from a few 10^{-5} down to $10^{-4}(\mathrm{GeV}/c^2)^2$.

Acknowledgement

This work is partly funded by the CNRS, the French Ile-de-France region in the framework of SESAME projects and the P2IO LabEx (ANR-10-LABX-0038) in the framework « Investissements d'Avenir » (ANR-11-IDEX-0003-01) managed by the French National Research Agency (ANR). Furthermore, we are benefiting from the research conducted in the scope of the IDEATE International Associate Laboratory (LIA) [30].

References

[1] http://clic-study.web.cern.ch/content/ctf3-general-description
[2] http://http://phil.lal.in2p3.fr/
[3] http://www.lal.in2p3.fr/ThomX
[4] M. El Khaldi, L. Garolfi, in *Proc. 6th International Particle Accelerator Conference* (Richmond, VA, USA, paper WEPMA008) 2757–2760. ISBN: 978-3-95450-168-7. http://jacow.org/IPAC2015/papers/wepma008.pdf (2015)
[5] D. Attie et al., e-print: arXiv:1601.04348
[6] X. Llopart, R. Ballabriga, M. Campbell, L. Tlustos, W. Wong, Nucl. Instrum. Methods Phys. Res. A **581**, 485-494 (2007). doi:10.1016/j.nima.2007.08.079
[7] D. Breton, E. Delagnes, and J. Maalmi, Nucl. Instrum. Methods Phys. Res. A **695**, 61–67 (2012). doi:10.1016/j.nima.2011.12.007
[8] X. Grave, R. Canedo, J.-F. Clavelin, S. Du, E. Legay, 14th IEEE-NPSS Real Time Conference, 119 (2005). doi:10.1109/RTC.2005.1547454
[9] A. Natochii et al., Technical report LAL/RT 15-04 (2015). Retrieved from http://hal.in2p3.fr/in2p3-01109588
[10] M. Assié et al., The European Physical Journal A **51(1)**, 11 (2015). doi:10.1140/epja/i2015-15011-6
[11] E. Niederlaender, KS-NK-06-010-EN-N, Technical report, Eurostat (2006)
[12] R. Orecchia, M. Krengli, B. A. Jereczek-Fossa, S. Franzetti, J. P. Gerard, Critical reviews in oncology/hematology **51(2)**, 81-90 (2004)

[13] M. Dosanjh, H. F. Hoffmann, G. Magrin, Nuclear Instruments and Methods in Physics Research Section A: Accelerators, Spectrometers, Detectors and Associated Equipment **571(1)**, 191-194 (2007)

[14] M. Lodge, M. Pijls-Johannesma, L. Stirk, A. J. Munro, D. De Ruysscher, T. Jefferson, Radiotherapy and Oncology **83(2)**, 110-122 (2007)

[15] M. Mohiuddin, J. H. Stevens, J. E. Reiff, M. S. Huq, and N. Suntharalingam, Radiat. Oncol. Invest. **4**, 41 (1996)

[16] H. J. Curtis, Radiation Research Supplement **7**, 250-257 (1967)

[17] A. Dilmanian et al., Proc. Natl. Acad. Sci. U. S. A. **103**, 9709 (2006)

[18] P. Regnard et al., Phys. Med. Biol. **53**, 861 (2008)

[19] Y. Prezado et al., J. Synchrotron Radiat. **19**, 60 (2012)

[20] A. Dilmanian et al., Neuro-Oncol. **4**, 26 (2002)

[21] I. Martinez-Rovira and Y. Prezado, Med. Phys. **42**, 685 (2015)

[22] C. Peucelle, C. Nauraye, A. Patriarca, E. Hierso, N. Fournier-Bidoz, I. Martínez-Rovira, Y. Prezado, Medical physics **42(12)**, 7108-7113 (2015)

[23] Y. Prezado et al., Med. Phys. **38**, 5012 (2011)

[24] J. C. Bernauer et al., Phys. Rev. C **90**, 015206 (2014)

[25] R. Pohl et al., Nature **466**, 213 (2010)

[26] J. C. Bernauer, R. Pohl, Scientific American **310**, 32 (2014)

[27] K. Griffioen, C. Carlson, S. Maddox, arXiv:nucl-ex:1509.06676

[28] D. W. Higinbotham, A. Amin Kabir, V. Lin, D. Meekins, B. Norum, B. Sawatzky, arXiv:nucl-ex:1510.01293

[29] M. O. Distler, T. Walcher, J. C. Bernauer, e-print arXiv:nucl-ex:1511.00479

[30] https://ideate.lal.in2p3.fr/en/home/

EPJ Web of Conferences **138**, 01013 (2017) DOI: 10.1051/epjconf/201713801013
Baldin ISHEPP XXIII

Minkowski space approach for the Bethe-Salpeter equation

Lauro Tomio[1,2,⋆], *Cristian* Gutierrez[1], and *Vitor* Gigante[3]

[1] *Instituto de Física Teórica, UNESP, 01140-070, São Paulo, SP, Brazil*
[2] *Instituto Tecnológico de Aeronáutica, DCTA, 12.228-900, São José dos Campos, SP, Brazil*
[3] *Laboratório de Física Teórica e Computacional - LFTC, Universidade Cruzeiro do Sul, 01506-000, São Paulo, SP, Brazil*

Abstract. The bound-state energy spectrum for two-boson system with a fixed exchanged massive boson is investigated by considering the scalar Bethe-Salpeter (BS) equation in Minkowski space within the ladder-approximation framework. The approach is extending to excited states a previous study performed for the ground state. The number of states in the spectrum is restricted by the coupling strength of the two-body interaction, as well as the mass of the exchanged boson. In the present approach, we use the Nakanishi Integral Representation and the formally exact null-plane projection. As usual a comparison between eigenvalues obtained in the Minkowski and Euclidean space is presented. Within such scheme, the light-front momentum-space valence wave functions are obtained for the ground and excited states, together with the corresponding quantities in the impact-parameter space. Among relevant features emerging from this study, we can point out the node structure of light-front wave function and the leading exponential fall-off of the valence wave function in the impact-parameter space.

1 Introduction

The study of the relativistic bound-state problem, by means of the fully covariant Bethe-Salpeter (BS) equation [1] directly in the Minkowski space, has been undertaken in the last two decades thanks to the Nakanishi integral representation (NIR) to represent the BS amplitude [2]. The NIR leads to an integral equation for the Nakanishi weight-function, which is equivalent to the initial BS equation. The first attempts to use the NIR in the non-perturbative regime were performed by K. Kusaka and A.G. Williams [3, 4], when studying a bound-state system composed by two scalar particles interacting by means of another scalar, in the ladder approximation. To treat the singular behavior of the integral equation a suitable numerical treatment was adopted. Within the same problem, a substantial simplification in the derivation of the equation for the Nakanishi weight-function was achieved in the work developed by J. Carbonell and V. Karmanov [5]. In addition of using the NIR for the BS amplitude, the light-front (LF) projection was implemented in a twofold in [5], in order to remove the singularities present in the integral equation of the Nakanishi weight-function and to obtain the LF valence wave function for the bound state. This method is shown to be valid for any irreducible kernel in the BS equation. Further simplifications to the integral equation derived in [5] were achieved by using the uniqueness conjecture of the Nakanishi weight-function, as shown in [6, 7] by considering

⋆e-mail: tomio@ift.unesp.br

the ladder kernel. The exploration of the method to include kernels beyond the ladder has successfully been performed in [8, 9]. Moreover, in [10], within the NIR the effect of the self-energy contribution to the binding energy of the ground state has been explored in the scalar theory. It is important to point out that the very relevant problem of two-fermion bound state has been investigated within the NIR framework in [11, 12], for the case $J^\pi = 0^+$, where J is the total spin and π the parity of the bound state.

The previous successful applications of the NIR for the BS equation in Minkowski space encouraged us to formulate an extended program to investigate other interesting problems, by considering the NIR together with LF projection methods. As a natural extension of the previous analysis, which was limited only for ground states, within a recent work done in [13], the s−wave spectrum of excited states for the scalar theory was explored in the ladder kernel approximation. This problem, as far as we know, was for the first time investigated in the Minkowski space, with relevant details appearing in [14]. In the present communication, we complement the results presented in [14] with some further analysis considering the results obtained in [13]. In addition of computing the binding energies, which are available in the Euclidean space, the method allows one to obtain the LF wave function, which should reveal the node structure typical for excited states. As usual, the calculation in the Euclidean space is used to verify the reliability of the method by comparing the eigenvalues in Minkowski and Euclidean spaces. In order to widen the applicability of the calculations in Minkowski space, we explore the structure of the valence wave function in the impact parameter (IP) space. This quantity is obtained as the Fourier transform of the transverse momentum of the LF wave function, while keeping the longitudinal component. Thus, it provides a correlation between the longitudinal momentum of the constituent with its transversal probability distribution. We were motivated to study this quantity because of the increasing interest in hadron physics, when investigating structural properties of hadrons, such as the parton transverse-momentum distributions [15, 16]. A detailed description of the wave function in the IP space is presented here for the first-excited state.

Next, in Sec. 2 we review the NIR method to study bound states, with particular focus on an extension of the approach to obtain the excited states. In Sec. 3, in order to enhance the confidence on the approach and appreciate its numerical performance, we include a comparison between the eigenvalues obtained in the Minkowski and Euclidean spaces. The consistency of the results are also verified for the eigenvalues in the non-relativistic limit. In Sec. 4, we discuss the results obtained for the LF momentum-space valence wave function, considering the ground and first-excited states. The valence wave function in the impact-parameter space is analysed in Sec. 5, presenting our results corresponding to the first excited state. Finally, in Sec. 6, we resume this report with a brief conclusion.

2 Minkowski space Bethe-Salpeter equation and bound-state spectrum

The BS equation for a bound state composed by two scalar particles of equal mass m in the ladder approximation, without considering vertex neither self-energy corrections, reads as [6, 7, 14]:

$$\varphi(k,p) = \frac{i}{(p/2+k)^2 - m^2 + i\epsilon}\,\frac{i}{(p/2-k)^2 - m^2 + i\epsilon}\int \frac{d^4k'}{(2\pi)^4}\,\frac{i(-ig)^2}{(k-k')^2 - \mu^2 + i\epsilon}\,\varphi(k',p), \quad (1)$$

where μ is the exchanged-scalar mass. The BS amplitude written in terms of the NIR reads as

$$\varphi(k,p) = -i\int_0^\infty d\gamma' \int_{-1}^{1} dz' \frac{g(\gamma',z')}{[\gamma' + \kappa^2 - k^2 - (p\cdot k)z' - i\epsilon]^3}, \quad (2)$$

where $\kappa^2 = m^2 - M^2/4 > 0$ for bound states, with M being the total energy (bound-state mass), which is usually related to the binding energy B, as $B = 2m - M$. Therefore, $B \in [0, 2m]$. By introducing the BS amplitude in terms of the NIR, (2) into (1), and performing the LF projection, we obtain the following integral equation for $g(\gamma', z')$:

$$\int_0^\infty d\gamma' \frac{g(\gamma', z)}{[\gamma' + \gamma + z^2 m^2 + (1 - z^2)\kappa^2]} = \int_0^\infty d\gamma' \int_{-1}^{1} dz' \, V(\gamma, z; \gamma', z') \, g(\gamma', z'), \tag{3}$$

where the expression for $V(\gamma, z; \gamma', z')$, with a detailed derivation using the ladder approximation, can be found in [7]. In order to solve (3), two basis expansions were introduced for the Nakanishi weight-function [7], in terms of Laguerre (γ–integrations) and Gegenbauer (z–integrations) polynomials. The success of this basis expansion when studying the ground state led us to consider the same polynomial basis in our numerical approach to study the excited states. In our numerical study, we found that the basis numbers for the γ and z integrations, when considering the excited states with good accuracy ($N_\gamma = 26$, $N_z = 10$), are not too much larger than the corresponding numbers for the ground-state ($N_\gamma = 14$, $N_z = 10$). In order to obtain the spectrum of the excited states in the numerical approach, one can slightly change the integral equation used for the ground state calculation in the following form

$$\lambda g = \left(\frac{1}{\alpha_{gr}}\right)^{-1} \mathcal{B}^{-1}(M) \, \mathcal{A}(M) \, g, \tag{4}$$

where $\mathcal{B}(M)$ and $\mathcal{A}(M)$ represent symbolically the operators in the left- and right-hand side of (3), respectively, where $\alpha_{gr} = g^2/(16\pi m^2)$ (positive defined) is obtained from the ground-state calculation. The (4) has a pivotal role in solving the BS equation for bound states: (i) First, it allows to obtain the coupling constant for the ground state, α_{gr}; (ii) next, by fixing α_{gr}, one can reach the corresponding excited-state values in the spectrum, by looking for values $M = 2m - B_{gr} > M_{gr}$ such that $\lambda = 1$.

By following the approach as explained above in our numerical investigation, we summarize our results for the behavior of the coupling constant as function of the binding energy in figure 1. In this figure, we present three panels corresponding to three different values of the mass of the exchanged boson ($\mu = 0.01\,m$, $0.1\,m$ and $0.5\,m$). In each panel our results are given for the ground and first two excited states. Once given μ, in order to identify the possible excited states, one should first obtain the coupling for the ground state and next look for the corresponding excited states. Therefore, the relevant parts of the panels are constrained by the corresponding maximum value of the ground-state energy ($B \leq 2\,m$). As in the non-relativistic case with Yukava potential, the number of excited states is finite for $\mu \neq 0$, increasing only when $\mu \ll m$ is approaching zero [17]. Theoretically, an infinite number of states is expected for $\mu = 0$ (Coulomb limit). However, as we can verify in the next section, this number increases slowly as μ is being reduced, with the corresponding coupling α_{gr} being adjusted by the ground-state. As shown in figure 1, the value of α_{gr} is restricted to a maximum value defined by $B \leq 2\,m$, such that in case that $\mu = 0.5\,m$ only one excited state can be found.

3 Comparison between the Minkowski and Euclidean eigenvalues

Here we provide the results we have obtained for a comparison between the eigenvalues in the Minkowski space, using the NIR, with those obtained by solving the Euclidean BS equation. With $B(n)$ ($n = 0, 1, 2, 3..$) defined as the binding energy of the ground and excited states, where $B(0) = B_{gr}$, we choose $B(0) = 1.9m$ for the binding energy of the ground state, using different mass values μ for the exchanged boson. This choice of a strongly-bound ground-state energy is motivated by the fact that the spectrum will be more sensitive to relativistic effects. Our main results are displayed in table 1 where we can observe a very satisfactory agreement between the values obtained in both Minkowski and Euclidean spaces, which provides a further reliability of the method. As an expected feature of

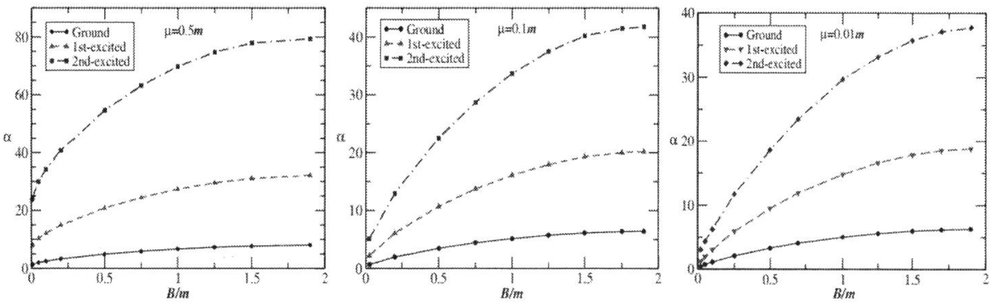

Figure 1. Behavior of the coupling constant α as function of the binding energy B, in units of m, for three different values of the mass μ of the exchanged boson, considering the ground and first two excited states.

the relativistic spectrum one can observe that, as μ increases the number of excited states decreases. On the other side, when μ is reduced approaching to zero, we can observe an increasing number of excited states (toward infinite for $\mu = 0$), in agreement with the non-relativistic spectrum [17].

Table 1. Comparison of the spectra obtained in the Euclidean and Minkowski spaces [14], for the binding energies $B(n)$ of two boson with mass m, by varying the mass μ of the exchanged boson and, consequently, the value of the coupling α_{gr}. In this case, the value of the ground-state binding energy was fixed to $B(0) = 1.9\,m$.

$(\mu/m, \alpha_{gr})$		**Minkowski**	**Euclidean**
(0.05, 6.324)	$B(1)/m$	0.259	0.258
	$B(2)/m$	0.090	0.090
(0.1, 6.437)	$B(1)/m$	0.221	0.220
	$B(2)/m$	0.050	0.051
(0.5, 8.047)	$B(1)/m$	0.0082	0.0082

The energy ratios $B(n)/B(0)$ are also displayed in figure 2, showing the corresponding behavior as we vary the mass values of the exchanged boson μ. We consider the first two excited states, with $B(0) = 0.3\,m$ (left panel) and $0.5\,m$ (right panel). The non-relativistic limits are also being presented for $\mu \to 0$ in order to verify the consistency of the relativistic results. For that, we consider the analytic expression obtained in the Euclidean space for the bound state of two spinless bosons exchanging a massless scalar boson. In lower orders of α, this expression is given by [18]

$$B(n) = \frac{m}{4} \frac{\alpha^2}{(n+1)^2} \left[1 + \frac{4\alpha}{\pi} \ln \alpha \right] + \dots \qquad (n \geq 0). \tag{5}$$

The first term corresponds to the non-relativistic limit and it is reproduced by the energy ratios displayed in figure 2 for the first- and second-excited states in the limit $\mu = 0$. Notice that the logarithmic corrections are canceled when computing the energy ratios. Moreover, the agreement of the NIR results with this expansion given by (5) can also be verified when considering smaller values of $B(0)$, as observed from table 2. In this table, we display a set of binding energies in order to verify the range the validity of the expansion (5). One should notice that the agreement of the NIR results with the analytical expression is better for $B(0) \leq 0.02\,m$, which is expected, since (5) was obtained by performing a standard perturbative expansion in α.

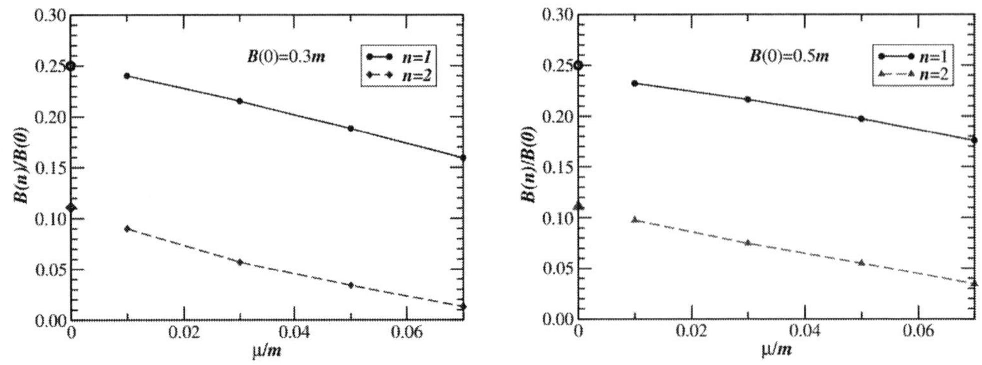

Figure 2. Energy ratios $B(n)/B(0)$ (where $B(0)$ is the ground-state energy and $n = 1, 2$) are shown for μ/m. As indicated in the panels, we have fixed $B(0)$ to 0.3 m (left panel) and 0.5 m (right panel), with the first excited state given with solid lines and second excited state with dashed lines. The symbols on the lines indicate the positions where we have the values obtained through (4). At the origin, as indicated inside both panels, we can reach the corresponding non-relativistic limits expressed by (5) for the ratio $B_{nr}(n)/B_{nr}(0) = 1/(n+1)^2$: $B_{nr}(1)/B_{nr}(0) = 0.25$ and $B_{nr}(2)/B_{nr}(0) = 0.11$. (Right panel extracted from [14]).

The results in the non-relativistic limit, given by the first term of (5) and displayed in the fourth column of table 2, show the obvious limitation of (5) to small values of α as compared with the third column, and considerable deviations from the results obtained with the NIR. Also, the relativistic results for the energy levels are consistently smaller than the corresponding non-relativistic ones, not only for small values of α but also for larger couplings, as in case of the extrapolated results shown in figure 2 for the ratios (where the α_{gr}'s are larger than one). As visually noticed, the extrapolated result is better verified in case that $B(0) = 0.3m$ than in the case that $B(0) = 0.5m$. The quite good results obtained for the extrapolated values of the energy ratios indicate that the expected deviations for the energies, verified in (5), are mainly canceled for the ratios in the limit $\mu = 0$. As in this limit the relativistic ratios between the excited bound-state energies of the spectrum $B(n)/B(0)$ are not larger than the corresponding non-relativistic ratios, being practically the same $B(n)/B(n-1) = (1 + 1/n)^2$, an infinite number of bound-state energies should also be verified for the relativistic case when $\mu = 0$, such that $\lim_{n\to\infty} B(n) = B(n-1)$, with a finite number given in case the exchange boson has $\mu \neq 0$.

4 Momentum space valence wave function

Once the Nakanishi weight-function is obtained by solving (3), we can obtain the corresponding LF wave function. In terms of the variables ξ and $\mathbf{k}_\perp$, where $\xi \equiv (1-z)/2$ and $k_\perp \equiv \sqrt{\gamma}$, the LF wave function reads,

$$\psi(\xi, \mathbf{k}_\perp) = \frac{1}{\sqrt{2}}\xi(1-\xi)\int_0^\infty d\gamma' \frac{g(\gamma', 1-2\xi)}{\left[\gamma' + k_\perp^2 + \kappa^2 + (1-2\xi)^2\frac{M^2}{4}\right]^2}. \tag{6}$$

In figure 3, we show the behavior of this LF wave function for the ground (two upper panels) and first excited (two lower panels) states. The ground-state wave function is given for $B(0) = 1.9\,m$, with $\mu = 0.1\,m$ and α_{gr}=6.437, with the corresponding first-excited state emerging at $B(1) = 0.22\,m$. The wave-function behaviors are presented in terms of the variable $k_\perp$ for some given values of ξ (two

Table 2. By considering the limit $\mu = 0$, the coupling α_{gr} (1st column) with the corresponding ground-state energies are shown by the results obtained from (i) the NIR approach (2nd column); (ii) the expansion (5) (3rd column); (iii) and the corresponding non-relativistic limits (4th column), obtained from the leading term of (5).

α_{gr}	$B(0)/m$ [NIR]	$B(0)/m$ [(5)]	$B_{nr}(0)/m$
0.272	0.010	0.010	0.018
0.344	0.015	0.016	0.030
0.401	0.020	0.021	0.040
0.460	0.025	0.029	0.053
0.519	0.030	0.038	0.067
0.621	0.040	0.060	0.096
0.711	0.050	0.087	0.126

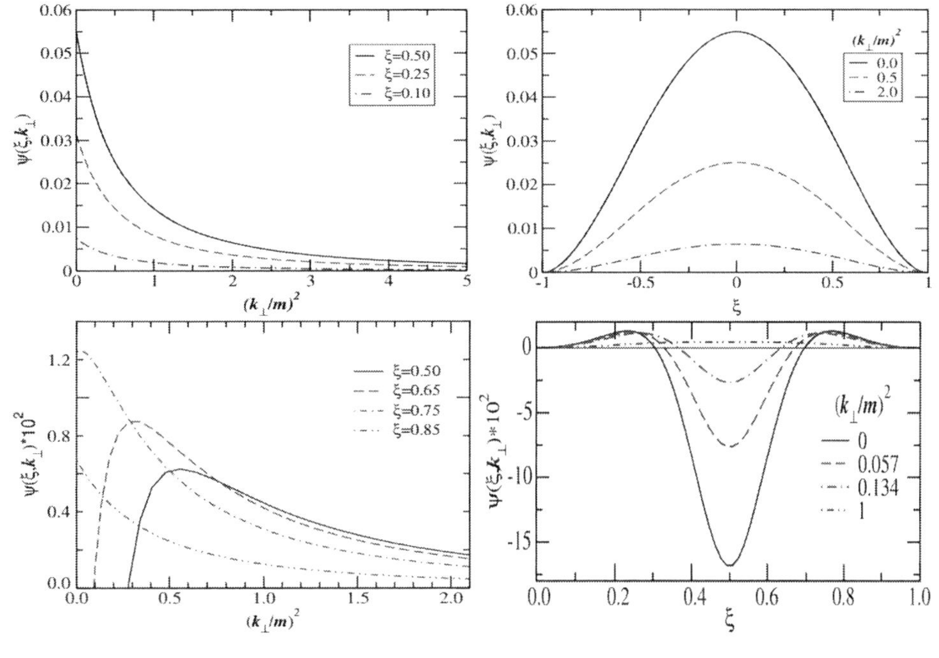

Figure 3. $\psi(\xi, \mathbf{k}_\perp)$ as function of $\mathbf{k}_\perp$ for fixed values of ξ (left panels) and as function of ξ for fixed values of $(\mathbf{k}_\perp/m)^2$ (right panels). For $\mu = 0.1\,m$, the plots are given for the ground state with $B(0) = 1.9\,m$ (upper frames) and for the first-excited state, $B(1) = 0.22\,m$ (lower frames). (For more details, see [14]).

left panels) and in terms of the variable ξ for some fixed values of $k_\perp$ (two right panels). >From the plotted results for the excited state (lower panels of figure 3), one can verify the expected node structure typical for an excited state. In addition, for the second-excited state, two nodes can be verified in the corresponding wave function (not shown here, but given in [14]).

By an inspection of the lower panels of figure 3, for the first-excited state the node structure is present for $(k_\perp/m)^2 < 1$ and $\xi < 0.75$, being symmetric with respect to $\xi = 1/2$. In particular, the node structure moves toward $\xi = 1/2$ as $k_\perp$ increases. Such behavior can be naively expected by the relation between the Cartesian variables for three-momenta with the corresponding LF ones:

$$M^2 = \left[\sqrt{\mathbf{p}_1^2 + m^2} + \sqrt{\mathbf{p}_1^2 + m^2}\right]^2 - (\mathbf{p}_1 + \mathbf{p}_2)^2 = p^+ \left[\frac{m^2 + \mathbf{p}_1^{\perp 2}}{p_1^+} + \frac{m^2 + \mathbf{p}_2^{\perp 2}}{p_2^+}\right] - \mathbf{p}_\perp^2. \tag{7}$$

In the rest frame, $\mathbf{p}_1 + \mathbf{p}_2$=0, and by introducing the relative coordinates we obtain

$$\mathbf{k}^2 = \frac{k_\perp^2 + m^2}{4\xi(1-\xi)} - m^2. \tag{8}$$

The first excited-state wave function has to display a node at a given fixed value of $\mathbf{k}^2$. Therefore, as shown by the lower panels of figure 3, by increasing the value of $k_\perp$, the corresponding variable ξ becomes constrained to approach 1/2 (i.e. the maximal value of $\xi(1-\xi)$).

5 Valence wave function in the impact-parameter space

A convenient and useful approach to study the valence amplitude is by considering the impact parameter (IP) space [16], in connection with the Generalized Parton Distributions. This is performed by keeping the longitudinal momentum and applying the Fourier transformation over the transverse momentum of the LF wave function. Such mixed representation shows the correlation between the momentum fraction carried by the constituent particle and the corresponding probability distribution in the transverse plane. This can be particularly interesting for the analysis of fermionic systems carrying both angular momentum and spin, with a distribution correlated to the longitudinal momentum, as a reflection of the internal dynamics.

The wave function in the IP-space can be used to compute the elastic electromagnetic form factor, generalizing the well-known relation of the non-relativistic form factor to a relativistic framework. In this case, the form factor is computed as the Fourier transform of the transverse charge density, as thoroughly discussed by Miller in [19]. For a composite bosonic state, the form factor can be written as

$$\mathcal{F}(Q^2 = -q^2) = \int d^2\mathbf{b}\, \rho(\mathbf{b})\, e^{-i\mathbf{b}\cdot\mathbf{q}_\perp}, \tag{9}$$

by considering the following points: (i) the momentum transfer q^μ is evaluated in the Breit frame with q^+= 0, (ii) $Q^2 = \mathbf{q}_\perp^2$, (iii) $\mathbf{b}$ belongs to the transverse plane or IP-space, and (iv) $\rho(\mathbf{b})$ is the impact parameter density, which is a sum of contributions from all Fock components of the LF wave function,

$$\rho(\mathbf{b}) = \rho_{\text{val}}(\mathbf{b}) + \text{ higher Fock states densities } \cdots . \tag{10}$$

The valence term is defined through the valence wave function in the IP-space, $\phi(\xi, \mathbf{b})$, as follows

$$\rho_{\text{val}}(\mathbf{b}) = \frac{1}{4\pi}\int_0^1 \frac{d\xi}{\xi(1-\xi)^3}\, |\phi(\xi, \mathbf{b}/(1-\xi))|^2. \tag{11}$$

In (11), the IP-space valence wave function is the 2D Fourier transform of $\psi(\xi, \mathbf{k}_\perp)$, given by

$$\phi(\xi, \mathbf{b}) = \int \frac{d^2\mathbf{k}_\perp}{(2\pi)^2}\, \psi(\xi, \mathbf{k}_\perp) e^{i\mathbf{k}_\perp\cdot\mathbf{b}}, \tag{12}$$

where $\phi(\xi, \mathbf{b})$ results to be symmetric with respect to $1 - 2\xi$, as a consequence of the symmetry of $g(\gamma, z)$ under the transformation $z \to -z$. By performing the 2D Fourier transformation, the IP-space valence wave function can be written within the NIR for the s-wave state as

$$\phi(\xi, \mathbf{b}) = \frac{\xi(1-\xi)}{4\pi\sqrt{2}} F(\xi, b), \tag{13}$$

where

$$F(\xi, b) = b \int_0^\infty d\gamma\, g(\gamma, 1-2\xi) \frac{K_1\left(b\sqrt{\gamma+\kappa^2+(1/2-\xi)^2M^2}\right)}{\sqrt{\gamma+\kappa^2+(1/2-\xi)^2M^2}}. \tag{14}$$

Therefore, the structure of the excited state was further investigated in our approach in the IP-space, with the corresponding results being presented in two panels given in figure 4. In the left panel, we show the function $F(\xi, b)$ in terms of the ξ variable, for fixed values of $m\,b$; and, in the right panel, in terms of $m\,b$, for fixed values of ξ. We consider the case that the binding energy is $B(1) = 0.22\,m$, obtained for $\mu = 0.1\,m$ and α_{gr}=6.437. The moving position of the node in the (ξ, b) plane, can be appreciated in both panels of the figure. As a general remark, the position of the node increases with $m\,b$ for ξ approaching the end-points and tends to be suppressed close to such boundaries. This is the counterpart of the previous observation done for the results shown in momentum space. In addition, as one can observe in the right panel of the figure 4, there is a fast fall-off of the wave function for large values of $m\,b$, which indeed can be verified to have an exponential behavior, as detailed below.

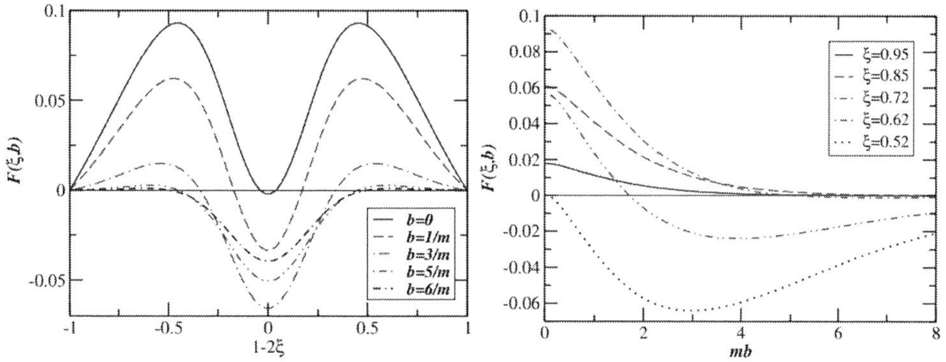

Figure 4. The first excited-state function $F(\xi, b)$, as given by (14), for $B(1) = 0.22\,m$, with $\mu = 0.1\,m$ and α_{gr}=6.437, is shown as function of ξ, for some given values of $m\,b$ (left panel); and, as function of $m\,b$, for given fixed values of ξ (right panel). In view of the symmetry, we should observe that all the plots in the right panel for fixed ξ are exactly the same for $1-\xi$. As evidenced in both the panels, the existence of zeros for $F(\xi, b)$ occurs only for values of $m\,b$ not larger than ~ 7.

The exponentially drops out behavior in $F(\xi, b)$ can be understood by a close analysis of (14). First, from a physically-motivated requirement [19], one should note that $\phi(\xi, b)$ is finite for $b \to 0$ (see also [7]), such that

$$\phi(\xi, b=0) = \frac{\xi\,(1-\xi)}{4\pi\sqrt{2}} \int_0^\infty d\gamma\, \frac{g(\gamma, 1-2\xi)}{\gamma+\kappa^2+(1/2-\xi)^2M^2} < \infty\ , \tag{15}$$

following that one can deduce that $g(\gamma, 1-2\xi)$ must vanish for $\gamma \to \infty$. Therefore, the relevant interval of γ in the integral given in (14) can be taken as effectively finite. Exploring such an observation, one can extract the driving exponential fall-off of $F(\xi, b)$ in the asymptotic limit $b \to \infty$. In this limit,

$$K_1(x)|_{x\to\infty} \to \left(\frac{\pi}{2x}\right)^{\frac{1}{2}} e^{-x}. \tag{16}$$

By considering the limit $b \to \infty$, we notice that the leading exponential behavior in the integral (14) is coming from values of $e^{-b\sqrt{\gamma+\kappa^2+(\xi-1/2)^2M^2}}$ with γ close to 0. Therefore, in such limit, we can single out the exponential fall-off behavior as

$$F(\xi, b)|_{b\to\infty} \to e^{-b\sqrt{\kappa^2+(\xi-1/2)^2M^2}} f(\xi, b), \tag{17}$$

where the reduced function $f(\xi, b)$ is expected to decrease more smoothly for large values of b. The feature shown in (17) is typical for the wave function in coordinate representation, i.e., exponential fall-off multiplied by a polynomial function.

6 Conclusions

In the present contribution, we are reporting some recent investigations on the spectrum of excited states of the Bethe-Salpeter equation in 3 + 1 space-time dimensions, considering the s-wave state in the ladder approximation, using the Nakanishi integral representation. We have found a finite number of excited states for non-zero exchanged-scalar mass, and we have successfully compared our results with the corresponding ones obtained in the Euclidean space, where obviously the Nakanishi integral representation is not assumed. Here we present some sample results obtained for the first two excited states; however the method has allowed us to obtain up to three excited states $B(n)(n = 1, 2...)$ for large enough ground-state binding energy $B(0)$ and small mass μ of the exchanged boson [13]. In order to verify the non-relativistic limit when the mass of the exchanged boson is approaching to zero, the ratios $B(n)/B(0)$ obtained through the Bethe-Salpeter approach are in agreement with those of the well known non-relativistic calculation. To compute these ratios for μ close to zero, an extrapolation was performed, due to numerical instabilities of the integral equation that occur for very small $B(0)$ and small mass of the exchanged boson. From our analysis of the valence wave function, which is more detailed in [13, 14], we have presented results for the expected node structure of the first-excited state when plotted as function of the LF momentum variables. The structure of the impact-parameter space valence wave function has also been explored, with our results presented for the first-excited state. In particular, for large transverse distances an exponential fall-off behavior was singled out in agreement with the non-relativistic case in the 3D Euclidean space. The success in applying the NIR to the BS formalism for the spectrum of bosonic systems, as outlined here and detailed in [13, 14], provides enough motivation for possible extensions, as to consider these studies beyond the ladder approximation [9]. Following previous investigations on two-fermion systems within the NIR [11, 12], it is quite relevant also to look for the corresponding excited bound-state spectrum.

Acknowledgments

The authors thank Profs. Tobias Frederico, Vladimir Karmanov, and Giovanni Salmè for clarifying discussions along the development of works related to this communication. We also acknowledge the partial financial support received from the Brazilian funding agencies CAPES, CNPq and FAPESP.

References

[1] E.E. Salpeter, H.A. Bethe, Phys. Rev. **84**, 1232 (1951)
[2] N. Nakanishi, Phys. Rev. **130**, 1230 (1963). *Graph Theory and Feynman Integrals* (Gordon and Breach, New York, 1971)
[3] K. Kusaka and A.G. Williams, Phys. Rev. D **51**, 7026 (1995)
[4] K. Kusaka, K. Simpson and A.G. Williams, Phys. Rev. D **56**, 5071 (1997)
[5] V. A. Karmanov and J. Carbonell, Eur. Phys. J. A **27**, 1 (2006)

 DOI: 10.1051/epjconf/201713801013

[6] T. Frederico, G. Salmè and M. Viviani, Phys. Rev. D **85**, 036009 (2012)
[7] T. Frederico, G. Salmè and M. Viviani, Phy. Rev. D **89**, 016010 (2014)
[8] J. Carbonell and V. A. Karmanov, Eur. Phys. J. A **27**, 11 (2006)
[9] V. Gigante, J.H.A. Nogueira, E. Ydrefors, C. Gutierrez, V.A. Karmanov and T. Frederico, arXiv:1611.03773 (2016)
[10] V. Sauli and J. Adam, Phys. Rev. D **67**, 085007 (2003)
[11] J. Carbonell, V. A. Karmanov, Eur. Phys. J. A **46**, 387 (2010)
[12] W. de Paula, T. Frederico, G. Salmè and M. Viviani, Phys. Rev. D **94**, 071901 (2016)
[13] C. Gutierrez, *Minkowski space Bethe-Salpeter equation within Nakanishi representation*, PhD thesis IFT-T.007/16 (Instituto de Física Teórica, Universidade Estadual Paulista, 2016)
[14] C. Gutierrez, V. Gigante, T. Frederico, G. Salmè, M. Viviani, L. Tomio, Phys. Lett. B **759**, 131 (2016)
[15] M. Burkardt, Int. J. Mod. Phys. A **18**, 173 (2003)
[16] M. Diehl, Phys. Rep. **388**, 41 (2003)
[17] E. R. Vrscay, Phys. Rev. A **33**, 1433 (1986)
[18] G. Feldman, T. Fulton and J. Townsend, Phys. Rev. D **7**, 1814 (1973)
[19] G. A. Miller, Annu. Rev. Nucl. Part. Sci. **60**, 25 (2010)

EPJ Web of Conferences **138**, 01014 (2017) DOI: 10.1051/epjconf/201713801014
Baldin ISHEPP XXIII

Thermal model for small systems

Jean Cleymans[1,**], *Boris* Hippolyte[2,***], *Helmut* Oeschler[3,****], *Krzysztof* Redlich[4,5,6,†], and *Natasha* Sharma[7,‡]

[1] *UCT-CERN Research Centre and Department of Physics, University of Cape Town, Rondebosch 7701, South Africa*
[2] *Institut Pluridisciplinaire Hubert Curien (IPHC), Université de Strasbourg, CNRS-IN2P3, Strasbourg, France*
[3] *Physikalisches Institut, Ruprecht-Karls-Universität Heidelberg, Heidelberg, Germany*
[4] *Institute of Theoretical Physics, University of Wrocław, Pl-45204 Wrocław, Poland*
[5] *ExtreMe Matter Institute EMMI, GSI, D-64291 Darmstadt, Germany*
[6] *Department of Physics, Duke University, Durham, NC 27708, USA*
[7] *Department of Physics, Panjab University, Chandigarh 160014, India*

Abstract. An analysis is presented of the expectations for particle production in collisions of small nuclei using the thermal model. The maxima observed in particle ratios of strange particles to pions as a function of beam energy in heavy ion collisions, are reduced when considering smaller nuclei. Of particular interest is the Λ/π^+ ratio which shows a strong maximum surviving even in collisions of small nuclei.

1 Introduction

A large effort is presently under way to study not only heavy- but also light-ion collisions. This is being motivated by the results obtained in heavy ion collisions like Pb-Pb and Au-Au, for the K^+/π^+, and also other particle ratios. It has been conjectured that these indicate a phase change in nuclear matter [1].

A consistent description of particle production in heavy-ion collisions, up to LHC energies, has emerged during the past two decades using a thermal-statistical model (referred to simply as thermal model in the remainder of this talk). It is based on the creation and subsequent decay of hadronic resonances produced in chemical equilibrium at a unique temperature and baryon chemical potential. According to this picture the hadronic resonances made up of the light flavors are produced in chemical equilibrium.

Indeed some particle ratios exhibit interesting features when studied as a function of the beam energy: (i) a maximum in the K^+/π^+ratio, (ii) a maximum in the Λ/π ratio, (iii) no maximum in the K^-/π^-ratio. The maxima occur at a center-of-mass energy of around 10 GeV [2–4]. It is interesting to note that the occurrence of these maxima happens in an energy regime where a maximum baryon

**e-mail: jean.cleymasn@uct.ac.za
***e-mail: hippolyt@in2p3.fr
****e-mail: h.oeschler@gsi.de
†e-mail: krzysztof.redlich@ift.uni.wroc.pl
‡e-mail: Natasha.Sharma@cern.ch

density occurs [5] and a transition from baryon-dominated freeze-out to a meson dominated one takes place [4]. An alternative interpretation is that these maxima reflect a phase change [1] to a deconfined state of matter.

The maxima mark a distinction between heavy-ion collisions and p-p collisions as they are not observed in the latter. This shows a clear difference between the two systems which is worthy of investigation.

It is the purpose of the present talk to report on an analysis [6] studying the transition from a small system like a p-p collision to a large system like a Pb-Pb or Au-Au collision and to follow the genesis of the maxima in certain particle ratios.

2 The model

A relativistic heavy-ion collision will go through several stages. At one of the later stages, the system will be dominated by hadronic resonances. The identifying feature of the thermal model is that all the resonances as listed by the Particle Data Group [7] are assumed to be in thermal and chemical equilibrium, an assumption which drastically reduces the number of free parameters and thus this stage is determined by just a few thermodynamic variables namely, the chemical freeze-out temperature T, the various chemical potentials μ determined by the conserved quantum numbers and by the volume V of the system. The latter plays no role when considering ratios of yields. It has been shown that this description is also the correct one [8–11] for a scaling expansion as first discussed by Bjorken [12].

In general, if the number of particles carrying quantum numbers related to a conservation law is small, then the grand-canonical description no longer holds. In such a case conservation of quantum numbers has to be implemented exactly in the canonical ensemble [13, 14]. In the case considered here there are two volume parameters: the overall volume of the system V, which determines the particle yields at fixed density and the strangeness correlation (cluster) volume V_c, which reflects the canonical suppression factor and reduces the densities of strange particles. The volume V_c is parameterized by the radius R_C which serves as a free parameter and defines the range of local strangeness equilibrium.

We start by presenting a brief reminder of the general concepts of the thermal model. In the grand-canonical ensemble, the volume V, temperature T and the chemical potentials $\vec{\mu}$ determine the partition function $Z(T, V, \vec{\mu})$. In the hadronic fireball of non-interacting hadrons, $\ln Z$ is the sum of the contributions of all i-particle species of energy E_i and spin-isospin degeneracy g_i by

$$\frac{1}{V}\ln Z(T, V, \vec{\mu}) = \sum_i Z_i^1(T, \vec{\mu}), \tag{1}$$

where $\vec{\mu} = (\mu_B, \mu_S, \mu_Q)$ are the chemical potentials related to the conservation of baryon number, strangeness and electric charge, respectively.

The one-particle partition function is given, in Boltzmann approximation, by

$$Z_i^1 \equiv \frac{g_i}{2\pi^2} m_i^2 T K_2\left(\frac{m_i}{T}\right)\exp(\mu). \tag{2}$$

The partition function contains all information needed to obtain the number density n_i of particle species i. Introducing the particle's specific chemical potential μ_i, one gets

$$n_i(T, \vec{\mu}) = \frac{1}{V}\left.\frac{\partial(T\ln Z)}{\partial\mu_i}\right|_{\mu_i=0}. \tag{3}$$

Any resonance that decays into species i contributes to the yields eventually measured. Therefore, the contributions from all heavier hadrons j that decay to hadron i with the branching fraction $\Gamma_{j\to i}$ are

$$n_i^{\rm decay} = \sum_j \Gamma_{j\to i}\, n_j. \tag{4}$$

Consequently, the final yield N_i of particle species i is the sum of the thermally produced particles and the decay products of resonances,

$$N_i = (n_i + n_i^{\rm decay})\, V. \tag{5}$$

From (3-5) it is clear that in the grand canonical ensemble the particle yields are determined by the volume of the fireball, its temperature and the chemical potentials.

In general, if the number of particles is small, then the grand-canonical description no longer holds. In such a case conservation laws have to be implemented exactly. We refer here only to strangeness conservation and consider charge and baryon number conservation to be fulfilled on the average. The density of strange particle i carrying strangeness s can be obtained from [13],

$$\begin{aligned} n_i^C &= \frac{Z_i^1}{Z_{S=0}^C} \sum_{k=-\infty}^{\infty} \sum_{p=-\infty}^{\infty} a_3^p a_2^k a_1^{-2k-3p-s} \\ & I_k(x_2) I_p(x_3) I_{-2k-3p-s}(x_1), \end{aligned} \tag{6}$$

where $Z_{S=0}^C$ is the canonical partition function

$$\begin{aligned} Z_{S=0}^C &= e^{S_0} \sum_{k=-\infty}^{\infty} \sum_{p=-\infty}^{\infty} a_3^p a_2^k a_1^{-2k-3p} \\ & I_k(x_2) I_p(x_3) I_{-2k-3p}(x_1), \end{aligned} \tag{7}$$

and Z_i^1 is the one-particle partition function (7) calculated for $\mu_S = 0$ in the Boltzmann approximation. The arguments of the Bessel functions $I_s(x)$ and the parameters a_i are introduced as,

$$a_s = \sqrt{S_s/S_{-s}}\ , \quad x_s = 2V\sqrt{S_s S_{-s}}, \tag{8}$$

where S_s is the sum of all $Z_k^1(\mu_S = 0)$ for particle species k carrying strangeness s.

In the limit where $x_n < 1$ (for $n = 1, 2$ and 3) the density of strange particles carrying strangeness s is well approximated by [13]

$$n_i^C \simeq n_i \frac{I_s(x_1)}{I_0(x_1)}. \tag{9}$$

From these equations it is clear that in the canonical ensemble the strange particle density depends explicitly on the volume through the arguments of the Bessel functions. This volume might be different from the overall volume V and is denoted as V_c [15, 16].

To account for the suppression beyond that expected in the canonical ensemble, it was assumed that exact strangeness conservation holds only in a small subvolume V_c of a system. The concept of a strangeness correlation volume V_c has been used in earlier studies [16–19].

3 Origin of the maxima

In the thermal model, the baryon chemical potential decreases continuously with increasing beam energy. At the same time the temperature increases rather quickly until it reaches a plateau. Following the rapid rise of the temperature at low beam energies, the Λ/π^+ and K^+/π^+also increase rapidly. This comes to a halt when the temperature reaches its limiting value. However, the baryon chemical potential keeps on decreasing. Consequently, the Λ/π and K^+/π^+ratios follow this decrease due to strangeness conservation as K^+ is produced in associated production together with a Λ. The two effects combined lead to maxima in both cases. For very high energies, the baryo-chemical potential no longer plays a role ($\mu_B \approx 0$) and the temperature is constant hence these ratios hardly vary [4].

To show this in more detail we present as an example in figure 1 lines where the $K^+/\pi^+$and the Λ/π^+ ratios remain constant in the $T - \mu_B$ plane. It should be noted that the maxima of these ratios do not occur in the same position, which remains to be confirmed experimentally. It is also worth noting that the maxima are not on but slightly above the freeze-out curve. The shaded regions indicate the values of these ratios in the $T - \mu_B$ plane.

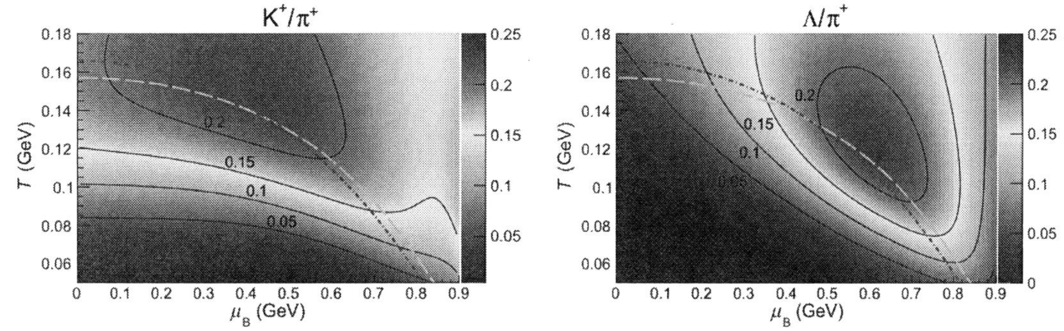

Figure 1. Values of the K^+/π^+(left-hand pane) and the Λ/π^+ (right-hand pane) ratios in the $T - \mu_B$ plane. Lines of constant values are indicated. The dashed-dotted line is the freeze-out curve obtained in [3] while the dashed line uses the parameterization given in [21]. Note that the maxima do not occur in the same position.

Similarly, in figure 2 we show in lines where the $\Xi^-/[(\pi^+ + \pi^-)/2]$ and the $\Omega^-/[(\pi^+ + \pi^-)/2]$ ratios remain constant in the $T - \mu_B$ plane. Again, it should be noted that the maxima of these ratios never occur in the same position. Again, it is also worth noting that the maxima are not on but slightly above the freeze-out curve. The shaded regions indicate the values of these ratios in the $T - \mu_B$ plane.

The differences in the positions of the maxima are caused by the different masses of the hadrons involved.

4 Particle ratios for small systems

To consider the case of the collisions of smaller nuclei we have to take into account the strangeness suppression according to the canonical model, i.e. the concept of strangeness correlation in clusters of a sub-volume $V_c \leq V$ [15, 16, 20].

In the following figures we show the trends of various particle ratios as a function of $\sqrt{s_{NN}}$. The dependence of T and μ_B on the beam energy is taken from heavy-ion collisions [3]. For p-p collisions slightly different parameters are more suited [22] but were not used. Therefore, the shown calculations give the general trend only. We have ignored the variations of parameters with system size.

We focus on the dependence of the thermal parameters with particular emphasis on the change in the strangeness correlation radius R_C. The parameters R = 10 fm (which is the value for central Pb-Pb collisions) and γ_S = 1 are kept fixed. The freeze-out values of T and μ_B will vary with the system size [20], however this has not been taken into account in the present work which aims to give a qualitative description of the effect only.

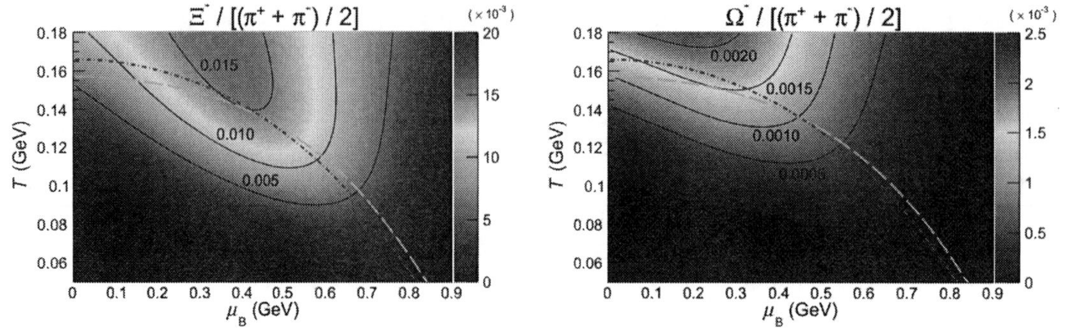

Figure 2. Values of the $\Xi^-/\frac{1}{2}[\pi^+ + \pi^-]$ (left-hand pane) and the $\Omega^-/\frac{1}{2}[\pi^+ + \pi^-]$ (right-hand pane) ratios in the $T - \mu_B$ plane. Lines of constant values are indicated. The dashed-dotted line is the freeze-out curve obtained in [3] while the dashed line uses the parameterization given in [21]. Note that the maxima do not occur in the same position.

The smaller system size is described by decreasing the value of the correlation radius R_C. This ensures that strangeness conservation is exact in R_C, and that strangeness production is suppressed with decreasing R_C.

In figure 3 we show the energy and system size dependence of two particle ratios calculated along the chemical freeze-out line. In figure 3 a maximum is seen in the K^+/π^+ ratio which gradually disappears when the correlation radius decreases. A different effect is seen in $\Lambda/\pi^\pm$ ratio. Here, the gradual decrease of the maximum is also seen but, contrarily to the K^+/π^+ ratio, it remains quite prominent even for a small correlation radius. Also, the maximum shifts, for smaller systems, towards higher $\sqrt{s_{NN}}$. For pp collisions which correspond to a R_C of about 1.5 fm [20], they will hardly be observed. It should also be noted that in the thermal model the maxima happen at different beam energies.

In figure 4 we plot the K^-/π^- ratio and $\bar{\Lambda}/\pi^-$ ratios as a function of beam energy for various values of the correlation radius. Due to the different signs of the chemical potentials entering this ratio, as compared to the particle ratios in figure 3 no maximum is obtained for any value of the beam energy or the correlation radius.

Again, we emphasize that the results presented here are of a qualitative nature. In particular there could be changes due to variations with the system size of the temperature and the baryon chemical potential. In addition the strangeness equilibration volume V_c could be energy and system size dependent.

The curves in figures 3 and 4 were calculated using THERMUS [23].

The particle ratio K^+/π^+ of hadrons produced in relativistic nucleus-nucleus collisions has also been studied in the context of the thermal model of hadrons with exact strangeness conservation in [24].

DOI: 10.1051/epjconf/201713801014

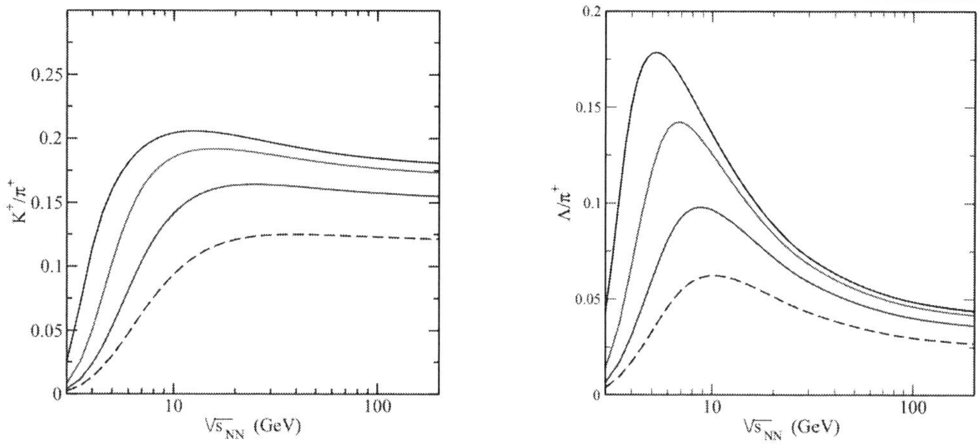

Figure 3. Values of the K^+/π^+(left-hand pane) and the Λ/π^+ (right-hand pane) ratios as a function of invariant beam energy for various strangeness correlation radii R_C, calculated using the thermal model [23]. The correlation radius is varied from 3.0 (top curve) to 2.5, 2.0, 1.5 and finally 1.0 fm (bottom curve). Note that the Λ/π^+ ratio is the ratio where the maximum stays most pronounced as the system size is reduced.

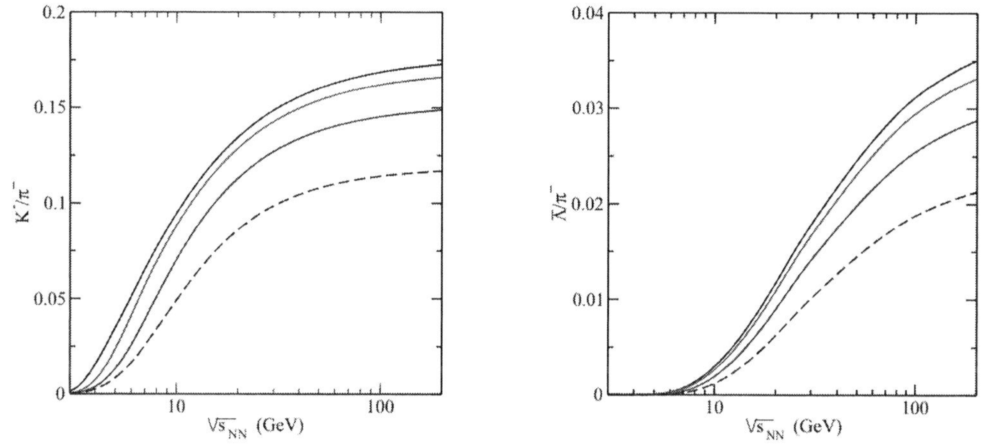

Figure 4. Values of the K^-/π^-(left-hand pane) and the $\bar{\Lambda}/\pi^-$ (right-hand pane) ratios as a function of invariant beam energy for various strangeness correlation radii R_C, calculated using the thermal model [23]. The correlation radius is varied from 3.0 (top curve) to 2.5, 2.0, 1.5 and finally 1.0 fm (bottom curve).

5 Conclusions

The thermal model qualitatively describes the presence of maxima in the $K^+/\pi^+$and the $\Lambda/\pi^\pm$ ratios at a beam energy of $\sqrt{s_{NN}}\approx$ 10 GeV. In this talk we have described what could possibly happen with different strange particles and pion yields in collisions of smaller systems due to constraints imposed by exact strangeness conservation. In particular, the Λ/π^+ ratio still shows a clear maximum even small systems. The pattern of these maxima is also quite special as they are not always at the same beam energy.

Acknowledgments
K.R. acknowledges the supported by the National Science Center, Poland under grant Maestro, DEC-2013/10/A/ST2/00106. N.S. acknowledges the support of DST-SERB Ramanujan Fellowship (D.O. No. SB/S2/RJN- 084/2015).

References

[1] M. Gazdzicki and M.I. Gorenstein, Acta Phys. Polon. B **30**, 2705 (1999)
[2] A. Andronic, P. Braun-Munzinger and J. Stachel, Nucl. Phys. A **772**, 167 (2006)
[3] J. Cleymans, H. Oeschler, K. Redlich, S. Wheaton, Phys. Rev. C **73**, 034905 (2006)
[4] J. Cleymans, H. Oeschler, K. Redlich, S. Wheaton, Phys. Lett. B **615**, 50 (2005)
[5] J. Randrup and J. Cleymans Eur. Phys. J. A **52**, 218 (2016)
[6] J. Cleymans, B. Hippolyte, H. Oeschler, K. Redlich, N. Sharma, arXiv:1603.0953[hep-ph]
[7] K.A. Olive et al., Particle Data Group, Chin. Phys. C **38**, 090001 (2014)
[8] J. Cleymans, Proceedings, 3rd International Conference on Physics andastrophysics of quark-gluon plasma (ICPA-QGP '97, 1998) 55
[9] J. Cleymans and K. Redlich, Phys. Rev. C **60**, 054908 (1999)
[10] S.V. Akkelin, P. Braun-Munzinger and Yu. M. Sinyukov, Nucl. Phys. A **710**, 439 (2002)
[11] W. Broniowski and W. Florkowski, Phys. Rev. Lett. **87**, 272302 (2001)
[12] J.D. Bjorken, Phys. Rev. D **27**, 140 (1983)
[13] P. Braun-Munzinger, J. Cleymans, H. Oeschler, K. Redlich, Nucl. Phys. A **697**, 902 (2002)
[14] P. Braun-Munzinger, K. Redlich, J. Stachel, in *Quark Gluon Plasma 3*, eds. R.C. Hwa and Xin-Nian Wang (World Scientific Publishing, 2003)
[15] J. Cleymans, H. Oeschler, K. Redlich, Phys. Rev. C **59**, 1663 (1999)
[16] S. Hamieh, K. Redlich, A. Tounsi, Phys. Lett. B **486**, 61 (2000)
[17] A. Tounsi, A. Mischke, K. Redlich, Nucl. Phys. A **715**, 565 (2003)
[18] H. Caines, J. Phys. G **32**, S171 (2006)
[19] C. Höhne, F. Pühlhofer, R. Stock, Phys. Lett. B **640**, 96 (2006)
[20] I. Kraus, J. Cleymans, H. Oeschler, K. Redlich, S. Wheaton, Phys. Rev. C **76**, 064903 (2007)
[21] V. Vovchenko, V.V. Begun, M.I. Gorenstein, Phys. Rev. C **93**, 064906 (2016)
[22] J. Cleymans, S. Kabana, I. Kraus, H. Oeschler, K. Redlich, N. Sharma, Phys. Rev. C **84**, 054916 (2011)
[23] S. Wheaton, J. Cleymans, M. Hauer, Comp. Phys. Comm. **180**, 84-106 (2009)
[24] V.D. Toneev and A.S. Parvan, J. Phys. G. **31**, 583 (2005)

Heavy flavor measurements at the STAR experiment

Pavol Federic[1,*]

[1] *Nuclear Physics Institute of the Czech Academy of Sciences, Rez near Prague, Czech Republic*

Abstract. In ultra-relativistic heavy-ion collisions at RHIC, a new state of nuclear matter with extreme properties is produced — the strongly interacting quark-gluon plasma. Heavy quarks, predominantly produced at early stages of collisions owing to their large masses, provide an exceptional probe for exploring the hot and dense medium created in such collisions. The Heavy Flavor Tracker and Muon Telescope Detector have been fully operational since 2014 and have significantly improved STAR's capabilities in measuring both open and hidden heavy flavor hadrons in heavy-ion collisions. We present an overview of recent heavy flavor results obtained at the STAR experiment.

1 Introduction

Heavy flavor quarks are predominantly produced early in heavy-ion collisions in hard scattering processes. They experience the full evolution of the system. These features make heavy quarks ideal probes for studying the properties of the strongly coupled Quark-Gluon Plasma (sQGP) at RHIC. Two new detectors dedicated to heavy flavor measurements, the Heavy Flavor Tracker (HFT) [1] and the Muon Telescope Detector (MTD) [2], were fully installed at the STAR experiment in the beginning of 2014. In this proceedings we will report recent results from STAR using the HFT and MTD.

2 STAR experiment

STAR is a large-acceptance, multi-purpose detector at RHIC. The main subsystems used in the following analyses are the Time Projection Chamber (TPC), Time Of Flight (TOF) detector, HFT and MTD.

The TPC reconstructs track trajectories, measures their momenta and provides particle identification via measuring ionization energy losses (dE/dx). The TOF greatly improves the identification capabilities, especially in low p_T region, through measuring the time-of-flight of charged particles.

The HFT consists of 3 sub-detector systems, the Silicon Pixel (PXL) detector utilizing the state-of-the-art Monolithic Active Pixel Sensor (MAPS), Intermediate Silicon Tracker (IST) and Silicon Strip Detector (SSD). The PXL detector is made of two layers of MAPS. The low material budget of the PXL detector (0.4-0.6% radiation length per layer) in combination with excellent position resolution provides precise measurements of track impact parameters. The track impact parameter resolution for charged kaons exceeds the requirement of less than 55 μm at transverse momentum (p_T) equal to 750 MeV/c.

*e-mail: pavolfederic@gmail.com

The MTD is made of multi-gap resistive plate chambers placed at the outermost layer of STAR behind the solenoidal magnet, which is used as a hadron absorber. It is designed for muon triggering and identification with timing resolution ~100 ps for $p_T > 1.2$ GeV/c. It has a geometrical acceptance of 45% in azimuth within $|\eta| < 0.5$.

3 Open heavy flavor measurements

The total charm production cross sections at mid-rapidity ($|\eta| < 1$) have been measured by STAR in p+p and Au+Au collisions at $\sqrt{s_{NN}} = 200$ GeV, and are found to scale with the number of binary nucleon+nucleon collisions [4]. This observation supports the expectation that heavy flavor quarks are mainly produced in initial hard parton scatterings at RHIC. Relatively long lifetimes of heavy flavor quarks enable them to experience the whole evolution of the system. Therefore they can serve as calibrated probes to study QGP.

Shown in figure 1 is the D^0 nuclear modification factor (R_{AA}) in 0-10% Au+Au collisions at $\sqrt{s_{NN}}$ = 200 GeV extracted from about 780 million Minimum Bias (MB) events with the HFT (black solid circles). Also shown are the STAR published results from 1.1 billion MB events recorded in 2010 and 2011 without the HFT (red open circles) [4]. The vertical bars (brackets) represent statistical (systematic) uncertainties, while the grey shaded areas indicate the systematic uncertainties from p+p reference for the R_{AA} obtained from the HFT data. The latter were not present for the published results due to cancellation of efficiency corrections between the p+p and Au+Au data. The boxes around unity denote the overall normalization uncertainties in the Au+Au and p+p data, respectively. The new R_{AA} measurement confirms the strong suppression of D^0 yields at high transverse momenta, which is comparable to the suppression observed for light hadrons and indicates strong charm interactions with the medium.

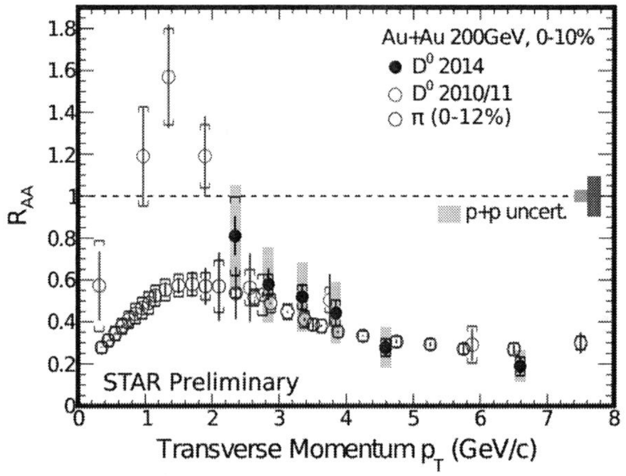

Figure 1. Measured D^0 R_{AA} for 0-10% central Au+Au collisions with (solid circles) and without (open circle) the HFT. The open blue circles show π R_{AA} [3] for comparison.

The HFT has also enabled the first measurements of elliptic flow (v_2) for open charm hadrons at RHIC. Significant non-zero D^0 v_2 is shown in the left panel of figure 2 for $p_T > 2$ GeV/c. The vertical bars (brackets) represent statistical (systematic) uncertainties, while the grey shaded areas show the estimate of contributions from non-flow effects. Shown in the right panel is the D^0 v_2/n_q as function of ($\sqrt{m_0^2 + p_T^2} - m_0)/n_q$, where m_0 is the hadron mass and n_q the number of constituent quarks. The D^0 v_2/n_q is below those of light hadrons [5], but still consistent within uncertainties, suggesting non-zero charm quark flow.

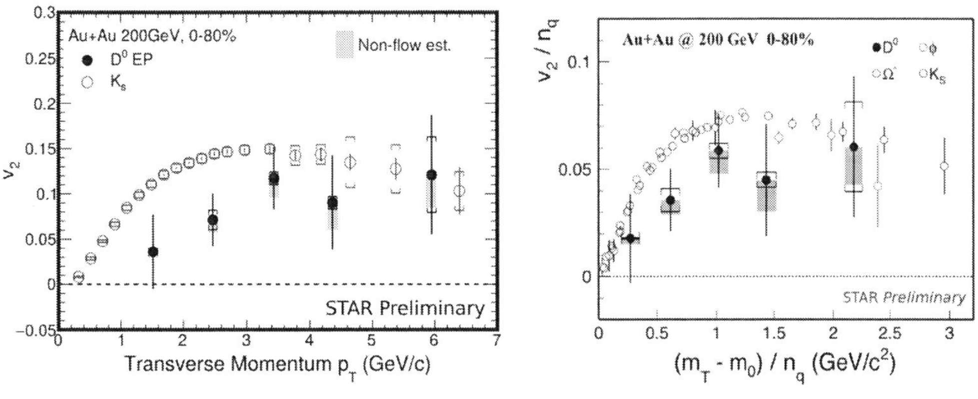

Figure 2. Left: measured D^0 v_2 compared to that of light hadrons; Right: v_2/n_q as function of ($\sqrt{m_0^2 + p_T^2} - m_0)/n_q$ for D^0 and other particle species.

The measurements of D^0 R_{AA} and v_2 are compared to theoretical calculations. As seen in figure 3, the TAMU [6] and SUBATECH [7] models describe both the D^0 R_{AA} and v_2. The DUKE [8] model describes the R_{AA} well but underestimates the v_2. These models use different approaches to describe interactions of charm quarks in the medium and allow to extract transport coefficients. The TAMU model uses a non-perturbative T-matrix approach with potentials obtained from Lattice QCD calculations to describe two-body interactions as a function of the transferred 4-momentum. The SUBATECH group employs a pQCD approach with the Hard Thermal Loop approximation for soft collisions. The charm quark diffusion coefficient $2\pi T D_s$ extracted from the TAMU (SUBATECH) model ranges between 2-~10 (2-4) for T/T_c = 1-2. The DUKE group uses a generalized Langevin approach with transport properties tuned to LHC data, i.e. $2\pi T D_s = 7$. Figure 4 shows the diffusion coefficient extracted from different model calculations [7, 9–14] compared to the yellow band inferred from the STAR data, which are consistent with the lattice QCD calculations [13, 14] within large uncertainties.

4 Quarkonium measurements with the MTD

While the heavy quark production can be calculated by pQCD, the evolution into quarkonia from quark-antiquark pair is a non-perturbative process. Precise measurements of quarkonium production in p+p collisions can help to understand quarkonium production mechanisms. The situation in heavy-ion collisions is more complicated. Quarkonia may dissociate owing to the color screening of the heavy quark-antiquark potential by the QGP. Such a dissociation depends on the temperature of the QGP and varies for different quarkonium states of different binding energies. On the other hand, cold nuclear matter effects (CNM), such as shadowing/anti-shadowing of parton distribution functions, initial-state parton energy loss and final-state nuclear absorption, along with regeneration via coalescence of deconfined charm quarks may also play a role. Therefore quarkonium production can be used to constrain the QGP temperature once other effects are understood.

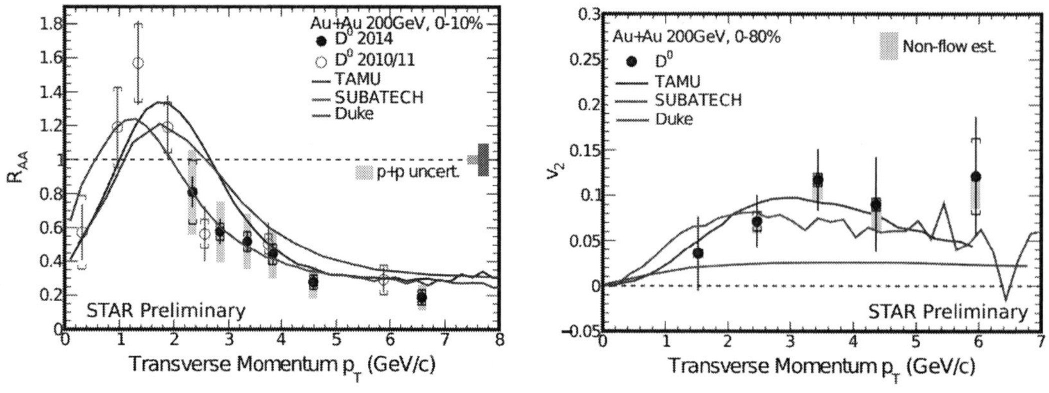

Figure 3. Left: D^0 R_{AA} in 0-10% centrality compared to model calculations; Right: D^0 v_2 in 0-80% centrality compared to model calculations.

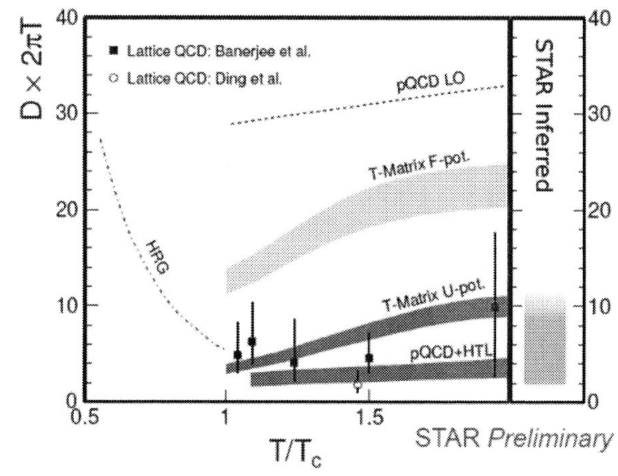

Figure 4. Charm quark diffusion coefficient from model calculations and the inferred range from STAR measurements. See text for details.

Figure 5 shows the results on J/ψ R_{AA} in Au+Au collisions at $\sqrt{s_{NN}}$ = 200 GeV in the di-electron and in di-muon channels for different collision centralities. The new MTD results from the di-muon decay channel are consistent with the di-electron channel results [15, 16] within uncertainties. The dependence of J/ψ R_{AA} on the number of participating nucleons (N_{part}) has also been measured. The left (right) panel of figure 6 shows the abovementioned di-muon results (red stars) for $p_T > 0$ (5) GeV/c compared to the results from PHENIX [17] in Au+Au collisions at the same centre-of-mass energy (black open circles) as well as the measurements from ALICE [18] and CMS [19] in Pb+Pb collisions at $\sqrt{s_{NN}} = 2.76$ TeV (blue squares). For $p_T > 0$ GeV/c, the STAR results are consistent with the PHENIX results, showing a decreasing R_{AA} towards more central collisions. On the other hand, the ALICE results are almost independent of centrality and significantly less suppressed in central collisions compared to STAR results. Such a difference could be explained by larger regeneration

 DOI: 10.1051/epjconf/201713801015

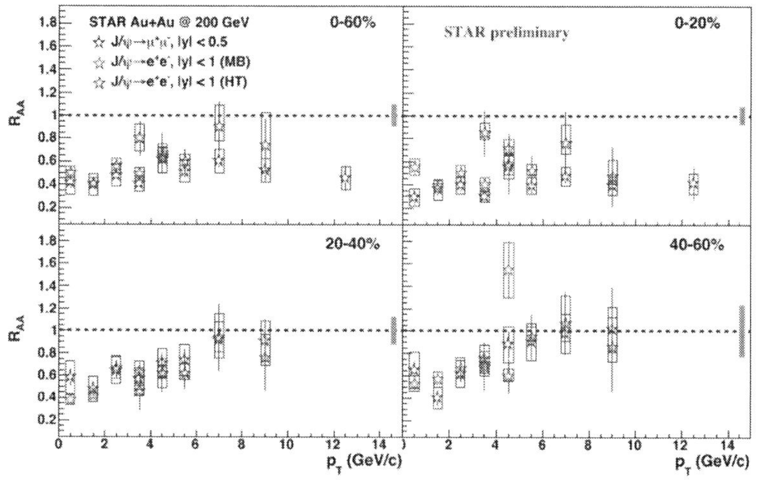

Figure 5. J/ψ R_{AA} in Au+Au collisions at $\sqrt{s_{NN}}$ = 200 GeV in four centrality classes.

contributions to low p_{T} J/ψ at the LHC with much higher $c\bar{c}$ yields. At high p_{T}, the R_{AA} at both RHIC and LHC have a decreasing trend towards more central collisions, and the LHC results are lower than those at RHIC. This is likely due to higher temperature reached at the LHC which leads to stronger J/ψ dissociation. Two different transport models [20, 21], both including dissociation and regeneration, can qualitatively describe the measured R_{AA}. However, there seems to be a tension between the model calculations and experimental data at high p_{T}.

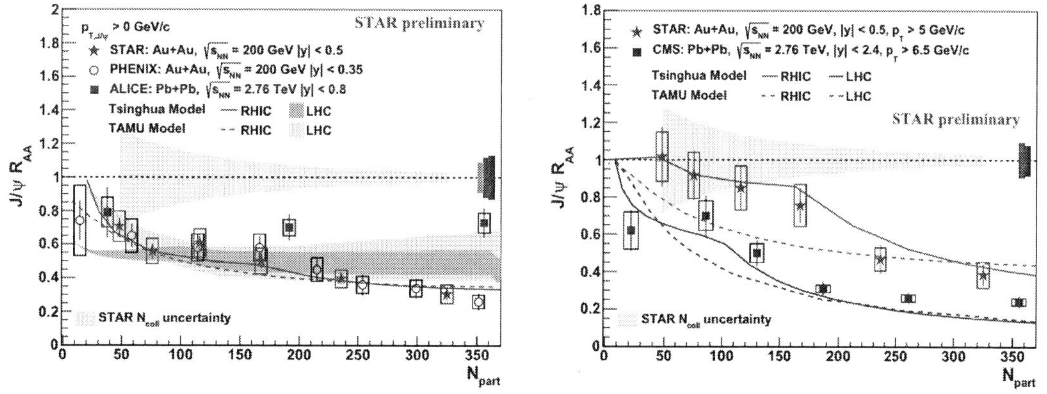

Figure 6. Left: J/ψ R_{AA} for $p_{\mathrm{T}} > 0$ GeV/c in Au+Au (Pb+Pb) collisions at $\sqrt{s_{NN}}$ = 200 GeV (2.76 TeV); Right: J/ψ R_{AA} for $p_{\mathrm{T}} > 5$ (6.5) GeV/c in Au+Au (Pb+Pb) collisions at $\sqrt{s_{NN}}$ = 200 GeV (2.76 TeV). The vertical bars (boxes) represent statistical (systematic) uncertainties. See text for details.

Shown in the left panel of figure 7 is the reconstructed Υ signal from the di-muon decay channel using the same MTD data described above. Compared to the di-electron channel, the di-muon channel has the advantage of better invariant mass resolution, thanks to the significantly suppressed bremsstrahlung radiation. The unlike-sign (red points) spectrum is fitted to the sum of Υ(1S), Υ(2S)

 DOI: 10.1051/epjconf/201713801015

and Υ(3S) signal distributions, $B\bar{B}$, Drell-Yan and combinatorial background distributions. The shapes of the signal distributions are Gaussian functions with means fixed to the PDG values, and widths determined from detector simulations, while the shape of $B\bar{B}$ and Drell-Yan background is determined from PYTHIA. Both the shape and normalization of the combinatorial background are obtained by fitting the like-sign (black points) spectrum. Shown in the right panel are the ratio of the excited Υ state yield with respect to the ground state yield, obtained from the abovementioned fit, the world-wide p+p data and CMS measurements [22, 23] in Pb+Pb collisions at $\sqrt{s_{NN}}$ = 2.76 TeV. There is an indication of larger Υ(2S + 3S)/Υ(1S) ratio at RHIC than that at the LHC.

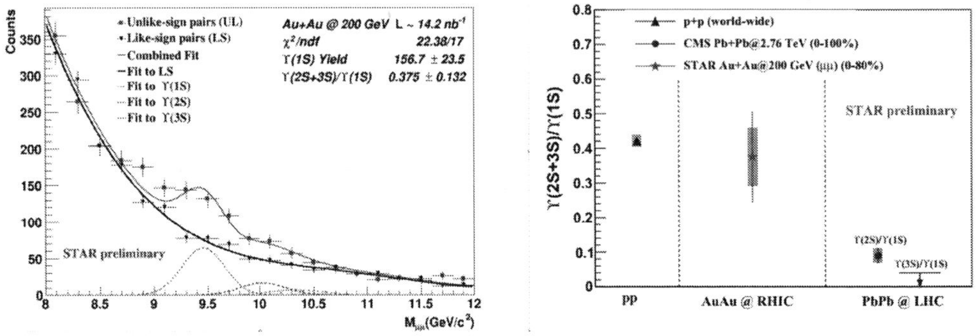

Figure 7. Left: invariant mass spectra of di-muon pairs in Au+Au collisions at $\sqrt{s_{NN}}$ = 200 GeV; Right: ratio of excited Υ states to the ground state measured in p+p collisions, Au+Au collisions at $\sqrt{s_{NN}}$ = 200 GeV and Pb+Pb collisions at $\sqrt{s_{NN}}$ = 2.76 TeV. See text for details.

5 Summary

STAR has carried out the first measurements with the recently installed HFT and MTD in Au+Au collisions. The measurements of D^0 R_{AA} and v_2 indicate strong interactions between charm quarks and the medium created in heavy-ion collisions. Model calculations which can simultaneously describe the R_{AA} and v_2 give a range of charm quark diffusion coefficients between 2 and ~12 for the QGP temperature between T_c and $2T_c$. The measurements of J/ψ with the MTD via the di-muon channel are consistent with the di-electron channel results and show a distinct rising trend in the R_{AA} with transverse momentum in 20-60% centrality. The suppression observed at high p_T in 0-20% central collisions is consistent with dissociation due to color screening. The measurements of high-p_T J/ψ R_{AA} and ratio of yields of different Υ states indicate a lower medium temperature reached at RHIC than at LHC.

Acknowledgements

The work has been supported by the grant 13-20841S of the Czech Science Foundation (GACR). The author's participation in the conference was supported by grant INGO LG14004207 of MSMT of the Czech Republic.

References

[1] Technical Design Report: The STAR Heavy Flavor Tracker (2011)
[2] L. Ruan et al., J. Phys. G: Nucl. Part. Phys. **36**, 095001 (2009)
[3] B. I. Abelev et al., Phys. Lett. B **655**, 104 (2007)
[4] L. Adamczyk et al. (STAR Collaboration), Phys. Rev. Lett. **113**, 142301 (2014)
[5] B. I. Abelev et al. (STAR Collaboration), Phys. Rev. C **77**, 054901 (2008)
[6] M. He, R. J. Fries and R. Rapp, Phys. Rev. C **86**, 014903 (2012)
[7] M. Nahrgang et. al., Phys. Rev. C **91**, 014904 (2015)
[8] W. Zha et. al., Phys. Rev. C **88**, 067901 (2013)
[9] A. Andronic et. al., The Eur. Phys. J. C **76(3)**, 107 (2016)
[10] O. Kaczmarek, Nuclear Physics A **931**, 633 (2014)
[11] F. Riek and R. Rapp, Phys. Rev. C **82**, 035201 (2010)
[12] B. Svetitsky, Phys. Rev. D **37**, 248 (1998)
[13] H. T. Ding et. al., Phys. Rev. D **86**, 014509 (2012)
[14] D. Banerjee, S. Datta, R. Gavai and P. Majumdar, Phys. Rev. D **85**, 014510 (2012)
[15] L. Adamczyk et. al. (STAR Collaboration), Phys. Lett. B **722**, 55 (2013)
[16] L. Adamczyk et. al. (STAR Collaboration), Phys. Rev. C **90**, 024906 (2014)
[17] A. Adare et al. (PHENIX collaboration), Phys. Rev. D **82**, 012001 (2010)
[18] B. B. Abelev et al. (ALICE Collaboration), Phys. Lett. B **734**, 314 (2014)
[19] S. Chatrchyan et al. (CMS Collaboration), JHEP **05**, 063 (2012)
[20] Y. Liu et. al., Phys. Let. B **678** 72 (2009). K. Zhou et. al., Phys. Rev. C **89**, 054911 (2014)
[21] X. Zhao and R. Rapp, Phys. Rev. C **82**, 064905 (2010)
[22] W. Zha et. al., Phys. Rev. C **88**, 067901 (2013)
[23] S. Chatrchyan et al. (CMS Collaboration), PRL **109**, 222301 (2012)

DOI: 10.1051/epjconf/201713801016

Recent STAR heavy-ion results

Mikhail Tokarev[1,*] (for the STAR Collaboration)

[1] *Joint Institute for Nuclear Research, 141980 Dubna, Russia*

Abstract. The recent STAR heavy-ion results obtained in the first phase of the RHIC Beam Energy Scan program are presented. The measurements of particle spectra have been performed over a wide range of collision energy $\sqrt{s_{NN}}$=7.7-200 GeV, centrality and transverse momentum of produced particles. The fixed target mode in heavy-ion collisions at the STAR experiment also extends considerably the range of search for the new physics. Heavy quarks provide an exceptional probe in understanding properties of the hot and dense medium created in such collisions. The Heavy Flavor Tracker (HFT) and Muon Telescope Detector (MTD) upgrades at the STAR experiment at RHIC significantly improved the experimental capabilities of TPC, ToF and EMC detectors in measuring both open and hidden heavy flavor hadrons in heavy-ion collisions.

1 Introduction

The main goal of the physics program at the Relativistic Heavy Ion Collider (RHIC) is to produce a new state of the nuclear matter in collisions of heavy ions and to investigate its properties over a wide range of energy, centrality with various probes and to determine the features of the phase diagram of the nuclear matter [1–4]. Experimental results from the RHIC support the hypothesis that a strongly-coupled nuclear medium with partonic degrees of freedom, namely the Quark-Gluon Plasma (QGP), is created in heavy-ion collisions at high energy. It is assumed that the medium produced in heavy-ion collisions is thermalized. Therefore the phase diagram is usually presented in the temperature-baryon chemical potential plane $\{T, \mu_B\}$. Both quantities can be varied by changing the energy and centrality of collision and the type and momentum of produced particle. The phase diagram should demonstrate a possible transition from a high energy density and high temperature phase in which quark and gluon degrees of freedom dominate to a phase where the relevant degrees of freedom are hadronic [5–9]. The new state at the top RHIC energy is found to be characterized by the suppression of the high transverse momentum (p_T) hadron production in Au+Au collisions relative to $p + p$ collisions [10], large elliptic flow (v_2) for hadrons with light and heavier strange valence quarks, differences between baryon and meson v_2 at intermediate p_T for Au+Au collisions [11, 12], enhanced correlated yields at large $\Delta\eta$ and $\Delta\phi \simeq 0$ [13]. These properties are associated with the existence of a quark-gluon phase of the nuclear matter.

The description of the experimentally found phase diagram of the nuclear matter in the framework of the Quantum Chromodynamics (QCD) is one of the important problems for the theory. It is the best

*e-mail: tokarev@jinr.ru

test of the QCD predictive force especially in the nonperturbative region. Lattice QCD calculations (see [14] and references therein) indicate that the system produced at $\mu_B = 0$ evolves through a rapid crossover in the quark-hadron phase transition. For collisions corresponding to large μ_B, the transition is expected to be first order. The point in the $\{T, \mu_B\}$ plane where the first order phase transition ends, is the QCD critical point. Searching for correspondence between the features (critical point, phase boundaries, type of phase transition) of QCD and experimentally found phase diagram of the nuclear matter is the current focus of the experimental and theory investigations [15]. Therefore the Beam Energy Scan (BES) Program planned at RHIC is aimed to clarify these problems [5–9]. The guiding idea is to vary the collision energy and look for the signatures of phase boundary and critical point i.e. to span the phase diagram from the top RHIC energy (lower μ_B) to the lowest possible energy (higher μ_B). We assume that the signatures of QGP created at $\sqrt{s_{NN}}$ = 200 GeV are established [1–4]. Therefore turn-off of these signatures at particular energy would suggest the crossing of the phase boundary. Similarly, near critical point, there would be enhanced fluctuations in multiplicity distributions of conserved quantities (net-charge, net-baryon) and correlations in distributions of correlation functions (HBT, forward-backward multiplicity). Large fluctuations or divergence of some observables at fixed kinematic and dynamical conditions is usually considered as a characteristic behavior of the system near a critical point. Before looking for these signatures, one should know the region ($\{T, \mu_B\}$) and mean transverse momentum ($< p_T >$) of the phase diagram which can be accessed.

The thermodynamic parameters T and μ_B can be obtained from the ratios and spectra of the produced particles assuming that the thermodynamical equilibrium of the produced nuclear system has been reached. The properties of bulk medium such as the multiplicity density, mean p_T may provide insight about the particle production mechanisms at these energies. The systematic study of these bulk properties with light and heavy probes may clarify the evolution and change in behavior of the system formed in heavy-ion collisions as a function of collision energy and centrality.

We will present the recent STAR heavy-ion results obtained in the first phase of the RHIC Beam Energy Scan program. The measurements of particle spectra have been performed over a wide range of collision energy $\sqrt{s_{NN}}$=7.7-200 GeV, centrality and transverse momentum of produced particles. The fixed target mode in heavy-ion collisions at the STAR experiment also extends considerably the range of search for the new physics. Heavy quarks ($D, J/\psi$) provide an exceptional probe in understanding properties of the hot and dense medium created in such collisions. The Heavy Flavor Tracker (HFT) and Muon Telescope Detector (MTD) upgrades at the STAR experiment at RHIC significantly improved the experimental capabilities of TPC, ToF and EMC detectors in measuring both open and hidden heavy flavor hadrons in heavy-ion collisions.

2 STAR detector

The main detectors of the STAR experiment used to obtain the results on p_T spectra, yields, and particle ratios for charged hadrons are the Time Projection Chamber (TPC) [18] and Time-Of- Flight detectors (ToF) [19]. The TPC is the primary tracking device at STAR. It is 4.2 m long and 4 m in diameter. Its uniform acceptance covers about ± 1 units of pseudorapidity (η) and the full azimuthal angle. TPC is placed in the uniform solenoidal magnetic field (0.5 T). The sensitive volume of the TPC contains P10 gas (10% methane, 90% argon) regulated at 2 mbar above atmospheric pressure. The TPC data are used to determine particle trajectories, momenta, and particle-type through ionization energy loss (dE/dx). The TOF is based on Multi-gap Resistive Plate Chamber (MRPC) technology and is very useful to identify the particles at relatively higher momentum range compared to TPC. The Multi-Gap Resistive Plate Chamber Time of Flight (MRPC ToF) [20] will completely surround the outer radius of the TPC, $|\eta| < 0.9$, $0 < \phi < 2\pi$. It consists of 23K channels from 120 modules. The details of the design and other characteristics of the STAR detectors can be found in [17].

3 Beam Energy Scan at STAR

The Beam Energy Scan program at RHIC is aimed to perform systematic study and data analysis of the particle production in Au+Au collisions over a range of collisions energy $\sqrt{s_{NN}}$ = 7.7 – 39 GeV and centrality [5–9]. It concerns as spectra of non- and identified particles, ratio of particle yields etc.

3.1 Nuclear modification factor

The nuclear modification factor, R_{CP}, is the binary collisions N_{bin} scaled ratio of central (C) to peripheral (P) transverse momentum distributions

$$R_{CP} = \frac{(N_{bin})_P}{(N_{bin})_C} \frac{(d^2N_A/dp_T d\eta)_C}{(d^2N_A/dp_T d\eta)_P}. \quad (1)$$

If hard processes were not affected by the presence of the medium, they should scale with the number of binary collisions and R_{CP} would equal unity. All high p_T particle species as shown in [21] are strongly suppressed by a factor of 5 in the central Au+Au collisions at $\sqrt{s_{NN}}$ = 200 GeV. The suppression at top RHIC energy provides evidence for partonic energy loss in QGP.

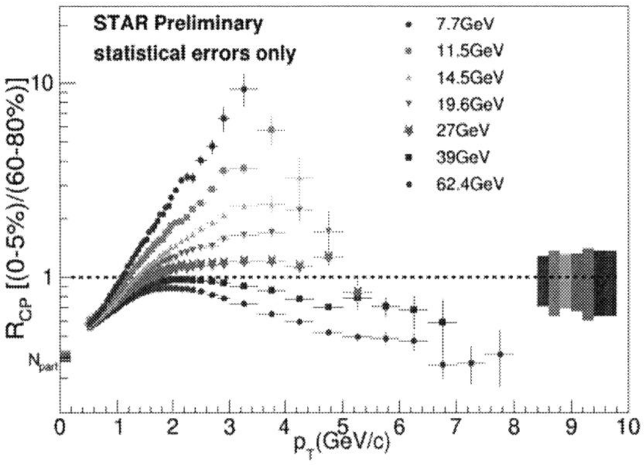

Figure 1. The nuclear modification factor R_{CP}: the ratio of the charged particle invariant yields in Au+Au collisions at the energy $\sqrt{s_{NN}}$ = 7.7 – 62.4 GeV in the range $|\eta| < 0.5$ for the centrality (0-5)% and (60-80)% versus transverse momentum p_T [22, 23].

Figure 1 shows the dependence of the ratio of the charged particle invariant yields in Au+Au collisions at the energy $\sqrt{s_{NN}}$ = 7.7, 11.5, 14.5, 19.6, 27, 39, and 62.4 GeV in the range $|\eta| < 0.5$ for the centrality (0-5)% and (60-80)% versus transverse momentum p_T [22, 23]. The ratio R_{CP} demonstrates strong enhancement of high-p_T charged hadrons for the lower collision energies in several ranges of p_T. For high p_T, a smooth transition from suppression to enhancement is observed with decreasing $\sqrt{s_{NN}}$.

3.2 Hadron spectra in Au+Au

The transverse momentum spectra of hadrons produced in high-energy collisions of heavy ions reflect features of constituent interactions in the nuclear medium. The medium modification is one of the effects (recombination, coalescence, energy loss, multiple scattering,...) that affects the shape of the spectrum. The properties of the created medium are experimentally studied by variation of the event centrality and collision energy. The particle spectra were found to reveal the exponential and power behavior at low- and high-p_T region, respectively.

Figure 2 shows the transverse momentum spectra of $\pi^+, K^+, p, \Xi^-, \Lambda$ and $\bar{p}$ particles produced in Au+Au collisions at $\sqrt{s_{NN}}$ = 14.5 GeV for different centralities and in the central rapidity range [24]. Errors shown are statistical only. The different fit functions shown by the dashed lines were used to obtain yields. These are Bose-Einstein function for π^+, $(m_T - m)$ exponential function for K^+, double exponential function for $p, \bar{p}$ and and Boltzmann function for Ξ^-, Λ. The spectra are measured using TPC+TOF up to 4 GeV/c for strange baryons and less than 2 GeV/c for $\pi, K, p, \bar{p}$. The similar spectra for identified particles at other BES energies were presented in [25–27].

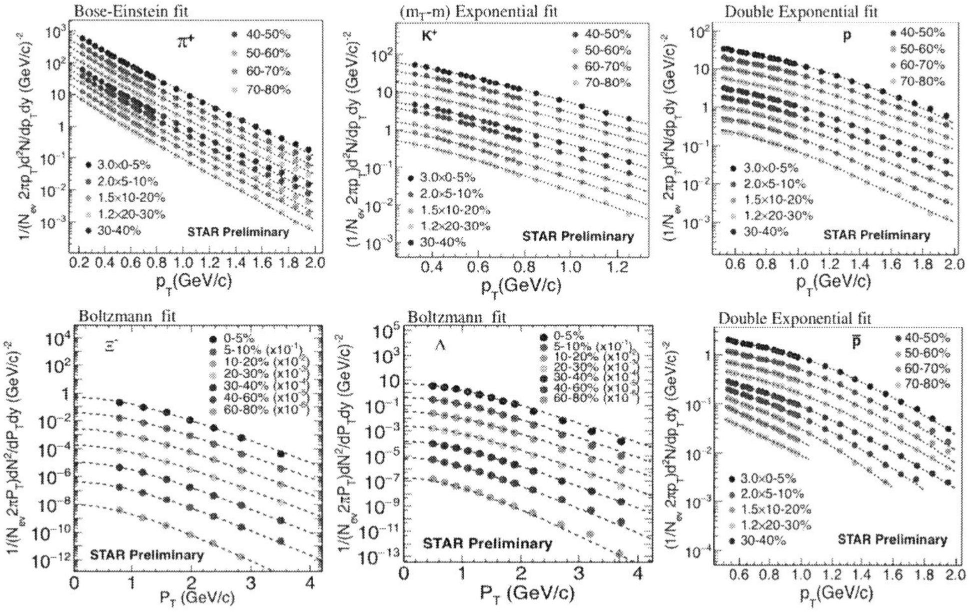

Figure 2. Mid-rapidity ($|y| < 0.5$) transverse momentum spectra for $\pi^+, K^+, p, \Xi^-, \Lambda$ and $\bar{p}$ in Au+Au collisions at $\sqrt{s_{NN}}$ = 14.5 GeV for different centralities. The curves represent Bose-Einstein function, $(m_T - m)$-exponential, Boltzmann and double-exponential function fits. The errors are statistical [24].

Multi-strange hadrons are important probes for the search of the phase boundaries of QGP (see [28] and references therein). The Ω hyperons and ϕ mesons are expected to have relatively small hadronic interaction cross sections. Therefore, they can carry the information directly from the chemical freeze-out stage with little or no distortion due to hadronic rescattering. Yields of these particles are minimally distorted from decay feed-down also. Therefore Ω hyperons and ϕ mesons offer a unique advantage in probing the transition from partonic to hadronic dynamics.

Figure 3 presents the first STAR measurements of mid-rapidity ($|y| < 0.5$) ϕ meson spectra with a broad p_T coverage for various collision centrality bins at selected BES energies $\sqrt{s_{NN}}$ = 7.7, 11.5, 19.6, 27, 39 GeV [28]. Similar spectra were also obtained for Ω hyperon production in Au+Au collisions. These results are used to examine features of formation mechanism in particle

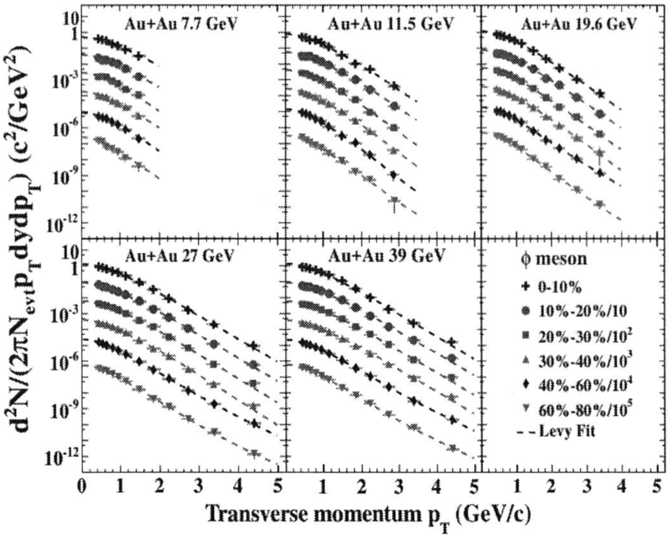

Figure 3. (Mid-rapidity ($|y| < 0.5$) transverse momentum spectra of ϕ mesons produced in Au+Au collisions at different centralities and energies ($\sqrt{s_{NN}} = 7.7, 11.5, 19.6, 27, 39$ GeV) [28]. The green bands represent systematic errors. The dashed curves represent fits to the experimental data with a Levy function [29].

production and explore possible changes from bulk partonic coalescence to hadron dominated dynamics as the colliding energy decreases. Spectra were corrected for reconstruction efficiency and geometrical acceptance. The systematic errors $(10 - 16)\%$ at $\sqrt{s_{NN}} = 11.5 - 39$ GeV and $(17 - 21)\%$ at $\sqrt{s_{NN}} = 7.7$ GeV are mainly due to the different signal extraction techniques, and the reconstruction efficiency corrections were also taken into account.

The production of light nuclei with small binding energy such as (anti)deuterons, can be used to study the freeze-out properties and local baryon density in high-energy nuclear collisions. The azimuthal anisotropic results of protons and deuterons have shown that the coalescence is the dominant process for the light nuclei production at later stage of the evolution. The coalescence parameter B_A for the production of nucleus A is expressed via the proton and neutron yields as follows:

$$\frac{d^3N_A}{dp_A^3} = B_A\left(\frac{d^3N_p}{dp_p^3}\right)^Z\left(\frac{d^3N_n}{dp_n^3}\right)^{(A-Z)} \tag{2}$$

Here Z, and $A - Z$ are the number of protons and neutrons, respectively. The parameter characterizes the measure of the phase space density for nucleons. It is defined in a similar way for produced anti-nuclei. A difference between $B_2(A)$ and $B_2(\bar{A})$ may imply that baryon and anti-baryon freeze-out at different densities.

Results of systematic study of colliding energy, centrality, transverse momentum dependence of mid-rapidity deuteron and anti-deuteron production measured by the STAR experiment from Au+Au collisions at $\sqrt{s_{NN}} = 7.7, 11.5, 14.5, 19.6, 27, 39$, and 200 GeV [31] are presented in figure 4. Deuterons, protons and their anti-particles were identified using the time projection chamber (TPC) and time-of-flight detector (TOF). Proton and anti-proton yields are corrected for weak decays. It was found that the baryon density at mid-rapidity decreases with increasing energy, yield of $\bar{d}$ in-

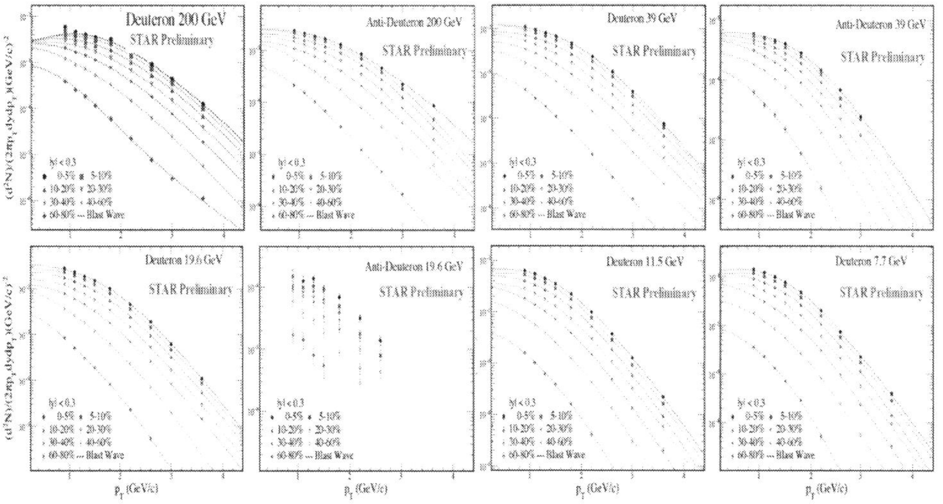

Figure 4. The transverse momentum distribution of d and $\bar{d}$ production in Au+Au collisions at different centralities and energies $\sqrt{s_{NN}}$ = 7.7, 11.5, 19.6, 39 GeV and 200 GeV [31].

creases with increasing energy, the mean p_T increases from peripheral to central collision and shows a weak energy dependence. The mean p_T difference between d and $\bar{d}$ was found to be small. The ratios $\bar{d}/d$ and $\bar{p}/p$ increase with increasing energy. The coalescence parameter $B_2(d)$ decreases as a collision energy increases, which shows that as energy decreases, the freeze-out volume for nucleons becomes smaller. The relative change $(B_2(d) - B_2(\bar{d}))/B_2(d)$ of the coalescence parameter shows little energy dependence, the average value is 0.29 ± 0.14, which might represent the iso-spin effect is not negligible in heavy-ion collision or particle and anti-particle have different freeze-out volume.

3.3 Nuclear flow

The medium created at top RHIC energies as reported in [1–4] passes through a partonic phase. Elliptic flow v_2 is one of the signal of collective phenomena of particle production

$$\frac{d^3N_A}{dp_A^3} = \frac{1}{2\pi p_T}\frac{dN_A}{dp_T}\left(1 + \sum_1^{\infty} 2v_n \cos(n(\phi - \Psi_r))\right). \tag{3}$$

Therefore it is of interest to study how the signal (v_2) evolves as a function of collision energy and centrality and type of produced particle when transverse particle densities change from low to high. Search for the switch from one to the other mechanism is one of the dramatic signatures of phase transition. If one scales the elliptic flow v_2/n_q and transverse kinetic energy E_T/n_q by the number n_q of constituent quarks in the hadron, all particles fall on a common curve. Such scaling are explained by quark coalescence, or recombination models, which provide an intriguing framework for hadronization of bulk partonic matter at RHIC. The essential degrees of freedom at the hadronization seem to be effective constituent quarks which have developed a collective elliptic flow during the partonic evolution. Elliptic flow v_2 values for identified particles and anti-particles at midrapidity in Au+Au collisions were measured by the STAR experiment in the Beam Energy Scan at RHIC at

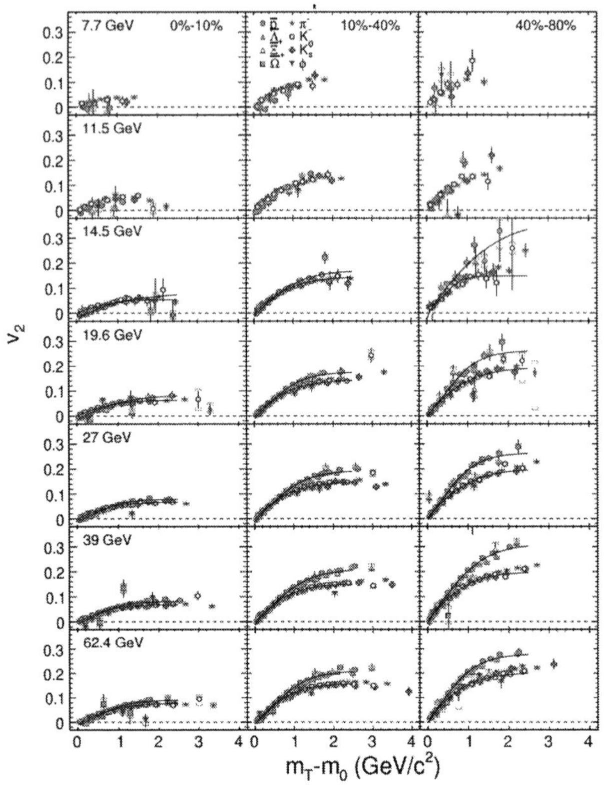

Figure 5. The elliptic flow v_2 of identified anti-particles $\pi^-, K^-, K^0, \bar{p}, \phi, \bar{\Lambda}, \bar{\Xi}^+, \bar{\Omega}^+$ as a function of transverse kinetic energy $E_T = m_T - m_0$, for $(0-10)\%$, $(10-40)\%$, and $(40-80)\%$ central Au+Au collisions at $\sqrt{s_{NN}} = 7.7, 11.5, 14.5, 19.6, 27, 39$, and 62.4 GeV [30]. The lines show simultaneous fits to baryons and mesons. The systematic errors are shown by the hooked error bars.

$\sqrt{s_{NN}} = 7.7, 11.5, 14.5, 19.6, 27, 39$ GeV and 62.4 GeV for three centrality classes [30]. A similar relative v_2 baryon-meson splitting for all centrality classes which is in agreement within 15% with the number-of-constituent quark scaling except at the lowest beam energies was observed. The larger v_2 for most particles relative to anti-particles shows a clear centrality dependence, with the largest difference for the most central collisions.

Figure 5 shows v_2 versus transverse kinetic energy $E_T = m_T - m_0$ of anti-particles $\pi^-, K^-, K^0, \bar{p}, \phi, \bar{\Lambda}, \bar{\Xi}^+, \bar{\Omega}^+$ for three centrality ranges of Au+Au collisions at $\sqrt{s_{NN}} = 7.7, 11.5, 14.5, 19.6, 27, 39$, and 62.4 GeV [30]. The splitting between baryons and mesons are found to be significant at 19.6 GeV and higher energies, and marginally significant at 14.5 GeV. There is no observed splitting for all centralities at 11.5 GeV and below. For these energies we are limited by the number of events and cannot draw a conclusion. For the ϕ meson at 14.5 GeV there were not enough events to plot the centrality dependence. For every particle species, energy, and centrality, v_2 increases with increasing E_T. At $E_T > 1$ GeV an onset of v_2 saturation can be observed. For the most central $(0-10)\%$ events the absolute baryon-meson splitting is significantly smaller compared to more peripheral collisions, partly because the values are smaller making the absolute difference smaller.

Note, that if there were a perfect NCQ scaling, the ratio $v_2(B)/v_2(M)$ should be equal to 1.5. It was found that the baryon-to-meson elliptic flow ratio for particles is higher than for anti-particles at all energies for $(0-10)\%$ and $(10-40)\%$ central collisions, but has no significant difference between particles and anti-particles for $(40-80)\%$. The ratio for anti-particles shows a centrality dependence which is increasing from central to peripheral collisions from about 1.3 to 1.6. But the ratio for particles does not show a significant centrality dependence. There is no significant beam energy dependence for the ratio of both particles and anti-particles for the points plotted, except for anti-particles at $(10-40)\%$ centrality. In addition, we can see from the ratio that NCQ scaling holds for particles at centralities of $(0-10)\%$ and $(10-40)\%$, but the ratio is slightly larger at $(40-80)\%$.

During the BES program, the STAR experiment has obtained results on v_2 at mid-rapidity ($|y| < 1.0$) for light nuclei $d, t, ^3He$ at $\sqrt{s_{NN}} = 7.7, 11.5, 19.6, 27, 39, 62.4$, and 200 GeV and anti-nuclei $\bar{d}$ at $\sqrt{s_{NN}} = 19.6, 27, 39, 62.4$, and 200 GeV and $^3\bar{He}$ at $\sqrt{s_{NN}} = 200$ GeV [32]. It was found that the light (anti-)nuclei $v_2(p_T)$ shows a monotonic rise with increasing p_T, mass ordering at low p_T and a reduction for more central collisions. It is observed that v_2 of nuclei and anti-nuclei are of similar magnitude at $\sqrt{s_{NN}} = 39$ GeV and above. The difference Δv_2 between d and $\bar{d}$ was found to follow the difference between p and $\bar{p}$ as a function of collision energy. The flow v_2 for ^3He and t nuclei shows similarity for all collision energies and in fact all the light-nuclei v_2 generally follow an atomic mass number scaling (v_2/A). The scaling indicates that the coalescence of nucleons might be the underlying mechanism of light-nuclei formation in high energy heavy-ion collisions. It is noted in [32] that light nuclei could have also formed via coalescence of quarks when v_2 and E_T are scaled by number of constituent quarks n_q instead of mass number A. Although this process seems physically acceptable, the survival of light nuclei, with their low binding energies (a few MeV), is highly unlikely under the high temperatures requisite for dissociating nucleons into quarks and gluons.

3.4 Thermodynamic parameters

The main aim of the RHIC BES program is to explore the phase diagram of nuclear matter created in heavy ion collisions [5]. It is assumed that the fireball produced in these collisions thermalizes rapidly leading to expansion and cooling of the nuclear system. Then the hadronization takes place and the particles get detected in the detectors.

Experimentally measured spectra of different particles over a wide range of centrality and collision energy can be used to determine thermodynamic parameters in the framework of the statistical thermal model [33–35] and the blast-wave model [36, 37]. The first model assumes that the produced system reached the thermodynamical equilibrium and statistical approach can be used to determine the chemical freeze-out temperature T_{ch} and baryon chemical potential μ_B. The second model considers that the local thermalization of particles at a kinetic freeze-out temperature T_{kin} and moving with a common radial flow velocity $<\beta>$. The kinetic freeze-out temperature and average collective velocity parameters are extracted from blast-wave fits to the identified hadron spectra. STAR spectra of particles π, K, p, Λ, Ξ and their anti-particles were used to determine the different particle ratios and to obtain the phase diagram in $\{T_{ch}, \mu_B\}$ and $\{T_{kin}, <\beta>\}$ planes [24–27].

Figure 6(a) demonstrates the dependence of T_{ch} and μ_B on centrality (number of participants N_{part}) at different energies $\sqrt{s_{NN}} = 7.7, 11.5, 14.5, 19.6, 27$, and 39 GeV. One can see from figure 6(a) that T_{ch} increases with the collision energy. The inverse behavior is observed for the chemical potential μ_B. The centrality dependence of μ_B is seen as well. Figure 6(b) shows the dependence of spectra of π^+, K^+, p particles and their anti-particles produced in Au+Au collisions on the transverse momentum at different collision energies and centralities. These data have been used to extract the kinetic freeze-out temperature T_{kin} and radial velocity $<\beta>$. One can see that $<\beta>$ decreases and T_{kin} increases from the central to peripheral collisions. An anti-correlation was observed between T_{kin} and $<\beta>$.

The largest value of the collective velocity and lowest temperature is reached in the most central collisions at the highest collision energy.

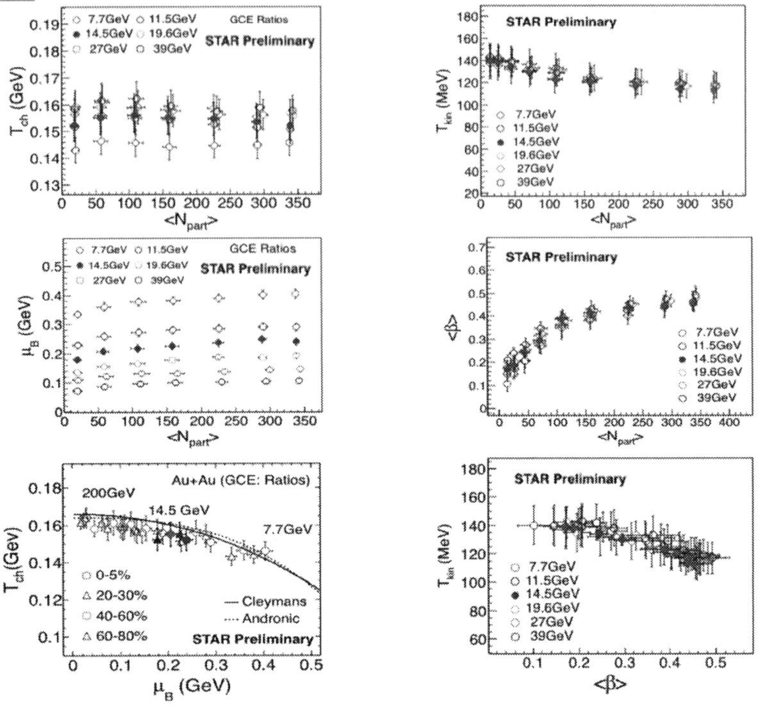

a) b)

Figure 6. Chemical $\{T_{ch}, \mu_B\}$ (a) and kinetic $\{T_{kin}, < \beta >\}$ (b) freeze-out thermodynamic parameters using the Grand-Canonical Ensemble of charged hadron production in Au+Au collisions for different collision energy and centrality [24, 38–40].

3.5 Self-similarity of hadron production in Au+Au collisions

The idea of the Beam Energy Scan programs at RHIC is to span the phase diagram from the top RHIC energy to the lowest possible energy. It is known that the study of critical phenomena is most preferable in terms of dimensionless variables. Therefore self-similarity of hadron interactions could be assumed to be a very fruitful concept to study collective phenomena in the hadron and nuclear matter. Important manifestation of this concept is a notion of scaling itself. Scaling in general means self-similarity at different scales. The approach based on z-scaling suggested in [41, 42] was treated as manifestation of the self-similarity property of the structure of colliding objects (hadrons, nuclei), the interaction mechanism of their constituents, and the process of constituent fragmentation into real hadrons. We use the approach for analysis of the spectra of charged hadrons obtained by STAR in BES program [43]. The approach [41, 42] relies on a hypothesis about self-similarity of hadron interactions at a constituent level. The assumption of self-similarity transforms to the requirement of simultaneous description of transverse momentum spectra corresponding to different collision energies, rapidities, and centralities by the same scaling function $\psi(z)$ depending on a similarity parameter z. The scaling function for the collision of nuclei is expressed in terms of the experimentally measured inclusive invariant cross section ($Ed^3\sigma/dp^3$), the multiplicity density ($dN/d\eta$), and the total inelastic cross section (σ_{in}), as follows:

$$\psi(z) = -\frac{\pi}{(dN/d\eta)\,\sigma_{\text{in}}} J^{-1} E \frac{d^3\sigma}{dp^3}. \tag{4}$$

Here J is the Jacobian for the transformation from $\{p_T^2, y\}$ to $\{z, \eta\}$. The function ψ is interpreted as a probability density to produce an inclusive particle with the corresponding value of z. The similarity parameter z is expressed via momentum fractions (x_1, x_2, y_a, y_b), multiplicity density, and parameters $\delta_{1,2}, \epsilon_{a,b}, c$, as follows:

$$z = z_0 \Omega^{-1}, \tag{5}$$

where $z_0 = \sqrt{s_\perp}/[(dN_{ch}/d\eta|_0)^c m_N]$, $\Omega = (1-x_1)^{\delta_1}(1-x_2)^{\delta_2}(1-y_a)^{\epsilon_a}(1-y_b)^{\epsilon_b}$. Here m_N is a nucleon mass. The quantity $\sqrt{s_\perp}$ is the transverse kinetic energy of the binary subprocess consumed on the production of the inclusive particle with mass m_1 and its counterpart with mass m_2. The constituents of the incoming nuclei carry fractions x_1, x_2 of their momenta. The inclusive particle carries the momentum fraction y_a of the scattered constituent. The parameters $\delta_{1,2}$ and $\epsilon_{a,b}$ describe structure of the colliding nuclei and fragmentation process, respectively. The parameter c is interpreted as a "specific heat" of the created medium. Simultaneous description of different spectra with the same $\psi(z)$ puts strong constraints on the values of these parameters, and thus allows for their determination. It was found that $\delta_{1,2}$ and c are constant at high energies, $\epsilon_{a,b}$ depends on multiplicity and $m_1 = m_2 \equiv m$. For the obtained values of $\delta_{1,2}, \epsilon_{a,b}$ and c, the momentum fractions are determined to minimize the resolution $\Omega^{-1}(x_1, x_2, y_a, y_b)$ with the constraint $(x_1P_1 + x_2P_2 - p/y_a)^2 = (x_1M_1 + x_2M_2 + m/y_b)^2$, which has sense of the momentum conservation law of a constituent subprocess. The fractal dimension of the nucleus is expressed in terms of the nucleon fractal dimension (δ) and the atomic number of nucleus (A) as follows, $\delta_A = A\delta$. The fragmentation fractal dimension in nucleus+nucleus collisions is parameterized in the form: $\epsilon_{AA} = \epsilon_0(dN/d\eta) + \epsilon_{pp}$. It depends on multiplicity density and nucleon fragmentation dimension ($\epsilon_{pp} = 0.2$). The negatively charged hadron spectra measured by STAR in Au+Au collisions at $\sqrt{s_{NN}}$ =7.7, 11.5, 19.6, 27, 62.4 and 200 GeV at mid-rapidity $|\eta| < 0.5$ as a function of transverse momentum p_T for different centrality classes of $(0-5)\%$, $(5-10)\%$,

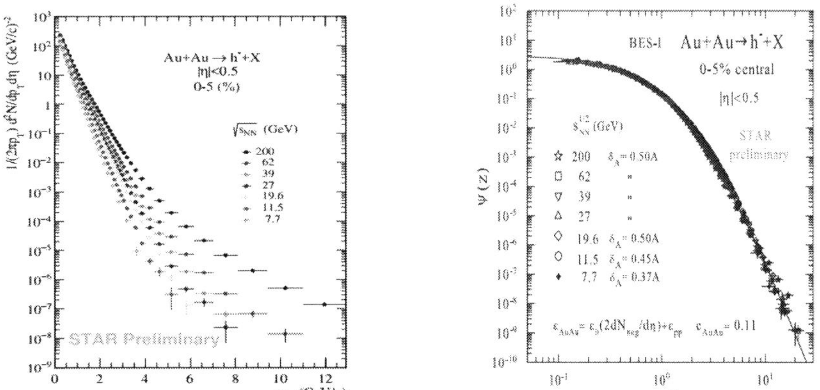

a)   b)

Figure 7. The dependence of the spectra of negatively charged particles produced in the most central Au+Au collisions at energy $\sqrt{s_{NN}} = 7.7, 11.5, 19.6, 27, 39, 62.4$, and 200 GeV in the range $|\eta| < 0.5$ [43] on the transverse momentum (a) and the corresponding dependence of scaling function $\psi(z)$ on the self-similarity parameter z (b).

(10 – 20)%, (20 – 40)%, (40 – 60)%, (60 – 80)% were presented in [43]. The transverse momentum spectra for these energies and for most central events (0% – 5%) are shown in figure 7(a). The distributions are measured over a wide momentum range $0.2 < p_T < 12.$ GeV/c. As we can see the data presentation in terms of dimensional quantities demonstrates a strong dependence on the scale. Figure 7(b) shows scaling function $\psi(z)$ as a function of self-similarity parameter z for the data presented in figure 7(a). We observe a "collapse" of the data onto a single curve. The dependence demonstrates the universality of the shape of the scaling curve, the constancy of the fractal dimensions of the nuclei and the fragmentation process in the range $\sqrt{s_{NN}} > 19.6$ GeV. The found regularity can be treated as manifestation of the self-similarity of the structure of colliding objects, interaction mechanism of their constituents, and processes of fragmentation of these constituents into real registered particles. It is assumed that the discontinuity of the model parameters ("heat capacity" and fractal dimensions) indicates on the existence of critical phenomena in a nuclear matter.

4 Heavy flavor

Charm and beauty quarks are created predominantly via initial hard scatterings in nucleon-nucleon collisions and the production rate is calculable with perturbative QCD techniques. The large masses are expected to be retained during the interactions with the nuclear medium. Heavy quarks are therefore predicted to be sensitive to transport and other properties of the early stages of the system when the QGP is expected to exist.

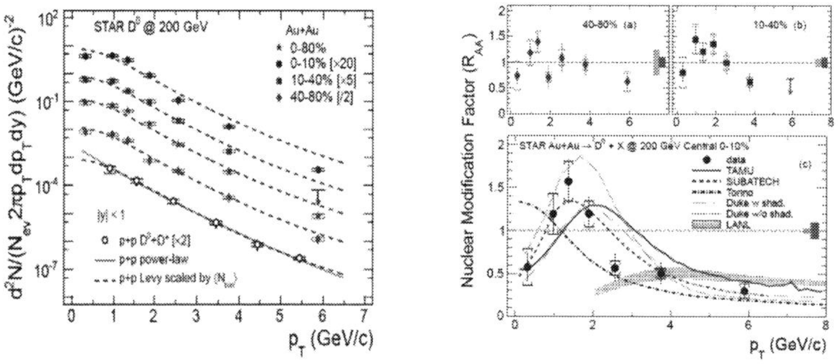

a) b)

Figure 8. The dependence of the differential invariant yield (a) and the nuclear modification factor R_{AA} (b) of D^0 meson production in Au+Au collisions at $\sqrt{s_{NN}} = 200$ GeV on the transverse momentum p_T for different centralities. The curves (left panel) are number-of-binary collisions scaled Levy functions from fitting to the p+p result (open circles) [51]. The arrow denotes the upper limit with 90% confidence level of the last data point for (10 – 40)% collisions. The systematic uncertainties are shown as square brackets. [50].

Energetic heavy quarks were predicted to lose less energy via gluon radiation than light quarks when they traverse the QGP. Initial RHIC and LHC measurements, however, show similar suppression at high transverse momentum, p_T, in central A+A collisions (see [50] and references therein). This has led to the reconsideration of the effects of heavy-quark collisional energy loss and requires follow-up measurements. Heavy-quark collective motion can provide experimental evidence for bulk medium thermalization. Model calculations show that interactions between heavy quarks and the QGP are sensitive to the drag/diffusion coefficients of the medium. These can be related to the shear viscosity-to-entropy ratio and other transport properties. Therefore, measurements of heavy-quark production at low and intermediate p_T are of particular relevance to these issues, and also for the interpretation of the

DOI: 10.1051/epjconf/201713801016

charmonia production in heavy-ion collisions. In elementary collisions, heavy quarks are expected to hadronize mainly through hard fragmentation. In high energy heavy-ion collisions, the large charm-pair abundance could increase the coalescence probability. The coalescence of charm with a light quark from the medium with a large radial flow may introduce a p_T-dependent modification to the observed charmed hadron spectrum compared to that from fragmentation. Furthermore, this may lead to a baryon-to-meson enhancement for charmed hadrons similar to that observed for light flavor hadrons.

Figure 8(a) shows the first measurement of the spectra (solid symbols) of D^0 meson production over a transverse momentum range $0.0 < p_T < 6.0$ GeV/c in Au+Au collisions at $\sqrt{s_{NN}} = 200$ GeV for different centrality bins [50]. The measurement was performed via invariant-mass reconstruction of the hadronic decay channel $D^0 \rightarrow K^- + \pi^+$ and its charge conjugate. The pp data, shown as open circles, contain D^0 data for $p_T < 2.0$ GeV/c and D^* data for $p_T > 2.0$ GeV/c. The main subsystems used for the D^0 analysis are the TPC and TOF. All measurements are presented as an average of D^0 and $\bar{D}^0$ yields at midrapidity ($|y| < 1$). The nuclear modification factor R_{AA} for D^0 meson production calculated as the ratio between the p_T spectrum in Au+Au collisions for the centrality bins of (40−80)%, (10−40)%, and (0−10)% to the Levy function fit to the p+p data is shown in figure 8(b). The statistical and systematic uncertainties are shown by the vertical lines and brackets, respectively. Strong suppression is observed in the most-central collisions for $p_T > 2.5$ GeV/c, while no evidence is found for suppression in peripheral collisions. In (0 − 10)% collisions, the suppression level is around 0.5 for $p_T > 3$ GeV/c, which is consistent with both the measurements of electrons from heavy-flavor hadron decays and the light hadrons. These results as noted in [50] indicate that charm quarks lose significant energy in the hot and dense medium produced in the central Au+Au collisions at RHIC and support the coalescence hadronization of D meson production.

The STAR experiment has completed data taking with two new detectors for heavy flavor studies, the Heavy Flavor Tracker (HFT) [44] and Muon Telescope Detector (MTD) [45–49].

Figure 9 demonstrates the transverse momentum spectra (blue symbols)(a) and nuclear modification factor (black symbols) (b) for D^0 meson production in Au+Au collisions obtained by using the TPC, TOF and HFT [52]. The spectra are measured up to 6.5 GeV/c and are in a good agreement with

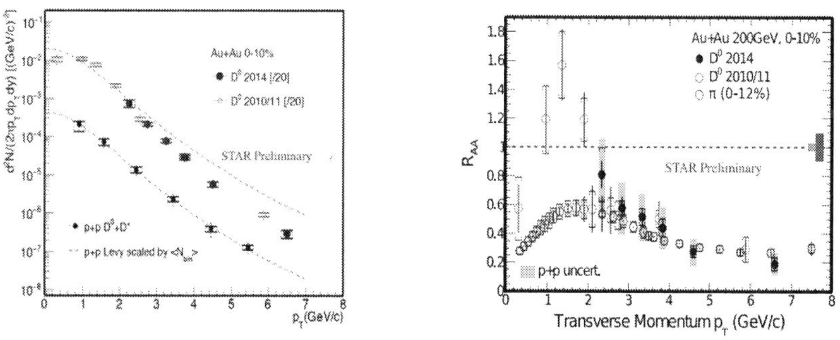

a) b)

Figure 9. The dependence of the differential invariant yield (a) and the nuclear modification factor R_{AA} (b) of D^0 meson production in Au+Au collisions at $\sqrt{s_{NN}} = 200$ GeV on the transverse momentum p_T for (0 − 10)% central collisions [52]. The curves (left panel) are number-of-binary collision scaled Levy functions from fitting to the p+p data (black symbols) [51]. Data taken with and without the HFT are shown by blue and red symbols on the left panel (a) and black and red symbols on the right panel (b). The systematic uncertainties are shown as square brackets.

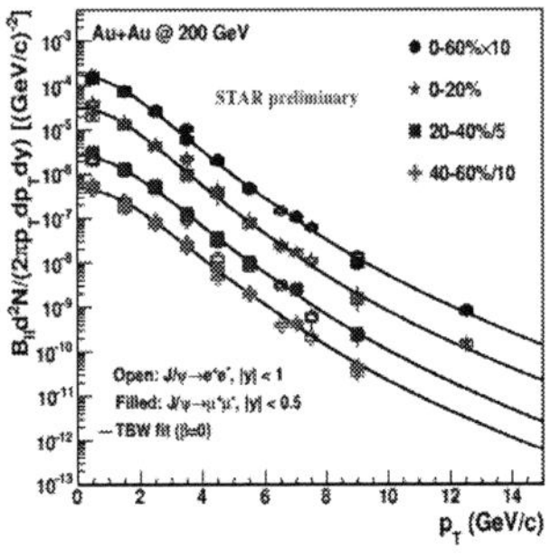

a)

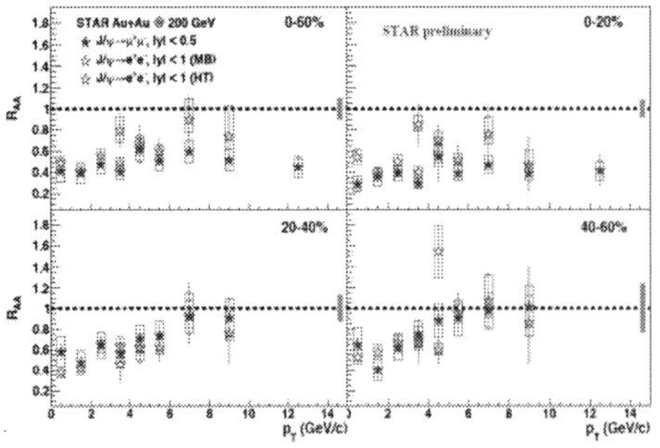

b)

Figure 10. The invariant yield of J/ψ scaled by the branching ratio B_{ll} (a) and the nuclear modification factor R_{AA} (b) as a function of p_T for different centralities in the di-muon (filled symbols) [53] and in the di-electron (open symbols) [56] channels.

the previous data [50] shown by red symbols. The value of R_{AA} is larger than 1 for $p_T \sim 1.5$ GeV/c. It indicates on charm coalescence with a radially flowing bulk medium. Significant suppression of D^0 production in most central $(0-10)\%$ Au+Au collisions for $p_T > 2.5$ GeV/c is observed. The value of R_{AA} at high p_T is found to be the same (~ 0.2) as for π mesons shown by blue symbols in figure 9(b). This result indicates on strong charm-medium interaction.

Using the MTD detector, STAR has studied J/ψ production in Au+Au collisions at $\sqrt{s_{NN}} =$ 200 GeV [53]. Figure 10(a) shows the invariant yield of J/ψ in Au+Au collisions at $\sqrt{s_{NN}} = 200$ GeV for different collision centralities. As seen from figure 10(a) the new results in the di-muon decay channel measured in the range $0 < p_T < 15$ GeV/c are consistent with previous results in the di-electron decay channel [54–56]. The Tsallis Blast-Wave fits to di-electron results shown by solid lines assume zero velocity for J/ψ.

Figure 10(b) shows the nuclear modification factor R_{AA} for di-muon (red star) and di-electron (open black star) channels as a function of transverse momentum for different centralities

$$R_{AA} = \frac{(\sigma_{inel})}{(N_{coll})} \frac{(d^2N_{AA}/dp_T dy)}{(d^2\sigma_{pp}/dp_T dy)}. \tag{6}$$

We see strong suppression of R_{AA} at low $p_T < 3-4$ GeV/c. The mechanism of dissociation plays a significant role in this p_T range. For middle centralities $(20-40)\%$ and $(40-60)\%$ the ratio R_{AA} increases with p_T and reaches unity at $p_T = 8-9$ GeV/c. For most central Au+Au collisions $(0-20)\%$ suppression is large up to $p_T \simeq 12$ GeV/c. The dependence of R_{AA} on transverse momentum is consistent with di-electron channel results for the entire p_T range within uncertainties in all centralities.

5 Fixed target mode at STAR

As it was noted above the RHIC BES-I program [5] was proposed to search for the possible critical point and to study the nature of the phase transition between hadronic and partonic phases of matter. The first goal of the BES-I program, namely the evidence for the disappearance of signatures of quark-gluon phase observed at the RHIC top energy, is already qualitatively reached with the results of BES-I. The original design of RHIC optimized the luminosity at $\sqrt{s_{NN}} = 200$ GeV. At lower energies, the beam is not as well-focused and the collision rate is low. The STAR BES Phase II White

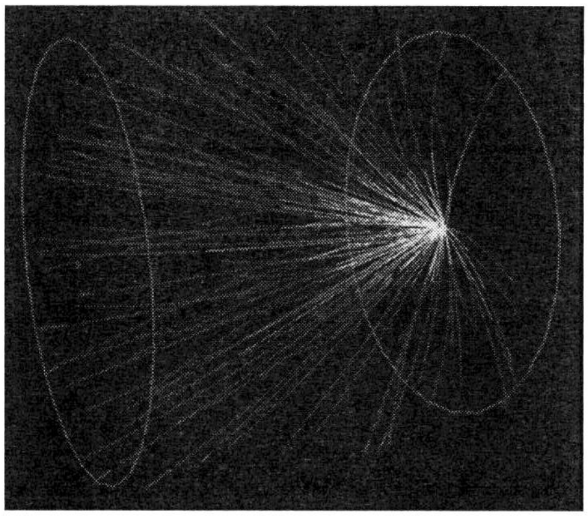

Figure 11. Central Au+Au collision event at $\sqrt{s_{NN}} = 3.9$ GeV in the fixed target mode. The reconstructed tracks from an interaction between a nucleus in the beam halo and a gold nucleus [59].

Paper details the motivation and the plans to return to energy scans at low RHIC energy in years 2018 and 2019 [16]. In order to provide adequate intensity for BES phase II, the RHIC Collider Accelerator Division (CAD) has been undertaking an upgrade with electron cooling [70–72]. It is expected that with electron cooling, the luminosity will increase by a factor of 2-5 at $\sqrt{s_{NN}} = 7.7$ GeV in the collider mode. Using interactions between beam nuclei and a stationary gold target [58–61] the feasibility has been demonstrated to study Au+Au events at center-of-mass energies below 7.7 GeV (up to 3.0 GeV) during the BES phase II. The results [57] obtained by NA49 Collaboration at CERN support the onset of deconfinement at a collision energy of 7.7 GeV. Therefore the enlarged statistics of BES phase II would allow us to improve on accuracy of earlier results, and to precisely determine at which collision energy QGP signatures disappear. The QGP signatures disappear at the low end of the BES-I energy range. An internal fixed-target mode will be used to access collision energies below 7.7 GeV.

Figure 11 shows the reconstructed tracks in Au+Au collision at $\sqrt{s_{NN}} = 3.9$ GeV from an event associated with the gold target. Figure 12 demonstrates the spectra (a) and ratio (b) of π^+ and π^- meson production in Au+Al collisions at energy $\sqrt{s_{NN}} = 3.0, 3.5, 4.0$ GeV.

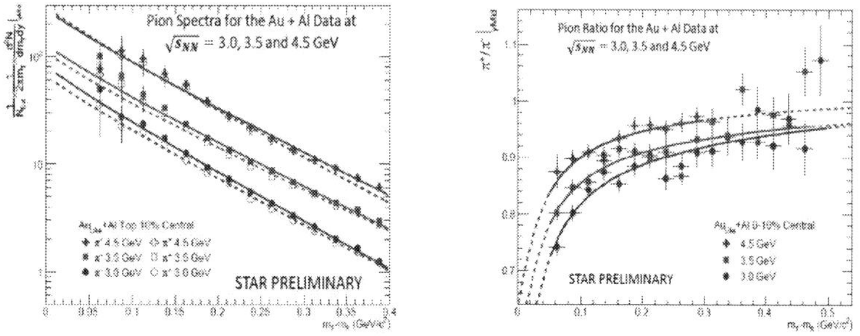

a) b)

Figure 12. Spectra (a) and ratio (b) of π^+ and π^- meson production in Au+Al collisions in fixed target mode at energy $\sqrt{s_{NN}} = 3.0, 3.5, 4.0$ GeV [58–61].

 DOI: 10.1051/epjconf/201713801016

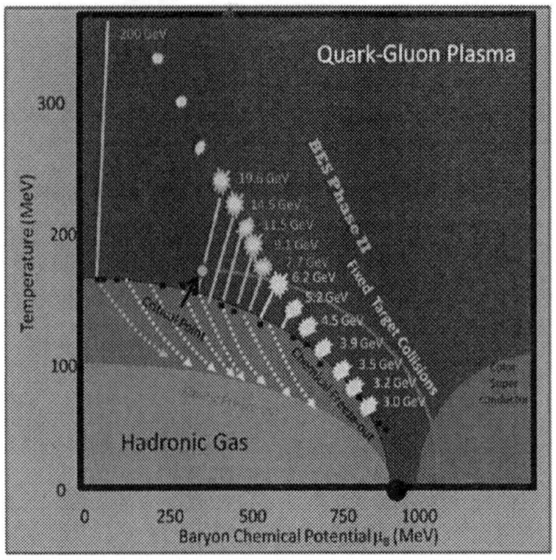

Figure 13. The supposed phase diagram of nuclear matter which is created in central Au+Au collisions at over a range $\sqrt{s_{NN}} = 3 - 200$ GeV in the (T_{ch}, μ_B) plane.

It is expected that the fixed-target program will enable STAR to make key measurements related to the phase diagram of nuclear matter below the reported onset of deconfinement at $\sqrt{s_{NN}} = 7.7$ GeV.

Figure 13 shows the reach of RHIC's BES program in the (T_{ch}, μ_B) plane at energies $\sqrt{s_{NN}} = 3.0 - 200$ GeV. This choice of energies provides almost uniform coverage of the unknown $\{T_{ch}, \mu_B\}$ region and may allow us to narrow down an area of interest for further study. The range of baryon chemical potential (μ_B) changes for BES-II energy in the collider and fixed target modes is shown in table 1. One can see that values of T_{ch} and μ_B cover a range from 420 MeV to 205 GeV at $\sqrt{s_{NN}} = 7.7$ and 20 GeV in the collider mode and from 750 MeV to 420 MeV at $\sqrt{s_{NN}} = 3.0$ and 7.7 GeV in the fixed target mode, respectively. Note, that this range is a larger range than at any other heavy-ion facility.

Table 1. The μ_B values are estimated from the systematics of central collisions at BES-II energies [16].

Collider energy (GeV)	Fixed Target energy (GeV)	Center-of-mass rapidity	Single beam energy (GeV)	Chemical potential μ_B (MeV) (collider mode)	Chemical potential μ_B (MeV) (fixed target mode)
19.6	4.47	1.52	8.87	205	589
17.2	4.21	1.46	7.67	230	608
14.5	3.90	1.37	6.32	250	633
13.0	3.72	1.32	5.57	288	649
11.5	3.53	1.25	4.82	315	666
9.1	3.19	1.13	3.62	370	699
7.7	2.98	1.05	2.92	420	721

6 STAR upgrades for BES-II

The limited event statistics from BES-I has already allowed significant progress to be made toward the goals that were established at the outset of this program [5]. There are clear indications that hadronic interactions dominate at the lower BES energies and several observables associated with the formation of a partonic phase at top RHIC energy show indications of turn-off. The successful performance of the BES-II program [16] and further understanding of the phase diagram of QCD matter is only possible with the STAR upgrade [62, 63], in combination with the higher RHIC luminosities due to electron cooling [70, 71]. New systems - the internal Time Projection Chamber (iTPC) and Event Plane Detector (EPD), will increase STAR's mid-rapidity acceptance, event plane resolution, and centrality determination. The BES phase II program requests Au+Au collisions at $\sqrt{s_{NN}} = 19.6, 14.6, 11.5$ and 7.7 GeV after electron cooling upgrade. The fixed gold target will allow collisions at low energy $\sqrt{s_{NN}} = 4.5, 4.0, 3.5$, and 3.0 GeV, respectively. These events can be triggered upon and recorded without reducing the number of collider events.

The inner TPC upgrade [64–66] will provide the increase of transverse momentum acceptance from $p_T > 125$ MeV to $p_T > 60$ MeV/c, improve dE/dx resolution from 7.5% to 6.2%, and increase acceptance from $|\eta| < 1$ to $|\eta| < 1.5$. The end cap Time-Of-Flight [67] will provide PID of $\pi/K/p$ for the whole iTPC acceptance in the collider and fixed target modes. The Event Plane Detector [68] will provide large η coverage $2.1 < |\eta| < 5.1$ compared to TPC ($|\eta| < 1.0$), high η (radial, 16) and azimuthal (24) segmentation, good timing resolution ($\sim$ 1 ns), and additional mid-rapidity independent event plane and centrality determination independent from TPC. It will be also used as a trigger detector for BES-II.

So the STAR upgrade (iTPC+eTOF+EPD) is crucial to extend energy reach to overlap/complementary AGS/FAIR/NICA/JPARC and provide the unprecedented coverage and PID for critical point search in BES-II.

7 Conclusions

The study of features of the phase diagram of the nuclear matter created in collisions of nuclei at RHIC is one of the major scientific tasks in heavy ion physics. These are location of a possible critical point and phase boundaries, determination of type of the phase transitions. The ability of RHIC to scan a wide range of beam energies ($\sqrt{s_{NN}} = 7.7 - 200$ GeV), beam and probe species provides a unique opportunity for the exploration of the phase diagram. RHIC has completed Phase-I of the beam energy scan program with center-of-mass beam energies $\sqrt{s_{NN}} = 7.7, 11.5, 14.5, 19.6, 27$ and 39 GeV. Some of resent results obtained by the STAR Collaboration were presented in the report. Among them is the dependence of the nuclear modification factor R_{CP} on transverse momentum p_T for different collision energy $\sqrt{s_{NN}}$. The results demonstrate the evolution of suppression of particle yields with $\sqrt{s_{NN}}$. It indicates on strong modification of a mechanism of the hadron production by the nuclear medium.

It was found that the spectra of different particle ($\pi, p, K, J/\psi, \phi, \Lambda, \Xi, \Omega, d, He, ...$) and anti-particle production at BES energy and different centrality of collisions over a wide range of p_T show strong sensitivity to medium environment and flavor content of produced particles. Indication of the nucleon coalescence mechanism for production of light nuclei ($d, {}^3He, {}^4He$) and their anti-nuclei was supported.

The dependence of elliptic flow on the transverse momentum p_T for particles $\pi, p, K, \phi, \Lambda, \Xi, \Omega$ and their anti-particles found at BES energies supports the hypothesis on parton collectivity and reveals features of the number constituent quark scaling in Au+Au collisions firstly observed at 200 GeV.

The spectra of strange and non-strange particles and anti-particles and their ratios measured by STAR over a wide range of centrality and collision energy were used to determine the thermodynamic parameters in the framework of the statistical thermal model and blast-wave model. The results allow us to scan the phase diagram of nuclear matter over a wide range in T_{ch}, μ_B and $T_{kin}, <\beta>$ planes and to verify a scenario of chemical and kinetic freeze-out.

Using new detectors - Heavy Flavor Tracker and Muon Telescope Detector, the STAR Collaboration has performed the measurements of D and J/ψ mesons production in Au+Au collisions at $\sqrt{s_{NN}} = 200$ GeV. The dependence of R_{AA} on the transverse momentum was found to be consistent with di-electron channel results and confirm strong suppression of particle yields for most central Au+Au collisions at large p_T as it has been observed for light flavor particles.

The STAR BES phase II sets out the physics case for an in-depth study of energies below 20 GeV with typically 20 times the statistics as in the same energy region in BES-I. It can be done in the collider and fixed target modes. Low-energy RHIC electron cooling is an essential point in the collider mode. The fixed target program will extend BES-II physics reach to the region below the onset of deconfinement down to energy $\sqrt{s_{NN}} = 3$ GeV.

In accordance to general theory of critical phenomena [73, 74] the "scaling" and "universality" are concepts developed to understand the behavior and properties of multiparticle systems. Scaling means that systems near critical point exhibiting self-similar properties are invariant under transformation of scale. According to universality, quite different systems behave in a remarkably similar fashion near the respective critical points. It is expected that experimental characteristics of heavy ion collisions such as fluctuation and correlation functions should reveal a power behavior, demonstrate a strong enhancement and be described by the critical exponents near critical point. Therefore search for new scaling laws of the particle production in Au+Au collisions is of interest to verify the fundamental physical principles such as self-similarity [75]. We consider that the collider and fixed target modes with new detector systems (HFT, MTD, iTPC, EPD, eTOF) significantly improve the capabilities of STAR for BES-II to detect key features of QCD phase diagram.

References

[1] J. Adams et al. (STAR collabotration), Nucl. Phys. A **757**, 102 (2005)
[2] K. Adcox et al. (PHENIX collabotration), Nucl. Phys. A **757**, 184 (2005)
[3] B. B. Back et al.(PHOBOS collabotration), Nucl. Phys. A **757**, 28 (2005)
[4] I. Arsene et al. (BRAHMS collabotration), Nucl. Phys. A **757**, 1 (2005)
[5] M. M. Aggarwal et al. (STAR Collaboration), "An Experimental Exploration of the QCD Phase Diagram: The Search for the Critical Point and the Onset of Deconfinement", arXiv:1007.2613
[6] L. Kumar (for STAR Collaboration), Nucl. Phys. A **830**, 275C (2009)
[7] L. Kumar (for STAR Collaboration), Nucl. Phys. A **862-863**, 125 (2011)
[8] B. Mohanty (for STAR Collaboration), Nucl. Phys. A **830**, 899C (2009)
[9] B.I. Abelev et al. (STAR Collaboration), Phys. Rev. C **81**, 024911 (2010)
[10] B.I. Abelev et al. (STAR Collaboration), Phys. Lett. B **655**, 104 (2007)
[11] J. Adams et al. (STAR Collaboration), Phys. Rev. Lett. **95**, 122301 (2005)
[12] A. Adare et al. (STAR Collaboration), Phys. Rev. Lett. **98**, 162301 (2007)
[13] B.I. Abelev et al., Phys. Rev. C **80**, 064912 (2009)
[14] Y. Aoki et al., Nature **443** 675, (2006). S. Ejiri, Phys. Rev. D **78**, 074507 (2008). E.S. Bowman and J. I. Kapusta, Phys. Rev. C **79**, 015202 (2009). M.A. Stephanov, Int. J. Mod. Phys. A **20**, 4387 (2005). S. Gupta et al., Science **332**, 1525 (2011)

[15] S.Sharma, "Recent Lattice QCD Results for RHIC BES", RHIC & AGS Annual Users' Meeting, Upton, BNL, January 7-10, 2016. https://www.bnl.gov/aum2016/
[16] STAR Collaboration, "Studying the Phase Diagram of QCD Matter at RHIC". https://drupal.star.bnl.gov/STAR/starnotes/public/sn0598
[17] K. Ackermann et al. (STAR Collaboration), Nucl. Instrum. Meth. A **499**, 624 (2003)
[18] M. Anderson et al. (STAR Collaboration), Nucl. Instrum. Meth. A **499**, 659 (2003)
[19] W. Llope et al. (STAR Collaboration), Nucl. Instrum. Meth. B **241**, 306 (2005)
[20] F. Geurts et al., Nucl. Instrum. Meth. A **533**, 60 (2004). Y. Wang et al., Nucl. Instrum. Meth. A **613**, 200 (2010)
[21] M.A.C. Lamont (for the STAR Collaboration), J. Phys. Conf. Ser. **50**, 192 (2006)
[22] S. Horvat (for the STAR Collaboration), arXiv:1303.7260, arXiv:1601.01644
[23] S. Horvat (for the STAR Collaboration), "Charged Hadron Nuclear Modification Factors in the Beam Energy Scan Data from STAR", 8th International Workshop on Critical Point and Onset of Deconfinement, PoS(CPOD 2013)002, March 11-15, 2013, Napa, California, USA
[24] V. Bairathi (for the STAR Collaboration), Nucl. Phys. A **956**, 292 (2016)
[25] L. Kumar (for the STAR Collaboration), Nucl. Phys. A **862-863**, 125 (2011)
[26] L. Kumar (for the STAR Collaboration), Nucl. Phys. A **904-905**, 256C (2013)
[27] L. Kumar (for the STAR Collaboration), Nucl. Phys. A **931**, 1114 (2014)
[28] L. Adamczyk et al. (STAR collabotration), Phys. Rev. C **93**, 21903 (2016)
[29] B.I. Abelev et al. (STAR Collaboration), Phys. Rev. C **79**, 064903 (2009)
[30] L. Adamczyk et al. (STAR collabotration), Phys. Rev. C **93**, 014907 (2016)
[31] N. Yu (for the STAR Collaboration), "Beam energy dependence of d and anti-d production in Au+Au collisions at RHIC", QM2015, September 27 - October 3, 2015 Kobe Fashion Mart, Kobe, Japan. http://qm2015.riken.jp/
[32] L. Adamczyk et al. (STAR collabotration), nucl-ex:1601.07052
[33] J. Cleymans et al., Phys. Rev. C **73**, 034905 (2006). S. Wheaton, J.Cleymans, Comput. Phys. Commun. **180**, 84 (2009)
[34] A. Andronic et al., Nucl. Phys. A **772**, 167 (2006). A. Andronic et al., Phys. Lett. B **675**, 312 (2009). A. Andronic et al., Nucl. Phys. A **834**, 237C (2010)
[35] F. Becattini et al., Phys. Rev. C **73**, 044905 (2006)
[36] E. Schnedermann, J. Sollfrank, U.W. Heinz, Phys. Rev. C **48**, 2462 (1993)
[37] B.I. Abelev et al. (STAR Collaboration), Phys. Rev. C **79**, 034909 (2009). B.I. Abelev et al. (STAR Collaboration), Phys. Rev. C **81**, 024911 (2010)
[38] L. Kumar (for the STAR Collaboration), Nucl. Phys. A **904**, 256c (2013)
[39] S. Das (for the STAR Collaboration), Nucl. Phys. A **904-905**, 891c (2013)
[40] B. Huang (for the STAR Collaboration), "Beam Energy Scan Results from STAR", 247 (2015), In: Proc. 50th Recontres de Moriond, "QCD and High Energy Interactions", La Thuile, Aosta Valley, Italy, March 21-28, 2015
[41] I. Zborovský, M. Tokarev, Phys. Rev. D **75**, 094008 (2007)
[42] I. Zborovský, M. Tokarev, Int. J. Mod. Phys. A **24**, 1417 (2009)
[43] M.V. Tokarev (for the STAR Collaboration), Int. J. Mod. Phys. Conf. Ser. **39**, 1560103 (2015)
[44] STAR Collaboration, Technical Design Report: The STAR Heavy Flavor Tracker (2011)
[45] L. Ruan et al. (for the STAR Collaboration), J. Phys. G: Nucl. Part. Phys. **36**, 095001 (2009)
[46] R. Ma (for the STAR Collaboration), "New STAR results", RHIC & AGS Annual Users' Meeting, BNL, USA, January 7-10, 2016. https://www.bnl.gov/aum2016/

[47] M.Lomnitz (for the STAR Collaboration), "Open heavy flavor measurements in Au+Au collisions at 200 GeV using the STAR Heavy Flavor Tracker", RHIC & AGS Annual Users' Meeting, BNL, USA, January 7-10, 2016. https://www.bnl.gov/aum2016/

[48] H. Qiu (for the STAR Collaboration), "Heavy flavor measurements at the STAR experiment", 38th International Conference on High Energy Physics, Chicago, USA, August 3-10, 2016. http://www.ichep2016.org/

[49] M. Lomnitz (for the STAR Collaboration), "Review of recent heavy flavor measurements from the STAR experiment" 5th International Conference on New Frontiers in Physics (ICNFP 2016), Kolymbari, Greece, July 6-14, 2016. http://indico.cern.ch/event/442094/overview

[50] L. Adamczyk et al. (STAR Collaboration), Phys. Rev. Lett. **113**, 142301 (2014)

[51] L. Adamczyk et al. (STAR Collaboration), Phys. Rev. D **86**, 072013 (2012)

[52] Guannan Xie (for the STAR Collaboration), Quark Matter 2015, Kobe, Japan. Rongrong Ma (for the STAR Collaboration), 2016 RHIC & AGS Annual Users Meeting, BNL, USA, June 7-10, 2015

[53] T. Todoroki (for the STAR collaboration) Strangeness in Quark Matter 2016, Berkeley, USA, June 27 - July 1, 2016. https://indico.cern.ch/event/403913/overview

[54] L. Adamczyk et al. (STAR Collaboration), Phys. Lett. B **722**, 55 (2013)

[55] L. Adamczyk et al. (STAR Collaboration), Phys. Rev. C **90**, 024906 (2014)

[56] D. Banerjee et al., Phys. Rev. D **85**, 014510 (2012). H. Ding et al., J. Phys. G **38**, 124070 (2010). B.I. Abelev et al. (STAR collaboration), Phys. Rev. D **83**, 052006 (2011)

[57] C. Alt et al. (NA49 Collaboration), Phys. Rev. C **77**, 024903 (2008)

[58] D. Sebra (for the STAR Collaboration), "The Beam Energy Scans at RHIC: Results from BES-I Prospects for BES-II", 34th Reimei Workshop "Physics of J-PARC Heavy-Ion Collisions", J-PARC, Tokai, Japan, August 8-9, 2016. http://asrc.jaea.go.jp/soshiki/gr/hadron/workshop/reimei-34th/

[59] K.Meehan (for the STAR Collaboration), "The Fixed-Target Experiment at STAR", RHIC & AGS Annual Users' Meeting, Upton, BNL, January 7-10, 2016. https://www.bnl.gov/aum2016/

[60] R.Reed (for the STAR Collaboration), "STAR Detector Upgrades" RHIC & AGS Annual Users' Meeting, Upton, BNL, January 7-10, 2016. https://www.bnl.gov/aum2016/

[61] K.Meehan (for the STAR Collaboration), "The Fixed-Target Experiment at STAR", 32nd Winter Workshop on Nuclear Dynamics (WWND2016) 28 February - 5 March, 2016 , Guadeloupe, French. https://indico.cern.ch/event/435555/

[62] A. Schmah (for the STAR Collaboration), "STAR Upgrades and Physics Program", Strangeness in Quark Matter 2016, Berkeley, USA, June 27 - July 1, 2016. https://indico.cern.ch/event/403913/overview

[63] R. Reed (for the STAR Collaboration), "STAR Detector Upgrades", RHIC & AGS Users' Meeting, 2016 RHIC & AGS Annual Users' Meeting, Upton, BNL, January 7-10, 2016. https://www.bnl.gov/aum2016/

[64] STAR Collaboration, Proposal for STAR inner TPC sector Upgrade (iTPC). http://drupal.star.bnl.gov/STAR/starnotes/public/sn0619

[65] STAR Collaboration, "Technical Design Report for the iTPC Upgrade". December 1, 2015. https://drupal.star.bnl.gov/STAR/starnotes/public/sn0644

[66] C. Yang (for the STAR Collaboration), "The iTPCupgrade at STAR", Strangeness in Quark Matter 2016, Berkeley, USA, June 27 - July 1, 2016. https://indico.cern.ch/event/403913/overview

[67] STAR Collaboration, eTOF Proposal, January, 2016

[68] STAR Collaboration, An Event Plane Detector for STAR, Proposal, May, 2016. https://drupal.star.bnl.gov/STAR/system/files/

[69] W. Fischer, "RHIC upgrades for the next decade", Satellite Workshop "Accelerators Revealing the QCD Secrets", XIIth Quark Confinement and Hadron Spectrum, Thessaloniki, Greece, 4 September, 2016

[70] White Paper "Low-Energy RHIC electron Cooler (LEReC)", September 19, 2013

[71] A. Fedotov, "LEReC overview: project goal and cooling approach", Low Energy RHIC electron Cooling (LEReC), MEIC Collaboration Meeting, 30-31 March, 2015

[72] D. Sebra (for the STAR Collaboration), "Exploring the QCD Phase Diagram: RHIC Beam Energy Scan II", Low Energy RHIC electron Cooling Review, BNL, USA, 13 August, 2013

[73] H.E. Stanley, *Introduction to Phase Transitions and Critical Phenomena* (Oxford, New York and Oxford, 1971)

[74] H.E. Stanley, Rev. Mod. Phys. **71**, S358 (1999)

[75] M. Tokarev, I. Zborovský, Nucl. Phys. Proc. Suppl. B **245**, 231 (2013)

EPJ Web of Conferences **138**, 01017 (2017) DOI: 10.1051/epjconf/201713801017
Baldin ISHEPP XXIII

J/ψ production at the STAR experiment

Pavla Federičová for the STAR collaboration[1,*]

[1] *Czech Technical University in Prague, Faculty of Nuclear Sciences and Physical Engineering, Czech Republic*

Abstract. The suppression of J/ψ production in heavy-ion collisions compared to p+p collisions, caused by the color screening in the deconfined medium, has been proposed as a signature of the formation of the Quark-Gluon Plasma. However, J/ψ production rates are also sensitive to other effects such as the cold nuclear matter effects and recombination. To understand the interplay of these mechanisms, it is important to study J/ψ production for different collision geometries.
In these proceedings, we present recent result on J/ψ production via the di-electron decay channel from the STAR experiment at RHIC in p+p collisions at $\sqrt{s_{NN}}$ = 200, 500 GeV, in d+Au collisions at $\sqrt{s_{NN}}$ = 200 GeV and in heavy-ion collisions, specifically Au+Au collisions at $\sqrt{s_{NN}}$ = 200 GeV and U+U collisions at $\sqrt{s_{NN}}$ = 193 GeV. We also show measurements on J/ψ production via the di-muon decay channel in p+p collisions at $\sqrt{s_{NN}}$ = 500 GeV and in Au+Au collisions at $\sqrt{s_{NN}}$ = 200 GeV using the newly installed Muon Telescope Detector.

1 Introduction

Quarkonium production is one of the essential probes to study the properties of the Quark-Gluon Plasma (QGP) formed in relativistic heavy-ion collisions. J/ψ is the lightest and most abundantly produced quarkonium state at the STAR experiment. The suppression of its production due to the color-screening effect in the medium has been proposed as a direct evidence of the QGP formation [1].

However, other effects can also affect the observed J/ψ yields. These effects include firstly Cold Nuclear Matter (CNM) effects such as initial state parton scattering, nuclear shadowing and nuclear absorption, and secondly regeneration via coalescence of deconfined charm quarks. To understand these different mechanisms, it is important to study the quarkonium production in different collision systems and at different collision energies.

The interpretation of medium-induced J/ψ modification requires a good understanding of its production mechanisms in p+p collisions, which include direct production via gluon fusion, feed-down from higher charmonium states and B-meson decay [2].

The modification of quarkonium production in nucleus+nucleus collisions (A+A) with respect to p+p collisions is usually quantified as the so-called nuclear modification factor (R_{AA}):

$$R_{AA} = \frac{1}{< N_{coll} >} \frac{d^2N_{AA}/dp_T dy}{d^2N_{pp}/dp_T dy}. \quad (1)$$

*e-mail: pavla.federicova@fjfi.cvut.cz

It is defined as the ratio of particle yields in A+A collisions to that in p+p collisions scaled by the average number of binary collisions $< N_{coll} >$. If there is no medium effect, the production in A+A collisions is a simple superposition of the nucleon-nucleon interactions and R_{AA} is equal to unity. If the production is suppressed (enhanced), then $R_{AA} < 1$ (> 1).

In these proceedings, we present the measurements of J/ψ production via the di-electron decay channel in p+p collisions at $\sqrt{s_{NN}}$ = 200, 500 GeV, in d+Au collisions at $\sqrt{s_{NN}}$ = 200 GeV and in heavy-ion collisions, such as Au+Au collisions at $\sqrt{s_{NN}}$ = 200 GeV and U+U collisions at $\sqrt{s_{NN}}$ = 193 GeV. We also report the first J/ψ measurements via the di-muon decay channel in p+p collisions at $\sqrt{s_{NN}}$ = 500 GeV and in Au+Au collisions at $\sqrt{s_{NN}}$ = 200 GeV using the newly installed Muon Telescope Detector (MTD) [3].

2 STAR experiment

The STAR experiment is a multi-purpose detector dedicated to the study of the strongly interacting matter at high temperature and high-energy density [4]. Its main purpose is to detect, track and identify charged particles at mid-rapidity. STAR consists of various subsystems which perform different tasks, such as triggering, particle identification or tracking. Enclosed in a 0.5 Tesla solenoidal magnet are a Time Projection Chamber (TPC) [5], a Time-Of-Flight detector (TOF) [6], and a Barrel Electro-Magnetic Calorimeter (BEMC) [7]. The TPC is the main tracking detector for STAR. It also provides particle identification capability through the measurement of specific energy loss (dE/dx). The TOF greatly improves particle identification at low transverse momenta (p_T). The BEMC is used to trigger on and identify high p_T electrons. These detectors cover full azimuth over a large pseudo-rapidity range ($|\eta| < 1$).

In 2014, the MTD [3] was fully installed outside of the solenoidal magnet at STAR. It is designed for triggering on and identifying muons with $p_T > 1.2$ GeV/c at mid-rapidity ($|y| < 0.5$). Compared to the di-electron decay channel, the di-muon decay channel is less affected by bremsstrahlung radiation, Dalitz decay, and γ conversion.

3 J/ψ measurement in p+p collisions

The differential cross section of J/ψ production as a function of p_T has been measured in p+p collisions. Figure 1 shows the STAR results (red circles) using the 2012 data and PHENIX results (black circles) both at $\sqrt{s}$ = 200 GeV via the di-electron channel. STAR and PHENIX results are consistent with each other. In figure 2, the results at $\sqrt{s}$ = 500 GeV from the di-muon (black circles) and di-electron channels (red circles) for $0 < p_T < 20$ GeV/c are shown. The CGC+NRQCD calculations (low p_T) [8] and NLO NRQCD calculations (high p_T) [9] can together describe the experimental data well both at 200 GeV and 500 GeV.

STAR also studied the dependence of J/ψ production on event activity in p+p collisions at $\sqrt{s}$ = 200 GeV and at $\sqrt{s}$ = 500 GeV. As can be seen in figure 3, a faster-than-linear growth of relative J/ψ yield versus event activity has been observed, which suggests correlation between soft and hard processes. Such an increase is found to be largely independent of collision energy when comparing the STAR results at $\sqrt{s}$ = 200 GeV and at $\sqrt{s}$ = 500 GeV with the LHC results at $\sqrt{s}$ = 7 TeV [10].

4 J/ψ measurement in d+Au collisions

In order to identify the modification to J/ψ production due to the presence of QGP, all other nuclear effects need to be understood and quantified. CNM effects [11] coming from the presence of ordinary

 DOI: 10.1051/epjconf/201713801017

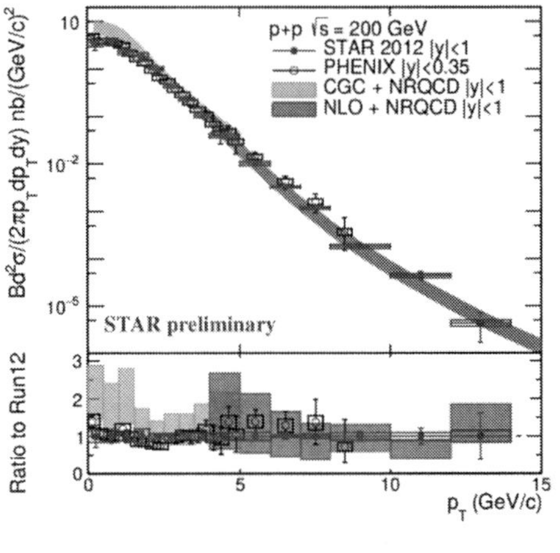

Figure 1. J/ψ production cross section as a function of p_T in p+p collision at $\sqrt{s}$ = 200 GeV measured via the di-electron decay channel by the STAR (closed circles) and PHENIX (open circles) experiments.

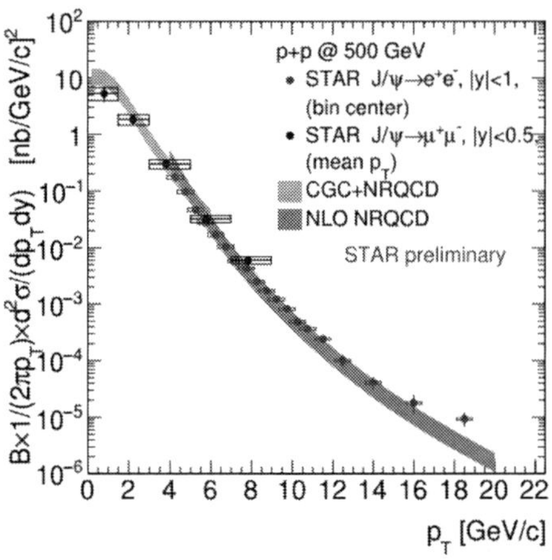

Figure 2. J/ψ cross section as a function of p_T in p+p collision at $\sqrt{s}$ = 500 GeV measured via the di-muon decay channel (black circle) and the di-electron decay channel (red circle).

nuclear matter in the collision can modify the production of J/ψ. These effects at RHIC are studied via d+Au collisions at $\sqrt{s_{NN}}$ = 200 GeV where the formation of the QGP is not expected, but nuclear matter is present.

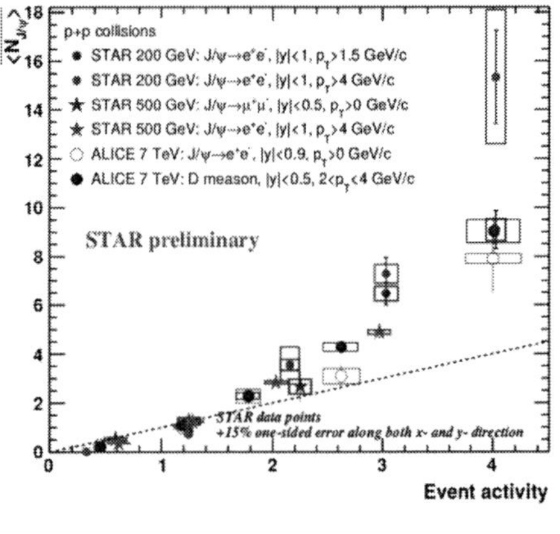

Figure 3. Relative J/ψ yield versus event activity in p+p collisions at $\sqrt{s}$ = 200 GeV and $\sqrt{s}$ = 500 GeV compared with the LHC results at $\sqrt{s}$ = 7 TeV.

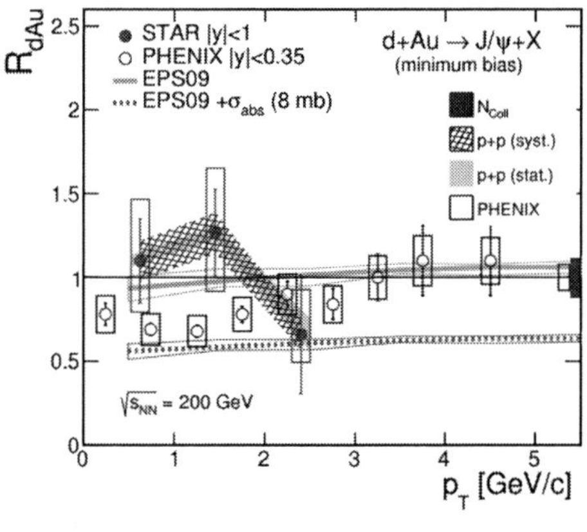

Figure 4. The J/ψ R_{AA} as a function of p_T in d+Au collisions (closed circles). The data are compared with PHENIX results (open circles) and EPS09 calculations.

The J/ψ nuclear modification factor measured in d+Au collisions as a function of p_T is shown in figure 4. The STAR results [12] are compared to the PHENIX data [13] and are in agreement within statistical and systematic uncertainties. The model calculations assuming only shadowing

(EPS09 [14]) as well as shadowing combined with nuclear absorption are also shown and both can qualitatively describe the data. At high p_T, the R_{dAu} is consistent with unity suggesting that the influence of CNM effects is small. At low p_T, the PHENIX results indicate sizeable CNM effects for $p_T < 2$ GeV/c.

5 J/ψ measurement in Au+Au collisions

J/ψ invariant yields in Au+Au collisions at $\sqrt{s_{NN}}$ = 200 GeV have been studied via both the di-electron and di-muon channels.

Figure 5 shows the J/ψ nuclear modification factor in the di-electron channel from STAR (solid and open circles) [15] and PHENIX (open squares) [16] as a function of p_T for different collision centralities. The J/ψ suppression decreases towards high p_T for all centralities and is consistent with unity at high p_T in peripheral and semi-peripheral collisions. The data are compared with theoretical predictions including color screening effect and regeneration of charm quarks by Zhao and Rapp (solid line) [17] and Liu et al. (dashed line) [18].

Centrality dependence of the J/ψ nuclear modification factor is shown in figure 6. The STAR data for $p_T < 5$ GeV/c (black circles) [15] and for $p_T > 5$ GeV/c (red circles) [19] indicate J/ψ suppression for central collisions. The suppression increases with collision centrality. The high p_T J/ψ mesons are less suppressed than the low p_T ones, which could be due to the less influence of recombination and CNM effects at high p_T. Theoretical models by Zhao and Rapp (solid line) [17] and Liu et al. (dashed line) [18] can qualitatively describe the data.

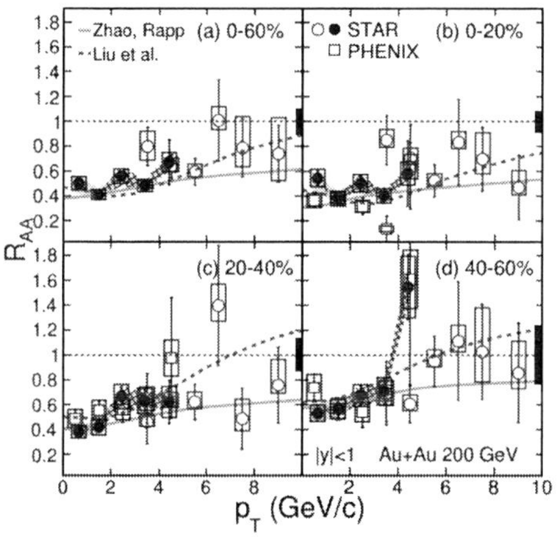

Figure 5. The J/ψ nuclear modification factor as a function of p_T for different centralities in Au+Au collisions.

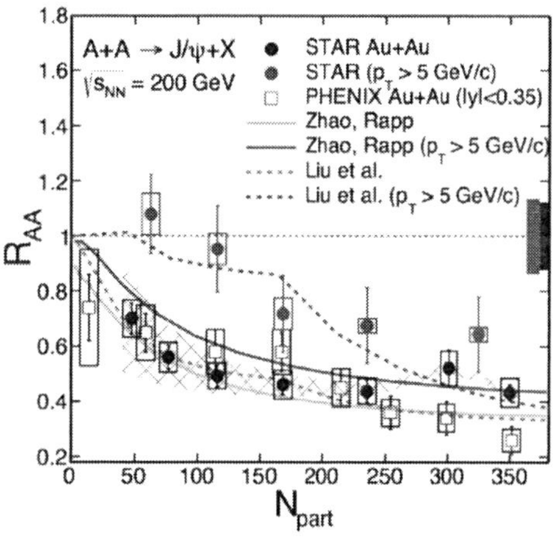

Figure 6. The J/ψ nuclear modification factor as a function of centrality for the STAR low (black circles) and high p_T data (red circles).

Figure 7 shows the invariant yield of J/ψ from the di-muon channel (solid points) as a function of p_T for different collision centralities. These new results are consistent with the STAR published results from the di-electron decay channel (open points) [20, 21] within uncertainties.

Centrality dependence of J/ψ R_{AA} from the MTD data (red stars) is shown in figure 8 for $p_T > 0$ GeV/c and in figure 9 for $p_T > 5$ GeV/c. Results from PHENIX in Au+Au collisions at $\sqrt{s_{NN}}$ = 200 GeV (black open circles) [16], as well as results from ALICE [22] and CMS [23] in

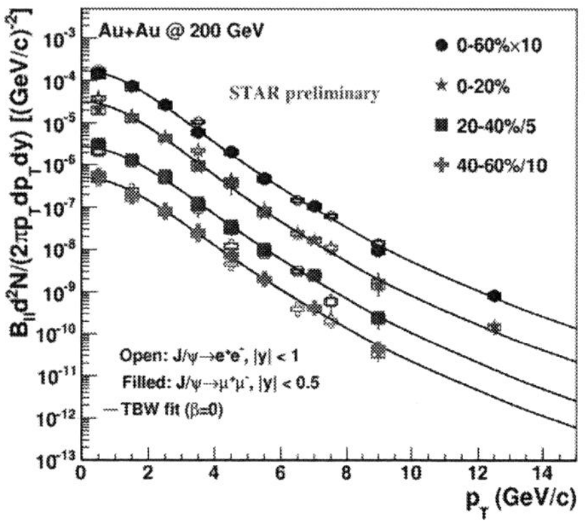

Figure 7. The J/ψ invariant yield as a function of p_T for different centralities in the di-muon decay channel (solid points) and in the di-electron decay channel (open points).

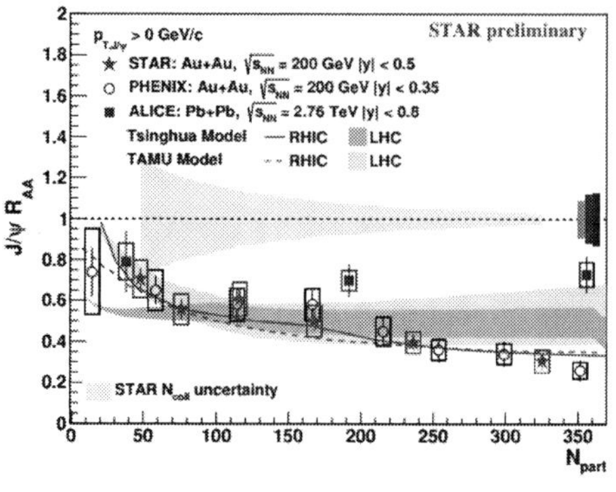

Figure 8. The J/ψ nuclear modification factor for $p_T > 0$ GeV/c in Au+Au collisions at $\sqrt{s_{NN}} = 200$ GeV and in Pb+Pb collisions at $\sqrt{s_{NN}} = 2.76$ TeV as a function of centrality.

Pb+Pb collisions at $\sqrt{s_{NN}} = 2.76$ TeV (blue solid circles) are shown in figure 8 and figure 9, respectively. For $p_T > 0$ GeV/c, the STAR results are consistent with the PHENIX results, both of which decrease with increasing centrality, while the ALICE results are almost independent of centrality and less suppressed than those at RHIC in central collisions. Such a difference can be explained by larger regeneration contributions to the low p_T J/ψ at the LHC due to higher $c\bar{c}$ yields. Both transport models from Tsinghua [18, 24] and Texas A&M University (TAMU) [17, 25], including dissociation and regeneration effects, can qualitatively describe the centrality dependence at RHIC, but tend to overestimate the suppression at the LHC.

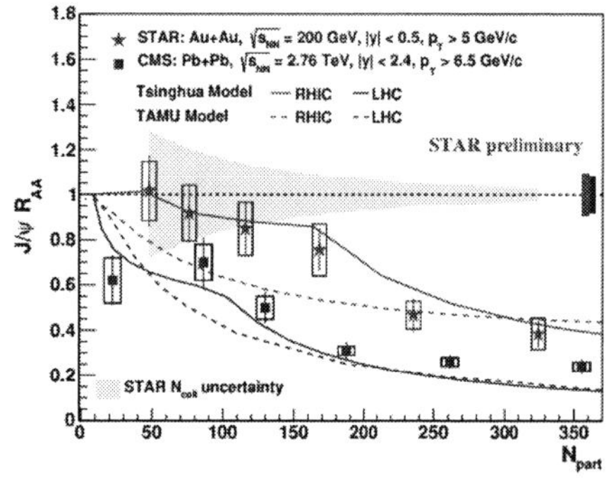

Figure 9. The J/ψ nuclear modification factor for $p_T > 5$ GeV/c in Au+Au collisions at $\sqrt{s_{NN}} = 200$ GeV and in Pb+Pb collisions at $\sqrt{s_{NN}} = 2.76$ TeV as a function of centrality.

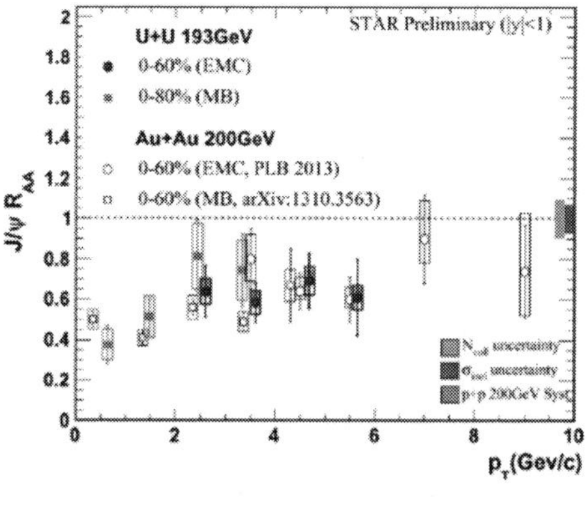

Figure 10. The J/ψ nuclear modification factor as a function of p_T in U+U collisions at $\sqrt{s_{NN}} = 193$ GeV compared with Au+Au collisions at $\sqrt{s_{NN}} = 200$ GeV.

For $p_T > 5$ GeV/c, the R_{AA} both at RHIC and the LHC have a decreasing trend towards the central collisions, but RHIC data exhibit less suppression than that at the LHC in all centralities. This is expected due to higher energy density and temperature reached at the LHC, leading to stronger J/ψ suppression. At high p_T, there is a tension among the transport models and experimental data.

6 J/ψ measurement in U+U collisions

STAR has also collected data for U+U collisions at $\sqrt{s_{NN}}$ = 193 GeV. Since uranium nuclei are larger than gold nuclei, it is expected that the energy density of the medium created in U+U collisions is higher than that in Au+Au collisions [26].

Figure 10 shows the STAR results on J/ψ nuclear modification factor in U+U collisions as a function of p_T. The suppression of J/ψ production is similar to that observed in Au+Au collisions at $\sqrt{s_{NN}}$ = 200 GeV [27].

7 Summary

We present the J/ψ measurements in the di-electron and the di-muon decay channels at STAR. In p+p collisions at $\sqrt{s}$ = 200 GeV and 500 GeV, inclusive J/ψ cross section can be described by CGC+NRQCD and NLO NRQCD model calculations. The modification of J/ψ production in d+Au collisions at $\sqrt{s_{NN}}$ = 200 GeV is consistent with no suppression at high p_T suggesting no CNM effects but indicating sizeable CNM effects at low p_T.

In Au+Au collisions at $\sqrt{s_{NN}}$ = 200 GeV, we observe clear evidence of J/ψ suppression indicating dissociation. The J/ψ R_{AA} can be qualitatively described by theoretical models including dissociation and regeneration. Suppression of J/ψ production in U+U collisions at $\sqrt{s_{NN}}$ = 193 GeV is similar to that observed in Au+Au collisions at $\sqrt{s_{NN}}$ = 200 GeV.

Acknowledgement

This work was supported by the grant No. 13-20841S of Grant Agency of the Czech Republic.

References

[1] T. Matsui and H. Satz, Phys.Lett. B **178**, 416-422 (1986)
[2] N. Brambilla et al., Eur. Phys. J. C **71**, 1534 (2011)
[3] L. Ruan et al., J. Phys. G: Nucl. Part. Phys. **36**, 095001 (2009)
[4] K. Ackermann et al. (STAR Collaboration), Nucl.Instrum.Meth. A **499**, 624-632 (2003)
[5] M. Anderson, J. Berkovitz, W. Betts, R. Bossingham, F. Bieser, et al., Nucl.Instrum.Meth. A **499**, 659-678 (2003)
[6] W. Llope (STAR Collaboration), Nucl.Instrum.Meth. A **661**, 110-113 (2012)
[7] M. Beddo (STAR Collaboration), Nucl.Instrum.Meth. A **499**, 725-739 (2003)
[8] Y.-Q. Ma, and R. Venugopalan, Phys. Rev. Lett. **113**, 192301 (2014)
[9] H. Shao et al., JHEP **05**, 103 (2015)
[10] J. Adam et al. (ALICE collaboration), J. High Energy Physics **09**, 148 (2015)
[11] R. Vogt, Phys. Rev. C **71**, 054902 (2005)
[12] L. Adamczyk et al. (STAR Collaboration), Phys. Rev. C **93**, 64904 (2016)
[13] A. Adare et al. (PHENIX Collaboration), Phys. Rev. C **87**, 034904 (2013)
[14] K. Eskola, H. Paukkunen, and C. Salgado, Nucl. Phys. A **830**, 599 (2009)
[15] L. Adamczyk et al. (STAR Collaboration), Phys. Rev. C **90**, 24906 (2014)

[16] A. Adare et al. (PHENIX Collaboration), Phys. Rev. Lett. **98**, 232301 (2007)
[17] X. Zhao and R. Rapp, Phys. Rev. C **82**, 064905 (2010)
[18] Y. Liu, Z. Qu, N. Xu, and P. Zhuang, Phys. Lett. B **678**, 72-76 (2009)
[19] L. Adamczyk et al. (STAR Collaboration), Phys. Lett. B **722**, 55 (2013)
[20] L. Adamczyk et al. (STAR Collaboration), Phys. Let. B **722**, 55-62 (2013)
[21] L. Adamczyk et al. (STAR Collaboration), Phys. Rev. C **90**, 024906 (2014)
[22] B. Abelev et al. (ALICE Collaboration), Phys. Let. B **734**, 314-327 (2014)
[23] S. Chatrchyan et al. (CMS Collaboration), JHEP **05**, 063 (2012)
[24] K. Zhou et al., Phys. Rev. C **89**, 054911 (2014)
[25] X. Zhao and R. Rapp, Nucl. Phys. A **859**, 114-125 (2011)
[26] D. Kikola, G. Odyniec, and R. Vogt, Phys. Rev. C **84**, 054907 (2011)
[27] W.M. Zha (STAR Collaboration), Nuclear Physics A **931**, 596-600 (2014)

DOI: 10.1051/epjconf/201713801018

Photo-absorption sum rules σ_{-1} and further questions about correlations of constituents inside bound hadrons

S.B. Gerasimov[1,*]

[1] *Joint Institute for Nuclear Research, Dubna, 141980, Russia*

Abstract. Within the non-relativistic approach to photo-absorption sum rules for the 3N(4N)-nuclei the new σ_{-1} sum rules are proposed which are based on general charge-symmetry (CS) consequences for the "CS-conjugated" triton and ^3He and the CS-self-conjugated ^4He. Combining the relativistic constituent quark model of the nucleon and the known spin-,isospin- and the dipole-moment fluctuation sum rules, the relevant moments of the quark coordinate distribution and correlation functions in the ground state of the nucleon are derived and discussed.

1 Introduction

The observed effects of the structure-dependent asymmetries in the static electromagnetic characteristics of hadrons, such as parameters of form-factors or coefficients of the electromagnetic polarizability, or in characteristics of processes induced by lepton- or photon-hadron interactions, e.g., total or partial cross sections, are known as very useful means to study and to understand the underlying dynamics.

The sum rules, in particular, have long served as a reliable constraints on, or relations between measured quantities, depending only on most general principles and statements of the theory.

In this work we shall concentrate mainly on two topics: the sum rules for static electromagnetic characteristics of hadrons and the role of the dynamical short-range correlations of nuclear constituents in description of the behavior of the total and polarized photonuclear or photon-nucleon cross section.

As it is known, the non-relativistic dipole sum rules continue to be a useful tool in the theory of the atomic and nuclear photo-effect.

$$\sigma_n(E1) = \int_{thr}^{\infty} d\omega\, \omega^n\, \sigma_{E1}(\omega).$$

Examples: $n = -2$ $\rightarrow$ Kramers-Heisenberg sum rule (SR) for static electric-dipole polarizability of a given quantum system;
$n = -1$ $\rightarrow$ the bremsstrahlung-weighted SR, dependent of charged-"parton" correlation in a given system;
$n = 0$ $\rightarrow$ the famous Thomas-Reiche-Kuhn SR, known as a precursor of the Quantum Mechanics. As far as the nucleon is the relativistic 3q system, the relativistic generalization of sum rules is needed, including the spin degrees of freedom and spin-dependent interactions and correlations of partons.

*e-mail: gerasb@theor.jinr.ru

2 The non-relativistic σ_{-1} sum rules for few-body nuclei

Besides sum rules including the measurable physical quantities and obtained on the base of the current algebras and the $p_z \to \infty$ techniques, the similar sum rules (but without the explicitly spin-dependent terms) were obtained in the non-relativistic (NR) nuclear physics context. In particular, the NR sum rules σ_{-1} for the s-shell nuclei ($A = Z + N \leq 4$) were earlier obtained from the assumption that the ground-state wave function is symmetric in the space coordinates of nucleons [1],[2] for each nucleus of the isotopic doublet ^3H and ^3He and for ^4He

$$4\pi^2\alpha\frac{NZ}{A-1}(\frac{1}{3})\cdot\langle r_{ch}^2\rangle_{NR} = \int\frac{d\nu}{\nu}\sigma_{E1}(\nu). \tag{1}$$

The present-day situation is that the accuracy of the theoretical results [3] is higher than the experimental uncertainties in the important energy range up to the pion production threshold [4] suggesting that new experimental data would be timely. The main objective of this section is a comparison of the Charge Independence (CI) and Charge Symmetry (CS) initial assumptions taken for the derivation of the σ_{-1} sum rules for the "mirror" ^{3}He -^{3}H and the "charge-self-conjugate" ^{4}He nuclei. The validity of the Foldy [1] and Khokhlov [2] sum rules assumes validity of the (nucleon) charge independence (CI) symmetry. Our objective in this section is to apply stronger charge symmetry (CS) requirements in derivation of the F-Kh-type sum rules. In accord with the CS-provisions for the mean values of the nucleon radii correlations in the mirror pair ^{3}H- and ^{3}He nuclei and the "self-conjugate" ^{4}He nucleus, we have get that the values of $< D^2 >_{^3He}$, $< D^2 >_{^3H}$, and $\langle r_{ch}^2\rangle_{^3H}$ can be expressed via one and the same linear combination of the introduced and undetermined parameters which is different from the analogues expression for $\langle r_{ch}^2\rangle_{^3He}$. Taken at face value this means that with our stricter assumption on the underlying symmetry i.e. the CS instead of CI-option, we get only one "mixed" sum rule instead of three (for ^{3}H,^{3}He and ^{4}He) earlier mentioned Foldy-Khokhlov sum rules:

$$\sigma_{-1}(^3H)[2.45mb] = \frac{4\pi^2\alpha}{3}\cdot\langle r_{pp}^2\rangle_{ch}^{^3H}[2.42mb] = \sigma_{-1}(^3He)[2.54\pm.09mb]. \tag{2}$$

The nuclei charge radii in the above relations and in what follows are understood as corresponding to the case of "point-charge-protons" ones which reached by subtracting from "physical" values standard corrections, e.g.[5].

The problem of the radii difference in the charge-symmetry-conjugated 3N-nuclei is of similar degree of complexity as the long-discussed binding energy difference of the ^{3}He and ^{3}H: $E_b(^3H) = 8.52MeV$ and $E_b(^3He) = 7.76MeV$, e.g. [6, 7]. For a qualitative understanding of the intended estimation, we resort to dimensional arguments and factors as follows:

$$\frac{E_b(^3He)}{E_b(^3H)}[.911] \cong \frac{\langle r_{ch}^2\rangle_{^3He}^{-1/2}}{\langle r_{ch}^2\rangle_{^3H}^{-1/2}}[0.899]. \tag{3}$$

The unexpected numerical closeness of two ratios written down on exclusively dimensional ground signals nevertheless that for its full dynamical explanation one should include into consideration not only the Coulomb interaction effects but all collection of the CS-breaking-effects as in [6, 7], that represent on the nuclei and hadronic scales the fundamental current u- and d-quark mass difference.

The modified $\sigma_{-1}(^4He)$ sum rule keeping the basic CS-symmetry limitations can be presented in the form

$$\sigma_{-1}(E1)^{^4He} = \frac{16}{9}\pi^2\alpha(\langle r_c{}^2\rangle_{^4He} + \beta_{pp} - \beta_{pn}). \tag{4}$$

The notations for the $\beta_{NN'}$ through the scalar products of the nucleon radial coordinates were introduced in the paper [1]. For the CS-self-conjugate $^4He_2$-nucleus all $\beta_{pp} = \beta_{nn} \neq \beta_{pn}$ can be computed if the explicit form of the ground state radial wave function is available.

3 Relativistic σ_{-1} sum rule and its application for one-electron atom

In what follows we consider relativistic dipole moment fluctuation sum rules in the "valence-parton" approximation, that is neglecting virtual particle-antiparticle configurations in the ground state of the considered systems or diffractively produced in the final states of photo-absorption reactions. The anomalous magnetic moment sum rules express a model-independent correspondence between static properties of a particle (or bound system of particles) and integrals over the photo-absorption spectrum. For particles with the spin $S = 1/2$ the sum rule for the anomalous magnetic moment κ (or GDH-sum rule) reads [8–10]

$$\frac{2\pi^2\alpha\kappa^2}{m^2} = \int_{thr}^{\infty} \frac{d\nu}{\nu}(\sigma_p(\nu) - \sigma_a(\nu)) \tag{5}$$

where $\sigma_{p(a)}$ refers to the cross-section with the parallel(anti-parallel) spins of the target and photon.

In QED,the validity of the sum rule for free electron was checked in two lowest orders of the perturbation theory [11, 12]. Later on, for the physical reasons, we shall replace κ^2 entering different sum rules just by its integral expression in the GDH sum rule. In particular, for the relativistic dipole moment fluctuation sum rule

$$4\pi^2\alpha[\frac{1}{3} < D^2 > -\frac{\kappa^2}{4m^2}] = \int_{thr}^{\infty} \frac{d\nu}{\nu}\sigma_{tot}(\nu), \tag{6}$$

we obtain another form to be used later

$$4\pi^2\alpha[\frac{1}{3} < D^2 >] = \int_{thr}^{\infty} \frac{d\nu}{\nu}(\sigma_p(\nu). \tag{7}$$

We apply first derived sum rule to the system of the highly ionized atom Pb^{81+}, initiated about half-century ago by J.S. Levinger and co-workers [13]. Using the form of the sum rule with our included term κ_{atom} we reduced deviation between left- and right-hand sides of the sum rule, discovered in earlier works, to one-half percent. Numerically:

$$4\pi^2\alpha\frac{1}{3} < D^2 > [937.2b] - 4\pi^2\alpha(\frac{\kappa}{2M})^2[67.9b] = \int_{thr}^{\infty} \frac{d\nu}{\nu}\sigma_{tot}(\nu)[874b], \tag{8}$$

where for the anomalous magnetic moment of the atom we used the total magnetic moment of electron bound in the lowest S-wave orbit [14].

4 Relativistic constituent quark model and nucleons

Following formally to the $p_z \to \infty$ techniques derivation of the Cabibbo-Radicati [15] or GDH sum rule [16], we can obtain the relation

$$4\pi^2\alpha(\frac{1}{3} < \vec{D}^2 >) = \int \frac{d\nu}{\nu}\sigma_p^{res}(\nu). \tag{9}$$

We use the definitions

$$\hat{D} = \int \vec{x}\hat{\rho}(\vec{x})d^3x = \sum_{j=1}^{3} Q_q(j)\vec{d}_j, \hat{r}_1^2 = \int \vec{x}^2\hat{\rho}(\vec{x})d^3x = \sum_{j=1}^{3} Q_q(j)\vec{d}_j^{\,2}. \quad (10)$$

The defined operators $Q_q(j)$ and $\vec{d}_j$ are the electric charges and configuration and spin variables of point-like interacting quarks in the infinite-momentum frame of the bound system.

Finally, we relate the electric dipole moment correlators consecutively for the proton, the neutron, the pure "isovector-nucleon" and the iso-vector part of the mean-squared radii operators, which all are sandwiched by the nucleon state vectors in the "infinite - momentum frame", with experimentally measurable data on the resonance parts of the photo-absorption cross sections on the proton and neutron presently known below $\sim$ 2 GeV [17]. The listed operator mean values are presented as follows

$$R_V = \frac{1}{2}(< r_1^2 >_P - < r_1^2 >_N) = \alpha - \frac{1}{2}\beta. \quad (11)$$

$$J_P = \frac{1}{3} < \hat{D}^2 >_P = \frac{8}{27}\alpha + \frac{1}{27}\beta + \frac{8}{27}\gamma - \frac{8}{27}\delta. \quad (12)$$

$$J_N = \frac{1}{3} < \hat{D}^2 >_N = \frac{2}{27}\alpha + \frac{4}{27}\beta + \frac{2}{27}\gamma - \frac{8}{27}\delta. \quad (13)$$

$$J_V = \frac{1}{3} < \hat{D}^2 >_V = \frac{2}{3}\alpha + \frac{1}{3}\beta + \frac{2}{3}\gamma - \frac{4}{3}\delta, \quad (14)$$

where $< \vec{d}_1^{\,2} > = < \vec{d}_2^{\,2} > = \alpha, < \vec{d}_3^{\,2} > = \beta, < \vec{d}_1 \cdot \vec{d}_2 > = \gamma, < \vec{d}_1 \cdot \vec{d}_3 > = < \vec{d}_2 \cdot \vec{d}_3 > = \delta$

indices "1" and "2" refer to the like quarks (i.e. to the $u(d)$- quarks inside the proton (neutron)), and "3" to the odd quark. Evaluation of the relativistic electric dipole moment fluctuation and the isovector charge radius sum rules for the nucleon was carried out with the available compilation [17] of the resonance pion-photoproduction data on the proton and neutron with the helicity $\lambda = 1/2(3/2)$ $A_{1/2}^{P(N)}$ and $A_{3/2}^{P(N)}$ and all integrals over photo-excited nucleon resonances were taken in the narrow resonance approximation:

$$J_{p(a)}^{res} \simeq \frac{4\pi m_n |A_{3/2(1/2)}^{res}|^2}{m_{res}^2 - m_n^2}, \quad (15)$$

where $m_{n(res)}$ is the nucleon (or resonance) mass. Solving the system of the linear equations and evaluating the $R_V, J_{P,N,V}$ with the help of experimentally known partial amplitudes of main photo-excited resonances, we find our final results for the numerical values α, β and the opening angle θ_{12} and θ_{13} between vectors $\vec{d}_1$ and $\vec{d}_2$ and vectors $\vec{d}_{1(2)}$ and $\vec{d}_3$:

$$\alpha^{1/2} = 0.75 \pm 0.06 f, \quad (16)$$

$$\beta^{1/2} = 0.77 \pm 0.12 f, \quad (17)$$

$$\theta_{12} \simeq 120^0, \quad (18)$$

$$\theta_{13} \simeq \theta_{23} \sim 120^0, \tag{19}$$

$$< {r_1}^2 >_V = 0.25 \pm 0.02 f^2 [exp : .29 f^2]. \tag{20}$$

Having got the values of the "global" parameters α,β, etc. defined in the infinite momentum frame, we put all three vectors with the pertinent numerical parameters into the plane transverse to $p_z \to \infty$ momentum and define parameters of the closed contour circumscribing their ends. The resulting curve is either ellipse with very small eccentricity or the circle. This fact is, in our opinion, at variance with rather popular "quark-diquark" models of the nucleon with, presumably, more noticeable asymmetry of the spatial correlations of the *uu*- and *ud* - quark pairs. It should be stressed that the accepted parametrization of the integrals $R_V \div J_V$ is corresponding to the CS-breaking option, that is, the cited pairwise closeness of the α and β, or γ and δ may be due to the approximate nature of our numerical estimates. Their actual difference may be connected, e.g., with the mass and coupling constant difference of the charged and neutral pion within the valence quarks of nucleons, that is with the mass-difference of the u- and d-quarks, as the fundamental reason of the charge-symmetry breaking.

5 Concluding remarks

1. The generalization of the non-relativistic Foldy-Khokhlov sum rules is proposed for the "CS-mirror" ^{3}He -^{3}H and the "CS-self-conjugate" ^{4}He nuclei. The introduced corrections replace the charge-independence, (CI)- approximations used earlier.

2. The relativistic current-algebra and dispersion relation based sum rule establishing new relation between characteristics of the electron bound in the hydrogen-like, highly-charged ions is checked numerically.

3. With the accuracy about 15%, the isovector charge radius of the nucleon was calculated via the Cabibbo-Radicati sum rule through the integral of total resonance photo-production of hadrons on the nucleons. The mentioned deficit should be referred to the "pion-sea" presence in the nucleon state vector.

4. Approximately derived values of the parameters defining quark correlations in the nucleon turn out qualitatively close to each other contrary to expectations based on the implementation of the CS-vs-CI-symmetry, as it is seen in pairwise comparison of α-vs-β or γ-vs-δ. Thus, these data don't seemingly be consistent with the asymmetric quark-diquark model of the nucleon.

References

[1] L.L. Foldy, Phys. Rev. **107**, 1303 (1957)
[2] Yu. K. Khokhlov, ZhETF **32**, 124 (1957)
[3] J. Golak, R. Skibinski, W. Glöckle, H. Kamada, A. Nogga, H. Witala, V. D. Efros, V. D. Leidemann, G. Orlandini, Nucl. Phys. A **707**, 365 (2002)
[4] V. N. Fetisov, A. N. Gorbunov, A. T. Varfolomeev, Nucl. Phys. **71**, 305 (1965)
[5] A. Ekström, G. R. Jansen, K. A. Wendt, G. Hagen, T. Papenbrock, B. D. Carlsson, C. Forssen, M. HjorTh-Jensen, P. Navratil, W.Nazarewicz, Phys. Rev. C **91**, 00301(R) (2015)
[6] R. A. Brandenburg, S. A. Coon, P. U. Sauer, Nucl. Phys. A **294**, 305 (1978)
[7] Y. Wu, S. Ishkawa, T. Sasakawa, Phys. Rev. Lett. **64**, 1875 (1990)
[8] S. B. Gerasimov, Sov. J. Nucl. Phys. **2**, 430 (1966)

[9] S. B. Gerasimov, Sov. J. Nucl.Phys. **5**, 902 (1967)
[10] S. D. Drell, A. C. Hearn, Phys. Rev. Lett. **16**, 908 (1966)
[11] S. B. Gerasimov, J. Moulin, Nucl. Phys. B **98**, 349 (1975)
[12] D. A. Dicus, R. Vega, Phys. Lett. B **501**, 44 (2001)
[13] W. B. Payne, J. S. Levinger, Phys. Rev. **101**, 1020 (1956)
[14] S. B. Gerasimov, Czech. J. Phys. **55**, A-2009 (2005)
[15] N. Cabibbo, L. A. Radicati, Phys. Lett. **19**, 687 (1966)
[16] M. Hosoda, K. Yamamoto, Prog. Theor. Phys. **36**, 425 (1966)
[17] K. A. Olive, et al. (Particle Data Group), Chin. Phys. C **38**, 090001 (2014)

EPJ Web of Conferences **138**, 01019 (2017) DOI: 10.1051/epjconf/201713801019
Baldin ISHEPP XXIII

Comparative analysis of the pion-nucleus scattering within the microscopic Folding and the local Kisslinger type potentials

Valery Lukyanov[1,★] *Elena* Zemlyanaya[1], *Konstantin* Lukyanov[1], and *Ibrahim* Abdul-Magead[2]

[1] *Joint Institute for Nuclear Research, Dubna, Russia*
[2] *Cairo University, Giza, Cairo, Egypt*

Abstract. Elastic scattering cross sections are calculated and compared to the data of $\pi^{\pm}$ + ^{28}Si and ^{40}Ca at energies 130, 180, and 230 MeV by using the both microscopic optical potentials (OP), the high-energy folding and the local Kisslinger type potentials. In the folding OP, we use the known nuclear density distributions while parameters of the elementary πN-scattering amplitude are fitted to the data with the aim to estimate the in-medium effect on pions scattered on bound nucleons. As to the local modified Kisslinger OP its parameters are known from the earlier studies. The cross sections for both OPs are calculated by solving the Klein-Gordon wave equation, and thus the relativistic and distortion effects on the process are accounted for exactly. A fairly well agreement with experimental data was obtained and the decisive role of a surface region of potentials was established where both OPs occur in close coincidence.

1 Introduction

In our theoretical study of π-nucleus scattering we apply two models. In *the first* one we use the optical potential (OP) from [1] where it was obtained as the local OP in a wave equation for the $\psi(r)$-function after the Krell-Ericson transformation [2] of the ordinary $\phi(r)$ wave function of an ordinary wave equation with the non-local Kisslinger OP $U = a(r) + \nabla b(r)\nabla$ [3]. The 12 parameters of this local OP were obtained in [1] by fitting the calculated π-A differential cross section to the data measured at energies in the region of the 3-3 pion-nucleon resonance. These $\phi(r)$ and $\psi(r)$ functions are shown to be phase equivalent solutions of the respective wave equations. *The second model* of OP [4] corresponds to the eikonal phase of the Glauber high-energy approximation [5] and, in fact, becomes the folding integral of the πN amplitude and form factor of the nuclear density distribution function. For the pion-nucleus scattering this OP was applied in [6], and unlike the Glauber approach for the scattering amplitude where one uses integration along the straight line trajectory of motion, in our calculations we solve the Klein-Gordon relativistic wave equation and thus relativistic and distortion effects are accounted for exactly. For both OPs, we compare the calculated differential cross sections with experimental data and thus obtain forms of respective potentials to establish the meaningful regions of nuclei that reveal themselves in the pion scattering.

The aims of our study are to explain the experimental data in the region of πN 3-3-resonance energies, to compare the calculated cross sections for both models, and to estimate the "in-medium"

★e-mail: vlukyanov@theor.jinr.ru

effect on the elementary πN scattering amplitude. The parameters of the latter are obtained by fitting the calculated πA differential cross sections to the respective experimental data on elastic scattering. The established best-fit "in-medium" πN parameters are compared with those for the corresponding "free" πN scattering amplitudes. Application is presented for elastic scattering of pions on nuclei ^{28}Si and ^{40}Ca at energies 130, 180, and 230 MeV.

2 Basic equations

The microscopic folding OP is taken as done in [4]

$$U(r) = -\frac{(\hbar c)\beta_c}{2(2\pi)^3}\sigma\,[i+\alpha]\cdot\int e^{-i\mathbf{qr}}\rho(q)\,f(q)\,d^3q, \tag{1}$$

where β_c is the speed of light, $f(q)$ – form factor of the πN-amplitude $F_{\pi N}(q)$, and $\rho(q)$ is the form factor of a nuclear density distribution

$$\rho(q) = \int e^{i\mathbf{qr}}\,\rho(r)\,d^3r. \tag{2}$$

Here we use the charge-independent principle which enables one to apply the averaged πN-amplitude

$$F_{\pi N}(q) = \frac{k}{4\pi}\sigma\,[i+\alpha]\cdot f(q), \qquad f(q) = e^{-\beta q^2/2}, \tag{3}$$

dependent on 3 parameters instead of 12 parameters needed for the separate $\pi^\pm p$- and $\pi^\pm n$- amplitudes. So, σ is the total cross section, α, the ratio of real to imaginary part of the amplitude at forward angles, and β, the slope parameter. In calculations of the folding OP (1) we take the nuclear density distribution $\rho(r)$ in the form of a symmetrized Fermi-function, for which the form factor $\rho(q)$ can be presented in an analytic form, and thus

$$\rho_{SF}(r) = \rho_0\frac{\sinh(R/a)}{\cosh(R/a)+\cosh(r/a)}, \quad \rho_0 = \frac{A}{1.25\pi R^3}\left[1+(\frac{\pi a}{R})^2\right]^{-1}, \tag{4}$$

$$\rho_{SF}(q) = -\rho_0\frac{4\pi^2 aR}{q}\frac{\cos qR}{\sinh(\pi a q)}\left[1-\left(\frac{\pi a}{R}\right)\coth(\pi a q)\tan qR\right]. \tag{5}$$

In our applications of the folding optical potential (1), the radius R and diffuseness parameters a of the target nuclei ^{32}Si and ^{40}Ca are taken from the respective electron-nucleus scattering data. In the case of the folding potential (1), when comparing calculated cross sections to experimental data we will fit three parameters σ, α, and β, and thus the obtained magnitudes of them will characterize the amplitude of scattering of pions on the bound nucleons, i.e. the so-called "in-medium" effect.

As to *the local model of Kisslinger – type OP* developed in [1] it was realized in the form

$$U(r) = \frac{(\hbar c)^2}{2E}\frac{1}{1-\alpha(r)}\Big\{q(r) - k^2\alpha(r) - (1/2)\nabla^2\alpha\Big[1+\frac{(1/2)\nabla^2\alpha}{1-\alpha}\Big]\Big\}, \tag{6}$$

where the first term arises from the s- and p-wave πN interaction, while the other smaller terms are from the p-wave alone, and

$$q(r) = -4\pi[\mathbf{b_0}\rho(r) \mp \mathbf{b_1}\Delta\rho(r)]/p_1 + \delta q, \tag{7}$$

$$\alpha(r) = 4\pi[\mathbf{c_0}\rho(r) \mp \mathbf{c_1}\Delta\rho(r)]/p_1 + 4\pi[\mathbf{C_0}\rho^2(r) \mp \mathbf{C_1}\rho(r)\Delta\rho(r)]/p_2, \tag{8}$$

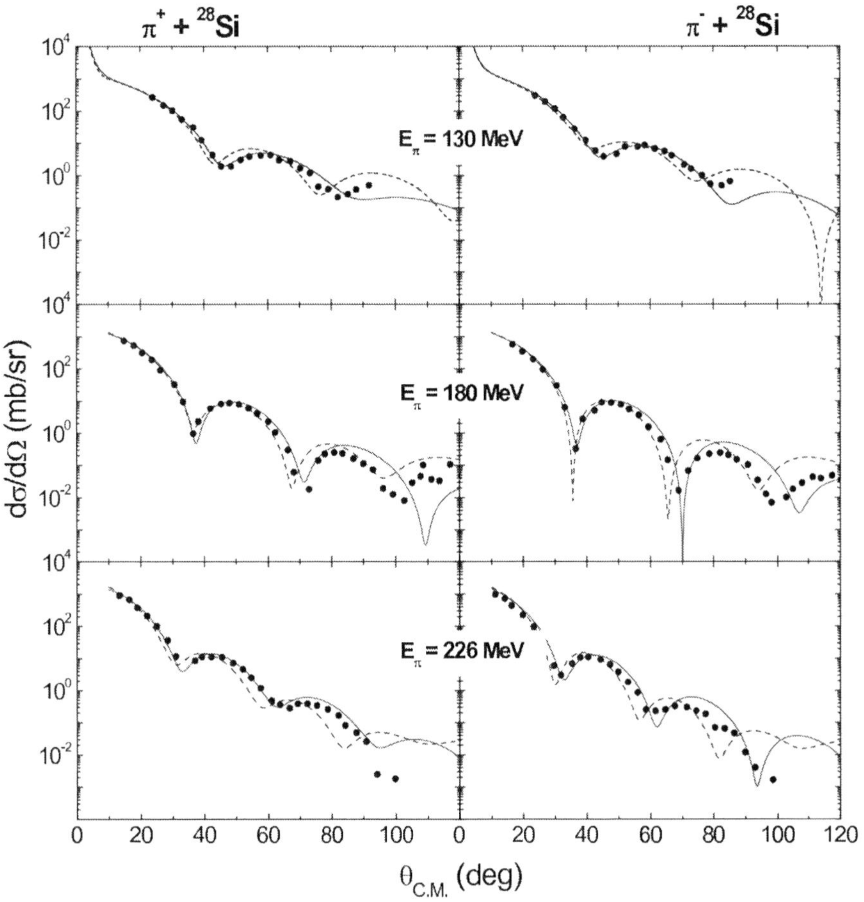

Figure 1. Comparisons of the calculated $\pi^{\pm} + {}^{28}Si$ elastic scattering cross sections at the pion energies 130, 180, and 230 MeV with experimental data from [7]. Solid (red) curves correspond to the folding OP, and dashed (blue) curves are for the local Kisslinger-type OP. The best fit "in-medium" parameters σ, α, β of the πN scattering amplitude are from table 1.

$$\delta q(r) = -2\pi\epsilon\nabla^2\{[\mathbf{c_0}\rho(r) \mp \mathbf{c_1}\Delta\rho(r)]/p_1 + \frac{1}{2}[\mathbf{C_0}\rho^2(r) \mp \mathbf{C_1}\rho(r)\Delta\rho(r)]/p_2]\}. \tag{9}$$

Here $\mathbf{b_{0,1}}$, $\mathbf{c_{0,1}}$, $\mathbf{C_{0,1}}$ are the complex parameters fitted, e.g. in [1] as functions of an energy. Then, E is the total pion energy in c.m system, $p_1 = 1 + \epsilon$, $p_2 = 1 + (1/2)\epsilon$ with $\epsilon = E/Mc^2$ and Mc^2 the nucleon mass.

The differential cross sections are calculated as done in [8] by solving the Klein-Gordon equation in its form at conditions $E \gg U$ (below $\hbar = c = 1$)

$$\left(\Delta + k^2\right)\psi(\vec{r}) = 2\bar{\mu}U(r)\psi(\vec{r}), \quad U(r) = U^H(r) + U_C(r). \tag{10}$$

Here k is the relativistic momentum of pions in center-of-mass (c.m.) system,

$$k = \frac{M_A k^{lab}}{\sqrt{(M_A + m_\pi)^2 + 2M_A T^{lab}}}, \quad k^{lab} = \sqrt{T^{lab}\,(T^{lab} + 2m_\pi)}, \tag{11}$$

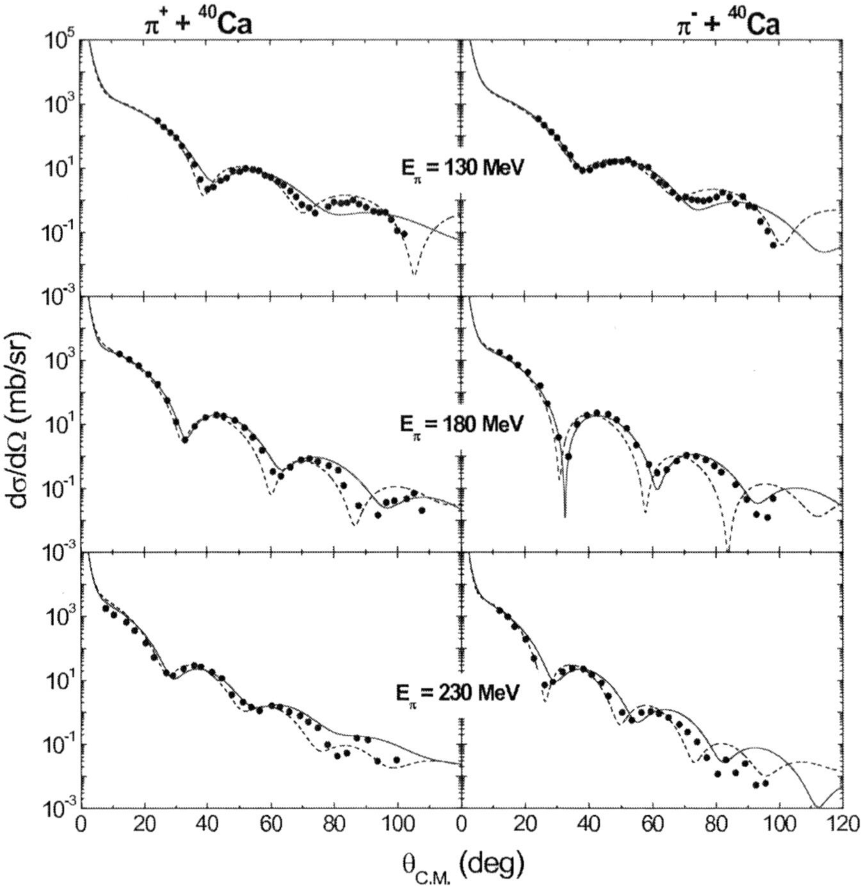

Figure 2. The same as in figure 1 but for the elastic scattering of $\pi^\pm$ + ^{40}Ca. Experimental data are from [9].

and $\bar{\mu} = \frac{EM_A}{E+M_A}$ is the relativistic reduced mass, $E = \sqrt{k^2 + m_\pi^2}$ – total energy, m_π and M_A are the pion and nucleus masses.

In the case of the folding potential (1), when comparing calculated pion-nucleus cross sections to experimental data, the fitted parameters σ, α, and β have a meaning of the so-called "in-medium" parameters for the amplitude of scattering of pions on the bound nucleons. The fitting procedure of their minimization was realized in [10] and [11] for the χ^2-function

$$\chi^2 = f(\sigma, \alpha, \beta) = \sum_i \frac{(y_i - \hat{y}_i(\sigma, \alpha, \beta))^2}{s_i^2}, \qquad (12)$$

where $y_i = \dfrac{d\sigma}{d\Omega}$ and $\hat{y}_i = \dfrac{d\sigma}{d\Omega}(\sigma, \alpha, \beta)$ are, respectively, experimental and theoretical differential cross sections, s_i are experimental errors.

3 Results of calculations

Figures 1 and 2 demonstrate a reasonable agreement of the calculated and experimental differential cross sections of elastic scattering of $\pi^\pm$-mesons on nuclei ^{32}Si and ^{40}Ca at energies 130, 180, and

Table 1. The best-fit parameters σ, α, β and respective χ^2/k quantities with k, the number of experimental points at $\Theta_{C.M.}$ between 0 and 80 degrees ([10]).

reaction	T^{lab}, MeV	σ, fm^2	α	β, fm^2	χ^2/k
$\pi^- + {}^{28}$Si	130	7.22	0.84	0.78	0.9
$\pi^+ + {}^{28}$Si		9.59	0.11	0.83	2.7
$\pi^- + {}^{40}$Ca		6.86	0.90	0.88	3.9
$\pi^+ + {}^{40}$Ca		8.21	0.10	0.76	14.4
$\pi^- + {}^{28}$Si	180	9.33	0.43	0.28	6.0
$\pi^+ + {}^{28}$Si		7.75	0.76	0.49	1.2
$\pi^- + {}^{40}$Ca		9.75	0.21	0.27	3.0
$\pi^+ + {}^{40}$Ca		5.75	1.09	0.69	2.9
$\pi^- + {}^{28}$Si	226	7.43	0.60	0.17	16.2
$\pi^+ + {}^{28}$Si		9.99	-0.13	0.12	21.5
$\pi^- + {}^{40}$Ca	230	5.34	0.79	0.23	6.4
$\pi^+ + {}^{40}$Ca		9.17	-0.10	0.24	9.6

230 MeV. Experimental data are taken from [7] and [9]. In the case of folding OP (solid red curves) we used the nuclear point-density-distributions (for centers of nucleons) in the form of a symmetrized fermi-function (4) with the radius R and diffuseness a parameters R = 3.134 fm and a = 0.477 fm for ^{28}Si, and R = 3.593 fm and a = 0.493 fm for ^{40}Ca (see [12]). As to the"in-medium" parameters σ, α, β of a πN scattering amplitude they have been fitted in [10] and [11], and the corresponding χ^2 magnitudes per one experimental point are presented in table 1 for the region of scattering angles up to 80 degrees. The "in-medium" effect is revealed more clearly when one compares the"in-medium" parameters to the respective parameters for the pion scattering on "free" nucleons done in [13]. Thus, to express the "in-medium" effect more markedly we show in figure 3 the differences of corresponding "in-medium" and "free" parameters as shadowed regions by blue color.

As for the cross sections calculated within the local Kisslinger-type optical potentials, they are shown in Figures 1 and 2 by dotted blue curves. The respective 12 parameters of OPs have already been fitted in [1] in dependence of an energy, and the full nuclear density parameters in the form of a symmetrized fermi-function (4) were taken as R = 3.134 fm and a = 0.537 fm for ^{28}Si from [14], and R = 3.53 fm and a = 0.562 fm for ^{40}Ca .Their applications for our case lead to about the same χ^2 magnitudes that were obtained for the folding OPs. Thus one can conclude that in spite of

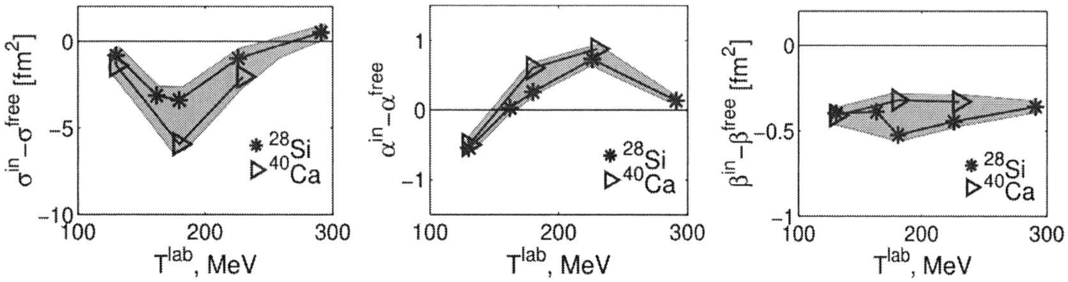

Figure 3. The behavior of differences of the fitted "in-medium" and "free" parameters of the πN scattering amplitude (3) at the energy region of 3-3-resonance.

DOI: 10.1051/epjconf/201713801019

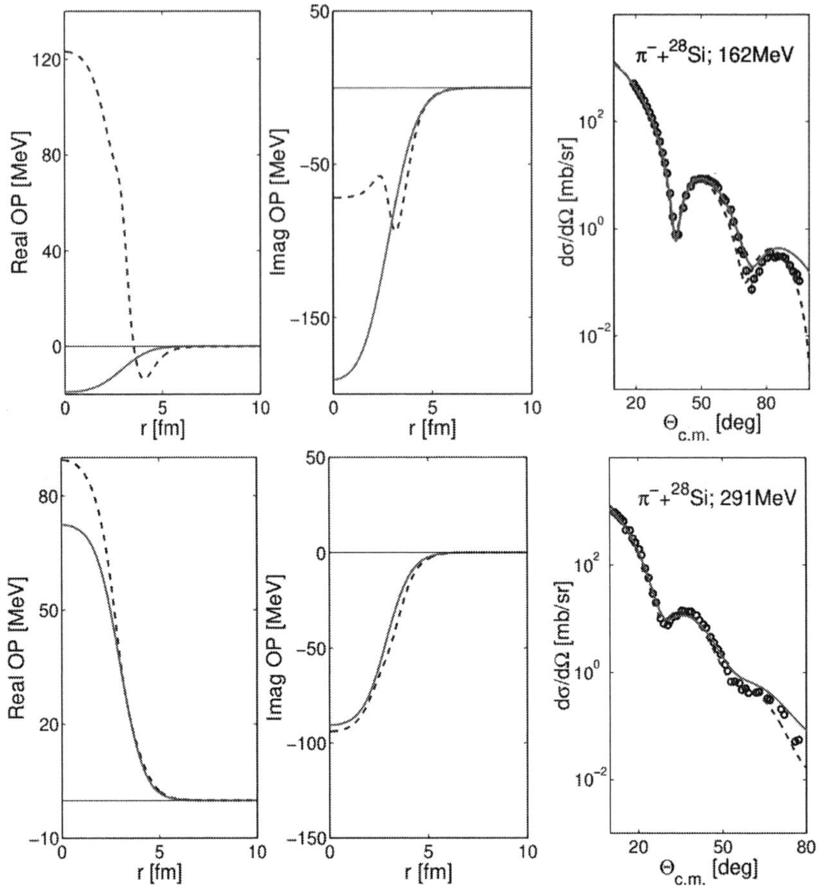

Figure 4. Comparisons of the folding (solid, red curves) and local Kisslinger-type (dashed, blue) optical potentials, and the corresponding cross sections for π^+-^{28}S scattering at T_{lab}=162 and 291 Mev.

the so different constructions of employed potentials and sometimes different behavior of them as function of r, both OPs yield cross sections in reasonably good agreement with the experimental data at energies of 3-3 resonance. However this does not mean that the forms of both OPs should be close to each other. Indeed, as an example, in figure 4 we show the optical potentials and corresponding cross sections at different energies 162 and 291 MeV for the π^- scattering on ^{32}Si. One can see that for each energy the calculated cross sections for both OPs are in a good coincidence. At the same time, at lower energy 162 MeV the real part of a local Kisslinger-type potential is repulsive in the inner part of a nucleus, while its imaginary part is negative and almost coincides to the folding OP in a surface region. On the other hand, at higher energy 291 MeV both OPs have similar forms for both real and imaginary parts, and have no any variations as function of r. So, one can conclude that behavior of the imaginary part of OP in its surface region plays a decisive role in explanation of experimental data at the first quarter of angles.

4 Summary

One can sum up that our approach based on 3-parameter OP, allows one to explain fairly well the experimental data on pion-nucleus elastic scattering. Moreover, we show that both microscopic folding and local Kisslinger-type optical potentials provide a good agreement with the experimental data of pion-nucleus elastic scattering at intermediate energies in the region of the πN 3-3 resonance between 130 and 230 MeV. Such an agreement takes place in spite of the fact that the Kisslinger-type potential has some variations in its inner part. At the same time both potentials are close to each other in the surface region and thus one can conclude that this region plays a decisive role in the mechanism of scattering. Comparisons of the πN cross section σ^{free} with σ^{in} at 3-3-resonance region show that the pion interaction with bound nucleons in nuclear matter is weaker than in the case of free nucleons. The behavior of the parameter α indicates that the refraction process increases at larger energies $T^{lab} \geq 170$ MeV.

In general, when modeling the elastic scattering data alone we face the ambiguity problem when several approaches explain the same amount of data. For example, in [15] just the same data as applied in figures 1,2 have been successfully explained basing on the optical potential that corresponds to the first and second order parts of multiple scattering theory and fitting the phenomenological ρ^2-dependent term added to account for, say, the pion collisions with the couples of nucleons in nuclei. In this connection, the data on inelastic πA-scattering may help us to reduce this ambiguity. Indeed, in the models of such processes one applies the derivative of the already fitted elastic scattering OP, that reveals itself mainly in the surface region. So, in [16] the data on the inelastic scattering cross sections with excitations of 2^+ and 3^- collective states of nuclei were explained fairly well basing on the known folding OPs and on the only additional parameter, the deformation of a nuclear surface.

References

[1] M.B. Johnson and G.R. Satchler, Ann. Phys. **248**, 134 (1996)
[2] M. Krell and T.E.O. Ericson, Nucl. Phys. B **11**, 521 (1969)
[3] L.S. Kisslinger, Phys. Rev. **98**, 761 (1955)
[4] V.K. Lukyanov, E.V. Zemlyanaya, K.V. Lukyanov, Phys. Atom. Nucl. **69**, 240 (2006)
[5] R.J. Glauber, *Lectures in Theoretical Physics* (Interscience, New York, 1959) 315
[6] V.K. Lukyanov, E.V. Zemlyanaya, K.V. Lukyanov, Phys. Atom. Nucl. **77**, 100 (2014)
[7] B.M. Preedom et al., Nucl. Phys. A **326**, 385 (1979)
[8] V.K. Lukyanov, E.V. Zemlyanaya, K.V. Lukyanov, K.M. Hanna, Phys. Atom. Nucl. **73**, 1443 (2010)
[9] P. Gretillat et al., Nucl. Phys. A **364**, 270 (1981)
[10] V.K. Lukyanov, E.V. Zemlyanaya, K.V. Lukyanov, E.I. Zhabitskaya, M.V. Zhabitsky, Preprint JINR P4-2012-105, Dubna, 2012
[11] V.K. Lukyanov, E.V. Zemlyanaya, K.V. Lukyanov, E.I. Zhabitskaya, M.V. Zhabitsky, Phys. Atom. Nucl. **77**, 100 (2014)
[12] V.K. Lukyanov, E.V. Zemlyanaya, B. Słowinski, Phys. Atom. Nucl. **67**, 1282 (2004)
[13] M.P. Locher, O. Steinmann, N. Straumann, Nucl. Phys. B **27**, 598 (1971)
[14] H. De Vries, C.W. De Jager, C. De Vries, Atomic Data and Nucl. Data Tables **36**, 495 (1987). Nucl. Phys. B **27**, 598 (1971)
[15] M. Gmitro, S.S. Kamalov, R. Mach, Phys. Rev. **36**, 1105 (1987)
[16] V.K. Lukyanov, E.V. Zemlyanaya, K.V. Lukyanov, I.A.M. Abdul-Magead, Phys. Atom. Nucl. **79**, 670 (2016)

Investigation of the compressed baryonic matter at the GSI accelerator complex*

V.P. Ladygin[1,**], *T.O.* Ablyazimov[1], *P.G.* Akishin[1], *E.P.* Akishina[1,2], *V.P.* Akishina[1], *M.I.* Baznat[1,3], *I.V.* Boguslavsky[1], *V.N.* Borshchov[4], *A.V.* Bychkov[1], *D.V.* Dementiev[1], *O.Yu.* Derenovskaya[1], *V.V.* Elsha[1], *K.K.* Gudima[1,3], *Yu.V.* Gusakov[1], *S.N.* Igolkin[5], *A.Yu.* Isupov[1], *V.V.* Ivanov[1], *G.D.* Kekelidze[1], *A.N.* Khrenov[1], *P.I.* Kisel[1,2], *M.G.* Korolev[6], *G.E.* Kozlov[1,7], *V.A.* Kramarenko[1], *P.K.* Kurilkin[1], *N.B.* Ladygina[1], *V.M.* Lysan[1], *A.I.* Malakhov[1], *I.P.* Martinovsky[8], *M.M.* Merkin[1,6], *Yu.A.* Murin[1], *S.S.* Parzhitsky[1], *V.A.* Penkin[1], *S.M.* Piyadin[1], *M.A.* Protsenko[1], *A.A.* Savenkov[1], *A.V.* Shabunov[1], *A.I.* Shafranovskaya[1], *A.D.* Sheremetiev[1], *O.G.* Tarasov[1], *I.T.* Tymchuk[4], *N.I.* Zamiatin[1], *A.I.* Zinchenko[1], and *E.V.* Zubarev[1]

[1]*Joint Institute for Nuclear Research, Dubna, Russian Federation*
[2]*Institut für Kernphysik, Goethe Universität Frankfurt, Frankfurt, Germany*
[3]*Institute of Applied Physics of ASM, Kishinev, Moldova*
[4]*National Science Center, Kharkov Institute of Physics and Technology, Kharkov, Ukraine*
[5]*State University of St.Petersburg,, St.Petersburg, Russian Federation*
[6]*Skobeltsyn Institute of Nuclear Physics of Moscow State University, Moscow, Russian Federation*
[7]*Frankfurt Institute for Advanced Studies, Goethe-Universität Frankfurt, Frankfurt, Germany*
[8]*State Enterprise PLANAR, Minsk, Belarus*

Abstract. The Compressed Baryonic Matter (CBM) experiment at FAIR will play a unique role in the exploration of the QCD phase diagram in the region of high net-baryon densities, because it is designed to run at unprecedented interaction rates. High-rate operation is the key prerequisite for high-precision measurements of multi-differential observables and of rare diagnostic probes which are sensitive to the dense phase of the nuclear fireball. The goal of the CBM experiment at SIS100 ($\sqrt{s_{NN}}$ = 2-4.9 GeV) is to discover fundamental properties of QCD matter, namely, the equation-of-state at high density as it is expected to occur in the core of neutron stars, effects of chiral symmetry, and the phase structure at large baryon-chemical potentials ($\mu_B \geq 500$ MeV).
We are focusing here on the contribution of JINR to the CBM experiment: design of the superconducting dipole magnet; manufacture of the straw and micro-strip silicon detectors, participation in the data taking and analysis algorithms and physics program.

1 Introduction

The investigation of nuclear matter at extreme conditions, i.e. at high temperatures and/or at high baryon densities, is one of the most challenging fields of modern physics. The major efforts are

*Dedicated to the memory of Prof. Yu.V. Zanevsky and Prof. V.D. Peshekhonov
**e-mail: vladygin@jinr.ru

devoted to the exploration of the phase diagram of strongly interacting matter using heavy-ion collisions at relativistic energies. While the experiments at the Relativistic Heavy Ion Collider (RHIC) at BNL and at the Large Hadron Collider (LHC) at CERN focus on the study of high temperatures, the Compressed Baryonic Matter (CBM) experiment at the future FAIR accelerator will concentrate on the investigation of highest baryon densities at still moderate temperatures. In particular, the CBM research program aims at the exploration of the structure of high density matter including the question of deconfinement and chiral phase transitions. The theoretical description of physics at high net baryon densities within the fundamental theory of strong interaction, Quantum Chromodynamics (QCD), is still strongly evolving and the scientific progress in "strong" nonperturbative QCD is driven by new experimental data. Figure 1 illustrates the phases of strongly interacting matter and their boundaries in a diagram of temperature versus baryon chemical potential [1]. The CBM experiment will enter a new era with diagnostic probes never measured before in the FAIR energy range, and thus has a unique research potential.

The high-intensity heavy-ion beams of the future FAIR accelerators, together with the planned Compressed Baryonic Matter (CBM) experiment, offer excellent possibilities to produce and to investigate baryonic matter at highest densities in the laboratory. The research program comprises the study of the structure and the equation-of-state of baryonic matter at densities comparable to the ones in the inner core of neutron stars. This includes the search for the phase boundary between hadronic and partonic matter, the critical endpoint, and the search for signatures for the onset of chiral symmetry restoration at high net-baryon densities [2, 3].

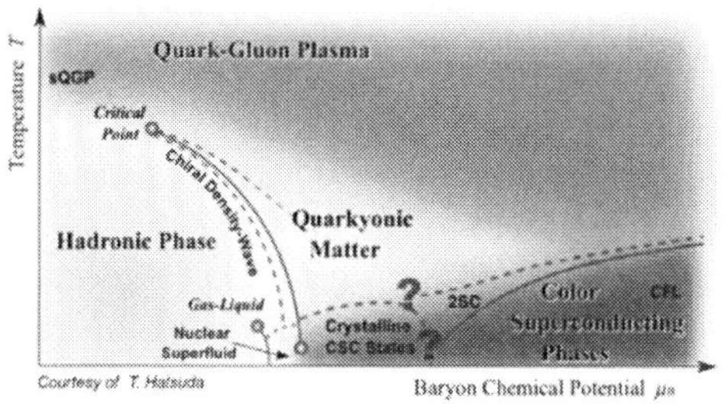

Figure 1. Sketch of the phase diagram for strongly interacting matter [1].

Figure 2. The CBM experimental setup together with HADES [4].

The most promising probes to study the high density QCD matter at SIS100 energies are the flow of identified particles, event-by-event fluctuations, strangeness production including hypernuclei and metastable strange objects, dileptons, charm production near threshold.

The collective flow of hadrons is driven by the pressure gradient created in the early fireball and provides information on the dense phase of the collision. The CBM experiment will therefore dramatically improve the data situation by measuring the the flow of identified particles in the FAIR energy range, including multi-strange hyperons and dileptons.

Event-by-event fluctuations of conserved quantities such as baryon number, strangeness and electrical charge can be to provide insight into the properties of matter created in high-energy nuclear collisions. Lattice QCD calculations suggest that higher moments of these distributions are more sensitive to the phase structure of the hot and dense matter created in such collisions.

The important probes of the excited medium created in heavy-ion collisions are the particles containing strange quarks. The expected particle yields are sufficient to study with unprecedented statistical significance the production and propagation of heavy strange and anti-strange baryons up to Ω^+ hyperons in dense nuclear matter. Also excited hyperon states can be identified. The CBM experiment at SIS100 will also measure hydrogen and helium hypernuclei with large statistics. The discovery of (double-) Lambda hypernuclei and the determination of their lifetimes will provide information on the hyperon-nucleon and hyperon-hyperon interactions, which are essential ingredients for the understanding of the nuclear matter equation-of-state at high densities, and, hence, of the structure of neutron stars.

Dileptons emitted in collisions of heavy ions offer the unique opportunity to investigate the microscopic properties of strongly interacting matter. Virtual photons are radiated off during the whole time evolution of a heavy- ion collision. Once produced, they decouple from the collision zone and materialize as muon or electron pairs. A very important part of the CBM research program is the possibility to study both di-electron and di-muon pairs production.

Charm production will be studied for the first time at beam energies close to production threshold. At these energies, the formation time of charmonium is small compared to the lifetime of strongly interacting system created in high-energy nuclear collisions. Therefore, CBM will study the interactions between fully formed J/Ψ and the dense medium with appropriate counting statistics and systematics. Systematic measurements of charmonium in p+A collisions with varying target mass number A at proton energies up to 30 GeV will shed light on the charmonium interaction with cold nuclear matter and constitute an important baseline for measurements in nuclear collisions.

2 CBM detector

The CBM detector has been designed as a multi- purpose device which will be capable to measure hadrons, electrons and muons in elementary nucleon and heavy-ion collisions over the full FAIR beam energy range. In order to perform high-precision multi-differential measurements of rare probes the experiment should run at event rates of 100 kHz up to 10 MHz for several months per year. To filter out weakly decaying particles like hyperons or D mesons, no simple trigger signal can be generated. Instead, the full events have to be reconstructed, and the decay topology has to be identified online by fast algorithms running on a high-performance computing farm hosted by the GSI Green IT cube. To utilize maximum rates, the data acquisition is based on self-triggered front-end electronics.

The CBM experimental setup together with HADES [4] is shown in figure 2. Each setup has its own target. During HADES operation, the beam will be stopped by a beam dump in front of the CBM detector, comprises the following components:

- a Superconducting Dipole Magnet,
- a Micro Vertex Detector (MVD) consisting of four layers of silicon monolithic active pixel sensors,
- a Silicon Tracking System (STS) based on double- sided silicon micro-strip sensors arranged in eight stations inside the magnetic field,
- a Time-of-Flight wall (TOF) based on Multi- Gap Resistive Plate Chambers (MRPC) with low-resistivity glass essential for high-rate operation,
- a Ring Imaging Cherenkov (RICH) detector comprising a CO_2 radiator, and a UV photon detector realized with multi-anode photomultipliers for electron identification,
- a Transition Radiation Detector (TRD) for pion suppression, particle tracking, and identification using specific energy loss,
- a Muon Chamber (MuCh) system for muon identification consisting of a set of gaseous micro-pattern chambers sandwiched between hadron absorber plates made of graphite and iron,

- a Projectile Spectator Detector (PSD) for event characterization,
- a First-Level-Event-Selection (FLES) system for online event reconstruction and selection.

For muon measurements, the RICH will be exchanged by the MuCh which is shown in a parking position to the right of the beam axis in figure 2. The preparation of the experiment is well advanced. According to the schedule, the CBM experiment will be ready to take the first beams from SIS100 in 2022.

3 JINR contribution to CBM

The contribution of JINR to the CBM experiment is the following: design of the superconducting dipole magnet; manufacture of the straw and micro-strip silicon detectors, participation in the data taking and analysis algorithms and physics program at FAIR [5].

3.1 Superconducting Dipole Magnet

The H-type superconducting magnet of the CBM experiment is designed to provide a vertical magnetic field with bending power of 1 T·m over a length of 1 m from the target. A perspective view of the magnet with the support is presented in figure 3. The magnet yoke consists of two conical poles and two horizontal and two vertical parts to close a flux return. Two cylindrical superconducting coils will be installed in their respective cryostats and will be cooled separately by helium bath based cooling system like the SAMURAI magnet at RIKEN [6]. The maximum energy storage is evaluated as 5.15 MJ when the operating current of the CBM magnet is rated at 686 A. The design of coil for the CBM magnet is based on the design of the Super-FRS dipole [7]. Each coil has 1749 turns and 53 layers. The superconducting wire is a Nb/Ti wire. The conductor insulation consists of 0.1 mm polyimide type and 0.2 mm glass fiber (G10) material. Each layer is additionally insulated from the others by three layers of 0.1 mm fiber glass fabric with epoxy resin. The detailed description of the CBM magnet can be found in [8].

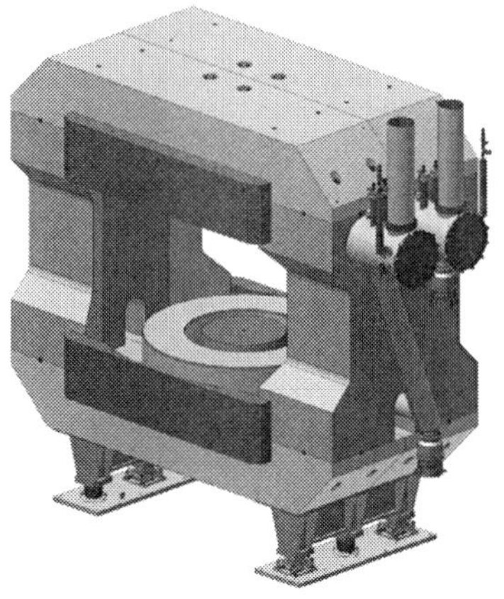

Figure 3. Perspective view of the CBM dipole magnet with the support.

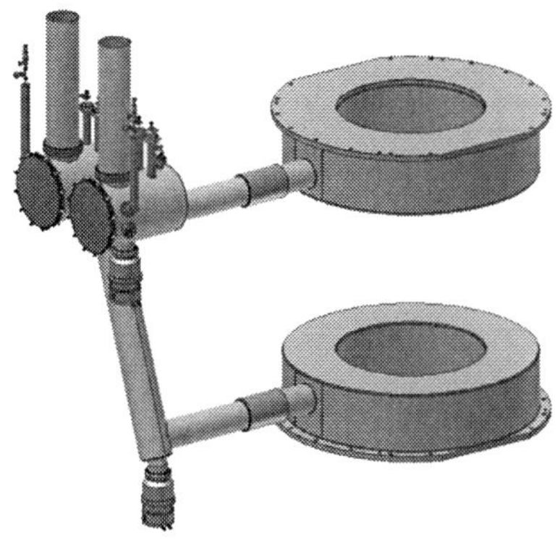

Figure 4. 3D model of cryostats with the feed boxes.

A full 3D model of CBM superconducting dipole magnet was completed by the feed boxes for the top and bottom superconducting coils presented in figure 4. Each of the sections includes an inner

 DOI: 10.1051/epjconf/201713801020

helium vessel, thermal shield and vacuum vessel. The vacuum vessel has flanges for connecting to the external cryogenic line, the turbo molecular pump, two 1000 A current leads, the multipin feedthrough with plug connectors, the safety valve of the vessel and the warm multi-purpose helium line (quench gas collection). Also, the vacuum vessel and the heat shield have a large diameter access hole to connect pipes of the external cryogenic line with helium pipes of coils. All these connections are made by means flexible metal hoses.

Internal helium vessel containing 10 l of liquid helium has four flanges. The high voltage feedthrough CF40HC8-CE-SST95 is connected to the top flange. It is necessary for passing superconducting wires with vaporized helium to current leads. The second flange is for the multi-pin feedthrough with plug connectors. The third flange is connected to the helium return line. The last flange is connected through built-in safety valve to the warm multi-purpose helium line (for quench gas collection) [9].

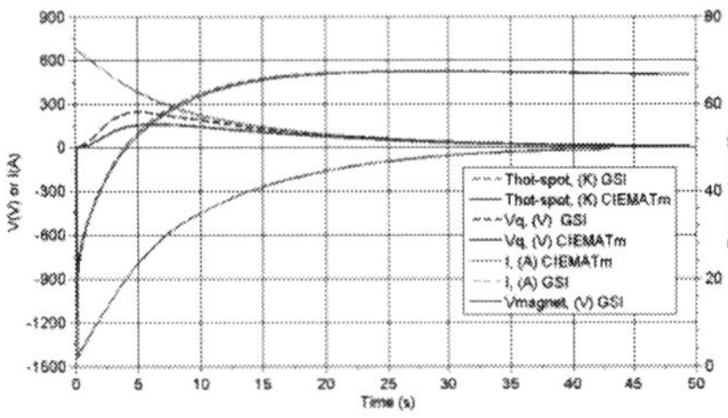

Figure 5. 3D quench simulation for the CBM magnet: magnet current, coil voltage, magnet voltage and "hot spot" temperature. ($R_d = 2.1\Omega$).

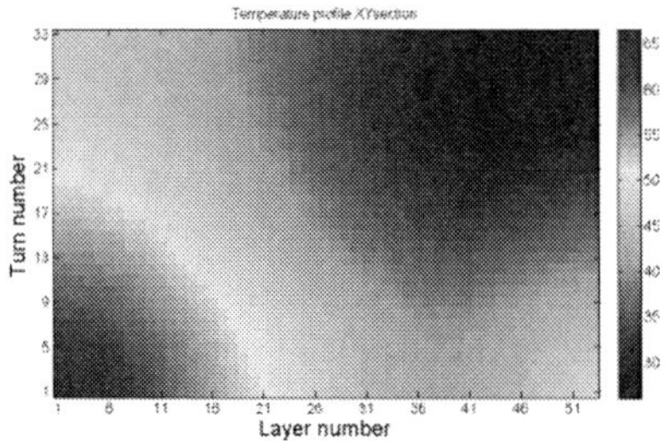

Figure 6. 3D quench simulation for the CBM magnet: temperature distribution in the coil cross section. ($R_d = 2.1\Omega$).

The 3D quench calculations have been performed for two quench protection schemes. One of them is based on the extraction of the energy stored in CBM magnet via external dump resistor of 1.9-2.1 Ohm. In this case around 80-86 % of the energy stored in the magnet (~5 MJ) will be dissipated outside of the cryostat [10, 11]. Figure 5 demonstrates magnet current, coil voltage, magnet voltage and "hot spot" temperature when the 2.1 Ohm dump resistor is used. The maximum "hot-spot" temperature is about 70 K. Figure 6 represents the temperature distribution in the cross section of the coil winding. The other quench protection scheme is based on the heating of the superconducting coil [12]. The quench protection system based on the energy extraction via dump resistor with resistance of 1.9-2.1 Ohm was chosen for CBM magnet due to the lower temperature gradient in the coil.

The electromagnetic calculation has been performed for the CBM dipole magnet with the distance between the poles of 1550 mm, what is 150 mm larger than required by TDR [8]. The bending power of enlarged magnet is about 0.88 T·m with the nominal operating current. The current must be increased by ~25% to reach the field integral of ~1 T·m [8]. However, in this case the reaction forces in the coil supports will increase by ~38%, what would require to change significantly the magnet design [13]. The electromagnetic calculation for the distance between the poles of 1440 mm is in progress.

3.2 Straw chambers for muon detector

The experimental challenge for muon measurements in heavy-ion collisions at FAIR energies is to identify low-momentum muons in an environment of high particle densities. The CBM concept is to track the particles through a hadron absorber system, and to perform a momentum- dependent muon identification.

This concept is realized by segmenting the hadron absorber in several layers, and placing triplets of tracking detector planes in the gaps between the absorber layers. The absorber/detector system is placed downstream of the Silicon Tracking System which determines the particle momentum. In order to reduce meson decays into muons the absorber/detector system has to be as compact as possible. The final design of the muon detector system for SIS300 consisting of 6 hadron absorber layers (60cm carbon and iron plates of 2×20 cm, 30 cm, 35 cm and 100 cm thickness) and 18 gaseous tracking chambers located in triplets behind each absorber slab is shown in figure 7. The identification of a muon depends on its momentum which varies with the mass of the vector mesons and with beam energy.

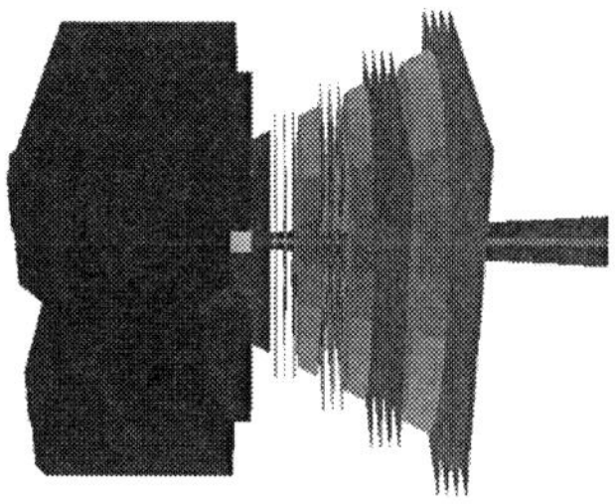

Figure 7. SIS100 CBM muon system configuration with straw tubes in two last detector stations.

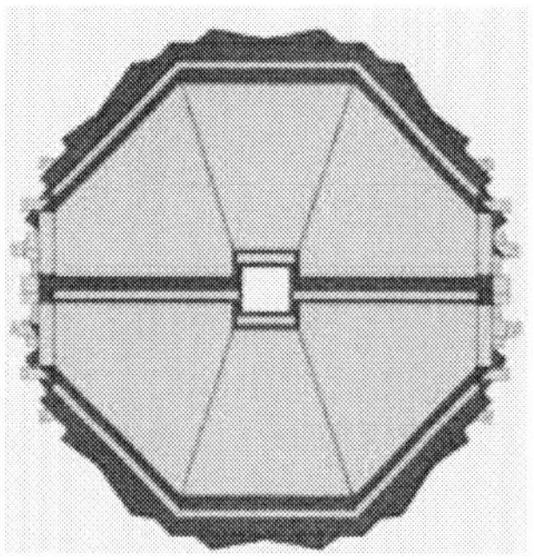

Figure 8. Schematic view of the MuCh station with three straw chambers.

The challenge for the muon chambers and for the track reconstruction algorithms is the very high particle density up to a maximum of 0.3 hits/cm^2 per central event in the first detector layers after the first absorber. For a reaction rate of 10 MHz this hit density translates into a hit rate of 3 MHz/cm^2. However, rather low particle multiplicities for the last detection layers (behind last muon absorbers) allows to use straw tubes. For MuCh use, an assembly of straw stations containing 12 double layers modules, each consisting of about 1184 straws, is in development. Each straw station contains three identical octagonal chambers measuring X and two rotated (+10°, -10°) coordinates of a passing charged particle. Each chamber consists of two identical modules with some overlap between them to avoid dead regions. The chambers are having inner holes for the beam pipe with a diameter of 43 cm [14]. The schematic view of the MuCh station with three straw chambers is shown in figure 8.

The main advantage of the straw tubes is the high speed with good spatial resolution from the point of view of the tracking of the detected particle. Sensitive time of the detector is not exceeding 80 ns and a spatial resolution is better than 200 μm at high performance up to the hit rate of 5-10 MHz/channel. The experience of the manufacture of the ATLAS TRT [15, 16], the COMPASS tracker [17] at CERN and the straw chambers for SVD-2 experiment [18] at U70 accelerator in Protvino confirms the high performance of the detectors.

The straw-chambers are satisfying to the following parameters: high uniformity of registration channels; long-term life in high radiation fields; good spatial and timing resolutions of the detector and low radiation length. The full size prototype of the octagonal straw detector for CBM [19] has been manufactured and tested. The straw detector with the rectangular shape [17, 18] is considered to reduce the cost and to improve the quality of the tracker.

3.3 Silicon Tracker System

The task of the STS [20] is to provide track reconstruction and momentum determination of charged particles. The multiplicity of charged particles is about 700 per event within the detector acceptance. The layers are located downstream of the target at distances between 30 cm and 100 cm inside the Superconducting Dipole Magnet with ~1 T·m magnetic field integral. The required momentum resolution is of the order of Δp/p~1%. This performance can be achieved with an ultra-low material budget of the stations, imposing particular restrictions on the location of power-dissipating front-end electronics in the fiducial volume. The concept of the STS tracking is based on silicon micro-strip sensors mounted onto lightweight mechanical support ladders [21]. The sensors will be read out through multi-line micro-cables with fast electronics at the periphery of the stations where cooling lines and other infrastructure can be placed. The micro-strip sensors will be double-sided with a stereo angle of 7.5°, a strip pitch of 60 μm, strip lengths between 20 and 60 mm, and a thickness of 300 μm of silicon. The micro-cables will be built from sandwiched polyimide-Aluminum layers of several 10 μm thickness [22]. 8 layer station will consist of 896 sensor modules from CiS (Germany) and Hamamatsu (Japan) manufactors. These modules will be assembled into 106 carbon ladders [21].

About third of the CBM STS sensor modules will be assembled to populate 60 ladders for the first four stations within the signed JINR-FAIR-CBM Collaboration Contract. JINR STS- team is working on testing of components of the modules, on the development of assembly methods, on the multiple fixtures and custom designed ladder assembly devices together with industrial partners in Ukraine and Belarus [23]. Module assembling site at JINR LHEP was fully equipped financed jointly by JINR and BMBF. Quality assurance group modified the standard automatic probe station developed for scanning the wafers with chips into a automatic device for testing double-sided micro-strip sensors. The first test station with a silicon micro-strip sensor for the in-beam tests at Nuclotron is designed, assembled and put to operation in the lab. It is ready for the installation in the STS test area at Nuclotron [24].

3.4 Methods, algorithms and software for fast event reconstruction

Event reconstruction, being a key part of success of the experiments, is the most complicated and time consuming task of the data analysis. In addition, in some cases a full on-line event reconstruction and selection must be done on-line at the high-level trigger stage. All of the above mentioned makes necessary to develop fast and efficient methods, algorithms and software for data analysis and to optimize them for running on an advanced high performance computing cluster.

JINR (LIT) takes part in the development of the algorithms and software for track and ring reconstruction in MuCh, TRD, RICH, MVD detectors as well as global track reconstruction. Track reconstruction methods are based on the track following and Kalman filter procedures. Ring reconstruction is based on the Hough Transform method. The clustering algorithms in MVD, STS and MuCh have been developed and included into simulation [25].

4D event reconstruction using the time slices information has been developed. The time-slice border problem effect of possible event splitting in the border region of neighbouring time-slices was investigated. Multi-vertex analysis reconstruction for the case of merged events which can not be resolved with the time measurement use only has been performed using the event topology [26].

First Level Event Selection software is under development using different manycore CPUs and GPUs platforms [27].

Two methods for the fast and reliable electron–positron identification using the energy losses of charged particles in the TRD (see figure 9) have been developed: an artificial neuron network (ANN) and a modified goodness-of-fit criterion ω_n^k (figure 10) [28].

DOI: 10.1051/epjconf/201713801020

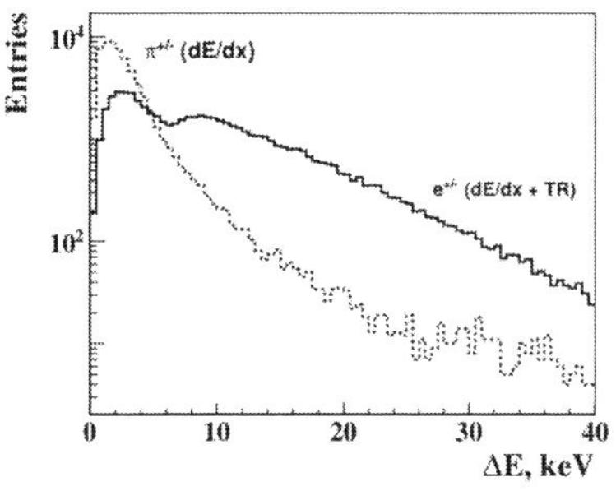

Figure 9. Energy loss distributions of pions (dashed line) and electrons (solid line) in the TRD

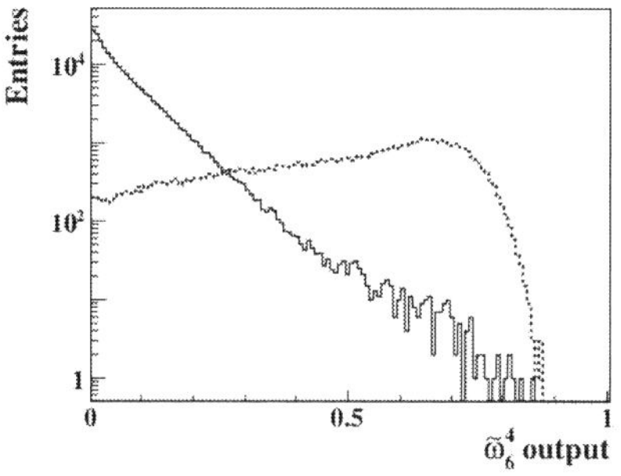

Figure 10. Distribution of ω_n^k random values for pions (solid line) and electrons (dashed line) [28].

The the vector finding method [29] is based on building vectors for each MuCh station separately and matching them to each other and to STS tracks through the absorbers. Presumably, such an approach should better handle a heterogeneous tracking environment (GEM and straw tube detectors) and offer higher flexibility with respect to algorithm tuning as compared with a track following method. Moreover, this algorithm is naturally parallelizable and can be used for the triggering application. The results were obtained for different SIS100 MuCh configurations (see figure 7) using the vector finding [30]. It has been shown good performance of the algorithm both for 4 double layers and 3 double layers of straw tubes for 2 last station of MuCh detector.

JINR contributes also to the CBMRoot and to the concept of CBM Databases [31] developments.

3.5 Simulation for physics at SIS100

. The measurements of the yields of light nuclei, which are formed by the coalescence of individual nucleons is an alternative to HBT interferometry method to extract the information on the source at freeze-out.

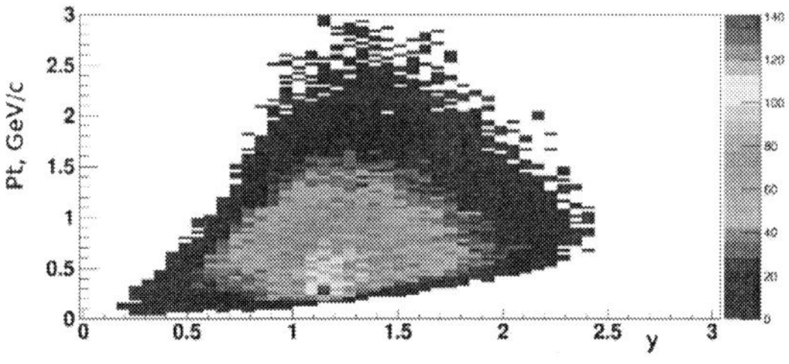

Figure 11. $y - p_T$ distribution for deuterons from $Au + Au$ central collisions at 4 A·GeV [32].

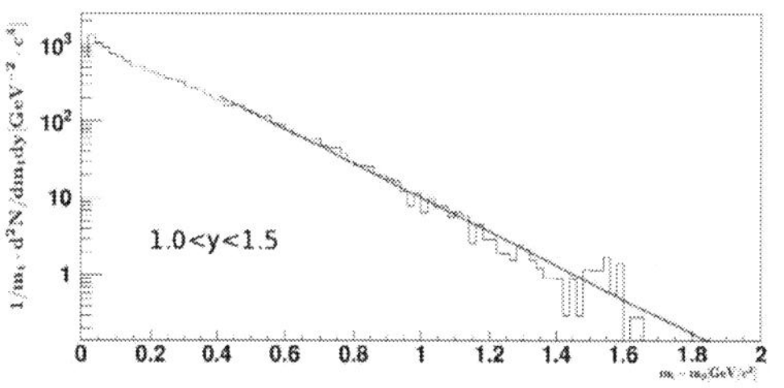

Figure 12. Transverse mass distribution for deuteron produced in the $Au+Au$ central collisions at 4 A·GeV [32].

The transverse momentum- rapidity, $y-p_T$, distribution for deuterons from the $Au+Au$ central collisions at 4 A·GeV [32] is presented in figure 11. The dominant reaction mechanism is a coalescence. The $1/m_t \cdot d^2N/dm_tdy$ distribution for deuterons from the $Au + Au$ central collisions at 4 A·GeV [32] is presented in figure 12. The deuterons and protons with the rapidity of $1.0\leq y \leq 1.5$ were selected for further analysis of the coalescence factor B_2. The calculated at $p_t = 0$ value of coalescence factor $B_2(m_t = m_0) = (3.1\pm0.2)\cdot 10^{-3}$ GeV^2 is larger by one order of magnitude than that measured in central $Pb + Pb$ collisions at SPS [33]. The B_2 factor increases exponentially as a function of the transverse mass difference $m_t - m_0$ with the inverse slope $T \approx 1.56$ GeV [34].

Feasibility studies of the $J/\Psi \to e^+e^-$ and $J/\Psi \to \mu^+\mu^-$ decays reconstruction have been performed [35]. It has been shown that one can expect ~60 and ~$5\cdot10^3$ $J/\Psi \to e^+e^-$ decays detected per hour at 10 MHz of the $Au + Au$ collisions at 10 and 25 A·GeV, respectively. The signal to background ratio is S/N 0.18 and 7.5 at 10 and 25 A·GeV, respectively. The feasibility to measure $J/\Psi \to e^+e^-$ in pC and pAu collisions has been also demonstrated.

3.6 Beam tests

JINR participate in the data analysis of the beam tests results for the RICH and TRD prototypes.

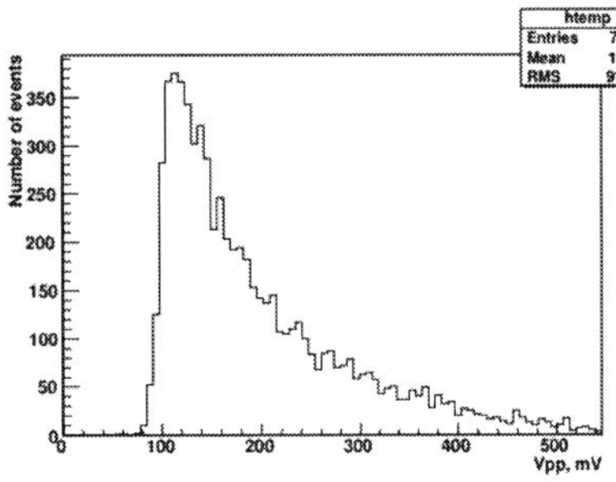

Figure 13. The amplitude of the scintillation detector irradiated by the neutron fluence of ~$7.5\cdot10^7$ n/cm^2.

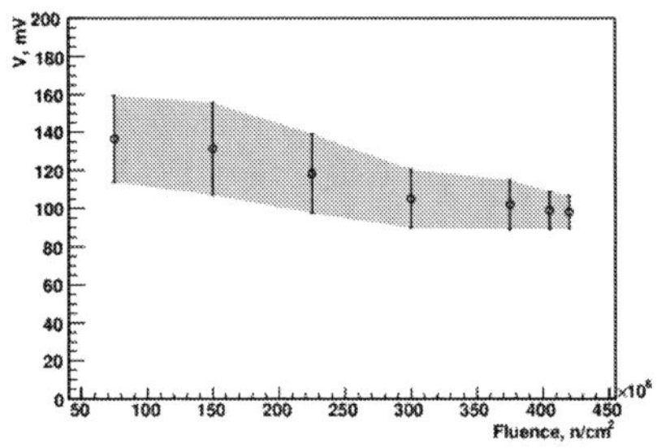

Figure 14. The dependence of the signal amplitude from the scintillation detector as a function of the neutron fluence.

The radiation hardness of APDs from different manufactures has been investigated at NPI in Řež [36] for PSD readout development. Further studies have been performed with a 10 channels scintillation detector prototype developed at LHEP JINR [37]. A prototype was irradiated using quasi-monoenergetic secondary neutron beam from cyclotron facility U120M at NPI in Řež. The signal from one of the scintillation detector in the beginning of the irradiation (after $7.5\cdot10^7$ n/cm^2 of the 1 MeV neutron fluence) is shown in figure 13. The decrease of the SiPM signal with the irradiation dose is demonstrated in figure 14. The signal decreases by a factor of ~1.4, while the intrinsic noise increases approximately twice after irradiation with the 1 MeV neutron fluence of $4.2\cdot10^8$ n/cm^2. The electronics demonstrated the stable operation during the irradiation.

4 Conclusions

- The CBM research program aims at the exploration of the structure of high density matter. For these purpose the advanced experimental setup will be build for high counting rate conditions expected at FAIR.
- The major goal for 2016-2020 is to construct CBM detector to be ready for data taking at SIS100.
- JINR participates in the CBM project very actively and its contribution is large.
- Experience of the design and construction of many elements of the CBM setup (the Superconducting Dipole Magnet [8], MuCh [14], STS [20]) is used for the BM@N at the external Nuclotron beams and for MPD at NICA.

References

[1] K. Fukushima and T. Hatsuda, Rept. Prog. Phys. **74**, 014001 (2011)
[2] B. Friman et al., Lect.Notes Phys. **814**, 1 (2011)
[3] S. Chattopadhyay et al., e-Print: arXiv:1607.01487 [nucl-ex]
[4] G. Agakishiev et al., Eur.Phys.J. A**41**, 243 (2009)
[5] P. Kurilkin et al., Int.J.Mod.Phys. Conf. Ser. **39**, 1560098 (2015)
[6] H. Sato et al., IEEE Trans. Appl. Supercond. **23**, 4500308 (2013)
[7] L.Z. Ma et al., IEEE Trans. Appl. Supercond. **22**, 9502104 (2012)
[8] The CBM Collaboration, *Technical Design Report for the CBM Superconducting Dipole Magnet*, GSI-2015-02000 (2013)
[9] Yu.V. Gusakov et al., *CBM Progress Report 2015* (GSI, Darmstadt, 2016) 9
[10] P. Kurilkin et al., *CBM Progress Report 2013* (GSI, Darmstadt, 2014) 8
[11] P. Kurilkin et al., J.Phys.Conf.Ser. **742**, 012017 (2016)
[12] P. Kurilkin et al., *CBM Progress Report 2015* (GSI, Darmstadt, 2016) 11
[13] Yu.V. Gusakov et al., *CBM Progress Report 2015* (GSI, Darmstadt, 2016) 8
[14] The CBM Collaboration, *Technical Design Report for the CBM Muon Chambers*, GSI-2015-02580 (2015)
[15] Yu.V. Gusakov et al., Phys.Part.Nucl. **41**, 1 (2010)
[16] A.S. Boldyrev et al., Instrum.Exp.Tech. **55**, 323 (2012)
[17] V.N. Bychkov et al., Nucl.Instrum.Meth. in Phys.Res. A **556**, 66 (2006)
[18] S.G. Basiladze et al., Instrum.Exp.Tech. **55**, 336 (2008)
[19] V.D. Peshekhonov et al., Phys.Part.Nucl.Lett. **9**, 172 (2012)
[20] The CBM Collaboration, *Technical Design Report for the CBM Silicon Tracking System*, GSI-2013-05499 (2013)
[21] S. Igolkin et al., *CBM Progress Report 2014* (GSI, Darmstadt, 2015) 48
[22] V.M. Borshchov et al., *CBM Progress Report 2015* (GSI, Darmstadt, 2016) 36
[23] A. Sheremetev et al., *CBM Progress Report 2015* (GSI, Darmstadt, 2016) 40
[24] D. Dementyev and Yu. Murin, *CBM Progress Report 2015* (GSI, Darmstadt, 2016) 31
[25] G. Kozlov, et al., *CBM Progress Report 2012* (GSI, Darmstadt, 2013) 91
[26] V. Akishina, et al., *CBM Progress Report 2012* (GSI, Darmstadt, 2013) 93
[27] V. Akishina, et al., *CBM Progress Report 2012* (GSI, Darmstadt, 2013) 92
[28] O.Yu. Derenovskaya and V.V. Ivanov, *CBM Progress Report 2012* (GSI, Darmstadt, 2013) 96
[29] P. Billoir et al., Nucl.Instrum.Meth. in Phys.Res. A **241**, 115 (1985)
[30] A. Zinchenko et al., *CBM Progress Report 2015* (GSI, Darmstadt, 2016) 121
[31] E.P. Akishina et al., *CBM Progress Report 2015* (GSI, Darmstadt, 2016) 123
[32] V.D. Toneev et al., Nucl.Phys. A **519**, 463C (1990)
[33] S.V. Afanasiev et al., Phys.Lett. B **486**, 22 (2000)
[34] V.P. Ladygin et al., *CBM Progress Report 2015* (GSI, Darmstadt, 2016) 127
[35] O.Yu. Derenovskaya et al., Phys.Part.Nucl.Lett. **11**, 560 (2014)
[36] V. Kushpil et al., Phys.Part.Nucl.Lett. **13**, 120 (2016)
[37] S.G. Reznikov et al., *CBM Progress Report 2015* (GSI, Darmstadt, 2016) 102

Overview of ALICE results in pp, pA and AA collisions

Rainer Schicker, for the ALICE Collaboration[1,★]

[1] *Phys. Inst., Im Neuenheimer Feld 226, 69120 Heidelberg, Germany*

Abstract. The ALICE experiment at the Large Hadron Collider (LHC) at CERN is optimized for recording events in the very large particle multiplicity environment of heavy-ion collisions at LHC energies. The ALICE collaboration has taken data in Pb-Pb collisions in Run I and Run II at nucleon-nucleon center-of-mass energies $\sqrt{s_{NN}}$ = 2.76 and 5.02 TeV, respectively, and in pp collisions at center-of-mass energies $\sqrt{s}$ = 0.9, 2.76, 5.02, 7, 8 and 13 TeV. The asymmetric system p-Pb was measured at a center-of-mass energy $\sqrt{s_{NN}}$ = 5.02 TeV. Selected physics results from the analysis of these data are presented, and an outline of the ALICE prospects for Run III is given.

1 Introduction

The ALICE experiment at the Large Hadron Collider (LHC) at CERN is a general purpose detector optimized for the measurement of high-energy Pb-Pb collisions [1]. The analysis of such collisions allows to study a variety of issues which are pertinent to an improved understanding of Quantum Chromodynamics (QCD), the theory of the strong interaction. At high temperatures and energy densities of the QCD medium, lattice QCD calculations predict a phase transition to a state of deconfined quarks and gluons [2]. High-energy nuclear collisions allow to reach such energy densities, however within a finite volume and for a limited time only. This phase transition is accompanied by the restoration of chiral symmetry, in which the quarks acquire their current mass. The understanding of the properties of the QCD medium created in heavy-ion collisions necessitates a good knowledge of the underlying collision dynamics. Intrinsic QCD-medium signals can be disentangled from initial cold-matter and final-state effects by comparing the heavy-ion observables to the corresponding quantities measured in pp and p-Pb collisions.

2 The ALICE Experiment

2.1 The History of the ALICE Experiment

The first conceptual ideas for experiments at the LHC were discussed in a workshop in 1990. Out of this workshop, an Expression of Interest, followed by a Letter of Intent and the Technical Proposal evolved. After approval of the ALICE central barrel system in 1997, the different detector systems were specified in Technical Design Reports. A muon spectrometer was later added, as well as a Transition Radiation Detector (TRD) and Electromagnetic Calorimeters EMCAL and DCAL in the central barrel.

★e-mail: schicker@physi.uni-heidelberg.de

2.2 The ALICE Physics Programme

2.2.1 The ALICE Physics Programme in pp-Collisions

The ALICE physics programme in pp collisions is complementary to the other LHC experiments due to the low p_T-threshold, and due to the excellent particle identification capability in the central barrel [3]. Besides being a reference for the analysis of Pb-Pb collisions, the pp collisions have interest in their own. The analysis of minimum-bias pp collisions addresses a plethora of issues relevant for the understanding of soft QCD processes, such as particle and multi-particle production in the non-perturbative QCD regime. The analysis of high-multiplicity pp collisions allows to study the possible onset of collectivity in small systems, such as evidenced in collective flow effects.

2.2.2 The ALICE Physics Programme in pA-Collisions

The analysis of data taken in p-Pb collisions, and its comparison to the corresponding quantity measured in Pb-Pb collisions, can be used to disentangle cold-matter initial and final-state effects from intrinsic QCD-medium effects. Of particular interest here are possible shadowing of nuclear Parton Distribution Functions (nPDFs), and the low-x behaviour of the gluon-PDF as signature of gluon saturation at low values of Bjorken-x.

2.2.3 The ALICE Physics Programme in AA-Collisions

The analysis of Pb-Pb collisions at LHC energies addresses a multitude of physics issues, above all the question of experimental observables to characterize the nature of the QCD phase diagram [4]. The observables derived in these analyses, and their correlations, can be used to study experimental signatures of the deconfinement and the chiral symmetry transition, and to examine the relationship between these two transitions.

2.3 The ALICE Detector Systems

In figure 1, the yoke of the ALICE magnet is shown in red. The innermost detector system, the Inner Tracking System (ITS), is zoomed in on the top right of the figure, showing the strip, drift and pixel subsystems. Radially outwards from the ITS, the Time Projection Chamber (TPC), the Transition Radiation Detector (TRD), and the Time-of-Flight detector (TOF) are shown. The V0 system serves as trigger detector, and its information is used to define event parameters such as high-multiplicity or rapidity-gap characteristics. In addition, there exist other small acceptance systems for particle identification and event characterization (FMD, HMPID, T0). The electromagnetic calorimeters EMCAL and PHOS are able to measure photons in a limited solid angle of the central barrel. A muon spectrometer complements the central barrel on one side within the pseudorapidity range $-4 < \eta < -2.5$.

2.4 The ALICE Performance

The ALICE detector systems shown in figure 1 provide the information for reconstructing the individual tracks, for fitting their momentum, and for identifying the particle type of the track.

The ALICE particle identification capability by the energy loss dE/dx in the TPC is shown on the left part, and by the Time-of-Flight as measured by the TOF detector on the right part of figure 2. The information from the TRD detector, located between the TPC and the TOF detector, is used to identify electrons. The separation of electrons from charged hadrons can in addition be achieved by the electromagnetic calorimeters.

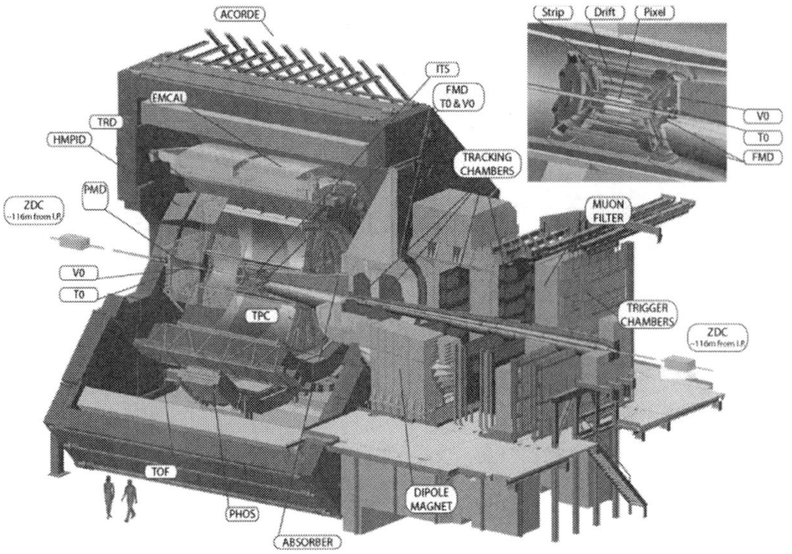

Figure 1: The detector systems of the ALICE central barrel, with the muon spectrometer shown on the right hand side of the central barrel.

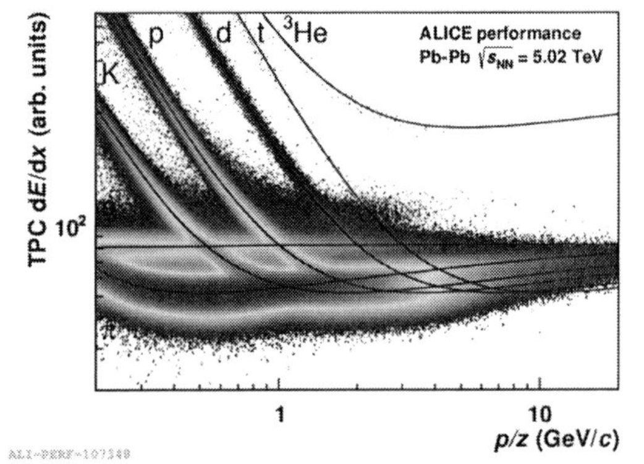

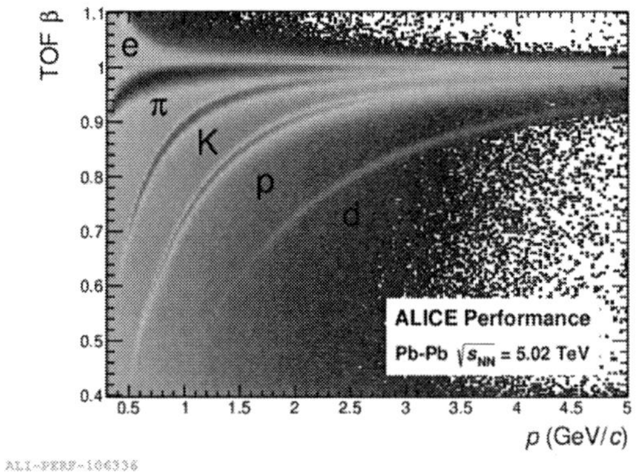

Figure 2: The particle identification by energy loss dE/dx in the TPC is shown on the left. The particle identification by TOF is shown on the right.

2.5 The Data Sample collected by ALICE

The statistics of the data samples recorded by the ALICE collaboration in Run I and Run II of the LHC is displayed in table 1.

In Run I and Run II, the ALICE collaboration has taken data in pp collisions at center-of-mass energies of $\sqrt{s}$ = 0.9, 2.76, 5.02, 7, 8 and 13 TeV. In Pb-Pb collisions, events were recorded in Run I and Run II at center-of-mass energies of $\sqrt{s_{NN}}$ = 2.76 and 5.02 TeV, respectively. In p-Pb collisions, data were recorded in Run I at the center-of-mass energy of $\sqrt{s_{NN}}$ = 5.02 TeV. During November and December of this year, data taking in p-Pb collisions is planned at $\sqrt{s_{NN}}$ = 5.02 and 8 TeV.

Table 1: LHC Runs I+II, 2009-2015:

system	run	year	$\sqrt{s_{NN}}$ (TeV)	int. lumi.
p-p	I	2009-2010	0.9	0.15 nb^{-1}
p-p	I	2011	2.76	1.1 nb^{-1}
p-p	I	2010-2011	7	4.8 pb^{-1}
p-p	I	2012	8	9.7 pb^{-1}
p-p	II	2015	5.02	2.5 nb^{-1}
p-p	II	2015	13	4.35 pb^{-1}
p-Pb	I	2013	5.02	30 nb^{-1}
Pb-Pb	I	2010,2011	2.76	0.1 nb^{-1}
Pb-Pb	II	2015	5.02	0.24 nb^{-1}

3 Proton-Proton Collisions

3.1 Multi-Parton Interactions

Particle production in high-energy hadronic collisions can get substantial contributions from Multi-Parton Interactions (MPI). Here, several partonic interactions can take place in a hadronic collision, resulting in a multiplicity dependence of observables such as particle yields and transverse momentum distributions. Analyzing MPI effects is of fundamental importance for studying the range of validity of Parton Distribution Functions (PDFs), an input when calculating perturbative QCD cross sections.

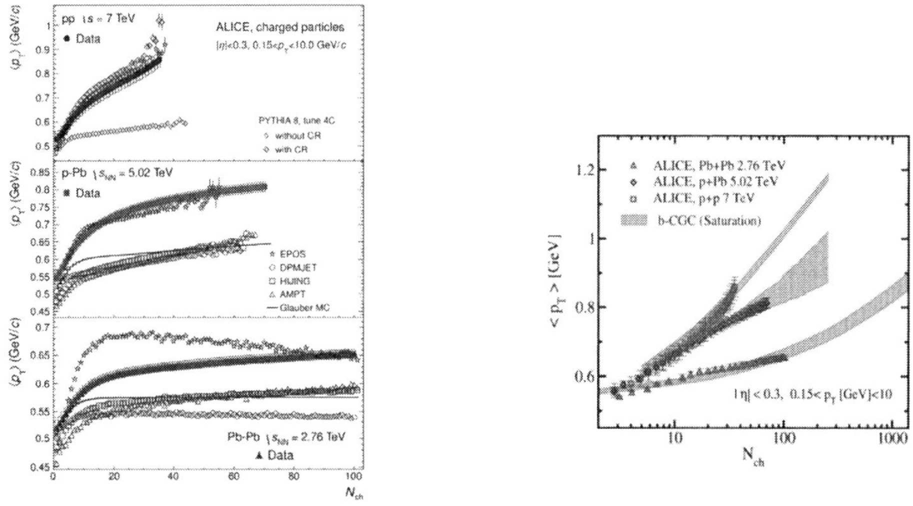

Figure 3: Average transverse momentum $<p_T>$ as function of charged particle multiplicity for pp, p-Pb and Pb-Pb collisions are shown on the left. The comparison of measured $<p_T>$ to the predictions of the Color Glass Condensate approach are shown on the right.

In the top left panel in figure 3, the ALICE measurement of the average transverse momentum in the range $|\eta| < 0.3$ for pp collisions at $\sqrt{s} = 7$ TeV is displayed as function of the charged particle multiplicity [5]. The results of the PYTHIA 8 calculations, shown by the open symbols, clearly show the importance of including colour reconnections. In the middle panel on the left, the corresponding ALICE data for p-Pb collisions are shown in purple, superimposed in black by the predictions of the EPOS, DPMJET, HIJING and AMPT models [6]. The EPOS model predicts the ALICE data reasonably well, even though it shows a different trend at multiplicities $N_{ch} < 20$ [7]. None of the other three models is able to describe the data. A Glauber MC calculation, shown by the solid black line, also fails to reproduce the data. In this Glauber MC approach, p-Pb collisions are assumed to consist of independent pp collisions, each contributing to the particle yield according to the yield as measured in pp collisions. In the bottom panel on the left, the ALICE data for Pb-Pb collisions are shown in red, superimposed in black by the predictions of the different model calculations. As for the p-Pb data, the DPMJET, HIJING and AMPT model as well as the MC Glauber calculation underpredict the measured $<p_T>$. The EPOS model overpredicts the data and shows a different trend.

On the right side of figure 3, the ALICE data are compared to Color Glass Condensate (CGC) model calculations [8]. The CGC approach, addressing the importance of the initial-state effect, is based on gluon saturation at low values of Bjorken-x, and Glasma physics. The CGC predictions are in much better agreement with the ALICE data than the model predictions shown on the left of figure 3.

3.2 Strangeness and Charm in Proton-Proton Collisions

The initial state of proton-proton collisions does not contain strange valence quarks. The measurement of strange hadrons hence allows the study of particle production in QCD both in the perturbative and non-perturbative sector. Strangeness can be created in hard perturbative $2 \to 2$ partonic scatterings by flavour creation ($gg \to s\bar{s}, q\bar{q} \to s\bar{s}$) and by flavour excitation ($gs \to gs, qs \to qs$). During partonic evolution, gluon splitting ($g \to s\bar{s}$) contributes to strangeness production. At low transverse momenta, strangeness production is dominated by non-perturbative processes such as string fragmentation.

On the left of figure 4, the p_T-integrated yield ratios of strange ($K^0_S, \Lambda, \overline{\Lambda}$) and multi-strange ($\Xi^-, \overline{\Xi^+}, \Omega, \overline{\Omega^+}$) hadrons measured in pp, p-Pb and Pb-Pb are shown [9]. These ratios represent a production rate which increases faster for strange hadrons than for non-strange hadrons. The strangeness enhancement increases with hadron strangeness, and with event activity defined by $dN_{ch}/d\eta$. This enhancement is not reflected by the PYTHIA generator shown in figure 4 by the solid lines. The EPOS generator describes the trend vs multiplicity, even though it fails for the triply strange Ω-Baryon. The Dipsy model describes the data best, but is also not doing well for the triply Ω-Baryon.

On the top right part of figure 4, the yield $dN_{J/\Psi}/dy$, normalized to $<dN_{J/\Psi}/dy>$, is shown as function of the normalized charged particle multiplicity $(dN_{ch}/d\eta)/<dN_{ch}/d\eta>$. This yield is shown in red at midrapidity $|y| < 0.9$, and in blue at forward rapidities $2.5 < |y| < 4.0$. An approximate linear correlation of these quantities is seen in figure 4. In the bottom right part, the normalized $dN_{J/\Psi}/dy$ yield as predicted by PYTHIA 6.4 in Perugia 2011 tune is shown. In this calculation, only J/Ψ's produced in hard scatterings via the NRQCD framework are calculated, whereas J/Ψ production from cluster formation during parton shower evolution or in MPI are not taken into account.

3.3 Charmed Mesons in Proton-Proton Collisions

The study of open-charm meson production in pp collisions at LHC energies allows to test predictions of perturbative QCD (pQCD) at the highest collider energies available. Such pQCD calculations are done in collinear factorization approach at next-to-leading order in the general-mass variable-flavour-number scheme (GM-VFNS), or at fixed order with next-to-leading-log resummation (FONLL). These calculations describe the p_T-differential production cross sections of D-mesons within uncertainties.

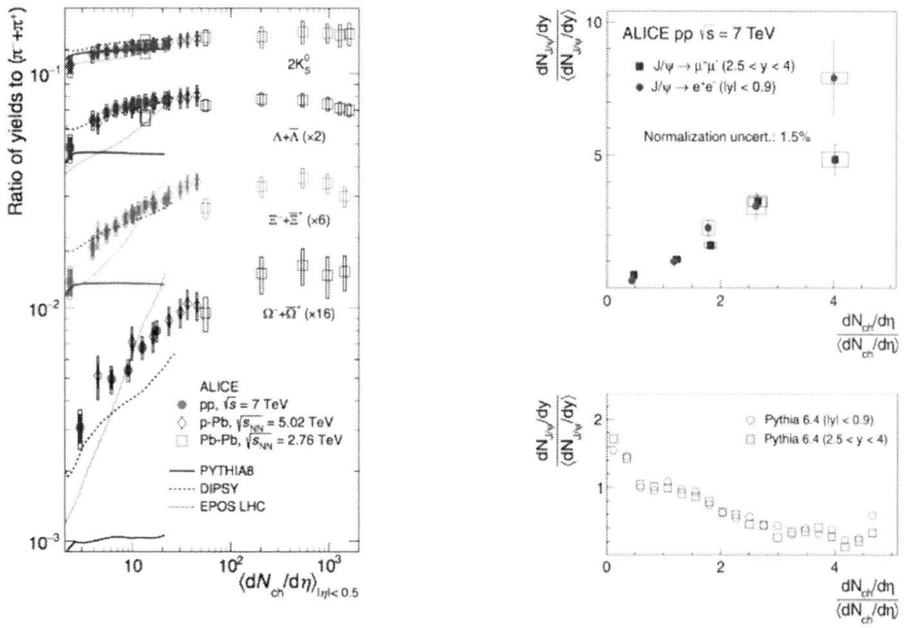

Figure 4: The yield ratio of strange and multi-strange hadrons to pions as function of $(dN/d\eta)$ is shown on the left. The normalized J/Ψ yield as function of normalized $dN_{ch}/d\eta$ is shown in the upper part on the right. The PYTHIA prediction of this normalized yield is shown in the lower part on the right.

The relative D-meson yield (average of D^0, D^+, D^{*+}) is shown in figure 5 in four p_T-intervals as function of the relative charged particle yield [10]. These measurements are compared to model predictions by PYTHIA 8 (red-dotted line), the EPOS models (green-dashed and green dot-dashed line), and the percolation model (blue dot-dashed line). In PYTHIA 8, heavy-flavour production is due to four mechanisms, the first hard 2→2 hard process, the subsequent hard process in MPI, gluon splitting from hard process, and from initial/final state gluon radiation (ISR/FSR). The dominating contribution in PYTHIA 8 at LHC energies is the ISR/FSR mechanism which contributes 62% and 40% for D and B-meson production, respectively. The EPOS model incorporates initial conditions according to the Gribov-Regge multiple scattering framework. Individual scatterings are initiated by Pomerons, and are associated to parton ladders. Each of these parton ladders represents a hard pQCD process with initial and final state radiation. A saturation scale is introduced for taking account of non-linear effects. With these initial conditions, a hydrodynamical evolution can be defined on the core of the collision. The EPOS calculation without hydro-evolution predicts an approximate linear increase of D-meson production as function of relative charged-particle multiplicity. The EPOS model with hydro-evolution results in a deviation from linear dependence. Within the percolation model, high-energy hadronic collisions are dominated by the exchange of colour sources of finite spatial dimension between projectile and target [11]. The percolation model predicts a faster than linear increase of the relative D-meson yield.

4 Proton-Lead Collisions

The analysis of the p-Pb collision system sheds light on the role of initial and final-state effects when studying differences between the pp and Pb-Pb systems [12]. The nuclear shadowing of PDFs, the

 DOI: 10.1051/epjconf/201713801021

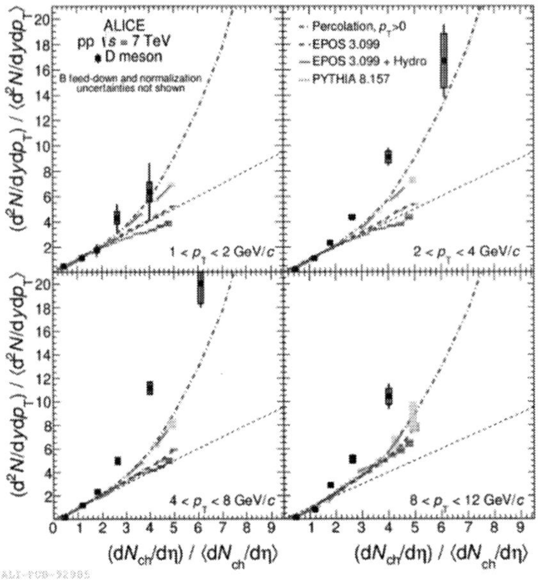

Figure 5: Average D-meson relative yield as function of relative charged-particle multiplicity at central rapidity in different p_T-intervals.

existence of CGC in the initial state and the hydrodynamic phase in the evolution of the system, for example, affect bulk observables quite differently in the pp, p-Pb and Pb-Pb collision systems.

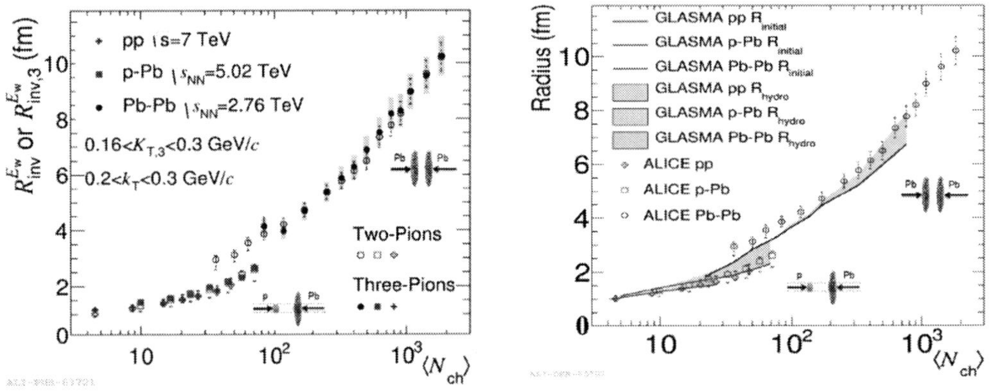

Figure 6: Pion source sizes are shown on the left, model calculations on the right.

The source sizes derived from pion correlations are shown on the left in figure 6 as a function of the charged particle multiplicity for the pp, p-Pb and Pb-Pb systems [13]. Here, R_{inv} expresses the invariant source radius derived by a Gaussian fit to the two-pion correlation function. The quantity $R_{inv,3}$ denotes the corresponding quantity as determined by analysis of three-pion cumulant distributions. At the same multiplicities within the range $60 < N_{ch} < 80$, the source radii in the pp and p-Pb systems

DOI: 10.1051/epjconf/201713801021

show similar values, with a significantly larger value in the Pb-Pb system reflecting the hydrodynamic expansion stage of a heavy-ion collision. These ALICE data are compared on the right of figure 6 to predictions by the CGC based IP-Glasma model [14].

5 Lead-Lead Collisions

5.1 Exploring the QCD Phase Diagram with Heavy-Ion Collisions

Collisions of heavy-ions at high energies produce QCD matter at extreme values of energy densities and temperatures [15]. Lattice gauge calculations indicate that a QCD phase transition takes place under such conditions. In lattice QCD approach, the partition function of the grand canonical ensemble is evaluated stochastically by Monte Carlo sampling of field configurations. Thermodynamic state functions such as pressure and energy densities can subsequently be evaluated. The transition from hadronic matter to a state of deconfined quarks and gluons, the Quark-Gluon Plasma (QGP) is seen in such calculations at a critical temperature of $T_{cr} \sim 160$ MeV. At this critical temperature, the chiral condensate $<\bar{q}q>$ is rapidly diminishing, thereby indicating the restoration of broken chiral symmetry [16]. A multitude of experimental observables are analyzed to characterize the nature of the QCD phase diagram, and to establish signals of partial chiral symmetry restoration. The present understanding of this deconfined state is a strongly coupled QGP which behaves as a liquid with a very low viscosity to entropy ratio [17].

5.2 Soft Probes in Heavy-Ion Collisions

The analysis of the azimuthal anisotropy of particle production in heavy-ion collisions reveals information on the initial geometry of the overlap zone of the colliding nuclei, on the equation of state of the produced QCD medium and its transport properties.

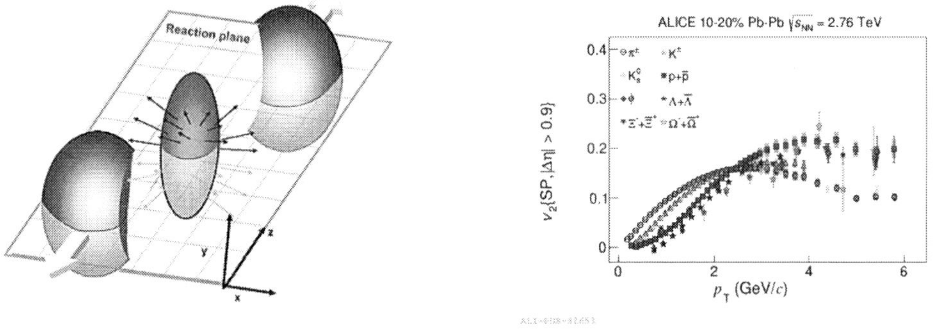

Figure 7: Heavy-ion collision geometry is shown on the left, v_2-flow coefficients on the right.

On the left side of figure 7, a heavy-ion collision geometry is shown. The azimuthal distribution of particles produced in the almond-shape overlap zone is analyzed with respect to the reaction plane by

$$\frac{d^2N}{p_T dp_T d\phi} \propto 1 + 2\sum_{n=1}^{\infty} v_n(p_T)\cos[n(\phi - \Psi_n)] \tag{1}$$

The azimuthal particle distribution of 1 is characterized by the symmetry planes Ψ_n and by the coefficients v_n. Such distributions test hydrodynamical models, and allow to extract properties of the QCD medium such as viscosity.

On the right side of figure 7, the v_2-coefficient of the Fourier decomposition of the azimuthal anisotropy is shown for the centrality class 10%-20% [18]. A distinctive mass ordering of the flow coefficient v_2 is seen at low values of transverse momentum p_T. This mass ordering is attributed to the interplay between elliptic and radial flow. Radial flow tends to deplete the spectrum at low values of p_T, which increases with increasing particle mass and transverse velocity. In a system with azimuthal anisotropy, this depletion is larger in-plane than out-of-plane, thereby reducing v_2. At large values of p_T, the flow coefficients v_2 show a tendency to order according to the particle type of baryon and meson. Such scaling was first observed at RHIC, and was interpreted as evidence that quark degrees of freedom dominate in the early stages of heavy-ion collisions when collective flow develops [19].

5.3 Hard Probes in Heavy-Ion Collisions

The analysis of jets produced in scatterings of highly virtual quarks or gluons provides an important tool to test the Standard Model in the perturbative region of QCD. In heavy-ion collisions, the analysis of jet quenchinq allows to study the QCD medium produced in the collision. In figure 8 on the left, a hard parton scattering is shown. The hadronization of the two partons after scattering results in two jets emerging on opposite sides in azimuthal angle. Of particular interest are hadron-photon and jet-photon correlations. The photon is expected to escape the interaction zone with much reduced interaction as compared to hadrons, and hence carries the information of the hard-scattering kinematics.

On the right of figure 8, a dijet event is shown with the leading jet of transverse momentum p_T = 205.1 GeV/c, and the subleading jet of p_T = 70.0 GeV/c at the opposite azimuthal angle [20].

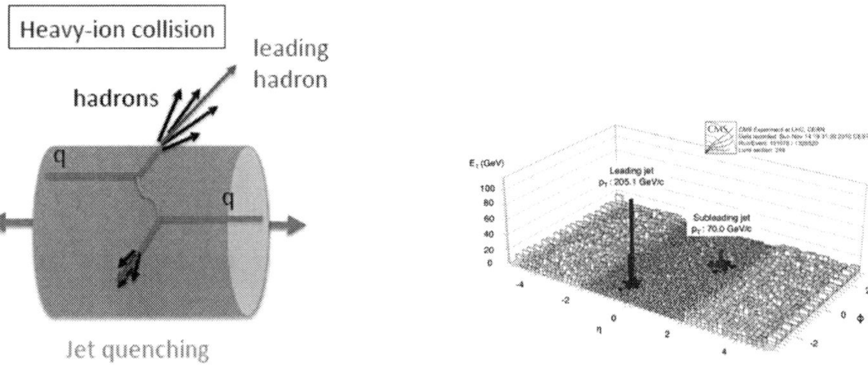

Figure 8: Hard parton scattering is shown on the left, a dijet event on the right (figure from [20]).

The energy of the partons of the jet is reduced in heavy-ion collisions as compared to pp collisions due to medium-induced gluon radiation and collisional energy loss. In perturbative QCD, the production cross section of the initial hard scattered parton can be calculated. The hadronization of the initial scattered parton can be calibrated by jet measurements in pp collisions. Jet quenching in heavy-ion collisions can be analyzed by studying whether these jet spectra can be understood as an incoherent superposition of nucleon-nucleon collisions. The analysis of ALICE data reveals that jets with transverse momenta $40 < p_{T,jet} < 120$ GeV/c in heavy-ion collisions are strongly suppressed in the 10% most central events [21].

5.4 Low Mass Dileptons

In the limit of vanishing quark masses, the QCD Lagrangian contains a symmetry due to the conserved right or left-handedness of the zero-mass spin-1/2 particles. This chiral symmetry is spontaneously broken, leading to a population of the QCD ground state by scalar quark-antiquark pairs [22]. In the QCD medium formed in heavy-ion collisions at high energies, the expected partial restoration of chiral symmetry is reflected by significant modifications of hadron properties. Of particular importance for the study of such in-medium modifications is the ρ-meson. Due to the short life-time of $c\tau$= 1.3 fm, ρ-decays occur predominantly within the QCD-medium formed in the heavy-ion collision. The measurement of the in-medium ρ-spectral function through the ρ-dilepton decays allows to test the nature of the formed QCD medium. Both an in-medium mass shift, as well as a broadening of the ρ-meson have been advocated [23, 24].

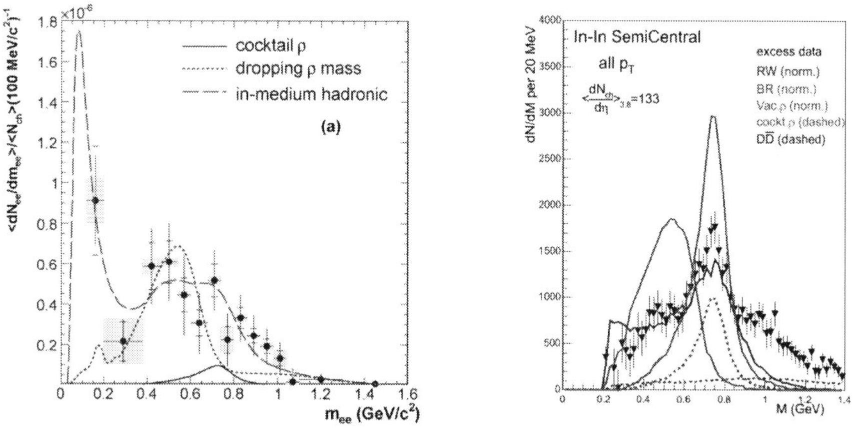

Figure 9: Invariant dilepton mass spectra measured by CERES (left) and NA60 (right).

In figure 9 on the left, the invariant mass distribution of e^+e^--pairs measured by the CERES collaboration in Pb-Au collisions at 158A GeV/c is shown [25]. The data points represent the e^+e^--pair yield after subtraction of the hadronic cocktail, i.e. the contribution to this yield from known sources such as π^0, η, η' and ω Dalitz-decays. In this figure, the dotted and long-dashed lines represent the expectations as derived from the dropping and broadening ρ-mass scenario, respectively. Within the statistics of these data, the dropping ρ-mass scenario is disfavored. On the right of figure 9, the excess of the $\mu^+\mu^-$-pair yield measured by the NA60 collaboration in In-In collisions is shown [26]. These data are compared to expectations from the dropping ρ-mass scenario (green line) and the broadening ρ-mass scenario (blue line). The solid red line represents the expectation of this yield in case of no changes to the ρ-spectral function. The dashed red line is shown for illustration, and indicates the expected contribution from the cocktail ρ bound by the ratio of ρ/ω = 1.2 measured at high p_T > 1.6 GeV/c. >From these data, an unmodified ρ is clearly ruled out, as well as the dropping mass scenario. The most realistic description of these data is within the broadening ρ-mass scenario.

The ALICE collaboration has measured the yield of low-mass dileptons at pseudorapidities $|\eta| < 0.8$ in Run I of the LHC in pp collisions at $\sqrt{s}$ = 7 TeV, and in p-Pb collisions at $\sqrt{s_{NN}}$ = 5.02 TeV.

The differential cross section for e^+e^--pair production in pp collisions is shown in figure 10 on the left. Superimposed to the data are the cocktail contributions expected from $\pi^0, \eta, \eta', \omega, \phi$ and J/Ψ Dalitz-decays, as well as from the two-body e^+e^--decays of ρ, ω, ϕ and J/Ψ. At pair masses

 DOI: 10.1051/epjconf/201713801021

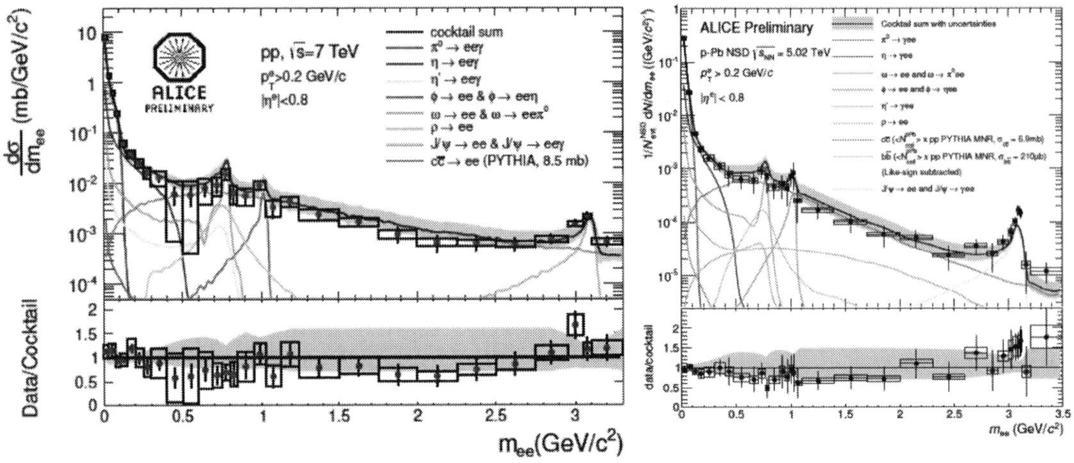

Figure 10: ALICE mass spectra of e^+e^--pairs in pp (left) and p-Pb collisions (right).

beyond the ϕ-mass, the spectrum is dominated by the contribution of $c\bar{c}$ decays. The expected cocktail contribution is consistent with the data up to a mass of m=3.5 GeV/c^2 as shown in the lower part on the left. The differential cross section for e^+e^--pair production in p-Pb collisions is shown in figure 10 on the right, together with the expected cocktail contributions. The expected cocktail contribution is consistent with the data up to a mass of m=3.5 GeV/c^2 as shown in the lower part on the right.

The dilepton spectra measured in pp and p-Pb collisions in ALICE serve as baseline to the corresponding measurements in Pb-Pb collisions. Besides the in-medium modification of low-mass vector mesons discussed above, the analysis of virtual direct photons in the low mass region and signatures of thermal radiation from the QGP in the intermediate mass region are of high interest. Such studies can be pursued in ALICE by analysing dielectrons in the central barrel as well as by examining dimuons in the ALICE muon spectrometer.

6 The Upgrade Programme of ALICE

The ALICE detector upgrade programme is driven by the future ALICE physics plans [27]. These plans center on rare physics probes where the ALICE capabilities on measuring low-p_T tracks, in conjunction with excellent particle identification, are unique. Such probes are, for example, heavy flavour hadrons and quarkonia at low p_T, low mass lepton pairs, jets, light nuclei and hypernuclei. These upgrade plans are formulated under the assumption that the LHC will increase the Pb-beam intensity such that an interaction rate of 50 kHz is reached in Run III, corresponding to an instantaneous luminosity of L = 6 x 10^{27}cm^{-2}s^{-1}. Standard trigger strategies cannot be applied in most cases of the above-mentioned rare probes. The ALICE detectors hence need to be modified such that all interactions can be inspected. Based on these considerations, the ALICE detector upgrade programme for Run III starting in the year 2021 consists of

- Upgrade of the TPC in order to operate in ungated mode to have a dead-time free readout
- A new high-resolution, thin-material Inner Tracking System (ITS). The resolution on distance of closest approach will be improved by a factor of about 3 as compared to the present performance
- Upgrade of the readout electronics of the TRD, TOF, PHOS detectors and the muon spectrometer for the future high-rate data taking

- Upgrade of the data acquisition and the high-level trigger system for the future high-rate data taking
- Upgrade of the offline processing software in order to handle the much larger number of events

7 Summary and Outlook

The ALICE collaboration has taken data in Run I and Run II of the LHC in pp, p-Pb and Pb-Pb collisions. A multitude of physics analyses of the Pb-Pb data addresses the nature of the QCD phase diagram. The study of pp collisions addresses a variety of questions relevant for an improved understanding of the non-perturbative sector of QCD. The analyses of p-Pb collisions reveal effects of cold-matter initial and final-state effects in heavy-ion induced reactions. The future of the ALICE physics programme is centered on measuring rare probes with much improved statistics as is presently available. The ALICE detectors are correspondingly upgraded in order to meet the increased data rate in the future.

Acknowledgements

This work is supported by the German Federal Ministry of Education and Research under promotional reference 05P15VHCA1.

References

[1] ALICE Collaboration (F. Carminati et al.), J. Phys. G **30**, 1517 (2004).
ALICE Collaboration (B. Alessandro et al.), J. Phys. G **32**, 1295 (2006)
[2] HotQCD Collaboration (A. Bazavov et al.), Phys. Rev. D **90**, 094503 (2014)
[3] ALICE Collaboration (B. Abelev et al.), Int. J. Mod. Phys. A **29**, 1430044 (2014)
[4] P. Braun-Munzinger and J. Wambach, Rev. Mod. Phys. **81**, 1031 (2009)
[5] ALICE Collaboration (B. Abelev et al.), Phys. Lett. B **727**, 371 (2013)
[6] S. Roesler, R. Engel and J. Ranft, DPMJET-III, (2000), arXiv:hep-ph/0012252.
X.-N. Wang and M. Gyulassy, Phys. Rev. D **44**, 3501 (1991).
Z.-W. Liu, C.M. Ko, B.-A. Li, B. Zhang and S. Pal, Phys. Rev. C **72**, 064901 (2005)
[7] T. Pierog, I. Karpenko, J. Katzy, E. Yatsenko and K. Werner, arXiv:1306.0121
[8] A.H. Rezaeian, Phys. Lett. B **727**, 218 (2013)
[9] ALICE Collaboration (J. Adam et al.), arXiv:1606.07424
[10] ALICE Collaboration (J. Adam et al.), JHEP **1509**, 148 (2015)
[11] E.G. Ferreiro and C. Pajares, arXiv:1501.03381
[12] ALICE Collaboration (J. Adam et al.), Phys. Rev. C **91(6)**, 064905 (2015)
[13] ALICE Collaboration (B. Abelev et al.), Phys. Lett. B **739**, 139 (2014)
[14] B. Schenke and R. Venugopalan, Phys. Rev. Lett. **113**, 102301 (2014)
[15] Y. Akiba et al., arXiv:1502.02730
[16] W. Weise, Nucl. Phys. A **690**, 98 (2001)
[17] H. Niemi, K.J. Eskola, R. Paatelainen and K. Tuominen, Nucl. Phys. A **956**, 312 (2016)
[18] ALICE Collaboration (B. Abelev et al.), JHEP **1506**, 190 (2015)
[19] STAR Collaboration (B.I. Abelev et al.), Phys. Rev. C **75**, 054906 (2007).
PHENIX Collaboration (A. Adare et al.,), Phys. Rev. Lett. **98**, 162301 (2007)
[20] CMS Collaboration (S. Chatrchyan et al.), Phys. Rev. C **84**, 024906 (2011)

[21] ALICE Collaboration (J. Adam et al.), Phys. Lett. B **746**, 1 (2015)
[22] A.W. Thomas and W. Weise, *The Structure of the Nucleon* (Wiley-VCH, Berlin, 2001) 149
[23] G.E. Brown and M. Rho, Phys. Rept. **363**, 85 (2002)
[24] R. Rapp and J. Wambach, Adv. Nucl. Phys. **25**, 1 (2000)
[25] CERES Collaboration (D. Adamova et al.), Phys. Lett. B **666**, 425 (2008)
[26] NA60 Collaboration (R. Arnaldi et al.), Phys. Rev. Lett. **96**, 162302 (2006)
[27] ALICE Collaboration (B. Abelev et al.), J. Phys. G **41**, 087001 (2014)

Hadronic resonance production with ALICE at the LHC

Sergey Kiselev for the ALICE collaboration[1,⋆]

[1] *Institute for Theoretical and Experimental Physics, Moscow, Russia*

Abstract. Recent results on short-lived hadronic resonances obtained by the ALICE experiment in pp, p-Pb and Pb-Pb collisions at the LHC energies are presented. Transverse momentum spectra, yields, ratios to long-lived particles and nuclear modification factors are discussed. The results are compared with model predictions and measurements at lower energies.

1 Introduction

Hadronic resonance production plays an important role both in elementary and in heavy-ion collisions. In heavy-ion collisions, the medium at high density and/or high temperature can modify the properties of short-lived resonances such as their masses, widths and spectral shapes [1]. Moreover, since the lifetimes of short-lived resonances are comparable with the time span of the late hadronic phase, regeneration and rescattering effects become important and resonance ratios to longer lived particles can be used to estimate the time interval between the chemical and kinetic freeze-out [2]. The measurements in pp and p-Pb collisions constitute a reference for nuclear collisions and provide information for tuning event generators inspired by Quantum Chromodynamics. Recent results on short-lived mesonic $\rho(770)^0$, $K^*(892)^0$, $\phi(1020)$ and baryonic $\Sigma(1385)^{*\pm}$, $\Xi(1530)^{*0}$ resonances (hereafter ρ^0, K^{*0}, ϕ, $\Sigma^{*\pm}$, Ξ^{*0}) obtained by the ALICE experiment are presented here. The ρ^0 has been measured in pp and Pb–Pb collisions at $\sqrt{s_{NN}} = 2.76$ TeV. The K^{*0} and ϕ have been measured in pp, p–Pb, and Pb–Pb collisions at various energies in different multiplicity or centrality intervals. The $\Sigma^{*\pm}$ and Ξ^{*0} have been measured in pp collisions at $\sqrt{s} = 7$ TeV and in p–Pb collisions at $\sqrt{s_{NN}} = 5.02$ TeV.

2 The ALICE detector

The ALICE detector [3, 4] at the LHC is designed to study Pb–Pb, p–Pb and pp collisions at TeV-scale center of mass energies. The components of the ALICE detector most directly related to the results presented here are described briefly. The Inner Tracking System (ITS) is a six-layers silicon-based detector that surrounds the interaction point and covers the pseudorapidity region $|\eta| < 0.9$. It is used to reconstruct the collision vertex and provides tracking and particle identification. Particle tracking is mainly provided by the Time Projection Chamber (TPC). The TPC also allows particles to be identified through their energy loss. The Time-of-Flight Detector (TOF) sits externally to the TPC and measures the time-of-flight of the particles, allowing for additional identification. Both the TPC and the TOF cover the central pseudorapidity region $|\eta| < 0.9$ and the full azimuthal angle. A pair

⋆e-mail: Sergey.Kiselev@cern.ch

of scintillation hodoscopes, covering $2.8 < \eta < 5.1$ (VZERO-A) and $-3.7 < \eta < -1.7$ (VZERO-B), were used for event triggering and the determination of the event multiplicity (pp, p–Pb) and centrality (Pb–Pb).

3 Analysis procedure

Resonances have been reconstructed via their main hadronic decay channels: $\rho^0 \rightarrow \pi^- + \pi^+$ (B.R.=1.00), $K^{*0} \rightarrow \pi^\pm + K^\mp$ (B.R.=0.67), $\phi \rightarrow K^+ + K^-$ (B.R.=0.49), $\Sigma^{*\pm} \rightarrow \Lambda + \pi^\pm$ (B.R.=0.87), $\Xi^{*0} \rightarrow \Xi^- + \pi^+$ (B.R.=0.67). For the baryonic resonances, the intermediate decay products, Λ and Ξ^-, are identified through selections based on their decay topology. Due to the short lifetimes of these resonances, their decay products cannot be distinguished from the particles coming from the primary vertex, and their yield can only be measured by first computing the invariant mass spectrum of all primary track pairs and then subtracting the combinatorial background. The combinatorial background is evaluated using the event-mixing technique or the like-sign technique. The resulting invariant mass distributions are then fitted. For the ρ^0 a cocktail is used, including a smooth function to describe the continuum residual background, plus peaks accounting for the contributions from K^0_S, $\omega(782)$, K^{*0}, $f_0(980)$ and $f_2(1270)$. The shape of the ρ^0 peak is described by the product of a p-wave relativistic Breit-Wigner function, a phase-space factor, a mass-dependent reconstruction efficiency and a Söding interference term [5]. For the other resonances, the background-subtracted invariant mass distributions are fitted using a Breit-Wigner function or a Voigtian function (convolution of Breit-Wigner and Gaussian which accounts for the detector resolution) plus a polynomial for the residual background, see [6–9] for details. After the signal is extracted, the raw yield is evaluated. The raw yields extracted in different p_T bins are corrected for efficiency and acceptance and the differential transverse momentum spectra are obtained. The corrected spectra are fitted with a Levy-Tsallis function [10] (pp and p–Pb collisions) or with a Blast Wave function [11] (Pb–Pb collisions), for extrapolation in the region where the spectra are not measured.

4 Results

Figure 1 presents the transverse momentum spectra for ρ^0 in pp and Pb–Pb collisions at $\sqrt{s_{NN}} = 2.76$ TeV, $\Sigma^{*\pm}$ and Ξ^{*0} in p–Pb collisions at $\sqrt{s_{NN}} = 5.02$ TeV. The transverse momentum spectrum of the ρ^0 meson in pp collisions has been compared to a number of PYTHIA [12] tunes and the PHOJET [13] event generators. None of them gives a fully satisfactory description of the data.

Figure 2 shows the $\langle p_T \rangle$ of K^{*0}, ϕ and p (which have similar masses) as a function of the cubic root of the average charged-particle multiplicity density, $\langle dN_{ch}/d\eta \rangle^{1/3}$, in the three collision systems pp, p–Pb and Pb–Pb [9]. The $\langle p_T \rangle$ values follow a similar trend with the system size in pp and p–Pb collisions, where they rise faster with multiplicity than in Pb–Pb collisions. An analogous behavior has been observed in [14] for charged particles and can be understood as the effect of color reconnection between strings produced in multi-parton interactions. In central Pb–Pb collisions, the $\langle p_T \rangle$ values for these three particles are consistent within uncertainties, as it is expected in presence of a common radial flow. However in pp and p–Pb collisions the $\langle p_T \rangle$ values for the K^{*0} and ϕ resonances are higher than for p. The mass ordering observed in central Pb–Pb collisions, where particle with similar mass has similar $\langle p_T \rangle$, is not observed in pp and p–Pb collisions.

Figure 3 presents the particle ratios ρ^0/π [15], K^{*0}/K and ϕ/K [7, 9] as a function of $\langle dN_{ch}/d\eta \rangle^{1/3}$ for various collision systems together with THERMUS [16] and grand-canonical [17] thermal models, and EPOS [18] calculations. The ρ^0/π and K^{*0}/K ratios exhibit a decreasing trend towards more central Pb–Pb collisions, where the measured ratios are about 60% of the thermal model predictions.

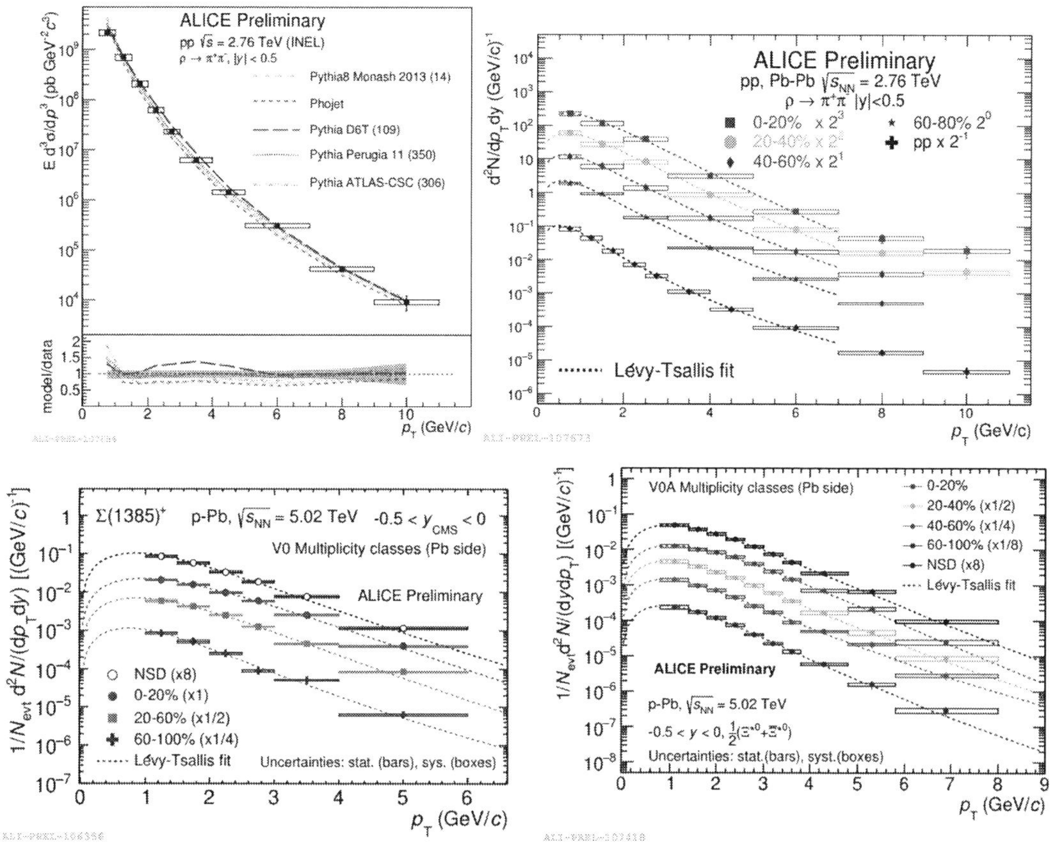

Figure 1. (color online) Transverse momentum spectra for ρ^0 in pp (top left) and Pb–Pb (top right) collisions at $\sqrt{s_{NN}} = 2.76$ TeV, $\Sigma(1385)^{\pm}$ (bottom left) and $\Xi(1530)^0$ (bottom right) in p–Pb collisions at $\sqrt{s_{NN}} = 5.02$ TeV.

This suppression is consistent with rescattering of ρ^0 and K^{*0} daughters in the hadronic phase of central collisions as the dominant effect and is at least qualitatively reproduced by calculations using the EPOS model with UrQMD [18]. As described in [7], the K^{*0}/K ratio measured in central Pb–Pb collisions can be used along with predictions from a thermal model and modelling of rescattering after hadronization [2] to obtain an estimate of 2 fm/c for the lower limit of the time between chemical and kinetic freeze-outs. Results on the K^{*0}/K ratio in pp and p–Pb collisions have large uncertainties, but they hint to a multiplicity-dependent suppression, which leaves the door open for the possibility of a hadron-gas phase with non-zero lifetime in high-multiplicity pp and p–Pb collisions. In Pb–Pb collisions, the ϕ/K ratio is nearly flat and agrees with the prediction of the thermal model. This suggests that rescattering effects are not important for ϕ, which has 10 times longer lifetime than K^{*0} and decays mainly after the end of the hadronic phase.

The particle ratios for baryonic resonances $\Sigma^{*\pm}/\Lambda$ and Ξ^{*0}/Ξ are shown in figure 4 for p–Pb collisions as a function of $\langle dN_{ch}/d\eta \rangle$ and for pp collisions. For the $\Sigma^{*\pm}/\Lambda$ ratio no strong dependence on energy or system size is observed. The measured $\Sigma^{*\pm}/\Lambda$ values are consistent with thermal model [16, 17] and PYTHIA8 [20] predictions but higher than the DPMJET [21] value for p–Pb collisions. The Ξ^{*0}/Ξ ratio also does not depend on multiplicity at the LHC. However the measured Ξ^{*0}/Ξ

DOI: 10.1051/epjconf/201713801022

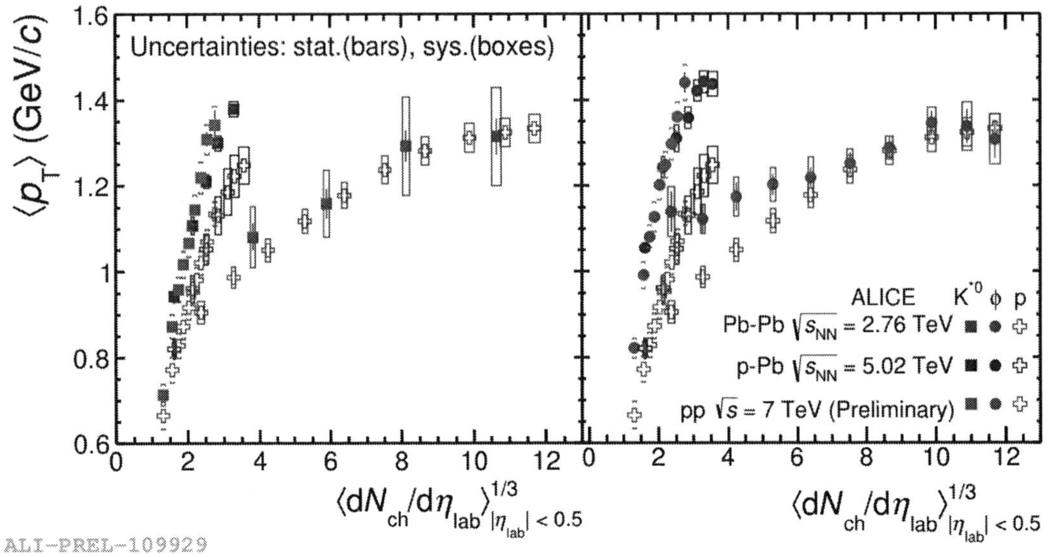

Figure 2. (color online) The mean transverse momentum as a function of $\langle dN_{ch}/d\eta\rangle^{1/3}$ for K^{*0} (left) and ϕ (right) compared to p (same data in both panels).

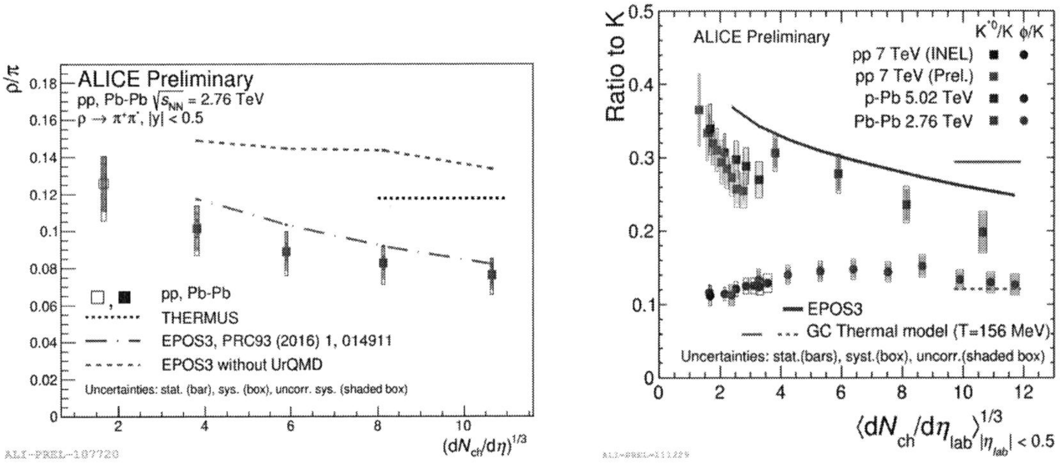

Figure 3. (color online) Ratios ρ^0/π (left), K^{*0}/K and ϕ/K (right) as a function of $\langle dN_{ch}/d\eta\rangle^{1/3}$ for various collision systems. The measurements are also compared to model [16–18] predictions.

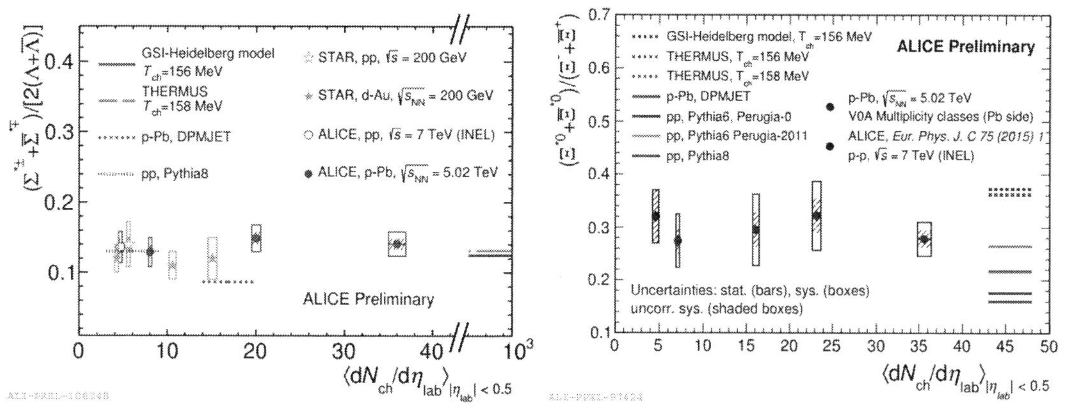

Figure 4. (color online) The particle ratios $\Sigma^{*\pm}/\Lambda$ (left) and Ξ^{*0}/Ξ (right) as a function of $\langle dN_{ch}/d\eta\rangle$ for various collision systems. STAR data are from [19].

values are lower than thermal model [16, 17] predictions and higher than the PYTHIA8 and DPMJET values. Since these ratios are flat, the $\Sigma^{*\pm}$ and Ξ^{*0} are enhanced relative to pions by the same amounts as the Λ and Ξ [22]. Therefore the enhancement of these strange baryons is driven by their strangeness content and not by their masses.

Figure 5 presents the nuclear modification factor of the ρ^0 resonance and stable hadrons as a function of p_T for central and peripheral Pb–Pb collisions at $\sqrt{s_{NN}}$ = 2.76 TeV. At high p_T, ρ^0 is

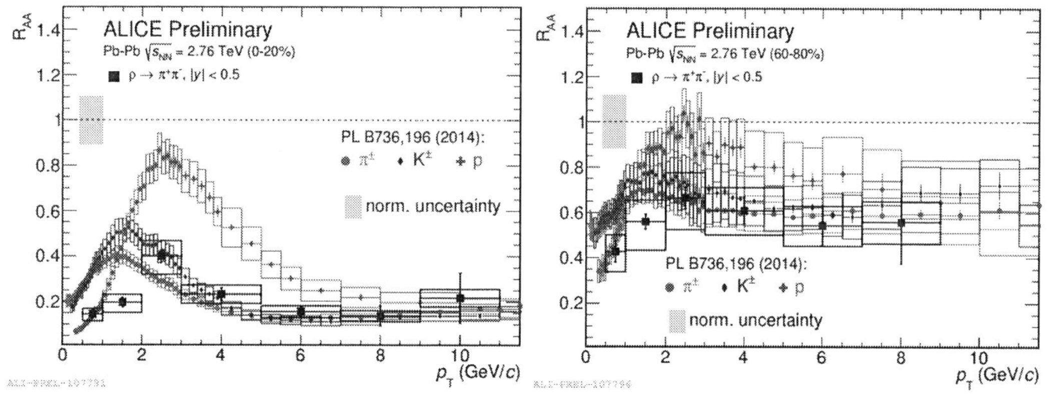

Figure 5. (color online) Nuclear modification factor of the ρ^0 resonance and stable hadrons as a function of transverse momentum for central (left) and peripheral (right) Pb–Pb collisions at $\sqrt{s_{NN}}$ = 2.76 TeV.

strongly suppressed for central collisions, $R_{AA} \approx 0.15$ - 0.2. The suppression is consistent with that measured for the stable hadrons. For peripheral collisions there is a moderate suppression close to the values for the stable hadrons. At low p_T, $p_T < 2$ GeV/c, the ρ^0 is more suppressed than charged hadrons, which would be consistent with the hypothesis that rescattering effects are important.

5 Conclusions

Recent results on short-lived hadronic resonances obtained by the ALICE experiment in pp, p-Pb and Pb-Pb collisions at the LHC energies have been presented. In pp and p–Pb collisions the $\langle p_T \rangle$ values for the K^{*0} and ϕ resonances rise faster with multiplicity than in Pb–Pb collisions. One possible explanation could be the effect of color reconnection between strings produced in multi-parton interactions. The mass ordering observed in central Pb–Pb collisions, where particles with similar masses (K^{*0}, ϕ and p) have similar $\langle p_T \rangle$, is not observed in pp and p–Pb collisions. The ρ^0/π and K^{*0}/K ratios exhibit a significant suppression going from peripheral to central Pb–Pb collisions, consistent with rescattering of the decay products of the short-lived ρ^0 and K^{*0} in the hadronic phase. The ϕ/K ratio is not suppressed due to the longer lifetime of the ϕ. The $\Sigma^{*\pm}/\Lambda$ ratio, measured in pp and p–Pb collisions, does not show a variation neither with the collision energy, nor with the system size. The Ξ^{*0}/Ξ ratios are higher than predicted by event generators in pp and p–Pb. In Pb–Pb collisions the nuclear modification factor of the ρ^0 resonance is consistent with that for other light hadrons at high p_T, where high momentum particle production is suppressed due to parton energy loss in the hot and dense medium created in the collision.

References

[1] G. E. Brown and M. Rho, Phys. Rev. Lett. **66**, 2720 (1991). R. Rapp, Nucl. Phys. A **725**, 254 (2003). E. V. Shuryak and G. Brown, Nucl. Phys. A **717**, 322 (2003)

[2] G. Torrieri and J. Rafelski, Phys. Lett. B **509**, 239 (2001). J. Rafelski, J. Letessier and G. Torrieri, Phys. Rev. C **64**, 054907 (1993). C. Markert, J. Rafelski and G. Torrieri, arXiv:0206260 [hep-ph]

[3] K. Aamodt et al., (ALICE Collaboration), JINST **3**, S08002 (2008)

[4] B. Abelev et al., (ALICE Collaboration), Int. J. Mod. Phys. A **29**, 1430044 (2014)

[5] A. Beddall et al., Acta Phys. Polon. B **39**, 173 (2008)

[6] K. Aamodt et al., (ALICE Collaboration), Eur. Phys. J. C **72**, 2183 (2012)

[7] B. Abelev et al., (ALICE Collaboration), Phys. Rev. C **91**, 024609 (2015)

[8] B. Abelev et al., (ALICE Collaboration), Eur. Phys. J. C **75**, 1 (2015)

[9] B. Abelev et al., (ALICE Collaboration), Eur. Phys. J. C **76**, 245 (2016)

[10] C. Tsallis, J. Stat. Phys. **52**, 479 (1988)

[11] E. Schnedermann, J. Sollfrank and U. Heinz, Phys. Rev. C **48**, 2462 (1993)

[12] T. Sjöstrand, S. Mrenna and P. Skands, J. High Energy Phys. **05**, 026 (2006)

[13] R. Engel et al., Z. Phys. C **66**, 203 (1995) R. Engel and J. Ranft, Phys. Rev. D **54**, 4244 (1996)

[14] B. Abelev et al., (ALICE Collaboration), Phys. Lett. B **727**, 371 (2013)

[15] A. G. Knospe, for the ALICE Collaboration, Conference Proceedings, SQM2016, Berkeley, California, USA, 27 June-1 July 2016, arXiv:1610.09529 [nucl-ex]

[16] S. Wheaton, J. Cleymans, and M. Hauer, Comp. Phys. Comm. **180**, 84 (2009)

[17] J. Stachel et al., J. Phys.: Conf. Ser. **509**, 012019 (2014)

[18] A.G. Knospe et al., Phys. Rev. C **93**, 014911 (2016)

[19] B. Abelev et al., (STAR Collaboration), Phys. Rev. C **78**, 044906 (2008)

[20] T. Sjöstrand, S. Mrenna and P. Skands, Comput. Phys. Comm. **178**, 852 (2008)

[21] S. Roesler, R. Engel and J. Ranft, Conference Proceedings, MC2000, Lisbon, Portugal, October 23-26 (2000) 1033, arXiv:0012252v1 [hep-ph]

[22] J. Adam et al., (ALICE Collaboration), Phys. Lett. B **758**, 389 (2016)

Neutron polarimetry in the momentum range 1-6 GeV

Ying Wang[1]*, *Dominique* Marchand[1], *Nikolay* Piskunov[2], and *Egle* Tomasi-Gustafsson[3]

[1] *Institut de Physique Nucléaire d'Orsay, CNRS/IN2P3 and Univ Paris-Sud, UMR 8608, Univ Paris-Saclay, 91406 Orsay, France*
[2] *Joint Institute for Nuclear Research, VBLHE, 141980 Dubna, Russia*
[3] *CEA, DRF, IRFU, SPhN, Saclay, 91191 Gif-sur-Yvette Cedex, France*

Abstract. In the framework of the ALPOM II experiment, we compare the figure of Merit of two processes for the neutron analyzing power measurement in the reactions $\vec{n} + p \to \vec{n} + p$ and $\vec{n} + p \to p + \vec{n}$ in the beam momentum range from 1 to 6 GeV. Based on a pole model calculation for the cross sections and existing data for analyzing powers, this study suggests that for neutron momentum larger than 3 GeV, the process of $\vec{n} + p \to p + \vec{n}$ is more effective for polarimetry.

1 Introduction

Since Akhiezer and Rekalo suggested the double polarization method for form factors (FFs) measurements in 1968, proton and neutron polarimetries have been revealed to be the most sensitive way to access the nucleon electric to magnetic FFs ratio [1, 2]. The ratio of the electric to magnetic FFs can be directly measured from the ratio of transverse to longitudinal polarization of the recoil nucleon in the eN scattering induced by linearly polarized electrons.

As the measurement of the longitudinal and transverse polarizations (directly related to GE/GM) are done simultaneously, measurements are affected by less systematic uncertainties. Moreover, as radiative corrections mostly cancel in the polarization ratio (at least the factorized terms), radiative corrections affect much less the results than for the Rosenbluth method, based on unpolarized cross section measurements [3].

At Jefferson Lab, the first double polarization experiment dedicated to measure the proton FF ratio was performed in 1998 at transfer momentum $Q^2 \simeq 3.5$ GeV^2 benefiting from the technological advances of the Continuous Electron Beam Accelerator Facility (CEBAF) providing a high current polarized beam [4].

Recently, measurements have been extended to Q^2 around 9 GeV^2 for the proton electromagnetic FFs ratio [5]. The planned approved experiments are expected to measure the proton FFs ratio up to Q^2 12 - 15 GeV^2 [6]. In the neutron case, it is mandatory to use the polarization method for neutron to extract the FFs ratio, due to the smallness of the neutron electric FF. Indeed, before it was possible to apply the polarization method, the data for the neutron FFs ratio could not have the same accuracy as for proton.

*e-mail: wangying@ipno.in2p3.fr

The principle of a polarization measurement is the precise determination of the azimuthal asymmetry of the particle. As the asymmetry becomes very small at high momentum, the polarization experiments can be very time consuming. In order to optimize the polarimetry, it is necessary to choose the most suitable reaction. The measurement of proton analyzing powers had been done by ALPOM at beam momentum 1.75 to 5.3 GeV [7]. To study the neutron analyzing power, elastic np scattering was used as polarizing reaction up to a momentum of 3 GeV [8]. It appears that the neutron analyzing power becomes smaller and smaller at forward angles with increasing energy, whereas it becomes large and negative when the outgoing proton is detected at backward angles [9]. The latter situation corresponds to charge exchange reactions (neutron backward, named CE) and gives a new opportunity to extend neutron polarimetry to high energies.

The purpose of this work is to give an estimation of the expected figure of Merit (FoM) of a polarimeter based on reactions $\vec{n} + p \rightarrow \vec{n} + p$ (neutron forward, named ZE) and $\vec{n} + p \rightarrow p + \vec{n}$ in the ALPOM II energy range with polarized proton and neutron beam momentum up to 7.5 GeV and 4.5 GeV respectively. The present results are based on existing data with the help of the calculation from [10]. The purpose of this paper is to determine at which incident momentum ZE and CE become competitive in terms of FoM.

The experiment ALPOM II is the upgrade of ALPOM [7], which will be carried out at the Veksler and Baldin Laboratory of High Energies (VBLHE) of Joint Institute for Nuclear Research (JINR) in Dubna, produced from the breakup of the deuteron beam accelerated at the Nuclotron. In order to do the measurement at high momentum, a hadron calorimeter has been added to the setup, which makes possible to select particles through their energy deposit.

2 Performance of the Polarimetry

The performance of a polarimeter is expressed in terms of the FoM, $\mathcal{F}$. For a given configuration, as target thickness, acceptance..., it is defined as

$$\mathcal{F}^2 = \int_\theta \varepsilon(\theta)\, A_y^2(\theta)\, d\theta. \tag{1}$$

The integration is performed over the angular range where we assume that the polarimeter is ideally efficient. In (1), $\varepsilon(\theta)$ is the differential efficiency of the process, defined as the differential cross section for the useful reaction (in our case the CE or ZE reaction) integrated over the azimuthal angle ϕ, normalized to the total cross section, σ_{tot}, in order to have a relative normalization. And the notation A_y is for the analyzing power, the single spin polarization observable when the beam is polarized. The total np cross section in the energy range of interest is quite constant and it is taken as $\sigma(N + N \rightarrow anything) \simeq 40$ mb. We can estimate the cross section on a CH2 target as follows:

$$\sigma_{tot}(N + CH2 \rightarrow anything) = 2\sigma(N + N) + A^{2/3}\sigma(N + N) \simeq (80 + 200)\, mb = 280\, mb. \tag{2}$$

The knowledge of FoM allows to estimate the number of incident events, N_{inc}, necessary to obtain a given error ΔP on the polarization measurement. It allows to estimate the duration of the experiment. The larger is the FoM, the shorter is the running time to get the same error on the polarization:

$$N_{inc} = \frac{2}{\Delta P^2 \mathcal{F}^2}. \tag{3}$$

2.1 Efficiency

The efficiencies for these two processes are calculated from the cross sections, assuming an ideal system with 100% experimental efficiency. The formula is,

$$\varepsilon(\cos\theta) = \frac{d\sigma/d\cos\theta}{\sigma_{tot}}. \tag{4}$$

The two diagrams of figure 1 are considered:

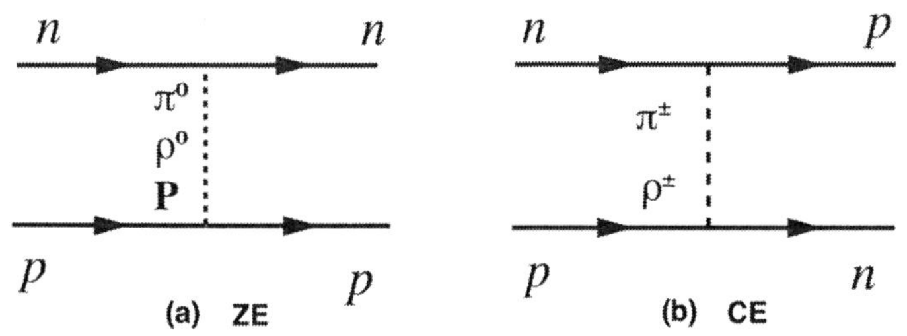

Figure 1. Diagrams of the pole model of elastic $np \to np(pn)$ scattering.

$$\begin{aligned} n(k_1) + p(k_2) &\to n(k_3) + p(k_4) \;\; t - channel\,(ZE), \\ n(k_1) + p(k_2) &\to p(k_4) + n(k_3) \;\; u - channel\,(CE). \end{aligned} \tag{5}$$

The Mandelstam variables are defined in terms of four-momenta as:

$$\begin{aligned} t &= (k_1 - k_3)^2 = (k_2 - k_4)^2, \\ u &= (k_1 - k_4)^2 = (k_2 - k_3)^2, \\ s &= (k_1 + k_2)^2 = (k_3 + k_4)^2. \end{aligned} \tag{6}$$

We use the same mass for neutron and proton $M_n = M_p = M$.

The formula and parameters of the pole model for t−dependent differential cross section are taken from [10],

$$\frac{d\sigma}{dt} = \frac{1}{64\pi s q^2}(|T_\pi(u) + T_\rho(u)|^2 + \frac{1}{4}|T_\pi(t) + T_\rho(t)|^2 + |T_P(t)|^2), \tag{7}$$

where $q^2 = s/4 - M^2$ and $T_{\pi/\rho,P}$ are amplitudes corresponding to π, ρ, pomeron exchanges respectively. The predictions following the (7), are illustrated in figure 2 for different values of the incident neutron beam momentum. The model gives a good description of the data for all values of the beam momentum at small t, and a small deviation at large values of t. However, the differences are quite small, always within a factor of 10. We plot in log-scale to highlight the small difference between data and model (figure 2). As expected, ZE is dominant at forward angles, while CE is dominant for backward angles. The same result can also be seen in figure 3 (left) where the s-dependence of the cross section is calculated for $t = 0$, in the forward region. The total cross sections, integrated over the full angular range, are shown in figure 3 (right). Both show a decreasing behavior when the momentum increases. The difference between CE and ZE contributions grows in the region from 0 to 2 GeV, and above 2 GeV, it becomes constant. According to the calculation of the efficiency, figure 3 shows that the ZE channel dominates the cross section. The other component of FoM formula, the analyzing power, is discussed in the next section.

DOI: 10.1051/epjconf/201713801023

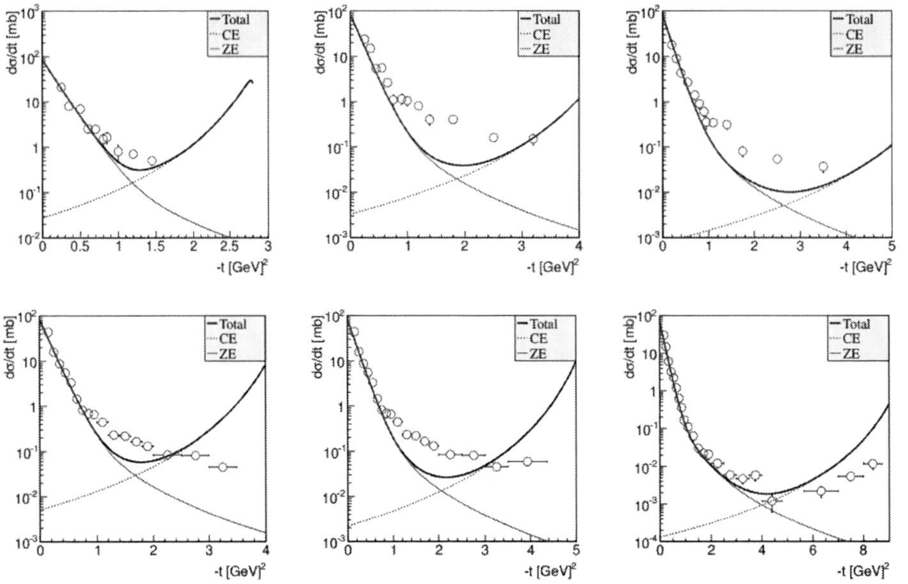

Figure 2. Top figures: data are from [11], corresponding to different incident beam momenta: P_{Lab}=2.241 GeV (left), 3.305 GeV (middle), and 4.337 GeV (right). Bottom figures: data are from [12] and correspond to beam momenta: P_{Lab}= 3.05 GeV (left), 3.570 GeV (middle), 6.120 GeV (right). The model prediction, [10], is also shown (black thick solid line) together with the ZE (red solid line) and CE (blue dotted line) contributions.

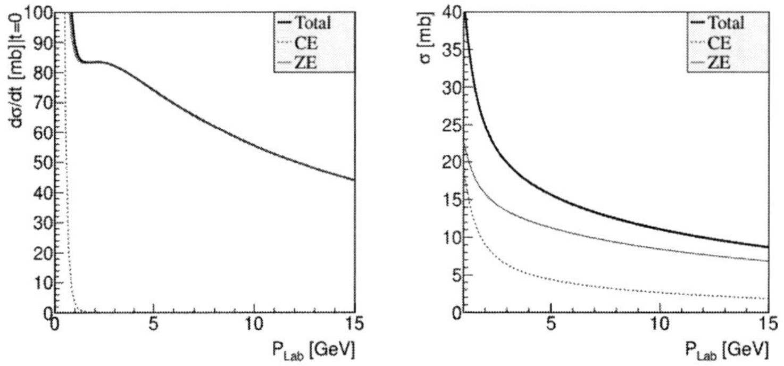

Figure 3. (Left: differential cross section as a function of P_{Lab}, according to (7) for t = 0. Reference [10], right: total cross section as a function of P_{Lab} (black solid thick line), ZE (thin red solid line), and CE (blue dotted line).

2.2 Analyzing Powers

To compare the analyzing powers of CE and ZE reactions, in figure 4 we plot the maximum absolute value of the polarization as a function of the incident momentum, with a linear function to extrapolate to the higher momentum region. The results have to be considered upper limits. With increasing momentum, the maximum absolute values of ZE and CE have an opposite behavior. While the ZE analyzing power decreases with increasing momentum, the CE analyzing power increases, which compensates the cross section contribution of CE reaction.

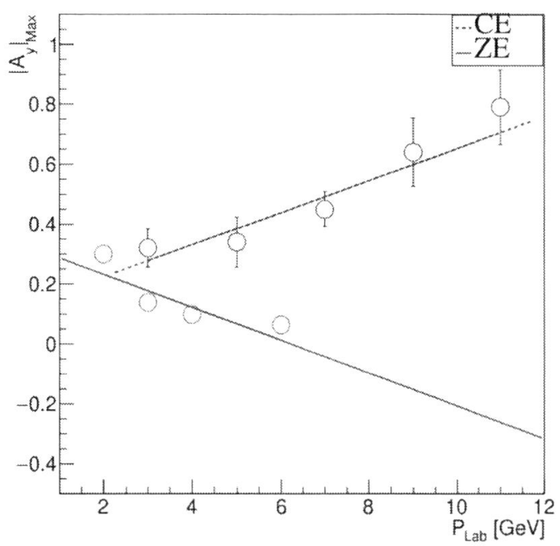

Figure 4. $|A_y|_{Max}$for $np \to np(pn)$ based on the polarization data from CE (blue dotted line) from [9] and ZE (red solid line) from [13] with the parameters as in table 1

Table 1. ZE parameters corresponding to figure 4.

CE	y = 0.1765 + 0.0425 x
ZE	y = 0.35 - 0.0536 x

2.3 Figure of Merit

The squared differential FoM (for CE and ZE) is defined as:

$$\mathcal{F}^2(\cos\theta) = A^2_{y(Max)} \cdot \varepsilon(\cos\theta). \tag{8}$$

Following this definition we calculate the integrated FoM (see figure 5) in the full angular range. As discussed above, the cross sections of both reactions decrease with increasing momentum, but CE is far smaller than ZE in the low momentum region, where the analyzing powers do not compensate this effect. So, in this momentum region, ZE has larger FoM. But for larger values of p_{Lab}, above 3 GeV, the analyzing power having a more important weight than the cross section in the figure of Merit, the CE reaction gives larger FoM, which is more effective for polarimetry. For the same running time, the measurement will be more precise, having smaller errors. In figure 5 the red empty circles are calculated from the Monte Carlo simulation of JLab proposal PR-11-009 on the basis of the ZE reaction [14]. This comparison is done for an ideal apparatus with 100% efficiency and acceptance. The interesting momentum range has to be validated by experimental data. The crossing point, attributing a factor of ten uncertainty in the cross section, may vary in the range 3 to 5 GeV.

3 Discussion and conclusion

In ALPOM II, at momenta of 3 GeV and higher, assuming 10^6 incident particles, the uncertainty of the polarization (ΔP) can be estimated to 0.045 (ZE) and 0.030 (CE), in spite of the fact that

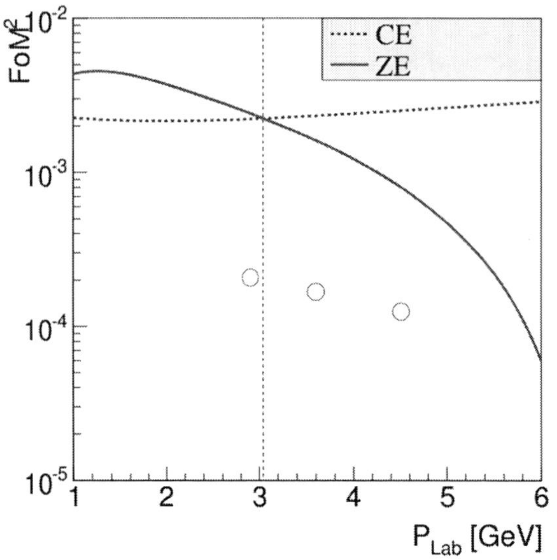

Figure 5. Integrated figure of Merit for $np \to np(pn)$ based on the fitting result from figure 4 (upper limit) for CE (blue dotted line) and ZE (red solid line). The red empty circles are calculated from the Monte Carlo simulation of JLab proposal PR-11-009.

the total cross section for ZE is larger than for CE. With increasing beam momentum, FoM of ZE decreases dramatically while CE stays about constant, then the CE becomes preferable, under the present assumptions of ideal experimental environment. However, our calculation can predict elastic scattering only up to 6 GeV, where analyzing powers have been measured. The model is not reliable for further extrapolation. Moreover, the uncertainty on the calculated figure of merit is essentially related to the uncertainty of the calculation and the lack of experimental data. Attributing a factor of 10 uncertainty on the CE cross section, affects the figure of merit (squared) by the same factor. The crossing is expected in a range of ΔP of 3 to 5 GeV.

In conclusion, we have shown the results of the comparison for FoMs of CE and ZE processes and we suggest to use the CE reaction for the neutron analyzing power measurement when the beam momentum is above 3 GeV. In ALPOM II experiment, the proton analyzing power will also be measured which is useful for the proton FF measurement [15].

References

[1] A. Akhiezer, M.P. Rekalo, Sov.Phys.Dokl. **13**, 572 (1968)

[2] A. Akhiezer, M.P. Rekalo, Sov.J.Part.Nucl. **4**, 277 (1974)

[3] M. Rosenbluth, Phys.Rev. **79**, 615 (1950)

[4] H. Anklin, L. deBever, K. Blomqvist, W. Boeglin, R. Böhm, M. Distler, R. Edelhoff, J. Friedrich, D. Fritschi, R. Geiges et al., Physics Letters B **428**, 248 (1998)

[5] A. Puckett, E. Brash, O. Gayou, M. Jones, L. Pentchev et al., Phys.Rev. C **85**, 045203 (2012). 1102.5737

[6] L.P. Pentchev, C.F. Perdrisat, C. Cisbani, V. Punjabi, B. Wojtsekhowski, JLab E12-07-109 (2007)

[7] L. Azhgirey, V. Arefev, S. Basilev, Y. Bushuev, V. Glagolev et al., Nucl.Instrum.Methods (2004)

[8] T. Eden, a. Madey, R. et, Phys. Rev. C **50**, R1749 (1994)

[9] M.A. Abolins, M.T. Lin, R.C. Ruchti, J.G. Horowitz, R.C. Kammerud, N.W. Reay, K. Reibel, N.R. Stanton, K.W. Edwards, D.G. Crabb et al., Phys. Rev. Lett. **30**, 1183 (1973)

[10] Y. Troyan, M. Anikina, A. Belyaev, A. Ierusalimov, A. Troyan, Physics of Particles and Nuclei Letters **11**, 101 (2014)

[11] M.N. Kreisler, F. Martin, M.L. Perl, M.J. Longo, S.T. Powell, Phys. Rev. Lett. **16**, 1217 (1966)

[12] B.G. Gibbard, L.W. Jones, M.J. Longo, J.R. O'Fallon, J. Cox, M.L. Perl, W.T. Toner, M.N. Kreisler, Phys. Rev. Lett. **24**, 22 (1970)

[13] R. Diebold, D.S. Ayres, S.L. Kramer, A.J. Pawlicki, A.B. Wicklund, Phys. Rev. Lett. **35**, 632 (1975)

[14] J. Arrington et al., JLab PR-11-009 (2011)

[15] N. Piskunov et al., *np charge exchange polarimetry in GeV region*. Report at XXIII International Baldin Seminar on High Energy Physics "Problems, Relativistic Nuclear Physics and Quantum Chromodynamics" (Dubna, Russia, September 19-24, 2016). http://relnp.jinr.ru/ishepp-xxiii/Presentations/Piskunov.pdf

EPJ Web of Conferences **138**, 01024 (2017) DOI: 10.1051/epjconf/201713801024
Baldin ISHEPP XXIII

η bound state: theory and experiment

Hartmut Machner[1,*]

[1] *Fakultät für Physik, Universität Duisburg-Essen, Lotharstr. 1, 47048 Duisburg, Germany*

Abstract. The search for a quasi bound η meson in atomic nuclei is reviewed. This tentative state is studied theoretically as well as experimentally.

1 Introduction

This report is based on a recent review by the author [1]. Other reviews covering the topic in part are in [2] and [3]. Atomic nuclei are built up by nucleons. The quark substructure is not visible. The nucleons are bound together by the strong force. The replacement of nucleons by Λ- or Σ-hyperons was successfully studied, by producing them via recoil-free kinematics, i. e. they are produced at rest. Bound system consisting of an atomic nucleus and another negatively charged particle, where binding appears due to the Coulomb interaction, are atoms. The successful method of producing pionic atoms was again recoil free kinematics. We will come back to this point.

Since the η is electrically neutral, binding can occur only via the strong interaction. The observation of an η bound state would be the first time that a boson is bound in a nucleus. Since a boson is not effected by the Pauli principle, it can be in a state where the nucleon density is maximal. Observation of such a state would allow to study the in medium properties of bound objects. The η is short lived ($\tau = 5 * 10^{-19}$ s) and therefore secondary beams of them are impossible.

It is well known that the Delta resonance $P_{33}(1232)$ plays a dominant role in case of pion production. It seems that the $S_{11}(1535)$ plays a similar role in case of η production although cross sections in this case are much smaller.

The scattering lengths of the pion-nucleon interaction is rather small and as a result the strong interaction shift in the $1s$ state of pionic atoms is repulsive. Contrary to this, the η-nucleon interaction at small momenta is attractive and rather strong. This was first pointed out by Bhalerao and Liu [4] and later applied by Haider and Liu [5] to predict quasi bound η mesons in atomic nuclei for mass numbers $A \geq 12$. In the following text we apply the standard sign convention in meson physics [6] for the s wave scattering parameters $p \cot \delta_0 = 1/a + 1/8(r_0 p^2)$ with p the η momentum, δ_o the s wave phase shift, a the scattering length and r_0 the effective range. For a real attractive potential $a_r < 0$ means binding. Contrary to the $\pi - N$ systems where the scattering length is real at very small energies here the $\eta N \rightarrow \pi N$ channel is always open and hence the scattering length is complex. From such large values for the scattering length $a(\eta N)$, Haider and Liu[5] have shown that η can be bound in nuclei with A $\geq$ 12. In the following text we frequently use the term bound state instead of the more strict quasi bound state. This is common in the literature.

*e-mail: hartmut.machner@uni-due.de

DOI: 10.1051/epjconf/201713801024

2 Theoretical Considerations

A state is called a bound state in the usual sense when the sum of its constituent masses is larger than the mass of the composite. In non-relativistic quantum mechanics binding is represented by an attractive potential and the state is a solution of the radial Klein-Gordon equation. These solutions lie on the imaginary axis in the momentum plane with $\mathrm{Im}(p) = p_i > 0$. However, a possible η bound state is not stable since always the interaction

$$\eta + N \to \pi + N' \tag{1}$$

with a nucleon N is possible. If the η bound state was in a s-state the energy of the final state is $m_\eta + m_N - B_\eta = m_\pi + m_{N'} + T_\pi + T_{N'}$ with T the kinetic energies in the final state and B_η the binding energy. Here we have neglected Fermi motion of the nucleon and the recoil of the residual nucleus. Assuming a binding energy $B_\eta = 10$ MeV, this leads to $T_\pi \approx 317$ MeV and $T_{N'} \approx 47.3$ MeV. These energies are clearly too large for the two final state particles to stay in the nucleus. Because of the possible decay of the state it is a quasi bound state and this fact is accounted for by a complex potential. The task is now to produce a complex potential for elastic scattering $\eta N \to \eta N$, construct from this a complex $\eta A \to \eta A$ potential and then search for poles in the upper part of the second quadrant in the complex plane.

The η-nucleon scattering length $a(\eta N)$ or more generally the matrix $T(\eta N \to \eta N)$ is quite poorly known. As stated above the lifetime of η's is short, so $a(\eta N)$ or $T(\eta N \to \eta N)$ has to be extracted in rather indirect ways. The inputs are production cross sections of $\pi^- p \to \eta n$ and $\gamma p \to \eta p$ reactions. Also decays into the channels γN, πN, $\pi\pi N$ and ηN were considered. The major mechanism that generates the imaginary part of $a(\eta A)$ is the reaction $\eta A_i \to N^*(A-1) \to \pi A_f$, where N^* is the nucleon resonance $N^*(1535)$ with a strong coupling to both the η and the pion [7], [8]. The deduced values for the scattering length range from $0.22 + i0.235$ fm to $1.14 + i0.31$ fm.

The standard approach is to construct from the η-nucleon scattering length an optical potential for the η-nucleus interaction with A the mass number of the nucleus, and then to solve a wave equation with this potential [5], [9], [8], [10]. The complex optical potential is given by

$$U_{\mathrm{opt}} = V + iW = -\frac{2\pi}{\mu} T(\eta N \to \eta N) A\rho(r) \tag{2}$$

with μ the reduced ηN mass, $T(\eta N \to \eta N)$ the η-nucleon transition matrix and ρ the nuclear density. We will call this relation as the $T\rho$ approximation. In the impulse approximation the relation

$$a(\eta N) = T(\eta N \to \eta N, \sqrt{s_0}) \tag{3}$$

with $\sqrt{s_0} = m_\eta + m_N$ holds. For a bound state one needs to know the T matrix at $\sqrt{s} = \sqrt{s_0} - B_\eta$ i.e. below threshold.

Different groups employed different wave equations to search for poles. This leads of course to different results. Another source of ambiguities are the different η nucleon scattering lengths.

Here we study the importance of the input on the final result. We compare the mass dependence of the binding energy and the width for the two extreme values of the scattering length. Such a comparison was made in [9]. The results with the smallest value of the scattering length is from [12] and the largest from [11]. The results are shown in figure 1. The larger real part of the scattering length leads to rather strong binding. However, the imaginary parts although are almost identical lead to strongly different widths. The Green-Wycech results give an almost mass independent width.

A possible way to extract the properties of a bound state is to extract the η nucleus scattering length from the final state interaction [6]. One has to measure an excitation function of a reaction

$${}^{Z_1}A_1 + {}^{Z_2}A_2 \to \{{}^{Z_1}A_1 + {}^{Z_2}A_2\}_{gs} + \eta \tag{4}$$

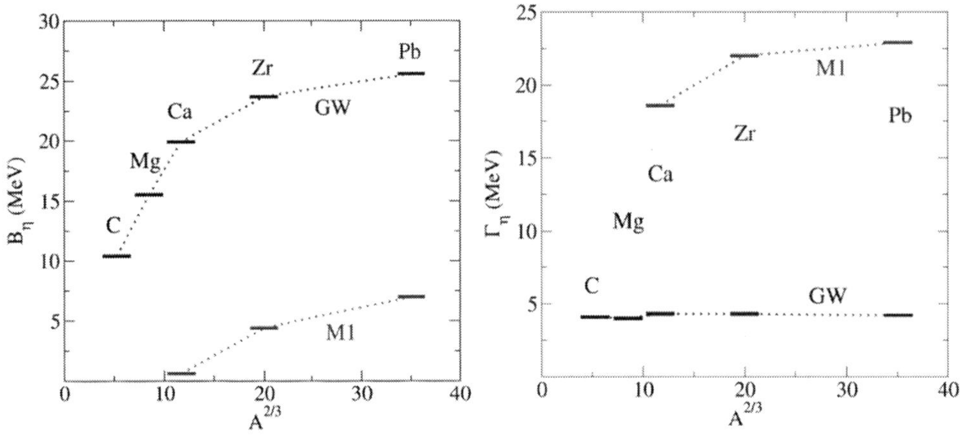

Figure 1. Theoretical calculations of the complex energies for η-bound nuclei in $1s$ state(from [9]). The results labeled with GW use the Green-Wycech [11] result of the scattering length $a(\eta N) = 0.97 + i0.26$ fm, while those labeled M1 use the result from Mai et al. [12] $a(\eta N) = 0.22 + i0.24$ fm. Left frame: the binding energy, right frame the width.

with $\{^{Z_1}A_1 +^{Z_2} A_2\}_{gs}$ the fused nuclear system in its ground state and the η relative to that in a s state. One can either measure the η or the fused nuclear system. The measurement of the decay of the η into photons does not allow the conclusion that the nuclear system is in its ground state due to the limited resolution in the two photon detection. Instead one measures the four-vector of the fused nucleus together with particle identification and reconstructs the properties of the η. This, however, limits the method to light nuclei.

The method is to extract the effective range parameters from the matrix element

$$|f_s|^2 = \frac{d\sigma_s}{d\Omega}\frac{p_i}{p_f} \tag{5}$$

with p the momenta in the incident and final state in the centre-of-mass (cm) system and $d\sigma_s/d\Omega$ the s wave part of the cross section, as it will be discussed in the next section. The scattering length a and effective range r_0 parameters have to be complex since always the channel $\eta + N \rightarrow \pi + N$ is open. Because the square of a is fitted to the data the sign of a_r cannot be found from such measurements. The case with more than one s wave will be discussed below.

One can naively assume that the s wave part of the cross section close to threshold is just $d\sigma_s/d\Omega = \sigma_{tot}/4\pi$. However, often other waves than just the s wave contribute to the total cross section even close to threshold. In this case the decomposition of the total cross section into partial waves is possible from the knowledge of spin observables in addition to cross sections. In the following paragraphs we will give some theoretical prerequisites allowing to extract the s wave contribution from measurements.

In the simplest approach the s wave amplitude is related to the scattering length via $f_s(p) = f_B/(1 - iap)$ with f_B the production amplitude. $f_s(p)$ has a pole in the complex plane that occurs for $p_0 = -i/a = -ia^*/|a|^2$ With $E = p_0^2/(2\mu_{\eta A})$ we find $B_\eta = (a_r^2 - a_i^2)/(2\mu_{\eta A}|a|^4)$ and $-\Gamma_\eta/2 = 2a_r a_i/(2\mu_{\eta A}|a|^4)$. For $B_\eta > 0$ follows $|a_r| > |a_i|$ and $a_r a_i < 0$. Unitarity requires $a_i > 0$ and therefore $a_r < 0$.

3 Experiments

The experimental searches for bound or quasi bound states is not a story of successes. Early experiments [13] were later shown not to be quasi free [14] and hence no effect could be seen. Other experiments [15], [16] and [17] were not conclusive. See [1] for a detailed discussion.

A photoproduction experiment was performed at the MAMI accelerator in Mainz making use of the TAPS spectrometer [18]. A tagged photon beam with 800 MeV maximum energy on a ^{3}He target was used. The reaction studied was

$$\gamma + {}^3\text{He} \rightarrow \pi^0 + p + X \tag{6}$$

as a function of W, which is the cm energy reduced by the deuteron mass and the ^{3}He binding energy. An enhancement was found in the difference of the spectra of the angular range 180° to 170° and 170° to 150°. The authors [18] claimed to have seen a bound η state which implies that the first occurred step $\gamma +^3$ He $\rightarrow \eta \otimes^3$ He was followed by $\eta + p \rightarrow N^{**} \rightarrow \pi^+ + p$.

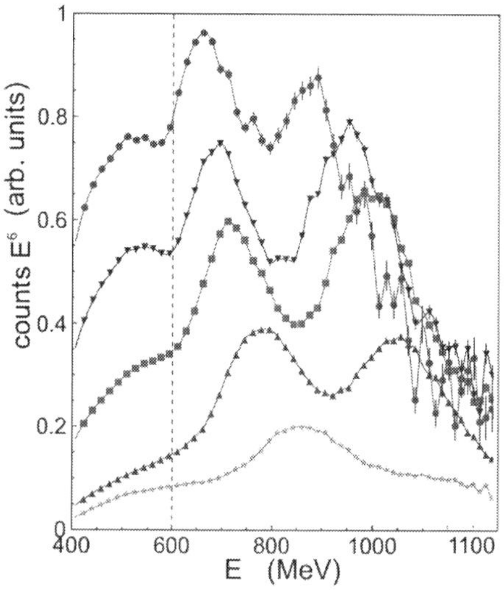

Figure 2. Excitation functions of $\pi^0 p$ back-to-back pairs for different ranges of the opening angle $\theta_{\pi^0} + \theta_p$ after removal of the overall energy dependence $\propto E_\gamma^{-6}$. From top to bottom opening angle ranges of: 165° - 180°, 150° - 165°, 140° - 150°, 130° - 140°, and 120° - 130°. The vertical line indicates the η-production threshold.

Almost the same group repeated the experiment with again the TAPS spectrometer plus the Crystal Ball detector [19]. The experiment benefitted not only from the now almost 4π acceptance but also from much higher statistics. The photon energies ranged from 0.45 GeV to 1.4 GeV. The result of this measurement is shown in figure 2. The strong rise of the $\pi^0 p$ cross section above the η production threshold is similar to the previous experiment and supports the possibility of a resonance in the threshold region. However, the structures visible at higher energies have not been seen in [18]. They are in the so-called second and third resonance region and their walk with angle is purely kinematical.

The WASA collaboration [20] studied the reaction $d + d \rightarrow \pi^- + p + {}^3$He. The idea is that an intermediate $\eta\alpha$ bound state might exist. The whole reaction chain is then

$$d + d \rightarrow \eta \otimes \alpha \rightarrow N^*(1535) + {}^3\text{He} \rightarrow (\pi^- + p) + {}^3\text{He}. \tag{7}$$

The deuteron beam momentum varied between 2.185 GeV/c and 2.400 GeV/c. No anomaly in the excitation function for beam momenta below and above threshold has been seen.

Experiments employing transfer reactions are favourable; in such experiments the whole beam momentum can be transferred to a nucleon or a cluster of nucleons. The remaining system then does not carry linear momentum and thus favours the probability that a produced η is bound to the residual nucleus. This method, originally developed in the production of hypernuclei [21], was successfully applied in the study of pionic atoms [22]. In order to transfer the beam momentum almost completely to the emerging particle it has to be emitted in the forward direction close at zero degree. One such experiment [23], [24] employed the GSI fragment separator. The search was done with the $d,^3he$ reaction. The spectrometer is flooded by break up protons having beam velocity and therefore the same magnetic rigidity p/Z as the ^{3}He particles of interest and are thus undistinguishable. So far no final result is published [25].

One nucleon transfer guarantees a rather large cross section. This is not the case for two nucleon transfer reaction $A + p \to (A-1) \bigotimes \eta + {}^3He$. However, it is just this experiment by the GEM collaboration [26] which claims to have observed an η mesic bound state with sufficient significance. We will therefore discuss this experiment in more detail. The experiment made use of two signatures simultaneously: transfer reaction with recoil free kinematics and back to back emission of a pion and a nucleon from a possible reaction chain $\eta + N \to N* \to N' + \pi$ with the N^* almost at rest. A proton beam from the COSY Jülich accelerator with momentum of 1745 MeV/c was used, where η mesic states with binding energies -30 MeV $\leq B_\eta \leq 0$ MeV can be produced with a momentum transfer $q \leq 30$ MeV/c. The high resolution magnetic spectrograph Big Karl [28] was used to identify ^{3}He

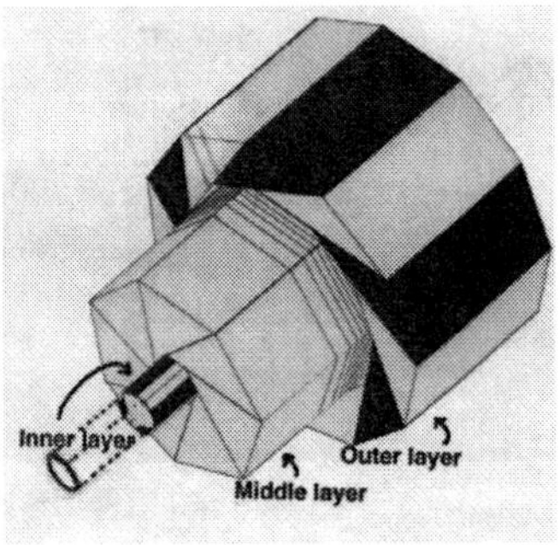

Figure 3. The ENSTAR detector [27] surrounding the target. It consists of wedges from scintillating material. Read out is performed by scintillating fibres collecting the light in grooves milled in the wedges and transporting it to photo tubes. One half of this detector is shown. The inner two layers are extruded for clarity.

ions and their momenta. The decay into a proton and π^- with the two final particles emitted almost back to back to each other was measured with a dedicated detector ENSTAR [27]. It surrounds the target and one half of it is shown in figure 3. By construction it is capable of determining azimuth and polar angle.

We want now to discuss the effect of the conditions applied to missing mass spectrum or binding energy spectrum. The momenta measured in the FP are shown in figure 4, converted to binding energy. This spectrum shows a peak on a continuum. This continuum was parameterized by a constant as well as polynomials while for the peak a Gaussian was assumed. In addition fits were performed applying Poisson statistics. The significance of the peak is around 5σ [26]. The centroid E_B and Gaussian width σ were found to be -12.0±2.2 MeV and 4.7±1.7 MeV.

The procedure applied by the GEM collaboration [26] to assume a further background below the peak was questioned by Haider and Liu [29]. The final state can also be reached by a non resonant reaction for which they used a microscopic-theory based nearly energy-independent amplitude. The need of adding non-resonant amplitude is further discussed in [30]. Then there will be an interference between this amplitude and the one for the resonant production. They fitted the corresponding amplitudes to the experimental data and found indeed a serious interference effect which shifts the calculated Breit-Wigner maximum towards the experimental maximum. The same is true for the width. One such fit is also shown in figure 5. For this curve the ηN scattering length is (0.250 + 0.123i) fm.

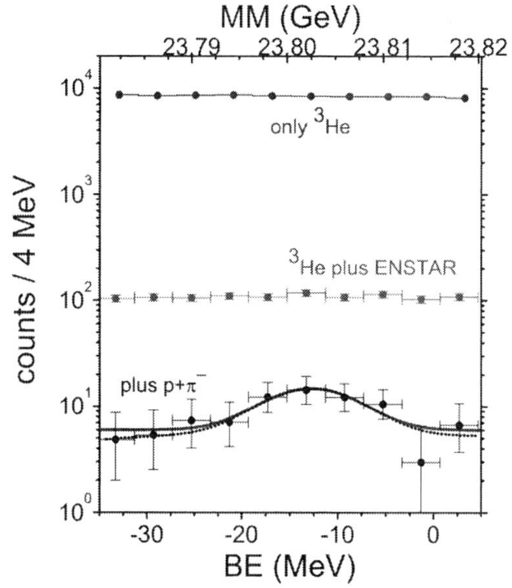

Figure 4. Binding energy spectra. Upper curve: data with the requirement of a ^{3}He in the focal plane. Middle curve: a coincidence between ^{3}He and the ENSTAR detector. Lower curve: coincidence between ^{3}He and a π^- and proton being back to back (bb) emitted recorded in the ENSTAR detector. The solid curve is a fitted Gaussian together with a constant, the dashed curve a gaussian plus a polynomial.

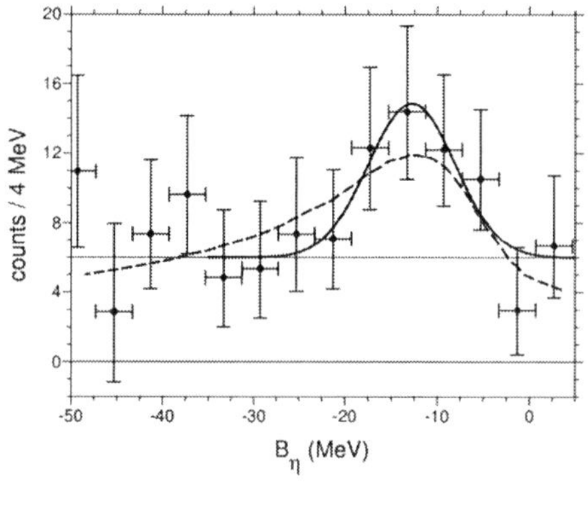

Figure 5. The final binding energy spectrum. Note the expanded view and the different binning compared to figure 4. The data are shown with Poisson error bars. A fit to the data with a Gaussian and a constant background are shown (solid curve). A fit with a Breit-Wigner form with a coherent non resonant fraction is shown as dashed curve (from [29]).

Another method proposed to search for η bound states is to study the final state interaction (FSI) between the η meson and a nucleus.

The most intensively system studied is the $d + p \rightarrow ^3$He$+\eta$ reaction. Data are from Refs. [31], [32], [33], [34], [35], [36], [37]. Those close to threshold are shown in figure 6. In this figure the

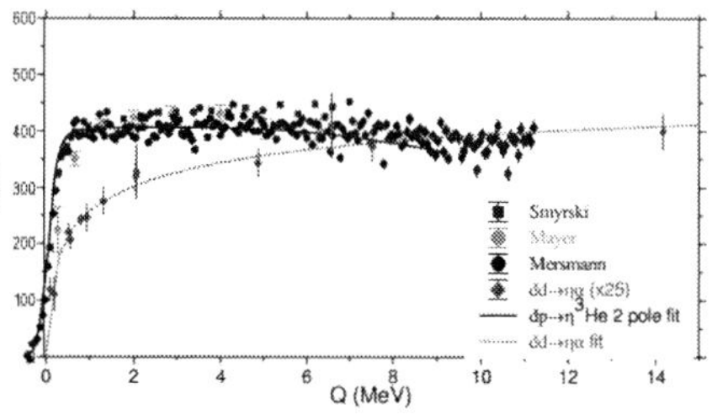

Figure 6. Comparison of the excitation functions for the two reactions $d + p \rightarrow \eta + ^3$He and $d + d \rightarrow \eta + ^4$He. The total cross sections for the former reaction are from [35] (diamonds), [36](squares), and[37] (dots, five points together). The solid curve is the two pole fit. Data for the latter reaction (triangles up) are from Frascaria et al. [38], Willis et al. [39], Wronska et al. [40], and the GEM collaboration Budzanowski et al. [41] The dotted curve is the scattering length fit to these data.

recently published data are compiled. Obviously there are differences between the different data sets. This leads of course to different results for the final state parameters. COSY 11 [36], quoted also as Smyrski, and ANKE [37], quoted also as Mersmann, were measured at COSY applying the internal deuteron beam. The momentum of the beam increased linearly with time. Data were taken continuously and later put into bins with widths ΔQ. Details of corrections for nonlinearity within bins and correction due to finite beam resolution are given in [36] and [42]. COSY 11 [36] applied (2) to their data and obtained

$$a_{\eta^3He} = \pm(2.9\pm2.7) + i(3.2\pm1.8)\text{ fm}. \tag{8}$$

This corresponds to a possible bound state at $B_\eta = -0.2 \pm 0.8$ MeV. So this result points more to a virtual than to a bound state. The half width is $\Gamma/2 = 1.9 \pm 0.4$ MeV. However, when we repeated the fit for the data from [36] we found different values and moreover they depend on the fit interval. While the value for a_i is quite stable, a_r varied from 0.0 ± 6000 fm, when the full data set is included in the fit, to 2.1 ± 2.7 fm, when the range is limited to 2.2 MeV. The large error is an indication that the option of fitting the full range is useless, because the assumption of pure s wave is wrong. The imaginary part is 3.6 ± 1.2 fm. These numbers are in agreement with the published values. This finding is an indication that already for excess energies above 2.2 MeV (2) is no more applicable and the effective range has to be considered in addition to the scattering length as it has been stressed in [43]. On the other hand the ANKE data show after the rapid rise a gentle decrease with increasing energy. Mersmann [42] has performed a corresponding fit to the ANKE data including the smearing as discussed above. This fit yielded $a_{\eta^3He} = [\pm(0.000 \pm 2.416) + i\cdot(6.572 \pm 0.501)]$ fm and $r_{0,\eta^3He} = [(0.000 \pm 2.416) + i\cdot(1.268 \pm 0.212)]$ fm. The scattering length and the effective range are thus determined by the imaginary parts alone. The fit results don't fulfill the criterion $|a_r| > |a_i|$. The ANKE collaboration [37] applied in addition another fitting form. They assumed a two pole representation of the final state interaction

$$f_s(p) = \frac{f_B}{(1-\frac{p}{p_1})(1-\frac{p}{p_2})} \tag{9}$$

with p_1 and p_2 two complex pole positions. From the position of the first pole one gets scattering length and effective range. which are$a_{\eta^3He} = \left[\pm\left(10.7 \pm 0.8^{+0.1}_{-0.5}\right) + i\cdot\left(1.5 \pm 2.6^{+1.0}_{-0.9}\right)\right]$ fm and $r_{0,\eta^3He} = \left[(1.9 \pm 0.1) + i\cdot\left(2.1 \pm 0.2^{+0.2}_{-0.0}\right)\right]$ fm. In obtaining these values a smearing of the energy scale due to a finite beam momentum distribution was applied. This results in a pole (if exists) at $B_\eta = 0.30 \pm 0.15 \pm 0.04$ MeV and $\Gamma_\eta/2 = 0.21 \pm 0.29 \pm 0.6$ MeV.

Although the FSI parameters differ drastically from those of the fit the two fit curves are practically indistinguishable especially in the strong rising part which is decisive for the scattering length. It is somewhat surprising that two fits with five parameters each and a one to one correspondence give so different results.

Measurements of reaction $d + d \to \eta + {}^4$He were reported in [38], [39], [40] and more recently in [41]. The cross section is much smaller than for the previously discussed reaction $d + p \to \eta + {}^3$He.

In a simultaneous analysis of the $d + p \to {}^3$He$+\eta$ reaction and the $d + d \to {}^4$He$+\eta$ reaction in terms of a simple optical model approach [39] it was found that for the possible binding energies the relation $B_\eta({}^3\text{He} + \eta) < B_\eta({}^4\text{He} + \eta)$ holds. However, in the measurements with a polarized beam on which the analysis was based, the full polar angle could not be measured. The s-wave cross section was extracted by assuming isotropic emission. This isn't true for the data measured at the higher momenta as can be seen by comparing to the data from [40]. The anisotropy could be either due to s waves plus p waves or to a $s-d$ wave interference. The problem could be only solved by applying polarized deuterons. Such an experiment was performed by the GEM collaboration at COSY Jülich [41] which will be discussed now in some detail.

The experiment was performed at a deuteron beam momentum of 2385.5 MeV/c corresponding to an excess energy of 16.6 MeV [41]. Recoiling α particles were identified and their four momentum vector measured with the magnetic spectrograph Big Karl [28]. The experiment made use of polarized as well as unpolarize deuteron beams. The experiment had certain acceptances so that the polarized cross section depends practically only on the analyzing power A_{xx}. The angular distribution of the

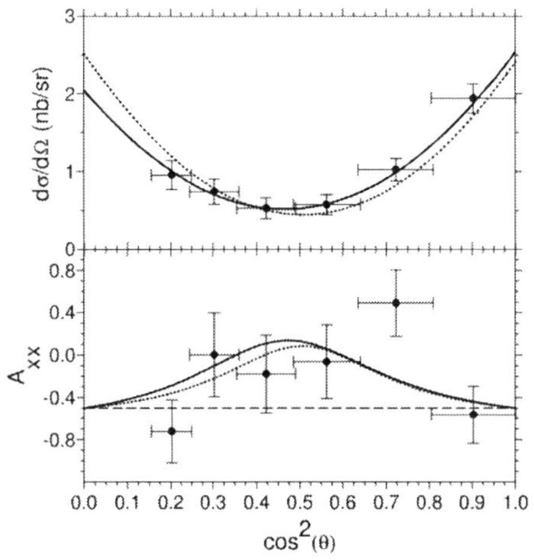

Figure 7. Angular distributions of the unpolarized cross section and the analyzing power A_{xx} (from [41]). The solid curves represent a fit with four partial waves, the dotted curves show fits with helicity amplitudes.

tensor analyzing power and of the unpolarized cross section are shown in figure 7. The s wave amplitude f_s can now be extracted. We find that $|f_s|^2 = 4.4 \pm 1.1$ nb/sr.

For the two highest energy measurements from [39] d wave contributions have to be considered. With the present result previous analysis could be corrected lading It is a good approximation to assume the d wave amplitudes A_2 and B to depend on the η momentum as p_η^2 and apply the results discussed here. This yields $|f_s|^2 = 13.8 \pm 1.2$ nb/sr and $|f_s|^2 = 10.6 \pm 1.3$ nb/sr for the momenta at 73 MeV/c and 91 MeV/c, respectively. For the Wronska result we find $|f_s|^2 = 14.3 \pm 2.4$ nb/sr at 86 MeV/c. We are now in a position to make a comparison of the world data for the s wave amplitude. This is done in figure 6. In a fit the production amplitude and the scattering length were fitted to the data yielding $a_{\eta\alpha} = [\pm(3.1 \pm 0.5) + i(0.0 \pm 0.5)]$ fm. This result corresponds to a bound state - if it exists - of $B_\eta = 3.71 \pm 0.09$ MeV and $\Gamma/2 = 0.0 \pm 0.2$ MeV. In this figure we compare the excitation functions for the present reaction with the one for $d + p \to \eta + {}^3\mathrm{He}$. The latter reaction shows a much more rapid rise than the former. This is an indication of the larger scattering length in case of the lighter system.

Two experiments have been reported leading to the mirror nucleus ^{7}Be. The experiments were performed at SATURNE Saclay [44] and COSY Jülich [45]. Both studies employed the reaction

$$p + {}^6\mathrm{Li} \to \eta + {}^7\mathrm{Be}\,. \tag{10}$$

At Saclay the η was measured through its two γ decay at a beam energy of 683 MeV corresponding to a beam momentum of 1322 MeV/c or to an excess energy of $Q = 19.13$ MeV. In total eight events were observed. Four excited states with $L = 1$ and $L = 3$ can contribute. The other experiment [45] was performed at a beam energy of 673.1 MeV, corresponding to 1310 MeV/c momentum or an excess energy of $Q = 11.28$ MeV. The recoiling ^{7}Be nuclei were detected in the spectrograph Big Karl. Since the $L = 3$ states are particle unstable only the two $L = 1$ states contribute. The standard detectors in the focal plane were not adequate for this experiment since the recoiling particles have rather low energies of $\approx$100 MeV. The MWDC's were replaced by multi-wire avalanche-chambers to measure the track, followed by two layers of scintillation detectors one meter apart. They allow particle identification via TOF measurement. All these devices were housed in a large vacuum box made of stainless steel.

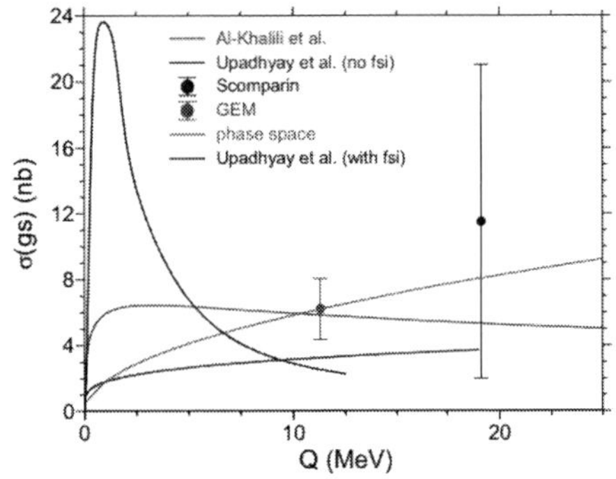

Figure 8. Excitation function for the $p + ^6$ Li $\rightarrow \eta + ^7$ Be(g.s.) reaction. The two data points are from Refs. [44] and [45]. The dashed curve is the Al-Khalili model [46] and the dashed-dotted is the phase space behaviour, both normalized to the GEM data point. The solid and the dotted curves are calculations [47] with and without final state interactions. The arrow indicates the region where only the ground state is involved in the reaction.

The η meson events were identified via the missing mass technique. Finally the counts were converted to a cross section. Assuming isotropic emission one gets the total cross section shown in figure 8. Together with the form factors from [46] and the amplitude $f(pd \rightarrow \eta^3\text{He})$ extracted from the two data sets discussed above, the cross section for the reaction leading to the ^{7}Be ground state could be extracted. The two data are shown in figure 8. Also shown is the energy dependence of the Al-Khalili model [46] normalized to the cross section of the GEM collaboration [45]. Also the normalized phase space dependence is shown. In addition model predictions [47] with and without FSI are shown. A measurement even closer to threshold preferably below the first excited state could distinguish between the different models and could answer whether strong FSI exists in this final channel. Upadhyay et al. [47] got from $a_{\eta N} = (0.88 + i0.41)$ fm a value $a_{\eta^7\text{Be}} = (-9.18 + i8.53)$ fm.

References

[1] H. Machner, J. Phys. G **42(4)**, 043001 (2015)
[2] N. G. Kelkar, K. P. Kemchandani, N. J. Upadhyay, and B. K. Jain, Rep. Progr. Phys. **76**, 066301 (2013)
[3] B. Krusche and C. Wilkin, Prog. Part. Nucl. Phys. **80**, 43–95 (2014)
[4] R. S. Bhalerao and L. C. Liu, Phys. Rev. Lett. **54**, 865 (1985)
[5] Q. Haider and L. C. Liu, Phys. Lett. B **172**, 257 (1986)
[6] M. K. Goldberger and K. M. Watson, *Collision Theory* (John Wiley and Sons, New York, London, Sidney, 1964)
[7] J. Behringer et al. (Particle Data Group), Phys. Rev. D **86**, 010001 (2012)
[8] S. Wycech, A. M. Green, and J. A. Niskanen, Phys. Rev. C **52**, 544 (1995)
[9] E. Friedman, A. Gal, and J. Mares, Phys. Lett. B **725**, 334 (2013)
[10] C. Wilkin, Phys. Rev. C **47**, R938 (1993)
[11] A. M. Green and S. Wycech, Phys. Rev. C **71**, 014001 (2005)
[12] M. Mai, P. C. Bruns, and U.-G. Meißner, Phys. Rev. D **86**, 094033 (2012)
[13] R. E. Chrien et al., Phys. Rev. Lett. **60**, 2595 (1988)
[14] H. Nagahiro, D. Jido, and S. Hirenzaki, Phys. Rev. C **80**, 025205 (2009)
[15] B. J. Lieb et al., in Karen Poelakker, editor, *Progress at LAMPF* (Los Alamos National Laboratory, 1988) LA-11670-PR, 52

[16] G. A. Sokol et al., Fizika B **8**, 85 (1999)
[17] V. A. Baskov et al., arXive nucl-ex:1212.6313 (2012)
[18] M. Pfeiffer et al., Phys. Rev. Lett. **92**, 252001 (2004)
[19] F. Pheron et al., Phys. Lett. B **709**, 21 (2012)
[20] P. Adlarson et al. (WASA collaboration), Phys. Rev. C **87**, 035204 (2013)
[21] W. Brückner et al., Phys. Lett. B **62**, 481 (1976)
[22] T. Yamazaki et al., Z. Physik A **355**, 219 (1996)
[23] R. S. Hayano and A. Gillitzer et al., GSI proposal S214 (1997)
[24] A. Gillitzer, Acta Physica Slovaka **56**, 269 (2006)
[25] A. Gillitzer, priv. communication (2014)
[26] A. Budzanowski et al. (COSY-GEM Collaboration), Phys. Rev. C **79**, 061001(R) (2009)
[27] M. Betigeri et al., Nucl. Instruments and Meth. in Phys. Res. A **578**, 198 (2007)
[28] M. Drochner et al., Nucl. Phys. A **643**, 55 (1998)
[29] Q. Haider and L. C. Liu, J. Phys. G **37**, 125104 (2010)
[30] Lon-Chang Liu and Q. Haider, Acta Physica Polonica **45**, 837 (2014)
[31] J. Berger et al., Phys. Rev. Lett. **61**, 919 (1988)
[32] M. Betigeri et al., Phys. Lett. B **472**, 267 (2000)
[33] H. H. Adam et al., Phys. Rev. C **75**, 014004 (2007)
[34] T. Rausmann et al., Phys. Rev. **80**, 017001 (2009)
[35] B. Mayer et al., Phys. Rev. C **53**, 2068 (1996)
[36] J. Smyrski et al., Phys. Lett. B **649**, 258 (2007)
[37] T. Mersmann et al., Phys. Rev. Lett. **98**, 242301 (2007)
[38] R. Frascaria et al., Phys. Rev. C **50**, R537 (1994)
[39] N. Willis et al., Phys. Lett. B **406**, 14 (1997)
[40] A. Wrońska et al., Eur. Phys. J. A **26**, 421 (2005)
[41] A. Budzanowski et al. (The GEM Collaboration), Nucl. Phys. A **821**, 193 (2009)
[42] T. Mersmann, PhD thesis, University Münster (2007)
[43] J. A. Niskanen and H. Machner, Nucl. Phys. A **902**, 40 (2013)
[44] E. Scomparin et al., J. Phys. G **19**, L51 (1993)
[45] A. Budzanowski et al., Phys. Rev. C **82**, 041001(R) (2010)
[46] J.S. Al-Khalili, M.B. Barbaro, and C. Wilkin, J. Phys. G **19**, 403 (1993)
[47] N. J. Upadhyay, N. G. Kelkar, and B. K. Jain, Nucl. Phys. A **824**, 70 (2009)

 DOI: 10.1051/epjconf/201713801025

Effect of the $\eta\eta$ channel and interference phenomena in the two-pion transitions of charmonia and bottomonia

Yu.S. Surovtsev[1,⋆], *P.* Bydžovský[2], *T.* Gutsche[3], *R.* Kamiński[4], *V.E.* Lyubovitskij[3,5,6], and *M.* Nagy[7]

[1] *Bogoliubov Laboratory of Theoretical Physics, Joint Institute for Nuclear Research, 141980 Dubna, Russia*
[2] *Nuclear Physics Institute of the CAS, 25068 Řež, Czech Republic*
[3] *Institut für Theoretische Physik, Universität Tübingen, Kepler Center for Astro and Particle Physics, Auf der Morgenstelle 14, D-72076 Tübingen, Germany*
[4] *Institute of Nuclear Physics of the PAN, Cracow 31342, Poland*
[5] *Department of Physics, Tomsk State University, 634050 Tomsk, Russia*
[6] *Mathematical Physics Department, Tomsk Polytechnic University, Lenin Avenue 30, 634050 Tomsk, Russia*
[7] *Institute of Physics, SAS, Bratislava 84511, Slovak Republic*

Abstract. The basic shape of di-pion mass spectra in the two-pion transitions of both charmonia and bottomonia states is explained by a unified mechanism based on contributions of the $\pi\pi$, $K\overline{K}$, and $\eta\eta$ coupled channels including their interference. The role of the individual f_0 resonances in shaping the di-pion mass distributions in the charmonia and bottomonia decays is considered.

1 Introduction

In this contribution we report on new results in continuation of our study of scalar meson properties analyzing jointly the data on the isoscalar S-wave processes $\pi\pi \to \pi\pi, K\overline{K}, \eta\eta$ and on the two-pion transitions of heavy mesons, when it is reasonable to consider that the two-pion pair is produced in the S-wave state and the final meson remains a spectator. We analyzed data on the charmonium decay processes — $J/\psi \to \phi\pi\pi$, $\psi(2S) \to J/\psi\pi\pi$ — from the Crystal Ball, DM2, Mark II, Mark III, and BES II Collaborations and practically all available data on two-pion transitions of the Υ mesons from the ARGUS, CLEO, CUSB, Crystal Ball, Belle, and *BaBar* Collaborations — $\Upsilon(mS) \to \Upsilon(nS)\pi\pi$ ($m > n$, $m = 2,3,4,5$, $n = 1,2,3$). Moreover, the contribution of multi-channel $\pi\pi$ scattering (namely, $\pi\pi \to \pi\pi, K\overline{K}, \eta\eta$) in the final-state interactions is considered. The multi-channel $\pi\pi$ scattering is described in our model-independent approach based on analyticity and unitarity and using a uniformization procedure. A novel feature in the analysis is accounting for effects from the $\eta\eta$ channel in the indicated two-pion transitions which is assumed not only kinematically, i.e., via including the channel threshold in the uniformizing variable, but also by adding the $\pi\pi \to \eta\eta$ amplitude in the corresponding formulae for the decays.

We showed that the experimentally observed interesting (even mysterious) behavior of the $\pi\pi$ spectra of the Υ-family decays, which starts to be apparent in the second radial excitation and is also seen for the higher states, — a bell-shaped form in the near-$\pi\pi$-threshold region, smooth dips at about

⋆e-mail: surovcev@theor.jinr.ru

0.6 GeV in the $\Upsilon(4S,5S) \to \Upsilon(1S)\pi^+\pi^-$, about 0.45 GeV in the $\Upsilon(4S,5S) \to \Upsilon(2S)\pi^+\pi^-$, and at about 0.7 GeV in the $\Upsilon(3S) \to \Upsilon(1S)(\pi^+\pi^-, \pi^0\pi^0)$, and also sharp dips near 1 GeV in the $\Upsilon(4S,5S) \to \Upsilon(1S)\pi^+\pi^-$ — can be explained by the interference between the $\pi\pi$ scattering amplitudes, $K\overline{K} \to \pi\pi$ and $\eta\eta \to \pi\pi$, in the final-state re-scattering (by the constructive interference in the near-$\pi\pi$-threshold region and by the destructive one in the dip regions). Note that in a number of works (see, e.g., [1] and the references therein) various assumptions were made to explain this observed behavior of the di-meson mass distributions.

We have explained the basic shape of di-pion mass spectra in the two-pion transitions of both charmonia and bottomonia on the basis of our previous conclusions on wide resonances without any additional assumptions. In works [2, 3] we have shown: If a wide resonance cannot decay into a channel which opens above its mass, but the resonance is strongly coupled to this channel, then one should consider this resonance as a multi-channel state allowing for this closed channel.

2 The effect of multi-channel $\pi\pi$ scattering in decays of the ψ- and Υ-meson families

We have considered the three-channel case of the multi-channel $\pi\pi$ scattering, i.e. the reactions $\pi\pi \to \pi\pi, K\overline{K}, \eta\eta$, because it was shown [2] that this is a minimal number of coupled channels needed for obtaining correct values of f_0-resonance parameters. In the combined analysis, the data for the multi-channel $\pi\pi$ scattering were taken from many papers (see in [3]). For the decay $J/\psi \to \phi\pi^+\pi^-$ data were taken from Mark III, DM2 and BES II Collaborations; for $\psi(2S) \to J/\psi(\pi^+\pi^-$ and $\pi^0\pi^0)$ — from Mark II and Crystal Ball(80) (see also in [3]). For $\Upsilon(2S) \to \Upsilon(1S)(\pi^+\pi^-$ and $\pi^0\pi^0)$ data were used from ARGUS [4], CLEO [5], CUSB [6], and Crystal Ball [7] Collaborations; for $\Upsilon(3S) \to \Upsilon(1S)(\pi^+\pi^-, \pi^0\pi^0)$ and $\Upsilon(3S) \to \Upsilon(2S)(\pi^+\pi^-, \pi^0\pi^0)$ — from CLEO [8, 9]; for $\Upsilon(4S) \to \Upsilon(1S,2S)\pi^+\pi^-$ — from *BaBar* [10] and Belle [11]; for $\Upsilon(5S) \to \Upsilon(1S,2S,3S)\pi^+\pi^-$ — from Belle Collaboration [11].

The di-meson mass distributions in the quarkonia decays are calculated using a formalism analogous to that proposed in [12] for the decays $J/\psi \to \phi(\pi\pi, K\overline{K})$ and $V' \to V\pi\pi$ $(V = \psi, \Upsilon)$ which is extended with allowing for amplitudes of transitions between the $\pi\pi$, $K\overline{K}$ and $\eta\eta$ channels in decay formulae. It is assumed that the pion pairs in the final state have zero isospin and spin. Only these pairs of pions undergo the final state interactions whereas the final $\Upsilon(nS)$ meson $(n < m)$ remains as a spectator. The decay amplitudes are related with the scattering amplitudes T_{ij} $(i, j = 1-\pi\pi, 2-K\overline{K}, 3-\eta\eta)$ as follows

$$F(J/\psi \to \phi\pi\pi) = c_1(s)T_{11} + \Big(\frac{\alpha_2}{s-\beta_2} + c_2(s)\Big)T_{21} + c_3(s)T_{31}, \quad (1)$$

$$F(\psi(2S) \to \psi(1S)\pi\pi) = d_1(s)T_{11} + d_2(s)T_{21} + d_3(s)T_{31}, \quad (2)$$

$$F(\Upsilon(mS) \to \Upsilon(nS)\pi\pi) = e_1^{(mn)}T_{11} + e_2^{(mn)}T_{21} + e_3^{(mn)}T_{31}, \quad (3)$$

$$m > n, \; m = 2,3,4,5, \; n = 1,2,3$$

where $c_i = \gamma_{i0} + \gamma_{i1}s$, $d_i = \delta_{i0} + \delta_{i1}s$ and $e_i^{(mn)} = \rho_{i0}^{(mn)} + \rho_{i1}^{(mn)}s$; indices m and n correspond to $\Upsilon(mS)$ and $\Upsilon(nS)$, respectively. The free parameters α_2, β_2, γ_{i0}, γ_{i1}, δ_{i0}, δ_{i1}, $\rho_{i0}^{(mn)}$ and $\rho_{i1}^{(mn)}$ depend on the couplings of J/ψ, $\psi(2S)$, and $\Upsilon(mS)$ to the channels $\pi\pi$, $K\overline{K}$ and $\eta\eta$. The pole term in (1) in front of T_{21} is an approximation of possible ϕK states, not forbidden by OZI rules.

The amplitudes T_{ij} are expressed through the S-matrix elements

$$S_{ij} = \delta_{ij} + 2i\sqrt{\rho_i\rho_j}T_{ij}, \quad (4)$$

where $\rho_i = \sqrt{1 - s_i/s}$ and s_i is the reaction threshold. The S-matrix elements are taken as the products

$$S = S_B S_{res} \tag{5}$$

where S_{res} represents the contribution of resonances, S_B is the background part. The S_{res}-matrix elements are parameterized on the uniformization plane of the $\pi\pi$-scattering S-matrix element by poles and zeros which represent resonances. The uniformization plane is obtained by a conformal map of the 8-sheeted Riemann surface, on which the three-channel S matrix is determined, onto the plane. In the uniformizing variable used [13]

$$w = \frac{\sqrt{(s-s_2)s_3} + \sqrt{(s-s_3)s_2}}{\sqrt{s(s_3-s_2)}} \quad (s_2 = 4m_K^2 \text{ and } s_3 = 4m_\eta^2) \tag{6}$$

we have neglected the $\pi\pi$-threshold branch point and allowed for the $K\overline{K}$- and $\eta\eta$-threshold branch points and left-hand branch point at $s = 0$ related to the crossed channels.

Resonance representations on the Riemann surface are obtained using formulae in table 1 [14], expressing analytic continuations of the S-matrix elements to all sheets in terms of those on the physical (I) sheet that have only the resonances zeros (beyond the real axis), at least, around the physical region. In table 1 the Roman numerals denote the Riemann-surface sheets, the superscript I

Table 1. Analytic continuations of the S-matrix elements

Process	I	II	III	IV	V	VI	VII	VIII
$1 \to 1$	S_{11}	$\frac{1}{S_{11}}$	$\frac{S_{22}}{D_{33}}$	$\frac{D_{33}}{S_{22}}$	$\frac{\det S}{D_{11}}$	$\frac{D_{11}}{\det S}$	$\frac{S_{33}}{D_{22}}$	$\frac{D_{22}}{S_{33}}$
$1 \to 2$	S_{12}	$\frac{iS_{12}}{S_{11}}$	$\frac{-S_{12}}{D_{33}}$	$\frac{iS_{12}}{S_{22}}$	$\frac{iD_{12}}{D_{11}}$	$\frac{-D_{12}}{\det S}$	$\frac{iD_{12}}{D_{22}}$	$\frac{D_{12}}{S_{33}}$
$2 \to 2$	S_{22}	$\frac{D_{33}}{S_{11}}$	$\frac{S_{11}}{D_{33}}$	$\frac{1}{S_{22}}$	$\frac{S_{33}}{D_{11}}$	$\frac{D_{22}}{\det S}$	$\frac{\det S}{D_{22}}$	$\frac{D_{11}}{S_{33}}$
$1 \to 3$	S_{13}	$\frac{iS_{13}}{S_{11}}$	$\frac{-iD_{13}}{D_{33}}$	$\frac{-D_{13}}{S_{22}}$	$\frac{-iD_{13}}{D_{11}}$	$\frac{D_{13}}{\det S}$	$\frac{-S_{13}}{D_{22}}$	$\frac{iS_{13}}{S_{33}}$
$2 \to 3$	S_{23}	$\frac{D_{23}}{S_{11}}$	$\frac{iD_{23}}{D_{33}}$	$\frac{iS_{23}}{S_{22}}$	$\frac{-S_{23}}{D_{11}}$	$\frac{-D_{23}}{\det S}$	$\frac{iD_{23}}{D_{22}}$	$\frac{iS_{23}}{S_{33}}$
$3 \to 3$	S_{33}	$\frac{D_{22}}{S_{11}}$	$\frac{\det S}{D_{33}}$	$\frac{D_{11}}{S_{22}}$	$\frac{S_{22}}{D_{11}}$	$\frac{D_{33}}{\det S}$	$\frac{S_{11}}{D_{22}}$	$\frac{1}{S_{33}}$

is omitted to simplify the notation, $\det S$ is the determinant of the 3×3 S-matrix on sheet I, $D_{\alpha\beta}$ is the minor of the element $S_{\alpha\beta}$, that is, $D_{11} = S_{22}S_{33} - S_{23}^2$, $D_{22} = S_{11}S_{33} - S_{13}^2$, $D_{33} = S_{11}S_{22} - S_{12}^2$, $D_{12} = S_{12}S_{33} - S_{13}S_{23}$, $D_{23} = S_{11}S_{23} - S_{12}S_{13}$, etc.

These formulae show how singularities and resonance poles and zeros are transferred from the matrix element S_{11} to matrix elements of coupled processes.

The background is introduced via the S_B-matrix elements in a natural way: on the threshold of each important channel there appears generally speaking a complex phase shift. It is important that we have obtained practically zero background of the $\pi\pi$ scattering in the scalar-isoscalar channel. First, this confirms well our assumption in (5). Second, this shows that the representation of multi-channel resonances by the pole and zeros on the uniformization plane given in table 1 is good and quite sufficient. This result is also a criterion for the correctness of the approach.

In table 2 we show the poles corresponding to f_0 resonances, obtained in the analysis. Generally, *the wide multi-channel states are most adequately represented by poles*, because the poles give the main model-independent effect of resonances and are rather stable characteristics for various models,

DOI: 10.1051/epjconf/201713801025

Table 2. The poles for resonances on the $\sqrt{s}$-plane. $\sqrt{s_r} = E_r - i\Gamma_r/2$.

Sheet		$f_0(500)$	$f_0(980)$	$f_0(1370)$	$f_0(1500)$	$f_0'(1500)$	$f_0(1710)$
II	E_r	521.6±12.4	1008.4±3.1			1512.4±4.9	
	$\Gamma_r/2$	467.3±5.9	33.5±1.5			287.2±12.9	
III	E_r	552.5±17.7	976.7±5.8	1387.2±24.4		1506.1±9.0	
	$\Gamma_r/2$	467.3±5.9	53.2±2.6	167.2±41.8		127.8±10.6	
IV	E_r			1387.2±24.4		1512.4±4.9	
	$\Gamma_r/2$			178.2±37.2		215.0±17.6	
V	E_r			1387.2±24.4	1493.9±3.1	1498.8±7.2	1732.8±43.2
	$\Gamma_r/2$			261.0±73.7	72.8±3.9	142.3±6.0	114.8±61.5
VI	E_r	573.4±29.1		1387.2±24.4	1493.9±5.6	1511.5±4.3	1732.8±43.2
	$\Gamma_r/2$	467.3±5.9		250.0±83.1	58.4±2.8	179.3±4.0	111.2±8.8
VII	E_r	542.5±25.5			1493.9±5.0	1500.4±9.3	1732.8±43.2
	$\Gamma_r/2$	467.3±5.9			47.8±9.3	99.9±18.0	55.2±38.0
VIII	E_r				1493.9±3.2	1512.4±4.9	1732.8±43.2
	$\Gamma_r/2$				62.2±9.2	298.4±14.5	58.8±16.4

whereas masses and total widths are very model-dependent for wide resonances [16]. The masses, widths, and the coupling constants of resonances should be calculated using the poles on sheets II, IV and VIII, because only on these sheets the analytic continuations have the forms (see table 1):

$$\propto 1/S_{11}^{\mathrm{I}}, \quad \propto 1/S_{22}^{\mathrm{I}} \text{ and } \propto 1/S_{33}^{\mathrm{I}},$$

respectively, i.e., the pole positions of resonances are at the same points of the complex-energy plane, as the resonance zeros on the physical sheet, and are not shifted due to the coupling of channels.

Further, since studying the decays of charmonia and bottomonia, we investigated the role of the individual f_0 resonances in contributing to the shape of the di-pion mass distributions in these decays, firstly we studied their role in forming the energy dependence of amplitudes of reactions $\pi\pi \to \pi\pi, K\overline{K}, \eta\eta$. In this case we switched off only those resonances [$f_0(500)$, $f_0(1370)$, $f_0(1500)$ and $f_0(1710)$], removal of which can be somehow compensated by correcting the background (maybe, with elements of the pseudo-background) to have the more-or-less acceptable description of the multi-channel $\pi\pi$ scattering. Below we therefore considered description of the multi-channel $\pi\pi$ scattering for two more cases:

- First, when leaving out a minimal set of the f_0 mesons consisting of the $f_0(500)$, $f_0(980)$, and $f_0'(1500)$, which is sufficient to achieve a description of the processes $\pi\pi \to \pi\pi, K\overline{K}, \eta\eta$ with a total $\chi^2/\mathrm{ndf} \approx 1.20$.
- Second, from the above-indicated three mesons only the $f_0(500)$ can be omitted while still obtaining a reasonable description of multi-channel $\pi\pi$ scattering (though with appearance of a pseudo-background) with the total $\chi^2/\mathrm{ndf} \approx 1.43$.

In figure 1 we show the obtained description of the processes $\pi\pi \to \pi\pi, K\overline{K}, \eta\eta$. The solid lines correspond to contribution of all relevant f_0-resonances; the dotted, of the $f_0(500)$, $f_0(980)$, and $f_0'(1500)$; the dashed, of the $f_0(980)$ and $f_0'(1500)$. One can see that the curves are quite similar in all three cases.

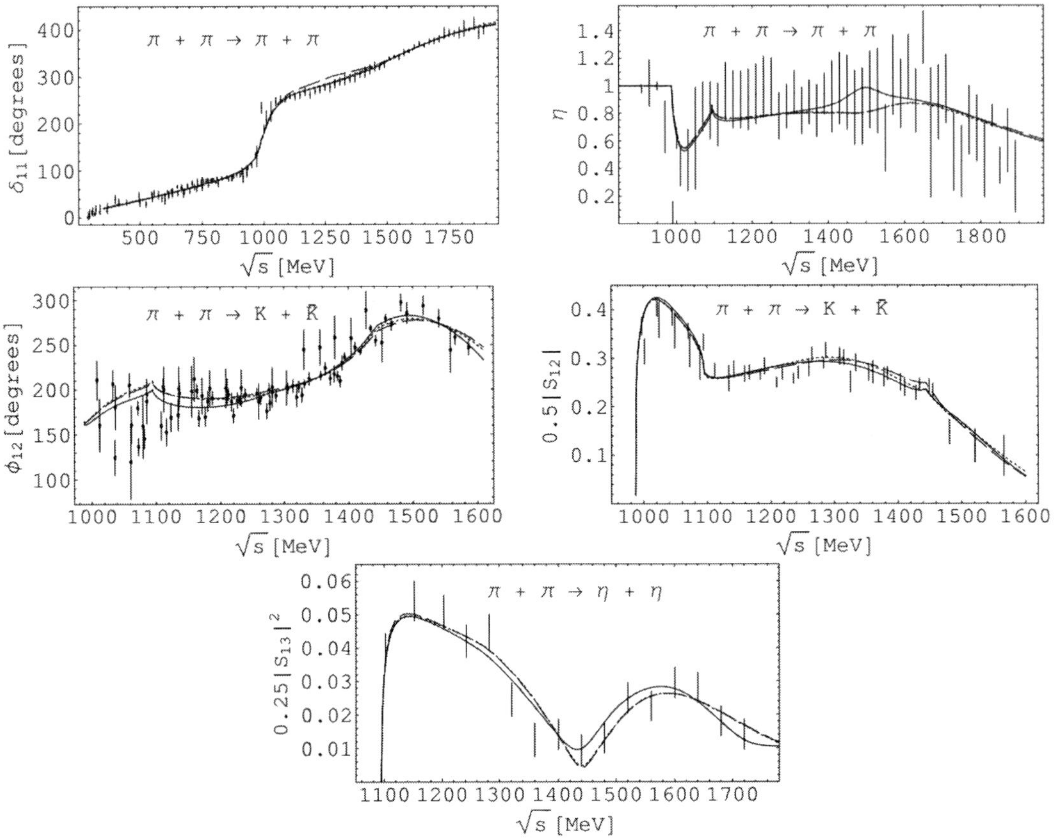

Figure 1. The phase shifts and moduli of the S-matrix element in the S-wave $\pi\pi$-scattering (upper panel), in $\pi\pi \to K\overline{K}$ (middle panel), and the squared modulus of the $\pi\pi \to \eta\eta$ S-matrix element (lower figure).

The di-meson mass distributions in the decay analysis were calculated using the relation

$$N|F|^2 \sqrt{(s-s_1)[m_\psi^2 - (\sqrt{s}-m_\phi)^2][m_\psi^2 - (\sqrt{s}+m_\phi)^2]} \quad (7)$$

for the decay $J/\psi \to \phi\pi\pi$ and with analogous relations for $\psi(2S) \to \psi(1S)\pi\pi$ and $\Upsilon(mS) \to \Upsilon(nS)\pi\pi$. The normalization to the experiment, N is: for $J/\psi \to \phi\pi\pi$ 0.5172 (Mark III), 0.1746 (DM 2) and 3.8 (BES II); for $\psi(2S) \to J/\psi\pi^+\pi^-$ 1.746 (Mark II); for $\psi(2S) \to J/\psi\pi^0\pi^0$ 1.6891 (Crystal Ball(80)); for $\Upsilon(2S) \to \Upsilon(1S)\pi^+\pi^-$ 4.1758 (ARGUS), 2.0445 (CLEO(94)) and 1.0782 (CUSB); for $\Upsilon(2S) \to \Upsilon(1S)\pi^0\pi^0$ 0.0761 (Crystal Ball(85)); for $\Upsilon(3S) \to \Upsilon(1S)(\pi^+\pi^-$ and $\pi^0\pi^0)$ 19.8825 and 4.622 (CLEO(07)); for $\Upsilon(3S) \to \Upsilon(2S)(\pi^+\pi^-$ and $\pi^0\pi^0)$ 1.6987 and 1.1803 (CLEO(94)); for $\Upsilon(4S) \to \Upsilon(1S)\pi^+\pi^-$ 4.6827 (*BaBar*(06)) and 0.3636 (Belle(07)); for $\Upsilon(4S) \to \Upsilon(2S)\pi^+\pi^-$, 37.9877 (*BaBar*(06)); for $\Upsilon(5S) \to \Upsilon(1S)\pi^+\pi^-$, $\Upsilon(5S) \to \Upsilon(2S)\pi^+\pi^-$ and $\Upsilon(5S) \to \Upsilon(3S)\pi^+\pi^-$, respectively 0.2047, 2.8376 and 6.9251 (Belle(12)).

A satisfactory description of all considered processes (including $\pi\pi \to \pi\pi, K\overline{K}, \eta\eta$) was obtained with the total $\chi^2/\text{ndf} = 736.457/(710 - 118) \approx 1.24$; for the $\pi\pi$ scattering, $\chi^2/\text{ndf} \approx 1.15$. Results for the distributions are shown in figures 2-4 with the same notation as in figure 1. Here the effects of omitting some resonance are more apparent than in figure 1.

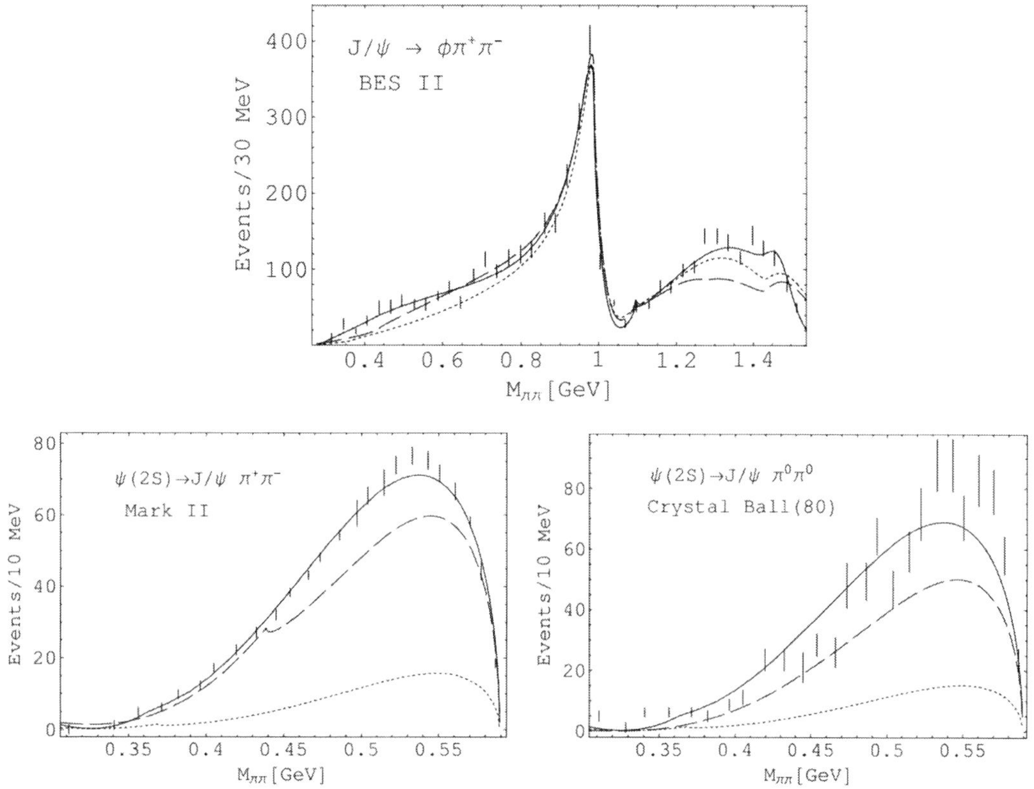

Figure 2. The decays $J/\psi \to \phi\pi\pi$ and $\psi(2S) \to J/\psi\pi\pi$. The solid lines correspond to contribution of all relevant f_0-resonances; the dotted, of the $f_0(500)$, $f_0(980)$, and $f_0'(1500)$; the dashed, of the $f_0(980)$ and $f_0'(1500)$.

3 Conclusions and Discussion

The combined analysis was performed for the data on isoscalar S-wave processes $\pi\pi \to \pi\pi, K\overline{K}, \eta\eta$ and on the decays of the charmonia — $J/\psi \to \phi\pi\pi$, $\psi(2S) \to J/\psi\pi\pi$ — and of the bottomonia — $\Upsilon(mS) \to \Upsilon(nS)\pi\pi$ ($m > n$, $m = 2,3,4,5$, $n = 1,2,3$) from the ARGUS, Crystal Ball, CLEO, CUSB, DM2, Mark II, Mark III, BES II, *BaBar*, and Belle Collaborations.

It is shown that the di-pion mass spectra in the above-indicated decays of charmonia and bottomonia are explained by the unified mechanism which is based on our previous conclusions on wide resonances [2, 3] and is related to contributions of the $\pi\pi$, $K\overline{K}$ and $\eta\eta$ coupled channels including their interference. It is shown that in the final states of these decays (except $\pi\pi$ scattering) the contribution of coupled processes, e.g., $K\overline{K}, \eta\eta \to \pi\pi$, is important even if these processes are energetically forbidden.

Accounting for the effect of the $\eta\eta$ channel in the considered decays, both kinematically (i.e. via the uniformizing variable) and also by adding the $\pi\pi \to \eta\eta$ amplitude in the formulas for the decays, permits us to eliminate nonphysical (i.e. those related with no channel thresholds) non-regularities in some $\pi\pi$ distributions, which are present without this extension of the description [15]. We obtained a reasonable and satisfactory description of all considered $\pi\pi$ spectra in the two-pion transitions of charmonium and bottomonium.

It was also very useful to consider the role of individual f_0 resonances in contributions to the di-pion mass distributions in the indicated decays. For example, it is seen that the sharp dips near 1 GeV in the $\Upsilon(4S, 5S) \to \Upsilon(1S)\pi^+\pi^-$ decays are related with the $f_0(500)$ contribution to the interfering amplitudes of $\pi\pi$ scattering and $K\overline{K}, \eta\eta \to \pi\pi$ processes. Namely consideration of this role of the $f_0(500)$ allows us to make a conclusion on existence of the sharp dip at about 1 GeV in the di-pion mass spectrum of the $\Upsilon(4S) \to \Upsilon(1S)\pi^+\pi^-$ decay where, unlike $\Upsilon(5S) \to \Upsilon(1S)\pi^+\pi^-$, the scarce data do not permit to draw such conclusions yet.

Also, a manifestation of the $f_0(1370)$ turned out to be interesting and unexpected. First, in the satisfactory description of the $\pi\pi$ spectrum of decay $J/\psi \to \phi\pi\pi$, the second large peak in the 1.4-GeV region can be naively explained as the contribution of the $f_0(1370)$. We have shown that this is not right – the constructive interference between the contributions of the $\eta\eta$ and $\pi\pi$ and $K\overline{K}$ channels plays the main role in formation of the 1.4-GeV peak. This is quite in agreement with our earlier conclusion that the $f_0(1370)$ has a dominant $s\bar{s}$ component [2].

On the other hand, it turned out that the $f_0(1370)$ contributes considerably in the near-$\pi\pi$-threshold region of many di-pion mass distributions, especially making the threshold bell-shaped form of the di-pion spectra in the decays $\Upsilon(mS) \to \Upsilon(nS)\pi\pi$ ($m > n, m = 3,4,5, n = 1,2,3$). This fact confirms, first, the existence of the $f_0(1370)$ (up to now there is no firm conviction if it exists or not). Second, that the exciting role of this meson in making the threshold bell-shaped form of the di-pion spectra can be explained as follows: the $f_0(1370)$, being predominantly the $s\bar{s}$ state [3] and practically not contributing to the $\pi\pi$-scattering amplitude, influences noticeably the $K\overline{K}$ scattering; e.g., it was shown that the $K\overline{K}$-scattering length is very sensitive to whether this state does exist or not [16]. The interference of contributions of the $\pi\pi$-scattering amplitude and the analytically-continued $\pi\pi \to K\overline{K}$ and $\pi\pi \to \eta\eta$ amplitudes lead to the observed results.

It is important that we have performed a combined analysis of available data on the processes $\pi\pi \to \pi\pi, K\overline{K}, \eta\eta$, on decays of charmonia $J/\psi \to \phi\pi\pi$, $\psi(2S) \to J/\psi(\pi\pi)$ and of bottomonia $\Upsilon(mS) \to \Upsilon(nS)\pi\pi$ ($m > n, m = 2,3,4,5, n = 1,2,3$) from the ARGUS, Crystal Ball, CLEO, CUSB, DM2, Mark II, Mark III, BES II, *BABAR*, and Belle Collaborations. The convincing description (including also the $\eta\eta$ channel) of practically all available data on two-pion transitions of the Ψ and the Υ mesons confirmed all our previous conclusions on the unified mechanism of formation of the basic di-pion spectra, which is based on our previous conclusions on wide resonances [2, 3] and is related to contributions of the $\pi\pi$, $K\overline{K}$ and $\eta\eta$ coupled channels including their interference. This also confirmed all our earlier results on the scalar mesons [3]; the most important results are:

1. Confirmation of the $f_0(500)$ with a mass of about 700 MeV and a width of 930 MeV (the pole position on sheet II is $514.5 \pm 12.4 - 465.6 \pm 5.9$ MeV).

2. An indication that the $f_0(980)$ (the pole on sheet II is $1008.1 \pm 3.1 - i(32.0 \pm 1.5)$ MeV) is neither a $q\bar{q}$ state nor the $K\overline{K}$ molecule.

3. An indication for the $f_0(1370)$ and $f_0(1710)$ to have a dominant $s\bar{s}$ component.

4. An indication for the existence of two states in the 1500-MeV region: the $f_0(1500)$ ($m_{res} \approx$ 1495 MeV, $\Gamma_{tot} \approx 124$ MeV) and the $f_0'(1500)$ ($m_{res} \approx 1539$ MeV, $\Gamma_{tot} \approx 574$ MeV).

This work was supported in part by the Heisenberg-Landau Program, by the Votruba-Blokhintsev Program for Cooperation of Czech Republic with JINR, by the Grant Agency of the Czech Republic (grant No. P203/15/04301), by the Grant Program of Plenipotentiary of Slovak Republic at JINR, by the Bogoliubov-Infeld Program for Cooperation of Poland with JINR, by the BMBF (Project 05P2015, BMBF-FSP 202), by Tomsk State University Competitiveness Improvement Program, the Russian Federation program "Nauka" (Contract No. 0.1526.2015, 3854), by Slovak Grant Agency VEGA under contract No.2/0197/14, and by the Polish National Science Center (NCN) grant DEC-2013/09/B/ST2/04382.

References

[1] Yu.A. Simonov and A.I. Veselov, Phys. Rev. D **79**, 034024 (2009)
[2] Yu.S. Surovtsev, P. Bydžovský, V.E. Lyubovitskij, R. Kamiński, and M. Nagy, J. Phys. G Nucl. Part. Phys. **41**, 025006 (2014). Phys. Rev. D **86**, 116002 (2012)
[3] Yu.S. Surovtsev, P. Bydžovský, V.E. Lyubovitskij, R. Kamiński, and M. Nagy, Phys. Rev. D **89**, 036010 (2014)
[4] H. Albrecht et al. (ARGUS Collaboration), Phys. Lett. **134B**, 137 (1984)
[5] D. Besson et al. (CLEO Collaboration), Phys. Rev. D **30**, 1433 (1984)
[6] V. Fonseca et al. (CUSB Collaboration), Nucl. Phys. B **242**, 31 (1984)
[7] D. Gelphman et al. (Crystal Ball Collaboration), Phys. Rev. D **32**, 2893 (1985)
[8] D. Cronin-Hennessy et al. (CLEO Collaboration), Phys. Rev. D **76**, 072001 (2007)
[9] F. Butler et al. (CLEO Collaboration), Phys. Rev. D **49**, 40 (1994)
[10] B. Aubert et al. (BaBar Collaboration), Phys. Rev. Lett. **96**, 232001 (2006)
[11] A. Sokolov et al. (Belle Collaboration), Phys. Rev. D **75**, 071103 (2007). A. Bondar et al. (Belle Collaboration), Phys. Rev. Lett. **108**, 122001 (2012)
[12] D. Morgan and M.R. Pennington, Phys. Rev. D **48**, 1185, 5422 (1993)
[13] Yu.S. Surovtsev, P. Bydžovský, and V.E. Lyubovitskij, Phys. Rev. D **85**, 036002 (2012)
[14] D. Krupa, V.A. Meshcheryakov and Yu.S. Surovtsev, Nuovo Cim. A **109**, 281 (1996)
[15] Yu.S. Surovtsev, P. Bydžovský, T. Gutsche, R. Kamiński, V.E. Lyubovitskij, and M. Nagy, Phys. Rev. D **92**, 036002 (2015). arXiv:1506.03023[hep-ph]
[16] Yu.S. Surovtsev, D. Krupa and M. Nagy, Eur. Phys. J. A **15**, 409 (2002). Czech. J. Phys. **56**, 807 (2006)

DOI: 10.1051/epjconf/201713801025

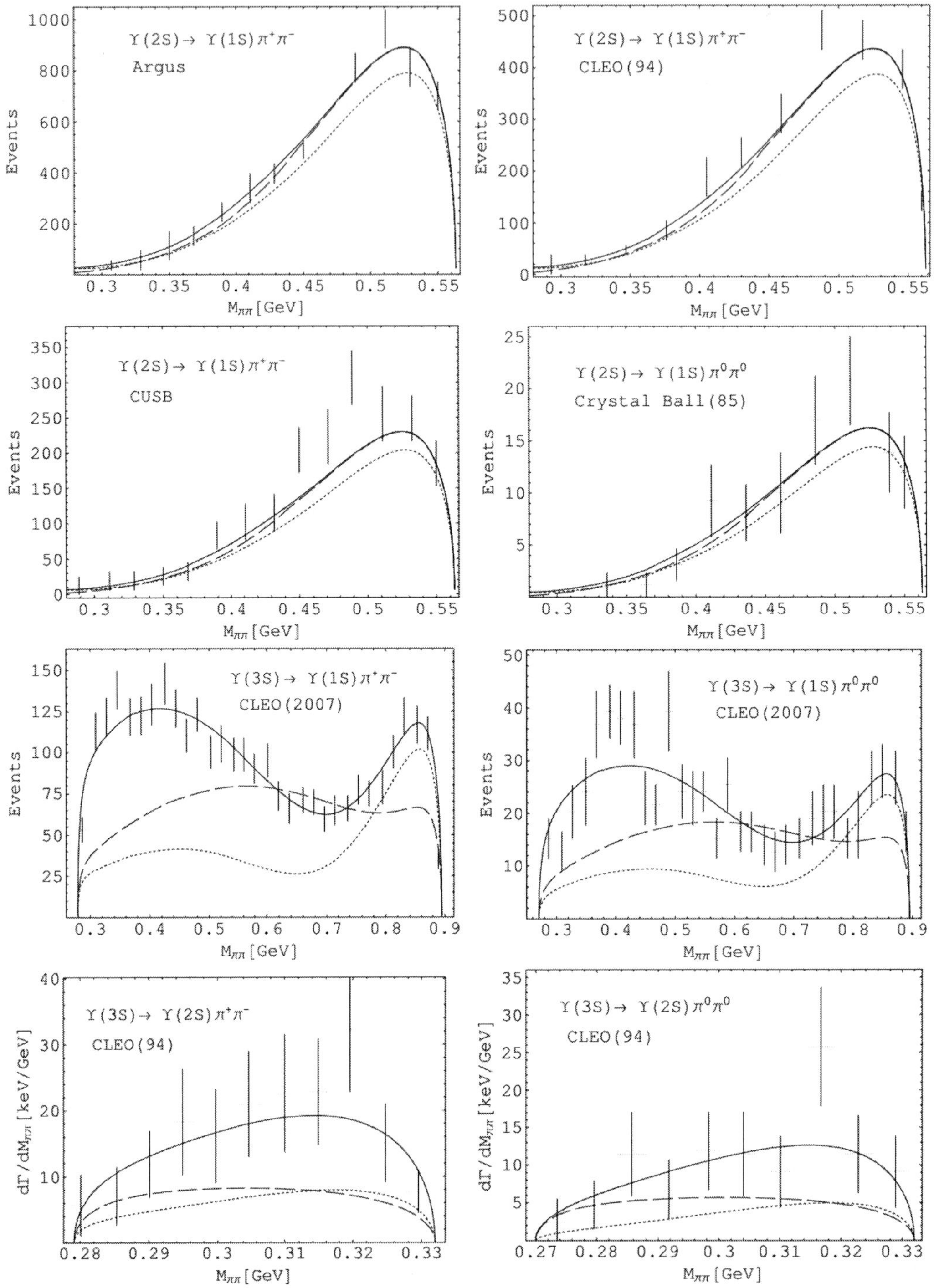

Figure 3. The decays $\Upsilon(2S) \to \Upsilon(1S)\pi\pi$ (two upper panels), $\Upsilon(3S) \to \Upsilon(1S)\pi\pi$ (middle panel) and $\Upsilon(3S) \to \Upsilon(2S)\pi\pi$ (lower panel). The solid lines correspond to contribution of all relevant f_0-resonances; the dotted, of the $f_0(500)$, $f_0(980)$, and $f_0'(1500)$; the dashed, of the $f_0(980)$ and $f_0'(1500)$.

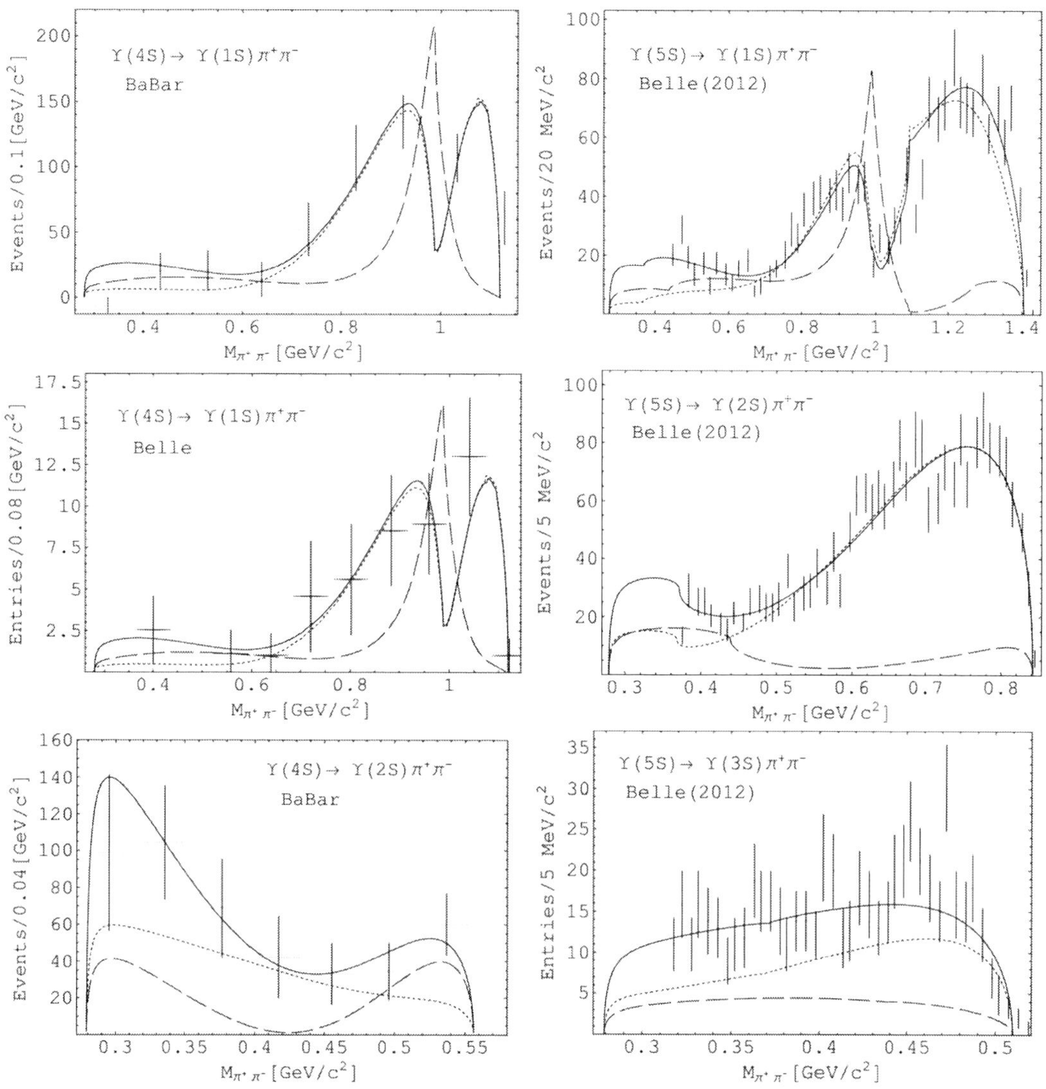

Figure 4. The decays $\Upsilon(4S) \to \Upsilon(1S, 2S)\pi^+\pi^-$ (left-hand) and $\Upsilon(5S) \to \Upsilon(ns)\pi^+\pi^-$ ($n = 1, 2, 3$) (right-hand). The solid lines correspond to contribution of all relevant f_0-resonances; the dotted, of the $f_0(500)$, $f_0(980)$, and $f_0'(1500)$; the dashed, of the $f_0(980)$ and $f_0'(1500)$.

EPJ Web of Conferences **138**, 01026 (2017) DOI: 10.1051/epjconf/201713801026
Baldin ISHEPP XXIII

A superconducting CW-LINAC for heavy ion acceleration at GSI

Winfried Barth[1,2,5*], *Kurt* Aulenbacher[2,4], *Markus* Basten[3], *Florian* Dziuba[2], *Viktor* Gettmann[2], *Maksym* Miski-Oglu[2], *Holger* Podlech[3] and *Stepan* Yaramyshev[1,5]

[1]GSI Helmholtzzentrum, 64291 Darmstadt, Germany
[2]HIM, Helmholtzinstitut, 55128 Mainz, Germany
[3]IAP, Goethe University, 60438 Frankfurt, Germany
[4]KPH, Johannes Gutenberg-University, 55099 Mainz, Germany
[5]MEPhI, National Research Nuclear University - Moscow Engineering Physics Institute, 115409 Moscow, Russia

Abstract. Recently the Universal Linear Accelerator (UNILAC) serves as a powerful high duty factor (25%) heavy ion beam accelerator for the ambitious experiment program at GSI. Beam time availability for SHE (Super Heavy Element)-research will be decreased due to the limitation of the UNILAC providing Uranium beams with an extremely high peak current for FAIR simultaneously. To keep the GSI-SHE program competitive on a high level and even beyond, a standalone superconducting continuous wave (100% duty factor) LINAC in combination with the upgraded GSI High Charge State injector is envisaged. In preparation for this, the first LINAC section (financed by HIM and GSI) will be tested with beam in 2017, demonstrating the future experimental capabilities. Further on the construction of an extended cryo module comprising two shorter Crossbar-H cavities is foreseen to test until end of 2017. As a final R&D step towards an entire LINAC three advanced cryo modules, each comprising two CH cavities, should be built until 2019, serving for first user experiments at the Coulomb barrier.

1 Introduction

Design and construction of cw high intensity proton and ion linacs is a crucial goal of worldwide accelerator technology development [1-4]. Especially compactness of an accelerator facility is a modern trend, which high intensity cw proton and ion linac development has to comply with [5-7]. In the low and medium-energy range cw-linacs can be used for several applications, as boron-neutron capture therapy (BNCT), high productivity isotope generation and material science. The high energy linac is an integrated and essential part of several large scale research facilities, as spallation neutron sources or accelerator driven systems. Therefore the study and investigation of the design, operation and optimization of cw-linacs, as well as progress in elaboration of the superconducting technology, is of superior interest for the accelerator community around the world.

* Corresponding author: w.barth@gsi.de

Providing heavy ion beams for the ambitious experiment program at GSI, the Universal Linear Accelerator (UNILAC) combined with the High Charge State Injector (HLI) served at last as a powerful high duty factor (25%) accelerator. An UNILAC upgrade program is ongoing, designated to prepare for high intensity high current synchrotron injector operation for FAIR (Facility for Antiproton and Ion Research) [8-10]. As a result beam time availability for SHE-research will be strongly diminished due to the duty factor limitation for the UNILAC FAIR injector operation. To keep the SHE program at GSI competitive [11], an upgrade program of the HLI was already initialized comprising a new 28 GHz ECR source and a cw capable RFQ and an IH-DTL [12]. A standalone sc cw-LINAC [13] in combination with the upgraded HLI is assumed to meet the demands of the experimental program at its best. With significantly higher beam intensity the SHE production rate will be increased as well.

2 General cw-LINAC layout

A conceptual layout [13] of a sc cw-LINAC was worked out, which allows the acceleration of highly charged ions with a mass to charge ratio of 6 at 1.4 MeV/u from the upgraded HLI. Nine superconducting CH cavities operated at 217 MHz accelerate the ions to energies between 3.5 MeV/u and 7.3 MeV/u, while the energy spread should be kept smaller than ±3keV/u. Superconducting solenoids mounted between the cavities provide for proper beam focusing. The general parameters are listed in table 1.

Table 1. Design parameters of the cw-LINAC.

Mass/Charge		6
Frequency	MHz	217
Max. beam current	mA	1
Injection Energy	AMeV	1.4
Output energy	AMeV	3.5 – 7.3
Output energy spread	AkeV	±3
Length of acceleration	m	12.7
Sc CH-cavities		9
Sc solenoids		7

R&D and prototyping (demonstrator project) [14, 15] in preparation of the proposed cw-LINAC is assigned to a collaboration of GSI, IAP and HIM, which was founded in 2009. The selected location to setup the demonstrator is in straightforward direction of the HLI at GSI (figure 1).

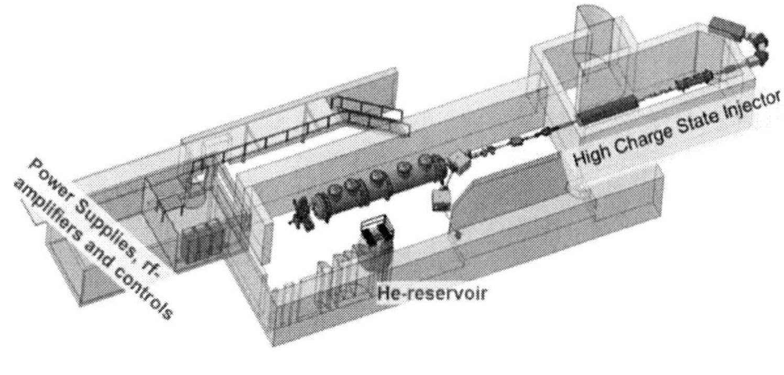

Figure 1. CH-multi cavity test environment@GSI.

Two existing experiments at the HLI had to move since the space is needed for the demonstrator test environment embedded in a new radiation protection cave. The liquid helium (LHe) supply is covered by a 3000 l tank. The consumed helium is collected in a 25 m3 recovery balloon and bottled by a compressor. For matching the beam from the HLI to the demonstrator a new rebuncher cavity as well as an existing rebuncher and an additional quadrupole doublet will be used. Moreover beam diagnostic devices, like SEM-profile grids and an emittance measurement devices and phase probes for beam energy measurements applying time of flight (TOF) has to be integrated in the beam line in front of and behind the demonstrator as well.

3 Status of the UNILAC High Current Performance

The GSI-UNILAC (figure 2) as well as the heavy ion synchrotron SIS18 will serve as a high current heavy ion injector for the new FAIR synchrotron SIS100.

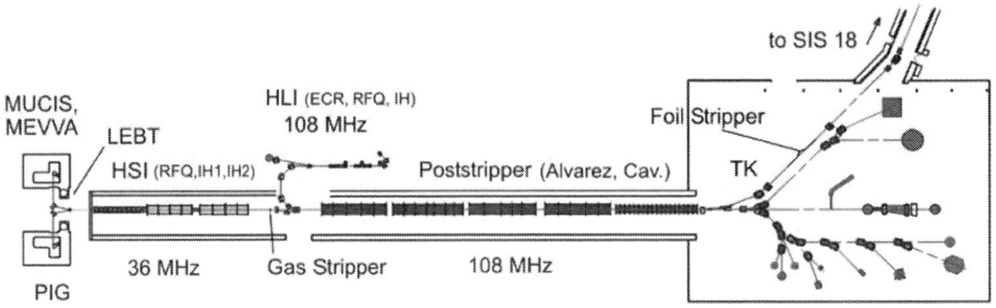

Figure 2. Layout of GSI UNILAC.

In the context of an advanced machine investigation program [16-18] in combination with the ongoing UNILAC upgrade program, a new uranium beam intensity record (11.5 emA, U^{29+}) at very high beam brilliance was achieved recently in a machine experiment campaign. This is an important step paving the way to fulfill the FAIR heavy ion high intensity beam requirements. The U^{4+}-beam current and brilliance was improved by applying a 7 hole-aperture extraction system at the VARIS ion source [19]. By optimization of the low energy beam transport and improved RFQ matching an RFQ transmission of 75% (11.3 emA) was achieved.

After advanced RF optimization and RF conditioning the HSI RFQ tank yields for reliable high-current uranium beam operation. Optimizing the MEBT (Medium Energy Beam Transport) between RFQ and IH DTL by increasing the transverse and longitudinal focusing the previously disturbing beam losses could be minimized significantly, resulting in a stable overall high current operation. Beam matching to the gas stripper by adapting the quadrupole channel resulted in a beam transmission of 90% in this section. For the first time an U^{4+} beam current of 7.6 emA was available for heavy ion stripping. Applying a H_2-stripper gas cell at an optimal target thickness ($\approx$14 μg/cm^2) and after re-optimization of the charge separation procedure under high current conditions, an increased stripping efficiency of about 21% was achieved. Finally the uranium beam intensity at injection into the poststripper accelerator was increased by almost a factor of 4 accomplished in a 22 months lasting machine optimizing program (figure 3).

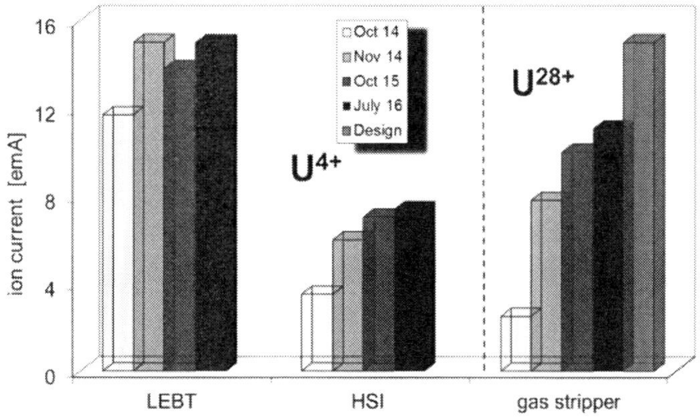

Figure 3. Measured uranium beam transmission at GSI-HSI and stripper section.

For this new record Uranium beam intensity (11.5 mA, U^{29+}) the beam emittance was measured carefully: The vertical beam emittance is significantly increased, while the horizontal emittance is decreased at higher beam current. For the high current beam dynamics layout of the gas stripper section, an enlarged vertical beam envelope in the interaction zone is foreseen, resulting also in an enhanced beam emittance growth due to strong particle straggling. Regarding high space charge power at higher ion beam current, the horizontal beam spot at the stripper has to be reduced as well. Thus horizontal beam brilliance at 1.4 MeV/u simply scales with the pulse current [9].

4 Status of cw-LINAC

The demonstrator (figure 4) [20] is the prototype of the first section of the proposed cw-LINAC, comprising a superconducting CH cavity embedded by two superconducting solenoids. A study was conducted, which showed an assembly concept for the cryostat comprising the solenoids and the cavity as well. It is intended to align the three components to the beam axis. The sc CH structure is the key component and offers a variety of research and development. The beam focusing solenoids provide maximum fields of 9.3 T at an overall length of 380 mm and a free beam aperture of 30 mm.

Figure 4. Cryostat with CH-cavity, high field solenoids, cold warm transitions and support system (Cryogenic Limited).

The demonstrator project kick-off took place in 2010. After complete fabrication all acceptance tests passed through, and all subcomponents are delivered. For the design of the support frame, which carries the cavity (embedded by the two sc solenoids), a suspension concept was applied (figure 5) [21, 22]. The support frame, as well the accelerator components, are suspended each by eight tie rods in a cross-like configuration balancing the mechanical stress during the cooldown and warm up (figure 6). As a consequence the suspended components may always stay within the tolerance limits related to the beam axis. Another important feature concerns the two sc solenoids: A configuration of one main coil out of NbSn and two compensation coils made from NbTi shields the maximum magnetic field of 9.3 T within a longitudinal distance of 10 cm down to 30 mT, acceptable at the position of the neighbored cavity (first RF-gap). The solenoids are connected to LHe ports inside the cryostat by copper tapes allowing dry cooling.

Figure 5. Sc cw-LINAC-demonstrator comprising a CH cavity embedded by two sc solenoids on a support frame, to be mounted inside the cryostat.

Figure 6. Detailed view of tie rods in a cross-like configuration.

The CH cavity is directly cooled with LHe supported by a helium jacket out of titanium. The vendor RI (Research Instruments GmbH, Germany) provided for sufficient cavity preparation. After final high pressure rinsing (HPR), the delivery of the cavity was accomplished in September 2015; a performance test at low temperature (4.2 K) and with low RF power was performed in October 2015 at Goethe University Frankfurt [23]. Three piezo frequency tuners are manufactured at GSI, tuning the resonance frequency, while RF-operation. Therefore a tuner dummy was already tested at Goethe University Frankfurt [24]. Commissioning of the demonstrator is planned in 2016 at the GSI HLI, which operates at 108 MHz (Figure 1). A new beam line in straightforward direction to the HLI, which transports the beam to the new radiation protection shelter locating the demonstrator, was designed regarding beam dynamical simulations. In June 2015 an emittance measurement at the future position of the demonstrator cryostat confirmed the latest beam dynamic simulations and the aspired positioning of the components in the beam transport line [25].

5 Advanced R&D

The successor of the demonstrator R&D is already the "Advanced Demonstrator" project. It is planned to build the first half of the future entire cw LINAC. Based on calculations of the beam dynamics [26, 27], the cw LINAC cryo module, comprising two CH cavities and one superconducting solenoid, was defined.

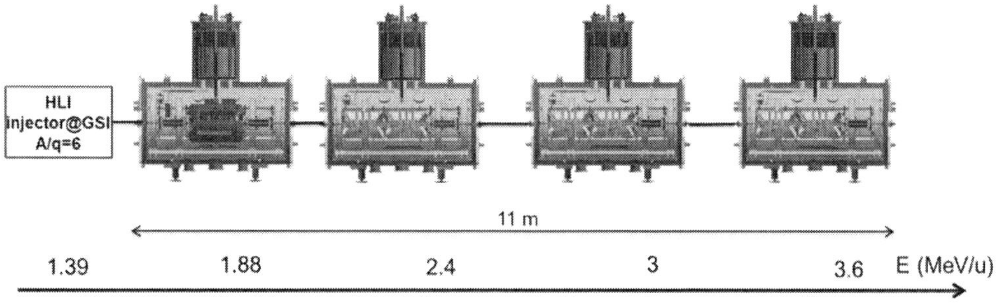

Figure 7. Layout of the Advanced Demonstrator.

Figure 7 shows the present layout based on the so called "short cavity" design. The usage of short CH cavities allows more flexibility in tuning the entire cw LINAC, maintaining the necessity of variable output energy and almost a neglect able emittance growth during the acceleration process. The design value of the accelerating gradient of $E_a = 5.5$ MV/m provides an effective accelerating voltage of 1.8 MV per (short) cavity. A sc solenoid with a maximum magnetic field of 9.3 T is sufficient to focus the heavy ion beam in both transversal planes. A cryo plant with a total power of 2.3 kW, currently used for the testing of superconducting dipole magnets for FAIR-SIS100, is foreseen to supply the Advanced Demonstrator by two cooling loops (4.4°K at 3.5 bar and 50°K at 11 bar respectively). The cryogenic power is delivered to the local helium distribution system via a double shielded transfer line. Each cryo module has its own distribution box with a Joule-Thompson expansion valve, producing liquid helium; a controlled valve maintains a constant pressure in the cavities within 10 mbar. The cryostat provides for the 4.2 K environment for solenoid as well as for the cavities with static losses of maximum 5 W. The accelerator components are mounted on a support frame by a so called Nuclotron suspension, which allows preserving the alignment of components during the cooldown within 0.2 mm (in transverse direction). The support frame itself is also suspended by a Nuclotron suspension, allowing for adjustment of the whole frame in situ under operational conditions.

6 Summary & Outlook

A UNILAC upgrade program is successfully ongoing, designated to prepare for high intensity high current synchrotron injector operation for FAIR. As a result beam time availability for SHE-research will be strongly diminished. A standalone sc cw-LINAC is envisaged to keep the GSI-SHE program competitive on a high level and even beyond. In preparation for this, the first LINAC section is already built as a demonstrator. After the successful RF-test of the demonstrator cavity further activities in preparation of the first beam test are conducted: Cavity treatments, pushing the accelerating gradient to the theoretical limit of about 25 MV/m, as well as ordering of components for the advanced demonstrator project. The beam test of the demonstrator cavity is scheduled for March 2017.

References

1. C. Prior, *Overview of High Intensity Accelerator Projects*, Proceedings of HB'10 (Morschach, Switzerland, 2010)
2. T.V. Kulevoy, S.M. Polozov, A.D. Fertman et al., Atomic Energy **117(5)**, 117:347 (2015)
3. P.N. Ostroumov, *Advances in CW Ion Linacs*, Proceedings of IPAC'15 (Richmond, VA, USA, 2015)
4. N. Solyak et al., *The Concept Design of the CW Linac of the Project X*, Proceedings of IPAC'10 (Kyoto, Japan, 2010)
5. S.M. Polozov et al., *Beam Dynamics Simulations and Code Comparison for a New CW RFQ Design*, Proceedings of HB'16 (Malmö, Sweden, 2016)
6. L. Weissman, D. Berkovits et al., Journal of Instrumentation **10**, T1004 (2015)
7. G.V. Trubnikov et al., Physics of Particles and Nuclei Letters **13(7)**, 915–918 (2016)
8. W. Barth et al., Phys. Rev. ST Accel. Beams **18(4)**, 040101 (2015)
9. W. Barth et al., *High Current Uranium Beam Measurements at GSI-UNILAC for FAIR*, Proceedings of HB'16 (Malmö, Sweden, 2016)
10. W. Barth et al., Phys. Rev. ST Accel. Beams **18(5)**, 050102 (2015)
11. S. Hofmann et al., *Heavy Element Research at GSI-SHIP*, Proceedings of EXON'04 (Peterhof, Russia, 2004)
12. W. Barth et al., *UNILAC-upgrade programme for the heavy element research at GSI-SHIP*, Proceedings of EPAC'06 (Edinburgh, Scotland, 2006)
13. S. Minaev et al., Phys. Rev. ST Accel. Beams **12**, 120101 (2009)
14. S. Mickat et al., *The sc cw-LINAC Demonstrator – SRF technology finds the way to GSI*, Proceedings of SRF'11 (Chicago, USA, 2011)
15. V. Gettmann et al., *The sc cw-LINAC Demonstrator – an R&D project for SRF technology*, Proceedings of SRF'11 (Chicago, USA, 2011)
16. W. Barth et al., *Upgrade program of the high current heavy ion UNILAC as an injector for FAIR*, NIM A **577(1-2)**, 211-214 (2007)
17. S. Yaramyshev et al., Phys. Rev. ST Accel. Beams **18(4)**, 050103 (2015)
18. S. Yaramyshev et al., *Development of the versatile multi particle code DYNAMIION*, NIM A **558(1)**, 90–94 (2006)
19. R. Hollinger et al., Nucl. Instrum. Meth. B **239**, 227 (2005)
20. S. Mickat, M. Amberg, K. Aulenbacher, W. Barth, F. Dziuba, V. Gettmann, S. Jacke, A. Orzhekhovsaya, H. Podlech, U. Ratzinger, K. Tinschert, *Status of the superconducting cw LINAC at GSI*. DOI: 10.1142/9789814508865_0044 (Vladivostok, Russia, 2013)

21. V. Gettmann et al., *The sc cw -LINAC Demonstrator - First section of a sc cw -LINAC*, Proceedings of SRF'11 (Chicago, Illinois, USA, 2011)
22. V. Gettmann et al., *Recent status new superconducting cw heavy ion LINAC@GSI*, Proceedings of SRF'15 (Whistler, Canada, 2015)
23. F. Dziuba et al., *First Performance Test on the Superconducting 217 MHz CH Cavity at 4.2 K*, Proceedings of LINAC'16 (East Lansing, MI, U.S.A., 2016)
24. M. Amberg et al., *The fast Piezo-based frequency tuner for sc CH-cavities*, Proceedings of LINAC'14 (Geneva, Switzerland, 2014)
25. M. Miski-Oglu, et al., *Steps towards superconducting cw-LINAC for heavy ions at GSI*, Proceedings of SRF'15 (Whistler, Canada, 2015)
26. M. Schwarz et al., *Beam Dynamics for the sc cw heavy ion LINAC at GSI*, Proceedings of IPAC'15 (Richmond, U.S.A., 2015)
27. S. Yaramyshev et al., *An Advanced Procedure for Longitudinal Beam Matching for SC CW Heavy Ion LINAC With Variable Output Energy*, Proceedings of HB'16 (Malmö, Sweden, 2016)

Status of the NICA project at JINR

Vladimir Kekelidze[1], *Alexandr* Kovalenko[1,⋆], *Rihard* Lednicky[1], *Viktor* Matveev[1], *Igor* Meshkov[1], *Alexandr* Sorin[1], and *Grigory* Trubnikov[1]

[1]*Joint Institute for Nuclear Research, Dubna, 141980 Russia*

Abstract. The NICA (Nuclotron-based Ion Collider fAcility) project is now under active realization at the Joint Institute for Nuclear Research (JINR, Dubna). The main goal of the project is a study of hot and dense strongly interacting matter in heavy-ion (up to Au) collisions at the center-of-mass energies up to 11 GeV per nucleon. Two modes of operation are foreseen, collider mode and extracted beams, with two detectors: MPD and BM@N. The both experiments are in preparation stage. An average luminosity in the collider mode is expected to be 10^{27} cm^{-2} s^{-1} for Au (79+). Extracted beams of various nuclei with maximum momenta of 13 GeV/c (for protons) will be available. A study of spin physics with extracted and colliding beams of polarized deuterons and protons at energies up to 27 GeV (for protons) is foreseen with the NICA facility. The proposed program allows one to search for possible signs of phase transitions and critical phenomena as well as to shed light on the problem of the nucleon spin structure.
Keywords: baryonic matter, heavy-ion interactions, fixed target and collider experiments, spin physics research

1 Introduction

The new research facility NICA aimed at study of heavy-ion and polarized proton and deuteron collisions is under design and construction at the Joint Institute for Nuclear Research (JINR) since 2010 [1]. The study of hot and dense baryonic matter should shed light on: in-medium properties of hadrons and the nuclear matter equation of state (EOS); the onset of deconfinement (OD) and/or chiral symmetry restoration (CSR); phase transition (PT), mixed phase and the critical end-point (CEP); possible local parity violation in strong interactions (LPV) [2–5]. It is indicated in a series of theoretical works, in particular, in [4] that heavy-ion collisions at $\sqrt{S}_{NN}$ 11 GeV allow attaining the highest baryon density. These calculations are illustrated in figure 1. The NICA research domain is attractive as being the expected region for searching of new phenomena at the maximum baryon density including possible phase transitions. The study of the nucleon spin content ("spin puzzle") and polarization phenomena in light- and heavy-ion interactions is another target of research at NICA. The high intensity and high polarization (>50%) of colliding beams could provide unique possibilities for this study.

The NICA construction plan up to 2023 foresees an essential development of the accelerator facility and construction of:

- the new spectrometer Baryonic Matter at Nuclotron (BM@N) in order to start the fixed target experiments with heavy ion beams extracted from the modernized Nuclotron;

⋆e-mail: kovalen@dubna.ru

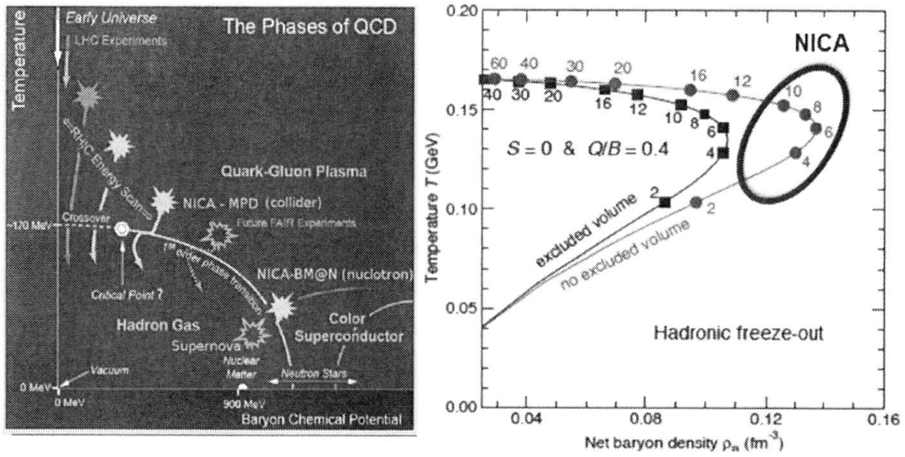

Figure 1. Freeze-out diagram for baryonic matter showing baryon density reachable at different facilities both in collider and fixed target experiments [4].

- the MultiPurpose Detector (MPD) at the NICA colliding beams with a primary goal to study heavy-ion collisions;
- preparation of the collider detector project "Spin Physics Detector" (SPD) aimed at a primary goal to study spin physics.

2 The NICA accelerator complex progress

The Nuclotron at Veksler and Baldin Laboratory of High Energy Physics (VBLHEP) of JINR was put into operation in 1993. It is based on the unique technology of superconducting fast cycling magnets developed at the Laboratory. The NICA facility (figure 2) includes: the injection complex, the booster, the upgraded Nuclotron and two storage rings with two interaction points IP1 and IP2 aimed at the MPD and SPD detectors respectively. The injection complex provides a wide set of ion species up to the heaviest one, Au, at an energy of 3.5 MeV/u with an expected intensity of $2{\cdot}10^9$ particles per cycle. The main efforts during the last few years were devoted to completion of the design, manufacturing and commissioning beam of both heavy ion, KRION-6T, and polarized proton and deuteron, SPI, sources aimed at reaching of the specified parameters, manufacturing of the HILAC and the new RFQ fore-injector that should replace the old high voltage pulsed transformer at the linac LU-20. The design and manufacturing of the equipment were carried out in collaboration with German and Russian research laboratories and companies [6]. The HILAC was manufactured at Bevatech company (Frankfurt, Germany) in accordance with technical specification developed by the LHEP experts. Completion of the injector's manufacturing and its tests at the accelerator together with the ion sources were being performed in 2015-2016. The source of polarized ions (SPI) with the new RFQ pre-accelerating section was assembled at the linac LU-20 in May 2016. The beam was accelerated in LU-20 and some days later, 12 of June, polarized deuteron beam from SPI was accelerated in the Nuclotron ring (figure 3). All equipment of HILAC was also assembled and tested with a beam. The specified technical parameters were reached. The illustrations below show a general view of the HILAC. The unique RF power supply system based on solid state transistors was designed, manufactured and shipped to Dubna. The booster synchrotron should accelerate ions up to 600 MeV/u. The magnetic ring of 211 m long is placed inside the window of the Synchrophasotron yoke. Technical

DOI: 10.1051/epjconf/201713801027

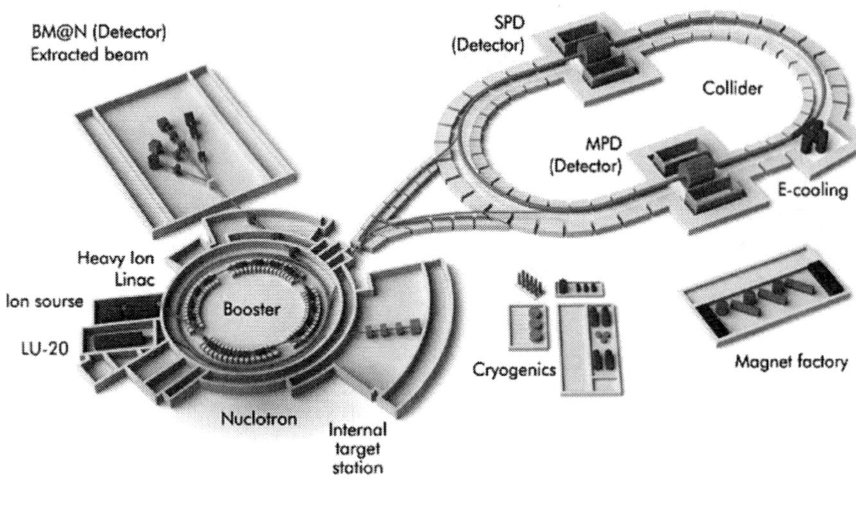

Figure 2. The NICA facility at the Laboratory of High Energy Physics of JINR.

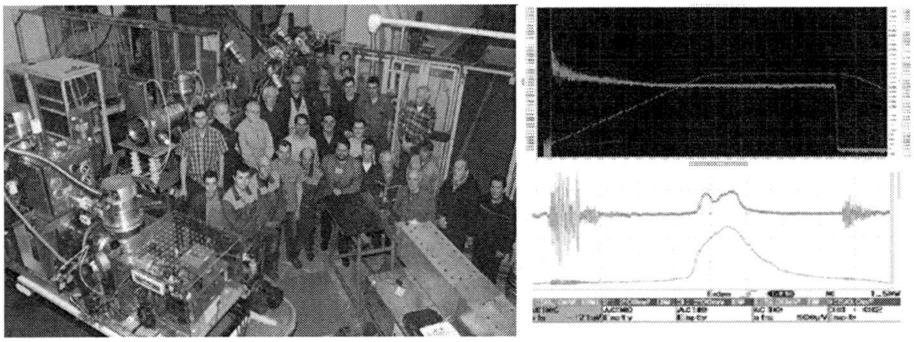

Figure 3. Basic equipment of the NICA polarized proton and deuteron fore-injector at the linac and the team.

design of the booster systems was performed. The most specific features of the booster are ultrahigh vacuum and the electron cooling system. The magnetic system elements were designed by the LHEP designers. The hardware parts like cryostats, basements etc. are under manufacturing at different factories out of Dubna. The most delicate part of the magnet production, namely: SC cable and coil fabrication, the cold mass assembling, magnetic measurements (warm and cold) and the bench tests are provided by the LHEP engineers and technicians at the Magnet Test Facility (figure 5). The serial fabrication and tests of the magnets for the NICA facility were started in 2016 [7].

The NICA cryogenics will be based on the modernized liquid helium plant that was built in the early 1990s for the Nuclotron. The main goals of the modernization are: increasing of the total refrigerator power from 4000 to 8000 W at 4.5 K, making a new distribution system of liquid helium and ensuring the shortest possible cool-down time. These goals will be achieved by means of an additional 1000 l/h helium liquefier and satellite refrigerators located near the accelerator rings. A Nitrogen system will be used for magnet thermal shield cooling at 77 K and at the first stage of cooling down all three rings of the Nuclotron/NICA with the total length of about 1 km and cold mass of 290 tons. Work is in a good progress.

DOI: 10.1051/epjconf/201713801027

Figure 4. Main parts of HILAC: left – RFQ, right - buncher and two DTL accelerating sections.

Figure 5. The NICA magnet test facility: left part – general view; right part – the booster dipole is being prepared for cold tests.

The two storage/collider rings will have two intersection points (IP). The major parameters of the NICA collider are the following: the maximum $B\rho$ = 45 Tm; vacuum in a beam chamber – 10-11 Torr; maximum dipole field 2 T; ion kinetic energy range from 1 GeV/u to 4.5 GeV/u for Au79+; zero beam crossing angle at IP; 9 m space for detector allocations at IP's; average luminosity L = 1027 $cm^{-2}s^{-1}$ for gold ion collisions at $\sqrt{s_{NN}}$ = 9 GeV. The collider ring 503.04 m long (twice as large as the Nuclotron ring) has a racetrack shape and is based on double-aperture (top-to-bottom) superferric magnets – dipoles and quadrupoles. The TDR of the collider has been prepared, nevertheless final design of some lattice elements are still necessary. Current construction program considers the collider rings commissioning time consistent with the MPD construction.

3 The BM@N experiment

The main research goals of the new setup – BM@N (Baryonic Matter at Nuclotron) at extracted Nuclotron beams are the following

1. Study of the properties of dense nuclear (dominantly baryonic) matter with strangeness in heavy-ion (A+A) collisions in particular:

• production mechanisms and modifications of hadron properties in dense nuclear matter ("in-medium effects") using different probes, namely: strange mesons, strange and multi-strange baryons; vector mesons via hadronic or dilepton/photon mode).

• study of the EoS with strangeness;

• hyper-matter production: search for light hypernuclei and multi-strange metastable objects.

2. Study of elementary reactions: pp, pn(d) as "reference" to pin down nuclear effects

3. Search for ‚cold' nuclear matter with pA – collisions.

BM@N at the first stage can study the in-medium effects on strangeness measuring of a variety of observables at different energies and centralities in heavy-ion collisions in order to find an "anomalous" behavior in comparison with theory. The observables sensitive to in-medium effects are the following: particle yields and ratios p_T-(m_T)-spectra, rapidity distributions, angular distributions, collective flow. The BM@N setup uses the existing wide aperture dipole magnet; a tracking system con-

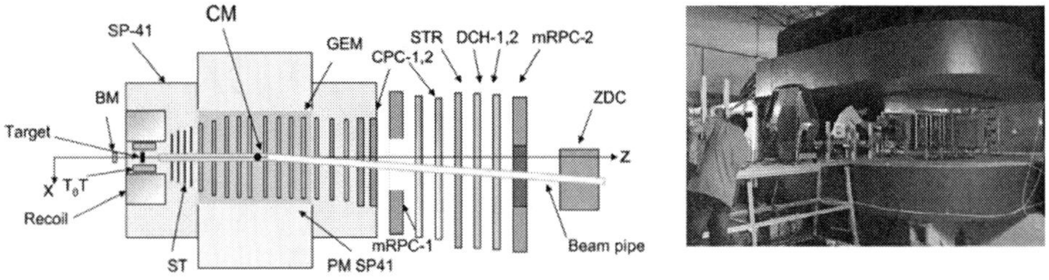

Figure 6. The BM@N scheme: GEM – central tracker inside analyzing magnet; DCH+Straw – outer tracker behind the magnet; mRPC+T0 – time-of-flight system; ZDC – centrality measurements.

sisting of 12 planes of high resolution gem chambers (GEM) and 8 planes of drift chambers (DCH); a time-of-flight (TOF) system based on the RPC chambers and a fast counter detector providing trigger signal; a zero degree hadron calorimeter (ZDC) for the reaction plane definition and estimation of the impact parameter. The main goal for the coming period is to start physics data taking at the set-up

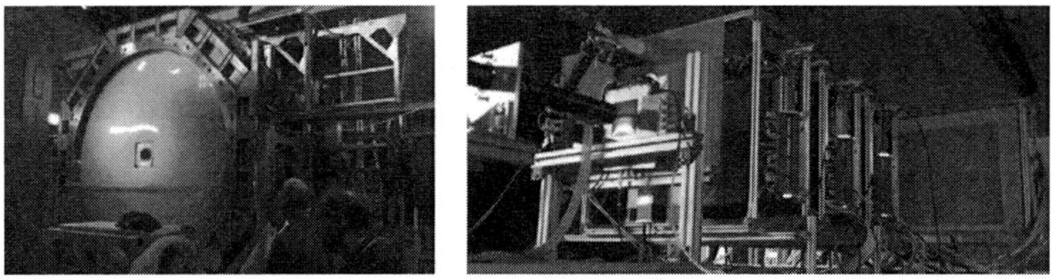

Figure 7. View of the assembled and tested BM@N detectors.

by the end of 2017. The further upgrade is foreseen at the second stage to equip the setup with a silicon vertex detector (in cooperation with GSI, Darmstadt), and with an electromagnetic calorimeter. The heavy-ion collision experiments at BM@N located in the fixed-target hall of the Nuclotron will provide a competitive research program focused on physics of dense nuclear and strange matter.

DOI: 10.1051/epjconf/201713801027

4 The MPD

The MPD experimental program is aimed to investigate both hot and dense baryonic matter and polarization phenomena. Preliminary list of the first priority physics tasks to be performed includes:

measurement of a large variety of signals at systematically changing conditions of collision (energy, centrality, system size) using as bulk observables 4π geometry particle yields (OD, EOS); multi-strange hyperon yields and spectra (OD, EOS); electromagnetic probes (CSR, OD); azimuthal charged-particle correlations (LPV); event-by-event fluctuation in hadron productions (CEP); correlations involving π, K, p, Λ (OD); directed and elliptic flows for identified hadron species (EOS,OD); reference data (i.e. p + p) will be taken at the same experimental conditions. Technical design of the

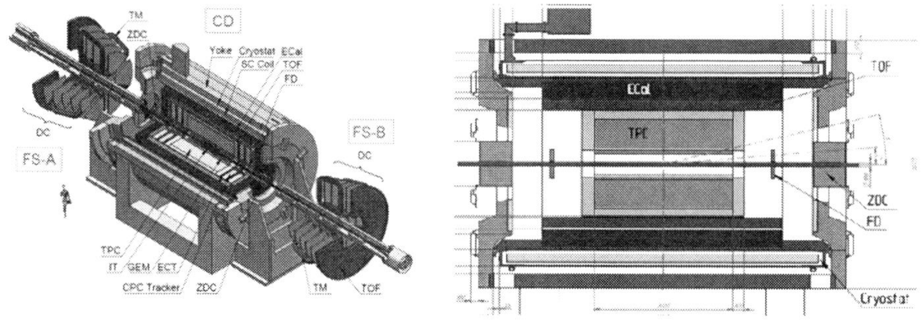

Figure 8. View of the MPD setup and the cross section of the central (barrel) part. The first stage detectors are: TPC, TOF, ECAL, ZDC, FD. Stage two includes IT and Endcaps.

detectors including beam tests is completed. The preparation for mass production is organized. Special problems, like the ECAL manufacturing and the design of this system with necessary capability for assembling inside the SC solenoid was also solved.

The status of MPD superconducting solenoid gives definite optimism: the contract with ASG company (Italy) was signed, works are in progress. The process of manufacturing heavy metallic parts of the solenoid yoke is illustrated at figure 9. The designed TPC assembly tools (see figure 10),

Figure 9. View of the solenoid yoke parts under manufacturing at the Vitkovice Heavy Machinery (CZ).

cooling and laser calibration system provides minimal TPC deformation (~32 μm) and is ready for production. The processes studied with the MPD were simulated using the dedicated software framework (MpdRoot). This software is based on the object - oriented framework FairRoot [8] and provides a powerful tool for detector performance studies, development of algorithms for reconstruction and physics analysis of the data. Based on the modeling, the MPD physics cases at the first stage (2023) are shown in table 1 below.

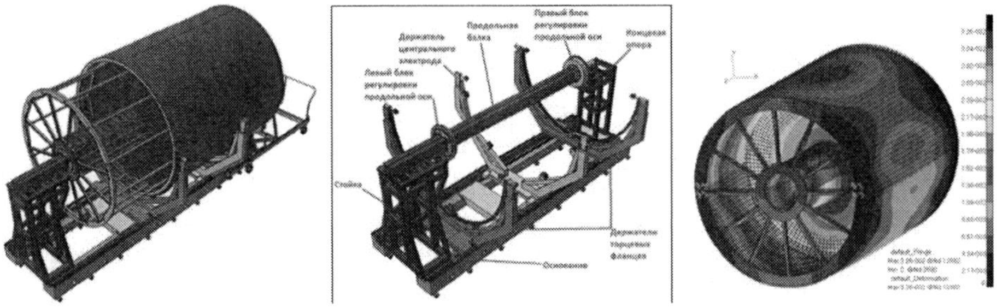

Figure 10. The TPC assembly tools.

Table 1. MPD physics cases at the first stage (2023).

Observable	Set-up	Coverage	New insights
Hadron yields & ratios	TPC, TOF, FHCAL,ECAL	$\lvert\eta\rvert < 1.5$ pT < 4 GeV/c	Data for $4<\sqrt{s}<7$ GeV, critical assessment of y-spectra and K/π-ratio
Hyperons: yields, flow, Polarization	TPC, TOF FHCAL	$\lvert\eta\rvert < 1.5$ pT < 4 GeV/c	New data on yields, flow and polarization at $\sqrt{s} < 7$ GeV.
Dileptons	TPC, TOF ECAL, FHCAL	$\lvert\eta\rvert < 1.2$ pT < 3 GeV/c	**low statistics data for comparison**
Fluctuations & Correlations	TPC, TOF ECAL, FHCAL	$\lvert\eta\rvert < 1.5$	New data on Ev-by-Ev fluct. for $\sqrt{s} > 4$ GeV
Chiral Magnetic & vortical effects	TPC, TOF FHCAL	$\lvert\eta\rvert < 1.5$ pT < 3 GeV/c	Data @ $\sqrt{s} < 7$GeV (CME) Vortical @ $4 < \sqrt{s} < 11$ GeV
(Hyper)Nuclei	TPC, TOF ZDC	$\lvert\eta\rvert < 1.5$ pT< 5 GeV/c	**low statistics data for comparison**

This stage is characterized by limited luminosity (to $5{\cdot}10^{25}$ cm^{-2}s^{-1}). The evaluated rate in Au + Au collisions at the maximum energy (10% central interactions) will be up to 7 kHz taking into account the designed luminosity of $L = 1{\cdot}10^{27}$ cm^{-2}s^{-1}. Groups from 12 institutions are working on the sub-detector prototyping and serial productions. More detailed information could be found in the conceptual design report [9] or at the NICA site: " nica.jinr.ru ". It was confirmed that the MPD is well optimized for studying in-medium effects caused by high baryon densities, such as: changing particle properties in hot and dense medium (broadening spectral functions etc.), event-by-event dynamical fluctuations of strange to non-strange particle ratios and others. There are optimal conditions at both BM@N and MPD detectors for observation and study of light hypernuclei. The simulations demonstrate enhanced yields of different single and double Λ light hydrogen and helium nuclei as a function of $\sqrt{s_{NN}}$. There are some novel proposals such as studying vorticity [10], helicity separation in heavy-ion collisions [11], and directed flow in asymmetric nuclear collisions [12] that could be studied at NICA at stage two.

The tests of the detector prototypes have been carried out at the Nuclotron test beams in several past years and will be continued in 2017 and after.

5 Status of the SPD preparation

The NICA program foresees that the SPD detector will be designed and installed at the IP-2 to study spin physics. The comparable parameters on measurements of different components of transverse quark-parton distribution functions are presented below. There is a number of processes besides the

Table 2. The comparable parameters on measurements of different components of transverse quark-parton distribution functions.

experiment	CERN, COMPASS-II	FAIR, PANDA	FNAL, E-906	RHIC, STAR	RHIC-PHENIX	NICA, SPD
mode	F.T.	F.T.	F.T.	collider	collider	collider
Beam/target	π-, p	anti-p, p	π-, p	pp	pp	pp, pd,dd
Polarization:b/t	0; 0.8	0; 0	0; 0	0.5	0.5	0.7
Luminosity	$2 \cdot 10^{33}$	$2 \cdot 10^{32}$	$3.5 \cdot 10^{35}$	$5 \cdot 10^{31}$	$5 \cdot 10^{31}$	10^{32}
$\sqrt{s}$, *GeV*	14	6	16	200, 500	200, 500	10 - 26
$x_{1(beam)}$ *range*	0.1-0.9	0.1-0.6	0.1-0.5	0.03-1.0	0.03-1.0	0.1-0.8
q_T, *GeV*	0.5 -4.0	0.5 -1.5	0.5 -3.0	1.0 -10.0	1.0 -10.0	0.5 -6.0
Lepton pairs,	μ-μ+	μ-μ+	μ-μ+	μ-μ+	μ-μ+	μ-μ+, e+e-
Data taking	2015	>2025	2013	>2016	>2016	>2020
Transversity	NO	NO	NO	YES	YES	YES
Boer-Mulders	YES	YES	YES	YES	YES	YES
Sivers	YES	YES	YES	YES	YES	YES
Pretzelosity	NO	NO	NO	NO	YES	YES
Worm Gear	NO	NO	NO	NO	NO	YES
Direct γ	NO	NO	NO	YES	YES	YES

mentioned, which could be studied with this detector and with the fixed target detectors at beams extracted from the upgraded Nuclotron, namely: DY processes with longitudinally and transversally polarized p and d beams; extraction of unknown (poorly known) parton distribution functions (PDF); PDFs from J/ψ production processes; spin effects in various exclusive and inclusive reactions; cross sections of diffractive processes; helicity amplitudes and double spin asymmetries (Krisch effect) in elastic reactions; spectroscopy of quarkoniums with any available decay modes. This can be done in the kinematic energy region which is not available at other experiments. The formation of the collaboration has been started. The analysis of the accelerator issues related to polarized proton and deuteron beams has been progressing during the last three years. Feasible schemes of the polarization control in the Nuclotron and collider have been proposed and analyzed [13, 14].

6 Status of the NICA site preparation

Preparation of the site for the collider and the detectors was started in 2013. The technical design project of the NICA collider building, the MPD and SPD halls and engineering infrastructure with the set of technical documentation was completed in 2014. The State expertise review has been successfully passed in accordance with the Russian regulations. A general contractor for civil construction of the NICA Complex is STRABAG Company. View of the site at the Conference date are shown in figure 11.

DOI: 10.1051/epjconf/201713801027

Figure 11. View of the NICA facility site in October 2016.

7 Concluding remarks

The NICA accelerator and experimental complex and the corresponding research program will provide a relevant research, which is both competitive and complementary to ones being carried out at other world leading scientific centres and facilities. The Agreement between the Government of the Russian Federation and the Joint Institute for Nuclear Research has been signed in June 2016. Additional special funds were allocated for the NICA Complex. Talking about NICA at the Academician A.M. Baldin Memorial Seminar, we would like to remind about the proposal of the universal collider design and construction – The Super-Nuclotron – the first collider project at JINR [15].

References

[1] V.D. Kekelidze et al., PoS (ICHEP2012) 411, (2013)
[2] A. Sissakian et al., Part. & Nucl Lett. **5**, 1 (2008)
[3] M. Gazdzicki et al., PL. B **585**, 115 (2004)
[4] J. Cleymans, and J. Randrup, Phys. Rev. C **74**, 04791 (2006)
[5] B. Mohanty, J. Phys. G: Nucl. & Part. Phys. **38**, 124023 (2011)
[6] A. Butenko et al., IPAC 2014, paper WEPRO067 (Dresden, June 2014)
[7] G. Trubnikov et al., ibid, paper TUPRO005 (Dresden, June 2014)
[8] D. Bertini et al., JoP,CS **119**, 032011, (2008). http://cbmroot.gsi.de
[9] MPD CDR v.1.2 (JINR, Dubna, 2010). http://www.nica.ru
[10] O. Rogachevsky et al., PR C **82**, 054910 (2010)
[11] M. Baznat et al., PR C **88**, 061901 (2013)
[12] V.D. Toneev, private communication 2014
[13] A.D. Kovalenko et al., Part. & Nucl **45(1)** (2014)
[14] I.A. Savin et al., arXiv:1408.3959 [hep-ex], 2014
[15] A.M. Baldin, A.D. Kovalenko, JINR Rapid Comm. JINR - 3(77)-96, Dubna 1996

Investigation of reactions using polarized and unpolarized deuteron beam at Nuclotron: current status and perspectives

M. Janek[1,⋆], *V.P.* Ladygin[2], *V.V.* Avdeichikov[2], *A.V.* Averyanov[2], *A.S.* Belov[3], *V.V.* Fimushkin[2], *Yu.V.* Gurchin[2], *A.Yu.* Isupov[2], *J-T.* Karachuk[2,4], *A.N.* Khrenov[2], *D.O.* Krivenkov[2], *A.K.* Kurilkin[2], *P.K.* Kurilkin[2], *N.B.* Ladygina[2], *A.N.* Livanov[2], *G.* Martinska[5], *S.M.* Piyadin[2], *S.G.* Reznikov[2], *G.* Tarjanyiova[1], *A.A.* Terekhin[2], *Ya.T.* Skhomenko[2,6], *I.E.* Vnukov[6], and *L.S.* Zolin[2]

[1]*Physics Department, University of Zilina, Univerzitna 1, 01001 Zilina, Slovakia*
[2]*Joint Institute for Nuclear Research, Joliot-Curie 6, 141980 Dubna, Moscow region, Russia*
[3]*Institute for Nuclear Research of Russian Academy of Sciences, Moscow, Russia*
[4]*Advanced Research Institute for Electrical Engineering, Splaiul Unirii 313, Bucharest, Romania*
[5]*Institute of Physics, University of P. J. Shafarik, Park Angelinum 9, 04001 Kosice, Slovakia*
[6]*Belgorod State University, Pobedy 85, 308015 Belgorod, Russia*

Abstract. Deuteron induced reactions at intermediate energies are investigated at Internal Target Station (ITS) of Nuclotron. The dp elastic scattering data have been accumulated in the angular range (70° - 120°) in c.m from 0.3 to 2.0 GeV of deuteron energy. Part of the obtained results of differential cross sections and analyzing powers of dp elastic scattering are compared with theoretical predictions based on relativistic multi-scattering model. The dp breakup data are obtained with the detection of two outgoing protons at the angles of (19° - 54°) in laboratory frame at the deuteron energies from 0.3 to 0.5 GeV in various detector configurations. One of the tools to study spin structure of short range correlations is the measurements of the cross section, tensor analyzing power T_{20} and spin correlation parameter $C_{y,y}$ in the $d^3He \to p(0°)^4He$ reaction at the energies between 1.0 and 1.75 GeV using polarized ^3He target and extracted polarized deuteron beam from new polarized ion source at Nuclotron. For this purposes the feasibility study has been performed. Proposed experiments on extracted deuteron beam of Nuclotron is aimed to investigate: non-nucleonic degrees of freedom and baryonic resonances properties in the d(A,d(0°))X and d(A,π-(0°))X reactions at different energies, non-nucleonic degrees of freedom via the measurements of the tensor A_{yy} and vector A_y analyzing powers in d(A,π-)X reaction and polarization properties of the baryonic resonances in the d(A,d)X reaction and others. First measurement of vector and tensor polarization of new polarized ion source has been performed in June-July of 2016y.

1 Introduction

It is natural to describe few nucleon systems based on nucleon-nucleon (NN) potentials. There are two approaches how the NN potentials are constructed. Realistic potentials as AV18 [1], CD-Bonn [2] and Nijm I, II and 93 [3] are based on meson-exchange picture. Effective field theories are based on

⋆e-mail: janek@fyzika.uniza.sk

spontaneously broken chiral symmetry [4]. These realistic NN potentials and effective field theories describe NN data with very high accuracy. Many experimental and theoretic studies are aimed on the lightest nuclei in the intermediate energy region below the pion threshold. The advantage of this energy region is that the exact calculations can be done. Nowadays, Faddeev equations for all NN potentials can be solved exactly. But these realistic NN are unable to describe binding energy of three [5] and four nucleon systems [6] and many scattering experiments. Cross sections and the tensor analyzing powers obtained from *Nd* scattering at energies below 30 MeV are well described with the corresponding theoretical calculations in which only the NN potentials have been used. In case of the vector analyzing power A_y such an agreement couldn't be achieved using any known NN potential. The inclusion of the three nucleon force (3NF) did not improve the situation leading to the well known A_y puzzle. The most used contribution in theoretical models which included 3NF are two pion exchange between all three nucleons. This contribution has been proposed long ago by Fujita-Miyazawa [7]. Idea of two pion exchange has been adopted and further developed in models of Urbana IX [8] and Tucson-Melbourne [9]. The success has been demonstrated in descriptions of binding energies of three and four nucleon systems. The cross section data of the *pd* elastic scattering obtained at 250 MeV cannot be reproduced by the Faddeev calculations [10]. Existence of new type of short range 3NF could be the reason for observed discrepancy. It has been found that in the vicinity of the Sagara discrepancy the currently known 3NF contribute by up to 30% to the *dp* elastic scattering cross section at intermediate energies [11]. The investigation of $dp \rightarrow ppn$ reaction at deuteron energy of 270 MeV at RIKEN [12] and IUCF [13] reveals variation in contribution of NN and 3NFs in polarization observables. Vector analyzing power A_y can be described using only NN forces. Similar picture has been observed in KVI in *dp* breakup reaction at energy of 130 MeV [14]. Inclusion of 3NF improves a part of the polarization data but breaks other. The spin structure of the *np* SRCs has been investigated at JINR via the measurements of the tensor analyzing power A_{yy} in deuteron inclusive breakup at different energies in the wide regions of the x_F and transverse proton momentum p_T [15]. Obtained results were compared with calculations using standard [16] and covariant deuteron wave functions [17], respectively. The A_{yy} data clearly demonstrate the dependence on two internal variables, x_F and p_T (or their combinations). However, the use of the deuteron structure function that depends on two variables [17] does not allow to describe the data. Few nucleon scattering has a rich phase space. Variety of polarization observables can be measured under different kinematic conditions. The comparison of polarization observables obtained in various parts of phase space with theory can give information about dynamics beyond pairwise potential, Coulomb and relativistic effects and their relative strength.

2 Experiments at ITS of Nuclotron

The goal of the Deuteron Spin Structure (DSS) experimental program is to obtain the information about two and 3NFs, including their spin dependent parts from *dp*-elastic scattering at the energies between 0.3 – 2.0 GeV and *dp*-breakup with registration of two protons at deuteron energies of 0.3 – 0.5 GeV [18]. Deuterons are accelerated by linear accelerator and Nuclotron up to required energy (see figure 1). Polyethylene and Carbon targets enclosed in a spherical hull of Internal Target Station (ITS) [19] were used. *Dp* effect is obtained by subtracting Carbon contribution from Polyethylene spectra. Different detector setups were used for *dp* breakup and elastic scattering reaction investigations.

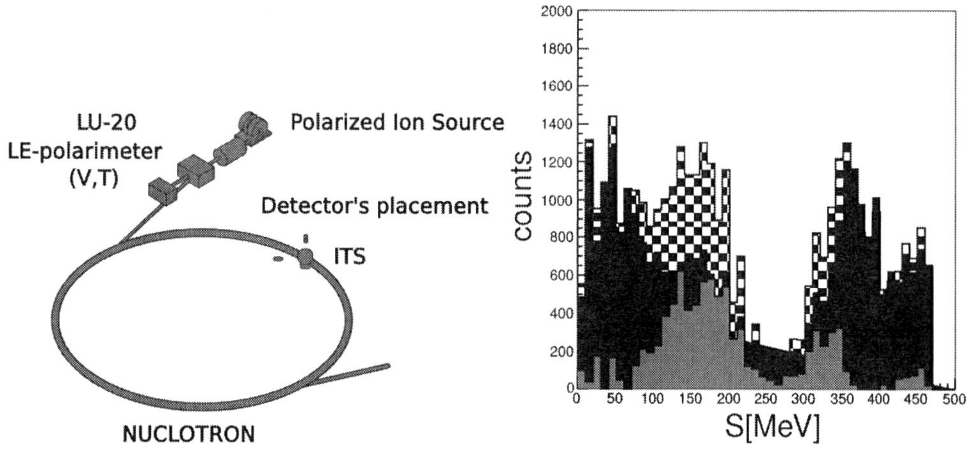

Figure 1. Left panel - Sketched view of Nuclotron accelerator with marked positions of ITS, detector setup for dp elastic and dp breakup reaction investigations, ion source and low energy polarimeter. Right panel - checked, black and grey histograms represent preliminary results of S distribution obtained on Polyethylene, Carbon and subtracted (dp) one for the detector arms placed at angles of $\Theta_1 = 35.0°$, $\Theta_2 = 43.0°$ and $\Phi_{12} = 180°$ at deuteron energy of 0.4 GeV.

2.1 Dp breakup reaction

The $dp \rightarrow ppn$ breakup reaction is investigated by $\Delta E - E$ scintillation detectors. The deuteron energies are 0.3, 0.4 and 0.5 GeV. Up to eight $\Delta E - E$ detectors can be used in experiment. Each detector consists of thin and thick scintillator. Both scintillators are of a tube shape, the thin one of a height of 1 cm, the thick one of a 20 cm height, with the diameter of the cross section 8 cm and 10 cm, respectively. Two Photomultiplier tubes PMTs-85 are positioned opposite to each other at the outside cylindrical surface of the thin scintillator. At the bottom end of the E scintillator a photomultiplier tube PMT-63 is positioned. Solid angle of each of detector is 4.6°. The details of the $\Delta E - E$ detector construction can be found in [20]. Energy spectra and missing mass distributions of dp breakup reaction were compared with GEANT4 simulation [21]. Reasonable agreement gives us opportunity to handle the signal and background in more precise way. The GEANT4 modelling also confirms that energy losses and scattering angles of the particles which passed through the spherical steel hull that surrounds the target are relatively small except of region of very low energy up to 20 MeV. But protons with this energy energy are cut by coincidence condition of $\Delta E - E$ detector.

Special setup of the detectors has been chosen in the experiment performed at 2014 to follow inverse kinematics of the reaction in which relativistic effects can play a role [22]. The kinematic variable S corresponds to the arc-length along the kinematic curve with zero point chosen at the minimal value of deposited energy in the second arm. Points of kinematic S-curve are calculated with energy step of 5 MeV in case of dp breakup reaction for the detector arms placed at polar angles of $\Theta_1 = 35.0°$, $\Theta_2 = 43.0°$ and $\Phi_{12} = 180°$ (azimuthal angle between arms) at deuteron energy of 0.4 GeV. The S is increasing in anti-clockwise direction. The number of breakup events in an interval $S - \Delta S/2$, and $S + \Delta S/2$ was obtained by projecting the events on a line perpendicular to the S-

DOI: 10.1051/epjconf/201713801028

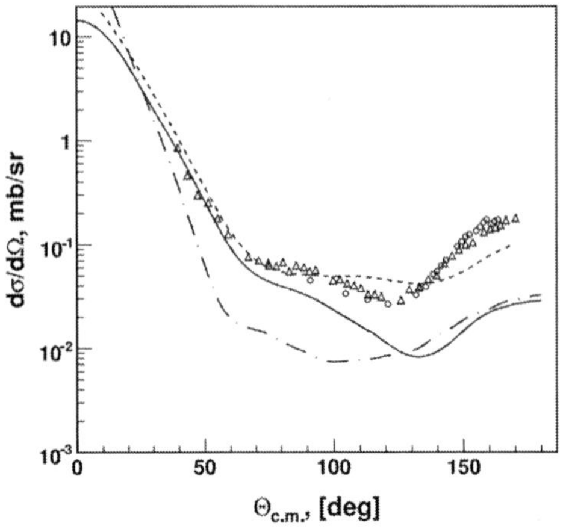

Figure 2. The angular dependence of the differential cross section obtained at 940 MeV [24] and 850 MeV [25] are represented by the open circles and triangles, respectively. The solid, dashed and dot-dashed lines are the results of the non-relativistic Faddeev calculations [26], relativistic multiple scattering model [27, 28] and optical potential approach [29].

curve. Preliminary result for energy step of $\Delta S = 5$ MeV and for the arm's polar angles of $\Theta_1 = 35°$, $\Theta_2 = 43°$ and azimuthal angle between them $\Phi_{12} = 180°$ for Polyethylene (checked spectra), Carbon (black spectra) and subtracted (grey spectra) at deuteron energy of 0.4 GeV is presented on the right panel in figure 1. Carbon normalization was performed on the parts of Polyethylene spectra where only Carbon content is assumed.

2.2 Dp elastic scattering

The results on the angular dependencies of the analyzing powers A_y, A_{yy} and A_{xx} of dp- elastic scattering have been obtained at Nuclotron at 880 MeV [23]. The differential cross section obtained at 940 MeV [24] and 850 MeV [25] in the earlier experiments are shown in figure 2 by the open circles and triangles, respectively. The solid, dashed and dot-dashed lines are the results of the non-relativistic Faddeev calculations [26], relativistic multiple scattering model [27, 28] and optical potential approach [29]. The Faddeev calculations without invoking $3N$ forces reproduce the behaviour of the analyzing powers, however, they fail to describe the cross section data at the scattering angles larger than 70° in the centre of mass system. The calculations performed within relativistic multiple scattering model [27, 28] describes the data on the vector analyzing power A_y and cross section. However, there are some problems in the description of the tensor analyzing powers at large angles in the centre of mass system. The optical potential approach fails to reproduce both cross section and analyzing powers. The observed deficiencies in the description of the differential cross section and deuteron analyzing powers at $T_d \sim$880 MeV obtained at quite large transverse momenta require the consideration of the additional mechanisms, for instance, $3N$ forces. Since present $3N$ forces models cannot improve the agreement with the data obtained even at lower energies, new models of $3N$ forces (including their short-range part) should be considered. For instance, the next step in the relativistic multiple scattering model [27, 28] development could be taking into account the explicit Δ- isobar excitation.

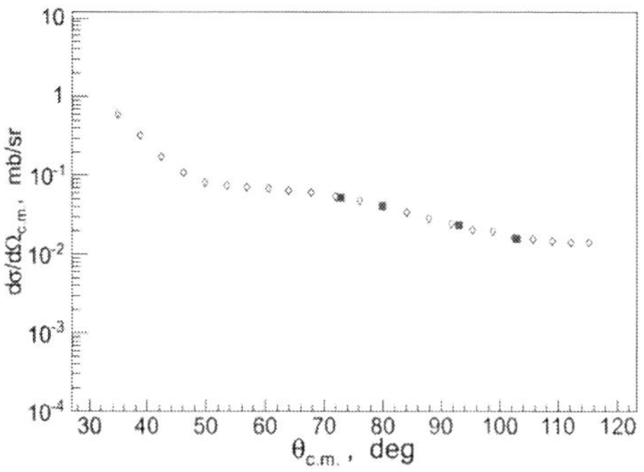

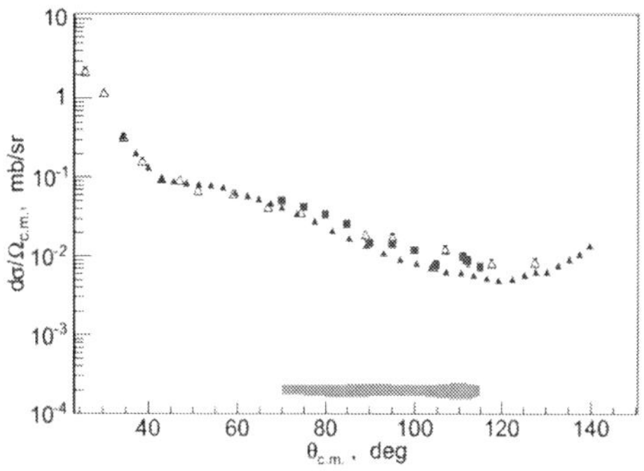

Figure 3. The angular dependence for differential cross section at 650 MeV/n (left panel) and 700 MeV/n (right panel). Rhombuses represent data obtained at 641.3 MeV/n [30]. Solid and open triangles represent data from 800 MeV/n [31] and 794 MeV/n [30], respectively. Red squares are data obtained at Nuclotron. Their systematic errors are shown by the solid blue band.

Dp elastic data have been obtained in the energy range from 0.3 to 2.0 GeV. Presented *dp* elastic data were obtained by high energy *dp* detector setup. Firstly, the normalization was performed at the energy of 650 MeV/n. The angular dependence of differential cross section [30] was approximated by the exponential function in angular range $71° < \theta_{c.m.} < 105°$. Obtained normalization was used to obtain the differential cross section for other energies. Angular dependence of differential cross section at 650 MeV/n and 700 MeV/n are shown on the left and right panels of figure 3. The systematic errors are shown by the solid blue band.

3 Experiments at extracted deuteron beam

Transportation line of the Nuclotron extracted beam to the start version of BM@N spectrometer is shown in figure 4. Spin physics with extracted polarized deuteron beam from new polarized deuteron source [32] can be performed with the start version of the BM@N setup [33]. The BM@N experimental setup will be installed at the renovated 6V beamline [34] in the fixed-target hall of the Nuclotron. The 6V beamline contains the quadrupole lenses doublet, two dipole magnets allowing to correct the beam position in the vertical and horizontal planes, and SP-41 dipole magnet with enlarged aperture for the momentum measurements [35].

Spin program realization requires to install the polarimeters (with CH_2-C targets) at F3 and F5 focuses for polarimetry of the deuteron (or proton) beam. Liquid hydrogen (deuterium) target, polarized $^3\vec{He}$ target [36] and set of nuclear targets is needed to place at F5 or F3. The spin physics with the target position at F5 (or F3) focus is related with the measurements of the polarization observables in the reactions with the emission of the detected particle at forward angles. The solid angle is defined by the lenses doublet 7k100–8k100 of the VP1 transportation beam line.

3.1 $d(^3He, p)^4He$ reaction

The goal of the ^{3}He(d, p)^{4}He reaction study at Nuclotron is to understand the reasons of strange structure in the behaviour of the tensor analyzing power T_{20} in dp- backward elastic scattering [38, 39]

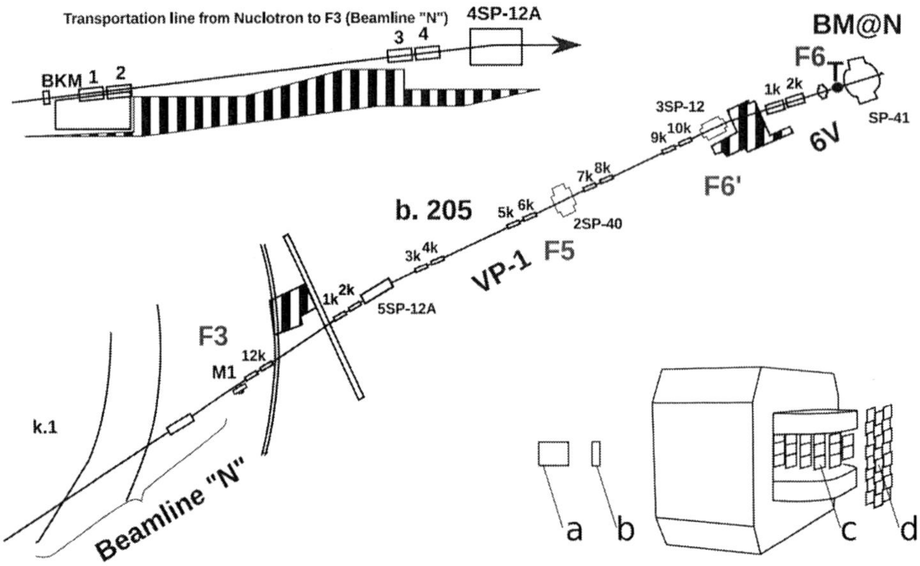

Figure 4. Transportation line of the Nuclotron extracted beam to the BM@N spectrometer with start version of BM@N setup (bottom right). Spin physics with BM@N can be associated with different target positions: at F5 (F3), F6' and F6. Schematic view of the part of BM@N setup used in simulation is shown in the right-down corner; a - ^3He target, b - start counter (quartz radiator), c - 12 GEM tracker stations, d - TOF wall.

at the internal momentum k ~0.3–0.5 GeV/c in the vicinity of the D- wave dominance. The experiments performed at RIKEN at the energies below 270 MeV have shown that the polarization correlation coefficient $C_{//}$ for the ^{3}He$(d,p)^4$He reaction may be a unique probe to the D-state admixture in deuteron [40] due to the strong spin-selectivity in neutron capture process by ^{3}He nucleus. In the one-nucleon exchange (ONE), the expression for $C_{//}$ is proportional to the D-state fraction in deuteron T_{20} and κ_0 for dp backward elastic scattering which include S- and D-state interference term together with a squared D contribution. Thus expected that $C_{//}$ may be a candidate to provide an information on the deuteron structure complementary to those from T_{20} and κ_0 obtained in dp- backward elastic scattering [38].

The main goal of the proposed experiment is to obtain the data on $C_{//}$ in the energy region of 1.0–1.75 GeV, where the contribution from the deuteron D-state is expected to reach a maximum in one-nucleon exchange approximation, to obtain new information on the strange structure observed in the behaviour of T_{20} in the dp- backward elastic scattering and to realize experiment on the full determination of the matrix element of the ^{3}He$(d,p)^4$He reaction in the model independent way. These data will help us also to understand the short-range spin structure of deuteron and effects of non-nucleonic degrees of freedom. For these purposes polarized deuteron beam from new PIS [32] and spin-exchange-type polarized ^{3}He target developed at CNS of Tokyo University [36] and modified for the experiment at Nuclotron can be used. Experiment is planned to be carried out with extracted deuteron beam of Nuclotron at the BM@N area at energy range of 1.0 - 1.75 GeV. The feasibility study of the $d^3He \to p^4He$ reaction have been investigated using GEANT4 and ROOT simulation package [37]. There are serious requests on beam size and emittance in the target crossover point. Deuteron beam hits the ^3He target which is placed in the one of the focuses of the beam transportation line of Nuclotron, F5 focus point is assumed in next. The length of polarized ^3He target is 30 cm. Scattered protons and deuterons, as well as primary deuterons pass along transportation beam line

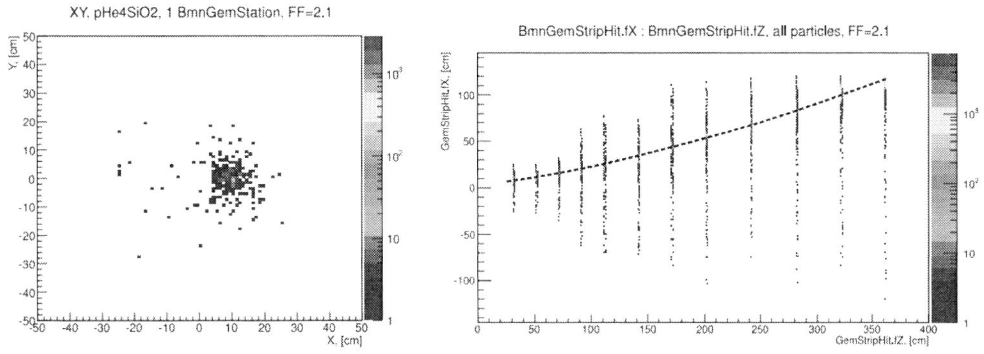

Figure 5. Left panel - the XY secondary beam profile at the 1-st GEM station for the magnetic field B_y of ~0.9 T. Right panel - the XZ profile of the secondary beam hits in the GEM tracker stations for the magnetic field B_y of ~0.9 T. Dashed line represents mean path of scattered particles.

with magnetic elements tuned for the optimal momentum of the secondary protons ~ 75% of the initial deuteron beam momentum. The simulation has been performed for the initial deuteron kinetic energy T_d = 1.5 GeV and 30 cm of the ^3He target cell using the GEANT4 toolkit. The software model reflects only substantial parts of BM@N setup (start version). In the first variant of simulation ^3He target has only been used while in the second one the quartz radiator of 2 mm thick was added as a TOF start counter after ^3He target. The hadronic interaction of the secondary particles with the target, air and start counter was simulated using GEANT4 with using QBBC physics list with the Binary Cascade model. Quartz radiator spreads the size and the number of particles in the secondary beam due to additional multiple scattering in comparison with the first variant, however, the number of these events is quite small for both cases. The XY secondary beam profile of the first GEM tracker station is presented on the left panel in figure 5. One can see that the X- position of the beam is ~10 cm from the central axis of the spectrometer. The XZ profile of the secondary beam hits in the GEM tracker stations is shown on the right panel in figure 5. The deflection angle is about ~320 mr. The time difference between mRPC and start detectors for protons and deuterons for the base line of ~24 m are separated by ~20 ns which is good enough to distinguish between them. The further simulation has been performed for the setup with quartz radiator. The momentum reconstruction of particles is based on the information derived from 12 GEM stations placed in the inhomogeneous magnetic field defined by a field map. The simulation has been performed at different field map scale factors: "half" of the magnetic field value and at maximal one. BM@N setup with 12 stations of GEM tracker situated in the magnet can provide the momentum resolution about few percents. The time difference between mRPC and start detector will be used to separate protons from the background deuterons. Additional improvement of the momentum resolution can be done by removing of the additional material along the beam line as well as by the increasing of the space resolution of the tracking detectors.

3.2 Other possible reaction at extracted deuteron beam

The polarization observables in the $p(d,p)d$ [38, 39] and $d(d,p)t$ [41–43] reactions in the collinear geometry can be measured using liquid hydrogen or deuterium target or CH_2 and CD_2 solid targets with Carbon background subtraction. The solid nuclear target can be placed inside the pole of the 3SP40 magnet. It can be varied from the beginning to the middle of the pole of the 3SP40 magnet. This configuration can be used for the measurements of the tensor A_{yy} and vector A_y analyzing powers (and, possibly, vector polarization transfer coefficient C_y^y) in inclusive deuteron breakup, $A(d,p)X$, at large transverse proton momenta [15] at the highest available energy at Nuclotron; of the tensor A_{yy} and vector A_y analyzing powers [44–46] (and, possibly, vector polarization transfer coefficient C_y^y [47]) in the inelastic deuteron scattering, $A(d,d')X$, in the vicinity of the baryonic resonances excitation; of the tensor A_{yy} and vector A_y analyzing powers in the inclusive pion production, $A(d,\pi^-)X$, [48] also at the highest available energy at Nuclotron. The measurements of the analyzing powers in inclusive deuteron breakup and the inelastic deuteron scattering in the vicinity of the baryonic resonances excitation can be performed simultaneously. All these experiments require the additional TOF detector placed between F5 and F6 focuses. The separation of the primary deuteron and secondary beams at focus point F6 is provided by the modernized analyzing magnet SP-41 [35]. The physics at F6 is related with the baryonic resonances spin properties studies at the energies between 2 and 6 GeV of the deuteron kinetic energy and includes the measurements of the tensor A_{yy} and vector A_y analyzing powers in quasi-elastic and inelastic $A(d,pp)X$ reaction; of the tensor A_{yy} and vector A_y analyzing powers in the inelastic deuteron scattering $A(d,d')X$ [47] and $A(d,d')\pi^{\pm}X$ reactions; investigation of the vector analyzing power in neutron induced reactions (with the proton spectator detection) like $np \to pn$, $np \to pp\pi^-$, $np \to np\pi^+\pi^-$ [49, 50] etc. For these experiments the full size RPC wall is required. In parallel, the measurement of the dp- elastic scattering [23, 51], $dd \to {}^3\mathrm{He}n$ [52–54] and $dA \to {}^3\mathrm{He}X$ [55] processes can be performed. The spin studies with BM@N require also the advanced deuteron beam polarimetry at Nuclotron discussed in [56, 57].

4 Deuteron polarization measurement

Measurements of the deuteron vector and tensor beam polarization were performed at the end of linear accelerator in June-July 2016 with new polarized ion source. Deuteron vector and tensor beam polarization at energy of 10 MeV measured with low energy polarimeter are shown in figure 6. Red symbols represent the 1-4 transition with ideal values of polarization (p_z,p_{zz}) = (+1/2,-1/2), blue symbols are for the case of 3-4 transition with ideal values of polarization (p_z,p_{zz}) = (-1/2,-1/2). The measurement of deuteron polarization with low energy polarimeter shows good values of vector and tensor polarizations.

5 Conclusion

Dp breakup and *dp* elastic data along with theoretical description based on relativistic multi-scattering model were discussed. Monte Carlo study of $^3He(d,p)^4He$ reaction at 1.5 GeV confirms suitability of experiment performed on extracted beam. New PIS can provide unique opportunity for the studies of the spin effects and polarization phenomena in few body systems at ITS and also on extracted beam with the BM@N setup. Commissioning of the upgraded DSS setup has been performed at 270 MeV using unpolarized deuteron beam in June-July 2016. The measurement with low energy polarimeter showed good values of vector and tensor polarizations. New data can be measured at ITS at the Nuclotron using new PIS, e.g. data on the analyzing powers A_y, A_{yy} and A_{xx} in dp- elastic scattering at various energies up to 2000 MeV, and the dp- nonmesonic breakup at the energies between 300 and 500 MeV for various kinematic configurations. Proposed experiments on extracted deuteron beam were discussed.

Acknowledgement

The authors thank P. N. Batyuk, A.D. Kovalenko, L.V. Kutuzova, V.I. Maximenkova, S. P. Merts, Yu.V. Prokofichev, V.B. Shutov, A.V. Turbabin and V.N. Zubets for their help in the preparation of new

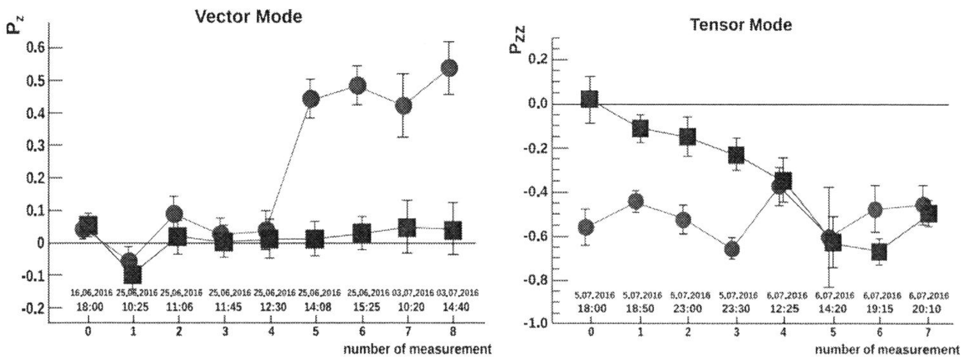

Figure 6. Left panel - deuteron vector beam polarization at energy of 10 MeV. Right panel - deuteron tensor beam polarization at energy of 10 MeV. Red symbols represent the 1-4 transition with ideal values of polarization (p_z,p_{zz}) = (+1/2,-1/2), blue symbols are for the case of 3-4 transition with ideal values of polarization (p_z,p_{zz}) = (-1/2,-1/2).

polarized ion source and experimental setup. The work has been supported in part by the RFBR under grant 16-02-00203a, JINR-Slovak Republic and JINR-Romania Scientific Cooperation programs in 2015-2016 years.

References

[1] S. A. Coon et al., Nucl. Phys. A **317**, 242 (1979)
[2] S. A. Coon, W. Glockle, Phys. Rev. C **23**, 1790 (1981)
[3] V. G. J. Stoks, R. A. M. Klomp, C. P. F. Terheggen, J. J. de Swart, Phys. Rev. C **49**, 2950 (1994)
[4] P.F. Bedanque, U. Kolck, Ann. Rev. Nucl. Part. Sci. **52**, 339 (2002)
[5] H. Setze et al., AIP Conf. Proc. **334**, 463 (1995)
[6] W. Tornow et al., Few-Body Syst. Suppl. **8**, 163 (1995)
[7] F. Correll et al., Nucl. Phys. A **475**, 407 (1998)
[8] G. Rauprich et al., Nucl. Phys. A **535**, 313 (1991)
[9] H. Patberg et al., Phys. Rev. C **53**, 1497 (1996)
[10] K. Hatanaka, Y. Shimizu et al., Phys.Rev. C **66**, 044002 (2002)
[11] N. Sakamoto et al., Phys. Lett. B **367**, 60 (1996)
[12] K. Sekiguchi et al., Phys. Rev. C **79**, 054008 (2009)
[13] H.O. Meyer et al., Phys. Rev. Lett. **93**, 112502 (2004)
[14] S. Kistryn et al., Phys. Rev. C **72**, 044006 (2005)
[15] L.S. Azhgirey et al., Phys.Lett. B **387**, 37 (1996)
[16] R. Machleidt, Phys.Rev. C **63**, 024001 (2001)
[17] V.A. Karmanov and A.V. Smirnov, Nucl.Phys. A **546**, 691 (1992)
[18] V.P. Ladygin et al., J.Phys.Conf.Ser. **295**, 012131 (2011)
[19] A.I. Malakhov et al., Nucl. Instr. Meth. in Phys. Res. A **440**, 320 (2000)
[20] S.M. Piyadin et al., Physics of Particles and Nuclei Letters **8(2)**, 107 (2011)

[21] M. Janek et al., Physics of Particles and Nuclei Letters, **11(4)**, 552 (2014)
[22] H. Witala et al., Few-Body Syst. **49**, 61 (2011)
[23] P.K. Kurilkin et al., Phys.Lett. B **715**, 61 (2012)
[24] J.C. Alder et al., Phys.Rev. C **6**, 2010 (1972)
[25] N.E. Booth et al., Phys.Rev. D **4**, 1261 (1971)
[26] W. Glöckle, H. Witala, D. Hüber, H. Kamada and J. Golak, Phys.Rept. **274**, 107 (1996)
[27] N.B. Ladygina, Phys.Atom.Nucl. **71**, 2039 (2008)
[28] N.B. Ladygina, Eur.Phys.J. A **42**, 91 (2009)
[29] M.A. Shikhalev, Phys.Atom.Nucl. **72**, 588 (2009)
[30] E. Culmez, Phys.Rev. C **43(5)** (1991)
[31] E. Winkelmann et al., Phys.Rev. C **21(6)**, 2535-2541 (1980)
[32] V.V. Fimushkin et al., Eur. Phys. J. ST **162**, 275 (2008)
[33] V. Ladygin et al., PoS(Baldin-ISHEPP-XXI)38 (2012)
[34] S.M. Piyadin et al., Phys. Part. Nucl. Lett. **9**, 589 (2012)
[35] P.G. Akishin et al., Phys.Part.Nucl.Lett. **12**, 305 (2015)
[36] T. Uesaka et al., Nucl.Instr. and Meth. in Phys.Res. A **402**, 212 (1998)
[37] M. Janek et al., Journal of physics: conference series **678**, 012032 (2016)
[38] V. Punjabi et al., Phys.Lett. B **350**, 178 (1995)
[39] L.S. Azhgirey et al., Phys.Lett. B **391**, 22 (1997)
[40] T. Uesaka et al., Phys.Lett. B **467**, 199 (1999)
[41] V.P. Ladygin et al., Phys.Lett. B **598**, 47 (2004)
[42] N.B. Ladygina, Few Body Syst. **53**, 253 (2012)
[43] N.B. Ladygina, Phys.Part.Nucl. **45**, 187 (2014)
[44] V.P. Ladygin et al., Eur.Phys.J. A **8**, 409 (2000)
[45] L.S. Azhgirey et al., Phys.Atom.Nucl. **68**, 991 (2005)
[46] V.P. Ladygin et al., Phys.Atom.Nucl. **69**, 852 (2006)
[47] V.P. Ladygin and N.B. Ladygina, Phys.Atom.Nucl. **65**, 182 (2002)
[48] L.S. Azhgirey et al., Phys.Atom.Nucl. **74**, 1392 (2011)
[49] G. Agakishiev et al., Phys.Lett. B **750**, 184 (2015)
[50] A. Jerusalimov et al., Eur.Phys.J. A **51**, 83 (2015)
[51] P.K. Kurilkin et al., Phys.Part.Nucl.Lett. **8**, 1081 (2011)
[52] M. Janek et al., Eur.Phys.J. A **33**, 39 (2007)
[53] A.K. Kurilkin et al., Int.J.Mod.Phys. A **24**, 526 (2009)
[54] A.K. Kurilkin et al., Phys.Rev. C **87**, 051001 (2013)
[55] V.P. Ladygin and N.B. Ladygina, Phys.Atom.Nucl. **58**, 1283 (1995)
[56] P.K. Kurilkin et al., Nucl.Instr.Meth. in Phys.Res. A **642**, 45 (2011)
[57] P.K. Kurilkin and V.P. Ladygin, Phys.Part.Nucl. **45**, 265 (2014)

GlueX overview: status and some future plans

Maria Patsyuk[1,★] on behalf of the GlueX Collaboration

[1] *Massachusetts Institute of Technology, Cambridge, MA, USA*

Abstract. The GlueX experiment at Thomas Jefferson National Accelerator Facility (Jefferson Lab) has started data taking in late 2014 with its first commissioning beam. All of the detector systems are now performing at or near design specifications and events are being fully reconstructed. Linearly-polarized photons were successfully produced through coherent bremsstrahlung. An upgrade of the particle identification (PID) system with a GlueX DIRC detector, planned for 2018, will allow identification of final state kaons. The construction of the GlueX DIRC has already started. One of the plans for GlueX is to study properties of short-range correlations (SRC) in nuclei, which will shed new light on the quark-gluon structure of bound nucleons.

1 Introduction

The GlueX Experiment [1] is a key element of the Jefferson Lab 12 GeV upgrade. The experiment is at the end of a new beamline from the Continuous Electron Beam Accelerator Facility (CEBAF) at Jefferson Lab, that will use 12 GeV electrons to deliver linearly-polarized photons to a new experimental area, Hall D. The primary physics goal of GlueX is to discover and study the properties of hybrid mesons - particles where the gluonic field contributes directly to the J^{PC} quantum numbers of the mesons [2]. Lattice QCD calculations indicate that several of the nonets of these hybrid mesons have exotic quantum numbers, forbidden J^{PC} for a simple fermion-antifermion system [3]. In addition, the expected masses for the lightest hybrids are well-matched to the energy and kinematics accessible to the GlueX experiment. Commissioning of the GlueX experiment started in late 2014 and continued through the spring of 2015. The first physics quality beam is expected in Fall 2016. In this paper we present some of the first results of the GlueX experiment, as well as describe the planned upgrade of the particle identification system (PID) and plans for studying of Short Range Correlations (SRC) using photonuclear reactions and GlueX spectrometer.

2 The Beamline

The Hall D is a photon-only area taking advantage of the 12 GeV electron beam from the upgraded CEBAF. After extraction, electrons pass through a bremsstrahlung radiator and then into the dipole magnet for tagging the energy of the scattered electrons (see figure 1 on the left). Those electrons not interacting in the bremsstrahlung radiator are deflected into an electron dump attached to the tagger hall. For the electrons producing a bremsstrahlung photon in the region between 25% and

★e-mail: mpatsyuk@mit.edu

98% of the primary electron beam energy, the scattered electrons are detected in a pair of tagger hodoscopes, thus tagging the energy of the photon. For 12 GeV/c electrons, the fine-grained *Tagger Microscope* (TAGM) has been placed to tag photons in the range between 8.2 GeV and 9.2 GeV, with each microscope element spanning 10 MeV in energy. For both higher- and lower-energy photons, the *Tagger Hodoscope* (TAGH) tags photons using elements about 30 MeV wide.

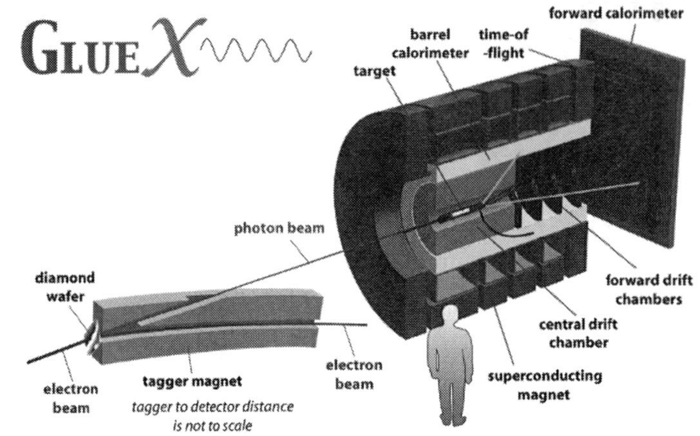

Figure 1. A schematic drawing of the photon tagger and the GlueX detector at Jefferson Lab.

To produce a linearly polarized coherent bremsstrahlung photon beam, a 20 μm diamond crystal[1] is utilized as a radiator. The photons in the coherent momentum range from 8.4 GeV/c to 9 GeV/c are linearly polarized relative to the crystal axes in the diamond. The peak polarization is expected to be 40%. To produce two perpendicular directions of polarization, the diamond can be rotated. Coherent bremsstrahlung photons are produced in the direction of the incident electrons, while incoherent bremsstrahlung photons are produced in a cone around this direction.

The photons travel through an 80 m long vacuum beamline before entering Hall D. Here, the off-axis photons are removed by passing the beam through a 3.4 mm diameter collimator[2] that has beed instrumented around the hole with an *Active Collimator* [4] that provides fast feedback on the beam positioning. After that photons pass through a *Triplet Polarimeter* that is used to monitor the linear polarization. Then photons enter a *Pair Spectrometer* system [6] that is used to monitor both the energy and intensity of the beam. The pair spectrometer is read out in coincidence with TAGM and TAGH, which allows determination and monitoring of the energy calibration of the tagger system. Finally photons travel into the GlueX spectrometer, where some of them interact with the liquid hydrogen target. The remainder pass into the photon beam dump at the back of Hall D.

Initial operation of the GlueX was with beam intensity of 10^7 photons per second in the coherent peak (photon energy in the range 8.4 – 9 GeV), with the ultimate rate limit expected to be an order of magnitude higher.

3 The GlueX Detector

The GlueX detector, schematically shown in figure 1 on the right, has near uniform acceptance for both charged and neutral particles in the polar angle range from 1° to 120°. The 30 cm long target

[1]During the spring 2015 running of GlueX at lower electron energies, 50 μm and 100 μm thick diamonds were used.
[2]During the spring 2015 run, a 5 mm diameter collimator was used.

filled with liquid hydrogen is located in the barrel part of the GlueX detector. The solenoid magnet provides a magnetic field of about 2 T along the direction of the beam. Particles produced in the interaction are first coming through a scintillator-based *Start Counter*, which surrounds the target and the beamline in the forward direction. The light signal is detected by silicon photomultipliers [7, 8] providing track timing with 300 ps resolution as well as some particle identification based on the energy loss measurement. The timing resolution is enough to identify the bucket from the electron beam 499 MHz radio frequency (RF) structure for each photon.

The *Central Drift Chamber* (CDC) [9] immediately surrounding the start counter is based on straw-tube technology and provides position measurement along the charged tracks with 150 μm accuracy in the $r - \phi$ plane and at the mm level along the beam axis. The CDC also provides dE/dx information for charged tracks separating kaons from pions up to a momentum of 1 GeV/c. The *forward drift chamber* (FDC) [10] complements the tracking in the forward region and performs PID by measuring dE/dx.

The *Barrel Calorimeter* (BCAL) [11–13] surrounds the tracking devices inside the solenoid. BCAL is covering the polar angle acceptance between 12° and 160° and reconstructs π^0 using decay photons of energy below 100 MeV. The energy resolution is σ_E/E $6\%/\sqrt{E} \oplus 2\%$. The timing information is used to provide time-of-flight for particles interacting in the BCAL, and is used in the global PID for both charged particles and photons.

The *Forward Calorimeter* (FCAL) [14] is located downstream of the solenoid magnet and covers the polar angles between 1° and 12°. FCAL is sensitive to photons with energy above 100 MeV and has similar energy resolution.

The *Time-Of-Flight* (TOF) wall located upstream of FCAL consists of two layers of scintillator bars, which are read out on both sides using photomultiplier tubes. The TOF provides both timing and energy-loss information. The timing resolution of the TOF system is 100 ps, which provides PID for forward going particles.

All of the detector systems are now performing at or near the design specifications. Figure 2 shows an illustration of the PID capabilities for CDC, TOF, and FCAL. The left plot shows the dE/dx measured by CDC for positively charged particles. Clear π/p separation is evident up to about 1 GeV/c. In the middle plot the calculated β of positively charged particles is shows as a function of momentum, and clear bands are seen for π^+, K^+, e^+, and protons, which clear proton identification up to 3 GeV/c and π/K separation up to about 2 GeV/c. The right-hand image in figure 2 shows separation of electrons/positrons from hadrons in the FCAL. The middle plot was made using the spring 2015 commissioning data, whereas the others are based on the fall 2014 commissioning data.

To fully use the potential of the GlueX detector several experiments are planned. The study of hybrid meson spectroscopy has started in 2016, the first data is sufficient to measure the beam asymmetry for a wide range of reactions ($\gamma p \to (\pi, \eta, \eta', \rho, \omega, \phi)p$). The polarization transfer to the ρ meson is described in detail in [18]. The physics data from 2017 and 2018 (low intensity mode with 10 MHz in the peak) are expected to allow measurements of cross sections, spin-density matrix elements, and start of exotic hybrids search and the partial wave analysis. The measurement of the charged pion polarizability based on the Primakoff effect is planned for 2018.

The GlueX strangeness program complements the mapping of hybrid spectrum by identifying kaon final states. The quark flavor composition of a meson can be experimentally inferred by comparing the branching ratios for strange and non-strange decays. For a systematic study of the kaon final states an upgrade of the PID system is planned for 2018. A new DIRC (Detection of Internally Reflected Cherenkov light) detector [5] will be installed in the forward region of the spectrometer. After that GlueX will run in a high intensity mode (50 MHz in the peak) focusing on identification of states with hidden strangeness and hyperon resonances.

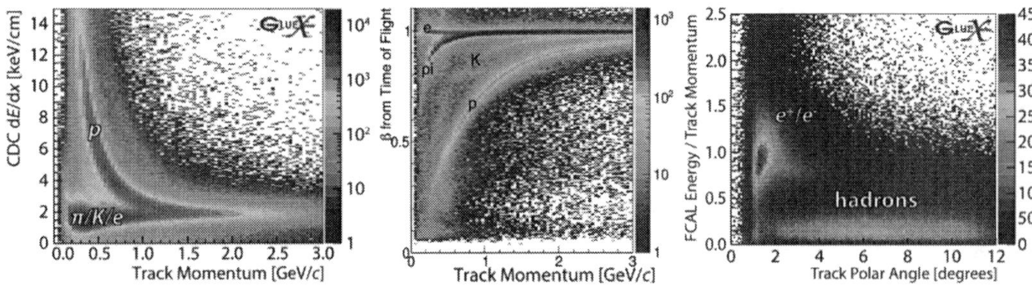

Figure 2. For positively charged particles: (Left) energy loss dE/dx in the CDC as a function of reconstructed particle momentum and (Middle) the β as determined from both fime-of-flight and path-length measurements in the detector versus the reconstructed particle momentum. The energy measured by the FCAL corresponding to the track momentum is plotted against the track polar angle on the right-hand side.

3.1 PID Upgrade

The proposed PID upgrade for GlueX is based on the DIRC (Detection of Internally Reflected Cherenkov light) principle [19]. The GlueX DIRC will reuse four existing BaBar DIRC [20] bar boxes filled with long and narrow radiators ($1.75 \times 3.5 \times 490$ cm^3) without any modifications. Each pair of boxes will be attached to a new compact photon camera, inspired by the SuperB FDIRC prototype[3] [21]. The GlueX DIRC design is described in detail in [22]. The current GlueX DIRC design is going to provide clean separation between kaons and pions with at least three standard deviations for momenta less than 4 GeV/c.

The GlueX DIRC is a thin wall of fused silica radiators located about 4 m away from the target, attached to photon cameras read out by an array of multi-anode photomultipliers (MaPMT). The GlueX DIRC covers the polar angle range from 2° to 11° and requires four unaltered BaBar bar boxes each containing 12 radiators. The bar boxes will be oriented horizontally and placed directly upstream of the TOF detector, with two bar boxes above and two below the beam. Each pair of the bar boxes will be attached to a photon camera, one of which will be located to the left, and another one to the right of the beam. The support structure of the DIRC will allow the pairs of bar boxes to slide vertically out of the active area of the detector for experiments requiring minimal material budget in front of the forward calorimeter.

The design of the photon camera is shown in figure 3. The FDIRC photon camera design is based on compact focusing blocks made of fused silica, one for each bar box in a barrel. For the GlueX planar orientation one common camera, filled with distilled water, will be used for two bar boxes together. This design with wider cameras reduces the fraction of photons reflecting on the sides, simplifying the hit pattern and removing ambiguities in the reconstruction.

Another important modification of the photon camera design is the approximation of a cylindrical mirror by three flat segments for simplicity in alignment and construction. The optimized radius of the mirror approximated by three flat segments allows using fewer photosensors, which is the main cost driver for GlueX DIRC. The resolution on the Cherenkov angle is approximately the same for the designs with the cylindrical and the three-segmented mirrors.

We plan to use H12700 [23] 64-channel MaPMTs from Hamamatsu for the read out system. The electronics design is based on the CLAS12 RICH [24], since it has similar specifications. The expected timing resolution is about 1 ns.

[3] FDIRC was designed to be successor of the BaBar DIRC, but the SuperB project was canceled in 2012

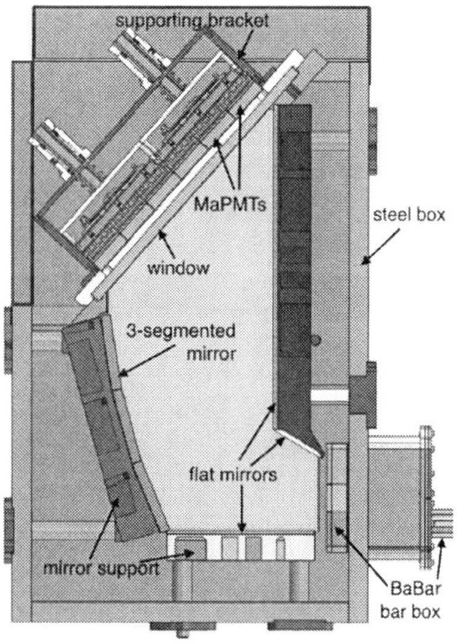

Figure 3. Schematic diagram of the cross section of the photon camera. The attached bar boxes are shown at the bottom right.

Construction has been started. The calibration system is being studied. Preparations for transporting the bar boxes from SLAC to Jefferson Lab are underway. An initial installation of two bar boxes and one photon camera in HallD is planned for 2017.

3.2 Short Range Correlations (SRC) at GlueX

Recent high-momentum transfer measurements [27–35] have shown that nucleons in nuclear ground states can form pairs with large relative momentum and small center-of-mass momentum of the pair because of short-range (scalar and tensor) components of the nucleon-nucleon interaction. These pairs are referred to as short-range correlated (SRC) pairs. As one form of cold dense nuclear matter fluctuations, SRC pairs are characterized by nucleonic degrees of freedom and medium densities of $\rho \approx 2 - 3\,\rho_0$, where ρ_0 is the nominal density of the nucleus. Another approach to the cold dense nuclear matter fluctuations at higher densities of $\rho \approx 5 - 10\,\rho_0$ involving quark degrees of freedom is being developed at JINR (Dubna) [36]. It was initially proposed by A. M. Baldin [37] in 1971.

Experimentally, SRC can be studied using a hard reaction between a high-momentum probe and a nucleon in a nucleus belonging to a SRC pair. Detection of the nucleon, its recoil partner and, if possible, the probe, sheds light on properties of SRC pairs, which define nucleon structure modifications in the nuclear medium. The current exclusive SRC studies at BNL [31–33] and JLab [35] provided statistics of tens to hundreds of events, but there are plans to perform a next generation exclusive SRC studies at GSI, JLab, and possibly, Nuclotron [38, 39], detecting tens of thousands of SRC events.

The GlueX spectrometer offers unique possibilities to measure Short Range Correlations (SRC) in exclusive photonuclear reactions using a set of thin foils instead for the liquid hydrogen target. The high energy photon beam complements the set of different probes of protons and electrons used so far and ensures hard knockout reactions. Another advantage of using a real photon is that it can interact with the neutron from a SRC pair leading to a charged final state ($\gamma n \to \pi^- p$). According to the cross section of the reaction $\gamma p \to \pi^+ n$ shown in figure 4, γp scattering selects mostly forward going high momentum proton in a nucleus. On the other hand, the GlueX detector seem to offer good neutron detection efficiency in BCAL. This allows probing *nn* SRC pairs to understand better the role of protons versus neutrons in a nucleus.

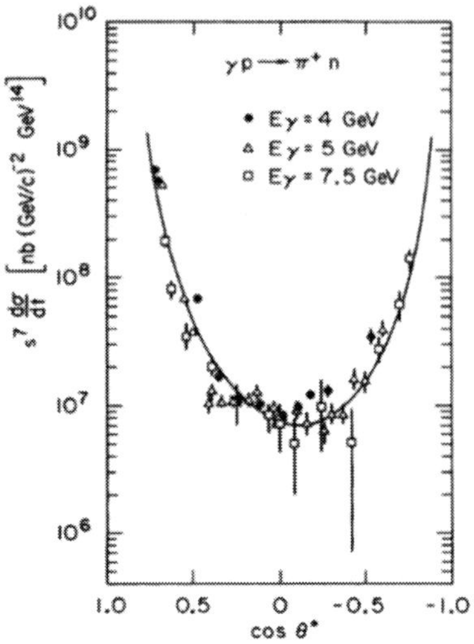

Figure 4. $s^7 d\sigma/dt$ versus cosine of the scattering angle in the center of mass (θ^*). The solid line shows the empirical function $(1-z)^{-5}(1+z)^{-4}$, where $z = \cos\theta^*$, which is an empirical fit to the angular distribution. Figure from [40].

Thin nuclear targets will be made of ^{2}H, ^{4}He, ^{12}C, ^{28}Si, ^{40}Ca, ^{48}Ca, ^{54}Fe, ^{93}Nb, ^{111}Ag, ^{208}Pb. GlueX will measure the branching ratio for many reactions including $\gamma p \to \pi^-\Delta^{++}, \rho^0 p, K^+\Lambda, K^+\Sigma^0$ and the equivalent reactions for neutrons.

3.3 Summary

The GlueX experiment has been successfully commissioned, and the first analysis based on the available data is in progress. All detector systems are near design specifications and are ready to start physics running soon. The first stage of GlueX running in 2016 – 2018 at low intensity (10 MHz in the peak) is devoted to measurements of cross sections, asymmetries, and spin-density matrix elements, as well as identification of known resonances and hybrids using partial wave analysis.

In 2018 it is planned to increase the beam intensity by a factor of 5 to cover all parts of the GlueX exotic hybrid program and upgrade the PID system by installing a DIRC detector. This will significantly improve the separation between kaons and pions and allow effective detection of strange final states.

For a more distant future there are plans to study short range correlations in exclusive photonuclear reactions using the GlueX spectrometer.

References

[1] The GlueX Collaboration, The GlueX Experiment in Hall D, 2010, Presentation to JLab PAC 36. Available at: http://www.gluex.org/docs/pac36_update.pdf

[2] C. A. Meyer and E.S. Swanson, Prog. Part. Nucl. Phys. **82**, 21-58 (2015)

[3] J. J. Dudek et al., Phys. Rev. D **88(9)**, 094505 (2013). [arXiv:1309.2608 [hep-lat]]
[4] G. Miller and D. R. Walz, Nucl. Instrum. Meth. **117**, 33 (1974)
[5] J. Stevens et al., JINST **11(07)**, C07010 (2016). [arXiv:1606.05645 [physics.ins-det]]
[6] F. Barbosa et al., Nucl. Instrum. Meth. A **795**, 376 (2015)
[7] F. Barbosa, et al., Nucl. Instrum. Meth. A **695**, 100 (2015)
[8] O. Soto, et al., Nucl. Instrum. Meth. A **732**, 431 (2013)
[9] Y. Van Haarlem et al., Nucl. Instrum. Meth. A **622**, 142 (2010). [arXiv:1004.3796 [nucl-ex]]
[10] Berdnikov, V. V. et al., Instr. Exp. Tech. **58**, 25 (2015)
[11] B.D. Leverington et al., Nucl. Instrum. Meth. A **596**, 327 (2008)
[12] Z. Papandreou et al., Nucl. Instrum. Meth. A **596**, 338 (2008)
[13] A.E. Baulin et al., Nucl. Instrum. Meth. A **715**, 48 (2013)
[14] K. Moriya et al., Nucl. Instrum. Meth. A **726**, 60 (2013)
[15] GlueX Collaboration, "Mapping the spectrum of light quark Mesons and gluonic excitations with linearly polarized protons", Jefferson Lab PAC 30 Proposal (2006). Available at: http://www.gluex.org/docs/pac30proposal.pdf
[16] A. AlekSejevs et al. [GlueX Collaboration], arXiv:1305.1523 [nucl-ex]
[17] M. Dugger et al. [GlueX Collaboration], arXiv:1408.0215 [physics.ins-det]
[18] H. Al Ghoul et al. [GlueX Collaboration], AIP Conf. Proc. **1735**, 020001 (2016). [arXiv:1512.03699 [nucl-ex]]
[19] P. Coyle et al., Nucl. Inst. Meth. A **343**, 292 (1994)
[20] I. Adam et al., Nucl. Inst. Meth. A **538**, 281 (2005)
[21] B. Dey et al., Nucl. Instrum. Meth. A **775**, 112 (2015). [arXiv:1410.0075 [physics.ins-det]]
[22] GlueX Collaboration, "GlueX DIRC Technical Design Report" (2015). Available at: http://argus.phys.uregina.argus.phys.uregina.ca/gluex/DocDB/0028/002809/003/dirc_tdr.pdf
[23] https://www.hamamatsu.com/resources/pdf/etd/H12700_TPMH1348E.pdf
[24] CLAS12 Collaboration, "CLAS12 RICH Technical Design Report" (2013). Available at: https://www.ge.infn.it/jlab12/files/RICH_TDR.pdf
[25] K. Nishimura et al., Nucl. Instr. and Meth. A **639**, 177 (2011)
[26] J. Hardin and M. Williams, arXiv:1608.01180 [physics.data-an]
[27] K. S. Egiyan et al., Phys. Rev. C Nucl. Phys. **68**, 014313 (2003)
[28] K. S. Egiyan et al., Phys. Rev. Lett. **96**, 082501 (2006)
[29] R. A. Niyazov et al., Phys. Rev. Lett. **92**, 052303 (2004)
[30] F. Benmokhtar et al., Phys. Rev. Lett. **94**, 082305 (2005)
[31] J. L. S. Aclander et al., Phys. Lett. B **453**, 211 (1999)
[32] A. Tang et al., Phys. Rev. Lett. **90**, 042301 (2003)
[33] A. Malki et al., Phys. Rev. C Nucl. Phys. **65**, 015207 (2002)
[34] E. Piasetzky et al., Phys. Rev. Lett. **97**, 162504 (2006)
[35] R. Shneor et al., Phys. Rev. Lett. **99**, 072501 (2007)
[36] S. S. Shimanskiy, arXiv:1411.7211 [nucl-th]
[37] A. M. Baldin, Bulletin of the Lebedev Physics Institute, LPI RAS **1**, 35 (1971)
[38] B. V. Vasilishin et al., JINR Preprint 9-86-512 (Dubna, 1986)
[39] NICA Conceptual Design Report, JINR
[40] R. L. Anderson et al., Phys. Rev. D **14(3)**, 13 (1976)

Progress of analysis of dissociation of ^{10}C, ^{10}B and ^{12}C nuclei in nuclear track emulsion

D. A. Artemenkov[1], *V.* Bradnova[1], *E.* Firu[2], *M.* Haiduc[2], *R. R.* Kattabekov[1], *S. P.* Kharlamov[3], *N. K.* Kornegrutsa[1], *K. Z.* Mamatkulov[1], *A.* Neagu[2], *P. A.* Rukoyatkin[1], *V. V.* Rusakova[1], *R.* Stanoeva[4], *A. A.* Zaitsev[1,3], *P. I.* Zarubin[1,3,*], and *I. G.* Zarubina[1]

[1] *Joint Institute for Nuclear Research, Dubna, Russia*
[2] *Institute of Space Science, Magurele, Romania*
[3] *P.N. Lebedev Physical Institute of the Russian Academy of Sciences, Moscow, Russia*
[4] *South-Western University, Blagoevgrad, Bulgaria*

Abstract. Recent findings related with the unstable nuclei ^{8}Be and ^{9}B in the coherent dissociation of relativistic nuclei ^{10}C, ^{10}B and ^{12}C in nuclear track emulsion ("white"stars) are highlighted. Selection of the ^{10}C "white"stars accompanied by ^{8}Be (^{9}B) leads to appearance of the distinct peak with a maximum at 4.1 ± 0.3 MeV in the excitation energy distribution of $2\alpha 2$p ensembles. The ^{8}Be$_{g.s.}$ nucleus is manifested in the coherent dissociation ^{10}B → 2He + H with a probability of 25 ± 5 % including 13 ± 3 % of ^{9}B decays. A probability ratio of the mirror channels ^{9}B + n and ^{9}Be + p is estimated to be 10 ± 1. Reanalysis of relativistic ^{12}C dissociation in lead enriched emulsion revealed 9 3α-events corresponding to the Hoyle state.

1 Introduction

The family of nuclei composing the beginning of the isotope table provides a wholesome "laboratory"which allows one to study the evolution from cluster to shell nuclear structure (figure 1). As "building blocks"the light nuclei include the lightest clusters having no excited states, namely, α-particles, tritons, ^{3}He nuclei, and deuterons, and nucleons which virtual associations coexist in dynamical equilibrium. A pair of α-particles can constitute the unstable ^{8}Be nucleus in the ground ^{8}Be$_{g.s.}$ or 1st exited ^{8}Be$_{2+}$ states. The stable ^{7}Be and ^{7}Li nuclei are important in the structure of the neutron deficient and neutron rich nuclei, respectively. When the nucleons or clusters are added to the ^{8}Be, ^{7}Be and ^{7}Li nuclei the last ones serve as cores in the subsequent stable and radioactive isotopes. Then, the unstable ^{9}B and stable ^{9}Be can possess core roles of an equal importance in heavier nuclear structures. A balanced superposition of the cores, clusters and nucleons in appropriate spin-parity states determine ground state parameters of a corresponding nucleus.

The cluster structure of light nuclei including radioactive ones is under study by means of 0.5 mm pellicles of nuclear track emulsion (NTE) exposed longwise to primary and secondary beams of relativistic nuclei of the JINR Nuclotron (reviewed in [1], [2]). The studies are focused on coherent dissociation events (called "white"stars) which do not feature either slow fragments or charged

*e-mail: zarubin@lhe.jinr.ru

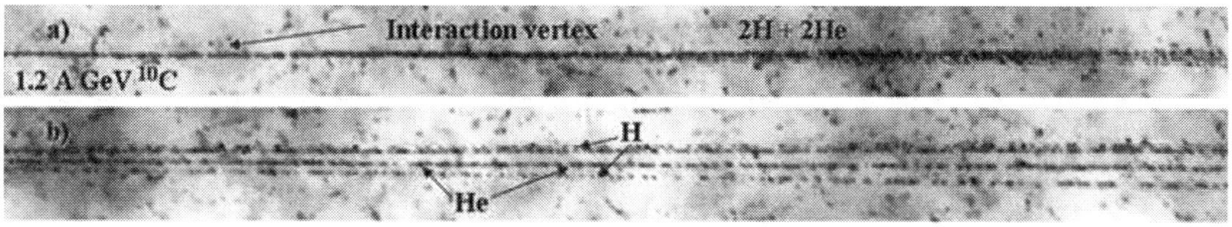

Figure 1. Macrophoto of the coherent dissociation event of 1.2 A GeV ^{10}C nucleus into pairs of He and H nuclei; a) primary track, approximate position of interaction vertex and appearance of fragment tracks and b) tracks of fragments are resolved; opening angles between tracks are Θ_{2He} = 5.9 mrad, Θ_{HeH} = 8.6, 16.6, 3.0, 17.6 mrad, Θ_{2p} = 20.1 mrad. Both 2HeH triples in the event correspond to ^{9}B decays.

mesons. This empirical feature allows one to assume a glancing character of such collisions and that excitations of relativistic nuclei under study are minimal. A main underlying mechanism of coherent dissociation is nuclear diffraction interaction processing without nuclear density overlap and angular momentum transfer.

Events of coherent dissociation are called "white"stars because of the absence of tracks of strongly ionizing particles (figure 1). The term "white"star reflects aptly a sharp "breakdown"of the ionization density at the interaction vertex upon going over from the primary-nucleus track to secondary tracks within a 6° cone at 1.2 A GeV. This special feature generates a fundamental problem for electronic methods because more difficulties should be overcome in detecting events where the degree of dissociation is higher. On the contrary, such events in NTE are observed and interpreted in the most straightforward way, and their distribution among interaction channels characterized by different compositions of charged fragments is determined exhaustively.

The probability distribution of the final configurations of fragments in "white"stars makes it possible to reveal their contributions to the structure of nuclei under consideration. In the case of dissociation, specific configurations arise at random without sampling and that the dissociation mechanism itself does not lead to the sampling of such states via angular-momentum or isospin exchange. By and large, available results confirm the assumption that the cluster features of light nuclei determine the pattern of their relativistic dissociation. At the same time, events that involve the dissociation of deeply bound cluster states and which cannot arise at low collision energy are detected.

Reconstruction of the decays of relativistic ^{8}Be and ^{9}B nuclei is possible by the energy variable Q = M* - M, where $M^{*2} = \Sigma(P_i \cdot P_k)$, M is the total mass of fragments, and $P_{i,k}$ are their 4-momenta defined under the assumption of conservation of an initial momentum per nucleon by fragments. When the identification of relativistic fragment can be reasonably supposed the quasi-invariant variable Q allows one to estimate the excitation energy of their complex ensembles uniting all angular measurements in an event. For the "white"stars of ^{9}Be and ^{10}C nuclei the assumption that He fragments correspond to ^{4}He nuclei (α), and H ones in ^{10}C – ^{1}H (p) is justified. Then ^{8}Be and ^{9}B identification is reduced to measurements of the opening angles between the directions of fragment emission. Recent and rather unexpected findings of these studies of nuclear clustering related with the unstable nuclei and obtained in relativistic experimental approach are highlighted below.

2 The unstable nuclei as cores

Contribution of the unstable nuclei ^{6}Be, ^{8}Be and ^{9}B into dissociation of relativistic nuclei ^{7,9}Be, ^{10}B and ^{10,11}C is under current study. The general pattern is as follows.

DOI: 10.1051/epjconf/201713801030

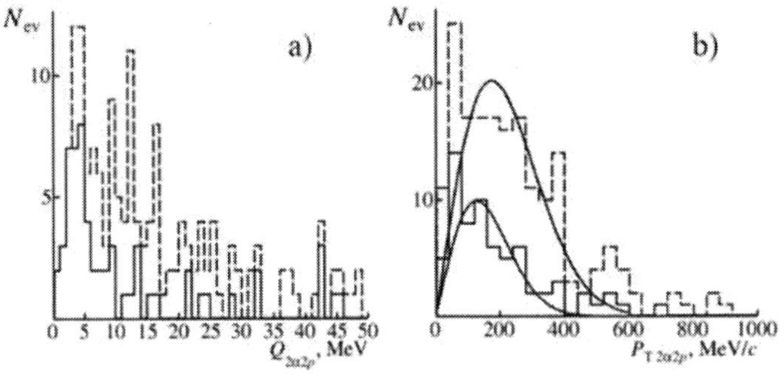

Figure 2. Distributions over energy $Q_{2\alpha 2p}$ (a) and total transverse momentum $P_{T2\alpha 2p}$ (b) of all "white"stars ^{10}C $\rightarrow$ 2He + 2H (dashed) and the ones with the presence of ^{9}B (solid).

Weights of the configuration ^{6}Be + n to the ^{7}Be structure is estimated at a level of 8 ± 1 % which is near the value of 5 ± 1 % for the configuration ^{6}Li + p.

Distributions over the opening angle of α-pairs indicate to a simultaneous presence of virtual ^{8}Be$_{g.s.}$ and ^{8}Be$_{2+}$ states in the ground states of the ^{9}Be and ^{10}C nuclei.

The core ^{9}B is manifested in the ^{10}C nucleus with a probability of 30 ± 4 %. ^{8}Be$_{g.s.}$ decays in ^{10}C "white"stars always arise through the ^{9}B decays. For ^{10}C "white"stars it has to be assumed that ^{6}Be and ^{8}Be$_{g.s.}$ are produced as interfering parts of 2α2p ensembles due to impossibility of separation of the ^{6}Be and ^{8}Be$_{g.s.}$ decays.

In a charge state distribution of fragments the share of the channel ^{10}B $\rightarrow$ 2He + H is 77%. On the basis of measurements of fragment emission angles it is determined that unstable nucleus ^{8}Be$_{g.s.}$ manifests itself with a probability of 25 ± 5 % where 13 ± 3 % of it occurs due to decays of the unstable nucleus ^{9}B. Channel Be + H appeared subdued accounting for about 2 % of "white"stars.

^{8}Be$_{g.s.}$ decays are presented in 24 ± 7 % of 2He + 2H and 27 ± 11 % of the 3He of the ^{11}C "white"stars. ^{9}B decays are identified in "white"stars ^{11}C $\rightarrow$ 2He + 2H constituting 14 % of the ^{11}C "white"stars. As in the ^{10}C case ^{8}Be$_{g.s.}$ decays in ^{11}C "white"stars almost always arise through ^{9}B decays. On this ground the channel ^{9}B + H amounts 14 ± 3 %.

3 Possibility of the 2α2p resonance around 4.1 MeV

An interesting feature is manifested in excitation energy distribution $Q_{2\alpha 2p}$ of 2α2p "quartets"provided by completeness of their observation. Figure 2a shows distribution $Q_{2\alpha 2p}$ of all "white"stars ^{10}C $\rightarrow$ 2He + 2H which appears at a first glance to be scattered. The distribution $Q_{2\alpha 2p}$ of the ^{10}C stars containing ^{9}B decays features the distinct peak with a maximum at 4.1 ± 0.3 MeV at RMS of 2.0 MeV. Its width is determined by the accepted momentum approximation. The peak statistics present 17 ± 4 % of the total number of the ^{10}C "white"stars or 65 ± 14 % of those containing ^{9}B decays.

Distribution over a total momentum $P_{T2\alpha 2p}$ of all 2α2p ensembles (figure 2b) is described by a Rayleigh function with the parameter σ = 175 ± 10 MeV/c while in the case of the presence of ^{8}Be$_{g.s.}$ (^{9}B) decays it is significantly less σ = 127 ± 16 MeV/c. Our observation of such a state manifesting in extra narrow 2α2p jets is grounded on selection of evidently glancing collisions which reduce dramatically a continuum contribution. It is worth noting the observation of a single "white"star 2α2p having $Q_{2\alpha 2p}$ equal to 0.77 MeV in which both 2αp triples correspond to ^{9}B decays with $Q_{2\alpha p}$ of 0.22 and 0.67 MeV, $Q_{2\alpha}$ of 0.14 MeV and $Q_{\alpha 2p}$ of 0.64 and 0.6 MeV (figure 1).

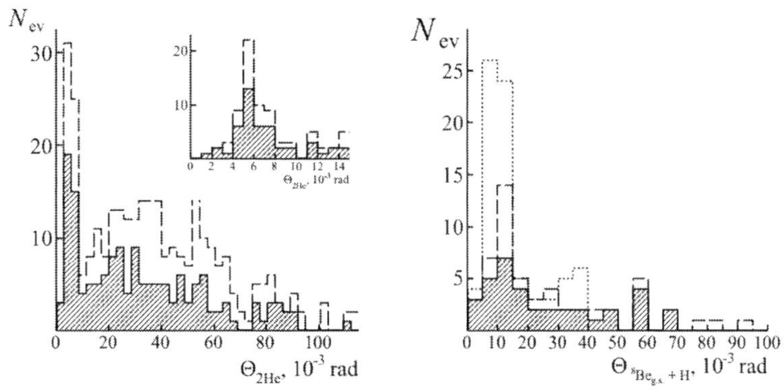

Figure 3. Left: distributions over opening angle Θ_{He} between directions of He fragments in all ^{10}B stars 2He + H (dashed) and "white"^{10}B stars 2He +H (hatched). Right: distributions over opening angle $\Theta_{8Be g.s.}$ + H between directions of fragments ^{8}Be$_{g.s.}$ and H fragments in ^{10}C "white"stars (dotted), all ^{10}B stars (right, dashed) and "white"^{10}B stars (right, hatched).

4 Asymmetry in mirror channels for the ^{10}B nucleus

The early analysis of the NTE exposured to 1 A GeV ^{10}B nuclei has pointed out that triples 2He + H constitute about 65 % among 50 "white"stars found to that time. However, origin of this effect has not been studied being in a "shadow"of emerging studies with radioactive nuclei. Meanwhile, the 2He + H triple dominance indicates the possible presence in ^{10}B of structures ^{9}B$_{g.s.}$ + n side by side with the mirror one ^{9}Be + p. It is interesting to verify whether they have equal contributions or not. Another opportunity is that the ^{10}B nucleus can incorporate the "dilute"^{9}Be cluster in the superpositions ^{8}Be$_{g.s.}$ + n and ^{8}Be$_{2+}$ + n. Both them lead to 3-prong "white"stars out of ^{9}B$_{g.s.}$ decays. Thus, a new round of the ^{10}B analysis has been recently started which progress is summarized below.

The distribution of 2He pairs over the opening angle Θ_{2He} in an interval $0 < \Theta_{2He} < 10.5$ mrad allows one to count 62 decays ^{8}Be$_{g.s.}$ in all found events ^{10}B $\to$ 2He + H including 39 in the "white"stars (figure 3, left). These numbers give 19 ± 3 % and 25 ± 5 % in the respective statistics. Then, the condition on the opening angle $\Theta(^8Be_{g.s.} + H)$ <25 mrad (figure 3, right) allows one to identify 32 decays ^{9}B in all found events and 21 in the "white"stars which constitute, respectively, 9 ± 2 % and 13 ± 3 % contributions of the subset ^{8}Be$_{g.s.}$ + H in the channel 2He + H. Thus, in the "white"star case decays ^{9}B explain just 52 ± 16 % of decays ^{8}Be$_{g.s.}$. This way, the idea about simultaneous coexistence in ^{10}B of superposition of cores ^{8}Be$_{g.s.}$ – ^{8}Be$_{2+}$ and ^{9}B obtains a ground.

Statistics of "white"stars found without sampling one allows to compare the probability of dissociation in the channels ^{9}B + n and ^{9}Be + p (just two events). Measurements of fragment emission angles became possible only in 82 of the 108 events 2He + H, which determines the reconstruction efficiency of the ten ^{9}B decays found among them. On this basis, a probability ratio of the mirror channels ^{9}B + n and ^{9}Be + p is estimated to be 10 ± 1. Accounting for observation efficiency of the "white"stars ^{9}Be + p does not affect qualitatively this ratio.

This fact is quite unexpected and even intriguing. Perhaps it points to the predominance of the ^{9}Be core in nuclear molecular form 2α + n appearing in the dissociation channels containing ^{8}Be$_{2+}$ or ^{8}Be$_{g.s.}$ without ^{9}B decays. The core ^{9}B represents such a structure originally. Another explanation may be based on a broader spatial distribution of neutrons in the ^{10}B compared to protons.

Identification of He and H isotopes by a multiple scattering method progressing now will promote the analysis. In particular, the cluster configuration involving the deuteron ^{8}Be$_{2+}$ + d can be a source of $^8Be_{2+}$ decays. Besides, since the channel $^{10}B \rightarrow ^6Li + \alpha$ is observed with 10% probability contribution of the "dilute"^{6}Li cluster into the 2α + p(d) channel can be expected. Thus, with attraction of existing knowledge on ^{9}Be and ^{6}Li the pattern ^{10}B dissociation via decays $^8Be_{g.s.}$, $^8Be_{2+}$ and ^{9}B can be disentangled step by step. If successful, it will lead to better understanding for the neighbouring nuclei ^{11}C and, then, ^{12}N.

5 Search for the Hoyle state

Search for production of alpha-particle triples in the second excited state 0_2^+ of the ^{12}C nucleus (the first unbound or Hoyle state) in dissociation of relativistic ^{12}C nuclei in nuclear track emulsion (NTE) has been recently suggested. Traditionally the nucleus ^{12}C is regarded as a "laboratory"for the development the α-particle clustering concepts. It is a possible that in the ground state of $^{12}C_{g.s.}$ there are two pairs of α-clusters with orbital angular momenta equal to 2 (D-wave). In this case the basic configurations are ^{8}Be nuclei in the first excited state 2^+. In a classical pattern one may imagine a rotation in opposite directions of two α-clusters with angular momenta equal to 2 around a common center represented by a third α-cluster. Then the remaining combination of two α-clusters should correspond to the ground state of the nucleus ^{8}Be with spin and parity 0^+ (S-wave). As a result the superposition of the pair states in the ensemble of three α-clusters leads to a zero spin in $^{12}C_{g.s.}$. Naturally, this simplified model requires a quantum-mechanical consideration. Nevertheless, its validity should be confirmed by an intensive formation of states $^8Be_{2+}$ and $^8Be_{g.s.}$ with a predominance of the former one in reactions of knocking out α-particles from ^{12}C nuclei.

Such a concept does not contradict the mechanism of the synthesis of the nucleus ^{12}C accepted in nuclear astrophysics. Fusion of a triple of α-particles occurs through its second excited state 0_2^+ (the Hoyle state) located at 290 keV above the breakup threshold $^{12}C \rightarrow 3\alpha$. Basically, each pair of α-particles in it corresponds to $^8Be_{g.s.}$. In the transition $0_2^+ \rightarrow 2_1^+$ with emission of a photon to the first excited state of ^{12}C, which is bound one, an α-pair in the D-wave should arise in a 3α ensemble in order to provide conservation of the angular momentum. The subsequent transition to $^{12}C_{g.s.}$, which is also accompanied by emission of a photon leads to the formation of another α-particle pair in the D-wave state. This pair should have an opposite angular momentum with respect to the first pair to ensure zero spin value of the ground state $^{12}C_{g.s.}$. Thus, the nucleus $^{12}C_{g.s.}$ does acquire polarization. Figuratively being expressed it does conserve an "invisible rotation". The ratio of the yields of α-particle pairs through the states $^8Be_{2+}$ and $^8Be_{g.s.}$ in disintegrations of nuclei ^{12}C not accompanied by a transfer of the angular momentum is a key parameter which should reflect the spin-cluster structure $^{12}C_{g.s.}$.

Importance of the discussed structure is determined not only by interest to describe $^{12}C_{g.s.}$, but also the fact that it is the starting configuration for the reverse process of generating 3α-particle ensembles in the Hoyle state. It is assumed that this state after $^8Be_{g.s.}$ is a Bose-Einstein condensate [3] consisting of α-particles with zero angular momentum. Its identification in breakups of ^{12}C allows one to advance to generation of condensate states of larger number of α-particles. Fundamental aspect seems to be related to the fact that in order to recreate the condensate it is necessary to "evacuate"two hidden rotations $^{12}C_{g.s.}$. The Coulomb dissociation of a nucleus on a heavy nucleus appears to be the most suitable process since few photon exchanges in it are possible. In principle, contribution of electromagnetic interaction can be enhanced by addition of the lead.

Repeated analysis of early exposures of lead enriched NTE to relativistic ^{12}C nuclei revealed the following. Fraction of 3α events containing ^{8}Be is 40 %. 9 events of 72 ones correspond to the Hoyle state in distribution over excitation energy $Q_{3\alpha}$ of α-triples (figure 4). This observation motivates the search for the Hoyle state on statistics of hundreds events of electromagnetic dissociation. It is assumed that this state is the alpha Bose-Einstein condensate composed of α-particles with zero

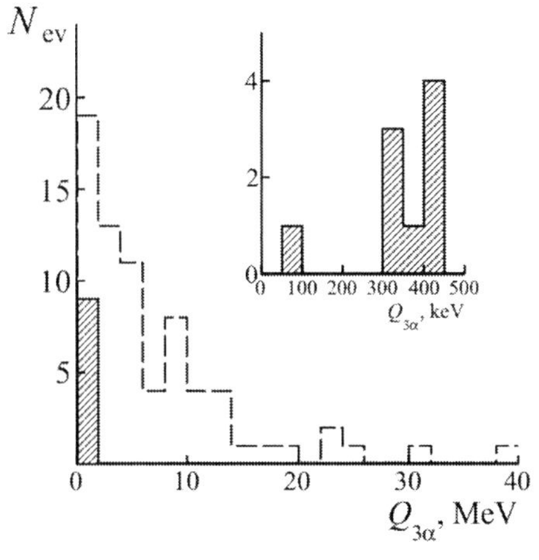

Figure 4. Distribution over excitation energy $Q_{3\alpha}$ of α-triples produced in coherent dissociation of ^{12}C of momentum of 4.5 A GeV/c.

angular momenta in which all pairs are close to the unstable ^{8}Be nucleus. Identification of the Hoyle state will open up perspective for search in relativistic approach of condensate states in dissociation of heavier nuclei.

Stacks of NTE pellicles exposed to ^{12}C of energy in the range from hundreds MeV to few tenths GeV per nucleon at the JINR Nuclotron and the IHEP accelerator complex (Protvino) will serve as the material for this study. At a sufficient statistics of the coherent dissociation events of ^{12}C nuclei emission angles of α-particle triples will be measured with resolution allowing reconstruction of decays of the unstable ^{8}Be nucleus. NTE pellicles will be enriched with lead because the Coulomb dissociation of a relativistic nucleus on a heavy target nucleus is relevant for the Hoyle state excitation.

References

[1] P. I. Zarubin, Lect. Notes Phys. **875**, Clusters in Nuclei **3**, 51—93 (2014). [arXiv: 1309.4881]

[2] D. A. Artemenkov, A. A. Zaitsev, P. I. Zarubin, Physics of Elementary Particles and Atomic Nuclei (in press) (2016). [arXiv: 1607.08020]

[3] T. Yamada, Y. Funaki, H. Horiuchi, G. Roepke, P. Schuck, A. Tohsaki, Lect. Notes Phys. **848**, Clusters in Nuclei **2**, 229–298, Springer International Publishing (2012). [arXiv: 1309.4881]

Development of the Baldin approach for the relativistic nuclear interactions

Denis Artemenkov[1,*], *Alexander* Malakhov[1,**], and *Gennady* Lykasov[1,***]

[1] *Joint Institute for nuclear research, Dubna, Moscow region, 141980, Russia*

Abstract. This article presents the development of the A.M.Baldin approach using a description of relativistic nuclear interactions in the four velocity space. The modification of this approach taking into account the energy dependence of inclusive spectra of hadron has allowed us to describe the differential cross section as a function of the transverse momentum and a satisfactory description of the slopes of the hadron spectra as functions of the energy, has been obtained. The hadron spectra as functions of the energy, has been obtained. Results of our calculations have coincided with experimental data in a wide energy range with high precision.

1 Introduction

A.M. Baldin was invited to consider relativistic nuclear interactions in the four dimension velocity space. The approach of studying relativistic nuclear interactions in the four dimension velocity space proved to be very fruitful. This approach allowed him to classify nuclear interactions, effectively highlight the four-dimensional hadron jets, and study the asymptotic properties of the nuclear interactions at high energies [1]-[3]. This article presents further development of this approach.

Let us put down the interaction of nucleus I with nucleus II resulting in the inclusive particle 1 production as follows:

$$I + II \rightarrow 1 + \ldots \tag{1}$$

According to Baldin's assumption more than one nucleon in the nucleus I can participate in the interaction (1). The value of $N_I = \lambda \cdot A_I$ is the efficient number of nucleons inside the nucleus I, participating in the interaction which is called the cumulative number. Its values lie in the region of $0 \leq N_I \leq A_I$. The cumulative area complies with $N_I > 1$. Of course, the same situation will be for the nucleus II, and you can enter the cumulative number of N_{II}.

For reaction (1) with the production of the inclusive particle 1, we can write the law of conservation of four-momentum in the following form:

$$(N_I P_I + N_{II} P_{II} - p_1)^2 = (N_I m_0 + N_{II} m_0 + M)^2, \tag{2}$$

where N_I and N_{II} are cumulative numbers (the number of nucleons involved in the interaction); P_I, P_{II}, p_1 are four-momenta of the nuclei I and II and particle 1, respectively; m_0 is the mass of the

*e-mail: artemenkov@lhe.jinr.ru
**e-mail: malakhov@lhe.jinr.ru
***e-mail: lykasov@jinr.ru

nucleon; M is the mass of the particle providing the conservation of the baryon number, strangeness, and other quantum numbers. For particles produced without accompanying antiparticles (π mesons, jets and others) $M = 0$. In [3] the parameter of self-similarity was introduced, that allowed one to describe the differential cross section of the yield of a large class of particles in relativistic nuclear collisions:

$$\Pi = \min 1/2[(u_I N_I + u_{II} N_{II})^2]^{1/2} \quad , \tag{3}$$

where u_I and u_{II} are four-velocities of the nuclei I and II.

In this case the invariant cross-sections of the output inclusive particles of different types at nucleus interactions with atomic numbers A_I and A_{II}, are described by the universal dependence in a wide energy range and different atomic numbers of the colliding nuclei:

$$E \cdot d^3\sigma/dp^3 = C_1 A_I^{\alpha(N_I)} \cdot A_{II}^{\alpha(N_{II})} \cdot \exp(-\Pi/C_2)\,, \tag{4}$$

where $\alpha(N_I) = 1/3 + N_I/3$, $\alpha(N_{II}) = 1/3 + N_{II}/3$, $C_1 = 1.9 \cdot 10^4$ mb $\cdot$ GeV$^{-2} \cdot c^3 \cdot$ st^{-1} and $C_2 = 0.125 \pm 0.002$.

2 Self-similarity parameter in the central rapidity region

In the central rapidity region it is possible to find the analytical expression for Π [4]. For this case N_I and N_{II} are equal to each other: $N_I = N_{II} = N$.

$$N = [1 + (1 + \Phi_\delta/\Phi^2)^{1/2}]\Phi \quad , \tag{5}$$

where

$$\Phi = 2m_0(m_{1t}\cosh Y + M)/\sinh^2 Y \quad , \tag{6}$$

$$\Phi_\delta = (M^2 - m_1^2)/(4m_0^2 \cdot \sinh^2 Y). \tag{7}$$

Here $m_{1t} = (m_1^2 + p_t^2)^{1/2}$ is the transverse mass of the particle 1, Y is the rapidity of the colliding nuclei. And then

$$\Pi = N \cosh Y. \tag{8}$$

3 Further development of Baldin's approach

For the inclusive cross-section of hadron production in the central rapidity region and according to AGK (Abramovsky, Gribov, Kanchelly) cancellation [5] we can write the following expression for the quark contribution:

$$E(d^3\sigma/dp^3)_q = \phi_q(y = 0, p_t) \cdot \Sigma[n\sigma_n(s)]. \tag{9}$$

According to [6] we write:

$$E(d^3\sigma/dp^3)_q = \phi_q(y = 0, p_t) \cdot \Sigma[n\sigma_n(s)] = \phi_q(y = 0, p_t) \cdot g \cdot (s/s_0)^\Delta \tag{10}$$

According to [7], [8] the inclusive cross-section of hadron production, taking into account the gluon contribution, can be written in the following form:

$$E(d^3\sigma/dp^3)_g = \phi_g(y = 0, p_t) \cdot \Sigma[n\sigma_n(s)] = \phi_g(y = 0, p_t) \cdot \Sigma[(n-1) \cdot \sigma_n(s)] =$$

$$\phi_g(y = 0, p_t) \cdot [g \cdot (s/s_0)^\Delta - \sigma_{nd}] \quad . \tag{11}$$

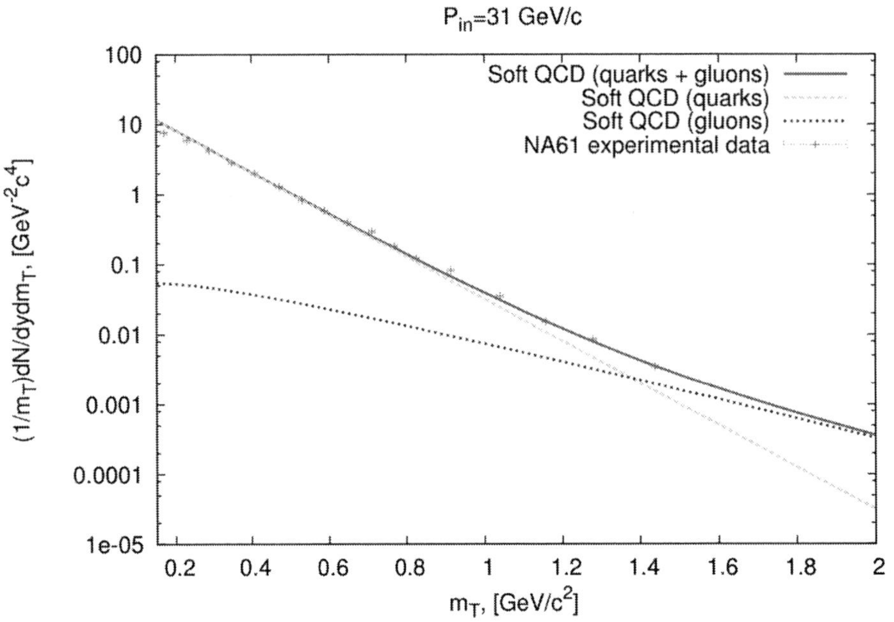

Figure 1. Results of the calculations of the inclusive cross-section of pion production in pp collisions as a function of the transverse mass at the initial momenta $P_{in} = 31$ GeV/c. They are compared to the NA61 experimental data [9].

Thus, taking into account the quark and gluon contributions we will obtain the following expression:

$$E(d^3\sigma/dp^3) = [\phi_q(y=0,p_t) + \phi_g(y=0,p_t) \cdot (1 - \sigma_{nd}/[g(s/s_0)^{\Delta})] \cdot g \cdot (s/s_0)^{\Delta} \quad . \tag{12}$$

In formulas (9)-(12) we apply the following symbols: σ_n - cross-section of hadron production by means of the n-pomeron exchange; $\phi = \phi(\Pi)$; g - constant ($\sim$ 20 mbarn), which is calculated within the "quasi-eikonal" approximation [6]; $S_0 = 1$ GeV^2; $\Delta = [\alpha_p(0) - 1] \sim 0.08$, where $\alpha_p(0)$ is the subcritical Pomeron intercept.

The first part of the inclusive spectrum (Soft QCD (quarks)) is related to the function $\phi_q(y = 0, \Pi)$, which is fitted by the following form:

$$\phi_q(y=0,\Pi) = A_q exp(-\Pi/C_q) \quad , \tag{13}$$

where $A_q = 3.68(\text{GeV}/c)^{-2}, C_q = 0.147$.

The function $\phi_q(y = 0, \Pi)$ related to the second part (Soft QCD (gluons)) of the spectrum is fitted by the following form:

$$\phi_g(y=0,\Pi) \;=\; A_g \sqrt{m_{1t}} exp(-\Pi/C_g)\,, \tag{14}$$

where $A_g = 1.7249$ $(\text{GeV}/c)^{-2}, C_g = 0.289$. Using (12) we can calculate the inclusive cross section of hadron production as a function of the transverse mass.

In figure 1 the inclusive spectrum $(1/m_{1t})d\sigma/dm_{1t}dy$ of π^--mesons produced in pp collisions at the initial momentum $P_{in} = 31$ GeV/c is presented versus their transverse mass m_t. Using only the first part of the spectrum $P_{in} = 158$ GeV/c, which is due to the quark contribution, the conventional string

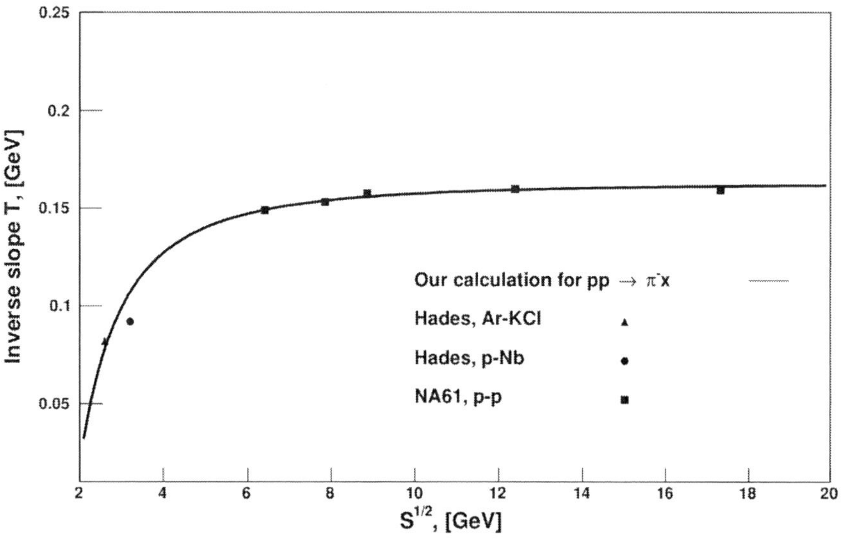

Figure 2. Results of calculations of the inverse slope parameter T on the energy dependence for the negative pion production in pp interactions. The experimental points are taken from [9, 10].

model, let us call it the Soft QCD (quarks), one can describe the NA61 data [9] rather satisfactorily at $p_{in} = 31$ GeV/c and $m_{1t} < 1$ GeV/c^2.

This part of the inclusive spectrum corresponds to the dashed line in figure 1. The inclusion of the second part of spectrum due to the contribution of gluons (Soft QCD (gluons)), the dotted line, allowed us to describe all the NA61 data up to $m_{1t} = 1.5$ GeV/c, see the solid line in figure 1 (Soft QCD (quarks + gluons)).

Figure 1 shows that the deviation of the simple exponential form of the pion spectrum (dashed line) from the total description of the NA61 data (solid line) appears only at $m_{1t} > 0.8$ GeV/c^2 and it is very small. Therefore, the spectrum approximately can be presented in the exponential form, which depends both on m_{1t} and s [13]:

$$\rho(s, m_{1t}) \simeq \phi_q(y = 0, \Pi) g(s/s_0)^{\Delta} = g(s/s_0)^{\Delta} A_q exp(-\frac{m_{1t}}{C'_q m_0(1 - 4m_0^2/s)}) =$$

$$g(s/s_0)^{\Delta} A_q exp(-m_{1t}/T), \tag{15}$$

were

$$T = C'_q m_0(1 - 4m_0^2/s) \tag{16}$$

is the inverse slope parameter, which is called sometimes as the thermal freeze-out temperature.

Thus the value of T given by (16) is in direct relation of the similarity parameter of Π to the Mandelstam relativistic invariant variables s, t, u.

One can see from (16) that this thermal freeze-out temperature depends on the initial energy of collided protons. That is the direct consequence of the self-similarity approach, which operates the four-momentum velocity formalism. This s-dependence of T is significant at low initial energies and at $s \gg m_0^2$ the slope parameter T becomes independent of s. We have calculated the inverse slope parameter T, as a function of the energy $\sqrt{s}$ given by (16) and presented in figure 2. There is a good agreement with the experimental data [9].

 DOI: 10.1051/epjconf/201713801031

In figure 3 we present the calculations of inclusive spectra of charged hadrons (mainly pions and kaons) produced in pp collision at $\sqrt{s}$ = 7 TeV performed by using (12) and the perturbative QCD (PQCD) within the leading order (LO) [7, 8] compared to the LHC data.

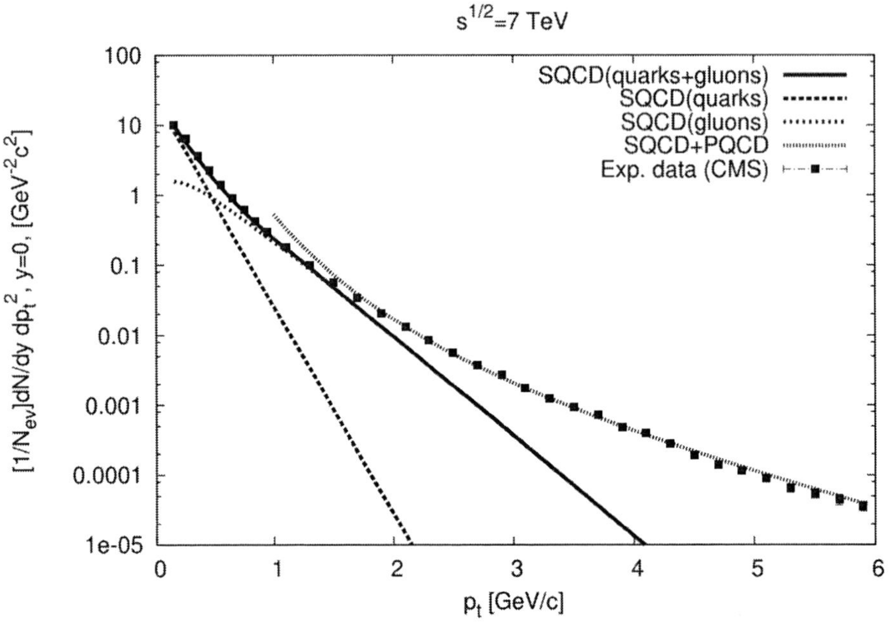

Figure 3. Results of the calculations of the inclusive cross-section of charge hadrons produced in pp collisions at the LHC energies as a function of their transverse momentum pt at $\sqrt{s}$ = 7 TeV. The points are the LHC experimental data [11, 12].

These spectra are the sums of inclusive spectra of pions and kaons, therefore they are presented as a function of the transverse momentum p_t instead of functions of the transverse mass m_{1t} because the masses of the pion and kaon are different. In addition to the part of spectrum, which corresponds to (12) , see the solid lines in these figures, we have also included the PQCD calculations, see the dotted lines. The PQCD calculation within the LO is divergent at low p_t, therefore, the dotted lines go up, when p_t decreases. The kinematical region about $p_t \simeq 1.8 - 2.2$ GeV$/c$ can be treated as the matching region of the nonpertubative QCD (Soft QCD) and the pertubative QCD PQCD. From figure 3, one can see that it is possible to describe rather well these inclusive spectra in a wide region of p_t at the LHC energies matching these two approaches. In [13] there are more details of the results of our calculations for the inclusive cross sections of charge hadrons produced in pp collisions at the SPS and LHC energies.

4 Conclusion

The inclusive hadron spectrum in the space of four velocities is presented within the self-similarity approach as a function of the similarity parameter Π.

We have shown that the energy dependence of the similarity parameter Π included within this approach is very significant at low energies, namely at $\sqrt{s}$ < 6 GeV, and rather well reproduces the experimental data on the inverse slope or the thermal freeze-out temperature of the inclusive spectrum of hadrons produced in pp collisions, it increases and saturates when $\sqrt{s}$ grows. This is very significant for a theoretical interpretation of future experimental data planned to obtain at FAIR, CBM (Darmstadt, Germany) and NICA (Dubna, Russia) projects.

However, we have also shown that the s-dependence of Π is not enough to describe the inclusive spectra of hadrons produced in the midrapidity region, for example, in pp collisions in the wide region of initial energy, especially at the LHC energies. Therefore, we modify the self-similarity approach using the quark gluon string model (QGSM) including the contribution of non-perturbative gluons, which are very significant to describe the experimental data on inclusive hadron spectra in the midrapidity region at the transverse momenta p_t up to $2-3$ GeV$/c$.

To describe the data in the midrapidity region and values of p_t up to $2-3$ GeV$/c$, we have modified the simple exponential form of the spectrum, as a function of Π, and presented it in two parts due to the contribution of quarks and gluons, each of them has different energy dependence. To extend the application of the suggested approach to analyze these inclusive p_t-spectra at large hadron transverse momenta, we have to include the perturbative QCD (PQCD) calculation, which results in the main contribution at $p_t > 2-3$ GeV$/c$.

References

[1] A. M. Baldin, L. A. Didenko, Fortsch.Phys. **38**, 261-332 (1990)
[2] A. M. Baldin, A. I. Malakhov, and A. N. Sissakian, Physics of Particles and Nuclei Suppl. **32**, S4-S30 (2001)
[3] A. M. Baldin, A. A. Baldin, Phys. Particles and Nuclei **29(3)**, 232 (1998)
[4] A. M. Baldin, A. I. Malakhov, JINR Rapid Communications **(87)-98(1)**, 5-12 (1998)
[5] V. Abramovsky, V. N. Gribov and O. Konchelli, Sov.J.Nucl.Phys. **18**, 308 (1973)
[6] K. A. Ter-Martirosyan, Sov.J.Nucl.Phys. **44**, 817 (1986)
[7] V. A. Bednyakov, A. A. Grinyuk, G. I. Lykasov, M. Pogosyan, Int.J.Mod.Phys. A **27**, 1250042 (2012)
[8] A. A. Grinyuk, G. I. Lykasov, A. V. Lipatov, N. P. Zotov, Phys.Rev. D **87**, 074017 (2013)
[9] A. A. Abgrall et al., Eur.Phys.J. C **74**, 2794 (2014)
[10] G. Agakishiev et al., arXiv:1512.07070, Eur.Phys.J. A **52**, 178 (2016)
[11] ATLAS Collab. (G. Aad et al.), New J.Phys. **13**, 053033 (2011)
[12] CMS Collab. (V. Khachatryan et al.), Phys. Rev. Lett. **105**, 022002 (2010)
[13] D. A. Artemenkov, G. I. Lykasov and A. I. Malakhov, Int.J.Mod.Phys. A **30**, 1550127 (2015)

Zero-momentum correlator in SU(2) Landau gauge gluodynamics under various boundary conditions

Igor Bogolubsky[1,*] and *Alla* Bogolubskaya[1,**]

[1] *Joint Institute for Nuclear Research, Dubna, Russia*

Abstract. We make new simulations of $SU(2)$ zero spatial momentum (ZM) gluon correlator in Landau gauge gluodynamics paying special attention to possible lattice artefacts and Gribov copy effect. In particular we started investigation of the correlator dependence on the choice of boundary conditions (BCs) by comparing results for periodic and zero-field (ZF) boundary conditions at various β values. Time behaviour of the ZM correlator for ZF BCs corresponds to approximately constant in time effective gluon mass thus providing additional evidence in favour of decoupling behaviour of momentum-dependent gluon propagator in the IR region. We have found that at fixed lattice sizes and β values the ZM gluon correlator and effective gluon mass for periodic and ZF boundary conditions can differ considerably.

1 Introduction

The only solution which has been found in lattice studies of momentum-dependent gluon and ghost propagators within Landau gauge gluodynamics on sufficiently large lattices

$$D^{ab}_{\mu\nu} = \delta^{ab}\left(\delta_{\mu\nu} - \frac{q_\mu q_\nu}{q^2}\right)\frac{Z(q^2)}{q^2}, \quad G^{ab} = \delta^{ab}\frac{J(q^2)}{q^2}$$

is *decoupling*, or *regular* one, which is characterized by the finite nonzero gluon propagator $D(p^2)$ and the finite ghost dressing function $J(q^2)$ for $q^2 \to 0$ (for details see [1–3]). This solution describes the "effectively massive" gluon and breaks the BRST symmetry contrary to the *scaling*, or *conformal* solution, corresponding to massless gluon, zero $D(0)$ and infinite $J(q^2 \to 0)$. After thorough checking of possible lattice artefacts due to finite-volume, finite-size and Gribov copy effects the *decoupling* solution has been widely accepted by the "community of nonperturbative QCD studies" as the only physical solution. However up to now dependence of the lattice propagators on the choice of boundary conditions has not been reported, to our knowledge (note, that in most lattice studies of gluodynamics periodic boundary conditions (PBCs) are traditionally used). Here we present results of the comparative study of the ZM gluon correlator in $SU(2)$ gluodynamics under periodic and zero-field (sometimes called "open") boundary conditions.

*e-mail: bogolubs@jinr.ru
**e-mail: abogol@jinr.ru

2 Zero-momentum gluon correlator

2.1 Definition of correlators

Lattice studies of the gauge-dependent "space",

$$S(t) = \sum_{\vec{x}} \sum_{i=1,2,3} \mathrm{Tr}\,\langle\, A_i(t,\vec{x})\, A_i(0,\vec{x})\,\rangle\,,$$

and "time"

$$T(t) = \sum_{\vec{x}} \mathrm{Tr}\,\langle\, A_0(t,\vec{x})\, A_0(0,\vec{x})\,\rangle\,,$$

ZM gluon correlators in the Landau gauge was started in [4] and continued in [5–7].

2.2 Landau gauge fixing

To fix the Landau gauge, we apply to all links a gauge transformation $g \in G$ ($G = SU(2)$) mapping $U_{x,\mu} \to {}^gU_{x,\mu} = g_x U_{x,\mu}\, g^{\dagger}_{x+\hat{\mu}}$, with the aim to maximize a gauge functional

$$F_U[g] = \frac{1}{8V} \sum_{x,\mu} \Re\mathrm{e}\,\mathrm{Tr}\,{}^gU_{x,\mu}\,, \tag{1}$$

or, more exactly, to find the *global* maximum of $F_U[g]$ [8]. Search for the global maximum in practice is a complicated problem which becomes exceedingly time consuming with increasing lattice volume. Starting from an initial random gauge transformation g_x one generally arrives at one of many *local* maxima of $F_U[g]$. The corresponding gauge-fixed configurations are called Gribov copies; they all satisfy the differential gauge condition $\partial_\mu A_\mu = 0$. To avoid appearance of lattice Gribov copies we successfully applied the Simulated Annealing (SA) method (see, e.g., [3] and references therein), finalized by overrelaxation (OR) iterations. The SA method – often called a "stochastic optimization method"– allows quasi-equilibrium tunneling through functional barriers, in the course of a "temperature" T decrease. It is proven that *infinitely slow* "cooling" of a system under consideration (decreasing of T) allows to reach the *global* maximum of its underlying functional. In practice, the greater is the number of SA "cooling" steps, the closer is the value of $F_U[g]$ local maxima reached in simulations to its global maximum. In this work we for the first time applied the SA technique to the study of the ZM gluon correlator which allowed to diminish dispersion of the correlator $S(t)$ values and thus raise accuracy of "effective" gluon mass calculation.

2.3 Periodic vs Zero-field BCs; ZM correlators

We accomplished high-statistic Monte-Carlo studies of ZM correlators on lattices $L_s^3 \times L_t$, $L_t = 2L_s$ (in most cases), with $L_s = 10, 12, 14, 16, 18$; our maximal lattice extension was $22^3 \times 30$.

The typical number of MC configurations was of order 10^4 with one gauge fixed copy obtained for each MC configurations. For gauge fixing we use $SA + OR$ method with rather slow cooling (large number of SA steps) thus reaching values of the gauge functional G_F very close to its global maximum. We have checked that $T(t)$ is constant in t with very high accuracy (in Landau gauge it should be t-independent [4]), which shows reliability of our simulations.

In the first series of simulations we use zero-field boundary conditions (ZF BCs). It means that link variables $U_{x,\mu}$ are given by unit $SU(2)$ matrix for all links out of the minimal hypercubic volume containing all $L_s^3 \times L_t$ lattice sites.

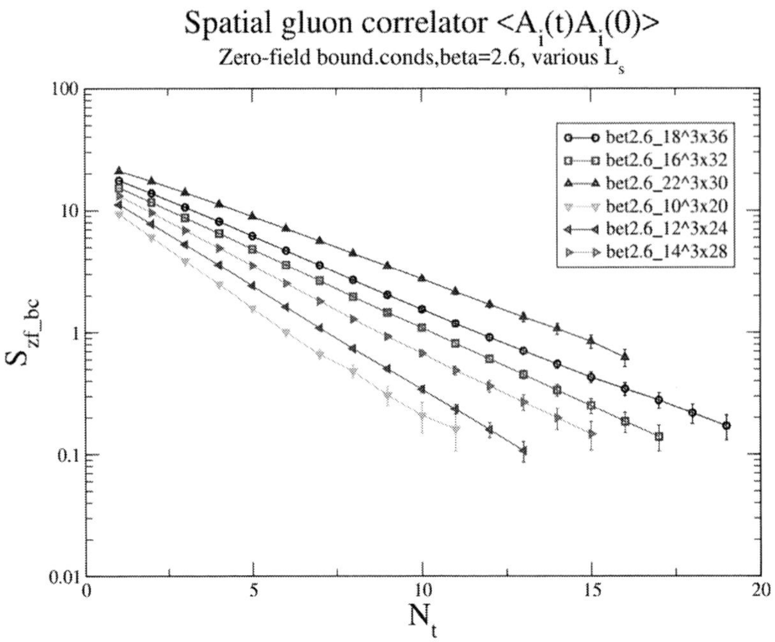

Figure 1. ZM gluon correlator $S(t)$ at $\beta = 2.6$ and various lattice extensions for ZF BCs.

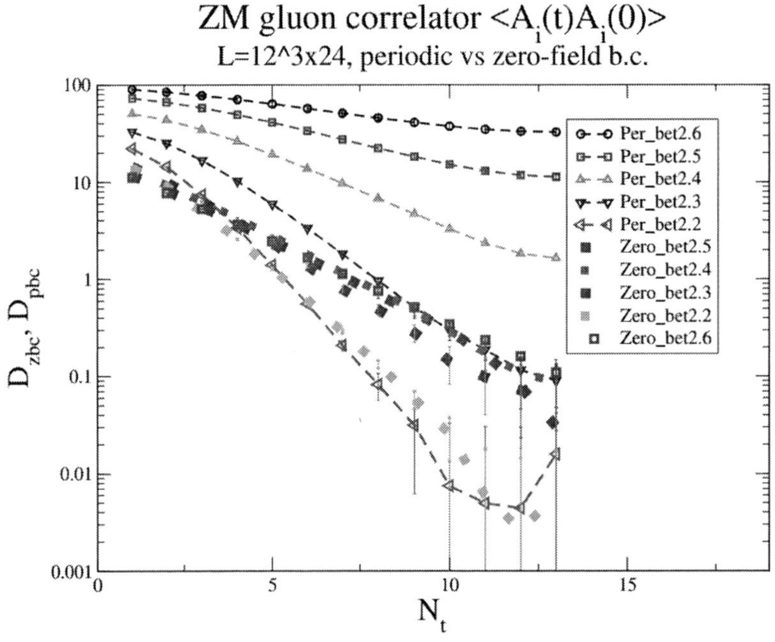

Figure 2. ZM gluon correlator $S(t)$, $L = 12^3 \times 24$, periodic vs zero-field BCs.

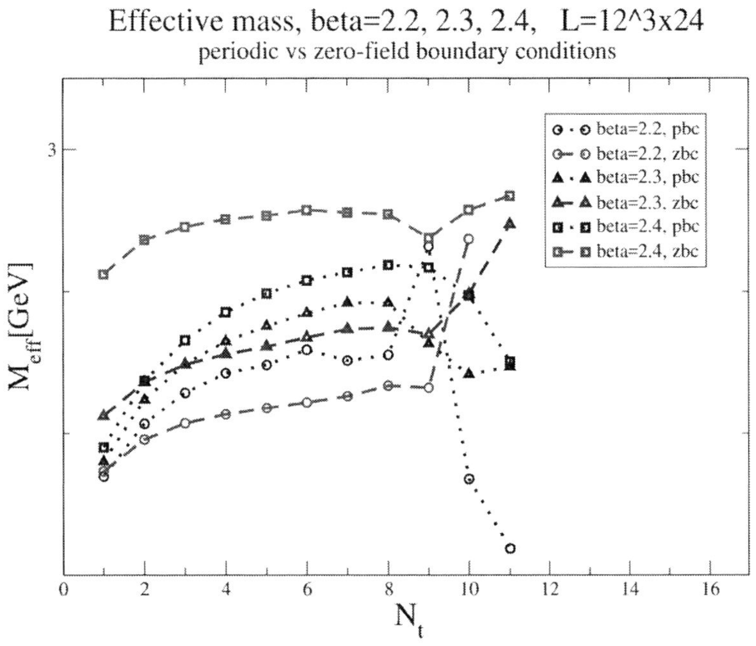

Figure 3. Comparison of dependencies of gluon effective masses on N_t under PBSc and ZF BCs at $L = 12^3 \times 24$ for $\beta = 2.2, 2.3, 2.4$.

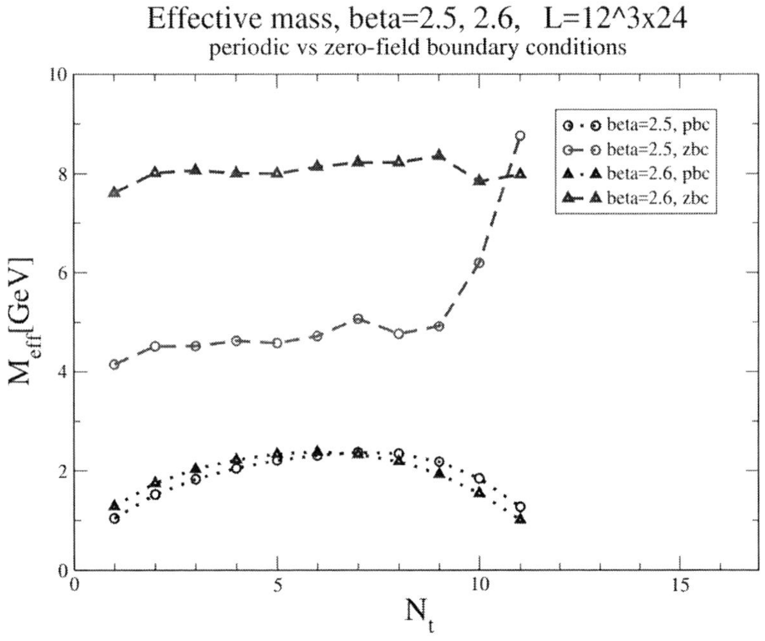

Figure 4. Comparison of dependencies of gluon effective masses on N_t under PBSc and ZF BCs at $L = 12^3 \times 24$ for $\beta = 2.5, 2.6$.

The varying correlator $S(t)$ obtained for $\beta = 2.6$ is plotted in figure 1. It worth noting that correlators $S(t)$ found for ZF BCs at various lattice volumes are presented in the logarithmic scale by nearly straight lines which correspond to gluon "effective" masses independent of N_t contrary to what was found for periodic boundary conditions (PBCs) in [4–7] (see also below).

In the second series of simulations we compare results obtained on the lattice $L = 12^3 \times 24$ for 2 types of boundary conditions: periodic (which are traditionally used in lattice experiments, PBCs) and zero-field (ZF BCs) ones; the results are presented in figure 2. One can see that the $S(t)$ correlators obtained for PBCs and ZF BCs can differ considerably.

Finite-volume effects (mainly) and insufficient statistics (number of MC configurations) can lead to distortion of awaited exponential decrease of $S(T)$ correlator; in particular, the decrease is in most cases smaller at small N_t, and relative statistical errors become progressively larger when $N_t \to L_t/2 + 1$. That is why the notion of N_t-dependent "effective" mass

$$M_{eff}(N_t) = -\ln \frac{S(N_{t+1})}{S(N_t)}$$

is introduced. In figure 3 and figure 4 we compare dependencies of "effective" masses for PBCs and ZF BCs on the lattice $L = 12^3 \times 24$ and various $\beta = \frac{2}{g^2}$ (figure 3 for $\beta = 2.2, 2.3, 2.4$ and figure 4 for $\beta = 2.5, 2.6$). One can see the quick increase of $M_{eff}(N_t)$ found under ZF BCs at larger $\beta = 2.5, 2.6$ for which correlation length becomes larger than the physical lattice extension $L_s \times a(\beta)$ due to small values of $a(\beta)$. One might assume that for larger β ZF BCs introduce stronger finite-volume effects, which however should become weaker with growing lattice extension L_s.

Our simulations of ZM gluon correlators at $\beta = 2.6$ and various $L = L_s^3 \times L_t$ with $L_s = 2L_t$ allow to find out whether it is really the case. In figure 5 it is clearly seen that the effective mass curves for

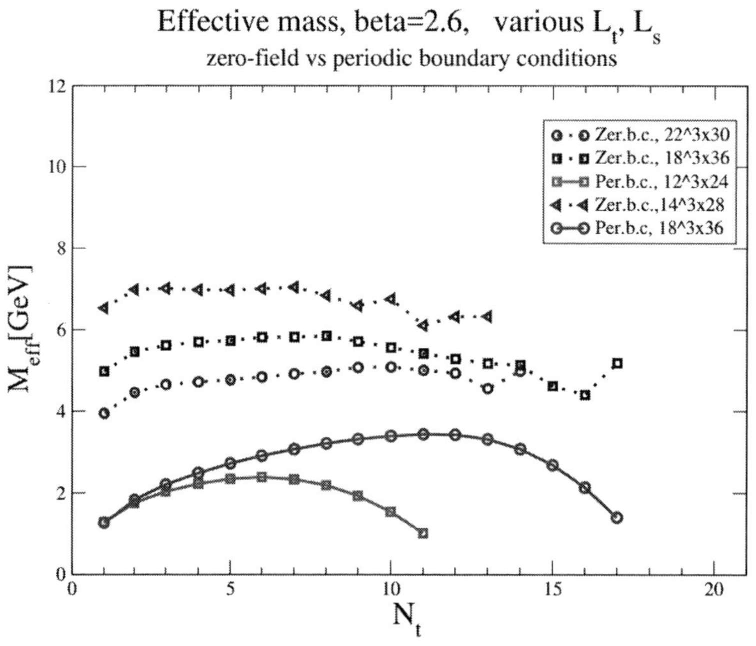

Figure 5. Convergence of dependencies of effective masses at $\beta = 2.6$ for (i) PBCs and (ii) ZF BCs when increasing lattice extensions

PBCs and ZF BCs mutually converge at growing lattice extensions, tending to some limiting curve which could be used for estimation of effective gluon mass being obtained from the ZM correlator $S(t)$.

Note here, that mass estimates for larger β's are physically preferable because for these β's finite-size effects are considerably lower. Relying on the simulations of the ZM correlator with $\beta = 2.6$ one might propose the following bounds for gluon mass M_g : $2MeV \leq M_g \leq 3.5MeV$.

3 Conclusions

We studied behaviour of zero-momentum gluon correlators having in mind existence of Gribov ambiguity which appears when Landau gauge fixing procedure is applied within Monte-Carlo lattice approach. We used one-copy $SA + OR$ technique with slow cooling which provides good practical solution of the Gribov copy problem. Numerical studies of ZM correlators were accomplished on lattices of various space and time extensions L_s and L_t and various β values.

We have studied dependence of ZM correlator $S(t)$ on the choice of boundary conditions. To this aim we have made simulations of the $S(t)$ ZM correlator both at periodic and zero-field boundary conditions and have studied behaviour of effective masses found from this correlator. In all cases we have found that effective $SU(2)$ gluon mass turns out to be nonzero (thus confirming qualitative results of [4–7] on this issue) and have got the following estimate of M_g in case of $SU(2)$ group: $2MeV \leq M_g \leq 3.5MeV$.

Appearance of nonzero effective gluon mass found when studying zero-momentum gluon correlator can be considered as additional argument in favour of physical importance of decoupling, or regular solution for gluon and ghost momentum-dependent propagators which has been found in Landau gauge gluodynamics in [3] and is characterized by nonzero effective gluon mass.

References

[1] A. Cucchieri, T. Mendes, PoS (LAT2007) 297 (2007). Phys. Rev. Lett. **100**, 241601 (2008)

[2] A. Sternbeck, L. von Smekal, D.B. Leinweber, A.G. Williams, PoS (LAT2007) 340 (2007)

[3] I.L. Bogolubsky, E.-M. Ilgenfritz, M. Müller-Preussker, A. Sternbeck, PoS (LAT2007) 290 (2007). Phys. Lett. B **676**, 69 (2009)

[4] J.E. Mandula and M. Ogilvie, Phys. Lett. B **185**, 127 (1987)

[5] R. Gupta, G. Guralnik, G. Kilcup, A. Patel, S. Sharpe and T. Warnock, Phys.Rev. D **36**, 2813 (1987)

[6] C. Bernard, C. Parrinello, A. Soni, Nucl. Phys. Proc. Suppl **30**, 535 (1993)

[7] J.E. Mandula, Phys. Rept **315**, 273 (1999)

[8] D. Zwanziger, Nucl. Phys. B **364**, 127 (1991)

Study of lattice QCD at finite chemical potential using the canonical ensemble approach

V. G. Bornyakov[1,3,4,⋆], *D. L.* Boyda[1,2,4], *V. A.* Goy[1,2,4], *A. V.* Molochkov[1,4], *Atsushi* Nakamura[1,5,6], *A. A.* Nikolaev[1,4], and *V. I.* Zakharov[1,4]

[1] *School of Biomedicine, Far Eastern Federal University, 690950 Vladivostok, Russia*
[2] *School of Natural Sciences, Far Eastern Federal University, 690950 Vladivostok, Russia*
[3] *Institute for High Energy Physics NRC Kurchatov Institute, 142281 Protvino, Russia*
[4] *Institute of Theoretical and Experimental Physics NRC Kurchatov Institute, 117218 Moscow, Russia*
[5] *Research Center for Nuclear Physics (RCNP), Osaka University, Ibaraki, Osaka, 567-0047, Japan*
[6] *Theoretical Research Division, Nishina Center, RIKEN, Wako 351-0198, Japan*

Abstract. A new approach to the computation of the canonical partition functions in $N_f = 2$ lattice QCD is presented. The results for the number density as function of imaginary chemical potential obtained in the confining and deconfining phases are used to determine the phase transition line at real chemical potential.

1 Introduction

Recent results of heavy ion collision experiments at RHIC [1] and LHC [2] shed some light on the properties of the quark gluon plasma and the position of the transition line in the temperature – baryon density plane. New experiments at lower energy will be carried out at FAIR (GSI) and NICA (JINR). To fully explore the phase diagram theoretically it is necessary to make computations at finite temperature and finite baryon chemical potential. For finite temperature, lattice QCD is the only ab-initio method available and many results had been obtained. However, for finite baryon density lattice QCD faces the so-called complex action problem (or sign problem) which prevents direct simulations. Various proposals exist at the moment to solve this problem see, e.g. reviews [3–5], and yet it is still very hard to get reliable results at $\mu_B/T > 1$. Our work is devoted to the further development of the canonical partition function approach.

The fermion determinant at nonzero baryon chemical potential μ_B, $\det\Delta(\mu_B)$, is in general not real. This makes it impossible to apply standard importance sampling techniques to systems described by the grand canonical partition function

$$Z_{GC}(\mu_q, T, V) = \int \mathcal{D}U (\det\Delta(\mu_q))^{N_f} e^{-S_G}, \tag{1}$$

where S_G is some gauge field action, $\mu_q = \mu_B/3$ is the quark chemical potential, $T = 1/(aN_t)$ is the temperature, $V = (aN_s)^3$ is the volume, a is the lattice spacing, and N_t, N_s - the numbers of lattice sites in time and space directions.

⋆e-mail: vitaly.bornyakov@ihep.ru

The canonical approach was studied in a number of papers [6–13]. It is based on the following relations. The representation of the grand canonical partition function $Z_{GC}(\mu_q, T, V)$ in terms of the canonical ones $Z_C(n, T, V)$ (where $\xi = e^{\mu_q/T}$ is the fugacity)

$$Z_{GC}(\mu, T, V) = \sum_{n=-\infty}^{\infty} Z_C(n, T, V)\xi^n, \tag{2}$$

is called fugacity expansion. The inverse of this equation can be written in the form [14]

$$Z_C\,(n, T, V) = \int_0^{2\pi} \frac{d\theta}{2\pi} e^{-in\theta} Z_{GC}(\theta, T, V). \tag{3}$$

$Z_{GC}(\theta, T, V)$ is the grand canonical partition function for imaginary chemical potential $\mu_q = i\mu_{qI} \equiv iT\theta$. Standard Monte Carlo simulations are able to represent the measure of this partition function since the fermionic determinant is real for imaginary μ_q.

At imaginary chemical potential, the QCD partition function Z_{GC} is a periodic function of θ: $Z_{GC}(\theta) = Z_{GC}(\theta + 2\pi/3)$. This symmetry is called Roberge-Weiss symmetry [15]. As a consequence of this periodicity the canonical partition functions $Z_C(n, T, V)$ are nonzero only for net quark numbers $n = 3k$. QCD reveals a rich phase structure at nonzero θ, depending on the number of flavors N_f and the (degenerate) quark mass m. This phase structure is shown in figure 1. T_c is the confinement/deconfinement crossover point at zero chemical potential. The line ($T \geq T_{RW}, \mu_I/T = \pi/3$) indicates the first of a series of first order phase transitions at $\mu_I/T = (2k + 1)\pi/3$. Along the curve between T_c and T_{RW}, the crossover is expected to change to a first order transition for small and large quark masses, see e.g. [16]. The quark number density n_q for N_f degenerate quark flavours is defined

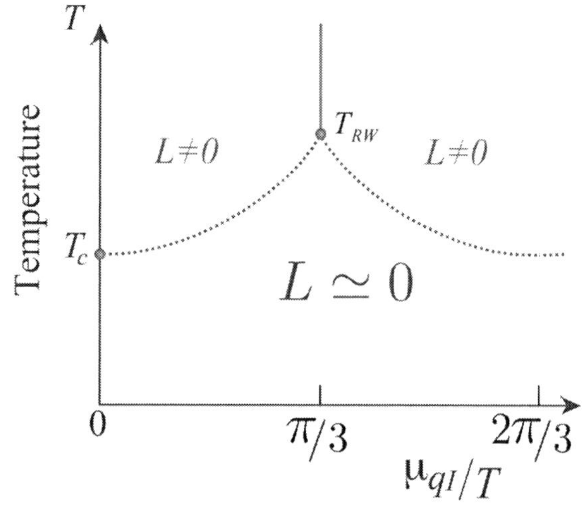

Figure 1. Schematic figure of Roberge-Weiss phase structure in the pure imaginary chemical potential regions.

by the following equation:

$$\frac{n_q}{T^3} = \frac{1}{VT^2}\frac{\partial}{\partial \mu_q} \ln Z_{GC} = \frac{N_f N_t^3}{N_s^3 Z_{GC}} \int \mathcal{D}U e^{-S_G} (\det \Delta(\mu_q))^{N_f} \mathrm{tr}\left[\Delta^{-1} \frac{\partial \Delta}{\partial \mu_q/T}\right]. \tag{4}$$

It can be computed numerically for imaginary chemical potential, and n_q turns out also purely imaginary: $n_q = in_{qI}$.

From (2) and (4) it follows that the densities n_q and n_{qI} are related to $Z_C(n,T,V)$ (in the following we will use the notation Z_n for the ratio $Z_C(n,T,V)/Z_C(0,T,V)$) by the equations

$$n_q/T^3 = \mathcal{N}\frac{2\sum_{n>0} nZ_n \sinh(n\theta)}{1+2\sum_{n>0} Z_n \cosh(n\theta)}, \quad n_{qI}/T^3 = \mathcal{N}\frac{2\sum_{n>0} nZ_n \sin(n\theta)}{1+2\sum_{n>0} Z_n \cos(n\theta)}. \tag{5}$$

$\mathcal{N} = \frac{N_t^3}{N_s^3}$ is a normalization constant. Our suggestion is to compute Z_n using equation (5) for n_{qI}.

One can compute $Z_{GC}(\theta,T,V)$ using numerical data for n_{qI}/T^3 via numerical integration

$$L_Z(\theta) \equiv \log\frac{Z_{GC}(\theta,T,V)}{Z_{GC}(0,T,V)} = -V\int_0^\theta d\tilde{\theta}\, n_{qI}(\tilde{\theta}), \tag{6}$$

where we have omitted the arguments T and V from the grand canonical partition function for brevity. Then the Z_n can be computed as

$$Z_n = \frac{\int_0^{2\pi}\frac{d\theta}{2\pi}e^{-in\theta}e^{L_Z(\theta)}}{\int_0^{2\pi}\frac{d\theta}{2\pi}e^{L_Z(\theta)}}. \tag{7}$$

In our work we use a modified version of this approach [17]. Instead of performing the numerical integration in (6) we fitted n_{qI}/T^3 to theoretically motivated functions of μ_{qI}. It is known that the density of a noninteracting quark gas is described by

$$n_q/T^3 = N_f\left(2\frac{\mu_q}{T} + \frac{2}{\pi^2}\left(\frac{\mu_q}{T}\right)^3\right). \tag{8}$$

We thus fitted the data for n_{qI} to a polynomial containing odd powers of θ

$$n_{qI}(\theta)/T^3 = \sum_{n=1}^{n_{max}} a_{2n-1}\theta^{2n-1} \tag{9}$$

in the deconfining phase. This type of the fit was also used in [18] and [19].

In the confining phase (below T_c) the hadron resonance gas model provides a good description of the chemical potential dependence of thermodynamic observables [20]. Thus, it is reasonable to fit the density in the confining phase by a Fourier expansion

$$n_{qI}(\theta)/T^3 = \sum_{n=1}^{n_{max}} f_{3n}\sin(3n\theta) \tag{10}$$

Again this type of the fit was used in [18], were the conclusion was that it works well.

To demonstrate our method we made simulations of lattice QCD with $N_f = 2$ clover improved Wilson quarks and Iwasaki improved gauge field action (for detailed definitions of the lattice action see [17]). We have simulated on $16^3 \times 4$ lattices at temperatures $T/T_c = 1.35, 1.20$ and 1.08 in the deconfinement phase and $T/T_c = 0.99, 0.93, 0.84$ in the confinement phase along the line of constant physics with $m_\pi/m_\rho = 0.8$. All parameters of the action, including the c_{SW} parameter, were taken from the WHOT-QCD collaboration paper [21]. We have computed the number density on samples of $N_{conf} = 1800$ configurations, using every 10-th trajectory produced by the Hybrid Monte Carlo algorithm.

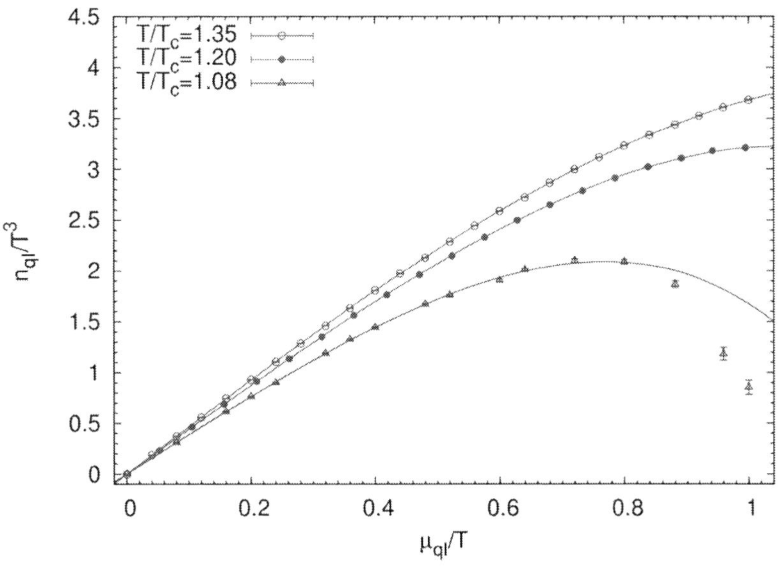

Figure 2. The quark density n_{ql} as function of θ in the deconfinement phase at temperatures $T/T_c = 1.35, 1.20, 1.08$. The curves show fits to the function (9).

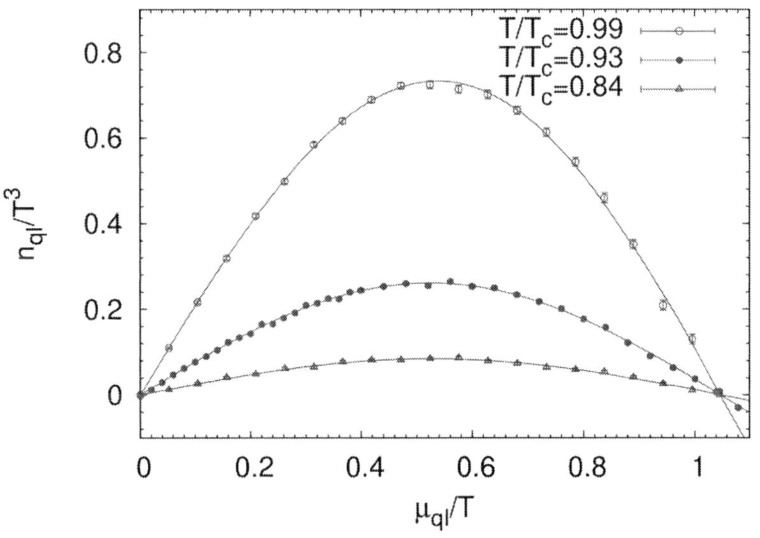

Figure 3. The quark density n_{ql} as function of θ in the confining phase. The curves show fits to (10) with $n_{max} = 1$ for $T/T_c = 0.84, 0.93$ and $n_{max} = 2$ for $T/T_c = 0.99$.

2 Results

In figure 2 and figure 3 we show numerical results for n_{ql} as a function of θ in the deconfining and confining phases, respectively [17]. Note, that the behavior of n_{ql} at $T = 1.08T_c$ is different from that at higher temperatures. This temperature is still below T_{RW}, and at $\theta = \pi/3$ there is no first order phase transition to be hit. Instead, n_{qI} is continuous and passes a crossover to the confinement phase at about $\theta = 0.92(2)$.

It is not yet clear how to fit the data over a range of μ_{qI} connecting pieces of deconfining and confining phase. Here we used a fit to the function (9) with $n_{max} = 3$ over the range $\theta \in [0, 0.8]$, i.e. including only the deconfining phase. In this case we should consider the fit as a Taylor expansion.

We have computed Z_n using the new procedure described in the previous section and compared it with results obtained with use of the hopping parameter expansion. We found good agreement between two methods indicating that the new method works well [17]. This allows us to proceed with an analytical continuation to the real values of μ_q beyond the Taylor expansion validity range (for all temperatures except $T/T_c = 1.08$). Respective results are shown in figure 4. Using the

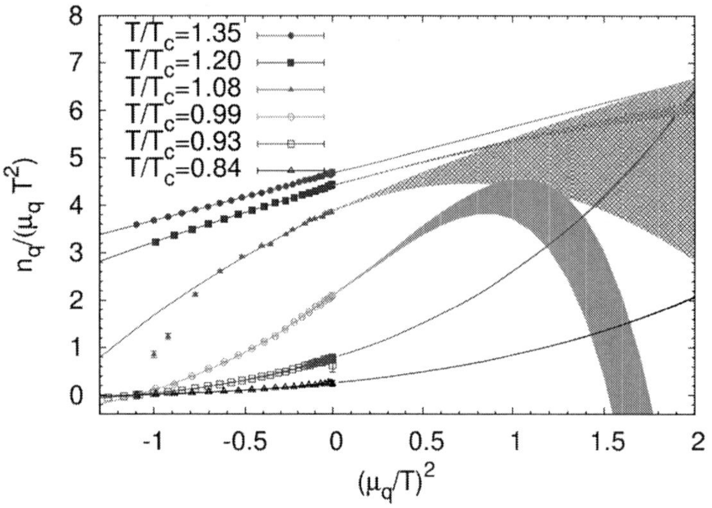

Figure 4. Analytical continuation for the number density vs. μ_q^2. The curves show respective fits (9) or (10). The width of the curves indicates the statistical error of the extrapolation to $\mu_q^2 > 0$.

results for the number density n_q we can compute the temperature of the transition from the hadron phase to the quark-gluon plasma phase using the following procedure. Our results for the number density at temperatures $T > T_c$ as function of the chemical potential μ_q are reliable even for large μ_q for $T > T_{RW}$, while in the temperature interval $T_c < T < T_{RW}$ they are presently reliable for small μ_q only. We can use these results to compute the pressure $\Delta P_{deconf}(T, \mu_q) = P(T, \mu_q) - P(T, 0)$ as function of μ_q and then extrapolate the pressure for fixed μ_q to temperatures $T < T_c$. To obtain a good extrapolation we need the pressure computed for more values of temperature than is available now. But three values which we have in this work are sufficient to demonstrate the idea. We then find the transition temperature $T_c(\mu_q)$ solving numerically the equation $\Delta P_{deconf}(T, \mu_q) = \Delta P_{conf}(T, \mu_q)$, where $\Delta P_{conf}(T, /mu_q)$ is the excess pressure obtained from the number density in the confined phase. In this paper we use an extrapolation for the coefficients in (9) rather than for the pressure itself. This extrapolation is shown in figure 5. We fitted the data for $a_i, i = 1, 3, 5$ by a polynomial $b_0 + b_1\, T/T_c + b_2(T/T_c)^2$ and then computed the extrapolated values of these parameters at $T/T_c = 0.93$ and 0.84. The extrapolated values were used to compute $\Delta P_{deconf}(T, \mu_q)/T^4$ as

$$\frac{\Delta P_{deconf}(T, \mu_q)}{T^4} = \frac{a_1}{2}\left(\frac{T}{T_c}\right)^2 - \frac{a_3}{4}\left(\frac{T}{T_c}\right)^4 + \frac{a_5}{6}\left(\frac{T}{T_c}\right)^6 \tag{11}$$

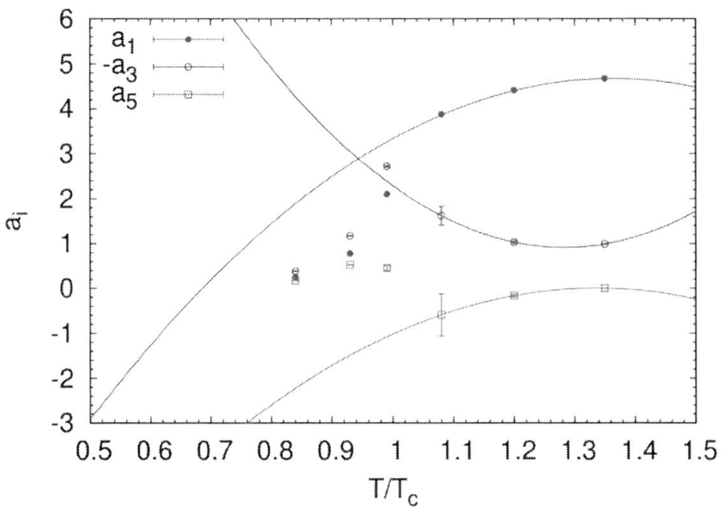

Figure 5. Constants a_i of (9) for six values of temperature. The curves show polynomial fits over $T > T_c$ range.

We show the results in figure 6 together with the pressure computed by integration of (10) for T/T_c =

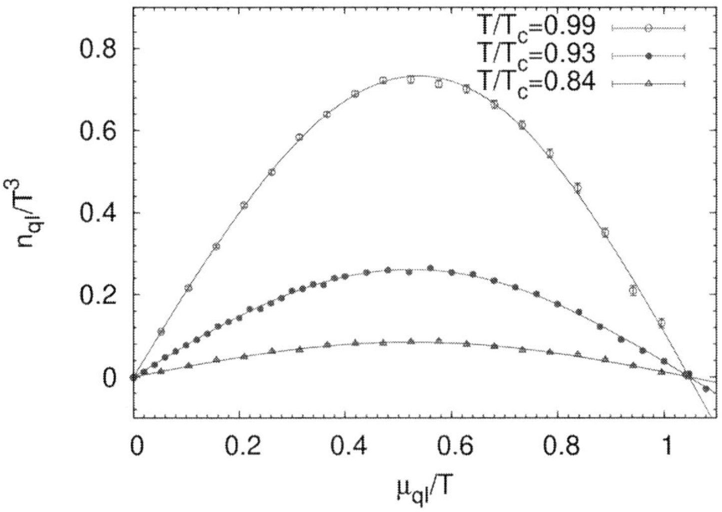

Figure 6. The pressure computed via (11) (denoted by 'deconf', thin curves) and by integration of (10) (denoted by 'conf', thick curves) for $T/T_c = 0.93, 0.84$.

0.93, 0.84. The crossing points determine the values of μ_q on the transition line at the respective values of temperature. We show the two crossing points in figure 7 together with a fit of the form

$$T_c(\mu_q)/T_c = 1 - C\left(\mu_q/T_c\right)^2 . \tag{12}$$

The result for the fit parameter is $C = 0.07$. We have to emphasize that in figures 6 and 7 we do not show errors since we are not yet able to compute them. For the time being, these figures are shown to

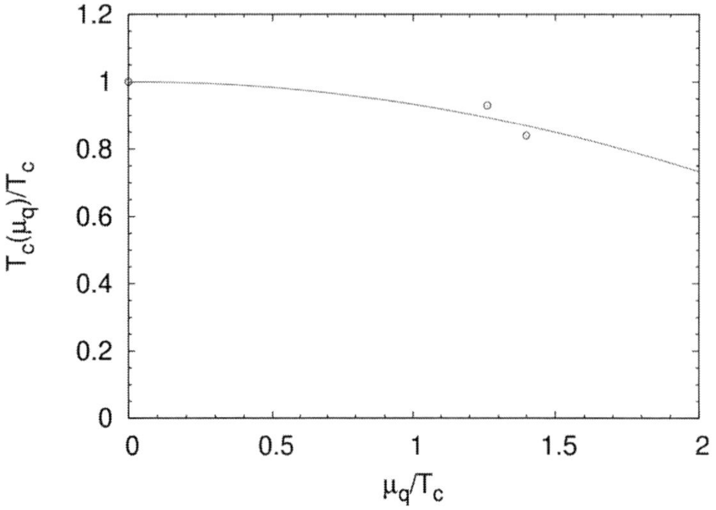

Figure 7. The transition line in the temperature – chemical potential plane. The curve shows a quadratic fit.

illustrate our idea about the method to find the shape of the transition curve. Still it is encouraging that the value of the coefficient C obtained is of correct order of magnitude. We can compare with [22, 23] where $C = 0.051(3)$ and $C = 0.065(7)$ respectively were found in lattice QCD with $N_f = 2$. In both papers the direct analytical continuation of $\frac{T_c(\mu_q)}{T_c}$ from imaginary μ_q was used.

Here we have presented a new method to compute the canonical partition functions Z_n. It is based on fitting n_{qI} for all values of μ_{qI} to the theoretically motivated fitting functions: a polynomial fit (9) in the deconfinement phase for T above T_{RW} and a Fourier-type fit (10) in the confinement phase. We have also explained how the transition line at real μ_q can be computed using the results which we obtained for the number density n_q. We are planning to enlarge the grid of temperature values to improve the method. It is worth to note that the method of direct analytical continuation for $\frac{T_c(\mu_q)}{T_c}$ from imaginary chemical potential can be applied to our data. This will help us to improve the method presented here.

Acknowledgments
This work was completed due to support by RSF grant under contract 15-12-20008. Computer simulations were performed on the FEFU GPU cluster Vostok-1 and MSU 'Lomonosov' supercomputer.

References

[1] J. Adams et al. (STAR), Nucl. Phys. A **757**, 102 (2005). nucl-ex/0501009
[2] K. Aamodt et al. (ALICE), JINST **3**, S08002 (2008)
[3] S. Muroya, A. Nakamura, C. Nonaka, T. Takaishi, Prog. Theor. Phys. **110**, 615 (2003). hep-lat/0306031
[4] O. Philipsen, PoS (LAT2005) 016 (2006), [PoSJHW2005,012(2006)]. hep-lat/0510077
[5] P. de Forcrand, PoS (LAT2009) 010 (2009). hep-lat/1005.0539
[6] P. de Forcrand, S. Kratochvila, Nucl. Phys. Proc. Suppl. **153**, 62 (2006). hep-lat/0602024
[7] S. Ejiri, Phys. Rev. D **78**, 074507 (2008). hep-lat/0804.3227
[8] A. Li, A. Alexandru, K.F. Liu, X. Meng, Phys. Rev. D **82**, 054502 (2010). hep-lat/1005.4158

[9] A. Li, A. Alexandru, K.F. Liu, Phys. Rev. D **84**, 071503 (2011). hep-ph/1103.3045
[10] J. Danzer, C. Gattringer, Phys. Rev. D **86**, 014502 (2012). hep-lat/1204.1020
[11] C. Gattringer, H.P. Schadler, Phys. Rev. D **91**, 074511 (2015). hep-lat/1411.5133
[12] R. Fukuda, A. Nakamura, S. Oka, Phys. Rev. D **93**, 094508 (2016). hep-lat/1504.06351
[13] A. Nakamura, S. Oka, Y. Taniguchi, JHEP **02**, 054 (2016). hep-lat/1504.04471
[14] A. Hasenfratz, D. Toussaint, Nucl. Phys. B **371**, 539 (1992)
[15] A. Roberge, N. Weiss, Nucl. Phys. B **275**, 734 (1986)
[16] C. Bonati, P. de Forcrand, M. D'Elia, O. Philipsen, F. Sanfilippo, Phys. Rev. D **90**, 074030 (2014). hep-lat/1408.5086
[17] V.G. Bornyakov, D.L. Boyda, V.A. Goy, A.V. Molochkov, A. Nakamura, A.A. Nikolaev, V.I. Zakharov (2016), hep-lat/1611.04229
[18] J. Takahashi, H. Kouno, M. Yahiro, Phys. Rev. D **91**, 014501 (2015). hep-lat/1410.7518
[19] J. Gunther, R. Bellwied, S. Borsanyi, Z. Fodor, S.D. Katz, A. Pasztor, C. Ratti (2016), hep-lat/1607.02493
[20] F. Karsch, K. Redlich, A. Tawfik, Phys. Lett. B **571**, 67 (2003). hep-ph/0306208
[21] S. Ejiri, Y. Maezawa, N. Ukita, S. Aoki, T. Hatsuda, N. Ishii, K. Kanaya, T. Umeda (WHOT-QCD), Phys. Rev. D **82**, 014508 (2010). hep-lat/0909.2121
[22] P. de Forcrand, O. Philipsen, Nucl. Phys. B **642**, 290 (2002). hep-lat/0205016
[23] L.K. Wu, X.Q. Luo, H.S. Chen, Phys. Rev. D **76**, 034505 (2007). hep-lat/0611035

Nonperturbative quantization 'a la Heisenberg for non-Abelian gauge theories: two-equation approximation. Applications: a flux tube solution and a scalar model of glueball

Vladimir Dzhunushaliev[1,2,⋆]

[1] *IETP, Al-Farabi KazNU, Almaty, 050040, Kazakhstan*
[2] *Dept. Theor. and Nucl. Phys., KazNU, Almaty, 050040, Kazakhstan*

Abstract. The nonperturbative quantization technique à la Heisenberg is applied for the SU(3) gauge theory. The operator Yang-Mills equation and corresponding infinite set of equations for all Green's functions are considered. Gauge degrees of freedom are splitted into two groups: (1) $A^a_\mu \in SU(2) \times U(1) \subset SU(3)$; (2) coset degrees of freedom $SU(3)/SU(2) \times U(1)$. Using some assumptions about 2- and 4-point Green's functions, the infinite set of equations is truncated to two equations. The first equation is the $SU(2)\times U(1)$ Yang-Mills equation, and the second equation describes a gluon condensate formed by coset fields. A flux tube solution describing longitudinal color electric fields stretched between quark and antiquark located at the ± infinities is obtained. It is shown that the dual Meissner effect appears in this solution: the electric field is pushed out from the gluon condensate.

1 Introduction

One of the most important problems in modern physics is the problem of nonperturbative quantization. For example, the problem appears when explaining confinement in quantum chromodynamics. The problem is that force lines between a quark and an antiquark are confined into a flux tube and the field strength decreases exponentially with distance from quarks. This situation is fundamentally different from the situation with positive and negative charges in electrodynamics where the field strength decreases as a power law. In the 1950's, W. Heisenberg offered the procedure of nonperturbative (NP) quantization for a nonlinear spinor field [1]. Here we apply Heisenberg's idea to nonperturbative quantization of SU(3) gauge fields. This talk is based on [2].

2 Nonperturbative quantization à la Heisenberg for non-Abelian gauge theories

According to Heisenberg, the SU(3) Yang-Mills operator equations can be written as follows

$$D_\nu \widehat{F}^{a\mu\nu} = 0, \tag{1}$$

⋆e-mail: v.dzhunushaliev@gmail.com

where $\hat{F}^B_{\mu\nu} = \partial_\mu \hat{A}^B_\nu - \partial_\nu \hat{A}^B_\mu + g f^{BCD} \hat{A}^B_\mu \hat{A}^D_\nu$ is the field strength operator; $\hat{A}^B_\mu$ is the gauge potential operator; $B, C, D = 1, \ldots, N$ are the SU(3) color indices; g is the coupling constant; f^{BCD} are the structure constants for the SU(3) gauge group.

In order to solve this equation, we have to write an infinite set of equations for all Green's functions

$$\left\langle D_\nu \widehat{F}^{A\mu\nu}(x) \right\rangle = 0, \tag{2}$$

$$\left\langle \hat{A}^{B_1}_{\alpha_1}(x_1) D_\nu \widehat{F}^{A\mu\nu}(x) \right\rangle = 0, \tag{3}$$

$$\left\langle \hat{A}^{B_1}_{\alpha_1}(x_1) \hat{A}^{B_2}_{\alpha_2}(x_2) D_\nu \widehat{F}^{A\mu\nu}(x) \right\rangle = 0, \tag{4}$$

$$\ldots = 0, \tag{5}$$

$$\left\langle \hat{A}^{B_1}_{\alpha_1}(x_1) \ldots \hat{A}^{B_n}_{\alpha_n}(x_n) D_\nu \widehat{F}^{A\mu\nu}(x) \right\rangle = 0 \tag{6}$$

$$\ldots = 0. \tag{7}$$

Here $\langle(\ldots)\rangle = \langle Q|(\ldots)|Q\rangle$ and $|Q\rangle$ is a quantum state of a given physical system. For brevity, we will write $\rangle$ instead of $|Q\rangle$.

The solution of the full set of equations gives us full information on the quantum state $|Q\rangle$ and field operators $\hat{A}^B_\mu$. In this sense, we can say that the solution of the full set of equations is the solution of the operator field equations. It is similar to such situation in probability theory: the knowledge of probability density is equivalent to the knowledge of all averages of a given random variable.

2.1 Two-equation approximation

It is not possible to solve the infinite set of equations (2)-(7). Practically we have to truncate the set of equations (2)-(7), and the truncated system will approximately describe the solution of the full system.

In order to do that, we will assume that in some physical situations all $SU(3)$ degrees of freedom can be decomposed into two groups. In the first group, the gauge fields $\hat{A}^a_\mu = \langle \hat{A}^a_\mu \rangle + i\delta\hat{A}^a_\mu \in SU(2) \times U(1) \subset SU(3)$, where $\langle \hat{A}^a_\mu \rangle = A^a_\mu$ will be treated as classical fields, and $\delta\hat{A}^a_\mu$ are quantum fluctuations around the classical field A^a_μ. In the second group, the gauge fields $A^m_\mu \in SU(3)/SU(2) \times U(1)$ are pure quantum ones in the sense that $\langle \hat{A}^m_\mu \rangle = 0$.

We will consider physical systems where the quantum average of odd degrees of the gauge field are zero,

$$\left\langle \left(\hat{A}^{m_1}_{\mu_1}(x_1) \ldots \hat{A}^{m_{2k+1}}_{\mu_{2k+1}}(x_{2k+1}) \right) \right\rangle = \left\langle \left(\delta\hat{A}^{a_1}_{\mu_1}(x_1) \ldots \delta\hat{A}^{a_{2k+1}}_{\mu_{2k+1}}(x_{2k+1}) \right) \right\rangle = 0. \tag{8}$$

The decomposition of field strengths $\hat{F}^a_{\mu\nu}$ and $\hat{F}^m_{\mu\nu}$ into $(\cdots)^a$ and $(\cdots)^m$ parts can be written as follows:

$$\hat{F}^a_{\mu\nu} = \hat{\mathcal{F}}^a_{\mu\nu} + g f^{amn} \hat{A}^m_\mu \hat{A}^n_\nu, \tag{9}$$

$$\hat{F}^m_{\mu\nu} = \partial_{[\mu} \hat{A}^m_{\nu]} - g f^{amn} \hat{A}^a_{[\mu} \hat{A}^n_{\nu]} + g f^{mpq} \hat{A}^p_\mu \hat{A}^q_\nu. \tag{10}$$

Here $\hat{\mathcal{F}}^a_{\mu\nu} = \partial_\mu \hat{A}^a_\nu - \partial_\nu \hat{A}^a_\mu + g f^{abc} \hat{A}^b_\mu \hat{A}^c_\nu$ is the field strength tensor of the subgroup $SU(2) \times U(1)$; $\hat{A}^B_{[\mu} \hat{A}^C_{\nu]} = \hat{A}^B_\mu \hat{A}^C_\nu - \hat{A}^B_\nu \hat{A}^C_\mu$ is the antisymmetrization procedure; $a, b, c, \ldots$ are the subgroup indices $SU(2) \times U(1)$, and $m, n, p, \ldots$ are the coset indices.

Let us consider the equation (2) for $A = a$. After algebraic manipulations, we have

$$\tilde{D}_\nu \mathcal{F}^{a\mu\nu} - \left(m^2\right)^{ab\mu\nu} A^b_\nu + \left(\mu^2\right)^{ab\mu\nu} A^b_\nu = j^{a\mu}, \tag{11}$$

where $\tilde{D}_\mu = \partial_\mu + g f^{abc} A^b_\mu$ is the covariant derivative in the subgroup $SU(2) \times U(1)$. 2-point Green's functions for the gauge fields $\delta\hat{A}^a_\mu \in SU(2) \times U(1)$ and for the coset $A^m_\mu \in SU(N)/SU(2) \times U(1)$ are defined as

$$G^{mn\mu\nu}(y,x) = \left\langle \hat{A}^{m\mu}(y)\hat{A}^{n\nu}(x)\right\rangle, \tag{12}$$

$$G^{ab\mu\nu}(y,x) = \left\langle \delta\hat{A}^{a\mu}(y)\delta\hat{A}^{b\nu}(x)\right\rangle. \tag{13}$$

Equation (2) for $A = m$ gives us

$$\left\langle D_\nu \widehat{F}^{m\mu\nu}\right\rangle = 0. \tag{14}$$

After tedious calculations, the equation (3) for $A = m, B_1 = r, \alpha_1 = \alpha$ gives us

$$\tilde{D}_\nu \mathcal{F}^{a\mu\nu} - \left[\left(m^2\right)^{ab\mu\nu} - \left(\mu^2\right)^{ab\mu\nu}\right] A^b_\nu = j^{a\mu}, \tag{15}$$

$$\begin{aligned}
&\left[\partial_{x^\nu}\partial^{x^\mu} G^{rm\alpha\nu}(y,x) - \Box_x G^{rm\alpha\mu}(y,x)\right]_{y=x} + \\
&g f^{amn}\Big\{-\partial_{x^\nu}\left[A^{a\mu}(x)G^{rn\alpha\nu}(y,x) - A^{a\nu}(x)G^{rn\alpha\mu}(y,x)\right]_{y=x} - \\
&A^a_\nu(x)\left[\partial^{x^\mu} G^{rn\alpha\nu}(y,x) - \partial^{x^\nu} G^{rn\alpha\mu}(y,x)\right]_{y=x} + \\
&G^{rn\alpha}{}_\nu(x,x)\mathcal{F}^{a\mu\nu}(x)\Big\} + g^2 f^{amn} f^{bnp} A^a_\nu(x)\Big[A^{b\mu}(x)G^{rp\alpha\nu}(x,x) - A^{b\nu}(x)G^{rp\alpha\mu}(x,x)\Big] + \\
&g^2 f^{amn} f^{apq} G^{rnpq\alpha}{}_\nu{}^{\mu\nu}(x,x,x,x) = 0,
\end{aligned} \tag{16}$$

where

$$\left(m^2\right)^{ab\mu\nu} = -g^2\left[f^{abc}f^{cpq}G^{pq\mu\nu} - f^{amn}f^{bnp}\left(\eta^{\mu\nu}G^{mp}{}_\alpha{}^\alpha - G^{mp\nu\mu}\right)\right], \tag{17}$$

$$j^{a\mu} = -g f^{amn}\left\{\partial_{x^\nu} G^{mn\mu\nu}(x,x) + \left[\partial^{x^\mu} G^{mn}{}_\nu{}^\nu(y,x) - \partial^{x^\nu} G^{mn}{}_\nu{}^\mu(y,x)\right]_{y=x}\right\}, \tag{18}$$

$$\left(\mu^2\right)^{ab\mu\nu} = -g^2\left(f^{abc}f^{cde}G^{de\mu\nu} + \eta^{\mu\nu}f^{adc}f^{cbe}G^{de}{}_\alpha{}^\alpha + f^{aec}f^{cdb}G^{ed\nu\mu}\right), \tag{19}$$

and the 4-point Green's function $G^{mnpq}{}_{\mu\nu\rho\sigma}(x,y,z,u) = \left\langle \hat{A}^m_\mu(x)\hat{A}^n_\nu(y)\hat{A}^p_\rho(z)\hat{A}^q_\sigma(u)\right\rangle$.

3 Scalar approximation for the condensate equation

Let us consider the case when all 2-point Green's functions for coset degrees of freedom have the same order and are approximately described by one function. In this case we can assume that every 2-point Green's function can be approximately presented as

$$G^{mn\mu\nu}(y,x) \approx C^{mn\mu\nu} D(y,x), \tag{20}$$

where $C^{mn\mu}{}_\mu$ is a constant and $D(x,x)$ is an averaged dispersion of quantum field $\hat{A}^m_\mu$. We use the following 𝔄𝔫𝔰𝔞𝔱𝔷 for $C^{mn\mu\nu}$

$$C^{mn\mu\nu} = \delta^{mn}\mathcal{A}^\mu\mathcal{A}^\nu, \tag{21}$$

$\mathcal{A}^\mu\mathcal{A}_\mu$ is a constant; $\mathcal{A}^\mu\partial_\mu = \partial_\mu\mathcal{A}^\mu = 0$. In order to close the equation (16), we assume that a 4-point Green's function is a bilinear combination of 2-point Green's functions:

$$G^{mnpq\alpha\beta\mu\nu}(x,x,x,x) \approx C^{mnpq\alpha\beta\mu\nu} D(x,x)\left[M^2 - D(x,x)\right], \tag{22}$$

where $C^{mnpq\alpha\beta\mu\nu}$ and M^2 are constants. After that point, we obtain the following equation for the averaged dispersion of the coset quantum fields $\hat{A}^m_\mu$:

$$\Box_x D(y,x)_{y=x} - \left(m^2_\phi\right)^{ab\mu\nu} A^a_\nu(x) A^b_\mu(x) D(x,x) - \lambda D(x,x)\left[M^2 - D(x,x)\right] = 0, \tag{23}$$

where

$$\left(m_\phi^2\right)^{ab\mu\nu} = g^2 f^{amn} f^{bnm} \frac{\mathcal{A}^\mu \mathcal{A}^\nu - \eta^{\mu\nu} \mathcal{A}^\alpha \mathcal{A}_\alpha}{3\mathcal{A}^\alpha \mathcal{A}_\alpha}, \tag{24}$$

$$\lambda = g^2 f^{amn} f^{apq} \frac{C^{mnpq\ \mu\nu}_{\mu\nu}}{3\mathcal{A}^\alpha \mathcal{A}_\alpha}. \tag{25}$$

The solution of this equation is sought in the form

$$D(y,x) = \phi(y)\phi(x). \tag{26}$$

Substituting (26) into (23), we have the following equation for the gluon condensate ϕ formed by the coset quantum gauge fields $A^m_\mu \in SU(3)/SU(2)\times U(1)$

$$\Box\phi - \left(m_\phi^2\right)^{ab\mu\nu} A^a_\nu A^b_\mu \phi - \lambda\phi\left(M^2 - \phi^2\right) = 0. \tag{27}$$

4 SU(3) flux tube

In order to consider a flux tube solution describing force lines between a quark and an antiquark, let us simplify the equation (29)

$$\tilde{D}_\nu \mathcal{F}^{a\mu\nu} - \left[\left(m^2\right)^{ab\mu\nu} - \left(\mu^2\right)^{ab\mu\nu}\right] A^b_\nu = j^{a\nu}, \tag{28}$$

$$\Box\phi - \left(m_\phi^2\right)^{ab\mu\nu} A^a_\nu A^b_\mu \phi - \lambda\phi\left(M^2 - \phi^2\right) = 0. \tag{29}$$

4.1 Evaluation of $\left(\mu^2\right)^{ab\mu\nu}$, $\left(m^2\right)^{ab\mu\nu}$, and $\left(m_\phi^2\right)^{ab\mu\nu}$ for the flux tube solution

We will consider a physical system where 2-point Green's functions can approximately be expressed as follows:

$$G^{ab\mu\nu}(y,x) \approx \Delta^{ab} \mathcal{B}^\mu \mathcal{B}^\nu, \tag{30}$$

where $\mathcal{B}_\mu \mathcal{B}^\mu$ is a constant and

$$\Delta^{ab} = \begin{pmatrix} \delta_1 & 0 & 0 \\ 0 & \delta_2 & 0 \\ 0 & 0 & \delta_3 \end{pmatrix}. \tag{31}$$

From (31) we see that our system has some color anisotropy in the $SU(2)$ subgroup: $G^{11\mu\nu} \neq G^{22\mu\nu} \neq G^{33\mu\nu}$. The vectors $\mathcal{A}^\mu$ and $\mathcal{B}^\mu$ are

$$\mathcal{A}^\mu = \left(0, 0, 0, \frac{\mathcal{A}_\varphi}{\rho}\right), \tag{32}$$

$$\mathcal{B}^\mu = \left(0, 0, \mathcal{B}_\rho, \frac{\mathcal{B}_\varphi}{\rho}\right). \tag{33}$$

With such choice of the vectors A^a_μ and $\mathcal{B}^\mu$, we have

$$\left(\mu^2\right)^{1bt\nu} A^b_\nu = g^2\left(\mathcal{B}_\rho^2 + \mathcal{B}_\varphi^2\right)(\delta_2 + \delta_3) A^1_t = \mu_1^2 A^1_t, \tag{34}$$

$$\left(\mu^2\right)^{2bz\nu} A^b_\nu = -g^2\left(\mathcal{B}_\rho^2 + \mathcal{B}_\varphi^2\right)(\delta_1 + \delta_3) A^2_z = -\mu_2^2 A^2_z, \tag{35}$$

$$\left(m^2\right)^{1bt\nu} A^b_\nu = \left(3g^2\mathcal{A}_\varphi^2\right) A^1_t = m^2 A^1_t, \tag{36}$$

$$\left(m^2\right)^{2bz\nu} A^b_\nu = -\left(3g^2\mathcal{A}_\varphi^2\right) A^2_z = -m^2 A^2_z, \tag{37}$$

$$\left(m_\phi^2\right)^{ab\mu\nu} A^a_\nu A^b_\mu = \frac{g^2}{4}\left[\left(A^1_t\right)^2 - \left(A^2_z\right)^2\right] = m_\phi^2\left[\left(A^1_t\right)^2 - \left(A^2_z\right)^2\right]. \tag{38}$$

Insertion of (20) and (21) into (18) yields

$$j^{a\mu} = 0. \tag{39}$$

4.2 Numerical solution for the flux tube

We choose the 4-potential in the form (in the cylindrical coordinate system t, z, ρ, φ)

$$A^1_t(\rho) = \frac{f(\rho)}{g}; \quad A^2_z(\rho) = \frac{v(\rho)}{g}; \quad \phi(\rho) = \phi(\rho). \tag{40}$$

The corresponding color electric and magnetic fields are then

$$E^3_z(\rho) = F^3_{tz} = \frac{f(\rho)v(\rho)}{g}, \tag{41}$$

$$E^1_\rho(\rho) = F^1_{t\rho} = -\frac{f'(\rho)}{g}, \tag{42}$$

$$H^2_\varphi(\rho) = \rho\epsilon_{\varphi\rho z}F^{2\rho z} = -\rho\frac{v'(\rho)}{g}. \tag{43}$$

Substituting (40) into the equations (28) and (29) (and after redefining $\phi = \tilde{\phi}/m$, $M = \tilde{M}/m$, and $\lambda = m^4\tilde{\lambda}/g$), we have

$$f'' + \frac{f'}{\rho} = f\left(-v^2 + \tilde{\phi}^2 - \mu_1^2\right), \tag{44}$$

$$v'' + \frac{v'}{\rho} = v\left(f^2 + \tilde{\phi}^2 - \mu_2^2\right), \tag{45}$$

$$\tilde{\phi}'' + \frac{\tilde{\phi}'}{\rho} = \frac{m_\phi^2}{g^2}\tilde{\phi}\left[-f^2 + v^2 + \tilde{\lambda}\left(\tilde{\phi}^2 - M^2\right)\right]. \tag{46}$$

Here the prime denotes differentiation with respect to ρ. Equations (44)-(46) are solved as a non-linear eigenvalue problem with the eigenvalues $\mu_{1,2}, M$ and the eigenfunctions $f, v, \tilde{\phi}$. The boundary conditions are

$$f(0) = 0.2, f'(0) = 0;$$
$$v(0) = 0.5, v'(0) = 0;$$
$$\phi(0) = 1.0, \phi'(0) = 0. \tag{47}$$

The results of calculations are presented in figures 1 and 2.

From (44)-(46) we can obtain the asymptotic behavior of the functions $f(\rho), v(x)$, and $\tilde{\phi}(\rho)$:

$$f(\rho) \approx f_0\frac{e^{-\rho\sqrt{M^2-\mu_1^2}}}{\sqrt{\rho}}, \tag{48}$$

$$v(\rho) \approx v_0\frac{e^{-\rho\sqrt{M^2-\mu_2^2}}}{\sqrt{\rho}}, \tag{49}$$

$$\tilde{\phi}(\rho) \approx M - \phi_0\frac{e^{-\rho\sqrt{2\lambda M^2}}}{\sqrt{\rho}}, \tag{50}$$

where f_0, v_0, and ϕ_0 are some constants. The flux of the longitudinal electric field is

$$\Phi = \int E^3_z ds = \frac{2\pi}{g}\int_0^\infty \rho f(\rho)v(\rho)d\rho < \infty. \tag{51}$$

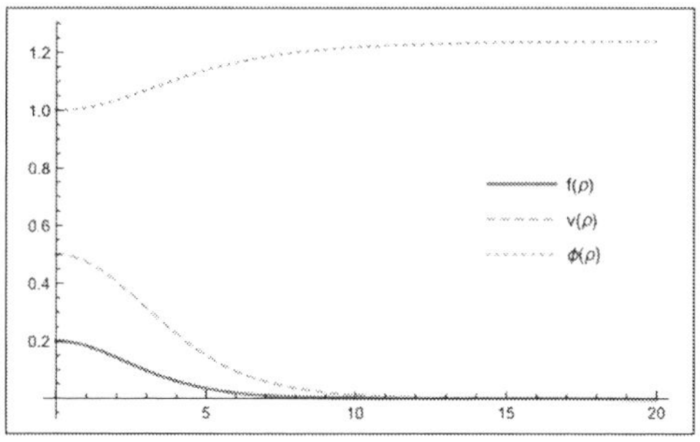

Figure 1. The functions $f(\rho), v(\rho), \tilde{\phi}(\rho)$. The solid curve is $f(\rho)$, the dashed curve is $v(\rho)$, the dotted curve is $\tilde{\phi}(\rho)$. $\mu_1 = 1.0956253, \mu_2 = 1.1494285, M = 1.19505, \tilde{\lambda} = 0.1$.

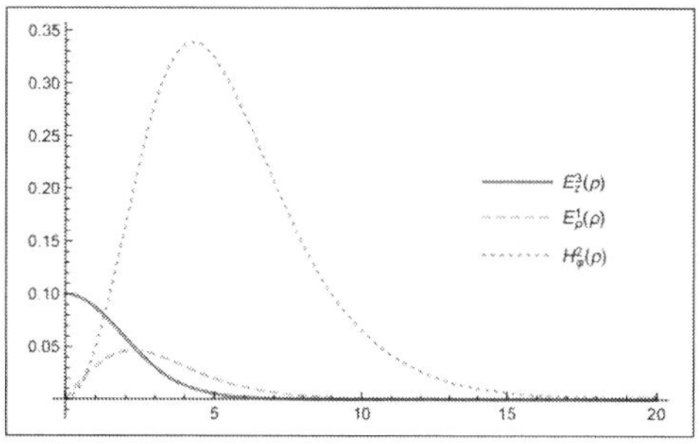

Figure 2. The chromoelectric and chromomagnetic fields $E_z^3(\rho), E_\rho^1(\rho), H_\varphi^2(\rho)$. The solid curve is $E_z^3(\rho)$, the dashed curve is $E_\rho^1(\rho)$, the dotted curve is $H_\varphi^2(\rho)$

5 Discussion and conclusions

- Nonperturbative quantization à la Heisenberg for QCD has been offered.
- After some simplified assumptions, the two-equation approximation has been obtained.
- The first equation is for gauge fields from a subgroup, the second one is for quantum coset fields.
- Applications: a flux tube solution, a scalar model of glueball, and the dual Meissner effect.

Acknowledgements

This work was supported by Grant Φ.0755 in fundamental research in natural sciences by the MES of RK. I am very grateful to N. Kochelev and V. Folomeev for fruitful discussions and comments.

References

[1] W. Heisenberg, *Introduction to the unified field theory of elementary particles.*, (Max - Planck - Institut für Physik und Astrophysik, Interscience Publishers London, New York, Sydney, 1966), 177 p.
W. Heisenberg, Nachr. Akad. Wiss. Göttingen **8**, 111 (1953).
W. Heisenberg, Zs. Naturforsch. **9a**, 292 (1954).
W. Heisenberg, F. Kortel und H. Mütter, Zs. Naturforsch. **10a**, 425 (1955).
W. Heisenberg, Zs. für Phys. **144**, 1 (1956).
P. Askali and W. Heisenberg, Zs. Naturforsch. **12a**, 177 (1957).
W. Heisenberg, Nucl. Phys. **4**, 532 (1957).
W. Heisenberg, Rev. Mod. Phys. **29**, 269 (1957)
[2] V. Dzhunushaliev, arXiv:1608.05662 [hep-ph]

EPJ Web of Conferences **138**, 02004 (2017) DOI: 10.1051/epjconf/201713802004
Baldin ISHEPP XXIII

Deviation pattern approach for optimizing perturbative terms of QCD renormalization group invariant observables

M.R. Khellat[1]* and *A.* Mirjalili[1]**

[1] *Physics Department, Yazd University, P.O.Box, 89195-741, Yazd, Iran*

Abstract. We first consider the idea of renormalization group-induced estimates, in the context of optimization procedures, for the Brodsky-Lepage-Mackenzie approach to generate higher-order contributions to QCD perturbative series. Secondly, we develop the deviation pattern approach (DPA) in which through a series of comparisons between lower-order RG-induced estimates and the corresponding analytical calculations, one could modify higher-order RG-induced estimates. Finally, using the normal estimation procedure and DPA, we get estimates of α_s^4 corrections for the Bjorken sum rule of polarized deep-inelastic scattering and for the non-singlet contribution to the Adler function.

1 Introduction

One of the main objectives of studying different ways of optimizing perturbative expansions for physical quantities is to disclose critical information about the sources of ambiguities that arise at different orders in perturbation theory studies. The limits and validity of these perturbative descriptions is another theoretical challenge which should be addressed particularly in a complete optimization prescription. Consequently, unambiguous determination of the factorization scale would be crucial. Light-front holographic formalism is an instance of a complete optimization prescription which defines an effective coupling for hadron dynamics at all momenta and takes advantage of the principle of maximum conformality (PMC) [1–3] to fix the renormalization scale ambiguity. A review of this program alongside its current status and existing debates, such as the alternatives to PMC, can be found in [4].

Moreover, important investigations have been going on for several years by the authors of [5–9] to understand the consequences of considering a conformal symmetry (CS) limit in QCD. The objective of these investigations involves: to find at first a plausible generalization of the Brodsky-Lepage-Mackenzie (BLM) approach which is capable of resumming charge renormalization contributions in QCD perturbative series at different orders, and secondly clarify whether PMC, in the CS limit, generates series with scheme-independent coefficients. As a result of these investigations, an extension of the BLM approach, called sequential-BLM (seBLM), has been developed and evaluated in [5, 7]. The procedure of seBLM is formulated based on the proposed $\{\beta\}$-expansion for QCD observables and resums the β-dependent terms into a single renormalization scale through the renormalization-group equation (RGE) by rearranging these terms in several stages. It should be mentioned that the original

* mkhellat@gmail.com
** a.mirjalili@yazd.ac.ir

version of PMC assumes a β-representation for the coefficients of QCD perturbative series which is different from the $\{\beta\}$-expansion of seBLM; however, PMC-II employs the same combination of β coefficients as seBLM and performs a resummation of the β-pattern into different scales at different orders [10, 11].

In addition, to resolve the issue of scheme-invariance, PMC-II suggests an analogy between the general structure of a QCD perturbative series and the induced terms in the structure of the series when the series is expressed in a $\mathcal{R}_\delta$ scheme [10], as a subclass of the minimal subtraction schemes. The characteristic of this class of MS-like schemes is that they are related to each other through scale transformations. In fact, scheme-dependence of the BLM approach, as the predecessor of PMC, has been previously discussed in several cases [12–14]. Probably, PMC-II follows the strategy of identifying a class of plausible schemes, which translate their renormalization scales into each other, as a remedy to demonstrate a relation between scheme and scale ambiguities. However, this point should be considered cautiously even within the $\mathcal{R}_\delta$ schemes.

The next important theoretical criteria regarding any formulation of an optimization prescription is whether that formulation respects the general structure of perturbative expansions and the renormalization-group or not. A carefully established estimation procedure is potentially capable of revealing such characteristics of an optimization formulation. Motivated by the renormalization group equation, the idea of RG-inspired estimates was developed in [15] to generate estimates for scheme-independent optimization prescriptions in QCD. We would devise a similar plan for the BLM scale-fixing procedure in which one would adopt a series of estimates for a renormalization group invariant observable at different orders up to $O(\alpha_s^n)$. After this stage, we would modify the $\sim \alpha_s^i, 2 \le i \le n$ estimate to produce an estimate at $\sim \alpha_s^{n+1}$. To generate an $(i+1)$-th order estimate, we start from the i-th order BLM scale-fixed series and evolve the series using the $(i+1)$-th order RGE.

2 The procedure of estimation

Consider a RG invariant perturbative expansion in an MS-like scheme as $R(Q^2) = r_0 + r_1 a_s(Q^2) + \sum_{i=2}^{\infty} r_i a_s{}^i(Q^2)$ for which the normalization $a_s = \alpha_s/\pi$ has been chosen. At any order, this series is supposed to represent a measurable quantity within that order of perturbative approximation. On the other hand, due to the renormalization procedure, the truncated series has become a function of the renormalized parameters. To resolve the issue, one could take advantage of the BLM proposal to assign a single renormalization scale or multiple scales at different orders for the series. Here, we follow the single scale extension of BLM proposed in [16] and extend the corresponding definition of the BLM scale to any order through the following recursive relation [17]

$$\ell^{(k)} \equiv \ln\left(\mu_{BLM}^2/\mu_0^2\right) = \ell^{(k-1)} + \sum_{i=0}^{k-2} c_{ki} a_p^{k-2}(\mu_{BLM}) \,. \tag{1}$$

In (1) the NLO and N^{2}LO BLM scales and coefficients, corresponding to $k = 2$ and $k = 3$, can be adopted from [17]. Taking the N$^{(k-1)}$LO BLM scale-fixed series

$$R_{BLM}^{(k)} = r_0 + r_1 a_p(\mu_{BLM}) + \bar{r}_2 a_p^2(\mu_{BLM}) + \ldots + \bar{r}_k a_p^k(\mu_{BLM}),$$

one could evolve the series using the RG equation at order $(k+1)$

$$\mu^2 \frac{\mathrm{d}a}{\mathrm{d}\mu^2} = -\sum_{i=0}^{k-1} \beta_i a(\mu)^{i+2},$$
$$a(\mu_0) = \exp\left(-\ell^{(k)} \beta(a) \partial_a\right) a \,|_{a=a(\mu_{BLM})} \,, \tag{2}$$

where the operator representation of [5] has been adopted to formulate the evolution of the renormalized coupling through the renormalization scales.

Here, we would formulate our procedures based on the n_f expansion $r_i = \sum_{k=0}^{i-1} r_{ik} n_f^k$ for the coefficients of the RG-invariant expansion. However, we should note that any implementation of BLM should obey its core principle to resum vacuum polarization contributions, which are accumulated in the $\{\beta\}$-coefficients of the renormalization scheme. The trick to take care of vacuum polarization insertions by resumming flavor-dependent terms would not work at N^{4}LO and beyond where n_f^1 contributions start to appear which are not related to charge renormalization. As a result, our proposed formulation is restricted to be valid up to N^{4}LO.

N^{2}LO and N^{3}LO estimates equivalent to $k = 3$ and $k = 4$ can be found in (1,8) from [17]. For instance, the analytical $SU(N_c)$ expression for the N^{2}LO estimate would be

$$r_3^{(est)} = -\frac{121}{16}\frac{C_A^2}{T_F^2}\frac{r_{21}^2}{r_1} - \frac{17}{8}\frac{C_A^2}{T_F}r_{21} - \frac{11}{2}\frac{C_A}{T_F}\frac{r_{20}r_{21}}{r_1} + \left(2\frac{r_{20}r_{21}}{r_1} + \frac{5}{4}r_{21}C_A + \frac{3}{4}r_{21}C_F\right)n_f + \frac{r_{21}^2}{r_1}n_f^2 . \quad (3)$$

Here $[T^a, T^a]_{ij} = C_F\delta_{ij}$ and $f^{acd}f^{bcd} = C_A\delta^{ab}$ are quadratic Casimir operators of the fundamental and adjoint representations of the color group $SU(N_c)$ and $\mathrm{tr}(T^aT^b) = T_F\delta^{ab}$ is the trace normalization of the fundamental representation; $\{C_F = \frac{N_c^2-1}{2N_c}, C_A = N_c, T_F = 1/2\}$. As it is also explained in the next section, $\sim a^3$ and $\sim a^4$ estimates in [17] are generated by the inverse of (2) as an operator equation at $O(\alpha_s^3)$ and $O(\alpha_s^4)$-levels incorporating $\ell^{(1)}$ and $\ell^{(2)}$, respectively. The third-order estimation is given by (3) in which the presence of $-\ell^{(1)} = -3r_{21}/T_F r_1$ is evident. It should also be noted that, in this framework, transition from n_f expansion to seBLM $\{\beta\}$-pattern would not be possible.

The most important characteristic of these N^{2}LO and N^{3}LO estimates is that they vanish completely at the perturbative quenched QED (pqQED) limit $\{C_F = 1, C_A = 0, T_F = 1, n_f = 0\}$. We refer the reader to a detailed study of the pqQED model, the specifications of the conformally invariant limit of the model. Its connection with the consideration of massless perturbation theory results, obtained in pqQED can be found in [18].

2.1 Deviation pattern approach

For generating the exact N^{k}LO results from a N^{k-1}LO BLM scale-fixed series, we need $\ell^{(k)}$ and $\bar{r}_k$; however, we just have access to $\ell^{(k-1)}$ and the estimates are generated on the basis of $\bar{r}_k = 0$. In other words, estimates are naturally deviated from the exact results. On the other hand, these deviations occur at all orders and, for purely numerical purposes, we can take advantage of our knowledge of the deviation of lower-order estimates from the exact results. A possible algorithm which could directly include deviations in the estimation procedure by performing numerical comparisons and modifications would be,

1. $k = 2$ estimation and modification:
   (a) generate the N^{2}LO estimate via (3)
   (b) make the comparison $r_3^{(est)} = r_3^{(exact)}$ which is equivalent to the following three equations

$$r_{30}^{(exact)} = r_{30}^{(est)} = -\frac{121}{16}\frac{C_A^2}{T_F^2}\frac{r_{21}^2}{r_1} - \frac{17}{8}\frac{C_A^2}{T_F}r_{21} - \frac{11}{2}\frac{C_A}{T_F}\frac{r_{20}r_{21}}{r_1},$$

$$r_{31}^{(exact)} = r_{31}^{(est)} = 2\frac{r_{20}r_{21}}{r_1} + \frac{5}{4}r_{21}C_A + \frac{3}{4}r_{21}C_F, \quad r_{32}^{(exact)} = r_{32}^{(est)} = \frac{r_{21}^2}{r_1} .$$

(c) solve the three equations for $\{r_1, r_{20}, r_{21}\}$ and substitute the solutions in the expression for $r_4^{(est)}$; the modification is like changing the weights of the constituents of $r_4^{(est)}$.

2. $k > 2$ estimations and modifications (suppose we have access to exact results up to N^{k}LO):

   (a) generate N^{2}LO to N^{k}LO estimates, i.e. $k-1$ estimates.

   (b) make the comparisons $\{r_j^{(est)} = r_j, 3 \leq j \leq k+1\}$ which are equivalent to $3+4+\ldots+k = (1/2)(k-1)(k+4)$ equations $\{r_{mn}^{(est)} = r_{mn}^{(exact)}\}$ for $3 \leq m \leq k+1$ and $0 \leq n < m$.

   (c) perform a selection of equations because the number of free parameters controlling the estimates is $(k/2)(k+1)$; therefore, $(k-2)$ comparisons should be left out (for an explanation of the point in 4th order estimations, see section 2.2 in [17]).

   (d) adapt $(k/2)(k+1)$ parameters obtained from the previous step to modify $r_{k+2}^{(est)}$.

3 Adler function and Bjorken sum rule

The primary building block of the Adler function and R-ratio is the vacuum polarization function $\Pi(Q^2)$ which is a scale-dependent object in QFT and in the $\overline{\text{MS}}$-scheme is a function of the renormalization scale μ and the running coupling a_s

$$(q_\mu q_\nu - q^2 g_{\mu\nu})\Pi(Q^2) = i\int d^4x e^{iq.x}\langle 0 \mid T[j_\mu(x)j_\nu(0)] \mid 0\rangle\,.$$

Here $Q^2 = -q^2$ and the time-dependent correlator is responsible for the production of the vacuum polarization. The Adler function and R-ratio are related to the vacuum polarization function and the hadronic EM vacuum polarization function as follow [19, 20]

$$\begin{aligned} D(Q^2) &= -12\pi^2 Q^2 \frac{d}{dQ^2}\Pi(Q^2)\,, \\ \tilde{R}(s) &= 6\pi\left(\Pi^{EM}(-s-i\epsilon)) - \Pi^{EM}(-s+i\epsilon)\right)\,. \end{aligned} \tag{4}$$

These two functions are related to each other through the well-known integral transformations

$$D(Q^2) = \int_0^\infty \frac{Q^2\tilde{R}(s)\,ds}{(s+Q^2)^2},\quad \tilde{R}(s) = \frac{i}{2\pi}\int_{s-i\epsilon}^{s+i\epsilon}\frac{dz}{z}D_{pt}(-z)\,. \tag{5}$$

The corresponding perturbative expansions $D_{pt}(Q^2) = \sum_n d_n^{pt}\alpha_s^n(Q^2)$ and $\tilde{R}(s) = \sum_n \tilde{r}_n\alpha_s^n(s)$ would also become connected by introducing appropriate analytical continuation procedures [21].

The Bjorken polarized sum rule is an integration over the difference between proton and neutron polarized structure functions[22]

$$\begin{aligned} \Gamma_1^{p-n} &= \int_0^1 dx\,\left[g_1^p(x,Q^2) - g_1^n(x,Q^2)\right] \\ &= \frac{g_A}{6}C^{Bjp}(Q^2) + \sum_{j=2}^{\infty}\frac{\mu_{2j}^{p-n}(Q^2)}{Q^{2j-2}}\,. \end{aligned}$$

g_A represents the charge of the axial vector current of the nucleon. The non-singlet coefficient function C_{ns}^{Bjp} and the non-singlet Adler function D^{ns} would have the following perturbation theory expansions

Table 1. C_{ns}^{Bjp} and $D^{ns} \sim a^4$ estimates

C^{Bjp}	n_f^0	n_f^1	n_f^2	n_f^3	D^{NS}	n_f^0	n_f^1	n_f^2	n_f^3
c_4^{est}	-265.4	95.26	-5.94	0.08	d_4^{est}	362.1	-99.08	5.04	-0.05
c_4^{est1}	-261.3	67.71	-3.62	0.04	d_4^{est1}	317.2	-93.26	4.76	-0.03
c_4^{exact}	-479.4	123.4	-7.69	0.10	d_4^{exact}	407.4	-103.3	5.63	-0.03

in the $\overline{\text{MS}}$-scheme,

$$D^{ns}(a_s) = 1 + d_1 a_s + d_2 a_s^2 + d_3 a_s^3 + d_4 a_s^4 + O(a_s^5)\,, \tag{6}$$

$$C_{ns}^{Bjp}(a_s) = 1 + c_1 a_s + c_2 a_s^2 + c_3 a_s^3 + c_4 a_s^4 + O(a_s^5)\,. \tag{7}$$

It would be convenient to compare the exact results for the color structures at N^{2}LO approximation with the color structures of the respective estimated contributions to these functions, i.e. d_3 and c_3 with d_3^{est} and c_3^{est}. We could write the corresponding differences as follows:

$$\begin{aligned} d_3 - d_3^{est} =& \underline{\underline{-\frac{69}{128}C_F^3}} + C_F^2 T_F n_f\left[\frac{15}{64} + \frac{17}{4}\zeta_3 - \underline{\underline{5\zeta_5}}\right] + C_F T_F^2 n_f^2\left[\frac{119}{432} + \frac{14}{9}\zeta_3 - \underline{\frac{4}{3}\zeta_3^2}\right] \\ &+ C_F^2 C_A\left[-\frac{133}{128} - \frac{77}{8}\zeta_3 + \underline{\underline{\frac{55}{4}\zeta_5}}\right] + C_F C_A T_F n_f\left[-\frac{924}{432} - \frac{329}{36}\zeta_3 + \underline{\frac{22}{3}\zeta_3^2} + \underline{\underline{\frac{5}{6}\zeta_5}}\right] \\ &+ C_F C_A^2\left[\frac{24569}{6912} + \frac{407}{36}\zeta_3 - \underline{\frac{121}{12}\zeta_3^2} - \underline{\underline{\frac{55}{24}\zeta_5}}\right], \end{aligned} \tag{8}$$

$$\begin{aligned} c_3 - c_3^{est} =& \underline{\underline{-\frac{3}{128}C_F^3}} + C_F^2 T_F n_f\left[\frac{155}{576} - \underline{\underline{\frac{5}{12}\zeta_3}}\right] + C_F T_F^2 n_f^2\left[-\frac{43}{216}\right] + C_F^2 C_A\left[-\frac{145}{576} - \underline{\underline{\frac{11}{12}\zeta_3}}\right] \\ &+ C_F C_A T_F n_f\left[\frac{1339}{864} + \underline{\underline{\frac{3}{4}\zeta_3 - \frac{5}{6}\zeta_5}}\right] + C_F C_A^2\left[-\frac{2143}{864} + \underline{\underline{\frac{55}{24}\zeta_5}}\right]. \end{aligned} \tag{9}$$

The terms which are double-underlined do exclusively belong to the exact analytical results d_3 and c_3 while the terms proportional to ζ_3^2 in (8) belong to d_3^{est}. Transcendental Riemann functions related to the Bjorken sum rule of the polarized lepton-hadron DIS C_{ns}^{Bjp} start to appear at $\sim a^3$ while for D^{ns} they exist at $\sim a^2$; this is the reason why there are no ζ_3 and ζ_5 contributions in c_3^{est}.

Having an input on the difference between N^{2}LO estimates of D^{ns} and C_{ns}^{Bjp} and the corresponding exact results, it would be possible to follow the estimation procedure of Sec. 2 and the deviation pattern approach in Sec. 2.1 to produce N^{3}LO estimates for these functions in table 1. The *est* superscript refers to the method of Sec. 2 and *est*1 superscript to the DPA estimates. As listed in the table, one can also find the exact numerical results for both $D^{ns}(Q^2)$ [20] and $C_{ns}^{Bjp}(Q^2)$ [23] at the fourth-order of perturbative approximation.

Acknowledgment. M.K thanks Andrei Kataev for the very useful discussions and his comments on several topics related to the work done here. He is also grateful for the financial support provided by JINR and the organization committee of the XXIII International Baldin Seminar on High Energy Physics Problems which was dedicated to the 90th anniversary of Academician A. M. Baldin. The authors acknowledge the comments of the respected referee which improved and clarified parts of the paper.

References

[1] S. J. Brodsky and L. Di Giustino, Phys. Rev. D **86**, 085026 (2012)
[2] S. J. Brodsky and X. G. Wu, Phys. Rev. D **85**, 034038 (2012). Erratum: [Phys. Rev. D **86**, 079903 (2012)]
[3] S. J. Brodsky, M. Mojaza and X. G. Wu, Phys. Rev. D **89**, 014027 (2014)
[4] S. J. Brodsky, J. Few-Body Syst. **57**, 703 (2016)
[5] S. V. Mikhailov, JHEP **0706**, 009 (2007)
[6] A. L. Kataev and S. V. Mikhailov, Theor. Math. Phys. **170**, 139 (2012)
[7] A. L. Kataev and S V. Mikhailov, Phys. Rev. D **91**, 014007 (2015)
[8] G. Cvetic and A. L. Kataev, Phys. Rev. D **94**, 014006 (2016)
[9] A. L. Kataev and S. V. Mikhailov, JHEP **1611**, 079 (2016)
[10] M. Mojaza, S. J. Brodsky, and Xing-Gang Wu, Phys. Rev. Lett. **110**, 192001 (2013)
[11] Hong-Hao Ma et al., Phys. Rev. D **91**, 094028 (2015)
[12] W. Celmaster and P. M. Stevenson, Phys. Lett. B **125**, 493 (1983)
[13] J. Chyla, Phys. Lett. B **356**, 341 (1995)
[14] S. J. Brodsky, V. S. Fadin, V. T. Kim, L. N. Lipatov, and G. B. Pivovarov, JETP Lett. **70**, 155 (1999)
[15] A. L. Kataev and V. V. Starshenko, Mod. Phys. Lett. A **10**, 235 (1995)
[16] G. Grunberg and A. L. Kataev, Phys. Lett. B **279**, 352 (1992)
[17] A. Mirjalili and M.R. Khellat, Int. J. Mod. Phys. A **29**, 1450178 (2014)
[18] A. L. Kataev, JHEP **1402**, 092 (2014)
[19] S. L. Adler, Phys. Rev. D **10**, 3714 (1974)
[20] P. A. Baikov, K. G. Chetyrkin and J. H. Kuhn, Phys. Rev. Lett. **101**, 012002 (2008)
[21] D. V. Shirkov, arXiv:hep-ph/0003242. D. V. Shirkov, Theor. Math. Phys. **127**, 409 (2001)
[22] R. S. Pasechnik, D. V. Shirkov and O. V. Teryaev, Phys. Rev. D **78**, 071902 (2008)
[23] P. A. Baikov, K. G. Chetyrkin and J. H. Kuhn, Phys. Rev. Lett. **104**, 132004 (2010)

Pion transition form factor in the domain model of QCD vacuum

Sergei Nedelko[1,⋆] and *Vladimir* Voronin[1]

[1] *Bogoliubov Laboratory of Theoretical Physics, JINR, 141980 Dubna, Russia*

Abstract. Domain model of QCD allows for description of wide range of meson observables, dynamical chiral symmetry breaking, and resolution of $U_A(1)$ and strong CP problems. The purpose of the present study is transition form factor of neutral pseudoscalar mesons, and how they are influenced by the typical vacuum configurations of the model – almost everywhere homogeneous Abelian (anti-)self-dual fields. Asymptotic behavior of the calculated pion form factor supports Belle trend, but the asymptotic value of $Q^2 F_{\pi\gamma^*\gamma}$ differs from the prediction of factorization theorems.

1 Introduction

Experimental data for pion transition form factor $F_{\pi\gamma^*\gamma}$ obtained by the BaBar collaboration [1] indicate growth of $Q^2 F_{\pi\gamma^*\gamma}$ at large Q^2 that is inconsistent with rigid prediction of QCD factorization theorems [2]

$$F_{\pi\gamma^*\gamma} \sim \frac{\sqrt{2} f_\pi}{Q^2}, \tag{1}$$

where $f_\pi = 131$ MeV. Results of experiments on pion transition form factor $F_{\pi\gamma^*\gamma}$ carried out by the Belle [3] collaboration published later demonstrate qualitatively different behavior at large momenta, though they still allow violation of bound (1).

Particular source of contributions that could explain BaBar trend is the QCD vacuum described by ensemble of almost everywhere homogeneous Abelian (anti-)self-dual fields with nonzero scalar condensate $\langle F^2 \rangle$ — the key feature of the domain model [4–8] — that can possibly mix short and long-range dynamics. Short-range fluctuations in the model are separated from long-range modes with scalar condensate and integrated out, while the latter are treated nonperturbatively during bosonization procedure. Chiral symmetry is spontaneously broken by vacuum field, there exists nonzero scalar quark condensate $\langle \bar{q}q \rangle$ (see [4] for details). The $U_A(1)$ problem is resolved without introducing the strong charge-parity (CP) violation. This approach demonstrated its power in description of wide range of meson phenomenology (masses of light, heavy-light and double-heavy mesons, leptonic decay constants, transition constants, all of the above including excited mesons). Overall precision of the approach in the lowest approximation is about 10 – 15% with a few exceptions. The model also allows consistent treatment of processes involving higher number of hadrons. Results of calculations of decay constants g_{VPP} are given below.

Propagators and meson-quark vertices in the vacuum field are not translation-invariant, therefore momentum is not conserved in every vertex but only in an entire diagram. The purpose of the present

⋆e-mail: nedelko@theor.jinr.ru

Table 1. Values of parameters used for calculations.

$m_{u/d}$(MeV)	m_s(MeV)	m_c(MeV)	m_b(MeV)	Λ(MeV)	α_s	R(fm)
145	376	1566	4879	416	3.45	1.12

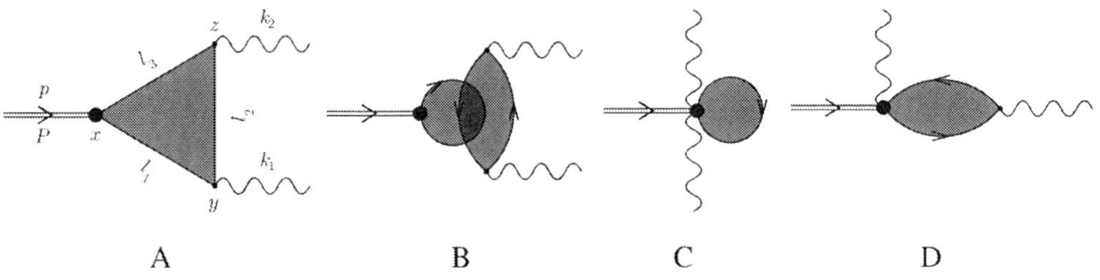

Figure 1. Possible contributions to transition form factor.

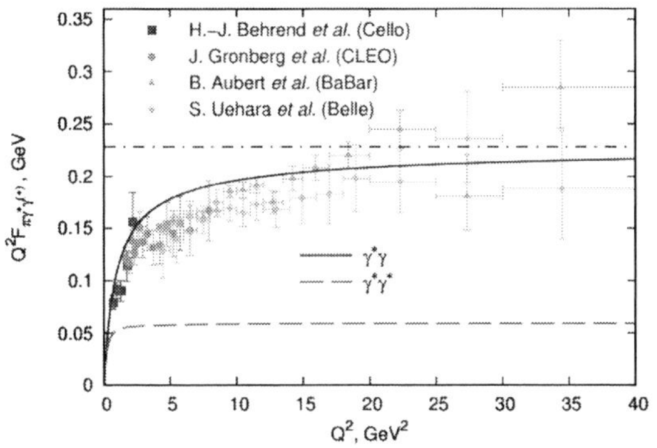

Figure 2. Pion transition form factor in asymmetric (solid line) and symmetric (dashed line) kinematics. $g_{\pi\gamma\gamma} = F_{\pi\gamma^*\gamma(0)} = 0.272\ \mathrm{GeV}^{-1}$. The data are taken from [1, 3, 9, 10].

paper is to investigate how this feature of the domain model affects transition form factors of pseudoscalar mesons. All calculations are performed with the same set of parameters that was used for calculation of meson spectra [4], see table 1.

2 Decay constants and transition form factor of pion

Interaction of a meson with two photons is described by the diagrams shown in figure 1. Currently, the inhomogeneity of the vacuum gluon ensemble is not taken into account completely. As a consequence, only the diagram A gives nonzero contribution. Result of computation of pion transition form factor in comparison with experimental data is shown in figure 2.

It turns out that nonconservation of momentum in every single vertex does not cause growth of $Q^2 F_{\pi\gamma^*\gamma}$ at large Q^2, and calculations show that at asymptotically large Q^2

$$F_{\pi\gamma^*\gamma} \sim \varkappa \frac{\sqrt{2} f_\pi}{Q^2}, \quad \varkappa \approx 1.24.$$

Qualitatively, asymptotic behavior of $Q^2 F_{\pi\gamma^*\gamma}$ at large Q^2 coincides with the factorization prediction, but the value of constant $\varkappa$ substantially differs from unity. This difference is a result of translational noninvariance of vertices. At the same time, this noninvariance does not affect asymptotic behavior

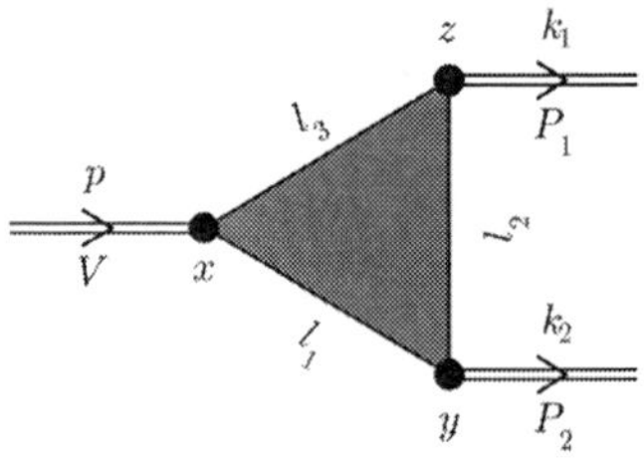

Figure 3. Diagram of decay of vector meson into a couple of pseudoscalar mesons.

Table 2. Results of calculation of g_{VPP} for various decays. g^*_{VPP} are results that one would obtain if local gauge invariance was neglected. Parameters of the model are given in table 1.

Decay	g_{VPP} [11]	g_{VPP}	g^*_{VPP}
$\rho^0 \to \pi^+\pi^-$	5.95	7.61	1.14
$\omega \to \pi^+\pi^-$	0.17	0	0
$K^{*\pm} \to K^\pm\pi^0$	3.23	3.56	0.65
$K^{*\pm} \to K^0\pi^\pm$	4.57	5.03	0.91
$\varphi \to K^+K^-$	4.47	5.69	1.11
$D^{*\pm} \to D^0\pi^\pm$	8.41	7.94	16.31
$D^{*\pm} \to D^\pm\pi^0$	5.66	5.62	11.53

of form factor in symmetric kinematics (two photons with equal virtuality Q^2). Thus, asymptotics calculated within the model

$$F_{\pi\gamma^*\gamma^*} \sim \varkappa^* \frac{\sqrt{2}f_\pi}{3Q^2}, \quad \varkappa^* = 1$$

matches factorization prediction

$$F_{\pi\gamma^*\gamma^*} \sim \frac{\sqrt{2}f_\pi}{3Q^2}. \tag{2}$$

The fact that prediction (2) is reproduced by the model, while prediction (1) is not, comes as no surprise because straightforward QCD factorization works well in the former case only, i.e. when both photons are highly virtual [12, 13].

3 Strong decays of vector mesons

The diagram describing process of decay of vector meson into a couple of pseudoscalar mesons is shown in figure 3. This process is treated consistently with previously obtained within the model meson observables. Calculated values of g_{VPP} are given in table 2.

Invariance of meson-meson amplitude under local background field gauge transformation turns out to be of crucial importance for correct description of the decays. If one reduces local gauge invariance to just the global one, decay constants change dramatically, especially $g_{\rho\pi\pi}$ (compare third and fourth column of table 2). This is to be compared with usually underestimated $g_{\rho\pi\pi}$ decay constant [14, 15]. Value of $g_{\omega\pi\pi}$ is exactly zero because of ideal mixing of ω and ϕ mesons and employed approximation of $SU(2)$ isospin symmetry ($m_u = m_d$).

4 Outlook

It is important to describe transition form factors of η, η', η_c simultaneously with pion. It was shown that in the instanton liquid model diagrams of type B in figure 1 contribute to transition form factors [16]. While this contribution is suppressed by difference of masses of up and down quarks and negligible in the present consideration, it should be taken into account in order to describe transition form factors of other mesons. However, incorporating this contribution into the domain model requires explicit accounting for inhomogeneity of the vacuum ensemble.

References

[1] B. Aubert et al. [BaBar Collaboration], Phys. Rev. D **80**, 052002 (2009). [arXiv:0905.4778 [hep-ex]]
[2] G. P. Lepage and S. J. Brodsky, Phys. Rev. D **22**, 2157 (1980)
[3] S. Uehara et al. [Belle Collaboration], Phys. Rev. D **86**, 092007 (2012). [arXiv:1205.3249 [hep-ex]]
[4] S. N. Nedelko and V. E. Voronin, Phys. Rev. D **93(9)**, 094010 (2016). [arXiv:1603.01447 [hep-ph]]
[5] A. C. Kalloniatis and S. N. Nedelko, Phys. Rev. D **64**, 114025 (2001)
[6] G. V. Efimov, and S. N. Nedelko, Phys. Rev. D **51**, 176 (1995)
[7] J. V. Burdanov, G. V. Efimov, S. N. Nedelko, S. A. Solunin, Phys. Rev. D **54**, 4483 (1996)
[8] A. C. Kalloniatis and S. N. Nedelko, Phys. Rev. D **69**, 074029 (2004)
[9] H. J. Behrend et al. [CELLO Collaboration], Z. Phys. C **49**, 401 (1991)
[10] J. Gronberg et al. [CLEO Collaboration], Phys. Rev. D **57**, 33 (1998). [hep-ex/9707031]
[11] K. A. Olive et al. [Particle Data Group Collaboration], Chin. Phys. C **38**, 090001 (2014)
[12] S. V. Mikhailov and N. G. Stefanis, Nucl. Phys. B **821**, 291 (2009). [arXiv:0905.4004 [hep-ph]]
[13] A. V. Radyushkin and R. T. Ruskov, Nucl. Phys. B **481**, 625 (1996). [hep-ph/9603408]
[14] V. Bernard, A. H. Blin, B. Hiller, U. G. Meissner and M. C. Ruivo, Phys. Lett. B **305**, 163 (1993). [hep-ph/9302245]
[15] H. B. Deng, X. L. Chen and W. Z. Deng, Chin. Phys. C **38(1)**, 013103 (2014). [arXiv:1304.5279 [hep-ph]]
[16] N. I. Kochelev and V. Vento, Phys. Rev. D **81**, 034009 (2010). [arXiv:0912.2172 [hep-ph]]

EPJ Web of Conferences **138**, 02006 (2017) DOI: 10.1051/epjconf/201713802006
Baldin ISHEPP XXIII

Search for short strings in e^+e^--annihilation

Marina Kozhevnikova[1,*], *Armen* Oganesian[1,2,**], and *Oleg* Teryaev[1,***]

[1] *Joint Institute for Nuclear Research, Dubna, Russia*
[2] *Institute of Theoretical and Experimental Physics, Moscow, Russia*

Abstract. In this work the behavior of power corrections to Adler function in operator product expansion (OPE) is studied, in particular the possible contribution of operator of dimension 2. The OPE terms of dimension 4 and 6 are taken into account. Various experimental data on reactions of e^+e^--annihilation to pion channels ($\pi^+\pi^-$, $2\pi^+2\pi^-$, $\pi^+\pi^-2\pi^0$, $3\pi^+3\pi^-$, $2\pi^+2\pi^-2\pi^0$) are used. The high-precision fits of the experimental data are obtained and used. The method based on Borel transformation of Adler function is applied. It is shown that the contribution of operator of dimension 2 is negative being compatible to zero at three standard deviations level. The strong (anti)correlation between short string and local gluon condensate is found.

1 Introduction

The problem of the existence of the operator with dimension 2, whose contribution to QCD sum rules [1] is proportional to $1/Q^2$, and search for relevant OPE corrections are performed already for quite a long time [2–4].

In the pioneering paper [2] a concept of short strings leading to corrections with dimension 2 was suggested. In Cornell potential [5]

$$V(r) \approx -\frac{4\alpha_s(r)}{3r} + kr$$

the term kr describes string potential (connected to the phenomenon of confinement) at short distances, leading[3] to the correction $\sim k/Q^2$. In paper [3] the correlation between short strings and perturbation theory order was established [4]. The contribution of the operator with dimension 2 to the e^+e^- data was studied later [4, 7] and it was found to be compatible to zero with large errors.

Our analysis is based on the use of Adler function being a simple two-point correlator which is convenient to compare to experimental data. We use a large number of them and perform an accurate numerical analysis. It consists of construction of the fitting model, calculating R-ratio and D-function in dispersion form, applying Borel transformation (BT), construction of sum rule and extraction of the OPE coefficients.

The main purpose is to verify whether the operator with dimension 2 exists.

*e-mail: m_kojevnikova@mail.ru
**e-mail: armen@itep.ru
***e-mail: teryaev@theor.jinr.ru

[3] Note that in OPE the first correction to e^+e^--annihilation cross section is $\sim \langle G_{\mu\nu}G^{\mu\nu}\rangle / Q^4$.

[4] It is analogous to the interplay of perturbation theory order and higher twist in deep inelastic scattering [6]

2 Fits

We limit ourselves by the data with isospin $I = 1$, thereby avoiding strange particles, which are difficult to measure. While the data are presented as sets of experimental points, for accurate analysis it is preferable to obtain analytic fits of experimental data which are convenient to integrate. Whereas there are different annihilation channels data obtained from different detectors, we fit every channel separately and find the formula for full cross section.

The analysis is performed using the data on e^+e^--annihilation to pion channels: $e^+e^- \to \pi^+\pi^-$ (CMD and OLYA detectors) [8], $e^+e^- \to 2\pi^+2\pi^-$ (M3N, GG2, DM1, CMD, OLYA, DM2, SND, CMD2, SND, BaBar) [9–19], $e^+e^- \to \pi^+\pi^-2\pi^0$ (OLYA, CMD2, SND, DM2, Frascati-ADONE-GAMMA-GAMMA-2) [20–25], $e^+e^- \to 3\pi^+3\pi^-$ (BaBar) [26], $e^+e^- \to 2\pi^+2\pi^-2\pi^0$ (BaBar) [26].

For description of the data on squared pion form factor $|F_\pi(s)|^2$ depending on energy $\sqrt{s}$ of reaction $e^+e^- \to \pi^+\pi^-$ we use the three-resonance (Gounaris-Sakurai) model [8], [27], where the form factor of each resonance V is calculated using Breit-Wigner formula:

$$F^{\rm BW}(s, m_V, \Gamma_V) = \frac{m_V^2(1 + d \cdot \Gamma_V/m_V)}{m_V^2 - s + f(s, m_V, \Gamma_V) - i\, m_V\, \Gamma_V(s)}, \text{ where } \Gamma_V(s) = \Gamma_V \left(\frac{k(s)}{k(m_V^2)}\right)^3,$$

$$k(s) = \frac{\sqrt{s - 4m_\pi^2}}{2}, \quad f(s, m_V, \Gamma_V) = \Gamma_V \frac{m_V^2}{k(m_V^2)^3}[k^2(s)(h(s) - h(m_V^2)) - (s - m_V^2)k^2(m_V^2)h'(m_V^2)],$$

$$h(s) = \frac{2}{\pi}\frac{k(s)}{\sqrt{s}} \ln\left(\frac{\sqrt{s} + 2m_\pi}{2m_\pi}\right), \quad h'(mV^2) = h'(s)\,|_{s=mV^2}, \quad F^{\rm BW}(0, m_V, \Gamma_V) = 1 \text{ automatically}.$$

The full pion form factor with resonances ρ, ω and ρ' has the following form:

$$F_\pi(s, \{\rho, \omega, \rho'\}\}) = \frac{F_\pi(s, \{\rho, \omega\}) + \alpha_{\rho'} F^{\rm BW}\left(s, m_{\rho'}, \Gamma_{\rho'}\right)}{1 + \alpha_{\rho'}},$$

$$\text{where} \quad F_\pi(s, \{\rho, \omega\}) = F^{\rm BW}(s, m_\rho, \Gamma_\rho)\, \frac{1 + \alpha_\omega F^{\rm BW}(s, m_\omega, \Gamma_\omega)}{1 + \alpha_\omega}.$$

Performing the χ^2-minimization we get the fit of the experimental data on $|F_\pi|^2(s)$ taken from work [8] characterized by $\chi^2_{\rm b.f.} = 0.82$ and the following resonances parameters values[5]:

$$\begin{aligned} m_\rho &= 0.770\text{ GeV (PDG)}, \quad \Gamma_\rho = 0.149 \pm 0.008\text{ GeV};\\ m_\omega &= 0.782\text{ GeV (PDG)}, \quad \Gamma_\omega = 0.009\text{ GeV (PDG)}, \quad \alpha_\omega = 0.0021 \pm 0.0017;\\ m_{\rho'} &= 1.354 \pm 0.116\text{ GeV}, \quad \Gamma_{\rho'} = 0.344 \pm 0.157\text{ GeV}, \quad \alpha_{\rho'} = -0.089 \pm 0.024;\\ d &= 0.384 \pm 0.200\text{ GeV}. \end{aligned}$$

For cross sections of the processes $e^+e^- \to 2\pi^+2\pi^-$, $e^+e^- \to \pi^+\pi^-2\pi^0$, $e^+e^- \to 2\pi^+2\pi^-2\pi^0$, and $e^+e^- \to 3\pi^+3\pi^-$ the description in the form of sum of three Gaussian curves, describing wide resonances, is assumed:

$$\sigma(s, \{M_i, \sigma_i, \alpha_i\}) = \sum_{i=1}^{3} \frac{\alpha_i^2}{\alpha_1^2 + \alpha_2^2 + \alpha_3^2}\, e^{-(\sqrt{s}-M_i)^2/(2\sigma_i^2)}.$$

The results are shown in table 1.

[5] We use values of m_ρ, m_ω and Γ_ω taken from PDG [28]

 DOI: 10.1051/epjconf/201713802006

Table 1. The fitting results for particular e^+e^--annihilation channels.

$e^+e^- \to 2\pi^+2\pi^-$	i	M_i, GeV	σ_i, GeV	α_i
	1	1.289 ± 0.034	0.213 ± 0.028	0.462 ± 0.058
$\chi^2_{\text{b.f.}} = 0.97$	2	1.573 ± 0.017	0.227 ± 0.011	1.522 ± 0.121
	3	2.524 ± 0.070	0.356 ± 0.037	2.427 ± 0.791
$e^+e^- \to \pi^+\pi^-2\pi^0$	i	M_i, GeV	σ_i, GeV	α_i
	1	1.304 ± 0.250	0.258 ± 0.066	1.024 ± 1.064
$\chi^2_{\text{b.f.}} = 1.12$	2	1.748 ± 0.411	0.283 ± 0.052	1.333 ± 0.275
	3	2.322 ± 0.201	0.194 ± 1.027	0.817 ± 2.826
$e^+e^- \to 3\pi^+3\pi^-$	i	M_i, GeV	σ_i, GeV	α_i
	1	1.811 ± 0.027	0.091 ± 0.023	0.183 ± 0.040
$\chi^2_{\text{b.f.}} = 0.58$	2	1.897 ± 0.034	0.237 ± 0.027	0.291 ± 0.029
	3	2.247 ± 0.042	0.134 ± 0.264	0.956 ± 0.534
$e^+e^- \to 2\pi^+2\pi^-2\pi^0$	i	M_i, GeV	σ_i, GeV	α_i
	1	1.740 ± 0.023	0.115 ± 0.019	0.529 ± 0.082
$\chi^2_{\text{b.f.}} = 0.75$	2	2.027 ± 0.080	0.287 ± 0.062	0.362 ± 0.082
	3	2.274 ± 0.042	0.248 ± 0.026	1.051 ± 0.255

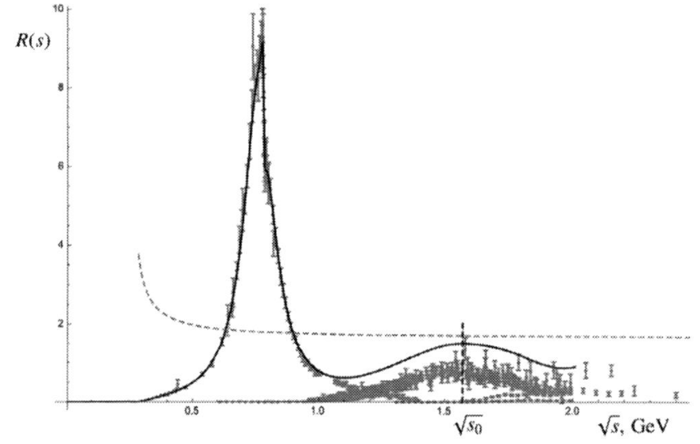

Figure 1. The full R-ratio in dependence on energy $\sqrt{s}$ at $\sqrt{s} \leq 2$ GeV and the theoretical representation of R-ratio. The data on particular channels of e^+e^--annihilation are presented by blue dots with errors.

3 R-ratio, Adler function and sum rule

The full R-ratio is the sum of R-ratios of particular processes. $R(s) = \sum\limits_i \dfrac{\sigma_{e^+e^-\to\text{hadrons of type } i}(s)}{\sigma_{e^+e^-\to\mu^+\mu^-}(s)}$.

Extracting the current with isospin $I = 1$ ($j_\mu = \frac{1}{2}(\bar{u}\gamma_\mu u - \bar{d}\gamma_\mu d)$) from full electromagnetic current, we obtain the formula for D-function with isospin $I = 1$ in the framework of OPE (for example, see work [29], (3.4)):

$$D_{\text{PT+OPE}}(Q^2) = \frac{3}{2}\left[1 + \frac{\alpha_s(Q^2)}{\pi} + \sum_{n\geq 1}\Gamma(n)\,\frac{a_{2n}}{Q^{2n}}\right], \tag{1}$$

where a_{2n} are OPE coefficients, we take into account three first power corrections.

The another form of D-function, dispersion, is the following:

$$D_{\rm exp}(Q^2) = Q^2 \int_{4m_\pi^2}^{\infty} \frac{R_{\text{exp-th}}(s)\,ds}{(s+Q^2)^2} = Q^2 \int_{4m_\pi^2}^{s_0} \frac{R_{\rm exp}(s)\,ds}{(s+Q^2)^2} + Q^2 \int_{s_0}^{\infty} \frac{R_{\rm th}(s)\,ds}{(s+Q^2)^2}, \tag{2}$$

$$\text{where} \quad R_{\text{exp-th}}(s) = R_{\rm exp}(s)\,\theta(s<s_0) + R_{\rm th}(s)\,\theta(s>s_0), \quad R_{\rm th}(s) = \frac{3}{2}\left[1 + \frac{\alpha_s(s)}{\pi}\right];$$

$R_{\rm exp}(s)$ is our fitting result (shown by black solid curve in figure 1) and $R_{\rm th}(s)$ is the one-loop approximation of perturbative QCD (shown by red dashed curve in figure 1).

The continuum threshold $\sqrt{s_0} = 1.57$ GeV is chosen to guarantee that $R_{\rm exp}(s)$ and $R_{\rm th}(s)$ and even their first derivatives take similar values (under the statistical uncertainties of experimental data), as it is seen on figure 1 (vertical dashed line). The function $R_{\rm exp}(s)$ decreases above $\sqrt{s_0} = 1.57$ GeV, which is explained by the absence of the data on e^+e^--annihilation to 8π channels.

Equating the both forms of Adler function: by OPE (1) and dispersion (2), and applying BT, we get the sum rule:

$$\Phi_{\rm exp}(M^2) = \Phi_{\rm PT+OPE}(M^2), \tag{3}$$

$$\text{where} \quad \Phi_{\rm exp}(M^2) = \int_0^{\infty} R_{\text{exp-th}}(s)\left(1 - \frac{s}{M^2}\right) e^{-s/M^2}\,\frac{ds}{M^2},$$

$$\Phi_{\rm PT+OPE}(M^2) = \frac{3}{2}\frac{\hat{B}_{Q^2\to M^2}[\alpha_s(Q^2)]}{\pi} + \frac{3}{2}\left(\frac{C_2}{M^2} + \frac{C_4}{M^4} + \frac{C_6}{M^6}\right), \quad C_{2n} = a_{2n},$$

$$\text{and} \quad \hat{B}_{Q^2\to M^2}\left[\alpha_s(Q^2)\right] = \frac{4\pi}{b_0}\left[\frac{1}{M^2}\int_0^{\infty}\frac{e^{-s/M^2}ds}{\ln^2(s/\Lambda^2)+\pi^2} + \frac{\Lambda^2}{M^2}e^{\Lambda^2/M^2}\right], \quad b_0 = 11 - 2N_f/3,$$

$\Lambda = \Lambda_{\rm QCD} = 0.25$ GeV is the QCD scale parameter. M^2 is the Borel mass, appearing in equation after Borel transformation.

4 The sum rule analysis

Finally, let us turn to the discussion of obtained sum rule (3). The Borel mass M^2 is varied in interval $0.75 \div 4$ GeV^2 which is divided to 20 points, then the coefficients C_2 and C_4 are determined by χ^2-minimization, and $C_6 = -\frac{448\pi^3}{27}\,\alpha_s\,\langle\bar{q}q\rangle^2 = -0.121$ GeV^6, which can be expressed in terms of quark condensate [29], is fixed [30]. The coefficient C_4 can be expressed in terms of gluon condensate $C_4 = \frac{2\pi^2}{3}\left\langle\frac{\alpha_s GG}{\pi}\right\rangle$ [1].

Our result is shown in figure 2; the regions corresponding to 1, 2 and 3 σ-levels are marked in red, blue and yellow. The minimal value is $\chi^2_{\rm min} = 0.483$, and corresponding gluon condensate and C_2 are found to be: $\langle(\alpha_s/\pi)\,GG\rangle = 0.025\,\text{GeV}^4$, $C_2 = -0.086\,\text{GeV}^2$. The allowed intervals for C_2 and gluon condensate at one σ-level are:

$$C_2 = -0.086 \pm 0.050\,\text{GeV}^2, \quad \left\langle\frac{\alpha_s GG}{\pi}\right\rangle = 0.025 \pm 0.012\,\text{GeV}^4.$$

At one σ-level our results do not contradict with the well known results [29–31], but at the same time C_2 is not compatible with zero. One can see that $C_2 = 0$ is possible at 3 σ-level.

 DOI: 10.1051/epjconf/201713802006

Furthermore, by the form of regions corresponding our results (see figure 2) we can suppose that there is some (anti)correlation between C_2 and C_4. A rough estimation of this connection is expressed by formula:

$$C_2 \approx -5\,\mathrm{GeV}^{-2}\left\langle\frac{\alpha_s GG}{\pi}\right\rangle + 0.025\,\mathrm{GeV}^2.$$

One can compare the received values with the existing values of the gluon condensate [29–31].

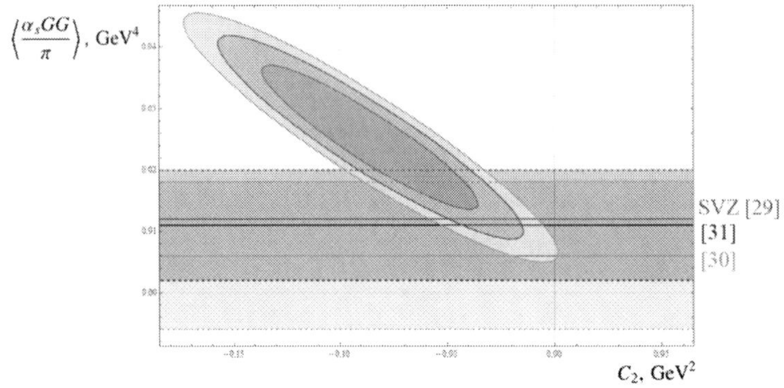

Figure 2. The allowed regions for C_2 and gluon condensate at 1, 2 and 3 σ-level, determined by $\chi^2 \le \chi^2_{\min} + 1$ (red), $\chi^2 \le \chi^2_{\min} + 2$ (blue) and $\chi^2 \le \chi^2_{\min} + 3$ (yellow). The regions of existing data on gluon condensate are shown for comparison. Central values are marked by solid horizontal lines, errors ranges are marked by dashed lines.

5 Conclusions and outlook

The resonance contribution fitting model is developed, the Adler function with Borelization is obtained and its precise numerical analysis is performed.

C_2 has negative sign and is compatible with zero only at 3 σ-level. At 1 σ-level: $C_2 = -0.086 \pm 0.050\,\mathrm{GeV}^2$, $\langle\alpha_s/\pi\, GG\rangle = 0.025 \pm 0.012\,\mathrm{GeV}^4$.

Strong (anti)correlation between short strings and local gluon condensate is found:

$$C_2 \approx -5\,\mathrm{GeV}^{-2}\left\langle\frac{\alpha_s GG}{\pi}\right\rangle + 0.025\,\mathrm{GeV}^2.$$

We plan to perform an analogous analysis in the framework of the analytic perturbation theory [32–34].

Acknowlegements.
We dedicate this paper to the memory of Alexander Petrovich Bakulev (1956-2012) with whom we started this work. We are grateful to A. L. Kataev, S. V. Mikhailov, O. P. Solovtsova for useful discussions and comments. M. K. is thankful to Yu. L. Ryzhykau for valuable advices and assistance in numerical analysis. The work is supported in part by RFBR Grant-140100647.

References

[1] M. A. Shifman, A. I. Vainshtein, V.I. Zakharov, Nucl. Phys. B **147**, 385-447 (1979)
[2] K.G. Chetyrkin, S. Narison, V.I. Zakharov, Nucl.Phys. B **550**, 353-374 (1999)

[3] S. Narison, V.I. Zakharov, Phys. Lett. B **679**, 355-361 (2009)
[4] S. Bodenstein, C.A. Dominguez, S.I. Eidelman, H. Spiesberger, K. Schilcher, JHEP **039**, 1201 (2012)
[5] E. Eichten, K. Gottfried, T. Kinoshita, J. B. Kogut, K. D. Lane and T. M. Yan, Phys. Rev. Lett. **34**, 369 (1975). Erratum: [Phys. Rev. Lett. **36**, 1276 (1976)]
[6] A. L. Kataev, G. Parente and A. V. Sidorov, Fiz. Elem. Chast. Atom. Yadra **31(7B)**, 29 (2000). [Phys. Part. Nucl. **31(7B)**, 29 (2000)]
[7] H. Spiesberger, Mod. Phys. Lett. A **28**, 1360009 (2013)
[8] L. M. Barkov et al., Nucl. Phys. B **256**, 365–384 (1985)
[9] G. Cosme, B. Dudelzak, B. Grelaud, B. Jean-Marie, S. Jullian et al., Nucl. Phys. B **152**, 215-231 (1979)
[10] C. Bacci, R. Baldini-Celio, G. Battistoni, G. Capon, R. Del Fabbro et al., Phys. Lett. B **95**, 139-142 (1980)
[11] A. Cordier, D. Bisello, J. Bizot, J. Buon, B. Delcourt et al., Phys. Lett. B **109**, 129-132 (1982)
[12] L. M. Barkov, I. B. Vasserman, P. V. Vorobev, P. M. Ivanov, G. Y. Kezerashvili et al., Sov. J. Nucl. Phys. **47**, 248–252 (1988)
[13] L. M. Kurdadze, et al., JETP Lett. **47**, 512–515 (1988)
[14] D. Bisello et al., Nucl. Phys. Proc. Suppl. **21**, 111–117 (1991)
[15] S. I. Dolinsky, V. P. Druzhinin, M. S. Dubrovin, V. B. Golubev, V. N. Ivanchenko et al., Phys.Rept. **202**, 99–170 (1991)
[16] R. R. Akhmetshin et al., Phys. Lett. B **475**, 190–197 (2000)
[17] M. N. Achasov, K. I. Beloborodov, A. V. Berdyugin, A. G. Bogdanchikov, A. V. Bozhenok et al., J. Exp. Theor. Phys. **96**, 789–800 (2003)
[18] R. R. Akhmetshin et al., Phys. Lett. B **595**, 101–108 (2004)
[19] B. Aubert et al., Phys. Rev. D **71**, 052001 (2005)
[20] M. R. Whalley, J. Phys. G: Nucl. Part. Phys. **29**, A1-A133 (2003)
[21] L. M. Kurdadze et al., J. Exp. Theor. Phys. Lett. **43**, 643-645 (1986)
[22] R. R. Akhmetshin et al., Phys. Lett. B **466**, 392-402 (1999)
[23] M. N. Achasov et al., Preprint BUDKER-INP-2001-34 (2001)
[24] B. Bisello et al., *DM2 results on e+ e- annihilation into multi - hadrons in the 1350 MeV - 2400 MeV energy range. High-energy physics. Proceedings* (25th International Conference, Singapore, August 2-8, 1990) Vol. I+II. Preprint LAL-90-35 (1990)
[25] C. Bacci et al., Nucl. Phys. B **184**, 31-39 (1980)
[26] B. Aubert et al., Phys. Rev. D **73**, 052003 (2006)
[27] G. J. Gounaris, J. J. Sakurai, Phys. Rev. Lett. **21**, 244–247 (1968)
[28] K. A. Olive et al., Particle Data Group. Chin. Phys. C **38**, 1025–1034 (2014 and 2015 update)
[29] M. A. Shifman, A. I. Vainshtein, V. I. Zakharov, Nucl. Phys. B **147**, 448-518 (1979)
[30] B. V. Geshkenbein, B. L. Ioffe, and K. N. Zyablyuk, Phys. Rev. D **64**, 093009 (2001)
[31] B. L. Ioffe, K. N. Zyablyuk, Eur. Phys. J. C **27**, 229-241 (2003)
[32] D. V. Shirkov and I. L. Solovtsov, Phys. Rev. Lett. **79**, 1209 (1997). Theor. Math. Phys. **150**, 132 (2007)
[33] A. P. Bakulev, S. V. Mikhailov and N. G. Stefanis, Phys. Rev. D **72**, 074014 (2005). [Phys. Rev. D **72**, 119908 (2005)]
[34] V. L. Khandramai, R. S. Pasechnik, D. V. Shirkov, O. P. Solovtsova and O. V. Teryaev, Phys. Lett. B **706**, 340 (2012)

Electroweak properties of ρ-meson in the instant form of relativistic quantum mechanics

A.F. Krutov[1,⋆], *R.G.* Polezhaev[1,⋆⋆], and *V.E.* Troitsky[2,⋆⋆⋆]

[1] *Samara University, 443086 Samara, Russia*
[2] *D.V. Skobeltsyn Institute of Nuclear Physics, Moscow State University, Moscow 119991, Russia*

Abstract. Charge and magnetic radii, magnetic and quadrupole moments of the ρ-meson are calculated in framework of the developed by authors version of the instant form of relativistic quantum mechanics with fixed number of particles (IF RQM). The calculations are performed with different wave functions of quarks in the ρ-meson and using the so- called modified impulse approximation (MIA). The electromagnetic characteristics of ρ-meson are obtained without fitting parameters. The value of the magnetic moment coincides with available experimental data: $\mu_\rho = 2.1 \pm 0.5\, e/2M_\rho$.

1 Introduction

In recent years there has been significant progress in the experimental study of ρ-meson: measurement of the lepton decay constant from the process $\tau \to \rho\nu_\tau$ [1, 2], extraction of the ρ-meson magnetic moment from processes $\gamma^* \to 4\pi$ [3], measurement of the magnetic moment of decay $\rho \to \pi\gamma^*$ [4], the expected measurement of the electromagnetic form factors of the ρ-meson from the reaction $e^+ + e^- \to \rho^+ + \rho^-$ [5]. To the moment, there are a large number of calculations of the rho meson electroweak properties in different approaches (see, e.g., [6–11]). The theoretical calculations of the electroweak properties of the ρ-meson by different methods have important implications for understanding the processes in the transition region between nonperturbative and perturbative quark dynamics, for example. A fairly large amount of experimental information enables to estimate these approaches from the point of view of their ability to self-consistent description of these data.

In the present work the some electroweak characteristics of the ρ-meson (magnetic and quadrupole moments, the charge and magnetic mean-square radii, the lepton decay constant) are calculated in the framework of instant form of relativistic quantum mechanics (IF RQM)(see, e.g., review [12]). Developed by authors formulation of IF RQM is based on a direct realization of the Poincaré algebra on the set of the dynamic observables of the composite system and dates back to the work of P. Dirac [13]. Distinctive features of our variant of IF RQM are the original procedure of construction of the electroweak currents of a composite system and new formulation of the impulse approximation - the so-called modified impulse approximation (MIA). Unlike conventional impulse approximation MIA is formulated in the terms of the reduced current matrix elements on Poincaré group (form factors) and

⋆e-mail: krutov@ssau.ru
⋆⋆e-mail: polezaev@list.ru
⋆⋆⋆e-mail: troitsky@theory.sinp.msu.ru

 DOI: 10.1051/epjconf/201713802007

does not lead to violation of Lorentz-covariance and conservation law for the currents of composite systems.

In this work the model parameters are fixed from the calculations of electroweak properties of pion [14] and from description of the lepton decay constant of ρ-meson [15]. So, our calculations of the magnetic moment and the charge radius are performed without fitting parameters and give the good agreement with experimental data.

2 Electroweak characteristics of ρ-meson and numerical calculations

The electromagnetic current matrix element of the ρ-meson can be written in terms of conventional Sachs form factors for the system with the total angular momentum equal 1. To do this let us write the parameterization of the electromagnetic current matrix element in the Breit frame (see, e.g., [16]):

$$\langle \vec{p}_\rho\,, m_J | j_\mu(0) | \vec{p}_\rho\,'\,, m'_J \rangle = G^\mu(Q^2)\,,$$

$$G^0(Q^2) = 2p_{\rho 0}\left\{(\vec{\xi}\,'\vec{\xi}^{\,*})\,G_C(Q^2) + \left[(\vec{\xi}^{\,*}\vec{Q})(\vec{\xi}\,'\vec{Q}) - \frac{1}{3}Q^2(\vec{\xi}\,'\vec{\xi}^{\,*})\right]\frac{G_Q(Q^2)}{2M_\rho^2}\right\}\,,$$

$$\vec{G}(Q^2) = \frac{p_{\rho 0}}{M_\rho}\left[\vec{\xi}'\,(\vec{\xi}^{\,*}\vec{Q}) - \vec{\xi}^{\,*}(\vec{\xi}\,'\vec{Q})\right]G_M(Q^2)\,. \qquad (1)$$

Here G_C, G_Q, G_M are the charge, quadrupole and magnetic form factors, respectively, M_ρ is the ρ-meson mass.

The polarization vector in the Breit frame has the following form:

$$\xi^\mu(\pm 1) = \frac{1}{\sqrt{2}}(0\,,\,\mp 1\,,\,-i\,,\,0)\,,\quad \xi^\mu(0) = (0\,,\,0\,,\,0\,,\,1)\,. \qquad (2)$$

The variables in ξ are total angular momentum projections.

In the Breit frame:

$$q^\mu = (p_\rho - p'_\rho)^\mu = (0\,,\,\vec{Q})\,,\quad p^\mu_\rho = (p_{\rho 0}\,,\,\frac{1}{2}\vec{Q})\,,\quad p'_\rho{}^\mu = (p_{\rho 0}\,,\,-\frac{1}{2}\vec{Q})\,, \qquad (3)$$

$$p_{\rho 0} = \sqrt{M_\rho^2 + \frac{1}{4}Q^2}\,,\quad \vec{Q} = (0\,,\,0\,,\,Q)\,.$$

Our approach gives following integral representation for the electromagnetic form factors of composite systems with total angular momentum equal 1 in MIA (see,e.g., [17]):

$$G_C(Q^2) = \int d\sqrt{s}\,d\sqrt{s'}\,\varphi(s)\,g_{0C}(s\,,Q^2\,,s')\,\varphi(s')\,,$$

$$G_Q(Q^2) = \frac{2\,M_\rho^2}{Q^2}\int d\sqrt{s}\,d\sqrt{s'}\,\varphi(s)\,g_{0Q}(s\,,Q^2\,,s')\,\varphi(s')\,, \qquad (4)$$

$$G_M(Q^2) = -\,M_\rho\int d\sqrt{s}\,d\sqrt{s'}\,\varphi(s)\,g_{0M}(s\,,Q^2\,,s')\,\varphi(s')\,.$$

Here $g_{0C}(Q^2)$, $g_{0Q}(Q^2)$, $g_{0M}(Q^2)$ are so-called free two-particle charge, quadrupole and magnetic form factors, respectively, $\varphi(s)$ is wave function of quarks in the ρ-meson in sense of RQM.

From physical point of view the free two-particle form factors describe the electromagnetic properties of system of the two non-interacting particle with discrete quantum numbers of the ρ-meson.

These form factors are reduced matrix elements of the electromagnetic current operator of the free two-particle system on the Poincaré group. Generally speaking, these form factors are regular generalized functions (distributions) (see [12] for details). These generalized functions can be calculated by the methods of relativistic kinematics and have the form given in [17]. Exactly the appearance of $g_{0C}(Q^2), g_{0Q}(Q^2), g_{0M}(Q^2)$ in the integral representation for the electromagnetic form factor of the ρ-meson is the essence of MIA.

The wave functions in sense of IF RQM in (4) at $J = S = 1$, $l = 0$ are defined by the following expressions (see, e.g., [12]):

$$\varphi(s) = \sqrt[4]{s}\,\psi(k)\,k\,, \quad k = \frac{1}{2}\sqrt{s - 4\,M^2}\,, \tag{5}$$

where M is mass of constituents quark.

Normalization is given by the following condition:

$$\int \psi^2(k)\,k^2\,dk = 1\,, \tag{6}$$

here $\psi(k)$ is a model wave function.

The magnetic moment μ_ρ and the quadrupole moment Q_ρ of ρ meson were calculated using the relations given in [16]:

$$G_M(0) = \frac{M_\rho}{M}\,\mu_\rho\,, \quad G_Q(0) = M_\rho^2\,Q_\rho\,. \tag{7}$$

The static limit in (4) gives the following relativistic expressions for moments:

$$\mu_\rho = \frac{1}{2}\int_{2M}^{\infty} d\sqrt{s}\,\frac{\varphi^2(s)}{\sqrt{s - 4\,M^2}}\left\{1 - L(s) + (\kappa_u + \kappa_{\bar{d}})\left[1 - \frac{1}{2}\,L(s)\right]\right\}\,, \tag{8}$$

$$Q_\rho = -\int_{2M}^{\infty} d\sqrt{s}\,\frac{\varphi^2(s)}{\sqrt{s}}\left[\frac{M}{\sqrt{s} + 2\,M} + \kappa_u + \kappa_{\bar{d}}\right]\frac{L(s)}{4M\sqrt{s - 4M^2}}\,, \tag{9}$$

$$L(s) = \frac{2\,M^2}{\sqrt{s - 4\,M^2}\,(\sqrt{s} + 2\,M)}\left[\frac{1}{2\,M^2}\sqrt{s\,(s - 4\,M^2)} + \ln\frac{\sqrt{s} - \sqrt{s - 4\,M^2}}{\sqrt{s} + \sqrt{s - 4\,M^2}}\right]\,,$$

where κ_q is the quark anomalous magnetic moment.

The ρ-meson charge ($\langle\, r_\rho^2 \rangle_C$) and magnetic ($\langle\, r_\rho^2 \rangle_C$) radii are calculated from relation:

$$\langle\, r_\rho^2 \rangle_C = -6\,G_C'(0)\,, \quad \langle\, r_\rho^2 \rangle_M = -6\,G_M'(0)\,. \tag{10}$$

The lepton decay constant of the ρ-meson, f_ρ is defined by the following matrix element of the electroweak current (see, e.g., [18]):

$$\langle 0|j_\mu^\rho(0)|\vec{P}_\rho, m_\rho\rangle = i\,\sqrt{2}\,f_\rho\,\xi_\mu(m_\rho)\frac{1}{(2\pi)^{3/2}}\,, \tag{11}$$

where $\vec{P}_\rho$ is the meson three-momentum, $m_\rho = -1, 0, 1$ is the spin projection, $\xi_\mu(m_\rho)$ is the polarization vector that in the Breit frame has the form (2).

For the calculation of the ρ-meson lepton decay constant we used the method of construction of current matrix element which is nondiagonal with respect to the total angular momentum [15]. Final expression for f_ρ in approximation of four-fermion interaction in our approach has form [15]:

$$f_\rho = \frac{\sqrt{3}}{\sqrt{2}\pi}\int_0^{\infty} dk\,k^2\,\psi(k)\frac{(\sqrt{k^2 + M^2} + M)}{(k^2 + M^2)^{3/4}}\left(1 + \frac{k^2}{3(\sqrt{k^2 + M^2} + M)^2}\right)\,. \tag{12}$$

For the description of the relative motion of quarks the following phenomenological wave functions are used:

1. A Gaussian or harmonic oscillator wave function (see, e.g., [19]):

$$\psi(k) = N_{HO}\exp\left(-k^2/2\,b^2\right). \tag{13}$$

2. A power-law wave function (see, e.g., [19]):

$$\psi(k) = N_{PL}\,(k^2/b^2 + 1)^{-n}\,, \quad n = 2\,,3\,. \tag{14}$$

For Sachs form factors of constituent quarks we used the following expressions:

$$G^q_E(Q^2) = e_q\,f_q(Q^2)\,, \quad G^q_M(Q^2) = (e_q + \kappa_q)\,f_q(Q^2)\,, \tag{15}$$

where e_q is the quark charge.

For $f_q(Q^2)$ the form proposed in [20] is chosen:

$$f_q(Q^2) = \frac{1}{1 + \ln(1 + \langle r^2_q\rangle Q^2/6)}\,, \tag{16}$$

here $\langle r^2_q\rangle$ is the MSR of constituent quark.

The choice of form factor of the constituent quark as (15), (16) due to the fact that the asymptotic of the electromagnetic pion form factor at $Q^2 \to \infty$ obtained in our nonperturbative approach with (16) does coincide with that predicted by QCD including the pre-asymptotic factor (see, e.g., [21] for details).

Parameters in our calculation are fixed from description of electroweak properties of pion [14]. Let us note that our RQM describes well the experimental data for the pion form factor including the recent points [22]. Let us emphasize that the parameters used in our calculations were obtained from the fitting to the experimental data up to $Q^2 \simeq 0.26$ GeV^2 [23]. At that time the data for higher Q^2 was not correlated in different experiments and had significant uncertainties. The later data for pion form factor in JLab experiments up to Q^2 =2.45 GeV^2 were obtained with rather good accuracy. All experimental points obtained in JLab up to now agree very well with our prediction of 1998.

So, by analogy with pion calculations from [14] we use the following set of parameters: M=0.22 GeV for the constituent quark mass; the quark anomalous magnetic moments enter our expressions through the sum $(\kappa_u + \kappa_{\bar{d}})$ and we take $\kappa_u + \kappa_{\bar{d}} = 0.0268$ in natural units; for the quark MSR we use the relation $\langle r^2_q\rangle \simeq 0.3/M^2$. The parameters of the wave functions b in (13), (14) were fixed from the requirements of the description of experimental values of ρ-meson lepton decay constants 152 ± 8 MeV [1]. The ρ-meson mass in our calculations is taken from [2] M_ρ = 775.5±0.4 MeV.

So, all parameters of our model are fixed. Results of numerical calculations are presented in table 1. Let us remark that all electroweak characteristics of the ρ-meson except the lepton decay constant are calculated without fitting parameters.

As it can be seen from table the calculated value of the magnetic moment is in agreement with experimental data. The values of $\langle r^2_\rho\rangle_C$, while not measured directly, are important for testing various conjectures about strongly interacting systems. One of the interesting related prediction was introduced as a consequence of the so-called Wu–Yang hypothesis [24] (see also [25]), though it is remarkable by itself. Namely, one may define the radius of a hadron either in terms of the electroweak interaction (the mean square charge radius, $\langle r^2_{\rm ch}\rangle$, calculated for the ρ meson in this paper) or in terms of the strong interaction (this radius, $\langle r^2_{\rm st}\rangle$, is defined by the slope of the cross section of

DOI: 10.1051/epjconf/201713802007

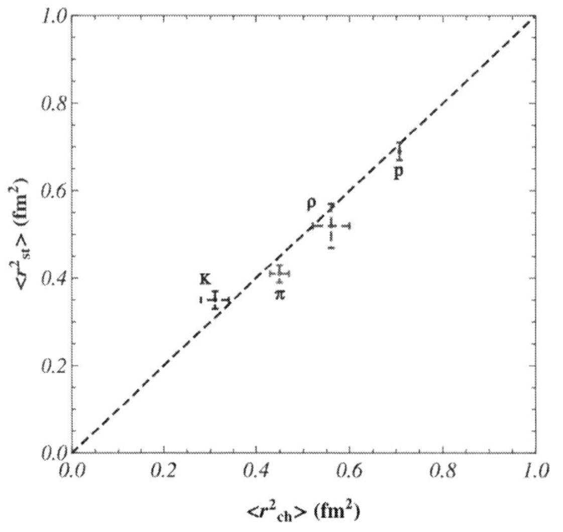

Figure 1. Relation between the strong-interaction hadronic radius $\langle r^2_{\rm st}\rangle$ and the charge radius $\langle r^2_{\rm ch}\rangle$ for light hadrons. Result for ρ-meson is obtained in our work with wave function (14) at $n = 3$.

Table 1. The ρ-meson electromagnetic moments and lepton decay constant obtained with the different model wave functions (13) – (14). $\langle r^2_\rho\rangle_C, \langle r^2_\rho\rangle_M$, are charge and magnetic MSR, respectively, in fm^2; μ_ρ is relativistic magnetic moment (8) in $e/2M_\rho$; Q_ρ is quadrupole moment (9) in fm^2. The parameters b in (13) and (14) are in GeV, f_ρ is ρ-meson lepton decay constant in MeV.

Wave functions	b	$\langle r^2_\rho\rangle_C$	$\langle r^2_\rho\rangle_M$	μ_ρ	Q_ρ	f_ρ
(13)	0.228	0.597	0.635	2.01	-0.0064	152.2
(14) n=2	0.217	0.579	0.532	2.12	-0.0064	153.6
(14) n=3	0.379	0.560	0.634	2.16	-0.0066	154.9
Experiment				2.1 ± 0.5 [3]		152 ± 8 [1]

hadron–proton scattering). The conjecture [25], which may be derived from, though not necessary implies, the hypothesis of [24], is the equality of the two radii,

$$\langle r^2_{\rm st}\rangle = \langle r^2_{\rm ch}\rangle\,. \tag{17}$$

This remarkable equality has been verified experimentally with a great degree of accuracy for the proton, π- and K- mesons. Even more demonstrative is figure 1, analogous to a figure from the paper [25], but presenting more recent data. We can see that the value of the ρ-meson charge radius obtained in this paper fits perfectly the conjecture (17).

3 Conclusions

In the present work the calculations of electroweak characteristics of the ρ-meson in the framework IF RQM are performed with different wave functions in modified impulse approximation (MIA). MIA does not violate the Lorentz covariance and conservation law for the electroweak currents in contrast with conventional impulse approximation. Other distinctive feature of our work is that all calculations were made without fitting parameters. The results for magnetic moment of the ρ-meson are consistent with recent experimental data [3]. The values of the charge radius satisfy the hypothesis of Wu-Yang that has been experimentally verified for a number of hadrons. One of the authors (VT) thanks S.Troitsky for interesting discussions. Authors (AK and RP) thank the Organizing committee for their kind invitation to XXIII International Baldin Seminar on High Energy Physics Problems and hospitality. This work was supported in part (AK and RP) by the Ministry of Education and Science of the Russian Federation (grant No. 1394, state task).

References

[1] C.S. Mello, A.N. da Silva, J.P.B.C. de Melo, T. Frederico, Few-Body Syst **56**, 509 (2015)
[2] K.A. Olive et al. (Particle Data Group), Chin. Phys. C **38**, 090001 (2014)
[3] D. Garcia Gudino, G.Toledo Sanchez, J.Mod. Phys. A **30**, 1550114 (2015)
[4] Yao W.N. et al. (Particle Data Group) J.Phys. G **33**, 1232 (2006)
[5] B. Aubert et al. (Babar Collaboration), Phys.Rev. D **78**, 071103 (2008)
[6] Jun He, B. Juliá-Díaz, and Yu-bing Dong, Phys.Lett. B **602**, 212 (2004)
[7] M.S. Bhagwat and P. Maris, Phys. Rev. C **77**, 025203 (2008)
[8] H.L.L. Roberts, A. Bashir, L.X. Gutierrez-Guerrero, C.D. Roberts and D.J. Wilson, Phys. Rev. C **83**, 065206 (2011)
[9] M. E. Carrillo-Serrano, W. Bentz, I. C. Cloët, and A. W. Thomas, Phys. Rev. C **92**, 015212 (2015)
[10] B. Owen, W. Kamel, D. Leinweber, B. Menadue, S. Mahbub, Phys. Rev. D **91**, 074503 (2015)
[11] B.L.G. Bakker, H.M. Choi and C.R. Ji, Phys. Rev. D **65**, 116001 (2002)
[12] A.F. Krutov and V.E. Troitsky, Phys. Part. Nucl. **40**, 136 (2009)
[13] P.A.M. Dirac, Rev.Mod.Phys. **21**, 392 (1949)
[14] A.F. Krutov and V.E. Troitsky, Eur. Phys. J. C **20**, 71 (2001)
[15] A.F. Krutov, R.G. Polezhaev, V.E. Troitsky, Phys. Rev. D **93**, 036007 (2016). arXiv:1602.00907 (2016)
[16] R.G. Arnold, C.E. Carlson, F. Gross, Phys. Rev. C **21**, 1426 (1980)
[17] A.F. Krutov, V.E. Troitsky, Phys. Rev. C **75**, 014001 (2007)
[18] W. Jaus, Phys.Rev. D **67**, 094010 (2003)
[19] F. Coester and W.N. Polyzou, Phys. Rev. C **71**, 028202 (2005)
[20] A.F. Krutov and V.E. Troitsky, Teor. Math. Phys. **116**, 907 (1998)
[21] S.V. Troitsky and V.E. Troitsky, Phys. Rev. D **88**, 093005 (2013)
[22] G.M. Hubert et.al. The Jefferson Lab F_π Collaboration, Phys.Rev. C **78**, 045203 (2008)
[23] S.R. Amendolia et al., Phys. Lett. B **146**, 116 (1984).
S.R. Amendolia et al., Nucl. Phys. B **277**, 168 (1986)
[24] T.T. Wu and C.N. Yang, Phys. Rev. B **137**, 708 (1965)
[25] B. Povh, J. Hüfner, Phys. Lett. B **245**, 653 (1990)

Sizes and distances in high energy physics

Vladimir Petrov[1,⋆]

[1] *Institute for High Energy Physics, NRC Kurchatov Institute, Protvino, Russia*

Abstract. This is a critical discussion of physical relevance of some space-time characteristics which are in current use in high energy physics.

1 Introduction

High-energy physics is in many ways a synonymous of relativistic physics. The latter circumstance introduces, as is well known, some peculiarities when one deals with coordinates, lengths, time intervals. The theoretical and interpretational toolkit of high-energy physics is dominated by energy-momentum characteristics while the spatio-temporal ones are directly unobservable and thus comparatively rare. Nonetheless, the latter are persistently present both in literature and physical discourses and sometimes play quite an important heuristic role. So Heisenberg's attempt in his S-matrix paradigm, developed later under the guise of the "analytic S-matrix approach" by G. Chew, to exorcise these entities from the particle physics as unobservable ones seems now failed. However, spatio-temporal parameters often are not clearly defined in quantum-field theoretical terms. Thereof, it necessary to at least reveal and clarify their true meaning. In this talk I will consider the following characteristics: charge radius, interaction range, interaction time and the size of the like-charge pion source.

2 Charge Radius

In PDG volumes one can find values of "charge radii" of hadrons. E. g. the proton charge radius is presented in two hypostasis: $r_{charge}(proton) = 0.84087 \pm 0.000039 fm$ (μp Lamb shift) and $r_{charge}(proton) = 0.8775 \pm 0.0051 fm$ (ep CODATA value).

How are these radii defined? In non-relativistic quantum mechanics when considering, say, the scattering of an electron off a composite system (atom) the relationship between the average radius of the scatterer and the derivative of its form factor $F(\mathbf{q}^2)$ is (with $\mathbf{q}$ the transferred 3-momentum)

$$\langle \mathbf{r}^2 \rangle = -6dF(\mathbf{q}^2)/d\mathbf{q}^2|_{\mathbf{q}^2=0}, \quad (1)$$

where

⋆e-mail: Vladimir.Petrov@ihep.ru

$$F(\mathbf{q}^2) = \int d\mathbf{r} e^{i\mathbf{q}\mathbf{r}} \rho(\mathbf{r}). \quad (2)$$

Here the average charge density is of the form

$$\rho(\mathbf{r}) = \sum_{k=1}^{N} e_k \rho_k(\mathbf{r}),$$

where

$$\rho_k(\mathbf{r}) = \int \prod_{j \neq k} d\mathbf{r}_j |\Psi(\{\mathbf{r}_j\}, \mathbf{r}_k = \mathbf{r})|^2. \quad (3)$$

So we get

$$\langle \mathbf{r}^2 \rangle = \sum_{k=1}^{N} e_k N_k \langle \mathbf{r}^2 \rangle_k,$$

where N_k is the number of constituents with charge e_k and

$$\sum_{k=1}^{N} e_k N_k$$

is the total charge of the system,

$$\sum_{k=1}^{N} N_k = N$$

while $\langle \mathbf{r}^2 \rangle_k$ stands for the average square of the distance of the k-th constituent from the center of mass of the system which is implicitly fixed in Eq.(3) with the condition

$$\delta(\sum_1^N \mathbf{r}_i m_i / \sum_1^N m_i).$$

From this expression it is evident that the "charge radius" generally does not give a clear idea of the geometrical extent of the system which is closer to the expression of the type

$$\langle \mathbf{r}^2 \rangle = \frac{1}{N} \sum_{k=1}^{N} \langle \mathbf{r}^2 \rangle_k.$$

The wave function is a solution of the Schrödinger equation with an appropriate instantaneous potential. When coming to the relativistic domain the use of the instantaneous potential becomes problematic because of the retardation effects. Nonetheless, Eq.(1) is being actively used for extracting from the data the "charge radius" of, say, the proton with further placing the result into the PDG Reviews as an important physical characteristic of the proton charge distribution. To avoid as much as possible the model dependence we first consider the formula (1) in the context of the general quantum-field theoretical expression via the Bogoliubov-LSZ reduction formalism. For simplicity but without loss of generality we consider the pion form factor $F(q^2)$ defined as follows

$$\langle p+q|J^\mu(0)|p\rangle = (2p^\mu + q^\mu)F(q^2),$$

where $J^\mu(x)$ is the electromagnetic current operator and $|p\rangle$ is a charged pion state. The function $F(q^2)$ is hermitian analytic in the complex q^2 plane with a cut $[4m_\pi^2, +\infty]$. In terms of the 4-momentum transfer $q^2 = q_0^2 - \mathbf{q}^2$ the "charge radius" of the pion is taken as

$$\langle \mathbf{r}^2\rangle = 6dF(q^2)/dq^2|_{q^2=0}. \tag{1*}$$

The only known model-independent way to introduce space-time coordinates is the use of the reduction formulae connecting Green functions (and hence scattering amplitudes) in momentum and configuration space. Using the reduction formalism [1] we get

$$F(q^2) = \tfrac{2p_\mu}{4m_\pi^2-q^2}\int d^4x e^{iqx}\langle 0|\tfrac{\delta J^\mu(x)}{\delta\phi^+(0)}|p\rangle.$$

Variation derivative is taken over the pion out-field. Up to a finite sum of quasi-local operators

$$\tfrac{\delta J^\mu(x)}{\delta\varphi^+(0)} = i\theta(-x)[J^\mu(x), I_\pi^+(0)].$$

Here $I_\pi^+(x) = i\frac{\delta S}{\delta\varphi_\pi^+(x)}S^+$ is the pion density operator. For definiteness let us take the laboratory frame where $\mathbf{p} = 0$. We get

$$F(q^2) = \tfrac{2m_\pi}{4m_\pi^2-q^2}\int d^4x \exp[-i\tfrac{q^2}{m_\pi} - i(\mathbf{xn})\sqrt{-q^2(1-q^2/m_\pi^2)}]\langle 0|\tfrac{\delta J^0(x)}{\delta\phi^+(0)}|\mathbf{p}=0\rangle.$$

Now, if we apply to this representation formula (1*) for the "charge radius" we have

$$\langle \mathbf{r}^2\rangle = \int d\mathbf{r}\mathbf{r}^2\rho_L(\mathbf{r}), \tag{4}$$

where

$$\rho_L(\mathbf{r}) = \frac{1}{2m_\pi}\int dx^0\langle 0|\frac{\delta J^0(x^0,\mathbf{r})}{\delta\phi^+(0,\mathbf{0})}|\mathbf{p}=0\rangle \tag{5}$$

has the meaning of the charge density inside the pion in rest.

When looking at Eq.(5) we notice that the distance $|\mathbf{r}|$ relates points taken at different times $x^0, 0$ and so the profile of the supposed charge distribution $\rho_L(\mathbf{r})$ does not give us an instantaneous snapshot of the charge distribution inside the pion but rather something smeared in time. One can prove (with use e.g. of the JLD representation for causal commutators) that in the non-relativistic limit

$$c\tfrac{1}{2m_\pi}\langle 0|\tfrac{\delta J^0(ct,\mathbf{r})}{\delta\phi^+(0,\mathbf{0})}|\mathbf{p}=0\rangle|_{c\longrightarrow\infty} = \delta(t)\Phi(\mathbf{r})$$

so we recover a NR quantum-mechanical expression like Eq.(3). On the other hand, one can argue that non-simultaneity in the definition of the charge radius (or a particle size) can be taken into account if we assume that arising uncertainty is given by the "retardation time" $\sim \langle r\rangle/c$ and may induce an uncertainty comparable with the very radius in question [2].

3 On "physical" proton radius

With all reservations stemming from the previous Section let us consider the charge radius of the nucleon from the point of its use in modelling practice. Quite often the value of the charge radius of the proton is taken in models of high-energy scattering to account for the size of the colliding nucleons. The fact of the matter is that the charge distribution may generally differ very much from the "matter distribution" due to the presence of the constituent charges making the distribution generally non-positive definite. E.g. the square of the charge radius of the neutron is **negative**, as follows from the definition (1*) and experimental data, and hence does not give us much to say about the neutron physical size. To make the point more concrete it is useful to express the charge radii of the proton and neutron in terms of the valence quarks. With account of the isotopic symmetry it is not difficult to obtain the following expression for the nucleon **physical** size

$$\left\langle r^2\right\rangle(nucleon) = r^2_{charge}(proton) + r^2_{charge}(neutron).$$

Let us take for definiteness the PDG values $r_{charge}(proton) = 0.707 fm^2$ (μp Lamb shift) and $r^2_{charge}(neutron) = -0.116 fm^2$.

Thereof we get for the physical radius of the nucleon

$$\sqrt{\langle r^2\rangle(nucleon)} = 0.769 fm$$

to compare with $r_{charge}(proton) = 0.841 fm$. At first sight the difference can seem insignificant but in cases where the energy dependence is logarithmic such difference leads to significant differences in the estimates of the relevant energies [2].

4 On the interaction range

The transverse interaction range $\left\langle b^2\right\rangle_{tot}$ (with **b** the impact parameter) is defined from the forward logarithmic slope $B(s)$

$$B(s) \equiv \left[\frac{\frac{\partial}{\partial t}\left(\frac{d\sigma}{dt}\right)}{\frac{d\sigma}{dt}}\right]_{t=0} = \partial ln[d\sigma/dt]/\partial t\ |_{t=0} = 2Re(\partial \ln(T(s,t))/\partial t)\ |_{t=0}$$

so that

$$2B(s) \approx \frac{\int db^2\, b^2\, \mathrm{Im}\,\tilde{\mathrm{T}}(s,b)}{\int db^2\, \mathrm{Im}\,\tilde{\mathrm{T}}(s,b)} =< b^2 >_{tot}, \qquad (6)$$

where $\tilde{T}(s,b) = \frac{1}{16\pi s}\int T(s,t)J_0(b\sqrt{-t})dt$.

It is a peculiar feature of the unitarity that the range of the transverse area where any process (*inelastic* included) can happen is defined by the quantity describing *elastic* scattering.

One can define the quantity $< b^2 >_{el}$ which seems to be more relevant for spatial description of elastic scattering only

$$\left(\int db^2 b^2 |\tilde{T}(s,b)|^2\right)/\left(\int db^2 |\tilde{T}(s,b)|^2\right) =< b^2 >_{el}.$$

However, it depends on the scattering phase which is very problematic as an observed quantity [3]:

$$< b^2 >_{el} =< (-t)B^2(s,t) + 4(-t)[\frac{\partial\Phi(s,t)}{\partial t}]^2 > .$$

Experiments show a very slow growth of $< b^2 >_{tot}$:

$$\sqrt{< b^2 >_{tot}}(0.01TeV@U-70, IHEP) \approx 0.9fm$$

while

$$\sqrt{< b^2 >_{tot}}(7TeV@LHC, CERN) \approx 1.3fm$$

only. If to take into account that physical values of the "valence" sizes of colliding nucleons $\sqrt{\frac{2}{3}\langle r^2\rangle (nucleon)}$ (seen in the impact parameter plane) are, as we have seen above, of order of 0.6 fm we conclude that colliding nucleons hardly cease to overlap even at the LHC energies. So even at the LHC we are very far from the "asymptopia" which implies the region of energies where

$$\sqrt{< b^2 >_{tot}} \gg \sqrt{\frac{2}{3}\langle r^2\rangle (nucleon)}.$$

Can one argue about the *longitudinal* range as well? Yes, but it is again related to the almost unobservable scattering phase:

$$\langle L^* \rangle = 4p^* \langle (\partial \Phi(s,t)/\partial t) \rangle ,$$

where asterisk means the center-of-mass (cms) values and p^* is the colliding hadrons' momentum. Here we can only say that existing models give a very fast grow of $\langle L^* \rangle$ proportional to the cms energy $\sqrt{s}$.

5 Time evolution of space-time correlation functions in hadron collisions

Let us consider the forward scattering amplitude. It can be cast in the following form

$$T(k,p) = \int d^4x \exp(ikx) \left\langle p | \frac{\delta I(-x/2)}{\delta \varphi(x/2)} | p \right\rangle . \tag{7}$$

For the sake of simplicity let us take the simple Regge pole model high-energy behaviour of this amplitude

$$T(k,p) \approx (2pk)^{\alpha(0)} \beta(k^2),$$

where we allow the projectile to be off-shell and consider the region $2pk \gg k^2$.

With a sufficient degree of rigour it was shown [4] that the correlator representing the amplitude in the configuration space behaves as follows:

$$\left\langle p | \frac{\delta I(-x/2)}{\delta \varphi(x/2)} | p \right\rangle \approx (px)^{\alpha(0)} g(x^2) \tag{8}$$

at

$$(px) \to \infty, x^2 \text{ fixed}.$$

Simple Regge-pole asymptotics can be generalized to more complicated cases. E.g. in the case of the functional saturation of the Froissart bound we would have

$$\left\langle p|\frac{\delta I(-x/2)}{\delta\varphi(x/2)}|p\right\rangle \approx (px)\ln^2(px)g(x^2). \tag{9}$$

In the laboratory frame and with account of the causality condition ($\delta I(-x/2)/\delta\varphi(x/2) = 0$ at $x\bar{\in}(x : x^0 \geq |\mathbf{x}|)$) we see that the correlation function $\langle p|\frac{\delta I(-x/2)}{\delta\varphi(x/2)}|p\rangle$ grows with the time t growth as $\sim t^{\alpha(0)}$ at $\alpha(0) > 0$ or as $\sim t\ln^2 t$ inside the upper light cone on any fixed hyperboloid $ct = \sqrt{x^2 + \mathbf{x}^2}$ (this corresponds to large values of one of the light-cone variables $x^{\pm}$). Such a behaviour of the correlation function is quite unusual because correlation functions we deal with in, say, statistical physics ordinarily die off at large times and distances.

We here should also mention the estimate of the interaction time which is being made supposing that it can be given by an "essential" space-time region where the exponential in Eq.(7) does not strongly oscillate. With such an assumption some authors (see e.g. [5]) obtained that the "effective interaction time" Δt_{eff} can be estimated (in the laboratory frame, e.g.) as

$$\Delta t_{eff} \approx E/m^2.$$

Accordingly the longitudinal distances can be estimated as $L_{eff} \sim \Delta t_{eff}$ so that at the LHC we would have the distances (reduced to the cms frame), where the scattering takes place, as large as 1500 fm. Sure, such an estimate does not take into account the correlator behaviour inside the "effective integration region" but Eqs.(8, 9) seem to support such a growth.

6 On the size of the like-charge pion source

Many years ago G. Goldhaber et al observed that the like charge pions significantly correlate in contrast to the unlike charge pions with this correlation seen in various correlation measures. With the obvious reason for like-charge pions to correlate due to the Bose-Einstein statistics the main aim of the further studies of these correlations was to extract from the data the spatial size of the region where these pions are being radiated from.

Generic designation of this correlation function is R(Q) where R can be defined in various ways dependent on the reference distribution exemplifying the uncorrelated samples. Figure 1 [6] represents a typical behaviour of R as function of $Q = \sqrt{-(k_1 - k_2)^2} = \sqrt{M_{\pi\pi}^2 - 4m_\pi^2}$.

Generic form (with insignificant deviations) of R is

$$R(Q) = 1 + \exp(-RQ). \tag{10}$$

And this is parameter R which is treated as the average size of the like-charge pion source. Such an interpretation stems from the conviction that the two-particle distribution can be represented as a Fourier transform of the modulus squared of some function in the configuration space as it takes place for the form factor in NR QM (see Eq.(2)). I did not manage to find out a consistent theoretical derivation of this elegant formula. If to accept the premises used in literature for the spatial distribution of the "source size" r then, in order to get $\exp(-QR)$, it should develop a singularity of the type $1/(r^2 + R^2)$ which can be at odds with analyticity properties of Wightman functions in configuration space [1]. Without going into the details of very interesting findings concerning the energy and multiplicity dependence of the parameter R (or even two such parameters [7]) I would like just to show the integral representation of the two-pion distribution function in terms of quantum-field theoretical space-time correlators.

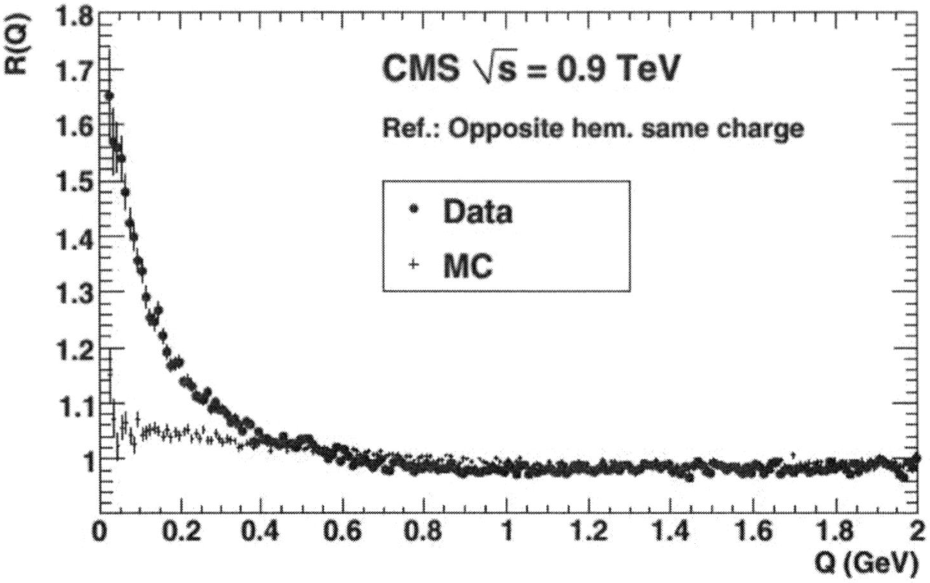

Figure 1. Typical BE correlation function as measured by CMS [6].

$$\frac{dN(N-1)}{d^3k_1d^3k_2} \sim \int d^4Xd^4\xi d^4\eta e^{iPX}e^{iq(\xi-\eta)}\left\langle pp, in|\bar{T}[I(X+\xi/2)I(X-\xi/2)]T[I^+(\eta/2)I^+(-\eta/2)]|pp, in\right\rangle$$

where $q = k_1 - k_2, P = (k_1 + k_2)$ and $I(x)$ is a pion density operator. First of all we see that in configuration space the integrand is non-diagonal which complicates its probabilistic interpretation. In any relevant channel the distance ξ refers to the 4-distance between the radiated pions. At the same time the momentum q related to the quantity Q is conjugated both to ξ and η. So it is difficult to relate the parameter R to the average source size. It remains to believe that we will finally eliminate these conceptual inconveniences though the problem seems quite a complicated from the formal theoretical viewpoints. I have, nonetheless, to stress that this statement in no way devalues intriguing regularities found in experimental and phenomenological studies for the parameter R. The problem is only what does it mean.

7 Conclusion

What was said in this talk is mostly setting of problems related to the lack of sound reasons for the standard physical interpretation of some space-time quantities, as they are normally defined, which I believe deserve the attention of the audience and hopefully will motivate further discussions and help to find acceptable solutions.

Acknowledgements

I am grateful to the organizers of the XXIII International Baldin Seminar on High Energy Physics Problems for invitation, hospitality and very good organization of the Seminar.

References

[1] N.N. Bogolubov, A.A. Logunov, A.I. Oksak, I. Todorov, *General Principles of Quantum Field Theory* (Kluwer Academic Publishers, 1990)

[2] V.A. Petrov, in *Proceedings of XXIXth International Workshop on High Energy Physics*, XXIXth International Workshop on High Energy Physics Protvino, Moscow Region, Russia, 26 – 28 June 2013

[3] J. Procházka, M.V. Lokajíček and V. Kundrát, Eur. Phys. J. Plus **131**, 147 (2016)

[4] R.A. Brandt, Nucl.Phys. B **83**, 60-92 (1974)

[5] V.N. Gribov, B.L. Ioffe and I.Ya. Pomeranchuk, Sov.J.Nucl.Phys. **2**, 549 (1966)

[6] CMS Collaboration, Phys. Rev. Lett. **105**, 032001 (2014)

[7] V.A. Khoze, A.D. Martin, M.G. Ryskin and V.A. Schegelsky, Eur.Phys. J. C **76(4)**, 193 (2016)

Recent results on light nuclei production from the NA49 experiment

V. I. Kolesnikov[1,★] for the NA49 Collaboration

[1]*Joint Institute for Nuclear Research, Dubna, Russia*

Abstract. Production of d and ^{3}He nuclei in central Pb+Pb interactions was studied at five collision energies ($\sqrt{s_{NN}}$ = 6.3, 7.6, 8.8, 12.3, and 17.3 GeV) with the NA49 detector at the CERN SPS. Transverse momentum spectra and rapidity distributions were measured. Yields are compared to predictions of statistical models. Phase-space distributions of light nuclei are discussed and compared to those of protons in the context of a coalescence approach. The coalescence parameters B_2 and B_3 for d and ^{3}He were determined at all energies.

1 Introduction

The main goal of the heavy-ion program at the CERN SPS is the study of the properties of hot and dense nuclear matter which can be probed experimentally by measuring particle composition, longitudinal and transverse momentum distributions of different species. The study of light nuclei production is of importance for several reasons. First of all, the mechanism of cluster formation in the interior of the fireball of a heavy-ion collision is not well understood and requires further quantitative investigations. It is likely that a significant fraction of light nuclei near the center-of-mass rapidity (midrapidity) are produced in a late stage of the reaction when the hadronic matter becomes diluted, thus probing the reaction dynamics at the time of the freeze-out. Secondly, in a conventional thermal model, particle multiplicities are predicted dependent on the bulk thermal parameters of the system - the chemical freeze-out temperature T, baryochemical potential μ_B and volume V. Though recent versions of statistical thermal models describe well hadron abundances in the range of heavy-ion collision energies from about 1 to several 10^3 GeV per nucleon [1, 2], the question was raised whether this approach is justified when applied to the production of nucleon clusters. Because thermal models consider particle yields integrated over the full phase space the lack of results on total multiplicities of light nuclei has up to now prevented a straightforward and quantitative test of the applicability of the statistical model approach to light nuclei production in the SPS energy range.

2 Data sets and analysis procedure

The data used in this analysis were collected in years 1996-2002. The NA49 experiment [3] utilized a ^{208}Pb beam at energies of 20A, 30A, 40A, 80A and 158A GeV impinging on a lead target of

★e-mail: Vadim.Kolesnikov@cern.ch

224 mg/cm^2 thickness corresponding to a 1% interaction probability. The interaction trigger selected the 12% most central collisions at 158*A* GeV, at other beam energies the data were recorded with a 7% central trigger. To ensure a good quality of event and track reconstruction, several cuts were applied to select events and tracks for further analysis. The identification of light nuclei (d,^{3}He) was based on momentum, dE/dx, and time-of-flight measurements. Deuteron candidates were required to have a TOF hit matched to the MTPC track, while the identification of ^{3}He candidates can rely completely on the specific energy loss measurement in the MTPC gas. The raw yields of clusters were corrected for geometrical acceptance, detector efficiency, and for the losses due to the PID selection criteria. The corrections were obtained from Monte Carlo (MC) simulations and from the data itself. More details about the analysis procedure can be found in [4].

3 Results

The invariant p_t spectra of identified ^{3}He nuclei at five collision energies in different rapidity intervals are shown in figure 1. The NA49 acceptance for light nuclei is sufficient to examine spectra in several rapidity intervals and their sizes are indicated in figure 1. The experimental distributions are scaled down successively for clarity of presentation and the same scaling factors were used for the rapidity slices located symmetrically with respect to the center of mass rapidity. In order to obtain particle rapidity density (dN/dy) we summed the measured points over the p_t distributions and added the extrapolations into unmeasured p_t regions exploiting information on the spectral shape. The extrapolation amounts to 3-7% of the dN/dy value for the ^{3}He spectra near mid-rapidity and increases to almost 40% for the results at the most forward rapidity. The overall systematic uncertainty for the yields of ^{3}He is of about 6% at mid-rapidity and approximately 10% for the most forward rapidity bin. For deuterons the systematic uncertainty varies from 5% to 15% depending on rapidity interval.

The cluster yields are plotted in figure 2 as a function of the normalized rapidity y/y_{beam}. All the rapidity distributions are concave and in order to quantify the shape all the data points were fitted with a parabola $a + b\,(y/y_{beam})^2$ (the fits are shown by dashed lines). The ratio of the fit parameters b/a (*relative concavity*) tends to increase with increasing beam momentum and cluster mass. Such a behavior of the invariant yields versus rapidity was earlier observed at AGS energies [5], where the relative concavity of the yield of clusters with atomic mass number from $A = 2$ to 4 progressively increases with A.

The total yield was then obtained by summing the measured values with the integral of the corresponding extrapolation function over the unmeasured region. To extrapolate the integral of dN/dy to full phase space, averaging of two different parameterizations of the rapidity spectra were employed as explained in [6]. The extrapolation accounts from 30% to 63% and from 20% to 85% of the total (4π) yield for ^{3}He and d, respectively, depending on the collision energy. The estimates for the 4π yield of deuterium and helium nuclei are plotted in figure 3 with red squares.

In the framework of a statistical thermal model the abundance N_C of a nucleon cluster of mass m, degeneracy factor g, charge q, and baryon number B is given by

$$N_C = \frac{gV}{\pi^2} m^2 T K_2(m/T) \exp\left(\frac{B\mu_B + q\mu_q}{T}\right), \quad (1)$$

where V, T, μ_B, μ_q, and K_2 are the source volume, temperature, baryochemical potential, charge potential, and Bessel function of the second kind. Such models have been able to reproduce the multiplicities of different types of particles in elementary and heavy-ion interactions. There are several parameterizations for the thermal fireball parameters T, μ_B, and V (or equivalently the fireball radius R) over a wide range of nuclear collision energies from AGS to LHC [7–10]. Using averages of these

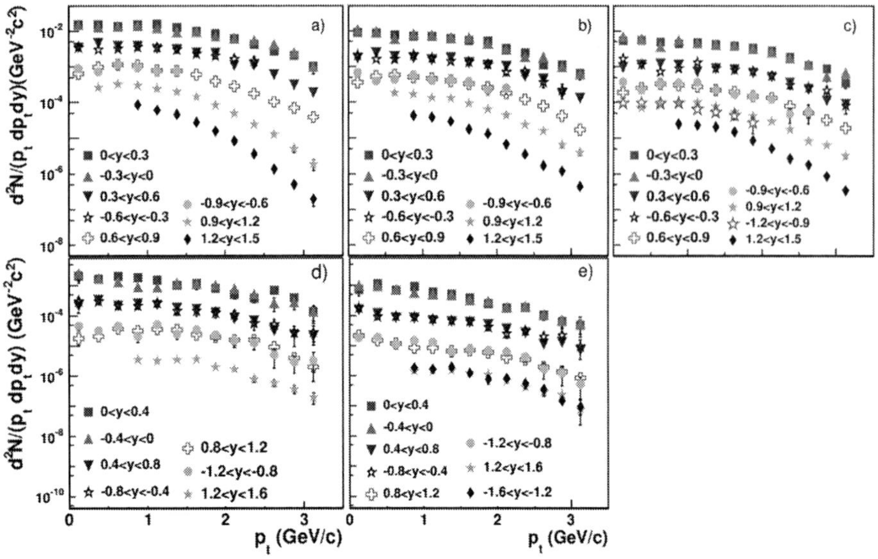

Figure 1. Invariant p_t spectra of ^{3}He at 20*A* (a), 30*A* (b), 40*A* (c), 80*A* (d), and 158*A* GeV (e). Only statistical errors are shown. The distributions near mid-rapidity are drawn to scale, other spectra are scaled down by successive powers of 5 for clarity. The same scaling factor is used for two rapidity slices that are symmetric about mid-rapidity ($y = 0$).

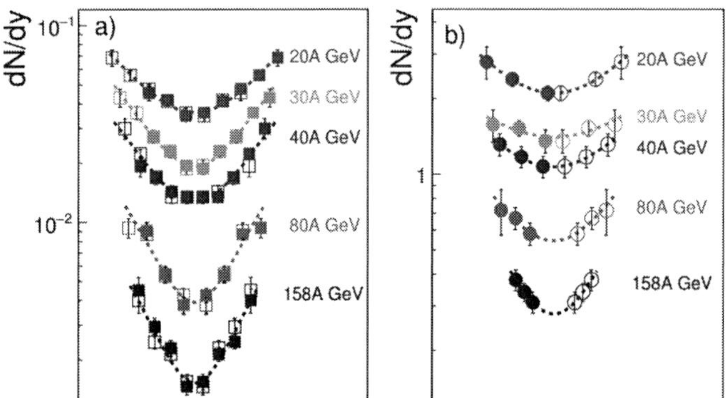

Figure 2. Rapidity distributions for ^{3}He (a) and deuterons (b) from central Pb+Pb collisions at 20*A*-158*A* GeV as a function of normalized rapidity y/y_{beam}. The solid symbols show the measurements and the open ones are the data points reflected around mid-rapidity. Dashed lines indicate parabolic fits to the rapidity spectra.

fireball parameters the mean multiplicities of d and ^{3}He were computed at all five collision energies according to 1. The results are plotted in figure 3 with blue circles. As it can be seen, thermal model calculations are capable of reproducing the energy dependence of the cluster multiplicities not only qualitatively but also quantitatively. The deviation of the calculations from the measured abundances does not exceed 2 standard deviations (see insets in figure 3). It appears at first glance that such a good agreement between experimental data on cluster production and thermodynamical model calculations

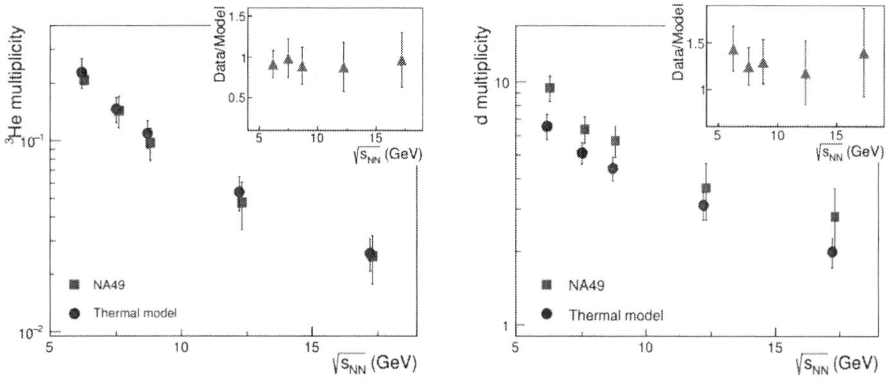

Figure 3. Total yield of ^{3}He (left panel) and deuterons (right panel) in central Pb+Pb collisions at 20A-158A GeV. The NA49 data are shown by red squares, thermal model calculations are indicated by blue circles. The insets show the ratio of the experimental data to the thermal model predictions.

looks surprising since one expects that chemical equilibrium of compound nucleon systems with the environment reaches later in time than the chemical freeze-out in the fireball has happened. The more diluted and colder matter is needed to make a net (fusion minus decomposition) rate positive for bound nucleon systems of several MeV binding energy. It has been recognized, however, that the light nuclei production process is basically determined by the entropy per baryon of the medium that is fixed at the chemical freeze-out [11, 12]. During further isentropic expansion stage a dynamic equilibrium has been reached for the light nuclei population: the loss due to disintegration are balanced by the gain in newly produced clusters.

In figure 4 (left panel) is plotted the average transverse energy $\langle m_t \rangle - m$ for ^{3}He as a function of normalized rapidity y/y_{beam} at five collision energies. As one can see, all the rapidity spectra follow a bell-like shape and we used Gaussian fits to get a mid-rapidity estimate for $\langle m_t \rangle - m$. In the right panel of figure 4 we present our results for the mid-rapidity $\langle m_t \rangle - m$ for protons and light nuclei from central Pb+Pb collisions at 20A-158A GeV. The data points for protons were taken from [13, 14], the values for d and ^{3}He were obtained in this study. Evidently, $\langle m_t \rangle - m$ rises approximately linearly with mass at all collision energies. These results may look surprising because it seems unlikely that objects of a few MeV binding energy per nucleon are participating in multiple thermalization collisions which generate the common velocity field inside fireballs of about 120-140 MeV temperature. On other hand, it was demonstrated in [15] that in the framework of the coalescence approach the choice of a suitable parameterizations for the spatial dependence of the single nucleon density can reproduce the observed mass dependence of the $\langle m_t \rangle - m$ of composites. For example, an interplay between a linear collective flow profile and a uniform density distribution gives an effective temperature rising linearly with mass.

In a coalescence approach [16, 17] the invariant yield N_A of clusters with charge Z and atomic mass number A is related to the product of the yields of protons N_{pr} and neutrons N_n through the coefficient B_A, the so-called coalescence parameter:

$$E_A \frac{d^3N_A}{d^3P_A} = B_A \left(E_{pr} \frac{d^3N_{pr}}{d^3p} \right)^Z \left(E_n \frac{d^3N_n}{d^3p} \right)^{A-Z}, \qquad (2)$$

where $p = P_A/A$. Assuming that the ratio of neutrons to protons is unity, B_A is then calculated by dividing the cluster yield at a given momentum P_A by the A-th power of the proton yields at

DOI: 10.1051/epjconf/201713803001

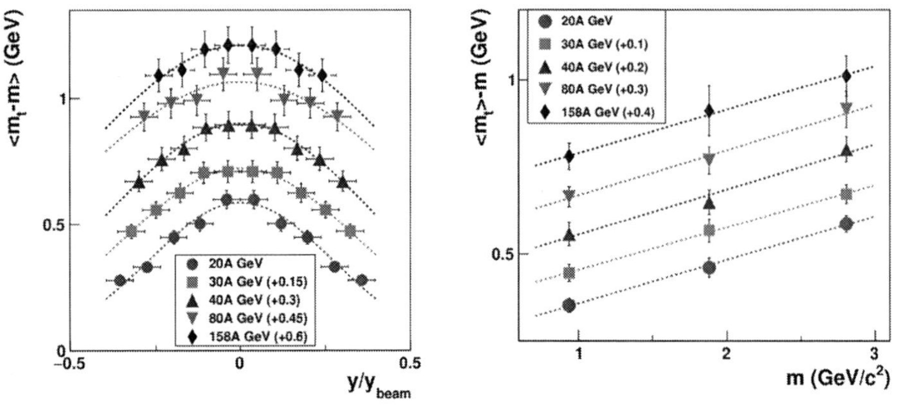

Figure 4. Left: mean transverse mass for ^{3}He as a function of y/y_{beam} from central Pb+Pb at 20A-158A GeV. Dashed lines indicate fits to a Gaussian. Right: mass dependence of $\langle m_t \rangle - m$ in central Pb+Pb collisions at 20A-158A GeV. Linear fits to the data points are indicated by dashed lines.

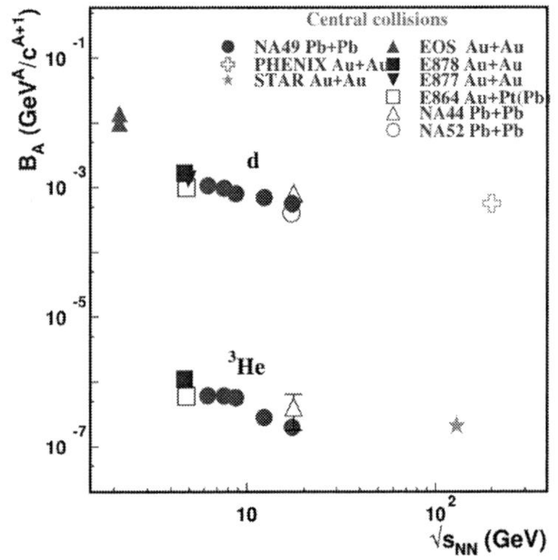

Figure 5. Coalescence parameters B_2 and B_3 from central A+A collisions.

P_A/A. Figure 5 compares the results for B_2 and B_3 at $p_t = 0$ from our study (red filled circles in figure 5) to experimental data from the Bevalac [18], AGS [5, 19], SPS [20, 21], and RHIC [22–24]. One concludes from this compilation that the coalescence parameters decrease only slowly with $\sqrt{s_{NN}}$ over a broad range of collision energies: within the SPS energy range the variation of the coalescence parameter is less than 40% and 60% for B_2 and B_3, respectively. In the framework of thermal models of cluster production [25, 26] the coalescence parameter is a measure of the source size: $B_A \approx (1/V)^{A-1}$. Thus, the observed energy dependence of B_A implies that the transverse size of the emitting source does not change much in this energy domain. This behavior is consistent with that found in two-pion interferometry measurements [27].

4 Summary

We present results on the production of d and ^{3}He nuclei in central Pb+Pb collisions at 20A-158A GeV recorded with the NA49 detector at the CERN SPS. The spectra for ^{3}He cover a wide range of rapidity and transverse momentum. Cluster yields were determined and exhibit a concave shape as function of rapidity with an increase of the degree of concavity for heavier systems. The yields of d and ^{3}He integrated over the full phase space agree with thermal model predictions at all collision energies. The transverse mass spectra of clusters were measured and the average values $<m_t>-m$ were found to increase linearly with the mass: this behavior favors a combination of a box density profile with a linear velocity profile in the source of the clusters. The coalescence parameters $B_{2,3}$ were derived showing a weak energy dependence. This observation suggests only a small increase of the freeze-out volume from AGS to RHIC energies.

References

[1] F. Becattini, J. Cleymans, A. Keränen, E. Suhonen, and K. Redlich, Phys. Rev. C **64**, 024901 (2001)
[2] A. Andronic, P. Braun-Munzinger, and J. Stachel, Nucl. Phys. A **772**, 167 (2006)
[3] S. V. Afanasiev et al. (NA49 Collaboration), Nucl. Instrum. Meth. A**430**, 210 (1999)
[4] T. Anticic et al. (NA49 Collaboration) Phys. Rev. C **94**, 044906 (2016)
[5] T. A. Armstrong et al. (E864 Collaboration), Phys. Rev. C **61**, 064908 (2000)
[6] T. Anticic et al. (NA49 Collaboration), Phys. Rev. C **85**, 044913 (2012)
[7] A. Andronic, P. Braun-Munzinger, and J. Stachel, Phys. Lett. B **673**, 142 (2009)
[8] J. Cleymans, H. Oeschler, K. Redlich, and S. Wheaton, Phys. Rev. C **73**, 034905 (2006)
[9] F. Becattini, M. Gaździcki, A. Keränen, J. Manninen, and R. Stock, Phys. Rev. C **69**, 024905 (2004)
[10] V. Vovchenko, V. V. Begun, and M. I. Gorenstein, arXiv:nucl-th/1512.08025v1
[11] P. J. Siemens and J. I. Kapusta, Phys. Rev. Lett. **43**, 1486 (1979)
[12] D. Hahn and H. Stöcker, Nucl. Phys. A **476**, 718 (1988)
[13] C. Alt et al. (NA49 Collaboration), Phys. Rev. C **73**, 044910 (2006)
[14] T. Anticic et al. (NA49 Collaboration), Phys. Rev. C **83**, 014901 (2011)
[15] A. Polleri, J. P. Bondorf, and I. N. Mishustin, Phys. Lett. B **419**, 19 (1998)
[16] S. T. Butler and C. A. Pearson, Phys. Rev. **129**, 836 (1963)
[17] A. Schwarzschild and C. Zupancic, Phys. Rev. **129**, 854 (1963)
[18] L. Ahle et al. (E802 Collaboration), Phys. Rev. C **60**, 064901 (1999)
[19] J. Barrette et al. (E877 Collaboration), Phys. Rev. C **61**, 044906 (2000)
[20] G. Ambrosini et al. (NA52 Collaboration), Phys. Lett. B **417**, 202 (1998)
[21] I. G. Bearden et al. (NA44 Collaboration), Eur. Phys. J. C **23**, 237 (2002)
[22] C. Adler et al. (STAR Collaboration) Phys. Rev. Lett. **87**, 262301 (2001)
[23] S. S. Adler et al. (PHENIX Collaboration) Phys. Rev. Lett. **94**, 122302 (2005)
[24] I. Arsene et al. Phys. Rev. C **83**, 044906 (2011)
[25] J. I. Kapusta, Phys. Rev. C **21**, 1301 (1980)
[26] A. Z. Mekjian, Phys. Rev. C **17**, 1051 (1978)
[27] C. Alt et al. (NA49 Collaboration), Phys. Rev. C **77**, 064908 (2008)

 DOI: 10.1051/epjconf/201713803002

Carbon fragmentation at 300 MeV/nucleon vs transport codes

B.M. Abramov[1,2], *P.N.* Alexeev[1], *Yu.A.* Borodin[1], *S.A.* Bulychjov[1], *I.A.* Dukhovskoy[1], *K.K.* Gudima[3], *A.I.* Khanov[1], *A.P.* Krutenkova[1], *V.V.* Kulikov[1,⋆], *M.A.* Martemianov[1], *S.G.* Mashnik[4,⋆⋆], *M.A.* Matsyuk[1], and *E.N.* Turdakina[1]

[1] *ITEP NSC "Kurchatov Institute", Moscow 117218, Russia*
[2] *MIPhT, Moscow Region 141700, Russia*
[3] *IAP, MD-2028 Chisinau, Moldova*
[4] *LANL, Los Alamos, NM 87545, USA*

Abstract. Momentum spectra of all long-lived fragments from ^{12}C fragmentation at 3.5^o have been measured. Differential cross sections span up to five orders of its magnitude. The preliminary results are compared with predictions of two transport codes: INCL++ and LAQGSM03.03. Both models give a good description of the data except few items which are discussed.

1 Introduction

The study of nucleus-nucleus interactions is one of the main aims of the modern nuclear physics. During last years, apart from an investigation of fundamental properties of these interactions, special attention has been paid to precise phenomenological description of these processes used in applications such as heavy-ion therapy where fragmentation is a main source of irradiation behind the Bragg peak. On this way a few simulation programs for nucleus-nucleus interactions has been created. They demand an experimental verification as well as refinement of their basic approaches. One of the aims of the FRAGM experiment performed at TWA (Tera Watt Accumulator) heavy-ion facility at ITEP was to obtain high precision data on nuclear fragmentation in the energy range accessible at this accelerator. In the framework of this experiment the data has been taken for carbon ion fragmentation in low-energy region. This work is a continuation of carbon fragmentation performed for higher energy [1] and for light fragments [2], [3]. In this publication we give preliminary results on all long-lived nuclear fragment emissions from 300 MeV incident carbon ion and compare them with the predictions of two transport codes.

2 Experiment

The experiment was carried out at the heavy-ion complex TWA at ITEP which includes an ion laser source, a linac, a booster and an accelerator-accumulator ring. Ions of 200-1000 MeV/n could be accumulated in this ring for successive use in experiments on the high-energy-density physics or

⋆e-mail: kulikov@itep.ru
⋆⋆deceased

DOI: 10.1051/epjconf/201713803002

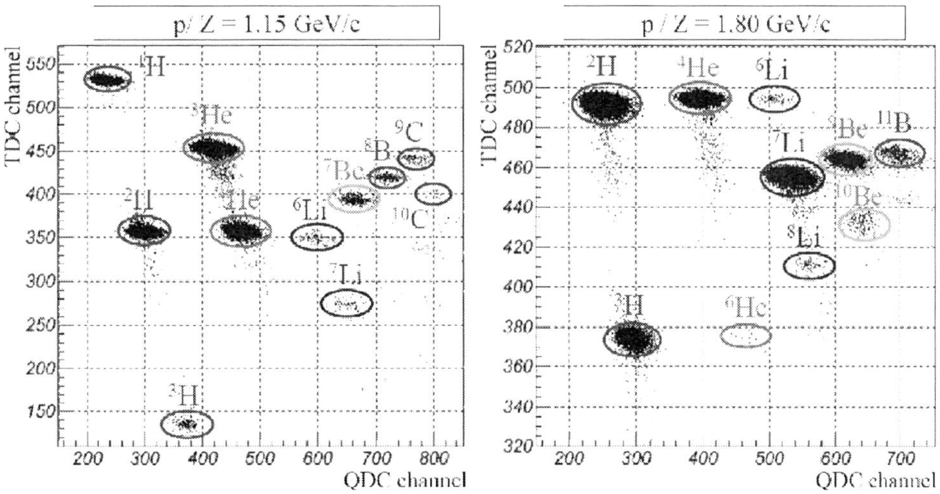

Figure 1. Two-dimensional plots QDC vs TOF channels for two rigidities of the beam-line spectrometer demonstrate good identification of the fragments.

accelerated to maximal energy of 4 GeV/nucleon. During our measurements each four seconds the carbon ions C^{+4} were accelerated in the booster up to 300 MeV/nucleon. Then while injection to the accelerator-accumulator ring they were totally stripped and captured in the ring. After that the beam was steered to the internal target of 50 μm Be foil strip providing the spill. It made simultaneously possible to have both a high luminosity due to multiple passage of the ions through the target and a small size of the source needed for a high momentum resolution of the subsequent magnetic analysis. The products of the carbon nucleus fragmentation outgoing at 3.5^{o} were momentum analyzed by the double-focus beam-line spectrometer of 42 meters long. Sets of few scintillation counters were placed at intermediate and final focuses for multiple measurements of ionization losses (dE/dx) and time of flight (TOF). The set up has been described in more details in [2]. Fragments with different charge and mass were unambiguously selected on two-dimensional plots dE/dx vs TOF. Two examples of this selection are given in figure 1 for different rigidities of the beam-line. The left is for 1.15 GeV/c that is near maximal emission of fragments with A/Z = 1.5 such as ^{3}He and ^{9}C. The right is for 1.80 GeV/c that is near the maximal yield of fragments with A/Z = 2.3 such as ^{7}Li, ^{9}Be and ^{11}B. The fragment momentum spectra were obtained by beam line energy scan in steps of 50 MeV/c. As a monitor we used a telescope of three scintillation counters that viewed the target at 2°. The preliminary results are presented in figure 2 normalized to INCL++ model predictions at the proton peak maximum.

3 Comparison of model predictions with experimental data

One of the aims of the FRAGM experiment is to test different models of ion-ion interactions. On this way it was possible for us to test three models from GEANT4 package supported by CERN: BC (Binary Cascade)[4], INCL++(C++ version of the Liege Intranuclear Cascade model)[5] and QMD (Quantum Molecular Dynamics)[6] as well as LAQGSM03.03 (the Los Alamos version of the Quark Gluon String Model)[7]. The last one is a main part of MCNP6 [8] transport code, supported and updated by LANL, USA. Only for two of them the results of the test will be given in this short

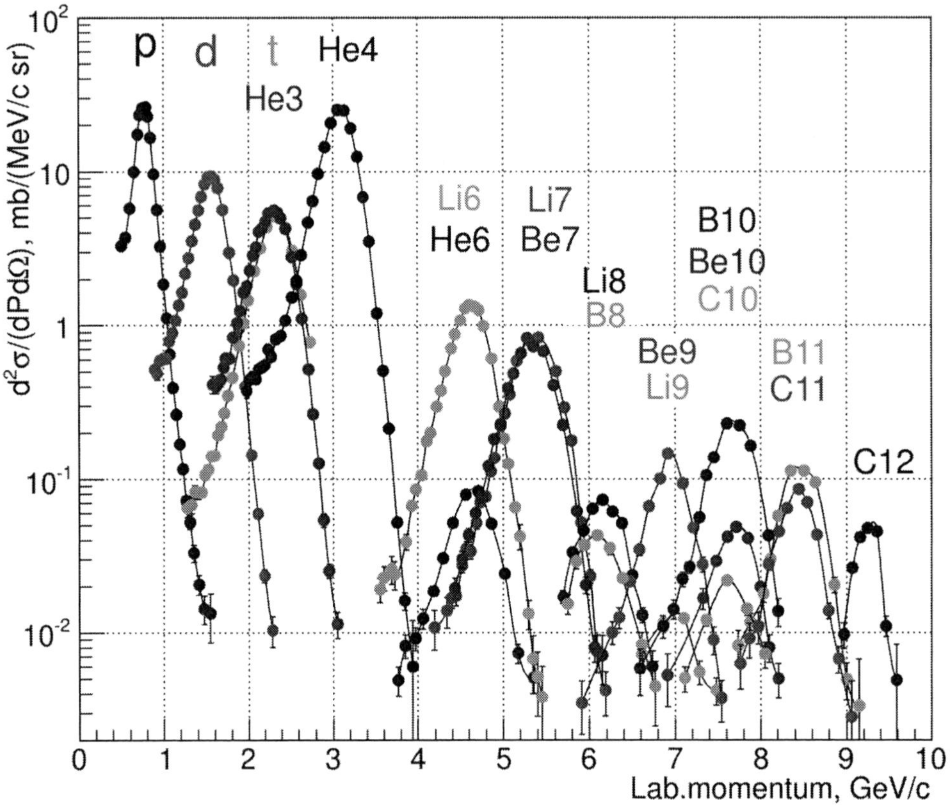

Figure 2. Laboratory momentum spectra of the fragments emitted at 3.5° from the carbon fragmentation at 300 MeV/nucleon. Smooth lines are drawn though the data points just to guide the eye.

publication. In figure 3 and figure 4 our data from figure 2 are compared with predictions of INCL++ and LAQGSM models, respectively, for fragments with the even atomic numbers and protons. The experimental data are given by red and magenta points and the model predictions by blue and light blue points. The agreement between INCL++ and data for protons, deuterons, ^{4}He, ^{6}Li and ^{10}B is perfect. The yields of ^{6}He,^{8}Li and ^{10}Be are overestimated by the model by a factor of 2-3 while the shapes of fragmentation peaks are reproduced well. INCL++ predicts very low cross section for ^{8}B emission and it cannot be seen in the scale of figure 3. The shape and value of the ^{12}C peak are different from the data. But it has to be mentioned here that ^{12}C emission can be connected with other mechanisms different from usually used in the fragmentation processes. The LAQGSM model (see figure 4) reproduces well the shapes of all fragmentation peaks, but positions of these peaks are slightly moved to higher momentum. The model predictions are in good agreement with the data for ^{8}B emission. This model does not give predictions for ^{12}C yield for above mentioned reasons.

4 Conclusion

The comparison performed shows a large potential of modern transport codes to describe the new high precision data on fragmentation in ion-ion interactions. The authors are grateful to operating

DOI: 10.1051/epjconf/201713803002

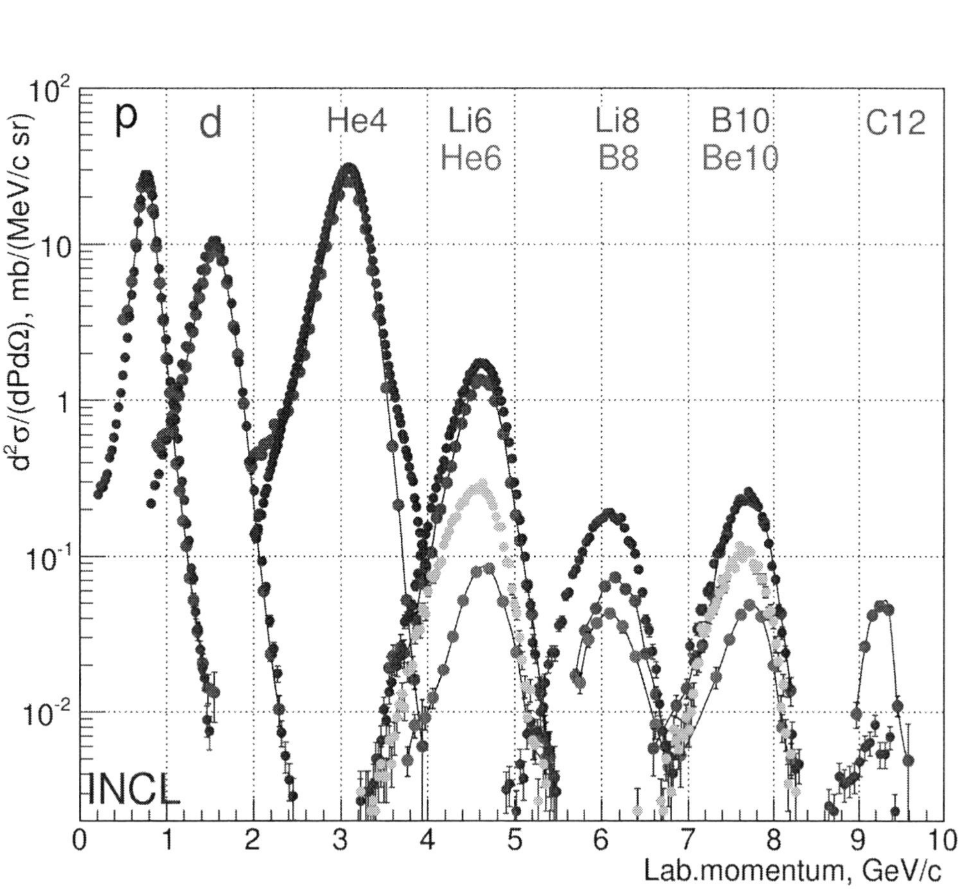

Figure 3. The same experimental data as in figure 2 but in comparison with the predictions of the INCL++ model.

personnel of ITEP TWA and to technicians of the FRAGM experiment for a large contribution in carrying out the measurements. Part of the work, performed at LANL, was carried out under DOE Contract No. DE-AC52-06NA25396. This work has been supported by Russian Foundation for Basic Research (grant RFBR 15-02-06308).

References

[1] B. M. Abramov et al., Yad.Fiz, **79**, 475 (2016). B. M. Abramov et al., Phys.Atom.Nucl. **79**, 700 (2016)

[2] B. M. Abramov et al., Pis'ma JETP **97**, 507 (2013). B. M. Abramov et al., JETP Lett. **97**, 439 (2013)

[3] B. M. Abramov et al., Yad.Fiz, **78**, 403 (2015). B. M. Abramov et al., Phys.Atom.Nucl. **78**, 373 (2015)

[4] G. Folger et al., Europ. Phys. J. A **21**, 407 (2004)

[5] D. Mancusi et al., Phys. Rev. C **90**, 054602 (2014)

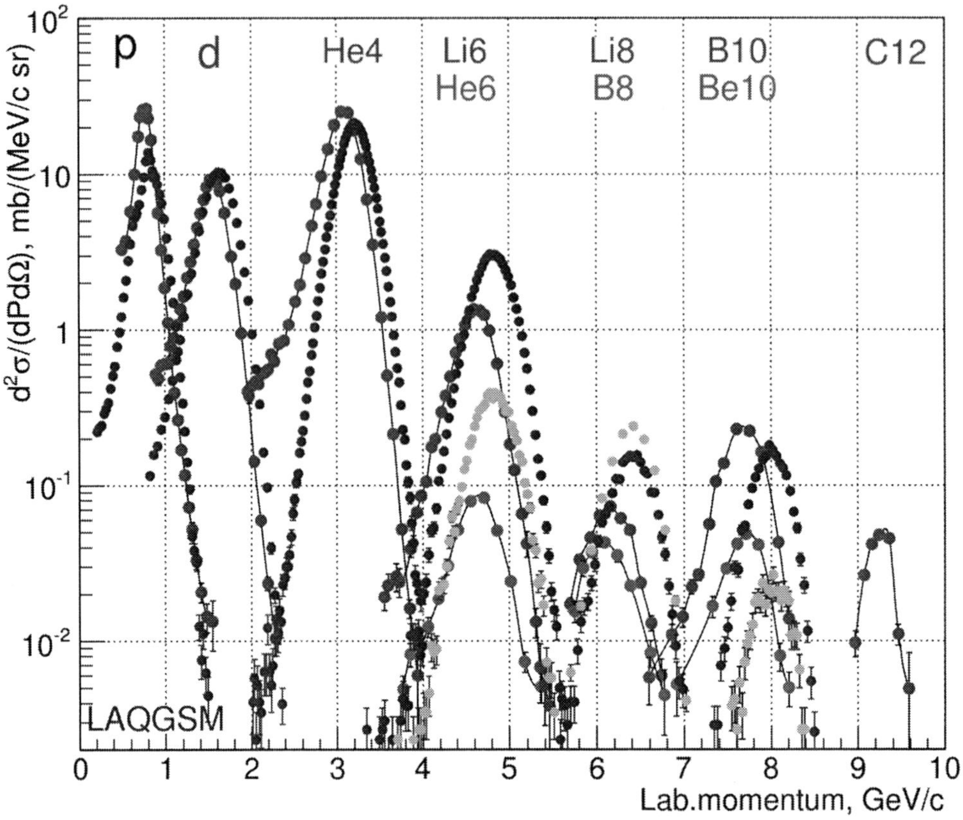

Figure 4. The same experimental data as in figure 2 but in comparison with the predictions of the LAQGSM03.03 model.

[6] T. Koi, http://geant4.cern.ch/results/papers/QMD-MC2010.pdf

[7] S G. Mashnik et al., LANL Report LA-UR-08-2931, Los Alamos (2008), arXiv:0805.0751. S G. Mashnik et al., LANL Report LA-UR-07-6198, Los Alamos (2007), arXiv:0709.1736. N. V. Mokhov, K. K. Gudima and S. I. Striganov, arXiv:1409.1086

[8] T. Goorley et al., Annals of Nuclear Energy **87**, 772 (2016)

Azimuthal anisotropies and initial-state fluctuations from SPS to LHC energies

Jovan Milosevic on behalf of the CERES/NA45 and CMS Collaboration[1,⋆]

[1] *Vinca Institute of Nuclear Sciences, M. Petrovica Alasa 12-14, 11001 Belgrade, Serbia*

Abstract. The v_3 coefficient, obtained using the PbAu data from the CERES detector at the top SPS energy of $\sqrt{s_{NN}}$ = 17.3 GeV, is presented. The v_2 is measured over a p_T range up to 100 GeV/c in PbPb collisions collected with the CMS detector. The $v_2\{2\}$ of charged and strange particles emitted in pp collisions shows a mass ordering effect. The $v_2\{4\}$ and $v_2\{6\}$ are comparable to the $v_2\{2\}$, and thus supports the collective nature of the long-range correlations in high-multiplicity pp collisions at 13 TeV. Principle Component Analysis (PCA) of two-particle harmonics ($V_{n\Delta}$) is studied in PbPb and high-multiplicity pPb collisions at the LHC. The factorization breaking of the $V_{n\Delta}$ can be attributed to the effect of initial-state fluctuations. Using a PCA, the $V_{n\Delta}$ are characterized through the leading and sub-leading modes. The leading modes are essentially equivalent to the $v_n\{2\}$. The sub-leading modes represent the largest sources of factorization breaking.

1 Introduction

One of main features of Quark Gluon Plasma is its hydrodynamic behavior characterized by v_n coefficients extracted using methods like two- and many-particle correlations [1], the scalar product (SP) [2, 3] and the Lee-Yang Zero (LYZ) [4, 5]. The CMS collaboration measured v_2 coefficient in PbPb collisions over a wide p_T range up to 100 GeV/c [6]. This measurement at high-p_T is complementary to the nuclear modification factor (R_{AA}) measurements. The long-range ($|\Delta\eta| > 2$) correlation known as the ridge is observed not only in PbPb collisions [7], but also in high-multiplicity pPb collisions [8] and even in smallest system formed in high-multiplicity pp collisions [9, 10] giving a hint that the ridge could have a hydrodynamic origin. To support it, the correlations among four or more charged particles are studied in pPb collisions at $\sqrt{s_{NN}}$ = 5.02 TeV [11] and in pp collisions at $\sqrt{s}$ = 5, 7 and 13 TeV [12]. The two-particle correlations between strange (K^0_S or $\Lambda/\bar{\Lambda}$) and charged particles [13] also revealed similar ridge structures, and the extracted v_2, scaled to the number of constituent quarks (n_q) shows that hydrodynamic behavior happens on the partonic level. It is shown in [14, 15] that even if the hydrodynamics is the only source of the long-range correlations, the initial-state fluctuations makes the event plane angle, Ψ_n, dependent on both, p_T and η. This leads to factorization breaking of the $V_{n\Delta}$ into a product of single-particle anisotropies v_n [16]. A PCA [17, 18] is applied on $V_{n\Delta}$ coefficients constructed from 2.76 TeV PbPb collisions and 5.02 TeV high-multiplicity pPb collisions [19]. Within the PCA approach, the $V_{n\Delta}$ coefficients are characterized through leading and sub-leading mode terms. The leading modes are essentially equivalent to the v_n harmonics

⋆e-mail: Jovan.Milosevic@cern.ch

extracted from two-particle correlations. The sub-leading modes represent the largest sources of factorization breaking. In the context of hydrodynamic models, they are a direct consequence of the initial-state fluctuations.

2 Experiment and data used

About 30 M of central PbAu collisions at the top SPS energy were collected. CERES detector worked in 0.5 T magnetic filed which enable precise p_T measurement. The η coverage close to mid-rapidity, together with a full azimuthal (ϕ) coverage enables flow study. Negative pions, identified using the differential energy loss dE/dx in the Time Projection Chamber (TPC) are used in this analysis.

The CMS tracker detector is surrounded with a super-conducting solenoid producing 3.8 T magnetic field which enable precise p_T measurements above 0.3 GeV/c [20]. The PbPb and pPb collisions at the LHC energies of $\sqrt{s_{NN}}$ = 2.76 TeV and 5.02 TeV with integrated luminosities of 160 μb^{-1} and 35 nb^{-1}, respectively, were collected. The data sets for pp collisions at $\sqrt{s}$ =5, 7 and 13 TeV have integrated luminosities of 1.0, 6.2 and 0.7 pb^{-1} respectively. A wide pseudorapidity coverage of $|\eta| < 2.5$ for the tracker and $2.9 < |\eta| < 5.2$ for the hadronic forward (HF) calorimeters together with a full azimuthal coverage excellently suites for studying the long-range flow correlations.

3 Results

In figure 1, the comparison between the hydrosolver+UrQMD [21] model predictions and CERES $v_3(p_T)$ measurement of negative pions in PbAu collisions at $\sqrt{s_{NN}}$ = 17.3 GeV [22] is shown. The predictions of this model are in a good agreement with the experimental results, except in the p_T region between 0.3 and 0.6 GeV/c where the model slightly underpredicts the experimental data. The $v_3(p_T)$ at the top SPS energy reach only about one half of the corresponding values at the top RHIC and LHC energies. The p_T integrated v_3 value is in an excellent agreement with the STAR v_3 value measured at 19.6 GeV [23]. Within the analyzed range, the CERES v_3 does not show centrality dependence [22].

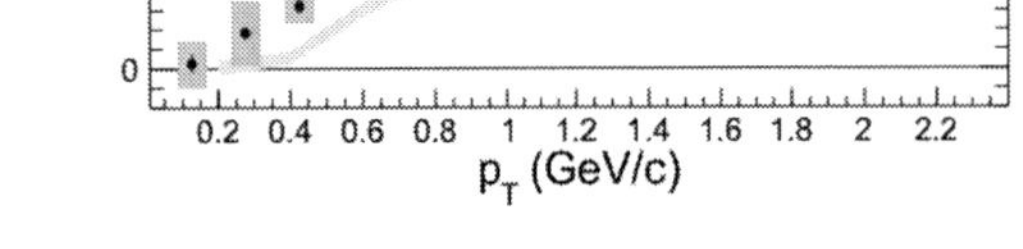

Figure 1. The negative pion $v_3(p_T)$ measured in PbAu collisions at $\sqrt{s_{NN}}$ = 17.3 GeV compared with hydrosolver+UrQMD model predictions [22]. Statistical uncertainties are represented with the error bars, while systematic ones are indicated by gray rectangles. Statistical errors of the model predictions are shown as yellow band.

The CMS showed in [6] that the v_3 is consistent with zero while the positive v_2 persists up to very high-p_T of nearly 100 GeV/c. To further explore the nature of the $v_2(p_T)$ dependence, the v_2 values are also calculated using the 4-, 6- and 8- particle cumulant analyses. The obtained results are shown in figure 2. In the low-p_T regions, the results show the usual hydrodynamic shape with $v_2\{SP\} > v_2\{4\} \approx v_2\{6\} \approx v_2\{8\}$. At high-$p_T$, results tend to converge to a unique distribution. Such observed collectivity at high-p_T is likely to be related to the jet quenching phenomena and the path-length dependence of the parton energy loss.

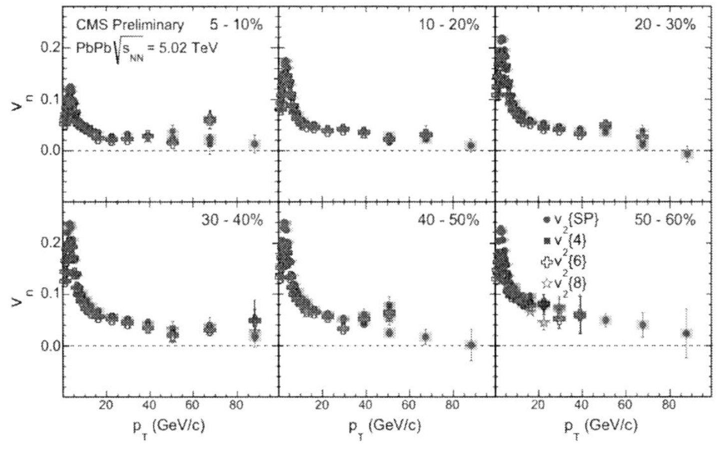

Figure 2. The SP v_2 results compared to those from cumulant method as a function of p_T over wide centrality range in 5.02 TeV PbPb collisions [6]. The systematic uncertainties are denoted with shaded boxes.

The v_2 coefficient in pp collisions at 13 TeV is studied using the two-particle correlation method as well as cumulant of fourth and sixth order [12]. The v_2, as a function of charged particle multiplicity, are shown in the left panel of figure 3. In the high multiplicity region, the results are consistent with each other. Also, the ordering of the pp results obtained from different methods is similar to what was observed in pPb (middle panel) and PbPb (right panel) collisions [11]: the relation $v_2\{2\} \geq v_2\{4\} \approx v_2\{6\}$ in all three colliding systems is an evidence of a collective behavior.

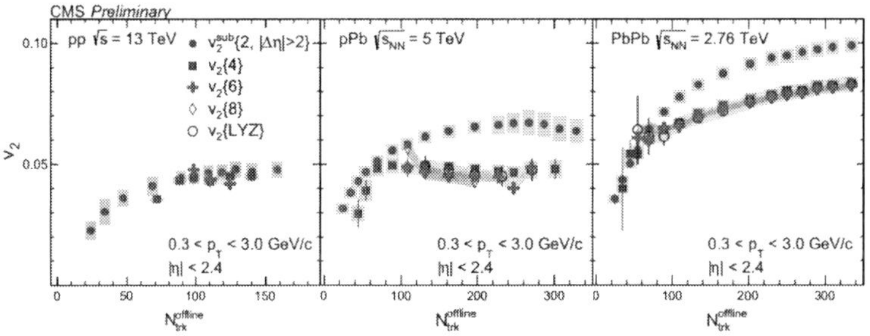

Figure 3. Left: The $v_2\{2\}$, $v_2\{4\}$ and $v_2\{6\}$ *vs* $N_{trk}^{offline}$ for charged particles at 13 TeV pp collisions [12]. Middle (right): The $v_2\{2\}$, $v_2\{4\}$, $v_2\{6\}$, $v_2\{8\}$, and $v_2\{LYZ\}$ at 5 TeV pPb (2.76 TeV PbPb) collisions [11]. The statistical (systematic) uncertainties are denoted with the error bars (shaded areas).

In order to further investigate possible collective behavior in pp collisions, the v_2 coefficients are also extracted from the correlations between the strange, K_S^0 and $\Lambda/\bar{\Lambda}$, and charged hadrons. The upper panel of figure 4 shows the v_2 values of charged hadrons, K_S^0 mesons and $\Lambda/\bar{\Lambda}$ hyperons as a function of p_T. At small-p_T, a mass ordering in the v_2 is seen in pp collisions. In the bottom panel of figure 4 are presented v_2 values scaled to the number of constituent quarks, n_q, plotted *vs* scaled kinetic transverse energy, KE_T/n_q, which shows that similarly as in the case of PbPb and pPb [13] collisions, a collective behavior on partonic level is present also in pp collisions at 13 TeV [12].

Due to the initial-state fluctuations, the factorization of the $V_{n\Delta}$ into a product of single-particle anisotropies does not hold precisely. A new observable, r_n (more details in [14–16]) is introduced. It is defined as $r_n = \frac{V_{n\Delta}(p_T^a, p_T^b)}{\sqrt{V_{n\Delta}(p_T^a, p_T^a)}\sqrt{V_{n\Delta}(p_T^b, p_T^b)}}$ which is approximately equal to the $\cos[n(\Psi_n(p_T^a) - \Psi_n(p_T^b))]$. If the r_n is smaller (equal) than 1 then factorization breaks (holds). The value greater than 1 means that there are unremoved non-flow effects. In figure 5 are shown magnitudes of the factorization breaking effect over a wide centrality/multiplicity range. In the case of the 2nd harmonic, the smallest effect is observed for semi-central collisions. Going to more central or more peripheral collisions the size of

 DOI: 10.1051/epjconf/201713803003

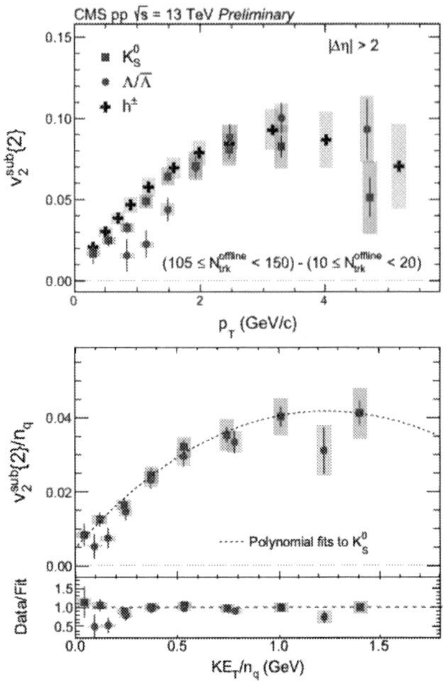

Figure 4. Top: the $v_2(p_T)$ for K_S^0 (blue squares), $\Lambda/\bar{\Lambda}$ (red circles), and charged particles (black crosses) for the multiplicity range $105 \leq N_{trk}^{offline} < 150$ in 13 TeV pp collisions. Middle: the n_q-scaled v_2 values of K_S^0 and $\Lambda/\bar{\Lambda}$ particles as a function of KE_T/n_q. Bottom: Ratios of v_2/n_q to a smooth fit function of v_2/n_q for K_S^0 *vs* KE_T/n_q [12]. The error bars correspond to statistical, while the shaded areas to systematic uncertainties.

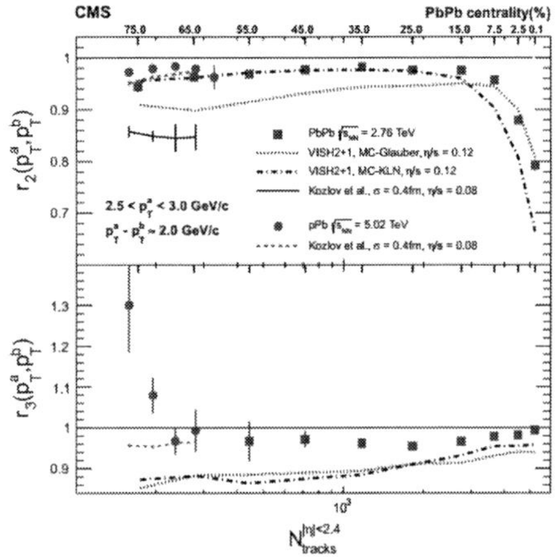

Figure 5. The factorization ratios, r_2 and r_3, as a function of event multiplicity in pPb and PbPb collisions [16]. The curves show few theoretical predictions for PbPb and pPb collisions. The horizontal solid lines denote the r_2 (top) and r_3 (bottom) value of unity. The error bars correspond to statistical uncertainties, while systematic uncertainties are negligible, and thus are not shown.

the effect increases. At the same multiplicity, the size of the effect is rather similar in the two colliding systems. In PbPb collisions, the size of the r_3 is small and nearly independent of centrality, while in pPb collisions the r_3 approaches to 1 and then goes significantly above 1 at lower multiplicities. The p_T-dependent data are qualitatively described by viscous hydrodynamic models with fluctuating initial-state conditions, while they are insensitive to the shear viscosity to entropy density ratio [16]. This promises of using the factorization data to disentangle contributions of the initial-state conditions and the medium's transport properties to the collective flow, and improving modeling of the evolution of the strongly-coupled quark gluon medium.

The PCA results from 5.02 TeV high-multiplicity pPb and 2.76 TeV PbPb collisions [19] are presented in figure 6 and 7 respectively. The extracted leading flow modes are essentially equal to the results [7, 24] obtained from two-particle correlations. A qualitatively new result obtained using the PCA method is the existence of the sub-leading mode, $v_n^{(2)}$, which is a consequence of the initial-state fluctuations. A non-zero $v_2^{(2)}$ is found both in pPb (left panel in figure 6) and in PbPb (upper

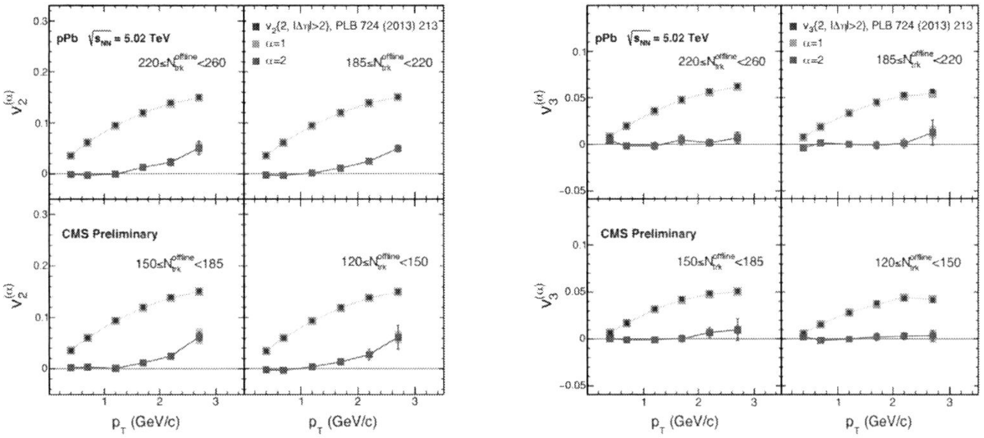

Figure 6. The first and the second mode of $v_2(p_T)$ (left) and $v_3(p_T)$ (right) from 5.02 TeV high-multiplicity pPb collisions [19]. The $v_2^{(1)}$ are compared to the v_n harmonics measured by the CMS collaboration using the two-particle correlation method [7]. The error bars correspond to statistical while shaded areas to systematic uncertainties.

row in figure 7) collisions. It is close to zero at small-p_T and achieves few percent at $p_T \approx 3$ GeV/c. While the strength of the effect does not depend on multiplicity in pPb collisions, there is a centrality dependence of the $v_2^{(2)}$ in PbPb collisions. The similar behavior is established in [16]. The $v_3^{(2)}$ is close to zero in both collision cases (right and bottom panel in figure 6 and 7) which is in an agreement with the results from [16].

4 Summary

In this contribution are presented results on the v_2 and v_3 measurements from the CERES at the top SPS energy, and from the CMS collaboration at the LHC energy. The CMS measured the v_2 up to 100 GeV/c. This result is complementary to the R_{AA} measurements. For the first time, the v_2 has been measured in pp collision at 13 TeV. This, together with the established n_q-scaling shows that collective behavior appears even in such small system as created in pp collision. Also, for the first time sub-leading flow modes, $v_n^{(2)}$, are measured in PbPb and high-multiplicity pPb collisions. The obtained results are in a qualitative agreement with factorization-breaking results.

References

[1] N. Borghini, P. M. Dinh, and J.-Y. Ollitrault, Phys. Rev C **64**, 054901 (2001)
[2] C. Adler et al., STAR Collaboration, Phys. Rev. C **66**, 034904 (2002)
[3] M. Luzum and J.-Y. Ollitrault, Phys. Rev. C **87**, 044907 (2013)
[4] R. S. Bhalerao, N. Borghini, and J.-Y. Ollitrault, Nucl. Phys. A **727**, 373 (2003)
[5] N. Borghini, R. S. Bhalerao, and J.-Y. Ollitrault, J. Phys. G **30**, S1213 (2004)
[6] CMS Collaboration, CMS-HIN-15-014 (2015)
[7] CMS Collaboration, Phys. Lett. B **724**, 213 (2013)

 DOI: 10.1051/epjconf/201713803003

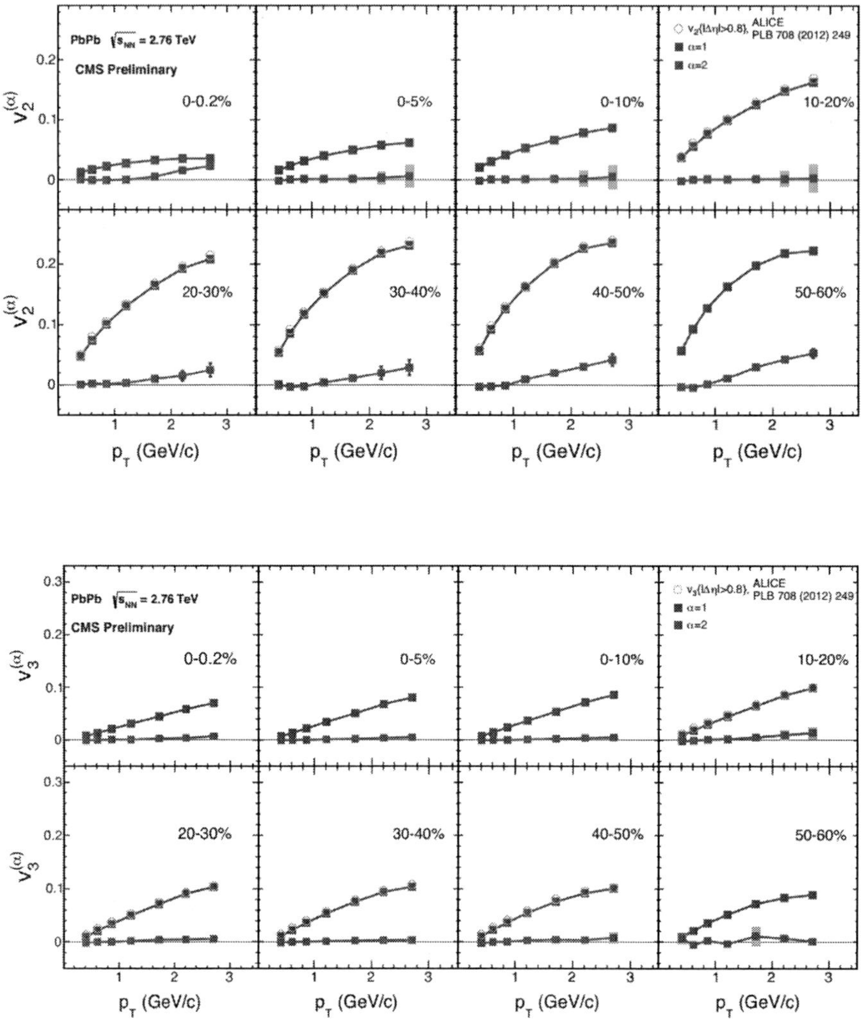

Figure 7. The first and the second mode of $v_2(p_T)$ (upper row) and $v_3(p_T)$ (bottom raw) from 2.76 TeV PbPb collisions [19]. The $v_2^{(1)}$ are compared to the v_n harmonics measured by the ALICE collaboration using the two-particle correlation method [24]. The error bars correspond to statistical while shaded areas to systematic uncertainties.

[8] CMS Collaboration, Phys. Lett. B **718**, 795 (2013)
[9] CMS Collaboration, JHEP **09**, 091 (2010)
[10] CMS Collaboration, Phys. Rev. Lett. **116**, 172301 (2016)
[11] CMS Collaboration, Phys. Rev. Lett. **115**, 012301 (2015)
[12] CMS Collaboration, arXiv:1606.06198, CMS-HIN-16-010 (2016)
[13] CMS Collaboration, Phys. Lett. B **742**, 200 (2015)
[14] F. G. Gardim et al., Phys. Rev. C **87**, 031901 (2013)
[15] U. Heinz et al., Phys. Rev. C **87**, 034913 (2013)
[16] CMS Collaboration, Phys. Rev. C **92**, 034911 (2015)

[17] R. S. Bhalerao, J.-Y. Ollitrault, S. Pal, and D. Teaney, Phys. Rev. Lett. **114**, 152301 (2015)
[18] A. Mazeliauskas, and D. Teaney, Phys. Rev. C **91**, 044902 (2015)
[19] CMS Collaboration, arXiv:1510.03373, CMS-HIN-15-010 (2015)
[20] CMS Collaboration, JINST **3**, S08004 (2008)
[21] I.A. Karpenko, P. Huovinen, H. Petersen, and M. Bleicher, Phys. Rev. C **91**, 064901 (2015)
[22] D. Adamova et al., CERES NA45 Collaboration, Nucl. Phys. A **957**, 99 (2017)
[23] L. Adamczyk et al., STAR Collaboration, Phys. Rev. Lett. **116**, 112302 (2016)
[24] ALICE Collaboration, Phys. Lett. B **708**, 249 (2012)

EPJ Web of Conferences **138**, 03004 (2017) DOI: 10.1051/epjconf/201713803004
Baldin ISHEPP XXIII

Study of clusters and hypernuclei production within PHSD+FRIGA model

Viktar Kireyeu[1,*], *Arnaud* Le Fèvre[2], and *Elena* Bratkovskaya[3]

[1] *Joint Institute for Nuclear Research, Dubna, Russia*
[2] *GSI Helmholtzzentrum für Schwerionenforschung, Darmstadt, Germany*
[3] *FIAS and ITP Frankfurt, Germany*

Abstract. We report on the results on the dynamical modelling of cluster formation with the new combined PHSD+FRIGA model at Nuclotron and NICA energies. The FRIGA clusterization algorithm, which can be applied to the transport models, is based on the simulated annealing technique to obtain the most bound configuration of fragments and nucleons. The PHSD+FRIGA model is able to predict isotope yields as well as hypernucleus production. Based on present predictions of the combined model we study the possibility to detect such clusters and hypernuclei in the BM@N and MPD/NICA detectors.

1 Introduction

Heavy-ion collisions provide the unique possibility to create and investigate hot and dense matter in the laboratory. At the initial stage of the reaction a QGP is formed, while the final stage is driven by the hadronization process and the formation of clusters. The capture of the produced hyperons by clusters of nucleons leads to the hypernuclei formation which is a very rare process at strangeness threshold energies. It is important to have the robust modelling of hypernuclei and cluster formation in order to study the detector replica and to have the possibility to optimize the experimental setup for the best efficiency.

Modelling the clusters formation is a complicated problem, and therefore in many approaches the fragment formation is simply omitted. This invalidates the prediction of single particle observables, because the cluster formation can modify the single particle spectra.

The simplest way to identify clusters is by employing coalescence or a minimum spanning tree procedure. The first needs a multitude of free parameters, whereas the second allows only for an identification at the end of the reaction which excludes any study on the physical origin [1].

*e-mail: vkireyeu@jinr.ru

 DOI: 10.1051/epjconf/201713803004

2 PHSD+FRIGA MODEL

The Parton-Hadron-String Dynamics (PHSD) [2] is a microscopic off-shell transport approach that describes the full evolution of a heavy-ion collision at relativistic energies from the initial hard scatterings and string formation through the dynamical deconfinement phase transition to the quark-gluon plasma as well as hadronization and to the subsequent interactions in the hadronic phase.

The Simulated Annealing Clusterization Algoritm (SACA) [3] and its new development and improved version "Fragment Recognition In General Application"; (FRIGA) [4] consists of the following steps: at first the algorithm takes the positions and momenta of all nucleons at time t to determine clusters within a phase space coalescence approach using the Minimum Spanning Tree technique (MST). In second step, the MST clusters and individual particles are recombined in all possible ways into fragments or left as single nucleons, such as to choose that configuration which has the highest binding energy [5]. This procedure is repeated many times (within a Metropolis procedure) and it automatically leads to the most bound configuration.

Clusters chosen that way at early times are the pre-fragments of the final state clusters, because fragments are not a random collection of nucleons at the end but the initial-final state correlations.

For our studies we use PHSD and FRIGA as one combined model: PHSD produces hadrons and stops at a certain time, which are input to the FRIGA code to determine clusters and hypernuclei. The time chosen for clustering can have a strong influence on the multiplicity of clusters.

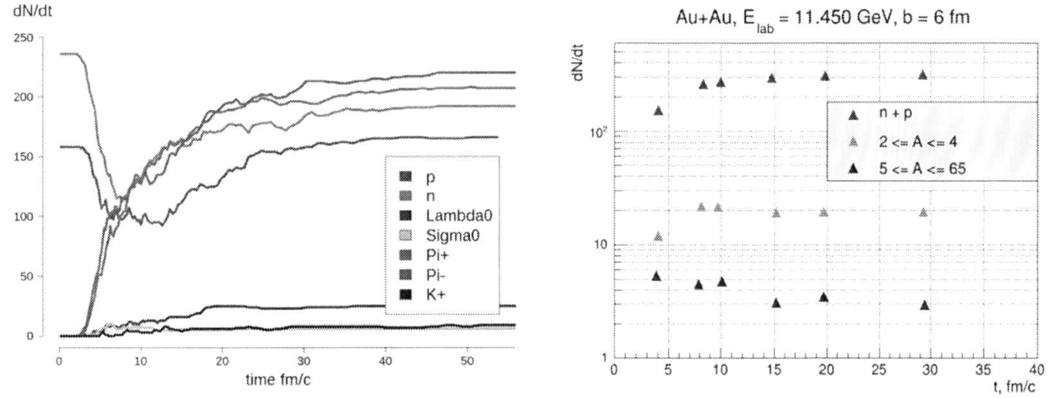

Figure 1. Time dependence of hadron (left) and fragment (right) multiplicities as predicted by the PHSD+FRIGA approach in Au+Au collisions at 11.45A GeV incident energy, 6 fm impact parameter.

Figure 1 illustrates such a time dependence in Au+Au semi-central (b=6 fm) collisions at 11.45A GeV incident energy ($\sqrt{s}$ = 5 GeV). There, the approach predicts a stabilization in the hadron and fragment yields around 15 fm/c. This earlier clusterization time as been chosen in the following for studying various fragments observables.

Figure 2 shows a comparison of PHSD+SACA predictions with AGS E-802 data [6] for central Au+Au collisions at 11.45 GeV, for the rapidity dependence of proton (left panel) and deuteron (right panel) yields. We observe there that the model reproduces well the experimental data.

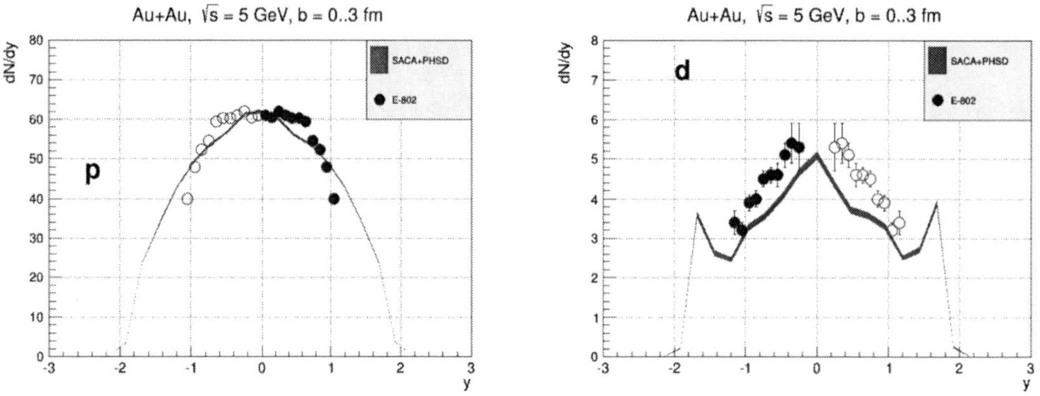

Figure 2. Yields of protons (left panel) and deuterons (right panel) as a function of rapidity (in the frame of the centre of the system) in central Au+Au collisions ($b < 3fm$) at 11.45A GeV. Red lines and black symbols show respectively the PHSD+FRIGA predictions and E-802 experimental mirrored about midrapidity (y=0) data.

3 Multi-Purpose Detector feasibility study

A new scientific program on heavy-ion physics NICA/MPD [7] has been launched at JINR (Dubna) for comprehensive exploration of the QCD phase diagram with ion species ranging from protons to Au^{79+} over the energy range $4 < \sqrt{s} < 11$ GeV. The major goal of the NICA/MPD project is the study of in-medium properties of hadrons and the nuclear matter equation of state, including a search for possible signals of deconfinement and/or chiral symmetry restoration, phase transitions and the QCD critical endpoint.

One of the tasks of the Multi-Purpose Detector [8] is to study the strangeness production. This task demands a good identification and reconstruction of heavy strange objects like hypernuclei and hyperons.

Figure 3 left panel shows the yields of hypernuclei and hyperons predicted by the PHSD+FRIGA approach in Au+Au collisions at 11.45A GeV. Preliminary results of a MPD feasibility study for hypertriton detection by the mean of reconstructed invariant mass distributions are shown on the right panel. There, the hypertriton signal is clearly viewed, with a good signal-to-background ratio of 3.3, and a detection efficiency of about 0.4%.

4 Conclusion

The PHSD+FRIGA model can produce clusters and hypernuclei and reproduce experimental data for 11.45 GeV beam energy ($\sqrt{s}$ = 5 GeV). These predictions have been already used for MPD performance studies. The new improved version of the clusterization algoritm FRIGA is still actively under development. It will be possible in a near future to obtain the fully dynamical cluster formation on each time step of the collisions.

DOI: 10.1051/epjconf/201713803004

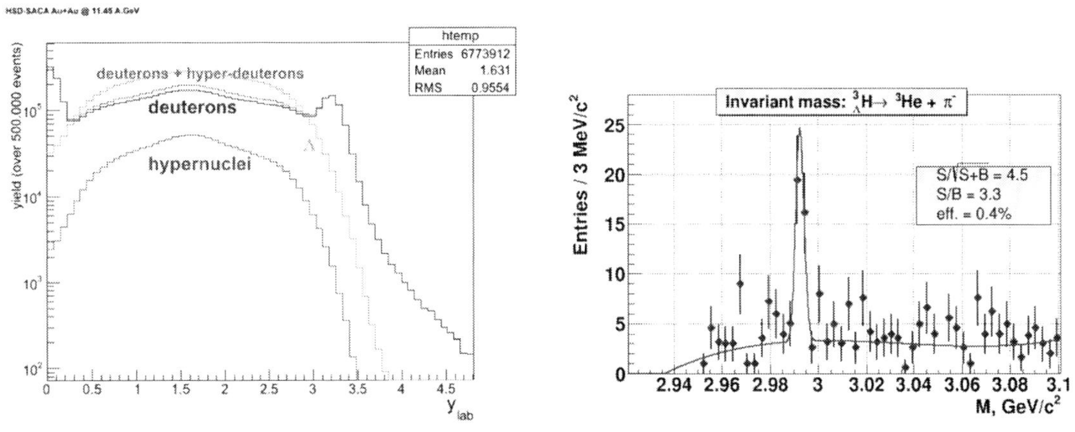

Figure 3. In Au+Au collisions at 11.45 GeV incident energy ($b < 3fm$): (left) rapidity (in the rest frame) dependence of yields of hypernuclei, deuterons and Lambda0 hyperons as predicted by PHSD+FRIGA; (right) invariant mass distribution of He3 and π^- vertexes as theoretically measured by the MPD set-up (preliminary results).

References

[1] P.B. Gossiaux, R. Puri, Ch. Hartnack, J. Aichelin, Nucl. Phys. A **619**, 379-390 (1997)
[2] E.L. Bratkovskaya, W. Cassing, Nucl.Phys. A **856**, 162-182 (2011)
[3] P.B. Gossiaux, R. Puri, Ch. Hartnack, J.Aichelin, Nuclear Physics A **619**, 379-390 (1997)
[4] A. Le Févre et al., J. Phys.: Conf. Ser. **668**, 012021 (2016)
[5] Dorso et al., Phys.Lett. B **301**, 328 (1993)
[6] L. Ahle et al., Phys. Rev. C **60**, 064901 (1999)
[7] A.S. Sorin et al., Nuclear Physics A **855**, 510-513 (2011)
[8] Kh.U. Abraamyan et al. (MPD Collaboration), NIM A **628**, 99-102 (2011)

 DOI: 10.1051/epjconf/201713803005

Hadron rapidity spectra within a hybrid model

A. S. Khvorostukhin[1,2,⋆] and *V. D.* Toneev[1,⋆⋆]

[1] *Joint Institute for Nuclear Research, 141980 Dubna, Russia*
[2] *Institute of Applied Physics, Moldova Academy of Science, MD-2028 Kishineu, Moldova*

Abstract. A multistage hybrid model is constructed what joins the initial non-equilibrium stage of interaction, described by the hadron string dynamics (HSD) model, to subsequent evolution of the expanding system treated within ideal hydrodynamics (the second stage). Particles can still rescatter after hydrodynamical expansion that is the third interaction stage. The developed hybrid model is assigned to describe heavy-ion collisions in the energy range of the NICA collider. Generally, the model is in reasonable agreement with the available data on proton rapidity spectra.

1 Introduction

Application of hydrodynamics to high-energy nuclear collisions has a long and vivid history which began almost 65 years ago with Landau's original work [1]. In this history a lot of papers have been written covering a wide spectrum of various and important problems of hydrodynamics. Modern phenomenological status of hydrodynamics is well reflected in recent review-articles (for example, see [2–4] and reference therein).

Hydrodynamics is a collective model of nuclear motion characterized by such physics parameters as temperature, pressure, equation of state (EoS), transport coefficients and so on. Unfortunately, the direct access to this information is impossible since the only available experimental information is contained in observable spectra of particles. Hydrodynamics gives the connection between observables and thermodynamic properties of excited nuclear matter. One has to keep in mind that hydrodynamics is applicable when the mean free path of a quasiparticle is smaller than the size of the system. In nuclear collisions this condition may be violated in rarefied matter at the initial stage of interaction, in peripheral collisions of nuclei and at the end of a hydrodynamical expansion. In this respect, the initial state in the hydrodynamic approach, i.e. space distributions of the energy density, charge density and velocity field, is postulated or calculated within other dynamical models. The subsequent second interaction stage is actually hydrodynamical expansion of dense matter with its EoS which can contain a phase transition (PT) of hadrons into a quark phase. Our model can also include post-hydrodynamic rescattering which is the final collision stage.

The ultimate aim of this work is the development of a multistage hybrid hydrokinetic approach to heavy-ion collisions in the range of moderate collision energies $\sqrt{s} \lesssim 10$ GeV, which are planned to be reached at the heavy-ion collider NICA (Dubna) [5] and heavy-ion accelerator FAIR (Darmstadt) [6].

⋆e-mail: hvorost@theor.jinr.ru
⋆⋆e-mail: toneev@theor.jinr.ru

The ideology of our model completely coincides with the hybrid hydro+UrQMD approach which has been successfully developed in recent years. The theoretical difficulty lies in the fact that one should construct a model describing available data in a unified way in the whole energy range considered and having a predictive power. So far the only hydrodynamic model, which well describes a variety of experimental data in this energy range is the three-fluid hydro model [7].

2 Hybrid model

2.1 Kinetic stage – hadron string dynamics model (HSD) [8]

Since the hydrodynamic equations are equations in partial derivatives, one needs to define initial conditions. In hybrid models these conditions are obtained from calculations in a kinetic model. In such way one takes into account the nonequilibrium evolution of the system at the beginning of the collision. As a model we choose the hadron string dynamics model [8], which describes a large variety of experimental data in the energy range of interest $E_{\rm lab} = 2 - 50$ AGeV.

In this paper, as a basis we consider both the AGS ($E_{\rm lab}$ = 6 and 10.7 AGeV), and SPS energy ($E_{\rm lab}$ = 40 AGeV, where $E_{\rm lab}$ is the kinetic energy of the bombarding nucleus in the laboratory system). To get smooth distributions in the density of energy and a baryon number in the initial state, the averaging over $5 \cdot 10^4$ events is used.

The transition from the kinetic description to the hydrodynamic one occurs at some time moment $t_{\rm start}$. It is assumed that at this time the excited system is close to an equilibrium state that may be characterized by conserving quantities, entropy or the ratio of entropy to the baryon charge [9]. As a criterium of the hydrodynamic stage start, we take the constancy of the ratio $S(t)/N_B(t)$, see [10] for details. Such approach to determination of the transition time to hydrodynamics allows one to easily take into account the effect of nuclear opacity. We also have to mention that only particles that have suffered interactions are taken into account to construct the initial state.

2.2 Hydrodynamic stage

The hydrodynamic equations express simply the conservation laws for the energy, the momentum and the baryon charge:

$$\partial_\mu T^{\mu\nu} = 0, \quad \partial_\mu J^\mu = 0, \tag{1}$$

The ideal hydrodynamics assumes that matter is in local equilibrium. Then the energy-momentum tensor $T^{\mu\nu}$ and the vector of the baryon current J^μ are

$$T^{\mu\nu} = (\varepsilon + P)\, u^\mu u^\nu - P g^{\mu\nu}, \quad J^\mu = n\, u^\mu. \tag{2}$$

Here $u^\mu = \gamma(1, \mathbf{v})$ is the 4-velocity of liquid, $\mathbf{v}$ is the 3-velocity, the Lorenz-factor is $\gamma = (1 - v^2)^{-1/2}$, quantities ε, n, P are the energy density, baryon density and pressure, and $g^{\mu\nu} = \mathrm{diag}(1, -1, -1, -1)$ is the metric tensor. Equations (1) should be completed by EoS $P = P(\varepsilon, n)$, then the system becomes closed.

Hydro equations (1) are solved by means of the SHASTA (the SHarp and Smooth Transport Algorithm) algorithm [11]. The details of our code can be found in [10]. We use the EoS for a hadron gas in the mean field proposed in [12] with included σ-meson.

2.3 Evaluation of observable: particlization procedure

A special task is the calculation of observables: rapidity distributions, transverse momentum spectra and azimuthal flows of particles. We use the approximation of "instantaneous freeze-out" which is frequently used in hydrodynamic models and assumes that on some space-time hypersurface there occurs an instantaneous transition from local equilibrium to collisionless particle expansion. In most cases it is postulated that the existence of some freezed-out cells do not essentially influence the dynamics of other parts of the system (as an example, see further isothermal and isoenergetic freeze-out). However, there are models where such influence is accounted, e.g., see [7].

The calculation can be finished when all cells are frozen. This is the short two-stage version of our model. However models of such kind neglect the third (again nonequilibrium) stage of nuclear collision which consists in possible particle rescatterings after fireball expansion. We construct also the full hybrid model including all three stages. But the main attention is paid to the short model since it gives better results.

To define an integration hypersurface on which the "instantaneous" transition from a liquid fluid to freely expanding particles occurs, we use the CORNELIUS algorithm [13], whose realization is available freely. We consider different freeze-out scenarios: 1) isochronous, when the calculation is completed at the given time moment; 2) isothermal (iso-T), when the cell is frozen if its temperature is less or equal the freeze-out temperature $T \leq T_{\rm frz}$, and 3) isoenergetic (iso-ε), which is entirely analogous to the isothermal one where the energy density plays the role of temperature. In the last two versions, the frozen cells are not excluded from the calculation after writing up them into a file and may influence the numerical solution, but this effect is nonessential.

The straightforward method to calculate observables is an application of the famous Cooper-Frye formulae [14]. Besides the "thermal" contribution calculated by it, resonance decays also have to be taken into account. However, below we shall use another approach allowing to speed up appreciably the numerical calculations. If the hypersurface is known, one may make the inverse transition from the fluid set to quasiparticles using the Monte-Carlo method – so called particlization. According to our algorithm, the number of particles in each cell is calculated following [15], while the 4-momentum is sampled according to [16]. The contribution of space-like cells is ignored.

To test our generator, the simulated distributions are compared to direct calculations of spectra, according to the Cooper-Frye formulae. We see a good agreement of the directly calculated spectra with the particlization results that allows us to apply the particlization procedure to calculation of observables.

Since our hydrodynamic equations do not include a separate equation for an electric charge current conservation, we use the symmetric EoS. Therefore, we need to estimate the particle fraction with the given electric charge among particles with the same mass (for example, the ratio of protons to the total number of nucleons). For the isochronous scenario, multiplying a number of nucleons by the isotopic factor Z/A we get dN/dy at $y = 0$ for protons essentially larger (closer to experiment) than for two others freeze-outs. So, an additional isotopic factor depends also on the freeze-out scenario. Thus, below we will compare the averaged experimental proton multiplicity with the calculated $(n_p + n_n)/2$ for the iso-ε and iso-T scenarios.

As to the pion yield, in all cases we take the isospin average $(n_{\pi^+} + n_{\pi^-} + n_{\pi^0})/3$. Our calculations show that such averaging is closer to the yield of n_{π^+}. One should note that the charge asymmetry is observed clearly in nuclear experiments in the NICA energy range. At energies of interest the ratio n_{π^-}/n_{π^+} exceeds remarkable 1. However, any version of the hadronic transport model HSD overestimates the pion multiplicity in the NICA energy range (see for example [17]). This overestimation of pions results in a too low kaon-to-pion ratio K^+/π^+. This problem is actively discussed in the last years and frequently associated with the signal of a possible quark-hadron phase transition.

In our analysis of central nucleus-nucleus collisions, by default, we use the impact parameter $b = 1$ fm at all energies under discussion since, as a rule, it gives results close to experiments.

3 Properties and results

3.1 The two-stage version

It is evident that the assumption on isochronous freeze-out is not realistic and can be used only for test calculations or as an intermediate stage. In realistic models where observable quantities are calculated on a "frozen" hypersurface, a constant temperature or the energy density scenarios are used. To construct a model having predictive power, one needs to choose a scenario and a method for calculating its parameters depending on collision energy.

Our calculations show that to reproduce the two-hump structure at $E_{\rm lab} = 40$ AGeV, one needs to take rather large values of the parameters $\varepsilon_{\rm frz}$ or $T_{\rm frz}$. It is interesting that in contrast with the kinetic HSD model, the hybrid model has no problem with the description of proton spectra at large rapidity (see figure 1).

The results [18] obtained in the model with isochronous freeze-out demonstrate the point that to get a two-hump structure, the two-phase EoS is necessary while our model shows that the choice of the initial state and method/parameter of freeze-out plays not less important role. Such ambiguity between the choice of a proper initial state or EoS was also noted in [19]. Since the model [18] with the two-phase EoS predicts the two-hump structure at $E_{\rm lab} = 10.7$ AGeV as well, one may conclude that the method used there for constructing the initial state is too simplified.

In our hybrid model, the two-hump structure arises also if we take later time of transition to hydrodynamics. This confirms the conclusion about an important role of the initial state.

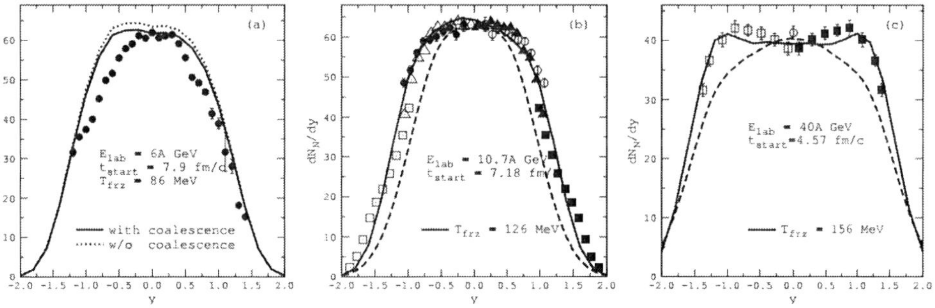

Figure 1. Ultimate proton spectra within the isothermal scenario at energies (*a*) $E_{\rm lab} = 6$ AGeV, (*b*) $E_{\rm lab} = 10.7$ AGeV and (*c*) $E_{\rm lab} = 40$ AGeV. The solid lines are our results, the ones dashed are the results with isochronous freeze-out from [18] (*b, c*). Experimental points for $E_{\rm lab} = 6$ AGeV are taken from [20], for $E_{\rm lab} = 10.7$ AGeV – [21] (triangles), [22] (squares) and [23] (circles), and for $E_{\rm lab} = 40$ AGeV – [24] (squares) and [25] (the circle).

To get the best description of the experiment, our model needs the freeze-out parameter depending on the collision energy. This dependence for freeze-out temperature $T_{\rm frz}$ can be obtained within the phenomenological statistical model of hadron and resonance production [26]. Thus, the solution is suggested to use as input data temperatures found from the data analysis within the statistical model. For all considered energies $E_{\rm lab} = 6, 10.7$, and 40 AGeV, a good agreement with the experiment for protons is achieved if we take respectively the values of $T_{\rm frz} = 86^{+13}_{-3}$, 124±7, and 156±11 MeV [26] (see figure 1).

The pion rapidity distribution height in the rapidity range $y \approx 0$ depends on both the freeze-out parameter and $t_{\rm start}$. However, we did not succeed in reproducing experimental pion spectra in either the isothermal or isoenergetic scenarios. The same behaviour is observed within the hybrid version of the UrQMD model with nonviscous hydrodynamics [27]. Our model differs from it only by the point that there the initial state is taken from the UrQMD model and hydrodynamics additionally includes the electric charge conservation. Hence the results should be quite similar. The hydro-UrQMD results [27] are also lower than experimental data for pion rapidity distributions.

As has been noted for the first time in [28] and clearly demonstrated in [27], the lack of pions is due to the absence in our model of dissipative effects which increase the entropy. In addition, one should remember that the SPS data are given for π^- pions, while the isospin averaged calculation results are closer to the number of less abundant π^+ pions.

Thus, our hybrid model has some difficulties with a simultaneous description of the AGS and SPS energy range. The model with isochronous freeze-out considered above [18] also has some problems in attempt to consider these both energies simultaneously because nobody succeeded in describing the shape of proton spectra within a single EoS.

3.2 The full version of the hybrid model

As is obvious, during the expansion at some time moment the system size becomes larger than the mean free path of particles and hydrodynamics breaks down. So the most justified approach has to include the posthydrodynamic rescattering. We realized this stage in our model and here we shortly mention main features of such full version of our code.

As a parameter of the back transition from fluids to particles, we choose the energy density in spirit of [27]. The particlization hypersurface is isochronous when the energy density of every cell becomes less the switching one.

Our results show that taking into account post-hydrodynamic rescattering decreases distributions both for protons and pions. The final rapidity spectra are worse than in the 2-stage version. The main reason for this is the isochronous transition since we obtain two-hump structure in proton distributions for $E_{\rm lab} = 10.7$ AGeV as in [18]. So we need more realistic procedure for switching to the kinetic description after the hydrodynamic expansion.

4 Conclusions

We propose the many-stage hybrid model to describe heavy-ion collisions in the energy range reachable at the heavy-ion collider NICA which is under construction in Dubna. The model can be realized in short and full versions.

The 2-stage model allows one to describe quantitatively the proton spectra within the isothermal freeze-out scenario if the temperatures $T_{\rm frz}$ are taken from the results of the statistical model processing the measured hadron yield. It is shown that within the hybrid model the parameters of the two-hump structure in the proton spectra may be obtained by either increasing the freeze-out temperature or energy density parameters or by transition to the hydrodynamical stage at a later time. The ideal hydrodynamics with the initial state calculated on the basis of the kinetic HSD model is not able to describe pion rapidity spectra simultaneously with those for protons. It is necessary to take into account the hadron matter viscosity at the hydrodynamic stage of the system evolution.

The importance of including dissipative effects in someway is demonstrated by all successful hydrodynamical models like 3-fluid hydro [7] or hydro+UrQMD [27].

Acknowledgements: We are grateful to D. Rischke and B. Betz for providing us the SHASTA code and E. Bratkovskaya for the HSD code. We are thankful to V. Voronyuk, Yu. Ivanov, Iu. Karpenko, A. Merdeev, L. Satarov and G. Sandukovskaya for useful discussions and constructive remarks.

References

[1] L. Landau, Izv. Akad. Nauk Ser.Fiz. **17**, 5164 (1953)
[2] U. W. Heinz and R. Snellings, Annu. Rev. Nucl. Part. Sci. **63**, 123 (2013)
[3] S. Jeon and U. Heinz, arXiv: 1503.03931 (2015)
[4] R. Derradi de Souza, T. Koide and T. Kodama, Prog. Part. Nucl. Phys. **86**, 35 (2016)
[5] A. N. Sissakian, A. S. Sorin and V. D. Toneev, Cong. Proc. C **060726**, 421 (2006)
[6] B. Friman et al., Lect. Notes Phys. **814**, 1 (2011)
[7] Yu. B. Ivanov, V. N. Russkikh, and V. D. Toneev, Phys. Rev. C **73**, 044904 (2006)
[8] W. Cassing and E. L. Bratkovskaya, Phys. Rept. **308**, 65 (1999)
[9] V. V. Skokov and V. D. Toneev, Yad. Phys. **70**, 114 (2007). [Phys. Atom. Nucl. **70**, 109 (2007)]
[10] A. S. Khvorostukhin and V. D. Toneev, arXiv: 1606.00987 [nucl-th]
[11] D. H. Rischke, S. Bernard, and J. A.Maruhn, Nucl. Phys. A **595**, 346 (1995)
[12] L. M. Satarov, M. N. Dmitriev, and I. N. Mishustin, Phys. Atom. Nucl. **72**, 1390 (2009)
[13] P. Huovinen and H. Petersen, Eur. Phys. J. A **48**, 171 (2012)
[14] F. Cooper and G. Frye, Phys. Rev. D **10**, 186 (1974)
[15] H. Petersen et al., Phys. Rev. C **78**, 044901 (2008)
[16] N. S. Amelin et al., Phys. Rev. C **74**, 064901 (2006)
[17] W. Cassing et al., arXiv: 1510.04120 (2015)
[18] A. V. Merdeev, *Hydrodynamic modelling of the quark-hadron phase transition, PhD thesisis* (Moscow, 2011, in russian). L. M. Satarov, private communication
[19] J. Sollfrank et al., Phys. Rev. C **55**, 392 (1997)
[20] J. L. Klay et al., Phys. Rev. Lett. **88**, 102301 (2002)
[21] L. Ahle et al., Phys. Rev. C **57**, R466 (1988)
[22] J. Barrette et al., Phys. Rev. C **62**, 024901 (2000)
[23] B. B. Back et al., Phys. Rev. Lett. **86**, 1970 (2001)
[24] T. Anticic et al., Phys. Rev. C **83**, 014901 (2011)
[25] T. Anticic et al., Phys. Rev. C **69**, 024902 (2004)
[26] A. Andronic, P. Braun-Munzinger, and J. Stachel, Nucl. Phys. A **772**, 167 (2006)
[27] Iu. A. Karpenko et al., Phys. Rev. C **91**, 064901 (2015)
[28] T. Hirano, J. Phys. G: Nucl. Part. Phys. **30**, S845 (2004)

EPJ Web of Conferences **138**, 03006 (2017) DOI: 10.1051/epjconf/201713803006
Baldin ISHEPP XXIII

Digamma diagnostics for the mixed-phase generation at NICA

V.I. Kukulin[1,⋆] and *M.N.* Platonova[1]

[1] *Skobeltsyn Institute of Nuclear Physics, Lomonosov Moscow State University, Moscow 119991, Russia*

Abstract. A novel type of diagnostics for dense and/or hot nuclear matter produced in heavy-ion collisions at NICA and similar future colliders (FAIR, etc.) is suggested. The diagnostics is based on an assumption (confirmed in many experiments worldwide) about intensive generation of light scalar mesons (σ) the consequent decay of which produces $\gamma\gamma$ pairs with the mass and width dependent upon density and temperature of the fireball produced in the collision process. Thus, measurements of the absolute yield, mass and width of the $\gamma\gamma$ signal carry valuable information about the state of fireball generated during the high-energy nuclear collision.

1 Introduction. Chiral symmetry restoration in hot and/or dense nuclear matter

One of the main goals of the future heavy-ion collider NICA in Dubna will be the search for signals of the phase transition between hadronic matter and quark-gluon plasma and search for new phases of baryonic matter, including the mixed phase. However, up to date one still has no full understanding for the state of dense and hot hadronic matter which is born in heavy-ion collisions, its phase and temperature distribution, etc. There is a key hypothesis about partial chiral symmetry restoration (CSR) in hot and/or dense nuclear matter inside the fireball, which is a consequence of basic principles of QCD. The CSR effect can be observed in two areas: (i) in a single hadron when it gets to be highly excited [1, 2] and (ii) in hot and/or dense nuclear matter, i.e., in medium [3–6]. In the latter case CSR occurs due to renormalization of hadron and σ-meson propagators in nuclear medium.

However the visible manifestation of both mechanisms seems to be the same: reduction of the σ-meson (and the valence quark) mass and the σ decay width into the $\pi\pi$ channel. In fact, the σ-meson mass and full width as taken from the current PDG tables [7] are M_σ = 400–550 MeV, Γ_σ = 400–700 MeV. This demonstrates a very large spread of the σ-meson basic parameters, which is highly non-typical for the light meson sector. E.g., the ρ-meson parameters are fixed with high accuracy: $M_\rho = 775.26 \pm 0.25$ MeV, $\Gamma_\rho = 149.1 \pm 0.8$ MeV. Thus, one can suggest the possible reason for such a large spread of σ-meson parameters to be not only big errors arising when extracting the σ mass and width from experimental data (due to a very large σ width) but the really different values of the σ basic parameters found in various hadronic processes. So, the σ-meson parameters as manifested in various hadronic processes depend upon the CSR degree and thus display the big variation non-typical for all other mesons. The full discussion of the σ-meson properties can be found in a recent review [8].

⋆e-mail: kukulin@nucl-th.sinp.msu.ru

The phase transition in hot matter and the temperature behavior of pseudoscalar as well as scalar meson parameters were investigated in, e.g., [3, 6]. Figure 1 illustrates the temperature dependence of M_π, M_σ and Γ_σ obtained in [6] in the framework of the effective nonlocal quark model. It is quite evident from figure 1 that the σ-meson mass and width drop strongly when the temperature approaches its critical value T_c.

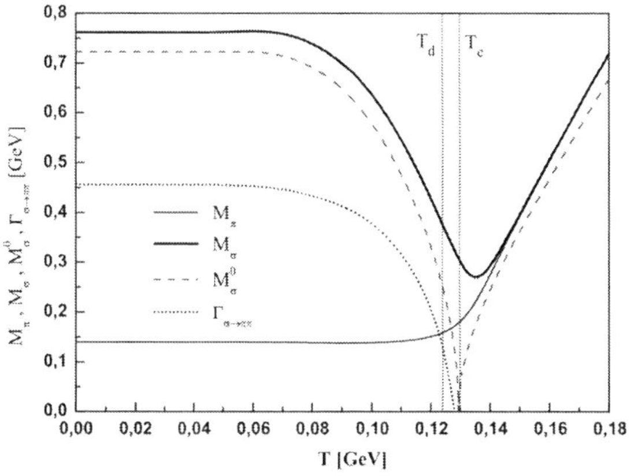

Figure 1. Temperature dependence of M_π, M_σ and Γ_σ [6]. M_σ^0 corresponds to M_σ in the chiral limit. The quark deconfinement temperature T_d and the critical temperature T_c of the chiral phase transition are also shown.

The density dependence of the σ-meson parameters has been studied in, e.g., [4, 5]. The density dependence of M_σ is essentially dictated by that of the condensate $\langle\sigma\rangle$. When the nuclear (quark) matter density becomes sufficiently high, the Pauli principle or temperature destroys the $q\bar{q}$ condensate, the valence quark motion is liberated from the condensate, and thus the CSR happens. This effect is rather similar to disappearance of superconductivity in electron gas in metals when the temperature is rising, since Cooper pairs are destroyed.

The in-medium σ-mass ansatz for the nuclear matter densities $\rho < \rho_0$ (ρ_0 being the nuclear saturated density) is $M_\sigma(\rho) = M_\sigma\,(1 - \alpha\rho/\rho_0)$, where M_σ is the vacuum σ-meson mass and the parameter α was found from QCD sum rules or model calculations to be $\alpha \simeq 0.2$–0.3 [4, 5]. From this ansatz, it is evident that the σ mass goes down when the matter density increases, and its width gets much reduced (due to the phase-space reduction). Thus, the σ-meson propagator in nuclear medium becomes much sharper than in vacuum due to CSR effects [5]. The overall importance of CSR can be concluded from the citation [3]: *"Observing the restoration of chiral symmetry at finite temperature is one of the central aims in the future relativistic heavy-ion experiments at the BNL RHIC and CERN LHC"*.

2 Sources for σ-meson production in hadronic collisions

Below we will review shortly some possible sources for the light scalar σ-meson production in hadronic collisions.

(i) *Dibaryon production in pn, pd and dd collisions at E ~ 1 GeV/A.* Recent experiments of the WASA-at-COSY Collaboration [9] on the basic 2π-production reaction $pn \to d + \pi^0\pi^0$ have shown a direct correlation of the old ABC-puzzle [10] with 2π emission from the $I(J^P) = 0(3^+)$ dibaryon state $D_{03}(2380)$. The same isoscalar resonance along with ABC enhancement have been observed also in $pd \to {}^3\mathrm{He} + \pi^0\pi^0$ and $dd \to {}^4\mathrm{He} + \pi^0\pi^0$ [11]. So, the authors [11] concluded that "the ABC effect in the double-pionic fusion to nuclei is traced back to a *pn resonance, which obviously is strong enough to survive even in the nuclear medium.*" However, the detailed mechanism of the D_{03} resonance decay which leads to the ABC enhancement is still not firmly established.

According to the dibaryon model for short-range NN interaction [12, 13], production of intermediate dibaryons in hadronic collisions is often associated with the scalar σ-meson production, with σ mass and width being renormalized strongly compared to those observed for σ mesons produced in vacuum (e.g., in $\pi\pi$ scattering). This renormalization happens due to CSR effects in the dibaryon state. Indeed, the $D_{03}(2380)$ dibaryon is a strongly excited hadron (with the excitation energy $E^* \simeq 500$ MeV), and the partial CSR was predicted for such states rather reliably [1]. Besides that, the $D_{03}(2380)$ dibaryon is a very dense object, since it has six quarks in a volume of one nucleon ($r(D_{03}) \simeq 0.7$–0.9 fm). So, the ABC effect can be interpreted as a consequence of the near-threshold σ-meson emission from the dibaryon, with the σ-meson parameters [14] $M_\sigma^{(\mathrm{ABC})} \simeq 300$ MeV, $\Gamma_\sigma^{(\mathrm{ABC})} \simeq 100$ MeV (cf. the σ-meson parameters found from $\pi\pi$ scattering in free space [8, 15]: $M_\sigma^{(\pi\pi)} = 441^{+16}_{-8}$ MeV, $\Gamma_\sigma^{(\pi\pi)} = 544^{+18}_{-25}$ MeV). The $\pi^0\pi^0$ invariant mass spectrum in the $pn \to d\pi^0\pi^0$ reaction at $\sqrt{s} = 2.38$ GeV calculated in the dibaryon model with the above (reduced) parameters of σ-meson is shown in figure 2 (a).

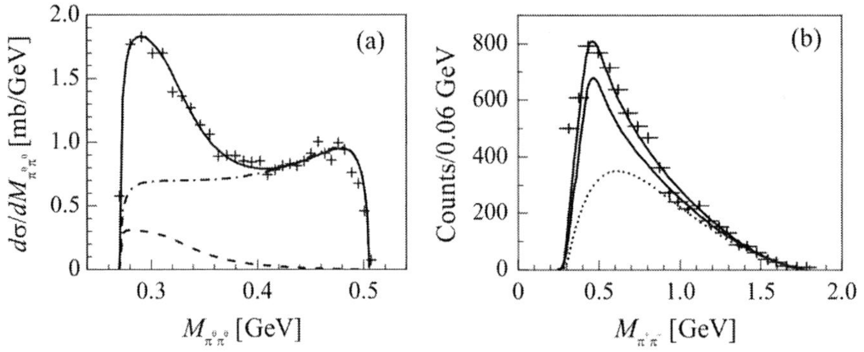

Figure 2. (a) Invariant $\pi^0\pi^0$ mass spectrum in $pn \to d\pi^0\pi^0$ at $T_p = 1.14$ GeV ($\sqrt{s} = 2.38$ GeV) [9] in comparison with the dibaryon model calculations [14] (solid line). Contributions of the near-threshold σ-meson production and the "background" process $D_{03}(2380) \to D_{12}(2150)\pi^0 \to d\pi^0\pi^0$ are shown by dashed and dash-dotted lines, respectively. (b) Invariant $\pi^+\pi^-$ mass spectrum in $pp \to pp\pi^+\pi^-$ at $p = 6.6$ GeV/c in comparison with a resonance fit (upper solid line). The 2π background without and with the contribution of the P_{11} at 1400 MeV is given by the dotted and lower solid line, respectively (for details, see [16]).

(ii) *Generation of the Roper resonance $N^*(1440)$ in NN, $N\alpha$, etc., collisions.* The authors [16] have analyzed carefully the $N^*(1440)$ generation in pp collisions at beam momenta 6–30 GeV/c and low momentum transfers $-t \leq 0.06$ GeV/c^2. The Roper resonance produced in such a way decays immediately (due to its very large width $\Gamma_{N^*} \simeq 350$ MeV) into $N\pi$ or $N\pi\pi$ channel, the latter decay going with a high probability via generation of an intermediate σ meson [7]. The $\pi^+\pi^-$ invariant mass distribution in the $pp \to pp\pi^+\pi^-$ process at 6.6 GeV/c is shown in figure 2 (b). It is clearly seen that the $\pi^+\pi^-$ production is dominated by the Roper resonance decay and is concentrated near $M_{\pi^+\pi^-} \simeq 400$ MeV corresponding to the σ-meson mass. So, the degree of CSR in the Roper resonance is less than in the dibaryon, which is likely due to the lower density of the Roper (three quarks instead of six).

(iii) *Single and double σ production from Pomeron.* It is well known that the high-energy pp scattering has a diffraction character and is governed by a Pomeron exchange. Furthermore, the Pomeron, being a scalar object, can produce σ mesons in high-energy pp collisions (see figure 3). The single and double σ emissions from Pomeron have been studied by Kisslinger et al. [17] who demonstrated that ca. 6–7% of the whole pp elastic cross section at 50 GeV goes to σ emission.

Experimental data of the GAMS Collaboration [18] on $2\pi^0$ production in high-energy pp collisions (at 450 GeV/c) revealed the high peak at $M_{\pi^0\pi^0} \simeq 500$ MeV and thus well confirmed the above Kisslinger et al. predictions (see figure 4).

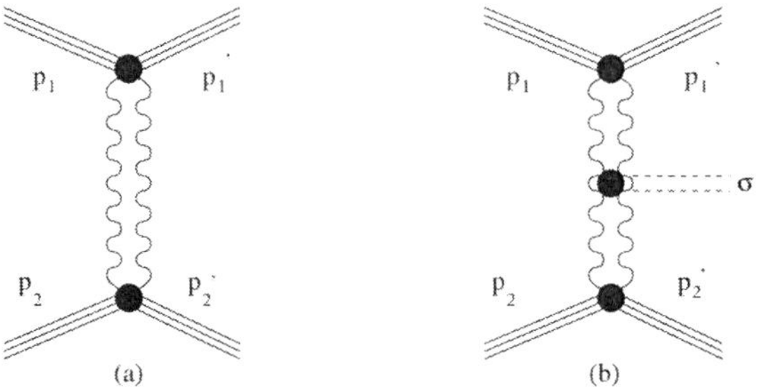

Figure 3. (a) Elastic pp scattering with Pomeron exchange. (b) Peripheral σ production with double Pomeron exchange [17].

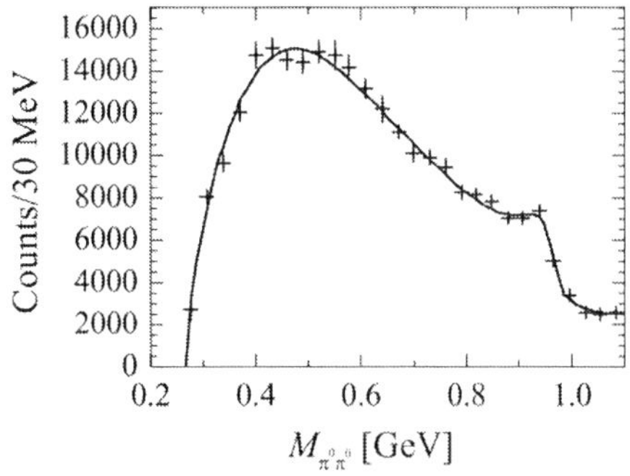

Figure 4. (a) Invariant $\pi^0\pi^0$ mass spectrum in $pp \to pp\pi^0\pi^0$ at 450 GeV/c (corrected for detection efficiency). The solid line represents the fit by an exponential-type function (for details, see [18]).

Thus, we demonstrated that in many hadronic collisions at intermediate and high energies there appears a rather high yield of light scalar σ mesons which mass and width values should carry information about the stage of CSR inside the fireball, or, in other words, about its temperature and density. In turn, this important information can be found by measuring the invariant-mass distributions of digamma signals from the σ-meson decays. Since the in-medium CSR effects lead to reduction of the σ mass and thus to suppression of the $\sigma \to \pi\pi$ decay mode, one should observe an increasing branching ratio for the $\sigma \to \gamma\gamma$ decay in hot and/or dense matter.

3 Experiments on digamma production in hadronic collisions

The initial $\gamma\gamma$-production experiments in intermediate-energy pp collisions were performed by the CELSIUS/WASA Collaboration [19]. They found a clear $\gamma\gamma$ signal at $M_{\gamma\gamma} \simeq$ 300–400 MeV (see figure 5 (a)), which turned out to be very stable against cuts. Furthermore, the model which incorporated very well the $\gamma\gamma$ events from π^0, η and $\pi^+\pi^-$ production gave practically no events in the area $M_{\gamma\gamma} \simeq$ 300–400 MeV, while the $\pi^0\pi^0$ production at $T_p = 1.36$ GeV gave about 15% of the observed counts between π^0 and η peaks. So, the authors [19] have come to the conclusion: "Since none of these simulated processes is able to account for the structure observed near the $\pi\pi$ threshold ..., we are led to consider seriously the possibility that the observed structure (at $M_{\gamma\gamma} \simeq$ 300–400 MeV) is real and might be due to the process $pp \to pp\sigma \to pp\gamma\gamma$, in particular also since $pp \to pp\pi^+\pi^-$ and $pp \to pp\pi^0\pi^0$ reactions are dominated by σ production".

Quite recently, there appeared new very encouraging $\gamma\gamma$ data with high statistics. In experiments done at the Dubna Nuclotron [20, 21], the authors analyzed the $\gamma\gamma$ spectra from pC and dC collisions at beam momenta 5.5 GeV/c (for protons) and 1.7–3.8 GeV/c per nucleon (for deuterons). The $\gamma\gamma$

DOI: 10.1051/epjconf/201713803006

yield from dC collisions at p = 2.75 GeV/cA and pC collisions at p = 5.5 GeV/c is depicted in figures 5 (b) and (c), respectively.

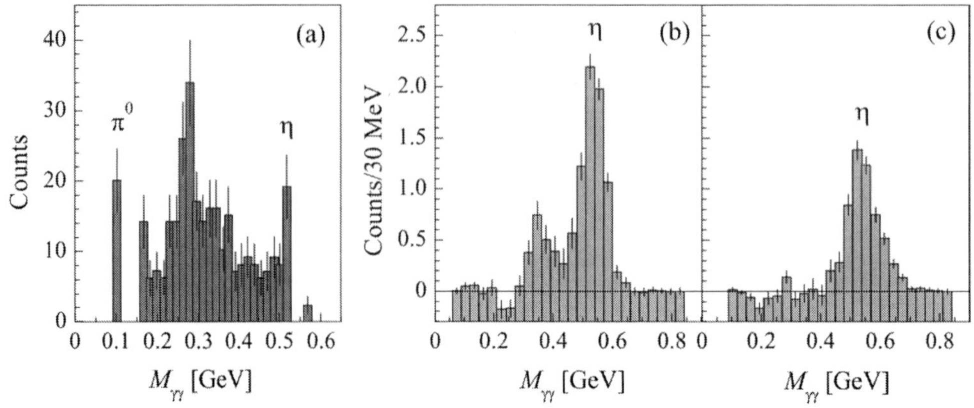

Figure 5. Invariant $\gamma\gamma$ mass spectra in (a) $pp \to pp + \gamma\gamma$ at T_p = 1.36 GeV [19], (b) dC $\to \gamma\gamma$ + X at p = 2.75 GeV/cA [20] and (c) pC $\to \gamma\gamma$ + X at p = 5.5 GeV/c [20]. The spectra in figures (b) and (c) were obtained after the background subtraction [20].

It is very interesting that the $\gamma\gamma$ signal at $M_{\gamma\gamma} \simeq$ 360 MeV with the width $\Gamma_{\gamma\gamma} \simeq$ 60 MeV clearly observed in dC collisions *is absent in pC collisions* (while the η meson is clearly seen in both cases). This enhancement at $M_{\gamma\gamma} \simeq 2 - 3M_\pi$ is quite similar to the ABC effect observed for near-threshold dipion production from the nucleon-nucleon and lightest nuclei collisions. So, it can be interpreted in terms of σ-meson production from the meson cloud surrounding the dibaryon component of the incident deuteron [13], see figure 6. In the kinematics of the Dubna experiment [20, 21], the possible interpretation of the above $\gamma\gamma$ signal can also be related to the well-known Landau–Pomeranchuk mechanism for bremsstrahlung in QED.

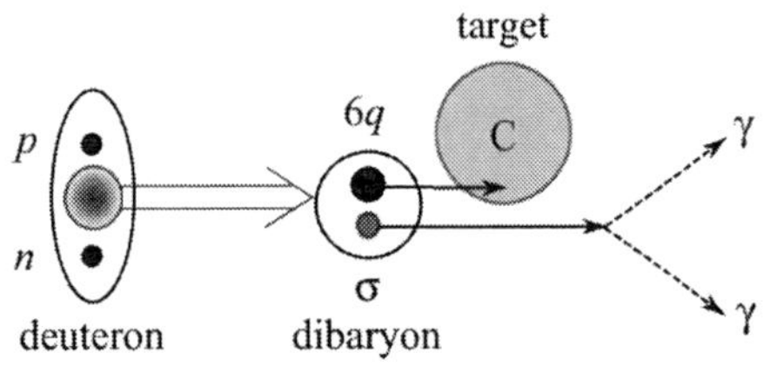

Figure 6. Dibaryon mechanism for $\gamma\gamma$ production in dC collisions.

According to Weldon [22], in hot and dense matter, due to removal of some important constraints valid in vacuum (Lorentz invariance, etc.), some processes forbidden in vacuum become possible, like the one-photon σ decay with a final dilepton production $\sigma \to \pi^+\pi^- \to \gamma \to e^+e^-$. Since the standard two-photon decay $\sigma \to \gamma\gamma$ is also valid in medium, one can relate the dilepton and digamma signals to each other. Thus, one gets two types of signals from the hot fireball, i.e., dilepton and digamma production, which carry valuable information about the state of hot matter produced in heavy-ion collisions. Recent results of the HADES Collaboration [23] seem to confirm this tight interrelation between dilepton and digamma production in heavy-ion collisions.

4 Conclusion

We have demonstrated above that in the intermediate- and high-energy collisions of nucleons and light or heavy ions a lot of light scalar σ mesons are emitted from different sources (excited hadrons, high-energy NN collisions, etc.). Masses and widths of these σ mesons will be dependent upon the hadron excitation energy (for excited hadrons) and the temperature and density of fireball arising in heavy-ion collisions. All these σ mesons have in general a very small branching ratio for the $\gamma\gamma$ decay ($\leq 10^{-5}$). So, it would be highly uneasy to observe these digamma signals at modest temperature and density of fireball.

When, however, the temperature and/or density become sufficiently high for restoration of chiral symmetry, M_σ and $\Gamma_{\sigma\to\pi\pi}$ drop down while the $\gamma\gamma$ signal gets much higher at $M_\sigma \simeq 2M_\pi$. When $M_\sigma \to 2M_\pi$ and $\Gamma_{\sigma\to\pi\pi} \to 0$, all numerous σ mesons produced in fireball will emit the $\gamma\gamma$ pairs. So, one will observe the large $\gamma\gamma$ burst from fireball with $M_{\gamma\gamma} \simeq 2M_\pi \simeq 280$ MeV. Simultaneously, observation of such a $\gamma\gamma$ burst will be a good signal for onset of the chiral phase transition in hot and/or dense nuclear matter. Thus, measuring the invariant mass and width of the $\gamma\gamma$ signal (together with the dilepton signal) will give unambiguous information about the chiral phase transition in hot and/or dense nuclear matter. Moreover, just the combination of the dilepton and digamma signals might be very advantageous in reducing the $\gamma\gamma$ background coming from $\pi^0 \to \gamma\gamma$ decay [24].

The authors acknowledge financial support from RFBR grants No. 16-02-00265, No. 16-02-00049 and No. 16-52-12005.

References

[1] L.Ya. Glozman, Phys. Lett. B **475**, 329 (2000)
[2] A.V. Nefediev, J.E.F.T. Ribeiro, and A.P. Szczepaniak, JETP Lett. **87**, 271 (2008)
[3] S. Chiku and T. Hatsuda, Phys. Rev. D **57**, 6 (1998)
[4] T. Hatsuda, T. Kunihiro, and H. Shimizu, Phys. Rev. Lett. **82**, 2840 (1999)
[5] Z. Aouissat, G. Chanfray, P. Schuck, and J. Wambach, Phys. Rev. C **61**, 012202 (1999). arXiv:nucl-th/9908076
[6] M.K. Volkov et al., Phys. Lett. B **424**, 235 (1998). D. Blaschke, Yu.L. Kalinovsky, A.E. Radzhabov, and M.K. Volkov, Phys. Part. Nucl. Lett. **3**, 327 (2006). arXiv:hep-ph/0508264
[7] C. Patrignani et al. (Particle Data Group), Chin. Phys. C **40**, 100001 (2016)
[8] J.R. Pelaez, Phys. Rep. **658**, 1 (2016). arxiv:hep-ph/1510.00653
[9] P. Adlarson et al., Phys. Rev. Lett. **106**, 242302 (2011). Phys. Lett. B **721**, 229 (2013)
[10] A. Abashian, N.E. Booth, and K.M. Crowe, Phys. Rev. Lett. **5**, 258 (1960)
[11] P. Adlarson et al., Phys. Rev. C **86**, 032201(R) (2012)
[12] V.I. Kukulin, I.T. Obukhovsky, V.N. Pomerantsev, and A. Faessler, J. Phys. G **27**, 1851 (2001)
[13] V.I. Kukulin et al., Ann. Phys. (N.Y.) **325**, 1173 (2010)
[14] M.N. Platonova and V.I. Kukulin, Phys. Rev. C **87**, 025202 (2013)
[15] I. Caprini, G. Colangelo, H. Leutwyler, Phys. Rev. Lett. **96**, 132001 (2006)
[16] H.P. Morsch and P. Zupranski, Phys. Rev. C **71**, 065203 (2005)
[17] L.S. Kisslinger, W. Ma, and P. Shen, Phys. Rev. D **71**, 094021 (2005)
[18] D. Alde et al., Phys. Lett. B **397**, 350 (1997)
[19] M. Bashkanov et al., Int. J. Mod. Phys. A **20**, 554 (2005). M. Bashkanov, arXiv:hep-ex/0406081
[20] Kh.U. Abraamyan et al., Phys. Rev. C **80**, 034001 (2009)
[21] Kh.U. Abraamyan et al., Eur. Phys. J. A **52**, 259 (2016); EPJ Web Conf., present volume
[22] H.A. Weldon, Phys. Lett. B **274**, 133 (1992)
[23] V.P. Ladygin, private communication
[24] P. Rehberg, Yu.L. Kalinovsky, and D. Blaschke, Nucl. Phys. A **622**, 478 (1997)

Hyperon production in HIC. The role of the initial state

Genis Musulmanbekov[1]*

[1] *JINR, Dubna, Russia*

Abstract. A new mechanism is proposed for interpretation of the non-monotonic, enhanced yield of hyperons and positive strange mesons (also known as "horn" effect) that was observed in central heavy ion collisions in the experiment NA49 at SPS. We argue that the data indicate the transition of the nucleons in the overlap region of colliding nuclei with high nuclear density into a certain hyperon phase. This nucleon-to-hyperon transition is accompanied by enhanced yields of positive kaons. The "horn"-effect is a consequence of an interplay between the energy/baryon density attained in the overlap region of colliding nuclei and the overlap time.

Understanding the behavior of nuclear matter under extreme conditions of high temperature and density is nowadays a challenge in physics of strong interactions. Many theoretical and experimental efforts have been dedicated to the physics of relativistic heavy–ion collisions looking for signatures of the quark gluon plasma (QGP). Restoration of chiral symmetry and deconfinement are expected to occur at high density and/or temperature of a hadronic matter. In heavy ion collisions even at low and intermediate energies the nuclear density increases more than 2–3 times. The question "How much of incident energy is converted into compression of nuclear matter?" has not found a definite answer yet. Observables indicating non-monotonic and unexpected (from theoretical point of view) behavior of emitted particles is particularly important. In this way the study of the strange particle production in heavy ion collisions is promising as they could serve a good diagnostic tool to investigate the properties of nuclear matter under extreme conditions. Strange particle yields have been thoroughly measured by many experiments with various colliding nuclei at different energies starting from kaon subthreshold production up to energies of SPS and RHIC [1–5]. The systematic study of hadron production in central Pb+Pb collisions at SPS performed by NA49 collaboration revealed a sharp structure in energy dependence of positive kaon to pion multiplicity ratio, K^+/π^+ (Left panel on figure 1) [4]. That peculiarity, called "horn"–effect, was later confirmed by Beam Energy Scan (BES) program of STAR collaboration at RHIC [5]. At the same time the ratio of negative mesons K^-/π^- increases monotonically tending to saturate at higher energies. Traditional transport models were unsuccessful in description of the data. In paper [7] authors claimed non–monotonic dependence of K^+/π^+ ratio as a signal of the onset of deconfinement. The idea that strangeness is a good signal of deconfinement was put forward by J. Rafelski in 1982 ([13]). The argument was the following: it is energetically favourable to produce $s\bar{s}$ - pairs in deconfined medium than a pairs strange hadrons in hadron gas. The former requires energy $\sim 300\ MeV$, while the strangeness production in hadronic reaction $N+N \longrightarrow N+Y+K^+$ requires $670\ MeV$ above threshold. Albeit a variety of models, statistical

*e-mail: genis@jinr.ru

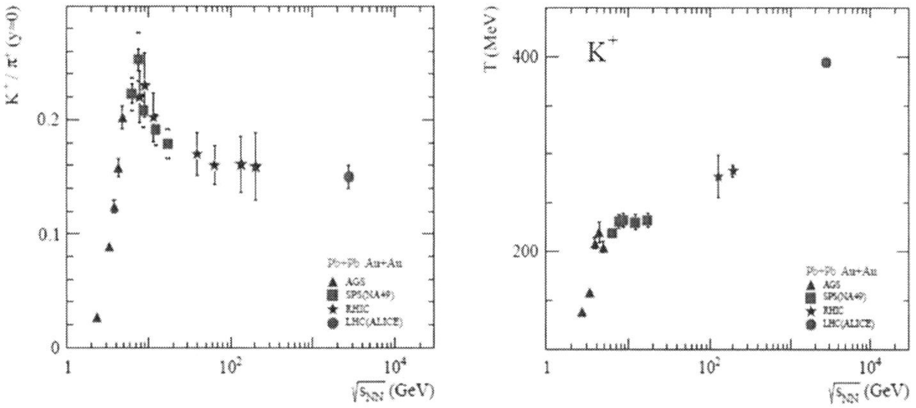

Figure 1. Left panel: K^+/π^+ ratio as a function of center of mass energy in central Au+Au and Pb+Pb collisions. Right panel: Inverse slope parameter of transverse momentum distributions of K^+s in these reactions. Data points are from AGS [1–3], SPS [4], RHIC [5] and LHC [6].

[8–10] and kinetic [11, 12] (with or without deconfinement) have been proposed for interpretation of "horn" structure a satisfactory understanding of it is still not complete.

We propose a new interpretation of the enhancement of the ratio K^+/π^+ and the "horn" structure as a consequence of transition of the nuclear matter in a highly compressed zone at the early stage of collision from a nucleonic to hyperonic state. Interpretation of the non–monotonic structure of K^+/π^+ has initiated intense theoretical activity. Several authors have attempted to reproduce the horn structure employing different approaches. "Horn" – like structure has been predicted in [7], as a manifestation of phase transition between thermalized hadronic and partonic phases. Several versions of Hadron Resonance Gas Model (HRGM) with and without thermal equilibrium describe more or less the effect [8–10]. In some of these models the origin of the horn is related to a transition of the hadron gas from a baryon–dominated to a meson–dominated gas at $\sqrt{s_{NN}} \approx 8\ GeV$ along the freeze-out curve. Authors in [12] elaborated the hadronic kinetic model in which the interplay between the available energy density and the lifetime of the fireball was able to describe data under the assumption that the total lifespan of the fireball is decreasing function of the collision energy for SPS energies above 30 $AGeV$.

Here we propose a new interpretation of non-monotonic behavior of K^+/π^+ ratio in heavy ion central collisions. At the initial phase of head–on collision the nuclear density ρ in the overlap/collision zone can reach several times the nuclear saturation density ρ_0. Since nucleons are extended objects of dimension ~ 1 fm each nucleon undergoes compression and occupies more than two times smaller volume than what was typical at the saturation density. Obviously, nucleons under such a strong compression/pressure have to modify their properties. We assume that under such conditions it is more preferable to nucleons to transit into a hyperonic state since hyperons containing one, two or three strange quarks occupy smaller volume than nucleons composed of light quarks. Thus, at the early stage of collision, before creation of a hot and dense fireball, a condensate of $s\bar{s}$ - pairs is created in the compressed overlap zone. Then $s-$quarks replace one (two, three) u or d quarks of nucleons, and $\bar{s}-$ quarks form with those replaced quarks strange mesons: $p, n \longrightarrow hyperons + kaons$. Therefore, in this "action–reaction" mechanism the transition of nucleons to hyperons is realized through the following channels:

$$
\begin{aligned}
p(uud) &\longrightarrow \Sigma^+(uus) + K^0(d\bar{s}), & n(udd) &\longrightarrow \Sigma^-(dds) + K^+(u\bar{s})\\
&\longrightarrow \Sigma^0(uds) + K^+(u\bar{s}), & &\longrightarrow \Sigma^0(uds) + K^0(d\bar{s})\\
&\longrightarrow \Lambda^0(uds) + K^+(u\bar{s}), & &\longrightarrow \Lambda^0(uds) + K^0(d\bar{s})\\
&\longrightarrow \Xi^-(dss) + 2K^+(u\bar{s}), & &\longrightarrow \Xi^0(uss) + 2K^0(d\bar{s})\\
&\longrightarrow \Xi^0(uss) + K^+(u\bar{s}) + K^0(d\bar{s}), & &\longrightarrow \Xi^-(dss) + K^+(u\bar{s}) + K^0(d\bar{s})\\
&\longrightarrow \Omega^-(sss) + 2K^+(u\bar{s}) + K^0(d\bar{s}), & &\longrightarrow \Omega^-(sss) + 2K^0(d\bar{s}) + K^+(u\bar{s})
\end{aligned}
$$

According to the transition channels the K^+s and K^0s, but any K^- are produced only. K^- can be produced in reactions $\pi Y \longrightarrow NK^-$ and $\phi \longrightarrow K^+K^-$. The first reaction should be suppressed because pions, composed of light quarks do not survive in a dense hadronic matter. ϕ–meson production in hadronic reactions are suppressed by OZI–rule. However, formation of a condensate of strange quarks in a dense matter can lead to weakening of OZI–rule, and the yield of K^-s at energies below "horn"–effect is due to ϕ–mesons decay. If this "action–reaction" mechanism is pertinent for the description of the "horn" structure, then the ratio $K^0/\ \pi^0$ should behave in the same way.

Another contribution to strange particle production comes from the kinetic mechanism of binary reactions. At low and intermediate energies the K^+s are produced predominantly in so–called "associated" reactions $N + N \longrightarrow N + Y + K^+$, $\pi + N \longrightarrow Y + K^+$ where N stands for nucleons and Y $(\Lambda, \Sigma, \Xi, \Omega)$ – for hyperons. K^-s can be produced on the next step, in reaction $\pi + Y \longrightarrow N + K^-$. As the collision energy increases, the contribution of these reactions to inelastic cross section decreases, due to increasing phase space that leads to increasing number of channels of strange particle production. Hence, one could argue that this (associated) production mechanism is sufficient to be consistent with horn–like non-monotonic K^+ spectrum. However, the results of transport models, URQMD and HSD, demonstrated that this mechanism was not able to describe quantitatively experimental data [14].

There are two factors providing the proposed transition mechanism: compression energy deposited into collision zone and time during which two colliding nuclei are overlapped – overlap time. Both are functions of collision energy. Which part of available energy is converted in creation/condensation of $s\bar{s}$ - pairs with subsequent formation of hyperons and kaons? This question addresses to compression of nuclear matter which could be derived from equation of state (EOS), relation between an energy density and baryonic density in the interaction zone. However, extraction of information on EOS both from experiment and theory is a matter of long debate. At high nuclear densities there are many uncertainties that make it difficult to extract the compression from experimental data. There are two hydrodynamical scenarios of initial conditions in hadron or heavy–ion collisions: *Landau solution* and *Bjorken solution*. In the Landau solution a compression in collision zone should be maximal because of the fully stopping power of hadronic matter. In contrast to the Landau scenario the core assumption of the Bjorken scenario is the boost–invariance that gives an accelerationless solution (zero stopping power) of the hydrodynamical equations. Truth – somewhere in the middle. For qualitative considerations we assume that the available energy to be converted into nuclear compression, E_{comp}, in the overlap zone is an increasing function of baryonic density.

On the other hand, the overlap time, τ_o, is decreasing function of the collision energy. Transition of nucleons to hyperons accompanied by strange meson production should take finite time which we denote by transition time, τ_{tr}. Naturally to expect that τ_{tr} is order of the nucleon size, $\approx 1fm$. Hence the proposed mechanism of transition can be realized if this transition time, τ_{tr}, is less than the overlap time, τ_o,when the net baryon density is maximal (on the average) at a given collision energy.

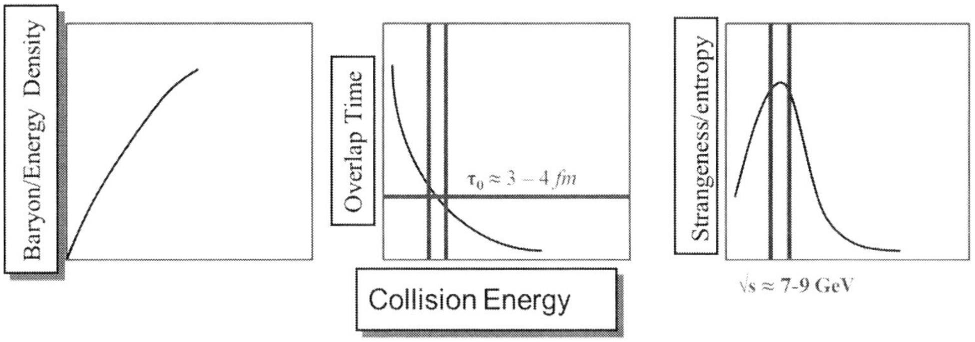

Figure 2. Interplay of compression energy increasing with collision energy (left panel) and overlap time decreasing with collision energy (middle panel) gives a peak of K^+/π^+ (right panel). Vertical lines outline location of the peak, and the horizontal line depicts the corresponding overlap time for gold–gold collisions.

Thus, the collision energy dependence of the probability of nucleon-to-hyperon transition is defined by interplay of the increasing compression energy and decreasing overlap time:

$$P_{N\to Y} \propto (\tau_o/\tau_{tr})^{\alpha} E(\rho/\rho_o)_{comp},$$

where α is the adjustable parameter and ρ_0 is the nuclear saturation density. This interplay results in the non-monotonic behaviour of yields of hyperons and K^+s. Besides energy dependence the overlap time depends on the sizes of colliding nuclei. For symmetrical nuclei in the center of mass system $\tau_o = [2R_A/(\gamma\beta)]b_{SP}$, where nuclear radii, $R_A = 1.2A^{1/3}$, γ – gamma-factor and β – velocity of colliding nuclei. Factor b_{SP} stands for taking into account the effect of stopping power. Following the Bjorken scenario we set the minimal value for the stopping power ($b_{SP} \sim 1$). Then the peak of K^+/π^+ for gold–gold collisions according to the data taking place at $\sqrt{s} = 7-9\ GeV$ corresponds to $\tau_o = 3-4\ fm$ (figure 2).

Additionally to the non-monotonic behaviour of the ratio K^+/π^+ the experiments observed at the same energies, so–called, "step" structure, the plateau in the inverse slope of transverse momentum distributions both of K^+ and K^- (Right panel on figure 1). According to our scenario this plateau emerges when kaon production via mechanism of binary interactions starts to dominate over the nucleon-to-hyperon transition mechanism, inasmuch as transverse momenta of kaons produced in binary interactions should be softer than those ones in transition mechanism. Up to the "horn" peak the K^-–mesons should be produced mainly in decays of ϕ–mesons which are created in the strange quark condensate, violating OZI suppression rule. Obviously, their transverse momentum spectra should be harder than those ones of kaons produced in binary interactions. These considerations are supported by observed in experiments strong enhancement of ϕ–mesons yield near and above threshold [15–17]. Also, compression of the nuclear matter in the overlap volume can lead to domination of vector meson production (ρ and ω) over pseudoscalar ones (pions). The enhanced yield of dileptons with invariant masses near the masses of vector mesons in comparison with the spectrum from elementary nucleon–nucleon reactions was observed by DLS and HADES [18, 19].

References

[1] L. Ahle et al., Phys. Lett. B **490**, 53 (2000)
[2] J. L. Klay et al., Phys. Rev. Lett. **88**, 102301 (2002)

[3] L. Ahle et al., Phys. Rev. C **60**, 064901 (1999)
[4] S. V. Afanasiev et al., Phys.Rev. C **66**, 054902 (2002). C. Alt et al., Phys. Rev. C **77**, 024903 (2008)
[5] L. Kumar (STAR Collaboration), J. Phys. G:Nucl. Part. Phys. **38**, 124145 (2011)
[6] M. Floris (ALICE Collaboration), J. Phys. G:Nucl. Part. Phys. **38**, 124025 (2011)
[7] M. Gazdzicki and M. Gorenstein, Acta. Phys. Pol. B **30**, 2705 (1999)
[8] J. Cleymans, H. Oeschler, K. Redlich, and S. Wheaton, Phys.Lett. B **615**, 50 (2005)
[9] A. Andronic, P. Braun-Munzinger and J. Stachel, Nucl.Phys. A **834**, 237C (2010)
[10] K.A. Bugaev et al., arXiv:1412.6571v2 (2014)
[11] J.K. Nayak et al., Acta Phys. Slov. **56**, 27 (2005)
[12] B. Tomasik and E.E. Kolomeitsev, Eur. Phys. J. C **49**, 115 (2007)
[13] J. Rafelski, Phys. Rep. **88**, 331 (1982)
[14] E. Bratkovskaya et al., Phys. Rev. C **69**, 054907 (2005)
[15] B. B. Back et al., Phys. Rev. C **69**, 054901 (2004)
[16] C. Alt et al., Phys. Rev. C **78**, 044907 (2008)
[17] G. Agakishiev et al., Phys. Rev. C **80**, 025209 (2009)
[18] R. G. Porter et al., Phys. Rev. Lett. **79**, 1229 (1997)
[19] A. Rustamov et al., Int. Jour. Mod. Phys. A **26**, 384 (2011)

Transverse momentum distributions of hadrons in the Tsallis-1 and Tsallis-2 statistics

A.S. Parvan[1,2,3,⋆]

[1]*Bogoliubov Laboratory of Theoretical Physics, Joint Institute for Nuclear Research, Dubna, Russia*
[2]*Department of Theoretical Physics, Horia Hulubei National Institute of Physics and Nuclear Engineering, Bucharest-Magurele, Romania*
[3]*Institute of Applied Physics, Moldova Academy of Sciences, Chisinau, Republic of Moldova*

Abstract. We considered the ultrarelativistic transverse momentum distributions of the Tsallis-1 and Tsallis-2 statistics using two regularization schemes. It was revealed that the cut-off parameter strongly influences the behavior of the transverse momentum distribution in both statistics. We have also found that the ultrarelativistic transverse momentum distribution of the Tsallis-1 statistics is transformed to the momentum distribution of the Tsallis-2 statistics by identifying $q \to 1/q_c$.

1 Introduction

The transverse momentum distributions of the Tsallis-factorized statistics [1, 2] are largely used to analyze the LHC and RHIC data on the transverse momentum distributions of hadrons created in the proton-proton and heavy-ion collisions. The Tsallis statistics [3, 4] was applied to treat the experimental transverse momentum distributions of hadrons only in the ultrarelativistic case and only for the first variant of the Tsallis statistics [5]. In [6] it was demonstrated that the Tsallis-factorized statistics is not equivalent to the Tsallis statistics. In [7] it was found out that the parameter q, which indicates the deviation of the Tsallis distribution from the usual Boltzmann distribution, effectively takes into account the hard processes of the partons reactions, which lead to the creation of energetically high-p_T hadrons. The main purpose of this study is to establish the relation between the cut-off parameter and the behavior of the transverse momentum distributions of the Tsallis-1 and Tsallis-2 statistics.

2 Tsallis-1 statistics for $q < 1$

In the Tsallis-1 statistics, the ensemble averages, the probabilities of microstates and the norm equation in the grand canonical ensemble are written as follows [8]:

$$\langle A \rangle = \sum_i A_i \left[1 + \frac{q-1}{q}\frac{\Lambda - E_i + \mu N_i}{T}\right]^{\frac{1}{q-1}}, \qquad p_i = \left[1 + \frac{q-1}{q}\frac{\Lambda - E_i + \mu N_i}{T}\right]^{\frac{1}{q-1}}, \tag{1}$$

$$1 = \sum_i \left[1 + \frac{q-1}{q}\frac{\Lambda - E_i + \mu N_i}{T}\right]^{\frac{1}{q-1}}, \tag{2}$$

⋆e-mail: parvan@theor.jinr.ru;parvan@theory.nipne.ro

where E_i and N_i are the energy and the number of particles in the i-th microscopic state of the system, respectively, Λ is a norm function and $q \in \mathbf{R}$ is a real parameter taking values $0 < q < \infty$.

The norm equation (2) for the Maxwell-Boltzmann ultrarelativistic ideal gas of the Tsallis-1 statistics in the grand canonical ensemble can be written as [6]

$$\sum_{\{n_{\vec{p}\sigma}\}} \frac{1}{\prod\limits_{\vec{p}\sigma} n_{\vec{p}\sigma}!} \left[1 + \frac{q-1}{q} \frac{\Lambda - \sum\limits_{\vec{p}\sigma} n_{\vec{p}\sigma}(\varepsilon_{\vec{p}} - \mu)}{T}\right]^{\frac{1}{q-1}} = 1, \tag{3}$$

where $\varepsilon_{\vec{p}} = |\vec{p}|$ is the one-particle energy and $n_{\vec{p}\sigma} = 0, 1, \ldots, \infty$ are the occupation numbers. Using the integral representations for the Gamma-function [9]

$$x^{-y} = \frac{1}{\Gamma(y)} \int\limits_0^\infty t^{y-1} e^{-tx} dt, \quad Re(x) > 0, Re(y) > 0 \tag{4}$$

we obtain

$$\frac{1}{\Gamma\left(\frac{1}{1-q}\right)} \int\limits_0^\infty t^{\frac{1}{1-q}-1} e^{-t\left[1+\frac{q-1}{q}\frac{\Lambda}{T}\right]} Z(\beta') dt = 1, \tag{5}$$

where $\beta' = -(q-1)t/(qT)$ and

$$Z(\beta') = \sum_{\{n_{\vec{p}\sigma}\}} \frac{1}{\prod\limits_{\vec{p}\sigma} n_{\vec{p}\sigma}!} e^{-\beta' \sum\limits_{\vec{p}\sigma} n_{\vec{p}\sigma}(\varepsilon_{\vec{p}} - \mu)} = \exp\left(\sum_{\vec{p}\sigma} e^{-\beta'(\varepsilon_{\vec{p}} - \mu)}\right). \tag{6}$$

For the ultrarelativistic ideal gas the partition function (6) can be rewritten as

$$Z(\beta') = \exp\left(\frac{gV}{\pi^2 \beta'^3} e^{\beta' \mu}\right) = \sum_{N=0}^\infty \frac{1}{N!} \left(\frac{gV}{\pi^2 \beta'^3}\right)^N e^{\beta' \mu N}, \tag{7}$$

where g is the spin degeneracy factor. Substituting (7) into (5) and using (4) we have [6]

$$\sum_{N=0}^{N_0} \frac{\tilde{\omega}^N}{N!} \frac{\Gamma\left(\frac{1}{1-q} - 3N\right)}{\left(\frac{1-q}{q}\right)^{3N} \Gamma\left(\frac{1}{1-q}\right)} \left[1 + \frac{q-1}{q} \frac{\Lambda + \mu N}{T}\right]^{\frac{1}{q-1}+3N} = 1, \tag{8}$$

where $\tilde{\omega} = gVT^3/\pi^2$ and $N_0 < 1/(3(1-q))$. The terms with $N > N_0$ in the series (8) are divergent and they should be excluded. Note that N is not the number of particles.

The mean occupation numbers for the Maxwell-Boltzmann ultrarelativistic ideal gas in the Tsallis-1 statistics can be written as [6]

$$\langle n_{\vec{p}\sigma} \rangle = \sum_{\{n_{\vec{p}\sigma}\}} n_{\vec{p}\sigma} \frac{1}{\prod\limits_{\vec{p}\sigma} n_{\vec{p}\sigma}!} \left[1 + \frac{q-1}{q} \frac{\Lambda - \sum\limits_{\vec{p}\sigma} n_{\vec{p}\sigma}(\varepsilon_{\vec{p}} - \mu)}{T}\right]^{\frac{1}{q-1}}. \tag{9}$$

Using (4), we obtain

$$\langle n_{\vec{p}\sigma} \rangle = \frac{1}{\Gamma\left(\frac{1}{1-q}\right)} \int\limits_0^\infty t^{\frac{1}{1-q}-1} e^{-t\left[1+\frac{q-1}{q}\frac{\Lambda}{T}\right]} Z(\beta') \langle n_{\vec{p}\sigma} \rangle_G dt, \tag{10}$$

where

$$\langle n_{\vec{p}\sigma}\rangle_G = \frac{1}{Z(\beta')}\sum_{\{n_{\vec{p}\sigma}\}} n_{\vec{p}\sigma}\frac{1}{\prod\limits_{\vec{p}\sigma} n_{\vec{p}\sigma}!} e^{-\beta'\sum\limits_{\vec{p}\sigma} n_{\vec{p}\sigma}(\varepsilon_{\vec{p}}-\mu)} = e^{-\beta'(\varepsilon_{\vec{p}}-\mu)}. \tag{11}$$

Substituting (11) and (7) into (10) and using (4), we obtain [6]

$$\langle n_{\vec{p}\sigma}\rangle = \sum_{N=0}^{N_0}\frac{\tilde{\omega}^N}{N!}\frac{\Gamma\left(\frac{1}{1-q}-3N\right)}{\left(\frac{1-q}{q}\right)^{3N}\Gamma\left(\frac{1}{1-q}\right)}\left[1+\frac{q-1}{q}\frac{\Lambda-\varepsilon_{\vec{p}}+\mu(N+1)}{T}\right]^{\frac{1}{q-1}+3N}, \tag{12}$$

where the upper bound of summation N_0 is the same as in (8). Then the transverse momentum distribution for the ultrarelativistic particles can be written as [6]

$$\frac{d^2N}{dp_T dy} = \frac{gV}{(2\pi)^2}p_T^2\cosh y\sum_{N=0}^{N_0}\frac{\tilde{\omega}^N}{N!}\frac{\Gamma\left(\frac{1}{1-q}-3N\right)}{\left(\frac{1-q}{q}\right)^{3N}\Gamma\left(\frac{1}{1-q}\right)}\left[1+\frac{q-1}{q}\frac{\Lambda-\varepsilon_{\vec{p}}+\mu(N+1)}{T}\right]^{\frac{1}{q-1}+3N}. \tag{13}$$

In the Gibbs limit $q \to 1$ (13) recovers the Maxwell-Boltzmann transverse momentum distribution of the Boltzmann-Gibbs statistics

$$\frac{d^2N}{dp_T dy} = \frac{gV}{(2\pi)^2}p_T^2\cosh y\, e^{-\frac{p_T\cosh y-\mu}{T}}. \tag{14}$$

Taking the zeroth term $N = 0$ in (13) and (8) and transforming $q \to 1/q_c$ [6], we exactly obtain the transverse momentum distribution of the Tsallis-factorized statistics [1, 2]

$$\frac{d^2N}{dp_T dy} = \frac{gVp_T^2\cosh y}{(2\pi)^2}\left[1+(q_c-1)\frac{p_T\cosh y-\mu}{T}\right]^{\frac{q_c}{1-q_c}}. \tag{15}$$

To find the upper bound of summation N_0, we rewrite (8) in the form $\sum_N \phi(N) = 1$, where

$$\phi(N) = \frac{\tilde{\omega}^N}{N!}\frac{\Gamma\left(\frac{1}{1-q}-3N\right)}{\left(\frac{1-q}{q}\right)^{3N}\Gamma\left(\frac{1}{1-q}\right)}\left[1+\frac{q-1}{q}\frac{\Lambda+\mu N}{T}\right]^{\frac{1}{q-1}+3N}. \tag{16}$$

In the Gibbs limit $q \to 1$ (16) recovers the Maxwell-Boltzmann function of the Boltzmann-Gibbs statistics $\phi(N) = \frac{\tilde{\omega}^N}{N!}e^{\frac{\Lambda+\mu N}{T}}$, where $\Lambda = -T\tilde{\omega}\, e^{\frac{\mu}{T}}$.

We have two possibilities to find the cut-off parameter N_0 in (8). See figure 1. In the first method of truncation of the series, which we denote as the Model A, we find N_0 from the local minimum of the function $\ln\phi(N)$:

$$\left.\frac{\partial\ln\phi(N)}{\partial N}\right|_{N=N_0} = 0. \tag{17}$$

However, in the second method of truncation of the series, which we denote as the Model B, we find N_0 from the inflection point of the function $\ln\phi(N)$ [6]:

$$\left.\frac{\partial^2\ln\phi(N)}{\partial N^2}\right|_{N=N_0} = 0. \tag{18}$$

The transverse momentum distribution of the Model A differs essentially from the transverse momentum distributions of the Model B and the Tsallis-factorized statistics due to the large values of the cut-off parameter N_0. However, the transverse momentum distribution of the Model B is close to the transverse momentum distribution of the Tsallis-factorized statistics, because the values of N_0 in the Model B are close to zero.

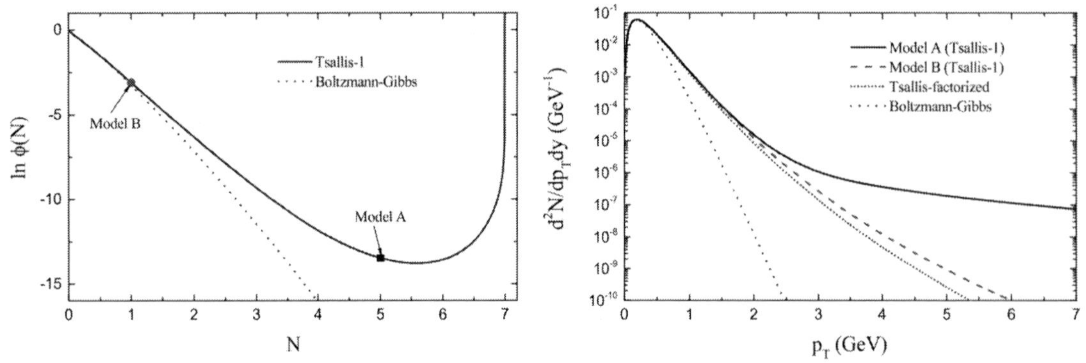

Figure 1. Two methods to fix the cut-off parameter N_0 in the Tsallis-1 statistics. The behavior of $\ln\phi(N)$ as a function of N (left panel) and the transverse momentum distribution (right panel) for the ultrarelativistic ideal gas of π^0 pions in the Tsallis-1 statistics in the volume $V = 1$ fm at the temperature $T = 90$ MeV, chemical potential $\mu = 0$, rapidity $y = 0$ and $q_c = 1.05$ ($q = 1/q_c$).

3 Tsallis-2 statistics for $q_c > 1$

In the Tsallis-2 statistics, the ensemble averages, the probabilities of microstates and the partition function in the grand canonical ensemble are written as follows [4, 6]:

$$\langle A\rangle = \frac{1}{Z^{q_c}}\sum_i A_i\left[1-(1-q_c)\frac{E_i-\mu N_i}{T}\right]^{\frac{q_c}{1-q_c}}, \qquad p_i=\frac{1}{Z}\left[1-(1-q_c)\frac{E_i-\mu N_i}{T}\right]^{\frac{1}{1-q_c}}, \tag{19}$$

$$Z = \sum_i\left[1-(1-q_c)\frac{E_i-\mu N_i}{T}\right]^{\frac{1}{1-q_c}}, \tag{20}$$

where Z is the partition function and $q_c \in \mathbf{R}$ is a real parameter taking values $0 < q_c < \infty$. Note that q_c is a finite number and it cannot be equal to infinity.

The partition function (20) for the Maxwell-Boltzmann ultrarelativistic ideal gas in the framework of the Tsallis-1 statistics in the grand canonical ensemble can be written as [6]

$$Z=\sum_{N=0}^{N_0}\frac{\tilde{\omega}^N}{N!}\frac{\Gamma\left(\frac{1}{q_c-1}-3N\right)}{(q_c-1)^{3N}\Gamma\left(\frac{1}{q_c-1}\right)}\left[1+(1-q_c)\frac{\mu N}{T}\right]^{\frac{1}{1-q_c}+3N}, \tag{21}$$

where $N_0 < 1/(3(q_c-1))$. The terms with $N > N_0$ in the series (21) are divergent and they should be excluded. Note that N is not the number of particles.

The mean occupation numbers for the Maxwell-Boltzmann ultrarelativistic ideal gas in the Tsallis-2 statistics in the grand canonical ensemble for $q_c > 1$ can be written as [6]

$$\langle n_{\vec{p}\sigma}\rangle=\frac{1}{Z^{q_c}}\sum_{N=0}^{N_0}\frac{\tilde{\omega}^N}{N!}\frac{\Gamma\left(\frac{q_c}{q_c-1}-3N\right)}{(q_c-1)^{3N}\Gamma\left(\frac{q_c}{q_c-1}\right)}\left[1+(q_c-1)\frac{\varepsilon_{\vec{p}}-\mu(N+1)}{T}\right]^{\frac{q_c}{1-q_c}+3N}, \tag{22}$$

where the upper bound of summation N_0 is the same as in (21). Then the transverse momentum distribution for the ultrarelativistic particles can be written as

$$\frac{d^2N}{dp_Tdy}=\frac{gV}{(2\pi)^2}p_T^2\cosh y\,\frac{1}{Z^{q_c}}\sum_{N=0}^{N_0}\frac{\tilde{\omega}^N}{N!}\frac{\Gamma\left(\frac{q_c}{q_c-1}-3N\right)}{(q_c-1)^{3N}\Gamma\left(\frac{q_c}{q_c-1}\right)}\left[1+(q_c-1)\frac{\varepsilon_{\vec{p}}-\mu(N+1)}{T}\right]^{\frac{q_c}{1-q_c}+3N}. \tag{23}$$

In the Gibbs limit $q_c \to 1$, (23) recovers the Maxwell-Boltzmann transverse momentum distribution (14) of the Boltzmann-Gibbs statistics. Taking the zeroth term $N = 0$ in (23) and (21), we exactly obtain the transverse momentum distribution (15) of the Tsallis-factorized statistics.

To find the upper bound of summation N_0, we rewrite the partition function (21) in the form $Z = \sum_N \phi(N)$, where

$$\phi(N) = \frac{\tilde{\omega}^N}{N!} \frac{\Gamma\left(\frac{1}{q_c-1} - 3N\right)}{(q_c-1)^{3N}\Gamma\left(\frac{1}{q_c-1}\right)} \left[1 + (1-q_c)\frac{\mu N}{T}\right]^{\frac{1}{1-q_c}+3N}. \tag{24}$$

In the Gibbs limit $q_c \to 1$ (24) recovers the Maxwell-Boltzmann function of the Boltzmann-Gibbs statistics $\phi(N) = \frac{\tilde{\omega}^N}{N!} e^{\frac{\mu N}{T}}$.

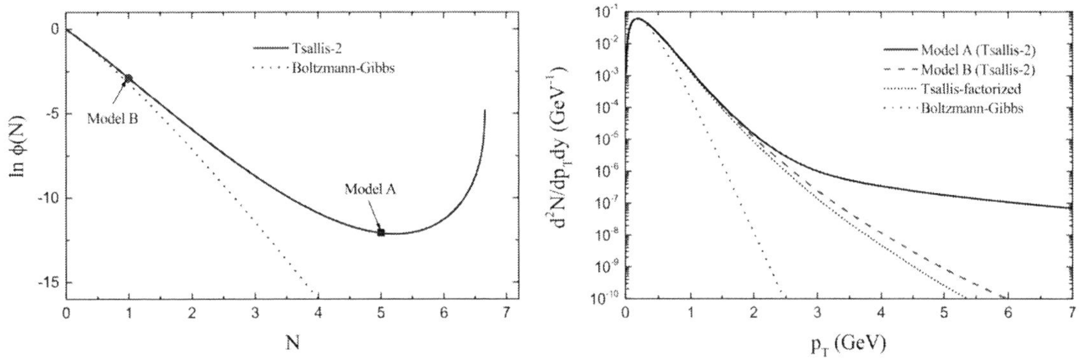

Figure 2. Two methods to fix the cut-off parameter N_0 in the Tsallis-2 statistics. The behavior of $\ln\phi(N)$ as a function of N (left panel) and the transverse momentum distribution (right panel) for the ultrarelativistic ideal gas of π^0 pions in the Tsallis-2 statistics. The values of the parameters are the same as in figure 1.

In the Model A, the cut-off parameter N_0 is found from the local minimum of the function $\ln\phi(N)$ substituting (24) into (17). However, in the Model B, the cut-off parameter N_0 is found from the inflection point of the function $\ln\phi(N)$ substituting (24) into (18). The results of these two methods are illustrated in figure 2. Thus, in the Tsallis-2 statistics the transverse momentum distribution of the Model A differs essentially from the transverse momentum distribution of the Model B and the Tsallis-factorized statistics due to the large values of the cut-off parameter N_0. However, the results of the Model B are close to the results of the Tsallis-factorized statistics, because the values of N_0 in the Model B are close to zero.

4 Comparison of the Tsallis-1 and Tsallis-2 statistics

In figure 3, we compare the transverse momentum distributions of the Tsallis-1 statistics with the transverse momentum distributions of the Tsallis-2 statistics in the Model A and the Model B. The parameter q of the Tsallis-1 statistics was obtained by the equality $q = 1/q_c$. We have obtained that the results of the Tsallis-1 statistics practically coincide with the results of the Tsallis-2 statistics. Thus, the ultrarelativistic transverse momentum distribution of the Tsallis-1 statistics is transformed to the ultrarelativistic transverse momentum distribution of the Tsallis-2 statistics by identifying $q \to 1/q_c$.

DOI: 10.1051/epjconf/201713803008

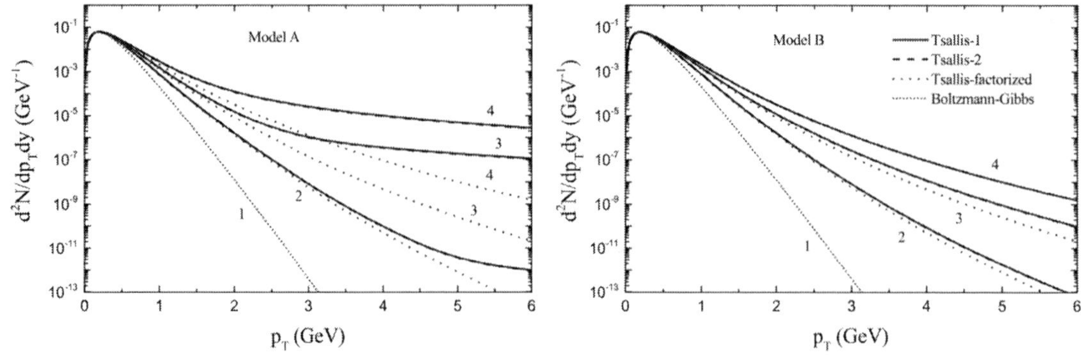

Figure 3. The transverse momentum distribution for the ultrarelativistic ideal gas of π^0 pions in the Model A (left panel) and the Model B (right panel) for the Tsallis-1 and Tsallis-2 statistics in the volume $V = 1$ fm at the temperature $T = 90$ MeV, chemical potential $\mu = 0$, rapidity $y = 0$ and different values of q_c $(q = 1/q_c)$. The lines $1, 2, 3$ and 4 correspond to $q_c = 1, 1.03, 1.05$ and 1.07, respectively.

5 Conclusions

The analytical formulae for the ultrarelativistic transverse momentum distributions of the Tsallis-1 and Tsallis-2 statistics were derived in detail. To regularize them, we use two cut-off prescriptions: the Model A and the Model B. It was revealed that the cut-off parameter strongly influences the behavior of the transverse momentum distribution in both the Tsallis-1 and Tsallis-2 statistics. In the Model A, in which the cut-off parameter is found from the local minimum of the function $\ln \phi(N)$, the transverse momentum distribution differs essentially from the transverse momentum distribution of the Tsallis-factorized statistics. However, in the Model B, in which the cut-off parameter is found from the inflection point of the function $\ln \phi(N)$, the transverse momentum distribution is very close to the transverse momentum distribution of the Tsallis-factorized statistics. We have also found that the ultrarelativistic transverse momentum distribution of the Tsallis-1 statistics is transformed to the momentum distribution of the Tsallis-2 statistics by identifying $q \to 1/q_c$.

Acknowledgments: This work was supported in part by the joint research project and grant of JINR and IFIN-HH (protocol N 4543).

References

[1] J. Cleymans, D. Worku, J. Phys. G: Nucl. Part. Phys. **39**, 025006 (2012)
[2] J. Cleymans, D. Worku, Eur. Phys. J. A **48**, 160 (2012)
[3] C. Tsallis, J. Stat. Phys. **52**, 479 (1988)
[4] C. Tsallis, R.S. Mendes, A.R. Plastino, Physica A **261**, 534 (1998)
[5] A.S. Parvan, Eur. Phys. J. A **52**, 355 (2016)
[6] A.S. Parvan, arXiv:1607.07670v2 [hep-ph]
[7] A.S. Parvan, O.V. Teryaev, J. Cleymans, arXiv:1607.01956v2 [nucl-th]
[8] A.S. Parvan, Eur. Phys. J. A **51**, 108 (2015)
[9] D. Prato, Phys. Lett. A **203**, 165 (1995)

Heavy-ion collisions in a fixed target mode at the LHC beams

Alexey Kurepin[1,*] and *Nataliya* Topilskaya[1,**]

[1] *Institute for Nuclear Research, RAS, Moscow, Russia*

Abstract. The interaction of high-energy LHC beams with fixed target, including polarized nuclei targets, can expand the range of fundamental physical investigations accessible at CERN. High-intensity beam of protons and lead ions accumulated in the LHC collider allows to apply the gas-target system like the LHCb SMOG or HERMES systems. It is also possible to use the beam halo by placing in the halo the fixed target in the form of thin ribbon or use a bend crystal to extract the beam. In the extracted beam it is possible to install a polarized target. Using the proton and ion beams of the LHC with fixed targets, the data in the energy interval between maximum energy of the SPS and the nominal RHIC energy in p-A and A-A collisions could be obtained. The fixed target mode allows the intensive study of rare processes, the study of polarization phenomena, the measurements of the parameters needed to analyze the data of cosmic rays and neutrino astrophysics, detailed study of the processes of quarkonia production and suppression. The high statistics data on quarkonium production at these energies will give the possibility to clarify the mechanism of production, to investigate the importance of recombination process and the energy dependence on the phase transition of nuclear matter to the quark-gluon phase. Also the physical program includes the study of the Drell-Yan process, D-meson production, flow and spin physics.

1 Introduction

At high temperature and at large energy density the existence of Quark-Gluon Plasma (QGP) is predicted by lattice Quantum Chromo Dynamics (QCD). But despite of big success in recent years at RHIC and LHC colliders the properties of nuclear matter at extreme conditions remain poorly understood. Progress in our understanding of the properties of nuclear matter requires systematic and high statistical measurements also in low energy region. Unfortunately, the data produced by the beam energy scan at RHIC have large statistical and systematical errors. The luminosity in collider experiments at RHIC decreases at low energy. The fixed target experiments at the LHC beams will have the number of advantages compared with the experiments on collider. Using the high-intensity LHC beams will significantly increase the statistical accuracy of the data and provide better acceptance and efficiency of measurements. It is possible to study p-A and Pb-A collisions with different nuclear targets and to use polarized targets. The fixed target experiments give possibility to access to high x_F Feynman domain by measuring the far backward region in the center-of-mass (c.m.) system. If the proton and ion beams would be used at the LHC with the fixed target, the energy interval between SPS

*e-mail: kurepin@inr.ru
**e-mail: topilska@inr.ru

 DOI: 10.1051/epjconf/201713803009

energy ($\sqrt{s_{NN}}$ = 29 GeV) and RHIC nominal energy ($\sqrt{s_{NN}}$ = 200 GeV) in p-A and A-A collisions could be covered. For proton beam with 450 GeV energy the energy in N-N c.m. is $\sqrt{s_{NN}}$ = 29.1 GeV. For pp and p-A collisions with the 7 TeV proton beam $\sqrt{s_{NN}}$ = 114.6 GeV with the rapidity shift 4.8 ($y_{c.m.} = 0 \rightarrow y_{lab} = 4.8$). For Pb-p and Pb-A collisions using lead beam with energy 2.75 TeV the c.m. N-N energy is $\sqrt{s_{NN}}$ = 71.8 GeV with the rapidity shift 4.3. Using LHC beams with reduced energy it would be possible to scan energy range between SPS and RHIC. This is a unique possibility to clarify the mechanism of charmonium, J/ψ and ψ(2S) production, to investigate the contribution of recombination process, since the probability of recombination decreases with decreasing energy of collision, and to search the critical point of the phase transition.

2 Quarkonium production at the CERN Large Hadron Collider LHC

At the LHC energies five quarkonium states from two families are under study, charmonia: J/ψ and ψ(2S) and bottomonia: Υ(1S), Υ(2S) and Υ(3S). In the color-screening model quarkonium states are suppressed in the QGP with different dissociation probabilities for the various mass states, depending on their binding energy [1]. The CMS Collaboration at the LHC reported the observation of the sequential suppression of bottomonium states in Pb-Pb collisions at $\sqrt{s_{NN}}$ = 2.76 TeV [2, 3]. However, other nuclear matter effects besides color screening, as well as cold nuclear matter (CNM) effects, regeneration and feed down from higher states and B-meson decay for J/ψ production should be taken into account.

The results of recent measurements by the ALICE Collaboration are compatible with a regeneration mechanism playing an important role in the production of J/ψ in Pb-Pb collisions at the LHC [4–6]. Additional J/ψ are expected to be produced from deconfined charm quarks by kinetic recombination in the QGP [7] or by statistical hadronization at the phase boundary [8]. Due to the lower production cross section of $b\bar{b}$ pairs compared to $c\bar{c}$ pairs, the regeneration of Υ(1S) is expected to be smaller than that of J/ψ [9].

On the other hand, effects related to the presence of CNM can also modify the production of quarkonia in nucleus-nucleus collisions. Cold nuclear matter effects can be separated into initial and final-state effects. Initial-state effects include shadowing and parton energy loss. The nuclear Parton Distribution Function (nPDF) is different from that of free nucleons (PDF). Since the gluon fusion mechanism dominates in heavy-quarks production, quarkonium production is particularly sensitive to the gluon nPDF, which is presently not well known. The heavy-quark pair can interact with the nuclear matter and eventually break up, that is normal nuclear absorption. The final-state resonance can also interact with surrounding comovers and lose energy or even break up. In four experiments ALICE, ATLAS, CMS and LHCb at the LHC, the quarkonium production was measured in different rapidity and transverse momentum ranges. The J/ψ production was measured in pp collisions at $\sqrt{s_{NN}}$ = 2.76, 7, 8 and 13 TeV and in Pb-Pb collisions at $\sqrt{s_{NN}}$ = 2.76 and 5.02 TeV. The quarkonium production in p-Pb collisions at $\sqrt{s_{NN}}$ = 5.02 TeV was measured for detailed investigation of the cold nuclear matter effects and investigation of production mechanism. In 2016 p-Pb collisions would be measured also at $\sqrt{s_{NN}}$ = 8 TeV.

The data of the J/ψ production in pp collision at the same energy as in Pb-Pb collision, gives the reference for extracting the nuclear modification factor R_{AA}. There is a good agreement for pp collision between the data obtained in ALICE, LHCb, CMS and ATLAS experiments in the same kinematical domains. The contribution of B-decay to J/ψ production cross section is approximately 10 % for p_T near 1.5 GeV/c [11]. It depends on rapidity and increases for larger J/ψ transverse momentum.

Charmonium production in Pb-Pb collisions at 2.76 and 5.02 TeV was measured in ALICE, CMS and ATLAS experiments. In ALICE experiment the transverse momentum of inclusive J/ψ mesons was measured from values near zero up to 12 GeV/c in rapidity range $|y| < 0.9$ (for J/ψ decay into two electrons) and $2.5 < y < 4$ (for muons channel). The comparison of PHENIX and ALICE R_{AA} results is shown in figure 1 [6, 12].

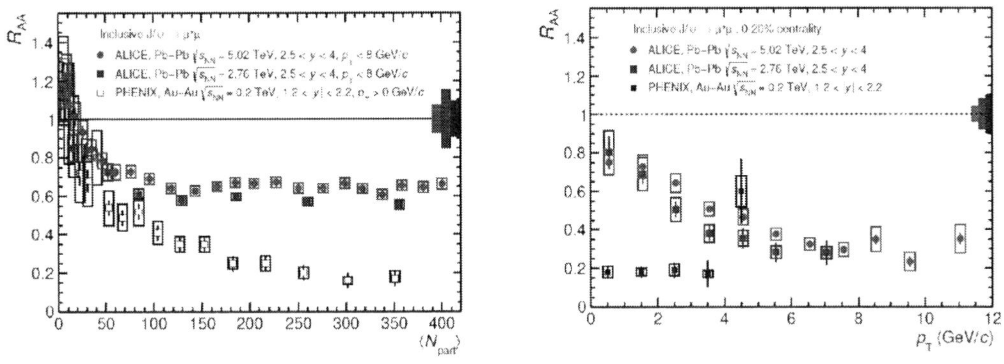

Figure 1. The comparison of PHENIX and ALICE data of R_{AA}. The left panel shows R_{AA} depending on centrality and the right panel on transverse momentum.

The ALICE data show less suppression compared to PHENIX results especially at low p_T [5]. At the LHC the charm quark density produced in the collisions increases with respect to RHIC energy. This may enhance the probability to create J/ψ from recombination of charm quarks. This contribution leads to an increase of the R_{AA} at the LHC with respect to RHIC and from forward- to mid-rapidity, where the density of charm quarks is higher. Moreover, in order to recombine, two charm quarks need to be close enough in phase space, so the effect will be larger at low transverse momentum of J/ψ. Comparison of the inclusive J/ψ production in Pb-Pb collisions at $\sqrt{s_{NN}} = 5.02$ TeV with $\sqrt{s_{NN}} =$ 2.76 TeV data shows the same behavior and values in spite of the expected larger values of the charm quark density. However the theoretical calculations by statistic and dynamical models provide a good description of the measurements within uncertainties [6].

Inclusive J/ψ and $\psi(2S)$ productions have been studied at the LHC in p-Pb collisions at energy $\sqrt{s_{NN}} = 5.02$ TeV. The ALICE and LHCb results are in good agreement. At ALICE the measurement is performed in the center of mass rapidity domains $2.03 < y < 3.53$ and $-4.46 < y < -2.96$ in muon channel and in electron channel in the center of mass rapidity domains $-1.37 < y < 0.43$ [13]. While at forward rapidity, corresponding to the proton direction, a suppression of the J/ψ yield with respect to binary-scaled pp collisions is observed, in the backward region no suppression is presented. Theoretical models including parton shadowing and coherent energy loss mechanism describe J/ψ production but could not reproduce additional suppression of $\psi(2S)$ [14].

The inclusive production cross sections of $\Upsilon(1S)$ and $\Upsilon(2S)$ as a function of p_T and rapidity have been measured using ALICE detector at forward rapidity in pp collisions at $\sqrt{s_{NN}} = 7$ TeV [15]. The results are in good agreement with measurements from LHCb experiment [16] over similar p_T and rapidity ranges. They are complemented by CMS measurements at mid-rapidity [17]. They are also in good agreement with NRQCD calculations.

The ALICE measurement of inclusive $\Upsilon(1S)$ production in Pb-Pb collisions was carried out at forward rapidity [18]. A strong suppression is observed with respect to pp collisions scaled by the number of independent nucleon-nucleon collisions. The ALICE results are compared with those of

CMS ($|y| < 2.4$. The observed suppression is stronger at forward than at mid-rapidity. The model [9] predicts a nearly constant R_{AA} which is in disagreement with ALICE and CMS data.

The production of inclusive Υ(1S) and Υ(2S) was measured also in p-Pb collisions at $\sqrt{s_{NN}}$ = 5.02 TeV in ALICE. Theoretical models with partonic energy loss plus shadowing could describe the data at forward rapidity but underestimate the suppression at backward rapidity.

Quarkonium production at LHC is a useful probe for the investigation of QGP formation and for testing pQCD models in pp scattering. For p-Pb collisions theoretical models including nuclear suppression, parton shadowing and coherent energy loss mechanisms could reproduce J/ψ production, but fail to describe additional suppression of ψ(2S) and underestimate the observed Υ(1S) suppression at forward rapidity. In Pb-Pb collisions the evidence for strong J/ψ suppression at large transverse momentum and additional J/ψ production from regeneration at low p_T was obtained. However, in order to get the ultimate data on the explanation of all observed effects in the anomalous charmonia states suppression due to the formation of Quark-Gluon Plasma we need to investigate the problem at intermediate energies.

3 Experiments in fixed target mode at the LHC beams

3.1 Quarkonium production at fixed target

Progress in our understanding of the properties of nuclear matter requires systematic and high statistical measurements also in low-energy region. The energy intervals between SPS, RHIC and LHC are very important to study the mechanism of quarkonium production and suppression, in order to investigate medium effects and conditions of quark-gluon plasma formation. Unfortunately, beam energy scan at RHIC produced data with low statistic and large systematic errors because the luminosity at RHIC decreases with energy reduction. It is important to study the mechanism of quarkonium production and suppression at low energies up to 35 GeV/per nucleon at the FAIR in CBM and at NICA collider in Dubna with high statistics. At CERN SPS the fixed target experiment CHIC for charmonium study at energy up to 20 GeV is under preparation. At RHIC there are experiments at fixed target at STAR installations. If the proton and ion beams would be used at the LHC with the fixed target, the energy interval between SPS energy ($\sqrt{s_{NN}}$ = 29 GeV) and RHIC nominal energy ($\sqrt{s_{NN}}$ = 200 GeV) in p-A and A-A collisions could be covered [19].

3.2 Three approaches for the fixed target experiments at the LHC. The possibility to use polarized target

There are three options of using high-intensity LHC beams for fixed target experiment AFTER (A Fixed Target ExpeRiment). The target in the form of thin ribbon could be placed around the main orbit of the LHC. The life time of the beam is determined by the beam-beam and beam-gas interactions. Therefore after some time the particles will leave the main orbit and interact with the target ribbon. So for fixed target measurements only halo of the beam will be used and the target can provide some extra collimation. Therefore no distortion of the main beam will be introduced. Since the target ribbon should not interfere during the beam formation and acceleration process it should be lifted in the working position after the tuning of the beam [19, 20]. This type of experiment could be carried out at the existing installations (such as ALICE or LHCb). The experiment with the target in the form of thin ribbon placed in the beam halo may be the first stage of the multi-purpose experiment AFTER [21].

The second option is the installation of the internal gas target like the LHCb SMOG or HERMES systems. The internal gas system SMOG with low gas density is currently used by LHCb Collaboration as a luminosity monitor and was initially designed for the monitoring the transverse size of the beam. It works during extended period of time without any influence on the other LHC experiments. In 2012 and 2013 LHCb measured p-Ne and Pb-Ne collisions with the c.m. NN energy $\sqrt{s_{\rm NN}}$ = 86.6 GeV for proton beam and 54.5 GeV for lead beam. During 2015 year the measuring of p-Ne, p-He and p-Ar collisions at $\sqrt{s_{\rm NN}}$ = 110.4 GeV and Pb-Ar collisions at $\sqrt{s_{\rm NN}}$ = 68.6 GeV were successfully performed. No distortion of the LHC performance was observed. Using the system like the HERMES experiment at HERA the higher gas pressure could be obtained with the possibility of the gas target polarization [22].

The third option consists of using the bent crystal placed in the halo of the LHC beams. Several protons or lead ions in the bunch could be channeled along the crystal curvature. The extracted beam can be used with an extra beamline and special installation for polarization of the target. This extraction allows to obtain very clean and collimated beam. This technique does not affect the LHC performance as it will use only halo of the beam. Similar technique is investigated and currently used by the UA9 Collaboration as a possibility to improve the LHC beam collimation. In the case of the gas target the polarization is possible. Using the target similar to HERMES system, the effective polarization of H or D can reach 0.8. Using extracted beam, different systems for target polarization could be used. In general, H or D gas targets provide good possibility for high value of polarization with similar data taking periods. The effective gas pressure could be similar to the HERMES and polarized target would have a length near one meter [22].

3.3 The proposed measurements

The experiment AFTER has a wide physical program, gives possibility to use different beams and targets of large thickness including the polarized ones. The physical program includes quarkonium production, the study of the Drell-Yan process, investigation of D-meson production, the study of flow, spin physics, energy scan for the search of the critical point of phase transition of nuclear matter to Quark-Gluon Plasma. Quarkonia are the important probes for investigation the deconfinement at ultra-relativistic heavy-ion collisions. But the mechanism of quarkonium production and suppression is not well understood in some energy and kinematical domain, especially at large Feynman $x_{\rm F}$, where cold nuclear matter effects expected to be large. The high statistical data in these domains give possibility to understand the mechanism of production and to estimate the contribution of recombination process. It would be possible to investigate production not only J/ψ, ψ(2S) and Υ(nS) mesons, but also η_c and χ_c states. These measurements require high statistics, large acceptance and good efficiency. Investigation of Drell-Yan process is very promising task in itself, but it is also background process in quarkonium production. The study of D-meson production is an important part of the measurement of charmonium production and dissociation.

Now RHIC is the only one collider of polarized protons with possibility to study spin physics at ultra-relativistic energies. There are no polarized beams at the Fermilab Tevatron or at the LHC. The polarized beams are planned to be build at the NICA collider in Dubna at low energy. But it becomes possible to perform new investigations of single transverse spin asymmetries (STSA), A_N, using a polarized fixed target at the LHC beams. It is connected with the Sivers effect [23] observed many years ago and provided the correlation of the transverse momentum of the partons, k_N to the nucleon spin. This effect depends on transverse momentum distribution (TMD). In the case of fixed target experiment AFTER at the LHC beams because of the boost between laboratory and centre-of-mass systems, measurements of the STSAs in the rapidity range of experiments ALICE or LHCb would investigate the region of large parton momentum fraction in the polarized hadron, $x^{\uparrow}$ [21]. This is the range, where asymmetries are expected to be the largest. In the case of Drell-Yan, the quark Sivers function is predicted to have an opposite sign to the Deep Inelastic Scattering (DIS). It was investigated in the study of the Drell-Yan process in COMPASS at CERN and in Fermilab experiments with polarized targets at low values of $x^{\uparrow}$. The experiment AFTER could provide measurement of STSAs of Drell-Yan over a wide range of $x^{\uparrow}$ in a single experiment.

Practically nothing is known about the gluon Sivers effect. The PHENIX collaboration was the first to investigate STSAs of gluon sensitive probes in J/ψ production. The value of A_N was found to be compatible with zero but with high errors, allowing the values as large as 20 %. In the experiment AFTER provided in the high luminosity at the LHC beams it would be possible to measure STSAs for bottomonia, spin-singlet states as η_c, to investigate gluon Sivers effect and importance of the gluon orbital angular momentum to solve the nucleon-spin problem.

4 Conclusions

The collisions of the high-energy LHC beams with fixed target, including polarized nuclei targets, can expand the range of fundamental physical investigations accessible at CERN. Using the proton and ion beams at the LHC with fixed targets, the data in the energy interval between maximum energy of the SPS and the nominal RHIC energy in p-A and A-A collisions could be obtained. The fixed target mode opens possibility for precise and unique measurements and advances our understanding in the fields of heavy-ion, hadron, spin and astroparticle physics, study of flow, energy scan for searching the critical point of the phase transition of matter to Quark-Gluon Plasma (QGP).

References

[1] T. Matsui and H. Satz, Phys. Lett. B **178**, 416 (1986)
[2] S. Chatrchyan et al., CMS Collaboration, Phys. Rev. Lett. **109**, 222301 (2012)
[3] S. Chatrchyan et al., CMS Collaboration, Phys. Rev. Lett. **107**, 052302 (2011)
[4] B. Abelev et al., ALICE collaboration, Phys. Rev. Lett. **109**, 072301 (2012)
[5] B. Abelev et al., ALICE collaboration, Phys. Lett. B **743**, 314 (2014)
[6] J. Adams et al., ALICE collaboration, arXiv:1606.08197 [nucl-ex]
[7] R. L. Thews et al., Eur. Phys. J C **43**, 97 (2005). R. L. Thews and M.I. Mangano, Phys. Rev. C **73**, 014904 (2006)
[8] A. Andronic, P. Braun-Munzinger, K. Redlich, and J. Stachel, Phys. Rev. Lett. **659**, 149 (2008)
[9] A. Emerick, X. Zhao and R. Rapp, Eur. Phys. J A **48**, 72 (2012)
[10] H. Fujii, K. Watanabe, Nucl. Phys. A **915**, 1 (2013)
[11] T. Dahms, CMS Collaboraion, J. Phys. G **38**, 124105 (2011)
[12] A. Adare et al., PHENIX Collaboration, Phys. Rev. C **84**, 054912 (2011)
[13] B. Abelev et al., ALICE collaboration, JHEP **02**, 073 (2014)
[14] B. Abelev et al., ALICE collaboration, JHEP **12**, 073 (2014)
[15] B. Abelev et al., ALICE collaboration, Eur. Phys. J C **73**, 2662 (2013)
[16] R. Aaij et al., LHCb collaboration, Eur. Phys. J C **72**, 2100 (2012)
[17] S. Chatrchyan et al., CMS Collaboration, Phys. Lett. B **727**, 101 (2013)
[18] B. Abelev et al., ALICE collaboration, Phys. Rev. Lett. B **738**, 361 (2014)
[19] A. B. Kurepin, N. S. Topilskaya and M. B. Golubeva, Phys. Atom. Nucl. **74**, 446 (2011)
[20] A. B. Kurepin and N. S. Topilskaya, Adv. High Energy Phys. **2015**, 760840 (2015)
[21] S. J.Brodsky, F. Fleuret, C. Hadjidakis, J. P. Lansberg, Phys. Rept. **522**, 239 (2013)
[22] J. P. Lansberg et al., PoS (PSTP2015) 042 (2015)
[23] D. W. Sivers, Phys. Rev. D **41**, 83 (1990)

Long-range rapidity correlations between mean transverse momenta

Vladimir Vechernin[1,★]

[1] *Saint Petersburg State University, 7/9 Universitetskaya nab., St.Petersburg, 199034 Russia*

Abstract. The forward-backward correlation strength between event-mean transverse momenta of particles produced in separated rapidity intervals in high energy hadronic collisions is analyzed in the simple model with the string fusion on a transverse lattice. The coefficient of long-range rapidity p_t-p_t correlation is obtained using the negative binomial distribution for the number of strings, which is closer to reality than the poissonian one. The results demonstrate good agreement with the asymptotes of the correlation coefficient analytically calculated earlier at large string density. It is observed that at LHC energy the p_t-p_t correlation coefficient reveals the drop for most central collisions. The model calculations show that the physical reason for this decrease is the attenuation of color field fluctuations due to the string fusion at large string density, whereas at RHIC energy the string density is not enough to provide a decline of the correlation coefficient for most central collisions.

1 Introduction

As expected, the experimental studies of long-range rapidity correlations can give us the information about the initial stage of high energy hadronic interactions [1]. In [2] the study of the long-range forward-backward (FB) correlations between multiplicities in two separated rapidity windows to find a signature of the string fusion and percolation phenomenon [3–5] in ultrarelativistic heavy ion collisions has been proposed. Later it was realized [6–12] that the investigations of the FB correlations involving intensive observables, such e.g. as the event-mean transverse momentum, enable to obtain clearer signal about the initial stage of hadronic interaction, including the process of string fusion, compared to usual FB multiplicity correlations.

In the present paper we consider the correlation between mean-event transverse momenta of charged particles in separated rapidity intervals [9, 11, 12], as an example of the correlation between intensive observables:

$$p_{tF} = \frac{1}{n_F}\sum_{i=1}^{n_F} |\mathbf{p}_{tF}^i| , \qquad p_{tB} = \frac{1}{n_B}\sum_{i=1}^{n_B} |\mathbf{p}_{tB}^i| . \tag{1}$$

We perform the MC calculations in the model, which allows to take into account the string fusion effects by introducing a lattice in transverse plane [7, 8]. We obtain the coefficient of long-range

★e-mail: v.vechernin@spbu.ru

rapidity p_t-p_t correlation, b_{pp}, using the negative binomial distribution (NBD) for the fluctuation in number of strings, which is closer to reality than the poissonian one, which was exploited in the previous works [9, 11]. The results demonstrate good agreement with the asymptotes of the correlation coefficient, analytically calculated at large string density in [12]. We also discuss the impact of the obtained results on the b_{pp} correlation coefficient centrality dependence at RHIC and LHC energies.

2 Definitions

To quantify the FB correlation between the event-mean transverse momenta, (1), measured event-by-event in rapidity intervals, $\delta\eta_F$ and $\delta\eta_B$, separated by some rapidity gap, $\Delta\eta$, one usually introduces the FB correlation function (the regression) [9, 11, 12]:

$$\langle p_B \rangle_{p_F} = f(p_F)\,, \tag{2}$$

which gives the dependence of the mean value of the variable p_B on the variable p_F. (From here we will omit the subscript t.) As usually we will characterize the strength of the FB correlation by the correlation coefficient b_{pp} [13, 14]:

$$b_{pp} = \frac{\langle p_F p_B \rangle - \langle p_F \rangle \langle p_B \rangle}{\langle p_F^2 \rangle - \langle p_F \rangle^2} = \frac{\mathrm{cov}(p_F, p_B)}{D_{p_F}}\,. \tag{3}$$

Another possible option [8, 9, 12]:

$$b'_{pp} = \left. \frac{d\langle p_B \rangle_{p_F}}{dp_F} \right|_{p_F=\langle p_F \rangle}\,. \tag{4}$$

These two definitions coincide, when the correlation function (2) is linear (a linear regression), yet may differ in a general case (see discussion in [15]).

Due to the locality of strong interaction in rapidity the FB correlation naturally divides into two parts: the short- and the long-range ones. The short-range correlation takes place only between particles produced from a same source (string). It goes to zero for large enough rapidity gaps, $\Delta\eta \gtrsim 2$. The origin of the long-range correlations at large $\Delta\eta$ is event-by-event variance in the number and properties of particle sources (strings [16], cut pomerons [17] and so on), formed in high energy hadronic interaction. In present paper we will consider only this long-range part of the correlation, as we are interested in the study of the variations in number and properties of the strings, produced at the initial stage of the hadronic interaction.

Moreover for the considered correlation between event-mean transverse momenta, compared to the correlation between multiplicities, we expect smaller influence on results of the trivial fluctuations in the number of sources (strings) - the so-called "volume" fluctuations. Being intensive variables, the p_F and p_B do not depend directly on the number of strings. They are sensitive to changes in the characteristics of the sources, e.g. due to the string fusion processes.

3 The model

We will take into account the string fusion effects in simplified form by introducing a lattice in the transverse plane. This discrete version of string fusion model [5] was introduced in [7] and then was exploited for a description of various phenomena (correlations, anisotropic azimuthal flows, the ridge) in high energy hadronic collisions [8–12, 18–20]. In this approach one splits the impact parameter plane into M cells with the area equal to the transverse area of a single string, σ_{str}, and supposes the fusion of all strings with the centers in a given cell. Then each string configuration can be completely defined by the set of integers:

$$C_\eta = \{\eta_1, ..., \eta_M\} \ , \tag{5}$$

where η_i is a number of initial strings fused in a given i-th cell . This leads to the splitting of the transverse area into domains with different, fluctuating values of color field within them.

We suppose that the η_i in each cell fluctuates around the same mean value η with a scaled variance ω_η. Then in accordance with the string fusion prescription [5] the mean number of charged particles in forward and backward observation windows, produced from the fragmentation of fused strings in the i-th cell, is given by the expressions

$$\overline{n}_i^F = \mu_F \sqrt{\eta_i} \ , \qquad \overline{n}_i^B = \mu_B \sqrt{\eta_i} \ , \tag{6}$$

where μ_F is a mean number of particles produced from decay of a single string in the forward window $\delta\eta_F$. It is clear that μ_F is proportional to the window width: $\mu_F = \mu_0 \, \delta\eta_F$ and the same for the backward window. Because we study only LRC, we also suppose the independent fragmentation of the i-th fused string in forward and backward windows. So the numbers of particles n_i^F and n_i^B fluctuates independently around the above mean values with some scaled variance ω_μ.

Finally, in accordance with the string fusion ideas [5] we suppose that the transverse momentum distribution of particles, produced from the η_i fused strings, has enlarged mean transverse momentum:

$$\overline{p}(\eta_i) = p_0 \sqrt[4]{\eta_i} \ , \tag{7}$$

where p_0 is a mean transverse momentum for particles produced from a decay of a single string.

4 Correlation between event-mean transverse momenta

In present work we are using the definition (3) for the MC calculation of the correlation coefficient b_{pp}. This approach requires the calculation only some averages, entering in (3). Whereas all our previous calculations of the correlation coefficients in string fusion model [7–12] were based on the approach in which we have calculated at first the FB correlation function (2) and only then the corresponding correlation coefficient by formula (4).

Basically to calculate b_{pp} by direct MC simulations in this model one has, along with the string configuration C_η, (5), to generate also the following configurations:

$$C_n^F = \left\{n_1^F, ..., n_M^F\right\} \ , \quad C_n^B = \left\{n_1^B, ..., n_M^B\right\} \ , \tag{8}$$

$$C_p^F = \left\{p_1^{1F}, ..., p_1^{n_1^F F}; ...; p_M^{1F}, ..., p_M^{n_M^F F}\right\} \ , \tag{9}$$

and the same for C_p^B. Here n_i^F and n_i^B are the numbers of charged particles, produced in forward and backward window from the fragmentation of fused string in i-th cell and p_i^{jF} and p_i^{jB} are the transverse momenta of all n_i^F and n_i^B particles produced from i-th cell in forward and backward window.

Yet such a direct MC calculation of the correlation coefficient is not optimal. The approach, in which part of the computations is performed analytically, proves to be more practical. It allows considerably reduce the amount of MC simulations to get the same accuracy.

As it is shown in [21] in the model under consideration the ingredients, entering the numerator and denominator in the formula (3), can be presented, after averaging over the configurations C_p^F and C_p^B, as follows

$$\langle p_F \rangle = \sum_i \langle \langle n_i^F / n_F \rangle_{C_n^F} \overline{p}(\eta_i) \rangle_{C_\eta} \ , \qquad \langle p_B \rangle = \sum_i \langle \langle n_i^B / n_B \rangle_{C_n^B} \overline{p}(\eta_i) \rangle_{C_\eta} \ , \tag{10}$$

DOI: 10.1051/epjconf/201713803010

$$\langle p_F p_B\rangle = \left\langle \left[\sum_i \langle n_i^F/n_F\rangle_{C_n^F} \overline{p}(\eta_i)\right]\left[\sum_k \langle n_i^B/n_B\rangle_{C_n^B} \overline{p}(\eta_k)\right]\right\rangle_{C_\eta} , \tag{11}$$

$$\langle p_F^2\rangle = \left\langle \sum_i \langle n_i^F/n_F^2\rangle_{C_n^F} d_{p_i}(\eta_i) + \sum_{i,k} \langle n_i^F n_k^F/n_F^2\rangle_{C_n^F} \overline{p}(\eta_i)\overline{p}(\eta_k)\right\rangle_{C_\eta} , \tag{12}$$

where $n_F = \sum_{i=1}^M n_i^F$ and $n_B = \sum_{i=1}^M n_i^B$. The $d_{p_i}(\eta_i) \equiv \overline{p^2}(\eta_i) - \overline{p}^2(\eta_i)$ is the variance of transverse momentum for particles produced from a decay of the fused string in the i-th cell.

For popular transverse momentum distributions with one-dimensional parameter $\tilde{p}$, due to dimensional reasons, we have from (7) $\tilde{p} \sim \sqrt[4]{\eta_i}$. This leads to the following connection:

$$d_{p_i}(\eta_i) = \gamma \overline{p}^2(\eta_i) , \tag{13}$$

where the dimensionless parameter γ depends only on the type of the distribution and is not changed by string fusion (see [11, 21]).

In paper [8] it was shown that for the poissonian distribution in n_i^B averaging of the n_i^B/n_B over configurations C_n^B reduces to the simple substitution:

$$n_i^B \to \overline{n}_i^B = \mu_B \overline{n}(\eta_i) = \mu_B \sqrt{\eta_i} . \tag{14}$$

In general case we can use this substitution for n_i^B and the similar one for n_i^F only as some approximation. The MC simulations show that this approximation works in some cases very well [22, 23]. In this approximation the formulae (10)-(12) are simplified to

$$\langle p_F\rangle = \langle p_B\rangle = \left\langle \frac{\sum_i \overline{n}(\eta_i)\overline{p}(\eta_i)}{\sum_k \overline{n}(\eta_k)}\right\rangle_{C_\eta} \qquad \langle p_F p_B\rangle = \left\langle \left[\frac{\sum_i \overline{n}(\eta_i)\overline{p}(\eta_i)}{\sum_k \overline{n}(\eta_k)}\right]^2\right\rangle_{C_\eta} ,$$

$$\langle p_F^2\rangle = \gamma \left\langle \frac{\sum_i \overline{n}(\eta_i)\overline{p}^2(\eta_i)}{[\sum_k \overline{n}(\eta_k)]^2}\right\rangle_{C_\eta} + \langle p_F p_B\rangle . \tag{15}$$

The remaining averaging over configurations C_η in (15) can be done easily by MC simulations.

Taking into account that the event-by-event fluctuation in the number of strings, $N = \sum_i \eta_i$, is non-poissonian in both pp and AA collisions (see e.g. [24]), in present work we have used the NBD as the distribution in the number of strings, which is closer to reality than the poissonian one.

In figure 1 we present the value of the correlation coefficient b_{pp}, (3), between event-mean transverse momenta as a function of η (the η/σ_{str} is a string density), calculated by MC simulations using the formulae (10)-(12) (the points) and by simplified MC simulations based on the formulae (15) (the solid curves). For comparison the results for both the NBD and the poisson distribution in a number of strings are presented (upper and lower curves). In the same figure the asymptotes of b_{pp} at large string density, analytically calculated by the formula

$$b_{pp} = \omega_\eta \mu_F / (\omega_\eta \mu_F + 16\gamma \sqrt{\eta}) , \tag{16}$$

obtained in [12], are also presented by the dashed curves.

We see in figure 1, comparing the points and the solid curves, that the simplified MC procedure (15), based on the substitution (14), works very well both for the NBD and the poisson distribution of the number of strings. We see also that the value of the correlation coefficient calculated using the NBD with $\omega_\eta = 2$ is approximately two times higher then for the poisson distribution, as we could expect from the asymptotic formula (16). For both distributions we observe also the good agreement

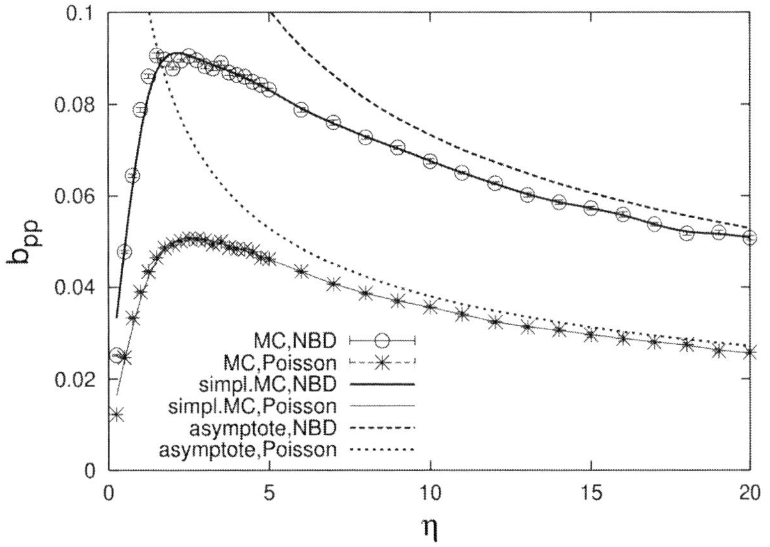

Figure 1. The coefficient b_{pp}, (3), of the long-range FB correlation between event-mean transverse momenta, arising due to string fusion processes, as a function of η (the η/σ_{str} is a string density), calculated for the NBD ($\omega_\eta = 2$) and the poisson distribution ($\omega_\eta = 1$) in a number of strings (upper and lower curves). The results are obtained by 100000 MC simulations, using (10)-(12), (the points) and by simplified MC simulations, using (15), (the solid curves). The dashed curves - the asymptotes of b_{pp}, (16), analytically calculated at large string density in [12]. The value of other parameters: $\mu_F = \mu_B = 1$ and $\omega_\mu = 1$ (the multiplicity and scaled variance in the observation window for a single string); $M = 64$ (the number of cells); γ=0.5, see the formula (13).

of the MC results with the analytical asymptotes, (16), found in [12] at large string density, yet this asymptote starts to work at larger string density η for the NBD, than for the poisson one.

Basing on the dependency, shown in figure 1, the general behavior of the p_t-p_t correlation coefficient, b_{pp}, as a function of the heavy ion collision energy and centrality can be qualitatively understood. We have to take into account that the maximum string density, achieved for 0-10% of the most central Au-Au collisions at energy 200 GeV/nucleon in RHIC, corresponds to the value η=2.88±0.09 and the one for 0-5% of the most central Pb-Pb collisions at energy 2.76 TeV/nucleon at LHC reaches η=10.56±1.05 [11, 25]. So, depending on the collision centrality, it varies approximately in the intervals (0, 3) for RHIC and (0, 11) for LHC.

In figure 1 we see that the increase of the string density η with centrality from 0 to 3 leads to the monotonic growth of the correlation coefficient, which at RHIC energies reaches the highest value in the most central collisions. Whereas at LHC energies the correlation coefficient reaches the highest value at some intermediate value of centrality, which corresponds to the string density $\eta \simeq 3$. The further increase of η with centrality from 3 to 11 leads to the decrease of the correlation coefficient, in contrast with the situation at RHIC energies. So at LHC energies the p_t-p_t correlation coefficient reaches the highest value in semicentral collisions at some intermediate value of centrality.

Note that this decrease of the correlation coefficient for the most central collisions at LHC energies corresponds to the decrease of the asymptotic (16) with the growth of η. It is clear from the figure 1 that the RHIC data, corresponding to the η region (0, 3), can not be described by the asymptotic formula (16), which starts to work at $\eta \gtrsim 7 \div 8$. This asymptotic decrease with string density, is reached only for central heavy ion collisions at LHC energies.

5 Concluding remarks

The forward-backward correlation strength between event-mean transverse momenta of particles produced in separated rapidity intervals in high energy hadronic collisions is analyzed in the simple model with the string fusion on a transverse lattice. It is found that at LHC energy the p_t-p_t correlation coefficient must reveal the drop for most central collisions. The model calculations show that the physical reason for this decrease is the attenuation of color field fluctuations due to the string fusion at large string density, whereas at RHIC energy the string density is not enough to provide a decline of the correlation coefficient even for most central collisions. This qualitative conclusion is confirmed by the results [26, 27], obtained in more realistic dipole-based MC string fusion model [28, 29].

The developed methods can be applied to the analysis of the long-range rapidity correlations involving heavy flavor particles. Assuming the sensitivity of the heavy flavor particle yields to the string fusion processes it seems fruitful to study the correlation between the event-mean transverse momentum and the relative yield of strange and charmed particles.

The research was funded by the grant of the Russian Science Foundation (project 16-12-10176).

References

[1] A. Dumitru, F. Gelis, L. McLerran, R. Venugopalan, Nucl. Phys. A **810**, 91 (2008)
[2] N.S. Amelin et al., Phys. Rev. Lett. **73**, 2813 (1994)
[3] T.S. Biro, H.B. Nielsen, J. Knoll, Nucl. Phys. B **245**, 449 (1984)
[4] A. Bialas, W. Czyz, Nucl. Phys. B **267**, 242 (1986)
[5] M.A. Braun, C. Pajares, Phys. Lett. B **287**, 154 (1992). M.A. Braun, C. Pajares, Nucl. Phys. B **390**, 542 (1993)
[6] M.A. Braun, C. Pajares, Phys. Rev. Lett. **85**, 4864 (2000)
[7] V.V. Vechernin, R.S. Kolevatov, Vestn. SPbU, Ser.4, No.4, 11 (2004). arXiv: hep-ph/0305136
[8] M.A. Braun, R.S. Kolevatov, C. Pajares, V.V. Vechernin, Eur. Phys. J. C **32**, 535 (2004)
[9] ALICE collaboration et al., J. Phys. G **32**, 1295 (2006), [Sect. 6.5.15]
[10] V.V. Vechernin, R.S. Kolevatov, Phys. Atom. Nucl. **70**, 1797 (2007)
[11] V.V. Vechernin, R.S. Kolevatov, Phys. Atom. Nucl. **70**, 1809 (2007)
[12] V.V. Vechernin, Theor. Math. Phys. **184**, 1271 (2015)
[13] M.A. Braun, C. Pajares, V.V. Vechernin, Phys. Lett. B **493**, **54** (2000)
[14] J. Adam et al. [ALICE Collab.], JHEP **05**, 097 (2015)
[15] V.V. Vechernin, in *Proc. of Baldin ISHEPP XX v.2* (JINR, Dubna, 2011) 10. arXiv:1012.0214
[16] V. Vechernin, Nucl. Phys. A **939**, 21 (2015)
[17] A. Capella, A. Krzywicki, Phys. Rev. D **18**, 4120 (1978)
[18] M.A. Braun, C. Pajares, Eur. Phys. J. C **71**, 1558 (2011)
[19] M.A. Braun, C. Pajares, V.V. Vechernin, Nucl. Phys. A **906**, 14 (2013)
[20] M.A. Braun, C. Pajares, V.V. Vechernin, Eur. Phys. J. A **51**, 44 (2015)
[21] V.V. Vechernin, Theor. Math. Phys. (2016) [in press]
[22] E. Andronov, V. Vechernin, PoS (QFTHEP 2013) 054 (2014)
[23] E.V. Andronov, Theor. Math. Phys. **185**, 1383 (2015)
[24] V.V. Vechernin, H.S. Nguyen, Phys. Rev. C **84**, 054909 (2011)
[25] J. Dias de Deuset al., Eur. Phys. J. C **72**, 2123 (2012)
[26] V. Kovalenko, V. Vechernin, EPJ Web of Conferences **66**, 04015 (2014)
[27] V. Kovalenko, V. Vechernin, arXiv:1611.07274 (2016)
[28] V.N. Kovalenko, Phys. Atom. Nucl. **76**, 1189 (2013)
[29] V. Kovalenko, V. Vechernin, PoS (Baldin ISHEPP XXI) 077 (2012)

Critical temperature in weakly interacting multicomponent field theory

V.I. Yukalov[1,*] and *E.P.* Yukalova[2,**]

[1] *Bogolubov Laboratory of Theoretical Physics, Joint Institute for Nuclear Research, Dubna 141980, Russia*
[2] *Laboratory of Information Technologies, Joint Institute for Nuclear Research, Dubna 141980, Russia*

Abstract. A method is suggested for calculating the critical temperature in multicomponent field theory with weak interactions. The method is based on self-similar approximation theory allowing for the extrapolation of series in powers of asymptotically small coupling to finite and even infinite couplings. The extrapolation for the critical temperature employs self-similar factor approximants. The found results are in perfect agreement with Monte Carlo simulations.

1 Introduction

Nuclear matter under extreme conditions, when varying baryon density or temperature, can experience different phase transitions [1–7]. The first thing one needs in studying a phase transition is to find out the related transition point, which is not always an easy task. In the present report, we explain typical problems arising in calculating transition points by the example of an effective multicomponent field theory. The critical temperature in this theory, being calculated by means of loop expansions, leads to a series whose terms are divergent. So that one has to sum a series of divergent terms, while the resulting temperature is, of course, finite. We suggest a method for overcoming this difficulty and illustrate it by calculating the critical temperature of an N-component field theory for different N. Comparing our approach with Monte Carlo simulations, when these are available, we find that our method provides the results that are in beautiful agreement with the Monte Carlo data.

We emphasize that our aim in this report is not to consider a particular phase transition that could occur in nuclear matter under varying parameters, but to develop a general strategy that would be valid for calculating critical temperatures in any of phase transitions of second-order type. As examples of such critical phase transitions in nuclear matter, occurring under varying barion density or temperature, it is possible to mention the transition at the critical point between hadron gas and hadron liquid, the transitions between hadron liquid and nuclear superfluid, between quark-gluon plasma and colour superconductor, Bose condensation of multiquark clusters, possible transitions between gluon and gluball phases, and some scenarios of deconfinement transition [1–7]. In the present report, we treat

*e-mail: yukalov@theor.jinr.ru
**e-mail: yukalova@theor.jinr.ru

a model possessing those problems that are typical in calculating critical temperatures in practically any continuous phase transition.

In what follows, we employ the system of units, where the Planck and Boltzmann constants are set to unity.

We start with a nonrelativistic 3 + 1 - dimensional field theory with the action

$$S[\psi] = \int L[\psi]\, dt\,, \qquad L[\psi] = \int \psi^\dagger i\, \frac{\partial}{\partial t}\, \psi\, dx \;-\; H[\psi]\,, \tag{1}$$

whose Lagrangian contains the energy part

$$H[\psi] = \int \left[\psi^\dagger \left(-\frac{\nabla^2}{2m} - \mu \right) \psi + \frac{\lambda}{2}\, |\psi|^4 \right] dx\,, \tag{2}$$

with the coupling parameter

$$\lambda = 4\pi\, \frac{a_s}{m} \tag{3}$$

expressed through the scattering length a_s. The term $|\psi|^4$ is understood as $(\psi^\dagger\psi)^2$. In the three-dimensional space, the variable $x = \{x_\alpha : \alpha = 1, 2, 3\}$.

Being interested in static quantities at finite temperature, we pass to the imaginary time formalism by accomplishing the Wick rotation [8]. Then the field ψ is periodic with respect to the imaginary time, $\psi(x, 0) = \psi(x, \beta)$, where $\beta \equiv 1/T$ is inverse temperature. Due to this periodicity, the field ψ may be decomposed into imaginary-time frequency modes with Matsubara frequencies $2\pi T j$, where the integer $j = 0, \pm 1, \pm 2, \ldots$. Near the transition point, the main contribution to the partition function is played by the zero Matsubara mode, while the nonzero Matsubara frequencies can be integrated out. To take into account N internal degrees of freedom, the term $|\psi|^2$ is interpreted as $\sum_{n=1}^{N} \varphi_n^2$, with the field φ_n for each of the components. In this way, we come to the partition function being the trace of the Boltzmann weight $\exp S[\varphi]$, with the effective action

$$S[\varphi] = \int \left[\frac{1}{2} \left(\frac{\partial \varphi}{\partial x} \right)^2 + \frac{m_{eff}^2}{2}\, \varphi^2 + \frac{\lambda_{eff}}{4!}\, \varphi^4 \right] dx\,, \tag{4}$$

corresponding to an N-component field theory, with the effective parameters

$$m_{eff}^2 = -2mT\mu\,, \qquad \lambda_{eff} = 48\pi m T a_s\,. \tag{5}$$

The details of the above procedure are well known and were described in many publications, e.g., in Refs. [8–12].

Thus, we need to study phase transitions in the N-component filed theory with action (4). This action represents a wide class of physical systems, depending on the number of components. For instance, $N = 0$ corresponds to dilute polymer solutions, $N = 1$, to the Ising-type models, $N = 2$, to the XY model and superfluids, and $N = 3$ corresponds to the Heisenberg model.

2 Phase transition temperature

Action (4) describes the $O(N)$-symmetric multicomponent filed theory with the N-component field $\varphi = \{\varphi_n(x) : n = 1, 2, \ldots, N\}$. The action is invariant under the inversion symmetry

$$\varphi_n \to -\varphi_n \qquad (n = 1, 2, \ldots, N)\,. \tag{6}$$

This implies that one of the solutions for the order parameter $\eta \equiv \langle\varphi\rangle$, which is the average of the field, is zero, $\langle\varphi\rangle = 0$.

But below the critical temperature T_c, the symmetry can become spontaneously broken, when a nonzero order parameter $\langle\varphi\rangle$ provides a lower free energy $F = -T \ln \mathrm{Tr} \exp(-S[\varphi])$, so that

$$F(\eta \neq 0) < F(\eta \equiv 0) . \tag{7}$$

One says that at the critical temperature T_c there occurs a phase transition, where the order parameter changes from zero to a nonzero value:

$$\eta \equiv 0 \qquad (T > T_c) ; \qquad \eta \neq 0 \qquad (T < T_c) . \tag{8}$$

In the considered case, the order parameter at T_c is continuous, which signifies the second order phase transition.

The transition temperature depends on the interaction strength that can be characterized by the gas parameter

$$\gamma \equiv \rho^{1/3} a_s , \tag{9}$$

in which ρ is a particle density and the scattering length is assumed to be positive, $a_s > 0$. For the ideal gas, with no interactions, the critical temperature is

$$T_0 \equiv T_c(0) = \frac{2\pi}{m} \left[\frac{\rho}{\zeta(3/2)} \right]^{2/3} , \tag{10}$$

which is evident from (1) and (2) under $\lambda = 0$. The problem is to find the dependence of the critical temperature $T_c(\gamma)$ on the gas parameter at small values of the latter.

One defines the relative critical temperature shift

$$\frac{\Delta T_c}{T_0} \equiv \frac{T_c - T_0}{T_0} , \tag{11}$$

which, at weak interactions, is linear in the gas parameter:

$$\frac{\Delta T_c}{T_0} \simeq c_1 \gamma \qquad (\gamma \to 0) , \tag{12}$$

as it has been proved in Refs. [9–12]. The difficulty in finding the coefficient c_1 is in the impossibility of employing perturbation theory with respect to γ, which, as explained below, looses its sense at the point T_c. Various ways of solving this problem have been reviewed in [13, 14].

One is able to calculate the shift coefficient c_1 involving the loop expansion [15], which yields [16–18] an expansion in powers of the variable

$$x = (N+2) \, \frac{\lambda_{eff}}{\sqrt{\mu_{eff}}} , \tag{13}$$

in which N is the number of the components, λ_{eff} is an effective coupling, defined in (5), and μ_{eff} is

$$\mu_{eff} = m_{eff}^2 - \Sigma(0,0) , \tag{14}$$

with $\Sigma(0,0)$ being the self-energy at zero limit of energy and momentum.

DOI: 10.1051/epjconf/201713803011

In seven loops, one comes to the asymptotic expansion

$$c_1(x) \simeq \sum_{n=1}^{5} a_n x^n \qquad (x \to 0) \; . \tag{15}$$

The problem arises, when one considers the above expansion at the critical temperature T_c, where μ_{eff} becomes zero, because of which the expansion variable tends to infinity, $x \to \infty$. Since this expansion has been derived for asymptotically small x, the tendency of the latter to infinity makes the expansion senseless. To make sense of such a difficulty, it is necessary to invent a way of finding an effective sum of the asymptotic series, derived for $x \to 0$, which would extrapolate the sum to finite values of x and including $x \to \infty$.

It is worth mentioning that the extrapolation of asymptotic series for a small variable to arbitrary values of the latter can be accomplished in the frame of optimized perturbation theory [19, 20]. Different variants of this theory have been employed for finding the shift coefficient c_1 by introducing control functions with a variable change [17, 18] or incorporating them into initial approximations [21–24]. However, the results of such calculations strongly depend on how the control functions are introduced, and also these calculations are rather cumbersome requiring the use of optimization conditions defining control functions at each approximation order.

Below we suggest a general uniquely defined method for extrapolating asymptotic series of a small variable to arbitrary values of the variable, including its infinite limit.

3 Self-similar approximation theory

A general method of extrapolation is based on self-similar approximation theory [25–28]. Although the final prescription that will be formulated below is rather simple, we briefly delineate the main steps of the theory to give the reader the impression of how the employed approximants have been obtained.

Self-similar approximation theory is based on the following important points. First, it is necessary to transform an initially given divergent sequence $\{f_k(x)\}$ of the series of order k for the sought function into a convergent sequence $\{F_k(x, u_k(x))\}$ by incorporating control functions $u_k(x)$. The latter can be introduced either in an initial approximation $f_0(x, u)$, or through a variable change $x = z(x, u)$, or by a sequence transformation $T(u) f_k(x) = F_k(x, u)$. A sequence is convergent if and only if, for a given positive ε, there exists k_ε such that for all $k > k_\varepsilon$ the Cauchy criterion of convergence is valid:

$$|F_{k+p}(x, u_{k+p}) - F_k(x, u_k)| < \varepsilon \; . \tag{16}$$

Control functions can be defined by minimizing the Cauchy cost functional

$$C[u] = \frac{1}{2} \sum_k |F_{k+p}(x, u_{k+p}) - F_k(x, u_k)|^2 \; . \tag{17}$$

At the next step, we consider the passage from an approximation F_k to F_{k+1} and so on as the motion in discrete time k. To formulate this in terminology of dynamical theory, we impose the reonomic constraint

$$F_0(x, u_k(x)) = f \; , \qquad x = x_k(f) \tag{18}$$

and introduce the endomorphism

$$y_k(f) \equiv F_k(x_k(f), u_k(x_k(f))) \; . \tag{19}$$

In order for the Cauchy cost functional (17) to have the absolute minimum zero, it is necessary that endomorphism (19) would satisfy the relation

$$y_{k+p}(f) = y_k(y_p(f)) \tag{20}$$

that is termed the property of *functional self-similarity*. Relation (20) defines a dynamical system in discrete time, that is, a cascade, whose trajectory is bijective to the approximation sequence $\{F_k\}$, because of which this dynamical system is called the *approximation cascade*.

It is more convenient to deal with a dynamical system in continuous time, i.e., a flow. For this purpose, it is possible to embed the approximation cascade into an *approximation flow*, which implies that the flow trajectory passes through all points of the cascade trajectory,

$$\{y_k(f):\ k \in \mathbb{Z}_+\} \subset \{y_t(f):\ t \in \mathbb{R}_+\} . \tag{21}$$

For the approximation flow, the self-similarity relation (20) can be rewritten as the Lie equation

$$\frac{\partial}{\partial t}\, y_t(f) = v(y_t(f)) , \qquad v(f) \equiv \left[\frac{\partial}{\partial t}\, y_t(f)\right]_{t=0} . \tag{22}$$

The Lie equation of motion, being integrated, yields the evolution integral

$$\int_{y_k}^{y_k^*} \frac{dy}{v_k(y)} = t_k , \tag{23}$$

in which v_k is the cascade velocity that is the Euler discretization of the flow velocity, y_k^* is a quasi-fixed point, and t_k is an effective time required for reaching y_k^* from y_k. The quasi-fixed point y_k^* represents the effective limit of the flow trajectory, hence, being bijective to the approximation sequence through notation (19), it represents the self-similar approximation $f_k^*(x)$ for the sought function $f(x)$.

The stability of the approximation procedure can be controlled by studying the local map multipliers

$$\mu_k(f) \equiv \frac{\delta y_k(f)}{\delta y_{k-1}(f)} . \tag{24}$$

The motion is locally stable provided that the absolute values of the map multipliers are not larger than one. Approaching a fixed point, one has $|\mu_k(f)| \to 1$ as $y_k \to y^*$. More details on the self-similar approximation theory can be found in the papers [25–28] and summarized in the review papers [29–31].

Let us concretize the approach by considering a function for which one can find only an asymptotic series at a small variable,

$$f(x) \simeq f_k(x) \qquad (x \to 0) , \tag{25}$$

with the k-th order expansion

$$f_k(x) = f_0(x)\left(1 + \sum_{n=1}^{k} a_n x^n\right) , \tag{26}$$

in which $f_0(x)$ is a known term. Applying a fractal transform [30] to series (26) and following the procedure delineated above, we come to the self-similar factor approximants

$$f_k^*(x) = f_0(x) \prod_{i=1}^{N_k} (1 + A_i x)^{n_i} , \tag{27}$$

where $N_k = k/2$ for even $k = 2, 4, \ldots$ and $N_k = (k+1)/2$ for odd $k = 3, 5, \ldots$. The parameters A_i and n_i are defined by the accuracy-through-order procedure $f_k^*(x) \simeq f_k(x)$ as $x \to 0$, with the scaling $A_1 = 1$ for odd approximants [32–35].

We aim at studying the limit of large x. Let the known term $f_0(x)$ behave as

$$f_0(x) \simeq Ax^{\alpha} \qquad (x \to \infty) . \tag{28}$$

If the large-variable limit of the sought function is

$$f(x) \simeq Bx^{\beta} \qquad (x \to \infty) , \tag{29}$$

then the powers of approximant (27) have to obey the constraint

$$\alpha + \sum_{i=1}^{N_k} n_i = \beta \qquad (\beta \neq 0) , \tag{30}$$

which, for the case of a finite large-variable limit, reduces to

$$\alpha + \sum_{i=1}^{N_k} n_i = 0 \qquad (\beta = 0) . \tag{31}$$

In this way, we get a sequence of self-similar factor approximants (27). If a number of terms in sum (26) are available, then all we need is to observe the numerical convergence of the sequence $\{f_k^*\}$. When there are a few of such terms, the convergence of self-similar approximants can be essentially accelerated by constructing a quadratic spline

$$q(x,t) = b_0(x) + b_1(x)t + b_2(x)t^2 , \tag{32}$$

whose coefficient functions are defined through the conditions

$$q(x,0) = f_{k-2}^*(x) , \qquad q(x,1) = f_{k-1}^*(x) , \qquad q(x,2) = f_k^*(x) . \tag{33}$$

The self-similar approximant for this spline reads as

$$q^*(x,t) = b_0(x)[1 + A(x)t]^{n(x)} . \tag{34}$$

The latter provides an extrapolation of the sequence f_{k-2}^*, f_{k-1}^*, and f_k^* to larger orders by setting $t \geq 2$. The minimal extrapolation involves $t = 2$ and $t = 3$ yielding the doubly renormalized approximant

$$f_k^{**}(x) = \frac{1}{2}\,[q^*(x,2) + q^*(x,3)] , \tag{35}$$

whose error value is given by the difference of $q^*(x,2)$ and $q^*(x,3)$.

4 Calculation of temperature shift

We accomplish this procedure for the critical temperature shift (12) using the fifth-order series (15). The coefficients a_n, calculated using the data of the seven-loop expansion by Kastening [17, 18], for different numbers of the components N, are presented in Table 1.

Note that in expansion (15) the variable x is assumed to be small, such that $x \to 0$, independently of the values of the parameters entering notation (13). But at the end, we have to find the limit $x \to \infty$, because of which the final answer for the coefficient c_1 is $c_1 = \lim_{x\to\infty} c_1^{**}(x)$.

Following the scheme of the previous section, we find the k-order factor approximants $f_k(x)^*$ for the quantity $c_1(x)$ in third order,

$$f_3^*(x) = a_1 x(1+x)^{n_1}(1+A_2x)^{n_2} ,$$

with $n_1 + n_2 = -1$, in fourth order,

$$f_4^*(x) = a_1 x(1+A_1x)^{n_1}(1+A_2x)^{n_2} ,$$

where $n_1 + n_2 = -1$, and in fifth order,

$$f_5^*(x) = a_1 x(1+x)^{n_1}(1+A_2x)^{n_2}(1+A_3x)^{n_3} ,$$

with $n_1 + n_2 + n_3 = -1$.

Since finally we need the limits $x \to \infty$, we find from the above approximants the limits

$$f_k^* \equiv \lim_{x\to\infty} f_k^*(x)$$

in the forms

$$f_3^* = a_1 A_2^{n_2} , \qquad f_4^* = a_1 A_1^{n_1} A_2^{n_2} , \qquad f_5^* = a_1 A_2^{n_2} A_3^{n_3} ,$$

whose values are given in Table 2. After taking the limit $x \to \infty$, the variable x does not enter the expressions below.

Accelerating the convergence by means of splines, as is explained above, we construct the quadratic polynomials

$$q(t) = b_0 + b_1 t + b_2 t^2 ,$$

with the coefficients defined by the equations

$$q(0) = f_3^* , \qquad q(1) = f_4^* , \qquad q(2) = f_5^* ,$$

which are listed in Table 3. The factor approximant for the spline $q(t)$ reads as

$$q^*(t) = b_0(1+At)^n ,$$

with the values A and n from Table 3. The values

$$q_2^* \equiv q^*(2) , \qquad q_3^* \equiv q^*(3)$$

are presented in Table 2.

Finally, the sought coefficient c_1, defined by formula (35), under $x \to \infty$ is given in Table 4. We also compare our calculations for c_1 with the available Monte Carlo simulations [36–39], as well as with the results obtained by optimized perturbation theory (OPT) [17, 18]. As it is seen, the values of c_1, derived in our approach, are in perfect agreement with the Monte Carlo simulations. The advantage of the self-similar factor approximants, we employed, is that they are uniquely defined and do not involve the use of control functions at the final step. While the results of optimized perturbation theory strongly depend on the choice of such control functions.

DOI: 10.1051/epjconf/201713803011

Table 1. The coefficients a_n of the asymptotic expansion for $c_1(x)$.

N	0	1	2	3	4
a_1	0.111643	0.111643	0.111643	0.111643	0.111643
a_2	−0.0264412	−0.0198309	−0.0165258	−0.0145427	−0.0132206
a_3	0.0086215	0.00480687	0.00330574	0.00253504	0.0020754
a_4	−0.0034786	−0.00143209	−0.000807353	−0.000536123	−0.000392939
a_5	0.00164029	0.00049561	0.000227835	0.000130398	0.0000852025

Table 2. Self-similar factor approximants $f_k^*(\infty)$, $k = 3, 4, 5$, and q_m^*, $m = 2, 3$.

N	0	1	2	3	4
f_3^*	0.548162	0.700831	0.823464	0.92360	1.00681
f_4^*	0.646479	0.858742	1.02476	1.15706	1.26426
f_5^*	0.682986	0.888388	1.05111	1.18101	1.28629
q_2^*	0.7347	1.008	1.219	1.384	1.516
q_3^*	0.7999	1.112	1.352	1.540	1.688

Table 3. Coefficients of polynomials $q(t) = b_0 + b_1 t + b_2 t^2$, and self-similar approximants $q^*(t) = b_0(1 + At)^n$.

N	0	1	2	3	4
b_0	0.548162	0.700831	0.823464	0.9236	1.00681
b_1	0.129222	0.222044	0.288772	0.338205	0.375156
b_2	−0.030905	−0.0641325	−0.087474	−0.104749	−0.11771
A	0.714061	0.894486	0.956514	0.985622	1.00014
n	0.330135	0.354202	0.366623	0.371523	0.372564

5 Discussion

We have suggested a method of extrapolating asymptotic series for a small variable $x \longrightarrow 0$ to any finite values of x, including the infinite limit $x \longrightarrow \infty$. The method is based on self-similar approximation theory. We have applied the method to calculating the critical temperature of weakly interacting $O(N)$ symmetric field theory. We have considered the cases with the number of the components $N = 0, 1, 2, 3, 4$, for which the seven-loop expansions are known. Note that the case of $N = \infty$ allows for an exact solution [10, 40] yielding

$$c_1(N = \infty) = \frac{8\pi}{3[\zeta(3/2)]^{4/3}} = 2.328473 .$$

As is evident from (11) and (12), knowing c_1 immediately gives the critical temperature

$$T_c \simeq T_0(1 + c_1\gamma).$$

Our results are in very good agreement with Monte Carlo simulations. The advantage of the used self-similar factor approximants is in their simplicity and unique definition, involving no control functions in the final calculations.

We are grateful to B. Kastening for discussions. Financial support from RFBR (grant #14 – 02 – 00723) is acknowledged.

Table 4. Results for the coefficient c_1 of the critical temperature shift obtained using self-similar factor approximants, compared to Monte Carlo simulations and optimized perturbation theory.

N	c_1	Monte Carlo	OPT
0	0.77± 0.03		0.81± 0.09
1	1.06± 0.05	1.09± 0.09 [39]	1.07± 0.10
2	1.29± 0.07	1.29± 0.05 [36] 1.32± 0.02 [37, 38]	1.27± 0.11
3	1.46± 0.08		1.43± 0.11
4	1.60± 0.09	1.60± 0.10 [39]	1.54± 0.11

References

[1] H. Satz, Phys. Rep. **88**, 349 (1982)
[2] R. Hagedorn, Riv. Nuovo Cimento **6**, 1 (1983)
[3] J. Cleymans, R. Gavai, and E. Suhonen, Phys. Rep. **130**, 217 (1986)
[4] H. Reeves, Phys. Rep. **201**, 335 (1991)
[5] V.I. Yukalov and E.P. Yukalova, Physica A **243**, 382 (1997)
[6] V.I. Yukalov and E.P. Yukalova, Phys. Part. Nucl. **28**, 37 (1997)
[7] V.I. Yukalov and E.P. Yukalova, POS (Baldin ISHEPP XXI) **2012**, 046 (2012)
[8] H. Kleinert, V. Schulde-Frohlinde, *Critical Properties of ϕ^4 Theories* (World Scientific, Singapore, 2001)
[9] G. Baym, J.P. Blaizot, M. Holzmann, F. Laloë, D. Vautherin, Phys. Rev. Lett. **83**, 1703 (1999)
[10] G. Baym, J.P. Balzot, J. Zinn-Justin, Eur. Phys. Lett. **49**, 150 (2000)
[11] M. Holzmann, G. Baym, J.P. Balizot, F. Laloë, Phys. Rev. Lett. **87**, 120413 (2001)
[12] J. Zinn-Justin, *Quantum Field Theory and Critical Phenomena* (Clarendon, Oxford, 2002)
[13] J.O. Andersen, Rev. Mod. Phys. **76**, 599 (2004)
[14] V.I. Yukalov, Laser Phys. Lett. **1**, 435 (2004)
[15] H. Kleinert, *Path Integrals* (World Scientific, Singapore, 2003)
[16] B. Kastening, Laser Phys. **14**, 586 (2004)
[17] B. Kastening, Phys. Rev. A **69**, 043613 (2004)
[18] B. Kastening, Phys. Rev. A **70**, 043621 (2004)
[19] V.I. Yukalov, Moscow Univ. Phys. Bull. **31**, 10 (1976)
[20] V.I. Yukalov, Theor. Math. Phys. **28**, 652 (1976)
[21] F.F. de Souza Cruz, M.B. Pinto, R.O. Ramos, Phys. Rev. B **64**, 014515 (2001)
[22] F.F. de Souza Cruz, M.B. Pinto, R.O. Ramos, P. Senna, Phys. Rev. A **65**, 053613 (2002)
[23] J.L. Kneur, M.B. Pinto, R.O. Ramos, Phys. Rev. A **68**, 043615 (2003)
[24] J.L. Kneur, A. Neveu, M.B. Pinto, Phys. Rev. A **69**, 053624 (2004)
[25] V.I. Yukalov, Physica A **167**, 833 (1990)
[26] V.I. Yukalov, Phys. Rev. A **42**, 3324 (1990)
[27] V.I. Yukalov, J. Math. Phys. **32**, 1235 (1991)
[28] V.I. Yukalov, J. Math. Phys. **33**, 3994 (1992)
[29] V.I. Yukalov and E.P. Yukalova, Ann. Phys. (N.Y.) **277**, 219 (1999)
[30] V.I. Yukalov and E.P. Yukalova, Chaos Solit. Fract. **14**, 839 (2002)
[31] V.I. Yukalov and E.P. Yukalova, POS (Baldin ISHEPP XXII) **2014**, 080 (2014)

DOI: 10.1051/epjconf/201713803011

[32] V.I. Yukalov, S. Gluzman, and D. Sornette, Physica A **328**, 409 (2003)
[33] S. Gluzman, V.I. Yukalov, and D. Sornette, Phys. Rev. E **67**, 026109 (2003)
[34] V.I. Yukalov and E.P. Yukalova, Phys. Lett. A **368**, 341 (2007)
[35] V.I. Yukalov and E.P. Yukalova, Eur. Phys. J. B **55**, 93 (2007)
[36] V.A. Kashurnikov, N. Prokofev, and B. Svistunov, Phys. Rev. Lett. **87**, 120402 (2001)
[37] P. Arnold and G. Moore, Phys. Rev. Lett. **87**, 120401 (2001)
[38] P. Arnold and G. Moore, Phys. Rev. E **64**, 066113 (2001)
[39] X. Sun, Phys. Rev. E **67**, 066702 (2003)
[40] P. Arnold and B. Tomašik, Phys. Rev. A **62**, 063604 (2000)

The differential cross section for the dp-elastic scattering at 500 to 900 MeV/nucleon

Arkadiy Terekhin[1,*], *Vladimir* Ladygin[1], *Yuriy* Gurchin[1], *Alexander* Isupov[1], *Marian* Janek[2], *Anatoliy* Khrenov[1], *Aleksey* Kurilkin[1], *Pavel* Kurilkin[1], *Nadeghda* Ladygina[1], *Semyon* Piyadin[1], *Sergey* Reznikov[1], and *Igor* Vnukov[3]

[1] *Joint Institute for Nuclear Researches, Dubna 141980, Russia*
[2] *Physics Department, University of Žilina, Žilina 010 26, Slovak Republic*
[3] *Belgorod State University, Belgorod, 308015, Russia*

Abstract. The results of the differential cross section measurements of dp-elastic scattering at energies from 500 to 900 MeV/nucleon at Nuclotron JINR are reported. The data were obtained for the angular range of 70-120° in the center-of-mass (c.m.s.). The results are compared with existing world data.

1 Introduction

The dp-elastic scattering is the simplest reaction to study the deuteron structure. The experimental material for this reaction covers the energy range from tens to thousands MeV/n [1]-[9]. The theoretical approaches based on the solution of the three-particle Schroedinger equation [10–12] and on the Faddeev calculations [13, 14] are described of the three-nucleon scattering with high accuracy at energies below 200 MeV/n [15, 16]. The theoretical calculations using not only 2N forces but also different 3N forces [17, 18] give the best agreement with experimental data. However, the discrepancy between the theory and experiment increases with the energy increasing. The Glauber scattering theory which takes both single and double interactions is a classic approach at the energies higher than 400 MeV/n. [19, 20]. The experimental data can be fitted with a relativistic multiple-scattering theory which uses off-mass-shell extrapolations of the nucleon-nucleon amplitudes suggested by the structure of derivative meson-nucleon couplings [21, 22].

The new differential cross section data were obtained at internal target station ITS [23] at Nuclotron at the energy of 1000 MeV/n [6]. Recently, the deuteron vector and tensor analyzing powers have been obtained at 440 MeV/n [25]. The preliminary data on the differential cross section for dp-elastic scattering at energies from 500 to 900 MeV/n, obtained at ITS at Nuclotron are presented in this paper.

2 Experiment and data analysis

The measurements were performed at ITS [23] at Nuclotron JINR by using new ITS DAQ system [26]. Two pairs of scintillation detectors placed symmetrically with respect to the beam direction

*e-mail: aterekhin@jinr.ru

 DOI: 10.1051/epjconf/201713803012

Table 1. Characteristics of detectors.

Detector	Size,mm^3	Distance from target,cm	Angular span,deg lab.sys.	c.m.s.
P	20x60x20	58	2	4
D	10x40x24	56	1	2
PP	50x50x10	56	5	10

were used to register the elastically-scattered deuterons and protons. Also two scintillation counters (PP-detectors) were used to register the quasi-elastically scattered protons. All counters based on the Hamamatsu H7416MOD PMT were used previously in the experiment [25]. The DP-detectors were rotated to give an angular range of the θ_{lab} = 19° to 50° ($\theta_{c.m.}$ = 70° to 120°). The PP-detectors were mounted at the angle corresponding to quasi-elastic scattering at $\theta_{c.m.}$ = 90° and remained stationary throughout the experiment at each energy. The size of the scintillation detectors, distances to the target, angular spans in the laboratory system and c.m.s. are shown in table 1. The layout of the counters with respect to the beam direction is shown in figure 1.

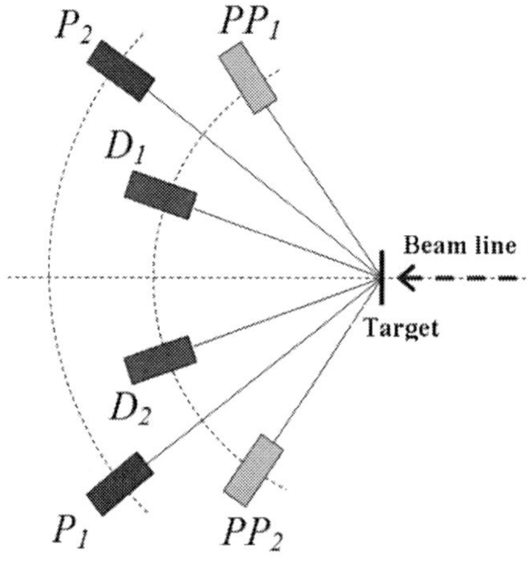

Figure 1. Layout of the counters with respect to the beam direction. $D_{1,2}$, $P_{1,2}$- deuteron and proton detectors, $PP_{1,2}$- detectors to register the pp-quasi-elastic scattering.

The procedure to obtain differential cross section data was realized by analysis of the amplitude and timing spectra. The estimation of the background in the amplitude data was performed by using the temporary gates on the deuteron and proton time difference spectra. The subtraction of the timing signal from deuteron- and proton- counters for 650 MeV/n is shown in figure 2. In this distribution the dp - elastic scattering events (I domain) and the background (II and III domains) are selected so that the widths of both domains are equal. The amplitude distribution for proton counter obtained by using these timing gates is shown in figure 3 A. The subtraction of the resulting spectra allows to reduce the background (figure 3 B).

The next stage is the CH_2-C subtraction procedure. The carbon background subtraction normalization coefficient k is deduced from the interval $a_{min} < a < a_{max}$, where a - channels of CH_2- and

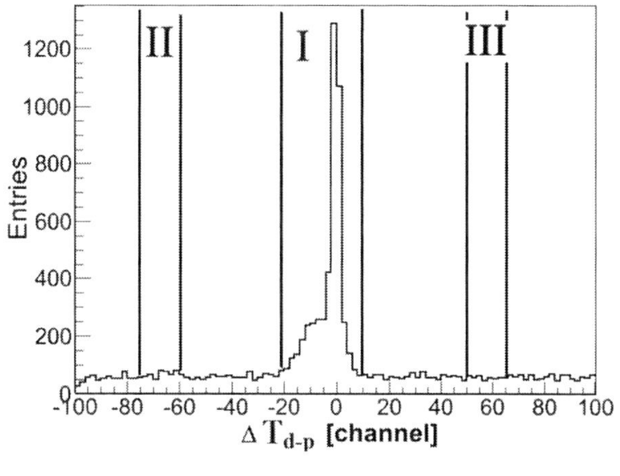

Figure 2. The time difference of signals arrival from the deuteron and proton counters for $\theta_{c.m.} = 73°$ at 650 MeV/n.

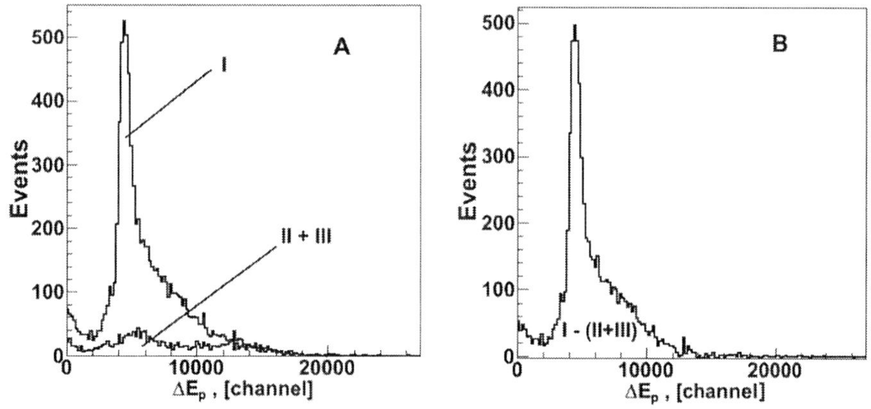

Figure 3. The background subtraction procedure for the amplitude spectrum of proton counter for $\theta_{c.m.} = 73°$ at 650 MeV/n.

C-amplitude distributions:

$$k = \frac{N_{CH_2}|_{a_{min}<a<a_{max}}}{N_C|_{a_{min}<a<a_{max}}}. \quad (1)$$

Here N_{CH_2} and N_C - CH_2 - and C - amplitude distributions integrals in a - interval within the window shown in figure 4 A by the solid lines. The carbon background can be then subtracted as:

$$N_{dp} = N_{CH_2} - kN_C, \quad (2)$$

where N_{dp} is the resulting dp - elastic scattering distribution, N_{CH_2} is the total CH_2 - distribution, kN_C is the normalized C - distribution within the window shown in figure 4 B by the dotted lines.

DOI: 10.1051/epjconf/201713803012

In figure 4 A the CH_2 - distribution is shown by the solid line. The normalized C - spectrum is shown by the dotted line. In figure 4 B the result of subtraction is demonstrated. Such procedure was performed for proton amplitude spectra for each $\theta_{c.m.}$.

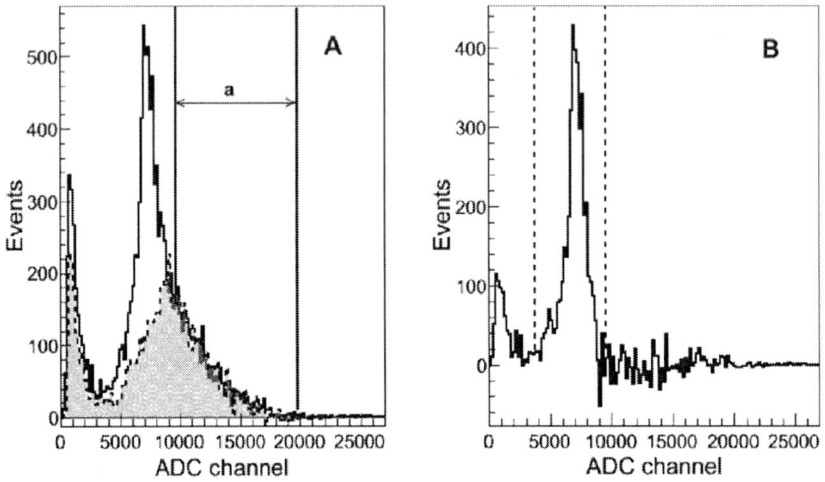

Figure 4. The procedure of CH_2 - C subtraction for $\theta_{c.m.} = 73°$ at 650MeV/n. A - is the CH_2- and normalized C- distributions given by the solid and dotted histograms, respectively, vertical solid lines - is the interval of the normalization. B - is the result of $CH_2 - C$ subtraction, vertical dashed lines are the gates indicating the domain of the *dp*-elastic scattering events.

3 Differential cross section

The experimental data for 650 MeV/n were normalized to the data [5]. The normalization factor for this energy was used to obtain the differential cross section for other energies. The differential cross section angular dependence [5] was approximated by the function $f(\theta_{c.m.}) = P_0 exp(P_1 \theta_{c.m.})$ in angular range $71° < \theta_{c.m.} < 108°$. The obtained parameters P_0 and P_1 were fixed. After this our data at 650 MeV/n was approximated by the function $C^{650}_{norm} f(\theta_{c.m.})$. The calculated value is equal to $C^{650}_{norm} = 79.98 \pm 5.29$ mb. The obtained differential cross section angular dependence at 650 MeV/n is shown in figure 5. The normalization factor C^{700}_{norm} at 700 MeV/n was calculated as composition $C^{650}_{norm} R$, where R - ratio of the pp – elastic cross section at 650 MeV and at 700 MeV in the PP – monitor counter angular span in the c.m.s. which is equal to $R = 1.22 \pm 0.05$. Analogous procedure was performed for 500 and 900 MeV/n. The differential cross section for 700,500 and 900 MeV/n are shown in figure 6,7,8.

One can see, that the new results are in agreement with the behavior of the world data.

4 Conclusion

The procedure on the dp-elastic scattering differential cross section at high energies at ITS at Nuclotron using CH2 - C subtraction is established.

The preliminary differential cross section data for dp - elastic scattering at 500, 700 and 900 MeV/n are obtained. New results are compared with the existing data for similar values of energies. The data are in reasonable agreement.

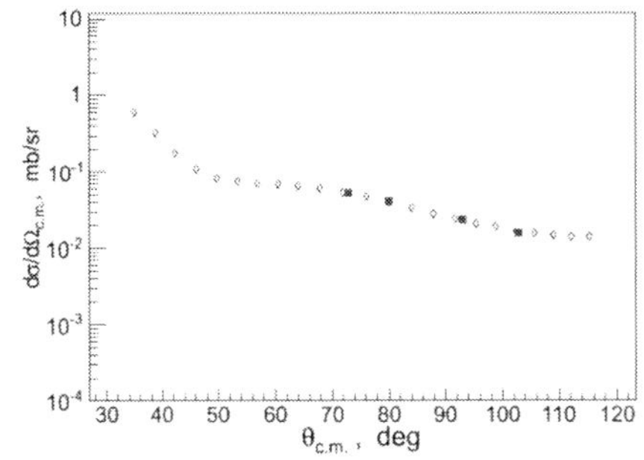

Figure 5. The differential cross section for dp-elastic scattering at 650 MeV/n. Squares - the results of this work, triangles - data from [5].

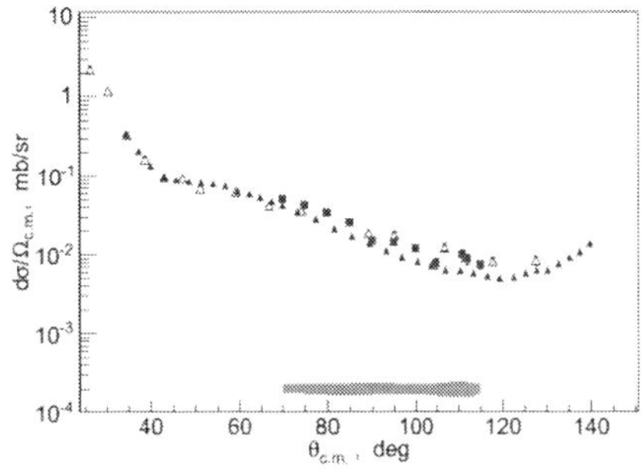

Figure 6. The differential cross section for dp-elastic scattering at 700 MeV/n. Squares - the results of this work, solid band - the systematic errors, open triangles - data from [5], solid triangles - data from [7].

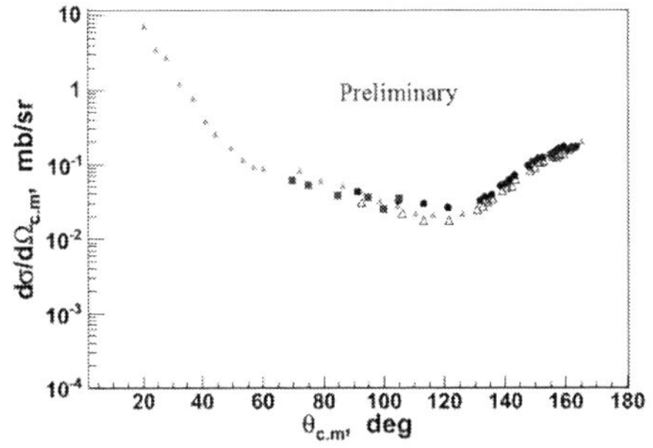

Figure 7. The differential cross section for dp-elastic scattering at 500 MeV/n. Squares - the results of this work, circles – data at 470 MeV/n [3], open triangles – data at 590 MeV/n [3], full triangles – data from 580 MeV/n [4].

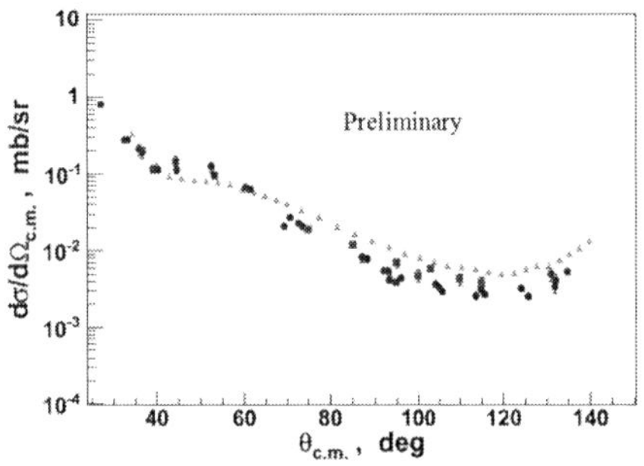

Figure 8. The differential cross section for dp-elastic scattering at 900 MeV/n. Squares - the results of this work, circles – data at 1000 MeV/n [8], triangles - data from 800 MeV/n [7].

The work has been supported in part by the RFBR grant 16-02-00203a and JINR - Slovak Republic Scientific Cooperation Program in 2015-2016.

References

[1] K. Ermisch et al., Phys. Rev. C **71**, 064004 (2005)
[2] K. Sekiguchi et al., Phys. Rev. C **65**, 034003 (2002)
[3] J.C. Alder et al., Phys. Rev. C **6(6)**, 2010 (1972)
[4] J.S. Vincent et al., Phys. Rev. Lett. **24(5)**, 236 (1970)
[5] E. Culmez et al., Phys. Rev. C **43(5)**, 2067 (1991)
[6] A.A. Terekhin et al., Eur. Phys. J. A **48**, 182 (2012)

[7] E. Winkelmann et al., Phys. Rev. C **21**, 2535 (1980)
[8] G.W. Bennet et al., Phys. Rev. Lett. **19**, 387 (1967)
[9] P.K. Kurilkin et al., PoS (Baldin ISHEPP-XXI) 040 (2012)
[10] A. Kievsky, M.Viviani, S.Rosati, Phys.Rev. C **64**, 024002 (2001)
[11] M. Viviani, A.Kievsky, S.Rosati, Few-Body Syst. **30**, 39 (2001)
[12] A. Deltuva et al., Phys.Rev. C **71**, 064003 (2005)
[13] H. Witala, Th.Cornelius, W.Glockle, Few-Body Systems **3**, 123 (1988)
[14] J.L. Friar, G.L.Payne, W.Glockle, D.Huber, H.Witala, Phys.Rev. C **51**, 2356 (1995)
[15] W. Glockle et al., Phys.Rep. **274**, 107 (1996)
[16] J. Kuros-Zolnierczuk, H.Witala et al., Phys.Rev. C **66**, 024004 (2002)
[17] S.A. Coon, H.K.Han, Few-Body Systems **30**, 131 (2001)
[18] B.S. Rudliner et al., Phys. Rev. C **56**, 1720 (1997)
[19] V. Franco, R.J.Glauber, Phys. Rev. Lett. **16**, 944 (1966)
[20] V. Franco, E.Coleman, Phys.Rev.Lett. **17**, 827 (1966)
[21] E.Bleszynski et al., AIP Conf. **150**, 1208 (1986)
[22] N.B. Ladygina et al., Eur. Phys. J. A **42**, 91 (2009)
[23] A.I. Malakhov et al., Phys. Res. A **440**, 320 (2000)
[24] A.A. Terekhin et al., Phys.Part.Nucl.Lett. **12(5)**, 695 (2015)
[25] P.K. Kurilkin et al., Phys. Lett. B **715**, 61 (2012)
[26] A.Yu. Isupov et al., Nucl. Instrum. Methods Phys. Res. A **698**, 127 (2013)

EPJ Web of Conferences **138**, 03013 (2017) DOI: 10.1051/epjconf/201713803013
Baldin ISHEPP XXIII

Beam and target dependencies of two-kaon femtoscopic correlations in SELEX

Grigory Nigmatkulov[1,⋆] (for the SELEX collaboration)

[1] *National Research Nuclear University MEPhI (Moscow Engineering Physics Institute), Kashirskoe highway 31, Moscow, 115409, Russia*

Abstract. The space and time extents of the particle emitting source created in collisions of particles with nuclei can be measured using a correlation femtoscopy method. In this paper we show the preliminary measurements of $K^{\pm}K^{\pm}$ correlation functions in 600 GeV/c Σ^-, π^- and 540 GeV/c p interactions with carbon and copper nuclei. The one-dimensional correlation functions are parameterized by a Gaussian form. The dependencies of the charged kaon emitting source radii on the different h+A collisions are presented.

1 Introduction

Multiparticle production in high-energy collisions of particles and/or nuclei is a basic property of the strong interaction which is described by the Quantum Chromodynamics (QCD). The formation of hadrons from quarks and gluons (hadronization) which are confined occurs on a distance scale of 1 fm. The multiparticle production is a dynamic process, evolving space and time. Correlations of particles with small relative momenta (also known as femtoscopy or HBT intensity interferometry) [1] are used to probe space-time structure of systems created in high-energy collisions. These correlations arise from Bose-Einstein enhancement of identical-boson production at low relative momenta. In this paper we report on the results of two-kaon femtoscopic analysis performed for 600 GeV/c Σ^-, π^- and 540 GeV/c p interactions with carbon and copper nuclei.

2 Data sample

The SELEX (E781) experiment is a three-stage magnetic spectrometer which was designed to study hadroproduction in the forward hemisphere ($x_F > 0.1$) [2]. Data were taken with 600 GeV/c negative hadron beam ($\simeq$50%Σ^- and $\simeq$50%π^-) and with 540 GeV/c positive hadron beam ($\simeq$92%p and $\simeq$8%π^+). The beam particle was identified by a transition radiation detector and tracked by 8 beam silicon strip detectors located before the charm production target. A segmented target consists of 2 copper and 3 diamond (carbon) sheets, separated by 1.5 cm clearance. The total target thickness was 5% on an interaction length for protons. Produced particles were tracked in a set of 20 vertex silicon strip detectors arranged in 4 sets of planes rotated by 45°. The single-track efficiency of each detector

⋆e-mail: nigmatkulov@gmail.com, ganigmatkulov@mephi.ru

was exceeding 98%. All strip detectors have a resolution of 6.5 μm (pitch/ $\sqrt{12}$). The average longitudinal vertex position resolution was 270 μm for primary vertex and 560 μm for secondary vertex. The particle momentum was measured by deflection of the track position by two magnets in a system of proportional wire chambers (PWC). In order to achieve high-resolution momentum measurements (up to full beam energy), the beam region of PWCs was covered with double-sided silicon detectors. Momentum resolution of 100 GeV/c track was $\sigma_p/p \approx 0.5\%$. Particles with momentum above 22 GeV/c traversed to a Ring Imaging Cherenkov detector (RICH) and identified as π, K or p. The kaon identification efficiency was over 90% above the kaon threshold ($\approx$43 GeV/c). The average number of tracks reaching the RICH was about 5 per event. The RICH detector provided 2σ K/π separation up to 165 GeV/c [4]. In order to achieve high kaon identification purity, only tracks with the momentum from 45 to 160 GeV/c were selected. Charged kaons were identified with the likelihood to be a kaon at least three times exceeding any other particle hypothesis [5]. The electrons were eliminated from the analysis using the transition radiation detector which was placed before RICH.

To measure the femtoscopic radii dependence on the nuclear mass A, the charged particles produced in the copper and carbon targets were analyzed separately. The reconstructed z-position of the primary vertex required to be in vicinity of a Cu or C target and not more than 0.05 mm from the edge of any target [3]. Primary tracks which have vertex silicon track segment matched with downstream segments in forward spectrometers used in the analysis. Each track was extrapolated to the primary vertex to determine its transverse miss distance. Only tracks with miss distance less than 20 μm were used. After applying the cuts 1065655, 249210, 53308 K^+K^+ and 3688541, 337034, 48298 K^-K^- pairs were selected for Σ^-, π^- and p beams.

3 Femtoscopy

The main tool of the correlation femtoscopy method is the correlation function $C(\mathbf{q})$ which is defined as a ratio $C(\mathbf{q}) = A(\mathbf{q})/B(\mathbf{q})$, where $A(\mathbf{q})$ is the distribution of two-particle relative momenta $\mathbf{q} = \mathbf{p}_1 - \mathbf{p}_2$, and $B(\mathbf{q})$ is a similar distribution in all aspects except the absence of the quantum statistical correlations. A is constructed from identical particles from the same event and B is constructed using the event mixing technique where particles of interest are combined from different events. The limited statistics available (about 50k identical-kaon pairs for p induced collisions) allowed us to perform the analysis using only one-dimensional correlation function $C_2(q_{inv})$. The q_{inv} is a one-dimensional Lorentz scalar $q_{inv} = \sqrt{(\mathbf{p}_1 - \mathbf{p}_2)^2 - (E_1 - E_2)^2}$. The momentum correlations of particles emitted at nuclear distances are also influenced by the effect of final-state interaction (FSI), Coulomb and strong interactions [6]. For identical kaons, the effect of strong interaction is negligible. In order to extract the emitting source radius, R, the correlation function can be fitted with the parameterization assuming that the emission function describes a static source and has the Gaussian profile:

$$C_2(q_{inv}) = N\left(1 - \lambda + \lambda K(q_{inv})\left(1 + \exp(-R^2 q_{inv}^2)\right)\right) D(q_{inv}), \tag{1}$$

where N is a normalization factor, λ is a correlation strength and the factor $D(q_{inv})$ is so-called baseline distribution that takes into account non-femtoscopic correlations. The factor $K(q_{inv})$ takes into account two charged kaon Coulomb repulsion [7, 8].

For elimination of possible biases because of the construction of reference sample, one may use Monte Carlo simulation where the quantum statistical effects are absent and build so-called double ratio:

$$C_2(q_{inv}) = \frac{A(q_{inv})/B(q_{inv})}{A^{MC}(q_{inv})/B^{MC}(q_{inv})}, \tag{2}$$

where $A^{MC}(q_{inv})$ and $B^{MC}(q_{inv})$ are the Monte Carlo relative momentum distributions from the same and mixed events, respectively.

4 Results and Discussions

The positive and negative kaon femtoscopic correlations are measured using the same detector setup, cuts and fitting procedures. This allows to compare the emitting region parameters for different h+A collisions. The analysis was performed separately for the copper and carbon targets and 600 GeV/c Σ^-, π^- and 540 GeV/c p beams. In order to construct the uncorrelated reference sample the event mixing technique was used. Kaons from adjacent events were grouped by the production target and beam particle. Relative momentum distributions were combined in the numerator and denominator separately for copper and carbon targets before constructing the ratios. The correlation functions were normalized such that the $C(q_{inv}) = 1$ for the $0.6 < q_{inv} < 0.8$ GeV/c region, where the influence of the non-femtoscopic correlations is small and the Bose-Einstein correlations are absent. The peak at $q_{inv} < 0.4$ GeV/c is due to the quantum statistics correlations.

Particle pairs with small relative momentum are spatially close to each other and affected by the track reconstruction inefficiencies such as track splitting and track merging. Track merging effects suppress the correlation function at low q_{inv}. To eliminate this effect, we require the tracks in a pair to be well separated (the average separation in the silicon strip detector region must be greater than 100 μm) in addition to the cuts used in the previous analysis [5]. The systematic uncertainty due to the two-particle effects was estimated by varying the cuts and does not exceed 8%.

The smearing of single particle momenta was studied by embedding simulated kaon tracks with known momenta through the detector. Using detailed studies of the momentum resolution, we correct the correlation functions with an iterative technique similar to that used by NA44 and STAR collaborations [9, 10]. Experimental correlation functions were corrected for momentum resolution using the expression:

$$C_2^{corr}(q_{inv}) = \frac{C_2^{uncorr}(q_{inv})C_2^{unsmeared}(q_{inv})}{C_2^{smeared}(q_{inv})}, \tag{3}$$

where $C_2^{corr}(q_{inv})$ is a corrected correlation function and $C_2^{uncorr}(q_{inv})$ represents a measured correlation function. The $C_2^{unsmeared}(q_{inv})$ and $C_2^{smeared}(q_{inv})$ are without and with taking into account the momentum resolution effect, respectively.

Figure 1 shows the correlation function before (black circles) and after (red circles) the correction on the momentum resolution. The momentum resolution correction affects the extracted source radii not more than on 1% and has negligible effect on the correlation strength.

In order to disentangle the correlation signals induced by non-femtoscopic correlations, the Monte Carlo (MC) generator PYTHIA-6.4.28 with Perugia 2011 tune is used [11, 12]. PYTHIA does not contain Bose-Einstein correlations and final-state interactions (Coulomb and strong) and thus will introduce only the correlations due to the energy-momentum conservation effects. In figure 2 the simulated correlation functions of positively and negatively charged kaons produced in the interactions of the negative pion beam with copper and carbon targets are compared to the measured ones. The source parameters were estimated in the following way. First, the simulated correlation functions were fitted with the first order polynomial $D(q_{inv}) = a + bq_{inv}$. Then the experimental correlation function is fitted by (1), taking the $D(q_{inv})$ from the PYTHIA fit. In order to estimate the systematic uncertainty, different function forms were used for the baseline function. The systematic uncertainty due to the baseline determination does not exceed 2%.

Figure 3 shows the correlation functions (double ratios) obtained according to (2) for positive (black solid circles) and negative (blue empty circles) kaon pairs. The correlation functions were fitted using (1) and the non-femtoscopic term was equal to unity. Solid red and dashed magenta lines represent the correlation function fits of positive and negative kaons, respectively.

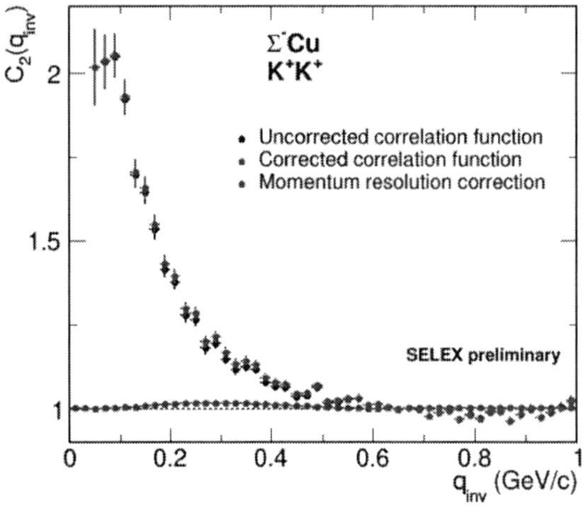

Figure 1. The positive kaon experimental correlation function measured for Σ^-Cu interactions before (black circles) and after (red circles) the momentum resolution correction. Magenta circles represent the correction factor due to the momentum resolution.

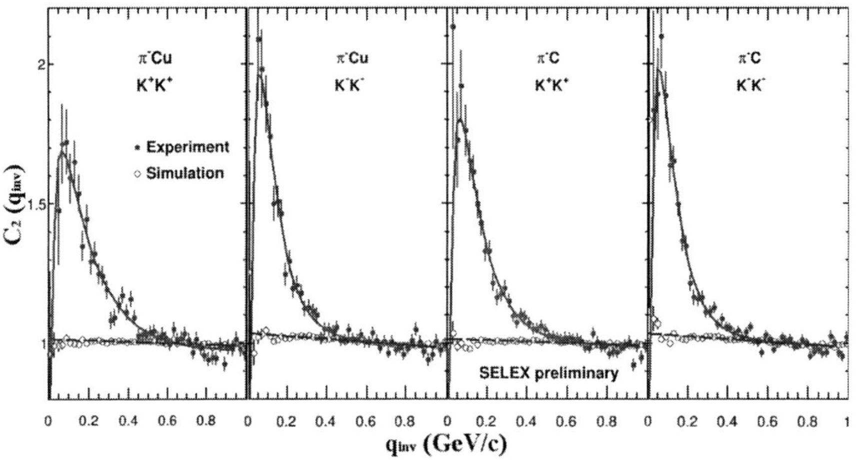

Figure 2. The experimental (red solid circles) and simulated (blue empty circles) correlation functions of positive and negative kaons obtained for $\pi^- Cu$ and $\pi^- C$ interactions. The blue dashed lines represent baseline fits. The red lines are the correlation function fits using (1).

The difference between the extracted source parameters obtained by previously discussed methods does not exceed 4%.

Figure 4 shows the measured source radii (left plot) and correlation strength parameters (right plot) of positive (red solid circles) and negative (blue empty circles) kaon pairs obtained for 600 GeV/c Σ^-, π^- and 540 GeV/c p interactions with carbon and copper nuclei. The uncertainty due to the fit range was estimated by varying the upper limit from q_{inv} = 1 GeV/c to q_{inv} = 0.6 GeV/c and lower limit from q_{inv} = 0 GeV/c to q_{inv} = 0.2 GeV/c, and did not exceed 6%.

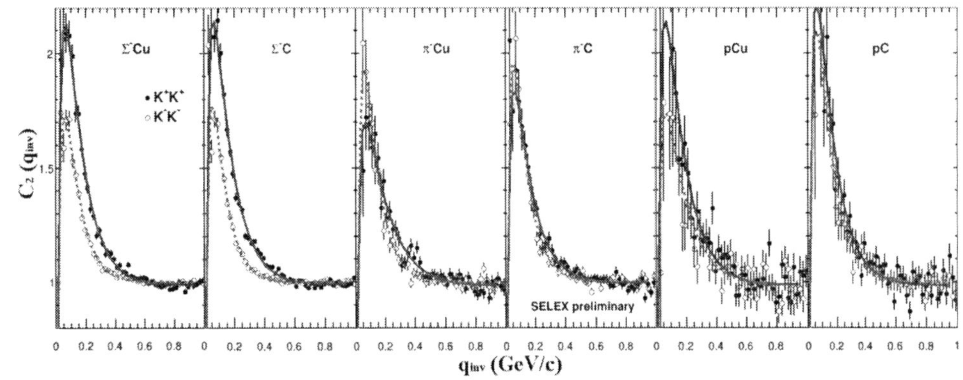

Figure 3. The positive (black solid circles) and negative (blue empty circles) kaon double ratios obtained for Σ^{-}Cu (a) and Σ^{-}C interactions. The solid red and dashed magenta lines represent fits performed using 1 of positively and negatively charged kaon pairs, respectively.

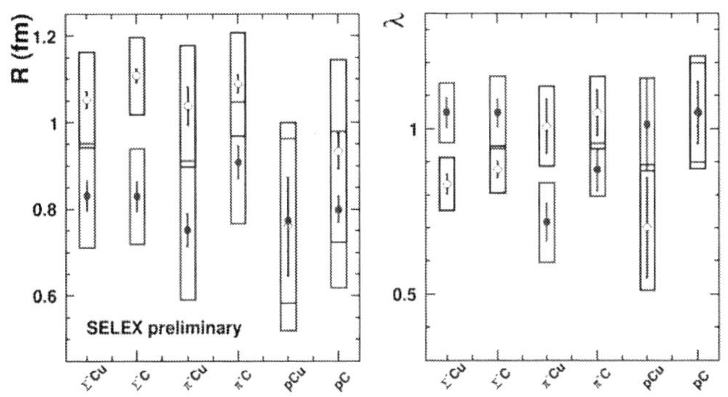

Figure 4. The dependence of the emitting source radii (left) and correlation strength parameter (right) on the type of the h+A collision. The red solid and blue empty circles represent the estimated source parameters for positive and negative kaons, respectively. Boxes represent systematic uncertainties on the measured values.

The systematic errors (boxes) on R and λ for each beam type and target material are taken as the root mean square spread of the values calculated for the different sources of the systematic uncertainty.

Due to the limited acceptance and statistics it is not possible to select events with different multiplicity (collision centrality) classes for each collision type. In the current analysis, the extracted source radii for 600 GeV/c Σ^{-}, π^{-} and 540 GeV/c p interactions with Cu and C nuclei are consistent within uncertainties and do not depend on the target material. The source radii of negatively charged kaons measured for Σ^{-} and π^{-} beams are generally larger than those for positive kaons. For Σ^{-} induced interactions the negative kaon correlation functions are smaller than for positive kaons at $q_{inv} \leq 0.6$ GeV/c. The λ parameters obtained for positive and negative kaons for different hadron beams are similar for carbon and copper nuclei.

DOI: 10.1051/epjconf/201713803013

5 Summary

We have presented the results from the SELEX experiment on positive and negative kaon femtoscopic measurements in 600 GeV/c Σ^-, π^- and 540 GeV/c p interactions with Cu and C nuclei. The one-dimensional source parameters were extracted for all h+A collision types. The measured sizes of the emitting region for Σ^-Cu(C), π^-Cu(C) and pCu(C) interactions are similar and consistent with each other within uncertainties. The source radii of negatively charged kaons measured for Σ^- and π^- beams are generally larger than those for positive kaons but consistent within the uncertainties. For Σ^-Cu(C) interactions the correlation functions of negative kaons are smaller than those for positive kaons at q_{inv} $\leq$0.6 GeV/c.

Acknowledgments

Authors would like to thank the support from National Research Nuclear University MEPhI in the framework of the Russian Academic Excellence Project (contract No. 02.a03.21.0005, 27.08.2013).

References

[1] G. Goldhaber et al., Phys. Rev. **120**, 300 (1960)
[2] J.S. Russ et al., Proc. Int. Conf. on High Energy Phys. **II**, 1259 (1998)
[3] U. Dersch et al., Phys. Lett. B **579**, 277 (2000)
[4] J. Engelfried et al., Nucl. Instr. and Meth. A **409**, 53 (1998)
[5] G.A. Nigmatkulov et al., Phys. Lett. B **753**, 458 (2016)
[6] R. Lednický et al., Phys. Lett. B **373**, 30 (1996)
[7] M.G. Bowler, Phys. Lett. B **270**, 69 (1991)
[8] Yu.M. Sinyukov et al., Phys. Lett. B **432**, 248 (1998)
[9] H. Bøggild et al., Phys. Lett. B **302**, 510 (1993)
[10] J. Adams et al., Phys. Rev. C **71**, 044096 (2005)
[11] T. Sjöstrand, S. Mrenna, P. Skands, JHEP **05**, 026 (2006)
[12] P. Skands, Phys. Rev. D **82**, 074018 (2010)
[13] K. Aamodt et al., Phys. Rev. D **82**, 052001 (2010)

EPJ Web of Conferences **138**, 04001 (2017) DOI: 10.1051/epjconf/201713804001
Baldin ISHEPP XXIII

Study of heavy exotic baryon resonances

Yubing Dong[1,2,*,**] and *Qi-Fang* Lü[1,***]

[1] *Institute of High Energy Physics, Chinese Academy of Sciences, Beijing 100049, P. R. China*
[2] *Theoretical Physics Center for Science Facilities (TPCSF), CAS, Beijing 100049, P. R. China*

Abstract. The possible production of the neutral $P_c(4380)$ and $P_c(4450)$ in the J-PARC experiment is discussed. Different spin-parity assignments of the P_c states are considered and differential cross sections of $\pi^- p \to J/\psi p$ are calculated. Moreover, the strong decays of $P_c \to J/\psi p$ are estimated based on a $\Sigma_c \bar{D}^*$ molecular scenario. Our results disfavor all the $P-$ wave $\Sigma_c \bar{D}^*$ assignments, while favor the $S-$ wave $\Sigma_c \bar{D}^*$ picture of the two P_c states comparing to the experimental data. In addition, we also analyze the possible $J^P = 3/2^-$ state in $\Sigma_c^* \bar{D}$ and $\Sigma_c^* \bar{D}^*$ pictures with heavy quark limit.

1 Introduction

It is well-known that two hidden charm pentaquark resonances $P_c^+(4380)$ and $P_c^+(4450)$ were reported by the LHCb Collaboration in the process of $\Lambda_b^0 \to J/\psi K^- p$ [1]. Their masses and widths are $(4380 \pm 8 \pm 29; 205 \pm 18 \pm 86)$ MeV for $P_c(4380)$ and $(4449.8 \pm 1.7 \pm 2.5; 39 \pm 5 \pm 19)$ MeV for $P_c(4450)$, respectively. Their isospins are $I = 1/2$. Till now, the spin-parities of the two states are not clear, and some possible spin-parity assignments of $(3/2^-, 5/2^+)$, $(3/2^+, 5/2^-)$, and $(5/2^+, 3/2^-)$ are suggested by the experimental analyses. Due to their strong decay mode of $J/\psi p$, the two states are expected to have five quark content $uudc\bar{c}$. The branching fractions $\mathcal{B}(\Lambda_b^0 \to P_c^+ K^-)\mathcal{B}(P_c^+ \to J/\psi p)$ of the decays $\Lambda_b^0 \to J/\psi K^- p$ are given by [2]. In addition, the two P_c states were also confirmed in the process of $\Lambda_b^0 \to J/\psi \pi^- p$ [3].

Since the observation of $X(3872)$, many exotic mesons XYZ particles have been reported. Those states cannot be easily understood as the $q\bar{q}$ mesons due to their narrow widths and their masses very near to the thresholds of the two mesons [4, 5]. Most of them, particularly the charged Z_c and Z_b, should have four-quark content. The newly observation of the two charged P_c states provides a unique perspective to the exotic baryon states. There are many interesting studies on these two P_c states in the literature. They may be interpreted as the compact pentaquark states, where the five quarks are tightly bounded and interact with each other via color magnetic interaction directly [6, 7]. They are also regarded as the loosely bound molecular states composed of one meson and one baryon [8–13]. Moreover, the kinematical effects due to the re-scattering processes on the two states have been carefully discussed [14, 15]. It should be stressed that besides their mass spectrum, their

*This work is supported by National Sciences Foundations of China No. 11475192 as well as supported, in part, by the DFG and the NSFC through funds (11621131001) provided to the Sino-German CRC 110 "Symmetries and the Emergence of Structure in QCD".
**e-mail: dongyb@ihep.ac.cn
***e-mail: lvqifang@ihep.ac.cn

DOI: 10.1051/epjconf/201713804001

production in other facilities and the decay behaviors can also provide information about their intrinsic structures. [12] has discussed the $P_c(4450) \to J/\psi\ p$ partial decay width, by regarding it as a bound state of charmonium $\psi(2S)$ and nucleon, and [16] performed a comprehensive calculation based on the heavy quark symmetry and spin rearrangement scheme, where the $\Sigma_c^*\bar{D}$, $\Sigma_c\bar{D}^*$, and $\Sigma_c^*\bar{D}^*$ molecular candidates are all taken into account simultaneously. In the $\Lambda_b^0 \to J/\psi K^- p$ decay process, only charged P_c states can be observed. The $\gamma p \to J/\psi p$ reaction with charged P_c production has been proposed for the future JLab. experiment [17–20]. In addition, the production of their neutral partners is expected to be observed in J-PARC experiment by our previous study [21].

This paper is organized as follows. In Sect. 2 we show the production of the neutral P_c states in the future J-PARC experiment. In Sect. 3, the strong decay widths $J/\psi p$ under a $\Sigma_c\bar{D}^*$ molecule assignment with spin-parity $J^P = 3/2^\pm$ or $5/2^\pm$ are estimated. In our calculation, we use a phenomenological effective Lagrangian approach. It should be noticed that this technique, for evaluating composite hadron systems, has been widely applied for various molecular states, where the compositeness condition corresponding to $Z = 0$ has been employed [22–24]. Finally, a short summary is given in Sec. 4.

2 Production of the two P_c states at J-PARC

We first study the $\pi^- p \to J/\psi n$ reaction within an effective Lagrangian approach [21]. The relevant Feynman diagrams are displayed in figure 1. The s-channel contributions from the P_c states with different spin-parity assumptions are taken into account. The u-channel contribution is negligible due to the highly off-shell intermediate P_c states. The background contributions from t-channel π and ρ meson exchanges are considered as well.

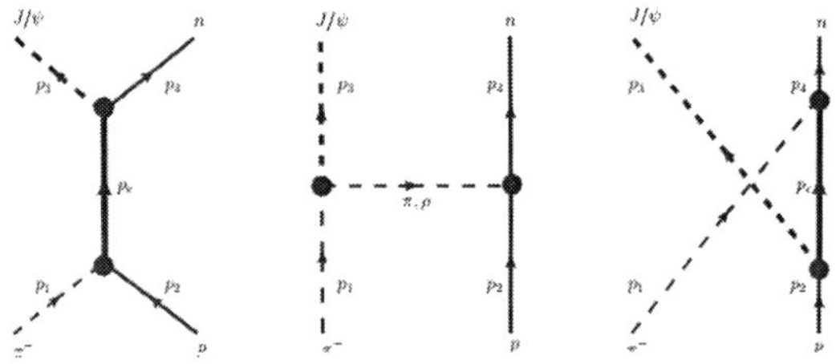

Figure 1. Feynman diagrams for $\pi^- p \to J/\psi n$ reaction.

The effective Lagrangians for $P_c N J/\psi$ couplings can be written as [25]:

$$\mathcal{L}^{3/2^\pm}_{P_c N J/\psi} = -\frac{ig_1}{2M_N}\bar{N}\Gamma^{(\pm)}_\nu \psi^{\mu\nu} P_{c\mu} + \text{H.c.}, \qquad \mathcal{L}^{5/2^\pm}_{P_c N J/\psi} = \frac{g_1}{(2M_N)^2}\bar{N}\Gamma^{(\mp)}_\nu \partial^\alpha \psi^{\mu\nu} P_{c\mu\alpha} + \text{H.c.}, \tag{1}$$

for $J^P = 3/2^\pm$, and $5/2^\pm$, respectively. In (1) the Γ matrices are defined as $\Gamma^+_\mu \equiv \gamma_\mu\gamma_5$ and $\Gamma^-_\mu \equiv \gamma_\mu$, respectively. In our calculation, the higher partial wave terms are neglected due to the small momentum of the final $J/\psi N$ states compared with nucleon mass and the lack of the experimental data [17]. The effective Lagrangians for $P_c N\pi$ vertexes are

$$\begin{aligned}
\mathcal{L}^{3/2^+}_{P_c N\pi} &= \frac{g_{P_c N\pi}}{m_\pi}\bar{N}\vec{\tau}\cdot\partial_\mu\vec{\pi}P_c^\mu + \text{H.c.}, \qquad \mathcal{L}^{3/2^-}_{P_c N\pi} = \frac{g_{P_c N\pi}}{m_\pi^2}\bar{N}\gamma_5\gamma_\mu\vec{\tau}\cdot\partial^\mu\partial_\nu\vec{\pi}P_c^\nu + \text{H.c.}, \\
\mathcal{L}^{5/2^+}_{P_c N\pi} &= \frac{g_{P_c N\pi}}{m_\pi^3}\bar{N}\gamma_5\gamma_\mu\vec{\tau}\cdot\partial^\mu\partial_\nu\partial_\lambda\vec{\pi}P_c^{\nu\lambda} + \text{H.c.}, \qquad \mathcal{L}^{5/2^-}_{P_c N\pi} = \frac{g_{P_c N\pi}}{m_\pi^2}\bar{N}\vec{\tau}\cdot\partial_\mu\partial_\nu\vec{\pi}P_c^{\mu\nu} + \text{H.c.},
\end{aligned} \tag{2}$$

DOI: 10.1051/epjconf/201713804001

Table 1. Coupling constants of P_cNJ/ψ and $P_cN\pi$ with different J^P assignments by assuming the branching ratios are 10% and 1%, respectively.

State	Channel	$3/2^+$	$3/2^-$	$5/2^+$	$5/2^-$
$P_c(4380)$	$J/\psi N$	1.09	0.49	2.17	5.13
	πN	8.56×10^{-3}	3.43×10^{-4}	3.59×10^{-5}	8.95×10^{-4}
$P_c(4450)$	$J/\psi N$	0.41	0.20	0.80	1.75
	πN	3.65×10^{-3}	1.43×10^{-4}	1.47×10^{-5}	3.75×10^{-4}

for $J^P = 3/2^\pm$ and $J^P = 5/2^\pm$, respectively. Furthermore, the couplings for t-channel diagram are also needed. We adopt the commonly used Lagrangian densities for $J/\psi\pi\pi$, $J/\psi\pi\rho$, πNN, and ρNN vertexes given by [26, 27].

The $P_c \to NJ/\psi$ and $P_c \to N\pi$ decay processes are calculated and the relevant coupling constants can be obtained from the partial decay widths with different J^P assignments. Assumed that the branching ratios are 10% and 1% for $P_c \to NJ/\psi$ and $P_c \to N\pi$, respectively, the obtained coupling constants are listed in table 1. To account for the fact that hadrons are not point-like particles, we also employ the form factors [28] in our calculations. For example, we have $F(q^2) = \frac{\Lambda_{P_c}^4}{\Lambda_{P_c}^4+(q^2-M_{P_c}^2)^2}$ for the intermediate P_c in the S-channel diagram with the cutoff parameter $\Lambda_{P_c} = 0.5$ GeV [25].

Then, the differential cross section is

$$\frac{d\sigma}{dcos\theta} = \frac{M_N^2}{16\pi s}\frac{|\vec{p}_3^{\,c.m.}|}{|\vec{p}_1^{\,c.m.}|}\left|\mathcal{M}_{\pi^-p\to J/\psi n}\right|, \tag{3}$$

in the center-of-mass (c.m.) frame with θ being the scattering angle of the outgoing J/ψ relative to the incoming pion beam, and $\vec{p}_1^{\,c.m.}$ and $\vec{p}_3^{\,c.m.}$ the three-momenta of π and J/ψ mesons in c.m. frame. Since the relative phase between different amplitudes is unknown, the interference terms between them are ignored here. This scheme avoids adding new free parameters and our calculation shows that the magnitudes of different contributions are much different, and hence the effects are expected to be small enough.

From our calculated total cross sections, we find that the π and ρ meson exchanges provide significant background contributions, while there are two narrow bump structures from $P_c^0(4380)$ and $P_c^0(4450)$ contributions. With the different J^P assignments, the divergences among the total cross sections are not clear, which can hardly be used to identify the spin-parity of the two P_c states. The peaks are in the magnitude of 1 μb at the c.m. energy W = 4.38 GeV and 4.45 GeV, which can be measured in the future J-PARC experiment. It should be stressed that the two P_c states are firstly observed in the $J/\psi p$ invariant mass, and this decay mode can occur via fallen apart mechanism. For the πN channel, although the decay width is suppressed by $c\bar{c}$ pair annihilation, this process is still an OZI-allowed two-body strong decay with large phase space. Here, we assume the branching ratios for the two P_c states to $J/\psi N$ and πN channels being 10% and 1%, respectively.

The estimated differential cross sections are shown in figure 2. It is seen that the t-channel meson exchanges provide forward contribution in the whole energy region. The differential cross sections at 4.38 GeV and 4.45 GeV are mainly from $P_c^0(4380)$ and $P_c^0(4450)$ contributions, respectively. The angular distributions of the two P_c states are obviously different with the forward background contribution and display significant different behaviors with different J^P assignments. These specific features may be observed by the future experiment at J-PARC and can help us to distinguish different spin-parity assumptions.

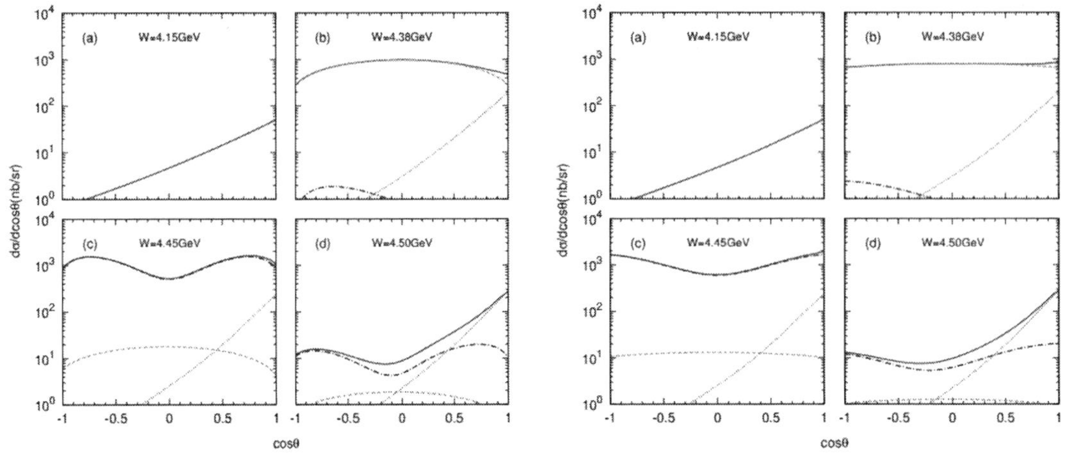

Figure 2. The differential cross sections for the $\pi^- p \to J/\psi n$ reaction at the c.m. energies W = 4.15 GeV, 4.38 GeV, 4.45 GeV, and 4.45 GeV, respectively. The $(P_c^0(4380), P_c^0(4450))$ respectively correspond to $(3/2^+, 5/2^-)$ (left panel) and $(3/2^-, 5/2^+)$ (right panel) assumptions. The red solid, green dashed, blue dot-dashed, and pink short dotted curves stand for the total, ($P_c^0(4380)$ and $P_c^0(4450)$), and the background contributions, respectively.

3 Strong decays of the two P_c states in a $\Sigma_c\bar{D}^*$ molecular scenario

In this section, we show a calculation for the strong decays of the two P_c states based on a $\Sigma_c\bar{D}^*$ molecule scenario [13]. The simplest effective Lagrangian describing the $P_c\Sigma_c\bar{D}^*$ couplings can be expressed as

$$\begin{aligned} \mathcal{L}^{3/2^\pm}_{P_c\Sigma_c\bar{D}^*}(x) &= -ig_{P_c\Sigma_c\bar{D}^*}\int d^4y\Phi(y^2)\bar{\Sigma}_c(x+\omega_{\bar{D}^*}y)\Gamma^{(\pm)}\bar{D}^*_\mu(x-\omega_{\Sigma_c}y)P^\mu_c(x) + \text{H.c.}, \\ \mathcal{L}^{5/2^\pm}_{P_c\Sigma_c\bar{D}^*}(x) &= g_{P_c\Sigma_c\bar{D}^*}\int d^4y\Phi(y^2)\bar{\Sigma}_c(x+\omega_{\bar{D}^*}y)\Gamma^{(\mp)}\partial_\mu\bar{D}^*_\nu(x-\omega_{\Sigma_c}y)P^{\mu\nu}_c(x) + \text{H.c.}, \end{aligned} \tag{4}$$

where $\omega_{\bar{D}^*} = m_{\bar{D}^*}/(m_{\Sigma_c} + m_{\bar{D}^*})$ and $\omega_{\Sigma_c} = m_{\Sigma_c}/(m_{\Sigma_c} + m_{\bar{D}^*})$. The $\Gamma^{(\pm)}$ matrices are defined as $\Gamma^+ = \gamma_5$ and $\Gamma^- = I$. The correlation function $\Phi(y^2)$ is chosen to describe the distribution of Σ_c and $\bar{D}^*$ in P_c states. In order to render the Feynman diagrams ultraviolet finite, its Fourier transform of $\Phi(y^2)$ should vanish fast in the ultraviolet region of the Euclidean space. Here, we adopt the commonly used Gaussian-type and thus, the Fourier transform can be written as $\tilde{\Phi}(p_E^2/\Lambda^2) \doteq \exp(-p_E^2/\Lambda^2)$, with p_E being the Euclidean Jacobi momentum and Λ a free size parameter characterizing the distribution of the two components in the molecule. This parameter is about 1 GeV and may vary in different systems. The coupling constant $g_{P_c\Sigma_c\bar{D}^*}$ can be determined by the compositeness condition [29]. This condition requires that the renormalization constant of the hadronic molecular wave function is equal to zero, $Z_{P_c} = 1 - \partial\Sigma^{T(\pm)}_{P_c}(p)/\partial\not{p}\big|_{\not{p}=m_{P_c}} \equiv 0$, where the $\Sigma^{T(\pm)}_{P_c}(p)$ is the transverse part of the mass operator. Then, the hadronic decay $P_c^+ \to J/\psi p$, shown in figure 3 with the exchanging of D^*, D, and Σ_c hadrons, can be estimated if the effective Lagrangians $\mathcal{L}_{\psi D^{(*)}D^{(*)}}$ and $\mathcal{L}_{\Sigma_c ND^*(D)}$ are known. Paper [13] gives their explicit expressions for the Lagrangians and the relevant coupling constants.

The calculated partial decay widths of $P_c^+ \to J/\psi p$ are presented in table 2 with $\Lambda = 1$ GeV. In our calculation, we found the partial decay widths increase for $J^P = 3/2^\pm$ assignment, while decrease for $J^P = 5/2^\pm$ when Λ increases. For the $P_c(4380)$ state, the predicted partial decay widths in all the

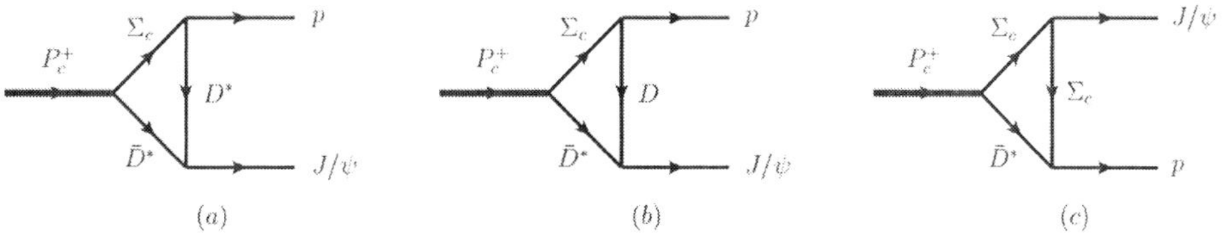

Figure 3. Feynman diagrams for the $P_c^+ \to J/\psi p$ decay process.

four cases are all smaller than the experimental total width within the errors. Since the $J^P = 5/2^-$ assignment is not favored by the experiment [1], and the $\Sigma_c\bar{D}^*$ molecule should be in D wave in this case, the $J^P = 5/2^-$ assumption for $P_c(4380)$ is ruled out, while others are consistent with the present data. For the $P_c(4450)$ state, since $J^P = 3/2^+$ assignment is not favored by the LHCb collaboration [1] and the calculated partial decay width is much larger than the experimental total width, this case can be completely excluded as well. Also, the $J^P = 5/2^+$ case is disfavored due to the large predicted width, and the D wave $\Sigma_c\bar{D}^*$ molecule with $J^P = 5/2^-$ can be hardly formed through the one boson exchange interaction. Hence, only the S wave $\Sigma_c\bar{D}^*$ molecule is left for the $P_c(4450)$ state.

Table 2. Partial decay widths of $P_c^+ \to J/\psi p$ with different J^P assignments with $\Lambda = 1$ GeV. The unit is in MeV.

State	$3/2^+$	$3/2^-$	$5/2^+$	$5/2^-$	Expt. [1]
$P_c(4380)$	173.12	38.12	169.51	4.96	$205 \pm 18 \pm 86$
$P_c(4450)$	369.82	25.00	76.15	3.39	$39 \pm 5 \pm 19$

Besides the $\Sigma_c\bar{D}^*$ molecular picture, the $\Sigma_c^*\bar{D}$ and $\Sigma_c^*\bar{D}^*$ scenarios for the two P_c states are also proposed in the literature. In the heavy quark limit, the S-wave D and D^* mesons can be categorized into a doublet as well as the heavy baryons Σ_c and Σ_c^*. Based on the symmetry and spin rearrangement scheme, [16] gave the ratios of the partial decay widths in the different molecular scenarios. This model independent result shows that for the three S-wave $\Sigma_c\bar{D}^*$, $\Sigma_c^*\bar{D}$, and $\Sigma_c^*\bar{D}^*$ molecules with $J^P = 3/2^-$, the ratios of their $J/\psi p$ decay widths satisfy the relation of $\Gamma[(\Sigma_c\bar{D}^*)] : \Gamma[(\Sigma_c^*\bar{D})] : \Gamma[(\Sigma_c^*\bar{D}^*)] = 1.0 : 2.7 : 5.4$. Simply employing those ratios and the results shown in table 2 with $\Lambda = 1$ *GeV*, we can estimate the $J/\psi\ p$ decay widths of the two P_c states in the $\Sigma_c^*\bar{D}$ and $\Sigma_c^*\bar{D}^*$ molecular scenarios with $J^P = 3/2^-$. The obtained results of $\Gamma^{3/2-}_{P_c(4380)}[(\Sigma_c^*\bar{D})]$, $\Gamma^{3/2-}_{P_c^+(4380)}[(\Sigma_c^*\bar{D}^*)]$ $\Gamma^{3/2^-}_{P_c^+(4450)}[(\Sigma_c^*\bar{D})]$, and $\Gamma^{3/2^-}_{P_c^+(4450)}[(\Sigma_c^*\bar{D}^*)]$ are 102.9, 205.9, 67.5, and 135 MeV, respectively. The much larger partial decay width $\Gamma^{3/2^-}_{P_c^+(4450)}[(\Sigma_c^*\bar{D}^*)]$, comparing to the data, excludes the possibility of $P_c(4450)$ as a S-wave $J^P = 3/2^-$ $\Sigma_c^*\bar{D}^*$ molecule. Also, the $P_c(4450)$ as the $\Sigma_c^*\bar{D}$ system is not favored due to its higher mass over the threshold and a slightly large partial decay width. The above discussion shows that if the $P_c(4450)$ state has the spin-parity $J^P = 3/2^-$, only $\Sigma_c\bar{D}^*$ system, out of the three molecular scenarios, is allowed.

4 Summary

In this paper we show our estimate for the production of the two neutral P_c states in the $\pi^- p \to J/\psi n$ process, which may be measured in the future experiment at J-PARC. The calculated differential cross sections are expected to give more information about the spin-parities of the two states. In the calculation, the partial branching ratios of $P_cJ/\psi N$ and $P_c\pi N$ are assumed to be 10% and 1%. Moreover, we also show a calculation for the strong decay of $P_c \rightarrow J/\psi p$ based on a $\Sigma_c\bar{D}^*$ molecule scenario. The estimated widths exclude all the P-wave $\Sigma_c\bar{D}^*$ assumptions in this case. For the $J^P = 3/2^-$ $\Sigma_c^*\bar{D}$ and $\Sigma_c^*\bar{D}^*$ scenarios, only $\Sigma_c^*\bar{D}$ for $P_c(4380)$ case might be possible. Particularly, the S-wave $\Sigma_c\bar{D}^*$ assignments for $P_c(4380)$ and $P_c(4450)$ may be both favored by the experimental data. It should be noted that this conclusion contradicts to the fact that the two P_c states have different spins and opposite parities. Since $P_c(4380)$ is broad while $P_c(4450)$ is narrow, and $P_c(4450)$ is much closer to the threshold of $\Sigma_c\bar{D}^*$, one should be careful to draw any conclusion for the $P_c(4380)$ state. This indicates an agreement with the interpretations of [8], in which the $P_c(4450)$ is a $J^P = 3/2^-$ $\Sigma_c\bar{D}^*$ resonance and $P_c(4380)$ may be not a genuine state. Theoretical investigations on other decay modes and further experimental information on the spin-parities and the partial decay widths will be helpful to understand the nature of the two P_c states.

References

[1] R. Aaij et al. [LHCb Collaboration], Phys. Rev. Lett. **115**, 072001 (2015)
[2] R. Aaij et al. [LHCb Collaboration], Chin. Phys. C **40**, 011001 (2016)
[3] R. Aaij et al. [LHCb Collaboration], Phys. Rev. Lett. **117**, 082003 (2016)
[4] X. Liu, Chin. Sci. Bull. **59**, 3815 (2014)
[5] H. X. Chen, X. Liu, and S. L. Zhu, Phys. Rept. **639**, 1-121 (2016)
[6] L. Maiani, A. D. Polosa, and V. Riquer, Phys. Lett. B **749**, 289 (2015)
[7] R. F. Lebed, Phys. Lett. B **749**, 454 (2015)
[8] M. Karliner and J. L. Rosner, Phys. Rev. Lett. **115**, 122001 (2015)
[9] R. Chen, X. Liu, X. Q. Li, and S. L. Zhu, Phys. Rev. Lett. **115**, 132002 (2015)
[10] J. He, Phys. Lett. B **753**, 547 (2016)
[11] U. -G. Meiner and J. A. Oller, Phys. Lett. B **751**, 59 (2015)
[12] M. I. Eides, V. I. Petrov, and M. V. Polyakov, Phys. Rev. D **93**, 054039 (2016)
[13] Q. F. Lü and Y. B. Dong, Phys. Rev. D **93**, 074020 (2016)
[14] F. K. Guo, U. -G. Meiner, W. Wang, and Q. Zhao, Phys. Rev. D **92**, 071502 (2015)
[15] X. H. Liu, Q. Wang, and Q. Zhao, Phys. Lett. B **757**, 231 (2016)
[16] G. J. Wang, L. Ma, X. Liu, and S. L. Zhu, Phys. Rev. D **93**, 034031 (2016)
[17] Q. Wang, X. H. Liu and Q. Zhao, Phys. Rev. D **92**, 034022 (2015)
[18] V. Kubarovsky and M. B. Voloshin, Phys. Rev. D **92**, 031502 (2015)
[19] M. Karliner and J. L. Rosner, Phys. Lett. B **752**, 329 (2016)
[20] A. N. H. Blin, C. Fernandez-Ramirez, A. Jackura, et al., Phys. Rev. D **94**, 034002 (2016)
[21] Q. F. Lü, X. Y. Wang, J. J. Xie, X. R. Chen, and Y. B. Dong, Phys. Rev. D **93**, 034009 (2016)
[22] Y. B. Dong, A. Faessler, T. Gutsche, and V. E. Lyubovitskij, Phys. Rev. D **77**, 094013 (2008)
[23] Y. B. Dong, A. Faessler, T. Gutsche, et al., Phys. Rev. D **79**, 094013 (2009)
[24] Y. B. Dong, A. Faessler, T. Gutsche, and V. E. Lyubovitskij, J. Phys. G **38**, 015001 (2011)
[25] S. H. Kim, S. I. Nam, Y. Oh and H. C. Kim, Phys. Rev. D **84**, 114023 (2011)
[26] Q. F. Lü, J. J. Xie and D. M. Li, Phys. Rev. C **90**, 034002 (2014)
[27] J. J. Wu and T.-S. H. Lee, Phys. Rev. C **88**, 015205 (2013)
[28] J. J. Xie, B. S. Zou and H. C. Chiang, Phys. Rev. C **77**, 015206 (2008)
[29] S. Weinberg, Phys. Rev. **130**, 776 (1963)

EPJ Web of Conferences **138**, 04002 (2017) DOI: 10.1051/epjconf/201713804002
Baldin ISHEPP XXIII

Modification of the strange meson properties in nuclear matter

Yury Kiselev[1,⋆], *Michael* Hartmann[2], *Eduard* Paryev[1,3], and *Andrey* Polyanskiy[1]

[1] *Institute for Theoretical and Experimental Physics, Moscow, Russia*
[2] *Institute for Nuclear Physics, FZ-Juelich, Juelich, Germany*
[3] *Institute for Nuclear Research, Russian Academy of Science, Moscow, Russia*

Abstract. To make progress in understanding the properties of the strange mesons with open (K^-) and hidden (ϕ) strangeness in the nuclear medium, the measurements of the production of K^+K^- pairs with invariant masses corresponding to both the ϕ and non-ϕ regions in proton collisions with C, Cu, Ag, and Au targets at an incident beam energy of 2.83 GeV were recently performed by the ANKE Collaboration at COSY. The data from the ϕ region of invariant masses provide information on the momentum dependence of the ϕ nuclear transparency ratio and the in-medium ϕ meson width over the momentum range of 0.6-1.6 GeV/c. Analysis of the data from the non-ϕ region of K^+K^- invariant masses, where differential cross sections for pair production on the four targets were obtained as functions of the laboratory K^- momentum, allows to draw the conclusion of the strength of the K^--nuclear potential.

1 Introduction

The study of hadron properties in a strongly interacting environment has been a very active research field over the last years, especially in connection with questions of the partial restoration of chiral symmetry in hot/dense nuclear matter and of the existence of a K^- condensate in neutron stars. Modification of hadron properties, mass and width, has been predicted by the QCD sum rules and the models with chiral Lagrangians. As the strength of the effects is expected to be proportional to the baryon density and/or the temperature, the main efforts over the last decades were focused at the quests for these effects in the relativistic heavy-ion collisions. However, the difficulties associated with the problem of the description of system with a drastic change in time its density and temperature did not allow to make the final conclusion about the existence of the modification effects. An alternative way to get the information on the hadron medium modification effects is the study of the reaction with elementary probes (photons, pions, and protons) on the nuclei. The advantage of this approach is connected to the fact that the process of hadron production proceeds in cold static nuclear matter of a well-defined density while the values of the effects are somewhat less but still comparable to those of heavy-ion collisions. At normal nuclear density, $\rho_0 = 0.16 fm^{-3}$, the in-medium modification of mesons can be expressed in terms of complex nuclear optical potential U. In this case the Breit-Wigner formula acquires two new terms: $S(M) = [\Gamma_0/2 - ImU]/[(M-(M_0+ReU))^2+(\Gamma_0/2-ImU)^2]$. The real part of the potential ReU determines a shift of the pole of the meson spectral function while

⋆e-mail: yurikis@itep.ru

the imaginary part ImU determines its broadening in nuclear matter. It is expected that both parts of the potential U depend on the meson momentum relative to the nuclear matter. Here we report on the results of the investigation of the in-medium properties of the mesons with hidden (ϕ) and open (K^-) strangeness produced by 2.83 GeV proton beam on nucleus targets. According to the theoretical predictions a decrease of the ϕ meson mass at normal nuclear density is small ($< 1\%$) and the main medium effect on the ϕ is substantial increase of its width compared to its vacuum magnitude of 4.43 MeV. This magnitude corresponds to the decay length $L = 46fm$ what significantly exceeds the size of nuclei. In spite of almost all produced ϕ's decay into K^+K^- pairs outside the nucleus, the study of the ϕ absorption inside the nuclei provides a possibility to derive the imaginary part of the nuclear optical potential and, hence, the ϕ width $\Gamma = -2ImU$ in nuclear matter [1]. The study of the non-resonant (direct) K^- mesons production gives the opportunity to deduce the magnitude of the real part of the K^--nuclear potential. In this case, in contrast with the ϕ, the K^- creation occurs inside nuclei and the surrounding nuclear medium affects the properties of produced mesons.

2 Experiment

The experiment was performed at the COoler SYnchrotron (COSY) of the Forschungszentrum Juelich (Germany) using the ANKE magnetic spectrometer that is located at an internal target station of the storage ring. ANKE contains three dipole magnets; D1 and D3 divert the circulating beam onto the target and back into the COSY ring, respectively, while D2 is the analyzing magnet [2]. A series of thin and narrow C, Cu, Ag, and Au targets was inserted in a circulating beam of 2.83 GeV protons in front of the main spectrometer magnet D2. The ANKE spectrometer has detection systems placed to the right and left of the beam to register positively and negatively charged ejectiles which are the K^+ and K^- in the case of the present study. Although only used here for efficiency studies, forward-going charged particles could also be measured in coincidence. The spectrometer is equipped by the proportional chambers for tracking, the scintillation counters for the time-of-flight and energy loss measurements, and Cherenkov counter for the pion suppression. The accessible ranges of the K^+ and K^- meson momenta were $0.2 < p < 0.6$ GeV/c and $0.2 < p < 1.0$ GeV/c, respectively. The polar production angle was restricted to 12^0 both for positively and negatively charged kaons. The initial proton kinetic energy of 2.83 GeV corresponds to the excess energy of 108 MeV above the threshold for the kaon pair creation in proton-nucleon collisions. The integrated luminosity for each target is calculated using the measured flux of π^+ mesons with momenta of 500 MeV/c produced at small laboratory angles. Knowing the cross section of this reaction [3] allowed calculation of the differential cross sections for the kaon pair production. The positively charged kaon was first selected using a dedicated detection system that can identify a K^+ against a pion and/or proton background that is 10^5 times more intense [4], [5]. The coincident K^- was subsequently identified from the time-of-flight difference between the stop counters in the negative and positive detector systems [6].

3 Analysis of data

The resulting invariant mass spectrum of the selected K^+K^- pair for C is given in figure 1. One can see that there is strong ϕ signal that sits on a broad distribution of non-resonant kaon pair production together with a relatively small number of misidentified events. To separate the non-resonant kaon pair production events from those arising from the ϕ resonance, a cut on the invariant mass of K^+K^- pairs, IM $< 1.005\ GeV/c^2$, was applied in the subsequent analysis. To determine the in-medium ϕ meson width a 3σ cut around the ϕ peak was made to select K^+K^- events steaming from the ϕ decay. For each of the four targets this part of mass spectrum was fitted by the incoherent sum of a Breit-Wigner function with the natural ϕ width, convoluted with Gaussian resolution function with σ=1

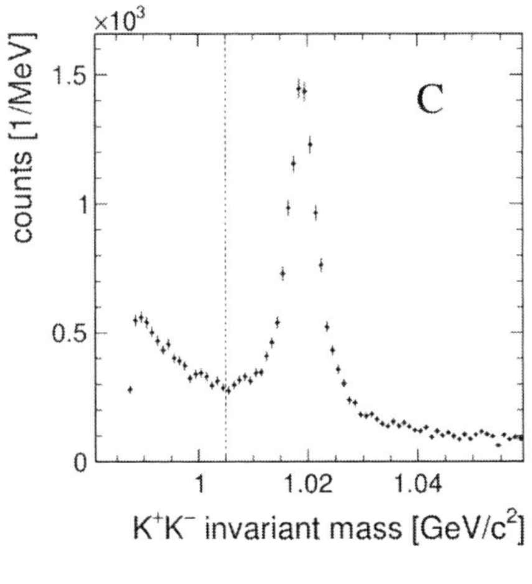

Figure 1. Invariant mass (IM) distribution for K^+K^- pair produced in pC collisions at 2.83 *GeV* beam energy. The vertical line indicates the cut IM=1.005 GeV/c^2 used for the separation of the non-resonant (left) and resonant (right) production processes.

MeV, and polynomial background function. Since the acceptance and efficiency corrections in ANKE are essentially target-independent, the ratio of the numbers of reconstructed ϕ mesons corresponds to the ratio of the cross sections for ϕ meson production in pA and pC collisions in the ANKE window. Values of the nuclear transparency ratio normalized to carbon, $R = (12/A)(\sigma_\phi^A/\sigma_\phi^C)$, averaged over the ϕ momentum range 0.6-1.6 GeV/c, are presented in our work [7]. The number of reconstructed ϕ mesons for each target was between 7000 and 10000.

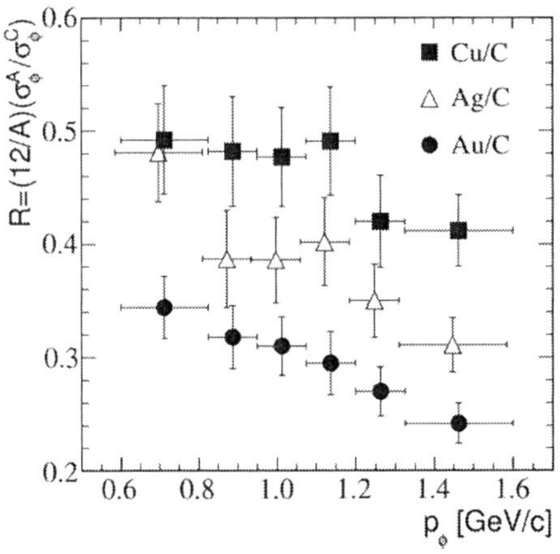

Figure 2. Momentum dependence of the transparency ratio R, normalized to carbon, for Cu, Ag, and Au targets.

These data allowed the momentum dependence of ϕ nuclear transparency, the in-medium ϕ width to be determined for four targets. The transparency ratio for Au, Ag, and Cu targets as a function of the ϕ laboratory momentum p_ϕ is shown in figure 2. The values of the transparency ratio can be measured more precisely than the cross sections themselves as the uncertainties in the absolute normalization and the acceptance calculations as well as the detector-dependent errors mostly cancel out in the ratio. The momentum dependence of the ϕ meson width has been extracted from the comparison of the measured transparency ratios with those calculated in frame of different approaches. We consider three models. The first model is based on the chiral invariant approach elaborated by the Valencia group [1], second one relayed on the BUU transport calculations of the Rossendorf group [8] and the third model is based on the spectral function approach developed by E. Paryev [9]. By comparing the calculated and measured values of the transparency ratio for three target combinations, it is possible

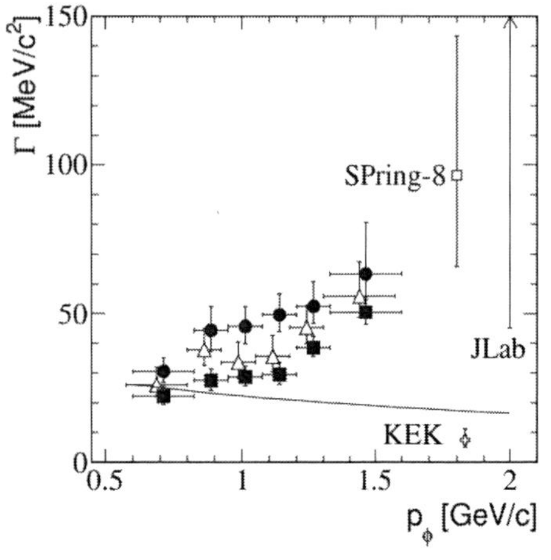

Figure 3. In-medium width of the ϕ meson in the nuclear rest frame at saturation density ρ_0 as a function of the ϕ momentum. The points have been evaluated by comparing the data of figure 2 with the results of three model calculations (model [1]-solid squares, model [9]-solid circles and model [8]-open triangles). Also shown are the results from the other experiments [11], [12], [13]. The solid line represents the Γ_ϕ calculated on the basis of the predicted ϕ self-energy in nuclear matter [14].

Table 1. The ϕ meson width Γ [MeV/c^2] at the laboratory momentum p_ϕ [GeV/c]. The results of the calculation assuming no medium effects on the ϕ width and the mean values of the ϕ meson width shown in figure 3 are presented in the second and third lines, respectively.

p_ϕ	0.71	0.89	1.01	1.14	1.26	1.42
Γ_{vac}	16.6	18.2	19.0	19.8	20.3	21.0
Γ	26	34	37	40	45	54

to determine the weighted average of the ϕ width Γ for each momentum bin in the nuclear rest frame for density ρ_0. The result of applying this procedure is presented in figure 3. As it can be seen from figure 3 the application of three models yields broadly consistent results. Observed significant increasing width with momentum does not depend on the applied model.

The total in-medium ϕ width Γ comprises decay and collision components. The value of both components can be modified due to the interaction of the ϕ with nuclear environment. In order to draw conclusion whether the observed ϕ width exhibits non-trivial medium effects, one has to compare this with the case of no influence of the surrounding matter on the ϕ properties. To do so we calculated $\Gamma_{vac} = \Gamma_{dec} + \Gamma_{col}$ for the center momentum of each momentum bin assuming no medium effects on the ϕ width. The calculations were performed using the vacuum value of the Γ_{dec}=4.43 MeV/c^2 from PDG and the value of the Γ_{col} deduced in the low-density approximation $\Gamma_{col} = \beta\rho_0\sigma_{vac}(\phi N)$. Here $\beta = p/E$ is the ϕ meson velocity relative to the nuclear matter having a density of ρ_0, while $\sigma_{vac}(\phi N)$ is the mean vacuum ϕ-nucleon cross section of 9 mb. The latter value is known from the analysis of the experiments on the ϕ photoproduction from proton and it is in line with the prediction of the additive quark model. The results of the calculations are presented in table 1 together with mean values of the width Γ shown in figure 3. Comparison of the second and third lines of the table provides the clear evidence for the modification of the ϕ width in strongly interacting environment already at normal nuclear density.

The question about the modification of the K^- meson properties in nuclear matter are still the subject of very intense debate. This is due to the complicated dynamics of antikaons inside nuclei, which leads to modifications of their in-medium properties. The calculations, based on chiral Lagrangians or on meson-exchange potentials, predict relatively shallow low-energy K^--nucleus potentials with central depths of the order of -50 to -80 MeV. On the other hand, fits to the K^- atomic data, in terms of phenomenological density dependent optical potentials or relativistic mean-field calculations, lead to much stronger potentials with depths of about -200 MeV at density ρ_0. Current experimental estimates of the potential are spread over the wide range from -30 MeV to -200 MeV. To make progress in understanding the strength of the K^- interaction in the nuclear medium we studied the properties of the K^- produced together with K^+ mesons on nuclear targets in the non-resonant process. Such events form low mass part of the invariant mass spectrum IM < 1.005 GeV/c^2 (see figure 1). In contrast to the inclusive measurements, this data set does not contain kaon pairs steaming from the ϕ decay outside the nuclei. For further interpretation of the data on non-resonant K^- meson production the corre-

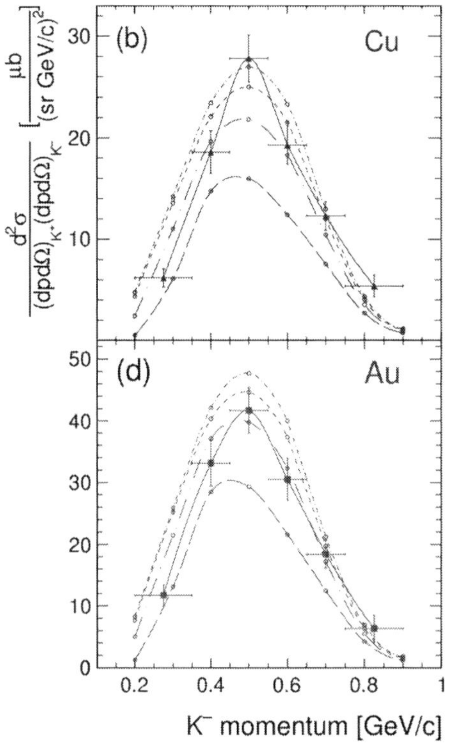

Figure 4. Double-differential cross sections for the production of non-resonant K^+K^- pairs on Cu, (b), and Au (d) targets as a function of the K^- laboratory momentum. The experimental data are averaged over K^+ momenta in the range $200 < p_{K^+} < 600$ MeV/c. The curves represent, from bottom to top, model calculations [10] for K^- potential depth U=0 MeV (long dashed), -60 MeV (dot long dashed), -126 MeV (short dashed), and -180 MeV (dot short dashed), respectively. The solid lines are simple spline functions through the experimental points.

sponding differential cross section were calculated in the frame of the collision model, based on the nuclear spectral function [10], assuming four options for the K^- nuclear potential depth at normal nuclear matter density, viz. U=0 MeV, U=-60 MeV U=-126 MeV, and U=-180 MeV. The calculations take into account the ANKE acceptance. An example of the performed calculations shown in figure 4 demonstrates rather strong sensitivity of the cross sections to the depth of the potential and, therefore, can be used for its determination.

In the following analysis of the data, aiming at the determination of the real part of the antikaon nuclear potential at saturation density, we make use of the cross sections integrated over the measured K^- momentum interval 0.2-1.0 GeV/c. Due to the increased number of counts, this approach has the advantage of decreasing significantly the statistical and partially systematic uncertainties. Evidently, the antikaon potential depth extracted in this way will correspond to an average K^- momentum of about 0.5 GeV/c, in the vicinity of which the main strength of the measured distributions is concentrated. The calculations of the corresponding cross sections were performed within the model for different potential strengths. The values of $\sigma_{exp}/\sigma_{cal}(U)$ are shown in figure 5 for U = 0 MeV, U = -60 MeV, U = -126 MeV, and U = -180 MeV. Also shown is a third-order polynomial fit to the complete data set of ratios. It is seen from the figure that the condition that $\sigma_{exp}/\sigma_{cal}(U) = 1$ is achieved if $U = -63^{+15}_{-12} MeV$. This means that the K^- effective mass in the nuclear matter of normal density is

 DOI: 10.1051/epjconf/201713804002

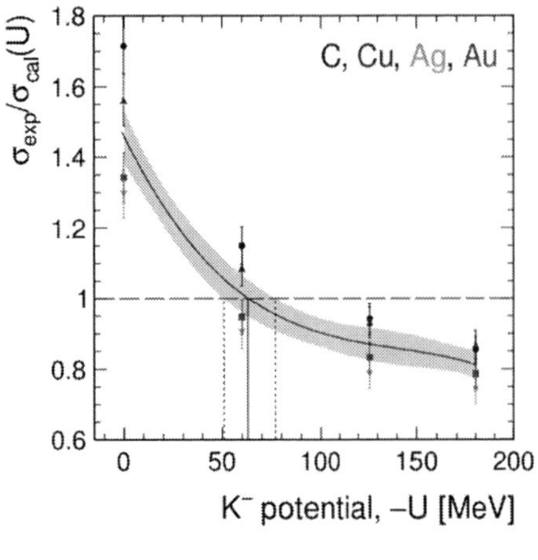

Figure 5. Ratio of the measured integrated cross section for the non-resonant K^+K^- production by protons on a given nucleus A to the corresponding cross sections, calculated within the adopted model and supposing four values for the K^- potential depth at nuclear matter density: U=0 MeV, -60 MeV, -126 MeV, -180 MeV. The curve represents a third-order polynomial fit of all ratios presented in the figure, with the shaded band indicating 1σ confidence interval.

less than its free mass by about of 60 MeV. Acting on the cautious side and using the maximum value of the final systematic uncertainty we came to the expanded error band of $U = -63^{+50}_{-31} MeV$. Within the uncertainties quoted, the value obtained for the potential depth is consistent with the moderate K^--nucleus potential that is predicted by calculations based on chiral Lagrangians or on meson-exchange potentials. However, it is hard to reconcile our value with the deep potential of order -200 MeV.

4 Conclusion

The data analysis of the performed experiment on the K^+K^- pair production off nuclei evidences for the properties of mesons with open (K^-) and hidden (ϕ) strangeness in the nuclear matter significantly differ from those in the free space.

References

[1] V. K. Magas et al., Phys. Rev. C **71**, 065202 (2005)
[2] S. Barsov et al., Nucl. Instrum. Methods Phys. Res., Sect. A **462**, 364 (2010)
[3] M. Hartmann et al., Phys. Rev. C **85**, 035206 (2012)
[4] M. Hartmann et al., Int. J. Mod. Phys. A **22**, 317 (2007)
[5] M. Buesher et al., Nucl. Instrum. Methods Phys. Res., Sect. A **481**, 378 (2002)
[6] Yu. T. Kiselev et al., Phys. Rev. C **92**, 065201 (2015)
[7] A. Polyanskiy et al., Phys. Lett. B **695**, 74 (2011)
[8] H. Schade, Ph.D. thesis, University of Dresden, 2010 (unpublished)
[9] E. Ya. Paryev, J. Phys. G **36**, 015103 (2009)
[10] E. Ya. Paryev, M. Hartmann, and Yu. T. Kiselev, J. Phys. G **42**, 0175107 (2015)
[11] R. Muto et al., Phys. Rev. Lett. **98**, 042501 (2007)
[12] T. Ishikawa et al., Phys. Lett. B **608**, 215 (2005)
[13] M. H. Wood et al., Phys. Rev. Lett. **105**, 112301 (2010)
[14] D. Cabrera et al., Nucl. Phys. A **733**, 130 (2004)

Electroproduction of lightest nucleon resonances up to Q*2=12 GeV*2 in quark models at light front

I.T. Obukhovsky[1,⋆], *A.* Faessler[2], *Th.* Gutsche[2], *D.K.* Fedorov[1], and *V.E.* Lyubovitskij[2,3,4]

[1] *Institute for Nuclear Physics, Moscow State University, 19991, Russia*
[2] *Institut für Theoretische Physik, Universität Tübingen, Kepler Center for Astro and Particle Physics, Auf der Morgenstelle 14, D-72076 Tübingen, Germany*
[3] *Department of Physics, Tomsk State University, 634050 Tomsk, Russia*
[4] *Laboratory of Particle Physics, Mathematical Physics Department, Tomsk Polytechnic University, 634050 Tomsk, Russia*

Abstract. The lightest nucleon resonances are described at light front as mixed states of the $3q$ cluster ("quark core") possessing a definite value of the inner orbital momentum $L = 0,1$ and a hadron molecular state, $N+\sigma$ or $\Lambda+K$. Helicity amplitudes of the resonance electroproduction off the proton are calculated at large Q^2 up to 12 GeV^2 and compared to the last CLAS data. At this basis we have estimated the probability of quark core in lightest nucleon resonances and predicted the high Q^2 behaviour of the resonance electrocoupling.

1 Introduction

The last decade has been marked by a significant progress in the experimental study of low-lying baryon resonances. Specifically new insights have been obtained in π and 2π electroproduction off the proton with the polarized electron beam at JLab (CLAS/CLAS12 Collaborations, see, e.g., recent overviews [1, 2]). In the context of projected extensive studies of baryons with $J^P = 1/2^{\pm}, 3/2^{\pm}, 5/2^{\pm}$, etc., there is an interest in calculation of electrocouplings of baryons at large Q^2.

There are many theoretical approaches to the problem, which start from the first principles, Lattice QCD, Dyson-Schwinger (DS)E or Bethe-Salpeter (BS) equations, Light-front QCD, AdS/QCD etc., but some rough estimates could be made on the basis of a light-front quark model. It implies the construction of a good basis of light-front quark configurations possessing a definite value of the orbital angular momentum (L) and satisfying the Pauli exclusion principle. Our approach is to fit parameters of light-front quark configurations to the elastic nucleon form factors and use them to construct a correct basis for excited ($L = 0,1$) nucleon states to calculate the transition form factors at large Q^2 up to 12 GeV^2.

Light-front quark wave functions were successfully used by many authors for description of nucleon form factors and transition amplitudes as before appearance of the polarized electron data [3–5] as well as after these (taking into account new high-quality data) [6–9]. However these works used the data up to $Q^2 \lesssim$ 3-4 GeV^2 which were only available. Now CLAS12 plans measurements up to 12

⋆e-mail: obukh@nucl-th.sinp.msu.ru

GeV^2 that initiates the theoretical study of the resonance electroproduction at higher Q^2 in terms of relativistic approaches.

In our recent work [10] we have generalized our earlier non-relativistic approach to the Roper resonance electroproduction at $Q^2 < 4\,\text{GeV}^2$ [11] by going to higher Q^2 in terms of light-front quark configurations. Solving this problem we have run into another difficulty: the contribution of excited relativistic quark configurations (at least for $L = 0,1$) overestimates the transition amplitudes $N+\gamma^* \to N^*$, while the elastic $N + \gamma^* \to N$ form factors calculated at the same basis are in a good agreement with the data. It follows from these results that, possibly, other (softer) degrees of freedom should be taken into consideration for excited states along with the quark core. Then at high Q^2 the contribution of such soft components should be quickly dying out, and only the quark core contribution will survive with a relatively small weight. Here we consider this possibility.

2 Nucleon and Roper resonance wave functions at light front

In the case of the Roper resonance the solution of the problem is almost evident: its inner structure cannot be adequately described in terms of constituent quark degrees of freedom only. As a result the quark model fails to explain the observed mass and decay width of the Roper resonance. Starting from this we have considered the Roper resonance $R = N_{1/2^+}(1440)$ as a mixed state of the radially excited quark configuration $3q^* = sp^2[3]_X$ and the "hadron molecule" (a loosely bound state of the nucleon and σ meson $(N\sigma)_{mol} = |N + \sigma\rangle$):

$$R = \cos\theta\,|3q^*\rangle + \sin\theta\,|N + \sigma\rangle \tag{1}$$

The parameter θ was adjusted to optimize the description of the helicity amplitude $A_{1/2}$ only. We found that at the value of $\cos\theta$ =0.7 - 0.8 this model correlates well with the recent CLAS data on the both $A_{1/2}$ and $S_{1/2}$ helicity amplitudes [11].

However, in [11] we have used non-relativistic h.o. quark configurations (i.e. the Gaussians) for the quark core of baryons, and thus such calculations would be senseless at $Q^2 \gtrsim$ 3-4 GeV^2. At high Q^2 the contribution of soft components of the baryon (the meson cloud, "molecular" admixtures, etc.) to transition form factors falls off in comparison to the quark core contribution. Hence only the quark contribution should be considered at high Q^2.

Unfortunately, the form factors defined with a Gaussian as a quark core wave function (with the scale parameter normalized on the nucleon radius) are also quickly dying out at $Q^2 \gtrsim$ 3-4 GeV^2. Possible alternatives to the Gaussian wave function have been considered in several works: a superposition of many Gaussians [6], a pole-like w.f. [3] and a model with the running quark mass [7]. We have chosen a pole-like form of the w.f. The pole-like form of the nucleon ground state wave function Φ_{0S}

$$\Phi_{0S}(\xi, \eta, k_\perp, K_\perp) = \frac{N_{0S}}{(1 + \mathcal{M}_0^2/\beta^2)^\gamma}, \tag{2}$$

where $\mathcal{M}_0^2 = \frac{M^2+k_\perp^2}{\eta\xi(1-\xi)} + \frac{\eta M^2+K_\perp^2}{\eta(1-\eta)}$, was firstly fitted to the elastic nucleon form factors by Schlumpf[3] with γ =3.5 and $\beta \approx$2M. Here $\boldsymbol{k}$, $\boldsymbol{K}$ are (relativistic) relative moments in quark pair "1-2" and in $2q$ cluster – third quark pair "(12)-3" respectively; the $\xi, \eta, \boldsymbol{k}_\perp, \boldsymbol{K}_\perp$ are the light-front variables with $x_1 = \xi\eta$, $x_2 = (1-\xi)\eta$, $x_3 = 1-\eta$, and $\mathcal{M}_0$ is the mass of free $3q$ system (M is the constituent quark mass).

Such form of Φ_{0S} is as yet unjustified, but it should be noticed that at least in the meson sector the pole-like form of the pion $\bar{q}q$ w.f. $f_\pi\varphi_\pi(x, k_\perp^2) = \frac{9}{4\pi^2}\left(1 + \frac{k_\perp^2}{4M^2x(1-x)}\right)^{-\kappa}$ with κ =1 was recently reconstructed [12] starting from the Bethe-Salpeter wave function projected onto the light front (there are also approximations with κ =1 - 2). The nucleon pole-like w.f. $\Phi_{0S}(\xi, \eta, k_\perp, K_\perp)$ looks like a generalization of the pion $\bar{q}q$ w.f. for the case of $3q$ system.

Starting from the "+" component of the quark current on light front

$$I^+ = \sum_{j=1}^{3} I_j^+, \quad I_j^+ = e_{qj}\left(f_{1j} + i\hat{n}_z \cdot [\sigma_j \times q_\perp] f_{2j}\right) \tag{3}$$

(without quark form factors f_i, but with a small anomalous quark magnetic moment $\varkappa_q$, i.e. $f_1 = 1$, $f_2 = \varkappa_q$), we have fitted free parameters of the model to the modern data on nucleon form factors within the full measured range $0 \leq Q^2 \leq 32$ GeV^2 including the electron polarization data on the ratio G_E/G_M at $Q^2 \lesssim 6$-8 GeV^2. With the values γ =3.51, M =0.251 GeV, $\varkappa_u$ =-0.0028, $\varkappa_d$ =0.0224, β_u =0.579 GeV, β_d =0.5 GeV we have obtained quiet good description [10] of all the nucleon data including the static electromagnetic properties of nucleons.

For the Roper resonance we used a light-front analogue of the radially excited quark configuration

$$\Phi_{2S} = N_{2S}\left(1 - c_R\frac{M_0^2}{\beta^2}\right)\Phi_{0S}, \qquad \langle\Phi_{2S}|\Phi_{0S}\rangle = 0 \tag{4}$$

and for the negative parity baryons, e.g. $N_{1/2^-}(1535)$, we used

$$\Phi_{1P} = N_{1P}|\boldsymbol{K}|\Phi_{0S} \quad \text{and} \quad N_{1P}|\boldsymbol{k}|\Phi_{0S} \tag{5}$$

After comparison with the data it turns out that the quark-core weight ($\cos\theta$) in the mixed model (1) should be reduced from $\cos\theta \simeq 0.7$-0.8 of the nonrelativistic (Gaussian) model to the value of $\cos\theta \lesssim 0.5$ for the relativistic (pole-like) model. Moreover the transition form factors for negative parity nucleon resonances (at least for $L = 1$, next Section) behave at large Q^2 similarly to the Roper one (see figure 1). At the same time the nucleon elastic form factors were successfully described without reducing the quark core weight.

3 High Q^2. Quark configurations and Melosh transformations

A good basis of relativistic quark configurations possessing the definite value of orbital momentum L and satisfying the Pauli exclusion principle is needed for baryon resonances with $J^P = 1/2^\pm, 3/2^\pm, 5/2^\pm$. We start from non-relativistic shell-model configurations and change the h.o. wave functions by the light-front ones (Gaussian or pole-like) dependent on the relativistic relative moments $\boldsymbol{k}$, $\boldsymbol{K}$ and expressed in light-front invariants ξ, η, $\lambda_\perp$, $\Lambda_\perp$. At this stage, as usual, there are problems with boosts (in the instant form of dynamics) or rotations (at the light front). In both cases generators of transformations depend on the dynamics, and thus such shell-model basis is only useful for transformations of kinematic subgroups of the full Lorentz group. Nevertheless some difficulties can be resolved going to the rest frame for definition of the total angular momentum $J = L + S$ or partial one $j = l + s$ for each two-body subsystem of the $3q$ bound state and going to the Breit frame for description of the ep collision. However such technique could only be considered as a reasonable approximation to "true" dynamical calculations.

Recall that the spin $\vec{s}_i$ of the particle is uniquely determined in its own rest frame, where $p_i^\nu = \overset{\circ}{p}{}^\nu_i \equiv \{M, 0, 0, 0\}$. In the moving frame the canonical spin state is determined by a rotationless Lorentz boost $\lambda(\boldsymbol{p}_i \leftarrow \overset{\circ}{p}_i)$ (we use notations of [13, 14]): $|\boldsymbol{p}_i; s_i\mu_i\rangle_c = U(\lambda(\boldsymbol{p}_i \leftarrow \overset{\circ}{p}_i))|\overset{\circ}{p}_i; s_i\mu_i\rangle$, while the light-front spin state is determined by another type of Lorentz transformation, which leads to the same momentum p_i^ν, but would be represented as a two-step process [13]: $l(\boldsymbol{p}_i \leftarrow \overset{\circ}{p}_i) = \lambda(\boldsymbol{p}_i \leftarrow \boldsymbol{p}_\infty)\lambda(\boldsymbol{p}_\infty \leftarrow \overset{\circ}{p}_i)$ where $\boldsymbol{p}_\infty = p_\infty \hat{\boldsymbol{n}}_z$ is the quark momentum at the infinite momentum frame, $|\boldsymbol{p}_i; s_i\mu_i\rangle_f = U(l(\boldsymbol{p}_i \leftarrow \overset{\circ}{p}_i))|\overset{\circ}{p}_i; s_i\mu_i\rangle$ [the full manifold of l transformations forms a subgroup of the Lorentz group, which is a kinematic subgroup for the light-front dynamics, — an analog of the rotational subgroup used in the instant form]. As a result, canonical and front states are related by a specific rotation

$$|\boldsymbol{p}_i, s_i\mu_i\rangle_c = \sum_{\mu_i'} |\boldsymbol{p}_i, s_i\mu_i'\rangle_f D^{s_i}_{\mu_i'\mu_i}[R(\boldsymbol{p}_i, M_i)], \quad R(\boldsymbol{p}_i, M_i) = \lambda(\overset{\circ}{p}_i \leftarrow p_i)\lambda(\boldsymbol{p}_i \leftarrow \boldsymbol{p}_\infty)\lambda(\boldsymbol{p}_\infty \leftarrow \overset{\circ}{p}_i), \tag{6}$$

which is known as the Melosh transformation. For the quark ($s_i = 1/2$) one can write: $D^{\frac{1}{2}}_{\mu_i'\mu_i}[R^{-1}(\boldsymbol{p}_i, M_i)] = \frac{M_i + p_i^+ + i\hat{n}_z\cdot[\sigma\times \boldsymbol{p}_{i\perp}]}{\sqrt{(M+p_i^+)^2 + p_{i\perp}^2}}$. The Lorentz transformation $l(\boldsymbol{p}_i' \leftarrow \boldsymbol{p}_i)$ does not rotate the front spin that is very convenient for description of the states in moving reference frames. But the front basis is not convenient for construction of rotationally covariant states. For the quark pair "12" a rotationally covariant state can be constructed in the proper rest frame of the pair (cm), where $(p_1 + p_2)^\nu_{cm} \equiv \mathcal{P}^\nu_{12cm} = \overset{\circ}{\mathcal{P}}{}^\nu{}_{12} = \{\mathcal{M}_{0(12)}, 0, 0, 0\}$:

$$|\overset{\circ}{\mathcal{P}}_{12}, j_{12}j^z_{12}(ls_{12})\rangle = \sum_{\mu_1\mu_2\mu_{12}m} (s_1\mu_1 s_2\mu_2|s_{12}\mu_{12})(lms_{12}\mu_{12}|j_{12}j^z_{12}) \int d^2\hat{k} Y_{l\mu_l}(\hat{\boldsymbol{k}})|\boldsymbol{p}_{1cm}\mu_1\rangle_c|\boldsymbol{p}_{2cm}\mu_2\rangle_c \tag{7}$$

It is well known [13–15] that in the case of free particles the Lorentz boost does not destroy this state. The right side of (7) is transforming as the state of an elementary particle with total spin j_{12} and mass $\mathcal{M}_{0(12)} = \sqrt{\frac{M^2 + k_\perp^2}{\xi(1-\xi)}}$. Hence, to construct a $3q$ state with the total angular momentum J, we should go to the center-of-mass frame of $3q$ system (CM) and use the same formula as (7) for another "two-body state": cluster 12+quark 3. This can be realized by substitution of responding CM moments, $\mathcal{P}_{12CM}$ and $\boldsymbol{p}_{3CM}$, to (7): $\boldsymbol{p}_{1cm} \to \boldsymbol{p}_{3CM}$, $\boldsymbol{p}_{2cm} \to \mathcal{P}_{12CM}$, $Y_{l\mu_l}(\hat{\boldsymbol{k}}) \to Y_{L\mu_L}(\hat{\boldsymbol{K}})$, where $\boldsymbol{p}_{3CM} = \boldsymbol{K}$, $\mathcal{P}_{12CM} = -\boldsymbol{K}$, $\boldsymbol{p}_{1cm} = \boldsymbol{k}$, $\boldsymbol{p}_{2cm} = -\boldsymbol{k}$.

Using the light-front boosts for the transition from frame $cm(12)$ to frame CM(12-3) (and further from frame CM(12-3) to the Breit frame) implies several Melosh rotations which change the wave functions of initial and final states. Fortunately the Melosh rotation of all spins does not violate the Pauli exclusion principle for the full $3q$ wave function (see, e.g., a discussion in [10]), though the calculation technique (fraction parentage coefficients, etc.) becomes more cumbersome, especially in the case of nonzero orbital moments L and l.

4 Wave functions and matrix elements

In the long run we have obtained the expressions for spin-orbital parts of searched wave functions $\Psi^{(0)}_{J(LS(s_{12}))\mu_J}$ (for $l = L = 0$), $\Psi^{(1)}_{J(LS(s_{12}))\mu_J}$ (for $L = 1$, $l = 0$) and $\Psi^{(2)}_{J(lS(s_{12}))\mu_J}$ (for $L = 0$, $l = 1$) used in the matrix element of current (3) for the third quark in the transition $N \to N^*$,

$$\langle\Psi^{(1)}_{J'(L'S'(s'_{12}))\mu'_J}|\hat{I}^+_3|\Psi^{(0)}_{J(LS(s_{12}))\mu_J}\rangle = \delta(\mathcal{P}'_B - \mathcal{P}_B - \boldsymbol{q}_\perp)\sum_{\{\mu_i\}} e_{q_3}\left(f_1\delta_{\mu'_3\mu_3} + f_2 q_\perp(-1)^{1/2-\mu_3}\delta_{-\mu'_3\mu_3}\right)$$

$$\times \int \frac{d^2\lambda_\perp d^2\Lambda_\perp d\xi d\eta}{\xi(1-\xi)\eta(1-\eta)} \Psi^{(1)}_{J'(L'S'(s'_{12}))\mu'_J}(\mathcal{P}'_B, \boldsymbol{K}', \boldsymbol{k}'; \mu_1, \mu_2, \mu'_3)\Psi^{(0)}_{J(LS(s_{12}))\mu_J}(\mathcal{P}_B, \boldsymbol{K}, \boldsymbol{k}; \mu_1, \mu_2, \mu_3) \tag{8}$$

Here $\Psi^{(0)}$ is the nucleon w.f. , $\Psi^{(1)}$ and $\Psi^{(2)}$ are wave functions of a $N^*_{J'^-}$ resonance with orbital momentum $L' = 1$ and $J' = 1/2$ or $3/2$. Superscripts 1 and 2 denote the type of permutational symmetry in the coordinate space for the first two quarks (1 for the symmetric and 2 for the antisymmetric state, i.e. Yamanouchi symbols 112 and 121 respectively). $\mathcal{P}_B$ and $\mathcal{P}'_B$ are moments of initial nucleon and outgoing resonance in the Breit frame respectively, $\mathcal{P}_B = -\alpha \boldsymbol{q}_\perp/2$, $\mathcal{P}'_B = (1-\alpha)\boldsymbol{q}_\perp/2$, $\alpha = \frac{q_\perp^2 + M_{N^*}^2 - M_N^2}{2q_\perp}$, $q_B^\nu = \{0, \boldsymbol{q}_\perp, 0\}$, $Q^2 = \boldsymbol{q}_\perp^2$.

$$\begin{aligned}\Psi^{(1)}_{J(LS(s_{12}))\mu_J}(\mathcal{P}, \boldsymbol{K}, \boldsymbol{k}; \mu_1, \mu_2, \mu_3) &= {}_f\langle \boldsymbol{p}_1\mu_1, \boldsymbol{p}_2\mu_2, \boldsymbol{p}_3\mu_3 | \mathcal{P}, J(LS(s_{12}))\mu_J\rangle_f \\ &= \delta(\mathcal{P} - \boldsymbol{p}_1 - \boldsymbol{p}_2 - \boldsymbol{p}_3) K \Phi_0(\mathcal{M}_0) \sum_{\mu_S \mu_L} Y_{L\mu_L}(\hat{K}) (L\mu_L S\mu_S | J\mu_J) \sum_{\{\bar\mu\}} (s_1\bar\mu_1 s_2\bar\mu_2 | s_{12}\bar\mu_{12})(s_{12}\bar\mu_{12} s_3\bar\mu_3) \\ &\times D^{s_{12}}_{\mu_{12}\bar\mu_{12}}[R^{-1}(-\boldsymbol{K}, \mathcal{M}_{0(12)})] D^{s_1}_{\mu_1\bar\mu_1}[\tilde{R}^{-1}(\boldsymbol{k}, M_1)] D^{s_2}_{\mu_2\bar\mu_2}[\tilde{R}^{-1}(-\boldsymbol{k}, M_2)] D^{s_3}_{\mu_3\bar\mu_1}[R^{-1}(\boldsymbol{K}, M_3)], \quad (9)\end{aligned}$$

$$\begin{aligned}\Psi^{(2)}_{J(lS(s_{12}))\mu_J}(\mathcal{P}, \boldsymbol{K}, \boldsymbol{k}; \mu_1, \mu_2, \mu_3) &= \sum_{j_{12}} (-1)^{s_{12}+s_3+S+j_{12}} \sqrt{(2S+1)(2j_{12}+1)} \begin{Bmatrix} l & J & S \\ s_3 & s_{12} & j_{12} \end{Bmatrix} \\ &\times \Psi^{(2)}_{J(j_{12}(ls_{12})s_3)\mu_J}(\mathcal{P}, \boldsymbol{K}, \boldsymbol{k}; \mu_1, \mu_2, \mu_3), \quad (10)\end{aligned}$$

$$\begin{aligned}\Psi^{(2)}_{J(j_{12}(ls_{12})s_3)\mu_J}(\mathcal{P}, \boldsymbol{K}, \boldsymbol{k}; \mu_1, \mu_2, \mu_3) &= {}_f\langle \boldsymbol{p}_1\mu_1, \boldsymbol{p}_2\mu_2, \boldsymbol{p}_3\mu_3 | \mathcal{P}, J(j_{12}(ls_{12})s_3)\mu_J\rangle_f \\ &= k\Phi_0(\mathcal{M}_0) \sum_{\mu_{12}\mu_l} Y_{l\mu_l}(\hat{k}) \sum_{\{\bar\mu\}} (s_1\bar\mu_1 s_2\bar\mu_2 | s_{12}\mu_{12})(l\mu_l s_{12}\mu_{12} | j_{12}\bar\mu_{j_{12}})(|j_{12}\bar\mu_{j_{12}} s_3\bar\mu_3 | J\mu_J) \\ &\times D^{j_{12}}_{\mu_{j_{12}}\bar\mu_{j_{12}}}[R^{-1}(-\boldsymbol{K}, \mathcal{M}_{0(12)})] D^{s_1}_{\mu_1\bar\mu_1}[\tilde{R}^{-1}(\boldsymbol{k}, M_1)] D^{s_2}_{\mu_2\bar\mu_2}[\tilde{R}^{-1}(-\boldsymbol{k}, M_2)] D^{s_3}_{\mu_3\bar\mu_1}[R^{-1}(\boldsymbol{K}, M_3)]. \quad (11)\end{aligned}$$

Jacobians, as evident factors, are omitted in (9) – (11) to simplify expressions. It should be noticed that the angular part of w.f.'s (9) – (11), $kY_{l\mu_l}(\hat{\boldsymbol{k}})$ and $KY_{L\mu_L}(\hat{\boldsymbol{K}})$, depends on the canonical 3-momenta $\boldsymbol{k}$ and $\boldsymbol{K}$

$$\boldsymbol{k} \equiv \boldsymbol{p}_{1cm} = \left\{\boldsymbol{\lambda}_\perp, -\frac{1-2\xi}{2}\mathcal{M}_{0(12)}\right\}, \quad \boldsymbol{K} \equiv \boldsymbol{p}_{3CM} = \left\{\boldsymbol{\Lambda}_\perp, \left(\frac{1-2\eta}{2}\mathcal{M}_0 + \frac{\mathcal{M}^2_{0(12)} - M^2}{2\mathcal{M}_0}\right)\right\}, \quad (12)$$

which are expressed through light-front invariants $\xi, \eta, \boldsymbol{\lambda}_\perp, \boldsymbol{\Lambda}_\perp, \mathcal{M}_0, \mathcal{M}_{0(12)}$, where

$$\boldsymbol{\lambda}_\perp = \frac{x_2\boldsymbol{p}_{1\perp} - x_1\boldsymbol{p}_{2\perp}}{x_1 + x_2}, \quad \boldsymbol{\Lambda}_\perp = \frac{(x_1 + x_2)\boldsymbol{p}_{3\perp} - x_3(\boldsymbol{p}_{1\perp} + \boldsymbol{p}_{2\perp})}{x_1 + x_2 + x_3}, \quad (13)$$

Spin-orbital basis (9) – (11) given in the coordinate-spin (XS) space was used to construct quark configurations satisfying the Pauli exclusion principle. In the product of XS and isospin (T) spaces (XST) the fully symmetric states with Young scheme $[3]_{XST}$ are represented by expressions

$$\Psi_N(56^+) = \sqrt{\frac{1}{2}}\left[\Psi^{(0)}_{s_{12}=1}|[21]_T\, y_T^{(1)}\rangle + \Psi^{(0)}_{s_{12}=0}|[21]_T\, y_T^{(2)}\rangle\right] \quad (14)$$

for the nucleon and

$$\Psi_{N^*}(70^-) = \frac{1}{2}\left[\Psi^{(1)}_{s_{12}=1} - \Psi^{(2)}_{s_{12}=0}\right]|[21]_T\, y_T^{(1)}\rangle + \frac{1}{2}\left[\Psi^{(1)}_{s_{12}=0} + \Psi^{(2)}_{s_{12}=1}\right]|[21]_T\, y_T^{(2)}\rangle. \quad (15)$$

for negative parity resonances $N^*_{J^-}$ ($J = 1/2, 3/2$) from the 70^- multiplet of $SU(6)$ group classification. Here the isospin w.f. for $T = 1/2$ are marked by the Young scheme (partition) $[21]_T$ and

Yamanouchi symbols $y_T^{(1)} = (112)$ and $y_T^{(2)} = (121)$. For the positive charge ($T_z = +1/2$) these w.f. are given by the well known expressions:

$$|[21]_T\, y_T^{(1)} T_z = +1/2\rangle = \sqrt{\frac{2}{3}} uud - \sqrt{\frac{1}{3}} \frac{ud + du}{\sqrt{2}} u, \quad |[21]_T\, y_T^{(2)} T_z = +1/2\rangle = \frac{ud - du}{\sqrt{2}} u \qquad (16)$$

Spin-orbital states $\Psi^{(n)}_{s_{12}}$ for $n = 1,2$ are given in (9)-(11) and for $n = 0$ one can use $\Psi^{(2)}_{s_{12}}$ with $L = 0$ and Φ_{0S} instead of Φ_{1P}. In (15) we omitted all subscripts at $\Psi^{(n)}$, except s_{12}, since the values of $s_{12} = 0$ or 1 define the Yamanouchi symbol in the spin space, $y_S^{(1)} = (112)$ or $y_S^{(2)} = (121)$ respectively (in the case of the total spin $S = 1/2$, i.e. for the Young scheme $[21]_S$). Therefore the relations

$$\begin{aligned} \Psi^{(1)}_{s_{12}=1} &= |[21]_X\, y_X^{(1)}\rangle|[21]_S\, y_S^{(1)}\rangle, \quad \Psi^{(1)}_{s_{12}=0} = |[21]_X\, y_X^{(1)}\rangle|[21]_S\, y_S^{(2)}\rangle, \\ \Psi^{(2)}_{s_{12}=1} &= |[21]_X\, y_X^{(2)}\rangle|[21]_S\, y_S^{(1)}\rangle, \quad \Psi^{(2)}_{s_{12}=0} = |[21]_X\, y_X^{(2)}\rangle|[21]_S\, y_S^{(2)}\rangle, \end{aligned} \qquad (17)$$

are implied in (14) - (15).

Matrix element (8) should be rewritten for physical initial and final states defined by (14) -(15) with covariant w.f.s (9)-(11) to calculate the amplitudes of resonance electroproduction (one can follow the technique developed in [10]).

5 Results and outlook

Matrix elements of the I^+ component of quark current (8) allowed us to calculate the transverse ($A_{1/2}$) and longitudinal ($S_{1/2}$) helicity amplitudes for electroproduction of the lightest nucleon resonances $N^*_{1/2^+}(1440)$ and $N^*_{1/2^-}(1535)$ in a large Q^2 interval up to 12 GeV^2. We used the same technique (fractional parentage coefficients etc.) as in our recent work [10] and the same parameters of the light-front quark model. For the $N^*(1535)$ we used a mixed model analogous to that of the Roper resonance:

$$N^*(1535) = \cos\theta^* |3q^*\rangle + \sin\theta^* |\Lambda + K^+\rangle \qquad (18)$$

with the $|3q^*\rangle$ defined by (15). Parameter of mixing θ^* was adjusted to optimize the description of the helicity amplitude $A_{1/2}$, and we have obtained $\cos\theta^* = 0.6$. We also updated the fit of parameter θ for the Roper resonance reducing $\cos\theta = 0.57$ of [16] to the new value $\cos\theta = 0.5$ which is closer to the CLAS data at $Q^2 \gtrsim 3$ - 4 GeV^2. The results are shown in figure 1.

In summary it can be concluded that

1. The lightest nucleon resonances are described at light front as mixed states of the $3q$ cluster possessing a definite value of the inner orbital momentum $L = 0,1$ and a hadron molecular state, $N + \sigma$ or $\Lambda + K$.
2. Nucleon elastic form factors are successfully described in a large interval of Q^2 by the lowest light-front quark configuration without any hadron-molecular admixtures.
3. But in the case of nucleon resonances the quark core overestimates the transition amplitudes at large Q^2, and thus the weight of the quark core in the resonance should be relatively small because of excitation of higher Fock states.

Acknowledgements

This work was supported by the RFBR-DFG Grant No 16-52-12019, by the DFG Grants No FA-67-42-1 and GU-267/3-1, by the German Bundesministerium für Bildung und Forschung (BMBF) under Project 05P2015 - ALICE at High Rate (BMBF-FSP 202) "Jet- and fragmentation processes at ALICE and the parton structure of nuclei and structure of heavy hadrons", by Tomsk State University Competitiveness Improvement Program and the Russian Federation program "Nauka" (Contract No. 0.1526.2015, 3854).

DOI: 10.1051/epjconf/201713804003

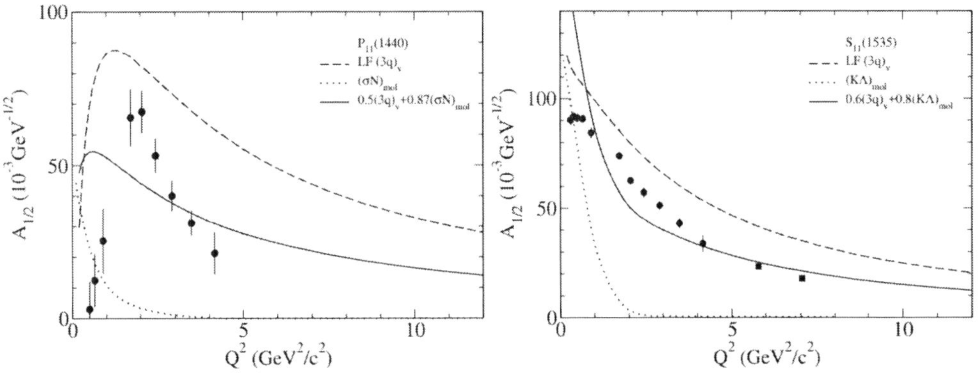

Figure 1. Transverse helicity amplitude $A_{1/2}$ for electroproduction of the Roper resonance (left panel) and the $N^*_{1/2^-}(1535)$ (right panel). Dashed lines are the 100% quark core contribution, dotted lines are the 100% hadron molecule contribution, solid lines are results for mixed models (1) and (18) with cosθ =0.5 and cosθ^* =0.6 respectively. The data are taken from [17, 18].

References

[1] I.G. Aznauryan and V.D. Burkert, Prog. Part. Nucl. Phys. **67**, 1 (2012)
[2] V.I. Mokeev, V.D.Burkert, D.S. Carman et al., Phys. Rev. C **93**, 025206 (2016)
[3] F. Schlumpf, Phys. Rev. D **47**, 4114 (1993)
[4] S. Capstick and B.D.Keister, Phys. Rev. D **51**, 3598 (1995)
[5] F. Cardarelli, E. Pace, G. Salme, and D. Simula, Nucl. Phys. A **623**, 361 (1996)
[6] S. Capstick, B.D. Keister, and D. Morel, J. Phys.: Conf. Ser. **69**, 012016 (2007)
[7] I.G. Aznauryan and V.D. Burkert, Phys. Rev. C **85**, 055202 (2012)
[8] G. Ramalho and K. Tsushima, Phys. Rev. C **81**, 074020 (2010)
[9] T. Gutsche, V.E Lyubovitskij, I. Schmidt, and A. Vega, Phys. Rev. D **89**, 036007 (2014)
[10] I.T. Obukhovsky, A. Faessler, T. Gutsche, and V.E Lyubovitskij, Phys. Rev. D **89**, 014032 (2014)
[11] I.T. Obukhovsky et al., Phys. Rev. D **84**, 014004 (2011)
[12] L. Chang, I. C. Cloet, J. J. Cobos-Martinez, C. D. Roberts et al., Phys. Rev. Lett. **110**, 132001 (2013)
[13] B. L. G. Bakker, L. A. Kondratyuk, and M. V. Terent'ev, Nucl. Phys. B **158**, 497 (1979)
[14] B.D. Keister and W.N. Polyzou, Adv. Nucl. Phys. **20**, 225 (1991)
[15] Iu. M. Shirokov, Soviet Physics JETP **8**, 703 (1959)
[16] I.T. Obukhovsky, A. Faessler, T. Gutsche, and V.E Lyubovitskij, J. Phys. G **41**, 095005 (2014)
[17] I.G. Aznauryan et al. CLAS Collaboration, Phys. Rev. C **78**, 045209 (2008). I.G. Aznauryan et al. CLAS Collaboration, Phys. Rev. C **80**, 055203 (2009)
[18] M.M. Dalton et al., Phys. Rev. C **80**, 015205 (2009)

EPJ Web of Conferences **138**, 04004 (2017) DOI: 10.1051/epjconf/201713804004
Baldin ISHEPP XXIII

The fermi coupling, mass spectrum and decay properties of mesons within the covariant confined quark model

Gurjav Ganbold[1,2,*]

[1] *Bogoliubov Laboratory of Theoretical Physics, JINR, Dubna 141980, Russia*
[2] *Institute of Physics and Technology, Ulaanbaatar 13330, Mongolia*

Abstract. Main properties of stable quark-antiquark bound states have been investigated by using a relativistic quark model with analytic confinement. A new insight into the problem of hadron mass generation is provided. It has been shown that the compositeness condition expressing the Yukawa coupling of the meson-quark interaction and the master equation relating the meson mass function to the Fermi coupling can together guarantee the equivalency of the both theories, thereby providing an interpretation of the meson field as a bound state of constituent quarks. A smooth behavior for the Fermi coupling $G(M)$ obeying reasonable accuracy for the meson estimated masses M has been obtained. We have also updated our previous numerical results for the weak-decay constants as well as the electromagnetic decay widths of mesons. The obtained estimates are found in reasonable agreement with the recent experimental data.

1 Introduction

Understanding of a number of phenomena in modern particle physics such as the quark confinement, hadronization, generation of hadron masses etc. requires correct description of hadron dynamics within theoretical models. Particularly, the Standard Model operates only fundamental particles (quarks, leptons, neutrinos), gauge and Higgs bosons, and it is not yet clear how to explain the appearance of the multitude of observed hadrons and elucidate the generation of their masses. The origin of the hadron masses is one of the puzzles of particle physics and the calculation of their mass spectra in a quality comparable to precise experimental data still remains a key problem in QCD.

Even before QCD was set up as the fundamental theory of strong interactions, it was understood that it is a difficult problem to describe a composite particle within QFT as based on the relativistic S-matrix. The original Lagrangian describes free fields and their interactions while the consideration of physical processes requires the renormalization, i.e. the transition from bare (unrenormalized) quantities to the physical (renormalized) ones. In particular, the bare field is related to the dressed one as $\phi_0 = Z^{1/2}\phi_r$ through the wave function renormalization constant Z. The bare field ϕ_0 may be eliminated from the Lagrangian by putting $Z = 0$, that is the *compositeness condition* (CC). The CC requirement $Z_H = 0$ is also one of the key ingredients in the *covariant confined quark model* (CCQM) development [1].

*e-mail: ganbold@theor.jinr.ru

Relativistic models with specific forms of analytically confined propagators have been developed to study some aspects of low-energy hadron physics [2]. Hereby, the role of analytic confinement in the formation of quark-antiquark bound states has been analyzed within Yukawa-type models. Additionally, the lowest mass of two-gluon bound state (the glueball) has been predicted in reasonable agreement with lattice estimations. A phenomenological model with an infrared confinement has been developed to study the dependence of the effective strong coupling α_s on the mass scale.

Below we apply the CCQM to the meson mass problem [3]. We use the CC requirement to determine the renormalized Yukawa coupling g_r as a function of model parameters and observed hadron masses m_H. Our constraint master equation expresses the Fermi coupling G as inversely proportional to the meson mass (self-energy) function that allows us to estimate the dependence $G(m_H)$ in a quite large region ranging from the π up to Υ meson. We also compared the obtained dependence of $G(m_H)$ with the α_s coupling [2] and found their nontrivial behaviors at large mass region.

2 Compositeness Condition and Mass Equation

In quantum field theory description of hadrons as composite particles was thoroughly studied (e.g., in papers [4]). It may be shown (on an example of mesons) that bound state structure of hadrons can be described equivalently by using phenomenological Lagrangians of either Yukawa-type:

$$\mathcal{L}^{Y}_{\rm int}(x) = g_H H(x) J_H(x) \tag{1}$$

or Fermi-type:

$$\mathcal{L}^{F}_{\rm int}(x) = \frac{G^2}{2} J^2_H(x)\,, \tag{2}$$

where $H(x)$ is a hadron field, $J_H(x) = \bar{q}(x)\Gamma_H q(x)$ is an interpolating current composed of constituent quarks, g_H and G are the renormalized Yukawa and Fermi couplings, respectively, Γ_H is a spin matrix. For such equivalence one should fulfill two conditions: the first one is the CC, which means that the renormalization constant of hadron field Z_H is equal to zero

$$Z_H = 1 - g^2_H\, \tilde{\Pi}'_H(m^2_H) = 0 \tag{3}$$

and the second one requires that the Fermi coupling G should obey the normalization condition

$$G\,\tilde{\Pi}(m^2_H) = 1\,, \tag{4}$$

where $\tilde{\Pi}_H(m^2_H)$ is the mass function of the hadron with mass m_H and $\tilde{\Pi}'_H(z) = d\tilde{\Pi}_H(z)/dz$.

Thus the vanishing of the wave function renormalization constant in the Yukawa theory may be interpreted as the elimination of any constituent bare field in the composite boson, while (4) determines the Fermi coupling.

The ideas proposed in [4] have been further pushed forward and successfully applied for study of different aspects of low-energy hadron physics: study of QCD bosonization and QCD vacuum, quark confinement and hadron phenomenology (form factors and decay properties of mesons, baryons and exotic states) [5]. We have recently proposed the CCQM model, where infrared quark confinement was introduced to guarantee the absence of all possible thresholds corresponding to quark production. The implementation of quark confinement allowed to use the same values for the constituent quark masses for the simplest quark-antiquark systems (mesons) and more complicated multiquark configurations (baryons, tetraquarks, etc.). It should be emphasized that the experimental values of hadron masses are used in all calculations performed in the CCQM. Up to now we have used only the compositeness condition in (3) to determine the renormalized Yukawa coupling g_H as a function of meson mass and model parameters. In paper [3] we used the normalization condition (4) to investigate the dependence of the coupling G on physical meson masses.

3 Meson Mass Function

The interaction of the ground-state (pseudoscalar and vector) mesons with their constituent quarks is described in the CCQM by a Lagrangian:

$$\mathcal{L}_{\rm int} = g_H H(x) J_H(x), \qquad J_H(x) = \int dx_1 \int dx_2 F_H(x; x_1, x_2)\, \bar{q}_2(x_2)\Gamma_H q_1(x_1). \tag{5}$$

Here, $\Gamma_P = i\gamma^5$ and $\Gamma_V^\mu = \gamma^\mu$ for the pseudoscalar and vector mesons, respectively.

For the vertex function F_H we use the translational invariant form

$$F_H(x, x_1, x_2) = \delta(x - w_1 x_1 - w_2 x_2)\, \Phi_H((x_1 - x_2)^2), \tag{6}$$

where $w_i = m_{q_i}/(m_{q_1} + m_{q_2})$. Its Fourier transform is chosen in a Gaussian form

$$\tilde{\Phi}_H(-p^2) = \int dx\, e^{ipx} \Phi_H(x^2) = e^{p^2/\Lambda_H^2}. \tag{7}$$

The size parameter Λ_H is an adjustable quantity. Since the calculation of the Feynman diagrams proceeds in the Euclidean region, where $p^2 = -p_E^2$, the vertex function decreases rapidly for $p_E^2 \to \infty$ and it provides ultraviolet convergence in the evaluation of any diagram.

By using (7) and the quark propagator in the Schwinger representation

$$S_q(x - y) = \int \frac{d^4k}{(2\pi)^4 i} \frac{e^{-ik(x-y)}}{m_q - \not{k}} \tag{8}$$

we finally rewrite the Fourier transform of the mass function as follows:

$$\begin{aligned} \tilde{\Pi}_H(p^2) &= \frac{3}{4\pi^2} \int\limits_0^\infty \frac{dt\, t}{a_H^2} \int\limits_0^1 ds\, e^{-t z_0 + z_H} \Big\{ \frac{n_H}{a_H} + m_{q_1} m_{q_2} + (w_1 - \frac{b}{a_H})(w_2 + \frac{b}{a_H}) p^2 \Big\}, \\ z_0 &= s m_{q_1}^2 + (1 - s)\, m_{q_2}^2 - s(1 - s) p^2, \qquad z_H = \frac{2 s_H t}{2 s_H + t} (s - w_2)^2 p^2, \\ a_H &= 2 s_H + t, \qquad b = (s - w_2)\, t. \end{aligned} \tag{9}$$

Here $n_P = 2$, $n_V = 1$ and $s_H = 1/\Lambda_H^2$. Note, a branching point appears in (9) at $p^2 = (m_{q_1} + m_{q_2})^2$. At this point the integral over t becomes divergent as $t \to \infty$ because of $z_0 = 0$ at $s = m_{q_2}/(m_{q_1} + m_{q_2})$. However, one can avoid the appearance of the threshold singularity by introducing an infrared cutoff $\tilde{\Pi}_H(p^2) = \int_0^\infty dt \ldots \to \tilde{\Pi}_H(p^2, \lambda) = \int_0^{1/\lambda^2} dt \ldots$ on the upper limit of the scale integration in (9). Hereby, the infrared cutoff parameter λ has a dimension of energy (mass).

Now we can investigate the dependence of the Fermi coupling G on the hadron masses, or estimate the mass by using (4) because the mass function $\tilde{\Pi}_H(p^2, \lambda)$ is well defined for finite $\lambda > 0$.

4 Meson Spectrum, Decay Constants and Widths

A given meson H is characterized by the coupling g_H, size parameter Λ_H (characterizes the distribution of the quarks inside the hadron), and two of the constituent quark masses m_q ($m_u = m_d$, m_s, m_c, m_b). The infrared confinement parameter λ is universal for all hadrons.

Parameters	$m_{u/d}$	m_s	m_c	m_b	λ
Values (GeV)	0.235	0.442	1.61	5.07	0.181

(10)

Table 1. The fitted values of the size parameters Λ_H (in GeV) [3].

π	K	D	D_s	B	B_s	B_c	η_c	η_b	
0.87	1.02	1.71	1.81	1.90	1.94	2.50	2.06	2.95	
ρ	ω	ϕ	J/Ψ	K^*	D^*	D_s^*	B^*	B_s^*	Υ
0.61	0.50	0.91	1.93	0.75	1.51	1.71	1.76	1.71	2.96

The model parameters have been determined by fitting meson leptonic decay constants and electromagnetic decay widths, known either from experiment or, from lattice simulations. The results of the fit for the model parameters are given in (10) and in table 1, respectively.

The obtained results of our (overconstrained) least-squares fit [3] are displayed in tables 2 and 3.

Table 2. Fit values for the leptonic decay constants f_H (in MeV) compared with experimental data.

	Fit Values	Data	Ref.		Fit Values	Data	Ref.
f_π	128.4	130.4 ± 0.2	[6, 7]	f_ρ	221.2	221 ± 1	[6]
f_K	156.0	156.1 ± 0.8	[6, 7]	f_ω	204.2	198 ± 2	[6]
f_D	206.7	206.7 ± 8.9	[6, 7]	f_ϕ	228.2	227 ± 2	[6]
f_{D_s}	257.5	257.5 ± 6.1	[6, 7]	$f_{J/\Psi}$	415.0	415 ± 7	[6]
f_B	189.7	192.8 ± 9.9	[8]	f_{K^*}	215.0	217 ± 7	[6]
f_{B_s}	235.3	238.8 ± 9.5	[8]	f_{D^*}	223.0	245 ± 20	[10]
f_{η_c}	386.6	438 ± 8	[9]	$f_{D_s^*}$	272.0	272 ± 26	[10]
f_{B_c}	445.6	489 ± 5	[9]	f_{B^*}	196.0	196 ± 44	[10]
f_{η_b}	609.1	801 ± 9	[9]	$f_{B_s^*}$	229.0	229 ± 46	[10]
				f_Υ	661.3	715 ± 5	[6]

Note, we did not include decay results for the $\eta(\eta')$-mesons because they have some additional features like the mixing angle and a possible significant gluon admixture to the conventional $q\bar{q}$-structure of the η'. The constituent quark masses and the values for the size parameters fall into

Table 3. Fit values for basic electromagnetic decay widths (in keV) compared with PDG-2012 data.

Process	Fit Values	Data [6]	Process	Fit Values	Data [6]
$\pi^0 \to \gamma\gamma$	5.07×10^{-3}	$(7.7 \pm 0.4) \times 10^{-3}$	$K^{*\pm} \to K^\pm\gamma$	57.7	50 ± 5
$\eta_c \to \gamma\gamma$	3.47	5.0 ± 0.4	$K^{*0} \to K^0\gamma$	129	116 ± 10
$\rho^\pm \to \pi^\pm\gamma$	76.3	67 ± 7	$D^{*\pm} \to D^\pm\gamma$	0.59	1.5 ± 0.5
$\omega \to \pi^0\gamma$	687	703 ± 25	$J/\Psi \to \eta_c\gamma$	1.90	1.58 ± 0.37

the expected range. The size parameters show the expected general pattern - the geometrical size of a meson, which is inversely proportional to Λ_{H_i}, decreases as the mass increases. The present numerical least-squares fit and the values for the model parameters supersede the results of our earlier analysis given in [1], where a different set of electromagnetic decays has been used. In the present fit we have also updated some of the theoretical/experimental input values.

Our prime goal is to study the behavior of the Fermi coupling G in (4) as a function of the hadron masses by keeping other parameters (infrared cutoff parameter λ, size parameters Λ_H and constituent quark masses m_q) fixed. The original dependence of G on the hadron mass is obtained by directly taking the physical values, and results in a decreasing but sawtooth-like behavior.

A smoothness criterion might be considered as a possibility, when values for the meson masses are computed through (4) as a function of the other model parameters. In table 4 we present our results for the dimensionless coupling $G\lambda^2$ in the case of exact fit (when the values of meson masses are taken from experimental data) and in the case of the smooth fit. The model meson masses estimated by using the smooth fit of G are shown in table 5.

Table 4. Estimations for dimensionless Fermi coupling $G\lambda^2$ in cases of exact and smooth fit [3].

	Exact fit	Smooth fit		Exact fit	Smooth fit
π	1.508	1.507	η_c	0.128	0.141
K	0.919	0.920	J/Ψ	0.129	0.139
ρ	0.571	0.560	B	0.215	0.125
ω	0.673	0.553	B^*	0.237	0.124
K^*	0.476	0.472	B_s	0.192	0.122
ϕ	0.377	0.400	B_s^*	0.232	0.121
D	0.224	0.195	B_c	0.0905	0.118
D_s	0.197	0.184	η_b	0.0612	0.0986
D^*	0.168	0.180	Υ	0.0600	0.0984
D_s^*	0.158	0.170			

Table 5. The fitted values for the meson masses (in MeV) [3].

	Model	Data [6]		Model	Data [6]
m_π	141.0	139.57018 ± 0.0003	m_{η_c}	2922.0	2983.7 ± 0.7
m_K	493.0	493.677 ± 0.016	$m_{J/\Psi}$	3067.0	3096.916 ± 0.011
m_ρ	778.0	775.26 ± 0.25	m_B	5425.0	5279.26 ± 0.17
m_ω	806.0	782.65 ± 0.12	m_{B^*}	5450.0	5325.2 ± 0.4
m_{K^*}	893.0	891.66 ± 0.26	m_{B_s}	5524.0	5366.77 ± 0.24
m_ϕ	1011.0	1019.45 ± 0.02	$m_{B_s^*}$	5566.0	5415.8 ± 1.5
m_D	1915.0	1869.62 ± 0.15	m_{B_c}	6041.0	6274.5 ± 1.8
m_{D_s}	1998.0	1968.50 ± 0.32	m_{η_b}	8806.0	9398.0 ± 3.2
m_{D^*}	2001.0	2010.29 ± 0.13	m_Υ	8880.0	9460.30 ± 0.26
$m_{D_s^*}$	2099.0	2112.3 ± 0.5			

5 Comparison of Fermi Coupling with Effective Strong Coupling

It might be interesting to compare the behavior of smooth G obtained in table 4 with the effective strong coupling α_s obtained in the QCD-inspired relativistic models with specific forms of analytically confined quark and gluon propagators [2]. In these models the nonlocal four-quark interaction is induced by one-gluon exchange between biquark currents, so, the resulting coupling α_s is dimensionless. The idea of such a comparison by re-scaling of one coupling to another, where both have a 'plateau' behavior, looks quite reasonable. Note that this plateau region is almost the same in both approaches. In figure 1 we compare the mass dependence of the re-scaled dimensionless Fermi coupling $\sim 1.74\,G\lambda^2$ (*solid line*) in [3] with the effective strong coupling α_s (*dashed line*) in [2]. One can see that these two curves are different in the region of small mass because of different confinement mechanisms realized in original models. However, these curves behave very similar to each other in the intermediate and heavy mass regions above ~ 2 GeV, that is a nontrivial observation and deserves additional consideration.

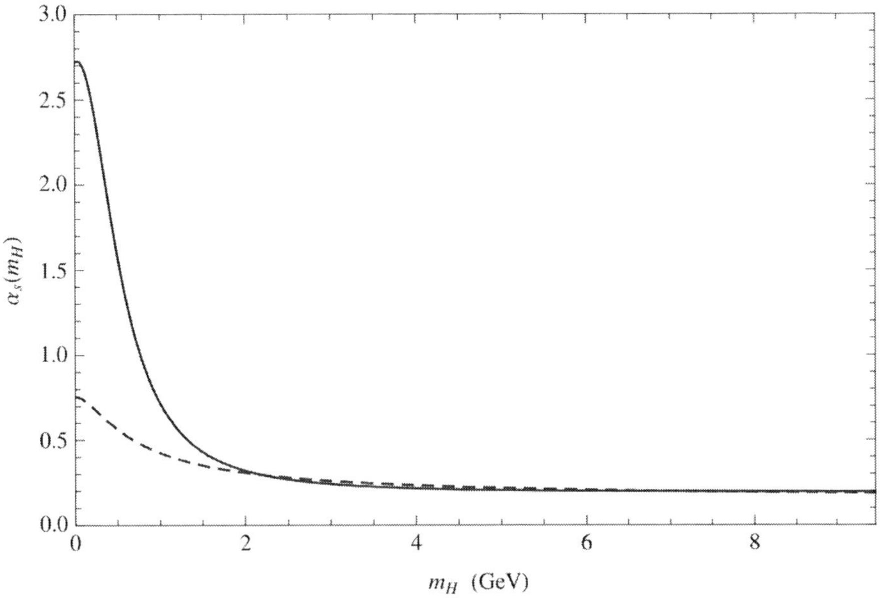

Figure 1. The dimensionless Fermi coupling $\lambda^2 G(m_H)$ (solid line) scaled in the region of large masses to the effective strong coupling $\alpha_s(m_H)$ obtained in [2] (dashed line).

References

[1] T. Branz et al., Phys. Rev. D **81**, 034010 (2010). M. A. Ivanov et al., Phys. Rev. D **85**, 034004 (2012)

[2] G. Ganbold, Phys. Rev. D **79**, 034034 (2009). Phys. Rev. D **81**, 094008 (2010). Phys. Part. Nucl. **43**, 79 (2012). Phys. Part. Nucl. **45**, 10 (2014)

[3] G. Ganbold, T. Gutsche, M. A. Ivanov, V. E. Lyubovitskij, J. Phys. G **42**, 075002 (2015)

[4] B. Jouvet, Nuovo Cim. **3**, 1133 (1956). A. Salam, Nuovo Cim. **25**, 224 (1962). M. A. Braun, Nucl. Phys. B **14**, 413 (1969). S. Weinberg, Phys. Rev. **130**, 776 (1963)

[5] G. V. Efimov and M. A. Ivanov, Int. J. Mod. Phys. A **4**, 2031 (1989). M. A. Ivanov et al., Phys. Rev. D **56**, 348 (1997). A. Faessler et al., Eur. Phys. J. C **4**, 18 (2002). G. V. Efimov and G. Ganbold, Phys. Rev. D **65**, 054012 (2002)

[6] J. Beringer et al. [Particle Data Group Collaboration], Phys. Rev. D **86**, 010001 (2012)

[7] J. L. Rosner, S. Stone, arXiv:1002.1655 [hep-ex]

[8] J. Laiho, E. Lunghi, R. S. Van de Water, Phys. Rev. D **81**, 034503 (2010)

[9] T. -W. Chiu et al. [TWQCD Collaboration], Phys. Lett. B **651**, 171 (2007)

[10] D. Becirevic et al., Phys. Rev. D **60**, 074501 (1999)

EPJ Web of Conferences **138**, 04005 (2017) DOI: 10.1051/epjconf/201713804005
Baldin ISHEPP XXIII

Kinematical fit and channel separation in np experiment

A.V. Belyaev[1]*, *A.P.* Jerusalimov[1], and *Yu.A.* Troyan[1]

[1] *Joint Institute for Nuclear Research, 141980 Dubna, Russia*

Abstract. The method of channels separation in particle-particle collisions is presented. The kinematical fit and weights for tracks combinations are used. The Kolmogorov criterion is applied to check agreement between experimental and theoretical χ^2 distributions. The investigations were made for data of np experiment in LHEP (JINR, Dubna) and for pp collisions in GSI (Darmstadt, Germany).

1 Introduction

The channel separation in many experiments in particle physics is often processed using value of a missing mass. In this case

1) kinematical parameters of particles are defined with a low precision
2) its values violate conservation laws
3) kinematical parameters of a neutral particle became inaccessible.

All this reasons make difficult an estimation of cross section errors. The kinematical fit and the proposed weight method allow to separate reaction channels eliminating reasons mentioned above.

2 Data and results

The neutron channel [1] of the 1m-H2-bubble chamber of the LHEP [2] worked in 70-90 years. Parameters of neutron beam are: momentum p_n $\sim$(1 - 5) GeV/c with relative error δp_n $\sim$ 2.5 % and with angle resolution Δ_n $\sim$ 0.5 mRad. Total $\sim 10^6$ events of the np collision were registered at stereo-photos using the bubble chamber.

Taking into account energy range of a np collision , one can get a list of hadrons, which may be produced in some channels:

positive charged particles:	p	d	π^+	K^+	Σ^+
negative charges particles:	π^-	K^-	Σ^-	Ξ^-	
neutral particles:	π^0	K^0	n	Λ	Σ^0

What is a problem? By some reasons data have no identification of particles, i.e. several hypotheses of a particle may correspond to every track which may be produced by a charged particle. Therefore a 3-prong np collision, for example, may be considered with a lot of combinations of track hypotheses like in the table 1.

*e-mail: belyaev.bljv@yandex.ru

Table 1.

Combination	Channel
$p\ p\ \pi^-$	$np \rightarrow pp\pi^-$
$p\ p\ \pi^-\ \pi^0$	$np \rightarrow pp\pi^-\pi^0$
$p\ p\ K^-\ \pi^0$	$np \rightarrow ppK^-K^0$
$p\ \pi^+\ \pi^-\ n$	$np \rightarrow p\pi^+\pi^-n$
$\pi^+\ p\ \pi^-\ n$	$np \rightarrow p\pi^+\pi^-n$
$p\ \pi^+\ \Sigma^-\ K^0$	$np \rightarrow p\pi^+\Sigma^-K^0$

It is necessary to select the true combination (channel) of a collision. Two instruments are represented here to solve this problem: 1) Kinematical fit; 2) "Weighting".

The kinematical fit [3] is the conditional minimization of the functional $\chi^2 = (\mathbf{x} - \mathbf{x}^m)\ \mathbf{G}^{-1}(\mathbf{x} - \mathbf{x}^m)$ with r constraint equations $F_r(x) = 0$ (r = 1,2,...) - number of degrees of freedom for χ^2),

where $\mathbf{x}^m$ - vector of measured kinematical parameters (conservation laws may be violated);

$\mathbf{G}^{-1}$ - error matrix for $\mathbf{x}^m$;

$\mathbf{x}$ - resulting vector of fitted kinematical parameters (conservation laws have no violations).

Constraint equations are usually based on conservation laws. Masses of secondary particles must be determined; they are used in constraint equations. Events which have only charged particles (r = 4) or (additionally) with a single neutral particle (r = 1) may be processed using the kinematical fit.

If the conditional minimization process has a success, the kinematical fit:
- fits a set of more precise kinematical parameters which keeps following conservation laws;
- gives less errors for fitted kinematical parameters;
- calculates kinematical parameters (with its errors) for unmeasured neutral particle (in the case of a channel with a single neutral, i.e. unmeasured, particle);
- gives the value of the χ^2 in the point of conditional minimum; the χ^2 determines a confidence probability for a channel of an interaction been represented in the $\mathbf{x}^m$.

If input data have no systematical errors then the experimental χ^2 distribution must be a standard one. This theoretical approach was tested and confirmed (new) with the Kolmogorov's criterion. For this purpose:
1) a set of ideal events (a model of phase volume for any np collision) was generated with the Monte Carlo generator;
2) ideal kinematical parameters were smeared using random gauss errors; 3) such model data were fitted with the kinematical fit procedure.

Then the Kolmogorov's criterion test demonstrated that model χ^2 distributions are in a fine agreement with standard ones. Conclusion (new): the kinematical fit procedure does not add systematical errors to resulting kinematical parameters (vector **x** above).
Note that an agreement with some fixed confidence level, obtained due to Kolmogorov's criterion, is a necessary condition of absence of systematical errors in resulting kinematical parameters.
But the kinematical fit can reach a conditional minimum for several various track combinations in the same event. Here an attempt to solve this problem with the **"weighting"** is presented.

It turned out that the conditional minimum in fitting process was reached only for 3 combinations. These combinations were supplied with the corresponding χ^2:

1	(7)	K^-	p	K^+	n	$\chi_1^2 = 2.45$
2	(15)	Σ^-	p	π^+	K^0	$\chi_2^2 = 0.05$
3	(16)	Σ^-	p	K^+		$\chi_3^2 = 3.11$

The weight of a combination is proposed in the following way (new approach)

1) Confidence probabilities $\mathbf{H}_j(\chi_j^2, r) = \int_{\chi_j^2}^{\infty} p(r, z)dz, \quad (j = 1, 2, 3)$, where $p(r, z) = \frac{z^{r/2-1} \cdot e^{-z/2}}{2^{r/2} \times \Gamma(r/2)}$ were calculated (figure 1).

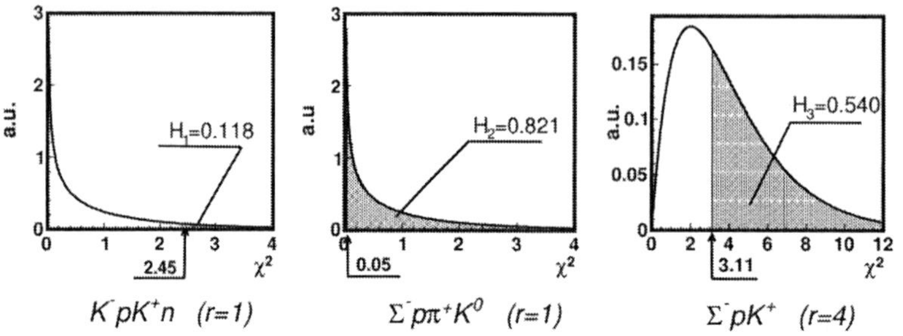

Figure 1. On the calculation of confidence probabilities of combinations.

2) the weight of $j-th$ combination is then defined as $w_j = \frac{H_j}{\sum_{i=1}^{3} H_i}, \quad (j = \overline{1,3})$

Results are combined in the table 2.

Table 2.

Combination	r	χ^2	$H(\chi_j^2, r)$	w	Channel
$K^- \ p \ K^+$	1	2.45	0.118	0.080	pK^+K^-n
$\Sigma^- \ p \ \pi^+ \ K^0$	1	0.005	0.821	0.555	$p\pi^+\Sigma^-K^0$
$\Sigma^- \ p \ K^+$	4	3.11	0.540	0.365	$pK^+\Sigma^-$

The "weight-method" differs from the method of "preference rules" which was used earlier. Using preference rules one selects only sole combination from the ones mentioned in example above. So, under the preference rules 1) combination with r = 4 has preference before all others with r = 1; 2) combination with minimal χ^2 has preference over all others (with the same value of r).

In the case of using the described weight-method the Kolmogorov's criterion produces **an agreement** (between experimental and standard χ^2 distributions), **which is better** than with some other methods that we used. This fact we can demonstrate with values of argument λ of the Kolmogorov's function $K(\lambda)$. In accordance with the Kolmogorov's criterion - smaller values of its argument λ correspond to bigger values of confidence probability $1 - K(\lambda)$, where the Kolmogorov's function is $K(\lambda) = \sum_{k=-\infty}^{\infty} (-1)^k e^{2k^2\lambda^2}$. Figures 2 and 3 shows differences between results of the preference rules and the weight-method. One can see that values of λ is much smaller for the weight-method.

Channel separation for all data was realized using the kinematical fit procedure and the described weight-method. Results of channel separation are given in tables 3 - 7.

The column "$\sum(w)$" contains sums of weights for all minimized combinations of given channel. It should be considered as effective number of events, corresponding to a given channel.

In cases when $\sum(w) > 1000$ the Kolmogorov's λ are shown.

One can see that the rounded value "$\sum\sum(w)$" equal to number of all events which have at least one minimized combination.

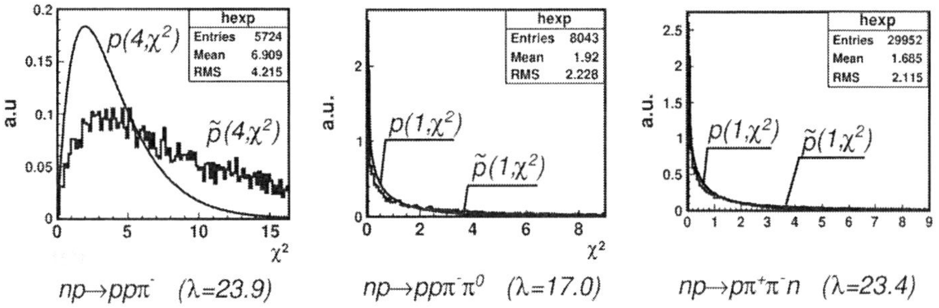

Figure 2. The preference rules: experimental functions $\tilde{p}(r,\chi^2)$ combined with theoretical ones $p(r,\chi^2)$.

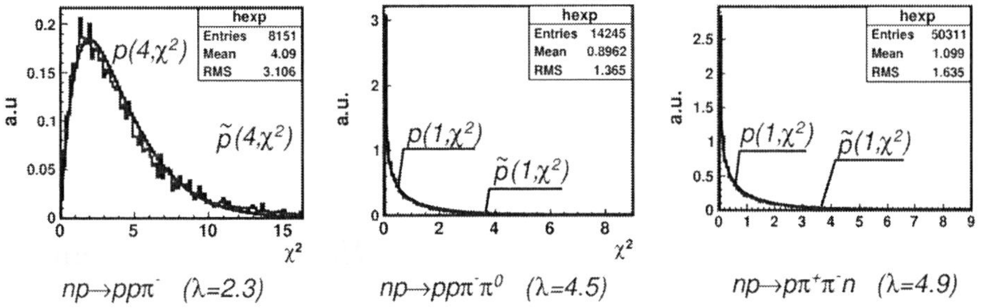

Figure 3. The weight method: experimental functions $\tilde{p}(r,\chi^2)$ combined with theoretical ones $p(r,\chi^2)$.

The contribution for a channel is calculated as $(\sum(w)/(\text{total events}))*100$.
For some channels of 1- and 3-prong events it was possible to compare branching ratios with ones published in [4] where they were calculated using preference rules and for smaller set of channels. Data shown in tables 3 - 7 should be considered as preliminary due to absence of the particle identification in events what makes values of the Kolmogorov's λ in some channels too big.

3 Conclusion

1. The method of channels separation in particle-particle collisions using the kinematical fit and weights for track combinations is presented.
2. The Kolmogorov's criterion was applied to check agreement between experimental and theoretical χ^2 distributions.
3. Channel separation using weights for 1-, 3- and 7-prong events of np collision (for various values of p_n) was realised.
4. The procedure of **Kinematical Fit** allowed to refine the parameters of the registered particles and to reconstruct the parameters of the neutral ones.

References

[1] Yu.D. Beznogikh et al., Prib.Tekh.Eksp. **4**, 202 (1969) (in Russian)
[2] V.V. Glagolev, http://lhe.jinr.ru/1mHBC/index.html
[3] J.P. Berge, F.T. Solmitz and H.D.Taft. Rev.Sci.instrum., **32**, 538-548 (1961)
[4] C. Besliu et al., Yad.Fiz. **43**, 888-892 (1986)

Table 3. 3-prong events at P_n=5.20 GeV/c

Channel	$\Sigma(w)$	Kolmogorov's λ	Contribution (%)
$d\,\pi^+\,K^-\,K^0$	1036.84	1.6	1.5±0.1
$p\,K^+\,K^-\,n$	2271.32	5.4	3.2±0.1
$d\,\pi^+\,\pi^-\,\pi^0$	3186.41	1.7	4.5±0.1
$p\,K^+\,\pi^-\,\Sigma^0$	3697.29	6.5	5.2±0.1
$p\,K^+\,\pi^-\,\Lambda$	4739.72	9.0	6.7±0.1
$p\,p\,\pi^-\,\pi^0$	6018.38	4.5	8.5±0.1
$p\,p\,\pi^-$	7018.73	2.3	9.9±0.1
$p\,\pi^+\,\pi^-\,n$	30695.38	4.9	43.1±0.2
$\Sigma(\Sigma(w))$	60140.97		84.7±0.3
Events with fit (M)	60141		84.7
Events with no fit (K)	10882		15.3
Total events (N)	71023		100.0

Table 4. 1-prong events at P_n=5.20 GeV/c

Channel	$\Sigma(w)$	Kolmogorov's λ	Contribution (%)
$d\,\pi^0$	311.77		2.9±0.2
$p\,n$	4465.24	1.9	42.2±0.6
$\Sigma(\Sigma(w))$	4777.00		45.1±0.7
Events with fit (K)	4777		45.1
Events with no fit (M)	5812		54.9
Total events (N)	10589		100.0

Table 5. 1-prong events at P_n=2.23 GeV/c

Channel	$\Sigma(w)$	Kolmogorov's λ	Contribution (%)
$d\,\pi^0$	443.48		3.8±0.2
$p\,n$	5476.57	2.4	47.3±0.6
$\Sigma(\Sigma(w))$	5920.00		51.1±0.7
Events with fit (K)	5920		51.1
Events with no fit (M)	5665		48.9
Total events (N)	11585		100.0

Table 6. 1-prong events at P_n=1.43 GeV/c

Channel	$\Sigma(w)$	Kolmogorov's λ	Contribution (%)
$d\,\pi^0$	411.72		6.7±0.3
$p\,n$	3995.28	1.5	64.6±1.0
$\Sigma(\Sigma(w))$	4407.00		71.3±1.1
Events with fit (K)	4407		71.3
Events with no fit (M)	1775		28.7
Total events (N)	6128		100.0

Table 7. 7-prong events at P_n=5.20 GeV/c

Channel	$\Sigma(w)$	Contribution (%)
$d\pi^+\pi^+\pi^+\pi^-\pi^-\pi^-$	2.55	<1.0
$d\pi^+\pi^+\pi^+\pi^-\pi^-\pi^-\pi^0$	7.07	2.0±0.8
$p\pi^+\pi^+\pi^+\pi^-\pi^-\pi^-n$	40.17	11.4±1.8
$pp\pi^+\pi^+\pi^-\pi^-\pi^-\pi^0$	48.44	13.7±2.0
$pp\pi^+\pi^+\pi^-\pi^-\pi^-$	189.78	53.8±3.9
$\Sigma(\Sigma(w))$	288.00	81.6±4.8
Events with fit (M)	288	81.6
Events with no fit (K)	65	18.4
Total events (N)	353	100.0

Photon and π^0 pairs production in proton-nucleus and deuteron-nucleus interactions. Results of experiments on internal beams of the Nuclotron

Khachik Abraamyan[1,2,⋆], *Alexander* Anisimov[1], *Mircea* Baznat[3], *Konstantin* Gudima[3], *Mikhail* Kozhin[1], *Vladimir* Kukulin[4,5], *Maxim* Nazarenko[6], *Sergey* Reznikov[1], and *Alexander* Sorin[1,7]

[1] *VBLHEP JINR, 141980 Dubna, Moscow region, Russia*
[2] *Yerevan State University, 0025, Yerevan, Armenia*
[3] *Institute of Applied Physics, MD-2028 Kishinev, Moldova*
[4] *Institute of Nuclear Physics, Moscow State University, 119992, Moscow, Russia*
[5] *Institut für Theoretische Physik, Universität Tübingen, D-72076 Tübingen, Germany*
[6] *Moscow State Institute of Radioengineering, Electronics and Automation, 119454 Moscow, Russia*
[7] *BLTP JINR, 141980 Dubna, Moscow region, Russia*

Abstract. The results of experiments on photon and π^0 pairs production in interactions of 5.5 GeV/c protons, 2.75 and 3.83 GeV/c per nucleon deuterons with carbon and copper nuclei are presented. As previously reported (see. [1] and references therein), along with π^0 and η mesons, a resonance structure in the invariant mass spectrum of two photons at $M_{\gamma\gamma} = 360 \pm 7 \pm 9$ MeV is observed in the reaction $dC \rightarrow \gamma + \gamma + X$ at momentum 2.75 GeV/c per nucleon. Here are some results of testing the signal and results of statistical analysis of the experimental data which allow to explain the lack of appearance of the signal in other experiments [2, 3] and make predictions for possible appearances of the signal in upcoming experiments.

1 Introduction

This paper is an extension of the research begun earlier (see. [1] and references therein). The first results on π^0 pairs production in interactions of 3.83 GeV/c per nucleon deuterons with copper nuclei are obtained. Also here are results of further verification of the signal detected in the spectrum of the invariant mass of photon pairs in the reaction $d + C \rightarrow \gamma + \gamma + X$ at 2.75 GeV/c per nucleon and results of statistical analysis of the experimental data: the possible influence of the internal beam characteristics on the properties of the events detected in different time intervals of the accelerator cycles and appearances of the signal in different parts of the collected statistics. These results allow to explain the lack of appearance of the signal in other experiments [2, 3] and make predictions for possible appearances of the signal in upcoming experiments.

⋆e-mail: abraam@bk.ru

2 π^0 pairs production in deuteron-copper interactions

Results of an analysis of the d(2.0 GeV/n)+C and p(4.6 GeV)+C events with 4 or more detected photons, are given in [1, 4]. Further analysis of the data for d(3.0 GeV/n)+Cu interactions obtained at lower thresholds of the discriminators (the discriminator thresholds were chosen to be 0.4 and 0.35 GeV for d + C and d + Cu reactions, respectively), shows that the structure, which observed in the d(2.0 GeV/n)+C and p(4.6 GeV)+C reactions after the background subtraction, is also observed in the d(3.0 GeV/n)+Cu reaction (see. figure 1).

3 Photon pair production

Below are the results of testing the signal detected in the spectrum of the invariant mass of photon pairs in the reaction $d + C \rightarrow \gamma + \gamma + X$ at 2.75 GeV/c per nucleon [5].

3.1 Another selection criteria

Another method to decrease the influence of possible violations of the energy-momentum conservation laws on the mixing-event background, is a restriction of energies of individual photons (instead of restriction of the sum of their energies [1, 5]). For a background suppression and minimization of systematical errors due to violation of conservation laws the following selection criteria were used:

(1b) the number of photons in an event, $N_\gamma \leq 3$ (the events of each type (one of the photons in the left arm and two on the right, etc.) were analysed separately, the results were summarized);

(2b) the energies of photons, $100 < E_\gamma < 750$ MeV;

The results of this analysis are given in figure 2.

3.2 The accuracy of the combinatorial background

To verify the accuracy of the combinatorial background we have used the method of "miscalibration" of modules of the γ-spectrometer by introduction of chaotic coefficients in the energy scale in each module of the spectrometer. The results are shown in figure 3.

3.3 Data simulation

In order to verify the experimental method of allocation of the observed signal there were simulations of studied reactions performed with accurate experimental measurement conditions, including the set of corresponding event class (see [5]). As the first step, any sources of the signal were not included in the model (see figure 4, left).

Then we have included the additional channel of two-γ creation by two-pion interaction with the observable structure formation. We assume that the resonance can be created by $\pi\pi$ interactions if the invariant mass of the pions obeys Breight-Wigner distribution with the observed parameters (see figure 4, right).

4 Statistical analysis of the experimental data

4.1 Different parts of the accelerator cycles

The purpose of this analysis is to investigate the possible influence of the internal beam characteristics on the properties of the events detected in different time intervals of the accelerator cycles. For this

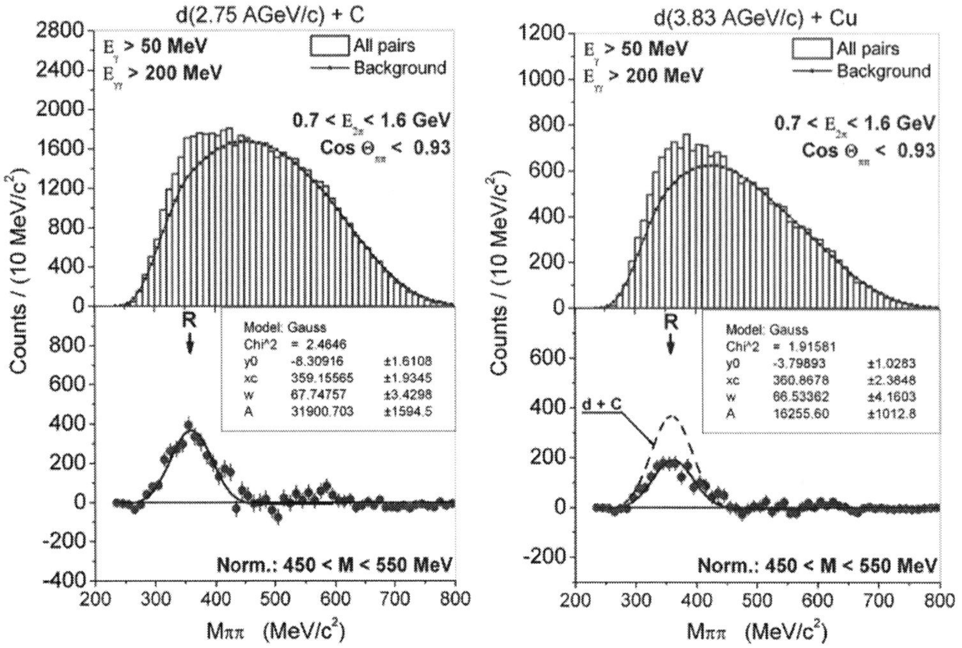

Figure 1. The invariant mass distributions of $\pi^0\pi^0$ pairs without (upper panels) and with (bottom panels) the background subtraction for the reactions $d+C$ at 2.75 GeV/c per nucleon [4] (left figure) and $d+Cu$ at 3.83 GeV/c per nucleon (right figure). The backgrounds are normalized to pair numbers in the interval: $450 < M_{\pi\pi} < 550$ MeV/c^2.

Figure 2. The invariant mass distributions (per 20 MeV/c^2) of $\gamma\gamma$ pairs satisfying criteria (1b)-(2b) in the reactions $d(2\ GeV/n)+C \rightarrow \gamma+\gamma+X$ (left figure) and $p(4.6\ GeV)+C \rightarrow \gamma+\gamma+X$ (right figure) after the background subtraction. The backgrounds are normalized to pair numbers in the interval: $200 < M_{\gamma\gamma} < 300$ MeV/c^2.

analysis we have used the data obtained on an internal beam of deuterons with momentum 2.75 GeV/c per nucleon [5]. The internal target consists of carbon wires with the diameter of 8 microns mounted in a rotatable frame. The overall construction is located in the vacuum shell of the accelerator. The duration of the irradiation cycle is 1 s. For the analysis, the data collected at the maximum beam intensity ($\sim 10^9$ per pulse), were selected (about 2/3 of all the data in the reaction d(2.0 GeV/n)+C [5]). The results of this analysis are shown in figures 5, 6.

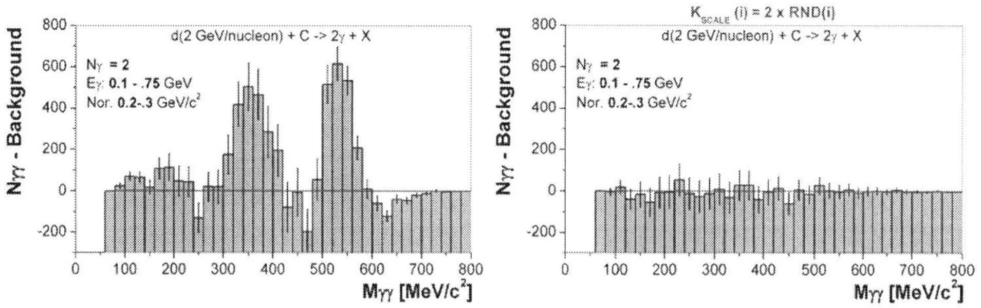

Figure 3. Invariant mass distributions (per 20 MeV$/c^2$) of $\gamma\gamma$ pairs in the reactions $d(2\ GeV/n) + C \rightarrow \gamma + \gamma + X$ after the background subtraction, at the number of photons in an event, $N_\gamma = 2$ and at the energies of photons, $100 < E_\gamma < 750$ MeV (left figure), and the same but after a "miscalibration" of the modules by the introduction of chaotic coefficients in the range $0 \div 2$: $E_i \rightarrow E_i \cdot 2 \cdot RND(i)$, (where E_i is the energy in the i-th module, $RND(i)$ are random numbers) (right figure). Here, the average of results obtained after the introduction of 6 different sets of random numbers is shown. The backgrounds are normalized to pair numbers in the interval: $200 < M_{\gamma\gamma} < 300$ MeV$/c^2$.

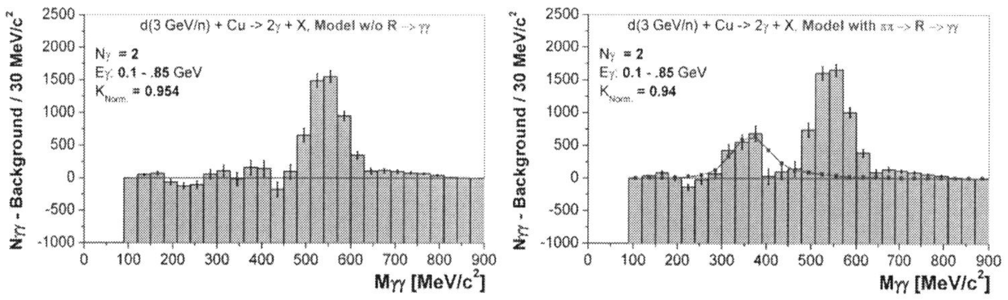

Figure 4. Invariant mass distributions of $\gamma\gamma$ pairs from the $d(3\ GeV/n)$+Cu reaction after the background subtraction, at the number of photons in an event, $N_\gamma = 2$ and at the energies of photons, $100 < E_\gamma < 850$ MeV, without (left figure) and with the reaction $\pi + \pi \rightarrow R \rightarrow \gamma + \gamma$ (right figure), under the real experimental conditions. The spectra are the result of processing the simulated data with the method which was applied to the experimental data. The contribution of photon pairs from the resonance decay (selected with the help of labels of photons applied in the model) is shown by the solid line. The auxiliary factors K_{Norm} of the backgrounds normalization (see the text, 4.2) are given in the figures.

4.2 Different parts of the collected statistics

The purpose of this analysis is to show different appearances of the signal at low (for a given signal-to-background ratio) statistics. In particular, this allows to explain the lack of appearance of the signal in other experiments [2, 3] and make predictions for possible appearances of the signal in upcoming experiments.

To create conditions, in terms of statistics similar to those of the experiment on TAPS [2] (see also our analysis of this experiment in [5]), we divided the statistics recorded in the reaction d(2.0 GeV/n)+C into 19 parts. Each part was analysed separately, under the conditions shown in figure 3. The spectrum in figure 3 contains a total of 189864 selected ($N_\gamma = 2$) events. The number of pairs after the background subtraction in the range of 300-420 MeV$/c^2$ (in the vicinity of the resonance

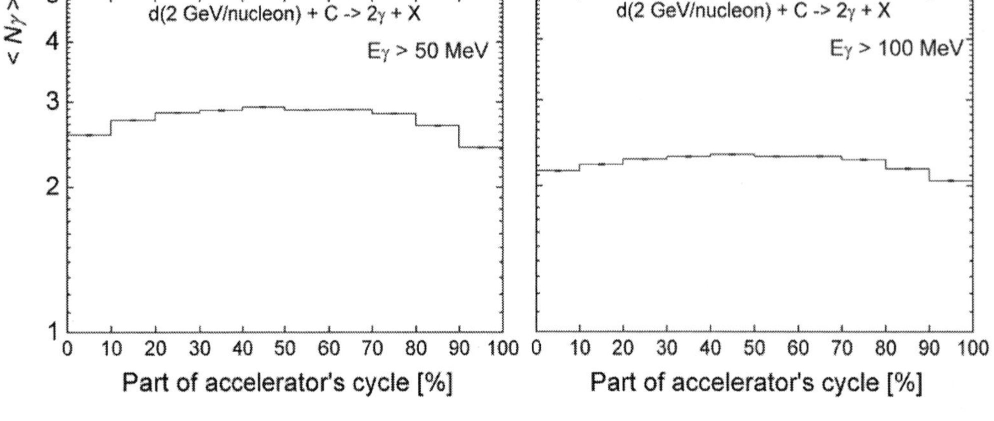

Figure 5. The mean number of detected photons in an event, recorded in different intervals of accelerator cycle, with a minimum energy of photons, $E_{\gamma min} = 50$ MeV (left) and $E_{\gamma min} = 100$ MeV (right).

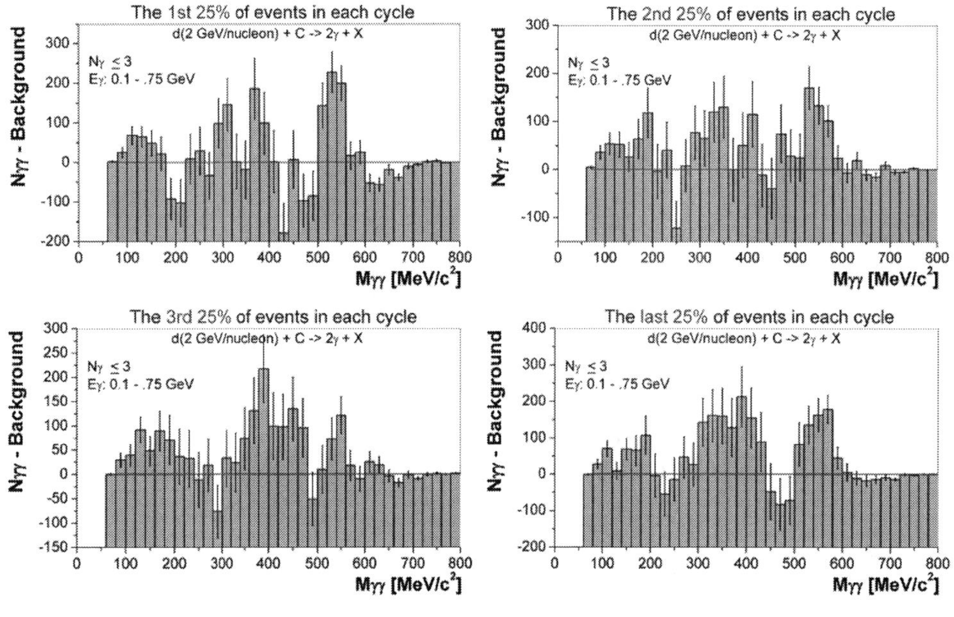

Figure 6. The invariant mass distributions of $\gamma\gamma$ pairs satisfying criteria (1b)-(2b) in the reactions $d(2\ GeV/n) + C \rightarrow \gamma + \gamma + X$ after the background subtraction, in the events recorded in different intervals of accelerator cycle. The backgrounds are normalized to pair numbers in the interval: $200 < M_{\gamma\gamma} < 300$ MeV/c^2.

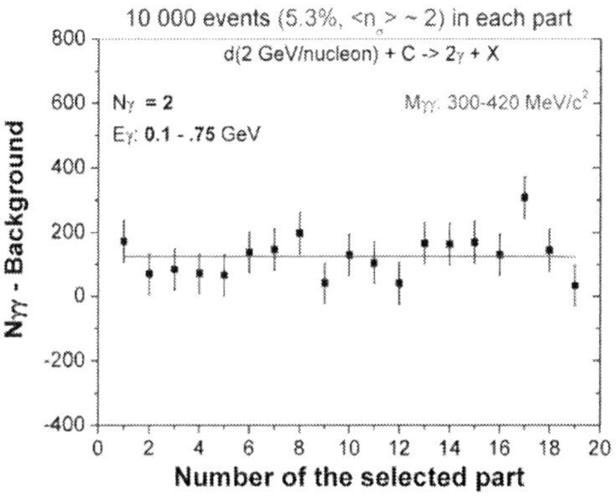

Figure 7. The number of pairs after the background subtraction in the range of 300-420 MeV/c^2 (in the vicinity of the resonance mass) for different parts of the statistics in the reaction $d(2\ GeV/n) + C \to \gamma + \gamma + X$, under the conditions: the number of photons in an event, $N_\gamma = 2$; the energies of photons, $100 < E_\gamma < 750$ MeV (see figure 3, left figure).

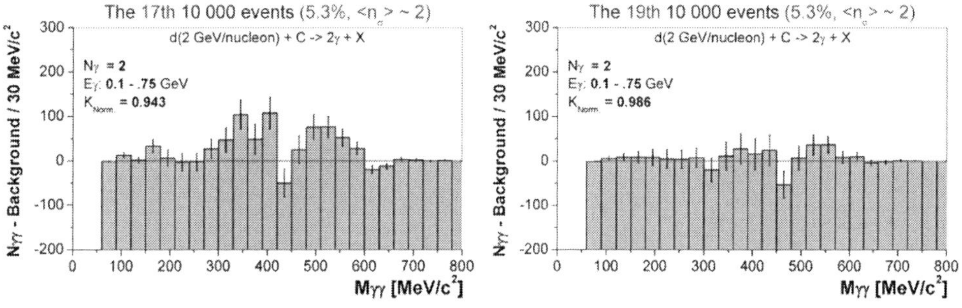

Figure 8. Invariant mass distributions (per 20 MeV/c^2) of $\gamma\gamma$ pairs in the reaction $d(2\ GeV/n) + C \to \gamma + \gamma + X$ after the background subtraction, under the same conditions as in figure 3, for different parts of the statistics collected in this experiment. The auxiliary normalization factors K_{Norm} (see the text) are given in the figures.

mass), is 2386 ± 280. Thus, the expected average number of pairs after the background subtraction in the above indicated range, for the each part is 126 ± 64, i.e. 2 standard deviations. In our experiment, each of these spectra is a result of $\sim 10^{11}$ d(2.0 GeV/n)+C interactions. Some results of this analysis are shown in figures 7, 8. The background normalization was carried out in two steps. First, the background is normalized to the total pair number. Naturally that at the event mixing the resonance maxima are not reproduced disturbing the overall normalization. At the second step this shortcoming of the background estimate is corrected by an auxiliary factor obtained by iterating treatment of the resonances contribution to the spectrum (in applying this method to the total spectrum (see figure 3) gives a result, which almost coincides with the result obtained after the normalization of the background to the number of pairs in the range 200-300 MeV/c^2).

References

[1] Kh.U. Abraamyan et al., Eur. Phys. J. A **52**, 259 (2016)
[2] R. Averbeck et al. (TAPS Collaboration), Z. Phys. A **359**, 65 (1997)
[3] WA80 Collaboration, Phys. Lett. B **361**, 14 (1995)
[4] Kh.U. Abraamyan et al., arXiv:1203.4198
[5] Kh.U. Abraamyan et al., Phys. Rev. C **80**, 034001 (2009)

Relativistically invariant self-similarity approach for description of collective phenomena

E. G. Baldina[1,2,*] and *A. A.* Baldin[1,2]

[1]Joint Institute for Nuclear Research, Dubna, 141980 Russia
[2]Institute for Advanced Studies "OMEGA", Dubna, 141986 Russia

Abstract. The self-similarity approach is applied to description of nuclear interactions. The obtained self-similarity solution quantitatively describes particle production in relativistic nuclear collisions. It is an appropriate tool for planning and optimization of experiments aimed at the search of new collective phenomena in highly excited nuclear matter, especially in the intermediate energy range.

Self-similarity is a special symmetry of solutions when the change in scales of independent variables can be compensated by the self-similarity transformation of other dynamic variables. This results in the reduction of the number of variables in any physical law. Self-similarity is an important tool for investigation of complex physical phenomena with the objective of finding regularities of general character coupling the multitude of experimentally measurable data.

Historically, self-similarity in elementary particle physics was introduced in relation with the deep inelastic electron scattering on protons, it is the well known Bjorken scaling [1] obtained directly from the conservation laws in the assumption since the squared momentum of particles participating in the reaction is much greater than the squared mass, the latter can be neglected.

Writing the momentum conservation law of the reaction, we have

$$P_1 + xP_2 = P_1' + \sum P_i', \tag{1}$$

where the subscript "1" denotes the electron, the subscript "2" the proton, and prime marks the quantities attributed to the particles after the interaction.

Introducing the so called Bjorken variable, $P_1 - P_1' = q$ and making simple transformations of (1) with account of the fact that the cross-products in the sum describing relative motion of all other particles produced in the reaction can be neglected, and finally, neglecting the squared masses of particles, as compared to the squared three-momenta, we get the known Bjorken scaling in the form

$$x = -\frac{q}{2P_2 q}. \tag{2}$$

The general self-similar solution [2, 3] including the Bjorken solution as a particular case, is obtained as follows.

* Corresponding author: e.baldina@mail.ru

We consider two interacting nuclei and assume that fractions X_1 and X_2 of masses (or fractions of momenta) of these nuclei participate in the interaction. The 4-momentum conservation is written as:

$$X_1P_1 + X_2P_2 = P_1' + P_2' + P_3 + \sum P_i' \tag{3}$$

Here, P_3 is the 4-momentum of the particle of interest produced in the reaction (note that conservation of quantum numbers should be taken into account explicitly by addition of corresponding terms to the right-hand side of (3)).

Again, we neglect terms of relative motion of produced particles and masses of all particles not registered, and seek the self-similar solution in the form

$$E\frac{d^3\sigma}{dp^3} = C_1 A_1^{\alpha(X_1)} A_2^{\alpha(X_2)} f(\Pi) \tag{4}$$

where $\Pi = \frac{1}{2}\left(X_1^2 + X_2^2 + 2X_1X_2\gamma_{12}\right)^{\frac{1}{2}}$ is the similarity parameter.

The fractions X_1 and X_2 are found by minimization of Π (figure 1) (see also [4]).

$$E\frac{d^3\sigma}{dp^3} = C_1 A_1^{\frac{1}{3}(1+X_1)} A_2^{\frac{1}{3}(1+X_2)} \exp(-\Pi/C_2) \tag{5}$$

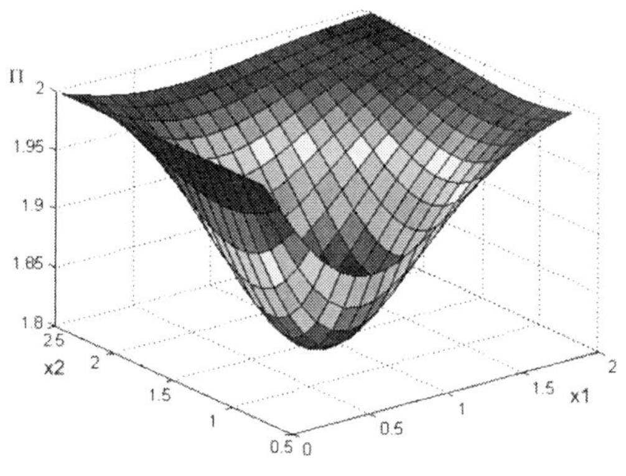

Figure 1. Minimum of Π in terms of X1, X2 coordinates.

This self-similar solution describes quantitatively a wide variety of experimental data on cumulative, twice cumulative particle production, and antimatter production [5-7] (figure 2).

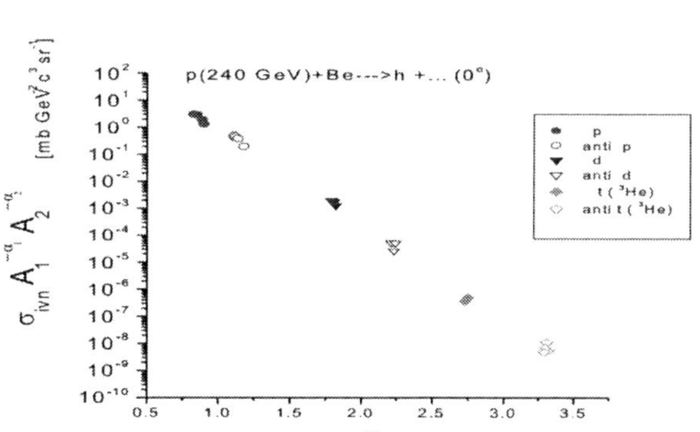

Figure 2. Matter and antimatter production: the self-similarity description.

It should be underlined that the widely used assumption $P^2 >> M^2$ is not valid even for very high momenta. It will be shown below that even a small mass difference essentially influences the reaction mechanisms.

The above self-similarity solution was found for reactions with low and intermediate transverse momenta (below 1 GeV/c). Recent experiment at the Institute for High Energy Physics (Protvino) [8], however, demonstrated that the A-dependence of the form $A^{(1+X_2)/3}$ well describes the dynamic dependence of the cross sections on X_2, while it fails to describe a stronger dependence on the mass number of the nucleus observed in this experiment (figure 3).

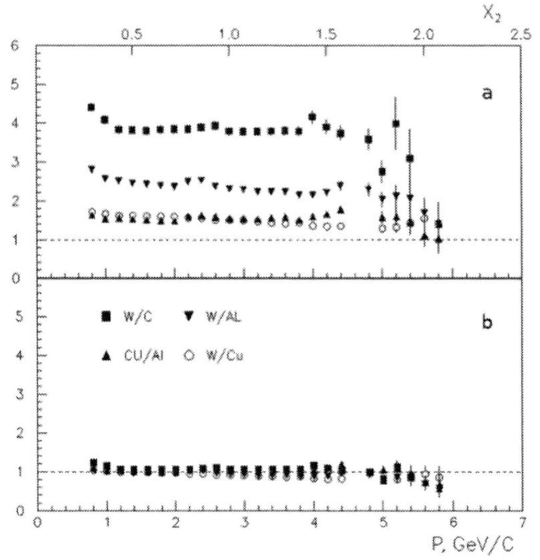

Figure 3. Ratio of cross sections of negative pion production on different nuclei multiplied by inverse A-dependence (see the text). The lower axis shows the momentum, the upper axis, X_2. (a) The ratios are obtained using the A-dependence in the form $A^{(1+X_2)/3}$, (b) the ratios are obtained using the A-dependence in the form $A^{(2.45+X_2)/3}$ [8].

It is possible to develop the self-similarity solution for description of high-P_t reactions; for this purpose the function $f(\Pi)$ in (4) should be added by a term depending on P_t.

The scientific program of research at the NICA collider [9], in the field of relativistic nuclear collision includes the study of the phenomenon of enhanced K^+ production in nucleon-nucleon collisions in the intermediate energy range, $\sqrt{S} \sim 3 \div 12\,GeV$ [10]. This phenomenon was observed in cumulative processes and is well described by the above self-similarity solution. The specific feature of this description is that it is necessary to take into account the channels with Λ Σ^0, Σ^+, and Σ^- hyperons for description of K meson production in the framework of self-similarity solution (4). Taking into account the contribution of all these channels yields the quantitative description of the enhanced K^+ production. This is a persuasive illustration that small mass corrections cannot be neglected in relativistic particle production.

Collective nuclear effects in the intermediate energy range have received great attention recently. Experiments in this field have been performed at SPS (CERN), are planned at NICA collider (JINR), FAIR (GSI), and RHIC (BNL). These experiments are aimed at investigation of exotic collective properties of matter in heavy ion interactions.

The self-similarity solution presented here well describes collective phenomena; therefore, it is reasonable to apply it for optimization of the future experiments at NICA aimed at the observation of these effects. Let us apply this self-similarity solution to predict the kinematic range where the particle production cross sections will be the largest or the most experimentally convenient.

Let us consider ϕ meson production in heavy ion collisions in the intermediate energy range.

Figure 4 shows the predicted invariant cross section of ϕ production and ϕ yield as a function of momentum in the laboratory reference frame in collision of 25 AGeV gold ions (we consider fixed target for convenience). Obviously, the particle yield changes by orders of magnitude with varying angle, which is extremely important from the experimental point of view. It should also be noted that the cross section maximum shifts toward the central region with increasing angle of emitted particle.

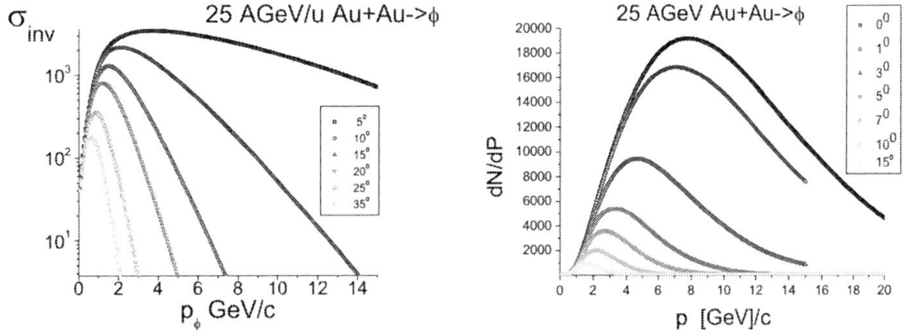

a) b)

Figure 4. (a) Predicted invariant cross section of ϕ production as a function of particle momentum for different angles; (b) ϕ meson yield as a function of momentum for different angles.

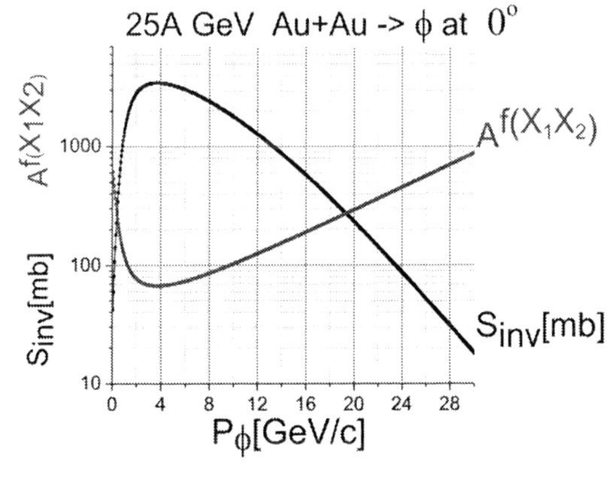

Figure 5. Invariant cross section and A-dependence of ϕ meson production in gold-gold collision at 25AGeV and zero angle.

Figure 5 demonstrates that the maximum cross section corresponds to the minimum A-dependence, which means minimum manifestation of collective phenomena. This is especially important for experiments in which the ratios of particle yields are studied.

Therefore, it is extremely important to take into account the acceptance of the particular experimental setup in data analysis of such experiments.

In conclusion, we state the following. The developed functional self-similarity solution quantitatively describes angular, energy and A-dependences of inclusive production cross sections of all hadrons with transverse momentum up to 1GeV/c. Higher transverse momenta require an additional mechanism to account for momentum dependence in the invariant cross section.

The analysis of inclusive spectra for the data selected in different ways shows that multiplicity in relativistic nuclear collisions has its origin basically in independent nucleon-nucleon interactions.

Interaction of very heavy nuclei (Au, Pb, etc.) may "entangle" the experimental picture of the reaction and complicate theoretical interpretation. Collective effects can be observed already for light and intermediate nuclei, although with lower multiplicity and combinatorial background.

Collective phenomena become extinct with increasing collision energy. Therefore, the energy range from hundreds MeV to tens of GeV is optimal for experimental observation of collective effects in nuclear collisions. Of interest is the study of the whole energy range available at the NICA collider, toward the lowest possible energies.

References

1. J. D. Bjorken, Phys. Rev. **179**, 1547 (1969)
2. A. A. Baldin, Phys. Part. Nucl. **56(3)**, 385 (1993)
3. A. M. Baldin and A. A. Baldin, Phys. Part. Nucl. **29(3)**, 577 (1998)
4. V. S. Stavinskii, in *Proc. of IX International Seminar on Problems of High Energy Physics* **1**, 190 (JINR, Dubna, 1988)
5. A. A. Baldin, E.G. Baldina, in *Selected Papers of the Seminar (2000-2005) Symmerties and Integrable Systems*, Ed. A. N. Sissakian, **1**, 41 (JINR, Dubna, 2006)
6. A. A. Baldin et al., JINR Rapid Comm. **3(92)**, 20 (1992)
7. A.A. Baldin, E.G. Baldina, E.N. Kladnitskaya, and O.V. Rogachevsky, Phys. Part. Nucl. Lett. **121**, 7 (2004)
8. V. V. Ammosov, et al., Phys. At. Nucl. **76**, 1213 (2013)
9. V. Kekelidze, R. Lednicky, V. Matveev, I. Meshkov, A. Sorin, and G. Trubnikov, Phys. Part. Nucl. Lett. **9(4)**, 313 (2012)
10. C. Blume, J. Phys. G **31**, S57 (2005)

EPJ Web of Conferences **138**, 05002 (2017)   DOI: 10.1051/epjconf/201713805002
Baldin ISHEPP XXIII

The hadron production in π^--C interaction at 40 GeV/c and QCD phase transition

B. Otgongerel[1,*], *Ts.* Baatar[1], *R.* Togoo[1], *A.I.* Malakhov[2], *G.* Sharkhuu[1], *M.* Sovd[1], and *T.* Tulgaa[1]

[1] *Institute of Physics and Technology, MAS, Enkhtaivan ave. 54B, Ulaanbaatar-51, Mongolia.*
[2] *Joint Institute for Nuclear Research, Dubna*

Abstract. In this paper, we proposed to study the phase transition process to use the new pair of variables, the temperature T and the cumulative number n_c (T,n_c). We considered the transverse energy spectra of protons and π^--mesons produced in π^--C interactions at 40 GeV/c as a function of cumulative number n_c (or four-dimensional momentum transfer t) and the baryonic chemical potential $\mu_b(\sqrt{t})$. Obtained results indicate the possible appearance of QCD phase transition of nuclear matter.

1 Introduction

Quantum chromodynamics (QCD) is the fundamental theory of the strong interaction and this is an asymptotically free theory [1,2] i. e. interactions between quarks and gluons become weaker as the mutual distance decreases or as exchanged momentum increases. Consequently at large temperatures and /or densities, the interactions which confine quarks and gluons inside hadrons should become sufficiently weak to release them [2].

There is hope that the QCD phase transition processes may be realized in the hadron-nucleus and nucleus-nucleus interactions at high energies and large momentum transfers and in the astrophysical objects of very high density such as neutron stars.

During the last years the collective phenomena such as the cumulative particle production [4], the production of nuclear matter with high densities, the phase transition from the hadronic matter to the quark-gluon plasma state and color-superconductivity is widely discussed in the literature [3-6, 11-17].

According to the different ideas and models, if these phenomena exist in the nature, then they will be observed in the hadron-nucleus and nucleus-nucleus interactions at high energies and large momentum transfers and should be influenced to the dynamics of interaction process and would be reflected in the angular and momentum characteristics of the reaction products.

The present status of the QCD phase diagram is shown schematically in figure 1. This figure was taken from [12].

Running coupling constant of QCD:

$$\alpha_s(Q^2) = \frac{1}{\beta_0 \ln \frac{Q^2}{\Lambda^2}}, \qquad \beta_0 = \frac{1}{4\pi}\left(11 - \frac{2}{3}n_f\right) \tag{1}$$

*e-mail: otgongerel_cs@yahoo.com

where Q^2 is momentum transfer square and Λ is the cut parameter. If parameter $T \geq \Lambda \approx 200$ MeV, QCD predicts the phase transition from hadronic state to QGP.

According to the theoretical investigations, the strongly interacting matter depending on the temperature and the density may occur in the next different phases: the hadronic phase, the mixed phase (h+QGP),the quark-gluon plasma (QGP) and color-superconducting quark matter [3, 5, 11, 12, 14-17].

For temperatures below $T_c \approx 200$ MeV and below nuclear matter density ρ, corresponding to net-baryon densities which are a few times greater than the ground state density of nuclear matter, the strongly interacting matter is in the hadronic phase. There is a line of the phase transition (the mixed phase) which separates the hadronic phase from the QGP. Finally, at large temperatures $T > T_c$ and densities the nuclear matter is in the QGP state. There is also the theoretical prediction about the existence of the color superconducting phase at comparatively small temperature and large baryon densities [14-17].

According to the color superconducting theory of QCD, the critical temperature T_c to color superconducting phase is described by the next formula [14-17]

$$T_c \cong 0.57 \cdot \Delta \cong 50MeV, \tag{2}$$

where Δ is the gap energy and the magnitude of this parameter, the fundamental energy scale that characterizes color superconductivity. The value of the color-superconducting gap parameter Δ is therefore of great importance in order to locate the transition line between the normal nuclear matter and the color-superconducting quark matter phases in the nuclear matter phase diagram.

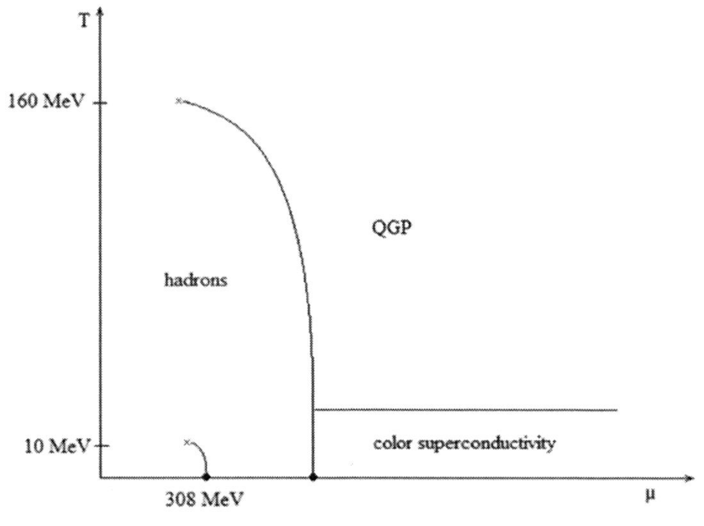

Figure 1. The phase diagram of strongly interacting matter (schematic).

The different model calculations [12, 14-17] yield predictions for the gap parameter Δ and critical temperature T_c. The energy gaps Δ are of order of 10 MeV to 100 MeV, with critical temperature about half as large ($T_c \sim 50MeV$).

The investigation of the multiparticle production process in hadron-nucleus and nucleus-nucleus interactions at high energies and large momentum transfers is very important for understanding the strong interaction mechanism and inner quark-gluon structure of nuclear matter.

In this paper we are considered the next reactions:

$$\pi^- C \rightarrow p + X \tag{3}$$

$$\pi^- C \to \pi^- + X. \tag{4}$$

This paper is a continuation of our previous publication [6,7].

2 Variables Used

2.1 Cumulative number

The cumulative number n_c in the fixed target experiment is determined by the next formula:

$$n_c = \frac{P_a \cdot P_c}{P_a \cdot P_b} \cong \frac{E_c - \beta_a p_{||}^c}{m_p}. \tag{5}$$

Here P_a, P_b and P_c are the four-momenta of the incident particle, target and the considered secondary particles, respectively. E_c is the energy and $p_{||}^c$ is the longitudinal momentum of the considering particle, $\beta_a = \frac{p_a}{E_a}$ is the incident particle velocity, m_p is the proton mass. From this formula we see that this variable is a **relativistic invariant** and n_c is the value of the target mass which is required for production of the secondary particle.

The connection of the variable n_c and momentum transfer t is determined by the next formula [7]:

$$t = -Q^2 = -(P_a - P_c)^2 \cong 2E_a \cdot \frac{m_p}{m_p} \cdot (E - p_{||}) - (m_a^2 + m_c^2) = S_{\pi^- p} \cdot n_c - (m_a^2 + m_c^2), \tag{6}$$

where $S_{\pi^- p} \simeq 2E_a \cdot m_p$ is the total energy squared of $\pi^- p$ interaction.

2.2 The baryonic chemical potential $\mu_b(\sqrt{t})$ as a function of the transferred momentum t

The paper [19] showed that the baryonic chemical potential μ_b(MeV) as a function of energy S_{NN} is parameterized by the following formula:

$$\mu_b(\sqrt{S_{hN}}) = \frac{1.303GeV}{1 + 0.286GeV^{-1}\sqrt{S_{hN}}}. \tag{7}$$

Then using the formula (6) which gives the connection between n_c, $S_{hN} \approx 2E_a m_p$ and transferred momentum t, we have a possibility to do the similar parameterization of the baryonic chemical potential $\mu_b(t)$ as a function of the transferred momentum t by the equation:

$$\mu_b(\sqrt{t}) = \frac{1.303GeV}{1 + 0.286GeV^{-1}\sqrt{t}} = \frac{1.303GeV}{1 + 0.286GeV^{-1}\sqrt{2E_a m_p n_c - (m_a^2 + m_c^2)}}, \tag{8}$$

where E_a and m_a are the incident particle energy and mass, n_c and m_c are the cumulative number and mass of the secondary particles.

Note that using formula (8) one can calculate values of $\mu_b(\sqrt{t})$ which belong to the secondary particles, in other words, this formula gives us a possibility to study the behaviour of the parameter T as a function of $\mu_b(\sqrt{t})$ using experimental data obtained in one experiment.

2.3 The effective temperature T

The transverse energy spectra of the secondary particles in the different n_c intervals are approximated by exponential function of the next form:

$$\frac{1}{2 \cdot E_t} \frac{\Delta N}{\Delta E_t} \sim e^{-bE_t}, \qquad E_t = \sqrt{p_t^2 + m^2}. \tag{9}$$

The effective temperature T is determined as the inverse slope parameter b:

$$T = \frac{1}{b}. \tag{10}$$

DOI: 10.1051/epjconf/201713805002

3 Experimental method

The experimental material was obtained with the help of the Dubna 2-meter propane (C_3H_8) bubble chamber exposed by π^-- mesons with momentum 40 GeV/c from Serpukhov accelerator. According to the advantage of the bubble chamber experiment all distributions in this paper are obtained under the condition of 4π geometry of secondary particles.

The average error of the momentum measurements is $\sim$ 12% and the average error of the angular measurements is $\sim 0.6^o$.

All secondary negative particles are taken as π^--mesons. The average boundary momentum from which π^--mesons were well identified in the propane bubble chamber is $\sim$ 40 MeV/c. In connection with the identification problem between energetic protons and π^+-mesons, protons with momentum more than $\sim$ 1 GeV/c are included to the π^+-mesons. The other experimental details are described in [8, 9].

4 Statistics

8791 $\pi^- C$ interactions are used in this analysis.

1. The total number of protons is 12441
2. The total number of π^- mesons is 30145.

5 Experimental results

The temperature (T) dependencies for secondary π^--mesons and protons from $\pi^- C$–interactions at 40 GeV/c as a function of the variable n_c (or t) are shown in figures 2, 3. From figure 2 we see that with increasing n_c the parameter T is increasing up to $n_c \cong 0.07$ and then in the $0.07 \leq n_c < 0.5$ interval the parameter T remains practically constant on the level $T \cong (0.233 \pm 0.004)$GeV and then again increases [13].

The corresponding dependence obtained for protons shows that the parameter T remains constant on the level $T_c \cong 50$ MeV in the $0.5 < n_c < 1.2$ interval and then increases.

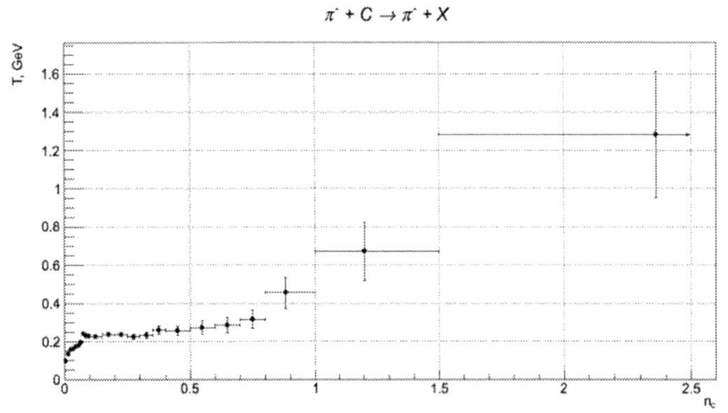

Figure 2. The dependence of the temperature T of π^-–mesons on cumulative number n_c from $\pi^- C$ interactions at 40 GeV/c

So we observe the essential change of the behavior of the parameter T of protons and π^-–mesons in $\pi^- C$ interactions at 40 GeV/c on the variable n_c in two different regions for protons and in three different regions for π^-–mesons. Such behavior may indicate the particle production different mechanism in these ranges of the variable n_c (or t) and temperature T. If so, the I region in which the

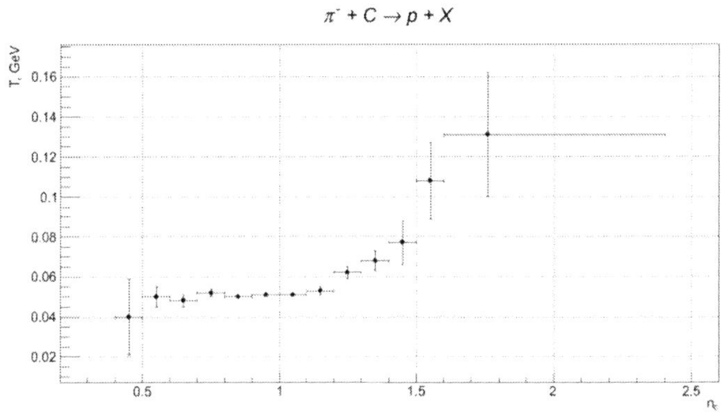

Figure 3. The dependence of the temperature T of protons on cumulative number n_c from $\pi^- C$ interactions at 40 GeV/c

temperature T increases for π^-–mesons may be connected with the thermalization of the strongly interacting objects. In this region the strongly interacting matter is in the hadronic phase.

In the II regions where the temperature T remains practically constant, the strongly interacting matter is in the thermodynamical equilibrium state (or the mixed phase).

In the III regions where the parameter T increases again, the strongly interacting matter is in the QGP phase.

We would like to note that this value of the critical temperature T_c, obtained for protons is in agreement with the theoretical prediction $T_c \cong 50$ MeV for the color-superconductivity of QCD [12, 14].

As mentioned above, the theoretical prediction of the critical temperature T_c for the color-superconductivity of QCD was obtained in the region of the phase diagram with high density and low temperature. Our experimental result also corresponds to this region.

From formula (8) we see that the baryonic chemical potential $\mu_b(\sqrt{t})$ is inversely proportional to the variable n_c and its numerical values are determined by n_c (or t).

So, the dependence of the parameter T on the variable $\mu_b(\sqrt{t})$ should give the inverse dependence (see figure 4, 5) in comparison with T dependencies as a function of n_c (see figure 2, 3).

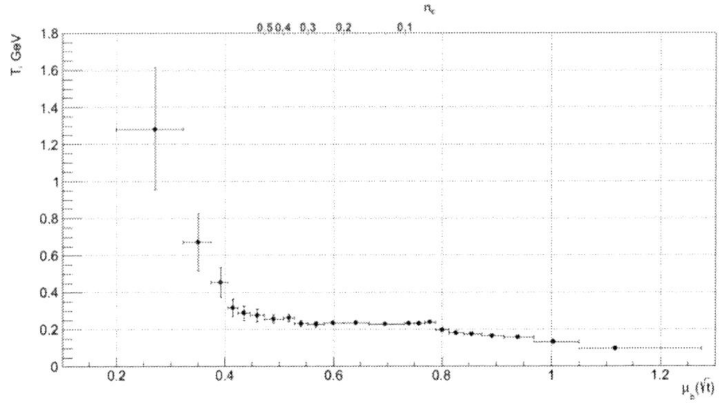

Figure 4. T dependence of π^--mesons as a function of the baryonic chemical potential $\mu_b(\sqrt{t})$ from $\pi^- C$ interactions at 40 GeV/c

Figure 6,7 show dependencies of the average values of the transverse momentum squared $\langle p_T^2 \rangle$ of π^--mesons and protons on the variable n_c. With increasing n_c the $\langle p_T^2 \rangle$ increases and then in the $0.07 < n_c < 0.5$ interval for π^--mesons and in the $0.5 < n_c < 1.2$ interval for protons remain practically constant and then again increases.

DOI: 10.1051/epjconf/201713805002

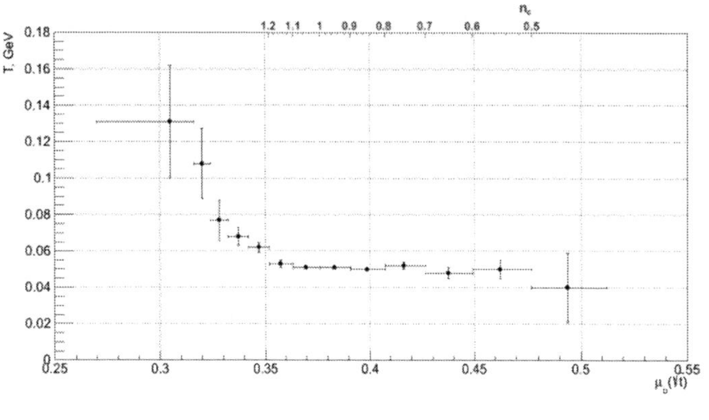

Figure 5. T dependence of protons as a function of the baryonic chemical potential $\mu_b(\sqrt{t})$ from π^-C interactions at 40 GeV/c

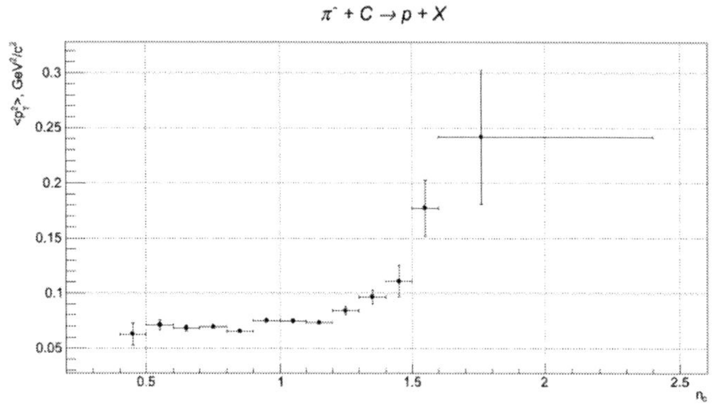

Figure 6. The dependence of the average values of the transverse momentum squared $\langle p_T^2 \rangle$ of protons on the variable n_c.

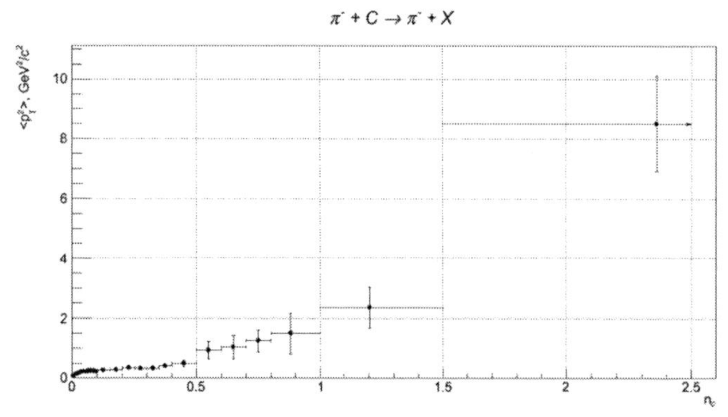

Figure 7. The dependence of the average values of the transverse momentum squared $\langle p_T^2 \rangle$ of π^--mesons on the variable n_c.

We would like to note that dependencies of the parameters T and $\langle p_T^2 \rangle$ as a function of cumulative number n_c of the secondary particles from π^-C interaction have very similar behaviour, in other words, these two dependencies have the n_c-regions with constant temperature. This similarity of these two dependencies may be connected with the saturation of the transverse momentum at equilibrium state.

In papers [22, 23], the dependence of $\langle p_T^2 \rangle$ on the cumulative number X of the secondary protons from p+A interactions at 10.14 GeV/c is considered. The result of this study is presented in figure 8. We note that the obtained dependence is in agreement with our result.

In papers [20, 21], the behavior of the phase transition process is determined in the following form:

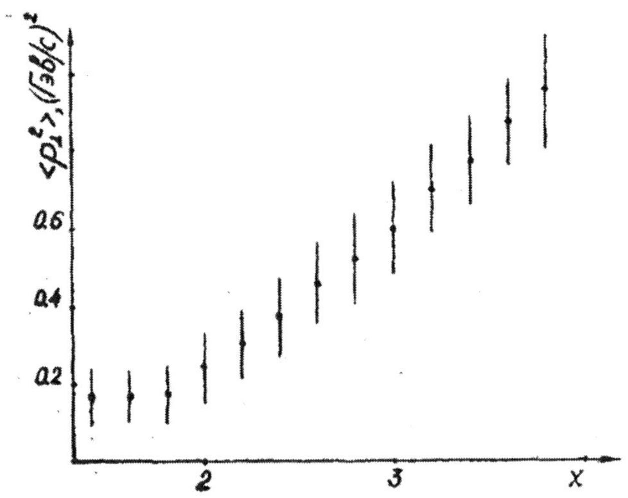

Figure 8. The dependence of $\langle p_T^2 \rangle$ on the cumulative number X [22, 23].

If a phase transition occurs (i.e. a rapid change of the number of degrees of freedom) one expects a monotonously rising curve interrupted by a plateau. This plateau is caused by the saturation of $\langle p_T \rangle$ in mixed phase. After the phase transition from e.g. color singlet states to colored constituents has been completed [20, 21] the mean transverse momentum rises again. Our experimental results are consistent with this statement.

6 Conclusions

- The analysis carried out in this paper gives us the possibility to study the phase transition process of nuclear matter.
- The T dependence of π^--mesons from $\pi^- C$ interactions at 40 GeV/c as a function of cumulative number n_c (or t) and the baryonic chemical potential $\mu_b(\sqrt{t})$ are in agreement with the QCD prediction on the phase transition from hadronic matter to QGP state.
- The baryonic chemical potential $\mu_b(\sqrt{t})$ is expressed by the variable n_c. So, these two variables $\mu_b(\sqrt{t})$ and n_c may be used with equal rights in order to study the phase transition process in hA and AA interactions at high energies.
- The experimental results obtained for the secondary protons produced in the target fragmentation region from $\pi^- C$ interactions at 40 GeV/c don't contradict to the prediction of $T_c \cong 50$ MeV for the color superconducting phase transition of QCD at large density and small temperature but the more detailed analysis is required for this case.

Acknowledgement

First of all we would like to express our thanks to the Dubna two meter propane bubble chamber collaboration members for their excellent experiment and to the Serpukhov accelerator staff for providing π^--meson beam at 40 GeV/c.

References

[1] D.J. Gross and F. Wilczek, Phys. Rev.Lett. **30**, 1343 (1973)
[2] J.C. Collins and M.J. Perry, Phys. Rev. Lett. **34**, 1353 (1975)
[3] A.M. Polyakov, Phys. Lett B **59**, 82 (1975)
[4] A.M. Baldin, Particles and Nuclei **8**, 429 (1977)
[5] R.V. Cavani, and H. Sats, Phys. Lett B **145**, 248 (1984)
[6] Ts. Baatar et. al., Preprint JINR p1-89424, Dubna (1989)
[7] Ts. Baatar et. al., Journal of Nuclear Physics **3(9)**, 52 (1990)
[8] O. Balea et al., Phys. Lett B **39**, 571 (1972)
[9] A.J. Abdurahmanov et al., Preprint JINR, p1-6937, Dubna (1973)
[10] N. Angelov et al., Preprint JINR, p1-9209, Dubna (1975)
[11] R. Hagedorn, Nouvo Cim. Suppl. **3**, 147 (1965)
[12] Dirk H. Rischke, arXiv:nucl-th/0305030v2 (2003)
[13] Ts. Baatar et. al., Preprint JINR E1-2012-13, Dubna (2012)
[14] K. Rajagopal, F. Wilczek, hep-ph/0011333 (2000)
[15] M.G. Alford, K. Rajagopal and F. Wilczek, Nucl. Phys. B **537**, 443 (1999)
[16] J.Barden, L.Cooper and Schriefer, Phys.Rev. **106**, 162 (1957)
[17] D.Sticlet, *Phase Transitions in the Early Universe-Electroweak and QCD Phase Transitions* (Master Program of Theoretical Physics)
[18] Ts. Baatar et. al., In Proceeding of XXI International Baldin Seminar on High Energy Physics, JINR, Dubna, Russia, September 10-15, 2012
[19] A. Andronic, P. Braun-Munzinger, J. Stachel, Nucl. Phys. A **772**, 167 (2006). [arxiv:nucl-th/0511071]
[20] L. Van Hove Phys. Lett. B **118**, 138 (1982)
[21] S.A. Bass, M. Gyulassy, H. Stocker and W. Greiner, arXiv: hep-ph/9810281 v2 (1998)
[22] A.M. Baldin et.al., Preprint JINR P1-83-432, Dubna (1983)
[23] S.V. Boyarinov et.al., Preprint ITEP 86-130, Moscow (1986)

π^0 production in dC- and dCu-interactions at a momentum of 4.5 GeV/c per nucleon

Khachik Abraamyan[1,2], *Mikhail* Kozhin[1,⋆], *Marlen* Khachaturyan[1], *Alfred* Khudaverdyan[2], *Georgy* Melkumov[1], and *Sergey* Reznikov[1]

[1] *VBLHEP JINR, 141980 Dubna, Moscow region, Russia*
[2] *Yerevan State University, 0025, Yerevan, Armenia*

Abstract. We report on measurements of neutral pion production in the inclusive reactions $d + C \rightarrow \pi^0 + x$ and $d + Cu \rightarrow \pi^0 + x$ at an incident momentum of 4.5 GeV/c per nucleon. The experiments were performed on the LHE 90-channel lead glass γ-spectrometer. The cross sections were measured over the kinematical region specified by the inequalities $\theta_\pi \leq 16^\circ$ and $E_\pi \geq 2$ GeV (in the laboratory frame). The cumulative number and transverse momentum dependencies of the exponent n in the invariant cross section parameterization $Ed^3\sigma/d^3p \sim A_t^n$ are investigated by comparing of the observed cross sections for π^0 production on carbon and copper targets in the intervals $0.6 \leq X \leq 1.8$ and $0.04 \leq P_T^2 \leq 0.40$ (GeV/c)^2. The double differential cross section for the reaction $d + C \rightarrow \pi^0 + x$ is measured using statistics of about $4.5 \cdot 10^4 \pi^0$ mesons. On the basis of these data we verified the so-called cluster mechanism of π^0 production. We have compared our data for the reaction $d + C \rightarrow \pi^0 + x$, extrapolated to $\theta_\pi = 0^\circ$, with the data from another experiments on π^- production: $d + C \rightarrow \pi^-(0^\circ) + x$ ($P = 1.75$ and 2.88 GeV/c per nucleon) [1]; $p + d \rightarrow \pi^-(180^\circ) + x$ and $p + d \rightarrow \pi^+(180^\circ) + x$ ($P = 8.9$ GeV/c per nucleon) [2]; $d + p \rightarrow \pi^-(0^\circ) + x$ ($P = 8.9$ GeV/c per nucleon) [3]. The invariant cross sections were approximated by an exponential function $Ed^3\sigma/d^3p \sim exp(-X/X_0)$. The slope parameter X_0 at different kinetic energies of the projectiles in the range of $1.05 \div 8.0$ GeV per nucleon is determined.

1 Introduction

We report here on measurements of inclusive π^0 production by relativistic deuterons on carbon and copper targets. The fragmentation of nuclei in high energy collisions into elementary particles with momenta far exceeding the average momentum per nucleon in the nucleus is one of the most interesting phenomena in high energy physics. This phenomenon, so-called cumulative production of particles was first observed by A.M. Baldin [4]. As in the previous investigations of our group [5–7], we aim at clarifying the mechanism of neutral pion emission near and beyond the kinematic limit for free nucleon-nucleon and nucleon-nucleus collisions. We have also investigated how π^0 production depends on the number of nucleons in the target and compared our results with quark-parton models of the nucleus [8].

⋆e-mail: jenaro@mail.ru

Our experiments allow to study the double differential cross section of the reaction $d + C \to \pi^0 + x$ and thus to study the properties of objects (hadronic clusters of fireballs, quarks, six-quark configurations) responsible for pion production in the above-mentioned kinematical region and to estimate the transverse component of valence quark momenta in protons and nuclei. Moreover, on the basis of this cross section we have been able to extrapolate our data to $\theta_\pi = 0°$, which allows us to compare them correctly with the data from another experiments on π^- production.

2 Experiment

The experiment was performed at the Synchrophasotron of the High Energy Laboratory using a 90-channel lead glass γ-spectrometer. Our experimental setup, described in detail in the [5, 9], allows to measure both the energies and emission angles of photons from π^0 decays. To monitor the number of beam particles hitting the target, a monitor telescope composed of three scintillation counters was used. For charge particle identifications 40 scintillator counter hodoscopes were used in front of the γ-spectrometer resulting in a 99% detection probability for charged particles.

Carbon and copper targets were exposed to a beam of deuterons with a momentum of 4.5 GeV/c, a momentum spread of $\Delta p/p = \pm 2\%$ and an intensity up to 10^5 particle per pulse. The distance along the beam between the target center and the upstream face of the spectrometer was 340 cm for the carbon target and 520 cm for the copper target. The corresponding acceptances in the laboratory emission angle for π^0 mesons were $\theta_\pi \leq 16°$ and 10°, respectively. About $2 \cdot 10^5$ triggers for the carbon target and $1.6 \cdot 10^5$ triggers for the copper target have been recorded during this experiment.

3 Event selection

The π^0 mesons are detected through their two-photon decay mode as a narrow peak in the invariant mass distribution. Photons are recognized as isolated and confined clusters (an area of adjacent modules in the γ-spectrometer with a signal exceeding the detection threshold). The photon energy is calculated from the energy of the cluster by applying a position-dependent leakage correction. Assuming that the photon originates at the target, its direction is determined from the geometrical position of the constituent crystals, weighted by the corresponding energy deposits.

Inclusive spectra of π^0 have been extracted from the raw data under the following selection criteria:

(1) $N_\gamma \geq 2$;

(2) $E_\gamma \geq 500$ MeV;

(3) $E_{\gamma\perp} \geq 120$ MeV;

where N_γ is the number of γ-quanta (clusters) in the event, E_γ is the energy of γ-quanta, $E_{\gamma\perp}$ is the transverse momentum of γ-quanta. $1.4 \cdot 10^5$ events have been recorded on the DST, satisfying this criteria.

4 Inclusive π^0 production spectrum

In order to identify π^0 mesons, all photon pair combinations are used to calculate the invariant mass in the each event. The combinatorial invariant mass spectra of the $\gamma\gamma$ combinations, selected according to the criteria (1) $\div$ (3) for the dC and dCu reactions, are shown in figure 1 (light color histograms). The dark color histograms in figure 1 represent the invariant mass distributions for $\gamma\gamma$ combinations selected accidentally from different events. These combinatorial distributions, so-called "events mixing" were used to determine the background. The dashed line on the dC histogram indicates the invariant mass spectrum of $\gamma\gamma$ pairs from a Monte Carlo simulation (for details, see [6]).

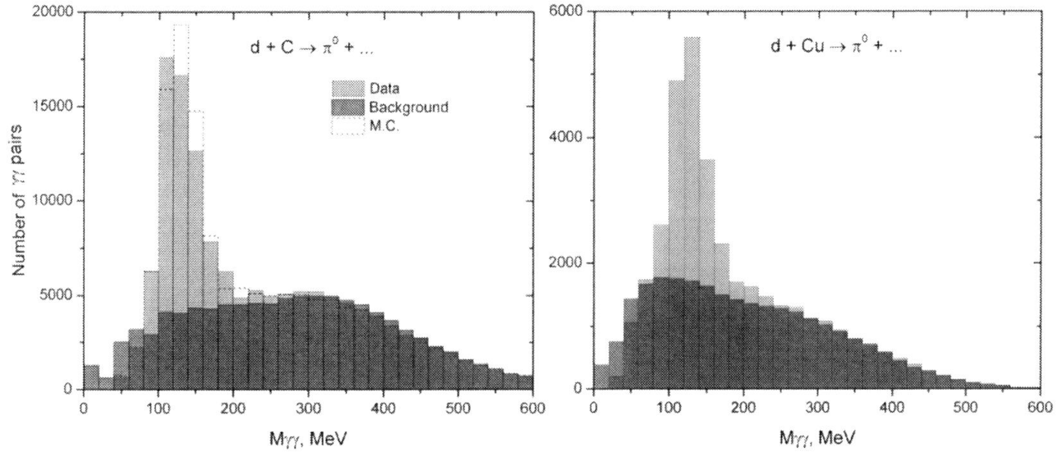

Figure 1. The combinatorial invariant mass spectra for the *dC* and *dCu* reactions.

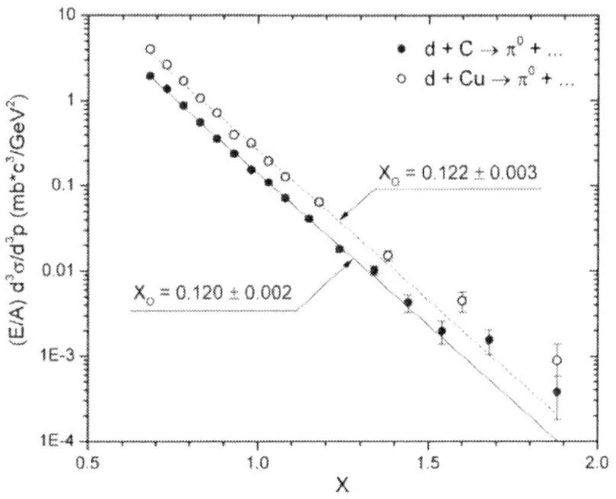

Figure 2. The invariant inclusive cross section of π^0 mesons for deuterons at 4.5 GeV/c per nucleon on C and Cu targets averaged in the range $\theta_\pi \leq 16^\circ$ and $E_\pi \geq 2$ GeV versus the cumulative number X. It can be parameterized by an exponential form: $Ed^3\sigma/d^3p \sim \exp(-X/X_0)$. The experimental values of slope parameters X_0, that characterize the quark-parton structure function of the fragmenting nucleus [10], are 0.120 ± 0.002 for C target and 0.122 ± 0.003 for Cu target and approximately equal within the errors.

In figure 2 we present invariant inclusive cross section of π^0 mesons for deuterons at 4.5 GeV/c per nucleon on C and Cu targets averaged in the range $\theta_\pi \leq 16^\circ$ and $E_\pi \geq 2$ GeV versus the cumulative number X. The X is determined by a minimum part of 4-momentum per nucleon in a fragmenting nucleus, necessary to pion production was kinematically allowed with the assumption that fragmentation occurs on one nucleon. It is equal to [3]:

$$X = \frac{P_r P_\pi - m_\pi^2/2}{(P_r P_f) - (P_f P_\pi) - m_N^2},$$

where P_r and P_f are four-momenta per nucleon of the nucleus, where fragmentation occurs and of the fragmented nucleus, respectively; m_N and m_π are the nucleon and pion mass; P_π is four-momenta of a pion. At the beam fragmentation (forward hemisphere) the formula becomes:

$$X = \frac{m_N E_\pi - m_\pi^2/2}{E_N m_N - E_N E_\pi - m_N^2 + P_N P_\pi \cos\theta_\pi},$$

 DOI: 10.1051/epjconf/201713805003

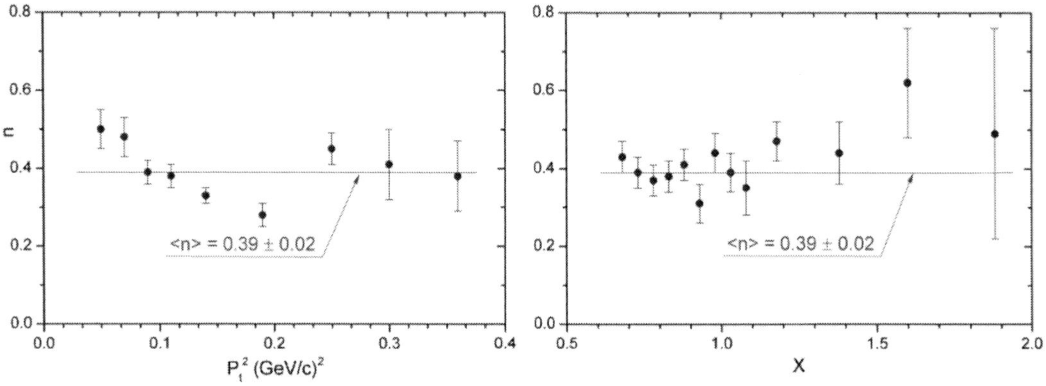

Figure 3. The target dependence parameter n in the parameterization $Ed^3\sigma/d^3p \sim A_t^n$ versus the transverse momenta P_T^2 and the cumulative number X for the reactions $d + A_t(C, Cu) \rightarrow \pi^0 + x$.

where P_N, P_π, E_N, E_π are the nucleon and pion momentum and energy, respectively; θ_π is the angle of pion emission (lab.syst.).

We present the target dependence in the form $Ed^3\sigma/d^3p \sim A_t^n$, where $A_t = 12$ and 63.5 is the nucleon number of the target. The dependence of the parameter n versus the transverse momenta P_T^2 and the cumulative number X is shown in figure 3.

The parton models [8] predict that n is independent of X and P_T at large X and small P_T. According to these models, an incident hadron fragments when one of its constituent quarks collides in the target. The spectator quarks that escape collisions and thus retain their original fraction X of the projectile momentum can fragment or recombine with a slow quark ($X \approx 0$) and form the large X and low P_T fragments. From the model we obtain a quantitative estimate $n \approx 0.38$, that is in good agreement with experimental data.

5 The double differential cross section

The double differential cross section of the reaction $d + C \rightarrow \pi^0 + x$ is first measured at 4.5 GeV/c per nucleon using statistics of about $4.5 \cdot 10^4$ π^0 mesons (figure 4, left).

The invariant cross section of this reaction versus the energy E_π and the emission angle θ_π of the π^0 in the lab. syst. is analyzed at $4° \leq \theta_\pi \leq 16°$ and $2.0 \leq E_\pi \leq 4.7$ GeV.

We have obtained the slope parameters in the energy spectra $Ed^3\sigma/d^3p$ (E_π, θ_π-fixed) $\sim exp(-E/T(\theta_\pi))$, for each interval of θ_π and expressed $T(\theta_\pi)$ as $T(\theta_\pi) = T_0(1-\beta^2)^{1/2}/(1-\beta \cdot \cos\theta_\pi)$.

Interpreting the parameter $T_0 \sim m_\pi$ as the universal hadron temperature [11]: $Ed^3\sigma/d^3p \sim exp(-E^*/T_0)$, we have obtained the values of β, the velocity of an intermediate cluster of nucleons which decays into a final pion (figure 4, right).

The obtained regularity of the $\beta(\theta_\pi)$ in the interval $\theta_\pi \leq 8°$ qualitatively agrees with the predictions of the cluster models ($\beta(\theta_\pi) \approx const$). However, for explanation of such small values of β one has to suppose second or more collisions in the target, which contradicts with observed target mass dependence.

Thus the obtained regularity of the $\beta(\theta_\pi)$ indicates the absence of intermediate clusters (clusters of 2, 3,... nucleons formed in collisions of incident nucleon with nucleons in the target nuclei) responsible for pion production in above-mentioned kinematical region.

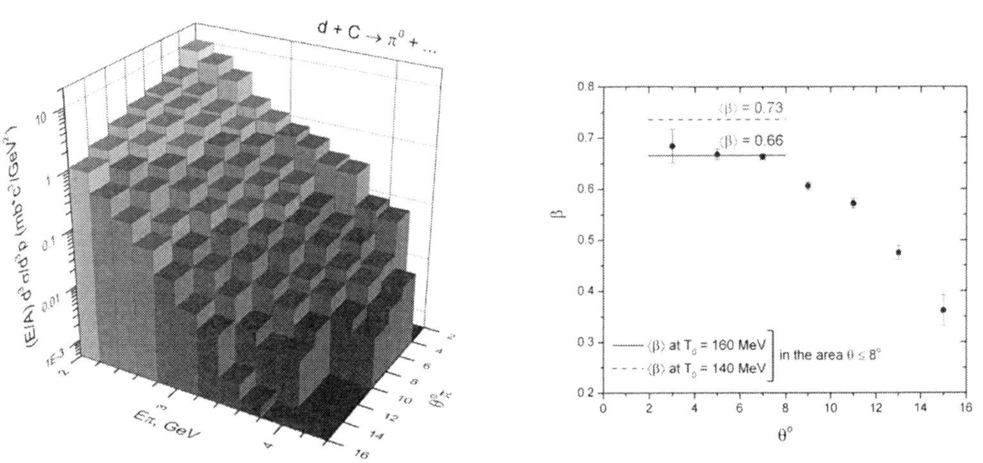

Figure 4. The invariant cross section $\sigma_{inv} = (E/2)d^3\sigma/d^3p$ of the reaction $d + C \rightarrow \pi^0 + x$ vs. the energy E and the emission angle θ of π^0 mesons (left), and the velocity β of a possible cluster vs. the emission angle θ (right).

6 Comparisons with the data from another experiments

We have compared the invariant cross section obtained from our data for the reaction $d + C \rightarrow \pi^0 + x$ with the data from another experiments on π^- production. For a more accurate comparison we made an extrapolation of our data to $\theta_\pi = 0°$ using results from the double differential cross section. In figure 5 (left) we present comparison with experiments, performed at the Bevalac of the Lawrence Berkeley Laboratory: $d + C \rightarrow \pi^-(0°) + x$ ($P = 1.75$ and 2.88 GeV/c per nucleon) [1]. In figure 5 (right) we present comparison with experiments, performed at the LHE Synchrophasotron, Dubna: $p + d \rightarrow \pi^-(180°) + x$ and $p + d \rightarrow \pi^+(180°) + x$ ($P = 8.9$ GeV/c, the DISK-2 setup) [2]; $d + p \rightarrow \pi^-(0°) + x$ ($P = 8.9$ GeV/c per nucleon, the SPHERE setup) [3]. In order to compare with proton data correctly, we have multiplied our data by $A^{-2/3}$ ($A = 12$). Parameter X_0 in the parameterization $(E/A_p)d^3\sigma/d^3p = A\exp(-X/X_0)$ have been extracted from all these data and shown in figure 6. The X_0 decreases at lower momenta of beam, and almost reaches a plateau about a momentum of 4.5 GeV/c per nucleon.

7 Conclusion

The invariant cross sections for the reactions $d + A_t(^{12}C, {}^{nat}Cu) \rightarrow \pi^0 + x$ were measured at the momentum 4.5 GeV/c per nucleon vs. cumulative number X within the range of $0.6 \leq X \leq 2.0$. The dependence of the cross sections on the mass of the target was presented in the form $Ed^3\sigma/d^3p \sim (A_t)^n$. Exponent $n = 0.39 \pm 0.02$ is independent on X within the kinematic region $X > 0.6$.

The predictions of the model of quark-parton recombination [8] are in good agreement with the observed A-dependence. The obtained regularity of the energy spectra of pions in different intervals of emission angles indicates the absence of intermediate clusters of nucleons [11], responsible for pion production in the kinematic region $\theta_\pi \leq 16°$ and $E_\pi \geq 2.0$ GeV.

We carried out comparisons of the invariant cross section of the reaction $d + C \rightarrow \pi^0 + x$, extrapolated to $\theta = 0°$, with the data from another experiments on π^- production. Parameter X_0 in the parameterization $(E/A_p)d^3\sigma/d^3p = A\exp(-X/X_0)$ decreases at lower momenta of beam, and almost reaches a plateau about a momentum of 4.5 GeV/c per nucleon.

 DOI: 10.1051/epjconf/201713805003

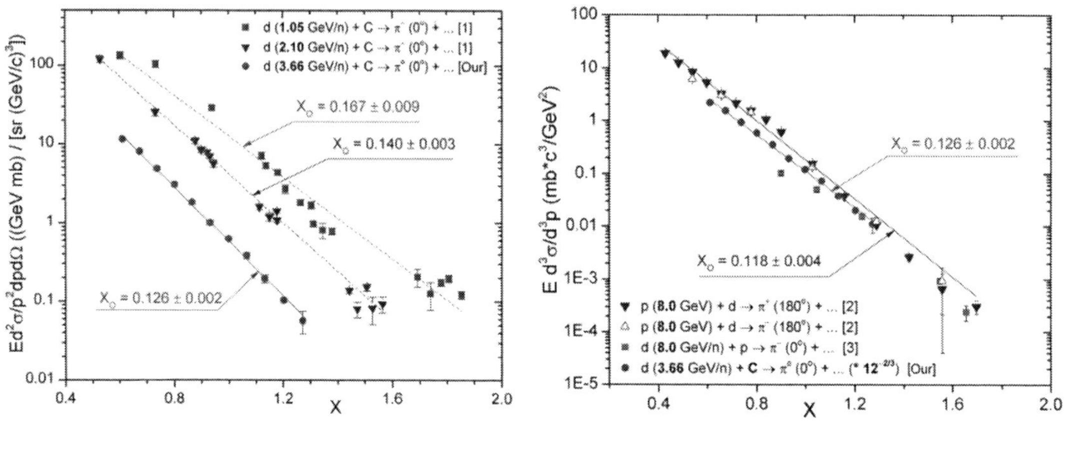

Figure 5. Comparisons of our data with the data from the experiments, performed at the Bevalac of the Lawrence Berkeley Laboratory (left), and at the LHE Synchrophasotron, Dubna (right).

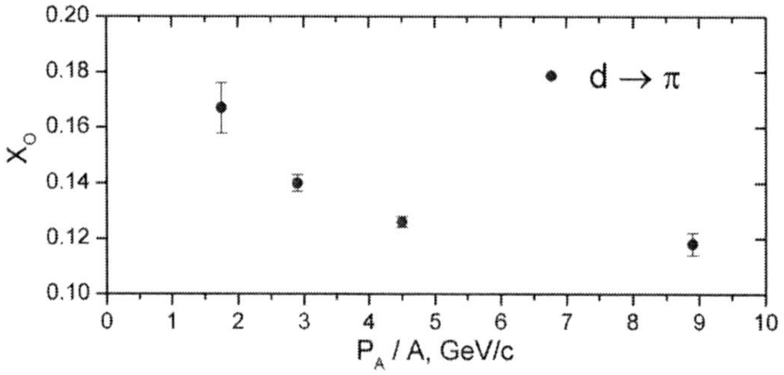

Figure 6. The slope parameter X_0 vs. momenta of deuterons.

References

[1] E. Moeller et al., Phys. Rev. C **28**, 1246 (1983)
[2] A.M. Baldin et al., JINR 1-82-28 (1982)
[3] Yu.S. Anisimov et al., Yad. Fiz. **60**, 1070 (1997)
[4] A.M. Baldin et al., Part. Nucl. **8**, 429 (1977)
[5] Kh.U. Abraamyan et al., Phys. Lett. B **323**, 1 (1994)
[6] Kh.U. Abraamyan et al., Yad. Fiz. **59**, 271 (1996)
[7] Kh.U. Abraamyan et al., Yad. Fiz. **60**, 2014 (1997)
[8] D. Berlad and A. Dar, Phys. Lett. B **102**, 385 (1981)
[9] R.G. Astvatsaturov et al., Nucl. Instr. and Meth. **163**, 343 (1979)
[10] A.M. Baldin et al., JINR D1 2-81-728 (1981)
[11] M.I Gorenshteyn, G.M. Zinovyev, V.P. Shelest, Yad. Fiz. **26**, 788 (1977)

NLO electroweak radiative corrections for four-fermionic process at Belle II

Aleksandrs Aleksejevs[1,⋆], *Svetlana* Barkanova[2,⋆⋆], and *Vladimir* Zykunov[3,⋆⋆⋆]

[1]*Memorial University of Newfoundland, Corner Brook, Canada*
[2]*Acadia University, Wolfville, Canada*
[3]*Belarusian State University of Transport, Gomel, Belarus; Joint Institute for Nuclear Research, Dubna, Russia*

Abstract. We discuss the next to the leading order (NLO) electroweak radiative corrections to the $e^-e^+ \to f^-f^+(\gamma)$ cross section asymmetry, for polarized and unpolarized beam scenario. The left-right and forward-backward amplitudes, with and without radiative corrections, are evaluated and compared for various kinematics. The hard bremsstrahlung is included for arbitrary energy cuts. The radiative corrections are shown to be significant and having a non-trivial dependency on the kinematic conditions. The calculations are relevant for the ultra-precise low-energy experiment Belle II planned at SuperKEKB.

1 Introduction

The precision electroweak physics will be accessed at low energy by the upcoming experiments such as Belle-II at SuperKEKB [1], MOLLER at JLab [2] and P2 at MESA [3] and play an important complimentary role to the direct new-physics search at the LHC. Belle II aims to determine the weak mixing angle at $\sqrt{s} = m(\Upsilon_{4S}) = 10.577$ GeV, with the same precision as the LEP/SLC measurements made at the Z-boson pole and for $e^-e^+ \to b\bar{b}$ but free from fragmentation uncertainties [4]. However, before the reliable information can be extracted from the experimental data, it is necessary to consider the higher-order effects, i.e. electroweak radiative corrections (EWC). The inclusion EWC is an indispensable part of any modern experiment, but will be of the most paramount importance for the upcoming ultra-precise measurements, and, in some cases, must include not only a full treatment of one-loop radiative corrections (NLO) but also the leading two-loop corrections (NNLO).

The significant theoretical effort has been dedicated to NLO EWC to electron-positron annihilation for more than three decades now, starting from [5] which discussed EWC for an arbitrary polarization. Later, the collaborations such as BHM and WOH [6, 7], LEPTOP [8], TOPAZ0 [9] and ZFITTER [10, 11], provided a new level of precision for EWC at the Z-pole required by the LEP and SLC colliders. More recent EWC of the post-LEP/SLC era are offered by KK [12] and SANC [13] codes. Although quick and accessible, these generic "ready-to-use" codes may not be optimal for the new

⋆e-mail: aaleksejevs@grenfell.mun.ca
⋆⋆e-mail: svetlana.barkanova@acadiau.ca
⋆⋆⋆e-mail: vladimir.zykunov@cern.ch

generation of precision experiments requiring a more custom approach. The previous work by our theory group [14] dedicated to the parity-violating (PV) tests of the Standard Model showed that the exact analytical one-loop calculations using the computer algebra approach did not only increase the theoretical precision dramatically, but also gave us an opportunity to verify previous calculations done in other formalisms [15]. A more detailed analysis of the low- and high-energy asymptotic behavior of electroweak radiative corrections in polarized $e^-e^+ \to \mu^-\mu^+$ process can be found in [16].

The main goal of this paper is to calculate a full gauge-invariant set of NLO EWC, both with computer algebra requiring no simplifications, using our semi-automatic approach (SAA) based on FeynArts, FormCalc and LoopTools [17], and analytically (on paper), in a compact form, using the asymptotic methods [6, 15], and to compare the results making sure that our calculations are error-free. A complete set of NNLO EWC is not yet available and not included in the current analysis, but the work is already in progress by several groups (for example, [18] and [19]) and planned to be completed in time for the upcoming experiments such as Belle II, MOLLER and P2.

2 NLO electroweak corrections

Let us start by defining the cross section for scattering of polarized electrons on unpolarized positrons,

$$e^-(p_1) + e^+(p_2) \to f^-(p_3) + f^+(p_4). \tag{1}$$

In the Born approximation (leading order) illustrated by the first diagram in figure 1, we have: $\sigma^0 \approx \frac{\pi^3}{2s}|M_0|^2$. Here, σ is the differential cross section $\sigma \equiv d\sigma/d(\cos\theta)$, θ is the scattering angle of the

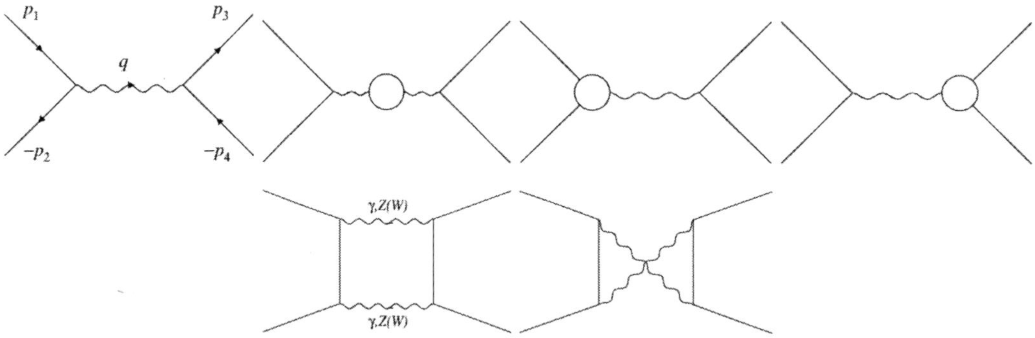

Figure 1. Feynman diagrams for the $e^-e^+ \to \mu^-\mu^+$ process in radiation-free kinematics

detected muon with 4-momentum p_3 in the center-of-mass system of the initial electron and positron, and M_0 is the Born ($O(\alpha)$) amplitude (matrix element). The 4-momenta of initial (p_1 and p_2) and final (p_3 and p_4) fermions generate a standard set of Mandelstam variables:

$$s = (p_1 + p_2)^2,\ t = (p_1 - p_3)^2,\ u = (p_1 - p_4)^2. \tag{2}$$

At the Next-to-Leading-Order ($O(\alpha^2)$), the differential cross section is given by the second term of the following expansion:

$$\sigma = \frac{\pi^3}{2s}|M_0 + M_1|^2 \approx \frac{\pi^3}{2s}(M_0 M_0^+ + 2\mathrm{Re}[M_1 M_0^+]), \tag{3}$$

where the one-loop amplitude M_1 is the sum of boson self-energy (BSE), vertex (Ver) and box diagrams (see figure 1):

$$M_1 = M_{\rm BSE} + M_{\rm Ver} + M_{\rm Box}. \tag{4}$$

There are no contributions from the electron self-energies in the on-shell renormalization scheme [20, 21]. The explicit form of the Born and one-loop amplitudes can be found in [22].

The bremsstrahlung diagrams corresponding to initial state radiation (ISR) and the final state radiation (FSR) are illustrated in figure 2. The differential cross section for the process

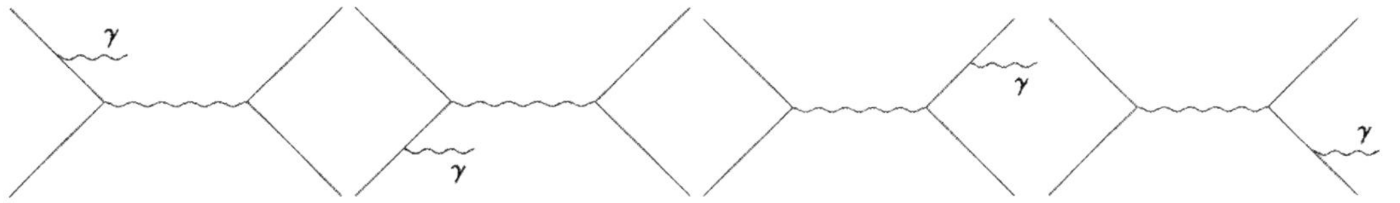

Figure 2. Diagrams with photon emission

$$e^-(p_1) + e^+(p_2) \to f^-(p_3) + f^+(p_4) + \gamma(p), \tag{5}$$

has a form

$$d\sigma_R = \frac{\alpha^3}{\pi^2 s} \sum |R|^2 d\Gamma_3. \tag{6}$$

The amplitude R, the phase space of emitted photon $d\Gamma_3$, and explanation of infrared divergence are given in [22].

3 Analysis

For input parameters, we used α, m_W, m_Z, m_H and fermionic masses from [23]. The light quark masses reproducing $\Delta\alpha^{(5)}_{had}(m_Z^2)$=0.02757 [24], are regulated by the hadronic vacuum polarization, which does not introduce a significant uncertainty to our results. Please see [14] for details.

Let us start with a comparison between the asymptotic and full calculations, using two relative corrections $\delta_+^{\rm NLO}$ and $\delta_-^{\rm NLO}$:

$$\delta_\pm^{\rm NLO} = \frac{\sigma_L^{\rm NLO} \pm \sigma_R^{\rm NLO}}{\sigma_L^0 \pm \sigma_R^0}. \tag{7}$$

They are additive, and allow to estimate the physical effect. The asymmetry can now be defined as:

$$\delta_A^{\rm NLO} = \frac{\delta_-^{\rm NLO} - \delta_+^{\rm NLO}}{1 + \delta_+^{\rm NLO}}. \tag{8}$$

The comparison between two methods is illustrated in figure 3. For now, we apply the same cut on the soft photon emission energy as in [25], $\omega = 0.05\sqrt{s}$, to be re-evaluated for each experiment specifically.

Let us denote the specific type of contribution in cross section or asymmetries by a subscript C. C can be 0 (Born contribution), 1 (one-loop EWC contribution), or 0+1 (both), i.e. $C = \{0, 1, 0{+}1\}$. The PV (left-right) asymmetry, the forward-backward asymmetry, and the left-right asymmetry constructed from the integrated cross sections are defined in the traditional way:

$$A^C_{LR} = \frac{\sigma^C_L - \sigma^C_R}{\sigma^C_L + \sigma^C_R}, \quad A^C_{FB} = \frac{\Sigma^C_F - \Sigma^C_B}{\Sigma^C_F + \Sigma^C_B}, \quad A^C_{LR\Sigma} = \frac{\Sigma^C_L - \Sigma^C_R}{\Sigma^C_L + \Sigma^C_R}. \tag{9}$$

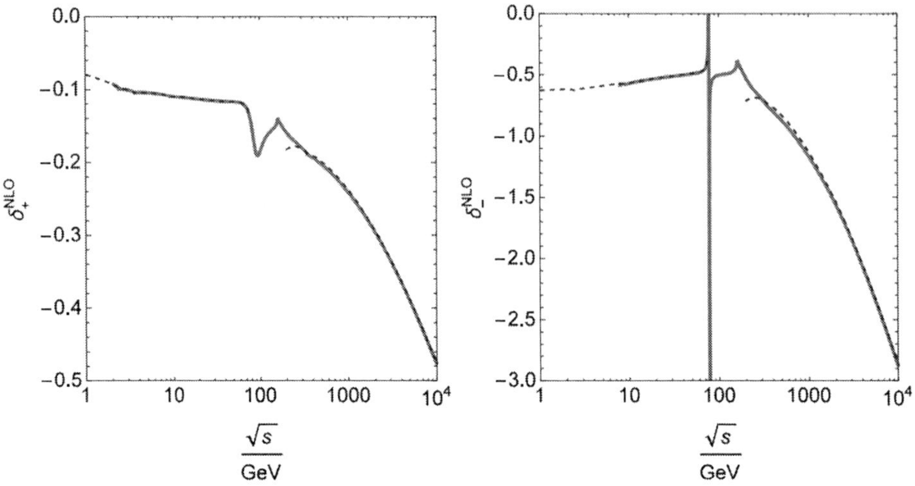

Figure 3. The relative NLO corrections at $\theta = 90°$ vs energy: solid line – computer algebra code (not working at very small $\sqrt{s}$), dashed line – asymptotic estimations, good for $\sqrt{s}$ below m_W and for $\sqrt{s}$ above m_Z.

The subscripts L and R on the cross sections $\sigma \equiv d\sigma/d(\cos\theta)$ correspond to the polarization degree of the electron $p_B = -1$ and $p_B = +1$ respectively.

The forward and backward cross sections defined as:

$$\Sigma_F^C = \int_0^{\cos a} \sigma_{00}^C \cdot d(\cos\theta), \quad \Sigma_B^C = \int_{-\cos a}^{0} \sigma_{00}^C \cdot d(\cos\theta).$$

and left and right integrated (over $a \le \theta \le b$ segment) cross sections are:

$$\Sigma_L^C = \int_{\cos b}^{\cos a} \sigma_L^C \cdot d(\cos\theta), \quad \Sigma_R^C = \int_{\cos b}^{\cos a} \sigma_R^C \cdot d(\cos\theta).$$

Figs. 4-6 illustrate these asymmetries with and without the radiative effects.

4 Conclusions

We compare the results for the full set of one-loop EWC to left-right and forward-backward asymmetries at energies relevant for Belle II at SuperKEKB obtained by two different methods: the exact (semi-automatic, with computer algebra) and approximate (asymptotic, on paper). Both methods have their advantages and limitations, but their combination proves that our work is error-free and provided more options to the experimental community. The bremsstrahlung process is fully controlled and the soft- and hard-photon approximations are compared. Although much improved in this work, the precision obtained with one-loop EWC is still insufficient for the upcoming experiments such as Belle II, MOLLER and P2 and will require at least the leading two-loop contributions.

Acknowledgments

We are grateful to the organizers of Baldin ISHEPP XXIII Seminar for hospitality and to Yu. Bystritskiy, and J.M. Roney for stimulating discussions. V.Z. thanks Harrison McCain Foundation for sup-

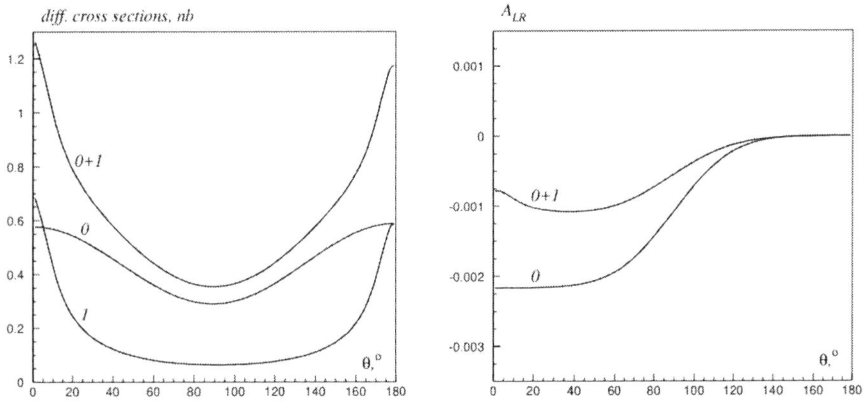

Figure 4. Left: unpolarized NLO corrected (0+1), Born (0), and their difference (1) differential cross sections vs scattering angle θ. Right: the polarization Born asymmetry (0) and asymmetry taking into account the NLO EWC (0+1) vs scattering angle θ. Note the large contribution of EWC to A_{LR} at small angles.

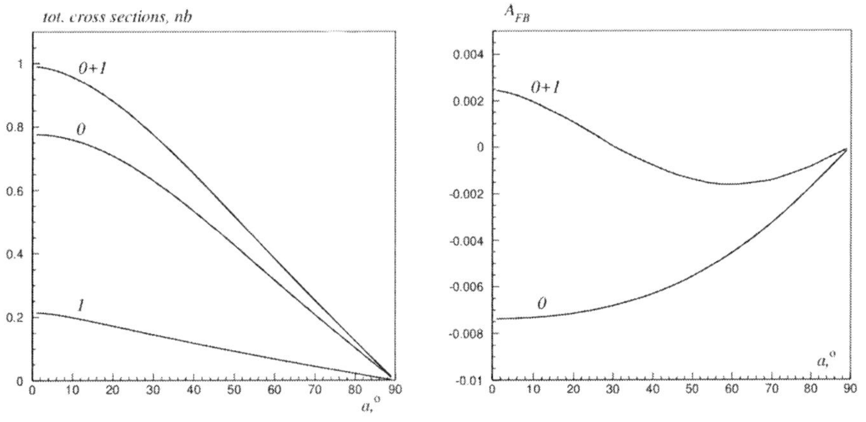

Figure 5. Left: unpolarized NLO corrected (0+1), Born (0), and their difference (1) total cross sections vs angle a. Right: the forward-backward Born asymmetry (0) and asymmetry taking into account the NLO EWC (0+1) vs angle a. Note the importance of EWC.

port and the Acadia University for hospitality in 2016. The work of A.A. and S.B. is supported by the Natural Science and Engineering Research Council of Canada (NSERC).

References

[1] Belle II Technical Design Report, Belle II Collaboration: T. Abe et al., arXiv:1011.0352 (2010)
[2] MOLLER Collab. (J. Benesch et al.) arXiv:1411.4088 [nucl-ex] JLAB-PHY-14-1986 (2014)
[3] N. Berger et al., Proceedings of the PhiPsi15, Sep. 23-26, 2015, Hefei, China (2015)

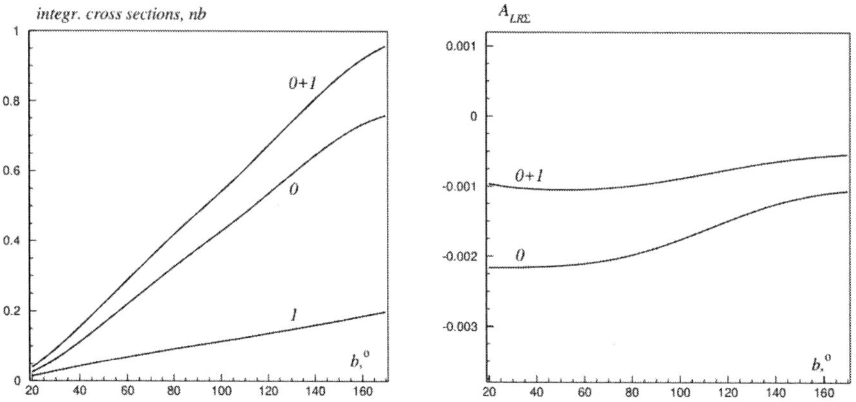

Figure 6. Left: unpolarized NLO corrected (0+1), Born (0), and their difference (1) integrated cross sections vs angle b at $a = 10°$. Right: the left-right integrated Born asymmetry (0) and asymmetry taking into account the NLO EWC (0+1) vs angle b at $a = 10°$.

[4] A. Bevan, arXiv:1112.1394v1 [hep-ex]
[5] M. Böhm and W. Hollik, Nucl. Phys. B **204**, 45 (1982)
[6] W. Hollik, Fortschr. Phys. **38**, 165 (1990)
[7] M. Bohm and W. Hollik, Z. Phys. C **23**, 31 (1984)
[8] V.A. Novikov, L.B. Okun, M.I. Vysotsky, Nucl. Phys. B **397**, 35 (1993)
[9] G. Montagna et al., Comput. Phys. Commun. **117**, 278 (1999)
[10] D. Bardin et al., Comput. Phys. Commun. **133**, 229 (2001)
[11] D.Yu. Bardin, S. Rieman, T. Rieman, Z. Phys. **32**, 121 (1986)
[12] S. Jadach, B.F.L. Ward and Z. Was, Phys. Rev. D **63**, 113009 (2001)
[13] A. Andonov et al., Phys. Part. Nucl. **34**, 1125 (2003)
[14] A. Aleksejevs et al., Phys. Rev. D **82**, 093013 (2010). A. Aleksejevs et al., Phys. Part. Nucl. **44**, 161 (2013)
[15] V.A. Zykunov, Phys. Rev. D **75**, 073019, (2007)
[16] A. Aleksejevs et al., arXiv:1609.08666 (2016)
[17] T. Hahn, M. Perez-Victoria, Comput. Phys. Commun. **118**, 153 (1999)
[18] A. Aleksejevs et al., Phys. Part. Nuclei Lett. **13**, 310 (2016)
[19] Richard J. Hill, arXiv:1605.02613 (2016)
[20] M. Böhm, H. Spiesberger, and W. Hollik, Fortschr. Phys. **34**, 687 (1986)
[21] A. Denner, Fortsch. Phys. **41**, 307 (1993)
[22] A.G. Aleksejevs, S.G. Barkanova, V.A. Zykunov, Phys. Atom. Nucl. **79**, 78 (2016)
[23] K. A. Olive et al. (Particle Data Group), Chin. Phys. C **38**, 090001 (2014)
[24] F. Jegerlehner, J. Phys. G. **29** 101 (2003). [hep-ph/0104304]
[25] A. Denner and S. Pozzorini, Eur. Phys. J. C **7**, 185 (1999)

DOI: 10.1051/epjconf/201713806002

Form factors and related quantities in clothed-particle representation

Alexander Shebeko[1,⋆] and *Adam* Arslanaliev[2]

[1]*Kharkov Institute of Physics & Technology, Ukraine*
[2]*The Karazin University, Kharkov, Ukraine*

Abstract. We show new applications of the notion of clothed particles in quantum field theory. Its realization by means of the clothing procedure put forward by Greenberg and Schweber allows one to express the total Hamiltonian H and other generators of the Poincaré group for a given system of interacting fields through the creation (annihilation) operators for the so-called clothed particles with physical (observed) properties. Here such a clothed particle representation is used to calculate the matrix elements (shortly, form factors) of the corresponding Nöther current operators sandwiched between the H eigenstates. Our calculations are performed with help of an iterative technique suggested by us earlier when constructing the $NN \to \pi NN$ transition operators. As an illustration, we outline some application of our approach in the spinor quantum electrodynamics.

1 Introduction

The method of unitary clothing transformations (UCTs) ([1]–[5]) allows us to develop an alternate approach for calculating the matrix elements $F^\mu(p',p) \equiv \langle p'; out|J^\mu(0)|p; in\rangle$ of the Nöther current density operator $J^\mu(0)$ sandwiched between the $in(out)$ states of interacting fields. Since every UCT $W(\alpha_c) = W(\alpha) = \exp R$, $R = -R^\dagger$ connects a primary set α of creation (annihilation) operators in the bare-particle representation (BPR) with new operators α_c in the clothed-particle representation (CPR) via similarity transformation $\alpha = W(\alpha_c)\alpha_c W^\dagger(\alpha_c)$, we consider the expansion

$$J^\mu(0) \quad = \quad WJ_c^\mu(0)W^\dagger = J_c^\mu(0) + [R, J_c^\mu(0)] + \frac{1}{2}[R,[R,J_c^\mu(0)]] + \ldots, \tag{1}$$

where $J_c^\mu(0)$ is the initial current in which the bare operators are replaced by their clothed counterparts. Along this guideline we have to handle R-commutators $[R,[R,\ldots[R,J]]] \equiv J^{[n]}$ with n brackets. As an illustration of the recursive procedure proposed in [3] for calculations with similar multiple commutators, we will show our results in case of the electromagnetic current operator of meson-nucleon system

$$J^\mu(0) = J_N^\mu(0) + J_\pi^\mu(0) + J_\rho^\mu(0) + J_\omega^\mu(0) + \ldots \tag{2}$$

and in quantum electrodynamics (QED) with

$$J^\mu(0) = J_e^\mu(0) = e\frac{1}{2}[\overline{\psi}(0), \gamma^\mu\psi(0)] = e : \overline{\psi}(0)\gamma^\mu\psi(0) :, \tag{3}$$

⋆e-mail: shebeko@kipt.kharkov.ua

where $\psi(0)$ is the electron-positron field $\psi(x)$ at $x = (0, \vec{0})$. In both cases the matrix elements of interest between the one-fermion states $|p; in\rangle = |p; out\rangle = b_c^{\dagger}(p)|\Omega\rangle$ with the 4-momentum $p = (p_0, \vec{p})$, $p_0 = \sqrt{\vec{p}^2 + m^2}$ are expressed through the Dirac $F_1(q^2)$ and Pauli $F_2(q^2)$ form factors (FFs) that depend on the 'transferred' 4-momentum $q = p' - p$ squared, viz.,

$$F^{\mu}(p', p) = e\bar{u}(p')\{F_1(q^2)\gamma^{\mu} + i\sigma^{\mu\nu}F_2(q^2)(p' - p)_{\nu}\}u(p). \quad (4)$$

Other notations embody the one-clothed fermion (electron, nucleon) creation operator $b_c^{\dagger}(p)$, fermion mass m and physical vacuum state $|\Omega\rangle$. All polarization indices are implied.

2 The UCT Method in Action

A key point of the clothing procedure in question is to remove the so-called bad terms from the total Hamiltonian for interacting particles

$$H \equiv H(\alpha) = H_F(\alpha) + H_I(\alpha) = W(\alpha_c)H(\alpha_c)W^{\dagger}(\alpha_c) \equiv K(\alpha_c).$$

By definition, such terms prevent the physical vacuum $|\Omega\rangle$ (the H lowest energy eigenstate) and the one-clothed-particle states $|n\rangle_c = a_c^{\dagger}(n)|\Omega\rangle$ to be the H eigenvectors for all necessary quantum numbers n included. Bad terms occur every time when any normally ordered product

$$a^{\dagger}(1')a^{\dagger}(2')...a^{\dagger}(n'_C)a(n_A)...a(2)a(1)$$

of class [C.A] embodies, at least, one substructure $\in [k.0]$ $(k = 1, 2...)$ or/and $[k.1]$ $(k = 2, 3, ...)$. In this context all primary Yukawa-type (trilinear) couplings should be eliminated from interaction $V(\alpha)$ that enters

$$H_I(\alpha) = V(\alpha) + \text{mass and vertex counterterms}.$$

It results in the form

$$H = K_F(\alpha_c) + K_I(\alpha_c) = K, \quad (5)$$

where free part $K_F(\alpha_c) = H_F(\alpha_c)$ while operator $K_I(\alpha_c)$ contains interactions between clothed particles. By construction, the latter has the property

$$K_I(\alpha_c)|\Omega\rangle = K_I(\alpha_c)|n\rangle_c \equiv 0.$$

For a boson-fermion (meson-nucleon, photon-electron) system
we have the decomposition

$$\begin{aligned} K_I(\alpha_c) = K(ff \to ff) + K(\bar{f}\bar{f} \to \bar{f}\bar{f}) + K(f\bar{f} \to f\bar{f}) \\ + K(bf \to bf) + K(b\bar{f} \to b\bar{f}) + K(f\bar{f} \leftrightarrow bb') \\ + K(ff \leftrightarrow bff) + K(f\bar{f} \leftrightarrow 3b) + K(3f \to 3f) + \cdots, \quad (6) \end{aligned}$$

where separate contributions are responsible for different physical processes so, for instance, operators $K(\gamma e \to \gamma e)$, $K(e\ e \leftrightarrow \gamma e\ e)$ and $K(3N \to 3N)$ can be used in describing the Compton scattering on electrons, electron-electron bremsstrahlung and modeling three-nucleon forces, respectively.
In particular, the fermion-fermion interaction operator in the CPR can be written as

$$K(ff \to ff) = \sum_b K_b(ff \to ff),$$

$$K_b(ff \to ff) = \int \sum_{\mu} d\vec{p}_1' d\vec{p}_2' d\vec{p}_1 d\vec{p}_2\, V_b(1', 2'; 1, 2) b_c^{\dagger}(1') b_c^{\dagger}(2') b_c(1) b_c(2),$$

where the symbol $\sum_{\mu}$ denotes the summation over fermion spin projections, $1 = \{\vec{p}_1, \mu_1\}$, etc.

3 The S-Matrix in SPR

Of interest is how the S–matrix for given interactions between "bare" particles can be expressed in terms of a new family of clothed-particle interactions ("quasipotentials")? As we have seen ([5] and references therein), such a reduction becomes possible provided that

$$\lim_{t \to \pm\infty} W_D(t) = 1, \tag{7}$$

this supplementary constraint imposed on UCTs in the Dirac (D) picture at the distant past and future. It leads to important links between *in(out)* states and clothed-particle ones. As well known, the former, being exact eigenstates of a total field Hamiltonian, are key elements of the collision theory in the Heisenberg (H) picture, e.g. within such a nonperturbative formalism by Lehmann, Symanzik and Zimmermann (LSZ), when evaluating the S–matrix $S_{fi} = \langle f; out \mid i; in \rangle$ with many-particle states

$$|p_1 \ldots p_n;\, in\,(out)\rangle = a^{\dagger}_{in\,(out)}(p_1) \ldots a^{\dagger}_{in\,(out)}(p_n)\,|\Omega\rangle. \tag{8}$$

Also let me recall that the creation (destruction) in(out) operators $a^{\dagger}_{in(out)}$ ($a_{in(out)}$) meet the same canonical commutation relations that $a_c^{\dagger}$ (a_c) do. We find for one-particle states,

$$|p;\, in\,(out)\rangle = a^{\dagger}_{in\,(out)}(p)|\Omega\rangle = a_c^{\dagger}(p)|\Omega\rangle \equiv |\, p;\, c\rangle. \tag{9}$$

Of course, (9) does not mean that $a_c^{\dagger}(p) = a^{\dagger}_{in\,(out)}(p)$. In this context, note links

$$|p_1\, p_2;\, in\rangle \equiv a^{\dagger}_{in}(p_1)a^{\dagger}_{in}(p_2)|\Omega\rangle = \Omega^{(+)}(p_1^0 + p_2^0)a_c^{\dagger}(p_1)a_c^{\dagger}(p_2|\Omega\rangle \tag{10}$$

and

$$|p_1\, p_2;\, out\rangle \equiv a^{\dagger}_{out}(p_1)a^{\dagger}_{out}(p_2)|\Omega\rangle = \Omega^{(-)}(p_1^0 + p_2^0)a_c^{\dagger}(p_1)a_c^{\dagger}(p_2|\Omega\rangle \tag{11}$$

with $\Omega^{(\pm)}(E) = \pm\iota \lim_{\epsilon\to+0} \epsilon G(E \pm \iota\epsilon)$, where $G(z) = (z - H)^{-1}$ is the Hamiltonian resolvent. Just such a transition has been used by us when describing properties of nucleon-nucleon collisions.
Now we employ such links to evaluate the matrix elements $\langle out|J|in\rangle$ of a current (source) J between initial and final states of leptons and hadrons. So for the electron-hadron scattering $e + h \to e' + h'$ one has to evaluate its amplitude

$$\langle e'h'; out|eh; in\rangle = 2\pi i \frac{m_e}{\sqrt{EE'}} e_0 \bar{u}_e(k')\gamma^{\mu} u_e(k) \frac{\langle h'|J_{\rho}(0)|h\rangle}{q^2} \delta(p' + k' - p - k), \tag{12}$$

where m_e is the mass of the physical electron, $q^2 = (k' - k)^2 = (p' - p)^2$, $u_e(k) \equiv u_e(k\sigma)$ the Dirac spinor for the electron with 4-momentum k and polarization σ.
In case of the one-hadron state $|h\rangle = |p\rangle = b_c^{\dagger}(p)|\Omega\rangle$ (other quantum numbers are suppressed) we use the expansion (1) so by introducing the contraction $J_c^{\mu}(0)l_{\mu} = J_c(0)$ with an arbitrary "polarization" 4-vector l we get the matrix elements

$$M_{OPEA} = \langle p'|J_c(0)|p\rangle \tag{13}$$

$$M^{[1]} = \langle p'|[R, J_c(0)]|p\rangle \tag{14}$$

$$M^{[2]} = \frac{1}{2!}\langle p'|[R, [R, J_c(0)]]|p\rangle \tag{15}$$

of successive R - commutators. Here the abbreviation OPEA is reminiscent of One-Photon-Exchange Approximation.

4 R-commutators in CPR

We have seen [3] how the commutators

$$[R, [R, \ldots [R, V]]] \equiv V^{[n]} \tag{16}$$

with n brackets ($n = 1, 2, \ldots$) and couplings V of the type $V = f * m + H.c.$, where $f * m$ is a polynomial composed of products of fermion and meson operators, become a spring of interactions between the clothed fermions and mesons, the mass and vertex contributions, etc. Let us consider the πN pseudoscalar coupling

$$V = \int \frac{d\vec{k}}{\omega_k} \hat{V}(k) a(k) + H.c., \quad R = \int \frac{d\vec{k}}{\omega_k} \hat{R}(k) a(k) - H.c., \tag{17}$$

where the generator R is determined by the condition $[H_F, R] = V^1$. Subsequent calculations are essentially simplified after introducing the following quantities $\hat{V}(k) =: F^\dagger V(k) F :$ and $\hat{R}(k) =: F^\dagger R(k) F :$ with the compact notation

$$F^\dagger X F \equiv \sum_{\mu'} \int \frac{d\vec{p}'}{E_{p'}} \sum_{\mu} \int \frac{d\vec{p}}{E_p} F^\dagger_{\epsilon'}(p'\mu') X_{\epsilon'\epsilon}(p'\mu', p\mu) F_\epsilon(p\mu) \tag{18}$$

for any 2×2 matrix

$$X = \begin{bmatrix} X_{11} & X_{12} \\ X_{21} & X_{22} \end{bmatrix}$$

and the column

$$F(p\mu) = \begin{bmatrix} F_1(p\mu) \\ F_2(p\mu) \end{bmatrix} = \begin{bmatrix} b(p\mu) \\ d^\dagger(p\mu) \end{bmatrix}.$$

In our case the matrices $V(k)$ and $R(k)$ can be written as

$$V_{\epsilon'\epsilon}(p'\mu', p\mu; \vec{k}) = \frac{igm}{\sqrt{2(2\pi)^3}} v_{\epsilon'\epsilon}(p'\mu', p\mu) \delta(\vec{k} + (-1)^{\epsilon'} \vec{p}' - (-1)^{\epsilon} \vec{p}) \tag{19}$$

$$v_{\epsilon'\epsilon}(p'\mu', p\mu) = \begin{bmatrix} \bar{u}(p'\mu')\gamma_5 u(p\mu) & \bar{u}(p'\mu')\gamma_5 v(p\mu) \\ \bar{v}(p'\mu')\gamma_5 u(p\mu) & \bar{v}(p'\mu')\gamma_5 v(p\mu) \end{bmatrix} \tag{20}$$

$$R_{\epsilon'\epsilon}(p'\mu', p\mu; \vec{k}) = -\frac{V_{\epsilon'\epsilon}(p'\mu', p\mu; \vec{k})}{\omega_{\vec{k}} + (-1)^{\epsilon'} E_{\vec{p}'} - (-1)^{\epsilon} E_{\vec{p}}} \tag{21}$$

with the sign-energy indices ($\epsilon', \epsilon = 1, 2$) and all necessary algebraic operations are based upon the covariant commutation rules

$$\{F_{\epsilon'}(p'\mu'), F^\dagger_\epsilon(p, \mu)\} = p_0 \delta(\vec{p}' - \vec{p}) \delta_{\mu'\mu} \delta_{\epsilon'\epsilon}. \tag{22}$$

In practical calculations it is convenient to use

$$V^{[n]} = \sum_{s=0}^{n} \frac{n!}{s!(n-s)!} f^{[n-s]} * m^{[s]} + H.c. \tag{23}$$

[1] In this case $m(k) = a_c(k)$ is the one-clothed-meson destruction operator and henceforth the label c will be omitted

and the recurrence relation

$$V^{[n]} = [R, V^{[n-1]}] = [\mathcal{R} - \mathcal{R}^{\dagger}, V^{[n-1]}] = [\mathcal{R}, V^{[n-1]}] + H.c. \quad (24)$$

as a test and a subsequent step when calculating more complex brackets.
Doing so, we find step by step

$$V^{[1]} = \int \frac{d\vec{k}}{\omega_k} \left[\{f(k)\}^{[1]} m(k) + f(k)\{m(k)\}^{[1]} \right] + H.c., \quad (25)$$

$$V^{[2]} = \int \frac{d\vec{k}}{\omega_k} \left[\{f(k)\}^{[2]} m(k) + 2\{f(k)\}^{[1]}\{m(k)\}^{[1]} f(k)\{m(k)\}^{[2]} \right] + H.c. \quad (26)$$

or

$$[R, V] = \int \frac{d\vec{k}_1}{\omega_{k_1}} \hat{V}(k_1)\hat{R}^{\dagger}(k_1) + \textit{terms quadratic in } a^{\dagger}(a) + H.c. \quad (27)$$

$$[R, [R, V]] = \int \frac{d\vec{k}_1}{\omega_{k_1}} \int \frac{d\vec{k}_2}{\omega_{k_2}} \{\hat{C}_{10}(k_1, k_2)a^{\dagger}(k_1) + \hat{C}_{01}(k_1, k_2)a(k_1)\} + \textit{terms cubic in } a^{\dagger}(a) + H.c. \quad (28)$$

to get the g^4-order interactions between the clothed fermions that enter the commutator

$$V^{[3]}_{ferm} = [R, V^{[2]}]_{ferm} = \int \frac{d\vec{k}_1}{\omega_{k_1}} \int \frac{d\vec{k}_2}{\omega_{k_2}} \hat{R}(k_1)\{\hat{C}_{10}(k_1, k_2) + \hat{C}^{\dagger}_{01}(k_1, k_2)\} + H.c. \quad (29)$$

along with the fermion mass renormalization terms in the same order. Here

$$\hat{C}_{10}(k_1, k_2) = -2[\hat{R}^{\dagger}(k_1), \hat{V}(k_2)]\hat{R}^{\dagger}(k_2) - \hat{V}(k_2)[\hat{R}^{\dagger}(k_1), \hat{R}^{\dagger}(k_2)] \quad (30)$$

$$\begin{aligned} \hat{C}_{01}(k_1, k_2) &= [\hat{R}(k_2), \hat{V}(k_1)]\hat{R}^{\dagger}(k_2) - [\hat{R}^{\dagger}(k_2), \hat{V}(k_1)]\hat{R}(k_2) \\ &- [\hat{R}(k_2), [\hat{R}^{\dagger}(k_2), \hat{V}(k_1)]] + 2[\hat{R}(k_1), \hat{V}(k_2)]\hat{R}^{\dagger}(k_2) \\ &+ \hat{V}(k_2)[\hat{R}(k_1), \hat{R}^{\dagger}(k_2)]. \end{aligned} \quad (31)$$

5 Calculations of typical matrix elements

Certainly, these ideas can be realized when calculating the meson and nucleon FFs for the density (2), where, e.g., the primary meson and nucleon currents are given by $J^{\mu}_{\pi}(\vec{x}) =: \vec{\pi}(\vec{x}) \times \partial^{\mu}\vec{\pi}(\vec{x})$: and $J^{\mu}_{N}(\vec{x}) =: \bar{\psi}(\vec{x})\gamma^{\mu}(1 \pm \tau_3)/2\psi(\vec{x})$:, and the electron (positron) FF in QED for the density (3). Let us remind that in the latter $\psi(0)$ is the electron-positron field in the Schrödinger S picture, while in the nucleon current density operator the field $\psi(0)$ has both nucleon and antinucleon components.

At this point, we would like to outline some application of our approach in QED. It is the case when the relevant R commutators should be calculated assuming that $V = V_{\gamma} + H.c.$, $V_{\gamma} = \int \frac{d\vec{k}}{k^0} : F^{\dagger}V(k, \sigma)F : c(k, \sigma)$ with the contraction $V(k, \sigma) = V_{\rho}(k)e^{\rho}(k, \sigma)$ where the c-number matrices $V_{\rho}(k)$ repeat the structure (19) after the replacement in the latter of ig by e and the matrix γ_5 by Dirac matrices γ_{ρ}. The corresponding generator is determined as it follows, viz., $R = R_{\gamma} - R^{\dagger}_{\gamma}$, $R_{\gamma} = \int \frac{d\vec{k}}{k^0} : F^{\dagger}R(k, \sigma)F : c(k, \sigma)$ with the contraction $R(k, \sigma) = R_{\rho}(k)e^{\rho}(k, \sigma)$ and matrix $R_{\rho}(k)$ composed of elements

$$R^{\rho}_{ij}(k) = -\frac{V^{\rho}_{ij}(k)}{(-1)^i E_{\vec{p}'} - (-1)^j E_{\vec{p}} + k^0} \quad (i, j = 1, 2). \quad (32)$$

Photon destruction operators $c(k,\sigma)$ enter the Fourier expansion of the e.m. field

$$A^{\rho}(\vec{x}) = \int \frac{d\vec{k}}{\sqrt{2(2\pi)^3k^0}} \sum_{\sigma} \left[e^{\rho}(k,\sigma)c(k,\sigma) + e^{\rho^*}(k_-,\sigma)c^{\dagger}(k_-,\sigma)\right] \exp(i\vec{k}\vec{x}), \tag{33}$$

together with photon polarization vectors $e^{\rho}(k,\sigma)$ that form the projection operator

$$\Pi_{\rho_1\rho_2}(k) = -g_{\rho_1\rho_2} + \frac{q^0(q_{\rho_1}n_{\rho_2} + q_{\rho_2}n_{\rho_1}) - q_{\rho_1}q_{\rho_2} - q^2 n_{\rho_1}n_{\rho_2}}{\vec{k}^2},$$

where $n^{\mu} = (1,0,0,0)$ is a fixed time-like vector, $q = (q_0, \vec{k})$, $q^2 = q_0^2 - \vec{k}^2$, but here q_0 being generally arbitrary is chosen to be equal to difference $q^0 = E_{\vec{p}'} - E_{\vec{p}}$ if once it appeared $\vec{k} = \vec{p}' - \vec{p}$ for two on-mass-shell fermions (electrons).
At last, the electron-positron current density in the CPR is given by $J_c(0) = N : F^{\dagger}jF : ((2\pi)^3 N = em_e)$ with the 2×2 matrix

$$j(p'\mu', p\mu) = \begin{bmatrix} \bar{u}(p'\mu')\hat{l}u(p\mu) & \bar{u}(p'\mu')\hat{l}v(p\mu) \\ \bar{v}(p'\mu')\hat{l}u(p\mu) & \bar{v}(p'\mu')\hat{l}v(p\mu) \end{bmatrix}. \tag{34}$$

Then, for example, we have

$$M^{[2]} = \frac{1}{2!}\langle p'|[R,[R,J]]|p\rangle, \tag{35}$$

$$J^{[1]} \equiv [R,J] = N\int \frac{d\vec{k}}{k^0} F^{\dagger}[R(k,\sigma), j]Fc(k,\sigma) + H.c. \tag{36}$$

so $J^{[2]} = [R,[R,J]] = [R, J^{[1]}]$ and the initial task reduces to finding the expectation value $\langle\Omega|b(p')[R,J^{[1]}]b^{\dagger}(p)|\Omega\rangle$.
Detailed calculations by these formulas and comparison with well known results are in preparation to be published somewhere else.

In our opinion, the approach exposed here can be helpful when calculating higher-order radiative corrections in mesodynamics, QED and other field models. In this context, our current explorations within the formalism of clothed particles are concentrated upon the three directions: i) test calculations of the Dirac and Pauli FFs in the e^2- and e^4 - orders to compared the standard ones; ii) calculations of the nucleon FFs in the g^2- and g^4- orders being aware of a decisive role of the normal ordering operation for separating the $b^{\dagger}b$ - contributions from the operators of fermion type $F^{\dagger}XFF^{\dagger}YF$; relying on our experience in evaluating the dipole magnetic and electric quadrupole deuteron moments in the CPR (see [5] and references therein) a further development of the theory of electromagnetic interactions with nuclei, in particular, including the construction of a new family of meson exchange two-body currents.

References

[1] A.V. Shebeko and M.I. Shirokov, Progr. Part. Nucl. Phys. **44**, 75 (2000)
[2] V.Yu. Korda and A.V. Shebeko, Phys. Rev. D **70**, 085011 (2004)
[3] V.Yu. Korda, L. Canton and A.V. Shebeko, Ann. Phys. **322**, 736 (2007)
[4] I. Dubovyk and A. Shebeko, Few Body Syst. **48**, 109 (2010)
[5] A. Shebeko, Chapter 1 In: *Advances in Quantum Field Theory*, ed. S. Ketov (InTech, 2012) 3-30

On the relativistic $^3D_1$ partial-wave contribution to the bound three-nucleon system

Sergey Bondarenko[1,*], *Valery* Burov[1,**], and *Sergey* Yurev[1,***]

[1] *Joint Institute for Nuclear Research, Dubna, 141980, Russia*

Abstract. The bound state of three nucleons is investigated using the Faddeev equations within the Bethe-Salpeter approach. The relativistic and nonrelativistic nucleon-nucleon interaction is chosen in a multirank separable form. The extension for partial-states with $L > 0$ is done. Three partial-wave states - $^1S_0$,$^3S_1$ and $^3D_1$ - are taken into account. The Gauss quadrature method is used to calculate the integrals and find the triton binding energy by iterations.

1 Introduction

Three-body calculations in nuclear physics are of great interest used for the description of three-nucleon bound states (^3He, T), processes of elastic, inelastic and deep inelastic scattering the leptons off light nuclei and also the hadron-deuteron reactions (for example, $pd \to pd$, $pd \to ppn$). The investigation of the nuclei ^3He and T is also interesting because it allows us to investigate further (in addition to the case of the deuteron) evolution of the bound nucleon thereby contributing to the explanation of so-called EMC-effect. In quantum mechanics the Faddeev equations are commonly used to describe the three-particle systems. The main feature of Faddeev equations is that all particles interact through a pair potential.

However at the high momentum transfer relativistic effects should be taken into account. The Bethe-Salpeter (BS) [1] equation is one of the most consistent approaches to describe the NN interaction. In this formalism, one has to deal with a system of nontrivial integral equations for both the NN scattered states and the bound state – the deuteron. To solve a system of integral equations, it is convenient to use a separable *Ansatz* [2] for the interaction kernel in the BS equation. In this case, one can transform integral equations into a system of algebraic linear ones which is easy to solve. Parameters of the interaction kernel are found from an analysis of the phase shifts and inelasticity, low-energy parameters and deuteron properties (binding energy, moments, etc.).

The relativistic three-particle systems are described by the Faddeev equations within the BS approach - so called Bethe-Salpeter-Faddeev equations. All nucleons have equal masses and the scalar propagators instead of spinor ones are used for simplicity. The spin-isospin structure of the nucleons is taken into account by using the so-called recoupling-coefficient matrix. The work mainly follows the ideas of the paper [3]. In the paper [4] only S partial-states were considered for the one-rank

*e-mail: bondarenko@jinr.ru
**e-mail: burov@jinr.ru
***e-mail: yurev@jinr.ru

kernel. In this paper the relativistic $^3D_1$ partial-wave is added into formalism using the three-rank Graz-II kernel [5].

The paper is organized as follows: in Sec. 2 the two-particle problem is considered, in Sec. 3 three-particle equations and partial-wave decomposition are described. In Sec. 4 the calculations and results are given. The summary is in the Sec. 5.

2 Two-particle case

Since the formalism of the Faddeev equations is based on the properties of the pair nucleon-nucleon interaction only some conclusions of two-body problem are given here.

The Bethe-Salpeter equation for the relativistic two-particle system is taken in the following form:

$$T(p, p'; s) = V(p, p') + \frac{i}{4\pi^3} \int d^4k\, V(p, k)\, G(k; s)\, T(k, p'; s) \tag{1}$$

where $T(p, p'; s)$ is the two-particle T matrix and $V(p, p')$ - kernel (potential) of the nucleon-nucleon interaction. The free two-particle Green function $G(k; s)$ is expressed, for simplicity, through the scalar propagator of the nucleons

$$G^{-1}(k; s) = [(P/2 + k)^2 - m_N^2 + i\epsilon][(P/2 - k)^2 - m_N^2 + i\epsilon]. \tag{2}$$

To solve equation (1) the separable *Ansatz* for the nucleon-nucleon potential $V(p, p')$ is used

$$V_{LL'}(p_0, p, p_0', p') = \sum_{n_1 n_2} \lambda_{n_1 n_2} g_{n_1}^{(L)}(p_0, p) g_{n_2}^{(L')}(p_0', p'). \tag{3}$$

In this case the two-particle T matrix has the following simple form:

$$T_{LL'}(p_0, p, p_0', p', s) = \sum_{n_1 n_2} \tau_{n_1 n_2}(s) g_{n_1}^{(L)}(p_0, p) g_{n_2}^{(L')}(p_0', p'). \tag{4}$$

where

$$[\tau^{-1}(s)]_{n_1 n_2} = [\lambda^{-1}]_{n_1 n_2} - \frac{i}{4\pi^3} \sum_{L=0,2} \int_{-\infty}^{\infty} dk_0 \int_0^{\infty} k^2\, dk\, g_{n_1}^{(L)}(k_0, k) g_{n_2}^{(L')}(k_0, k)\, G(k_0, k; s). \tag{5}$$

3 Three-particle case

Neglecting the three-particles interaction we write the equations for three-particle amplitude $T = \sum_{i=1}^{3} T^{(i)}$ in the following form:

$$\begin{bmatrix} T^{(1)} \\ T^{(2)} \\ T^{(3)} \end{bmatrix} = \begin{bmatrix} T_1 \\ T_2 \\ T_3 \end{bmatrix} - \begin{bmatrix} 0 & T_1G_1 & T_1G_1 \\ T_2G_2 & 0 & T_2G_2 \\ T_3G_3 & T_3G_3 & 0 \end{bmatrix} \begin{bmatrix} T^{(1)} \\ T^{(2)} \\ T^{(3)} \end{bmatrix} \tag{6}$$

where G_i is the free two-particle (j and n) Green function (ijn is cyclic permutation of (1,2,3)):

$$G_i(k_j, k_n) = S^{(j)}(k_j) S^{(n)}(k_n). \tag{7}$$

The two-particle amplitude T_i satisfies the Bethe-Salpeter equation (see previous section).

For the system of equal-mass particles the Jacobi momenta can be written in the following form:

$$p_i = \frac{1}{2}(k_j - k_n),\ q_i = \frac{1}{3}K - k_i,\ K = k_1 + k_2 + k_3. \tag{8}$$

Using expressions (8) the amplitude T (6) can be rewritten as

$$T^{(i)}(p_i, q_i; p_i', q_i'; K) = (2\pi)^4\delta^{(4)}(q_i - q_i')T_i(p_i; p_i'; P) \tag{9}$$
$$-i\int \frac{dp_i''}{(2\pi)^4} T_i(p_i; p_i''; P)G_i(k_j'', k_n'')[T^{(j)}(p_j'', q_i''; p_i', q_i'; K) + T^{(n)}(p_i'', q_i''; p_i, q_i'; K)],$$

where P is the total two-body momentum.

For the bound state the total momentum squared $s = K^2$ is fixed at the mass of the bound state (triton) $M_B = \sqrt{s} = 3m_N - E_B$. The equation (9) becomes homogeneous for the amplitude $\Psi^{(i)}(p_i, q_i; s)$

$$\Psi^{(i)}(p_i, q_i; s) = \langle p_i, q_i | T^{(i)} | M_B \rangle \equiv \Psi_{LM}(p, q; s), \tag{10}$$

For the equal-mass case all $\Psi^{(i)}$ functions are equal to each other and we can write

$$\Psi(p, q; K) = -i\int \frac{dq'}{(2\pi)^4}\Big[T(p, -\frac{q}{2} - q'; P)S(\frac{K}{3} + q + q')S(\frac{K}{3} - q')\Psi(q + \frac{q'}{2}, q'; K)$$
$$+T(p, \frac{q}{2} + q'; P)S(\frac{K}{3} + q + q')S(\frac{K}{3} - q')\Psi(-q - \frac{q'}{2}, q'; K)\Big]. \tag{11}$$

To perform the partial-wave decomposition several steps should be made:

1. construct two-body state in its rest frame;
2. boost two-body state into the three-particle rest frame;
3. construct three-body state.

The details of the partial-wave decomposition ideology for the quasi-potential equation can be found in the paper [6].

All steps should be made if the particles are considered as spin-one-half with corresponding spinor propagators $S^{(i)}$. However, as it was said in previous section we simplify the problem and consider the nucleons with scalar propagators (2). In this case the spin-isospin structure of the system can be presented as a simple multiplicative coefficient.

The total orbital angular momentum of the triton then can be written as $\boldsymbol{L} = \boldsymbol{l} + \boldsymbol{\lambda}$, where $\boldsymbol{l}$ is the angular momentum corresponding to a nucleon pair with relative momentum $\boldsymbol{p}$ and λ is the angular momentum corresponding to relative momentum $\boldsymbol{q}$.

To separate the angular dependence the amplitude can be written in the three-particle rest frame in the following form:

$$\Psi_{LM}(p, q; s) = \sum_{a\lambda} \Psi^{(a)}_{\lambda L}(p_0, |\mathbf{p}|, q_0, |\mathbf{q}|; s)\mathcal{Y}^{(a)}_{\lambda LM}(\hat{\mathbf{p}}, \hat{\mathbf{q}}), \tag{12}$$

with the angular part

$$\mathcal{Y}^{(a)}_{\lambda LM}(\hat{\mathbf{p}}, \hat{\mathbf{q}}) = \sum_{m\mu} C^{LM}_{lm\lambda\mu} Y_{lm}(\hat{\mathbf{p}})Y_{\lambda\mu}(\hat{\mathbf{q}}), \tag{13}$$

where the two-nucleon state with spin s, angular momentum l and total momentum j ($a \equiv {}^{2s+1}l_j$) is introduced, C are the Clebsch-Gordan coefficients and Y are the spherical harmonics.

If one considers two-nucleon separable interaction (with n_i being the separable index) the amplitude $\Psi^{(a)}_{\lambda L}$ can be written as

$$\Psi^{(a)}_{\lambda L}(p_0, p, q_0, q; s) = \sum_{n_1 n_2} g^{(a)}_{n_1}(p_0, p)\tau^{(a)}_{n_1 n_2}\Phi^{(a)}_{\lambda L n_2}(q_0, q; s), \tag{14}$$

where functions $\Phi^{(a)}_{\lambda Li}$ satisfy the following system of integral equations:

$$\Phi^{(a)}_{\lambda L n_2}(q_0, q; s) = \frac{i}{4\pi^3}\sum_{a'\lambda'}\sum_{n_3 n_4}\int_{-\infty}^{\infty} dq'_0 \int_0^{\infty} q'^2 dq' \times$$
$$Z^{(aa')}_{\lambda\lambda' n_2 n_3}(q_0, q; q'_0, q'; s)\frac{\tau^{(a')}_{n_3 n_4}[(\frac{2}{3}\sqrt{s}+q'_0)^2 - q'^2]}{(\frac{1}{3}\sqrt{s}-q'_0)^2 - q'^2 - m_N^2 + i\epsilon}\Phi^{(a')}_{\lambda' n4}(q'_0, q'; s), \tag{15}$$

with

$$Z^{(aa')}_{\lambda\lambda' n_2 n_3}(q_0, q; q'_0, q'; s) = C_{(aa')}\int d\cos\vartheta_{qq'} K^{(aa')}_{\lambda\lambda' L}(q, q', \cos\vartheta_{qq'})$$
$$\frac{g^{(a)}_{n_2}(-\frac{1}{2}q_0 - q'_0, |\mathbf{q}/2 + \mathbf{q}'|)g^{(a')}_{n_3}(q_0 + \frac{1}{2}q'_0, |\mathbf{q} + \mathbf{q}'/2|)}{(\frac{1}{3}\sqrt{s} + q_0 + q'_0)^2 - (\mathbf{q} + \mathbf{q}')^2 - m_N^2 + i\epsilon} \tag{16}$$

and

$$K^{(aa')}_{\lambda\lambda' L}(q, q', \cos\vartheta_{qq'}) = (4\pi)^{3/2}\frac{\sqrt{2\lambda + 1}}{2L + 1}$$
$$\sum_{mm'} C^{Lm}_{lm\lambda 0} C^{Lm}_{l'm'\lambda' m-m'} Y^*_{lm}(\vartheta, 0) Y_{l'm'}(\vartheta', 0) Y_{\lambda' m-m'}(\vartheta_{qq'}, 0) \tag{17}$$

where

$$\cos\vartheta = (\frac{q}{2} + q'\cos\vartheta_{qq'})/|\frac{\mathbf{q}}{2} + \mathbf{q}'|, \qquad \cos\vartheta' = (q + \frac{q'}{2}\cos\vartheta_{qq'})/|\mathbf{q} + \frac{\mathbf{q}'}{2}|$$

and

$$C_{(aa')} = \begin{bmatrix} \frac{1}{4} & -\frac{3}{4} & -\frac{3}{4} \\ -\frac{3}{4} & \frac{1}{4} & \frac{1}{4} \\ -\frac{3}{4} & \frac{1}{4} & \frac{1}{4} \end{bmatrix} \tag{18}$$

is the spin-isospin recoupling-coefficient matrix with $(a) = {}^1S_0, {}^3S_1, {}^3D_1$.

In the case of $L = 0$ we have $l = \lambda = 0$ or $l = \lambda = 2$ and expressions for non-zero K-functions are:

$$\begin{aligned} K_{00} &= 1 \\ K_{02} &= 4\pi A \\ K_{20} &= \sqrt{4\pi}Y^*_{20}(\vartheta, 0) \\ K_{22} &= \sqrt{(4\pi)^3}Y^*_{20}(\vartheta, 0)A \\ A &= \sum_{m'} C^{00}_{2m'2-m'} Y_{2m'}(\vartheta', 0) Y_{2-m'}(\vartheta_{qq'}, 0), \end{aligned} \tag{19}$$

and the K-matrix is:

$$K_{(aa')} = \begin{bmatrix} K_{00} & K_{00} & K_{02} \\ K_{00} & K_{00} & K_{02} \\ K_{20} & K_{20} & K_{22} \end{bmatrix} \tag{20}$$

In above formulae the Lorenz transformation from the rest system of two particles to the rest system of three particles is omitted. Since the g-functions depend on invariant p^2 momentum squared and for S partial-wave states the angular part is constant the only variable which should be transformed is the argument of spherical harmonics for $^3D_1$ partial-wave state. We believe that such transformation gives small contribution to the binding energy of three-nucleon system. However it should be done for nucleons with spinor propagators.

The system of integral equations (15-17) has several singularities in q_0' complex plain. However in the case of the bound three-particle system ($\sqrt{s} < 3m_N$) all singularities do not cross the path of integration on q_0' and thus do not affect to the Wick-rotation procedure $q_0' \to iq_4'$.

The system (15-17) after the Wick-rotation procedure is well analytically defined and can be solved using various numerical methods. One of them is discussed in the next section.

4 Solution and results

In the paper the iterations method is used to solve the system of integral equations. The mappings for the variable of integration on q $[0, \infty)$ and q_4 $(-\infty, \infty)$ to $[-1, 1]$ interval are introduced.

As it is shown in [7] the binding energy of the three-nucleon system can be found using the following condition for iterated solutions (see details in [7]):

$$\lim_{n\to\infty} \frac{\Phi_n(s)}{\Phi_{n-1}(s)}\bigg|_{s=M_B^2} = 1 \tag{21}$$

where n is a number of iterations.

The Graz-II kernels of interaction [5] with three probabilities ($p_D = 4, 5, 6\%$) of the $^3D_1$-state are used for calculations. The obtained results are shown in table 1 both for nonrelativistic and relativistic cases and for S-states and $(S + D)$ partial-wave states. The experimental value is 8.48 MeV.

Table 1. Three-nucleon binding energy calculated with Graz-II separable NN kernel.

Potential	Nonrelativistic		Relativistic	
	$^1S_0, ^3S_1$	$^1S_0, ^3S_1, ^3D_1$	$^1S_0, ^3S_1$	$^1S_0, ^3S_1, ^3D_1$
GRAZ-II, $p_D = 4$ %	8.372	8.334	8.628	8.617
GRAZ-II, $p_D = 5$ %	7.964	7.934	8.223	8.217
GRAZ-II, $p_D = 6$ %	7.569	7.548	7.832	7.831

As it is seen from the table 1 kernels with different p_D can give results for binding energy both bigger and smaller than the experimental value. The relativistic results are systematically higher then nonrelativistic ones to about 3%. Also the $^3D_1$ partial-state gives rather small contribution to the three-nucleon binding energy - about −0.5%.

Summary

In the paper three-body system is investigated using Bethe-Salpeter-Faddeev equations. In the calculations the three-rank Graz-II potential is used. The BSF integral equations are solved using the iterations method. The binding energy of the triton and amplitudes of the $^1S_0$, $^3S_1$ and $^3D_1$ partial-wave states of the triton are obtained.

It was shown that kernels with different p_D can give results for binding energy both bigger and smaller than the experimental value. The relativistic results are systematically higher then nonrelativistic ones to about 3% and the $^3D_1$ partial-state gives rather small contribution to the three-nucleon binding energy - about −0.5%.

Acknowledgements. This work was partially supported by the Russian Foundation for Basic Research grant № 16-02-00898.

References

[1] E.E. Salpeter, H.A. Bethe, Phys. Rev. **84**, 1232 (1951)
[2] S.G. Bondarenko, V.V. Burov, A.V. Molochkov, G.I. Smirnov, H. Toki, Prog. Part. Nucl. Phys. **48**, 449 (2002)
[3] G. Rupp and J. A. Tjon, Phys. Rev. C **37**, 1729 (1988)
[4] S. Bondarenko, V. Burov, S. Yurev, EPJ Web of Conferences **108**, 02015 (2016)
[5] L. Mathelitsch, W. Plessas, W. Schweiger, Phys. Rev. C **26**, 65 (1982). G. Rupp, J.A. Tjon, Phys. Rev. C. **45**, 2133 (1992)
[6] A. Stadler, F. Gross, M. Frank, Phys. Rev C **56**, 2365 (1997)
[7] R. A. Malfliet and J. A. Tjon, Nucl. Phys. A **127**,161-168 (1969)

Residual currents generated from vacuum by an electric field pulse in 2+1 dimensional QED models

S.A. Smolyansky[1,⋆], *D.V.* Churochkin[1], *V.V.* Dmitriev[1], *A.D.* Panferov[1], and *B.* Kämpfer[2]

[1] *Saratov State University, RU-410026, Saratov, Russia*
[2] *Helmholtz-Zentrum Dresden-Rossendolf, D-01314, Dresden, Germany*

Abstract. In the framework of strong field QED, the generation of a residual alternating polarization current is demonstrated, which remains after switching off an external field pulse. This effect is stipulated by inertial properties of the physical vacuum. In the standard vacuum $D = 2 + 1$ QED, this current is rapidly damped fast but can be available, apparently, for observation in the graphene, where the Fermi velocity $v_F \ll c$ plays an analogous role as the light velocity.

1 Introduction

We address the question whether there are residual electric currents which remain after the strong electromagnetic impulse passage. The total current is composed of the conductivity and polarization currents, and the residual conductivity current is constant. However, surprisingly, the residual polarization current turns out to be fast oscillating and damped. This is a peculiar inertial effect of the vacuum oscillations excited by an external field. Thus, according to this prediction, some alternate component will appear in the total residual current.

We will consider this effect in a rather general case for different variants of spinor QED models in the external spatially-homogeneous, time-dependent electric field where the spin effects can be separated from the basic effect of vacuum creation of the electron-positron plasma (EPP). We consider two such systems: $D = 2 + 1$ massless QED and graphene. The last model will be considered rather in detail. In conclusion, we discuss a manifestation of this effect in some other systems.

The discussed effect has also another interesting feature in the case of models with a running coupling constant. It turns out that the polarization current is dominating in the area of small coupling constant. However, this situation turns into the opposite for a large coupling constant, as in some QCD models.

We are going to consider firstly a general kinetic theory which unifies the mentioned QED models (Sect. 2). Such an analysis establishes the necessary prerequisites for a discussion of the residual currents. After that, as a concrete example, the residual currents will be considered in graphene on the basic of a nonperturbative kinetic approach (Sect. 3). A discussion of the effect will be given in Sect. 4.

⋆e-mail: smol@sgu.ru

2 Kinetic theory

The kinetic equation (KE) for the description of the EPP vacuum creation in a strong homogeneous time dependent electric field can be cast into the form of an integro-differential equation for several QED models at once [1]

$$\dot{f}(t) = \frac{1}{2}\lambda(t)\int_{t_0}^{t} dt'\lambda(t')[1-2f(t')]\cos\left[2\int_{t'}^{t} d\tau\varepsilon(\vec{p},\tau)\right]. \qquad (1)$$

The distribution function is defined in the momentum space, $f(t) = f(\vec{p}, t)$. Quasienergy $\varepsilon(\vec{p}, t)$ and the amplitude of the vacuum transition $\lambda(\vec{p}, t)$ have different form for different QED models.

Firstly, the KE (1) was obtained in $D = 3 + 1$ QED for a linearly polarized electric field in the framework of the quasiparticle representation [2, 3]. It is easy to show that the KE conserves its form in the two above-mentioned cases of $D = 2 + 1$ massless QED with a general Hamiltonian. In the case of graphene, it corresponds to a low-energy model of excitations around the K point in the Brillouin zone for clean graphene (e.g., [4]), Thus,

$$H(t) = V\int d^2x\Psi^{\dagger}(x)\hat{\vec{P}}\vec{\sigma}\Psi(x), \qquad (2)$$

where $\hat{P}_k = -i\hbar\nabla_k - (e/c)A_k(t)$, $k = 1,2$, $A_k(t) = \left(0, A^1(t), A^2(t), 0\right)$ in the Hamilton gauge; σ_k is the Pauli matrix; $V = c$ in the vacuum QED and $V = v_F \simeq 10^8$cm/s is the Fermi velocity in graphene. The KE (1) conserves its form also in the tight-binding model of nearest neighbor interactions [5]. Here, primary the Fock representation is defined by the stationary in-vacuum state |in>. The transition to the quasiparticle representation for graphene has been executed in [4] with the help of a unitary transformation which can be realized in explicit form. In this representation, the Hamiltonian (2) acquires the diagonal form

$$H(t) = \int \frac{d^2p}{(2\pi\hbar)^2}\varepsilon(\vec{p},t)\left[a^+(\vec{p},t)a(\vec{p},t) - b(-\vec{p},t)b^+(-\vec{p},t)\right] \qquad (3)$$

with the quasienergy $\varepsilon(\vec{p}, t) = V\sqrt{P^2(t)}$. The creation and annihilation operators of the quasi-electrons and quasi-holes are defined in the nonstationary vacuum |t> and are subject to the standard anti-commutation relations. They satisfy the equations of motion [4]

$$\begin{aligned}\dot{a}(\vec{p},t) &= \tfrac{i}{\hbar}\left[H(t), a(\vec{p},t)\right] + i\lambda(\vec{p},t)b^+(-\vec{p},t),\\ \dot{b}(-\vec{p},t) &= \tfrac{i}{\hbar}\left[H(t), b(-\vec{p},t)\right] - i\lambda(\vec{p},t)a^+(\vec{p},t),\end{aligned} \qquad (4)$$

where the transition amplitude is equal to

$$\lambda(\vec{p},t) = \left\{eV^2/2\varepsilon^2\right\}\left[E_1P_2 - E_2P_1\right] \qquad (5)$$

when exploiting $E(t) = -(1/c)\dot{A}(t)$.

In the quasiparticle representation, the electron and hole (positron) distribution functions are defined as

$$\begin{aligned}f^e(\vec{p},t) &=< \mathrm{in}|a^+(\vec{p},t)a^-(\vec{p},t)|\mathrm{in} >,\\ f^h(\vec{p},t) &=< \mathrm{in}|b^+(-\vec{p},t)b^-(-\vec{p},t)|\mathrm{in} > .\end{aligned} \qquad (6)$$

Their equality $f^e = f^h = f$ is ensured by the electroneutrality condition. Using then the standard procedure [3], definitions (6) and equations of motion (4), it is easy to obtain the KE (1).

It is useful to introduce the auxiliary functions

$$\left.\begin{array}{c} u(t) \\ v(t) \end{array}\right\} = \int_{t'}^{t} dt' \lambda(t') \left[1 - 2f(t')\right] \left\{ \begin{array}{c} \cos\theta(t,t') \\ \sin\theta(t,t') \end{array} \right. . \tag{7}$$

They can be rewritten as the anomalous averages

$$\begin{array}{c} f^{(+)}(\vec{p},t) =< \mathrm{in}|a^{+}(\vec{p},t)b^{+}(-\vec{p},t)|\mathrm{in} >, \\ f^{(-)}(\vec{p},t) =< \mathrm{in}|b^{-}(-\vec{p},t)a^{-}(\vec{p},t)|\mathrm{in} > \end{array} \tag{8}$$

by means of the relations

$$u = 2\Re f^{(+)} = 2\Re f^{(-)}, \quad v = 2\Im f^{(+)} = -2\Im f^{(-)}. \tag{9}$$

These functions describe the vacuum polarization (see below) and are useful in order to write the KE (1) in the form of a system of three ordinary differential equations:

$$\dot{f} = \frac{1}{2}\lambda u, \quad \dot{u} = \lambda(1-2f) - 2\varepsilon v, \quad \dot{v} = 2\varepsilon u. \tag{10}$$

It is natural to select the zero initial conditions $f(t_0) = u(t_0) = v(t_0) = 0$. The first equality means the absence of e-h excitations at $t = t_0$. The two last equalities can be interpreted as absence of the anomalous correlations (8) at $t = t_0$ what corresponds to the in-vacuum state in the case of vacuum QED. These properties can be correlated with the adiabatic hypothesis at $t \to -\infty$.

However, the opposite asymptotics of the u and v functions are the undamped oscillations after the shutoff of the external field (see below). It corresponds to the long action vacuum Zitterbewegung. It allows to introduce the peculiar time order parameter $\Phi(t) = u(t) + iv(t)$ which satisfies to the equation of motion

$$\dot{\Phi} - 2i\varepsilon\Phi = \lambda(1-2f). \tag{11}$$

The solution of this equation with the zero initial condition is

$$\Phi(t) = \int_{-\infty}^{t} dt' \lambda(t') \left[1 - 2f(t')\right] \exp\left[2i \int_{t'}^{t} d\tau \varepsilon(\tau)\right]. \tag{12}$$

Let us consider now a finite electric field which is shut down at the point of time t_{off}, i.e. $E(t > t_{off}) = 0$ and hence $\lambda(t > t_{off}) = 0$. Then, for $t > t_{off}$ it follows from (12) that

$$\Phi(t > t_{off}) = \Phi_{out} \exp\left[2i\varepsilon(t - t_{off})\right], \tag{13}$$

where the asymptotical value of the quasienergy is equal to

$$\varepsilon_{out} = \varepsilon(t \to \infty) = V\left[\left(\vec{p} - (e/c)\vec{A}_{out}\right)^2\right]^{1/2}, \tag{14}$$

$A_{out} = \lim_{t\to\infty} A(t)$. The momentum dependent amplitude

$$\Phi_{out}(\vec{p}) = \int_{-\infty}^{t_{off}} dt' \lambda(t') \left[1 - 2f(t')\right] \exp\left[2i \int_{t'}^{t_{off}} d\tau \varepsilon(\tau)\right] \tag{15}$$

is defined by the time structure of the external field pulse.

The presence of such undamped oscillations of the order parameter is a prerequisite of the discussed effect in the polarized current.

Thus, $|\Phi_{out}(\vec{p})|^2 = const$ after switching off the external field, i.e. the long-range order is formed. Such situation is typical for a phase transition in systems with broken symmetry. This conclusion follows also from the equation for the modulus of the order parameter

$$\frac{d}{dt}|\Phi|^2 = 2\lambda(1-2f)u \tag{16}$$

which can be obtained on the basis of (11).

3 Residual currents

The current density in the quasiparticle representation can be obtained from the Hamiltonian (2) and consists of conductivity and polarization currents,

$$j_k(t) = c\delta H/\delta A_k(t) = j_k^{cond}(t) + j_k^{pol}(t), \tag{17}$$

$$j_k^{cond}(t) = -2gecV^2 \int \frac{d^2p}{(2\pi\hbar)^2}\frac{P_k}{\varepsilon} f, \tag{18}$$

$$j_k^{pol}(t) = gecV^2 \int \frac{d^2p}{(2\pi\hbar)^2}\frac{P_k}{\varepsilon} u. \tag{19}$$

Let us note that these formulas are nonperturbative. On the r.h.s. of (18), the vacuum unit in the statistical factor $1 + 2f$ drops out. The integration is carried out within the Brillouin zone, and $g = 4$ is the degeneracy factor (two Dirac points and two pseudo-spin degrees of freedom).

Now we can investigate the evolution of the currents after shutdown of the external field at $t > t_{off}$. According to the KE (1) for $t > t_{off}$ $\dot{f}(t) = 0$, i.e. $f(t > t_{off}) = const$ and the residual conductivity current (18) will be constant. However, the residual polarization current (19) survives according to (13)-(15) and has a nontrivial time dependence,

$$j_k^{pol} = (t > t_{off}) = gecV^2 \int \frac{d^2p}{(2\pi\hbar)^2}\frac{P_k}{\varepsilon}\Phi_{out}\cos\left[2\varepsilon_{out}(t - t_{off})\right]. \tag{20}$$

The integration over the Brillouin zone leads to a smoothing of the undamped oscillations of the order parameter (13).

For an illustration let us consider a supergaussian electric field model with sharp fronts. figure 1 (left) represents the form of the field pulse

$$E(t) = E_0 \exp\left[-(t-t_m)^n/2\tau^n\right], \; n > 2 \tag{21}$$

for $n = 4$. The time dependence of the polarization current density during the whole period of evolution is shown on the r.h.s. picture.

From this picture follows that the lifetime T of the residual polarization current in graphene by far exceeds the field pulse duration, $T \gg \tau$.

On a qualitative level the similar picture arises also in $D = 2+1$ massless QED with the shortened lifetime in ratio $v_F/c \simeq 10^{-2} - 10^{-3}$.

It is interesting to note also that in the case of a small coupling constant the polarization current dominates, $|j^{pol}| \gg |j^{cond}|$ in the leading approximation. However, in the case of the running coupling constant depending on the mean energy (as in QCD) this inequality can be reversed.

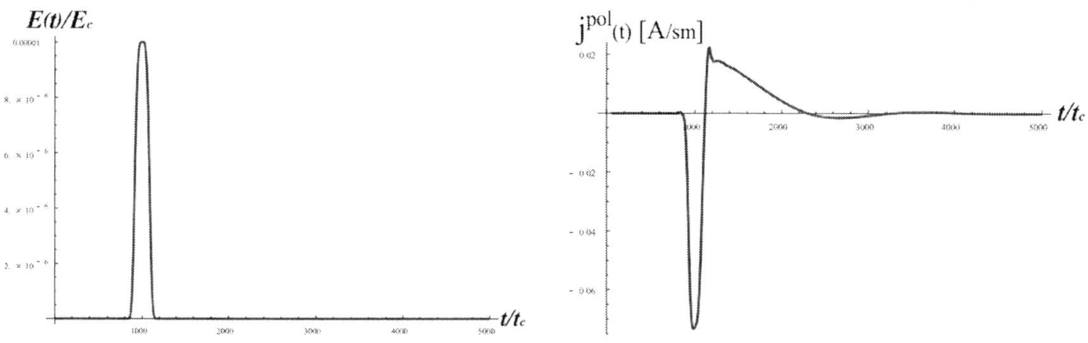

Figure 1. The electric field and the time scale normalized to the characteristic values $E_c = \hbar v_F / ea^2 = 1.088 \times 10^8 V/sm$ and $t_c = a/v_f = 2.46 \times 10^{-16} s$ ($a = 2.46\mathring{A}$ is the lattice spacing). **Left**: Supergaussian electric field (21) $n = 4$, $t_m = 2.46 \times 10^{-13} s$, $\tau = 2 \times 10^{-14} s$ and $E_0 = 1.088 \times 10^3 V/sm$. **Right**: The density of the polarization current.

4 Conclusion

The considered vacuum polarization effect is stipulated by inertial properties of the physical quantum vacuum or the vacuumlike state of graphene. In the last case the effect, apparently, can be available for an experimental test on the background of the constant residual conductivity current. However, this potential possibility needs a more detailed investigation of the dependence on external field parameters. On the other hand, it would be interesting to investigate this effect in more realistic massive QED models.

References

[1] S.A. Smolyansky, A.D. Panferov, D.B. Blaschke, L. Juchnowski, B. Kämpfer, A. Otto, arXiv: 1607.08775 [hep-ph]

[2] A.A. Grib, S.G. Mamaev, V.M. Mostepanenko, *Vacuum Quantum Effects in Strong External Fields* (Friedmann Laboratory Publishing, St.-Peterburg, 1994)

[3] S.M. Schmidt, D. Blaschke, G. Roepke, S.A. Smolyansky, A.V. Prozorkevich, and V.D. Toneev, Int. J. Mod. Phys. E **7**, 709 (1998)

[4] B. Dora, and R. Moessner, Phys. Rev. B **81**, 165431 (2010)

[5] H.C. Kao, M. Lewkowicz, and B. Rosenstein, Phys. Rev. B **82**, 035406 (2010)

Critical behaviour of (2+1)-dimensional QED: $1/N$-corrections

Anatoly V. Kotikov[1,⋆] and *Sofian* Teber[2,3,⋆⋆]

[1] *Bogoliubov Laboratory of Theoretical Physics, Joint Institute for Nuclear Research, 141980 Dubna, Russia*
[2] *Sorbonne Universités, UPMC Univ Paris 06, UMR 7589, LPTHE, F-75005, Paris, France and CNRS, UMR 7589, LPTHE, F-75005, Paris, France*
[3] *CNRS, UMR 7589, LPTHE, F-75005, Paris, France*

Abstract. We present recently obtained results for dynamical chiral symmetry breaking studied within $(2+1)$-dimensional QED with N four-component fermions. The leading and next-to-leading orders of the $1/N$ expansion are computed exactly in an arbitrary non-local gauge.

1 Introduction

In these Proceedings we present the results of our recent papers [1, 2], where the critical behavior of Quantum Electrodynamics in 2 + 1 dimensions (QED$_3$) has been studied. QED$_3$ is described by the Lagrangian:

$$L = \overline{\Psi}(i\hat{\partial} - e\hat{A})\Psi - \frac{1}{4}F^2_{\mu\nu}\,, \tag{1}$$

where Ψ is taken to be a four component complex spinor. In the presence of N fermion flavours, the model has a $U(2N)$ symmetry. A (parity-invariant) fermion mass term, $m\overline{\Psi}\Psi$, breaks this symmetry to $U(N)\times U(N)$. In the massless case, loop expansions are plagued by infrared divergences. The latter soften upon analyzing the model in a $1/N$ expansion [3]. Since the theory is super-renormalizable, the mass scale is then given by the dimensionless coupling constant: $a = Ne^2/8$, which is kept fixed as $N \to \infty$. Early studies of this model [4, 5] suggested that the physics is rapidly damped at momentum scales $p \gg a$ and that a fermion mass term breaking the flavour symmetry is dynamically generated at scales which are orders of magnitude smaller than the intrinsic scale a. Since then, dynamical chiral symmetry breaking (DχSB) in QED$_3$ and the dependence of the dynamical fermion mass on N have been the subject of extensive studies, see, *e.g.*, [1, 2, 4–13].

One of the central issue is related to the value of the critical fermion number, N_c, which is such that DχSB takes place only for $N < N_c$. An accurate determination of N_c is of crucial importance to understand the phase structure of QED$_3$ with far reaching implications from particle physics to planar condensed matter physics systems having relativistic-like low-energy excitations [14]. It turns out that the values that can be found in the literature vary from $N_c \to \infty$ [4, 6] corresponding to DχSB for all values of N, all the way to $N_c \to 0$ in the case where no sign of DχSB is found [7].

⋆e-mail: kotikov@theor.jinr.ru
⋆⋆e-mail: teber@lpthe.jussieu.fr

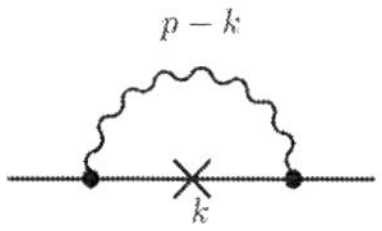

Figure 1. LO diagram to the dynamically generated mass $\Sigma(p)$. The crossed line denotes mass insertion.

Of importance to us in the following, is the approach of Appelquist et al. [5] who found that $N_c = 32/\pi^2 \approx 3.24$ by solving the Schwinger-Dyson (SD) gap equation using a leading order (LO) $1/N$-expansion. Lattice simulations in agreement with a finite non-zero value of N_c can be found in [8]. Soon after the analysis of [5], Nash approximately included next-to-leading order (NLO) corrections and performed a partial resummation of the wave-function renormalization constant at the level of the gap equation; he found [9]: $N_c \approx 3.28$.

Recently, in [1], the NLO corrections could be computed exactly in the Landau gauge upon refining the analysis of [10]. This led to $N_c \approx 3.29$, a value which is surprisingly close to the one of Nash in [9]. More recently, in [2], the results of [1] were generalized to an arbitrary non-local gauge [15]. Moreover, it was shown in [2] that a resummation of the wave-function renormalization yields a strong suppression of the gauge dependence of the critical fermion flavour number, $N_c(\xi)$ where ξ is the gauge fixing parameter, which is such that DχSB takes place for $N < N_c(\xi)$. Neglecting the gauge-dependent terms yields $N_c = 2.8469$, that coincides with results in [11]. In the general case, it is found that: $N_c(1) = 3.0084$ in the Feynman gauge, $N_c(0) = 3.0844$ in the Landau gauge and $N_c(2/3) = 3.0377$ in the $\xi = 2/3$ gauge where the leading order fermion wave function is finite. These results suggest that DχSB should take place for integer values $N \leq 3$. Using a very different method, Herbut obtained [12] a close value: $N_c \approx 2.89$.

It is the purpose of this work to review some of the basic steps of papers [1, 2] which represent an essential improvement with respect to Nash's approximate NLO results derived some 30 years ago.

2 Schwinger-Dyson equations

With the conventions of [1], the inverse fermion propagator is defined as: $S^{-1}(p) = [1 + A(p)]\,(i\hat{p} + \Sigma(p))$ where $A(p)$ is the fermion wave function and $\Sigma(p)$ is the dynamically generated parity-conserving mass which is taken to be the same for all the fermions. The SD equation for the fermion propagator may be decomposed into scalar and vector components as follows:

$$\tilde{\Sigma}(p) = \frac{2a}{N}\,\mathrm{Tr}\int \frac{d^3k}{(2\pi)^3}\,\frac{\gamma^\mu D_{\mu\nu}(p-k)\Sigma(k)\Gamma^\nu(p,k)}{[1+A(k)]\,(k^2+\Sigma^2(k))}\,, \quad A(p)p^2 = -\frac{2a}{N}\,\mathrm{Tr}\int \frac{d^3k}{(2\pi)^3}\,\frac{D_{\mu\nu}(p-k)\hat{p}\gamma^\mu\hat{k}\Gamma^\nu(p,k)}{[1+A(k)]\,(k^2+\Sigma^2(k))}\,, \tag{2}$$

where $\tilde{\Sigma}(p) = \Sigma(p)[1 + A(p)]$, $D_{\mu\nu}(p)$ is the photon propagator in the non-local ξ-gauge:

$$D_{\mu\nu}(p) = \frac{P^\xi_{\mu\nu}(p)}{p^2\,[1+\Pi(p)]}, \quad P^\xi_{\mu\nu}(p) = g_{\mu\nu} - (1-\xi)\frac{p_\mu p_\nu}{p^2}\,, \tag{3}$$

$\Pi(p)$ is the polarization operator and $\Gamma^\nu(p,k)$ is the vertex function. In the following, (2) will be studied for an arbitrary value of the gauge-fixing parameter ξ. All calculations will be performed with the help of the standard rules of perturbation theory for massless Feynman diagrams as in [16], see also the recent short review [17]. For the most complicated diagrams, the Gegenbauer polynomial technique will be used following [18].

3 Gap equation at leading order

The LO approximations in the $1/N$ expansion are given by: $A(p) = 0$, $\Pi(p) = a/|p|$ and $\Gamma^\nu(p,k) = \gamma^\nu$, where the fermion mass has been neglected in the calculation of $\Pi(p)$. A single diagram contributes to the gap equation (2) at LO, see figure 1, and the latter reads:

$$\Sigma(p) = \frac{8(2+\xi)a}{N} \int \frac{d^3k}{(2\pi)^3} \frac{\Sigma(k)}{(k^2+\Sigma^2(k))\,[(p-k)^2 + a\,|p-k|]} \,. \tag{4}$$

Following [5], we consider the limit of large a and linearize (4) which yields:

$$\Sigma(p) = \frac{8(2+\xi)}{N} \int \frac{d^3k}{(2\pi)^3} \frac{\Sigma(k)}{k^2\,|p-k|} \,. \tag{5}$$

The mass function may then be parameterized as [5]: $\Sigma(k) = B\,(k^2)^{-\alpha}$, where B is arbitrary and the index α has to be self-consistently determined. Using this Ansatz, (5) leads to the LO gap equation: ($\beta^{-1} = \alpha(1/2-\alpha)$ and $L \equiv \pi^2 N$)

$$1 = \frac{(2+\xi)\beta}{L} \quad \text{and} \quad \alpha_\pm = \frac{1}{4}\left(1 \pm \sqrt{1 - \frac{16(2+\xi)}{L}}\right), \tag{6}$$

which reproduces the solution given by Appelquist et al. [5]. The gauge-dependent critical number of fermions: $N_c \equiv N_c(\xi) = 16(2+\xi)/\pi^2$, is such that $\Sigma(p) = 0$ for $N > N_c$ and $\Sigma(0) \simeq \exp[-2\pi/(N_c/N - 1)^{1/2}]$, for $N < N_c$. Thus, DχSB occurs when α becomes complex, that is for $N < N_c$.

The gauge-dependent fermion wave function may be computed in a similar way. At LO, (2) simplifies as:

$$A(p)p^2 = -\frac{2a}{N}\mathrm{Tr}\int \frac{d^Dk}{(2\pi)^D} \frac{P^\xi_{\mu\nu}(p-k)\hat{p}\gamma^\mu\hat{k}\gamma^\nu}{k^2|p-k|}\,, \tag{7}$$

where the integral has been dimensionally regularized with $D = 3-2\varepsilon$. Taking the trace and computing the integral on the r.h.s. yields:

$$A(p) = \frac{\overline{\mu}^{2\varepsilon}}{p^{2\varepsilon}}\,C_1(\xi) + \mathrm{O}(\varepsilon), \quad C_1(\xi) = +\frac{2}{3\pi^2 N}\left((2-3\xi)\left[\frac{1}{\varepsilon} - 2\ln 2\right] + \frac{14}{3} - 6\xi\right), \tag{8}$$

where the $\overline{MS}$ parameter $\overline{\mu}$ has the standard form $\overline{\mu}^2 = 4\pi e^{-\gamma_E}\mu^2$ with the Euler constant γ_E. We note that in the $\xi = 2/3$-gauge, the value of $A(p)$ is finite and $C_1(\xi = 2/3) = +4/(9\pi^2 N)$. From (8), the LO wave-function renormalization constant may be extracted: $\lambda_A = \mu(d/d\mu)A(p) = 4(2-3\xi)/(3\pi^2 N)$ a result which coincides with the one of [19].

4 Next-to-leading order

We now consider the NLO contributions and parametrize them as:

$$\Sigma^{(\mathrm{NLO})}(p) = \left(\frac{8}{N}\right)^2 B\,\frac{(p^2)^{-\alpha}}{(4\pi)^3}\,(\Sigma_A + \Sigma_1 + 2\,\Sigma_2 + \Sigma_3)\,, \tag{9}$$

where each contribution to the linearized gap equation is represented graphically in figure 2. The gap equation has the following general form:

$$1 = \frac{(2+\xi)\beta}{L} + \frac{\overline{\Sigma}_A(\xi) + \overline{\Sigma}_1(\xi) + 2\,\overline{\Sigma}_2(\xi) + \overline{\Sigma}_3(\xi)}{L^2}\,, \quad \overline{\Sigma}_i = \pi\Sigma_i, \;\; (i = 1,2,3.A) \tag{10}$$

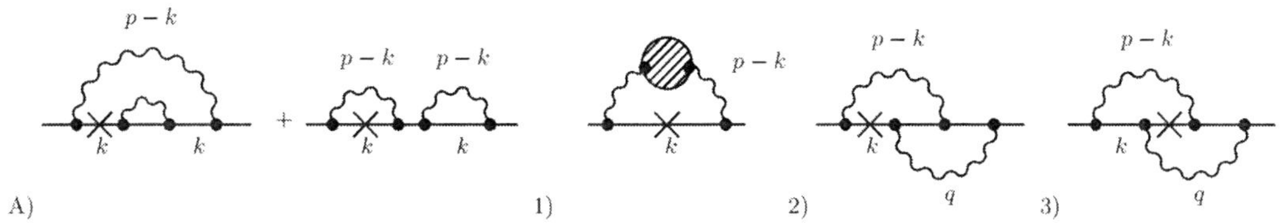

Figure 2. NLO diagrams to the dynamically generated mass $\Sigma(p)$. The shaded blob defines the two-loop polarization operator, see [1, 2] for details.

Performing the calculation of the diagrams shown in figure 2 (see [1, 2]), the gap equation (10) may be written in an explicit form as:

$$1 = \frac{(2+\xi)\beta}{L} + \frac{1}{L^2}\left[8S(\alpha,\xi) - 2(2+\xi)\hat{\Pi}\beta + \left(-\frac{5}{3} + \frac{26}{3}\xi - 3\xi^2\right)\beta^2 - 8\beta\left(\frac{2}{3}(1-\xi) - \xi^2\right)\right], \qquad (11)$$

where $\hat{\Pi} = 92/9 - \pi^2$ arises from the two-loop polarization operator in dimension $D = 3$ [20, 21] and $S(\alpha,\xi)$ contains the contributions of complicated diagrams. Considering (11) directly at the critical point $\alpha = 1/4$, *i.e.*, at $\beta = 16$, we have

$$L_c^2 - 16(2+\xi)L_c - 8\left[S(\xi) - 4(2+\xi)\hat{\Pi} - 16\left(4 - 50\xi/3 + 5\xi^2\right)\right] = 0, \qquad (12)$$

where $S(\xi) = S(\alpha = 1/4, \xi)$ and

$$8S(\xi) = 8(1-\xi)R_1 + (\xi^2-1)R_2 - (7+16\xi-3\xi^2)\frac{P_2}{16}, \quad R_1 = 163.7428, \ R_2 = 209.175, \ P_2 = 1260.720 \qquad (13)$$

Solving (12), we have two standard solutions:

$$L_{c,\pm} = 8(2+\xi) \pm \sqrt{d_1(\xi)}, \quad d_1(\xi) = 8\left[S(\xi) - 8\left(4 - \frac{112}{3}\xi + 9\xi^2 + \frac{2+\xi}{2}\hat{\Pi}\right)\right]. \qquad (14)$$

Combining these values with the one of $\hat{\Pi}$, yields: $N_c(\xi = 0) = 3.29$, $N_c(\xi = 2/3) = 3.09$, where "−" solutions are unphysical and there is no solution in the Feynman gauge. The range of ξ-values for which there is a solution corresponds to $\xi_- \leq \xi \leq \xi_+$, where $\xi_+ = 0.88$ and $\xi_- = -2.36$.

5 Resummation

Following [9], we would like to resum the LO term together with part of the NLO corrections containing terms $\sim \beta^2$. In order to do so, we will now rewrite the gap equation (11) in a form which is suitable for resummation. This amounts to extract the terms $\sim \beta$ and $\sim \beta^2$ from the complicated part of the fermion self-energy, $S(\alpha,\xi)$, yielding:

$$S(\alpha,\xi) = \frac{1}{4}(1-\xi)\beta(3\beta - 8) - \frac{1}{2}\xi(4+\xi)\beta + \tilde{S}(\alpha,\xi). \qquad (15)$$

At the critical point $\alpha = 1/4$ ($\beta = 16$), $\tilde{S}(\xi) = \tilde{S}(\alpha = 1/4, \xi)$ has the following form:

$$8\tilde{S}(\xi) = 8(1-\xi)\tilde{R}_1 + (\xi^2-1)\tilde{R}_2 - (7+16\xi-3\xi^2)\frac{\tilde{P}_2}{16}, \quad \tilde{R}_1 = 3.7428, \ \tilde{R}_2 = 1.175, \ \tilde{P}_2 = -19.28. \qquad (16)$$

With the help of the results (16), the gap equation (11) may be written as:

$$1 = \frac{(2+\xi)\beta}{L} + \frac{1}{L^2}\Big[8\tilde{S}(\alpha,\xi) - 2(2+\xi)\hat{\Pi}\beta + \left(\frac{2}{3} - \xi\right)(2+\xi)\beta^2 + 4\beta\left(\xi^2 - \frac{4}{3}\xi - \frac{16}{3}\right)\Big]. \tag{17}$$

At this point (11) and (17) are strictly equivalent to each other and yield the same values for $N_c(\xi)$. Equation (17) is the convenient starting point to perform a resummation of the wave function renormalization constant. To do it (see details in [2]) (17) can now be expressed as:

$$1 = \frac{8\beta}{3L} + \frac{1}{L^2}\left[8\tilde{S}(\alpha,\xi) - \frac{16}{3}\beta\left(\frac{40}{9} + \hat{\Pi}\right)\right], \tag{18}$$

which displays a strong suppression of the gauge dependence even at NLO as ξ-dependent terms do exist but they enter the gap equation only through the rest, $\tilde{S}$, which is very small numerically.

We now consider (18) at the critical point, $\alpha = 1/4$ ($\beta = 16$), which yields:

$$L_c^2 - \frac{128}{3}L_c - \left[8\tilde{S}(\xi) - \frac{256}{3}\left(\frac{40}{9} + \hat{\Pi}\right)\right] = 0. \tag{19}$$

Solving (19), we have two standard solutions:

$$L_{c,\pm} = \frac{64}{3} \pm \sqrt{d_2(\xi)}, \quad d_2(\xi) = \left(\frac{64}{3}\right)^2 + \left[8\tilde{S}(\xi) - \frac{256}{3}\left(\frac{40}{9} + \hat{\Pi}\right)\right]. \tag{20}$$

In order to provide a numerical estimate for N_c, we have used the values of $\tilde{R}_1$, $\tilde{R}_2$ and $\tilde{P}_2$ of (16). Combining these values together with the value of $\hat{\Pi}$, yields, for $N_c(\xi)$ ("$-$" solutions being unphysical):

$$N_c(0) = 3.08, \qquad N_c(2/3) = 3.04, \qquad N_c(1) = 3.01. \tag{21}$$

Actually, solutions exist for a broad range of values of ξ: $\tilde{\xi}_- \leq \xi \leq \tilde{\xi}_+$, where $\tilde{\xi}_+ = 4.042$ and $\tilde{\xi}_- = -8.412$; this is consistent with the weak ξ-dependence of the gap equation. Moreover, following [22], we think that the "right(est)" gauge choice is one close to $\xi = 2/3$ where the LO fermion wave function is finite. Indeed, upon resumming the theory, the value of $N_c(\xi)$ increases (decreases) for small (large) values of ξ. For $\xi = 2/3$, the value of N_c is very stable, decreasing only by 1-2% during resummation. Finally, if we neglect the rest, *i.e.*, $\tilde{S}(\xi) = 0$ in (19), the gap equation becomes ξ-independent and we have: $\overline{L}_c = 28.0981$ and therefore: $\overline{N}_c = 2.85$, a value that coincides with the one in [11].

6 Conclusion

We have presented the studies [1, 2] of DχSB in QED$_3$ by including $1/N^2$ corrections to the SD equation exactly and taking into account the full ξ-dependence of the gap equation. Following Nash, the wave function renormalization constant has been resummed at the level of the gap equation leading to a very weak gauge-variance of the critical fermion number N_c. The value obtained for the latter, (21), suggests that DχSB takes place for integer values $N \leq 3$ in QED$_3$.

Notice that the large-N limit of the photon propagator in QED$_3$ has precisely the same momentum dependence as the one in the so-called reduced QED, see [22]. One difference is that the gauge fixing parameter in reduced QED is twice less than the one in QED$_3$. Such a difference can be taken into account with the help of our present results for QED$_3$ together with the multi-loop results obtained in [20, 23]. The case of reduced QED, and its relation with dynamical gap generation in graphene which is the subject of active ongoing research, see, *e.g.*, the reviews [24], was considered in our paper [25].

References

[1] A. V. Kotikov, V. I. Shilin and S. Teber, Phys. Rev. D **94**, 056009 (2016)

[2] A. V. Kotikov and S. Teber, arXiv:1609.06912 [hep-th]

[3] T. Appelquist and R. Pisarski, Phys. Rev. D **23**, 2305 (1981). R. Jackiw and S. Templeton, Phys. Rev. D **23**, 2291 (1981). T. Appelquist and U. Heinz, Phys. Rev. D **24**, 2169 (1981)

[4] R. Pisarski, Phys. Rev. D **29**, 2423 (1984)

[5] T. Appelquist, D. Nash and L. C. R. Wijewardhana, Phys. Rev. Lett. **60**, 2575 (1988)

[6] M. R. Pennington and D. Walsh, Phys. Lett. **253B**, 246 (1991). D. C. Curtis, M. R. Pennington and D. Walsh, Phys. Lett. **295B**, 313 (1992). R. Pisarski, Phys. Rev. D **44**, 1866 (1991). V. Azcoiti and X. Q. Luo, Nucl. Phys. Proc. Suppl. **30**, 741 (1993). V. Azcoiti, V. Laliena and X. Q. Luo, Nucl. Phys. Proc. Suppl. **47**, 565 (1996)

[7] D. Atkinson, P. W. Johnson and P. Maris, Phys. Rev. D **42**, 602 (1990). N. Karthik and R. Narayanan, Phys. Rev. D **93**, 045020 (2016). Phys. Rev. D **94**, 065026 (2016)

[8] E. Dagotto, A. Kocic and J. B. Kogut, Phys. Rev. Lett. **62**, 1083 (1989). Nucl. Phys. B **334**, 279 (1990). S. J. Hands, J. B. Kogut, L. Scorzato and C. G. Strouthos, Phys. Rev. B **70**, 104501 (2004). C. Strouthos and J. B. Kogut, PoS (LAT 2007) 278 (2007)

[9] D. Nash, Phys. Rev. Lett. **62**, 3024 (1989)

[10] A. V. Kotikov, JETP Lett. **58**, 734 (1993). Phys. Atom. Nucl. **75**, 890 (2012)

[11] V. P. Gusynin and P. K. Pyatkovskiy, arXiv:1607.08582 [hep-ph]

[12] I. F. Herbut, Phys. Rev. D **94**, 025036 (2016)

[13] A. Bashir, A. Raya, S. Sanchez-Madrigal and C. D. Roberts, Few Body Syst. **46**, 229 (2009)

[14] J. B. Marston and I. Affleck, Phys. Rev. B **16**, 11538 (1989). L. B. Ioffe and A. I. Larkin, Phys. Rev. B **13**, 8988 (1989). G. W. Semenoff, Phys. Rev. Lett. **53**, 2449 (1984). P. R. Wallace, Phys. Rev. **71**, 622 (1947)

[15] E. H. Simmons, Phys. Rev. D **42**, 2933 (1990). T. Kugo and M. G. Mitchard, Phys. Lett. B **282**, 162 (1992)

[16] D. I. Kazakov, Phys. Lett. **133B**, 406 (1983). "Analytical Methods For Multiloop Calculations: Two Lectures On The Method Of Uniqueness" - 1984. Preprint JINR E2-84-410, Dubna

[17] S. Teber and A. V. Kotikov, arXiv:1602.01962 [hep-th]

[18] A. V. Kotikov, Phys. Lett. B **375**, 240 (1996)

[19] J. A. Gracey, Nucl. Phys. B **414**, 614 (1994)

[20] S. Teber, Phys. Rev. D **86**, 025005 (2012). A. V. Kotikov and S. Teber, Phys. Rev. D **87**, 087701 (2013)

[21] V. P. Gusynin, A. H. Hams and M. Reenders, Phys. Rev. D **63**, 045025 (2001). J.A. Gracey, Phys. Lett. B **317**, 415 (1993)

[22] E. V. Gorbar, V. P. Gusynin and V. P. Miransky, Phys. Rev. D **64**, 105028 (2001). E. C. Marino, Nucl. Phys. B **408**, 551 (1993). N. Dorey and N. E. Mavromatos, Nucl. Phys. B **386**, 614 (1992). A. Kovner and B. Rosenstein, Phys. Rev. D **42**, 4748 (1990)

[23] A. V. Kotikov and S. Teber, Phys. Rev. D **89**, 065038 (2014). S. Teber, Phys. Rev. D **89**, 067702 (2014)

[24] V. N. Kotov, B. Uchoa, V. M. Pereira, F. Guinea and A. H. Castro Neto, Rev. Mod. Rev. **84**, 1067 (2012). V. A. Miransky and I. A. Shovkovy, Phys. Rept. **576**, 1 (2015). V. P. Gusynin, Prob. Atomic Sci. Technol. **2013N3**, 29 (2013)

[25] A. V. Kotikov and S. Teber, arXiv:1610.00934 [hep-th]

EPJ Web of Conferences **138**, 07001 (2017) DOI: 10.1051/epjconf/201713807001
Baldin ISHEPP XXIII

Physics of double parton interactions

Recent results from DØ experiment

Georgy Golovanov[1,⋆]

[1] *Joint institute for nuclear research, Dubna, Russia*

Abstract. A study of processes with double parton interactions (DPI) in $p\bar{p}$ collisions collected by DØ detector at the Tevatron collider at $\sqrt{s}$ = 1.96 TeV is presented. The study includes measurements of DPI event fraction and the effective cross section, a process-independent parameter related to the effective interaction region inside a hadron. The measurements are done using processes with photon and hadronic jets in different parton-parton scatterings: photon+jet + dijet, photon+b/c + dijet and diphoton + dijet. The measured effective cross section is applied in order to estimate background from DPI processes to WH production at the Tevatron energy.

1 Introduction

High energy inelastic hadron-hadron collisions are generally described as a hard $2 \to 2$ partonic scattering with soft gluon radiation in initial and final states with account of fragmentation and hadronization effects. However, there is a possibility to find more than one hard parton-parton scattering within the same hadron-hadron collision. The cross section of a process with double parton interaction is proportional to cross sections of two partonic scatterings A and B: $\sigma_{\rm DPI} \equiv \sigma^{\rm A}\sigma^{\rm B}/\sigma_{\rm eff}$. The scaling parameter, $\sigma_{\rm eff}$, historically called the *effective cross section*, having a unit of cross section, characterizes the size of the effective interaction region within a hadron. The effective cross section is directly related to the spatial distribution of partons within a hadron and being a phenomenological parameter strongly needs experimental input. Moreover, some final states being produced by DPI mechanism can mimic rare processes produced by single parton-parton interaction (SPI). A study of DPI at high-$p_{\rm T}$ regime provides the important information about hadron structure and crucial for precise estimates of background for many rare processes.

2 Study of DPI kinematics in final states containing photons and jets

The DØ detector is a general purpose detector described in [1]. The events should pass triggers based on the identification of high-$p_{\rm T}$ cluster in the electromagnetic calorimeter with loose shower shape requirements for photons. Photon candidates are selected using calorimeter cells by creating clusters inside a cone with radii $\mathcal{R} \equiv [\Delta\eta^2 + \Delta\phi^2]^{1/2} = 0.2$ around a cell-initiator. A photon candidate in an event should be isolated in a ring $\mathcal{R} = 0.2$, the electromagnetic fraction of the total photon energy

⋆e-mail: golovang@fnal.gov

disposition should be > 96% and the probability to have a track associated to the calorimeter cluster to be < 0.1%.

Jets are reconstructed using the DØ Run II iterative midpoint cone algorithm with a cone size $\mathcal{R} = 0.5$. Measured jet energy does not necessarily match the energy of the parent parton caused by the parton-to-hadron fragmentation and interaction of hadrons with calorimeter material. Jet energy scale procedure is applied in order to correct measured jet energy to the true energy sum of all particles in a jet [2].

The technique used for $\sigma_{\rm eff}$ measurement is based on the difference of double parton-parton interaction and double $p\bar{p}$ collision event rates [3]. Assuming that scatterings in the two parton-parton hard processes are uncorrelated, DPI and double $p\bar{p}$ collision events should be kinematically identical. The effective cross section is extracted from the ratio of observed DPI and double $p\bar{p}$ collision event rates.

2.1 DPI in $p\bar{p} \to \gamma$+jet + dijet process

The associate production of photons and hadronic jets in leading order is based on Compton-like $qg \to q\gamma$ and annihilation $q\bar{q} \to g\gamma$ processes. Additional hadronic jets in an event are caused by gluon radiation in initial and final states that also leads to $p_{\rm T}$-imbalance between the photon and the leading jet. The main source of background to γ + jets production comes from processes with photons produced by hadron decays (π^0, η, etc.) and fragmentation photons, that mostly suppressed by the photon isolation requirement. The $qg \to q\gamma$ process dominates in broad kinematic region with the photon transverse momentum $p_{\rm T}^{\gamma} < 150$ GeV and its contribution to the total γ+ jets cross section becomes comparable to the annihilation process at $p_{\rm T}^{\gamma} \simeq 150$ GeV. The fraction of $qg \to q\gamma$ relative to the total γ + jets event cross section is shown in figure 1(a).

The DP γ+jet + dijet event sample is selected from DØ data with a single $p\bar{p}$ collision vertex [3]. The collected statistics of the sample corresponds to the integrated luminosity of 8.7 fb^{-1}. Each event must contain at least one photon in the pseudorapidity region $|\eta| < 1.0$ or $1.5 < |\eta| < 2.5$ and at least three jets with $|\eta| < 2.5$. Events are selected with the photon transverse momentum $p_{\rm T}^{\gamma} > 26$ GeV, leading (in $p_{\rm T}$) jet $p_{\rm T}^{jet\ 1} > 15$ GeV, while the next-to-leading (second) and third jets must have $15 < p_{\rm T}^{jet\ 2,3} < 30$ GeV. To avoid photon/jets overlapping the topology cuts requiring all object pairs (photon-jet or jet-jet) to be separated in $(\eta - \phi)$ space by distance $\Delta R(\gamma, \text{jet}) > 0.7$ and $\Delta R(\text{jet}, \text{jet}) > 1.0$ are applied.

Any event in γ+jet + dijet sample can be produced by two independent $2 \to 2$ parton-parton scatterings or by a single parton-parton $2 \to 4$ scattering with gluon radiation in an initial or final state as well. To identify the events with two independent parton-parton scatterings that produce γ+jet + dijet final state, an angular distribution sensitive to the kinematics of the DPI events is used. Among others, one can define a variable:

$$\Delta S \equiv \Delta\phi(\vec{q}_{\rm T}^{\,1}, \vec{q}_{\rm T}^{\,2}), \tag{1}$$

where $\Delta\phi$ is an azimuthal angle between the imbalance vectors of photon-leading jet ($\vec{q}_{\rm T}^{\,1}$) and dijet ($\vec{q}_{\rm T}^{\,2}$) pairs in γ+jet + dijet events. This angle is schematically shown in figure 2(a). The distribution of ΔS variable reflects angular properties of a mixture of γ+jet + dijet events containing both single and double parton-parton scatterings. The contribution from SPI events to the ΔS tends to shift distribution towards π because of the well-balanced in $p_{\rm T}$ photon and jets in a $2 \to 4$ process. The contribution from the DPI events should be flat since the ΔS angle between $\vec{q}_{\rm T}$ vectors of two independent $2 \to 2$ scatterings is arbitrary.

To study properties of DPI events and calculate their fractions in a single $p\bar{p}$ collision γ+jet + dijet data sample, one needs to construct a model that correctly reflects properties of DPI kinematics. The

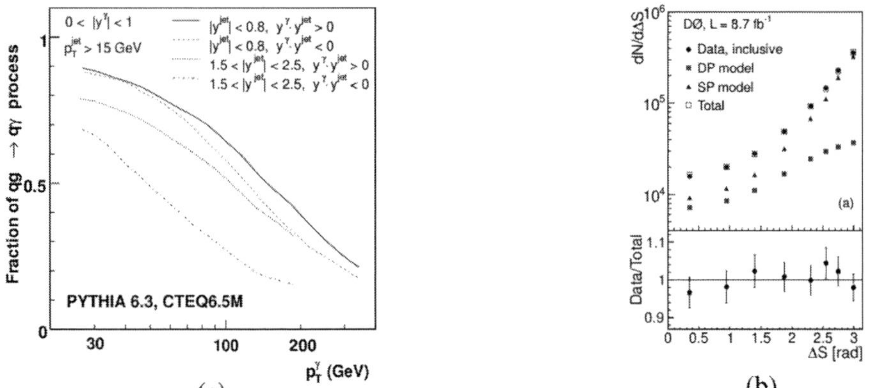

(a) (b)

Figure 1. (a) The fraction of events produced via the $qg \to q\gamma$ subprocess relative to the total associated photon and jet production for various photon and jet rapidity regions. (b) The ΔS distribution in the data, DPI, SPI models, and the sum of the DPI and SPI contributions weighted with their fractions ("Total") in γ+jet + dijet final state.

(a) (b)

Figure 2. A possible orientation of photon and jets transverse momenta vectors. Vectors $\vec{q}_T^{\,1}$ and $\vec{q}_T^{\,2}$ are the p_T-imbalance vectors of (a) γ+jet and dijet pairs in γ+jet + dijet events and (b) diphoton and dijet pairs in $\gamma\gamma$ + dijet events.

DPI model is composed by overlaying in a single event one event of an inclusive sample of γ + 1 jet events and one event of a sample of inelastic non-diffractive events selected with the minimum bias trigger and a requirement of at least one jet. Both samples contain only single $p\bar{p}$ collision events. The resulting mixed events, with jets re-ordered in p_T, are required to pass the γ+jet + dijet event selections described above. This model of DP events, called MIXDP, assumes independent parton scatterings with γ + jets and dijet final states by construction.

The DPI event fraction, $f_{\rm DPI}$, is found using a maximum likelihood fit of the ΔS distribution of the data to signal DPI and background SPI templates that are taken to be shapes of the ΔS distribution in the MIXDP model and SPI model simulated by SHERPA, respectively. Signal and background samples used as templates satisfy all the selection criteria applied to the data sample. The result of the fit is shown in figure 1(b). The found fraction of DPI events with total uncertainty is $f_{\rm DPI}^{\gamma jjj} = 0.206 \pm 0.007(stat) \pm 0.004(syst)$. Heaving measured number of DPI events, corresponding selection efficiencies and geometrical acceptances one can calculate effective cross section, which is found to be $\sigma_{\rm DPI}^{\gamma jjj} = 12.7 \pm 0.2(stat) \pm 1.3(syst)$ mb.

DOI: 10.1051/epjconf/201713807001

2.2 DPI in $p\bar{p} \to \gamma$+b/c + dijet process

The evolution of the study described in section 2.1 is a first measurement of the DPI rates and $\sigma_{\rm eff}$ involving heavy flavor leading jet using the γ+b/c + dijet final state and direct comparison to the results obtained with γ+jet + dijet events. The γ+b/c + dijet production is mainly caused by $b(c)g \to b(c)\gamma$ and $q\bar{q} \to g\gamma$ with $g \to Q\bar{Q}$, where $Q = b(c)$ [3]. Figure 3(a) shows the fractions of qg and bg subprocesses in events with γ+jet and $\gamma + b$-jet final states, calculated using default PYTHIA 6.4 settings and the CTEQ 6.1L parton distribution function. At $p_T^\gamma \simeq 30$ GeV the Compton-like scattering dominates over the annihilation process, contributing about (85 − 88)% of events. Since the initial quarks in the Compton-like scattering for inclusive γ+jet and $\gamma + b$-jet production are typically light and b/c quarks, respectively, the difference between effective cross sections measured in the two processes should be sensitive to the difference between light quark and heavy quark transverse spatial distributions in a hadron.

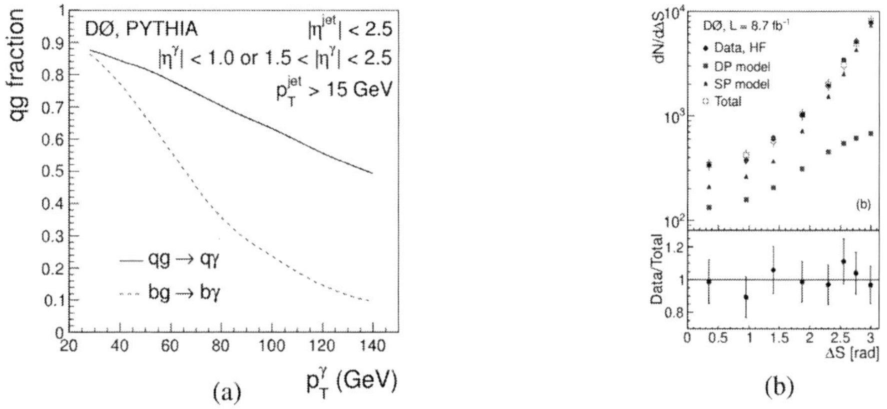

Figure 3. (a) The fractions of events produced via the $qg \to q\gamma$ and $bg \to b\gamma$ subprocesses. (b) The ΔS distribution in the data, DPI, SPI models, and the sum of the DPI and SPI contributions weighted with their fractions ("Total") in γ+b/c + dijet final state.

Events undergo the same selection criteria as described in Section 2.1 with additional leading jet requirements which ensure that there is sufficient information to identify the leading jet as a heavy flavor candidate and have typical efficiency of about 90%. The technique of the $f_{\rm DPI}$ and $\sigma_{\rm eff}$ extraction is similar to the previous measurements and the found fraction is $f_{\rm DPI}^{\gamma bjj} = 0.173 \pm 0.020(stat) \pm 0.002(syst)$, that gives the effective cross section $\sigma_{\rm DPI}^{\gamma bjj} = 14.6 \pm 0.6(stat) \pm 3.2(syst)$ mb. Figure 3(b) illustrates the results of the fitting procedure. Within uncertainties, the effective cross section in the γ+jet + dijet event sample is consistent with that in the event sample with identified heavy flavor jets, allowing to conclude that there is no evidence of $\sigma_{\rm eff}$ dependence on the initial parton flavor.

2.3 DPI in $p\bar{p} \to \gamma\gamma$ + dijet process

The goal of this study is to obtain the DPI rate and the effective cross section in the $\gamma\gamma$ + dijet final state [4]. The main contributions to diphoton production at the Tevatron are from the $q\bar{q} \to \gamma\gamma$ and $gg \to \gamma\gamma$ via direct $2 \to 2$ partonic processes, as well as from bremsstrahlung processes with a single and double parton-to-photon fragmentations. Figure 4(a) shows the relative fraction of the $gg \to \gamma\gamma$

contribution to the total diphoton cross section. In this analysis which restricts the transverse momenta of each of the two leading jets to the range of (15 – 40) GeV and the transverse momenta of each of the two leading photons to be above 15 GeV, the $q\bar{q}$ scattering significantly dominates the gg process, with $q\bar{q}$ fraction of about (70 – 80)%.

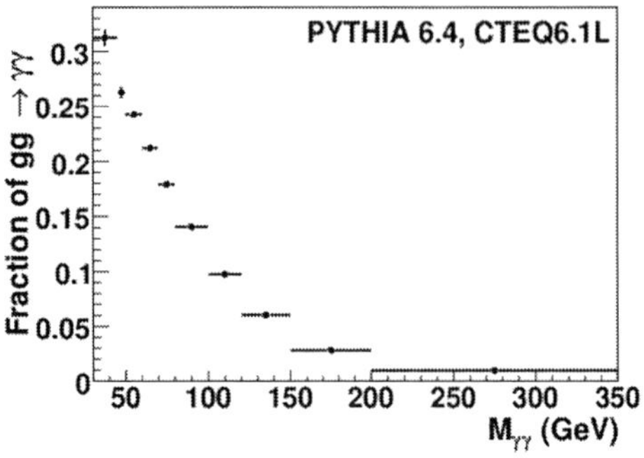

(a)

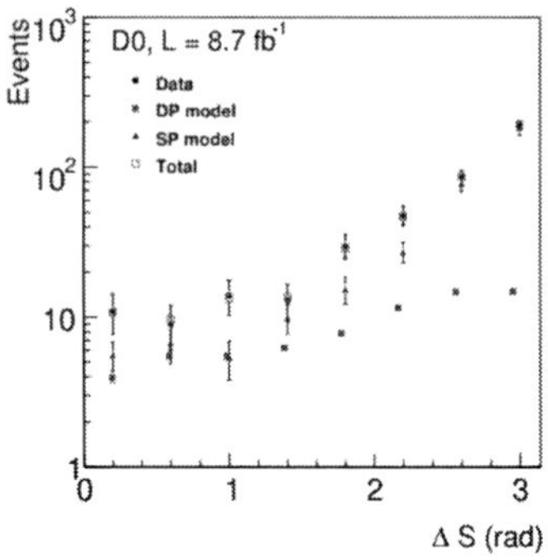

(b)

Figure 4. (a) The fraction of events produced via the $gg \to \gamma\gamma$ subprocess relative to the total diphoton cross section. (b) The ΔS distribution in the data, DPI, SPI models, and the sum of the DPI and SPI contributions weighted with their fractions ("Total") in $\gamma\gamma$ + dijet final state.

The DPI event fraction is found by fitting signal (MIXDP model constructed similarly to those described in sections 2.1 and 2.2) and background (SHERPA SPI sample) event models to the data using ΔS variable which is shown in figure 2(b). The result of the fit is presented in figure 4(b) and the found fraction of DPI events is $f_{\rm DPI}^{\gamma\gamma jj} = 0.213 \pm 0.061(stat) \pm 0.028(syst)$. Along with the measured selection efficiencies and geometric acceptances it gives the effective cross section $\sigma_{\rm DPI}^{\gamma\gamma jj} = 19.3 \pm 1.4(stat) \pm 7.8(syst)$ mb, which is in agreement with the previous measurements within uncertainties.

3 DPI as a background to WH production at the Tevatron

As it was mentioned above a distinctive feature of DPI events is an interaction of two parton-parton pairs within the same $p\bar{p}$ collision. The measured effective cross section, which is directly related to the spatial distribution of partons within a hadron, is used to estimate the rate of DPI events at the Tevatron energy as a background to associate WH production with $H \to b\bar{b}$ decay mode.

The HW production channel is simulated with Higgs boson masses of $m_{\rm H}$ = 115 GeV and $m_{\rm H}$ = 150 GeV. The DPI scattering is simulated by PYTHIA 8 as inclusive $q\bar{q} \to W + X$ production in the first parton process and inclusive QCD dijet production in the second process. The cross sections of the simulated events are normalized to either experimentally measured cross sections or to theoretical NNLO predictions. The differential cross sections $d\sigma/dM_{\rm jj}$ are calculated for the HW and DPI (W + dijet) events. To match the detector resolution, the jet transverse momenta are smeared in order to approximately reproduce the jet $p_{\rm T}$ resolution of the DØ detector. It is shown that the DPI cross section dominates the HW signal by more than two orders of magnitude, and the DPI cross section is caused mainly by the W+2 light jets (stemming from $u/d/s$-quarks or gluons) production [5].

In the signal HW events there are two b-jets in the final state. Since the leading DPI background is caused by the W + 2 light jet events, one should expect a significant reduction after requiring the jet b-tagging. To check this numerically a specific b-tagging requirement to the HW and DPI events is applied. Figure 5(a) shows the ratios of the HW event yield to the inclusive DPI W + dijet one in the

 DOI: 10.1051/epjconf/201713807001

dijet mass $M_{\rm jj}$ bins for the events selected by b-tagging. The uncertainty in each bin is caused by the normalization factors and the effective cross section. One can see that the Higgs boson events with $M_{\rm H}$ = 115 GeV are expected to be suppressed by about a factor 3 ($S/B \simeq 0.35$) in the peak position, while the signal events with $M_{\rm H}$ = 150 GeV are suppressed by about a factor 7. A set of variables sensitive to DPI kinematics is used as an input to a dedicated artificial neural network (ANN) to separate the HW from the DPI events. The new signal-to-background ratios are shown in figure 5(b). This ratio reaches about 2.2 at $M_{\rm jj} \simeq$ 115 GeV and about 2.7 at $M_{\rm jj} \simeq$ 150 GeV.

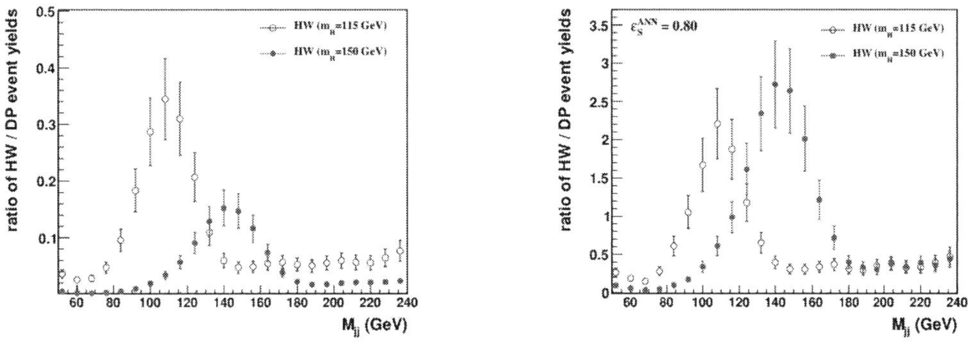

(a) (b)

Figure 5. The ratio of HW signal to DPI background event yields (a) with the b-tagging and (b) with account of the ANN selection efficiency taken for the HW events to be 80%.

4 Conclusion

Kinematic features of processes with double parton-parton scatterings have been studied by DØ experiment at the Tevatron. The effective cross section, a process independent parameter directly related to the size of the interaction region within a hadron, has been measured in processes with various final states containing isolated photons and hadronic jets and found to be $\sigma_{\rm DPI}^{\gamma jjj}$ = 12.7 ± 0.2($stat$) ± 1.3($syst$) mb, $\sigma_{\rm DPI}^{\gamma bjj}$ = 14.6 ± 0.6($stat$) ± 3.2($syst$) mb and $\sigma_{\rm DPI}^{\gamma\gamma jj}$ = 19.3 ± 1.4($stat$) ± 7.8($syst$) mb. All measured values are in agreement within uncertainties with each other as well as with most of the $\sigma_{\rm eff}$ measurements performed at the Tevatron and LHC. It is also shown that DPI production can be a significant background to many rare processes, especially with multi-jet final state. A set of variables allowing to reduce the DPI background is suggested.

References

[1] V. M. Abazov et al. [D0 Collaboration], Nucl. Instrum. Meth. A **565**, 463 (2006)
[2] V. M. Abazov et al. [D0 Collaboration], Nucl. Instrum. Meth. A **763**, 442 (2014)
[3] V. M. Abazov et al. [D0 Collaboration], Phys. Rev. D **89(7)**, 072006 (2014)
[4] V. M. Abazov et al. [D0 Collaboration], Phys. Rev. D **93(5)**, 052008 (2016)
[5] D. Bandurin, G. Golovanov and N. Skachkov, JHEP **1104**, 054 (2011)

EPJ Web of Conferences **138**, 07002 (2017) DOI: 10.1051/epjconf/201713807002
Baldin ISHEPP XXIII

Renormdynamics and production of fractals, fluctons, solitons, strings, unparticles, ...

Nugzar Makhaldiani[1,*]

[1]*Joint Institute for Nuclear Research, Dubna, Moscow Region, Russia*

Abstract. Concise introduction in renormdynamics with solvable models of (non)perturbative QCD and exotic states of fractals, fluctons, solitons, strings, unparticles...

We say that we find **New Physics** (NP) when either we find a phenomenon which is forbidden by SM in principle - this is the qualitative level of NP - or we find a significant deviation between precision calculations in SM of an observable quantity and a corresponding experimental value.

In QFT existence of a given theory means, that we can control its behavior at some scales (short or large distances) by renormalization theory [3]. If the theory exists, then we want to solve it, which means to determine what happens on other (large or short) scales. This is the problem (and content) of Renormdynamics. The result of the Renormdynamics, the solution of its discrete or continuous motion equations, is the effective QFT on a given scale (different from the initial one).

We will call Renormdynamics Functions (RDF) functions $g_n = f_n(t)$ which are solutions of the RD motion equations

$$\dot{g}_n = \beta_n(g), 1 \leq n \leq N. \tag{1}$$

In the simplest case of one coupling constant the function $g = f(t)$ is constant, $g = g_c$ when $\beta(g_c) = 0$, or is invertible (monotone). Indeed,

$$\dot{g} = f'(t) = f'(f^{-1}(g)) = \beta(g). \tag{2}$$

Each monotone interval ends by UV and IR fixed points and describes corresponding phase of the system. Note that the simplest case of the classical dynamics, the Hamiltonian system with one degree of freedom, is already two-dimensional, so we have no analog of one charge renormdynamics.

In the string theory, the connection between conformal invariance of the effective theory on the parametric world sheet and the motion equations of the fields on the embedding space is well known [8]. A more recent topic in this direction is AdS/CFT Duality [10]. In this approach for QCD coupling constant the following expression was obtained [2]

$$\alpha_{AdS}(Q^2) = \alpha(0)e^{-Q^2/4k^2}. \tag{3}$$

A corresponding β-function is

$$\beta(\alpha_{AdS}) = \frac{d\alpha_{AdS}}{d\ln Q^2} = -\frac{Q^2}{4k^2}\alpha_{AdS}(Q^2) = \alpha_{AdS}(Q^2)\ln\frac{\alpha_{AdS}(Q^2)}{\alpha(0)} \leq 0 \tag{4}$$

*e-mail: mnv@jinr.ru

So, this renormdynamics of QCD interpolates between the IR fixed point $\alpha(0)$, which we take as $\alpha(0) = 2$, and the UV fixed point $\alpha(\infty) = 0$. For the QCD running coupling considered in [4]

$$\alpha(q^2) = \frac{4\pi}{9\ln(\frac{q^2+m_g^2}{\Lambda^2})}, \tag{5}$$

where $m_g = 0.88GeV$, $\Lambda = 0.28GeV$, the β–function of renormdynamics is

$$\beta(\alpha) = -\frac{\alpha^2}{k}(1 - c\exp(-\frac{k}{\alpha})),\ k = \frac{4\pi}{9} = 1.40,\ c = \frac{m_g^2}{\Lambda^2} = (3.143)^2 = 9.88, \tag{6}$$

for a nontrivial (IR) fixed point we have $\alpha_{IR} = k/\ln c = 0.61$. For $\alpha(m) = 2$, at valence quark scale m we predict the gluon (or valence quark) mass as

$$m_g = \Lambda e^{\frac{k}{2\alpha(m)}} = 1.42\Lambda = m_N/3,\ \Lambda = 220MeV. \tag{7}$$

It is nice to have a nonperturbative β–function like (6), but it is more important to see which kind of nonperturbative corrections we need to have a phenomenological coupling constant dynamics. It was noted [18] that in valence quark parameterization $\alpha_s(m) = 2$, at a valence quark scale m.

In the 1870's G. J. Stoney [17], the physicist who coined the term "electron" and measured the value of elementary charge e, introduced as universal units of Nature for L, T, M :

$$l_S = \frac{e}{c^2}\sqrt{G},\ t_S = \frac{e}{c^3}\sqrt{G},\ m_S = \frac{e}{\sqrt{G}} \tag{8}$$

M. Planck introduced [15] as universal units of Nature for L, T, M:

$$m_P = \sqrt{\frac{hc}{G}} = \frac{m_S}{\sqrt{\alpha}},\ l_P = \frac{h}{cm_P} = \frac{l_S}{\sqrt{\alpha}} = 11.7l_S,\ t_P = \frac{l_P}{c} = \frac{t_S}{\sqrt{\alpha}} \tag{9}$$

Stoney's fundamental constants are more fundamental just because they are less than Planck's constants :) Due to the value of $\alpha^{-1} = 137$, we can consider relativity theory and quantum mechanics as deformations of the classical mechanics when deformation parameter $c = 137$ (in units $e = 1, \hbar = 1$) and $\hbar = 137$ (in units $e = 1, c = 1$), correspondingly. These deformations have an analytic sense of p-adic convergent series. The number 137 has a very interesting geometric sense, $137 = 11^2 + 4^2$. So, $\sqrt{137}$ is the hypotenuse length of a triangle with other sides of lengths 11 and 4. The Babylonians used a base 60 number system which is still used for measuring time - 60 seconds in a minute, 60 minutes in an hour - and for measuring angle - 360 degrees in a full turn. The base 60 number system has its origin in the ratio of the Sumerian mina (m) and Akkadian shekel (s), $m/s \simeq 60 = 3 \cdot 4 \cdot 5$. We also can consider base 137 system for fundamental theories. For the nuclear physics strong coupling phenomena description we may take as a base $p = 13$. For the hadronic physics, valence scale QCD, and graphen strong coupling phenomena description we may take as a base $p = 2$. For the weak coupling physics SM m_Z scale and MSSM unification scale phenomena description we may take as a base $p = 29$. There are different opinions about the number of fundamental constants [5]. According to Okun, there are three fundamental dimensionless constants in Nature: Planck's constant, $\hbar$; the velocity of light, c; and Newton's constant, G. According to Veneziano, there are only two: the string length L_s and c. According to Duff, there are no fundamental constants at all.

Usually $L_s = l_p$, so, the fundamental area is $L_s^2 = 137l_s^2$. The value $s_s = l_s^2$– Stoney area, is more like on a fundamental area :) In mathematics we have two kind of structures, discrete and continuous.

If a physical quantity has discrete values, it might have no dimension. If the values are continuous - the quantity might have a dimension, a unit of measure. These structures may depend on scale, e.g. on macroscopic scale condensed state of matter (and time) is well described as continuous medium, so we use dimensional units of length (and time). On the scale of atoms, the matter has a discrete structure, so we may count lattice sites and may not use a unit of length. If at small (e.g. at Plank) scale space (and/or time) is discrete, then we do not need a unit of length (time) for measuring, there is a fundamental length and we can just count.

Let us consider the following **discrete dynamics**:

$$S_{n+1} + S_{n-1} = \Phi(S_n), \tag{10}$$

which is obviously a (discrete) time (n) invertible in this implicit form. In the explicit form

$$S_{n+1} = F(S_n, S_{n-1}) = \Phi(S_n) - S_{n-1} \tag{11}$$

it is not obviously time invertible. If we take two step time lattice-make simplest discrete RD step and from one component-scalar S(n) construct two component-spinor $\Psi(n)$, we obtain explicit time invertible dynamics

$$\Psi_{n+1} = \Omega(\Psi_n),\ \Psi_{n+1} = \begin{pmatrix} S_{n+2} \\ S_{n+1} \end{pmatrix},\ \Psi_n = \begin{pmatrix} S_n \\ S_{n-1} \end{pmatrix}. \tag{12}$$

This **dynamical mechanism of origin spin** which connects time inversion symmetry and spin was invented when the theory of quanputers [12] was constructed. This mechanism indicates that with time inversion symmetry we can have only composed scalar fields. With the discovery of the Higgs particle with mass 125 *GeV*, a nice number $m_W/m_H \simeq 80/120 = 2/3$ appears, which, at least for me, indicates for composed nature of W and H, with the same mass of about 40 *GeV* two and three valence constituents respectively. The fermion constituents ψ_n^a of W and scalar constituents φ_n^a of H compose scalar super multiplet (φ_n^a, ψ_n^a) with a flavor index n and color index a. Another notation is (h, sh)-(He, She:). The 40 GeV constituents may be a good candidates for dark matter particles. Coupling constant unification at $\alpha_u^{-1} = 29.0$ and scale $10^{16} GeV$ in MSSM [13] has a relict on the SM scale: $\alpha_2^{-1}(m) = 29.0$ at $m = 41 GeV$. If we extrapolate the SM value of $\alpha^{-1}(m_Z)$ to electron mass scale, we find $\alpha^{-1}(m_e) = 137.0$ Recent discovery of the second Higgs particle with mass $M_H = 750 GeV$ indicates an interesting structure. It is curious that $M_H/m_h = 750/125 = 6$!

In the Standard Model of Particle Physics (SM), the values of the coupling constants and masses of particles depend on the scale according to the renormdynamic motion equations. One charge a, one mass m RD equations are

$$\dot{\alpha} = \beta(\alpha),\ \dot{m} = \gamma(\alpha) m. \tag{13}$$

For the electron and nucleon masses, electrodynamic and pion-nucleon fine structure constants we have an empirical relation: $m_e/\alpha \simeq m_N/\alpha_{\pi N}$. We take the relation $m/\alpha = const$ as an integral of renormdynamics motion equations for m and α, find exact form of the β function in the minimal mass parameterization

$$\begin{aligned} &\gamma(\alpha) = \gamma_1\alpha + \gamma_2\alpha^2 + ... = \gamma_1 A, \\ &A = f^{-1}(\alpha) = \alpha + \gamma_2/\gamma_1\alpha^2 + \gamma_3/\gamma_1\alpha^3 + ..., \\ &\alpha = f(A) = A + f_2 A^2 + f_3 A^3 + ... \end{aligned} \tag{14}$$

From the integral of motion, in the minimal mass parameterization: $\gamma(\alpha) = \gamma_1\alpha$, we obtain

$$(\ln\alpha)^{\cdot} = (\ln m)^{\cdot} \Rightarrow \beta(\alpha)/\alpha = \gamma(\alpha) = \gamma_1\alpha \Rightarrow \beta(\alpha) = \beta_2\alpha^2,\ \beta_2 = \gamma_1 \tag{15}$$

DOI: 10.1051/epjconf/201713807002

so, we have the following algebraic-diofant equations for the flavor and color content of the theory

$$\beta_2 = \gamma_1, \ \beta_n = 0, \ n \geq 3, \tag{16}$$

and prediction for the dimension of space-time: $D = 4$. Solutions of the motion equations are

$$\alpha(t) = \frac{\alpha_0}{1 - \alpha_0\beta_2 t}, \ m(t) = m_0|\alpha_0^{-1} - \beta_2 t|^{-\gamma_1/\beta_2} = \frac{m_0}{\alpha_0}\alpha(t). \tag{17}$$

In the multidimensional renormdynamics, when we have several (N) coupling constants and masses, we assume that there is maximal number ($N - 1$) of integrals of motion H_n. If the number of integrals is N, we do not have dynamics, we have only statics - finite field theory, $\alpha_n = const, \ n = 1, ..., N$. The idea of reduction to the one dimensional renormdynamics is simple:

$$\begin{aligned} &\frac{d\alpha_n}{dt} = \beta_n(\alpha_1, ..., \alpha_{(N-1)}, \alpha_N) \Rightarrow \frac{d\alpha_n}{d\alpha} = B_n(\alpha_1, ..., \alpha_{(N-1)}, \alpha), \ \alpha = \alpha_N, \\ &B_n(\alpha_1, ..., \alpha_{(N-1)}, \alpha) = \beta_n(\alpha_1, ..., \alpha_{(N-1)}, \alpha)/\beta_N(\alpha_1, ..., \alpha_{(N-1)}, \alpha), \\ &\alpha_n = \sum_{k\geq 1} f_{nk}\alpha^k, \ \ n = 1, 2, ..., N-1. \end{aligned} \tag{18}$$

Generally speaking, the integrals of motion may explicitly depend on time, $H_n = H_n(\alpha, t), \ 1 \leq n \leq N$. Introducing extra coupling constant $\alpha_{N+1} = t, \ \beta_{N+1} = 1$, we may reduce N-dimensional case with time dependent integrals, to the $N + 1$–dimensional case of time independent integrals. The motion equations can be derived from the integrals of motion:

$$0 = \dot{H}_n = \partial_t H_n + \frac{\partial H_n}{\partial \alpha_m}\dot{\alpha}_m \Rightarrow \dot{\alpha}_n = \beta_n(\alpha) = \frac{\det A_n}{\det A}, \ A^{-1}_{nm} = \frac{\partial H_n}{\partial \alpha_m}. \tag{19}$$

When the number of time dependent integrals are $M < N$, we may derive for M coupling constants RD equations and the remaining $N - M$ coupling constants maybe considered as control parameters. Defining some $N - M$ criteria of optimality, we may find corresponding optimal trajectories.

Solitons are particlelike states, solutions of motion equations and they quantum extensions. Examples are solitons of the sine-Gordon motion equation or barions-skirmions of the Skyrme model [16]. In particle theory, the skyrmion was described by Tony Skyrme in 1962 and consists of a quantum superposition of baryons and resonance states. Skyrmions as topological objects are important in solid state physics. Researchers could read and write skyrmions using scanning tunneling microscopy. The topological charge, representing the existence and non-existence of skyrmions, can represent the bit states "1" and "0".

QCD consists of quarks and gluons. Quarks possess both color (r, g, b) and flavor $(u, d, s$, etc.), while gluons possess color (r, g, b) and anti-color $(\bar{r}, \bar{g}, \bar{b})$, but not flavor. An open **string** (a string with two endpoints) is ideally suited to account for such quantum numbers at its two ends. For quarks, one end represents color and the other end - flavor. For gluons, one end represents color and the other - anti-color. In string theory, there are **branes** (higher dimensional extended objects that are generalized membranes) to which the endpoints of an open string are confined. Applying this idea to QCD, we introduce N_c colored branes and N_f flavored branes at which open strings corresponding to quarks and gluons terminate. The energy of a string is given by the sum of the classical energy stored inside the string and the excitation energies of vibration and rotation. Because the classical energy of a string is proportional to its length and because gluons are massless, N_c colored branes should lie on top of each other. On the other hand, quarks possess intrinsic masses, and therefore the endpoints of a quark string, namely, a flavored brane and a colored brane should be separated from each other by a nonvanishing distance U. Then, the intrinsic quark mass m_q can be represented as $m_q = U\times$(string tension), where the string tension is the energy stored inside a unit length of string and is represented: string tension=$1/(2\pi\alpha')$ in terms of α' , historically called the Regge slope. To describe QCD, we have to prepare Dp-branes and Dq-branes with p, q≥3 for colored branes and flavored branes, respectively, and these branes should be located in the space of more than five dimensions. To evaluate the amplitude for a certain process to occur in the above picture, we have to sum up all the possible two-dimensional world sheets with the weight exp(iS), where the action S is given by S=(energy)×(time)=(area of the string's world sheet)/$2\pi\alpha'$, following the Feynman path integral formulation.

Cumulative Effect: Production of particles from nuclei in a region, kinematically forbidden for reactions with free nucleons is connected to the existence of **Fluctons** - droplets of dense cold nuclear matter. The size of fluctons is the same as nucleon one. The density of fluctons is several times more than the nucleon one. So, it seems that there are violation of the baryon number conservation and Pauli principle inside fluctons. The Pauli principle can be saved (small violation remains) if we suppose that small nucleons inside fluctons have a small color charge as indicate numerical experiments [1]: α_s beyond the critical size (and value) decreases toward zero - IR freedom.

Classical fields have canonical, rational for integer D, (mass-)dimensions. E.g. in electrodynamics

$$L = \int d^D x(\bar{\psi}(\gamma(\partial - eA) - m)\psi - \frac{1}{4}F^2),$$
$$d_\psi = [\psi] = (D-1)/2,\ d_A = (D-2)/2,\ d_e = (4-D)/2. \tag{20}$$

Quantum corrections introduce (anomaly) corrections to the canonical dimensions, so the fields and coupling constants become **fractals** [11]. At fixed points of RD, the fractals are self similar and their compositions present at low energy **unparticles,** [7]. We will illustrate the qualitative picture of the (un)particle(like) objects with the simplest model of a scalar field given by the following Lagrangian

$$L = L(\Phi, M, \lambda) = \frac{1}{2}(\partial_\mu \Phi) - \frac{1}{2}M^2\Phi^2 - V(\Phi),\ \mu = 0, 1, 2, ..., D \tag{21}$$

where we usually take the self interaction in the form

$$V(\Phi) = \lambda\Phi^n,\ n = -2, 1, 2, 3, 4, 6. \tag{22}$$

In renormalizable case,

$$n = \frac{2D}{D-2} = 2 + \varepsilon(D),\ \varepsilon(D) = \frac{4}{D-2},\ D = \frac{2n}{n-2} = 2 + \varepsilon(n),\ \varepsilon(n) = \frac{4}{n-2}, \tag{23}$$

sometimes we consider also intermediate values of n and D and other forms of V.

In the free (self non interacting) field (particle) approximation: $\lambda = 0$, but in external gravitational field we have

$$L = \sqrt{-g}L(\Phi, M, 0),\ g = detg_{\mu\nu}(x). \tag{24}$$

Now we will see a nice composite particle mechanism :) Let us take a substitute: $\Phi = \varphi^k$, than we find

$$L(g, \Phi, M) = L((k\varphi^{k-1})^4 g, \varphi, M/k),\ g_{\mu\nu}(x) \Rightarrow (k\varphi^{k-1})^{4/D}g_{\mu\nu}(x). \tag{25}$$

Indeed

$$L(g, \Phi, M) = \sqrt{-g}(k^2\varphi^{2(k-1)}\frac{1}{2}(\partial_\mu\varphi)^2 - \frac{1}{2}M^2\varphi^{2k}) = \sqrt{-g(k\varphi^{k-1})^4}(\frac{1}{2}(\partial_\mu\varphi)^2) - \frac{1}{2}(\frac{M}{k})^2\varphi^2). \tag{26}$$

Now, having an experience with constituent - composite particle relation, we turn attention on the self-interaction term,

$$L = \sqrt{-g(k\varphi^{k-1})^4}(... - \frac{\lambda}{k^2}\varphi^N),\ N = kn - 2(k-1). \quad (27)$$

Most natural value of n for stable systems ($1+1 \to 1+1$, $2 \to 2$) is $n = 4$. In this case, $N = 2k+2$ and only natural value of constituents for which we have a renormalizable interaction is $k = 2 \Rightarrow N = 6$ with corresponding space-time dimension $D = 3$. The most natural value for fission-fusion interaction ($1 \leftrightarrow 2$) is $n = 3 \Rightarrow N = k+2$, for which we have realistic values $k = 2$ and $N = 4, D = 4$:) Other interesting values of naturally interpretable monomial (polynomial) interactions generally corresponds to the non-integer, fractional-fractal dimensions of space(time) D, with fractal-flucton-unparticle interpretations of the corresponding states of matter.

The size of particle-like states (solutions of the motion equations) is defined as $l \sim M^{-1}$, because at the boundary region, the linear part of the motion equations dominates and the Yukawa-like asymptotic $\Phi(r) \sim e^{-Mr}$ acts. In a pion-nucleon model for nucleon size we have $l_N \sim m_\pi^{-1} \simeq 1.43$ fm. The amplitude of the state (at maximum) $A \sim \lambda^{-\alpha}$, $\alpha = 1/(n-2)$. Indeed, the motion equation does not contain the coupling constant after a scaling substitution $\Phi = \lambda^{-\alpha}\phi$, so a particle-like solution ϕ does not contain λ and corresponding solution $\Phi = \lambda^{-\alpha}\phi \sim \lambda^{-\alpha}$,

$$\triangle\Phi + M^2\Phi + \lambda n\Phi^{n-1} = \lambda^{-\alpha}(\triangle\phi + M^2\phi + \lambda^{1-(n-2)\alpha}n\phi^{n-1}) = 0. \quad (28)$$

At not so low energies from string theory we may extract the following scalar field theory

$$L = \sqrt{-g}(\frac{1}{2}(\partial_\mu\Phi)^2 - \frac{1}{2}M^2\Phi^2 - \lambda\Phi^3),\ \mu = 0, 1, ..., D-1,\ D = 6+\varepsilon, \quad (29)$$

where $\varepsilon \in [0, 20]$. The one-loop β-function is $\beta(a) = (D-6)a - \beta_2 a^2$, $a \sim \lambda^2$ and it has stable UV fixed point at $a = (D-6)/\beta_2$ and IR fixed point $a = 0$. Beyond this point we have an unparticle $\Phi = \phi^2$ with Lagrangian

$$L = \sqrt{-g'}(\frac{1}{2}(\partial_\mu\phi)^2 - \frac{1}{2}(\frac{M}{2})^2\phi^2 - \frac{\lambda}{4}\phi^4),\ \mu = 0, 1, ..., d-1,\ d = 4-\varepsilon,\ \varepsilon \in [0, 1]. \quad (30)$$

The one-loop β-function is $\beta(\lambda) = (d-4)\lambda + b\lambda^2$ and it has stable IR fixed point at $\lambda = (4-d)/b$. The UV fixed point is $\lambda = 0$. At this point we have reduction from higher dimensional Φ^3 to lower dimensional ϕ^4. Another possibilities is an unparticle $\Phi = \varphi^4$ with Lagrangian

$$L = \sqrt{-g''}(\frac{1}{2}(\partial_\mu\varphi)^2 - \frac{1}{2}(\frac{M}{4})^2\varphi^2 - \frac{\lambda}{6}\varphi^6),\ \mu = 0, 1, ..., d-1. \quad (31)$$

The one loop β-function is $\beta(\lambda) = (d-3)\lambda + c\lambda^2$, $d = 3-\varepsilon$, $\varepsilon \in [0, 2]$. The IR fixed point is $\lambda = \varepsilon/c$. UV fixed point is $\lambda = 0$. Similar consideration gives reduction from higher energy ϕ^4 model to lower energy φ^6 one. Some technical questions remain. One of them concern to the substitution $\Phi = \phi^2$. It restricts Φ as $\Phi \geq 0$. Formulation of positivity condition is not so easy. We will take another path, we define the interaction as $\Phi^3 = (\Phi^2)^{3/2} \geq 0$. Then the substitution $\Phi^2 = \phi^4$ will work. By the way, by this definition we made also another improvement: the potential becomes bounded from below. For the reduction the substitution $\Phi^2 = \phi^4$ also works,

$$L = \sqrt{-g}(\frac{1}{8\Phi^2})(\partial_\mu\Phi^2)^2 - \frac{1}{2}M^2\Phi^2 - \lambda(\Phi^2)^{n/2},\ n = 3, 4. \quad (32)$$

Note that by substitution

$$(\frac{\Phi}{\Phi_0})^2 = \phi^{2k},\ \phi^2 = \exp(\ln(\Phi^2/\Phi_0^2)/k) = 1 + \frac{1}{k}\ln(\frac{\Phi}{\Phi_0})^2 + O(K^{-2}),\ \phi = \pm 1 + O(k^{-1}) \quad (33)$$

we reduce the field theory to a discrete theory, to a system of bits. Also, changing dimension of space D and nonlinearity n restricted by condition

$$n = \frac{2D}{D-2},\ D = \frac{2n}{n-2},\ \frac{1}{n} + \frac{1}{D} = \frac{1}{2} \tag{34}$$

we assume that they are functions of scale or coupling constant, due to monotonic property of the coupling constant. We have the following relation

$$\beta_n = -\frac{4}{(D-2)^2}\beta_D,\ \beta_n = \mu\frac{dn}{d\mu} = \frac{dn}{d\lambda}\beta_\lambda,\ \beta_D = \mu\frac{dD}{d\mu} = \frac{dD}{d\lambda}\beta_\lambda. \tag{35}$$

References

[1] I.L. Bogolubsky, E.M. Ilgenfritz, M. Müller-Preussker, A. Sternbeck, Phys. Lett. B **676**, 69 (2009)
[2] S.J. Brodsky, G.F. de Tèramond, A. Deur, Phys. Rev. D **81**, 096010 (2010)
[3] J.C. Collins, *Renormalization* (Cambrige Univ. Press, London, 1984)
[4] D. Diakonov, Prog. Par. Nucl. Phys. **51**, 173 (2003)
[5] M. Duff, L. Okun, G. Veneziano, J. High Energy Phys. **0203**, 023 (2002)
[6] L.D. Faddeev and L.A. Takhtajan, *Hamiltonian methods in the theory of solitons* (Springer, Berlin, 1990)
[7] H. Georgi, Phys. Rev. Lett. **98**, 221601 (2007)
[8] S.V. Ketov, *Quantum Non-linear Sigma-models* (Springer, Berlin, 2000)
[9] N. Koblitz, *p-adic numbers, p-adic analysis, and Zeta-functions* (Springer, Berlin, 1977)
[10] J.M. Maldacena, Adv. Theor. Math. Phys. **2**, 231 (1988)
[11] N. Makhaldiani, Physics of Particles and Nuclei Letters **8**, 325 (2011)
[12] N. Makhaldiani, Atomic Nuclei **74**, 1040 (2011)
[13] N. Makhaldiani, Physics of Particles and Nuclei Letters **11**, 1019 (2014)
[14] L. Okun, Sov. Phys. Usp. **34**, 818 (1991)
[15] M. Planck, Ann. d. Phys. **1**, 69 (1900)
[16] T. Skyrme, Nuclear Physics **31**, 556 (1962)
[17] G. Stoney, Phil.Mag. **11**, 381-391 (1881)
[18] M.B. Voloshin, K.A. Ter-Martyrosian, *Gauge Theory of Elementary Particles* (Atomizdat, Moscow, 1984)

EPJ Web of Conferences **138**, 07003 (2017) DOI: 10.1051/epjconf/201713807003
Baldin ISHEPP XXIII

Extension of the Schrodinger equation

Vyacheslav Somsikov*

[1] *Institute of Ionosphere, Kamenskoe Plato, 050020, Kazakhstan*

Abstract. Extension of the Schrodinger equation is submitted by removing its limitations appearing due to the limitations of the formalism of Hamilton, based on which this equation was obtained. For this purpose the problems of quantum mechanics arising from the limitations of classical mechanics are discussed. These limitations, in particular, preclude the use of the Schrodinger equation to describe the time symmetry violation. The extension of the Schrodinger equation is realized based on the principle of duality symmetry. According to this principle the dynamics of the systems is determined by the symmetry of the system and by the symmetry of the space. The extension of the Schrodinger equation was obtained from the dual expression of energy, represented in operator form. For this purpose the independent micro - and macro-variables that determine respectively the dynamics of quantum particle system relative to its center of mass and the movement of the center of mass in space are used. The solution of the extended Schrodinger equation for the system near equilibrium is submitted. The main advantage of the extended Schrodinger equation is that it is applicable to describe the interaction and evolution of quantum systems in inhomogeneous field of external forces.

1 Introduction

The constructions of any physical theory are performed based on the basic provisions: models, hypotheses, axioms. As a rule, these basic provisions are taken in a simplified way, to allow us to describe the basic regularities, observed in experiments. For example the structureless material of point (MP) was used as a body model for construction of Newtonian mechanics [1]. This simplification has resulted in a serious contradiction between nature and its physical picture. Indeed, in according with the formalisms of classical mechanics all processes in nature are reversible. In reality we are witnessing of the irreversible picture. This contradiction was classified as a one of the chief problem of the physics [2, 3].

The explanation of the mechanism of irreversibility in the frames of the laws of the classical mechanics was found by the expansion of classical mechanics [4-6]. Expansion was achieved by replacing the material point, which used in Newtonian motion equation, on the particle, which has a structure. As a result, Newtonian mechanics became a special case of the new enhanced mechanics. The replacement MP on structural particle (SP) allowed us to remove the above mentioned contradiction between the reversible theoretical picture and the irreversible picture of the real nature [6].

The Schrodinger equation is the basis of quantum mechanics [7-10]. And like in the case of classical mechanics, it is faced with some difficulties, in particular in the description of the symmetry

*e-mail: vmsoms@rambler.ru

breaking and irreversible processes [11, 12]. We shall not discuss them. It is enough to say that classical mechanics is the limiting case of quantum mechanics. Therefore, the proof of existence of irreversibility in the frame of the laws of classical mechanics is a proof of the existence of the irreversible dynamics in the quantum mechanics. And we can suggest that the approach to the removal of the restrictions in quantum mechanics will be similar to the approach that has been used for the expansion of classical mechanics.

In this paper we consider the following questions: what restrictions of the Schrodinger equation follow from the restrictions of classical mechanics; how to remove these restrictions from the Schrodinger equation; how to obtain the extended Schrodinger equation.

2 The basic ideas of the quantum mechanics

Quantum mechanics was created as a result of solving problems to explain of the experimentally observed effects: quantization of energy, wave-particle duality. These properties characterize the behavior of quantum particles. It was found that the quantum particles combine the properties of conventional particles and waves. These properties, unusual for the classical mechanics, determine dynamics of the quantum particles. The quantization is associated with the observed discrete states of atomic systems [9-11]. The wave-particle duality manifested itself in that the separate particle during its registration, for example, by means of photographic plates, falls into a separate point on the plate, as it takes place for ordinary classical particles. At the same time, this point is not determined by the equations of motion or electrodynamics equations. The experiments showed that the spatial distribution of these particles was described by the wave equation [7]. According to experimental data, the energy and momentum with the defined frequency and wavelength can be associated with each micro-particle [2-5]: $E = h\nu = \hbar\omega, p = h\nu/c$. Here $\hbar = h/2\pi$ is a Planck's constant [7]; c is a velocity of light; $\omega = 2\pi\nu$ is an angular frequency; λ is a de Broglie wavelength. This allowed us to put the wave function into conformity with to the quantum particle. As it was found, the wave function of a free particle was a plane mono-chromatic de Broglie wave [9-11]: $\psi(r,t) = A\exp(i(kr - \omega t)/h)$, where $k = p/h$. As it can be seen from the experiments, the consistent scattering of particles, for example, on the crystal lattice, is equivalent to the scattering of the simultaneous flow of independents particles. In accordance with this result, the wave function of the aggregate of particles must satisfy the principle of superposition. Analysis of the experimental results of research of the dynamic properties of particles led Schrodinger to the idea that their dynamics was described by the wave equation but not by the Newton equation of motion [7-10]. The solution of the wave equation is the wave function, which is defined by the points in the configuration space. So, due to the quantum-wave dualism, particles can be associated with a wave field, which determines their spatial distribution. The amplitude of this field depends on the coordinates and time. Therefore this function is expressed through the so-called wave function $\psi(r,t) \equiv \psi(x,y,z,t)$. The value, $|\psi(x,y,z,t)|^2 dV$, means the probability of finding a particle at a time t in the volume element dV. It can be written as: $dW = |\psi(x,y,z,t)|^2 dV$. Based on the Hamilton formalism, Schrodinger found the equation for the wave function of the particles that described the experimental data [7]. The key idea, which helped him to assign the classical mechanics with the quantum mechanics was that the action in classical mechanics, up to a constant factor, was equal to Planck's constant, and coincided with the phase of the wave. There are several ways to obtain the Schrodinger equation. But the most logical way, which helps us to understand the relationship of this equation with the classical mechanics, belongs to Schrodinger [7]. Let us briefly describe the path that led Schrodinger to his equation.

3 How the Schrodinger equation was built

Schrodinger drew attention to the importance of conformity of the Hamilton-Jacobi equations [1] and the equations of geometrical optics. It consists in the fact that the action $S = \int_{t_0}^{t}(T - U)dt$ is equivalent to the phase of the wave (see Hamilton-Jacobi equation [1]). Here T is a kinetic energy, U-is a potential energy, t-is a time. Based on this, he decided to use the fact that for the wave function, the phase of which up to a constant factor coincides with the action, the following equation holds [9]: $u^2\nabla^2\psi(r,t) - \partial^2\psi(r,t)/\partial t^2 = 0$. If we substitute into this equation the value: $u = E/\sqrt{2m(E-U)}$, where m is a mass of the particle, and if we take into account that $\partial^2\psi(r,t)/\partial t^2 = -\omega^2\psi(r,t), E\psi(r,t) = i\hbar\partial\psi(r,t)/\partial t$, we get to the following equation [8-10]:

$$i\hbar\partial\psi(r,t)/\partial t = \{-\hbar^2\nabla^2/2m + U(r,t)\}\psi \tag{1}$$

It is the Schrodinger equation for one particle. For him the superposition principle holds. The Schrodinger equation is a first order in time. That is, the wave function for particles is uniquely determined by its initial value. The Schrodinger equation for a system can be obtained from the energy expression for the total energy of the particles. If we replace the energy and the momentum by the relevant operators, the Schrodinger equation for a system will have the form [9]:

$$\{i\hbar\partial/\partial t + \sum_{i=1}^{N}[\hbar^2\nabla^2/2m_i - U(r_i,t)] - W_{int}(r_1,r_2...r_N)\}\psi(r_1,r_2...r_N,t) = 0 \tag{2}$$

Here $W_{int}(r_1,r_2...r_N)$ - is the energy of interaction of particles, depending on the distance between them; $i = 1,2...N$; $U(r_i)$ is a potential energy of the i -th particle in the field of external forces.

We emphasize that the eq. (2) is applicable to systems of interacting particles. If the interactions are absent, for the stationary case the solution of eq. (2) can be represented as the product of the wave functions for each particle: $\psi = \psi_1(r_1),\psi_2(r_2)...\psi_N(r_N)$. This is equivalent to the sum of independent solutions for each particle.

In the general case the variables of the eq. (2) are not decoupled. The nature of the engagement variables in the eq. (2) is similar to the nature of the engagement of the variables for MP system which moves in a non-uniform force field. The variables in the laboratory coordinate are not orthogonal because they belong to the different symmetry groups [4, 6].

4 Why it is necessary to expanded the Schrodinger equation

We will proceed from the fact that to describe a system of interacting particles in quantum mechanics, like in classical mechanics, the eq. (2) shall be converted in accordance with the principle of duality symmetries (PDS) [4, 5]. The PDS means that the dynamic of the system is determined by the internal symmetry of the system and symmetry of space. That is, the energy of a system must be submitted as a sum of the motion energy and internal energy of the system. Then the nature of the violation of the time symmetry is connected with the transformation of the motion energy of the system into the internal energy [5]. The reasons for using the PDS are following. According to the laws of classical mechanics, matter is divisible to infinity. Therefore all the elements, including elementary particles, are structured. Note that there are no experiments, which would be contrary to this fact. Therefore, we assume that in quantum mechanics all particles have structure. Indeed, according to quantum mechanics, firstly, the energy of the ground state of the particles is different from zero, and secondly, the energy of the ground state of a quantum oscillator is not equal to zero. At the same time, the energy of the ground state is determined by the uncertainty principle. These facts agree with each other if we assume that all particles are structured and therefore have the internal energy. The minimum of this energy is determined by the Planck's constant. In turn, the forces determined by internal energy and the energy of motion, respectively. Therefore if the energy is not present in the dual form, it becomes possible to describe the violation of time symmetry when they move in inhomogeneous fields of external forces because the violation of time symmetry is caused by the transformation of the motion energy of the system in its internal energy. An example that confirms this view is the problem of the quantum oscillator [10].

There are also arguments, which are in favor of expansion of the Schrodinger equation. Due to the wave properties of the dynamics of the particles, the so-called Heisenberg's uncertainty principle takes place: $\Delta p \Delta r \geq \hbar$. This gives rise to the following questions. How is the violation of time symmetry connected with the uncertainty principle? How is it related to the infinite divisibility of matter? How the principle of uncertainty is associated with the fact that the accuracy of determining the dynamics of the particles cannot exceed the accuracy of determining the energy of motion at each point of the phase space, which is limited to changes of the internal energy of the system? Let us explain that the answers to these and many other questions must be sought on the basis of *PDS* [5,6].

There are at least two approaches to the interpretation of the uncertainty principle. Bohr argued that the uncertainty principle should be taken as a real manifestation of nature, without trying to look for an explanation. Einstein saw it as a restriction of the theory itself. From the standpoint of structuring matter at all hierarchical levels [8], we can propose an explanation of the uncertainty principle, that corresponds to the point of view of Einstein.

Let us take a system of interacting particles. For example, the scattering of the of the high energy electron flux on nucleons. Then assume that new particles appear as a result. This means that there must be a change in the internal energy of the reaction products while maintaining the total energy of the interacting systems. Since in this case the energy of motion of the system is reduced due to its conversion into internal energy, the violation of the symmetry of the time takes place. It means that $\delta A = A^d \neq 0$, where A^d is a right hand part in the variation of the least of action for any systems moving in inhomogeneous space [4, 5]. Then the nature of the uncertainty principle can be explained by the fact that the motion energy transforms into the internal energy resulting in strong particles interaction or motion systems in inhomogeneous space and we will have: $A^d \geq h$ or the A^d never equal to zero [4].

Transformation of the motion energy of the system into the internal energy, in classical mechanics determines the dissipative processes, which constitute the essence of the second law of thermodynamics for sufficiently large systems. It is natural to assume that for the quantum mechanics it will be similar. In extension of the Schrodinger equation, taking *PDS* into account, we replace the independent variables by the micro-and macro-variables. These variables create two independent groups of variables. Micro-variables determine the motion of the particle relative to the center of mass of the system. Macro-variables determine the motion of the center of mass of the system in the space [4,5].

5 The expansion of the Schrodinger equation for the system

Now let us show how to obtain the extended Schrodinger equation in connection with *PDS* for the system of two particles in the stationary case when the interaction between particles depends on the distance between them. In this case from eq. (2) we will have [9, 10]:

$$\{E + [\hbar^2\nabla_R/2M - U(r,R)] + [\hbar^2\nabla_r^2/2\mu - W_{int}(r)]\}\psi(r,R) = 0 \qquad (3)$$

Here $M = m_1 + m_2$ is a the total mass of the system, $\mu = m_1m_2/(m_1+m_2)$, $R = (m_1r_1+m_2r_2)/(m_1+m_2)$ are the coordinates of the center of mass (CM) of system, $r = r_1 - r_2$, E_{int} is an internal energy of the system, E_{cm} is energy of motion of the system CM, and $E_{int} + E_{cm} = E$. Eq. (3) is obtained by transition to the micro- and macro-variables. They characterize the movement of the CM of oscillator and the motion of each particle with respect to the CM. If $U(r, R) = W(r) + V(R)$, the variables are separated, and the wave function can be written as: $\psi(r, R) = \psi(r)\psi(R)$. In this case, eq. (3) is divided into two independent equations:

$$\{E_{int} + \hbar^2\nabla_r^2/2\mu - W(r) - W_{int}(r)\}\psi(r) = 0(a); \{E_{cm} + \hbar^2\nabla_R/2M - V(R)\}\psi(R) = 0\ (b) \quad (4)$$

Eq. (4 a) defines the relative movement of the particles in the CM system. Its solution is consistent with the one of a quantum oscillator in the case when the Hamiltonian has the form [8]: $\hat{H} = 1/2(\hat{p}^2/\mu + \mu\omega^2\hat{r})$, where $\hat{p}, \hat{r}$ are the operators of momentum and coordinate, ω is a the natural frequency of the oscillator.

Eq. (4 b) determines the motion in the space of the CM of oscillator. In the absence of an external field of forces, its solution has the form: $\psi(R) = exp(i/\hbar PR)$, where $P = \sqrt{2ME_{cm}}$. Micro- and macro-variables form two groups of independent variables. For the interacting systems, or when the system is moving in non-uniform field forces, these variables are connected with each other. As a result, as in classical mechanics, invariant is a sum of the energy of motion of the system and its internal energy.

The nature of the symmetry breaking is completely determined by the nonlinear terms, containing simultaneously the micro-and the macro-variables and describing the mutual transformation of the motion energy of the system and of the internal energy while maintaining their sum [4-6]. This means that in the general case the principle of superposition of the solutions for particles is not realized.

Thus, to describe the processes of the system dynamics and its evolution, in general, when the violation of time symmetries takes place, we need to use the total energy recorded in the micro- and macro-variables. For this purpose, the energy system of particles must be written as a sum of the energy of motion and the internal energy in the form of the operator. For quantum mechanics, this means that the Hamiltonian operator must be constructed as a sum of operators which describe the internal dynamics of the system and operators which determine the motion of the system in a heterogeneous space. Replacing the energy and momentum of the system by operators, we obtain from eq. (2) for stationary case:

$$\{E + \hbar^2\nabla_R^2/2M - U(\tilde{r}_1, \tilde{r}_2...\tilde{r}_N, R) + \sum_{i=1}^{N}[\hbar^2\nabla_{\tilde{r}_i}^2/2m - W_{int}(\tilde{r}_i)]\}\psi(\tilde{r}_1, \tilde{r}_2...\tilde{r}_N, R) = 0 \quad (5)$$

Here R is a CM coordinate system, $\tilde{r}_i$ are the coordinates of the i-th particle relative to CM of the system. We call this equation the extended Schrodinger equation. Contrary to the classical equation, it allows to take into account the transformation of the energy of motion of a quantum system in its internal energy. This transformation takes place when the system is moving in an inhomogeneous field of external forces. Without such transformation it is impossible to describe the formation of new attractors or quantum systems.

In the absence of engagement micro-variables with macro-variables, the total energy is a sum of the energy of motion of the whole system and the internal energy, respectively, i.e., $E = E_{int} + E_{cm}$. In this case the eq. (5) is divided into two equations:

$$\{E_{cm} + \hbar^2\nabla_R^2/2M - V(R)\}\psi(R) = 0 \quad (6)$$

$$\{E_{int} + \sum_{i=1}^{N}[\hbar^2\nabla_{\tilde{r}_i}^2/2m - W_{int}(\tilde{r}_i) - W(\tilde{r}_1, \tilde{r}_2...\tilde{r}_N)\}\psi(\tilde{r}_1, \tilde{r}_2...\tilde{r}_N) = 0 \quad (7)$$

where $\psi(\tilde{r}_1, \tilde{r}_2...\tilde{r}_N, R) = \phi(R)\varphi(\tilde{r}_1, \tilde{r}_2...\tilde{r}_N)$, $U(\tilde{r}_1, \tilde{r}_2...\tilde{r}_N) = W(\tilde{r}_1, \tilde{r}_2...\tilde{r}_N) + V(R)$.

6 Solutions of the Schrodinger equation for systems near equilibrium

Let us consider the case when the system moving in the field of external forces near equilibrium. In this case the potential energy of the system in external field can be written as: $U_N = (\alpha \sum_{i=1}^{N} r_i^2)/2$, where α is a positive coefficient (the value of the second derivative $\partial^2 U/\partial^2 r$ when $r = r_0$). Here all particles are identical. The quadratic functions of the potential energy of the system in an external field can be written in terms of the quadratic function, in which the arguments are the coordinates of the particles with respect to the CM and the CM coordinate system [4-6]. It follows from the equation: $N \sum_{i=1}^{N} r_i^2 = (\sum_{i=1}^{N} r_i)^2 + \sum_{i=1}^{N-1} \sum_{j=i+1}^{N} r_{ij}^2$, where $r_i - r_j = r_{ij}$. After completing in this equality the replacement $R_N = (\sum_{i=1}^{N} r_i)/N$, we will have for potential energy: $U_N = \alpha[R_N^2 + 1/N \sum_{i=1}^{N-1} \sum_{j=i+1}^{N} r_{ij}^2]/2$. Here the first term in the right hand side is the potential energy of the motion of the system in the field of external forces. The second term is determined by the differences between the coordinates of the micro-particles.

Let us transform the energy by replacing: $r_i = R_N + \tilde{r}_i$, where $\tilde{r}_i$ are coordinate of the micro-particles relative to CM. Because $\sum_{i=1}^{N} \tilde{r}_i = 0$ we obtain $U_N = \alpha[R_N^2 + \sum_{i=1}^{N} \tilde{r}_i^2]/2$. Here the second term is equal to the energy of motion of particles relative to CM. Hence, the potential energy of the system in the external field near the equilibrium point is divided into the potential energy of motion of its CM and the amount of potential energy of motion of all particles relative the CM: $U_N = U_N^{tr} + U_N^{ins}$, where $U_N^{tr} = \alpha R_N^2/2$, $U_N^{ins} = \alpha \sum_{i=1}^{N} \tilde{r}_i^2/2$. I.e., the internal energy of the system can be written in the coordinates and velocities of the particles with respect to the CM. In this case the eqs. (6, 7) can be written as:

$$\{E_{cm} + \hbar^2 \nabla_R^2/2M - \alpha R_N^2/2\}\phi(R) = 0 \tag{8}$$

$$\{E_{int} + \sum_{i=1}^{N} [\hbar^2 \nabla_{\tilde{r}_i}^2/2m - W_{int}(\tilde{r}_i) - \alpha \tilde{r}_i^2/2]\}\varphi(\tilde{r}_1, \tilde{r}_2 ... \tilde{r}_N) = 0 \tag{9}$$

where $\psi(\tilde{r}_1, \tilde{r}_2 ... \tilde{r}_N, R) = \phi(R)\varphi(\tilde{r}_1, \tilde{r}_2 ... \tilde{r}_N)$.

If the condition $W_{int}(\tilde{r}_i) = \beta \sum_{i=1}^{N} \tilde{r}_i^2/2$ is satisfied, then we have: $\varphi(\tilde{r}_1, \tilde{r}_2 ... \tilde{r}_N) = \Pi_{i=1}^{N} \varphi_i(\tilde{r}_i)$. In this case the eq. (9) splits into independent equations for oscillators:

$$\{E_i + [\hbar^2 \nabla_{\tilde{r}_i}^2/2m - (\alpha + \beta)\tilde{r}_i^2/2]\}\varphi(\tilde{r}_i) = 0 \tag{10}$$

when condition $E_{int} = \sum_{i=1}^{N} E_i$ takes place. Here E_i is kinetic energy of motion of each particle of the system relative to its CM. This is the equation for the independent quantum oscillators. For the wave function $\varphi(\tilde{r}_i)$ the energy levels are given by [5]: $\varepsilon_{n_i} = \hbar\omega_i(n_i + 1/2)$, where $\omega_i = (\sqrt{(\alpha + \beta)})/m$. The index i is shown here only in order to emphasize the affiliation of the energy levels of the particle.

Thus, near equilibrium in the external and internal fields, the eqs. (6, 7) separate into independent oscillator equations for each particle, as well as for the system as a whole. The integrability of the equations follows from the fact that near the equilibrium the system is described by the Hamiltonian formalism. The characteristic frequency of these oscillators is defined as the forces of interaction of micro-particles as well as of the external forces. The external field of forces plays the role of the additive for independent vibrations of the atoms near their equilibrium. But the oscillation of the system in an external field of force is independent from the internal forces. The same is the case for the system of MP classical mechanics near the equilibrium of the internal and external forces fields [6]. In the general case of an inhomogeneous field of external forces, the variables in the eqs. (6, 7) are engaged. This leads to violation of time symmetry.

7 Conclusion

Thus, the restrictions of the classical mechanics which led to the reversibility of the hamiltonian systems, lead to restrictions of the Schrodinger equation. To remove these restrictions, it is necessary to write the Schrodinger equation in accordance with the *PDS* by using energy operators written in the micro- and macro-variables. In this case, the Hamiltonian splits into two parts. Micro-variables determine the dynamics of particle system with respect to the CM, and macro-variables determine the motion of a system in space as a whole. This equation can describe nonlinear processes symmetry breaking for the system moving in inhomogeneous fields of forces. Symmetry breaking determined by the nonlinear terms in the expansion of the forces, which depend simultaneously on micro- and macro-variables. These nonlinear terms, as in the case of classical mechanics, associated with the D-entropy of quantum systems [5, 6]. The D-entropy determines the part of the motion energy of the system which is transformed into the internal energy.

If the system is close to equilibrium, the Schrodinger equation is splitted into independent equations of the oscillations of the entire system as a whole, and each of the particles relative to the center of mass. The dynamic of the each particle is determined by the field of external forces and the forces of their interaction. The dynamics of the system is determined by the external forces only.

The main advantage of the extended Schrodinger equation is that it is applicable to describe the interaction and evolution of quantum systems.

References

[1] K.Lanczos, *Variation principles of mechanics* (Mir, Moscow, 1962) 408
[2] V.L.Ginzburg, UFN, **177**, 346 (2007)
[3] I.Prigogine, *From Being to Becoming* (Nauka, Moscow, 1980) 342
[4] V.M.Somsikov, PoS (Baldin ISHEPP XXII) 047 (2014)
[5] V.M.Somsikov, Modern Physics Letters B **4**, 1-11 (2016)
[6] V.M.Somsikov, *To the basics of evolution of the physics* (Sci., Almaty, 2016) 306
[7] A.Schrodinger, Phys.Rev. **28**, 1049-1070 (1926)
[8] L.D.Landau, E.M.Lifshitz, *Quantum mechanics* (Sci., Moscow, 1989) 767
[9] V.G.Levich, Y.Vdovin, V.A. Myamlin, *Course of Theoretical Physics II* (Sci., Moscow, 1962) 820
[10] V.G.Zelevinskii, *Lectures on quantum mechanics* (Siberian university, Novosibirsk, 2002) 504
[11] B.B.Kadomtsev, UFN **165**, 967-973 (1995)
[12] J.Greenstein, A.Zajonc, *Quantum challenge. Modern research foundations of quantum mechanics* (Intelligence, Dolgoprudny, 2012) 432
[13] M.J.Klein, UFN **92**, 679-700 (1967)
[14] V.M.Somsikov, Intern. Journal of Sci. **4**, 30-36 (2015)

DOI: 10.1051/epjconf/201713807004

Field induced phase transition in the few photon regime

A. D. Panferov[1,⋆], *S. A.* Smolyansky[1], *A. I.* Titov[2], *B.* Kämpfer[3,4], *A.* Otto[3,4], *D. B.* Blaschke[2,5,6], and *L.* Juchnowski[5]

[1] *Saratov State University, Saratov, Russia*
[2] *Joint Institute for Nuclear Research, Dubna, Russia*
[3] *Helmholtz-Zentrum Dresden-Rossendorf, Germany*
[4] *Institut für Theoretische Physik, Dresden, Germany*
[5] *University of Wroclaw, Wroclaw, Poland*
[6] *National Research Nuclear University (MEPhI), Moscow, Russia*

Abstract. Some features of the field induced phase transition accompanied by the vacuum creation of an electron-positron plasma (EPP) in strong time-dependent electric fields have been discussed in the work [1] in the domain of the tunneling mechanism ($\omega \ll m$, where ω is the characteristic frequency of the external field and m is the electron mass). In the present contribution the features of this process will be considered in the few photon domain where $\omega \sim m$. We observe a narrowing of the transient domain of the quickly oscillations and, mainly, a considerable growth of the effectiveness of the EPP production. Under these circumstances, we see an increase of the effectiveness of the EPP creation in the particular case of a bifrequent excitation, where both mechanisms (tunneling and few photon) act simultaneously [2, 3].

1 Introduction

The present note is devoted to the investigation of some features of the dynamical Schwinger effect which concerns the creation of an electron-positron plasma (EPP) from the vacuum in fast varying spatially homogeneous external electric fields. Initially, the Schwinger effect was predicted for a constant electric field [4–6]. According to these estimates a noticeable vacuum creation of electron-positron pairs in a strong electric field is possible only near the very high critical value of the field strength $E_c = m^2/e = 1.3 \times 10^{16}$ V/cm. For this reason, the interest in this problem is linked to a practical basis provided, e.g., by the further development of ultra-high intensity laser systems. The characteristic frequency of the pulse for a laser in the optical range is considerably smaller than the electron mass (using natural units $\hbar = c = 1$). A virtual pair must accumulate the energy of a very large number of photons of the external field before being "created". The dependence of the efficiency of the pair production on the field strength has approximately the same character as in the case of a constant field. Therefore the hope for detecting the creation of an EPP created from vacuum by the dynamical Schwinger effect can be realized only for field strengths in the immediate vicinity of E_c.

Encouraging predictions of enhanced efficiency of EPP vacuum excitation for the assisted Schwinger effect using the bifrequent external field of two lasers stimulate the interest to study the

⋆e-mail: panferovad@info.sgu.ru

high frequency domain (few photon area) in the single-frequency regime. Below we discuss some preliminary results of research on the field induced phase transition in this domain using the kinetic equation (KE) approach [7] as the basis. A brief description of this tool is given in Sect. 2. Some features of the phase transition under the influence of a quickly alternating external field from an initial vacuum state to the out-state of residual EPP are presented in Sect. 3. We put emphasis on comparison with the tunneling domain. On the examples of the simplest models for pulsed fields we demonstrate a significant increase of the efficiency of the EPP creation in the few photon domain for both, the distribution function and the observable pair density in the final state. The reason for this enhancement is the Breit-Wheeler mechanism. A second feature is the significant narrowing of the transient domain of fast oscillations. A discussion of the results is given in Sect. 4.

2 Kinetic equation

The creation of the EPP in a homogeneous, linearly polarized electric field $E(t) = -\dot{A}(t)$ with the vector potential (in Hamiltonian gauge) $A^{\mu}(t) = (0, 0, 0, A(t))$ is described by the KE [7] for the distribution function

$$\dot{f}(\mathbf{p}, t) = \frac{1}{2}\lambda(\mathbf{p}, t) \int\limits_{t_0}^{t} dt' \lambda(\mathbf{p}, t')[1 - 2f(\mathbf{p}, t')] \cos\theta(t, t'), \tag{1}$$

where

$$\lambda(\mathbf{p}, t) = eE(t)\varepsilon_{\perp}/\varepsilon^2(\mathbf{p}, t), \quad \theta(t, t') = 2\int_{t'}^{t} d\tau\, \varepsilon(\mathbf{p}, \tau).$$

Here λ is the amplitude of the vacuum transitions, and θ is the high-frequency phase. The quasienergy $\varepsilon(\mathbf{p}, \tau)$, the transverse energy $\varepsilon_{\perp}$ and the longitudinal quasimomentum $P(t)$ are defined as

$$\varepsilon(\mathbf{p}, t) = \sqrt{\varepsilon_{\perp}^2(\mathbf{p}) + P^2(t)}, \quad \varepsilon_{\perp} = \sqrt{m^2 + p_{\perp}^2}, \quad P(t) = p_{\parallel} - eA(t).$$

Here $p_{\perp}$ and $p_{\parallel}$ are the momentum components perpendicular and parallel to the field vector.

The integro-differential equation (1) is equivalent to a system of three time-local ordinary differential equations

$$\dot{f}(\mathbf{p}, t) = \frac{1}{2}\lambda u(\mathbf{p}, t), \quad \dot{u}(\mathbf{p}, t) = \lambda(1 - 2f(\mathbf{p}, t)) - 2\varepsilon v(\mathbf{p}, t), \quad \dot{v}(\mathbf{p}, t) = 2\varepsilon u(\mathbf{p}, t), \tag{2}$$

where $u(\mathbf{p}, t), v(\mathbf{p}, t)$ are auxiliary functions describing vacuum polarization effects.

The distribution function $f(\mathbf{p}, t)$ is zero in the in-vacuum state where the external field is absent. The KE (1) allows to describe the EPP evolution and to get the final distribution function $f_{\text{out}}(\mathbf{p})$ in the out-state. The duration of the action of the external field is limited, its time dependence is arbitrary. The in- and out-vacuum states are different although $E_{\text{out}} = E_{\text{in}} = 0$ due to the fact that $A_{\text{out}} \neq A_{\text{in}}$.

The total pair number density in the out-state is defined as

$$n_{\text{out}} = 2\int \frac{d\mathbf{p}}{(2\pi)^3} f_{\text{out}}(\mathbf{p}). \tag{3}$$

3 EPP creation in the few photon area

We will investigate the KE (1) numerically for the simplest Eckart-Sauter model of the one-sheeted electric field with characteristic duration T of its action defined by

$$E(t) = E_0 \cosh^{-2}(t/T), \quad A(t) = -TE_0 \tanh(t/T). \tag{4}$$

A more realistic laser pulse model is the one with a Gaussian envelope [8]

$$E(t) = E_0 \cos(\omega t)\, e^{-t^2/2\tau^2}, \quad A(t) = -\sqrt{\frac{\pi}{8}} E_0 \tau \exp(-\sigma^2/2)\, \mathrm{erf}\left(\frac{t}{\sqrt{2}\tau} - i\frac{\sigma}{\sqrt{2}}\right) + c.c., \tag{5}$$

where $\sigma = \omega\tau$ is a dimensionless measure for the characteristic duration of the pulse τ connected with the number of oscillations of the carrier field.

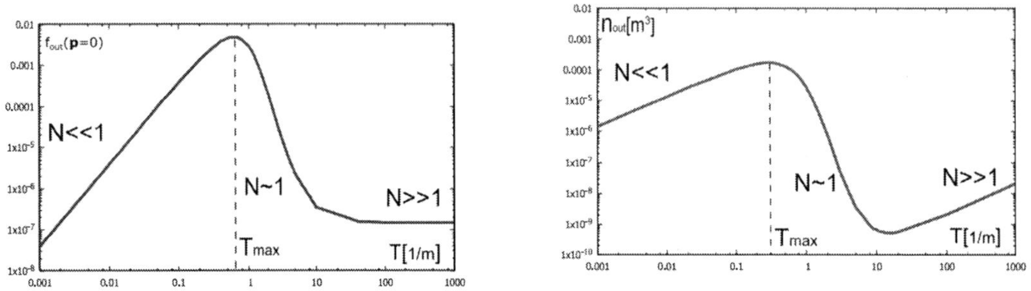

Figure 1. Behavior $f_{out}(\mathbf{p} = 0)$ (Left panel) and density (3) (Right panel) in the tunneling area ($N \gg 1$), the few photon regime ($N \sim 1$) and the domain of high-energy photons ($N \ll 1$) for the field model (4) with $E_0 = 0.2\, E_c$.

In figure 1 we demonstrate the resonant behaviour of the residual distribution function $f_{out}(\mathbf{p})$ in the point $\mathbf{p} = 0$ (left panel) and the total residual EPP number density n_{out} (3) (right panel) as functions of the field pulse duration T of the field model (4). The tendency of the function f_{out} to assume a constant value in the tunneling area $T \gg 1$ can be associated with the accumulation effect $n_{out} \sim T$ (right panel). This effect is well known from the exact solution of the problem [9]. A sharp decrease of the EPP creation efficiency is observed in the region of very small duration $T\,m \ll 1$. Analogous results can be obtained for the field model (5).

It would be natural to interpret these results in terms of the minimal photon number N required to overcome the energy gap and to create an electron-positron pair. The field models (4) and (5) themselves do not allow this because of the non-monochromaticity of these fields. However, on the qualitative level one can connect the value $T_{max}\, m \simeq 0.5$ of the maximal efficiency of EPP creation with the Breit-Wheeler process of electron-positron pair creation by a pair of colliding photons. Thus, $T_{max}\, m \sim N_{max} = 2$. When extrapolating this relation to both sides, the large and small durations, we obtain $N = 4T\, m$ for the field model (4). So, we can mark out on figure 1 and next figures the border between few photon and tunneling areas.

A similar correspondence can be introduced for the field model (5) in the case of a long pulse $\sigma \gg 1$, where $N = 2\, m/\omega$.

The existence of the maximum in the two photon domain correlates with the result of standard perturbation theory for $E_0/E_c \ll 1$ in the few photon region where this process is dominant [11].

The field strength dependence of the efficiency of EPP creation is depicted in figure 2. The left panel shows that sizeable values of the distribution function $f_{out}(\mathbf{p} = 0)$ can be reached for moderate field strengths when entering the few photon domain.

The right panel of figure 2 shows the dependence of the created EPP density on the field strength E_0 in the few photon regime at $T\, m = 5$. In the immediate neighborhood of the critical field $0.05\, E_c \lesssim$

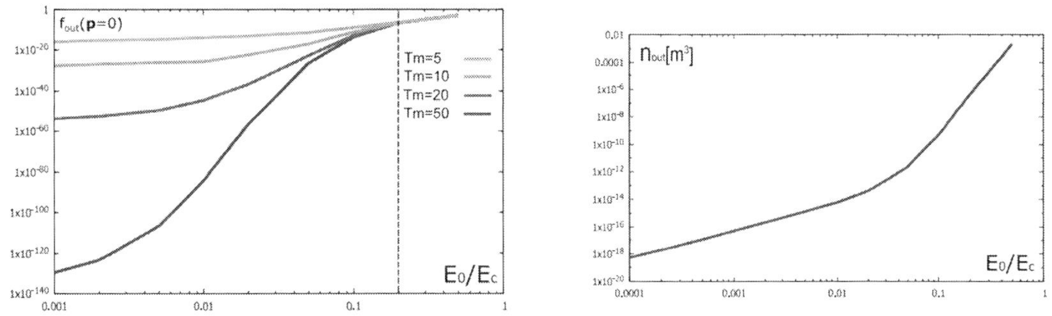

Figure 2. Left panel: Relation between $f_{\rm out}(\mathbf{p} = 0)$ and the amplitude of the external field E_0 for different values of pulse duration T. **Right panel:** The dependence of the density $n_{\rm out}$ on the external field E_0 in the few photon regime for $T\,m = 5$.

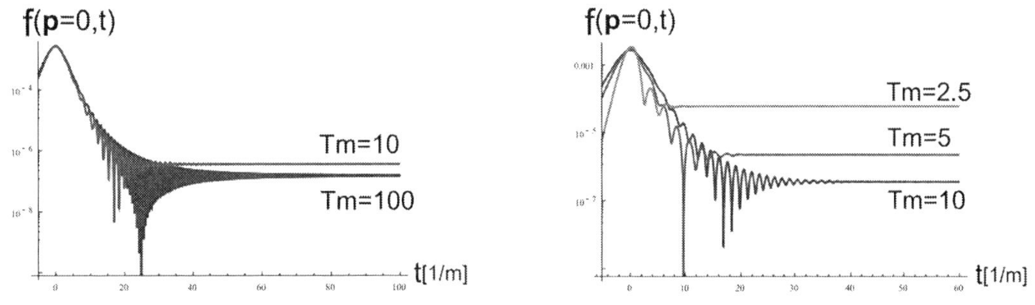

Figure 3. The transition from the quasiparticle EPP to the final out-state for the Eckart-Sauter pulse (4) with $E_0 = 0.2\,E_c$ at different values of T and $\mathbf{p} = 0$. **Left panel:** Transfer from the tunneling ($T\,m = 100$) to the few photon area ($T\,m = 10$). **Right panel:** Few photon area at different T.

$E_0 \lesssim 0.5\,E_c$ the efficiency of the pair creation process decreases very quickly (exponentially, as in the tunneling area). But at $E_0 \lesssim 0.02\,E_c$ and smaller it decreases as E_0^2. This promises an observable density of EPP even at field strengths several orders of magnitude smaller than E_c.

The left panel of the figure 3 shows a behavior of the distribution function that is typical for the phase transition from the vacuum to a final state of the EPP in the tunneling region ($T\,m = 100$) for a one-sheeted external electric field (4).

These are the three stages of evolution [1]: (i) quasiparticle EPP in the region of the maximal values of the external field, (ii) the transition region with rapid oscillations of the distribution function and (iii) the final state approaching a constant residual value $f_{\rm out}$.

Moreover, in the tunneling domain the saturation of the distribution function for a fixed point in momentum space appears. For example, the value of $f_{\rm out}$ at a given point $\mathbf{p} = 0$ does not depend on T and is defined by the field amplitude E_0 only. The left panel of figure 3 shows the deviation from this rule for $T\,m = 10$. The right panel shows a further sharp increase of $f_{\rm out}$ at the same E_0 only due to the lowering of T. One observes a gradual narrowing and a disappearance of the fluctuations in the

DOI: 10.1051/epjconf/201713807004

transient domain. This is due to the fact that the role of the vacuum polarization effects decreases with decreasing T.

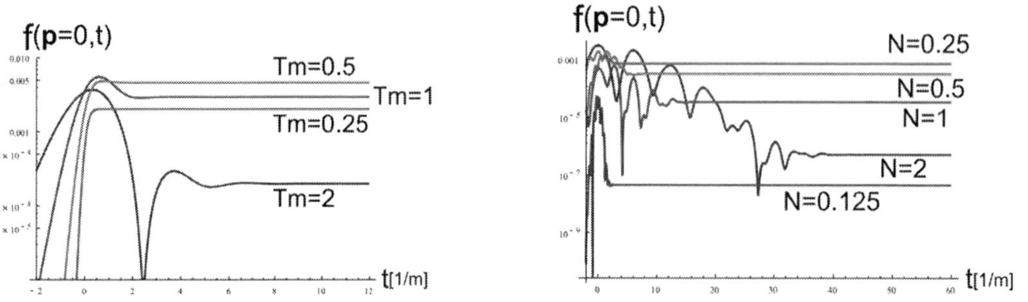

Figure 4. At a fixed electric field E_0 the maximum of f_{out} ($\mathbf{p}$ = 0) occurs in the vicinity of $T(N) \sim 1/m$. Entering further into the region with high photon energies results in a reduction of the efficiency of the pair creation process. **Left panel:** Features of the evolution of the distribution function in the vicinity of $T \sim 1/m$ for the field (4). **Right panel:** Similar results for field (5) in the vicinity $N \sim 1$ and $\sigma = 5$.

Figure 4 shows that both field models (4) and (5) exhibit a rapid growth of f_{out} in the few photon domain. However, near $T(N) \sim 1/m$ the growth stops and turns to a decrease for further increasing energy of the photons of the external field. There is a complication of the character of the evolution of the distribution function in the case of a more complex time dependence of the field (5) (right panel of figure 4). Here one observes also a gradual attenuation of oscillations in the transient region.

4 Conclusion

The results obtained in this contribution show that in the transition from the regime of tunneling (constant field and a field with a relatively slow time-dependence) to the few photon regime (quickly alternating fields with a characteristic frequency comparable to the Compton frequency) there are qualitative changes in the characteristics of the process of EPP creation from vacuum under the action of an external electric field. The evolution of the distribution function from the initial vacuum state to the final state loses a specific transient region of strong fluctuations. The transition to the few photon region is accompanied also by a considerable growth of the pair production effectiveness. So for the field strength $E_0 = 0.2\,E_c$ the pair density in the out state increases 3.34×10^5 times. But more importantly, the effect of pair production in the few photon domain can still be detected for much weaker fields, with field strengths several orders of magnitude smaller than E_c.

In contrast to the works on the bifrequent mechanism of vacuum EPP excitation [2, 3] the present investigation has shown that the transition in the few photon area ensures in itself a high intensity of the EPP creation process without participation of the tunneling mechanism. This is the main result of this work. It would be highly topical to perform a detailed comparison of the results of these two approaches.

Furthermore, it would be interesting to compare the results of the kinetic approach with the approach based on direct calculations of the probability of the Breit-Wheeler process in the framework of the S-matrix formalism [10].

Acknowledgements

S.A.S. and A.D.P. are grateful to the organizers of the XXIII Baldin Seminar for the support of their participation at this conference. The work of S.A.S., D.B.B. and L.J. was supported in part by the Polish Narodowe Centrum Nauki under grant No. UMO-2014/15/B/ST2/03752. D.B.B. was supported by the MEPhI Academic Excellence Project under contract No. 02.a03.21.0005. L.J. acknowledges support from the Bogoliubov-Infeld program for his visits to JINR Dubna in 2016.

References

[1] S. A. Smolyansky, A. D. Panferov, D. B. Blaschke, L. Juchnowski, B. Kämpfer and A. Otto, arXiv:1607.08775 [hep-ph]

[2] A. D. Panferov, S. A. Smolyansky, A. Otto, B. Kämpfer, D. B. Blaschke and L. Juchnowski, Eur. Phys. J. D **70(3)**, 56 (2016)

[3] A. Otto, T. Nousch, D. Seipt, B. Kämpfer, D. Blaschke, A. D. Panferov, S. A. Smolyansky and A. I. Titov, J. Plasma Phys. **82**, 655820301 (2016)

[4] F. Sauter, Z. Phys. **69**, 742 (1931)

[5] W. Heisenberg and H. Euler, Z. Phys. **98**, 714 (1936)

[6] J. S. Schwinger, Phys. Rev. **82**, 664 (1951)

[7] S. M. Schmidt, D. Blaschke, G. Röpke, S. A. Smolyansky, A. V. Prozorkevich and V. D. Toneev, Int. J. Mod. Phys. E **7**, 709 (1998)

[8] F. Hebenstreit, R. Alkofer, G. V. Dunne and H. Gies, Phys. Rev. Lett. **102**, 150404 (2009)

[9] N. B. Narozhny, and A. I. Nikishov, Sov. J. Nucl. Phys. **11**, 1072 (1970)

[10] A. I. Titov, B. Kämpfer, A. Hosaka and H. Takabe, Phys. Part. Nucl. **47(3)**, 456 (2016)

[11] D. B. Blaschke, B. Kämpfer, S. M. Schmidt, A. D. Panferov, A. V. Prozorkevich and S. A. Smolyansky, Phys. Rev. D **88(4)**, 045017 (2013)

Probing the low and high density nuclear matter by hadron scattering

S.M. Eliseev[1], *O.S.* Kosmachev[1]

[1]*Joint Institute for Nuclear Research, 141980, Dubna, Moscow Region, Russia*

Abstract. The subject of this report is connected with the widely discussed manifestation of in-medium effect in nuclear collisions. At present, the medium modification of hadron properties in the normal and in a dense nuclear matter is one of the fundamental questions of strong interaction physics. A new Glauber Monte Carlo model for hadron–nuclei interaction at intermediate energy is proposed. We utilized the principal assumptions as in the approaches of other authors describing nuclear collisions at high energy in the framework of the models without QGP. Yet, a number of new ingredients (noneikonal corrections, correlations of nucleons, in the nuclei, the nuclear Fermi motion, etc.) are introduced.

1 Introduction

Whether or not hadrons really change their properties in a nuclear medium, has been one of the long-standing problems in nuclear physics [1–4]. One of the essential assumption of long-familiar nuclear physics is that the protons and neutrons do not significantly change their properties and structure when they are in a nucleus.

Relativistic heavy-ion collisions are considered to be a favorable situation for the study of a large number of different phenomena which are expected to identify new properties of nuclear matter and nuclear interactions predicted by QCD. From QCD we know that at a sufficiently density, which may be obtained in high energy nucleus-nucleus collisions, color screening will lead to quark deconfinement. One of the most important subject discussed at present, are the difficulties related to the long distances in QCD. However, nucleons can significantly change their properties even in the *ground state of nuclei* [3, 4]. As it is seen from figure 1, the structure of nucleons is modified inside the nuclear medium (in-medium effect) at density $\rho \geq 0.5\rho_0$, ρ_0 is normal nuclear matter density. Inclusion of such unconventional EMC-type effects draw greater attention in nuclear physics [1, 2].

The renormalization of mass in the nuclear medium should be taken as a signature of partial restoration of chiral symmetry in *ground state nuclei*, even below the critical temperature and density. Such observations can direct to a some bridge between the investigations at intermediate and high energy. "This would represent a new paradigm for nuclear physics and help to build a bridge between QCD and nuclei "[1, 2].

In the last years, a lot of theoretical and experimental work has been devoted to the search of signals of such anomalous properties of hadrons in the nuclear environment *in-medium* effects [1].

Hadron modification has been one of the most debated topics in nuclear and particle physics conferences and attracted a great attention of scientists in different branches of particle, nuclear physics and astrophysics [5, 6].

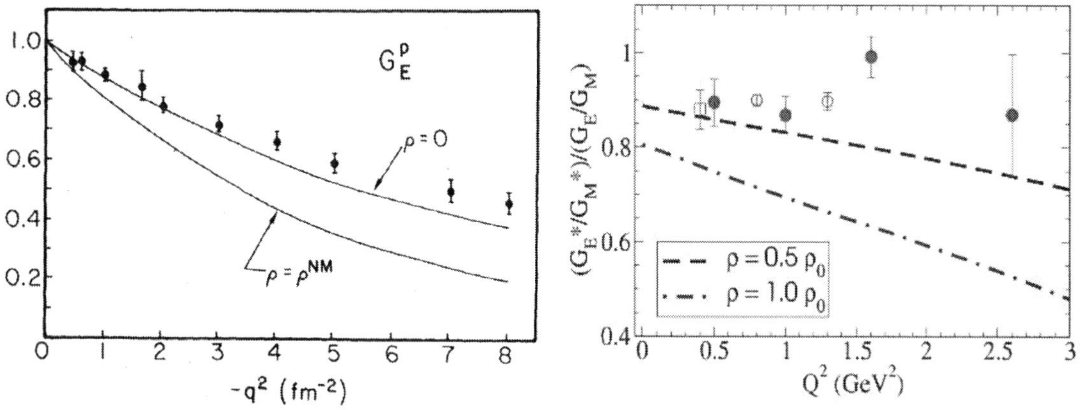

Figure 1. Left: calculated values of $G_E^p(q^2, \rho/\rho^{NM})$ as a function of $-q^2$ for $\rho/\rho^{NM} = 0$ (nucleon in vacuum) and for $\rho/\rho^{NM} = 1$ (nucleon in nuclear matter). Figure is taken from[4], figure 5. Right: double ratio for the proton electric/magnetic form factors in ^{4}He, normalized by the ratio in the vacuum versus the momentum transfer squared, Q^2. Data are from Mainz (square data point) and Jefferson Lab (circular data points). Predictions are given for normal nuclear matter density ($\rho = 1.0\rho_0$) and half normal nuclear density ($\rho = 0.5\rho_0$). Note the 10% reduction for $\rho = 0.5\rho_0$ and the 20% reduction for $\rho = 1.0\rho_0$ at $Q^2 = 0$. Figure is taken from [3], figure 14.

2 Glauber model

Presently, partly due to our lack of knowledge about the details of the confined mechanism, we are far away from a possibility to resolve the problem of *in-medium* effects from *the "first principles"*. Every attempt to extract some information concerning these effects from data will have (in part) the phenomenology.

Although the Glauber theory [7] was developed initially for high-energy projectiles, yet it is also employed at medium energies successfully [8, 9].

Glauber model became quite standard in the theoretical physics community, for details we refer to [10] and references therein. It is used in various theoretical models where it is essential to calculate the propagation of a particle through the medium. On the basis of this model several experimental data were interpreted, candidates for the signals of a phenomenon beyond the standard scenario: J/Ψ suppression, the phenomenon of color transparency, modeling in high energy nuclear collisions [10]. Therefore it is very tempting to use the same theoretical approach to recognize whether experiment on K^+–nuclei provides us with a hint for some new physics.

Kaons play a unique role in strong interaction physics. To examine the nuclear interior, the K^+–meson is regarded as a unique probe due to its long mean free path in the nuclear matter. Among all the hadrons, this makes the K^+ the weakest strongly interacting particle. This also means that K^+ can penetrate into the interior of a nucleus, thus probing the nucleus in a region where the density is high [11, 12].

According to the Glauber approximation the amplitude of a projectile–target elastic scattering may be written in the general form:

$$f(\vec{q}) = \frac{ik}{2\pi}\int e^{i\vec{q}\cdot\vec{b}}\Gamma(\vec{b})d\vec{b}, \text{ and } \Gamma(\vec{b}) = 1 - e^{i\chi(\vec{b})} = \frac{1}{2\pi ik}\int e^{-i\vec{q}\cdot\vec{b}}f(\vec{q})d\vec{q}, \tag{1}$$

where $\vec{b}$ is the impact parameter, χ is the corresponding phase shift function, and $f(\vec{q})$ is the elastic scattering amplitude of kaon–nucleon interaction.

Further, given the corresponding projectile–target nucleon amplitudes $f(q)$, one can express the above projectile–target nucleus amplitude in the **parameter–free** way. For projectile-nucleus scattering equation can be cast into the form:

$$F(\vec{Q}) = \frac{ik}{2\pi}\int e^{i\vec{Q}\cdot\vec{b}} < [1 - e^{i\chi(\vec{b},\vec{s}_1,\cdots,\vec{s}_A)}] > d\vec{b}, \tag{2}$$

where $\vec{s}_j$ is a component of the radius–vector $\vec{r}_j$ of the jth target–nucleon in the direction perpendicular to the incident momentum $\vec{k}$, while the brackets $< \ >$ denote the target ground–state average.

We used corrections to the Glauber's approximation which have been obtained by Wallace [13]:

$$\chi_j(b) = \sum_{n=0}^{j}\chi^{(n)}(b),\ \chi^{(n)}(b) = -\frac{\mu^{n+1}}{k(n+1)!}\left(\frac{b}{k^2}\frac{\partial}{\partial b} - \frac{\partial}{\partial k}\frac{1}{k}\right)^n \int_{-\infty}^{\infty} V^{n+1}(r)dz,\ V(r) = \frac{2\pi i}{\mu}f(0)\rho(r), \tag{3}$$

where μ is a reduced mass, and k is the momentum in the c.m.s. ($\hbar = c = 1$). The zeroth–order term in the above expansion represents the Glauber eikonal phase shift function, whereas higher–order terms, noneikonal effects. In this work, we truncate the tedious expansion of (3) and include, into our calculations, noneikonal corrections up to 3rd order. Then, the nuclear phase shift function can be written as follows,

$$\chi(b) = \chi_0(b) + \chi_{NE}(b), \qquad \chi_0(b) = \frac{2\pi}{k}\int_{-\infty}^{\infty} f(0)\rho(r)dz = -\frac{\mu}{k}\int_{-\infty}^{\infty} V(r)dz, \tag{4}$$

where $f(0)$ is the elementary forward scattering amplitude, $\rho(r)$ is the corresponding nuclear density. Also, one needs to point to the fact that the ^{28}Si nucleus is significantly clustered. Details of the nuclear structure is important for calculating differential cross sections [9]. In the report, we calculate only the total cross sections (see figure 2), and in this case it is necessary to know only the average density distribution of the nucleus [7, 8].

The parameters of kaon–nucleon amplitudes was taken from the partial wave analysis of K^+ nucleon scattering (for details we refer to [9] and references therein). Having thus the amplitudes $f(I)$ for isospins $I = 0$ and $I = 1$ we can readily find the K^+–proton amplitude as well as K^+–neutron amplitude

$$f(K^+p \to K^+p) = f(I=1), \qquad f(K^+n \to K^+n) = \frac{1}{2}\big[f(I=0) + f(I=1)\big].$$

For N = Z nuclei we used the average K^+–N amplitude

$$f(K^+N) = \frac{1}{2}\big[f(K^+p) + f(K^+n)\big].$$

 DOI: 10.1051/epjconf/201713807005

Further, given the corresponding projectile–target nucleon amplitudes $f(q)$, one can express the above projectile–target nucleus amplitude in the **parameter–free** way. It is then a straightforward matter to determine the total cross section for the case of K^+ – target nucleus scattering according to the optical theorem. The total and reaction cross sections may then be approximated by

$$\sigma_t = 4\pi \int_0^\infty \mathrm{Re}[1 - e^{i\chi(b)}]bdb, \qquad \sigma_r = 2\pi \int_0^\infty [1 - e^{-2\mathrm{Im}\chi(b)}]bdb, \tag{5}$$

where $\chi(b)$ is the nuclear phase shift function.

The detailed Fermi motion effect (using Monte Carlo simulation) are applied in the Glauber approach for nuclear collisions. In the present paper we consider a Gaussian parameterization of the momentum distribution of nucleons in nucleus used in analysis of high energy heavy-ion collisions and EMC effect [15]:

$$\rho(p) = N(e^{-p^2/p_0^2} + \epsilon_0\, e^{-p^2/q_0^2}) \tag{6}$$

where $p_0 = \sqrt{2/5}k_F$, $\epsilon_0 = 0.03$, $q_0 = \sqrt{3}p_0$, $k_F = 221$ MeV/c and N is a normalization constant.

3 Results and discussion

In figure 2 we show the dependence of the calculated and experimental cross sections of K^+–^{28}Si interactions *versus* kaon momentum. We can see that there is a universal discrepancy between the theoretical models and data for particles (K^+–mesons) interacting with ^{28}Si nuclei. In addition, optical potential models fail to describe the experimental data on K^+-nuclei interaction [12]. In comparison with previous works [12, 16–19], our approach improves the agreement between experimental data and theory. In addition, it is worth to emphasize that our results have been obtained without fitting any new parameters. Furthermore, K^+-meson probes the nucleus in a region of **high nuclear density** [12, 16–19]. At the same time, Glauber model gives an adequate description of collisions of proton with targets in the case of peripheral interactions, i.e. in the domain of **low nuclear density** [9].

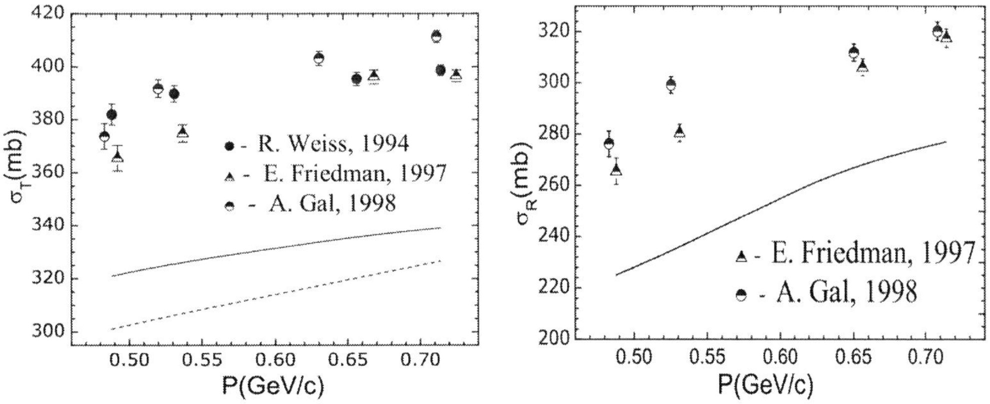

Figure 2. The calculated and experimental [16–18] total (σ_T) and reaction (σ_R) cross sections for K^+- meson interaction with ^{28}Si nuclei versus kaon momentum. The solid lines denote the prediction from our model. The dotted line demonstrates the results from the models used to calculate kaon-nucleus scattering: the covariant momentum-space optical potential model and the eikonal model (see table II, page 862 from [19]).

It is interesting to note - in both cases (electron-nucleon scattering and kaon-nucleus scattering) in-medium effect appears at **dense nuclear matter**, see figure 1 and figure 2. Comparison of the experimental data with the results of our calculations clearly demonstrates that the deeply penetrating hadron (K^+–meson) can "see" in the nucleus some "exotics" (beyond habitual mechanism of nuclear reactions). In our opinion the differences between models and data are more than just an artifact of the phenomenological models and that the introduction of some new in-medium effects is really required (see also [1–3, 5, 6]). In conclusion, investigation at intermediate energy of such type *"constitutes an important step for the future studies of baryons in high density matter with a strong magnetic field such as in a neutron star (or a magnetar), and in heavy ion collision experiment"* [1–3].

References

[1] A.W.Thomas, EPJ Web Conf. **123**, 01003 (2016)
[2] C. Cloet, Wolfgang Bentz, and Anthony W. Thomas, Phys. Rev. Lett. **116**, 032701 (2016)
[3] G. Ramalho, K. Tsushima, and A.W. Thomas, J. Phys. G **40**, 015102 (2013)
[4] L. S. Celenza et al., Phys. Rev. C **31**, 232 (1985)
[5] D. N. Voskresensky, D. Blaschke, Eur. Phys. J. A **52**, 67 (2016)
[6] Wei-Zhou Jiang, Bao-An Li, F. J. Fattoyev, Eur. Phys. J. A **51**, 119 (2015)
[7] R.G. Glauber, *"Lectures in Theoretical Physics"*, edited by W.E. Brittin and L.G. Dunham (Interscience, New York, 1959), vol. 1, p. 315
[8] R.J. Glauber, G. Matthiae, Nucl. Phys. B **21**, 135 (1970)
[9] R.H. Bassel and C. Wilkin. Phys. Rev. **174**, 1179 (1968) and references therein
[10] Michael L. Miller et al., Ann. Rev. Nucl. Part. Sci. **57**, (2007)
[11] G.E. Brown, C.B. Dover, P.B. Siegel, and W.Weise, Phys. Rev. Lett. **60**, 2723 (1988)
[12] E. Friedman, A. Gal, Phys. Rept. **452**, 89 (2007)
[13] S.J. Wallace, Ann. Phys. (N.Y.) **78**, 190 (1973)
[14] J.S. Hyslop, R.A. Arndt, L.D. Roper, R.L. Workman, Phys. Rev. D **46**, 961 (1992)
[15] Fujita T. and Hüfner J., Nucl. Phys. A **343**, 493 (1980)
[16] R. Weiss et al., Phys. Rev. C **49**, 2569 (1994)
[17] E. Friedman et al., Phys. Rev. C **55**, 1304 (1997)
[18] A. Gal, Nucl. Phys. A **639**, 485c (1998)
[19] M.F. Jiang, D.J. Ernst and C.M. Chen, Phys. Rev. C **51**, 857 (1995)

Notes on the problems of relativistic nuclear physics

S.M. Eliseev[1] and *O.S.* Kosmachev[1]

[1] *Joint institute for nuclear research, 141980, Dubna, Moscow Region, Russia*

Abstract. This article presents some arguments in favor of the necessity and opportunity to describe leptons and quarks on the united relativistic basis. The necessity is caused by experimental results of the relativistic nuclear physics. The opportunity has become really attainable taking into account the structure of the lepton equation groups and all their substructures obtained during the latest years.

Recently elementary particle physics has faced two challenges: dark matter and matter under extreme conditions. The second of them originally appeared as a product of relativistic nuclear physics. Both of these new facts are not described by existing theoretical models. The reason is that here we meet new physics, not just new events in physics which can be obtained in the framework of the already known, established notions.

It may be stressed that in the description of the observable ("bright") matter there is a gap between the lepton and hadron sectors. Unitary symmetry is an essential component of the Standard Model (SM). Its success is related to the fact that it successfully referred hard fixed substructures to hadrons - the quarks. But SM has not received them from the general principles. The lepton sector fell out completely from this structural classification of hadrons. It also, like quarks, turned out to be structureless. CM assigns sets of individual properties to leptons and quarks in the form of quantum numbers, but it can not name the reasons of their origin. This situation contradicts to the principle of the cause-and-effect relationship.

This fact is reflected in the conclusions of the O'Raifertaigh theorem [1]: the unification of the Lorentz group and internal symmetries of hadrons in the framework of a single group is possible only in the form of a direct product. As a consequence, we get a conclusion: all the components of the hadron multiplet have equal masses — that is not true.

On the other hand, a holistic description of the lepton sector has been recently developed by us [2]-[5]. A number of strict mathematical facts has been found and among them are: the presence of the individual structure of each lepton equation and the analogue of the confinement. It is detected clearly in the structure of both stable and unstable equations and occurs without any ad hoc assumptions. This structural confinement is necessary to understand the presence of substructures in the composition of the observed particles which do not exist in free states. In this case they are maximum invariant subgroups on which the expressions for the currents are formed. These subgroups appear automatically, i.e. their existence has not been assumed in advance.

The variety of the structures of different lepton equations is the basis for describing various properties of the leptons observed in the experiment. It turns out that in the description of the leptons presented in the classical works by Dirac [6], Pauli [7] and Majorana [8], this requirement is fulfilled if the structure is understood as a group structure of each wave equation obtained in these three articles. The group structures arise in the equations at the moment when you are writing them. However, until recently they have not been the subject of the study.

Indivisible structural components of all lepton equations are the four connectivity components of the Lorentz group. In this sense, relativism is the basis for the lepton equations. The four components of connectivity regulate their transformation properties relatively the discrete transformations: (R) is spatial inversion; (T) is the change of the time direction; (RT) - is the combined effect of the two transformations. In the lepton sector we have got a quite clear evidence of the fundamentality of lepton equation grounds. They fit into five assumptions which are necessary and sufficient conditions to formulate the lepton equations. The full set of equations admissible by five assumptions includes the massive and massless, charged and neutral, stable and unstable leptons having spin of 1/2.

The above mentioned holistic description of the lepton sector differs from the situation observed in the hadron sector on the basis of the quark model. As already mentioned above the quarks have a set of quantum numbers (including electric charge) without substructures, which would act as carriers of the declared characteristics. The proclamation of the hard quark structure is not related to the requirements of relativity. Subsequent various ways of quark relativization differ from what can be observed in the lepton sector. This structure of the lepton equations is the consequence of the relativity requirements and not vice versa. Such a permutation in the above procedure, as the practice of the latest 50 years has shown, leads to the differences which are impossible to avoid [9].

However, it becomes clear that there are no contra-indications to compose structures of hadrons in the same algorithmic sequence, as for leptons [5]. But it is necessary for this goal to add some complementary symmetries. In this case, relativism will be the basis of hadron substructures, which play the role of quarks. Structural components of the equations for stable leptons are four connectivity components of the homogeneous Lorentz group in the form of infinitesimal operators. Each of the connectivity components in the group of stable lepton equations is a subgroup of the order of 16. They are maximal subgroups which are the basis to form expressions for currents. We introduce the notations for them: group $\mathbf{d}_\gamma$ — is related with proper orthochronous representation; group $\mathbf{f}_\gamma$ — is related with the improper orthochronous or (*P*)- conjugate representation; group $\mathbf{b}_\gamma$ — is related with the proper antichronous or (*T*) — conjugate representation; group $\mathbf{c}_\gamma$ — is related with the antichronous improper or (PT)- conjugate representation.

The stable lepton groups have the following content:
the group of Dirac equation : $\mathbf{D}_\gamma(\mathbf{II}) = \{\mathbf{d}_\gamma, \mathbf{b}_\gamma, \mathbf{f}_\gamma\}$;
the group of Pauli equation for massless neutrinos: $\mathbf{D}_\gamma(\mathbf{III}) = \{\mathbf{d}_\gamma, \mathbf{b}_\gamma, \mathbf{c}_\gamma\, \mathbf{f}_\gamma\}$;
the group of Majorana equation for massive neutrinos: $\mathbf{D}_\gamma(\mathbf{I}) = \{\mathbf{d}_\gamma, \mathbf{c}_\gamma, \mathbf{f}_\gamma\}$.

By analogy with the Pauli equation group $\mathbf{D}_\gamma(\mathbf{III})$ it turned out to be possible to build additionally two groups with similar characteristics: $\mathbf{D}_\gamma(\mathbf{IV})$ and $\mathbf{D}_\gamma(\mathbf{V})$. Their peculiarity lies in the fact that each of them contains the connectivity components of one type only. In this case, they are candidates for massless particles, which have no antiparticles. However, it turns out that the both groups of the particles do not have the necessary elements to form the 4-vector current. Thus these particles can not exist in the vacuum (i.e. in the free states). Then it becomes clear that these groups appear automatically as substructures in groups of unstable leptons. Their structural content is the following:
the group for the massless (T)-"singlet": $\mathbf{D}_\gamma(\mathbf{IV}) = \{\mathbf{b}_\gamma\}$;
the group for the massless (PT)-"singlet": $\mathbf{D}_\gamma(\mathbf{V}) = \{\mathbf{c}_\gamma\}$.

For the unstable leptons the maximal invariant subgroups are subgroups of stable leptons which are part of each group of unstable leptons in various combinations.

The structures of the unstable lepton groups are as follows:
group $\mathbf{\Delta_1}$ is related with μ-lepton, $\mathbf{\Delta_1} = \{\mathbf{D}_\gamma(\mathbf{II}), \mathbf{D}_\gamma(\mathbf{III}), \mathbf{D}_\gamma(\mathbf{IV})\}$;
group $\mathbf{\Delta_3}$ is related with τ-lepton, $\mathbf{\Delta_3} = \{\mathbf{D}_\gamma(\mathbf{II}), \mathbf{D}_\gamma(\mathbf{I}), \mathbf{D}_\gamma(\mathbf{III})\}$;

group $\mathbf{\Delta_2}$ is related with unstable neutrino, $\mathbf{\Delta_2} = \{\mathbf{D}_\gamma(\mathbf{I}), \mathbf{D}_\gamma(\mathbf{III}), \mathbf{D}_\gamma(\mathbf{V})\}$.
Every group related with the corresponding equation has an original composition.

Some of the results of the holistic approach to describe the lepton sector are given below:

1. We have found an opportunity to list all types of lepton equations in the framework of the initial assumptions;
2. Structural individuality of each lepton equations has been determined;
3. The primary structural classification of the leptons has been obtained;
4. A simple and natural (without additional assumptions) opportunity has been found to describe the particles which do not exist in free states but occur in bounded states. It is the analogue of the confinement;
5. Basics of relativistic description of unstable leptons have been obtained. The structure of their equations indicates the internal self-consistency of the whole lepton sector in the framework of the proposed algorithm;
6. The existence of massive unstable neutrino has been predicted;
7. The existence of "doublets" of $\tau^{\pm}$-leptons has been predicted, i.e. $(\tau^{*})^{\pm}$-leptons. These four leptons form a quartet state on the basis of a united group.

It is evident that taking into account the structure of the leptons leads to non-trivial consequences for the whole sector. It is clear that the analogous consequences for hadrons will arise in a greater quantity because of their more complex structure. Thus, in the lepton sector the relativism (or more exactly all four connectivity components of the Lorentz group) is the initial structure-forming factor. We can not say the same about quarks. They were postulated as the constituent elements of hadrons and then relativized. The consequent decades have not confirmed the equivalence of this sequence of actions in comparison with the Dirac algorithm. This is probably the case when the struggle with the consequences without considering the initial reasons does not lead to the success.

All unstable elementary particles finally decay into stable leptons and two non-lepton exceptions — photons and protons. This situation naturally raises questions on the possibility of the reverse processes associated with the formation of the observed particles on the basis mentioned above. A reasonable answer to this obvious question is not possible beyond the methods and opportunities of the relativistic nuclear physics. One of the consequences of solving this large scientific problem will be formation of the elementary particles kinetics.

The main sign of the formation of matter under extreme conditions is an unusual (exaggerated) multiplicity of secondary particles. Such processes are not conceivable outside the consistent relativity because the ratio $E = mc^2$ there has no alternatives. The initial phase of extreme states of matter (let's call it protoplasma) can not be in the equilibrium condition because further self-organization (i.e., processes which are inverse to decays) occurs in the result of formation and decays of intermediate different unstable particles. The Lagrange's method is not applicable in these nonholonomic systems and the models based on it become inefficient.

The relativistic nuclear physics clearly defined the importance of two processes that play a key role for the high multiplicity of secondary particles: maximum destructurization, i.e., the formation of the most simple up to indivisible components in the collision of relativistic nuclei and self-organization i.e. formation of the complex by means of more simple elements in the environment called the matter under extreme conditions. There is no sense to speak about complete destruction without considering the structure of colliding particles. The question for today is as follows: how complete and close to reality is our knowledge about the structure of nuclei, hadrons and leptons?

The option of the minimal components as a result of the complete destruction of the colliding nuclei becomes the first priority for understanding the matter under extreme conditions. Another option is how the observed particles are formed in this unusual environment.

The structures of the stable leptons are so simple that they can not decay in the vacuum. It follows that the structure of stable leptons is simpler than the structure of quarks. From the formulas of structural composition it is seen that the components of the leptons can not play a role of quarks. Therefore, the study of matter under extreme conditions puts forward the problem of hierarchy of the complexity. It is necessary to answer the following question - what is the content of the initial environment (protoplasma), which is equally suitable to form leptons and hadrons in the finite state as well as many of the resonances in the intermediate states, including baryonic clusters observed only in the collisions of relativistic nuclei [10].

Taking into account the situation, when the complex is formed on the basis of the simple, we can call the initial phase of matter under extreme conditions as **photon-lepton protoplasma**. It is clear that it does not exclude the formation and continued existence of other components including the type, which is now called the quark-gluon plasma.

Unstable particles can be considered as one of manifestations of the matter under extreme conditions limited by lifetime of the particle and finite spatial volume. Therefore, the completed and detailed study of their decays, taking into account their structure and the structure of decay products is the most simple and informative way for finding the self-organization puzzle. From this point of view a comprehensive study of the decays of unstable leptons seems to be promising at this stage.

In this case the confinement and deconfinement of particles to the above processes becomes significant for both vacuum and photon-lepton protoplasma. It becomes clear that the mechanism adopted today for confinement of quarks is not universal. It will require revision taking into account the influence of the lepton sector composition of matter in the extreme state. In particular, traditional confinement has no place in stable leptons, where substructures exist, but quarks do not. A similar situation is observed with unstable leptons.

Experimental relativistic nuclear physics has identified a number of problems, which will be updated and refined in the course of their study. The following problems are considered to be the first priority and play the key role:

1. It is necessary to find the description of leptons and quarks on the unified relativistic basis.
2. Search for the quark structure must be extended by using the Dirac's algorithm i.e., similarly how the structure of electrons and other leptons was understood. And taking into account additional symmetries beyond the minimum required for the lepton sector.

The difficulty and subtlety of the above is that it is necessary strongly to differentiate the properties of substructures and those characteristics which are the consequence of the interactions between the observed objects containing these substructures. The so-called quantum numbers are introduced only on the results of the interactions and then attributed to the particles in free states, and even to the particles which do not exist in the free states. The negative role of the above factors and difficulties will reduce while increasing the number of the described particles on the common basis. Changes in quark phenomenology are needed in order to gradually study the new challenges for it must be internally self-consistent and independent of unnecessary assumptions.

References

[1] L. O'Raifertaigh, Phys. Rev. **139**, B1052-B1062 (1964)
[2] A. A. Gusev, O. S. Kosmachev, Phys. Part. Nucl. Lett. **5**, 67-71 (2008)
[3] O. Kosmachev, Phys. Part. Nucl. Lett. **7**, 149-174 (2010)
[4] O. Kosmachev, Phys. of At. Nucl. **74(6)**, 930-935 (2011)
[5] O. Kosmachev, Phys. of At. Nucl. **75(7)**, 885-887 (2012)
[6] P. Dirac. Proc. Roy. Soc. A, **117**, 610-624 (1928)
[7] W. Pauli, *Handbuch der Physik* (Springer, Berlin, 1933) 226
[8] E. Majorana, Il Nuovo Cimento **14**, 171-184 (1937)
[9] A. Pais, Rev. Mod. Phys. **38(2)**, 215-255 (1966)
[10] A. M. Baldin et al., Phys. of Nucl. **52(5)**, 1427-1436 (1990)

EPJ Web of Conferences **138**, 07007 (2017) DOI: 10.1051/epjconf/201713807007
Baldin ISHEPP XXIII

Gluon dominance model

Elena Kokoulina[1,2,⋆] and *Andrey* Kutov[3,⋆⋆]

[1] *Joint Institute for Nuclear Research, Dubna, Moscow region, 141980 Russia*
[2] *P.O.Sukhoi Gomel State Technical University, Gomel, 246000 Belarus*
[3] *Department of Mathematics, Ural Division, RAS, Syktyvkar, 167982 Russia*

Abstract. Study of multi-particle production has longer than the semi-centennial history. As it is known, with the growth of energy of accelerators, the new channels of reaction are being opened, the average number of secondary particles is increasing. Physicists are able to accelerate stable particles, such as electrons, positrons, protons, antiprotons, ions (both light and heavy). Rarely, they accelerate kaons and pions. The obtained experimental material stimulates the development of the different theoretical approaches. Since appearance of the modern theory of strong interactions, quantum chromodynamics (QCD), our understanding of multi-particle production is advanced significantly. The language of quarks and gluons is basic one at the explanation of observable phenomena. This review is devoted to the history of appearance and the following development of the gluon dominance model. This model is based on the pQCD and the phenomenological description of the hadronization stage. It permits to describe multiplicity distributions both for lepton and hadron interactions, especially in the high multiplicity region.

1 Introduction

Study of the multi-particle processes had become very popular in the middle of seventies [1, 2]. It has been found that in a collision of two relativistic particles a and b, three and more hadrons can be produced: $a+b \to c_1+c_2+\cdots+c_n$. The number of secondary particles, n, is called multiplicity. In the experiments, we distinguish multiplicity of charged, n_{ch}, and neutral, n_0, particles. Sum of charged and neutral particles is the total multiplicity: $n_{\rm tot} = n_{\rm ch} + n_0$.

For the description of multi-particle processes, the language of the mathematical statistics is applied [3]. The multiplicity distribution (MD), P_n, or the probability of production of n particles is the ratio $P_n = \sigma_n/\sigma_{inel}$, where σ_n is the topological cross section, $\sigma_{inel} = \sum_n \sigma_n$. The kth moment of P_n is $< n^k > = \overline{n^k} = \sum_n n^k P_n$. The moment of the first order is a mean multiplicity (an average value), $\overline{n}$. The kth central moment is $< (n- < n >)^k > = \overline{(n-\overline{n})^k}$. The central moment of the second order is a variance, $D_2 = \overline{n^2} - \overline{n}^2 = \sigma^2$, where the standard deviation σ is its square-root.

The number of secondary particles varies from event to event. But on average, the mean multiplicity is staying constant at the given energy and for the given interaction. The events in which the number of secondary particles considerably exceeds mean multiplicity are called the high multiplicity

⋆e-mail: kokoulina@jinr.ru
⋆⋆e-mail: akutov@rambler.ru

(HM) events and this region – high multiplicity region (HMR). Since last years, the HM event study has drawn a high attention that is connected with discovery of the collective behaviour of secondaries in nuclear and hadron interactions (ridges, flows, shock waves) [4–6].

Our SVD-2 Collaboration studies the HM events in a proton-proton, a proton-nucleus and a nucleus-nucleus interactions at the U-70 accelerator (IHEP, Protvino) and at Nuclotron (JINR, Dubna). Before beginning of these experiments, to predict the topological cross sections at the HMR, the gluon dominance model [7] has been developed. It evolved from the two stage model (TSM) [8] worked out for the description of MD in e^+e^- annihilation at high energies. Both models are based on the convolution of two main stages. In accordance of QCD, on the first stage, a quark-gluon cascade is developed. The second stage is hadronization. For this stage a phenomenological scheme is applied. Comparison of these models with experimental data on MD and other values shows good agreement.

Study of MD is important for understanding of the multi-particle production mechanism. TSM confirms the a fragmentation mechanism of hadronization in e^+e^- annihilation and GDM – a recombination mechanism in hadron interactions. Both models evidence an active role of gluons and a passivity of valency quarks at the multi-particle production.

This paper is organized as follows. In section 2, we describe the construction of TSM and its main results. In section 3, we develop this scheme to study hadron interactions and we call it GDM. It has been shown that the region of HM is stipulated for a gluon fission which can explain the tail of HM and even the ridge phenomenon [9]. The search for collective phenomena in pp interactions in HMR is carried out by the SVD-2 Collaboration at U-70 and Nuclotron. The growth of a scaled variance in HMR of the total multiplicity can be the indication the pion (Bose-Einstein) condensate formation [10]. Then we inform about our searching for the possible connection of the pion condensate and an excess yield of soft photons (whose energy is smaller than 50 MeV) in AA interactions at Nuclotron [11]. We round off with our future plans and conclusions in section 4.

2 TSM and e^+e^- annihilation

TSM has been constructed in the end of seventies [14]. It was aimed at the MD description of hadrons in the e^+e^- annihilation at high energies. In accordance with QCD, this process is realized through γ or Z^0-boson with formation of a quark-antiquark pair, its branching and a conversion of the initial and the newly born quarks and gluons to the observable hadrons: $e^+e^- \to q\overline{q} \to (q, \overline{q}, g) \to h_1 + h_2 + \cdots + h_n$. If quarks have sufficiently high energy, they can develop quark-gluon cascade $(q, \overline{q}, g)$. In [12, 13] three elementary processes are used in this description: 1) $q \to q+g$, 2) $g \to g+g$ and 3) $g \to q+\overline{q}$. The first two from them are predominated over the third one until the cascade is developed. Giovannini [12] has constructed and decided the system of the differential equations for the Markov branching process of quark and gluon. He writes them on the language of the generating functions (GF), $Q(z) = \sum_{n=0}^{\infty} P_n z^n$, where z – the auxiliary parameter. GF $Q(z)$ is connected with MD P_n:

$$P_n = \frac{1}{n!}\frac{\partial^n}{\partial z^n} Q(z)|_{z=0}\,.$$

The kth moment of MD is also calculated through the differentiation of GF: $\overline{n(n-1)\dots(n-k+1)} = \partial^n Q(z)\ /\partial z^n|_{z=1}$. For the quark jet, the qg-cascade leads to the negative binomial distribution (NBD):

$$P_m^q = \frac{k_p(k_p+1)\dots(k_p+m-1}{m!}\left(\frac{\overline{m}}{\overline{m}+k_p}\right)^m\left(\frac{k_p}{\overline{m}+k_p}\right)^{k_p} \quad (1)$$

with GF

$$Q^q(z) = \left[1 + \frac{\overline{m}}{k_p}(1-z)\right]^{-k_p}. \quad (2)$$

where k_p is the ratio of probabilities of two elementary processes 1) and 2): a quark bremsstrahlung of a gluon to a gluon fission, $\overline{m}$ – a mean multiplicity of gluons.

MD for a g-jet is the Farry distribution with GF $Q^g(z)$

$$P_m^g = \frac{1}{\overline{m}}\left(1 - \frac{1}{\overline{m}}\right)^{m-1}, \quad Q^g(z) = \frac{z}{\overline{m}}\left[1 - z\left(1 - \frac{1}{\overline{m}}\right)\right]^{-1}. \tag{3}$$

To get MD of hadrons, it is necessary to take into account a hadronization. From experimental data, it is known that the second correlative moment, $f_2 = \overline{n(n-1)} - \overline{n}^2$, for e^+e^- annihilation is negative at low energy. With growth of energy, it changes sign from "minus" to "plus" [15]. That behaviour is typical for a multi-particle production. Obviously, at the low energy, the qg-cascade is not enough developed. In this case the hadronization carries in MD the determinative contribution. As MD at hadronization stage has the negative second correlative moment, a binomial distribution (BD) has been chosen for this stage

$$P_n^h = \binom{N_P}{n}\left(\frac{\overline{n}_P^h}{N_P}\right)^n\left(1 - \frac{\overline{n}_P^h}{N_P}\right)^{N_P-n}, \tag{4}$$

where $\overline{n}_P^h$ and N_P are parameters having the sense of the mean multiplicity and maximum number of hadrons formed from a single parton $P = q, g$ (quark or gluon) through its passing of the hadronization stage. The second correlative moment for BD is equal to $f_2 = -(\overline{n}_P^h)^2/N_P < 0$. In the case of independent particle production which is described by Poisson distribution, $P_n = e^{-\overline{n}}\overline{n}^n/n!$ with GF $Q(z) = e^{\overline{n}(z-1)}$, $f_2 \equiv 0$. NBD gives positive $f_2 = \overline{m}^2/k_p$.

The convolution of this two stages is based on the hypothesis of the soft discolouration [14] and an equality of the hadron production probabilities from quark and gluon at the second stage. It leads to final expression for MD of hadrons in e^+e^- annihilation [8]

$$P_n = \Omega \sum_{m=0}^{M_g} P_m^q \binom{(2+\alpha m)N}{n}\left(\frac{\overline{n}^h}{N}\right)^n\left(1 - \frac{\overline{n}^h}{N}\right)^{(2+\alpha m)N-n}, \tag{5}$$

where Ω is a normalization factor, $N = N_q$, $\alpha = N_g/N_q$, M_g is the maximum number of active gluons (they can fragment to hadrons). TSM gives good description of MD in e^+e^- annihilation from 9 up to 189 GeV [8, 16]. In figure 1, on the left panel, the description of MD at 160 GeV [17] by TSM is shown. The mean multiplicity of gluons after the completion of the qg-cascade is described well by the logarithmic dependence with energy. The parameter of hadronization for a quark, $\overline{n}_q^h$ varies in the wide limits. In contrast to the quark parameter, the gluon parameter, $\overline{n}_g^h$ has a value close to one, that is confirmed by fitting: $\overline{n}_g^h = 0.947 \pm 0.002$ at χ^2/ndf = 34/17 = 2 (figure 1, the right panel). The TSM confirms predicted by I. Dremin [3] oscillations of the ratio of cumulative factorial moment to factorial moment as function of their rank, the growth of the period of oscillations at energy higher 90 GeV and changing of the sign of the second correlative moment with energy [8, 14]

$$f_2 = \left[\alpha^2\frac{\overline{m}^2}{k_p} + \alpha^2\overline{m} - \frac{2+\alpha\overline{m}}{N}\right](\overline{n}^h)^2.$$

The characteristics of the gluon jet obtained with TSM have been used in the description of MD in the tree-gluon decay of quarkonia, Υ(9.46) and Υ (10.02) [18]

$$P_n(m_\Upsilon) = \sum_{m'=0} \frac{(m'-1)(m'-2)}{2(\overline{m}/3)^3}\left(1 - \frac{1}{\overline{m}/3}\right)^{m'}\binom{(3+m')N_g}{n}\left(\frac{\overline{n}_g^h}{N_g}\right)^n\left(1 - \frac{\overline{n}_g^h}{N_g}\right)^{(3+m')N_g-n}.$$

 DOI: 10.1051/epjconf/201713807007

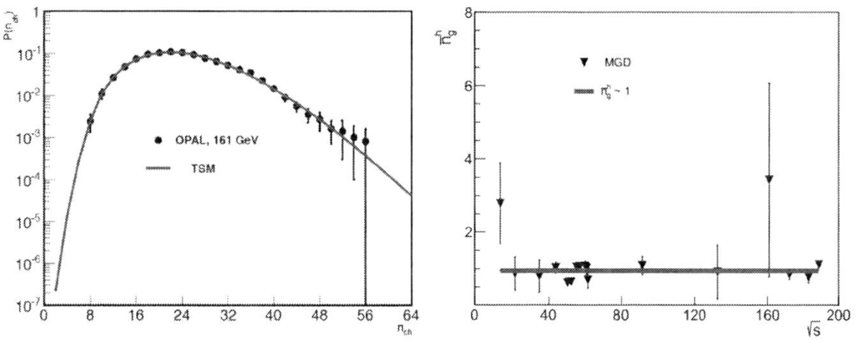

Figure 1. Left panel: The description of MD by TSM in e^+e^- annihilation at 161 GeV [17]. Right panel: the gluon parameter of hadronization versus energy.

The differences between mean multiplicity of hadrons in the region with two jet formation close to quarkonium and multiplicity of the three-gluon decay calculated in TSM is identical to the experimental value.

Using TSM we estimate values of mean multiplicity for e^+e^- annihilation at the energies of future ILC and CLIC accelerators: $\overline{n}(500\text{ GeV}) \approx 30 \pm 5$, $\overline{n}(1\text{ TeV}) \approx 50 \pm 10$.

3 GDM and hadron interactions

The "Thermalization" project was aimed at the collective phenomena searching for in HM region in pp interactions at the U-70 accelerator in IHEP (Protvino). The energy of a proton beam was 50 GeV. We have manufactured a HM-trigger to suppress registration of events with a small multiplicity. This allowed us to go down on the topological cross sections by three orders reaching max $n_{ch} = 24$ (at 50 GeV the kinematical limit is equal to 59 pions). At this energy, HM $n >> \overline{n} \approx 5$ for charged particles.

Before beginning of the "Thermalization" project, our SVD-2 Collaboration has found out that the Monte Carlo simulation underestimates the experimental data of the Mirabelle Collaboration [19] for topological cross section at the maximum value of the measured multiplicity, $n_{ch} = 18$, two orders in pp interactions at 70 GeV proton beam [20]. The phenomenological models gave them the different predictions too. One of them is MGD [7], called a gluon model, second one – quark model [21].

To modify TSM for hadron interactions, we have supposed that in hadron interactions, a valence quarks and the nascent gluons develop branching in accordance to QCD elementary processes.

The valency quark inclusion led us to values of $\overline{n}^h \ll 1$. With decrease of the number of valence quarks, this parameter begins to grow. Only completely an excluding of the valence quark (they are staying in the leading particles), we achieve the considerable improvement of χ^2, at that, the hadronization parameters become rather more than 1 as in the e^+e^- annihilation. It indicates the change of mechanism of hadronization: from fragmentation to recombination one [7].

This result has permitted us to formulate a gluon dominance model (GDM). According to this model, the basic sources of secondary hadrons are gluons, we call them active ones. As in TSM, the first stage is a formation of qg-medium and its branching, the second stage is hadronization of gluons. GDM exists in two schemes, with a fission of gluon and without of it. The first scheme testifies to half of the nascent gluons fragment in hadrons. The rest of them are caching up by quarks and reradiating can be sources of the soft photons (SP). MD of hadron in the second scheme presents the convolution

DOI: 10.1051/epjconf/201713807007

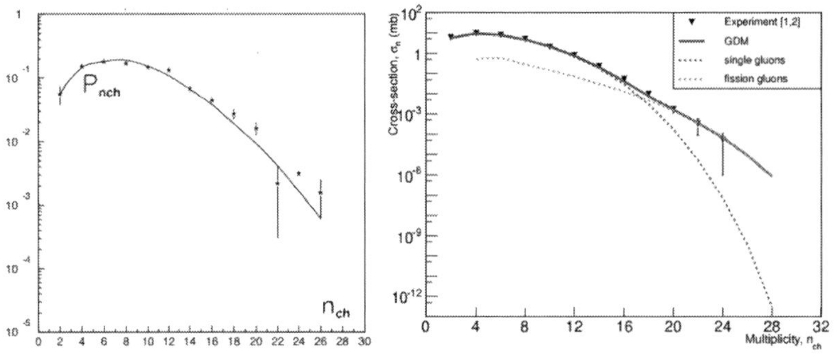

Figure 2. The description of MD by GDM in pp interactions. Left panel: at 300 GeV/c [22]; Right panel: at 50 GeV/c with taking into account the gluonic fission [23].

of two stages in the case of the independent appearance of active gluons

$$P_n = \sum_{m=1}^{M} \frac{\overline{m}^m e^{-\overline{m}}}{m!} \binom{mN}{n-2} \left(\frac{\overline{n}^h}{N}\right)^{n-2} \left(1 - \frac{\overline{n}^h}{N}\right)^{mN-(n-2)} . \tag{6}$$

In figure 2, left panel the description of MD at 300 GeV [22] by GDM is offered. In [9] it has been shown that with increasing of energy, a gluon fission becomes predominant in comparison to quark bremsstrahlung. In figure 2, right panel, the convolution of two contributions (without and with gluon splitting) describes well MD, including HM region.

The differences of topological cross sections $p\overline{p}$ annihilation and pp collisions at the same energy reveal two local maximums [24]. This two-humped structure can be described in the framework of GDM. For this purpose, some intermediate charged topologies determined by the valency quark combination have been introduced [25]. In the HM region, the "4"-topology (participate valency and sea quarks) carries noticeable contribution in these differences in HMR.

4 Pionic condensate and soft photons. Conclusion.

The SVD-2 Collaboration has received indication at the formation of the Bose-Einstein (pionic) condensate (BEC) in the high total multiplicity [26]. Begun and Gorenstein [27] within the framework of the ideal hadron gas model have shown that the sharp growth of the scaled variance, ω^0 (the ratio of a variance to mean multiplicity of the neutral pions at a given total multiplicity) is the signal of BEC formation. Our experiment testifies that the region of high total multiplicity is a very interesting one. The two-particle Bose-Einstein correlations in pp collisions at $\sqrt{s} = 0.9$ and 7 TeV measured with the ATLAS detector in the HM region show that the behaviour of their parameters testifies about approaching to the BEC ($\lambda \to 0$ – fully coherent emission of identical bosons, the effective radius R is staying constant or its measurement is difficult) [28].

Registration and research of the SP yield in hadron and nuclear interaction were the tasks standing before our experiment. There are some theoretical models which explain the excess of their yield by the BEC formation. That is why we have designed and manufactured an electromagnetic calorimeter with the low energy threshold smaller than 50 GeV [29]. Our first measurements confirmed the excess yield of SP in interactions of deuterium and lithium beams with a carbon target at 5A GeV/c

at Nuclotron. In the nearest future we plan studying: a connection between BEC formation and an excess of SP yield; a search of new resonances in the system of two γ-quanta ($f_0(500)$ or σ-meson, etc.); a study of increased yield of η^0-mesons in AA-interactions; research of pair photon interference (γ-femtometry); a search of P-parity violation effect in the events with high p_T in the BM@N setup.

References

[1] R. Hagedorn, Nuovo Cim. Suppl. **3**, 147-186 (1965)
[2] V.S. Murzin and L.I. Sarycheva, *Physics of hadron processes* (Energoatomizdat, Moscow, 1986, in Rissian) 280 p.
[3] I.M. Dremin, Phys. Lett. B **341**, 95-98 (1994)
[4] J.Adams et al. STAR Collaboration, Nucl. Phys. A **757**, 102-183 (2005)
[5] K.Adcox et al. PHENIX Collaboration, Nucl. Phys. A **757**, 184-283 (2005)
[6] S.Chatrchyan et al. JHEP **9**, 111-132 (2012)
[7] E. Kokoulina, A. Kutov, and V. Nikitin, Braz. J. Phys. **37**, 2C, 785-787 (2007)
[8] E.C. Kokoulina, in the proc. of XXXII ISMD, Alushta, Ukraine (World Scientific, 2002) 340
[9] E.A. Kuraev, E. A. S. Bakmaev and E.S. Kokoulina, Nucl. Phys. B **851**, 551-564 (2011)
[10] A.G. Afonin et al. SVD-2 Collaboration, EPJ Web of Conferences **37**, 06002 (2013)
[11] E. Ardashev et al. Eur.Phys.J. A **52**, 261 (2016)
[12] A. Giovannini, Nucl. Phys. B **161**, 429-448 (1979)
[13] K. Konishi, A. Ukawa and G. Veneziano, Nucl. Phys. B **157**, 45-107 (1979)
[14] V.I. Kuvshinov and E.S. Kokoulina, Acta Phys.Pol. B **13**, 533 (1982)
[15] J.G. Rushbrooke and B.R. Webber, Phys. Rep. **44**, 1-92 (1978)
[16] E.C. Kokoulina, In proceedings of11th Annual Seminar on Nonlinear Phenomena in Complex Systems: Chaos, Fractals, Phase Transitions, Selforganization. Minsk, Belarus, 13-16 May 2002, **11**, 23-136 (2003)
[17] K.Ackerstaff et al., The OPAL Collaboration, Z.Physics C **75**, 193-207 (1997)
[18] E.S. Kokoulina and V.I. Kuvshinov, Sov.Phys.J. **28**, 757-760 (1985)
[19] V.V.Ammosov et al., Phys. Let. B **42**, 519-521 (1972)
[20] E. Kokoulina, Acta Phys. Polon. **35**, 295-302 (2004)
[21] O.G. Chikilev and V.P. Chlyapnikov, Phys. Atom. Nucl. **55**, 779-783 (1992)
[22] A. Firestone et al., Phys. Rev. D **10**, 2080-2083 (1974)
[23] E.S. Kokoulina, Phys.Part.Nucl.Lett. **13(1)**, 74-78 (2016)
[24] J.G. Rushbrooke and B.R. Webber, Phys. Rep. **44**, 1-92 (1978)
[25] E.C. Kokoulina, American Institute of Physics. Conference proceedings **828**, 81-86 (2006)
[26] E. Kokoulina, Progr. of Theor. Phys. Suppl. **193**, 306-309 (2012)
[27] V.V. Begun and M.I. Gorenstein, Phys. Rev. C **77**, 064903 (2008)
[28] ATLAS Collaboration, Eur. Phys. J. C **75**, 466 (2015)
[29] E. Ardashev et al., Eur. Phys. J. A **52(8)**, 261 (2016)

 DOI: 10.1051/epjconf/201713807008

3π and 4π meson production in np interactions at intermediate energies

A.P. Jerusalimov[1]★, *A.V.* Belyaev[1], *V.P.* Ladygin[1], *V.N.* Pechenov[2], and *Yu.A.* Troyan[1]

[1] *Joint Institute for Nuclear Research, 141980 Dubna, Russia*
[2] *GSI Helmholtzzentrum fur Schwerionenforschung GmbH, 64291 Darmstadt, Germany*

Abstract. The study of 3π and 4π meson production in np interactions was carried out at the momenta of incident neutrons P_0=3.83, 4.42 and 5.20 GeV/c. The characteristics of the reactions were satisfactorily described by OPER model. For the better description of the reaction $np \to pp\pi^+\pi^-\pi^-\pi^0$ it was necessary to take into account the production of η^0 and ω^0 mesons.

1 Introduction

The study of multipion production in NN collisions is one way to obtain information about the NN, πN and $\pi\pi$ states, including:

dibaryons (including I=2 in ppπ^+),
dipions (narrow σ^0 meson, $\pi\pi$ state with I=2),
pentaquarks (I=5/2, S=+1),
missing resonances, etc.

Also the important task is the test of various models of pion production in NN interaction, such as, for example, Valencia model [1], Xu Cao model [2] and OPER model [3, 4].

2 Experiment

The neutron-proton interactions were studied using neutron beam and liquid hydrogen bubble (target) at the JINR Synchrophasotron [5]. The unique in fullness and precision data were obtained. It allowed to carry out the detailed study of inelastic np interactions in in a wide range of energies using the quasimonochromatic neutrons with P_0<2.5% under condition of 4π geometry.

The following reactions with 3 and 4 π mesons in the final states were studied:

$np \to pp\pi^+\pi^-\pi^-$,
$np \to pp\pi^+\pi^-\pi^-\pi^0$,
$np \to np\pi^+\pi^+\pi^-\pi^-$

at the momenta of P_0=3.83, 4.42 and 5.20 GeV/c (see figure 1).

The accuracy of the momentum and scattering angle reconstruction for the secondary charged particles was $\sigma_p/p \sim 2\%$ and $\sigma_\Theta \sim 10$ mrad, respectively. The separation of the reaction channels

★e-mail: jerus@jinr.ru

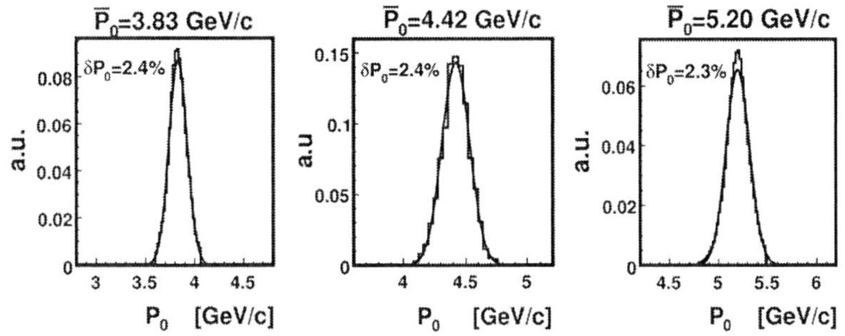

Figure 1. The momentum distributions of the incident neutrons at P_0= 3.83, 4.42 and 5.20 GeV/c

were carried out by the standard χ^2 procedure using corresponding constraint equations [6]. Finally, the numbers of events that were selected for the further study:

	P_0=3.83 GeV/c	P_0=4.42 GeV/c	P_0=5.20 GeV/c
$np \to pp\pi^+\pi^-\pi^-$	390 events	743 events	5790 events
$np \to pp\pi^+\pi^-\pi^-\pi^0$	66 events	215 events	2666 events
$np \to np\pi^+\pi^+\pi^-\pi^-$	83 events	344 events	4530 events

The cross sections of the considered reactions were presented in [7].

3 Data analysis

This reaction is characterized by:
- plentiful production of the Δ resonance, both from direct production and from Δ^* and N^* decays through the mode $\Delta\pi$,
- large peripherality of the secondary nucleons.

The model of reggeized π meson exchange suggested in ITEP [3] was taken to describe the experimental distributions of the considered reactions. The advantages of this model are:

- small number of free parameters (3 in our case),
- wide range of the described energies (2÷200 GeV),
- calculated values are automatically normalized to the reaction cross section.

The following diagrams were taken into account to calculate the characteristics of the reactions $np \to NN\,3\pi$ and $np \to NN\,4\pi$:

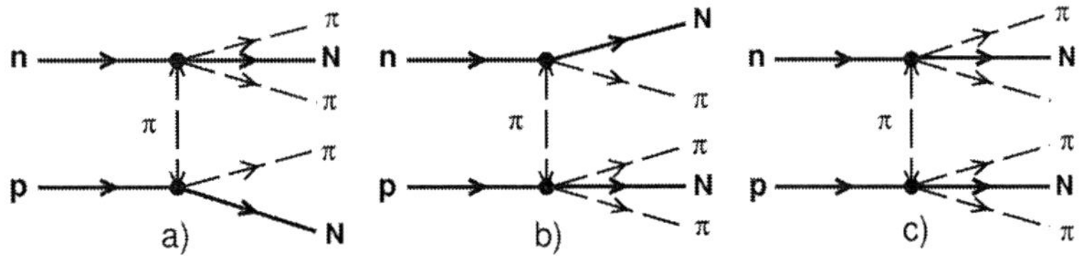

Figure 2. Diagrams of OPER model for the reactions of 3π and 4π meson production in np interactions.

3.1 Reaction $np \rightarrow pp\pi^+\pi^-\pi^-$

The results of the calculations using OPER model for the reaction $np \rightarrow pp\pi^+\pi^-\pi^-$ are shown in figure 3 for the data at P_0 = 5.2 GeV/c and in figure 4 for the data at P_0 = 4.42 GeV/c and P_0 = 3.83 GeV/c. One can see a good agreement between the experimental data and theoretical calculations.

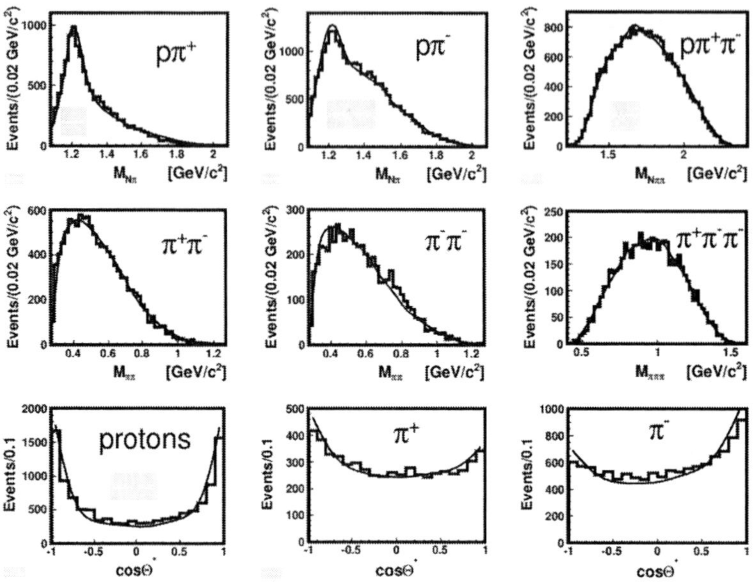

Figure 3. The distributions for the reaction $np \rightarrow pp\pi^+\pi^-\pi^-$ at P_0=5.20 GeV/c. Solid line - calculations using OPER model.

3.2 Reaction $np \rightarrow pp\pi^+\pi^-\pi^-\pi^0$

The results of the calculations using OPER model for the reaction $np \rightarrow pp\pi^+\pi^-\pi^-\pi^0$ are shown in figure 5 for the data at P_0 = 5.20 GeV/c.

One can see a good agreement between the experimental data and theoretical calculations except the region of η^0 and ω^0 mesons at the masses of $\pi^+\pi^-\pi^0$ combinations. As far as it concerned the description of η^0 and ω^0 mesons it is necessary to take into account the diagrams of the type that are showed in figure 6.

The diagrams 6a and 6b describe η^0-meson production through the production of N^*_{1535} resonance in πN interaction with the consequent decay $N^*_{1535} \rightarrow N\eta^0$. The "hanged" diagrams 6c and 6d describe η^0 meson production due to a-π or σ-η interaction.

The results at P_0=4.42 GeV/c and P_0=3.83 GeV/c are not presented due to a small statistics. But OPER model also described satisfactorily the experimental distribution and the signal of η^0 and ω^0 mesons production was also observed at these energies.

3.3 Reaction $np \rightarrow np\pi^+\pi^+\pi^-\pi^-$

The results of the calculations using OPER model for the reaction $np \rightarrow np\pi^+\pi^+\pi^-\pi^-$ are shown in figure 7 for the data at P_0 = 5.2 GeV/c and in figure 8 for the data at P_0 = 4.42 GeV/c and P_0 = 3.83 GeV/c.

One can see a good agreement between the experimental data and theoretical calculations.

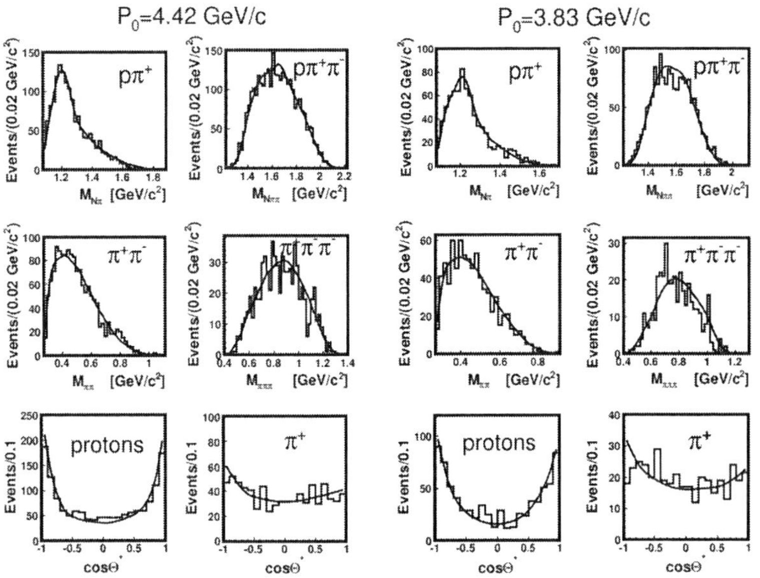

Figure 4. The distributions for the reaction $np \to pp\pi^+\pi^-\pi^-$ at P$_0$=4.42 GeV/c and P$_0$=3.83 GeV/c. Solid line - calculations using OPER model.

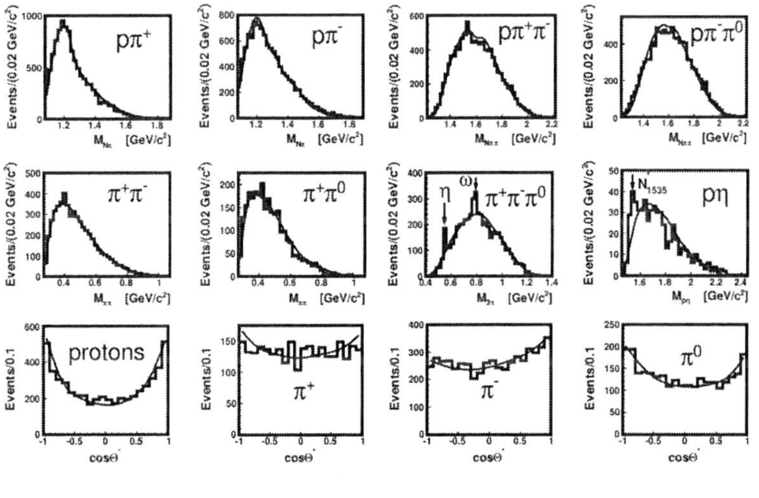

Figure 5. The distributions for the reaction $np \to pp\pi^+\pi^-\pi^-\pi^0$ at P$_0$=5.20 GeV/c. Solid line - calculations using OPER model.

4 Conclusion

Multi π mesons production in np interaction is provided by the excitation and $N\pi$ and $N\pi\pi$ decays of Δ^* and N^* resonances (taken from PWA and GIM).

The large peripherality of the secondary hadrons leads to the idea to use some exchange models (π, P etc. exchange).

It was shown that there are no noticeable signal of ρ meson production in the considered reactions.

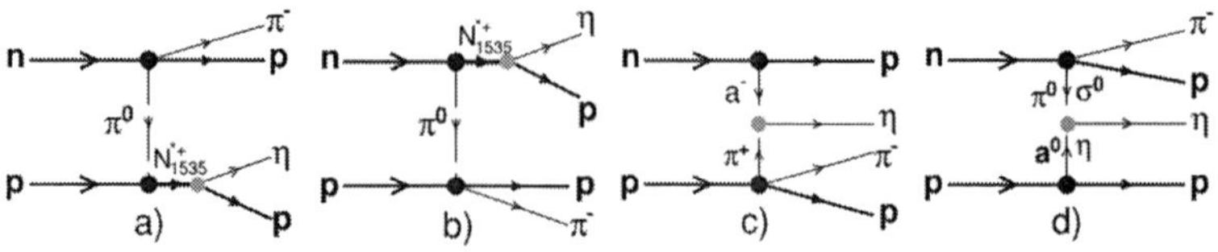

Figure 6. Diagrams of OPER model for the reaction $np \rightarrow pp\pi^-\eta^0$.

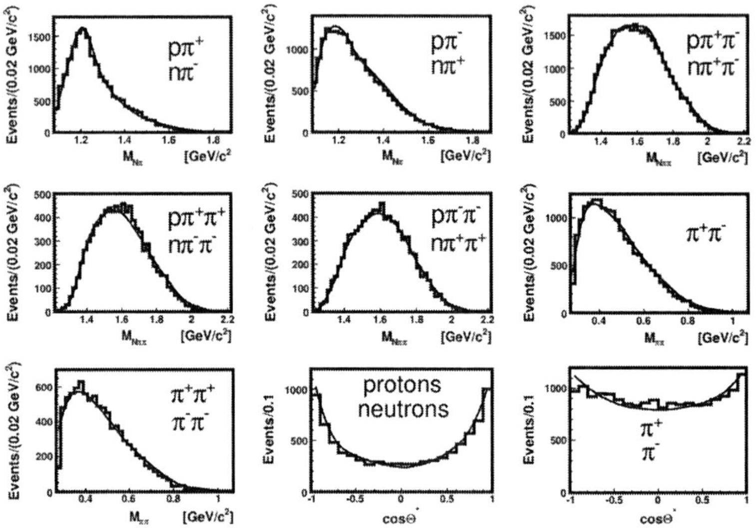

Figure 7. The distributions for the reaction $np \rightarrow np\pi^+\pi^+\pi^-\pi^-$ at P_0=5.20 GeV/c. Solid line - calculations using OPER model.

OPER model allows to obtain a good description of the characteristics of 3 and 4 pions production in np interactions.

To get a better description of the reaction $np \rightarrow pp\pi^+\pi^-\pi^-\pi^0$ it is necessary take into account η^0 and ω^0 production.

References

[1] L. Alvarez-Ruso, E. Oset, E. Hernandez, NP A **633**, 519 (1998)
[2] Xu Cao, Bing-Song Zou and Hu-Shan Xu, PR C **81**, 065201 (2010)
[3] L. Ponomarev, Part. and Nucl. **7**, 186 (1976) (in Russian)
[4] A.P. Jerusalimov et al., http://arxiv.org/pdf/1203.3330.pdf
[5] A.P. Gasparian et al., Prib.Tekh.Eksp. **2**, 37 (1977) (in Russian)
[6] J.P. Berge, F.T. Solmitz and H.D. Taft, Rev.Sci.Instrum. **32**, 338 (1961)
[7] C. Besliu et al., Sov.J.Nucl.Phys. **43**, 565 (1986)

DOI: 10.1051/epjconf/201713807008

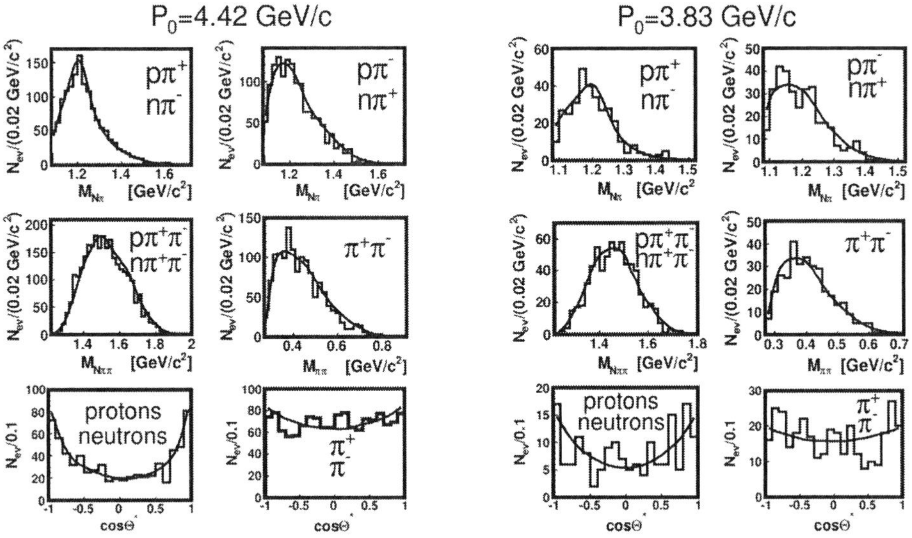

Figure 8. The distributions for the reaction $np \to np\pi^+\pi^+\pi^-\pi^-$ at P_0=4.42 GeV/c and P_0=3.83 GeV/c. Solid line - calculations using OPER model.

Analysis of fractal events with background

Tatiana Dedovich[1,*] and *Mikhail* Tokarev[1,**]

[1] *Joint Institute for Nuclear Research, Dubna, Russia*

Abstract. Properties of self-similarity and fractality in the processes of interactions of hadrons and nuclei at high energies are discussed. Different methods of fractal analysis (the box counting BC, system of the equations of p-adic coverage $SePaC$ methods) are described. Fractal analysis of mixed events was carried out by BC and $SePaC$ methods. The procedure of separation of fractals and background, estimation of the number of fractals in the original data set and contamination of the extracted data was proposed. The dependence of event contamination on multiplicity and background is studied. Reconstruction of the spectrum of fractal dimensions is found to depend on the method and background.

1 Introduction

The concept of the fractal is widely used in high-energy physics. Among them there are the fractal solution for the Schrödinger equation [1–3], the fractal space-time as a new method of regularization in the quantum field theory [4–7], the topological charge density distribution in the lattice gluodynamics as fractals [8], the intermittency in the spectra of secondary particles [9–11], the fractal structures of thermodynamical functions [12], the fractality of the momentum space as a signal of criticality in nuclear collisions [13], and the fractal properties of nuclei and events [14, 15].

A fractal is defined as a self-similar object characterized by the fractal dimension D_F which is more than the topological one D_T for the same object. The topological dimension determines the measure $M = \lim_{l\to 0} \sum_{i=1}^{N} l^{D_T}$, where N is a number of probes, covering the object with a size l. For the euclidean object this measure is independent of the scale, has a finite limit and determines the size of the object (length, square, volume). For a fractal the measure M tends to infinity, because N increases faster than a size l is reduced. In order to evaluate the rate of the growth the Hausdorff measure $M_H = \lim_{\delta\to 0} \sum_{i=1}^{N} l_i^d$ is introduced. Here N is a number of probes with a size $l < \delta$. The fractal dimension D_F is the value of d which provides the finite value of the measure M_H. Every fractal is characterized by the fractal dimension D_F and the box dimension $D_b = \lim_{\delta\to 0} lnN/lnl$. There are various types of fractals. Among them are fractals with independent, dependent and combined partition. They are analysed by the $SePaC$, BC and PaC methods in [16–18]. It was found, in particular, that D_F and D_b can be different for some types of fractals.

[*] e-mail: tdedovich@yandex.ru
[**] e-mail: tokarev@jinr.ru

2 z-Scaling

The description of inclusive cross sections of hadron production in collisions of protons and nuclei at high energies based on the principles of locality, self-similarity, fractality in the framework of the z-scaling concept was proposed in [19]. Authors have concluded that the structure of hadrons, mechanism of constituent interactions and hadronization process reveal self-similar properties over a wide scale range. In this approach an inclusive cross section is described in terms of the dimensionless function $\psi(z)$ depending on the self-similarity parameter z. The fractality is reflected in definition of $z = z_0(1 - x_1)^{-\delta_1}(1 - x_2)^{-\delta_2}$. Here z_0 is a slowly changing function of the kinematic variables, x_1 and x_2 are the scale-invariant fractions of incoming four-momenta P_1, P_2 of colliding objects and δ_1, δ_2 are their fractal dimensions. This methods has been applied for analysis of the inclusive spectra of hadrons and jets production at Tevatron and LHC [20, 21]. The dependence of the scaling function ψ on z was found to be independent of the collision energy and angle of produced particle. It is also described by the power law $\psi(z) \sim z^{-\beta}$ at high z. The value of the slope parameter β is independent of kinematic variables over a wide z-range. It confirms self-similarity and fractality of hadron production at various scales. Discontinuity of dimensions δ_1, δ_2 is assumed to be an indication of phase transition which is related to the change of physical sub-processes underlying inclusive particles production. Fractal dimensions have been determined from data analysis of inclusive cross sections and then verified in analysis of other data sets. These dimensions are fundamental characteristics of particles and processes and should be determined from the fractal analysis of the experimental data.

3 Fractal analysis

The fractal analysis is aimed to determine distributions characterizing the fractal properties of particles and processes. First of all the phase space in which the object reveals its fractal properties and an adequate method of analysis should be found. The original data set should be also described.

In the present work we use BC [16] and $SePaC$ [18] methods to study mixed events $N_{ev} = N_{frac} + N_{bg}$ consisting of fractals N_{frac} and background N_{bg}. The fractals ($N_{frac} = 1857$) of various types are generated by using the procedure of independent, dependent and combined partition. The background is chosen as to be random data set. Portion of background is changed from 0.05 to 0.95. The preliminary analysis includes estimation of number of fractals and extraction of data set of reconstructed fractals. Further the extracted data set is analysed to determine of spectrum of fractal dimensions and contamination.

4 *BC* method

In the section we briefly describe the BC method [16]. It consists of the following steps:
1. Reading out the data X.
2. Construction of the set of distributions of X. The number of bins M_i in distributions changes as multiplicatively $M_i = (P)^i$ or additively $M_{i+1} = M_i + step$.
3. Counting the number of non-zero N bins for each distribution.
4. Finding the value of the slope parameter D_b of the power function $N \sim M^{D_F}$ and corresponding χ^2 of the linear approximation of this function in a double-log scale.
6. Verification of the accuracy condition: if $\chi^2 < \chi^2_{lim}$ then the set of particles is considered to be a fractal.

Notice, that the BC method determines box dimension D_b and has one parameter χ^2_{lim}. The dimension errors are strongly dependent of the set of distributions and region of linear approximation.

Further we present the results of analysis of various events. The dependence of the portion *Prtn* of the events reconstructed as fractals on χ^2 for fractal, background and mixed events with different multiplicity $N = 64,\ 27$ is shown in figure 1.

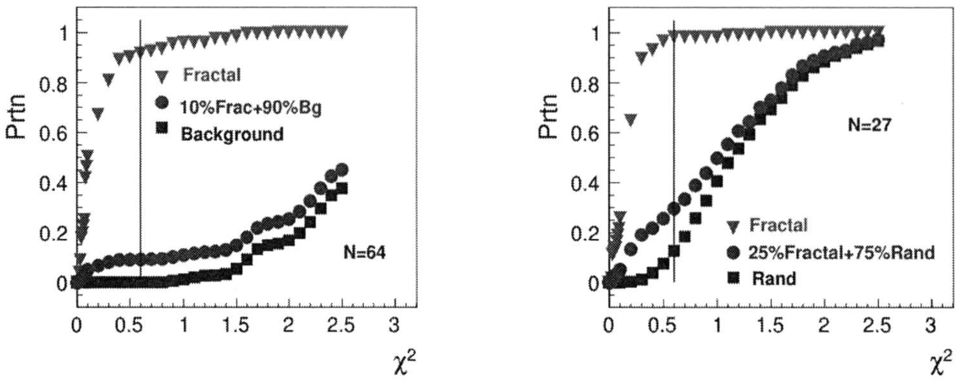

(a) (b)

Figure 1. *BC*: the dependence of the portion *Prtn* of the events reconstructed as fractals on χ^2_{lim} for fractal, background and mixed events with different multiplicity $N = 64$ (a), $N = 27$ (b).

One can see from figure 1 that for fractals (red symbols) the value of $min\chi^2$ at which $Prtn \approx 0.01$ does not exceed 0.1. The portion of all fractals which are reconstructed as fractals at $\chi^2 = 0.6$ reaches 0.9-0.96. For background (black symbols) the value of $min\chi^2$ increases and *Prtn* at $\chi^2 = 0.6$ decreases with increasing N. Notice, that fractals and background with multiplicity $N = 8$ are found to be not separable. The condition $min\chi^2 \leq 0.1$ corresponds to the existence of fractals in the original data set. The value 0.6 of χ^2 was used to extract fractals from the mixed events. We found that the extracted data $N_{ext} = P_{frac} \cdot N_{frac} + P_{bg} \cdot N_{bg}$ contain a significant portion of fractals ($P_{frac} = 0.9 \div 0.98$). For multiplicity $N \geq 64$ there is no background ($P_{bg} = 0$). For other multiplicities the extracted data have some portion ($P_{bg} \neq 0$) of the background. Thus, we can estimate N_{frac} in the original data set and contamination *Cont* of the extracted data as follows:

$$N_{frac} = \frac{N_{ext} - P_{bg} \cdot N_{ev}}{P_{frac} - P_{bg}}, \qquad Cont = \frac{P_{bg}}{N_{ext}} \cdot \frac{P_{frac} \cdot N_{ev} - N_{ext}}{P_{frac} - P_{bg}}. \tag{1}$$

5 *SePaC* method

In the section we briefly describe the *SePaC* method [18]. It consists of the following steps:
1. Reading out the data X.
2. Construction of P-adic coverages: $P = 3, ..P_{Max}$. Each coverage is a set of distributions of variable X. The number of bins M_i in distributions changes as a degree of basis P: $M_i = (P)^i$.
3. Counting the number of non-zero bins $N(lev, P)$.
4. Analysis of the system of the equations for verification of hypothesis of independent/dependent partition:
a. Construction of the system of the equations for all levels: $\sum_{i=1}^{N_{lev}} (d_{lev})^{D_F^{lev}} = 1$. Here N_{lev} is a number of levels, and d_{lev} is the length of non-zero ranges for each level.
b. Finding the solution D_F^{lev} of equations for each level by using the dichotomy method.

c. Determination of the averaged value $< D_F^{lev} >$ and deviation ΔD_F^{lev}.
d. Verification of the accuracy condition: if $\Delta D_F^{lev} < Dev$, then the set of particles is considered to be a fractal (P is the base and $D_F =< D_F^{lev} >$ is the fractal dimension).

Notice, that the $SePaC$ method determines fractal dimension D_F and has two parameters: maximal base P_{Max} and deviation Dev.

Further we present the results of analysis of various events. The dependence of the portion $Prtn$ of the events reconstructed as fractals on P_{Max} at $Dev = 0.18$ for fractal, background and mixed events with different multiplicity $N = 64, 27$ is shown in figure 2. One can see from figure 2, that for

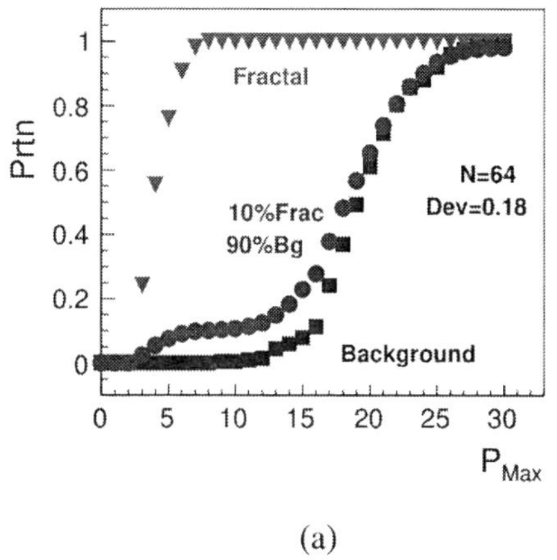

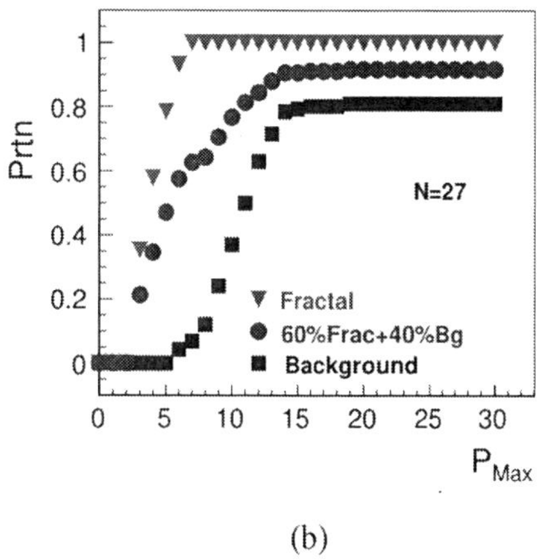

Figure 2. $SePaC$: the dependence of the portion $Prtn$ of the events reconstructed as fractal on Dev for fractal, background and mixed events with different multiplicity $N = 64$ (a), $N = 27$ (b).

fractals (red symbols) the value $minP_{Max}$ at which $Prtn \neq 0$ is equal to 3. Full reconstruction of the fractals takes place when P_{Max} is equal to the minimal value on the plateau. For background (black symbols) the value of $minP_{Max}$ increases with increasing N. The condition $minP_{Max} = 3$ corresponds to the existence of fractals in the original data set. The difference in the values of $minP_{Max}$ for fractals and background was used to extract fractals from the mixed events. The value of P_{Max} is taken to be minimal value at the first plateau or the value on the shoulder. We found that the extracted data $N_{ext} = N_{frac} + P_{bg} \cdot N_{bg}$ contain all fractals. For multiplicity $N \geq 64$ there are not background ($P_{bg} = 0$). For other multiplicities the extracted data have some portion ($P_{bg} \neq 0$) of the background. Thus, we can determine N_{frac} in the original data set and contamination $Cont$ of the extracted data as follows:

$$N_{frac} = \frac{N_{ext} - P_{bg} \cdot N_{ev}}{1 - P_{bg}}, \qquad Cont = \frac{P_{bg}}{N_{ext}} \cdot \frac{P_{frac} \cdot N_{ev} - N_{ext}}{1 - P_{bg}}. \tag{2}$$

6 Fractal dimension

We found that the $SePaC$-method separates the fractals and the background for all multiplicities and BC - only for events with $N \geq 16$. For high multiplicity events ($N \geq 64$) both methods completely suppress background. For events with smaller multiplicities contamination $Cont$ is found to be approximately equal for both methods. For mixed events with multiplicity $N = 32, 27$ and background less than 75% the value of $Cont$ for extracted data is estimated to be less than 30%. The similar result is obtained for events with $N = 16, 8$ and background less than 50%.

DOI: 10.1051/epjconf/201713807009

We would like to notice that the distribution of fractal dimension D_F for the extracted data is distorted due to systematic errors of the method and the presence of background events. Figure 3 shows the distribution of the error Er_{D_F} of fractal dimension found by the *SePaC* and *BC* methods for fractals with independent, dependent and combined partition.

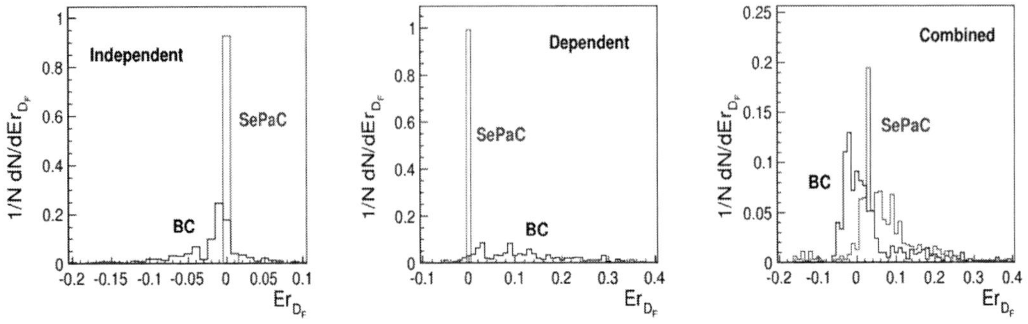

(a) (b) (c)

Figure 3. The distribution of the error Er_{D_F} of fractal dimension found by *SePaC* (red line) and *BC* (black line) methods for fractals with independent (a), dependent (b) and combine (c) partition.

One can see from figure 3, that the *SePaC* method exactly reconstructs dimension D_F for fractals with independent and dependent partition. For fractals with combined partition the distribution is found to be wide and shifted to positive values of Er_{D_F}. The distributions of errors obtained by the *BC* method demonstrate the asymmetric shape for all types of fractals. They are shifted to negative (a), positive (b) values of Er_{D_F} for fractals with independent and dependent partition, respectively. For fractals with combined partition the distribution (c) has positive and negative values of Er_{D_F}.

The distribution of fractal dimension D_F for background events with different multiplicity is shown in figure 4.

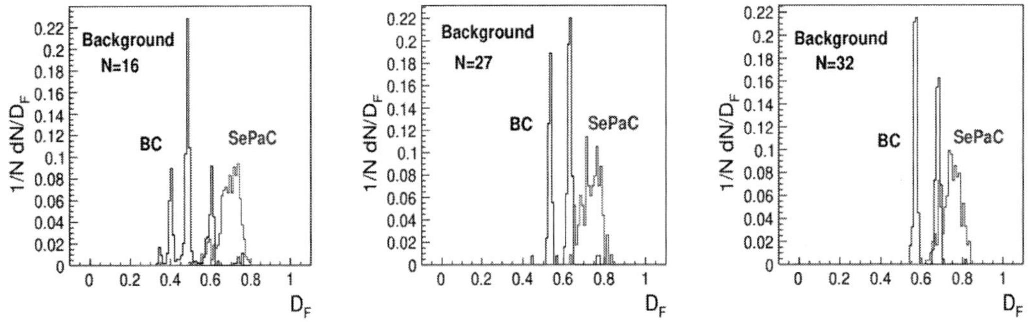

(a) (b) (c)

Figure 4. The distribution of fractal dimension D_F for background events with multiplicity $N = 16$ (a), 27 (b), 32 (c).

As it is seen from figure 4 the spectra of fractal dimensions reconstructed by the *BC* and *SePaC* methods from the background events are different. Notice that D_F changes in the range 0.6-0.8 for distribution obtained by the *SePaC* method. The spectrum of D_F reconstructed by the *BC* method has a more complex structure.

DOI: 10.1051/epjconf/201713807009

7 Conclusions

Fractal analysis of mixed events was carried out by the *BC* and *SePaC* methods. The procedure of separation of fractals and background, estimation of the number of fractals (N_{frac}) in the original data set and contamination (*Cont*) in the extracted data was proposed. We found that the *SePaC*-method separates the fractals and the background for all multiplicities and *BC* - only for events with $N \geq 16$. For high multiplicity events ($N \geq 64$) both methods completely suppress background. For events with smaller multiplicities contamination *Cont* was found to be approximately equal for both methods. The value of *Cont* depends on event multiplicity and background. It was shown that the distribution of fractal dimension D_F for the extracted data is distorted due to systematic errors of the used method and the presence of background events.

References

[1] M.V. Berry, J.Phys.A: Math.Gen. **29**, 6617 (1996)
[2] D. Wojcik et al., Phys.Rev.Lett. **85**, 5022 (2000)
[3] E. Akkermans et al., Phys.Rev.Lett. **105**, 230407 (2010)
[4] K. Svozil, J. Phys. A: Math. Gen. **20**, 3861
[5] A. Kar, S. Rajeev, Annals Phys. **327**, 102 (2012)
[6] G. Calcagni, Phys.Rev. D **84**, 061501 (2011)
[7] D. Moore et al., Phys.Rev. D **90**, 024075 (2014)
[8] P. Buividovich et al., Phys.Rev. D **86**, 074511 (2012)
[9] I. Dremin, Surveys High Energ.Phys. **6**, 141 (1992)
[10] M.K. Ghosh et al., DAE Symp.Nucl.Phys. **54**, 590 (2009)
[11] M. Rasool, S. Ahmad, Chaos Solitons Fractals **84**, 58 (2016)
[12] A. Deppman, Phys.Rev. D **93**, 054001 (2016)
[13] N.G. Antoniou, Phys.Rev. C **93**, 014908 (2016)
[14] Wei-Hu Ma et al., Chin.Phys. C **39**, 104101 (2015)
[15] I. Bunzarov et al., Phys.Part.Nucl.Lett. **11**, 404 (2014)
[16] B.B. Mandelbrot, The Fractal Geometry of Nature (Freeman, San Francisco, 1982)
[17] T.G. Dedovich, M.V. Tokarev, Phys. Part. Nucl. Lett. **8**, 521 (2011)
[18] T.G. Dedovich, M.V. Tokarev, Phys. Part. Nucl. Lett. **9**, 552 (2011)
[19] I. Zborovský, Yu.A. Panebratsev, M.V. Tokarev, G. Škoro, Phys. Rev. D **54**, 5548 (1996)
[20] M.V. Tokarev, T.G. Dedovich, Int. J. Mod. Phys. A **15**, 3495 (2000)
[21] M.V. Tokarev, T.G. Dedovich, and I.Zborovský, Int. J. Mod. Phys. A **27**, 1250115 (2012)

DOI: 10.1051/epjconf/201713808001

Search for T-invariance violation in pd-scattering

Yuriy Uzikov[1,2,3,⋆]

[1]*Joint Institute for Nuclear Research, Dubna 141980, Russia*
[2]*Dubna State University, Dubna 141980, Russia*
[3]*Department of Physics, M.V. Lomonosov Moscow State University, Moscow 11991, Russia*

Abstract. Scattering of protons with transversal polarization P_y^p on deuterons with tensor polarization P_{xz} provides a null-test signal for time reversal invariance violating but P-parity conserving effects. The dedicated experiment will be done at COSY facility at proton beam energy 135 MeV using internal deuteron target. I give a brief review of results of theoretical calculations of the null-test observable within the spin-dependent Glauber theory at beam energies 100-1000 MeV considering phenomenological T-violating P-conserving NN-interactions.

1 Introduction

CP-invariance violation (or T-invariance violation under CPT-symmetry) is required to account for the baryon asymmetry of the universe [1]. Since within the standard model the CP-violation observed in physics of kaons and B-mesons is by far not sufficient to explain this asymmetry [2], other sources of CP-violation have to be found. Possible measurements are classified according to whether or not measured observables violate P-parity as well as time reversal invariance. Various efforts are undertaken for measurements of the electric dipole moments (EDM) of elementary particles and atoms which are both T-violating and P-violating. The other type of observables which are T-violating but P-conserving (TVPC) received much less attention. The reason why TVPC effects are interesting is that experimental limits on them are still quite weak, in particular considerably weaker as compared to EDM.

Integrated cross section of scattering of protons with transversal polarization p_y^p on deuterons with tensor polarization P_{xz} provides a null-test signal for time-reversal invariance violating but parity conserving effects [3]. Since the intensity of TVPC interactions within the standard model are extremely small [4], observation of this effect at present level of experiments would directly indicate to physics beyond the standard model. A measurement of this observable is planned at the COSY accelerator in Jülich at 135 MeV in the TRIC experiment [5]. Theoretical study of the expected signal was performed at energies of the planned experiment [6–12]. Furthermore, certain relations between differential observables of the elastic pd-scattering caused by T-invariance requirement were obtained and degree of their violation by TVPC NN forces was studied [13, 14]. Below I will give a short review of the obtained results.

⋆e-mail: uzikov@jinr.ru

2 Null-test signal for T-invariance violation

The total cross section of the pd scattering with TVPC forces included can be written as [8]

$$\sigma_{tot} = \sigma_0^t + \sigma_1^t \mathbf{p}^p \cdot \mathbf{p}^d + \sigma_2^t (\mathbf{p}^p \cdot \mathbf{m})(\mathbf{p}^d \cdot \mathbf{m}) + \sigma_3^t P_{zz} + \widetilde{\sigma} p_y^p P_{xz}^d, \quad (1)$$

where $\mathbf{p}^p$ ($\mathbf{p}^d$) is the vector polarization of the initial proton (deuteron) and P_{zz} and P_{xz} are the tensor polarizations of the deuteron. The OZ axis is directed along the beam direction $\mathbf{m}$, the OY axis is directed along the vector polarization of the proton beam $\mathbf{p}^p$ and the OX axis is chosen to form the right-hand reference frame. The integrated cross sections σ_i^t ($i = 0, 1, 2, 3$) are caused by T-even P-even interaction, and the last term $\widetilde{\sigma}$ appears only in presence of the TVPC interactions and constitutes the TVPC null-test signal.

Calculation of the integrated cross sections σ_i^t and $\widetilde{\sigma}$ at beam energies > 100 MeV can be done on the basis of the spin-dependent Glauber theory of pd-scattering [15]. Indeed, as shown in [16], this theory allows one to describe well available data on differential spin observables at 135-200 MeV in forward hemisphere. In the Glauber theory one has to use the on-shell elastic NN scattering amplitudes as input. Hadronic amplitudes of the pN scattering are taken in the form of [15]

$$M_N(\mathbf{p}, \mathbf{q}; \sigma, \sigma_N) = A_N + C_N \sigma\hat{\mathbf{n}} + C_N' \sigma_N \hat{\mathbf{n}} + B_N(\sigma\hat{\mathbf{k}})(\sigma_N\hat{\mathbf{k}}) + \\ +(G_N + H_N)(\sigma\hat{\mathbf{q}})(\sigma_N\hat{\mathbf{q}}) + (G_N - H_N)(\sigma\hat{\mathbf{n}})(\sigma_N\hat{\mathbf{n}}), \quad (2)$$

where $\hat{\mathbf{q}}$, $\hat{\mathbf{k}}$ and $\hat{\mathbf{n}}$ are defined as unit vectors along the vectors $\mathbf{q} = (\mathbf{p}-\mathbf{p}')$, $\mathbf{k} = (\mathbf{p}+\mathbf{p}')$ and $\mathbf{n} = [\mathbf{k}\times\mathbf{q}]$, respectively; $\mathbf{p}$ ($\mathbf{p}'$) is the initial (final) proton momentum.

In general case TVPC NN interaction contains 18 different terms [17]. We consider here only the following terms of the t-matrix of the elastic pN scattering investigated in [6]:

$$t_{pN} = f_N(\sigma\cdot\sigma_N)(\mathbf{q}\cdot\mathbf{k})/m_p^2 + h_N[(\sigma\cdot\mathbf{k})(\sigma_N\cdot\mathbf{q}) + (\sigma_N\cdot\mathbf{k})(\sigma\cdot\mathbf{q}) - \frac{2}{3}(\sigma_N\cdot\sigma)(\mathbf{k}\cdot\mathbf{q})]/m_p^2 + \\ +g_N[\sigma\times\sigma_N]\cdot[\mathbf{q}\times\mathbf{k}]/m_p^2 + g_N'(\sigma-\sigma_N)\cdot i\,[\mathbf{q}\times\mathbf{k}][\tau\times\tau_N]_z/m_p^2. \quad (3)$$

Here σ (σ_N) is the Pauli matrix acting on the spin state of the proton (nucleon $N = p, n$), τ (τ_N) is the isospin matrix acting on the isospin state of the proton (nucleon), the momenta $\mathbf{q}$ and $\mathbf{k}$ are defined below (2) In the framework of the phenomenological meson exchange interaction the term g_N' corresponds to the $\rho-$meson exchange, and the h_N-term is caused by the axial h_1-meson exchange. As shown in [18], the contribution of the π- and σ-meson to TVPC NN interactions is excluded.

2.1 TVPC amplitude of the forward pd-scattering

One can write the forward elastic pd-scattering amplitude in a general form taking into account the TVPC interaction [8, 16] and then apply the generalized optical theorem to derive the formula (1) for the total cross section of the pd-scattering. In notations of [8] the integrated cross section $\widetilde{\sigma}$ is related to the TVPC term $\widetilde{g}$ of the forward pd elastic scattering amplitude as

$$\widetilde{\sigma} = -4\sqrt{\pi} Im \frac{2}{3}\widetilde{g}. \quad (4)$$

Furthermore, the forward TVPC amplitude $\widetilde{g}$ can be found within the Glauber theory [8]. Since the g'-term gives zero contribution to $\widetilde{g}$ [8], we consider the h and g terms and take into account both the

S and D wave of the deuteron. The following result was found for the forward TVPC amplitude on this way [12]

$$\widetilde{g} = \frac{i}{4\pi m_p}\int_0^{\infty} dqq^2\Big[S_0^{(0)}(q) - \sqrt{8}S_2^{(1)}(q) - 4S_0^{(2)}(q) + \sqrt{2}\frac{4}{3}S_2^{(2)}(q) + 9S_1^{(2)}(q)\Big] \qquad (5)$$
$$[-C_n'(q)h_p + C_p'(q)(g_n - h_n)],$$

where $S_i^{(j)}$ are the elastic form factors of the deuteron defined in [12]. The first term in the (big) squared brackets in (5), $S_0^{(0)}(q)$, corresponds to the S-wave approximation, the second term, $S_2^{(1)}(q)$, accounts for the S-D interference, and the last three terms contain the pure D-wave. The amplitude $\widetilde{g}$ does not contain a sum of one strong NN amplitude and one weak TVPC NN amplitude. Only products of these amplitudes appear in $\widetilde{g}$. This means that the strong (T-even P-even) background is excluded from the null-test observable $\widetilde{\sigma}$. Furthermore, available experimental uncertainties of the strong amplitudes of the NN-scattering do not affect the final result for $\widetilde{\sigma}$ drastically. That is why we believe that the accuracy of the Glauber calculations of the TVPC null-test signal is of the same order as for the pure hadronic total cross sections σ_i^t ($i = 0, 1, 2, 3$).

2.2 Why the g' term vanishes?

It was expected that the lowest mass meson exchange allowed in the phenomenological TVPC NN-interaction, i.e. the ρ-meson, dominates in the null-test signal of the TVPC effects and, consequently, the strength of these effects is measured usually in units of the corresponding coupling constant $\bar{g}_\rho$ [19]. However, in pd scattering considered in the framework of the Glauber theory the contribution of the g'-term to the null-test signal disappears. For the first time it was shown in [8]. Some qualitative explanations of the origin of this result were done in [10]. As it was shown in [8, 10], the crucial point is that within the Glauber theory of the multi-step scattering the highest order of rescattering is equal to A, where A is the number of the nucleons in the target nucleus. Therefore, for pd scattering ($A = 2$) only single and double scattering mechanisms are allowed. Contributions of the higher-order rescatterings are compensated by off-shell effects in the pN-scattering amplitudes which, therefore, are taken on-shell [20]. The single scattering contribution to the TVPC amplitude $\widetilde{g}$ is zero at zero transferred momentum $\mathbf{q} = 0$, as it follows from (3). For the double scattering mechanism there are two terms which differ by the order of the pure strong and TVPC NN-scattering. For the g'-type NN-interactions these two terms have opposite signs for the isospin matrix elements of the C-odd operator $[\tau \times \tau_N]_z$:

$$< n, p|[\tau \times \tau_N]_z|p, n >= -i2, \quad < p, n|[\tau \times \tau_N]_z|n, p >= i2 \qquad (6)$$

and this fact is the second crucial point in this consideration. Indeed, the linear dependence of the g'-term on $[\mathbf{q} \times \mathbf{k}]$ in (3) leads to cancellation of the strong A_N and B_N -terms from (2) in the final formula for the TVPC charge-exchange Glauber operator of the double scattering [10]. Only the C_N and C_N' strong amplitudes may contribute together with the g'-term. However, the sum of C_n and C_p terms is identically zero [10]. Furthermore, owing to (6) the rest term in T-operator associated with the C_N'-term is diagonal over the incoming (outcoming) proton spin states and therefore does not contribute to $\widetilde{g}$. Thus, this specific combination of isospin, spin and momentum dependencies of the g'-term provides its vanishing contribution to the null-test signal in $pd-$ scattering within the Glauber theory. Corrections to the Glauber theory may give a non-zero g'-contribution, but as usually, such corrections are small as compared to the leading Glauber terms.

One should note that g'-term gives contribution to the differential observables of the pd elastic scattering and violates the relations caused by T-invariance [14]. It is assumed [8], that this interaction can give a contribution to the null-test signal if it will be included into the deuteron wave function.

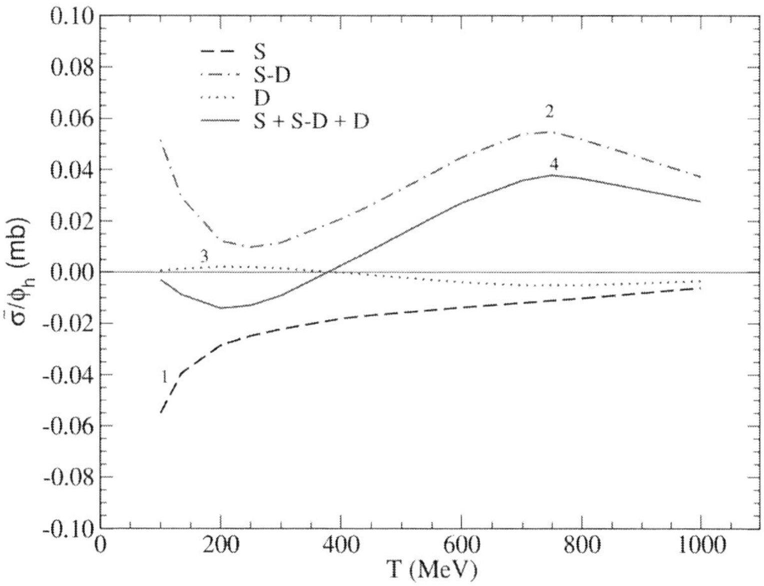

Figure 1. The TVPC signal $\widetilde{\sigma}$ for the h-term in units of the ratio (ϕ_h) of the TVPC and the strong h_1NN constants versus the proton beam energy T. The results of calculations [12] accounting for different terms in (5) with the CD Bonn deuteron wave function and hadronic pN- amplitudes from [22] are shown by the curves: S-wave (curve 1), S-D interference (2), D-wave (3), total result (4).

2.3 Coulomb interaction

In the first theoretical work [6] where the null-test signal was calculated within the impulse approximation, the Coulomb interaction was not considered. In [7] the Faddeev calculations were performed only for nd-scattering at rather low energies $\sim$ 100 keV.

Straightforward calculation on the basis of the optical theorem with the Coulomb amplitude included is rather questionable. On the other hand, being a T-even P-even interaction the Coulomb pp scattering cannot generate the TVPC amplitude $\widetilde{g}$, therefore its contribution to $\widetilde{g}$ is zero within the single scattering Glauber mechanism. In order to include the Coulomb interaction within the double scattering mechanism of the Glauber theory one should replace the pure hadronic T-even P-even pp amplitude M_p given in (2) by the sum $M_p + f_{pp}^C$, where f_{pp}^C is the Coulomb pp-scattering amplitude. Since the Coulomb pp-amplitude consists of spin-independent term and the $\boldsymbol{\sigma} \cdot \boldsymbol{\sigma}_p$ term caused by the Fermi-statistics, it is evident that the Coulomb term is added to the spin-independent term A_p, $A_p \to A_p + f_{pp}^C$ and double spin terms B_p, G_p, H_p but does not enter into the single spin terms C_p and C'_p. However, all amplitudes A_p, B_p, G_p, H_p are excluded from the TVPC amplitude (5) due to the specific spin structure of the TVPC transition operator of the forward pd scattering [8]. The only factor in (5) which contains the Coulomb effects is $C'_p(q)$. This is because the Coulomb scattering pp amplitude enters into the spin-independent term A_p, which, in turn, enters into the amplitude C'_p with the relativistic correction factor $q/2m_p$ [21] (m_p is the nucleon mass) $C'_p = C_p + iA_p\frac{q}{2m_p}$. When substituting this amplitude C'_p into (5) and making integration over the transferred momentum q, one can see that due to the presence of the factor $\sim q^3$ in the integrand the q^{-2} singularity of the Coulomb amplitude at $\mathbf{q} = 0$ does not lead to divergence of the $\widetilde{g}$ amplitude. As it was shown numerically [8], the Coulomb term makes rather weak influence on energy dependence of the null-test signal.

2.4 The deuteron D-wave contribution

The results obtained in [8] were restricted by the deuteron S-wave component. The maximum of the null-test signal was found in this approximation at lower energies, just corresponding to the TRIC experiment (curve 1 in figure 1). In [9, 12] the calculations were completed taking into account the D-wave of the deuteron. It was shown for the case of the h- and g-type of interaction that the deuteron D-wave gives very important contribution to the null-test signal due to interference between the deuteron S- and D- waves, whereas the pure D-component is negligible. First, one can see (figure 1), with the D-wave included a zero crossing of $\widetilde{\sigma}$ is possible even when the TVPC interaction itself is nonzero. In calculation [12] this occurs at beam energies about 400 MeV. Therefore, it is necessary to perform measurements at two or possibly more energies in order to achieve conclusive results. Second, according to predictions [9, 12], energies around 700-800 MeV are most promising for finding a signal (see curve 4 in figure 1). Some sensitivity to model dependent variations in the D-wave component is not too dramatic.

2.5 Possible spurious-effect

Let us consider possible spurious effects that could mimic a TVPC signal. One source of spurious signal is associated with a nonzero deuteron vector polarization $p_d^y \neq 0$ (in the direction of the incident-proton-beam polarization $\mathbf{p}^p$). In this case, the term $\sigma_1 P_y^p p_y^d$ in formula (1) contributes to the asymmetry corresponding to the difference of the event counting rates for the cases of $p_y^p P_{xz} > 0$ and $p_y^p P_{xz} < 0$ (with the fixed sign of P_{xz}), which is planned to be measured in the TRIC experiment [5]. At the beam energy of 135 MeV, the integrated hadronic cross sections calculated in [16] on the basis of the optical theorem take the values of $\sigma_0 = 78.5$ mb, $\sigma_1 = 3.7$ mb, $\sigma_2 = 17.4$ mb, $\sigma_3 = -1.1$ mb. This gives the ratio $r = \sigma_1/\sigma_0$ about 0.05. The goal of the TRIC experiment [5] is to measure the ratio $R = \widetilde{\sigma}/\sigma_0$ with an error not greater than 10^{-6}. Assuming that the signal-to-background ratio is $\widetilde{\sigma}/p_y^d\sigma_1 \sim 10$, one can see from the result obtained for r that the vector polarization of the deuteron should be smaller than $\sim 2 \times 10^{-6}$. Similar estimations can be done at 700-800 MeV. A possible way to avoid the contribution from the deuteron vector polarization is to measure the above mentioned asymmetry for opposite signs of the tensor polarization P_{xz} but with the fixed direction of the vector polarization of the proton.

3 Concluding remarks

Theoretical study of the null-test signal in pd-scattering performed on the basis of the spin-dependent Glauber theory demonstrates several unexpected effects. Among them are (i) disappearance of the contribution of the lowest mass meson exchange (ρ–meson) in the TVPC NN-interaction, caused by its specific isospin, spin and momentum dependence; (ii) a strong impact of the deuteron D-wave on the null-test signal arising due to destructive interference between the S- and D-wave in spite of zero transferred 3-momentum; (iii) oscillating behaviour of the null-test signal as a function of the beam energy, i.e. disappearance of the signal at some points in energy is possible even when the TVPC interaction itself is nonzero; (iv) very minor influence of the Coulomb interaction on the TVPC term of the forward scattering pd-amplitude $\widetilde{g}$. Only the h- and g-type terms are accounted in (5). The remaining terms of the TVPC NN-interaction listed in [17] are either off-shell type and, therefore, do not contribute in the Glauber theory of the pd-scattering and presumably suppressed beyond this theory, or of the h-type. In the latter case the h-terms differ only in their isospin dependence but have the same spin-momentum structure. Therefore, these terms can modify (5) with regard of the relative weight of the C'_p and C'_n amplitudes which one could examine in the future.

References

[1] A.D. Sakharov, Pis'ma v ZHETF **5**, 32 (1967)
[2] A. Riotto, M. Troden, Ann. Rev. Nucl. Part. Sci **49**, 35 (1999)
[3] H.E. Conzett, Phys. Rev. C **48**, 423 (1993)
[4] R.S. Conti, I.B. Khriplovich, Phys. Rev. Lett. **68**, 3262 (1992)
[5] COSY Proposal N 215, Spokespersons: P.D. Eversheim, B. Lorentz and Yu. Valdau, http://apps.fz-juelich.de/pax/paxwiki/images/8/8c/215-TRI_Prop_sum.pdf
[6] M. Beyer, Nucl. Phys. A **560**, 895 (1993)
[7] Y.-H. Song, R. Lazauskas, V. Gudkov, Phys. Rev. C **84**, 025501 (2011)
[8] Yu.N. Uzikov, A.A. Temerbayev, Phys. Rev. C **92**, 014002 (2015)
[9] Yu. N. Uzikov, Jour. Phys.: Conf. Ser. **678**, 012020 (2016)
[10] Yu.N. Uzikov, Eur. Phys. J.: Web of Conf. **113**, 04027 (2016)
[11] Yu.N. Uzikov, A.A. Temerbayev, Int. Jour. Mod. Phys.: Conf. Series **40**, 1660080 (2016)
[12] Yu.N. Uzikov, J. Haidenbauer, Phys. Rev. C **94**, 035501 (2016)
[13] A.A. Temerbayev, Yu. N. Uzikov, Izv. Ross. Akad. Nauk Ser. Fiz. **79**, 535 (2015). [Bull. Russ. Acad. Sci. Phys. **79**, 493 (2015)]
[14] A.A. Temerbayev, Yu. N. Uzikov, Izv. Ross. Akad. Nauk Ser. Fiz. **80**, 271 (2016). [Bull. Russ. Acad. Sci. Phys. **80(4)**, 242 (2016)]
[15] M.N. Platonova, V.I. Kukulin, Phys. Rev. C **81**, 014004 (2010)
[16] A.A. Temerbayev, Yu.N. Uzikov, Yad. Fiz. **78**, 38 (2015). [Phys. At. Nucl. 78, 35 (2015)]
[17] P. Herczeg, Nucl. Phys. **75**, 655 (1966)
[18] M. Simonius, Phys. Lett. B **78**, 147 (1975)
[19] J. Engel, C.R. Gould, V. Hnizdo, Phys. Rev. Lett. **73**, 3508 (1994)
[20] T.A. Osborn, Ann. Phys. **58**, 417 (1970)
[21] C. Sorensen, Phys.Rev. D **19**, 1444 (1979)
[22] R.A. Arndt, W.J. Briscoe, I.I. Strakovsky, R.L. Workman, Phys. Rev. C **76**, 025209 (2007). http://gwdac.phys.gwu.edu

Phenomenological description of spin effects in electromagnetic and strong interactions of quarks

Alexander J. Silenko[1,2,⋆] and *Oleg V.* Teryaev[1,3,⋆⋆]

[1] *Bogoliubov Laboratory of Theoretical Physics, Joint Institute for Nuclear Research, Dubna 141980, Russia*
[2] *Research Institute for Nuclear Problems, Belarusian State University, Minsk 220030, Belarus*
[3] *National Research Nuclear University "MEPhI" (Moscow Engineering Physics Institute), 115409 Moscow, Russia*

Abstract. Phenomenological description of interactions of relativistic quarks by the Dirac equation with the Cornell potential is given. The general form of the initial equation containing the vector and scalar parts of the Cornell potential is used at the arbitrary connection between these parts. The Hamiltonian in the Foldy-Wouthuysen representation is derived in the general form with allowance for the electromagnetic interactions. Unlike precedent investigations, it is relativistic and exact for terms of the zeroth and first powers in the Planck constant and also for such terms of the second power which describe contact interactions. General quantum mechanical equations of motion for the momentum and the spin are derived and the classical limit of the Hamiltonian and the equations of motion are found for the first time. A connection between the angular velocity of the quark spin precession and the force acting on it is determined. The energy of the spin-orbit interaction is rather high (of the order of 100 MeV). The terms describing the spin-orbit and contact interactions have opposite signs for the scalar and the vector parts of the Cornell potential. The evolution of the quark helicity and the spin-spin interaction of the quarks are also calculated.

1 Introduction

The Cornell potential finds numerous applications in the heavy quark physics. Among these applications, we can mention collisions of heavy ions which manifest new properties of quark-gluon plasma. It is naturally to use the conventional Cornell potential for a phenomenological description of these properties in the first approximation. The Cornell potential can be included into the Dirac equation in two possible ways. It is the most natural to consider the quark interaction like an electromagnetic interaction because the quanta of these interactions, photon and gluon, have the same spin. In this case, the Cornell potential is included into the four-potential of the Dirac equation. Another possibility is an inclusion of the Cornell potential into the scalar potential [1]. One often realizes the both possibilities and introduces the Cornell potential in the two (vector and scalar) forms [2–6]. Certainly, the vector Cornell potential is a four-vector. The use of traditional quantum-mechanical approaches allows one

⋆e-mail: alsilenko@mail.ru
⋆⋆e-mail: teryaev@theor.jinr.ru

to solve the resulting equation only in some specific case, when the parameters of the scalar and vector potentials are very specific [3–6].

In the present work, we use the both scalar and vector Cornell potentials and solve the more general problem of a strongly interacting Dirac particle (quark). Unlike other investigations [1–6], we use the approach based on the relativistic Foldy-Wouthuysen (FW) transformation and find a general solution of the problem for arbitrary parameters of the scalar and vector potentials. A possibility to restrict ourselves to a consideration of local interactions (because of short-range nuclear forces) allows us to realize advantages of this approach. Developed methods of the relativistic FW transformation [7–9] (see also [10–12] and references therein) ensure fulfilling this transformation in arbitrarily strong external fields. While this approach has not been used for strong interactions, it has proven itself as a powerful tool for a description of electromagnetic [11, 13, 14], gravitational [15–21], and weak [22, 23] interactions of single particles.

We use the system of units $\hbar = 1$, $c = 1$. Nevertheless, we include the Planck constant into several formulae.

2 Cornell potential and its inclusion into the Dirac equation

The standard approach in phenomenological description of interaction of quarks consists in the use of the Cornell potential having the form

$$\mathfrak{V}(r) = -\frac{a(r)}{r} + br. \tag{1}$$

In the first approximation, we can suppose that a and b are constant. In this case, $a \approx 100$ MeV·fm, $b \approx 400$ MeV·fm^{-1}. However, the second term in the Cornell potential is proportional to r^2 at condition that r is small [24].

For a phenomenological description of interaction of quarks, the Cornell potential should be included into the Dirac equation. This can be made in different ways. There are some similarities between the strong and electromagnetic interactions of quarks. Therefore, an inclusion of the potential (1) into the four-vector potential $\mathcal{A}^\mu = (\mathbb{F}, \mathcal{A})$ similarly to the Coulomb interaction is the most natural. In this case, $\mathbb{F}$ and $\mathcal{A}$ are the scalar and vector potentials of the source (quark). Their dependence on the velocity of the field source $\boldsymbol{V}$ is defined by the Lorentz transformations:

$$\mathbb{F} = \gamma\mathbb{F}_0, \qquad \mathcal{A} = \gamma\mathbb{F}_0\frac{\boldsymbol{V}}{c}, \tag{2}$$

where $\gamma = (1 - V^2/c^2)^{-1/2}$, and $\mathbb{F}_0$ is the potential of the source at rest. If the scalar Cornell potential is not used, $\mathbb{F}_0 = \mathfrak{V}(r)$.

Alternatively, the Cornell potential can be included into the Dirac equation as a scalar potential. In this case, the Cornell potential is multiplied by the matrix β, and its inclusion means the effective modification of the quark mass. The scalar and vector Cornell potentials are additive in the static limit. Therefore, the general Dirac Hamiltonian takes the form

$$\mathcal{H} = \beta[mc^2 + C_0\mathfrak{V}(r)] + e\Phi + \mathbb{F} + \boldsymbol{\alpha}\cdot\boldsymbol{\pi}, \qquad \boldsymbol{\pi} = \boldsymbol{p} - e\boldsymbol{A} - \mathcal{A}, \tag{3}$$

where $\boldsymbol{p} = -i\hbar\nabla$ is the momentum operator. The electromagnetic interaction is characterized by the scalar (Φ) and vector ($\boldsymbol{A}$) potentials. The potentials of the strong interaction are defined by (2) and (3) where

$$\mathbb{F}_0 = C_1\mathfrak{V}(r), \qquad C_0 + C_1 = 1. \tag{4}$$

DOI: 10.1051/epjconf/201713808002

One mostly incorporated both the scalar and vector potentials into the Dirac equation [3–6]. The case of specific relations between the two incorporated potentials was the only case investigated in these works. This equality does not have any physical substantiation but it allows one to obtain exact solutions [3–6]. In fact, the scalar and vector Cornell potentials present two different independent approaches to a description of the strong interaction. In [1], the vector potential has not been used. In this work, the FW transformation was performed only for a nonrelativistic Dirac particle in the weak field approximation. In the present work, the much more general problem of a high-precision description of a relativistic quark in an arbitrarily strong field is solved thanks to the use of a more universal method of the FW transformation. This allows us to calculate the effect of the strong spin-orbit interaction of relativistic quarks.

3 Phenomenological description of strong interaction with the vector Cornell potential

When the vector Cornell potential is used and the electromagnetic interaction of quarks is taken into account, the Dirac Hamiltonian takes the form

$$\left(i\hbar\gamma^\mu D_\mu - mc^2 + \frac{\mu'}{2}\sigma^{\mu\nu}F_{\mu\nu} + \frac{\mathfrak{m}'}{2}\sigma^{\mu\nu}\mathcal{G}_{\mu\nu}\right)\Psi = 0, \qquad D_\mu = \partial_\mu + ieA_\mu + i\mathcal{A}_\mu. \tag{5}$$

Here μ' and $\mathfrak{m}'$ are the anomalous magnetic and chromomagnetic moments, $F_{\mu\nu} = \partial_\mu A_\nu - \partial_\nu A_\mu$ and $\mathcal{G}_{\mu\nu} = \partial_\mu\mathcal{A}_\nu - \partial_\nu\mathcal{A}_\mu$ are the tensors of the electromagnetic and chromoelectromagnetic fields and D_μ is the covariant derivative. This equation generalizes the Dirac-Pauli one (see [13] and references therein). The strong interaction is here similar to the electromagnetic one.

In this case, the Hamiltonian in the Dirac representation takes the form

$$\mathcal{H} = \beta mc^2 + e\Phi + \mathbb{F} + \boldsymbol{\alpha}\cdot\boldsymbol{\pi} - \boldsymbol{\Pi}\cdot\boldsymbol{\mathcal{M}} + i\boldsymbol{\gamma}\cdot\boldsymbol{\mathcal{P}}, \tag{6}$$

where

$$\begin{array}{cc} \boldsymbol{\mathcal{M}} = \mu'\boldsymbol{B} + \mathfrak{m}'\boldsymbol{\mathcal{R}}, & \boldsymbol{\mathcal{P}} = \mu'\boldsymbol{E} + \mathfrak{m}'\boldsymbol{Q}, \\ \boldsymbol{Q} = -\nabla\mathbb{F} - \frac{\partial\boldsymbol{\mathcal{A}}}{\partial t}, & \boldsymbol{\mathcal{R}} = \nabla\times\boldsymbol{\mathcal{A}}. \end{array} \tag{7}$$

Here the chromoelectric and chromomagnetic fields, $\boldsymbol{Q}$ and $\boldsymbol{\mathcal{R}}$, are introduced.

Now we fulfill the transformation to the FW representation [25] with the method developed in [8, 9]. This representation holds a special place in quantum mechanics due to some unique properties. In this representation, quantum mechanical operators for relativistic particles in an external field have the same form as in the nonrelativistic quantum theory. In particular, the position operator [26] and momentum operator are equal to $\boldsymbol{r}$ and $\boldsymbol{p} = -i\hbar\nabla$, respectively, and the polarization operator for spin-1/2 particles is expressed by the Dirac matrix $\boldsymbol{\Pi}$. In other representations, these operators are given by much more cumbersome formulae (see [7, 25]). The relations between the operators in the FW representation are analogous to the relations between the corresponding classical quantities. The simple form of operators corresponding to classical observables is a great advantage of this representation. The above-mentioned properties of the FW representation allow one to use it successfully for passing to the semiclassical approximation and to the classical limit of relativistic quantum mechanics [25, 27]. We note that the Hamiltonian and all other operators are diagonal in two spinors (block-diagonal) in this representation.

When the FW representation is used, the passage to the classical limit is usually accomplished by simply replacing the operators in expressions for the Hamiltonian and in operator equations of motion with the corresponding classical quantities. The possibility of such a replacement, explicitly or implicitly used in practically all works devoted to the relativistic FW transformation, was rigorously proven in [10]. This possibility radically simplifies an interpretation of the basic quantum mechanical equations, especially in the relativistic case.

Initial Hamiltonian (6) can be presented in the following general form:

$$\mathcal{H} = \beta m + \mathcal{E} + \mathcal{O}, \quad \beta\mathcal{E} = \mathcal{E}\beta, \quad \beta\mathcal{O} = -\mathcal{O}\beta, \tag{8}$$

where $\mathcal{E}$ and $\mathcal{O}$ are even and odd (diagonal and off-diagonal in two spinors) operators, respectively.

The Hamiltonian transformed to the FW representation is exact for terms proportional to the zeroth and first powers of the Planck constant and also for terms describing contact interactions and proportional to $\hbar^2$. It is given by [8, 11]

$$\begin{gathered}\mathcal{H}_{FW} = \beta\epsilon + \mathcal{E} - \frac{1}{8}\left\{\frac{1}{\epsilon(\epsilon+m)}, [\mathcal{O}, [\mathcal{O}, \mathcal{Z}]]\right\},\\ \mathcal{Z} = \mathcal{E} - i\frac{\partial}{\partial t}, \quad \epsilon = \sqrt{m^2 + \mathcal{O}^2}.\end{gathered} \tag{9}$$

In the considered case, this Hamiltonian has the form

$$\begin{gathered}\mathcal{H}_{FW} = \beta\epsilon' + e\Phi + \mathbb{F} + \frac{\hbar}{8m^2}\left\{\frac{1}{\gamma(\gamma+1)}, \Big(\boldsymbol{\Sigma}\cdot\big[\boldsymbol{\pi}\times(e\boldsymbol{E} + \boldsymbol{Q})\\ -(e\boldsymbol{E} + \boldsymbol{Q})\times\boldsymbol{\pi}\big] - \hbar\nabla\cdot(e\boldsymbol{E} + \boldsymbol{Q})\Big)\right\}\\ + \frac{\hbar}{8m^2}\left\{\frac{1}{\gamma}, \Big(\boldsymbol{\Sigma}\cdot\big[\boldsymbol{\pi}\times(eG\boldsymbol{E} + G_c\boldsymbol{Q})\\ -(eG\boldsymbol{E} + G_c\boldsymbol{Q})\times\boldsymbol{\pi}\big] - \hbar\nabla\cdot(eG\boldsymbol{E} + G_c\boldsymbol{Q})\Big)\right\}\\ -\frac{\hbar}{4m}\left\{\frac{1}{\gamma}, \Big(\boldsymbol{\Pi}\cdot\big[e\boldsymbol{B} + \boldsymbol{\mathcal{R}}\big]\Big)\right\} - \frac{\hbar}{2m}\Big(\boldsymbol{\Pi}\cdot\big[eG\boldsymbol{B} + G_c\boldsymbol{\mathcal{R}}\big]\Big)\\ +\beta\frac{\hbar}{8m^3}\left\{\frac{1}{\gamma(\gamma+1)}, \Big[([eG\boldsymbol{B} + G_c\boldsymbol{\mathcal{R}}]\cdot\boldsymbol{\pi})(\boldsymbol{\Sigma}\cdot\boldsymbol{\pi}) + (\boldsymbol{\Sigma}\cdot\boldsymbol{\pi})(\boldsymbol{\pi}\cdot[eG\boldsymbol{B}\\ +G_c\boldsymbol{\mathcal{R}}]) + 2\pi\hbar(\boldsymbol{\pi}\cdot[eG\boldsymbol{j} + G_c\boldsymbol{\mathcal{I}}] + [eG\boldsymbol{j} + G_c\boldsymbol{\mathcal{I}}]\cdot\boldsymbol{\pi})\Big]\right\},\end{gathered} \tag{10}$$

where $\gamma = \epsilon'/m = \sqrt{m^2 + \pi^2}/m$ is the Lorentz factor, $G = (g-2)/2 = 2mc\mu'/(e\hbar)$, $G_c = (g_c - 2)/2 = 2mcm'/\hbar$, g_c is the chromomagnetic g factor similar to the electromagnetic one, $\boldsymbol{j} = [\nabla\times\boldsymbol{B} - \partial\boldsymbol{E}/(\partial t)]/(4\pi)$, and $\boldsymbol{\mathcal{I}} = [\nabla\times\boldsymbol{\mathcal{R}} - \partial\boldsymbol{Q}/(\partial t)]/(4\pi)$.

In the classical limit, the operator of the angular velocity of the spin precession is given by

$$\begin{gathered}\boldsymbol{\Omega} = \frac{1}{mc}\left\{\frac{1}{\gamma+1}\boldsymbol{\beta}\times(e\boldsymbol{E} + \boldsymbol{Q}) + \boldsymbol{\beta}\times(eG\boldsymbol{E} + G_c\boldsymbol{Q})\\ -\frac{1}{\gamma}(e\boldsymbol{B} + \boldsymbol{\mathcal{R}}) - (eG\boldsymbol{B} + G_c\boldsymbol{\mathcal{R}}) + \frac{\gamma}{\gamma+1}\left[\boldsymbol{\beta}\cdot(eG\boldsymbol{B} + G_c\boldsymbol{\mathcal{R}})\right]\boldsymbol{\beta}\right\}.\end{gathered} \tag{11}$$

The velocity of light c is explicitly included into this equation. In the classical limit, the force acting on the quark is similar to the Lorentz force:

$$\boldsymbol{F} = \frac{d\boldsymbol{\pi}}{dt} = e\boldsymbol{E} + \boldsymbol{Q} + \boldsymbol{\beta}\times(e\boldsymbol{B} + \boldsymbol{\mathcal{R}}), \tag{12}$$

where $\boldsymbol{\pi}$ is the kinetic momentum.

It is interesting to determine a change of the direction of particle motion in external fields. The equation of motion for the unit vector of the kinetic momentum direction, $\boldsymbol{N} = \boldsymbol{\pi}/|\boldsymbol{\pi}| = \boldsymbol{v}/|\boldsymbol{v}|$, has the form

$$\frac{d\boldsymbol{N}}{dt} = \boldsymbol{\omega}\times\boldsymbol{N}, \qquad \boldsymbol{\omega} = -\frac{1}{\gamma mc}\left[e\boldsymbol{B} + \boldsymbol{\mathcal{R}} - \frac{\boldsymbol{N}\times(e\boldsymbol{E} + \boldsymbol{Q})}{\beta}\right]. \tag{13}$$

The angular velocity of spin rotation relative to the kinetic momentum direction is equal to

$$o = \boldsymbol{\Omega} - \omega = \frac{1}{mc}\Big\{-(eG\boldsymbol{B} + G_c\boldsymbol{\mathcal{R}}) + \frac{\gamma}{\gamma+1}\boldsymbol{\beta}(\boldsymbol{\beta}\cdot[eG\boldsymbol{B} + G_c\boldsymbol{\mathcal{R}}]) \\ -\frac{1}{\gamma^2-1}\left[\boldsymbol{\beta}\times(e\boldsymbol{E}+\boldsymbol{Q})\right] + \boldsymbol{\beta}\times(eG\boldsymbol{E} + G_c\boldsymbol{Q})\Big\}. \tag{14}$$

Evidently, the quark helicity (the spin projection onto the momentum direction) is changed in magnetic and chromomagnetic fields only when the quark has the anomalous magnetic and chromomagnetic moments, respectively. The change of the quark helicity in electric and chromoelectric fields depends on both the normal and anomalous magnetic and chromomagnetic moments, respectively.

Let us consider the interaction of two quarks. Let $\boldsymbol{\mathcal{V}}$ be the velocity of the quark which is the source of the strong and electromagnetic fields. If we neglect the spin-spin interaction of the quarks in the first approximation, we obtain the following relations

$$\boldsymbol{B} = \frac{\boldsymbol{\mathcal{V}}}{c}\times\boldsymbol{E}, \quad \boldsymbol{\mathcal{R}} = \frac{\boldsymbol{\mathcal{V}}}{c}\times\boldsymbol{Q}. \tag{15}$$

When the interacting quarks belong to different nuclei, the vectors $\boldsymbol{v}$ and $\boldsymbol{\mathcal{V}}$ have opposite directions in the laboratory frame. At collisions of high energy nuclei, it can be often supposed that $|\boldsymbol{v}| \approx |\boldsymbol{\mathcal{V}}| \approx c$. An opposite situation takes place when the vectors $\boldsymbol{v}$ and $\boldsymbol{\mathcal{V}}$ are mainly co-directed. This situation can probably be realized in quark-gluon plasma due to a predominance of local interactions.

Let us use the center-of-mass system and consider the simplest (but practically very important) case of a collision of two nucleons. In this case, $\boldsymbol{\mathcal{V}} = -\boldsymbol{v}$. Let us determine the Hamiltonian of the first quark in the field of the second one. We restrict ourselves by a phenomenological description of main effects under peripheral collisions of relativistic nuclei. To perform a more precise quantitative analysis, one needs to take into account effects of retardation and to use the Breit equation as an initial one. Nevertheless, the approach used gives us a reasonable description of the force acting on the quark and of the torque acting on its spin.

When the electromagnetic interactions are neglected, the classical limit of the Hamiltonian (10) is given by

$$\mathcal{H} = \epsilon' + \mathbb{F} + \frac{1}{mc}\left(\frac{1}{\gamma} + \frac{1}{\gamma+1} + 2G_c\right)\boldsymbol{s}\cdot(\boldsymbol{\beta}\times\boldsymbol{Q}). \tag{16}$$

Here $\boldsymbol{s}$ is the classical spin vector. In this approximation, the angular velocity of the spin precession is equal to

$$\boldsymbol{\Omega} = \frac{1}{mc}\left(\frac{1}{\gamma} + \frac{1}{\gamma+1} + 2G_c\right)\boldsymbol{\beta}\times\boldsymbol{Q}. \tag{17}$$

We take into account that the contribution of the strong interaction into the spin precession is predominant for any energy.

It is rather convenient to express the angular velocity of the spin precession $\boldsymbol{\Omega}$ in the lab frame (more precisely, in the center-of-mass one) in terms of the force (12) in this frame. In the considered case, the force is given by

$$\boldsymbol{F} = \left(2 - \frac{1}{\gamma^2}\right)(e\boldsymbol{E} + \boldsymbol{Q}) - \boldsymbol{\beta}(\boldsymbol{\beta}\cdot[e\boldsymbol{E} + \boldsymbol{Q}]). \tag{18}$$

When only the strong interaction is taken into account, the angular velocity of spin precession is therefore expressed as follows:

$$\boldsymbol{\Omega} = \frac{1}{mc}\left(\frac{1}{\gamma} + \frac{1}{\gamma+1} + 2G_c\right)\frac{\gamma^2}{2\gamma^2-1}\boldsymbol{\beta}\times\boldsymbol{F}. \tag{19}$$

Since the force $\boldsymbol{F}$ defines the change of the momentum, it may be in principle extracted from experimental data. Thus, (19) is convenient for an analysis of experimental effects.

The sign of the projection of $\boldsymbol{\Omega}$ onto the normal to the scattering plane depends on the sign of G_c. If $G_c < 0$ (see [28, 29]) and the Lorentz factor is large enough, the vectors $\boldsymbol{\Omega}$ and $\boldsymbol{\beta}\times\boldsymbol{F}$ are oppositely directed.

Equation (19) defines the rather large angular velocity of the spin precession which corresponds to the rather high energy of the spin-orbit interaction (of the order of 100 MeV). The angular velocity of spin precession is of the same order of magnitude as the vorticity calculated in [30].

It is important to give a semiclassical description of evolution of the quark helicity. When we consider the case of a collision of two nucleons and neglect the spin-spin interaction of the quarks, we can use the relations (15) and $\mathcal{V} = -\boldsymbol{v}$. In this case, (14) takes the form

$$\boldsymbol{o} = \frac{1}{mc}\left[e\left(2G - \frac{1}{\gamma^2-1}\right)\boldsymbol{\beta}\times\boldsymbol{E} + \left(2G_c - \frac{1}{\gamma^2-1}\right)\boldsymbol{\beta}\times\boldsymbol{Q}\right]. \tag{20}$$

Taking into account the spin-spin interaction can also be necessary in some cases. In electrodynamics, the field of the immobile magnetic dipole $\tilde{\boldsymbol{\mu}}$ is given by

$$\boldsymbol{B} = \frac{3\boldsymbol{n}(\tilde{\boldsymbol{\mu}}\cdot\boldsymbol{n}) - \tilde{\boldsymbol{\mu}}}{r^3}.$$

The field of a moving magnetic dipole is defined by the Lorentz transformations.

The Hamiltonian of interaction of the immobile magnetic dipole $\boldsymbol{\mu}$ with the magnetic field $\boldsymbol{B}$ reads

$$\mathcal{H}_{int} = -\boldsymbol{\mu}\cdot\boldsymbol{B} = -\frac{3(\boldsymbol{\mu}\cdot\boldsymbol{n})(\tilde{\boldsymbol{\mu}}\cdot\boldsymbol{n}) - (\boldsymbol{\mu}\cdot\tilde{\boldsymbol{\mu}})}{r^3}. \tag{21}$$

Since the magnetic dipole moment is proportional to the spin [$\boldsymbol{\mu} = eg\boldsymbol{s}/(2mc)$], the Hamiltonian of interaction (21) corresponds to the Thomas-Bargmann-Michel-Telegdi equation [31, 32] (see also [13, 33]) in the general case. This equation describes the spin motion in electromagnetic fields:

$$\begin{gathered}\mathcal{H}_{int} = -\boldsymbol{\Omega}\cdot\boldsymbol{s}, \qquad \boldsymbol{\Omega} = -\frac{e}{mc}\left[\left(G + \frac{1}{\gamma}\right)\boldsymbol{B}\right. \\ \left. - \frac{\gamma G}{\gamma+1}(\boldsymbol{\beta}\cdot\boldsymbol{B})\boldsymbol{\beta} - \left(G + \frac{1}{\gamma+1}\right)\boldsymbol{\beta}\times\boldsymbol{E}\right],\end{gathered} \tag{22}$$

where $\boldsymbol{B}$ and $\boldsymbol{E}$ are the fields created by the magnetic dipole $\tilde{\boldsymbol{\mu}}$. The spin-spin interaction of quarks conditioned by the vector Cornell potential has similar properties.

Thus, the strong interaction of quarks leads to the spin-orbit interaction which results in a spin rotation and in a preferable orientation of the spin. The latter effect is similar to a spin polarization in targets placed in a magnetic field. In the case of strange quarks, this effect can be an additional explanation of hyperon polarization [30]. We can also state the existence of the spin-spin interaction and the possibility of a significant change of the quark helicity.

4 Dirac equation with the scalar and vector Cornell potentials

Let us consider the general case when the Dirac equation contains both the scalar and vector Cornell potentials. This equation was investigated in the recent works [2–6]. In these works, an analysis has been made only for specific values of C_0 and C_1. These values have been chosen by reason of symmetry. However, it is difficult to substantiate the corresponding symmetry of the Dirac equation by any physical arguments. The method of the FW transformation used in the present work allows us to derive the relativistic Hamiltonian for *arbitrary* C_0 and C_1. Certainly, these coefficients should satisfy the relation (4).

 DOI: 10.1051/epjconf/201713808002

With the use of (8) and (9), we obtain the following result:

$$\begin{aligned}
\mathcal{H}_{FW} = \beta\epsilon' + e\Phi + \mathbb{F} + \frac{\hbar}{8m^2}\Bigg\{\frac{1}{\gamma(\gamma+1)}, \Big(\boldsymbol{\Sigma}\cdot[\boldsymbol{\pi}\times(e\boldsymbol{E}+\boldsymbol{Q}) \\
-(e\boldsymbol{E}+\boldsymbol{Q})\times\boldsymbol{\pi}] - \boldsymbol{\Pi}\cdot(\boldsymbol{\pi}\times\mathfrak{E} - \mathfrak{E}\times\boldsymbol{\pi}) - \hbar\nabla\cdot(e\boldsymbol{E}+\boldsymbol{Q}) \\
+\beta\hbar\nabla\cdot\mathfrak{E}\Big)\Bigg\} + \frac{\hbar}{8m^2}\Bigg\{\frac{1}{\gamma}, \Big(\boldsymbol{\Sigma}\cdot[\boldsymbol{\pi}\times(eG\boldsymbol{E}+G_c\boldsymbol{Q}) \\
-(eG\boldsymbol{E}+G_c\boldsymbol{Q})\times\boldsymbol{\pi}] - \hbar\nabla\cdot(eG\boldsymbol{E}+G_c\boldsymbol{Q})\Big)\Bigg\} \\
-\frac{\hbar}{4m}\Bigg\{\frac{1}{\gamma}, \Big(\boldsymbol{\Pi}\cdot[e\boldsymbol{B}+\mathcal{R}]\Big)\Bigg\} - \frac{\hbar}{2m}\Big(\boldsymbol{\Pi}\cdot[eG\boldsymbol{B}+G_c\mathcal{R}]\Big) \\
+\beta\frac{\hbar}{8m^3}\Bigg\{\frac{1}{\gamma(\gamma+1)}, \Big[([eG\boldsymbol{B}+G_c\mathcal{R}]\cdot\boldsymbol{\pi})(\boldsymbol{\Sigma}\cdot\boldsymbol{\pi}) \\
+(\boldsymbol{\Sigma}\cdot\boldsymbol{\pi})(\boldsymbol{\pi}\cdot[eG\boldsymbol{B}+G_c\mathcal{R}]) \\
+2\pi\hbar(\boldsymbol{\pi}\cdot[eG\boldsymbol{j}+G_c\mathcal{I}] + [eG\boldsymbol{j}+G_c\mathcal{I}]\cdot\boldsymbol{\pi})\Big]\Bigg\}.
\end{aligned} \tag{23}$$

The analysis of the Hamiltonian (23) brings the following conclusions. The terms describing the spin-orbit and contact interactions have opposite signs for the scalar and vector Cornell potentials. The energy of the spin-orbit interaction is rather high and is of the order of 100 MeV. We can also mention that the transition to the classical limit of (23) reduces to the replacement of the operators in this equation with the corresponding classical quantities. In particular, the spin operators $\hbar\boldsymbol{\Sigma}$ and $\hbar\boldsymbol{\Pi}$ are substituted by the doubled classical spin vector, $2\boldsymbol{s}$.

5 Summary

We have considered the strong and electromagnetic interactions of Dirac particles (quarks). We have used the scalar and vector Cornell potentials for a phenomenological description of the strong interaction. The vector Cornell potential is similar to the four-vector potential of the electromagnetic field. Since the strong interaction is much closer to the electromagnetic interaction than to the scalar one, the use of the vector Cornell potential is more natural. Unlike previous investigations [1, 3, 5, 6], we have used the method of the relativistic FW transformation [7–9] applicable in arbitrarily strong external fields and have determined the relativistic FW Hamiltonian. In this Hamiltonian, the terms of the zeroth and first orders in the Planck constant and such terms of the order of $\hbar^2$ which describe contact interactions are exact. We have found the exact classical limit of this Hamiltonian. The adequate and exact classical limit of the strong interaction of quarks characterized by the Cornell potential has been determined for the first time. We also have taken into account the electromagnetic interaction which may not be negligible. Effective fields acting on the particle and spin have been found. General formulae for the angular velocity of spin precession have been derived. These formulae correspond to the high energy of the spin-orbit interaction (of the order of 100 MeV) for both the vector and scalar Cornell potentials. The terms describing the spin-orbit and contact interactions have opposite signs for the scalar and vector Cornell potentials. Dynamics of the quark helicity has been calculated and the spin-spin interaction of quarks has been considered. We have determined the relations between the angular velocity of spin precession and the force acting on the quark. Since this force defines the change of the momentum, it can be in principle extracted from experimental data.

Acknowledgments

This work was supported in part by the Belarusian Republican Foundation for Fundamental Research (Grant No. Ф16D-004) and by the Heisenberg-Landau program of the German Ministry for Science and Technology (Bundesministerium für Bildung und Forschung).

References

[1] U.D. Jentschura, J.H. Noble, J. Phys. A: Math. Theor. **47**, 045402 (2014)
[2] V.Yu. Lazur, O.K. Reity, V.V. Rubish, Phys. Rev. D **3**, 076003 (2011)
[3] H. Hassanabadi, E. Maghsoodi, S. Zarrinkamar, Ann. Phys. (Berlin) **525**, 944 (2013)
[4] H. Hassanabadi, E. Maghsoodi, S. Zarrinkamar, H. Rahimov, Adv. High Energy Phys. **2012**, 707041 (2012)
[5] M. Hamzavi, A.A. Rajabi, Ann. Phys. (New York) **334**, 316 (2013). M. Hamzavi, A.A. Rajabi, Chin. Phys. B **22**, 090301 (2013). M. Hamzavi, A.A. Rajabi, Chin. Phys. C **37**, 103102 (2013). M. Hamzavi, A.A. Rajabi, Few-Body Syst. **54**, 2067 (2013)
[6] L.A. Trevisan, C. Mirez, F.M. Andrade, Few-Body Syst. **55**, 1055 (2014)
[7] A.J. Silenko, J. Math. Phys. **44**, 2952 (2003)
[8] A.J. Silenko, Phys. Rev. A **77**, 012116 (2008)
[9] A.J. Silenko, Phys. Rev. A **91**, 022103 (2015)
[10] A.J. Silenko, Phys. Part. Nucl. Lett. **10**, 91 (2013)
[11] A.J. Silenko, Theor. Math. Phys. **176**, 987 (2013)
[12] A.J. Silenko, Phys. Rev. A **91**, 012111 (2015)
[13] A.J. Silenko, Theor. Math. Phys. **105**, 1224 (1995)
[14] A.J. Silenko, Russ. Phys. J. **48**, 788 (2005)
[15] A.J. Silenko, O.V. Teryaev, Phys. Rev. D **71**, 064016 (2005)
[16] A.J. Silenko, O.V. Teryaev, Phys. Rev. D **76**, 061101(R) (2007)
[17] Yu.N. Obukhov, A.J. Silenko, O.V. Teryaev, Phys. Rev. D **80**, 064044 (2009)
[18] Yu.N. Obukhov, A.J. Silenko, O.V. Teryaev, Phys. Rev. D **84**, 024025 (2011)
[19] Yu.N. Obukhov, A.J. Silenko, O.V. Teryaev, Phys. Rev. D **88**, 084014 (2013)
[20] Yu.N. Obukhov, A.J. Silenko, O.V. Teryaev, Phys. Rev. D **90**, 124068 (2014)
[21] Yu.N. Obukhov, A.J. Silenko, O.V. Teryaev, Phys. Rev. D **94**, 044019 (2016)
[22] A.J. Silenko, Nuclear Instrum. Methods in Phys. Res. B **114**, 259 (1996)
[23] A.J. Silenko, Theor. Math. Phys. **112**, 922 (1997)
[24] Ya.Ya. Balitskii, Nucl. Phys. B **254**, 166 (1983)
[25] L.L. Foldy, S.A. Wouthuysen, Phys. Rev. **78**, 29 (1950)
[26] T.D. Newton, E.P. Wigner, Rev. Mod. Phys. **21**, 400 (1949)
[27] J.P. Costella, B.H.J. McKellar, Am. J. Phys. **63**, 1119 (1995)
[28] D. Diakonov, Prog. Part. Nucl. Phys. **51**, 173 (2003)
[29] N. Kochelev, N. Korchagin, Phys. Lett. B **729**, 117 (2014)
[30] M. Baznat, K. Gudima, A. Sorin, O. Teryaev, Phys. Rev. C **88**, 061901 (2013)
[31] L.H. Thomas, Nature (London) **117**, 514 (1926). L.H. Thomas, Philos. Mag. **3**, 1 (1927)
[32] V. Bargmann, L. Michel and V.L. Telegdi, Phys. Rev. Lett. **2**, 435 (1959)
[33] A. J. Silenko, Phys. Scripta **90**, 065303 (2015)

EPJ Web of Conferences **138**, 08003 (2017) DOI: 10.1051/epjconf/201713808003
Baldin ISHEPP XXIII

Lambda polarization feasibility study at BM@N

Dilyna Suvarieva[1,2,⋆], *Konstantin* Gudima[3,⋆⋆], and *Alexander* Zinchenko[1,⋆⋆⋆]

[1] *Joint Institute for Nuclear Research, Dubna, Moscow region, Russia*
[2] *Plovdiv University "Paisii Hilendarski", Plovdiv, Bulgaria*
[3] *IAP AS, Chisinau, Moldova*

Abstract. Heavy strange objects (hyperons) could provide essential signatures of the excited and compressed baryonic matter. At NICA, it is planned to study hyperons both in the collider mode (MPD detector) and the fixed-target one (BM@N setup).
Measurements of strange hyperons polarization could give additional information on the strong interaction mechanisms. In heavy-ion collisions, such measurements are even more valuable since the polarization is expected to be sensitive to characteristics of the QCD medium (vorticity, hydrodynamic helicity) and to QCD anomalous transport.
In this analysis, the possibility to measure at BM@N the polarization of the lightest strange hyperon Λ is studied in Monte Carlo event samples produced with the DCM-QGSM generator. It is shown that the detector will allow to measure Λ polarization with a precision required to check the model predictions.

1 Introduction

The parity-preserving polarization of strange hyperons in hadronic interactions is transverse to scattering plane. It is large and sensitive to the strong interactions mechanisms. In the heavy-ion collisions this polarization can be diluted due to rescattering in the QCD medium, while the polarization transverse to reaction plane may emerge. It is sensitive to parity-odd characteristics of QCD medium (vorticity, hydrodynamic helicity) and to QCD anomalous transport.

In this analysis, the polarization of the lightest strange hyperon Λ, is studied. Its decay $\Lambda \to p + \pi^-$ is fully reconstructible and the polarization can be extracted from the angular distribution of the final state particles. The polarization is measured with respect to the production plane of the Λ hyperon: $\vec{n} = \hat{p}_{beam} \times \vec{p}$, where $\hat{p}_{beam}$ is aligned with the direction of the incoming beam and $\vec{p}$ is the Λ momentum.

In the Λ rest frame, the angle θ between the decay proton and the analyzing direction $\hat{n}$ will follow the probability distribution:

$$w(\cos\theta) = \frac{1}{2}(1 + \alpha P \cos\theta), \tag{1}$$

where $\alpha = 0.642$ is the decay asymmetry of the parity violating weak decay $\Lambda \to p + \pi^-$ and P is polarization in the direction of $\hat{n}$.

⋆e-mail: DilyanaSuvarieva@mail.bg
⋆⋆e-mail: gudima@cc.acad.md
⋆⋆⋆e-mail: Alexander.Zinchenko@jinr.ru

2 BM@N detector

A successful operation of the NICA complex [1] will require the existing machine Nuclotron to be upgraded in order to accelerate Au nuclei. After that, the improved Nuclotron beams will also be used to run a fixed target experiment BM@N [2].

The detector will allow to study $A+A$ collisions by measuring a variety of observables. Particle yields, ratios, transverse momentum spectra, rapidity and angular distributions, as well as fluctuations and correlations of hadrons will be studied as a function of the collision energy and centrality.

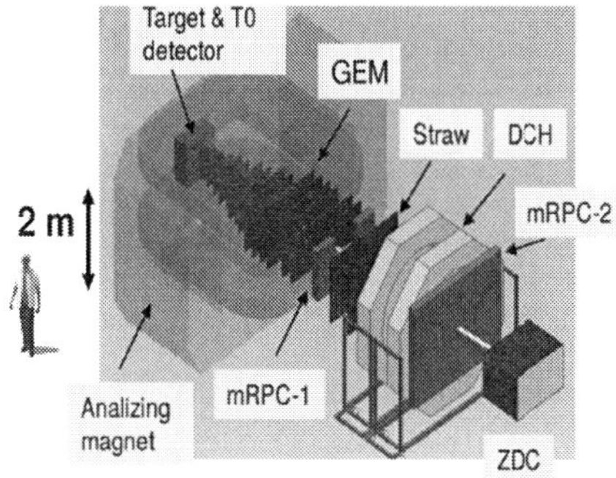

Figure 1. BM@N detector

A sketch of the proposed experimental setup is shown in figure 1. It combines high precision track measurements with time-of-flight information for particle identification and total energy measurements for event characterization. The charged track momenta and multiplicity will be measured with the set of GEM (Gas Electron Multipliers) detectors located downstream of the target inside the analyzing magnet of 0.8 T and drift chambers (Straw, DCH) situated outside the magnetic field. Design parameters of the time-of-flight detectors based on multigap Resistive Plate Chambers (mRPC-1, -2) with a strip read-out allow efficient discrimination between particle species with momentum up to a few GeV/c. The Zero Degree Calorimeter (ZDC) is designed for the collision centrality analysis by measuring the energy of forward going particles.

3 Analysis method

3.1 Signal extraction

The topology of Λ hyperon decays is shown in figure 2. The event topology (decay of a relatively long-lived particle into two particles) defines the selection criteria: relatively large distance of the closest approach (DCA) to the primary vertex of decay products, small track-to-track separation in the decay vertex, relatively large decay length of the mother particle. For higher efficiency, both the DCA and two-track separation cuts were applied in χ^2 - space, i.e normalized to their respective errors.

The exact values of selection cuts were found by performing a multidimensional scan over the whole set of selection criteria with a requirement to maximize the invariant mass peak significance, defined as $S/\sqrt{S+B}$, where S and B are total numbers of signal (described by the gaussian) and background (polynomial function) combinations inside $\pm 2\sigma$ interval around the peak position. While different physics analyses might prefer different selection quality criteria, the significance looks convenient to quantitatively evaluate effect of different factors on the reconstruction quality.

The corresponding scan procedure was realized as follows: during the particle combinations the parameters which have been chosen to serve as selection criteria (see above) were recorded along with the invariant mass value. Later, multiple loops over those variables were performed in some steps and their values were used as low or high thresholds, yielding the invariant mass peak significance for each set of selection cut values. Then, the maximum value was taken along with the corresponding set of selection parameters.

In order to subtract the background contribution from the measured $\cos\theta$-distribution, a so-called sideband approach was used. Its main idea is to subtract the distribution obtained for events well outside of the peak region of the Λ invariant mass distribution (in sidebands - figure 2) dominated by the background $p\pi^-$ combinations from the one in the peak region (dominated by signal combinations) according to the the following formula:

$$\cos\theta^S = \frac{(S+B)}{S}\cos\theta^{S+B} - \frac{B}{S}\cos\theta^B, \tag{2}$$

where S is a signal, B is a background, $\cos\theta^{S+B}$ is the distribution for the signal peak region and $\cos\theta^B$ for the sidebands (figure 3 left).

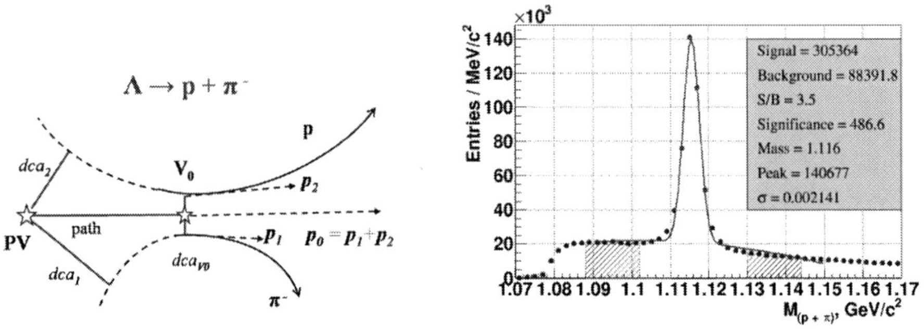

Figure 2. Left - event topology of a neutral particle two-prong decay; right - reconstructed $p\pi^-$ invariant mass distribution and sidebands used for background subtraction (hatched areas).

3.2 Polarization extraction

As it can be seen in figure 3 (filled circles on the right), the reconstructed $\cos\theta$-distribution can not be directly described by the expression (1) due to detector acceptance effects which distort the original linear dependence. In order to correct for detector effects, one can try to apply a simple approach based on the Monte Carlo simulation of the detector acceptance using events without the effect under study, i.e. with zero lambda polarization. The "unpolarized" $\cos\theta$-distribution is shown in figure 3 (empty circles). Dividing the "polarized" distribution by the "unpolarized" one it is possible to observe the net effect due to polarization (figure 5).

4 Results

The study was done for 500 thousand central *Au+Au* collisions (0-3.8 fm) with the beam kinetic energy $T_0 = 4.5$ GeV/nucleon ($\sqrt{s} = 3.5$A GeV). The event sample was created with the DCM-QGSM (Dubna Cascade Model - Quark-Gluon String Model) [3–7] event generator. Lambda polarization was calculated according to the model in [8], taking into account hyperon rescattering in the nuclear medium. The event generator output was used to produce two event samples: with and without polarization effects activated by switching on and off lambda anisotropical decay according to (1) during the particle transport with the GEANT3 package.

Figure 4 shows the model predictions for lambda polarization as functions of its transverse momentum and Feynman variable $x_F = p_\Lambda / p_{beam}$. Polarization increases with transverse momentum and reaches a maximum for some intermediate values of the Feynman variable. One can see also some fraction of zero-polarization lambdas which are the secondary ones (mostly from Σ^0 decays).

DOI: 10.1051/epjconf/201713808003

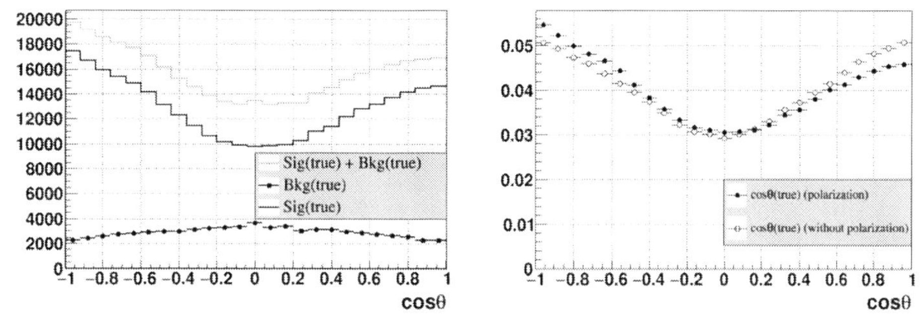

Figure 3. Left - $\cos\theta$-distributions for $p\pi^-$ combinations from the peak region of the invariant mass distribution (blue line - S+B), from the sidebands (circles - B) and background subtracted one (black line - S); right - background subtracted $\cos\theta$-distributions with (filled circles) and without (empty circles) polarization effects taken into account.

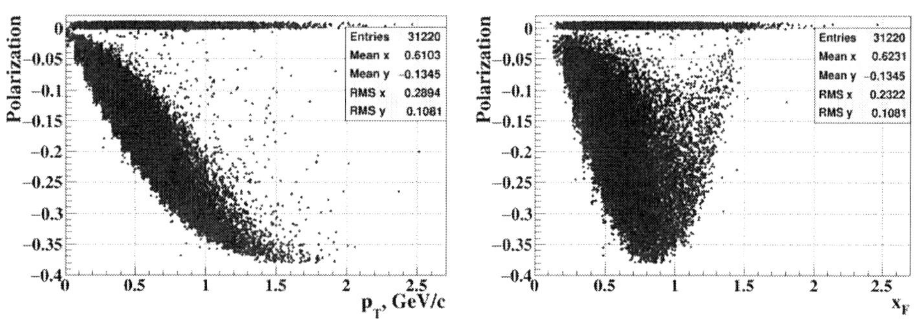

Figure 4. Left - lambda polarization vs p_T; right - vs x_F.

The extracted polarization value in the full phase space covered by the detector can be seen in figure 5 (left) as the slope parameter of the fitted straight line. For comparison, on the right plot of figure 5 one can see the true polarization distribution obtained for selected lambda-hyperons. There is a good agreement between the measured value (-0.1456) and the mean value of the true distribution (-0.1378).

In order to check the sensitivity of the method in different regions of the phase space, two intervals of p_T ($p_T < 0.6$ and $p_T \geq 0.6$ GeV/c) and x_F ($x_F < 0.5$ and $x_F = 0.5 - 1.0$) were chosen. One can see from figures 6 and 7 that for these intervals the measured and true polarization values are close to each other as well.

5 Conclusions

The analysis method and procedure for lambda polarization study have been developed and tested on simulated central $Au + Au$ events at T0 = 4.5A GeV in BM@N detector. The obtained results demonstrate quite good sensitivity of the detector to polarization variables. However, secondary lambda hyperons should be taken into account when interpreting the results.

As a continuation of this work, it is planned to do more detailed study of $\cos(\theta)$ - resolution dependence on lambda parameters and to investigate possible factors biasing the measurement results.

 DOI: 10.1051/epjconf/201713808003

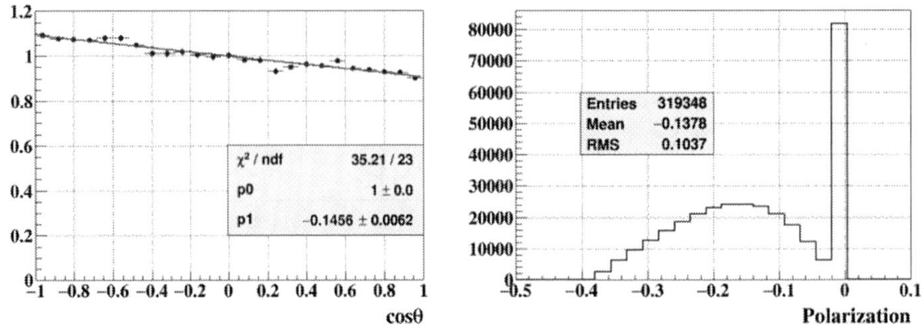

Figure 5. Left) reconstructed cos θ-distribution corrected for detector acceptance fitted to the straight line; right) true lambda polarization distribution for reconstructed events. Zero polarization corresponds to secondary lambdas.

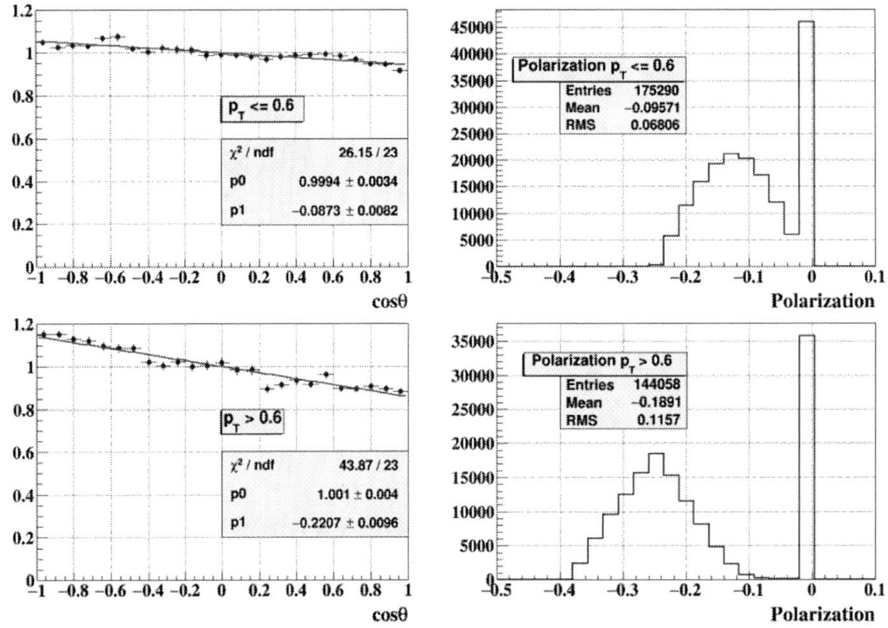

Figure 6. The same as in figure 5 for samples with $p_T < 0.6$ and $p_T \geq 0.6$ GeV/c.

The extension of the method to some other polarization studies (polarization transverse to reaction plane) will be evaluated.

Acknowledgements

We would like to thank Dr. O.Teryaev for attracting our attention to this physics topic and N.Geraksiev for helping with the background subtraction procedure.

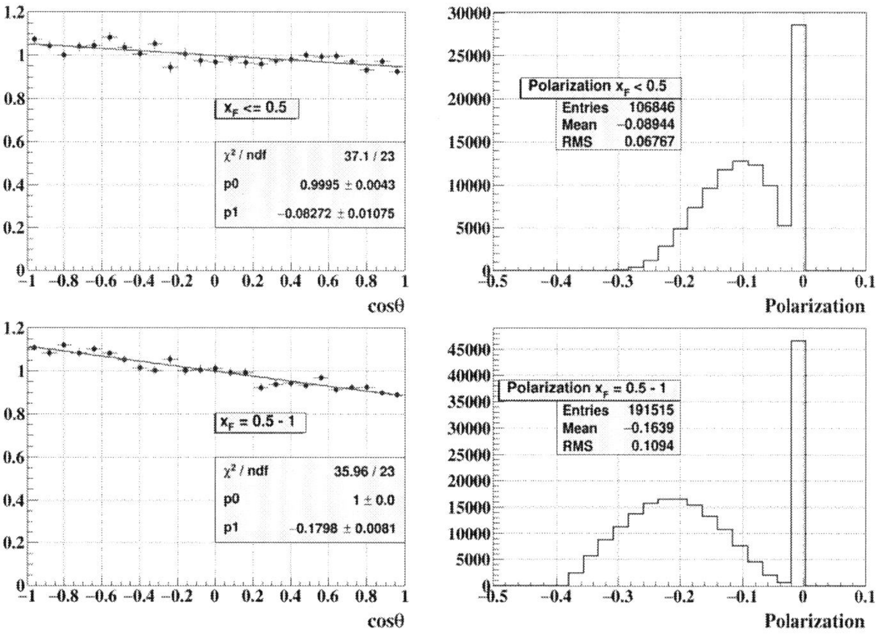

Figure 7. Left) reconstructed cos θ-distribution corrected for detector acceptance fitted to the straight line; right) true lambda polarization distribution for reconstructed events. Upper low - event sample with $x_F < 0.5$, lower row - with $x_F = 0.5 - 1.0$.

References

[1] http://nica.jinr.ru
[2] http://nica.jinr.ru/files/BM@N/BMN_CDR.pdf
[3] V.D. Toneev and K.K. Gudima, Nucl. Phys. A **400**, 173 (1983)
[4] N.S. Amelin, K.K. Gudima, and V.D. Toneev, Sov. Journ. of Nuclear Phys. **51**, 1730 (1990)
[5] N.S. Amelin et al., Phys. Rev. C **47**, 2299 (1993)
[6] K.K. Gudima, S.G. Mashnik, and A.J. Sierk, LANL Report LA-UR01-6804 (2001). https://mcnp.lanl.gov/
[7] S.G. Mashnik et al., LANL Report LA-UR-07-6198 (2007). arXiv:0709.173
[8] T. A. Degrand, J. Markkanen and H. I. Miettinen, Phys. Rev. D **32**, 2445 (1985)

Vector mesons polarization versus color transparency

Sergey Gevorkyan[1,⋆]

[1] *LHEP, Joint Institute for Nuclear Research, Dubna, Russia*

Abstract. Vector mesons $V = \rho, \omega, \varphi$ can be transversely (helicity $\lambda = \pm 1$) or longitudinally ($\lambda = 0$) polarized. Can vector mesons polarization influence on their interaction with nucleons and nuclei?
The meson-nucleon total cross sections can be extracted by measuring the absorption of mesons in production off nuclei as the nuclear absorption depends on the meson-nucleon total cross section and consequently on the vector meson polarization. Latter on I will discuss the possibility of difference in interaction of transverse and longitudinally polarized vector mesons with nucleons and experiments in which such difference can be detected.

1 Introduction

For many years the vector mesons $V(\rho, \omega, \varphi)$ electroproduction off nuclei has been considered as an effective tool to study the effect of color transparency (CT) (see e.g. [1]). The idea of CT is that hadron produced in certain hard-scattering processes will have a reduced probability of interaction in nuclear matter due to his smaller size compared with the physical hadron. Manifestation of the CT would be an increase in the value of the nuclear transparency defined as $Tr = \frac{d\sigma_A}{Ad\sigma_N}$, where $d\sigma_A(d\sigma_N)$ is the cross section for the process from nuclei target (free nucleon) in production. The CT effect can be understood within the dipole approach [2, 3] , where it is a result of small quark-antiquark dipole interaction with matter. In the hadronic basis this effect is the consequence of Gribov inelastic screening leading to weakening of interaction in nuclei [4].
On the other hand there is another mechanism of weakening of strong interaction as a result of its possible dependence on the polarization of vector meson. The vector meson scattering amplitude off nucleon at zero angle averaged over nucleon spin is determined by two quantities: $\sigma'_T = \sigma_T(1 - i\alpha_T)$ and $\sigma'_L = \sigma_T(1 - i\alpha_L)$, where $\sigma_{T(L)}$ is the total cross section for interaction of a transversely (longitudinally) polarized vector meson with nucleon and $\alpha_{T(L)} = Ref_{T(L)}(0)/Imf_{T(L)}(0)$ is the ratio of the real to imaginary part of the corresponding amplitudes at zero angle. The knowledge of these quantities is crucial for different models. For instance the naive quark model predicts $\sigma_T = \sigma_L$, whereas as it has been discussed later there are experimental indications and theoretical calculations which show that strong interaction of longitudinally polarized vector mesons with nucleons is much weaker than interaction of transversely polarized vector mesons.
In the figure 1 the dependence of nuclear transparency on the virtuality of photon Q^2 in ρ mesons electroproduction on two nuclei is depicted. With increasing of Q^2 the mesons absorption in nuclei decreases which usually has been interpreted as manifestation of CT effect.
On the other hand as it is seen from figure 2 with increasing Q^2 the fraction of longitudinally polarized

⋆e-mail: gevs@jinr.ru

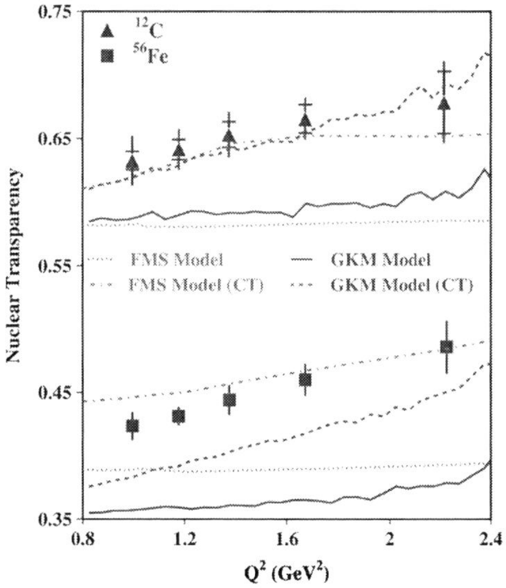

Figure 1. Nuclear transparency as a function of Q^2. Experimental data are from CLAS, JLab [5].

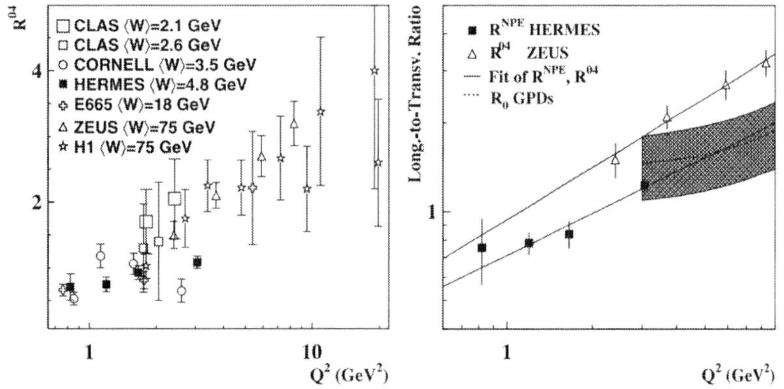

Figure 2. Q^2 dependence of the ratio $R = \sigma_L(\gamma^* p \rightarrow \rho p)/\sigma_T(\gamma^* p \rightarrow \rho p)$ for exclusive ρ^0 production on proton [6].

vector mesons also grows and if for instance $\sigma_T(VN) \gg \sigma_L(VN)$ the effect of absorption weakening with Q^2 is not a result of CT but at least partially is due to the difference between interaction of longitudinal and transverse vector mesons with nucleons.

2 The vector mesons polarization impact on their interaction with nucleons.

The first indication that interactions of vector mesons with nucleon can depend on its polarization come from the ρ and φ electroproduction on protons [7]. The ratio of the production cross section can be represented as $R = \frac{\sigma(\gamma_L p \rightarrow Vp)}{\sigma(\gamma_T p \rightarrow Vp)} = \xi^2 \frac{Q^2}{m_\rho^2}$, where the parameter ξ is the ratio of longitudinal to

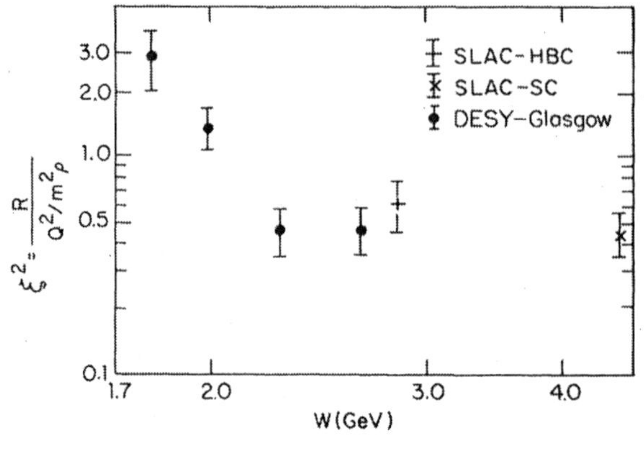

Figure 3. The dependence of the ξ^2 on invariant energy $W = \sqrt{s}$

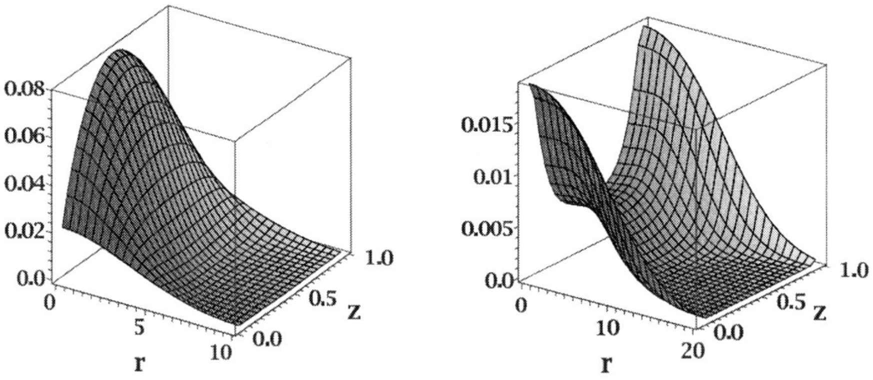

Figure 4. The two dimensional plots in z and r. Quarks distribution in longitudinally polarized ρ^0 meson (left) In transversely polarized one (right).

transverse ρ^0 total cross sections $\xi = \frac{\sigma_L(\rho p)}{\sigma_T(\rho p)}$. In the figure 3 the dependence of this ratio is shown and as it seen above the resonance region $\xi \approx 0.7$. The same quantity for φ meson: $\xi^2 = 0.33 \pm 0.08$. From theoretical point of view the possible difference between transverse and longitudinal cross sections is the result of different distribution of quarks in vector mesons depending on their polarizations. This effect takes place both for valence quarks distribution [8] and the distribution of constituent quarks [9, 10]. The distributions of constituent quarks in ρ^0 mesons for transverse and longitudinal polarizations as a function of transverse distance r between two quarks and z -the share of the vector meson light-cone momentum carried by quark are shown in the figure 4. One can see that quarks distribution in vector meson is very different for different polarizations [9, 10].

3 ω mesons photoproduction of nuclei.

Recently we proposed [11, 12] to measure the ω mesons photoproduction on a set of nuclei at photons energies $5GeV \leq E_\gamma \leq 10GeV$ available at Jefferson Lab. The deal is that incoherent photoproduction

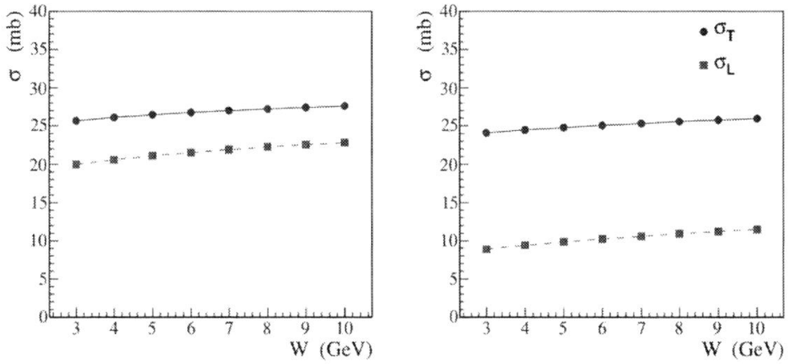

Figure 5. ρ meson total cross section with nucleon in two different parameterizations for vector meson wave function. a) Boosted Gaussian [13] (left) b) Light-cone relativistic model [9](right).

of ω mesons off nuclear targets is a unique way to get information on the possible dependence of strong interaction on the polarization of vector particle.
Unlike the photoproduction of ρ, φ mesons, which at moderate transfer momenta produced mainly transversely polarized (S-channel helicity conservation) the essential part of ω's in photoproduction are longitudinally polarized due to the significant contribution of pion-exchange in considered energy range [7].
From coherent photoproduction one can extract only the transverse cross section $\sigma'_T(\omega N) = \sigma_T(1 - i\alpha_T)$ as the pion exchange contributions from different nucleons cancel each other leading to photoproduction of only transversely polarized mesons. On the other hand in the incoherent photoproduction off nuclei where the ω mesons of both polarizations can be produced one can obtain the important information on longitudinal cross section.
The current quark models [13], [9] predict the considerable difference between total cross sections $\sigma_L(VN), \sigma_T(VN)$ whose dependence on invariant energy $W = \sqrt{s}$ is depicted in the figure 5.
As it was mentioned above the decrease of vector mesons absorption with rising Q^2 in electroproduction off nuclei is commonly interpreted as CT manifestation [5] can be a result of diminished cross section for longitudinally polarized vector mesons as their yield grows with photon virtuality Q^2. On the other hand the effect of color transparency is absent in vector mesons photoproduction at moderate transfer momenta. As a result the vector mesons production by real photons has an essential advantage compared to production by virtual photons.
I thank my coauthors E.Chudakov and A. Somov for fruitful collaboration and acknowledge the Jefferson Lab for permanent support.

References

[1] D. Dutta, K. Hafidi, M. Strikman, Color Transparency: past, present and future, arXiv:1211.2826 [nucl-th]
[2] S. R. Gevorkyan, A. M. Kotzinian, V. M. Jaloyan, Phys. Lett. B **212**, 251 (1988)
[3] N. N. Nikolaev, B. G. Zakharov, Z. Phys. C **49** 607 (1991)
[4] B.Z. Kopeliovich, J. Nemchik, I. Schmidt, Color Transparency at Low Energies: Predictions for JLAB, Phys.Rev. C **76**, 025210 (2007)
[5] L. El Fassi et al., Phys. Lett. B **712**, 325 (2012)

[6] A. Airapetian et al., Eur.Phys. J. C **62**, 659 (2009)
[7] T. H. Bauer, R. D. Spital, D. R. Yennie, F. M. Pipkin, Rev. Mod. Phys. **50**, 261 (1978)
[8] B. L. Ioffe, A. G. Oganesian, Phys. Rev. D **63**, 096006 (2001)
[9] J.R. Forshaw, R. Sandapen, JHEP **11**, 037 (2010)
[10] J.R. Forshaw, R. Sandapen, Phys. Rev. Lett. **109**, 081601 (2012)
[11] E. Chudakov, S. R. Gevorkyan, A. Somov, A Letter of Intend to Jefferson Lab PAC-43 (2015)
[12] E. Chudakov, S. R. Gevorkyan, A. Somov, Phys. Rev. C **93**, 015203 (2016)
[13] B. Z. Kopeliovich, J. Nemchik, I. Schmidt, Phys. Rev. C **76**, 015205 (2007)

Ratios of helicity amplitudes for exclusive ρ^0 electroproduction on transversely polarized proton

Serguei Manaenkov on behalf of the HERMES collaboration[1,*]

[1] *National Research Centre "Kurchatov Institute", Petersburg Nuclear Physics Institute, Russia*

Abstract. Exclusive ρ^0-meson electroproduction is studied by the HERMES experiment, using the 27.6 GeV longitudinally polarized electron/positron beam of HERA and a transversely polarized hydrogen target, in the kinematic region 1.0 GeV^2 < Q^2 < 7.0 GeV^2, 3.0 GeV < W < 6.3 GeV, and $-t'$ < 0.4 GeV^2. Using an unbinned maximum-likelihood method, 25 parameters are extracted. They determine the real and imaginary parts of the ratios of certain helicity amplitudes describing ρ^0-meson production by a virtual photon, where the denominator is the dominant amplitude $F_{0\frac{1}{2}0\frac{1}{2}}$. The latter is the nucleon-helicity-non-flip amplitude, which describes the production of a longitudinal ρ^0 meson by a longitudinal virtual photon. The transverse target polarization allows for the first time the extraction of ratios of a number of nucleon-helicity-flip amplitudes to $F_{0\frac{1}{2}0\frac{1}{2}}$. The ratios of nucleon-helicity-non-flip amplitudes are found to be in good agreement with those from the previous HERMES analysis.

1 Introduction

Exclusive electroproduction of vector mesons (V) on nucleons (N) gives information both on the reaction mechanisms and the nucleon structure [1]. Electroproduction at high energies can be considered to consist of three subprocesses: i) the incident lepton emits a virtual photon γ^*, which dissociates into a quark-antiquark pair; ii) this $q\bar{q}$ pair interacts strongly with the nucleon; iii) the observed vector meson is formed from the scattered $q\bar{q}$ pair. In Regge phenomenology, the interaction of the $q\bar{q}$ pair with the nucleon proceeds through the exchange of a pomeron or/and exchange of secondary reggeons. If the quantum numbers of the particle lying on the Regge trajectory are $J^P = 0^+,\ 1^-$, etc. (pomeron, ρ, f_2, ...), the process is denoted Natural Parity Exchange (NPE). Alternatively, the case of $J^P = 0^-,\ 1^+$, etc. (π, a_1, ...) corresponds to Unnatural Parity Exchange (UPE). In the framework of perturbative quantum chromodynamics valid at large photon virtuality Q^2 and high photon-nucleon center-of-mass (CM) energy W, the nucleon structure can also be studied through hard exclusive meson production, as the process amplitude contains Generalized Parton Distributions (GPDs) (see review [1]). However, the factorization property that permits to extract GPDs is rigorously proved only for the amplitudes $F_{0\frac{1}{2}0\pm\frac{1}{2}}$ of longitudinal vector meson production by longitudinal virtual photons [2]. In the Goloskokov-Kroll (GK) model (see [3] and references therein), the validity of factorization is assumed for some other amplitudes in addition to $F_{0\frac{1}{2}0\pm\frac{1}{2}}$ and this assumption is justified with a good description of the existing data. The presented HERMES data can be well described in the GK

*e-mail: manaenkov_si@pnpi.nrcki.ru

model if the pion exchange is taken into account (see [4] and references therein). This means that the GPD-based approach should be modified at intermediate energies and Q^2.

All observables in vector-meson electroproduction can be expressed in terms of the helicity amplitudes in the CM system describing ρ^0-meson production by a virtual photon. In particular, Spin-Density-Matrix Elements (SDMEs) are functions of the helicity amplitude ratios (HARs). Therefore HARs can be extracted from the data as was shown in [5]. For the first time, HARs with nucleon-helicity flip are obtained in the present analysis since the data on ρ^0-meson production by the longitudinally polarized electrons/positrons on a transversely polarized hydrogen target are accumulated.

2 Spin-density-matrix elements and helicity amplitudes

The SDMEs are the Fourier coefficients in angular distribution of the final-state particles. In the present paper, the formalism proposed in [6] for SDMEs is used. Any SDME can be expressed as a ratio of two sums of bilinear products of the helicity amplitudes $F_{\lambda_V\lambda'_N\lambda_\gamma\lambda_N}$ and their complex conjugate quantities $F^*_{\lambda_V\lambda'_N\lambda_\gamma\lambda_N}$. Here, $F_{\lambda_V\lambda'_N\lambda_\gamma\lambda_N}$ is the helicity amplitude of the process $\gamma^*(\lambda_\gamma) + N(\lambda_N) \to V(\lambda_V) + N(\lambda'_N)$, where the particle helicities are given in parentheses. The helicity amplitudes depend on W, Q^2, and $t' = t - t_{min}$, where t is the Mandelstam variable and $-t_{min}$ represents the smallest kinematically allowed value of $-t$ at fixed W and Q^2.

Any helicity amplitude can be decomposed into a sum of a NPE ($T_{\lambda_V\lambda'_N\lambda_\gamma\lambda_N}$) and an UPE ($U_{\lambda_V\lambda'_N\lambda_\gamma\lambda_N}$) amplitude: $F_{\lambda_V\lambda'_N\lambda_\gamma\lambda_N} = T_{\lambda_V\lambda'_N\lambda_\gamma\lambda_N} + U_{\lambda_V\lambda'_N\lambda_\gamma\lambda_N}$, for details see [6, 7]. The amplitudes obey the symmetry relations that hold because of parity conservation (see, e.g., [6, 7])

$$T_{\lambda_V\lambda'_N\lambda_\gamma\lambda_N} = (-1)^{-\lambda_V+\lambda_\gamma} T_{-\lambda_V\lambda'_N-\lambda_\gamma\lambda_N} = (-1)^{-\lambda_N+\lambda'_N} T_{\lambda_V-\lambda'_N\lambda_\gamma-\lambda_N}, \quad (1)$$

$$U_{\lambda_V\lambda'_N\lambda_\gamma\lambda_N} = -(-1)^{-\lambda_V+\lambda_\gamma} U_{-\lambda_V\lambda'_N-\lambda_\gamma\lambda_N} = -(-1)^{-\lambda_N+\lambda'_N} U_{\lambda_V-\lambda'_N\lambda_\gamma-\lambda_N}. \quad (2)$$

These symmetry properties of the helicity amplitudes permit to introduce the abbreviated notations:

$$T^{(1)}_{\lambda_V\lambda_\gamma} \equiv T_{\lambda_V\frac{1}{2}\lambda_\gamma\frac{1}{2}}, \quad U^{(1)}_{\lambda_V\lambda_\gamma} \equiv U_{\lambda_V\frac{1}{2}\lambda_\gamma\frac{1}{2}}, \quad T^{(2)}_{\lambda_V\lambda_\gamma} \equiv T_{\lambda_V\frac{1}{2}\lambda_\gamma-\frac{1}{2}}, \quad U^{(2)}_{\lambda_V\lambda_\gamma} \equiv U_{\lambda_V\frac{1}{2}\lambda_\gamma-\frac{1}{2}}. \quad (3)$$

All other amplitudes can be obtained from the symmetry relations given by (1 − 2). The HARs extracted in the present analysis are defined for $n = 1, 2$ as $t^{(n)}_{\lambda_V\lambda_\gamma} = T^{(n)}_{\lambda_V\lambda_\gamma}/T^{(1)}_{00}$, $u^{(n)}_{\lambda_V\lambda_\gamma} = U^{(n)}_{\lambda_V\lambda_\gamma}/T^{(1)}_{00}$.

3 Experiment and data analysis

3.1 Experiment

The data were accumulated with the HERMES spectrometer using the 27.6 GeV longitudinally polarized electron or positron beam of HERA and a gaseous hydrogen target. The HERMES setup included a forward spectrometer described in [8], in which the scattered lepton and the produced hadrons were detected within an angular acceptance of ±170 mrad horizontally and ±(40−140) mrad vertically. The tracking system had a momentum resolution of about 1.5% and an angular resolution of about 1 mrad. Lepton identification was accomplished using a transition-radiation detector, a preshower scintillator counter, and an electromagnetic calorimeter. The particle-identification system included also a dual-radiator ring-imaging Cherenkov detector [9] to identify hadrons. Combining the responses of the detectors in a likelihood method leads to an average lepton-identification efficiency of 98%, with a hadron contamination of less than 1%. The helicity of the beam was typically reversed every two months. The beam polarization was continuously measured by two Compton polarimeters [10, 11]. The average value of the beam polarization for the events used in the analysis is ±0.30 with a 0.03 uncertainty. A small part of the data were collected with an unpolarized target and the main part with a transversely polarized target [12], for which the polarization direction was reversed every 60 s to 180 s. The measured mean value of the target polarization is $\langle|P_T|\rangle = 0.72 \pm 0.06$ [13].

3.2 Data analysis

The ρ^0 mesons are produced and decay in the following exclusive reactions: $e + p \to e + p + \rho^0$, and $\rho^0 \to \pi^+ + \pi^-$. The event sample used in this analysis is practically the same as that used in [13]. The most important improvement is the application of a new tracking algorithm, which is based on a Kalman filter [14]. For the present analysis, the data are required to fulfill the following criteria: i) The scattered lepton has to have an energy larger than 3.5 GeV in order to not introduce effects from varying trigger thresholds. ii) The longitudinal beam polarization is restricted to the interval $15\% < |P_B| < 80\%$. iii) Events with exactly two oppositely charged hadrons and one lepton with the same charge as the beam lepton are selected. All tracks are required to originate from the same vertex. iv) No calorimeter clusters were observed (no π^0 is detected). v) The two-pion invariant mass is required to obey 0.6 GeV$\leq$ M($\pi^+\pi^-$)$\leq$ 1.0 GeV. vi) The recoiling proton was not detected, but instead reconstructed through the missing energy. Taking into account the spectrometer resolution, the missing energy ΔE has to lie in the interval -1.0 GeV $< \Delta E <$ 0.8 GeV. Here, $\Delta E = \frac{M_X^2 - M_p^2}{2M_p}$, with M_p being the proton mass and $M_X^2 = (p + q - p_{\pi^+} - p_{\pi^-})^2$ the missing mass squared, where p, q, p_{π^+}, and p_{π^-} are the four-momenta of target nucleon, virtual photon, and each of the two pions respectively. The distribution of missing energy ΔE, shown in [13], exhibits a clearly visible exclusive peak. vii) The kinematic constraints $Q^2 >$ 1.0 GeV^2, 6.3 GeV $> W >$ 3 GeV, $-t' <$ 0.4 GeV^2 are applied.

After application of all these requirements, the data sample contains 8741 events. These data are referred to in the following as data in the "entire kinematic region". The applied constraints do not fully suppress the background contribution. The exclusive sample contains contributions from double-diffractive processes, which should be negligible in the low ΔE region, from non-resonant $\pi^+\pi^-$ pair production, which is of the order of $1 - 2\%$ [13], from Δ-isobar excitation, which contribution in the HERMES acceptance is less than 7% [15] and from semi-inclusive deep-inelastic scattering (SIDIS) events. The presented results are not corrected for the former three processes, while a correction is applied for SIDIS background. The shaded histogram of missing energy ΔE (presented in [13]) shows SIDIS background obtained from a PYTHIA Monte Carlo (MC) simulation. It is normalized to the data in the region of 2 GeV $< \Delta E <$ 20 GeV. The simulation is used to determine the fraction of SIDIS background under the exclusive peak. This fraction f_{bg} increases from 7% to 23% for increasing $-t'$ and amounts on average to 11%. A fit of the angular distribution of the SIDIS MC events under the exclusive peak is performed. The obtained function W_{bg} having no free parameters and the angular distribution W_ρ of the exclusive ρ^0 production are combined to give the total angular distribution $W_{tot} = (1 - f_{bg})W_\rho + f_{bg}W_{bg}$. The function W_{tot} is used in an unbinned maximum-likelihood method, the detector properties are taken into account with using MC simulation. The contribution of background from SIDIS events to the obtained amplitude ratios is considered as one of the main sources of systematic uncertainty. This uncertainty is calculated as difference between HARs obtained for $f_{bg} = 0$ and HARs extracted when the background corrections are taken into account.

4 Results

The HARs obtained from the 25-parameter fit in the entire kinematic region ($\langle W \rangle$ = 4.73 GeV, $\langle Q^2 \rangle$ = 1.93 GeV^2, $\langle -t' \rangle$ = 0.132 GeV^2) are shown in figure 1. While the phase of $u_{11}^{(1)}$ is fixed according to the results of [16, 17], its modulus is fit so that the two crosses represent the results of fitting one free parameter. The value of Im[$t_{11}^{(1)}$] (open square) represents the result of [5]; its error bar shows the total uncertainty. The shadowed area corresponds to results obtained here, but also obtained in [5], while all other points are calculated for the first time. As is seen from figure 1, all HARs except $t_{11}^{(1)}$, $t_{01}^{(1)}$, and $u_{11}^{(1)}$, are compatible with zero within their total uncertainties. The HAR $t_{01}^{(1)}$ is responsible for the

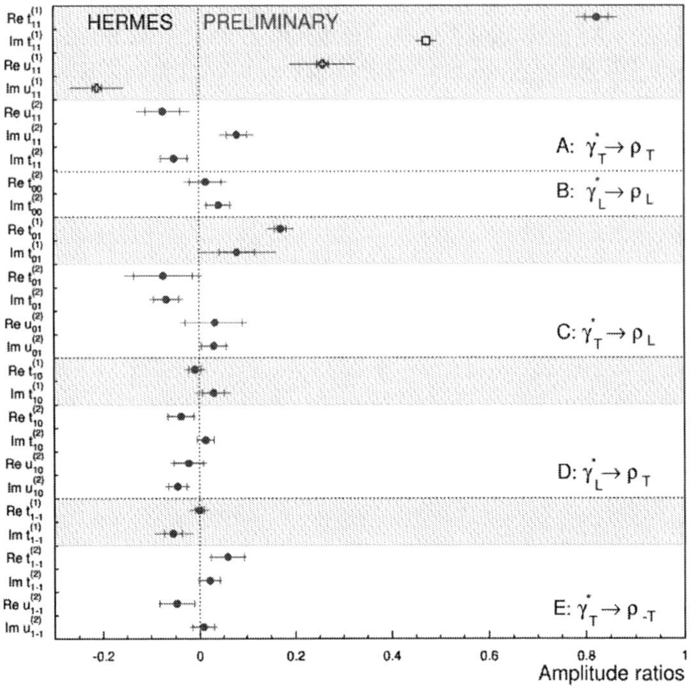

Figure 1. Helicity-amplitude ratios in the entire kinematic region. The phase of $u_{11}^{(1)}$ (open crosses) is fixed according to the results of [16, 17], its modulus is fit. The value of Im$[t_{11}^{(1)}]$ (open square) represents the result of [5]; its error bar shows the total uncertainty. For all other points the inner error bars represent the statistical uncertainty, while the outer ones represent statistical and systematic uncertainties added in quadrature.

violation of the s-channel helicity conservation approximation, while $u_{11}^{(1)}$ shows the role of UPE in ρ^0-meson production. Note that even the HARs compatible with zero are informative and can be used to fix parameters of theoretical models (see, e.g., the paper by S.V. Goloskokov in these proceedings).

A comparison of the SDMEs directly obtained from the HERMES data in [7] and [13] to those calculated with the HARs extracted in the present analysis is shown in figures 2-4. The correlation matrix for the 25 parameters is taken into account for the calculation of the statistical uncertainties of the SDMEs in the Diehl representation [6] $u_{\lambda_\gamma\lambda'_\gamma}^{\lambda_V\lambda'_V}$, $n_{\lambda_\gamma\lambda'_\gamma}^{\lambda_V\lambda'_V}$, and $s_{\lambda_\gamma\lambda'_\gamma}^{\lambda_V\lambda'_V}$ obtained from the HARs. The SDMEs $n_{\lambda_\gamma\lambda'_\gamma}^{\lambda_V\lambda'_V}$ and $s_{\lambda_\gamma\lambda'_\gamma}^{\lambda_V\lambda'_V}$, presented in figures 3 and 4, have linear contributions of the small HARs $t_{\lambda_V\lambda_\gamma}^{(2)}$ and $u_{\lambda_V\lambda_\gamma}^{(2)}$, respectively. These SDMEs can only be extracted from measurements with a transversely polarized target so that the nucleon-helicity-flip amplitude ratios $t_{\lambda_V\lambda_\gamma}^{(2)}$ and $u_{\lambda_V\lambda_\gamma}^{(2)}$ are extracted in this paper for the first time. The total uncertainty is the sum in quadrature of the statistical and the total systematic uncertainties. The SDMEs in figures 3-4 are reordered according to the SDME classes proposed in [7, 13]. Those SDMEs that can be extracted only from data taken with a longitudinally polarized lepton beam are shown in shaded areas.

Figure 2 shows that for each SDME $u_{\lambda_\gamma\lambda'_\gamma}^{\lambda_V\lambda'_V}$ determined from our present results for HARs, there exists an SDME $u_{\lambda_\gamma\lambda'_\gamma}^{\lambda_V\lambda'_V}$ published in [7]. However, figures 3 and 4 show that for some of the calculated SDMEs $n_{\lambda_\gamma\lambda'_\gamma}^{\lambda_V\lambda'_V}$ and $s_{\lambda_\gamma\lambda'_\gamma}^{\lambda_V\lambda'_V}$ no published results from [13] exist, because the beam polarization was not exploited in the analysis performed in [13]. While in [7] and [13] a total of 53 SDMEs could be extracted, the amplitude method presented here allows for the calculation of 71 SDMEs based on the

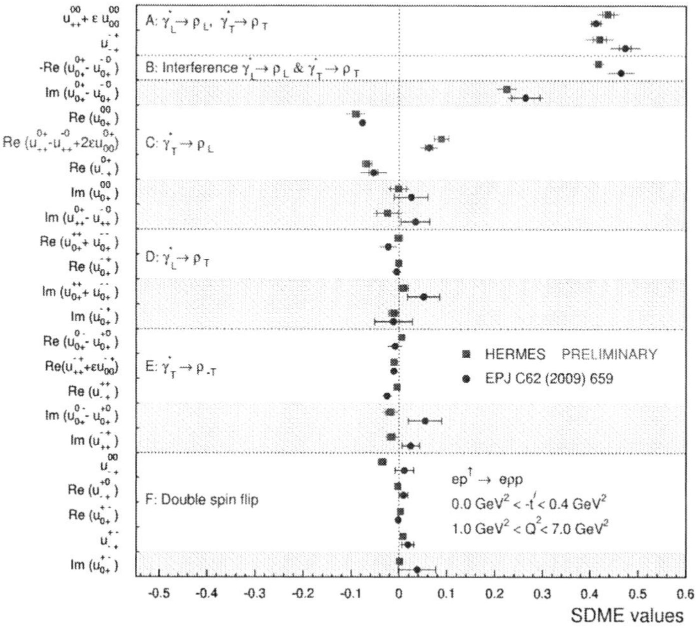

Figure 2. Comparison of the Diehl SDMEs [6] $u^{\lambda_V\lambda'_V}_{\lambda_\gamma\lambda'_\gamma}$ calculated from the helicity-amplitude ratios (red squares) and the SDMEs (blue circles) directly extracted in [7] in the entire kinematic region. For the first case, a 25-parameter fit is used. The points in the shaded area show SDMEs that can be obtained only if the beam is longitudinally polarized. The inner (outer) error bars represent the statistical (total) uncertainty.

extraction of 25 parameters. As seen from the figures, there is reasonable agreement between SDMEs directly extracted in [7, 13] and those calculated from the HARs.

References

[1] M. Diehl, Phys. Rep. **388**, 41 (2003)
[2] J.C. Collins, L. Frankfurt, M.S. Strikman, Phys. Rev. D **56**, 2982 (1997)
[3] S.V. Goloskokov and P. Kroll, Eur. Phys. J. C **53**, 367 (2008)
[4] S.V. Goloskokov and P. Kroll, Eur. Phys. J. A **50**, 146 (2014)
[5] A. Airapetian et al. (HERMES Collaboration), Eur. Phys. J. C **71**, 1609 (2011)
[6] M. Diehl, JHEP **0709**, 064 (2007)
[7] A. Airapetian et al. (HERMES Collaboration), Eur. Phys. J. C **62**, 659 (2009)
[8] K. Ackerstaff et al. (HERMES Collaboration), Nucl. Instr. and Meth. A **417**, 230 (1998)
[9] N. Akopov et al. (HERMES Collaboration), Nucl. Instr. and Meth. A **479**, 511 (2002)
[10] D.P. Barber et al. (HERMES Collaboration), Nucl. Instr. and Meth. A **329**, 79 (1993)
[11] M. Beckmann et al. (HERMES Collaboration), Nucl. Instr. and Meth. A **479**, 334 (2002)
[12] A. Airapetian et al. (HERMES Collaboration), Nucl. Instr. and Meth. A **540**, 68 (2005)
[13] A. Airapetian et al. (HERMES Collaboration), Phys. Lett. B **679**, 100 (2009)
[14] R. Fruhwirth, Nucl. Instr. and Meth. A **262**, 444 (1987)
[15] M. Tytgat, DESY-THESIS-2001-014 (2001)
[16] A. Airapetian et al. (HERMES Collaboration), Phys. Lett. B **513**, 301 (2001)
[17] A. Airapetian et al. (HERMES Collaboration), Eur. Phys. J. C **29**, 171 (2003)

DOI: 10.1051/epjconf/201713808005

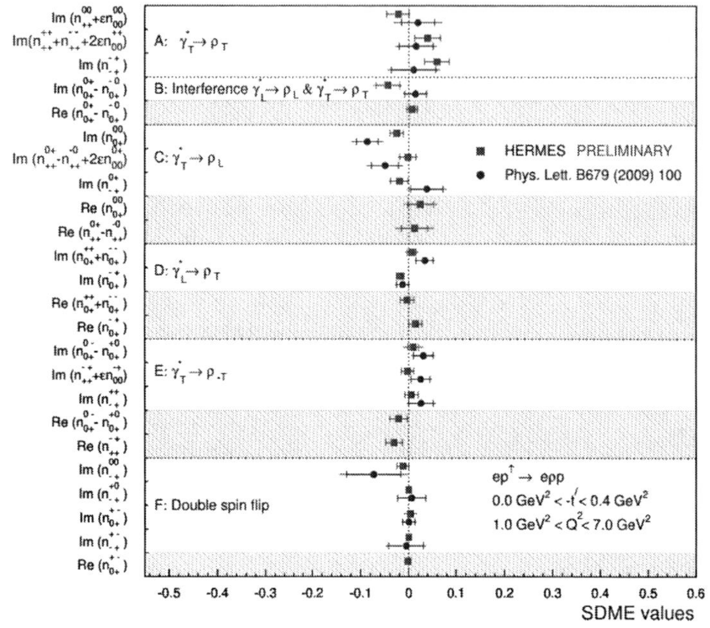

Figure 3. Comparison of the Diehl SDMEs [6] $n^{\lambda_V \lambda'_V}_{\lambda_\gamma \lambda'_\gamma}$ calculated from the helicity-amplitude ratios (red squares) and the SDMEs (blue circles) directly extracted in [13] in the entire kinematic region. For the first case, a 25-parameter fit is used. The points in the shaded area show SDMEs that can be obtained only if the beam is longitudinally polarized. The inner (outer) error bars represent the statistical (total) uncertainty.

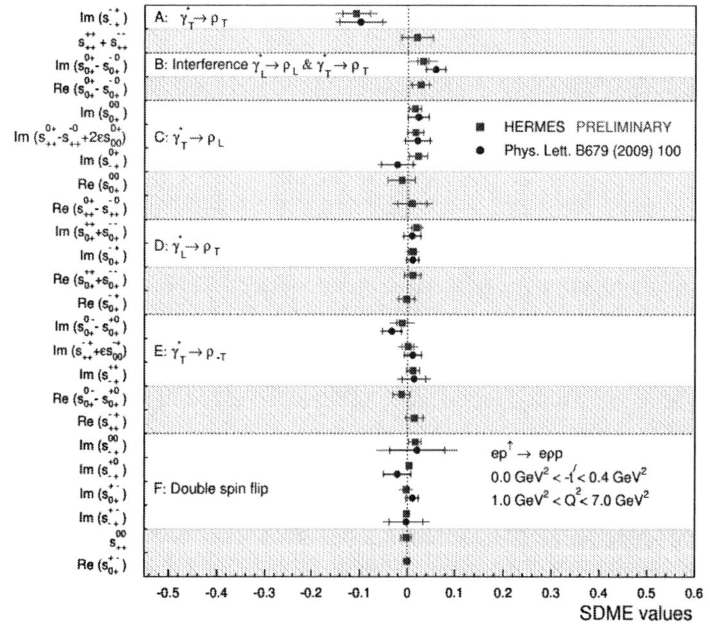

Figure 4. Comparison of the Diehl SDMEs [6] $s^{\lambda_V \lambda'_V}_{\lambda_\gamma \lambda'_\gamma}$ calculated from the helicity-amplitude ratios (red squares) and the SDMEs (blue circles) directly extracted in [13] in the entire kinematic region. For the first case, a 25-parameter fit is used. The points in the shaded area show SDMEs that can be obtained only if the beam is longitudinally polarized. The inner (outer) error bars represent the statistical (total) uncertainty.

Amplitude ratios in ρ^0 leptoproductions and GPDs

S.V. Goloskokov[1,★]

[1] *Bogoliubov Laboratory of Theoretical Physics, Joint Institute for Nuclear Research, Dubna 141980, Moscow region, Russia*

Abstract. We investigate exclusive leptoproduction of ρ^0 meson. These reactions were analyzed within the factorizing handbag approach. In our model good agreement of observables for light meson production with experimental data in a wide energy range was found.
Using the model results we calculate the ratio of different helicity amplitudes for a transversely polarized proton target to the leading twist longitudinal amplitude. Our results are close to the amplitude ratios measured by HERMES.

1 Introduction

In this report, investigation of ρ^0 meson leptoproduction is based on the handbag approach. Here the amplitudes at high Q^2 factorize into hard meson electroproduction off partons and the Generalized Parton Distributions (GPDs) [1]. Good agreement of our results for the cross sections and spin observables, expressed in terms of GPDs H, E, with experimental data were obtained in [2, 3]. We consider transversity effects H_T, $\bar{E}_T$, that have a twist-3 character [4]. This gives us a possibility to describe spin observables that are equal to zero without twist-3 contributions. Essential contributions from unnatural parity exchanges were found by HERMES [5] in the ω production. It was shown that the pion pole (PP) contributions [6] are significant for explanation of the large unnatural-parity effects at HERMES in ω and ρ^0 production. We discuss the amplitude properties and physical observables in section 2.

In section 3, using the model results [2–4, 6], we calculate the ratio of different helicity amplitudes to the leading twist longitudinal amplitude and compare the results with preliminary HERMES data [7].

2 ρ^0 meson leptoproduction. Physical observables

We can define Natural and Unnatural parity NP (UP) amplitudes as

$$T(U)_{\lambda\nu',\mu\nu} = \frac{1}{2}\,[M_{\lambda\nu',\mu\nu} \pm (-1)^{\mu-\lambda}\, M_{-\lambda\nu',-\mu\nu}]. \qquad (1)$$

★e-mail: goloskkv@theor.jinr.ru

Here ν, ν' are the initial and final proton helicities and μ, λ are the helicities of the photon and final ρ^0 meson. In what follows, according to [8], we use the notations for the proton spin-non flip and spin-flip amplitudes, respectively,

$$T(U)^{(1)}_{\lambda,\mu} = N(U)_{\lambda+,\mu+}; \quad T(U)^{(2)}_{\lambda,\mu} = N(U)_{\lambda-,\mu+}. \tag{2}$$

We calculate ρ^0 leptoproduction off proton within the handbag approach where the leading amplitude at high photon virtuality Q^2 can be represented in a factorized form [1] as a convolution of a hard meson subprocess amplitude off partons $\mathcal{H}^a_{\mu'+,\mu+}$, which is calculated perturbatively, and GPDs as

$$T^{(1)}_{\lambda\mu} \propto \int_{-1}^{1} d\overline{x}\,\mathcal{H}^a_{\lambda+,\mu+} F^a(\overline{x},\xi,t); \; T^{(2)}_{\lambda\mu} \propto -\frac{\sqrt{-t'}}{2m} \int_{-1}^{1} d\overline{x}\,\mathcal{H}^a_{\lambda+,\mu+} E^a(\overline{x},\xi,t). \tag{3}$$

Here a is a flavor factor. Generally, factorization is not valid for the not leading amplitudes. These problems can be solved in our model [2, 3] where subprocesses are calculated within the modified perturbative approach [9] in which quark transverse degrees of freedom accompanied by Sudakov suppressions are considered. The quark transverse momentum regularizes the end-point singularities in the TT amplitude thus it can be calculated.

GPDs contain broad information on the hadron structure. With the help of sum rules GPDs are connected with the hadron form factors, and information on the parton angular momenta can be extracted. In the forward limit $t = 0$ and zero skewness $\xi = 0$ GPDs are equivalent to ordinary Parton Distribution Functions (PDFs). The GPDs are estimated by us using the double distribution representation [10], which connects GPDs with PDFs through the double distribution function ω. For the valence quark contribution ω looks like

$$\omega_i(x,y,t) = h_i(x,t)\,\frac{3}{4}\,\frac{[(1-|x|)^2 - y^2]}{(1-|x|)^3}. \tag{4}$$

The functions h in (4) are parameterized as PDFs in the form

$$h(x,t) = N\,e^{b_0 t}\,x^{-\alpha(t)}\,(1-x)^n, \tag{5}$$

with the t- dependence which is considered in a Regge form, and $\alpha(t)$ is the corresponding Regge trajectory. The parameters for PDFs h in (5) are obtained from the analysis [11]; information about PDFs e is taken from [12]. The handbag approach describes successfully the light meson leptoproduction at HERMES, COMPASS and HERA energies [2, 3]. It can be seen from figure 1, (left) that we describe properly the energy dependence of the ρ^0 cross section from HERMES to HERA energies. Here GPDs H are essential. At $W < 5$GeV the cross section grows unexpectedly. This effect is not understood.

In figure 1, (right) we show the $A_{UT}^{\sin(\phi-\phi_s)}$ asymmetry for ρ^0 production at COMPASS energy which is determined mainly by the interference of GPDs H, E

$$A_{UT}^{\sin(\phi-\phi_s)} \sim \mathrm{Im}[< E >^* < H >]. \tag{6}$$

We describe properly t dependence of $A_{UT}^{\sin(\phi-\phi_s)}$ asymmetry at COMPASS.

Unfortunately, some asymmetries and Spin Density Matrix Elements (SPMEs), which are not small experimentally, are equal to zero in the leading-twist approximation of our GPD model. To describe the experimental data on the meson electroproduction at low Q^2, we consider the amplitudes $T_{01}^{(1,2)}$, which are determined in terms of the transversity GPDs H_T and $\bar{E}_T$. Within the handbag

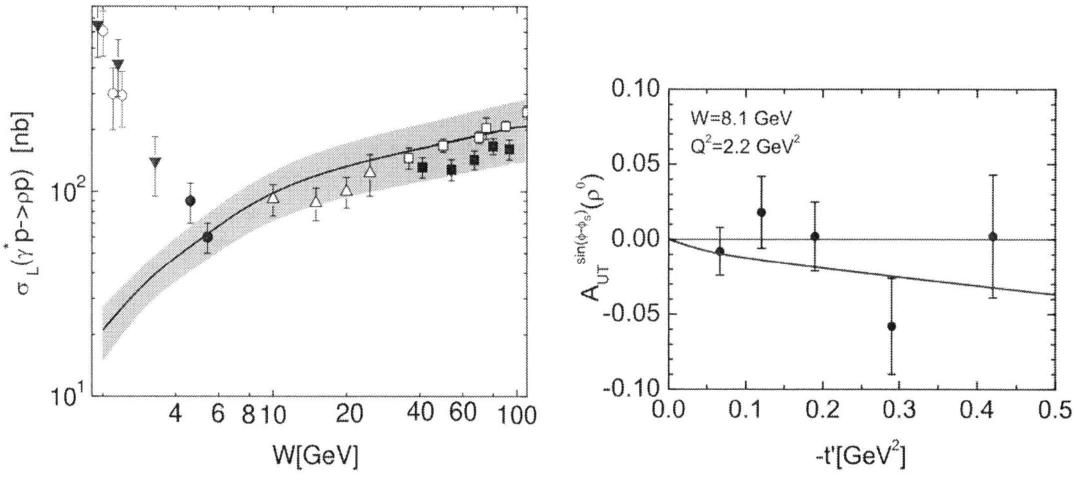

Figure 1. Left: Longitudinal ρ^0 cross section at $Q^2 = 4.0\,\mathrm{GeV}^2$. HERMES (solid circle), ZEUS (open square), H1 (solid square), E665 (open triangle), open circles- CLAS, CORNEL -solid triangle. **Right**: Model results for $A_{UT}^{\sin(\phi-\phi_s)}$ asymmetry for ρ^0 production with COMPASS data [13].

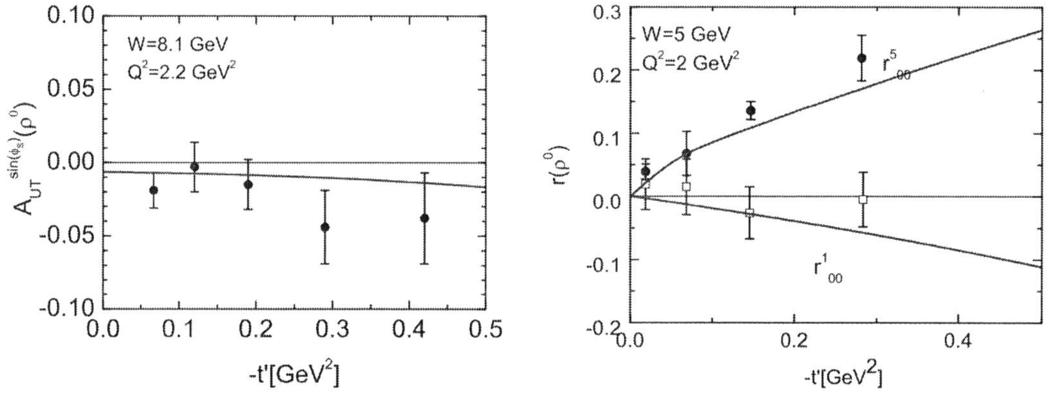

Figure 2. Left: Model results for $A_{UT}^{\sin(\phi_s)}$ asymmetry of ρ^0 production with COMPASS data [13]. **Right**: $\bar{E}_T$ effects in SDMEs of ρ^0 production. HERMES data are shown [14].

approach the transversity GPDs are accompanied by a twist-3 meson wave function in the hard subprocess amplitude $\mathcal{H}$ [4], which is the same for both the $M_{01}^{(1,2)}$ amplitudes:

$$T_{01}^{(1)} \propto \frac{\sqrt{-t'}}{4m} \int_{-1}^{1} d\bar{x}\mathcal{H}_{0-,++}\, \bar{E}_T(\bar{x},\xi,t);\; M_{01}^{(2)} \propto \int_{-1}^{1} d\bar{x}\mathcal{H}_{0-,++}\, H_T(\bar{x},\xi,t). \tag{7}$$

The H_T GPDs in the forward limit and $\xi = 0$ are equal to transversity PDFs δ and are parameterized by using the model [15]. Information on $\bar{E}_T$ is obtained now only from the lattice QCD [16]. We estimate the corresponding $\bar{e}_T$ PDFs from the lattice results using form (5). The double distribution is used to calculate transversity GPDs as before. Note that the amplitude $M_{01}^{(2)}$ has no definite parity. Really, $M_{01}^{(2)} = M_{0-++} = -T_{01}^{(2)} + U_{01}^{(2)}$ and $M_{0--+} = -T_{01}^{(2)} - U_{01}^{(2)}$. The last amplitude is equal to zero in the model and $T^{(2)}_{01} = -U^{(2)}_{01}$. Thus NP and UP contributions to the $M^{(2)}_{01}$ amplitude have the same value.

In figure 2, we show transversity effects in spin observables. In figure 2, (left) the $A_{UT}^{\sin(\phi_s)}$ asymmetry is presented together with COMPASS data. This asymmetry is determined by the H, H_T interference

$$A_{UT}^{\sin(\phi_s)} \sim \text{Im}[< H_T >^* < H >]. \tag{8}$$

The data are described quite well by the model results. In figure 2, (right) the SDMEs r^5_{00} and r^1_{00} are shown. The first SDME is determined by the H, $\bar{E}_T$ interference and the second one by the $\bar{E}_T$ contribution only

$$r^5_{00} \sim \text{Re}[< H >^* < \bar{E}_T >]; \quad r^1_{00} \sim - < \bar{E}_T > |^2. \tag{9}$$

We find that we describe properly both SDMEs at HERMES.

Now we shall discuss UP effects. The UP contribution to the $U^{(1)}_{11}$ amplitude is determined by the polarized $\tilde{H}$ quarks GPDs and can be observed in the A_{LL} asymmetry [3]. It was found that the $\tilde{H}$ effects provide the A_{LL} asymmetry that is a little bit smaller with respect to experiment [3].

The HERMES data for the ω production indicate the strong contributions from UP effects [5]. This can be seen, e.g., from the ratio of the unnatural and natural parity cross sections, which was found to be larger than unity. This effect can be caused [6] by the large PP contribution to this process.

The UP helicity amplitudes determined by PP contribution to ρ^0 leptoproduction looks as follows [6]. The dominant contributions are:

$$U^{(1)}_{11} \sim \frac{\rho_{\pi\rho}}{t - m_\pi^2} \frac{m \xi Q^2}{\sqrt{1-\xi^2}}, \quad U^{(2)}_{11} \sim -\frac{\rho_{\pi\rho}}{t - m_\pi^2} \frac{\sqrt{-t'}\, Q^2}{2}. \tag{10}$$

Here $\rho_{\pi\rho}$ is proportional to the $\pi\rho$ transition form factor $g_{\pi\rho}$. It can be determined from the ρ^0 meson radiative decay. For ρ^0 production the absolute value of the $\pi\rho$ transition form factor (FF) at zero photon virtuality can be estimated as [6]

$$\Gamma(\rho \to \pi\gamma) \sim \frac{\alpha_{elm}}{24} |g_{\pi\rho}(0)|^2 M_V^3; \quad |g_{\pi\rho}(0)| = .85\text{GeV}^{-1}. \tag{11}$$

This value of $g_{\pi\rho}$ was used in [6] in analysis of PP effects in ρ^0 production.

3 Amplitude ratios of ρ^0 meson leptoproduction.

Our GPD model with NP amplitudes determined by H, E, H_T, $\bar{E}_T$ GPDs and UP effects caused by $\tilde{H}$ and PP contributions describes well physical observables [2–4, 6]. However, observables in terms of GPDs have usually a quite complicated form. More direct information about GPDs can be found from the ratio of different helicity amplitudes on transversely polarized proton target to the leading twist longitudinal amplitude

$$t^{(i)}_{\lambda\mu} = T^{(i)}_{\lambda\mu}/T_{00}, \quad u^{(i)}_{\lambda\mu} = U^{(i)}_{\lambda\mu}/T_{00}, \tag{12}$$

which can be calculated using the model results [2–4, 6].

To be consistent with the HERMES analysis, we study the ratios $t(u)^{(i)}_{\lambda\mu}$ in the handbag approach using the HERMES definitions of the amplitude signs. The UP amplitudes (10) are dependent on the $g_{\pi\rho}$ sign. The model results for positive $g_{\pi\rho}$ transition FF (squares) and negative FF (triangles) together with HERMES data are shown in figure 3. We present here only the amplitude ratios which are not zero in the model.

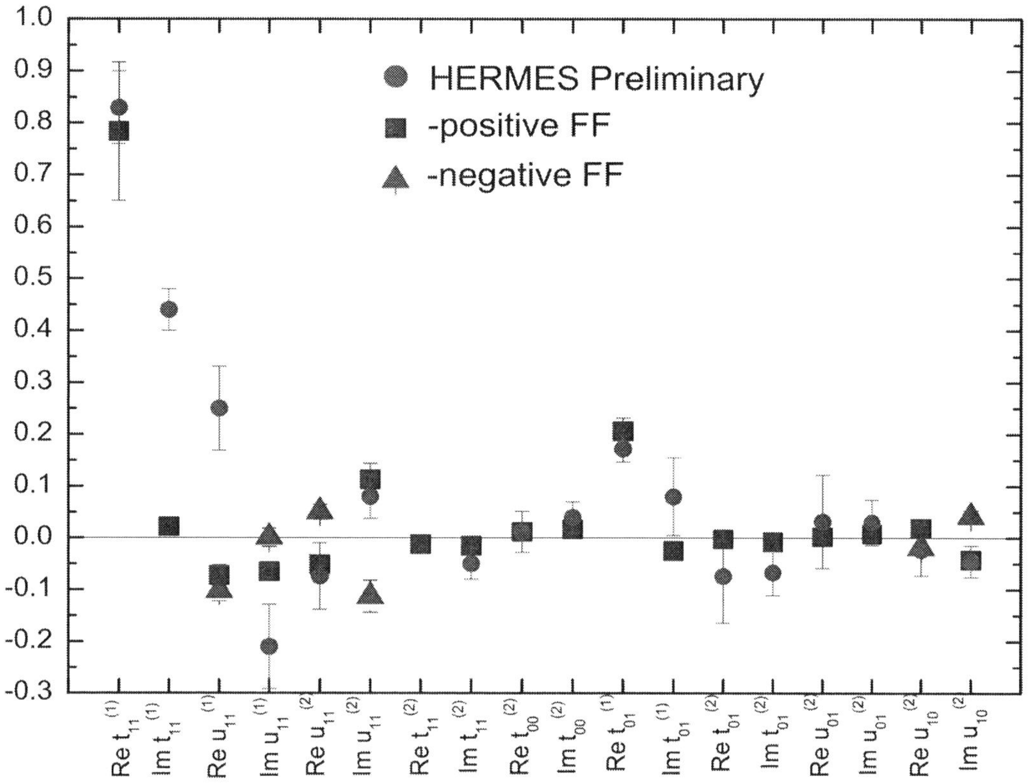

Figure 3. Model results for amplitude ratios of ρ^0 meson leptoproduction for positive and negative $\pi\rho$ transition FF together with HERMES preliminary data [7].

- t^1_{11} is the ratio of transverse to longitudinal amplitudes dominated by H GPDs. The amplitudes are mainly imaginary that give large Ret^1_{11}, which is consistent with data. Imt^1_{11} is rather small in the model. That is caused by the small relative phase between transverse to longitudinal amplitudes [3].
- Small value of u^1_{11} can be related to not so large value of $\tilde{H}$ and the corresponding PP contribution. See the discussion of the A_{LL} asymmetry.
- u^2_{11} is determined by PP effects. We have found a good result for the positive $\pi\rho$ transition FF.
- t^2_{11} is connected with the E GPDs contribution to the proton spin-flip amplitude for the transversely polarized photon and meson. The result is good for the imaginary part of the ratio. There are no experimental data for the real part.
- t^2_{00}- is determined by the E GPDs contribution to the longitudinal amplitude. The model results are not far from experiment.
- t^1_{01} is connected with the twist-3 transversity $\bar{E}_T$ effects. Our results are consistent with HERMES data.
- t^2_{01} and u^2_{01} are determined by the twist-3 transversity H_T effects. It was mentioned that we have connection $t^2_{01} = -u^2_{01}$ for these amplitudes. The amplitude ratios are in agreement with the model results, but they may be a little bit larger.

- u^2_{10}- is determined by PP contribution. The obtained results are consistent with experiment for the positive $\pi\rho$ transition FF.

We have analyzed meson electroproduction within the handbag approach. Modified perturbative approach was used to calculate the hard subprocess amplitude. GPDs were calculated using PDFs on the basis of the double distribution representation. Good description of different spin observables was found in the model [2–4, 6]. The ratio of different helicity amplitudes on the transversely polarized proton target to the leading twist longitudinal amplitude was calculated. The PP contribution to the UP amplitude ratios is dependent on the sign of the $\pi\rho$ transition FF. The model results for the positive sign of the $\pi\rho$ transition FF are compatible with HERMES data. For the negative sign of FF the results are worse. We can conclude that the positive sign of transition FF is preferable. This is consistent with conclusion done in HERMES and COMPASS papers [5, 17].

The work was supported in part by the Heisenberg-Landau program.

References

[1] X. Ji, Phys. Rev. D **55**, 7114 (1997).
A.V. Radyushkin, Phys. Lett. B **380**, 417 (1996).
J.C. Collins et al., Phys. Rev. D **56**, 2982 (1997)
[2] S.V. Goloskokov, P. Kroll, Euro. Phys. J. C **50**, 829 (2007)
[3] S.V. Goloskokov, P. Kroll, Euro. Phys. J. C **53**, 367 (2008)
[4] S.V. Goloskokov, P. Kroll, Euro. Phys. J. A **47**, 112 (2011)
[5] A. Airapetian et al. (HERMES Collab.), Euro. Phys. J. C **74**, 3110 (2014)
[6] S.V. Goloskokov, P. Kroll, Euro. Phys. J. A **50**, 146 (2014)
[7] S. Manaenkov (HERMES Collab.), Conference: C16-04-11 Proceedings. e-Print: arXiv:1607.02344 [hep-ph] (2016)
[8] A. Airapetian et al. (HERMES Collab.), Eur.Phys.J. C **71**, 1609 (2011)
[9] J. Botts, Sterman, Nucl. Phys. B **325**, 62 (1989)
[10] I.V. Musatov, A.V. Radyushkin, Phys. Rev. D **61**, 074027 (2000)
[11] J. Pumplin et al., JHEP **0207**, 012 (2002)
[12] M. Diehl, T. Feldmann, R. Jakob, P. Kroll, Euro. Phys. J. C **39**, 1 (2005).
M. Diehl, P. Kroll, Euro. Phys. J. C **73**, 2397 (2013)
[13] C. Adolph, et al. (COMPASS Collab.), Phys.Lett. B **731**, 19 (2014)
[14] A. Airapetian et al. (HERMES Collab.), Euro. Phys. J. C **62**, 659 (2009)
[15] M. Anselmino, M. Boglione, U. D'Alesio, A. Kotzinian, F. Murgia, A. Prokudin, S. Melis, Nucl. Phys. Proc. Suppl. **191**, 98 (2009)
[16] M. Gockeler et al. (QCDSF Collab. and UKQCD Collab.), Phys. Rev. Lett. **98**, 222001 (2007)
[17] C. Adolph et al. (COMPASS Collab.), e-Print: arXiv:1606.03725 [hep-ex] (2016)

Delta excitation in dp-elastic scattering at intermediate energies

Nadezhda Ladygina[1,*]

[1] *Joint Institute for Nuclear Research*

Abstract. In this report the deuteron-proton elastic scattering reaction is studied in the range of the deuteron kinetic energy from 500 MeV to 1300 MeV. dp-elastic scattering is considered in the relativistic multiple scattering expansion framework since the Faddeev calculation technique is not applicable at these energies. We start from the AGS-equations and iterate them up to a second-order of the nucleon-nucleon interaction. The four contributions are included into consideration: one-nucleon exchange (ONE), single scattering, double scattering, and the term corresponding the delta excitation in the intermediate state. The calculations are performed for four deuteron energies: 500, 880, 1200, and 1300 MeV. The results are presented in comparison with the experimental data on the dp-elastic scattering cross section.

1 Introduction

Elastic deuteron-proton scattering is the simplest example of the hadron nucleus collision, because a deuteron is the simplest nucleus containing only one proton and one neutron. In the previous papers [1],[2] we considered this reaction in the multiple scattering expansion formalism. Three contributions, one-nucleon-exchange (ONE), single-scattering (SS), and double scattering (DS), were taken into account. We got a reasonable agreement for the differential cross sections between the theoretical predictions and the experimental data almost in a whole angular range. However, the rise of the differential cross sections at the scattering angle larger than 140° was not described in the approach [1],[2].

In 1969 A.Kerman and L.Kisslinger supposed that resonances can play an important role in deuteron-proton backward elastic scattering [3]. Later the double-scattering diagram with Δ-isobar in the intermediate state was taken into account in dp backward scattering. The significant contribution of this diagram to the reaction amplitude was demonstrated in [4]-[6], [7]. However, the double scattering with nucleon in an intermediate state was not considered in these papers. Perhaps, it was the reason why the description of the differential cross sections energy dependence was not good enough.

The effort to take the Δ-isobar into account in order to describe dp-elastic scattering was also done in [8], [9]. In these papers deuteron-proton scattering was considered in a whole angular range, not only at $\theta^* = 180°$. Unfortunately, the process was studied at low energies, $T_d < 200$ MeV, where the Δ-isobar excitation effects are negligible.

In this paper we keep on considering deuteron-proton elastic scattering in the framework offered in papers [1],[2]. Here we add the Δ-isobar contribution in our approach. The detail of the model can be found in [10]. In this paper we present only basic points and obtained results.

*e-mail: nladygina@jinr.ru

 DOI: 10.1051/epjconf/201713808007

2 General formalism

According to the three-body collision theory, the amplitude of deuteron-proton elastic scattering $\mathcal{J}$ is defined by the matrix element of the transition operator U_{11}:

$$U_{dp\to dp} = \delta(E_d + E_p - E'_d - E'_p)\mathcal{J} =< 1(23)|[1 - P_{12} - P_{13}]U_{11}|1(23) > \quad . \tag{1}$$

Here the state $|1(23) >$ corresponds to the configuration, when nucleons 2 and 3 form the deuteron state and nucleon 1 is free. Appearance of the permutation operators for two nucleons P_{ij} reflects the fact that the initial and final states are antisymmetric due to an exchange of the two particles.

Following the Alt-Grassberger-Sandhas formalism [11],[12] we write the transition operator $U_{\beta\alpha}$ as:

$$U_{\beta\alpha} = -(1 - \delta_{\beta\alpha})(H_0 - z) - \sum_{\delta\neq\alpha} U_{\beta\delta}g_0t_\delta. \tag{2}$$

This transition operator depends on the potential through the channel transition-operator t_α of the two-particle subsystems:

$$t_\alpha = V_\alpha - V_\alpha g_\alpha V_\alpha \quad t_\alpha = V_\alpha - V_\alpha g_0 t_\alpha,$$

where g_0 is a free three-particle propagator and g_α is a two-particle resolvent in the three-particle space.

Iterating (2) up to the second order terms of t_i we can present the reaction amplitude as a sum of the following four contributions: one nucleon exchange, single scattering and double scattering with the nucleon and delta in the intermediate state, –

$$\mathcal{J}_{dp\to dp} = \mathcal{J}_{ONE} + \mathcal{J}_{SS} + \mathcal{J}_{DS} + \mathcal{J}_\Delta. \tag{3}$$

Schematically this sequence can be presented by the four diagrams figure (1). The terms in the dp-elastic scattering amplitude $\mathcal{J}$ in (3) correspond to: the one-nucleon-exchange (ONE) figure (1a), the single nucleon-nucleon scattering (SS) figure (1b), the double nucleon-nucleon scattering (DS) figure (1c), and scattering with Δ-isobar in intermedite state figure (1d).

3 Results and discussions.

We applied the presented above model to describe the differential cross sections. The formal definition of the total cross section of deuteron-proton elastic scattering is given by

$$\sigma(dp \to dp) = (2\pi)^4\frac{1}{6} \int \frac{d\vec{P}'_d}{E'_d}\frac{d\vec{p}'}{E'_p}\delta(E_d + E_p - E'_d - E'_p)\delta(\vec{P}_d + \vec{p} - \vec{P}'_d - \vec{p}') \cdot \frac{|T|^2}{\sqrt{(P_dp)^2 - P_d^2p^2}}, \tag{4}$$

where $|T|^2$ is the squared invariant amplitude defined as

$$|T|^2 = |\sqrt{E'_dE'_p}\,\mathcal{J}_{dp\to dp}\,\sqrt{E_dE_p}|^2 \equiv inv. \tag{5}$$

The expression for the differential cross section in the center-of-mass can be obtained through the scattering amplitude in the Breit frame as:

$$\frac{d\sigma}{dcos\theta^*} = \frac{(2\pi)^5}{24s}\left(s - m_N^2 - M_d^2 + \frac{t}{2}\right)^2 |\mathcal{J}^{(Breit)}_{dp\to dp}(s,t)|^2, \tag{6}$$

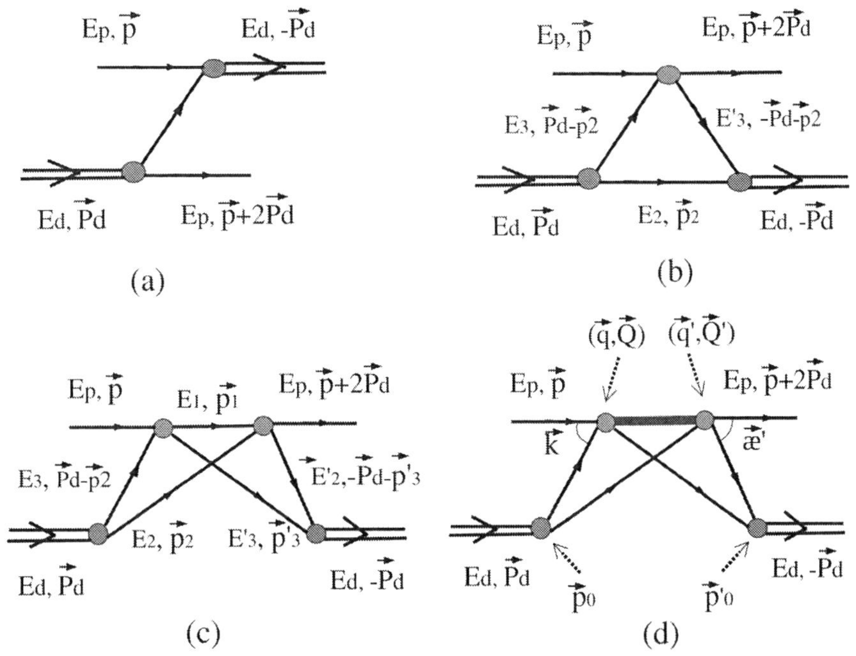

Figure 1. The diagrams included into consideration: (a) the one-nucleon exchange diagram; (b) the single scattering diagram; (c) the double scattering diagram with a nucleon in the intermediate state; (d) the double scattering diagram with Δ-isobar in the intermediate state.

where $s = (P_d + p)^2$ and $t = (p - p')^2$ are invariant Mandelstam variables.

We evaluated the dp-elastic differential cross section at four energies in a wide range. The results are presented for three cases: 1) only one-nucleon exchange (ONE) and single-scattering (SS) diagrams taken into account, 2) the double-scattering (DS) term is added into the consideration, and 3) the Δ-isobar excitation is included.

At first, the deuteron kinetic energy equal to 500 MeV was considered (figure 2). One can see, the SS-diagram plays an important role up to the scattering angle equal to about 60°. The DS-contribution is significant at the scattering angle between 60° and 160°. The difference between the results taking into account the double scattering and without it reaches 2-3 times at $\theta^* = 120°$.

This energy is not high enough for the Δ-isobar manifestation. Therefore, the inclusion of the Δ-contribution term into the scattering amplitude does not have to influence remarkably to the result. In fact, the results obtained with and without the Δ-isobar term are indistinguishable in a whole angular range.

The second energy is equal to 880 MeV (figure 3). One can see, the contribution of the double-scattering is very significant at this energy. The inclusion of the DS-term into consideration allows to describe the behaviour of the differential cross section at the scattering angle range 60° – 140°. The difference between ONE+SS curve and ONE+SS+DS one reaches about 15 times. Note, the Δ excitation begins to manifest itself at the angle equal to about 120° and describes the behaviour of the experimental data at the angle above 140° where the differential cross section sharply increases.

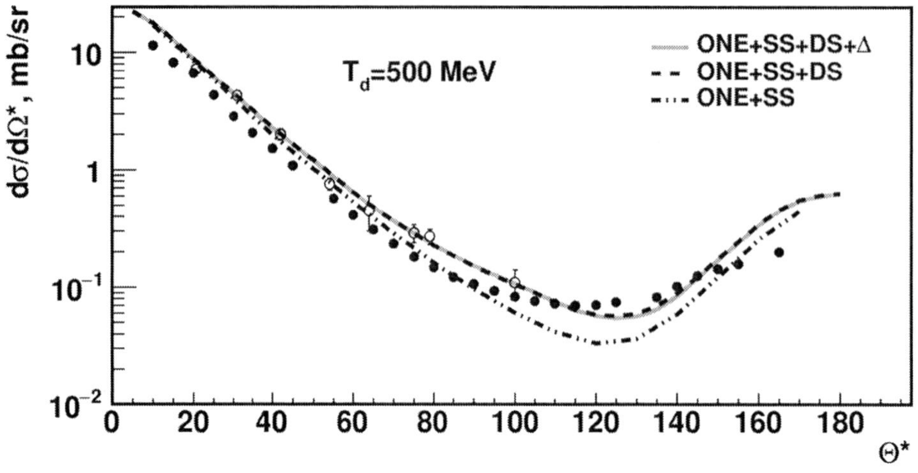

Figure 2. The differential cross section at the deuteron kinetic energy of 500 MeV as a function of the c.m. scattering angle. The data are taken from [13] (○), and [14] (•).

The dp-elastic differential cross sections at energies equal to T_d = 1200 MeV and T_d = 1300 MeV are presented in figure 4 and figure 5, respectively. Here we can observe further enhancement of the double scattering term influence on the result. Also the contribution of the Δ - isobar increases significantly at the scattering angle above 140°.

Analysing the obtained results one can conclude the following. The differential cross sections are described quite well up to the scattering angles equal to 60° taking into account only the one-nucleon-exchange and single-scattering terms. If we consider the dp-elastic scattering at the angles large than 60°, it is necessary to include the double scattering term into consideration. It should be noted the double-scattering contribution into the reaction amplitude increases with the deuteron energy growing and may change the value of the differential cross section on a few orders in comparison with the result obtained without inclusion of the DS-term.

The sharp rise of the differential cross section at $\theta^* \geq 140°$, which is observed in the experimental data, points at the appearance of the addition reaction mechanism. The inclusion of the Δ-excitation term into consideration allowed to describe this rise of the differential cross section at the large scattering angles. The contribution of the Δ-isobar mechanism grows with the initial deuteron energy. It is negligible at T_d = 500 MeV and significant at T_d = 1300 MeV.

In such a way we got a good description of the differential cross section of deuteron-proton elastic scattering in a whole angular range, from 0° to 180° of the scattering angle, in a wide deuteron energy interval, between 500 MeV and 1300 MeV.

Acknowledgements: The author is grateful to Dr. V.P. Ladygin for fruitful discussions and interest in this problem. This work has been supported by the Russian Foundation for Basic Research under grant № No.16-02-00203a.

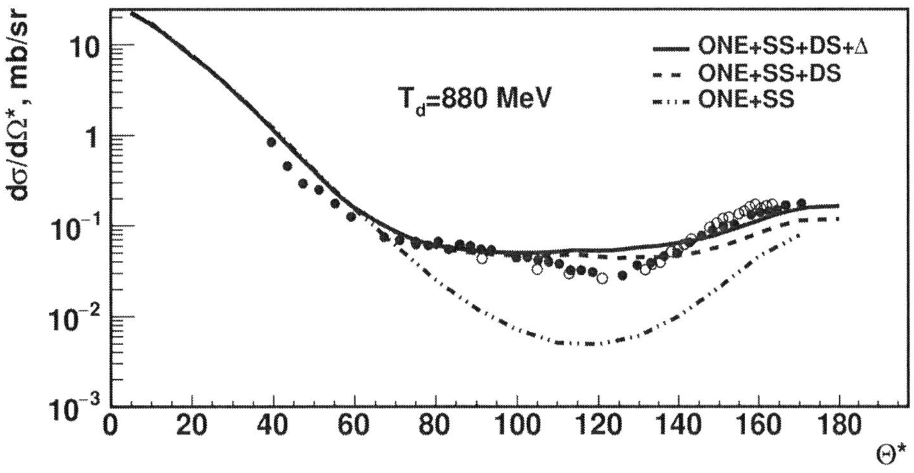

Figure 3. The differential cross section at the deuteron kinetic energy of 880 MeV as a function of the c.m. scattering angle. The data are taken from [15] (•), and [16] (∘).

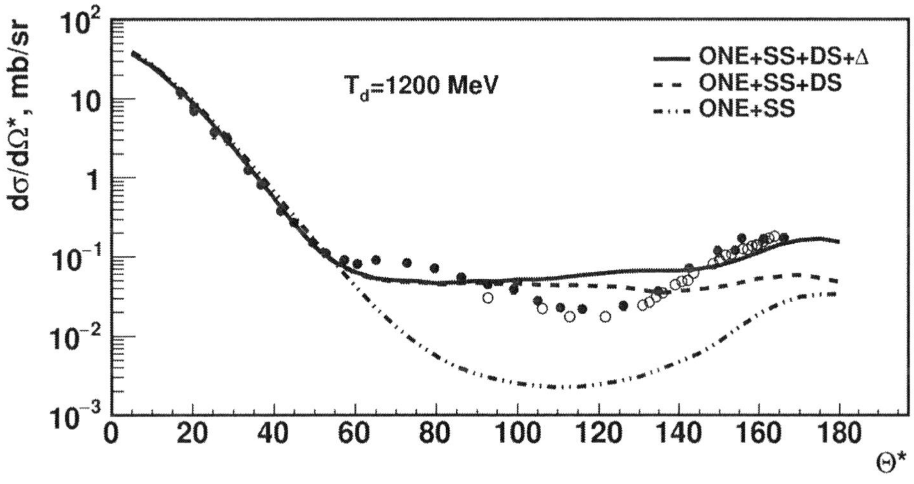

Figure 4. The differential cross section at the deuteron kinetic energy of 1200 MeV as a function of the c.m. scattering angle. The data are taken from [16] (∘), and [17] (•).

References

[1] N.B. Ladygina, Phys. At. Nucl. **71**, 2039 (2008)
[2] N.B. Ladygina, Eur. J. Phys. **A 42**, 91 (2009)
[3] A.K. Kerman, L.S. Kisslinger, Phys. Rev. **180**, 1483 (1969)
[4] L.A. Kondratyuk, F.M. Lev, Sov. J. Nucl. Phys. **26**, 153 (1977)

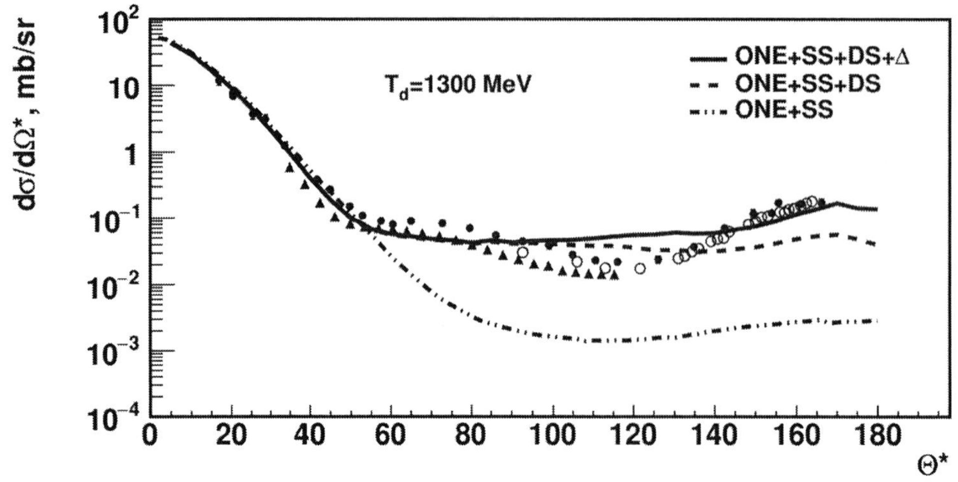

Figure 5. The differential cross section at the deuteron kinetic energy of 1300 MeV as a function of the c.m. scattering angle. The data are taken for T_d = 1200 MeV from T_d = 1180 MeV [16] (○), [17](●), and T_d = 1282 MeV [18] (▲).

[5] L.A. Kondratyuk, F.M. Lev, L.V. Shevchenko, Sov. J. Nucl. Phys. **29**, 558 (1979)
[6] L.A. Kondratyuk, F.M. Lev, L.V. Shevchenko, Yad. Fiz. **33**, 1208 (1981)
[7] Yu.N. Uzikov, Fiz.Elem.Chast.Atom.Yadra **29**, 1405 (1998)
[8] A. Deltuva, K. Chmielewski, P.U. Sauer, Phys. Rev. **C 67**, 034001 (2003)
[9] A. Deltuva, R. Machleidt, P.U. Sauer, Phys. Rev. **C 68**, 024005 (2003)
[10] N.B.Ladygina, Eur. J. Phys. **A 52**, 199 (2016)
[11] E.O. Alt, P. Grassberger, W. Sandhas, Nucl. Phys. **B 2**, 167 (1967)
[12] E. Schmid, H. Ziegelmann, *The Quantum Mechanical Three-Body Problem* (Oxford, Pergamon Press, 1974)
[13] R.D. Shamberger, Phys. Rev. **85**, 424 (1952)
[14] K. Hatanaka et al., Phys. Rev. **C 66**, 044002 (2002)
[15] N.E. Booth et al., Phys. Rev. **D 4**, 1261 (1971)
[16] J.C. Alder et al., Phys. Rev. **C 6**, 2010 (1972)
[17] E.T. Boschitz et al., Phys. Rev. **C 6**, 457 (1972)
[18] E. Gülmez et al., Phys. Rev. **C 42**, 2067 (1991)

Foundation of a unified physics

Ivanhoe B. Pestov

[1] *Joint institute for nuclear research, 141980, Dubna, Moscow Region, Russia*

Abstract. Since searches of new physics is not successful till now we need to have new approaches to the fundamental physics as a whole. In this report we exhibit new system of fundamental equations of physics (derived from the first principle) which gives rise to new physics and provides a basis for comparison with current one.

1 Introduction

To formulate new unified approach and understand why nature is just the way it is we abstract away from the known concepts and laws (a writer must not share your point of view) and put forward the grand relation: fundamental physical ideas, symmetries and equations are tightly connected with the idea of space and, hence, they have the geometrical origin and are formulated in the form which is completely independent of any outer and a priori conditions (everything is the concept of space and the concept of space is everything). Thus, the radically new idea of a unified physics is the most possible generality and the outlined first principle. The presented model is the only one unique logically possible physical theory since it gives the solution of the most difficult and long-standing problem in the history of science - the problem of time. When we discover connections of time with other objects, the reason of the existence of time becomes clear. The main body of references:[1]-[8].

2 General characteristic of fields and time

The most general realization of the idea of space is the concept of physical manifold. The physical manifold is a smooth manifold plus its connections with physical fields that transfer energy and momentum. The geometrical structure of the physical manifold (points, curves, congruences of curves, families of curves) determines a very restricted set of really geometrical quantities and along with that geometrical internal symmetry that makes these quantities variable and forms from them the fundamental physical fields: the positively defined Riemann metric g_{ij} which is the potential of the gravitational field providing connection of the physical fields with the physical manifold (it is characterized by the reparameterization invariance);

the linear (affine) connection P^i_{jk} which is the potential of the generalized electromagnetic field (the group of geometrical internal symmetry is a general linear group);

the scalar and covariant vector fields, and antisymmetric covariant tensor fields which are connected by the geometrical internal symmetry (spin symmetry) into the real spin field

$$\mathbf{A} = (a, a_i, a_{ij}, a_{ijk}, a_{ijkl}), \quad i, j, k, l = 1, 2, 3, 4.$$

The intrinsic space-time structure of the physical manifold is the temporal field (the natural clock) and bilateral symmetry defined by the arrow of time. The temporal field with respect to the coordinate system u^1, u^2, u^3, u^4 is denoted by $f(u) = f(u^1, u^2, u^3, u^4)$. The reading t of the natural clock is defined by the formula $t = f(u) = f(u^1, u^2, u^3, u^4)$ for any point of the physical manifold. The two events at the points u^1, u^2, u^3, u^4 and v^1, v^2, v^3, v^4 go on concurrently if $f(u^1, u^2, u^3, u^4) = f(v^1, v^2, v^3, v^4)$. The gradient of the temporal field (the arrow of time) is the vector field $\mathbf{t}$ with the components $t^i = (\nabla f)^i = g^{ij}\partial_j f = g^{ij}t_j$, where g^{ij} are the contravariant components of the Riemann metric g_{ij}. The construction of natural clock is defined by the fundamental equation :

$$D_{\mathbf{t}} f = (\nabla f)^2 = g^{ij}\frac{\partial f}{\partial u^i}\frac{\partial f}{\partial u^j} = 1, \tag{1}$$

which means that time flows equably. This equation has a general solution and a special solution known as the function of geodesic distance.

The arrow of time defines the fundamental discrete internal symmetry- bilateral symmetry. A pair of vector fields $\mathbf{v}$ and $\overline{\mathbf{v}}$ has bilateral symmetry if the sum of these fields is collinear to the gradient of a temporal field and their difference is orthogonal to it, $\overline{\mathbf{v}} + \mathbf{v} = \lambda\mathbf{t}, \quad (\overline{\mathbf{v}}, \mathbf{t}) = (\mathbf{v}, \mathbf{t})$, where $(\mathbf{v}, \mathbf{w}) = g_{ij}v^i w^j = v^i w_i$ is a scalar product. The bilateral symmetry may be represented as a linear transformation (reflection) $\overline{v}^i = R^i_j v^j$, where $R^i_j = 2t^i t_j - \delta^i_j$.

The bilateral symmetry defines the auxiliary metric $\overline{g}_{ij} = 2t_i t_j - g_{ij} = g_{ik}R^k_j, \quad \overline{g}^{ij} = 2t^i t^j - g^{ij}$, which defines time evolution of the physical fields.

In the geometrical (coordinate independent) form the time reversal invariance means that the theory is invariant with respect to the transformations $T: \quad t^i \to -t^i$.

3 Equations of Gravidynamics

The dynamical equations of the gravitational field and the law of energy conservation are represented in this section.

We denote

$$\Gamma^i_{jk} = \frac{1}{2}g^{il}(\partial_j g_{kl} + \partial_k g_{jl} - \partial_l g_{jk}), \quad R_{ijk}{}^l = \partial_i \Gamma^l_{jk} - \partial_j \Gamma^l_{ik} + \Gamma^l_{im}\Gamma^m_{jk} - \Gamma^l_{jm}\Gamma^m_{ik}$$

and get the Ricci tensor $R_{jk} = R_{ljk}{}^l$ and covariant derivative ∇_i with respect to Γ^i_{jk}. With this we introduce momentum of the gravitational field

$$P^i_j = \frac{1}{2}g^{ik}D_{\mathbf{t}}g_{jk} = \frac{1}{2}g^{ik}(\nabla_j t_k + \nabla_k t_j) = g^{ik}\nabla_j t_k = \nabla_j t^i,$$

where $D_{\mathbf{t}}$ is the Lie derivative along the arrow of time. The stress tensor of the gravitational field can be written in the following form:

$$S^i_j = h^i_k R^k_l h^l_j + D_{\mathbf{t}} P^i_j + \varphi P^i_j,$$

where $\varphi = \nabla_i t^i$ and $h^i_j = \delta^i_j - t_j t^i$. For a density of kinetic and potential energy of the gravitational field we have the following expressions:

$$T = \frac{1}{2}(P^i_i)^2 - \frac{1}{2}P^i_j P^j_i = \frac{1}{2}(TrP)^2 - \frac{1}{2}Tr(P^2), \quad U = \frac{1}{2}S = \frac{1}{2}S^l_l.$$

The first group of equations of Gravidynamics reads

$$D_t P^i_{\ j} + \varphi P^i_j + S^{\ i}_j + N^i_j = \frac{1}{2}\varepsilon h^{\ i}_j, \tag{2}$$

where N^i_j is the stress tensor of the generalized electromagnetic field and spin field, ε is the energy density of the gravitational field and other fields $\varepsilon = \varepsilon_g + \varepsilon_m$. The second group of equations of Gravidynamics reads

$$G_j = \Pi_j, \tag{3}$$

where

$$G_j = h^{\ i}_j(\nabla_l P^l_{\ i} - \partial_i P^l_{\ l})$$

is the energy flow vector of the gravitational field and Π_i is the energy flow vector of the other fields.

The law of energy conservation reads

$$D_t(\sqrt{g}\varepsilon) = 0, \quad \varepsilon = \varepsilon_g + \varepsilon_m = T + U + \varepsilon_m.$$

Since the temporal field enters into the Lagrangians of the physical fields in the form of the gradient of the scalar field $t_i = \partial_i f(u)$, the laws of a unified physics are invariant with respect to transformations of the form $f(u) \Rightarrow f(u) + a$, where a is a constant. This is internal symmetry that defines the law of energy conservation as a fundamental physical law of the nature which is true in all cases.

4 Equations of Generalized Electromagnetic Field

Here we establish dynamical equations of the generalized electromagnetic field (shortly gef) which are defined by gefsymmetry and describe a new form of matter which interacts only gravitationally and by this reason can be identified with the so called dark matter.

The potential of generalized electromagnetic field $P^i_{\ jk}$ is characterized by the internal symmetry $GL(4,\mathbf{R})$ with the law of transformation $\overline{P}^i_{\ jk} = U^i_{\ m} P^m_{\ jn} V^n_{\ k} + U^i_{\ m}\partial_j V^m_{\ k}$, where V^i_j are the components of the operator U^{-1} inverse to the operator U, $U^i_{\ k} V^k_{\ j} = \delta^i_{\ j}$. For brevity, we use the matrix notation $\mathbf{U} = (U^k_l)$, $\mathbf{P}_i = (P^k_{\ il})$, $\mathbf{E} = (\delta^k_l)$, $\mathbf{H}_{ij} = (H_{ijl}{}^k)$, $\mathrm{Tr}\,\mathbf{U} = U^k_{\ k}$. The transformations of gefsymmetry take the form $\overline{\mathbf{P}}_i = \mathbf{U}\mathbf{P}_i\mathbf{U}^{-1} + \mathbf{U}\partial_i\mathbf{U}^{-1} = \mathbf{P}_i + \mathbf{U}\mathrm{D}_i\mathbf{U}^{-1}$, where D_i is the natural differential operator associated with gefsymmetry only $\mathrm{D}_i\mathbf{U} = \partial_i\mathbf{U} + \mathbf{P}_i\mathbf{U} - \mathbf{U}\mathbf{P}_i = \partial_i\mathbf{U} + [\mathbf{P}_i, \mathbf{U}]$.

The Riemann tensor $\mathbf{B}_{ij} = \partial_i\mathbf{P}_j - \partial_j\mathbf{P}_i + [\mathbf{P}_i, \mathbf{P}_j]$ of $P^i_{\ jk}$ is reducible with respect to the transformations $\overline{\mathbf{B}}_{ij} = \mathbf{U}\mathbf{B}_{ij}\mathbf{U}^{-1}$, since $\mathbf{B}_{ij} = (\mathbf{B}_{ij} - \frac{1}{4}\mathrm{Tr}(\mathbf{B}_{ij})\mathbf{E}) + \frac{1}{4}\mathrm{Tr}(\mathbf{B}_{ij})\mathbf{E}$. Hence, the strength tensor of the generalized electromagnetic field is given by the formula

$$\mathbf{H}_{ij} = \mathbf{B}_{ij} - \frac{1}{4}\mathrm{Tr}(\mathbf{B}_{ij})\mathbf{E}, \quad \mathrm{Tr}(\mathbf{H}_{ij}) = 0$$

and the singlet state of the gef defines the tensor of the electromagnetic field $F_{ij} = \mathrm{Tr}(\mathbf{B}_{ij})$.

The ground state of gef is defined by the equation $\mathbf{B}_{ij} = 0$. We give a general solution of this equation. Let's consider four linear independent vector fields E^i_μ and one can construct purely algebraic components of the four covector fields E^μ_i, so that $E^i_\mu E^\mu_j = \delta^i_j$ holds valid. Setting $P^i_{\ jk} = L^i_{\ jk}$, where $L^i_{\ jk} = E^i_\mu \partial_j E^\mu_k$, we get a general solution of the equation in question. Let us introduce a tensor field $Q^i_{\ jk} = P^i_{\ jk} - L^i_{\ jk}$ and an irreducible deviation tensor.

$$T^i_{\ jk} = Q^i_{\ jk} - \frac{1}{4}Q^l_{\ jl}\delta^i_k, \quad \mathbf{T}_j = \mathbf{Q}_j - \frac{1}{4}\mathrm{Tr}(\mathbf{Q}_j)\mathbf{E}.$$

Furthermore, we introduce electric strength and magnetic strength of the generalized electromagnetic field by the direct and inverse mapping

$$\mathbf{J}_i = t^k \mathbf{H}_{ik}, \quad \mathbf{M}_i = \frac{1}{2} e_{ikjl} t^k \mathbf{H}^{jl} = t^k \overset{*}{\mathbf{H}}_{ik}, \quad \mathbf{H}_{ik} = -t_i \mathbf{J}_k + t_k \mathbf{J}_i - e_{ikjl} t^j \mathbf{M}^l.$$

The Lagrangian of gefdynamics takes the form

$$\mathcal{L}_P = -\frac{1}{4}\mathrm{Tr}(\mathbf{H}_{ij}\widetilde{\mathbf{H}}^{ij}) - \frac{\mu^2}{2}\mathrm{Tr}\,(\mathbf{T}_i\widetilde{\mathbf{T}}^i), \tag{4}$$

where μ is a constant of dimension of cm^{-1}, $\widetilde{\mathbf{H}}^{ij} = \overline{g}^{ik}\overline{g}^{jl}\mathbf{H}_{kl} = 2t^i\mathbf{J}^j - 2t^j\mathbf{J}^i + \mathbf{H}^{ij}$, $\widetilde{\mathbf{T}}^i = \overline{g}^{ik}\mathbf{T}_k$.

The first pair of equations of gefdynamics can be represented in the following form:

$$t^k \mathrm{D}_k \mathbf{M}^i - \mathbf{M}^k \partial_k t^i + \varphi \mathbf{M}^i = -e^{ijkl} t_j \mathrm{D}_k \mathbf{J}_l, \qquad \frac{1}{\sqrt{g}}\mathrm{D}_i(\sqrt{g}\mathbf{M}^i) = 0,$$

where $\varphi = \nabla_i t^i$.

The second pair of equations of gefdynamics reads

$$t^k D_k \mathbf{J}^i - \mathbf{J}^k \partial_k t^i + \varphi \mathbf{J}^i = e^{ijkl} t_j \mathrm{D}_k \mathbf{M}_l + \mu^2 \mathbf{S}^i, \qquad \frac{1}{\sqrt{g}}\mathrm{D}_i(\sqrt{g}\mathbf{J}^i) = \mu^2 \mathbf{S},$$

where $\mathbf{S}_i = h_i^j \mathbf{T}_j$, $\quad \mathbf{S} = t^k \mathbf{T}_k$.

Since $\mathrm{Tr}(\mathbf{J}_i) = \mathrm{Tr}(\mathbf{M}_i) = \mathrm{Tr}(\mathbf{T}_i) = 0$, the system of equations in question is simultaneous. This system of equations describes a new form of matter which is the structure element of the unified physics and has the characteristics of an entity known from the observations as Dark Matter.

5 Equations of Spindynamics

In a Unified Physics the Spindynamics includes the region of physical phenomena which are now investigated in the Standard Model. Thus, spin symmetry is geometrical internal symmetry which provides new and complete understanding of spin and, hence, theory of elementary particles as well. Here we present equations of Spindynamics and introduce operators of electrical charge and neutrino charge and discuss some questions connected with this innovation.

The spin symmetry group is a general linear group $GL(16, \mathbf{R})$ that acts in the linear space that represents the spin field. We constructed a natural general covariant basis in the Lie algebra $gl(2^n, \mathbf{R})$ of $GL(2^n, \mathbf{R})$ and describe spin as a bipolar structure on the group of spin symmetry $GL(2^n, \mathbf{R})$. The bipolar structure means that the two dual sets of commuting operators exist which define the Lie algebra of two dual groups S and $\widetilde{S}$. This result expresses in the mathematical form that the spin field can be put in correspondence to the physical top and that this field is the space of the two-valued representation of the dual groups S and $\widetilde{S}$.

To write Lagrangians of Spindynamics we define the positive definite scalar product and the auxiliary scalar product in the linear space in question as follows:

$$(\mathbf{A}|\mathbf{B}) = \sum\nolimits_{p=0}^{4} \frac{1}{p!} a_{i_1\cdots i_p} b_{j_1\cdots j_p} g^{i_1 j_1}\cdots g^{i_p j_p}, \quad \langle \mathbf{A}|\mathbf{B}\rangle = \sum\nolimits_{p=0}^{4} \frac{1}{p!} a_{i_1\cdots i_p} b_{j_1\cdots j_p} \overline{g}^{i_1 j_1}\cdots \overline{g}^{i_p j_p}.$$

The relation $\langle \mathbf{A}|\mathbf{B}\rangle = (\mathbf{A}|\mathrm{R}\mathbf{B})$ holds valid for these two scalar products, where $\mathrm{R} = -Q_{\mathrm{t}}\widetilde{Q}_{\mathrm{t}} = Q_{-\mathrm{t}}\widetilde{Q}_{\mathrm{t}} = Q_{\mathrm{t}}\widetilde{Q}_{-\mathrm{t}}$ and

$$Q_{\mathrm{t}} : \bar{a}_{i_1\cdots i_p} = p\, t_{[i_1} a_{i_2\cdots i_p]} - t^k a_{k i_1\cdots i_p}, \quad \widetilde{Q}_{\mathrm{t}} : \bar{a}_{i_1\cdots i_p} = (-1)^{p+1}(p\, t_{[i_1} a_{i_2\cdots i_p]} + t^k a_{k i_1\cdots i_p}).$$

DOI: 10.1051/epjconf/201713808008

The square brackets $[\cdots]$ denote the process of alternation, $t^i = g^{ij}t_j$ and $p = 0, 1, 2, 3, 4$. The operator R gives the straightforward representation of the bilateral symmetry and shows that this representation opens a new possibility to interpret the bilateral symmetry in the case of the spin field.

The Lagrangian of Spindynamics associated with this new representation of bilateral symmetry can be written only for the complex spin field and has the form:

$$\mathcal{L}_t = -\frac{i}{2}\langle \widetilde{Q}_t \overset{*}{\Psi} | \Pi \Psi \rangle + \frac{i}{2}\langle \widetilde{Q}_t \Psi | \Pi \overset{*}{\Psi} \rangle + im \langle \widetilde{Q}_t \overset{*}{\Psi} | \Psi \rangle. \tag{5}$$

It is not invariant with respect to time reversal $T: \quad t_i \to -t_i$, since $\mathcal{L}_{-t} = -\mathcal{L}_t$. The main differential operator of Spindynamics Π is defined by the new geometrical object, fundamental connection of Spindynamics

$$\overline{\Gamma}^i_{jk} = \Gamma^i_{jk} + t^i \nabla_j t_k - t_k \nabla_j t^i, \tag{6}$$

which is characterized by the equations $\overline{\nabla}_i t_j = 0, \quad \overline{\nabla}_i g_{jk} = 0, \quad \overline{\nabla}_i g^{jk} = 0$, where $\overline{\nabla}_i$ is the covariant derivative with respect to the connection (6). The operator Π is defined as follows:

$$(\Pi \mathbf{A})_{i_1 \cdots i_p} = p \overline{\nabla}_{[i_1} a_{i_2 \cdots i_p]} + \overline{\nabla}^k a_{k i_1 \cdots i_p} - 2 t^k \overline{\nabla}_t a_{k i_1 \cdots i_p}.$$

The Lagrangian of Spindynamics associated with the customary representation of bilateral symmetry can be written for the real spin field and takes the following form:

$$\mathcal{L}_t = \frac{1}{2}\langle \mathbf{A} | \Pi \mathbf{A} \rangle - \frac{m}{2}\langle \mathbf{A} | \mathbf{A} \rangle. \tag{7}$$

It is invariant with respect to the time reversal.

Equations for the free complex and real spin fields have the same form :

$$\Pi \Psi + \frac{1}{2}\varphi\, Q_t \Psi = m\Psi, \quad \Pi \mathbf{A} + \frac{1}{2}\varphi\, Q_t \mathbf{A} = m\mathbf{A}, \tag{8}$$

where $\varphi = \nabla_i t^i$, but there is a principal difference between the Lagrangians (5) and (7).

Since $\Pi \widetilde{Q}_t - \widetilde{Q}_t \Pi = 0, \quad \langle \widetilde{Q}_t \mathbf{A} | \mathbf{B} \rangle = -\langle \mathbf{A} | \widetilde{Q}_t \mathbf{B} \rangle$, the Lagrangian (7) is invariant with respect to the transformations $\widetilde{\mathbf{A}} = exp(\alpha \widetilde{Q}_t)\mathbf{A}$. For the density of the corresponding conserved current C^k one can find the expression $\rho = t_i C^i = \langle Q_t \mathbf{A} | \widetilde{Q}_t \mathbf{A} \rangle = (\mathbf{A} | \mathbf{A})$ which shows that there is a natural probability measure in the space of solution of equation (8) for the real spin field. This fundamental result gives us an opportunity to equate the operator $\widetilde{Q}_t$ with an operator of the electrical charge $\widetilde{Q}_t = Q_e$.

The operator of the neutrino charge Q_ν should be connected with the parity nonconservation and, hence, it is defined by the orientation of the physical manifold. The orientation is the antisymmetrical tensor e_{ijkl} ("element of volume") normalized as $e_{1234} = \sqrt{g}$, where $g = \mathrm{Det}(g_{ij})$.

We define the operator $\widetilde{H}$ in the component form

$$(\widetilde{H}\mathbf{A})_{i_1 \cdots i_p} = \frac{1}{(4-p)!}(-1)^{\frac{p(p+1)}{2}} e_{i_1 \cdots i_p j_1 \dots j_{4-p}} a^{j_1 \dots j_{4-p}}$$

and derive the following relations:

$$\widetilde{H}^2 = E, \quad \widetilde{H}\widetilde{Q}_t + \widetilde{Q}_t \widetilde{H} = 0, \quad \widetilde{H} Q_t = Q_t \widetilde{H} = 0.$$

The operator of the neutrino charge Q_ν takes the form $Q_\nu = \widetilde{Q}_t \widetilde{H} Q_t$. The operators Q_e and Q_ν anticommute

$$Q_e Q_\nu + Q_\nu Q_e = 0, \quad Q_e^2 = Q_\nu^2 = -E.$$

The operators Q_e, Q_ν and $Q_{e\nu} = Q_e Q_\nu$ define the representation of $SU(2)$ group and the quaternion structure.

The Lagrangian of Spindynamics (5) is invariant with respect to the transformations

$$\overline{\mathbf{\Psi}} = exp(\alpha Q_e)\mathbf{\Psi}, \quad \overline{\mathbf{\Psi}} = exp(\beta Q_\nu)\mathbf{\Psi}, \quad \overline{\mathbf{\Psi}} = exp(\gamma Q_{e\nu})\mathbf{\Psi},$$

but it is remarkable that the Lagrangian (7) is invariant only with respect to the transformation generated by the operator of the electric charge. We see that in one case the spin field carries only the electrical charge but in the second case it carries the electrical, neutrino and pseudo-charge. In both cases the spin field carries the internal spin which is generated by the geometrical internal symmetry defined by the antisymmetrical tensor field S_{ij} obeying the equation $\overline{\nabla}_i S_{jk} = 0$. We have no possibility to discuss here this important symmetry in detail.

6 Conclusion

Let us mention some evidences of the New Physics. Of course, the principal result is the dynamical equations of the gravitational field, the spin field and the generalized electromagnetic field which form the nucleus of the Unified Physics as a system and this system will reveal its internal content in the process of its exploitation.

It is discovered that in the Universe there are two space-time structures (the two natural clocks). One of them is defined by a general solution of the equation for temporal field (1). Another solution and a new space-time structure emerge as a special solution of this equation and corresponds to the rotating matter. The new space-time structure naturally explains the confinement and the baryon number conservation. Indeed, we cannot invoke the artificial concept of force to explain the confinement because for any force there is a more powerful one. But the new space-time structure is something else again. The baryon number conservation simply expresses in the symbolic form the existence of the second space-time structure and nothing else. The quark - lepton symmetry means that leptons live in the area of the usual space-time structure and quarks live in the area of the new one.

Spindynamics can be considered as a new purely rational realization of the Standard Model derived from the first principle only. The observed generations of quarks and leptons are different states of the complex spin field and the number of these generations equals four. The spin field describes the physical entity retaining the elementarity and is considered to be characterized solely by equations (8), the electrical charge and the internal spin in one case and the pseudocharge, electric charge, neutrino charge and the internal spin in the other case.

The problem of Dark Energy is solved. Dark Energy is simply the energy of the gravitational field. The gravitational field is physical one.

The problem of Dark Matter is solved. The Dark Matter is heavy light that interacts only gravitationally. Heavy photons are quants of the generalized electromagnetic field. Massless photons represent a singlet state of this field.

The most general and universal law of energy conservation is discovered it is valid in all cases and is an opportunity to be a new contributor of energy.

References

[1] I.B. Pestov. In: *Horizons in World Physics, Vol.248*, Ed. by A. Reimer, (Nova Science, New York, 2005), 1. Preprint No.E2–2004–105 JINR (Dubna, 2004). gr-qc/0507131

[2] I.B. Pestov, *New concept of time and gravity*, In: *Symmetries and Intergrable Systems, Vol. II*, Ed. by A.N. Sissakian. Preprint No. E2-96-424, JINR (Dubna, 1996)

[3] I.B. Pestov, J.Phys. A: Math.Gen. **33**, 3027 (2000)
[4] I.B. Pestov. In: *Dark Matter: New Research*, Ed. by Val Blain (Nova Science, New York 2006), 77. Preprint No. E2–2005–51, JINR (Dubna, 2005). gr-qc/0412096
[5] I.B. Pestov. Physics of Particles and Nuclei **44(3)**, 442 (2013)
[6] I.B. Pestov, Theor. and Math. Phys. **34(1)**, 48 (1978)
[7] I.B. Pestov, Physics of Atomic Nuclei **74(7)**, 1084 (2011). Preprint No. E2-2008-93, JINR (Dubna, 2008)
[8] I.B. Pestov, J.Phys.Math. **6**, 135 (2015). doi: 10.4172/2090-0902.1000135

On possible contribution of partons' orbital momentum to the proton spin

Boris Kostenko[1,⋆]

[1] *Joint Institute for Nuclear Research, Dubna, Russian Federation*

Abstract. Virtual parton excitations of proton achieve status of real particles after transition into the infinite momentum frame. In particular, QCD evolution of them should respect the conservation laws which in turn should be a necessary consequence of symmetry properties. I argue that these simple reasons may cast light on the so-called proton spin puzzle and upon an orbital momentum contribution of quarks and gluons to the proton spin. An already fulfilled experiment with a hint on possible observation of orbital excitations of gluons is pointed out.

1 Introduction

The relativistic covariance of the Dirac equation means that a fermion seen from a moving frame acquires an admixture of its antiparticles, exactly as the electric field acquires the magnetic field component if you look at it from a moving, $v \neq 0$, frame. For example, the Dirac field of a quark polarized at rest along z-axis is of the form [1]

$$\sqrt{\frac{E+m}{2m}}\left(1,0,0,\frac{p}{E+m}\right)^T$$

in a frame moving along z. It is easy to see that the electric charge and z-component of spin are conserved in the moving frame despite an increase of total intensity of the $q-\bar{q}$ field for a moving observer. Besides valence quarks, there are virtual quarks and gluons in proton, which may live very long in a fast moving frame, even forever if $v \longrightarrow 1$. They describe virtual excitations of nucleon, which may be considered as real ones in the infinite momentum frame (IMF). Just as excited states of hydrogen atom are described with taking into account corresponding physical symmetries, almost real excitations of nucleon may be naturally treated in the same way.

E. Fermi was the first who understood that electric field surrounding a charged particle and, as we understand it now, corresponding to virtual photons turns into real electromagnetic waves after transition into the high-speed reference frame [2]. Therefore, it was already shown by Fermi and developed further by Weizsäcker [3] and Williams [4] that physical picture of the world may undergo qualitative changes after transition into IMF[1]. On this evidence, it is not strange that the constituent

⋆e-mail: bkostenko@jinr.ru

[1]For a modern exposition of the subject see [5]. It is even more relevant to our study an investigation of gluon production in ultrarelativistic collisions of two nuclei considered as sources of non-Abelian Weizsäcker-Williams fields, see [6, 7].

quark model defined in the proton rest frame is not compatible with the helicity content for the current quarks and gluons defined in the above mentioned limit. Furthermore, the total contribution of quark helicities to the proton spin is small and consistent with zero within the range of experimental errors according to the EMC measurement of polarized structure functions [8]. Therefore it is natural to suggest on basis of the angular momentum conservation reason that the deficiency, which persists till now, is hidden in orbital angular momentum of quarks and gluons.

Now let us try to understand why gluons should indeed carry nonzero orbital momentum in the experiments of the EMC type.

2 Holographic principle and gluons

Three-dimensional nucleon turns into a two-dimensional object after transition into IMF. As it was explained above, nucleon's content may appear very different in 3-D and 2-D pictures as well as an appearance of a holographic plate differs from an image of a physical object embodied in it. Nevertheless, total information about nucleon spin should be the same in the both. We may drastically simplify the color-Maxwell equations,

$$\partial_\mu F^{\mu\nu} = gJ^\nu, \tag{1}$$

if take into account quasi-free character of knockout of partons from nucleon in deep inelastic scattering processes. This means that $g - q$ and $g - g$ interactions are negligibly rare, at least for very short intervals of time, and the color-Maxwell current may be omitted in the right hand side of (1). Eventually we obtain 8 color-Maxwell equations, one for each color, which coincides with the Maxwell equations for the free electromagnetic field.

To be explicit, let us introduce the color vector potential A^μ in a form as follows (hereafter the color index attributed to A^μ is omitted):

$$\mathbf{A} = \mathbf{e}_x u(x, y, z, t),$$

which satisfies to the wave equation,

$$u_{xx} + u_{yy} + u_{zz} - u_{tt} = 0.$$

After separation of time,

$$u(x, y, z, t) = e^{-i\omega t} v(x, y, z, \omega),$$

we obtain the Helmholtz equation with the wavenumber $k = \omega$,

$$v_{xx} + v_{yy} + v_{zz} + k^2 v = 0.$$

Now we use an ansatz

$$v(x, y, z, \omega) = e^{ikz} w(x, y, z, \omega)$$

with the view of taking advantage of a special role of z-direction and a high value of gluons' momentum in our picture. Because w depends on z much weaker than v, we neglect the term w_{xx} in the exact equation and arrive to

$$w_{xx} + w_{yy} + 2ikw_z = 0. \tag{2}$$

The well-known solution of (2)

$$w = (1/z)e^{ik(x^2+y^2)/2z}$$

may be transformed into a solution describing gluons confined in a small transverse region using a shift along z axis by a complex constant

$$z \to z_* = z - i\varepsilon, \qquad \varepsilon > 0.$$

Thus we acquire the solution of (2) in the form

$$g = (1/z_*)e^{ik(x^2+y^2)/2z_*} = (1/z_*)e^{ikz\psi}e^{-\rho^2/2\Delta^2(z_*)}, \tag{3}$$

where $\rho = \sqrt{x^2+y^2}$, $\psi = (\rho/|z_*|)^2/2$, $\Delta(z_*) = \sqrt{z^2+\varepsilon^2}/\sqrt{k\varepsilon}$. We are inside 2-D nucleon at $z = 0$, where $\Delta_0 \equiv \Delta(-i\varepsilon) = \sqrt{\varepsilon/k}$ and $\psi = \rho^2/2\varepsilon^2$. It is evident that localization corresponding to the gluon confinement is strong if $\Delta(-i\varepsilon)/\lambda = \sqrt{k\varepsilon} \gg 1$. According to the modern beliefs based on QCD calculations, gluon locations are classified according to the portion x of the total (nucleon) momentum it carries (see, e.g. [9]). Thus, it is usually suggested that for $0.1 < x < 1$ gluon is inside a valence quark, and it is centrally symmetric distributed inside nucleon at smaller x (with increasing Δ if x decrease).

Function (3) corresponds to the normal state of the two-dimensional harmonic oscillator. It describes the unexcited (ground) state of virtual gluons in nucleon. Besides, there are many excited virtual states which may live long if nucleon moves with a high velocity. For example, there are solutions of (2) describing excitations of two independent oscillators directed along x and y axes [10], which do not carry orbital momentum:

$$h = z_*^{-(n+m)/2}A_{nm}H_n(\mu x/\sqrt{z_*})H_m(\mu y/\sqrt{z_*})g,$$

where H_n, H_m are the Hermite polynomials, and $\mu = e^{i\pi/4}\sqrt{k/2}$, $A_{nm} = (-\mu)^{n+m}$. Of course, solutions of this type can hardly be used for description of virtual excitations of gluons because they do not reflect a true symmetry of the problem.

In the cylindrical coordinate system, equation (2) has solutions inside 2-D nucleon [11]

$$w_{p,m}(\rho,\phi,0) = \sqrt{\frac{2^{|m|+1}p!}{\pi(p+|m|)!}}\rho^{|m|}e^{-\rho^2}L_p^{|m|}(2\rho^2), \tag{4}$$

where $L_p^{|m|}$ for natural m are the generalized Laguerre polynomials. For short, we have changed notation $\rho \to \rho/\Delta_0$. The solutions (4) are known as Laguerre-Gauss (LG) modes which carry orbital angular momentum m directed along z-axis, and index p enumerates different radial excitation levels of virtual gluons. Thus, we may conclude that the virtual excitations of the gluon field should have nonzero orbital momentum by only symmetry reasons concerning its geometrical form.

3 A possibility to observe gluons' orbital excitations

The virtual electromagnetic excitations considered by Fermi turn into the real bremsstrahlung field if a fast electron carrying them undergo a blow [5]. Similarly, we may expect that hard collisions of nucleons may reveal the virtual orbital excitations of the gluon field. In experiments with a large momentum transfer, external impact may transform only the transverse components of virtual gluons into real ones and excitations of such a type may appear as narrow baryon resonances. Results of experiment [12, 13] fulfilled already in 1993 give a possible candidate for the phenomenon. A detailed analysis of the papers [12, 13] may be found in [14]. Below we sketch a probable theoretical approach to description of these data.

It is possible to find the dynamical symmetry lying behind the radial nodes of LG modes and construct generalized coherent states based on corresponding algebra for fixed m (see [11]). It may be shown that the radial nodes ladder operators,

$$\hat{\mathcal{P}}_+ |p,m\rangle = \mathcal{P}_+ |p+1,m\rangle, \qquad \hat{\mathcal{P}}_- |p,m\rangle = \mathcal{P}_- |p-1,m\rangle,$$

where $\hat{\mathcal{P}}_\pm = \hat{Q} \mp \left(\frac{|m|}{2} - \rho^2\right)$, $\hat{Q} = \frac{1}{2}\rho\partial_\rho + 2\rho^2 - (\hat{p} + |m| + 1)$ are the raising and lowering operators with eigenvalues $\mathcal{P}_+ = \sqrt{(p+1)(p+|m|+1)}$ and $\mathcal{P}_- = \sqrt{p(p+|m|)}$ correspondingly, and

$$\hat{p} = -\frac{1}{8\rho}\partial_\rho(\rho\partial_\rho) + \left(\frac{|m|^2}{8\rho^2} + \frac{\rho^2}{2} - \frac{|m|+1}{2}\right).$$

The ladder operators obey the $su(1,1)$ Lie algebra [15]

$$[\hat{\mathcal{P}}_+, \hat{\mathcal{P}}_-] = -2\hat{\mathcal{P}}_0, \qquad [\hat{\mathcal{P}}_0, \hat{\mathcal{P}}_\pm] = \pm\hat{\mathcal{P}}_\pm,$$

where $\hat{\mathcal{P}}_0 = (2\hat{p} + |m| + 1)/2$. The radial nodes at fixed m arise now as eigenvectors of the Casimir operator $\hat{C} = \hat{\mathcal{P}}_0^2 - 1/2(\hat{\mathcal{P}}_+\hat{\mathcal{P}}_- + \hat{\mathcal{P}}_-\hat{\mathcal{P}}_+)$: $\hat{C}|p,m\rangle = (m^2-1)/4\,|p,m\rangle$. The generalized coherent states corresponding to the $su(1,1)$ Lie group result from the displacement of the ground state $|0,m\rangle$:

$$|\alpha\rangle_m = e^{\alpha\hat{\mathcal{P}}_+ - \alpha^*\hat{\mathcal{P}}_-}|0,m\rangle = (1-|\alpha|^2)^{\frac{|m|+1}{2}} \sum_{p=0}^{\infty} \sqrt{\frac{(p+|m|)!}{|m|!p!}}\alpha^p |p,m\rangle, \tag{5}$$

where $\alpha = \tanh(\xi/2)e^{i\theta}$ is a complex number defined in terms of the two real quantities ξ and θ. The position representation of this coherent states in the two-dimensional world has the form [11]

$$\langle\rho,\phi,0\mid\alpha\rangle_m = \sqrt{\frac{2^{|m|+1}}{\pi\,|m|!}}\left(\frac{1-\alpha^*}{1-\alpha}\right)^{\frac{|m|+1}{2}} e^{im\phi - \frac{1+\alpha}{1-\alpha}\rho^2}\rho^{|m|}.$$

The factor $e^{im\phi}$ in the last expression says us that gluons in $su(1,1)$ coherent state have well-defined orbital angular momentum value of m per gluon. A similar formula with the same interpretation exists for photons in the coherent laser beams [16].

4 Quark confinement and the proton spin puzzle

We established in section 2 that the confinement of gluons in restricted domains and the cylindrical symmetry are sufficient grounds for their virtual excited states carry nonzero orbital momenta. Here we consider a general model which demonstrates that the confinement stimulates also quarks to change their projection of spin with the simultaneous alteration of their orbital angular momentum. This observation may be considered as a solution to the proton spin puzzle.

As the first step, let us establish whether free quarks can be confined by exponent in small transverse regions by analogy with (4). Using the fact that γ-matrices transforms as vectors under transformations of coordinates [17], $\gamma^s = \gamma^i\partial x^s/\partial x^i$, it is possible to obtain the Dirac equation in the cylindrical coordinate system:

$$\left[i\left(\gamma^\rho\frac{\partial}{\partial\rho} + \gamma^\varphi\frac{\partial}{\partial\varphi} + \gamma^z\frac{\partial}{\partial z}\right) + \gamma^0 E - M\right]\psi(\rho,\varphi,z;E) = 0, \tag{6}$$

DOI: 10.1051/epjconf/201713808009

where

$$\gamma^\rho = \begin{pmatrix} 0 & \sigma^\rho \\ -\sigma^\rho & 0 \end{pmatrix}, \quad \gamma^\varphi = \begin{pmatrix} 0 & \sigma^\varphi \\ -\sigma^\varphi & 0 \end{pmatrix}, \quad \gamma^z = \begin{pmatrix} 0 & \sigma^z \\ -\sigma^z & 0 \end{pmatrix} \quad \gamma^0 = \begin{pmatrix} I & 0 \\ 0 & -I \end{pmatrix}$$

and

$$\sigma^\rho = \begin{pmatrix} 0 & e^{-i\varphi} \\ e^{i\varphi} & 0 \end{pmatrix}, \quad \sigma^\varphi = \begin{pmatrix} 0 & -ie^{-i\varphi} \\ ie^{i\varphi} & 0 \end{pmatrix}, \quad \sigma^z = \begin{pmatrix} 1 & 0 \\ 0 & -1 \end{pmatrix}, \quad I = \begin{pmatrix} 1 & 0 \\ 0 & 1 \end{pmatrix}.$$

Then substitution $\psi = (u(\rho,\varphi), v(\rho,\varphi))^T e^{ip_z z}$ separates quarks and antiquarks and introduces the longitudinal component of momentum in explicit form. The substitution makes it possible to change from Dirac's equation to the system of two equations for quarks and antiquarks as follows:

$$(E - M_q)u + \left[i\left(\sigma^\rho \frac{\partial}{\partial\rho} + \frac{\sigma^\varphi}{\rho}\frac{\partial}{\partial\varphi}\right) - \sigma^z p_z\right] v = 0, \tag{7}$$

$$(E + M_{\bar{q}})v + \left[i\left(\sigma^\rho \frac{\partial}{\partial\rho} + \frac{\sigma^\varphi}{\rho}\frac{\partial}{\partial\varphi}\right) - \sigma^z p_z\right] u = 0. \tag{8}$$

Using (8), it is possible to express v in terms of u and substitute it in (7). As a result, we obtain

$$\left(\frac{\partial^2}{\partial\rho^2} + \frac{1}{\rho}\frac{\partial}{\partial\rho} + \frac{1}{\rho^2}\frac{\partial^2}{\partial\varphi^2}\right)u + p_\perp^2 u = 0, \tag{9}$$

where we have used the commutation relations $[\sigma^\alpha, \sigma^\beta] = 2i\varepsilon^{\alpha\beta\delta}\sigma^\delta$ with ranking (ρ, φ, z). Now we can perform a separation of variables by means of the substitution $u(\rho,\varphi) = w(\rho)e^{im\varphi}$ and acquire

$$\rho^2 \frac{d^2 w}{d\rho^2} + \rho\frac{dw}{d\rho} - m^2 w + p_\perp^2 \rho^2 w = 0, \tag{10}$$

where $m = 0, \pm1, \pm2, ...$ due to the single-valuedness of $u(\rho,\varphi)$. Equation (10) is Bessel's differential equation which gives the Bessel functions of the first kind $J_m(p_\perp\rho)$ as its solutions. The wave functions of quarks $u(\rho,\varphi)$ are the eigenstates of z component of orbital momentum operator $L_z = -i\partial/\partial\varphi$, which are localized in the vicinity of the point $\rho = 0$. However, this localization is insufficient for description of the confinement of quarks. Indeed, it is well known that $J_m(r)$ at long distances, $r \equiv p_\perp\rho \gg 1$, are oscillating functions damping as $1/\sqrt{r}$.

Here we consider a model describing the quark confinement using external static potential $V(\rho)$, which is chosen thereby to be cylindrically symmetric. For this purpose, let us insert in the Dirac equation (6) the expression $E - V(\rho)$ instead of E. Then equations (7) and (8) will undergo the corresponding change too, and instead of (9) we find

$$\left(\frac{\partial^2}{\partial\rho^2} + \frac{1}{\rho}\frac{\partial}{\partial\rho} + \frac{1}{\rho^2}\frac{\partial^2}{\partial\varphi^2}\right)u + [(E-V)^2 - (p_z^2 + M_q^2)]u - i\sigma^\rho \frac{dV/d\rho}{E - V + M_q}\left[i\left(\sigma^\rho\frac{\partial}{\partial\rho} + \frac{\sigma^\varphi}{\rho}\frac{\partial}{\partial\varphi}\right) - \sigma^z p_z\right]u = 0. \tag{11}$$

Now let us pay attention to the presence of the term $\sigma^\rho\sigma^z = \sigma^\varphi$ in (11). It destroys the initial polarization of quarks. Indeed, even initially the quark field was totaly polarized along z axis,

$$u(\rho,\varphi) = \begin{pmatrix} u_1 \\ 0 \end{pmatrix},$$

it obtains an admixture of states with the opposite polarization due to σ^{φ} term because

$$i\sigma^{\varphi}u(\rho,\varphi) = \begin{pmatrix} 0 \\ -u_1 e^{i\varphi} \end{pmatrix}.$$

It is also evident that the total angular momentum J_z, is conserved,

$$J_z = s_z + L_z = \frac{1}{2},$$

as it should be because of the cylindrically symmetric ambient conditions.

Of course, formula (11) says more than only about the influence of the confinement on the quark spin. Any external potential changes z projection of spin with the intensity proportional to the force $dV/d\rho$.

Author would like to thank anonymous referees for valuable comments.

References

[1] J.D. Bjorken, S.D. Drell, *Relativistic Quantum Mechanics* (McGraw-Hill, New York, 1964) 29
[2] E. Fermi, Z. Phys. **29**, 315 (1924)
[3] C. Weizsäcker, Z. Phys. **88**, 612 (1934)
[4] E.J. Williams, Phys. Rev. **45**, 729 (1934)
[5] M.E. Peskin, D.V. Schroeder, *Quantum Field Theory* (Addison-Wesley, N.Y., 1996) 209 – 210
[6] A. Kovner, L. McLerran, H. Weiget, Phys. Rev. D **52**, 6231 (1995)
[7] A. Kovner, L. McLerran, H. Weiget, Phys. Rev. D **52**, 2935 (1995)
[8] EMC, J. Ashman et al., Phys. Lett. B **206**, 364 (1988)
[9] C. Weiss, *Nuclear Physics with JLab12 and EIC*, (EIC Meeting, Stony Brook, 2014)
[10] A.P. Kiselev, Opt. Spectrosk. **102**, 603 (2007)
[11] E. Karimi, E. Santamato, Opt. Lett. **37**, 2484 (2012)
[12] Yu.A. Troyan, V.N. Pechenov, Yad. Fiz. **56**, 201 (1993)
[13] Yu.A. Troyan, Fiz. Elem. Chastits At. Yadra **24**, 683 (1993)
[14] B. Kostenko, J. Pribiš, Possible Observation of Phase Transitions in Two-Nucleon Systems, PoS(Baldin ISHEPP XXII) 122
[15] A. Perelomov, *Generalized Coherent States and Their Applications* (Springer, Berlin, 1986) 70
[16] L. Allen, M.W. Beijersbergen, R.J.C. Spreeuw, J.P. Woerdman, Phys. Rev. A **45**, 8185 (1992)
[17] A.H. Cook, Proc. R. Soc. Lond. A **383**, 247 (1982)

Parameters monitoring system for the experimental setup SCAN-3

Dmitriy Sakulin[1,*]

[1] *Joint Institute for Nuclear Research, 141980 Dubna, Moscow region, Russia*

Abstract. This overview begins with the presentation of General ideas about the physical parameters of SCAN-3. In addition, the basic principles of the monitoring system. The configuration of the web server of the monitoring system. In conclusion, the testing of the monitoring system on modules E-cal for the spectrometer BM@N.

1 Introduction

To perform the modern research is unthinkable without the use of computers. It is an important part of communication between the detectors and researchers. The development of computer technology and communications have allowed direct participants not to remain in the experiment, but to be active participants of these studies. This ideology is central in the modern experiment. System with this ideology will be presented in this overview.

2 The model of the monitoring system

2.1 Experimental setup SCAN-3

SCAN-3 is a three-arm Magnetic Spectrometer on the Nuclotron internal target for precision measurements of charged particles and neutrons. [1] It includes the following elements shown in figure 1. Required parameters of SCAN-3:

- Achieving the required precision is not possible without correction setup systems;
- The required accuracy of registration of the particle momentum is 1-2 percent;
- The accuracy of measurement depends on external parameters.

*e-mail: facolco@gmail.com

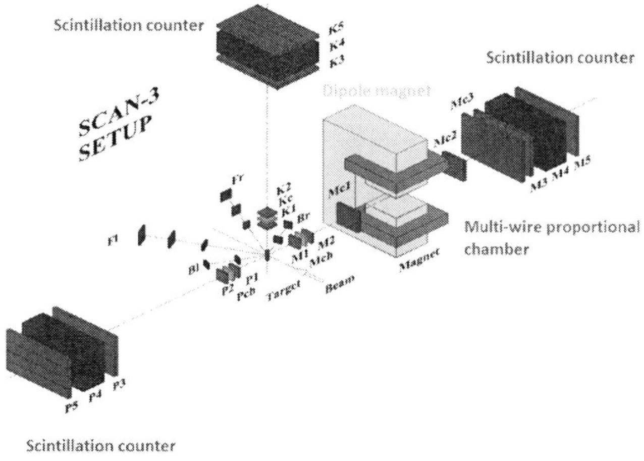

Figure 1. Experimental setup of SCAN-3.

2.2 Main tasks of the monitoring system

The considered system provides an opportunity:

- Collect, process, present and visualize data;
- Multisensory processing;
- Remote monitoring and control for the processes in the setup;
- Calibration and testing of individual subsystems.

Also, this work can be used for educational purposes.

2.3 Basic principles of the monitoring system

- Sensors are placed directly on the elements of the experiment;
- Using hardware and software for collection data (Arduino UNO platform);
- Using open-source relational database management system for storage data (MySQL);
- It is possible to use "raw" data;
- User-friendly interface through Highstock.

Arduino is an open-source electronics platform based on easy-to-use hardware and software. It is intended for anyone making interactive projects.

2.4 Block diagram of the monitoring system

Data from the sensors are transmitted to the database through internet connection (figure 2). The local server is located remotely. The clients access the server for the necessary information.

DOI: 10.1051/epjconf/201713809001

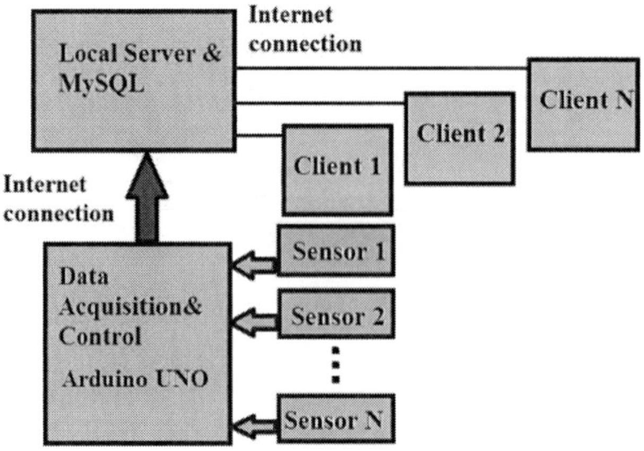

Figure 2. Block diagram of the monitoring system.

2.5 Realization of the system

[2]Arduino module with connected sensors looks as follows (figure 3). And it has also two sensors - humidity and magnetic field.

Figure 3. Arduino module with connected sensors.

3 Application of the monitoring system during testing of ECAL module for BMaN spectrometer

This monitoring system has been tested with ECAL module for BMaN spectrometer. In the test we considered an electromagnetic calorimeter designed to measure energy of particles which interact with matter primarily through the electromagnetic interaction (photons, charged leptons). The photos show the location of the sensors (figure 4).

 DOI: 10.1051/epjconf/201713809001

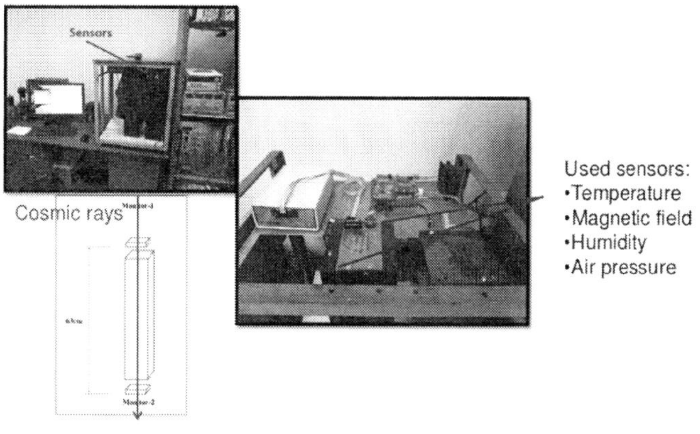

Figure 4. E-cal modules.

3.1 Testing monitoring of triggers

[3]The graph shows the monitoring of triggers from cosmic rays (figure 5). All days of experiment are accumulated and visualized in one day. Highstock in pure JavaScript, including sophisticated navigation options like a small navigator series, preset date ranges, date picker, scrolling and panning.

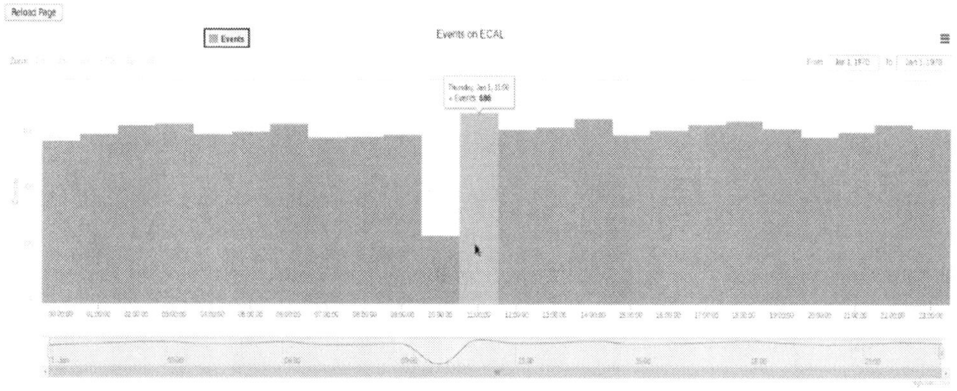

Figure 5. Triggers.

3.2 Visualization of data

This graph visually shows the values of the measuring sensors (figure 6). Highstock runs on the client side and it does not matter how the server is configured. This means that if the server is running PHP and MySQL or any other type of server in conjunction with the SQL, you can dynamically create the HTML and JavaScript with required Highstock.

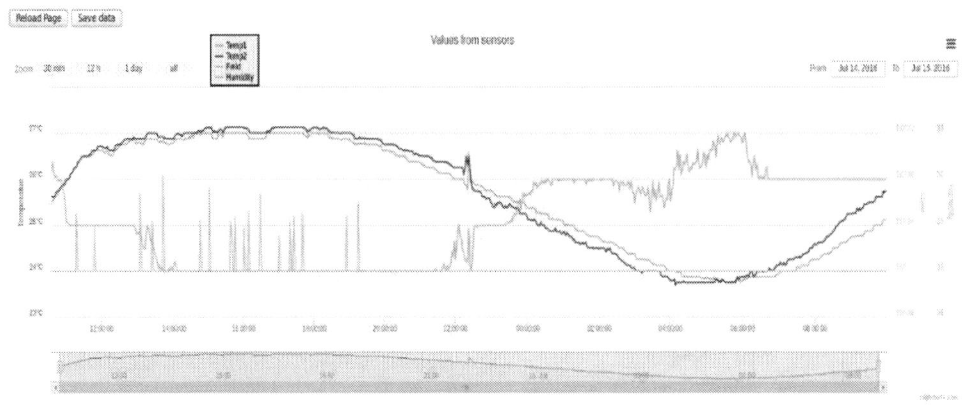

Figure 6. Visualization of data from sensors.

3.3 Temperature dependence

Significant changes in the readings of the magnetic field and the humidity are not detected. Therefore, we considered the temperature only. The graphs show the temperature dependence of the amplitudes of the signals (figure 7). The dependence for each channel was different. So after a temperature compensation for each channel we obtained the following results.

Temperature dependence of the amplitude of the signal.
Thermal compensation is turned **OFF**. Module N17 7 Jul 2016.

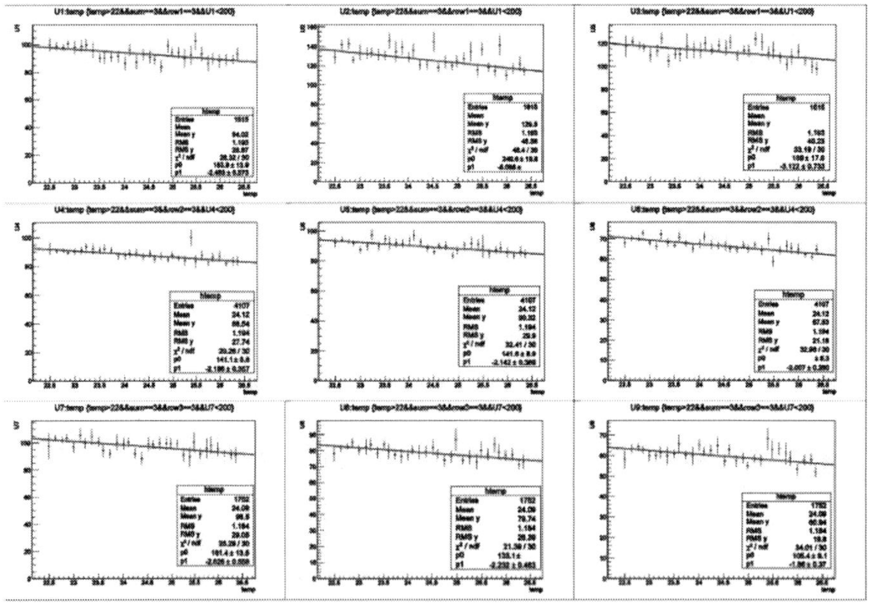

Figure 7. Temperature dependence.

DOI: 10.1051/epjconf/201713809001

4 Results

The graph shows the amplitudes of the 9 ADC channels (figure 8). The top graph without temperature compensation depicts the variation of the amplitude of the signal. The variation of signal disappears after the temperature compensation. That allows to achieve the required accuracy.

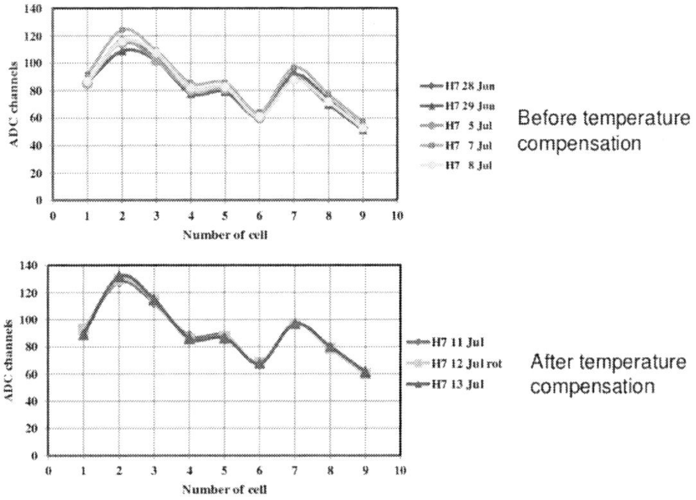

Figure 8. Graph measuring the amplitudes.

5 Conclusion

This monitoring system will be used in the system of slow control in the installation of the SCAN-3 on the Nuclotron, and also in the BM@N project. The system can be modified for any setups.

References

[1] New status of the project "n-nuclei" at the NUCLOTRON (2015), PoS (Baldin ISHEPP XXII) 125
[2] https://www.arduino.cc/
[3] http://www.highcharts.com/products/highstock

Creation of the precision magnetic spectrometer SCAN-3

S.V. Afanasiev[1,⋆], *Yu.S.* Anisimov[1], *A.A.* Baldin[1,2], *A.I.* Berlev[1,2], *D.K.* Dryablov[1,⋆⋆], *B.V.* Dubinchik[1], *A.F.* Elishev[1], *O.V.* Fateev,[1], *Z.A.* Igamkulov[1], *Yu.F.* Krechetov[1], *I.V.* Kudashkin[1,2], *S.N.* Kuznechov[1], *A.I.* Malakhov[1], *V.A.* Smirnov[1], *S.S.* Shimansky[1], *J.* Kliman[3], *V.* Matousek[3], *S.* Gmutsa[3], *I.* Turzo[3], *I.* Cruceru[4], *M.* Cruceru[4], *F.* Constantin[4], *G.* Niolescu[4], *L.* Ciolacu[4], *M.* Paraipan[5], *S.* Vokál,[6], *J.* Vrláková[6], *V.A.* Baskov,[7], *A.I.* Lebedev[7], *A.I.* L'vov[7], *L.N.* Pavlyuchenko[7], *V.V.* Polyansky[7], *E.V.* Rzhanov[7], *S.S.* Sidorin[7], *G.A.* Sokol[7], *I.V.* Glavanakov[8], *A.N.* Tabachenko[8], *D.M.* Jomurodov[9], *R.N.* Bekmirzaev[9], *R.M.* Ibadov[10], and *M.U.* Sultanov[10]

[1] *Joint Institute for Nuclear Research, Dubna, Russia*
[2] *Institute advanced studies «OMEGA», Dubna, Russia*
[3] *Institute of Physics, Slovak Academy of Sciences, Slovak republic*
[4] *Horia Hulubei National Institute of R&D for Physics and Nuclear engineering, IFIN-HH Bucharest, Romania*
[5] *Institute for Space Science, Bucharest-Magurele, Romania*
[6] *Faculty of Science, University of P.J. Šafárik, Košice, Slovak republic*
[7] *Lebedev Physical Institute, Moscow, Russia*
[8] *Tomsk Polytechnic University, Tomsk, Russia*
[9] *Jizzakh State Pedagogical Institute, Uzbekistan*
[10] *Samarkand State University, Uzbekistan*

Abstract. The new JINR project [1] is aimed at studies of highly excited nuclear matter created in nuclei by a high-energy deuteron beam. The matter is studied through observation of its particular decay products - pairs of energetic particles with a wide opening angle, close to 180°. The new precision hybrid magnetic spectrometer SCAN-3 is to be built for detecting charged ($\pi^\pm$, $K^\pm$, p) and neutral (n) particles produced at the JINR Nuclotron internal target in dA collisions. One of the main and complex tasks is a study of low-energy ηA interaction and a search for η-bound states (η-mesic nuclei). Basic elements of the spectrometer and its characteristics are discussed in the article.

1 Introduction

Among the lightest pseudoscalar mesons the eta (and eta-prime) are very interesting objects because their properties are strongly influenced by a flavour SU(3) singlet component, presumably made of gluons, as the axial U(1) anomaly suggests. Masses and decay widths of these mesons, as well as their interactions with nucleons and nuclei, crucially depend on the singlet component that is affected by the nuclear environment. In this way studies of ηN and ηA interactions become related with very fundamental aspects of nuclear physics and QCD [2].

⋆e-mail: afanasev@lhe.jinr.ru
⋆⋆e-mail: dryablov@lhe.jinr.ru

One of the main aims of the experiment SCAN-3 at the internal deuteron beam of the JINR Nuclotron is directed to studies of the ηA system. More specifically, we will study quasi-bound states of the η in nuclei first predicted in theoretical works by Haider an Liu [3], who discovered that ηN and ηA interaction is attractive at low energies.

The experiment SCAN-3 is aimed at investigations of dA collisions and observation of the fast decay products of the produced bound η in nuclei, πN and NN pairs, caused by the internal annihilation-at-rest reactions ηN$\rightarrow$ πN and ηNN$\rightarrow$NN. Information on the binding of η and eventually the low-energy ηA interaction can be inferred from precise measurements of the energy spectra of the ηN and NN pairs. Principles of this approach are explained in more detail [4].

In 2006-2010 a set of experimental data at two arms time-of-flight setup SCAN-2 and the internal d-beam of Nuclotron was obtained. The JINR collaboration performed the search for back-to-back πp pairs related to the η-mesic bound states in $d + ^{12}C \rightarrow \pi + p + X$ process [5]. An observation of the πp back-to-back correlation as well as the resonance like structure below η production threshold could be associated with the two-body N* resonance decay related with formation of an η-mesic nucleus. The new project SCAN-3 provides a significant increase of the energy resolution of the setup by including a magnetic spectrometer in the experimental installation and by increasing the time-of-flight base.

2 Requirements of the experiment

2.1 What we are looking for?

The interaction of the η-meson with a nucleon near threshold is mainly determined by the S_{11}, $J^P=\frac{1}{2}^-$ resonance N*(1535), which is just 49 MeV above the ηN threshold (1486 MeV) and has a width Γ=150 MeV, thus covering the whole low energy region of the ηN interaction. The S_{11} resonance also decays to πN, γN and $\pi\pi$N channels. It is important to recognize that if we register πN pair with approximately equal but opposite momentum components, even with some suitable total energy E_π + $E_p \approx m_\eta + m_N$ = 1486 MeV, it does not necessarily mean that we have registered the decay products of the η-mesic nucleus.

Generally [4], observation of a relatively narrow resonance peak in the spectrum of E_η in the subthreshold region $E_\eta<m_\eta$ is mandatory for claiming an observation of η-mesic nuclei at all.

2.2 Requirements to precisions

Theoretical predictions for the binding energy and width of the η-nuclear levels are highly dependent on the assumed η-nuclear potential and nucleus size. Apparently, future experiments should assume that the peak width will be about 10 MeV, and therefore they should provide accurate measurements of particle energies not worse than ~ 3-5 MeV, so that the accuracy of the total energy of the pair will be at least 5-7 MeV.

In our experiment, the requirement of accuracy of energy determination is weakened (approximately in ~1.5 times). It is caused by effects of a broadening of observable peak in a total energy spectrum of the correlated pair, caused by energy dispersion of intra-nuclear nucleons on which an annihilation of bounded meson is realized. This dispersion additionally increases observable width of peak by ~20 MeV. Thus the requirement to accuracy at level of 5-7 MeV is superfluous. That reduces the accuracy to the level of 10 MeV.

If we consider the process η+N_i $\rightarrow$ π+N with initial particles at rest, the kinetic energy and momentum of the secondary particles must be equal to (if we ignore the effect of binding and Fermi motion): E_π=313 MeV, E_N=94 MeV, p_π=p_N=431 MeV/c. Accordingly, experiment performance demands measurement precisions of kinetic energies at level of 1% for such of pions and 3% for such of nucleons. If we consider the process η+N_i+N_j $\rightarrow$ N_1+N_2, the kinetic energy and momentum are about E_1=E_2=273 MeV, p_1=p_2=767 MeV/c. Accordingly, experiment performance at the declared level demands accuracy in measuring of kinetic energies of such nucleons at level of 1 %.

3 Layout of the setup

We propose to extend a setup used at the Nuclotron in our previous studies in order to achieve the energy resolution for both charged and neutral components of the correlated πN and NN pairs of $\sigma \approx$ 5 MeV. To this aim we plan to construct a magnetic spectrometer (M arm) and two sectioned neutron detectors with a TOF base up to 6 m (P and K arms), see figure 1. The TOF arm (P-arm) is supposed to be used primarily for registration of protons and neutrons with the desired characteristics. The TOF technique for protons is well-proven in the previous experimental setup SCAN-2. Magnetic arm (M-arm) will be used for registration of pions. Technical parameters of the detectors, including their space and time resolution, have been specified using GEANT simulations [6]. Data acquisition system is to be built using VME modules produced in JINR.

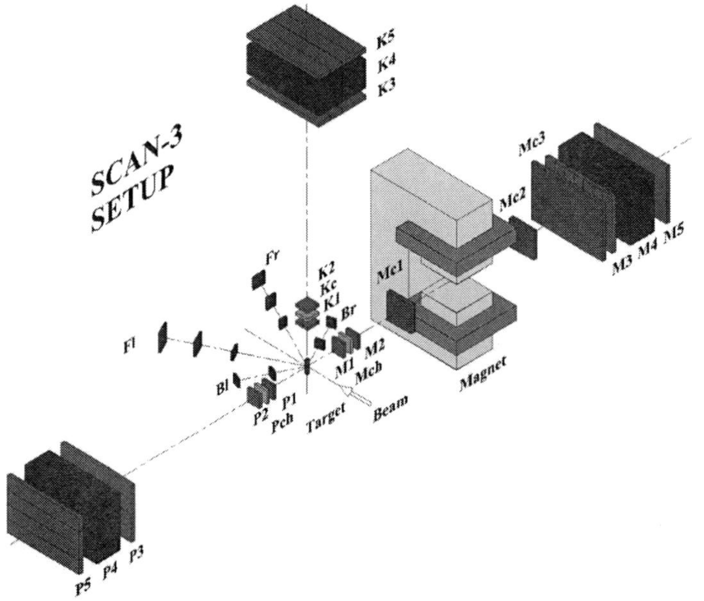

Figure 1. Layout of the SCAN-3 setup. Fr, Fl, Br, Bl – scintillator monitor counters. P1-P3, K1-K3, M1-M3 – trigger time-of-flight scintillator detectors. Pch, Kch, Mch – threshold Cherenkov detectors. Mc1, Mc2 – drift chambers. Mc3 – multiwire proportional chamber. M4 –scintillator total absorption detector. P4, K4 – two sets of scintillator neutron counters. P5, K5, M5 – veto detectors.

4 Magnetic system

At the moment, multiwire proportional chamber (Mc3) is ready for use, two biplanar drift chambers (Mc1 and Mc2) are simulated, the magnet (B_{max}=7 kG) is installed in the area of the Nuclotron internal target.

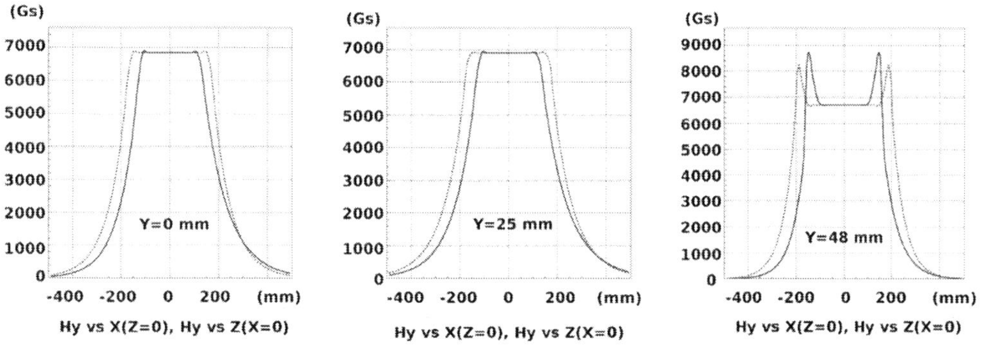

Figure 2. The magnetic field of the SP-46. The dependence of the vertical component of the magnetic field Hy (in Gs) on the coordinates X (axis is perpendicular to the average momentum of the analyzed particles) and Z (axis is parallel to the average momentum of the analyzed particles) in the planes: Y=0 mm (left figure), Y=25 mm (middle figure), Y=48 mm (right figure).

The use of the magnet SP-46 is a good offer for a magnetic spectrometer. Parameters of the magnet are: the transverse gap - 100 (vertical direction) x 300 (horizontal direction) mm^2, magnetic field - 7 kGs at the current of 20 A, the length of the magnetic track - 420mm. The vertical gap of SP-46 can be very easily increased from 100 mm to 170 mm by removing the 35mm flat parts of the magnet (the field will drop to ~5 kGs). The magnetic field was measured in the case of series-connected coils (total resistance of 2.5 Ohm) and current I = 20 A. Some results of magnetic field measurements are shown in the figure 2.

We have analyzed the accuracy of determining of the deflection angle of π and p in the magnetic field depending on the properties of the detectors, the geometry and materials. In this case we have not considered the boundary perturbations of the magnetic field. It is necessary for further simulations. Data analysis for pions has shown that the accuracy of the angular resolution could be better, then 1.4% (it corresponds to the pion kinetic energy of 4.4 MeV). Data analysis for protons has shown that the accuracy of the angular resolution could be better than 2.4% (it corresponds to the proton kinetic energy of 6.5 MeV).

The figure 3 shows the normalized momentum distributions of protons and pions in the M-arm. Reducing the particle detection efficiency at low momenta is justified by the fact that particles do not reach the detectors P3 or K3, which are included in the trigger.

5 Neutron detectors

P1-P3, M1-M3 time-of-flight scintillator detectors need to be calibrated. Prototype of a neutron detector is tested. The figure 4 shows geometry of the neutron detector.

The energy of the neutron is determined by time of flight on the basis of up to 6 m. The neutron detector will consist of several elements. Each element is an assembly of four long plastic scintillator with a thickness of 3 cm. The ends of each element are connected to a fast PMT with a photocathode diameter of about 100 mm. The ends of each of the four plates are further connected to the photomultiplier with a diameter of about 2 cm to identify the plate number. Simulation showed that the

DOI: 10.1051/epjconf/201713809002

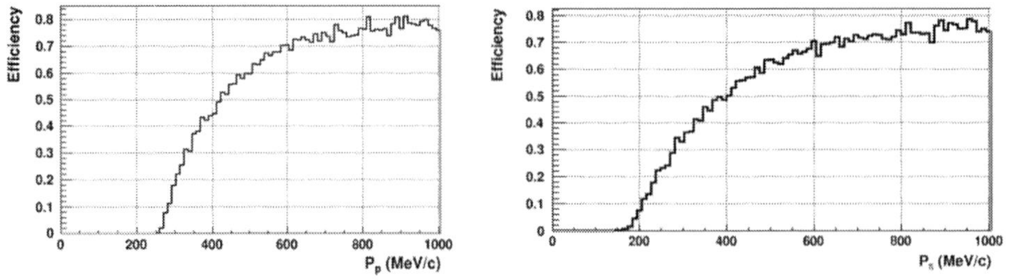

Figure 3. Momentum efficiency of protons (left) and pions (right) in the M-arm of the SCAN3.

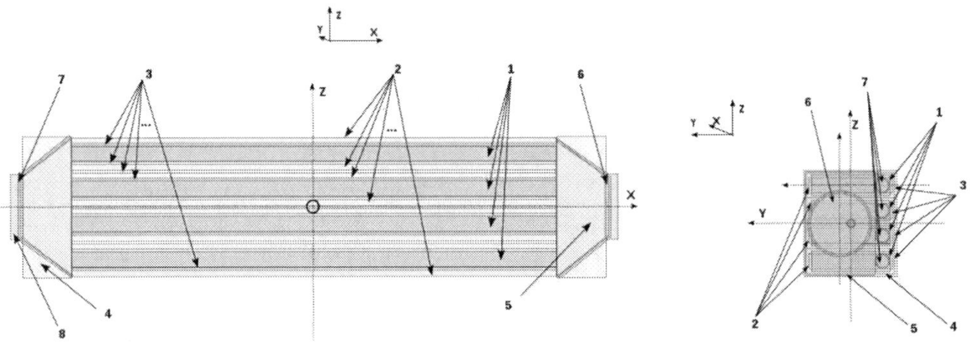

Figure 4. Geometry of the neutron detector: 1 - plates of organic scintillator (NE110 type, optics), 2 - air gap, 3 - wrap material for plates (synthetic paper tyvek, optics), 4 - air volume of the fiber (optics), 5 - reflecting wall of the optical fiber (aluminized Mylar 0.3 mm, optics), 6 - photocathode of PMT (XP2041 type), 7 - photocathode of PMT (PMT-85 type).

required energy resolution of neutron (about 5 MeV) is achieved with a time resolution of the neutron detector of 200 ps. Experimental study showed that the photomultiplier XP2020 is the most suitable, it provides a time resolution of about 250-280 ps.

To improve time resolution, it was proposed to use a PMT from the system identification plate. The researchers have studied the time characteristics of the FEU-85, FEU-87 with different voltage dividers. FEU-87 showed approximately 20% better time resolution than PMT-85, however, the light collection of PMT-85 about 30% higher.

It was found that the simultaneous use of the information from XP2020 and PMT-87 (or PMT-85) allows us to reach the required level of temporary and, therefore, the energy resolution of neutron detector. Testing of the detector continues.

6 Conclusions

The first main aims are:
- building of a magnetic spectrometer for detection of charged components of the correlated pairs with

the energy resolution of 4-5 MeV;
- building of a precision time-of-flight neutron detector with a similar energy resolution;
SCAN-3 will allow to carry out the following studies:
- studies of the $S_{11}(1535)$ and $\Delta(1232)$ resonances in the nuclear matter;
- a search for η-nuclei and Δ-nuclei in dA-collisions through narrow resonance peaks in the total energy spectra of the correlated pairs;
- determination of binding energies and widths of η in nuclei;
- measurements of the cross section $\sigma({}_{\eta}A)$ of η-nuclei formation in dA collisions; measurements of the A-dependence of $\sigma({}_{\eta}A)$;
- measurements of relative rates of πp and pN events;
- measurements of the cross section of Δ-nuclei formation in dA collisions;
- determination of binding energies and widths of quasi-bound states of Δ in nuclei;
- measurements of relative rates of π^+p, π^-p and pN events.

References

[1] 45th meeting of the JINR Programme Advisory Committees for Particle Physics: http://indico.jinr.ru/materialDisplay.py?materialId=3&confId=1589
[2] S.D. Bass and A.W. Thomas, Acta Phys. Pol. **45**, 627 (2014)
[3] Q. Haider and L.C. Liu, Phys. Lett. B **172**, 257 (1986)
[4] V.A.Baskov et al., e-Print: arXiv:1212.6313 [nucl-ex]
[5] S. V. Afanasiev et al., Nucl. Phys. B (Proc. Suppl.) **219-220**, 255 (2011)
[6] S.V. Afanasiev et al., Int. J. Mod. Phys.: Conf. Series **39**, 156009 (2015)

Light ion beams for ADS energy production

A. A. Baldin[1,2,*], *A. I.* Berlev[1,2], *M.* Paraipan[1,3], and *S. I.* Tyutyunnikov[1]

[1]Joint Institute for Nuclear Research, Dubna, 141980 Russia
[2]Institute for Advanced Studies "OMEGA", Dubna, 141986 Russia
[3]Institute of Space Science, Bucharest-Magurele, 077125 Romania

Abstract. The comparative analysis and optimization of the energy efficiency for proton and ion beams in ADS systems is performed via GEANT4 simulation with account of energy consumption for different accelerator types. It is demonstrated that for light nuclei, beginning from ^{7}Li, with energies above 1 GeV/nucleon ion beams are substantially more efficient than 1–3 GeV proton beams.

1 Introduction

The application of accelerated beams for power production has been discussed since 1950[se] [1]. The concept of an energy amplifier implies neutron flux generation at interaction of an extracted beam with a target and energy production based on ^{238}U, thorium ^{232}Th, and other heavy elements fission.

ADS systems possess a number of advantages, such as the possibility of afterburning (transmutation) of radioactive waste (minor actinides), the shorter, as compared to fast reactors, fissile materials production cycle, and so on [2–5]. With increasing beam energy, and correspondingly, hardening the secondary neutron spectrum, it may become possible to use spent nuclear fuel as the fission source [6, 7].

Long-term theoretical and experimental studies formed a widely spread opinion that ADS-optimal beams are protons with an energy from 500 MeV to 1.5 GeV [4, 5, 9]. The commonly considered concept of ADS energy production [10] assumes that a high current proton beam with an energy of ~1 GeV and a target station of heavy elements (natural uranium, thorium) provides an energy amplification factor of 30–40, which makes the development of an energy producing installation realistic. Recent experimental studies [7, 14] with deuteron beams, however, as well as the results of simulation demonstrate that the proton beam energy for ADS is higher, in a range of 2–4 GeV [12]. Since the energy required for accelerating a beam in an accelerator grows quadratically with the beam energy, it is necessary to optimize the beam taking into account the energy spent for the accelerator maintenance.

The possibility of replacing proton beams by light ion beams has been studied insufficiently. Thus, it was stated in [10, 11] that ions with a mass higher than that of tritium are less efficient for ADS. In [8], on the contrary, it was demonstrated that the most

* Corresponding author: an.baldin@mail.ru

efficient is the relativistic ^{20}Ne nuclear beam with an energy of ~3 GeV/nucleon, which agrees with our preliminary results [16].

2 Simulation

The interaction of protons and ions (deuterons, tritium nuclei, ^{7}Li, ^{9}Be, ^{11}B, ^{12}C, ^{14}N, ^{20}Ne, ^{24}Mg, ^{32}S, and ^{40}Ca) with a quasi-infinite target was simulated using GEANT4. In all calculations the cylindrical target from natural uranium with a length of 160 cm and a radius of 60 cm was considered. The simulation was performed for beam energies from 0.3 to 10 GeV/nucleon. The well studied part of the neutron spectrum below 20 MeV was simulated using the parameterizations based on the experimental data ENDF (Evaluated Nuclear Data File). Special attention was paid to the simulation of inelastic hadron interactions.

It was demonstrated in [17] that GEANT4 is applicable for quantitative description of interaction of beams with intermediate energies (from hundreds MeV to several GeV) with various heavy metal targets with a precision of 30%. The comparative analysis of calculations using different codes, namely, GEANT4, MCNPX, and SHIELD [18] also demonstrated satisfactory agreement on a level of 30% for ion beam interaction with thick targets. Experimental data on distribution of fission reactions inside the extended uranium targets using 0.5–4 GeV/nucleon deuteron beams [7, 14] agree within 30% with the results of GEANT4 simulation.

All ionization losses (including ionization losses of δ electrons and gamma conversion) for all charged particles, including primary beam particles and all secondary particles from successive nuclear reactions inside the target, namely, fission fragments, particles produced in multifragmentation, all produced particles and decaying charged particles until their complete stopping in the target (or until they reach the boundary and leave the bulk of the target), were summed for calculation of the energy deposited in the target E_{dep}.

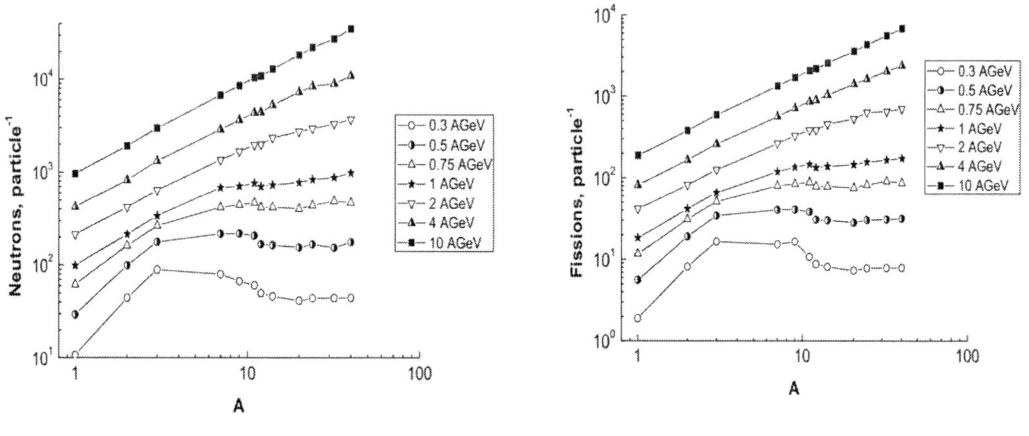

(a) (b)

Figure 1. (a) Neutron yield and (b) number of fissions as functions of the beam ion mass number.

Figure 1 shows the simulated (a) total number of neutrons produced in the target and (b) the number of fissions in natural uranium normalized per one beam ion as functions of the mass number of beam ions for the set of energies in a range of 0.3–10 GeV/nucleon.

Figure 2 shows the integral energy deposited Edep in the considered target for different beam energies as a function of mass number of the beam particles.

The energy deposition in the quasi-infinite target for 10 GeV/nucleon approaches the asymptotic dependence on the mass number of beam ions from protons to ^{40}Ca. Note that the transition to the asymptotic (integral deposited energy proportional to the ion mass number) depends on the beam ion energy. For 1 GeV/nucleon this asymptotic is applicable for light nuclei from protons to lithium, and for 2 GeV/nucleon, up to carbon ions.

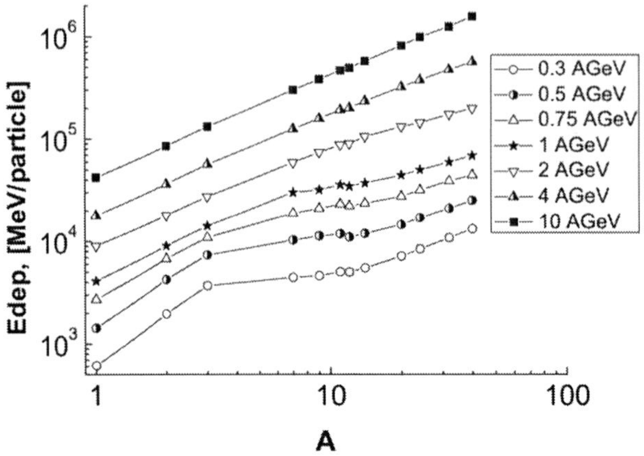

Figure 2. Integral energy deposition per one beam ion as a function of mass number of the incident ion.

The uranium fission channel makes the main contribution into the energy amplification factor due to nuclear transformations. It was demonstrated in [15] that for uranium and other actinides the ratio of the fission cross section and the total inelastic cross section is ~90% for incident proton energy below 1 GeV, and this ratio smoothly tends to 40% for proton energies above 10 GeV. Taking into account the known energy of one uranium nucleus fission event equal to 190 MeV, the relative contribution into the energy deposition in the target from fission reactions as a function of energy for protons and several light nuclei is shown in figure 3.

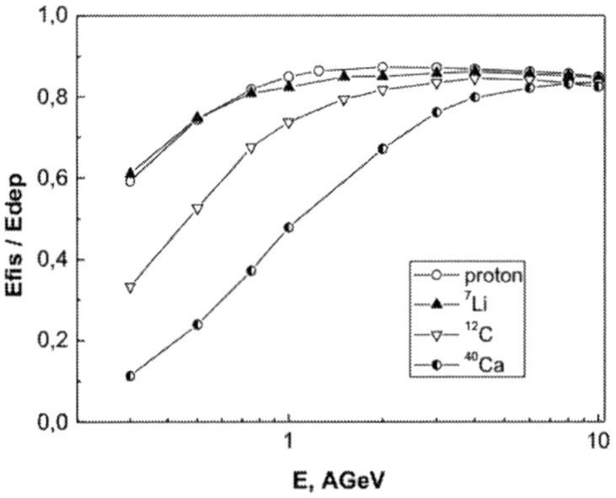

Figure 3. Relative contribution of energy deposited in the target from uranium fission reactions as a function of kinetic energy per nucleon of incident ions.

The simulations shown in figure 2 demonstrate that the 0.5 GeV/nucleon deuteron beam has the same energy deposition as the 1 GeV proton beam, and the 0.5 GeV/nucleon ^{7}Li beam is equivalent to the 3 GeV proton beam.

In this study, for the purpose of comparative analysis of energy consumption in different types of accelerators and accelerated ions, we assume that the energy spent for maintaining the magnetic field per unit length is the same for all cyclic accelerators. The energy spent for maintaining the magnetic field in a synchrotron is $E_{spent} \sim AP/Z$, where Z is the charge, A is the mass number, and P is the momentum of the accelerated ion. For an isochronous cyclotron, we have $E_{spent} \sim (AP/Z)^2$. We assume that for a linear accelerator the spent energy is $E_{spent} \sim AE_k/Z$, where E_k is the kinetic energy per nucleon of the accelerated ion.

Let us consider the relative efficiency ε_r for ion 2, in comparison with ion 1 for different accelerator types,

$$\varepsilon_r = \frac{E_{dep2}}{E_{dep1}} \frac{E_{spent1}}{E_{spent2}}. \quad (1)$$

Here, E_{dep2} and E_{dep1} are the energies deposited in the target for ions 1 and 2 (see figure 2), and E_{spent1} and E_{spent2} are the energy expenses according to the above assumptions for each accelerator type.

Figure 4 shows the relative efficiencies ε_r of energy deposition in the quasi-infinite uranium target as functions of energy and type of accelerated ions. All dependences are normalized to the 1 GeV proton beam (ion 1).

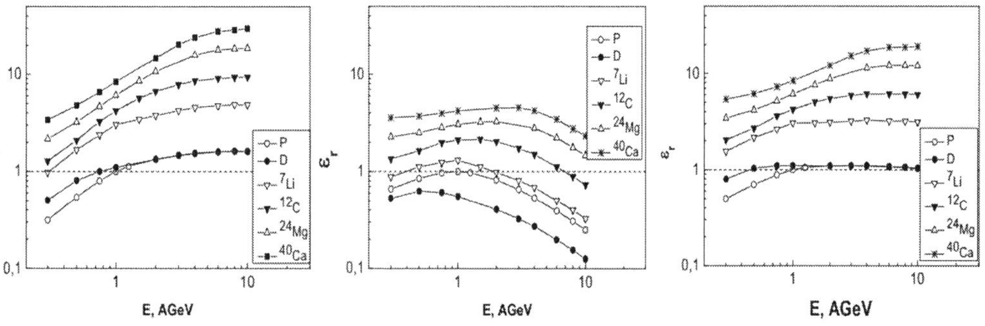

a) b) c)

Figure 4. Relative efficiency of ion energy deposition with respect to 1 GeV protons for beams accelerated in a (a) synchrotron, (b) cyclotron, and (c) linear accelerator.

It can be seen from figure 4 that the optimal proton energy for a synchrotron is ~ 3 GeV, for a linear accelerator, ~ 1–1.5 GeV, and for a cyclotron, ~ 1 GeV. The energy efficiency of light ions is considerably higher than that of protons. Noting this energy efficiency, the optimal energy range shifts from 1 GeV/nucleon for ^{7}Li to 4–6 GeV/nucleon for ^{40}Ca.

It should also be noted that the 0.5 GeV/nucleon ^{7}Li beam is equivalent from the point of view of energy deposition to the 2 GeV proton beam (or 0.75 GeV/nucleon ^{7}Li beam is equivalent to the 3 GeV proton beam). This allows one to obtain the equivalent energy with the same or higher efficiency using an accelerator with the twice as small size using the ^{7}Li beam, as compared to the use of the proton beam.

3 Conclusions

The optimal proton energy for a synchrotron is ~3 GeV, for a linear accelerator, ~1–1.5 GeV, and for a cyclotron, ~1 GeV.

The optimal accelerator energy depends on the type of accelerated ion; thus, for carbon the optimal energy is 2 GeV/nucleon, and for ^{40}Ca ions, 4 GeV/nucleon.

In terms of energy deposition, 0.5 GeV/nucleon ^{7}Li beam is equivalent to 2 GeV protons. This is especially topical for the choice and optimization of the accelerator parameters for an accelerator driven fast neutron reactor.

The experiments with carbon beams [14] indicate that it is necessary to develop and specify the models for description of the experimentally observed enhancement of the hard part of the neutron spectrum in an energy range of 0.5–4 GeV/nucleon. The forthcoming experiments with light ion beams at Nuclotron are topical.

We thank V. I. Yurevich, N. M. Sobolevskii, A. I. Malakhov, A. D. Kovalenko, and A. V. Butenko for helpful discussions.

References

1. G. A. Bartolomeev, *Intense Neutron Source, CONF-660*, 925 (Santa Fe, 1963)
2. V. P. Dmitrievskii, Phys. Part. Nucl. **28(3)**, 815 (1997)
3. V. N. Mikhailov, *JINR Rapid Commun. 6[80]-96*, 17 (1996)
4. C. Rubbia et al., *CERN/AT/93-47* (1993)
5. K. Ismailov, M. Saito, H. Sagara, K. Nishihara, Progress in Nucl. Energy **53**, 925 (2011)
6. J. Adam, et al. Collaboration E&T-RAW (Energy and Transmutation RAW) *JINR Preprint, E1-2010-61* (2010)
7. A. A. Baldin et al., Collaboration E&T-RAW (Energy and Transmutation RAW), Phys. Part. Nucl. Lett. **8(6)**, 606 (2011)
8. D. G. Koshkarev, N. M. Sobolevskii, A. V. Barkhudaryan, Atomic Energy, **105**(3), 225 (2008)
9. V. F. Batyaev et al., Atomic Energy, **104(4)**, 319 (2008)
10. S. R. Hashemi-Nezhad, W. Westmeier, M. Zamani-Valasiadou, B. Thomauske, R. Brandt, Annals of Nuclear Energy **38**, 1144 (2011)
11. V. I. Yurevich, Phys. Part. Nucl. **41(5)**, 778 (2010)
12. V. Pronskikh, N. V. Mokhov, I. Novitski, and S. I. Tyutyunikov, *12th Meeting on Shielding Aspects of Accelerators, Targets and Irradiation Facilities (SATIF-12)* (FNAL, Batavia, Illinois, USA, 2014)
13. W. Furman et al. Collaboration E&T-RAW (Energy and Transmutation RAW), *XXI International Baldin Seminar on High Energy Physics Problems*, (JINR, Dubna, Russia, 2012)
14. M. Artiushenko, A. Baldin, V. Chilap, A. Chinenov, W. Furman, K. Husak, Yu. Petrusenko, V. Sotnikov, S. Tyutyunnikov, V. Voronko, and I. Zhuk, *XXII International Baldin Seminar of High Energy Physics Problem*, (JINR, Dubna, Russia, 2014)
15. V. I. Yurevich, V. A. Nikolaev, R. M. Yakovlev, and A. N. Sosnin, Phys. At. Nucl. **65(8)**, 1383 (2002)
16. M. Paraipan, A. A. Baldin, M. G. Kadykov, and S. I. Tyutyunikov, *XXI International Baldin Seminar on High Energy Physics Problems,* (JINR, Dubna, Russia, 2012)
17. A. A. Baldin, A. I. Berlev, I. V. Kudashkin, G. Mogildea, M. Mogildea, M. Paraipan, and S. I. Tyutyunnikov, Phys. Part. Nucl. Lett. **13(2)**, 249 (2016)
18. A. Krylov, M. Paraipan, N. Sobolevsky, G. Timoshenko, and V. Tretyakov, Phys. Part. Nucl. Lett. **11(4)**, 549 (2014)

 DOI: 10.1051/epjconf/201713810002

Measurements and usage of cross sections of various (n,xn) threshold reactions

P. Chudoba[1,2,⋆], *J.* Vrzalová[1], *O.* Svoboda[1], *A.* Krása[1], *A.* Kugler[1], *M.* Majerle[1], *M.* Suchopár[1,3], and *V.* Wagner[1]

[1]*Nuclear Physics Institute of the CAS,250 68 Řež near Prague,Czech Republic*
[2]*Charles University, Faculty of Mathematics and Physics, Ke Karlovu 3, 121 16 Praha 2*
[3]*Faculty of Nuclear Sciences and Physical Engineering, Czech Technical University in Prague, Břehová 7, 115 19 Prague, Czech Republic*

Abstract. Current trend in nuclear reactor physics is a transition from technologies using thermal neutrons to technologies utilizing fast neutrons. Unfortunately focus was put mainly on the thermal neutrons for a long time and lead to very good knowledge about this low energy region, but very scarce coverage of the high energy region. This means that there is a gap in the knowledge of excitation functions for higher energies. This gap spreads from 20 MeV up to 1 GeV and higher. This is exactly the energy region needed for description of advanced nuclear systems such as accelerator driven systems (ADS). Our group from Nuclear Physics Institute (NPI) of the CAS is a member of an international collaboration Energy & Transmutation of Radioactive Waste (E&T RAW). This collaboration focuses on ADS for many years. In order to measure neutron field within ADS models it is necessary to know excitation functions of reactions used to monitor the neutron field. In many cases there are almost no experimental data for suitable reactions. Worse and quite common case is that there are no data at all. Therefore we are also focusing on measurements of these data in order to fill the databases as well as to allow further improvements of codes for nuclear data calculations.

⋆e-mail: chudoba@ujf.cas.cz

1 Introduction

Neutron field monitoring is crucial for many devices and systems such as spallation neutron sources, accelerator driven systems and other kinds of advanced nuclear systems. Since neutrons are electrically neutral hadrons it is difficult to detect them. The neutron spectroscopy is even more difficult. If it is not necessary to measure neutron flux on-line it is possible to take advantages of nuclear reactions and delayed measurements of a number of produced nuclei. One of possibilities is to utilize (n,xn) reactions and γ-spectroscopy methods.

Unfortunately experimental coverage and the knowledge of excitation functions are sufficient only for energies approximately up to 20 MeV. For higher energies the data are very scarce and as the consequence the parameters for the corresponding excitation functions are based rather on nuclear models than on experimental data. Our group from NPI focuses on cross section measurements of (n,xn) reactions on various materials. The purpose of using of various materials is to measure the neutron spectrum at some particular point based on the unfolding with help of the (n,xn) threshold reactions.

This article describes the procedure of calculation of cross section and possible problems which are necessary to be kept under the eye. Particularly it will be shown on the ^{197}Au(n,8n)^{190}Au reaction instance. Therefore its main purpose is to show the procedure, its peculiarities and the consequences arising from the insufficient amount of experimental data in higher energy region. Our group has long tradition in measurement of cross sections in various energy regions e.g. [1].

2 Cross section measurements

The road-plan of the measurement is quite straightforward. Going from the backwards the γ-spectroscopy methods are used to obtain the number of nuclei produced by the investigated reaction. In order to produce the nuclei it is necessary to irradiate samples with a viable neutron beam. The samples must be in form which allows good handling and measurement and must have mass which is optimized accordingly to the neutron source intensity. We are using thin metallic foils with dimensions in order of several squared centimeters and a mass of $\approx$0.5g.

2.1 Neutron source

For the cross sections measurements we take advantage of an opportunity to use a quasi-monoenergetic (QM) neutron source. In fact we are able to use two similar sources based on reaction ^{7}Li(p,n)^{7}Be. One is driven by the U-120M cyclotron at the NPI [2] and the other is driven by the Gustaf Werner Cyclotron at The Svedberg Laboratory (TSL) [3] of the Uppsala University. The ^{197}Au(n,8n)^{190}Au reaction was measured using the neutron source at TSL. The measurement was carried out within the EFNUDAT program.

2.2 γ-spectroscopy

After the irradiation and necessary security checks the samples are moved from irradiation housing to a spectroscopy laboratory. For the measurement a spectrometer based on HPGe detector is used. The obtained spectrum is then evaluated in a spectroscopy program. The evaluation consists of fitting the peaks which correspond to the γ-quanta emitted by the produced nuclei. The results from evaluation are in form of areas of the peaks which correspond to number of detected decays with respective errors originating from the fitting procedure. Once the areas are evaluated then the spectroscopy equation (1) [4] takes place. Various spectroscopic corrections are involved within discussed equation.

$$N_{yield} = \frac{S_{peak} \cdot C_{abs}(E)}{I_\gamma \cdot \varepsilon_p(E) \cdot COI(E) \cdot C_{area}} \frac{t_{real}}{t_{live}} \frac{e^{\lambda \cdot t_0}}{1 - e^{-\lambda \cdot t_{real}}} \frac{\lambda \cdot t_{irr}}{1 - e^{-\lambda \cdot t_{irr}}}. \tag{1}$$

The variables and parameters in the equation (1) are S_{peak} - peak area, $C_{abs}(E)$ - self absorption correction, I_γ - γ-quantum intensity, $\varepsilon_p(E)$ - peak efficiency, $COI(E)$ - coincidence correction coefficient, C_{area} - square emitter correction, t_{real} - real time, t_{live} - live time, t_0 - time between end of irradiation and start of the measurement, λ - decay constant, t_{irr} - time of the irradiation.

Once the yield is calculated it is possible to calculate the cross section for a particular reaction. Integral number of neutrons is necessary for the calculation. The calculation is described by a forthcoming equation (2)

$$\sigma = \frac{N_{yield} \cdot S \cdot A \cdot B_a}{N_n \cdot N_A \cdot m}. \tag{2}$$

The variables are S - sample area, A - atomic number, B_a - beam instability correction coefficient, N_n - number of neutrons, N_A - Avogadro's number and m - mass of the sample.

2.3 Background subtraction

Since the used neutron source is not monoenergetic the neutron spectrum consists of the QM peak and the lower energy continuum, it is necessary to perform a subtraction of the nuclei produced by the lower energy continuum. The correction can be quite significant. Its magnitude depends on the excitation function shape and reaction threshold energy. The neutron spectrum and one type of excitation functions is shown in figure 1(a). From this figure it is obvious how the low energy neutrons induce the investigated reaction.

In order to correct the number of produced nuclei N_{yield} we use a folding of neutron spectra and the excitation function. Particularly the ratio between the folding within the QM peak energy interval and the folding in the full neutron energy region. Since we have the spectra and excitation functions in form of bins, it is necessary to move from continuous integration into discrete summation. The procedure is expressed via equation (3).

$$C_{bgr} = \frac{\int_{Peak} \sigma(E) \cdot N(E) \mathrm{d}E}{\int_{Spectrum} \sigma(E) \cdot N(E) \mathrm{d}E} \longrightarrow C_{bgr} = \frac{\sum_{i \in Peak} \sigma_i \cdot N_i}{\sum_i \sigma_i \cdot N_i} \tag{3}$$

The variables stands for σ - excitation function and $N(E)$ for neutron spectrum. The number of nuclei produced by the QM peak is then

$$N_{yield}^{QM} = C_{bgr} \cdot N_{yield}. \tag{4}$$

3 Results and their development

Following the steps described earlier we have arrived to a result shown in figure 1(a). It is possible to see, that the agreement with TALYS 1.8 [5] calculation is excellent. The excitation function was obtained using default parameters with level density described using Constant Temperature Formula (CTF) and Back-shifted Fermi Gas Model (FGM). Since TALYS 1.8 provides a set of parameters based on the experimental data we have made another calculation using this parameterization called "'Best parameters"'. The results together with the default model are shown in figure 1(b). Once again it is possible to see excellent agreement, which suggest strong model dependency. Figures 2(a) and (b) show the excitation functions calculated with TALYS 1.8 and different level density parameters used

 DOI: 10.1051/epjconf/201713810002

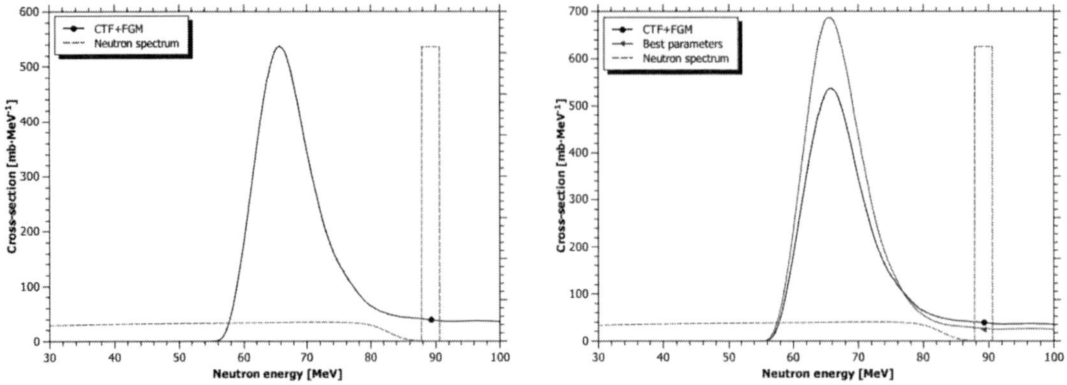

(a) (b)

Figure 1. (a) Comparison between calculated excitation function with default parameters and experimental point; (b) Comparison of two parameterization of excitation function showing strong model dependency. Calculated excitation functions were used for background subtraction.

for the background subtraction procedure and corresponding results of cross section. It is possible to see, that the model dependence is more or less significant for all excitation functions.

First motivation for the previous calculations was to recalculate older results with the newest version of TALYS. Due to obtained results, a deeper study was conducted. Using script for fitting of the yields of two different isotopes with overlapping γ-lines the yields and cross sections for lower energies were obtained.

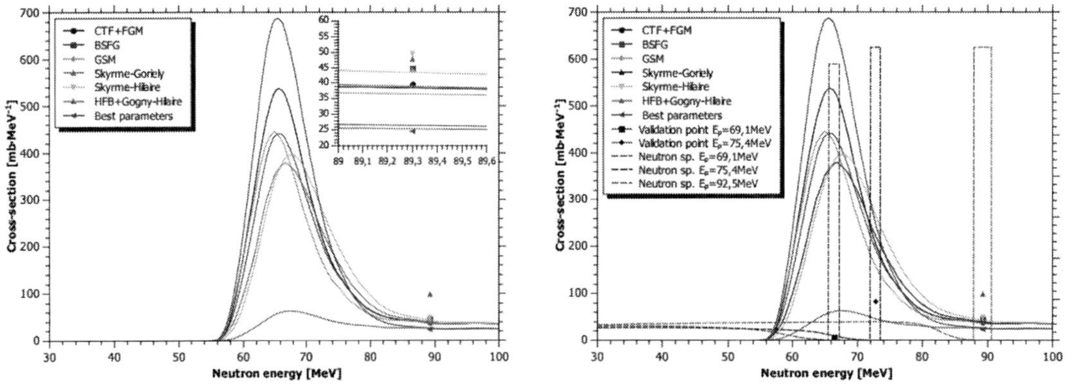

(a) (b)

Figure 2. (a) Visualization of behavior of excitation functions with different models of level density and "best parameterization" and its effect on experimental values; (b) Introduction of additional experimental points in order to determine the most accurate excitation function.

DOI: 10.1051/epjconf/201713810002

Figure 2(b) shows that the excitation functions are highly overestimated and shifted to a lower energy. The calculations for the validation points were done without background subtraction procedure. However from the figure 2(b) it can be seen that those points would be influenced only slightly. This is proven in the next figure.

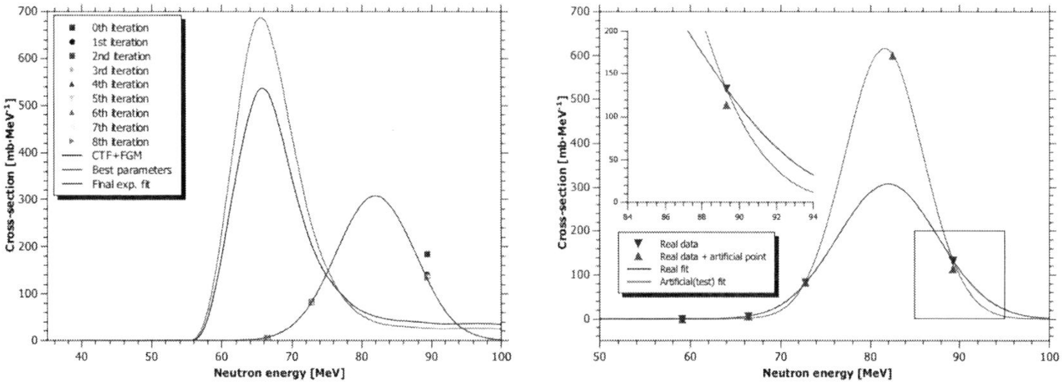

(a) (b)

Figure 3. (a) Evolution of experimental points during self-consistent iteration of the excitation function; (b) Test of correctness of the excitation function maximum.

Figure 3(a) shows the iteration of the cross section points. It can be seen that first two points are almost without influence of the low energy neutrons. The third point has dropped after first step and then the changes were only minimal. The iteration was stopped when the relative changes were lower than 0.01%. Since there are no experimental data for the maximum of the fitted peak an artificial point was introduced into data-set to increase the maximum to similar value as the calculated excitation function. Figure 3(b) shows that the excitation function with the artificial point is overestimated. It is propagating as undershooting of the fitted excitation function if it is taken as the parameter for the background subtraction procedure. We can see no such undershooting for the excitation function without the artificial point.

4 Conclusion

We have seen that the calculated excitation functions and the corresponding cross sections calculated with their help do not have to be accurate even if there is good agreement between the calculation and the experiment. Furthermore it has shown up that the excitation function can be even shifted in energy. Our fit of the excitation function is performed with Gaussian as the fitted function. There is no reason to doubt that the calculated shape is completely wrong and the Gaussian agrees with this shape very well in the peak region. The disagreement in shape has only negligible influence on the results. Since we have no data in the tail region for higher energies we can use the Gaussian for the iterative background subtraction. The data are still in preliminary phase and that is the reason why there are no numerical data or the error bars. However there is not going to be any significant change of the outcome with the finalization.

Acknowledgments

The experiment was supported by the European Union's program EFNUDAT.

References

[1] J. Vrzalová, O. Svoboda, A. Krása, A. Kugler, M. Majerle, M. Suchopár, V. Wagner, Nuclear Instruments and Methods in Physics Research Section A: Accelerators, Spectrometers, Detectors and Associated Equipment **726**, 84 (2013)

[2] P. Bém, V. Burjan, J. Dobeš, U. Fischer, M. Götz, M. Honusek, V. Kroha, J. Novák, S.P. Simakov, E. Šimečková, *The NPI cyclotron-based fast neutron facility*, in *ND2007* (EDP Sciences, Les Ulis, France, 2007). http://nd2007.edpsciences.org/10.1051/ndata:07598

[3] A.V. Prokofiev, J. Blomgren, O. Bystrom, C. Ekstrom, S. Pomp, U. Tippawan, V. Ziemann, M. Osterlund, Radiation Protection Dosimetry **126**, 18 (2007)

[4] V. Wagner, O. Svoboda, J. Vrzalová, M. Suchopár, B. Geier, A. Kugler, M. Honusek, Journal of Physics: Conference Series **366**, 012047 (2012)

[5] A. Koning, D. Rochman, Nuclear Data Sheets **113**, 2841 (2012)

Monte carlo simulations of Yttrium reaction rates in Quinta uranium target

M. Suchopár[1,2,★], *V.* Wagner[1,2], *O.* Svoboda[1], *J.* Vrzalová[1,2,4], *P.* Chudoba[1,3], *P.* Tichý[1,2,4], *A.* Kugler[1], *J.* Adam[1,4], *L.* Závorka[2,4], *A.* Baldin[4,5], *W.* Furman[4], *M.* Kadykov[4,†], *J.* Khushvaktov[4], *A.* Solnyshkin[4], *V.* Tsoupko-Sitnikov[4], *S.* Tyutyunnikov[4], *M.* Bielewicz[4,6], *S.* Kilim[6], *E.* Strugalska-Gola[6], and *M.* Szuta[6]

[1]*Nuclear Physics Institute of the ASCR PRI, Hlavní 130, 250 68 Řež, Czech Republic*
[2]*Faculty of Nuclear Sciences and Physical Engineering, Czech Technical University in Prague, Břehová 7, 115 19 Prague 1, Czech Republic*
[3]*Faculty of Mathematics and Physics, Charles University in Prague, Ke Karlovu 3, 121 16 Prague 2, Czech Republic*
[4]*Joint Institute for Nuclear Research, Joliot-Curie 6, 141 980 Dubna, Russia*
[5]*Institute for Advanced Studies "OMEGA", Universitetskaya 19, 141 980 Dubna, Russia*
[6]*National Centre for Nuclear Research, 05-400 Otwock-Świerk, Poland*

Abstract. The international collaboration Energy and Transmutation of Radioactive Waste (E&T RAW) performed intensive studies of several simple accelerator-driven system (ADS) setups consisting of lead, uranium and graphite which were irradiated by relativistic proton and deuteron beams in the past years at the Joint Institute for Nuclear Research (JINR) in Dubna, Russia. The most recent setup called Quinta, consisting of natural uranium target-blanket and lead shielding, was irradiated by deuteron beams in the energy range between 1 and 8 GeV in three accelerator runs at JINR Nuclotron in 2011 and 2012 with yttrium samples among others inserted inside the setup to measure the neutron flux in various places. Suitable activation detectors serve as one of possible tools for monitoring of proton and deuteron beams and for measurements of neutron field distribution in ADS studies. Yttrium is one of such suitable materials for monitoring of high energy neutrons. Various threshold reactions can be observed in yttrium samples. The yields of isotopes produced in the samples were determined using the activation method. Monte Carlo simulations of the reaction rates leading to production of different isotopes were performed in the MCNPX transport code and compared with the experimental results obtained from the yttrium samples.

1 Introduction

Yttrium as an activation material has several advantages when investigating fast neutron fluxes produced in spallation reactions of relativistic light ions impacting on thick heavy metal targets used in

★e-mail: suchopar@ujf.cas.cz
†deceased

ADS studies [1–4]. Yttrium in natural occurrence is monoisotopic with one stable isotope ^{89}Y. Products of (n,xn) threshold reactions on yttrium are easily identifiable. Half-lives of the products have suitable duration for gamma spectroscopy. Gamma transitions are intensive enough for detection and they are also well separated from each other. The yttrium activation detectors were irradiated during experiments with the Quinta uranium target [5, 6] at the Joint Institute for Nuclear Research (JINR) in Dubna, Russia. The deuteron beams with energies ranging from 1 GeV up to 8 GeV were produced by JINR Nuclotron synchrotron. Residual radionuclides in the samples were measured by the gamma spectrometry.

2 Experimental method and simulation procedure

The samples had a form of pills made of compressed yttrium powder with a diameter of 10 mm, a thickness of 1.5 mm, and a weight of ~0.6-0.8 g [9]. The purity of used ^{89}Y was 99.99%. These samples were placed on the front side of the detector plates in two positions at radial distances of 4 and 8 cm from the target axis (5 and 8 cm in case of the first detector plate with a hole for the beam entrance). The aluminium detector plates were located in the gaps between the setup sections at longitudinal distances of 0, 12.9, 26.2, 39.1, 52.4, and 65.3 cm from the target beginning. Schema of the experimental setup and locations of yttrium activation detectors in the Quinta assembly are depicted in figure 1. Twelve yttrium samples in total were used in each irradiation of the Quinta setup in 2, 4 and 6 GeV deuteron experiments carried out in March 2011 (without lead shielding) [7], 1 and 4 GeV deuteron experiments in December 2011 (with lead shielding) [9], and 2, 4 and 8 GeV deuteron experiments in December 2012 (again with lead shielding) [8]. Products of ^{89}Y(n,2n)^{88}Y, ^{89}Y(n,3n)^{87}Y, ^{89}Y(n,4n)^{86}Y, and ^{89}Y(n,5n)^{85}Y reactions with threshold energies 11.6, 21.1, 33.0, and 42.6 MeV respectively were detected using gamma spectroscopy and their yields per gram of activation material per beam deuteron were determined. The acquired gamma spectra were analysed using the Deimos32 spectroscopy software [10] with all necessary spectroscopic corrections applied [11]. The yield of each isotope was then calculated as a weighted average of at least two gamma lines.

The beam integral was determined using an aluminium foil located at sufficient distance from the setup (about 3 m) so that the measurement was not affected by backscattered neutrons from the setup and reaction rates from the activation foils were normalized to this beam monitor. The beam integral was determined with around 10% accuracy due to the fit of the excitation function of the ^{27}Al(d,3p2n)^{24}Na monitoring reaction as only 3 values for this reaction cross section are present in the EXFOR database in the GeV energy range [12, 13]. The beam position and profile were determined using the solid state nuclear track detectors (SSNTD) by means of the measurement of distribution of induced fission rates by the beam in natural lead samples [14, 15]. A fit of the experimental data by Gaussian distribution in the horizontal and vertical direction was used as input parameters for the particle source in the Monte Carlo simulations.

We simulated the isotope production in threshold (n,xn) reactions in yttrium samples and compared them with the experimental results obtained by the group from the National Centre for Nuclear Research in Otwock-Świerk, Poland [7–9]. The Monte Carlo simulations of the yields of activation reactions in the samples were performed with the MCNPX code version 2.7.0 [16]. Standard cross section libraries ENDF/B-VII.1 [17] included in the MCNPX code package were used for the simulations coupled with the INCL intranuclear cascade and ABLA fission-evaporation models for the primary beam interaction and high energy particle transport. Microscopic cross sections of ^{89}Y(n,xn) threshold reactions were calculated by the Talys code version 1.6 [18] with default parameters of the level density models. The simulated yields were calculated by folding of the reaction cross sections calculated by Talys (up to 200 MeV) and MCNPX (up to the beam energy) with neutron, proton and deuteron spectra computed by MCNPX. The contribution to the total reaction yield is produced

DOI: 10.1051/epjconf/201713810003

mainly by neutrons, protons and deuterons (near the beam axis), contribution by other particles (such as charged pions) can be neglected being on the level of the simulation uncertainties.

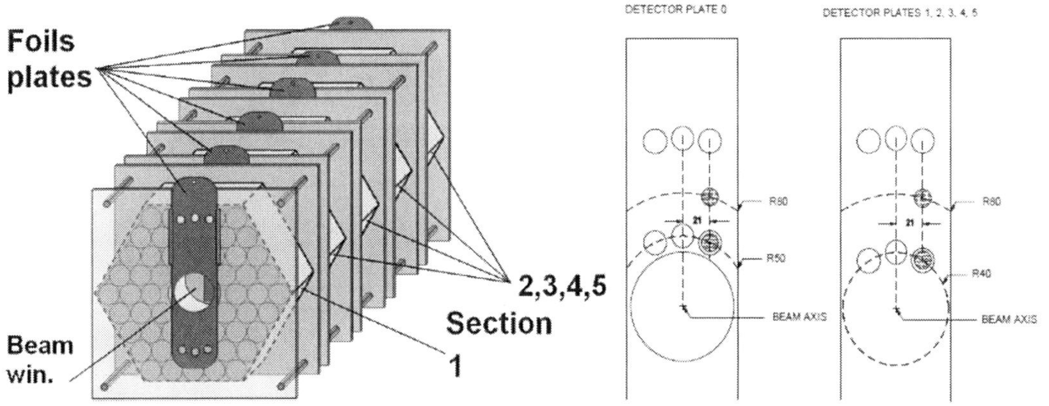

Figure 1. Schema of the Quinta setup (left). Locations of yttrium activation samples on the detector plates (right) [8]. Dimensions are in millimeters.

3 Discussion of results

After proper normalization of the data, we could compare the experimental spatial distributions of the reaction rates with our Monte Carlo calculations in the MCNPX code. The absolute and relative comparisons of experimental and simulated data in the axial direction in case of the three above mentioned 4 GeV experiments are shown in figure 2-4. The ratio of experimental to simulated yields varies between 0.5 and 1.5 in most cases and confirms relatively good agreement between the experiment and simulation, although the simulation generally underestimates the experimental data. The only apparent discrepancy appears at the maximum of isotopes production which is located at about 26 cm from the target beginning where the simulation underestimates the experimental data by a factor of two for the position at 4 cm in the 2, 4 and 6 GeV experiments performed without lead shielding but not in the rest of experiments where the Quinta setup was shielded with lead. This may be caused by a slight misplacement of the samples from their anticipated positions on the third plate due to the presence of other groups' samples rather than an incorrect beam displacement input in the simulation as the other foils near the beam axis are not influenced by this effect. Another discrepancy was observed in case of the 1 GeV experiment but only for the front detector plate. It could be caused by inaccurate beam position determination on the front side of the target or inaccurate placement of the samples on the first detector plate. Therefore, it has to be emphasized that the results strongly depend on precise beam parameters determination and are very sensitive to accurate activation foils placement in the setup.

4 Conclusion

Measurements of fast neutron field in the Quinta assembly were performed by means of yttrium activation detectors irradiated with 1, 2, 4, 6 and 8 GeV deuteron beams from the Nuclotron accelerator at JINR, Dubna. Experimental data on (n,xn) reaction rate distributions provided by the group from

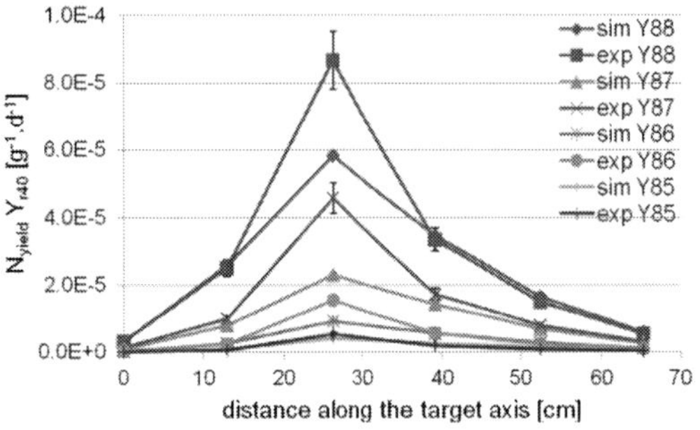

(a) radius 40 mm

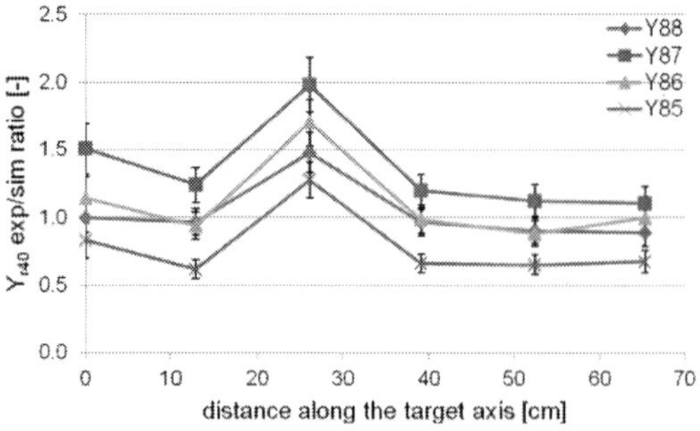

(b) radius 40 mm, exp/sim ratio

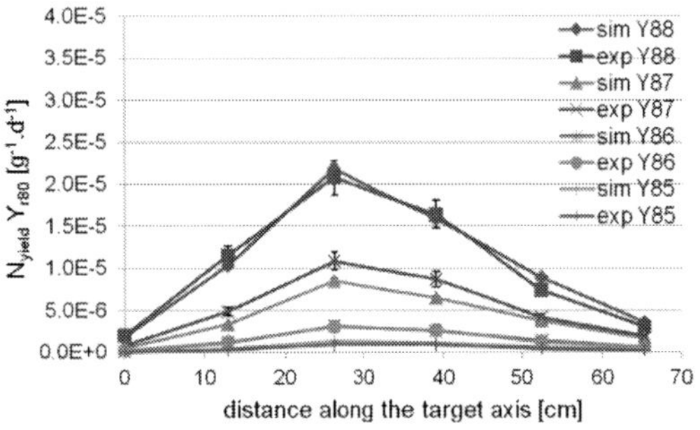

(c) radius 80 mm

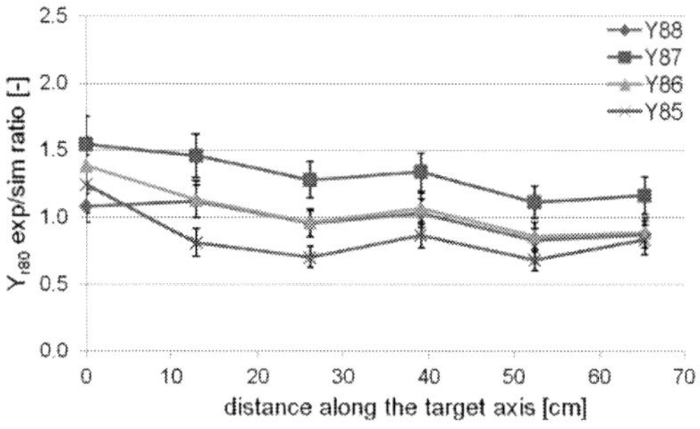

(d) radius 80 mm, exp/sim ratio

Figure 2. Measured and calculated yields of observed isotopes in yttrium detectors in longitudinal direction at 40 mm (a) and 80 mm (c) above the target axis and relative comparison of experimental and simulated yields at 40 mm (b) and 80 mm (d) above the target axis, 4 GeV experiment in March 2011.

the National Centre for Nuclear Research in Otwock-Świerk, Poland were compared with reaction rates calculated by the MCNPX code. The experimental spatial distribution of ^{88}Y, ^{87}Y, ^{86}Y and ^{85}Y production as well as the simulated one has a maximum at about 26 cm from the front of the target assembly as it was expected. The shape of reaction rate distribution is generally similar for all beam energies but differs in absolute value, which depends also on impacting beam position on the target. The ratio of experimental and simulated yields varies in the interval 0.5-2.0 in most cases and confirms relatively good agreement between the two, although the simulation generally underestimates the experimental data. Some discrepancy was observed at the maximum of isotopes production, where the simulation apparently underestimates the experimental data for the position at 4 cm in the 2, 4 and 6 GeV experiments performed without lead shielding but not in other experiments with lead shielding in place.

References

[1] V. Wagner et al., PoS (Baldin ISHEPP XXI) 090 (2012)
[2] V. Wagner et al., Journal of Physics: Conference Series **366**, 012047 (2012)
[3] P. Chudoba et al., PoS (Baldin ISHEPP XXII) 054 (2014)
[4] P. Chudoba et al., Physics Procedia **59**, 114–118 (2014)

DOI: 10.1051/epjconf/201713810003

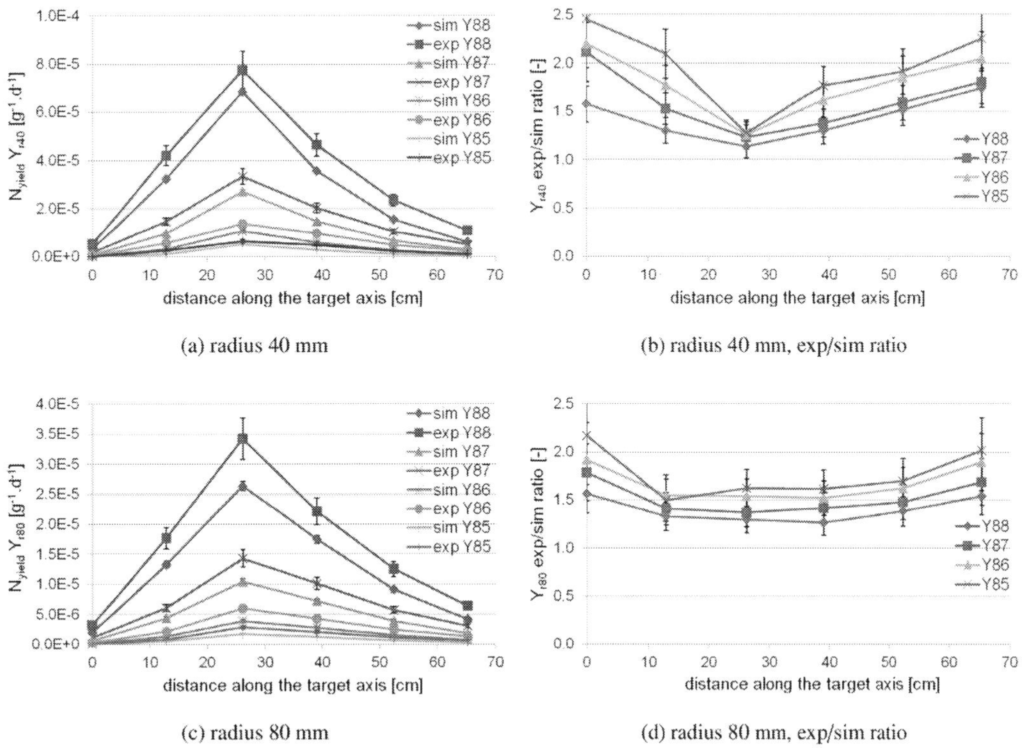

(a) radius 40 mm

(b) radius 40 mm, exp/sim ratio

(c) radius 80 mm

(d) radius 80 mm, exp/sim ratio

Figure 3. Measured and calculated yields of observed isotopes in yttrium detectors in longitudinal direction at 40 mm (a) and 80 mm (c) above the target axis and relative comparison of experimental and simulated yields at 40 mm (b) and 80 mm (d) above the target axis, 4 GeV experiment in December 2011.

[5] W. Furman et al., PoS (Baldin ISHEPP XXI) 086 (2012)
[6] M. Suchopár et al., PoS (Baldin ISHEPP XXI) 091 (2012)
[7] M.M. Bielewicz et al., PoS (Baldin ISHEPP XXI) 092 (2012)
[8] M.M. Bielewicz et al., PoS (Baldin ISHEPP XXII) 052 (2014)
[9] M. Szuta et al., PoS (Baldin ISHEPP XXII) 060 (2014)
[10] J. Frána, Journal of Radioanalytical and Nuclear Chemistry **257**, 583-587 (2003)
[11] M. Majerle et al., JINR Preprint E11-2009-178 (Dubna, Russia, 2009)
[12] J. Banaigs et al., Nuclear Instruments and Methods in Physics Research **95**, 307-311 (1971)
[13] P. Kozma and V.V. Yanovski, Czech Journal of Physics **40**, 393 (1990)
[14] S.R. Hashemi-Nezhad et al., Nuclear Instruments and Methods in Physics Research A **664**, 154-160 (2012)
[15] V. Bukhal et al., PoS (Baldin ISHEPP XXII) 053 (2014)
[16] D.B. Pelowitz, editor, *MCNPX User's Manual, Version 2.7.0*, LANL report LA-CP-11-00438 (Los Alamos, USA, 2011)
[17] M.B. Chadwick et al., Nuclear Data Sheets **112**, 2887–2996 (2011)
[18] A.J. Koning et al., *TALYS-1.6, A nuclear reaction program, User manual* (NRG Petten, The Netherlands, 2013)

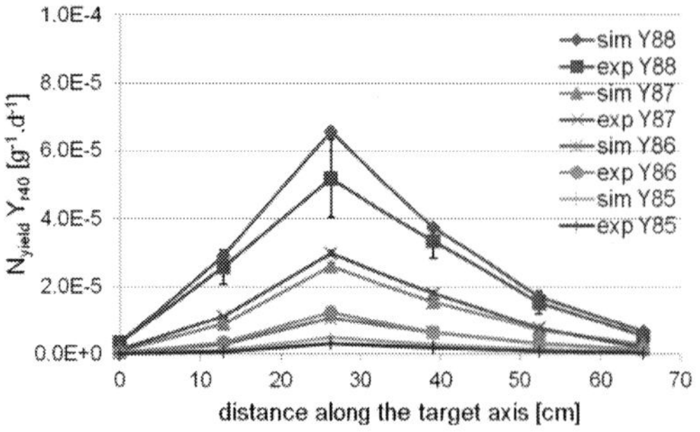

(a) radius 40 mm

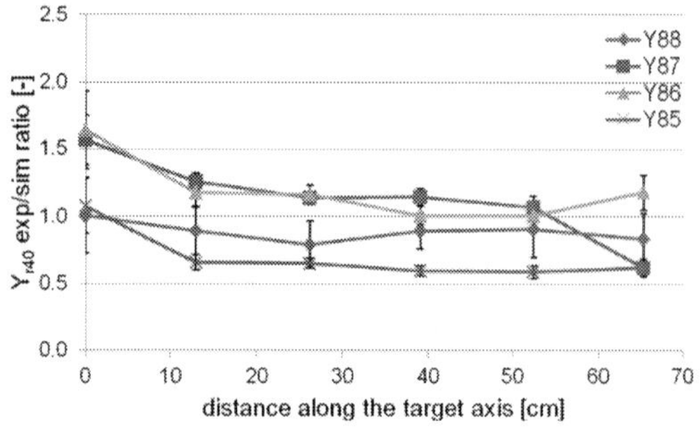

(b) radius 40 mm, exp/sim ratio

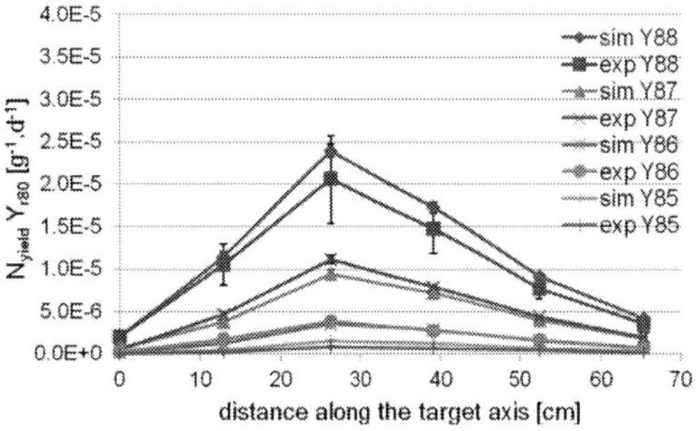

(c) radius 80 mm

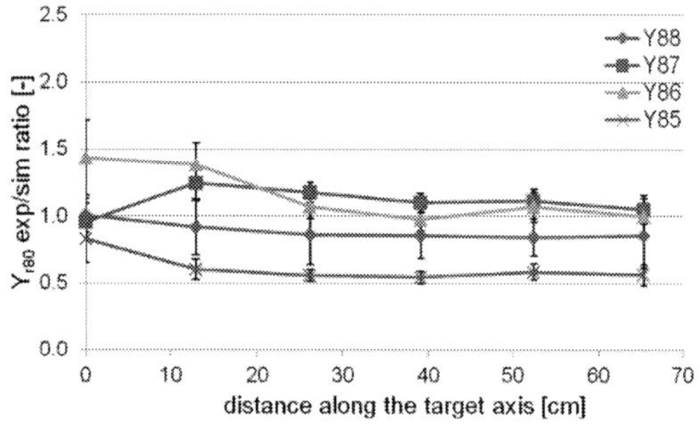

(d) radius 80 mm, exp/sim ratio

Figure 4. Measured and calculated yields of observed isotopes in yttrium detectors in longitudinal direction at 40 mm (a) and 80 mm (c) above the target axis and relative comparison of experimental and simulated yields at 40 mm (b) and 80 mm (d) above the target axis, 4 GeV experiment in December 2012.

Comparison of neutron induced fission and capture in Np-237 and Pu-239 irradiated in QUINTA assembly with 660 MeV proton beam

Stanislaw Kilim[1,⋆], *Elzbieta* Strugalska-Gola[1,⋆⋆], *Marcin* Szuta[1,⋆⋆⋆], *Marcin* Bielewicz[1,2,⋆⋆⋆⋆], *Sergey* Tyutyunnikov[2,†], and *Vladimir* Stegailov[2,‡]

[1]*National Centre for Nuclear Research ul. Andrzeja Soltana 7, 05-400 Otwock, Poland*
[2]*Joint Institute for Nuclear Research Joliot Curie str. 6, 141980 Dubna, Russia*

Abstract. Two Np-237 samples and one Pu-239 were irradiated in spallation neutrons produced in ADS setup QUINTA. The accelerated beam consisted of protons of energy 660 MeV. The method was based on gamma-ray spectrometry measurement. During analysis of the spectra several fission products and one actinide were identified. Fission product activities gave the number of fissions. The actinide (Np-238), a result of neutron capture by Np-237 gave the number of captures. In a similar manner the number of fissions in Pu-239 was determined. The Pu-240, a product of neutron capture by Pu-239, activity was impossible to measure.

1 Introduction

Np-237 and Pu-239 are produced artificially in a power reactor as a byproduct of energy production. They interact with reactor neutrons in two ways either their nuclei are split in two parts, fission products or they capture the neutrons resulting in production of other actinides. About 10-20% of fissionable isotopes are transmuted in reactor this way. As they are long lived and in a typical nuclear power reactor the parasitic neutron capture prevails, it is difficult to incinerate them. They accumulate. The only way to neutralise them is their fission.

Figure 1 shows Np-237 and Pu-239 fission product yield (the upper part) and fission and neutron capture cross section dependence on energy. It is seen that only for neutrons of energy above 1 MeV the fission prevails the capture. It means that either fast reactor or Accelerator Driven System (ADS) provide neutrons of the most appropriate energy to incinerate the actinides. In an international research project "Energy & Transmutation of Radio-Active Waste" (E&T RAW) QUINTA setup is a kind of ADS-type spallation neutron source made of natural uranium surrounded by lead shield. It is located in Dubna, Russia in Joint Institute for Nuclear Research.

⋆e-mail: stanislaw.kilim@ncbj.gov.pl
⋆⋆e-mail: elzbieta.strugalska-gola@ncbj.gov.pl
⋆⋆⋆e-mail: marcin.szuta@ncbj.gov.pl
⋆⋆⋆⋆e-mail: marcin.bielewicz@ncbj.gov.pl
†e-mail: tsi210647@yandex.ru
‡e-mail: stegajlov2013@yandex.ru

 DOI: 10.1051/epjconf/201713810004

This work shows the results of Np-237 and Pu-239 incineration measurements. Proton beam accelerated in a PHASOTRON impinged on QUINTA core producing fast neutrons, spallation neutron field. Np-237 and Pu-239 samples were irradiated in it and resulting gamma activity was measured then.

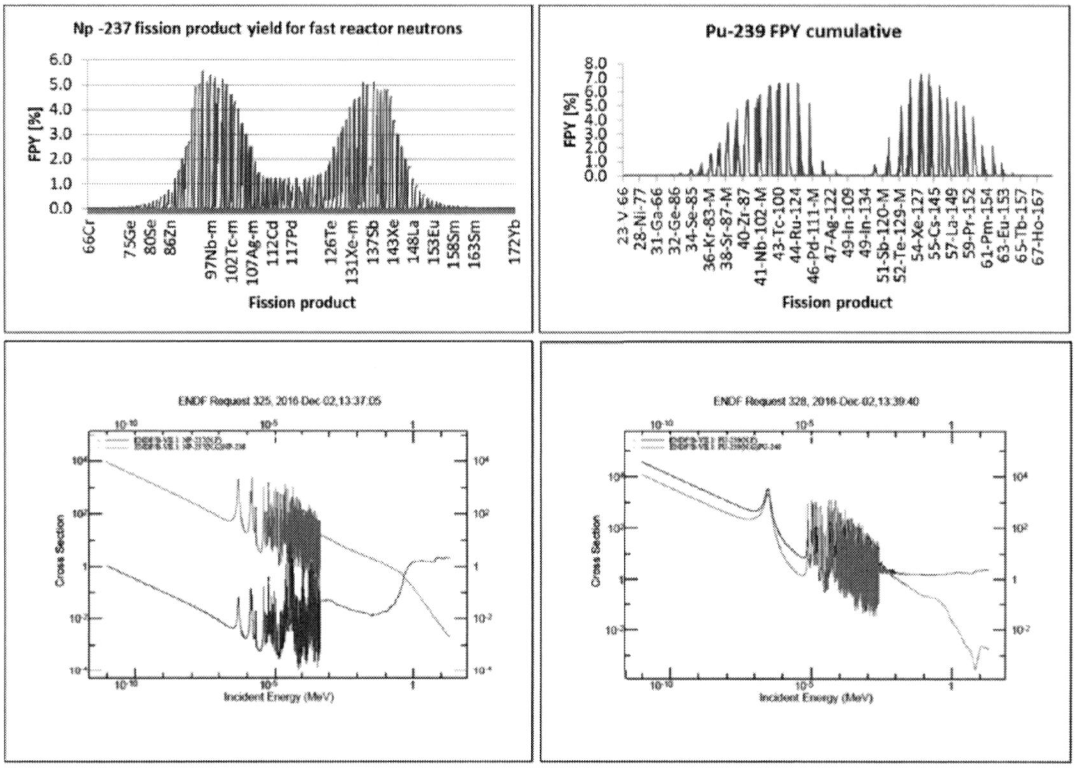

Figure 1. Np-237 (left) and Pu-239 (right) fission product yield (FPY) - upper view and (n,f) and (n,γ) cross section bottom view.

2 Experiment and work out description

The QUINTA setup is shown below - figure 2. Its core contains 500 kg of natural uranium elements cladded in aluminium. More details on QUINTA can be found in [1]. Actinide samples were located in the "window marked" on red, seen on the right drawing.

 DOI: 10.1051/epjconf/201713810004

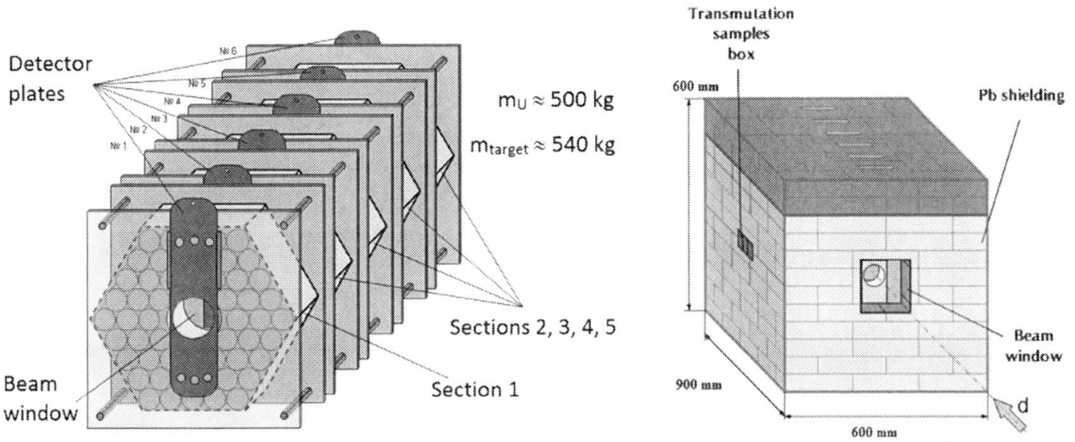

Figure 2. QUINTA setup view - internal core and front view.

Two experiments were carried out, see table 1. HPGE detector was used for gamma spectra registration then. To protect the detector against overload by X and low-energy gammas special filters were applied:

- 1mm Cu, 1 mm Cd and 12 mm Pb for Np-237 sample measurements.
- 1mm Cu, 1 mm Cd and 8 mm Pb for Pu-239 sample measurements.

The spectra were analyzed then with GammaMCA program to get a peak energy and a peak area. The peak energy lets us identify the parent isotope. There are two classes of the parent isotopes, fission products and absorption products.

The found peak areas were corrected for gamma line intensity, detector efficiency, fission product yield, cooling time and detector dead time formula 1. Some isotopes are chained, i.e. the decaying predecessor in chain gives additional rise to its quantity. More complicated formula 2 was used then. The final result was either fission rate, i.e. number off fissions per second, per sample gram and per one proton or capture rate. As concerns Pu-239 both fission and capture take place in neutron field. The main difference from Np-237 is in proportions. The fission prevails capture in Pu-239. In addition, neutron capture product Pu-240 undergoes α decay to U-236 with no gamma emission in 72.8% or with 45.242 keV X-radiation emission in 27.1%. Because of used Pb-filters and low activity caused by long (6563y) half life time it was impossible to determine number of captures in Pu-239.

Table 1. Experiments data

Experiment date	Irradiation duration	Beam	Energy	Fluence	Samples
2014-11-08	6h	Protons	660 MeV	$4.5x10^{15}$	Np-237-4
2015-12-04	5h 43m	Protons	660 MeV	$8.64x10^{14}$	Np-237-4 Pu-239-3

$$R_{f1g} = \frac{A_{1g}}{m\phi\gamma\epsilon_{pg}I_g}\frac{\lambda_1 t_{irr}}{(1-e^{-\lambda_1 t_{irr}})}\frac{e^{\lambda_1 t_+}}{(1-e^{-\lambda_1 t_{real}})}\frac{t_{real}}{t_{live}} \quad (1)$$

$$R_{f2g} = \frac{A_{2g}}{m\phi\epsilon_g I_g}\frac{t_{real}}{t_{live}}\frac{t_{irr}}{[c_1(1-e^{-\lambda_1 t_{real}})e^{-\lambda_1 t_+} + c_2(1-e^{-\lambda_2 t_{real}})e^{-\lambda_2 t_+}]}$$

$$c_1 = \frac{\gamma_1}{\lambda_1}(1-e^{-\lambda_1 t_{irr}})$$

$$c_2 = \frac{\gamma_2}{\lambda_2}(1-e^{-\lambda_2 t_{irr}}) \qquad (2)$$

R_{fg} - actinide fission rate, per deuteron and per gram
g - gamma line index
f - reaction index (f = fission)
Ag - gamma peak area
γ - isotope production yield [%]
m - activation sample mass [g]
ϵ_{pg} - gamma spectrometer efficiency
I_g - gamma line intensity [%]
ϕ - deuteron integral number
λ - isotope decay constant
t_+ - cooling time
t_{ir} - irradiation time
t_{real} - real time of measurement
t_{live} - live time of measurement

To get capture rate a very similar formula was used except the γ_f correction omitted. Isotope data are from table of Isotopes, 8E [2]. Fission product yield data come from ENDF/BVII.1 [3].

3 Results

The Np-237 results are both in graphical, see figure 3, and tabular forms below. Please, look at the γ_2[%] column in table 2. Case of γ_2 greater than zero means that the identified isotope has a decaying predecessor and formula 2 was used for reaction rate determination. Table 3 and table 4 present the summary results for each experiment.

The Pu-239 results are limited to graphical form only as their values are still preliminary - figure 4 shows identified fission products and determined preliminary fission rate - number of fissions per second, per sample gram and per one proton.

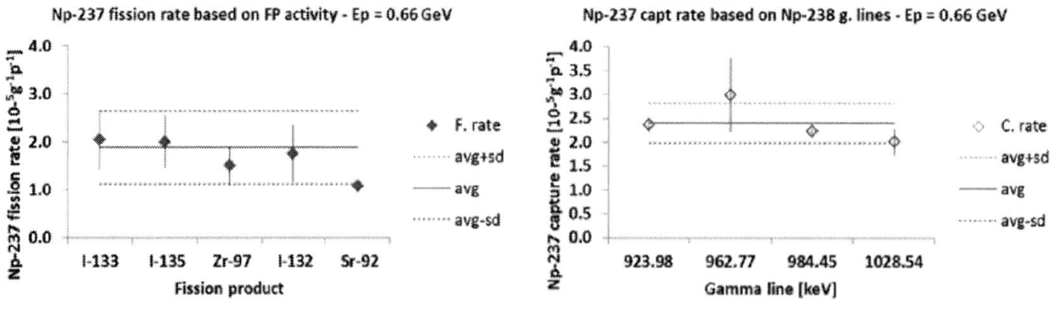

Figure 3. Np-237 fission (left) and capture (right) rates based on experiment 2014-12.

Table 2. Gamma peaks identified in Np-237 spectra.

$E_g[keV]$	$Isotope$	$T_{1/2}$	$I_g[\%]$	$\gamma_1[\%]$	$\gamma_2[\%]$
529.87	133I	20.87h	87	4.45	
555.57	91Sr	9.63h	95	2.67	
630.19	132I	2.295h	13.3	3.98	0.0407
641.28	142La	91.1 min	47.4	4.5	
657.94	97Nb	72.1m	98.23	5.35	0.031
661.66	137Cs	30.05y	85.1	5.11	
667.71	132I	2.295h	98.7	3.98	0.0407
743.36	97Zr	16.744h	93.6	5.35	
749.80	91Sr	9.63h	23.61	2.67	
756.72	95Zr	64.032d	54.38	5.54	
772.60	132I	2.295h	75.6	3.98	0.0407
954.55	132I	2.295h	17.6	3.98	0.0407
1038.76	135I	6.57h	8.01	4.16	
1131.51	135I	6.57h	22.6	4.16	
1260.41	135I	6.57h	28.7	4.16	
1383.93	92Sr	2.66h	90	4.01	
1457.56	135I	6.57h	8.73	4.16	
1678.03	135I	6.57h	9.62	4.16	
1791.20	135I	6.57h	7.77	4.16	
923.98	238Np	2.117d	2.869	100	
962.77	238Np	2.117d	0.702	100	
984.45	238Np	2.117d	27.8	100	
1028.54	238Np	2.117d	20.38	100	

Table 3. Experiment 2014-12-08 results summary for Np-237 sample.

	Avg	**Std. dev.**	**Std. dev. [%]**
Fission rate $[g^{-1}p^{-1}]$	1.67E-05	3.53E-06	21.10
Capture rate $[g^{-1}p^{-1}]$	2.40E-05	4.17E-06	17.40
fission/capture	0.70	0.19	27.35

Table 4. Experiment 2015-12-04 results summary for Np-237 sample - preliminary.

	Avg	**Std. dev.**	**Std. dev. [%]**
Fission rate $[g^{-1}p^{-1}]$	8.62E-06	2.17E-06	25.24
Capture rate $[g^{-1}p^{-1}]$	1.49E-05	1.65E-06	11.07
fission/capture	0.58	0.16	27.55

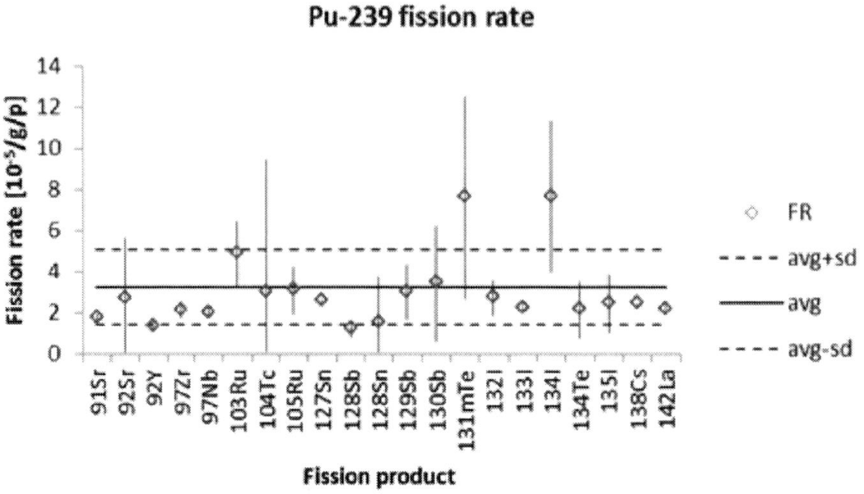

Figure 4. Pu-239 fission rate based on fission product activities - preliminary.

4 Conclusions

Np-237 and Pu-239 samples were irradiated in spallation neutron field to search for incineration feasibility. Both Np-237 fission and neutron capture products were identified. The fission to capture ratio was found to be below 1 which means that neutron capture still prevails in QUINTA neutron field.

Pu-239 fission products were identified as well but the fission rate value is still preliminary. The applied method did not allow to determine capture rate.

Acknowledgments

The authors would like to thank the staff of the PHASOTRON accelerator of the Joint Institute for Nuclear Research, Dubna for providing access to the research facilities used in these experiments.

References

[1] W. Furman et al., PoS (Baldin ISHEPP XXI) 086

[2] TABLE OF ISOTOPES, 8E.
http://www.wiley.com/legacy/products/subject/physics/toi/

[3] Evaluated Nuclear Data File (ENDF).
https://www-nds.iaea.org/exfor/servlet/E4sGetIntSection?SectID=2297333&req=1124&e4up=0
https://www-nds.iaea.org/exfor/servlet/E4sGetIntSection?SectID=2297342&req=1123&e4up=0

DOI: 10.1051/epjconf/201713810005

Study of neutron spectra in extended uranium target. New experimental data

M. Paraipan[1,2,⋆] and E& T Collaboration

[1] *Joint Institute for Nuclear Research,6, Joliot Curie St., 141980 Dubna, Moscow Region, Russia*
[2] *Institute of Space Science,409, Atomistilor St., Măgurele - Bucharest, 077125, Romania*

Abstract. The spatial distribution of neutron fluences in the extended uranium target ("Quinta" assembly) irradiated with 0.66 GeV proton, 4 AGeV deuteron and carbon beams was studied using the reactions with different threshold energy (E_{th}). The data sets were obtained with ^{59}Co samples. The accumulation rates for the following isotopes: ^{60}Co (E_{th} 0 MeV), ^{59}Fe (E_{th} 3 MeV), ^{58}Co (E_{th} 10 MeV), ^{57}Co (E_{th} 20 MeV), ^{56}Co (E_{th} 32 MeV), ^{47}Sc (E_{th} 55 MeV) , and ^{48}V (E_{th} 70 MeV) were measured with HPGe spectrometer. The experimental accumulation rates were compared with the predictions of the simulations with Geant4 code. Substantial difference between the reconstructed and the simulated data for the hard part of the neutron spectrum was analyzed.

1 Introduction

The use of proton beams for energy production in accelerator-driven systems (ADS) was extensively investigated [1]-[3] and the general opinion is that the optimal energy of proton beam lies in the range 1-3 GeV. The possibility to use heavy-ion beams for ADS was less analyzed and the conclusions of authors are contradictory [4]-[5]. Simulations performed using the code Geant4 in a quasi-infinite uranium target predict that one cannot improve the energetic efficiency of proton or deuteron beams by increasing the beam energy above 2-3 AGeV, but one can get a higher efficiency by accelerating heavier ions [6]. Based on this fact, a series of experiments with ion beams started at JINR. A more detailed analysis of particle distributions and spectra was realized by measuring the accumulation rates of isotopes with threshold energies from 0 to 70 MeV, using probes of ^{59}Co placed at different positions inside the target. In this study the distribution of the accumulation rates of these isotopes in the uranium target "Quinta" irradiated by 0.66 GeV proton, 4 AGeV deuteron and carbon beams is presented. The experimental data are compared with the predictions obtained using Geant4.10.1 [7] simulation code.

2 Experimental setup and method

"Quinta" is an extended ^{nat}U target composed from 5 section. Each section is an assembly of rods with diameter 36 mm and length 104 mm placed in a hexagonally shaped Al container. There are gaps of 17 mm between sections. In these gaps we introduce Al plates with probes from various

⋆e-mail: mparaipan@sunhe.jinr.ru

materials, placed at different radial distances from the beam axis. The first section has a beam window with diameter 144 mm. The entire assembly is surrounded with 10 cm thick Pb blanket with external dimensions 300 mm × 300 mm × 900 mm and beam window 150 mm × 150 mm × 100 mm. The scheme of the target is presented in figure 1. The longitudinal distances are measured from the surface of the blanket in the beam direction.

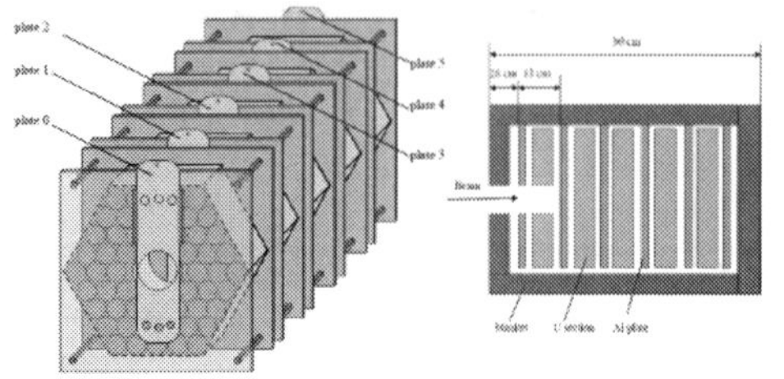

Figure 1. The scheme of "Quinta" setup.

Cobalt was chosen as material for the samples because is mono-isotopic (^{59}Co) and because there are sufficient experimental data on the excitation function under irradiation by neutrons and protons for the following isotopes covering a wide neutron energy range: ^{60}Co (threshold energy E_{th} = 0 MeV), ^{59}Fe (E_{th} =3 MeV), ^{58}Co (E_{th} =10 MeV), ^{57}Co (E_{th} =15 MeV), ^{56}Co (E_{th} =25 MeV), ^{47}Sc (E_{th} =55 MeV),and ^{48}V (E_{th} =70 MeV). Available experimental data on an isotope cross section is an important condition for using the isotope for the comparison with the simulated spectra and for the reconstruction of the energy spectrum of the particles. For neutrons the experimental data on the excitation functions in Co are available for energies until 180 MeV. For higher energies we use the values obtained with Geant4 and the binary cascade model (which provides the predictions closest to the experimental data for both protons and neutrons) properly adjusted to agree with the experimental data. An example of the simulated neutron spectrum registered at a longitudinal position of 39 cm and a radius of 8 cm is shown in figure 1a. This figure also shows the threshold energy for the isotopes. The probabilities of isotope production corresponding to the spectrum in figure 2a are shown in figure 2b, as functions of the neutron energy. This figure illustrates the energy interval in which the accumulation rate of a given isotope provides information on the neutron fluence. These energy intervals are approximately: 3 MeV - 40 MeV for ^{59}Fe, 10 MeV - 40 MeV for ^{58}Co, 20 MeV - 200 MeV for ^{57}Co, 30 MeV - 300 MeV for ^{56}Co, >60 MeV for ^{47}Sc, and > 80 MeV for ^{48}V.

The target was irradiated by protons with an energy of 0.66 GeV, deuteron and carbon beams with an energy of 4 AGeV. The beam position was monitored using the multiwire and pad ionization chambers. Solid state track detectors placed on the front side of the target were also used. The total beam intensity was measured with ionization chambers (for deuteron and carbon beams) and by activation of Al foil for the proton beam (reaction $^{27}Al(p,x)^{24}Na$). Detailed description of the beam monitoring system can be found in [8]. The values of the integral beam intensity were $2.5 \cdot 10^{15}$ for

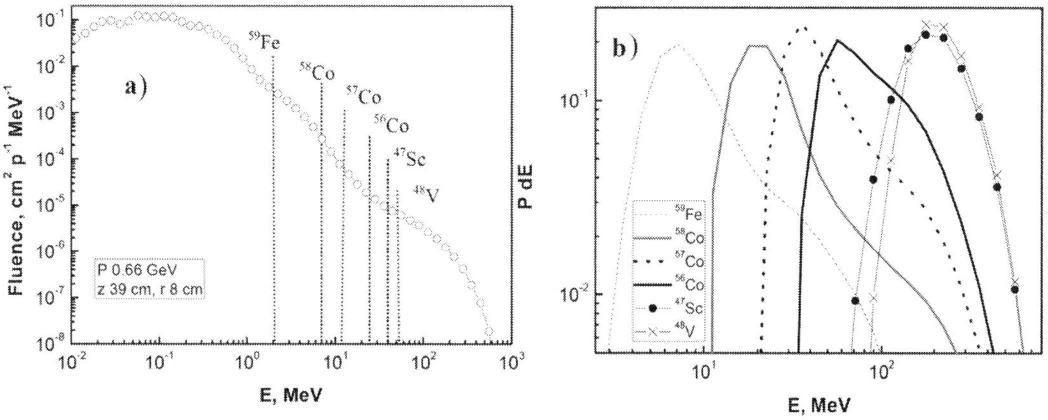

Figure 2. Simulated neutron spectrum at z= 39 cm and r= 8 cm (a); corresponding probabilities of isotope production (b).

protons 0.66 GeV,6.2 · 10^{12} for deuteron 4 AGeV beam, and 4.8 · 10^{10} for carbon 4 AGeV beam, measured with a relative error of 10 %. The accumulation rates of the isotopes in each sample were determined by measuring the γ-activities using the HPGe γ-spectrometers. The spectrometers were calibrated with a set of standard γ-sources. The reported error for the activity of γ-standards is 2 %.

The simulations with Geant4 were performed with two combinations of models for the inelastic hadron interaction. In the variant labeled "BC-Bert" we used the Bertini cascade model for hadrons and Geant4 binary cascade model for ions. The deexcitation stage of the interaction was described by the standard Geant4 model for deexcitation. In the variant labeled "INCL" the Liege cascade model was used for both hadrons and ions, together with the ABLA model for deexcitation. For the energy below 50 AMeV the inelastic interaction was modeled using the precompound exciton model. The interactions of neutrons with the energy below 20 MeV are modeled using the high precision experimental neutron database from ENDF/B VII-1 library.

3 Results and discussion

The spatial distribution of accumulated isotopes depends on the projectile type, initial energy and the threshold energy for the isotopes. The comparison between the radial distribution of ^{59}Fe, ^{58}Co and ^{48}V in the case of irradiation by 0.66 GeV protons and 4 AGeV deuterons is shown in figure 3. The accumulation rates of the isotopes N_{isot} are plotted as functions of radius, for z= 39 cm (after the second section, figure 3a) and z= 78 cm (after the fifth section, figure 3b). The data are scaled by the factors shown in the figure, in order to present all three isotopes in the same figure.

The distribution of an isotope reflects the distribution of particles with the energy above the isotope threshold energy. The distributions can be well parameterized with one (for ^{60}Co, ^{59}Fe) or two exponentials (for the isotopes with higher energy threshold). The corresponding fits are shown in figure 3. Due to the fit simplicity, we choose it to present the distributions for calculating the values of isotopes at a given distance from the beam axis.

The comparison between the accumulation rates of isotopes under irradiation by proton 0.66 GeV, deuteron 4 AGeV and carbon 4 AGeV beams is given in table 1. In the first column of table 1 the longitudinal (z) and radial (r) positions of the samples are given. The second column contains infor-

 DOI: 10.1051/epjconf/201713810005

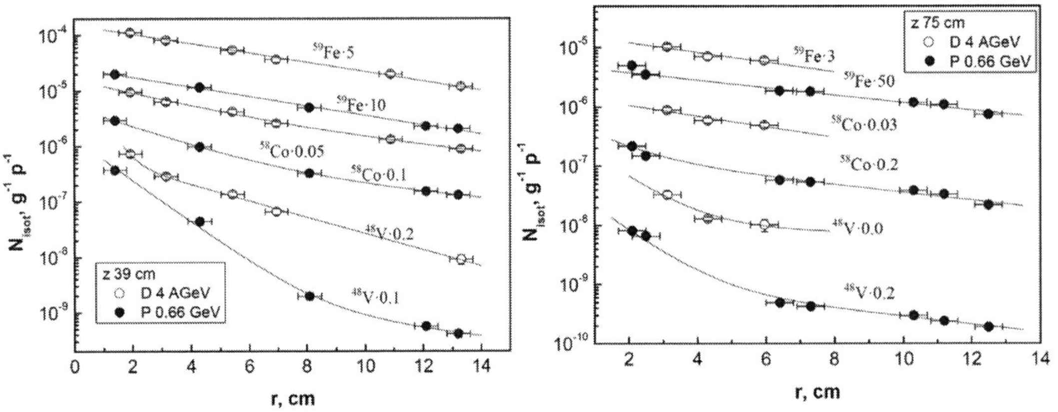

Figure 3. Radial distributions of ^{59}Fe, ^{58}Co, ^{48}V at (a) z =39 cm and (b) z= 78 cm under irradiation by 0.66 GeV protons and 4 AGeV deuterons. Lines show the exponential fit (see the text).

mation on the projectile and its energy per nucleon. The experimental and simulated using Geant4 code values for the accumulation rates of the isotopes are given in the remaining columns.

Let us analyze the change of the particle spectrum at a given position, under irradiation of the target with proton 0.66 GeV, deuteron 4 AGeV, and carbon 4AGeV beams. The coefficient $R_{i,j}$ defined as the ratio of the accumulation rate for the isotope i and the accumulation rate for the isotope with the lowest energy threshold (in our case ^{60}Co) under the target irradiation by the beam j is used as a measure of the spectrum hardness:

$$R_{i,j} = N_{i,j}/N_{^{60}Co,j} \tag{1}$$

The values of $R_{i,j}$ are determined with higher precision than the accumulation rate $N_{i,j}$ because the errors for $R_{i,j}$ do not include the errors from the beam intensity measurements. The ratios $R_{i,D4AGeV}/R_{i,P0.66GeV}$ as functions of the isotope threshold energy are shown in figure 4a for different positions inside the target. It can be seen that the hardness of the particle spectra under irradiation by 4 AGeV deuterons increases with increasing radial and longitudinal distance inside the target. The growth is more pronounced for particles with the energy above 70 MeV. After the first 10 cm (after section2 – z= 39 cm) the ratio $R_{i,D4AGeV}/R_{i,P0.66GeV}$ for ^{48}V is 1 at a radius of 6 cm, and increases to 1.3 at a radius of 13.9 cm. After the fifth section (40 cm in uranium) the ratio reaches a value of 6. These values reflect the generation of secondaries with higher energy, and with a more pronounced forward peak during the irradiation by the deuteron beam with an energy of 4 AGeV in comparison with the 0.66 GeV proton beam.

An important conclusion made from the comparison between spectra obtained under the target irradiation by carbon and deuteron beams with the same energy per nucleon (4 AGeV) illustrated in figure 4b is that in all positions the spectra obtained with the carbon beam are harder than the spectra obtained with the deuteron beam and the difference increase with increasing threshold energy of the isotope. The ratio $R_{i,C4AGeV}/R_{i,D4AGeV}$ for ^{48}V is 1.4 at the second section and a radius of 6 cm, and 1.55 at the fifth section. The simulation poorly reproduces the hardening of the spectrum. The simulation with the BC-Bert model reproduces the trend, but significantly underestimates the effect. The simulation with the INCL model predicts the spectra with similar hardness for carbon and deuteron beams.

Table 1. Experimental and simulated accumulation rates of the isotopes, under irradiation by proton 0.66 GeV, deuteron 4 AGeV and carbon 4 AGeV beams.

z/R, cm	Part/ Energy	Exp/ Sim	^{60}Co $g^{-1}p^{-1}$	^{59}Fe $g^{-1}p^{-1}$	^{58}Co $g^{-1}p^{-1}$	^{57}Co $g^{-1}p^{-1}$	^{56}Co $g^{-1}p^{-1}$	^{48}V $g^{-1}p^{-1}$
39/6.1	P/0.66	exp	(8.1±0.8)e-6	(7.9±0.8)e-7	(5.5±0.6)e-6	(1.8±0.2)e-6	(2.7±0.3)e-7	(8.6±0.9)e-8
		BC-Bert	9e-6	1e-6	6e-6	1.7e-6	2.7e-7	6.7e-8
		INCL	8.2e-6	1e-6	5.55e-6	1.7e-6	2.65e-7	9e-8
	D 4	exp	(1.1±0.1)e-4	(9.1±1)e-6	(6.7±0.7)e-5	(1.9±0.2)e-5	(2.5±0.3)e-6	(9.5±1)e-7
		BC-Bert	1.14e-4	9.5e-6	6.7e-5	1.9e-5	2.46e-6	9.3e-7
		INCL	8.1e-5	1e-5	5.9e-5	1.5e-5	2.1e-6	5.9e-7
	C/4	exp	(5.5±0.6)e-4	(4.8±0.5)e-5	(3.9±0.4)e-4	(1.2±0.13)e-4	(1.8±0.2)e-5	(7.1±0.8)e-6
		BC-Bert	5.5e-4	7e-5	4.7e-4	1.15e-4	1.6e-5	3.56e-6
		INCL	4.3e-4	5.5e-5	3.3e-4	8.5e-5	1.25e-5	3.9e-6
39/13.9	P/0.66	exp	(4.3±0.4)e-6	(1.7±0.2)e-7	(1.2±0.1)e-6	(3.5±0.35)e-7	(4.3±0.4)e-8	(3.9±0.4)e-9
		BC-Bert	3.4e-6	1.9e-7	1.2e-6	2.9e-7	3.6e-8	3.1e-9
		INCL	3.2e-6	1.8e-7	1.1e-6	3e-7	3.6e-8	3.2e-9
	D/4	exp	(6±0.6)e-5	(2.1±0.2)e-6	(1.64±0.17)e-5	(4.9±0.5)e-6	(6.2±0.7)e-7	(7.2±0.8)e-8
		BC-Bert	4.36e-5	2.76e-6	1.7e-5	4.2e-6	5.1e-7	5.4e-8
		INCL	3.3e-5	2.2e-6	1.25e-5	3e-6	3.8e-7	5.9e-8
	C/4	exp	(2.7±0.3)e-4	(1.1±0.1)e-5	(8.6±0.9)e-5	(2.6±0.3)e-5	(3.7±0.5)e-6	-
		BC-Bert	2.54e-4	1.45e-5	8.64e-5	2e-5	2.4e-6	
		INCL	1.77e-4	1.23e-5	8.2e-5	1.75e-5	2.2e-6	
65/6.5	P/0.66	exp	(2.9±0.3)e-6	(1±0.1)e-7	(7.8±0.8)e-7	(2.44±0.24)e-7	(3.6±0.4)e-8	(8.7±0.9)e-9
		BC-Bert	1.9e-6	7.53e-8	4.86e-7	1.45e-7	2.34e-8	5.4e-9
		INCL	1.7e-6	7.3e-8	4.74e-7	1.54e-7	2.3e-8	5.16e-9
	D/4	exp	(6.3±0.65)e-5	(4.1±0.4)e-6	(2.9±0.3)e-5	(8.7±0.9)e-6	(1.26±0.13)e-6	(5±0.5)e-7
		BC-Bert	4.8e-5	4.08e-6	2.68e-5	7.07e-6	1.04e-6	2.77e-7
		INCL	4.65e-5	4.57e-6	2.73e-5	7.24e-6	1.09e-6	3.64e-7
	C/4	exp	(3.1±0.3)e-4	(2.3±0.2)e-5	(1.7±0.2)e-4	(5.6±0.6)e-5	(8.4±1)e-6	(3.6±0.5)e-6
		BC-Bert	2.73e-4	2.24e-5	1.49e-4	3.85e-5	5.8e-6	1.69e-6
		INCL	2.5e-4	2.36e-5	1.49e-4	3.83e-5	5.8e-6	2.02e-6
78/6.9	P/0.66	exp	(1.8±0.2)e-6	(3.9±0.4)e-8	(2.8±0.3)e-7	(8.95±0.9)e-8	(1.2±0.12)e-8	(2.6±0.3)e-9
		BC-Bert	6.94e-7	1.84e-8	1.2e-7	3.73e-8	6.1e-9	1.3e-9
		INCL	6.25e-7	1.8e-8	1.26e-7	4e-8	5.56e-9	1.14e-9
	D/4	exp	(3±0.3)e-5	(2±0.2)e-6	(1.6±0.2)e-5	(5±0.5)e-6	(6.4±0.7)e-7	(2.6±0.3)e-7
		BC-Bert	2e-5	1.54e-6	1.13e-5	2.83e-6	4.4e-7	1.24e-7
		INCL	2e-5	1.96e-6	1.35e-5	3.8e-6	5.66e-7	2.07e-7
	C/4	exp	(1.4±0.14)e-4	(1.1±0.1)e-5	(9.1±0.9)e-5	(3.1±0.3)e-5	(4.5±0.6)e-6	(1.9±0.2)e-6
		BC-Bert	1.06e-4	8.4e-6	6e-5	1.6e-5	2.5e-6	7.95e-7
		INCL	1e-4	9.7e-6	6.7e-5	1.9e-5	2.83e-6	1.07e-6

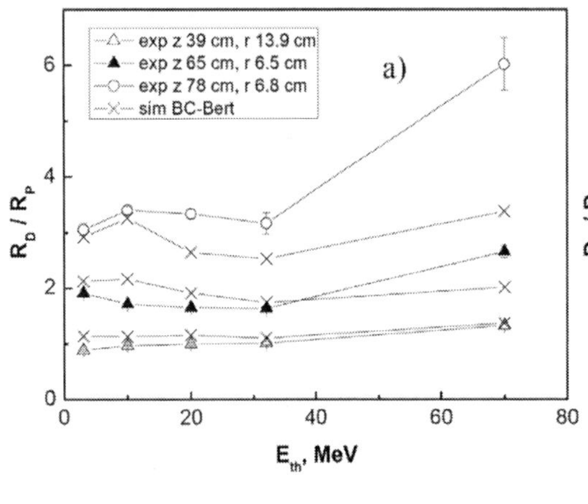

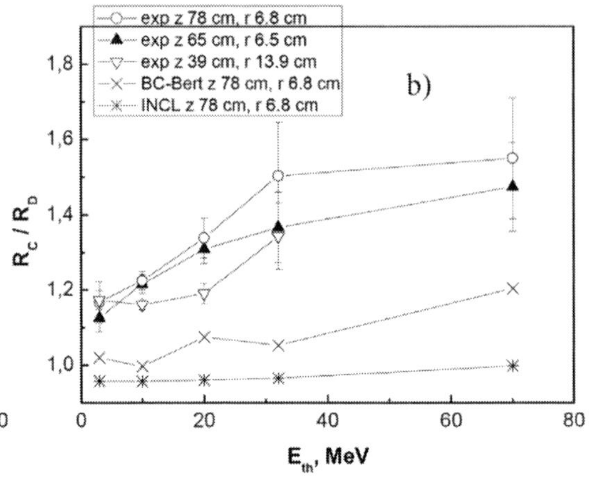

Figure 4. Experimental and simulated ratios (a) $R_{i,D4AGeV}/R_{i,P0.66GeV}$ and (b) $R_{i,C4AGeV}/R_{i,D4AGeV}$ as functions of the threshold energy of the isotopes, at different positions inside the target.

4 Conclusion

The particle spectra and spatial distributions inside the extended uranium target "Quinta" were studied by measuring the accumulation rate of isotopes in samples of ^{59}Co, under target irradiation with proton beam with energy 0.66 GeV, deuteron beam with energy 4 AGeV, and carbon beam with energy 4 AGeV. A set of 7 isotopes with threshold energy production from 0 to 70 MeV was measured. The experimental data show that the particle spectra become harder under irradiation by carbon in comparison with deuteron beam with the same energy per nucleon.

The experimental results were compared with the simulation performed using the code Geant4. The simulations show good agreement with the experimental data in the proximity of the shower maximum, but underestimate the production of isotopes 1.5-2 times at the end of the target. The simulation does not reproduce correctly the hardening of the spectrum under irradiation with carbon beam in comparison with deuteron beam with the same energy per nucleon.

The obtained results demonstrate that it is necessary to improve the simulation codes and to continue the experimental investigations with the extended target irradiated by light ion beams in an energy range of 1- 4 AGeV.

References

[1] R.G. Vasil'kov et al., Sov. J. At. Energy **44(4)**, 329-335 (1978)

[2] R.G. Alsmiller Jr. et al., Nucl. Sci. Eng. **79**, 162-166 (1981)

[3] C. Rubbia et al., preprint CERN/AT/95-44 (ET) (1995)

[4] D. G. Koshkarev, Technical Physics **74**, 137-139 (2004)

[5] S.R. Hashemi-Nezhad, W. Westmeier, M. Zamani-Valasiadou, B. Thomauske, R. Brandt, Ann. Nucl. Energy **38**, 1144–1155 (2011)

[6] M. Paraipan, A.A. Baldin, M.G. Kadykov, S.I. Tyutyunikov, XXI Baldin ISHEPP Dubna, Russia, September 10-15 (2012)

[7] S. Agostinelli et al., Nucl. Instr. Meth. A **506**, 250- 303 (2003)

[8] A.A. Baldin et al., Phys. Part. Nucl. Lett. **13**, 210- 217 (2016)

Comparison of two fast neutron fluence measurement methods based on Np-237 fission-to-capture ratio measurement (spectral index) and a reverse dark current measurement in a planar silicon detector

Marcin Szuta[1,*], *Stanislaw* Kilim[1], *Elzbieta* Strugalska-Gola[1], *Marcin* Bielewicz[1,2], *Nikolay* Zamyatin[2], *Anastasia* Shafronovskaia[2], and *Sergey* Tyutyunnikov[2]

[1]*National Centre for Nuclear Research, 05-400 Otwock-Świerk, Poland*
[2]*Joint Institute for Nuclear Research, 141980 Dubna, Russia*

Abstract. The idea of the first method is to search the neutron energy for the ratio of fission cross section to capture cross section of the selected actinide isotope from the nuclear data base that is equal to the measured ratio of the fissioned and captured actinide isotope Np-237.
The idea of the second method consists in the measurement of the reverse dark current increase, which is linearly proportional to neutron fluence, induced by the fast neutron irradiation in planar silicon detectors.
Np-237 samples and planar silicon detectors were placed inside a subcritical assembly (the Quinta assembly at the Joint Institute for Nuclear Research, Russia) very close to each other assuming that in both samples the same neutron fluence should pass.
We concluded that minor actinide samples and planar silicon detectors can be used as neutron fluence detectors especially in the high neutron energy range, where measurements are difficult.
Considering the importance of high energy neutron measurement in the ADS (Accelerator Driven System), actinide and silicon detectors could be a very useful tool.

1 Introduction

Minor actinides (MA) (Np, Am, Cm) are generated in nuclear fuel during irradiation in a reactor. These elements significantly contribute in radio-toxicity and heat generation of spent nuclear fuel. The key issue for reducing the MA radio-toxicity is the partitioning and transmutation of MA in fast neutron reactors (FR) or accelerator driven systems (ADS).

The problem is to determine the optimal place in the ADS for incineration of actinides. This can be done by application of actinides and silicon semiconductors as detectors of fast neutron fluence.

The Quinta experimental assembly simulating a deeply sub-critical assembly, located at the Joint Institute for Nuclear Research (JINR), Dubna, Russia, was used for preliminary application of the fast neutron spectrum to the actinides and silicon semiconductors to study the neutron fluence. The Quinta target was irradiated with a pulsed proton extracted from the Phasotron accelerator.

*e-mail: marcin.szuta@ncbj.gov.pl

2 Fast neutron fluence measurement methods

2.1 Actinide fast neutron fluence measurement method

Since the measured spectral indexes are defined as the ratio of average fission and capture cross sections, the retrieved distinct fission and capture cross sections for the distinct neutron energy from the nuclear data base describe the average values.

Having the average fission and capture cross section values, we can evaluate the average neutron flux in the location of the actinide sample using the measured amount of fissioned and captured actinide isotopes. Two different equations for fissioned (N_{yf}) and captured (N_{yc}) actinide isotopes should give the same average neutron flux value, that is a proof for correct measurement.

The number of neutron induced fissioned (N_{yf}) and neutron captured actinide isotopes (N_{yc}) in the actinide sample of volume V_p can be expressed as follows:

$$N_{yf} = V_p\bar{\Phi}N\bar{\sigma}_f t \quad ; \quad N_{yc} = V_p\bar{\Phi}N\bar{\sigma}_c t \quad , \tag{1}$$

where

$\bar{\Phi}$ - average neutron flux in the place of the actinide sample location [n/cm$^2 \cdot s$],
N - number of actinide isotopes in a volume unit [cm^{-3}],
$\bar{\sigma}_f$, $\bar{\sigma}_c$ - average microscopic cross section for the reactions (n, f) or (n, γ) $[barns]$,
t - irradiation time.

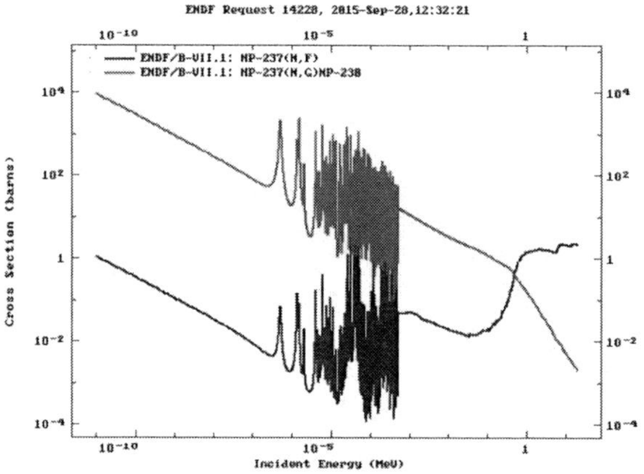

Figure 1. Cross sections of Np-237(n,g)Np-238 and Np-237(n,f) reactions.

The fission/absorption ratios are consistently higher for the fast spectrum. Thus, in a fast spectrum, actinides are preferentially fissioned, not transmuted into higher actinides (see figure 1).

2.2 Silicon semiconductor fast neutron fluence measurement method

Fast neutrons with an energy of $en > 100$ keV produce damage in the volume of the silicon semiconductor in the form of vacancies in the crystal lattice (knocking-out the primary atom from the lattice site).

The silicon detector changes its electrical parameters (characteristics) under neutron irradiation due to radiation defects in the crystal lattice. The radiation defects lead, among others, to reverse current growth.

The reverse dark current of the detector grows linearly [1] with increasing fast neutron fluence:

$$\Delta I = \alpha_I \times V \times \Phi \quad , \tag{2}$$

where:

$\Delta I = (I_{irrad} - I_{nonirrad})$, $[A]$, increment detector current,

V - volume of the detector, $[cm^{-3}]$,

Φ - neutron fluence, $[cm^{-2}]$, - equivalent for silicon damage fluence of fast neutrons with an energy of 1 MeV,

$\alpha_I = (5 \pm 10\%)\ 10^{-17}$ $[\ A\ cm^{-1}]$, constant of radiation.

Neutron radiation induced damage in silicon detectors is presented in figure 2. The values of neutron silicon detector damage cross section as a function of neutron energy presented in figure 2 are extracted from the figure described "Function of defect formation of D(E) in MeV $\times$ mb, (Ougouag) and the function NIEL-FN-522 in keV $\times$ cm^2/g", presented in reference [2].

The damage cross section in function of energy starts with the threshold mode at the energy of 170 keV [2].

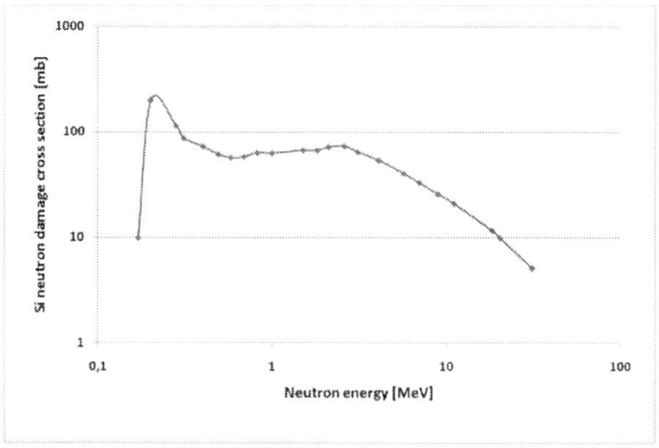

Figure 2. Neutron silicon detector damage cross section as a function of neutron energy [2].

We can roughly assume that in the neutron energy range from 0.4 MeV to 4 MeV the silicon detector damage cross section is constant which indicates that in this neutron energy range the precision of measured neutron fluence is defined by the changes of the damage cross section in the range, which is not greater than 25 %. This in turn implies that the calibration of the silicon detector with a neutron flux of any energy from this range is fulfilled within accuracy of 25%. That is why the measured neutron fluence can be termed as an equivalent for silicon detector damage by fast neutrons with an energy of 1 MeV.

3 Motivation of comparison of the two fast neutron fluence measurement methods

Performing an experimental study of transmutation of actinides in a deeply sub-critical assembly controlled by an external neutron source (ADS) - the Quinta assembly, we proposed minor actinide application as a neutron fluence and average neutron energy detector in the place of their location in order to determine the optimal place in the ADS for the incineration of the actinides [3].

Independently, within the project "Study of deep subcritical electronuclear systems and feasibility of their application for energy production and radioactive waste transmutation - E&T - RAW", a special JINR group concentrated on measuring fast neutron fluence using planar silicon detectors.

Accidentally, since location of some silicon detectors was very close to the location of the actinide Np-237 samples during certain experimental session and their results were quite comparable, we undertook to continue experimentally the comparison.

The relationship between the neutron fluences measured for three different energies of deuteron beams (2, 4 and 8 *GeV*) by the actinide detector and by the silicon detector is very alike. The values of the neutron fluencies measured by the actinide detectors are about 2.5 times higher than that measured by the silicon detectors. This is understandable because the silicon detectors measured the same neutron field, as the actinide detector, but weakened by 10 cm of a lead reflector. This observation motivated us for the comparison of the two fast neutron fluence measurement methods.

4 Experimental study on (sounds strange, as comparison is a way of studying) the comparison of the two fast neutron fluence measurement methods

4.1 The Quinta assembly

The Quinta assembly, located at the Joint Institute for Nuclear Research (JINR), Dubna, Russia, was used in the study of the actinide incineration for application of a fast neutron spectrum. It is an assembly of a massive uranium target and lead shielding (see figure 3). The Quinta assembly consists of a total of 512 *kg* of natural uranium. It is composed of five sections, each being 114 *mm* long and separated by a 17 *mm* air gap, which allows the placement of samples mounted onto special plates. The uranium exists as many cylindrical rods, where each rod is 36 *mm* in diameter, 104 *mm* in length and 1.72 *kg* in mass. The five sections are surrounded by 100 *mm* thick lead bricks on all six sides of a total weight of 1780 *kg*. One of the sides of the lead shield-reflector has a rectangular window for placing the actinide samples (see figure 3) [4, 5].

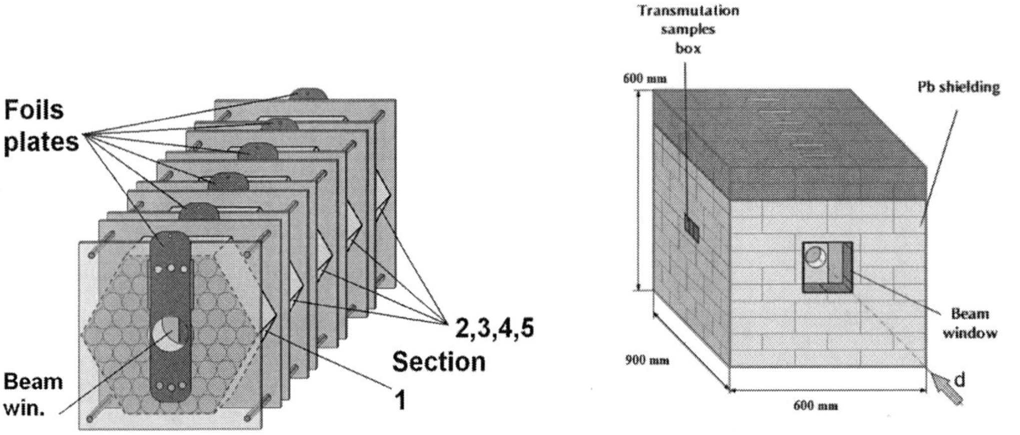

Figure 3. Scheme of the Quinta assembly. On the left there is a view of the uranium target with supporting structures and plastics used for sample placement (detector's plates), on the right there is a view of the lead shielding enfolding the target with transmutation samples box (left window) for the actinide sample placing in the shielding.

4.2 Principles of the measurements

After irradiation, the actinide samples were removed and transported away to be analyzed with gamma spectrometry. The actinide samples were measured using an HPGe n-type coaxial detector manufactured by CANBERRA. The measurements began 1.5 *h* after irradiation had stopped, continuing for up to 6 days afterwards. The spectra collection times ranged from 15 *min* to just over 3 *h*. All spectra were analyzed using the DEIMOS [6] program. Spectroscopic corrections make about 1% uncertainty. Statistical uncertainty of the Gauss fit of the gamma peaks in the DEIMOS code depends on intensity and its value is usually between 1% and 10%. Final uncertainty is set as a square root of the second powers of particular statistical and systematic uncertainties.

The total yield of the observed isotopes per one gram of activated material and one deuteron was calculated N_{yield}.

The procedure of evaluation of the actinide fission rate per deuteron and per gram ($I_{f\gamma}$) consists in the calculation of the number of the produced nuclei N_{yield} for each isotope and then dividing it by the isotope production efficiency per 1 fission (γ_f):

$$I_{f\gamma} = N_{yield}/\gamma_f, \tag{3}$$

where

$I_{f\gamma}$ - actinide fission rate per deuteron and per gram.

The gamma-ray spectrum analysis was carried out in a well established manner. The measured activities of the considered isotopes at the end of irradiation (EOI) were corrected for the decay during the irradiation according to a known deuteron beam profile, for the decay between EOI and start of counting (cooling time) and for the decay during the counting time and also not forgetting about dead time correction (t_{real}/t_{live}). Thus, one can calculate the relative number of the considered produced isotopes at the end of irradiation for each measurement.

4.3 Irradiation details

In order to compare the two methods, we performed an experiment where the silicon detector was placed very close to the ^{237}Np sample in the rectangular window in the lead shielding (see figure 3) of the QUINTA assembly assuming that both detectors were irradiated by the same neutron flux.

The Quinta target was irradiated with a pulsed proton beam of energies of 0.66 GeV extracted from the Phasotron accelerator. The total number of protons is equal to 8.64(30) × 10^{14} during the time of irradiation equal to 20580 s.

For the Np-237 sample of 1.147 g weight, the number of Np-237 fissions per sample (N_{yf}) was equal to 1.458 × 10^{10} and the number of captures per sample (N_{yc}) was equal to 2.096 × 10^{10}. The measured ratio of fission to capture for the proton beam of energy of 0.66 GeV is equal to 0.695±0.195.

Applying the try-and-error method, we find the neutron energy for which the ratio of fission cross section to capture cross section of the actinide Np-237 from the nuclear data base (ENDF/B-VII.1) is equal to the measured ratio of the fissioned and captured actinide isotope. The retrieved distinct fission and capture cross sections for the distinct neutron energy from the nuclear data base, which describe the average values, let us evaluate the neutron fluencies using the equations (1).

Adding the values of neutron fluencies measured by the silicon detector method we can compare the two methods (see table 1).

Table 1. Compilation of neutron fluencies measured by the two methods

Deuteron beam energy/year	0.66 GeV/2014
Neutron fluence using for evaluation the number of fissions in the sample [n/cm^2] ×10^{13}	1.67
Neutron fluence using for evaluation the number of captures in the sample [n/cm^2] ×10^{13}	1.64
Neutron fluence using silicon detectors [n/cm^2] ×10^{13}	1.11

5 Discussion

The mathematical structures of the equations describing the neutron fluence of the two methods are very alike. In both methods the values which are proportional to the neutron fluence depend on the cross section of the neutron inducing damage in the silicon detector and cross sections of the neutron inducing fission and capture on the Np-237 minor actinide (see figure 1 and figure 2). The cross sections used in both methods depending on the neutron energy are the primary reasons of the final result (I and N_{yf} or N_{yc}) which let us evaluate the searched neutron fluence. The further similarity is that the volumes of the detectors in both methods are taken into account.

In the research, the fluence was estimated to be about 1013 n/cm^2. The fluence measured by the silicon detector method was about 35% less than the fluence estimated by the actinide method. This could be expected because the silicon detector method allows effective measuring of the fast neutrons of an energy higher than 170 keV. So, the difference between the two fast neutron fluence measurements gives estimation of the neutron fluence for a neutron energy below 170 keV. This is an additional approach.

6 Conclusion

Minor actinide samples and planar silicon detectors can be used as neutron fluence detectors especially in the high neutron energy range where the measurements are impeded.

Given the importance of high-energy neutron measurements in the ADS (Accelerator Driven System), actinide and silicon detectors could be a very useful tool.

The difference between the two fast neutron fluence measurements of the two methods presented above gives, in addition, an estimation of the neutron fluence at neutron energy below 170 keV - the additional approach.

References

[1] I.A. Golutvin et al., preprint E14-95-97, Dubna (1995)

[2] A. Van Ginneken, *Non ionizing energy deposition in silicon for radiation damage studies* (Fermi Nat. Accelerator Lab. Report FN-522, 1989)

[3] M. Szuta, S. Kilim, E. Strugalska-Gola, M. Bielewicz, S. Tyutyunnikov, *Feasibility study of minor actinide application as a neutron fluency and average neutron energy detector in the place of their location*, Fifth Meeting of the Expert Group on Reactor Physics and Advanced Nuclear System (EGRPANS) of Working Party on Scientific Issues of Reactor Systems (WPRS) (NEA Headquarters, Paris, France, 19 February 2015)

[4] "E&T RAW" Collab. (I. Adam et al.) E1-2010-61, Dubna (2010)

[5] N.L. Asquith et al. Annals of Nuclear Energy **63**, 742-750 (2014)

[6] J. Frana Radioanal. and Nucl. Chem. **257**, 583 (2003)

DOI: 10.1051/epjconf/201713810007

Average fast neutron flux in three energy ranges in the Quinta assembly irradiated by two types of beams

Elzbieta Strugalska-Gola[1,⋆], *Marcin* Bielewicz[1,2,⋆⋆], *Stanislaw* Kilim[1], *Marcin* Szuta[1], and *Sergey* Tyutyunnikov[2]

[1]*National Centre for Nuclear Research, 05-400 Otwock-Świerk, Poland*
[2]*Joint Institute for Nuclear Research, 141980 Dubna, Russia*

Abstract. This work was performed within the international project "Energy plus Transmutation of Radioactive Wastes" (E&T - RAW) for investigations of energy production and transmutation of radioactive waste of the nuclear power industry. ^{89}Y (Yttrium 89) samples were located in the Quinta assembly in order to measure an average high neutron flux density in three different energy ranges using deuteron and proton beams from Dubna accelerators. Our analysis showed that the neutron density flux for the neutron energy range 20.8 - 32.7 MeV is higher than for the neutron energy range 11.5 - 20.8 MeV both for protons with an energy of 0.66 GeV and deuterons with an energy of 2 GeV, while for deuteron beams of 4 and 6 GeV we did not observe this.

1 Introduction - Motivation

Up to now, we have been concentrated on measuring the neutron flux distribution in the deeply subcritical Quinta assembly versus the axis and radius of the assembly for the neutron energy above 10 MeV applying proton and deuteron beams of an energy from 1 GeV to 8 GeV extracted from the NUCLOTRON accelerator. We have applied a proton beam of an energy of 0.66 GeV recently extracted from the PHASOTRON accelerator. Nuclear data handling at the experiment session have turned our attention to the neutron flux density measured on five foil plates in terms of three different neutron energies: 11.5 - 20.8 MeV, 20.8 - 32.7 MeV and 32.7 - 100 MeV. The neutron density flux for the neutron energy range 20.8 - 32.7 MeV being higher than for the range 11.5 - 20.8 MeV has appeared an unexpected feature of the measurement. Repeating in the same way data handling for the other planes as in collecting them in one figure it can be noticed that the same effect is observed for planes 2 - 5 but except plane 1, which describes the average neutron flux density in the first section where process of spallation begins. This unexpected feature of the measurement is presented in figure 1.

Finding this unexpected feature of an average neutron flux in the experiment with the proton beam of an energy of 0.66 GeV motivated us to make an overview of our experiments performed earlier for the deuteron beam of an energy of 1 - 8 GeV.

⋆e-mail: elzbieta.strugalska-gola@ncbj.gov.pl
⋆⋆e-mail: marcin.bielewicz@ncbj.gov.pl

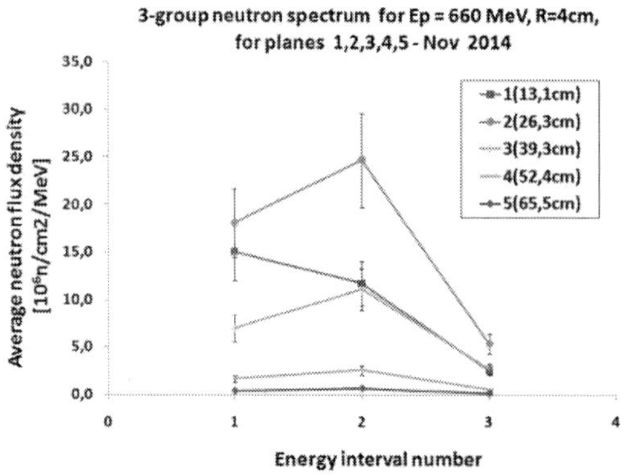

Figure 1. Average neutron flux density versus three different energy ranges on the X axis: 1 - (11.5-20.8 MeV), 2 - (20.8-32 MeV), 3 - (32-100 MeV) on five planes (legend on the right side of the figure); R=4 cm, proton beam energy 0.66 GeV, Experiment: Nov 2014.

2 Principles of the measurement

To study the spectrum of high-energy neutrons, we have used threshold ^{89}Y detectors which have the following advantages: one stable isotope, several threshold reaction channels, several resulting isotopes with a half life time long enough - longer than 12 hours. The first threshold energy of the reaction (n, 2n) giving ^{88}Y has been for an energy of neutrons equal to 11.5 MeV. The next possible threshold energies 20.8, 32.7, 42.1 and 54.4 MeV have been for the reactions (n, 3n) ^{87}Y, (n, 4n) ^{86}Y, (n, 5n) ^{85}Y and (n, 6n) ^{84}Y respectively [1]. The neutron field has been determined with a certain number of ^{89}Y foils placed in specified positions (given by the radial and axial distance) inside the experimental facility (figure 2). After neutron irradiation, the gamma activity of ^{89}Y foils has been measured with an HPGe spectrometer. Taking into account necessary corrections [1] and spectra analysis by the DEIMOS program [2], we have determined isotope production per one gram of a sample and per one beam deuteron/proton at specified positions of the Quinta assembly.

Having isotope production determined inside of the experimental facility for the three isotopes ^{88}Y, ^{87}Y and ^{86}Y, we have been able to evaluate three average high-energy neutron fluxes in each ^{89}Y foil location for certain energy ranges [3]. The energy ranges have been roughly determined by the microscopic cross section in function of the energy for the (n, xn) reactions of the three isotopes ^{88}Y, ^{87}Y and ^{86}Y. The following three threshold energies 11.5, 20.8 and 32.7 MeV for the reactions (n, 2n)^{89}Y ^{88}Y, ^{89}Y (n, 3n)^{87}Y and ^{89}Y (n, 4n)^{86}Y have appointed the first two energy ranges (11.5 - 20.8 MeV) and (20.8 - 32.7 MeV) of the neutron fluxes $\bar{\Phi}_1$ and $\bar{\Phi}_2$. The third energy range $\bar{\Phi}_3$ has begun at 32.7 MeV and ends at 100 MeV, when the microscopic cross section has been low comparatively to the maximum cross section.

The Quinta assembly, located at the Joint Institute for Nuclear Research (JINR), Dubna, Russia is an assembly of a massive uranium target and lead shielding (figure 2). The Quinta assembly consists of a total of 512 kg of natural uranium. It is composed of five sections, each being 114 mm long and separated by a 17 mm air gap, which allows the placement of samples mounted onto special plates. The uranium exists as many cylindrical rods, where each rod is 36 mm in diameter, 104 mm in length and 1.72 kg in mass. The first section contains only 54 rods after removal of the central 7 rods for creating a beam window. This beam window is 80 mm in diameter and serves to reduce the loss of backward emitted/scattered neutrons. The front and back of each section are bounded then by additional aluminum plates 350 × 350 × 5 mm. The five sections are mounted onto a single slab of

DOI: 10.1051/epjconf/201713810007

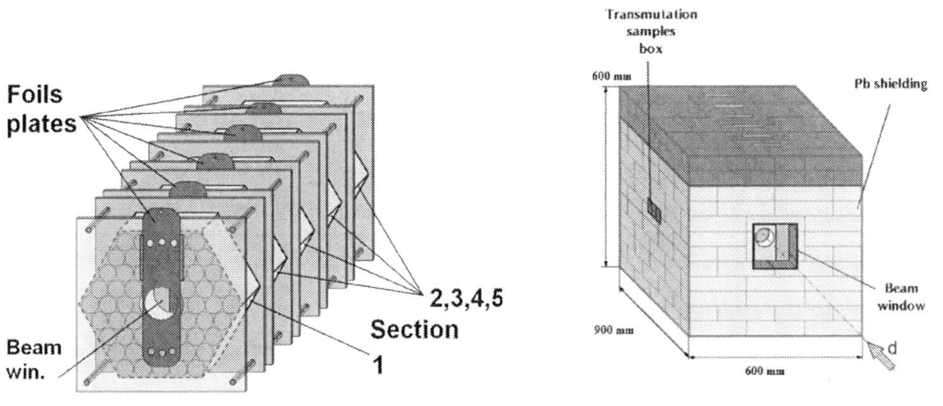

Figure 2. Scheme of the Quinta assembly. On the left, there is a view of the uranium target with supporting structures and plastics used for sample placing (foils plates), on the right, there is a view of the lead shielding enfolding the target.

aluminum with a thickness of 25 mm and surrounded by 100 mm thick lead bricks (shielding) on all six sides of a total weight of 1780 kg. The sample plates are labeled 1-6, starting from the direction of the incident beam [4]-[5].

3 Overview of experiments for the deuteron beam with an energy of 2 - 6 GeV

Repeating data handling in the same way as in figure 1 for the deuteron beam with an energy of 2 GeV shows (figure 3) that the same effect is observed for planes 2, 4 and 5 and is not clear for planes 1 and 3 which describes the average neutron flux density in the first and third sections.

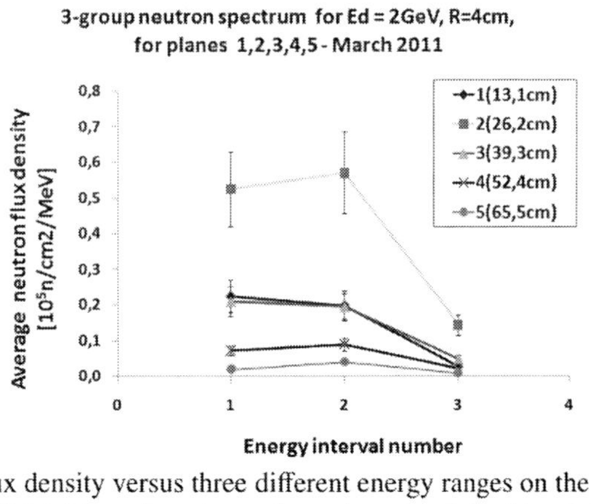

Figure 3. Average neutron flux density versus three different energy ranges on the X axis: 1 - (11.5-20.8 MeV), 2 - (20.8-32 MeV), 3 - (32-100 MeV) on five planes (legend on the right side of the figure); R=4 cm, deuteron beam energy 2 GeV, experiment III. 2011.

However the average neutron flux density versus three different energy ranges: 1 - (11.5-20.8 MeV), 2 - (20.8-32 MeV), 3 - (32-100 MeV) for R=4 cm and five planes for the deuteron beam

DOI: 10.1051/epjconf/201713810007

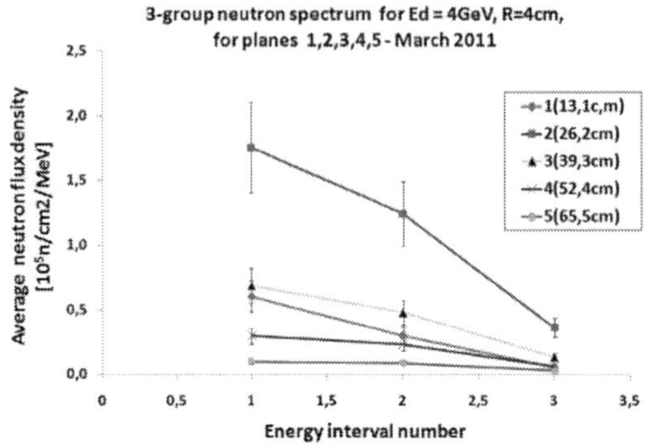

Figure 4. Average neutron flux density versus three different energy ranges on the X axis: 1- (11.5-20.8 MeV), 2- (20.8-32 MeV), 3- (32-100MeV) on five planes (legend on the right side of the figure); R=4 cm, deuteron beam energy 4 GeV, experiment III. 2011.

with an energy of 4 GeV behaves as expected (figure 4). In this case, the explanation like for the experimental data in figure 1, 3 is not sufficient.

Also the average neutron flux density versus three different energy ranges: 1 - (11.5-20.8 MeV), 2 - (20.8-32 MeV), 3 - (32-100 MeV) for R=4 cm and five planes for the deuteron beam with an energy of 6 GeV behaves as expected (figure 5).

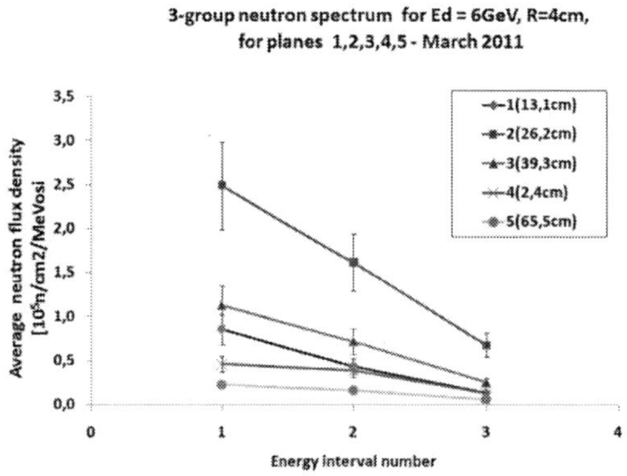

Figure 5. Average neutron flux density versus three different energy ranges on the X axis: 1- (11.5-20.8 MeV), 2- (20.8-32 MeV), 3- (32-100 MeV) on five planes (legend on the right side of the figure); R=4 cm, deuteron beam energy 6 GeV, experiment III. 2011.

4 Discussion

Such an unexpected behavior of the average neutron flux density versus three different energy ranges can be explained to some extent by the dependence of the micro cross section in the considered neutron energy range shown in figure 1 and figure 3. The threshold neutron energies of yttrium (n,xn) reactions determine the considered energy ranges.

In the first energy range 11.5-20.8 MeV, the area under the cross section curve of $^{89}Y(n,2n)^{88}Y$ is smaller than in the second energy range of the same reaction. Moreover there is also an area under the curve of $^{89}Y(n,3n)^{87}Y$ (figure 6). Since the measured amount of neutrons is proportional to the area under the cross section curves, the results in figure 1 and figure 3 are understandable.

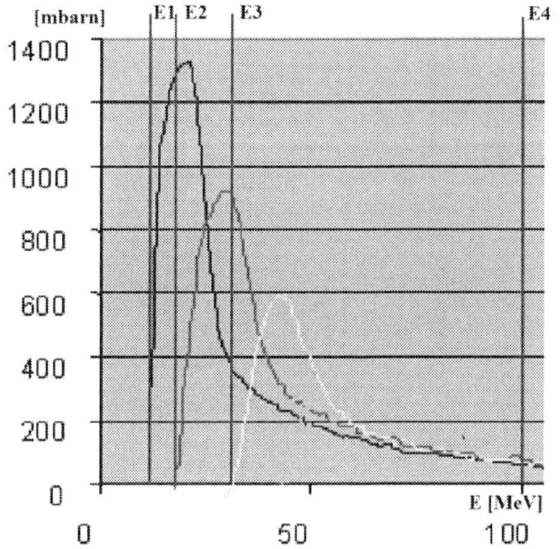

Figure 6. Cross sections of the three yttrium (n,xn) reactions - threshold energies: E1 = 11.5 MeV ^{88}Y, E2 = 20.8 MeV ^{87}Y, E3 = 32.7 MeV ^{86}Y. TALYS calculation [6].

The above explanation for the experimental data presented in figure 4 and figure 5 is not sufficient. Calculations using the Monte Carlo method [1] show that in the energy range (20.8-32 MeV) the neutron fluence is higher than in the range 11.5-20.8 MeV for the energy of the deuteron beam equal to 1 GeV (figure 7).

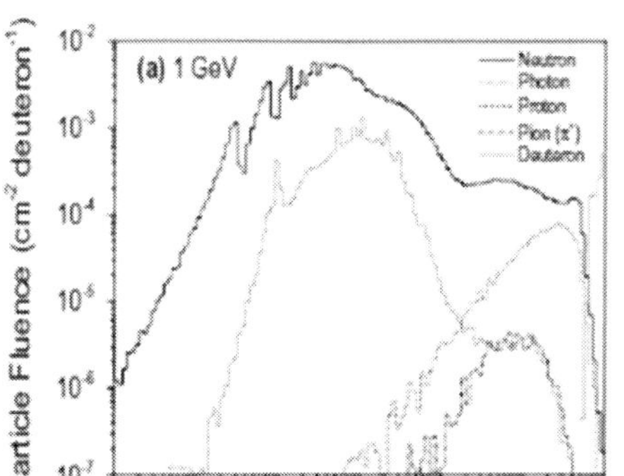

Figure 7. Neutron spectrum as a function of energy (black solid line) for the deuteron beam with an energy of 1 GeV [7].

Follow the same article [7]. Calculations using the Monte Carlo method show that in the three energy ranges (11.5-20.8 MeV, 20.8-32 MeV and 32-100 MeV) the neutron fluence is continuously decreasing with the neutron energy for the energy of the deuteron beam equal to 4 GeV (figure 8). This explains the behavior of the neutron flux density presented in figure 4 and figure 5.

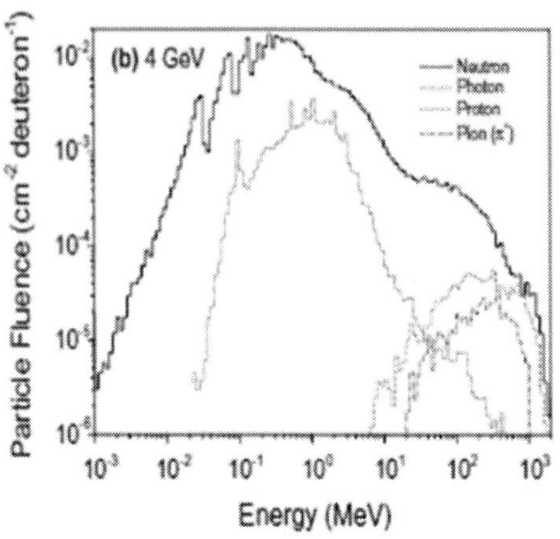

Figure 8. Neutron spectrum as a function of energy (black solid line) for the deuteron beam with an energy of 4 GeV [7].

5 Conclusions

This brief analysis carried out above let us to infer that the neutron flux density in the range above 10 MeV should be our concern in the future experimental research.

Theoretical analysis carried out by the authors [7] using MC method explains to some extent our experimental observation.

The tendencies presented above are vague and need to be checked in next experiments because we do not see clearly this effect in all our experimental results.

Though the measured amount of neutrons proportional to the area under the cross section curves in the energy range 11.5 - 20.8 MeV explains the results for proton or deuteron beams in the range from 0.66 to 2 GeV, it does not explain the behavior of the average neutron flux for an energy of different beams higher than 2 GeV.

The measurement of the high neutron flux density and explanation of the results still need to be developed by an additional experiments and calculation.

Acknowledgments: We would like to thank the personnel of Dubna laboratories and "E+T RAW" collaboration.

References

[1] M. Bielewicz et al. Phys. Soc. **59(2)**, 2014 (2011)
[2] J. Frana Radioanal. and Nucl. Chem. **257**, 583 (2003)
[3] M. Bielewicz et al. Nuclear Data Sheets **119**, 296-298 (2014)
[4] E&T RAW - Collaboration, Communication of the JINR, E1-2010-61 (2010)
[5] W. Furman et al., PoS (Baldin ISHEPP XXI) 086. www.pos.sissa.it (2013)
[6] A. J. Koning, S. Hilaire, M. C. Duijvestijn, TALYS Santa Fe, AIP Conference Proceedings **769**, 1154-9 (2005)
[7] N. L. Asquith et al., Annals of Nuclear Energy **63**, 742-750 (2014)

Study of nuclear fragmentation at MPD/NICA

M.B. Golubeva[1,*], *A.P.* Ivashkin[1,**], and *A.B.* Kurepin[1,***]

[1] *Institute for Nuclear Research RAS, Moscow, Russia*

Abstract. Due to the much lower beam energy of NICA compared to the RHIC and LHC hadron colliders and the fixed target experiments at SPS the role and performance of the forward detectors of NICA are quite different. The Neutron Zero Degree Calorimeter could be used for the measurement and monitoring of luminosity, however with lower efficiency of neutron detection produced in ultra-peripheral collisions. The use of Forward Hadron Calorimeter for the determination of centrality is impossible by simply counting the number of spectators because of the ambiguity of the impact parameter dependence. This ambiguity could be removed if the angular distribution of the spectators will be taken into account. It is shown by the simulation with LAQGSM model that the forward multiplicity detector like V0 of ALICE could not be used for the determination of centrality. However it could provide the valuable information on the nuclear fragmentation of heavy ions.

Collider NICA will be the first collider working at much lower energy than RHIC and LHC. This fact should be taken into account by construction of the detectors systems of NICA. The barrel detectors like TPC, TOF, TRD and ECAL could be very similar due to the central geometry. But the performance of forward detectors is different because of the important contribution of nuclear fragments in the yield of produced particles in the collision. Therefore the specific role of forward detectors in the MPD/NICA project should be considered. From the experience of ALICE at LHC and STAR and PHENIX at RHIC it could be seen that forward detectors in the nucleus - nucleus collisions are mostly used for trigger and zero time formation and for centrality and reaction plane determination.

The most familiar method for the classification of the events by centrality is the use of the multiplicity forward detector, like V0 of ALICE [1]. The multiplicity measured by V0 detector is matched to the centrality classes corresponding to the gaps in percents of the impact parameter of collision [2]. The multiplicity distribution is rather smooth except of very peripheral events, which are accounted for by electromagnetic interaction with larger cross section.

Another method of centrality determination, used mostly for the experiments with the fixed target, is the measurements of the amplitude of the Zero Degree Calorimeter (ZDC), which is proportional to the number of the spectators emitted very close to the zero angle. The same method for collider is used with the Neutron ZDC [3] since charged fragments are deflected by the magnets in the beam line. For MPD/NICA the NZDC could be used for the measurement and monitoring of luminosity, however with lower efficiency of neutron detection produced in ultra-peripheral collisions [4]. The

* e-mail: marina@inr.ru
** e-mail: ivashkin@inr.ru
*** e-mail: kurepin@inr.ru

DOI: 10.1051/epjconf/201713811001

Forward Hadron Calorimeter (FHCAL), proposed for MPD/NICA, could not be used for determination of centrality by simply counting the number of spectators because of the ambiguity of the impact parameter dependence [5]. This ambiguity could be removed by the measurement of the asymmetry of signal amplitudes between the outer and inner modules of the calorimeter [6]. In this article the possibility to determine the centrality with the forward multiplicity detector (FMD) for MPD/NICA is investigated. This scintillator detector is segmented in several rings in the radial direction and divided in several sectors in the azimuthal direction with internal diameter 10 cm and external diameter 130 cm. The thickness of scintillator is 2 cm.

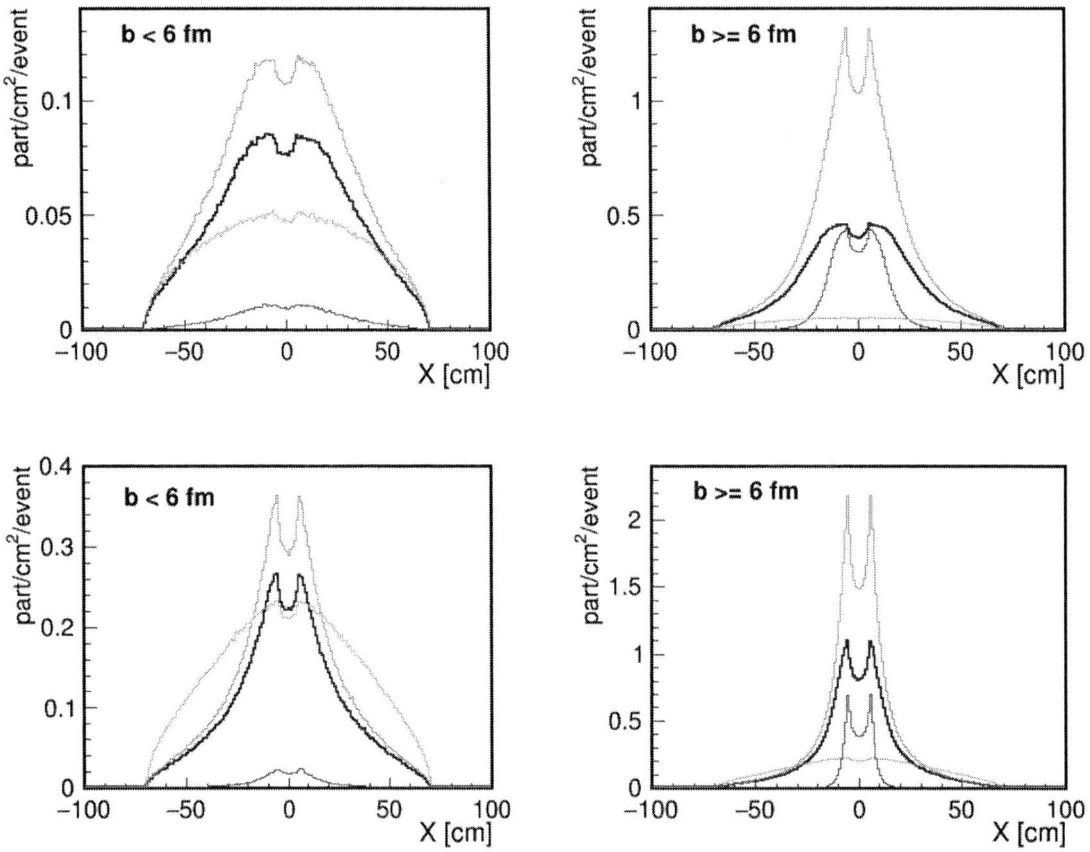

Figure 1. Horizontal distributions of protons (blue), neutrons (green), pions (light-blue) and fragments (red) produced in Au-Au collisions at $\sqrt{s} = 5$ GeV (top) and $\sqrt{s} = 11$ GeV (bottom) and impact parameters less than 6 fm (left) and larger than 6 fm (right).

For the simulation of Au-Au collisions the LAQGSM model [7] was used for the center of mass energies 5 and 11 GeV. In figure 1 the linear transverse spatial distributions are shown for protons, neutrons, pions and fragments (A > 1) for collisions with the impact parameter less 6 fm and larger 6 fm. The minimum at the centre is the result of the beam hole.

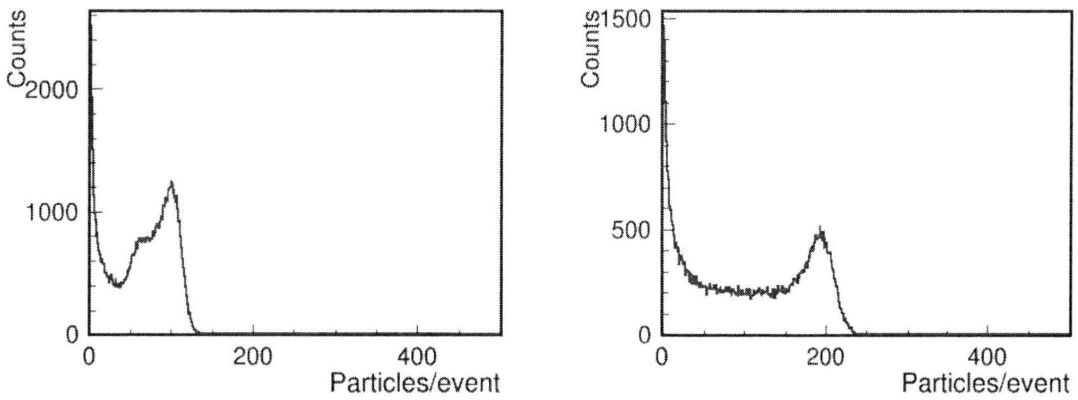

Figure 2. Charged particle multiplicity distributions at $\sqrt{s}$ = 5 GeV (left) and $\sqrt{s}$ = 11 GeV (right).

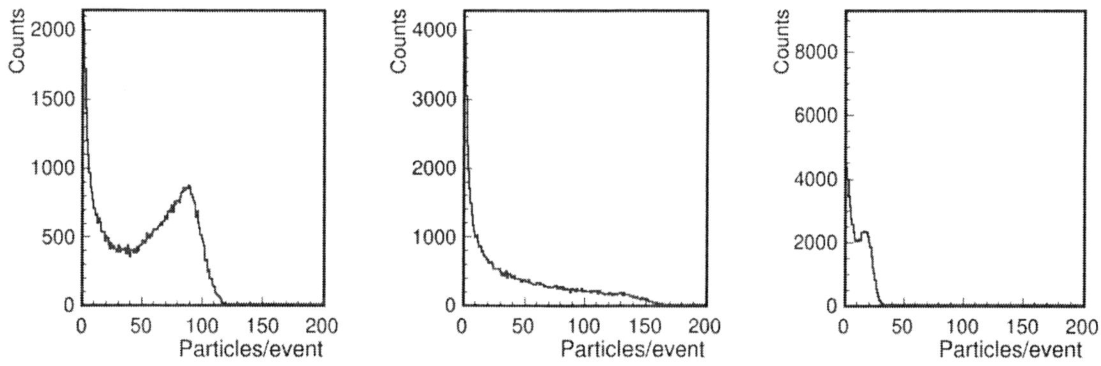

Figure 3. Multiplicity distributions of protons (left), pions (middle) and fragments (right) produced in Au-Au collisions at $\sqrt{s}$ = 11 GeV.

Calculated multiplicity distributions are shown in figure 2. The form of these distributions is different from that at high energy of ALICE. There are two maxima: one at small multiplicity and the second close to the limit of the number of particles per event.

It is seen from the calculations of multiplicity distributions made separately for protons, pions and fragments (figure 3) that the maximum at high multiplicity is due to protons, which could be considered as spectators from Au-Au collisions, arising at medium impact parameter. The same result was obtained for neutrons. This fact could change the common picture of the centrality determination in nucleus-nucleus collisions at intermediate energies. Because of Fermi motion of particle spectators they are deflected from the beam line and hit the surface of the FMD detector. However the emission of proton spectators is not uniform in the dependence on the impact parameter.

In the case of super-high energy of LHC in ALICE installation the spectators with high momenta do not hit the surface of V0 detector but pass inside the beam pipe.

 DOI: 10.1051/epjconf/201713811001

The mechanism of nuclear fragmentation in nucleus - nucleus collisions cannot be explained by simple geometrical picture. The study of nuclear fragmentation could be done at MPD/NICA with FHCAL, NZDC and possible FMD detectors in order to check different models of fragment production like Fermi break-up, preequilibrium emission and coalescence of produced particles. However for this purpose the FMD detector should be able to identify different fragments. This could be made with high granularity of FMD, energy loss and time-of-flight measurements. New information on fragment production will provide valuable information not only from scientific point of view, but also for many applications.

In conclusion we can note that the determination of the centrality of the collision in the case of relatively low beam energy of the NICA is not a simple problem. The spectators produced with sufficient transverse Fermi momentum are only partly hitting the detectors placed around the beam pipe, the heavier fragments are passing into the beam hole. Therefore the joint analysis of FHCAL, possible FMD and information from TPC and Neutron Zero Degree Calorimeter should be taken into account.

References

[1] The ALICE Collaboration, JINST **8**, P10016 (2013)
[2] The ALICE Collaboration, Phys. Rev. Lett. **106**, 032301 (2011). [arXiv, nucl-ex/1012,1657]
[3] The ALICE Collaboration, JHEP **1511**, 127 (2015). [arXiv, nucl-ex/1506, 08808]
[4] A.G. Litvinenko, this volume
[5] M. Golubeva et al., Yad. Fiz. **76**, 2 (2013)
[6] A. Ivashkin et al., MPD Forward Hadron Calorimeter (FHCAL), Technical Design Report, 1 (2016)
[7] S.G. Mashnik et al., arXiv:1607.02506 [nucl-th]

Selection of the elastic scattering events in interactions of the NICA colliding proton (deuteron) beams

Vasily Sharov[1,★]

[1] *Veksler and Baldin Laboratory of High Energy Physics, Joint Institute for Nuclear Research, Dubna, Russia*

Abstract. The features of the kinematics of elastic pp (dd) scattering in the collider system, as well as some issues concerning registration and selection of elastic scattering events in the *NICA* colliding beams are considered. Equality and the opposite direction of the scattered particle momenta provide a powerful selection criterion for elastic collisions. Variants of the organization of the trigger signal for recording tracks of secondary particles and *DAQ* system are given. The estimates of the characteristics of elastic NN processes are obtained from available $d\sigma/d\Omega_{CM}$ data for the elastic pp and np scattering. The paper presents examples of simulations using the Monte-Carlo of elastic pp scattering in the colliding proton beams and quasi-elastic np scattering in the colliding deuteron beams and evaluates the outputs of these processes at the *NICA* collider.

1 Introduction

The *NICA* collider [1] will provide the colliding beams of polarized protons and deuterons with the energy range of $T_p = 5 \div 12.6\ GeV$ ($\sqrt{s_{pp}} = 12 \div 27\ GeV$) for protons and $T_d = 2 \div 5.5\ GeV$ ($\sqrt{s_{NN}} = 4 \div 13\ GeV$) for deuterons. Obtaining new data on spin-dependent observables in elastic NN and dd scattering at the colliding beams energies $T_{p,d}$ above a few GeV could be the priority area of $VBLHEP$ research at the *NICA* collider.

In our previous reports and publications [2–4] we have considered opportunities of studying the spin 1/2 and 1 observables in the elastic interactions of the colliding proton and deuteron beams of the *NICA* collider. It has been shown that in elastic interactions of colliding proton beams one can measure energy and angular dependencies of the analyzing powers of the reactions with one polarized beam A_{oono} or A_{ooon}, and the spin correlation parameters A_{ookk} and A_{oonn} in collisions of both polarized proton beams. In elastic interactions of the colliding deuteron beams it is possible to obtain a complete set of energy and angular dependencies of components of the vector and tensor analyzing powers A_y, A_{yy}, and A_{xx} for the dN and dd reactions. The planned luminosity of the polarized colliding pp and dd beams will allow one to reach a sufficiently high counting rate of the elastic events in these measurements. The use of the colliding polarized pp and dd beams for the spin-dependent observables research will have a number of significant advantages in comparison with the experiments on the "fixed" target. The angular acceptance of the collider detector covers the full solid angle 4π radians when the wide ranges of the energies of the pp, dd interactions and the 4-momentum transfer squared are available.

★e-mail: vsharov41@mail.ru

To obtain the values of the spin-dependent pp or dd observables, it is necessary to measure the asymmetry of the outputs of the processes under study, i.e. - the ratio of the difference to the sum of the elastic differential cross sections with opposite signs of polarization of the colliding particles (or right-left asymmetry of the outputs for the same sign of the polarization of the beams).

In this article we discuss the kinematics of the elastic pp and dd scattering processes in the collider system and some of the issues concerning the registration and selection of elastic events. The kinematics features are discussed in the 2-nd section of the paper. The 3-rd section presents the Spin Physics Detector (SPD) elements and Monte Carlo (MC) simulations of scattering events. The fourth section gives the examples of the MC simulations for elastic events in the collider system using the existing experimental data on $d\sigma/d\Omega$ of elastic pp and np collisions.

2 The kinematics of elastic pp and dd scattering in the collider system

To determine possible criteria for the selection of elastic events in interactions of colliding beams, we consider the features of the two-particle $1 + 2 \rightarrow 3 + 4$ kinematics in the collider system. The coordinate system of the collider is simultaneously the laboratory system and Centre-of-Mass system of the colliding particles. The total momentum of the colliding particles and the total momentum of the reaction products are equal to zero. In this case, the momenta of incoming and outgoing particles are equal to each other $|\mathbf{p}_1| = |\mathbf{p}_2| = |\mathbf{p}_3| = |\mathbf{p}_4| = |\mathbf{p}_{beam}|$ and are oriented in the opposite directions $\mathbf{p}_1 \leftrightarrows \mathbf{p}_2$ and $\mathbf{p}_3 \leftrightarrows \mathbf{p}_4$. The equality and opposite direction of the outgoing particle momenta provide a powerful criterion to select elastic pp and dd events in the coordinate system of the collider. A

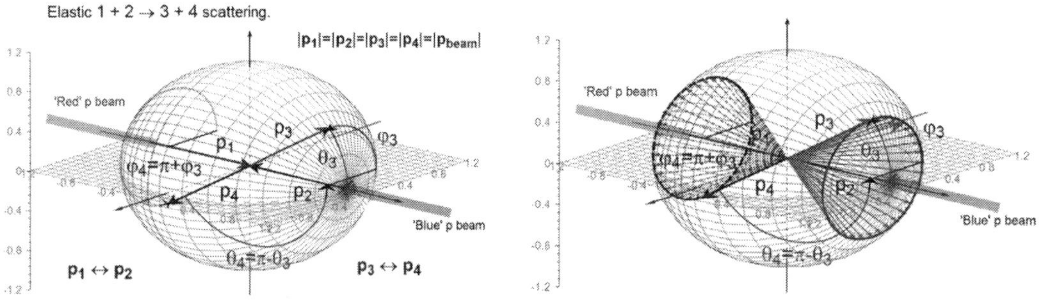

(a) **(b)**

Figure 1. **(a)** Three-dimensional region of possible values and orientations of the particle momenta in the elastic pp and dd collisions at the $NICA$ collider. **(b)** The distributions of the outgoing particles momenta are isotropic in the azimuthal direction.

three-dimensional range of possible values and orientation of the momentum of the particles in elastic pp and dd collisions in the collider represent a sphere with the radius value of $R = |\mathbf{p}_{beam}|$. Figure 1a. The direction of the momenta of secondary particles is determined by the values of polar θ_3 and θ_4 and azimuthal φ_3 and φ_4 angles of scattering. >From the condition $\mathbf{p}_3 \rightleftarrows \mathbf{p}_4$ it follows that the relation between the polar and azimuthal angles of the outgoing particles should be following: $\theta_4 = \pi - \theta_3$ and $\varphi_4 = \pi - \varphi_3$. The probability of the certain orientation of the $\mathbf{p}_3$ and $\mathbf{p}_4$ vectors is determined by the value of 4-momentum transfer squared $t_{1,3}$ from particle 1 to 3 or from particle 2 to 4. The value of $t_{1,3}$ for pp collisions in the CM system is defined by the following expression: $t_{1,3} = 2m_p^2 - 2E_p^2 + 2p_p^2 cos(\theta_{1,3})$. It is obvious that for the fixed value of $t_{1,3}$, i.e. at the fixed angle θ, the output of the elastic processes will be the same at any value of the azimuthal angle. In the collisions of unpolarized particles the spatial distribution of the outgoing particle momenta is isotropic in azimuthal direction, i.e., for each value of the polar angle θ the value of the azimuthal angle φ is distributed uniformly in the interval $0 \div 2\pi$. See figure 1b.

 DOI: 10.1051/epjconf/201713811002

>From the above kinematics, it follows that to select elastic events, you have to register two tracks of secondary particles and be sure that these tracks are oppositely directed. This condition specifies requirements to find the proper detectors for registration and selection of elastic processes in studies with colliding polarized *pp* and *dd* beams of the *NICA* collider.

3 Spin Physics Detector and Monte Carlo simulations of scattering events

The concept of the detector for spin physics studies at the *NICA* collider has been presented in [5]. The required set of devices for the Spin Physics Detector described in this publication can be also successfully used to register the elastic processes. The scheme of the detectors arrangement to register and select the elastic *pp*, *np* and *dd* events is shown in figure 2a. Two-coordinate detectors - internal (IT) and external trackers, determine the tracks of secondary particles, i.e., the directions of the scattered reaction products. Hodoscopes of scintillating counters (HSC) and resistive plate chambers (RPC) are proposed as trigger detectors for the data acquisition system (DAQ).

The use of a solenoid magnet in the SPD detector provides another powerful criterion for the selection of elastic events. The magnetic field of this magnet shifts the azimuthal φ coordinates of secondary particles that allows one to carry out off-line data processing to determine and compare their momenta. The solenoid magnetic field does not affect θ polar coordinates of secondary particles and so there are no problems in setting the θ trigger.

To register and select of the elastic *neutron-proton* (*np*) events we can use a Range System (RS) (particle range detectors) provided by the concept of SPD. The use of these detectors will allow us to estimate the energy value of the scattered *neutrons* and determine its approximate θ and φ coordinates. The signal from the selected part of the range detectors will allow one to select and register the elastic *np* events included in the trigger condition.

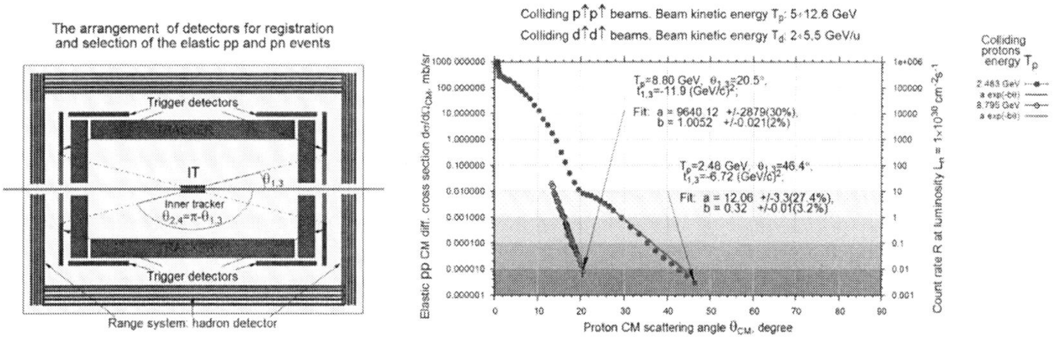

(a) (b)

Figure 2. **(a)** The arrangement of detectors in the SPD module for registration and selection of the elastic *pp*, *np* and *dd* events. **(b)** Fit of the experimental data sets on the elastic $d\sigma/d\Omega(pp)$ at the p beam energies of $T_p = 2.48$ and $8.8\ GeV$.

Angular θ, φ dependencies of the elastic events detected by SPD can be visualized by Monte Carlo modeling and used while preparing and carrying out these studies. The values of the parameters required for Monte Carlo simulations can be taken from available sets of experimental data $d\sigma/d\Omega(NN)$ in the Center of Mass (see the compilation [6]). Examples of the experimental $d\sigma/d\Omega(pp)$ and $d\sigma/d\Omega(np)$ data from [6] are shown in figures 2b and 5c.

The scattering events registered by the SPD coordinate detectors can be represented by a set of points — the coordinates of particles 3 and 4 in the θ, φ plane. Distributions of elastically scattered particles 3 and 4 in the θ coordinate can be approximated by exponential dependencies taken from the data [6] shown in figures 2b and 5c. The angular φ distribution of the elastic events can be uniform.

DOI: 10.1051/epjconf/201713811002

For inelastic scattered particles the θ and φ angular distributions can also be uniform. Under the above assumptions generated by the procedure, the Monte Carlo angular θ, φ distributions of elastic and inelastic events without any selection criteria are shown in figure 3a. Here the black dots represent the inelastic events, red – elastic scattering of the proton from the red oncoming beam, and the blue dots - elastic scattering of the proton from the blue beam.

The counting rate R of scattering events in the collider is proportional to the interaction cross section σ_{int} and luminosity of the collider L: $R = \sigma_{int} \cdot L$. The total pp cross sections in the energy range of the $NICA$ collider are following: $\sigma_{tot} \approx 40\ mb$, $\sigma_{tot,el} \approx 10\ mb$ and $\sigma_{tot,in} \approx 30\ mb$. The estimated luminosity of the colliding beams of polarized protons will reach $L_{pp} \approx 1 \cdot 10^{32}\ cm^{-2}s^{-1}$ at beam energies more than $T_p \approx 11\ GeV$ [7]. At $T_p \approx 3\ GeV$ the luminosity is expected to be equal to $L_{pp} \approx 1 \cdot 10^{30}\ cm^{-2}s^{-1}$. In this paper, we consider the examples of nucleon-nucleon collisions at energies of colliding nucleons $T_p = 2.48\ CeV$, $T_p = 8.8\ GeV$ and $T_n \approx 1.43\ GeV$. Due to these estimates at $T_p = 8.8\ GeV$ we have used the value $L_{pp} \approx 1 \cdot 10^{32}\ cm^{-2}s^{-1}$, and for the other energies the values of $L_{pp} \approx 1 \cdot 10^{30}\ cm^{-2}s^{-1}$. Then the expected counting rates will be $R_{el} \approx 1 \cdot 10^4 s^{-1}$ and

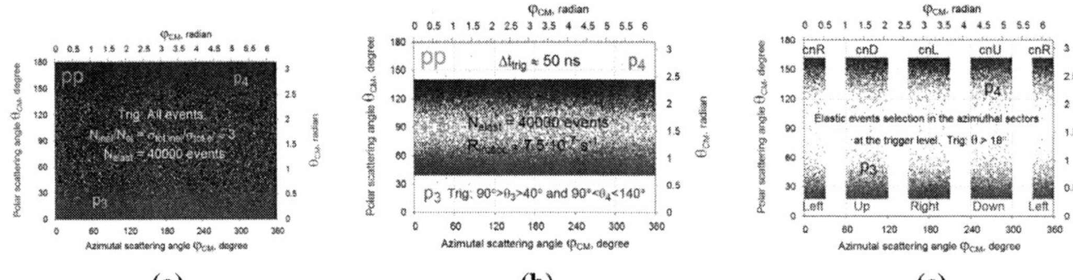

(a) **(b)** **(c)**

Figure 3. **(a)** Example of a Monte Carlo simulation for all the events without selection. **(b)** Simulation of elastic pp events with two trigger conditions. **(c)** The left-right and up-down events selection in the azimuthal sectors at the trigger level.

$R_{in} \approx 3 \cdot 10^4 s^{-1}$. When the trigger signal duration is $\Delta t \sim 50\ ns$, the probability of the inelastic event presence in this interval is $p_{in} \sim R_{in} \cdot 5 \cdot 10^{-8} = 1.5 \cdot 10^{-3}$ and of the elastic ones is $p_{el} \sim 5 \cdot 10^{-4}$. The counting rate of accidental coincidences, i.e. the probability of detecting the inelastic events along with the elastic ones with trigger conditions for the given scattering angle $C3$: $90^\circ \geq \theta_3 \geq C_3^\circ$ and $90^\circ \leq \theta_4 \leq \pi - C_3^\circ$, will be $R_{acc} \approx p_{el} \cdot R_{in} \approx 7.5 \cdot 10^{-7} s^{-}1$. This level of accidentals leads to almost complete absence of inelastic events in the following figures for the results of the MC simulations.

Figures 3b and 3c show examples of using different types of trigger conditions for the elastic pp events selected in the θ, φ plane. Selection of the elastic events in the given range only of the polar angles $\theta_{3,4}$ (fig. 3b) provides the maximum event rate. In $\vec{d}\vec{d}$ collisions to measure all the components of polarization observables, it is required to measure the left-right and up-down asymmetries of the outputs of the elastic process. In this case, the trigger detectors must identify separately $\Delta\varphi$ zones of scattering particles of the "red" beam to the left, right, up and down in the forward hemisphere ($0 < \theta < 90^\circ$) and the corresponding $\Delta\varphi$ areas of conjugated particles of the "blue" beam scattered in the back hemisphere ($90^\circ < \theta < 180^\circ$).

4 Examples of Monte Carlo simulation for pp and np collisions

By fitting the experimental θ angular dependencies of $d\sigma/d\Omega(pp)$ and $d\sigma/d\Omega(np)$ [6] from figure 2b and 5c, we get the values of the slope parameters of exponents to use Monte Carlo visualization for these processes. The values of $d\sigma/d\Omega$ from these data at the maximal angle θ_{max} allow us to estimate the event counting rate at this angle. The results of Monte Carlo simulations for the energies of colliding protons of $T_p = 2.48\ GeV$ and $T_p = 8.8\ GeV$ are given in figure 4 and figure 5a. Figure 4a shows the dependence of $t_{1,3}(\theta)$ in the range of $0 \leq \theta \leq 90^\circ$ for $d\sigma/d\Omega(pp)$ data at

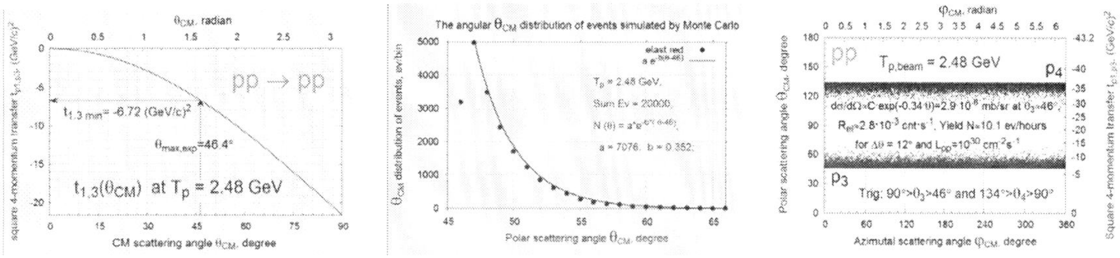

(a) (b) (c)

Figure 4. (**a**) The range of $t_{1,3}(\theta_3)$ values when $0 \leq \theta_3 \leq 90°$ for pp at $T_p = 2.48\ GeV$. (**b**) The θ_{CM} distribution of the events simulated by MC procedure at $T_p = 2.48\ GeV$. (**c**) Results of the MC simulations for pp at $T_p = 2.48\ GeV$.

$T_p = 2.48\ GeV$. The maximal angle and the corresponding $t_{1,3}(\theta_{max})$ value are indicated. Generated by the MC procedure the θ distributions of elastic and quasi-elastic events are shown in figure 4b for pp collisions at $T_p = 2.48\ GeV$, and in figure 6a for np collisions at $T_n = 1.427\ GeV$. Figs. 4c, 5a, 6b, and 6c show the MC generated θ, φ distributions of the elastic events and the expected counting rates $R(\theta) = L \int \int d\sigma/d\Omega(\theta){\cdot}sin\theta{\cdot}d\theta d\varphi \simeq 2\pi L d\sigma/d\Omega(\theta){\cdot}sin\theta{\cdot}\Delta\theta\ [s^{-1}]$ and yields $Y(\theta) = R(\theta){\cdot}3.6{\cdot}10^3\ [h^{-1}]$ in the specified band $[\theta_{max} \div \theta_{max} + \Delta\theta]$ at the number of T_{beam} values.

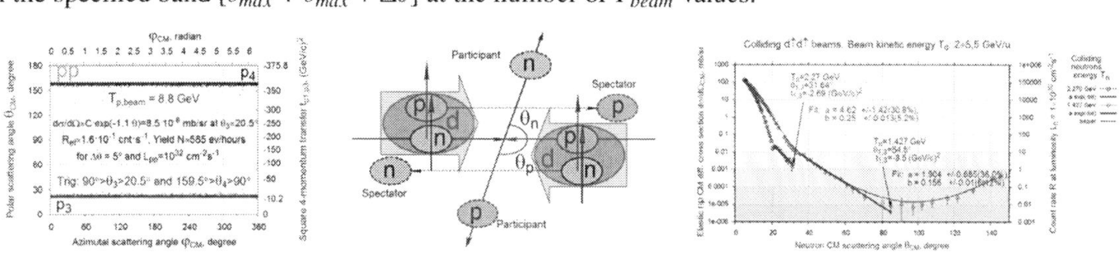

(a) (b) (c)

Figure 5. (**a**) Results of the Monte Carlo simulations for pp collisions at $T_p = 8.8\ GeV$. (**b**) The quasi-elastic $\vec{N}\vec{N}$ scattering in the polarized deuteron $\vec{d}\vec{d}$ collisions. (**c**) Fit of the experimental data sets on $d\sigma/d\Omega$ for elastic np scattering at the energies of the colliding dd beams at $T_d = 2.85$ and $4.54\ GeV$.

Assuming that the deuteron is a weakly bounded system, one can apply the formalism of elastic pp scattering to describe a quasi-elastic NN scattering independently on each nucleon in the deuteron. See figure 5b. The nucleons in the deuteron have the same value and orientation of polarization as the

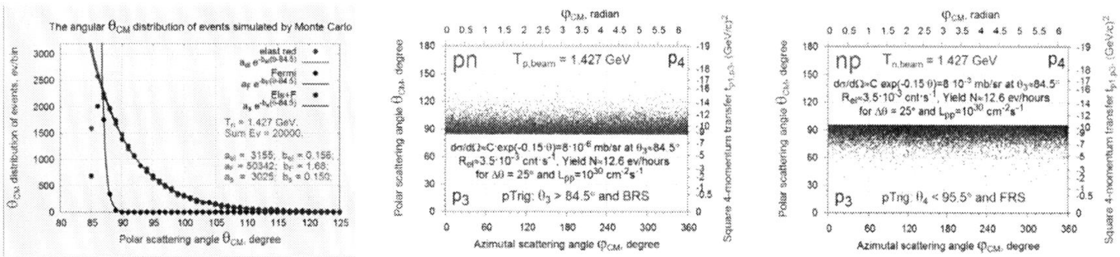

(a) (b) (c)

Figure 6. (**a**) The θ_{CM} distribution of events simulated by the MC procedure at $T_n = 1.427\ GeV$. (**b**) Results of the MC simulations for pn collisions at $T_n = 1.427\ GeV$. (**c**) The same but for np collisions.

deuteron itself $\vec{P}_N = \vec{P}_d$ and half deuteron kinetic energy $T_N = 0.5\ T_d$. The Fermi motion of nucleons in the deuteron influences weakly on the angular distribution of scattering products $\Delta\theta_{Fm} < 22\ mrad$ at $p_p \sim 2.2\ GeV/c$. The Gaussian momentum spread of the nucleon in the deuteron is also small $\sigma_p \approx 2.5\%$. Results of the MC simulations for pn and np collisions at $T_n = 1.427\ GeV$ are presented in figure 6. Figure 6a shows the θ distributions simulated by the MC procedure at $T_n = 1.427\ GeV$

for: - elastic np scattering (red dots and curve); - due to the Fermi motion of the nucleons in the deuteron (black dots and curve); and - elastic scattering + Fermi motion (blue dots and curve). The effect of Fermi motion is noticeable only within the limits of its θ distribution. Trigger conditions for the distributions of figure 6b and 6c include HSC or/and RPC signal $\theta_3 > 84.5°$ from the pn scattered proton and the signal of the scattered neutron from the corresponding backward part of the RS detectors - BRS and vice versa for the np scattering.

5 Conclusion

The features of the kinematics of the processes of elastic pp and dd scattering in the collider system have been considered. The equality and opposite direction of the momenta of the scattered particles provide a powerful selection criterion for elastic collisions in this system. Possible choices of the trigger signal to register the tracks of secondary particles and data acquisition system have been given.

According to available $d\sigma/d\Omega_{CM}(pp)$ and $d\sigma/d\Omega_{CM}(np)$ data in the energy range of the $NICA$ collider we have estimated the parameters of the angular distributions for the Monte Carlo simulations of these processes. The results of MC simulations of the elastic and quasi-elastic pp and np processes and estimations of counting rates and outputs of these reactions at specified scattering angles have been given. The presented examples of the MC simulations of elastic events can be used in the research of elastic pp, np, and dd interactions in the colliding beams at the $NICA$ collider. The obtained results have shown that in the colliding pp and dd beams at the $NICA$ collider can obtain a sufficiently high counting rate and outputs of the elastic collision events.

Acknowledgements

The author is grateful to his colleague E.V. Chernykh for his help to prepare the talk and this contribution.

References

[1] NICA - Nuclotron - based Ion Collider fAcility - a Base for Fundamental Research and Innovation Development, (JINR, Dubna 2014). http://nica.jinr.ru/files/Booklet-NICA.pdf

[2] V.I. Sharov, *Proceedings of "XV Advanced Research Workshop on High Energy Spin Physics. (DSPIN-13)", Dubna, October 8-12, 2013*, edited by A.V. Efremov, S.V. Goloskokov, JINR, E1,2-2014-12, (JINR, Dubna, 2014), 367-370

[3] V.I. Sharov, *Proceedings of "XXII International Baldin Seminar on High Energy Physics Problems. Baldin ISHEPP XXII", Dubna, September 15-20, 2014*, PoS (Baldin ISHEPP XXII) 102. http://pos.sissa

[4] V.I. Sharov, *Proceedings of "XVI Workshop on High Energy Spin Physics (D-SPIN2015)". Dubna, September 15-20, 2014*, Journal of Physics: Conference Series **678**, 012011 (2016)

[5] I.A. Savin, A.V. Efremov et al., *Spin Physics Experiments at NICA-SPD with polarized proton and deuteron beams*, Letter of Intent. (2014)

[6] J. Bystricky and F. Lehar, *Nucleon-Nucleon Scattering Data*, editors H. Behrens and G. Ebel (Fachinformationszentrum Karlsruhe, 1981 Edition, Nr. 11-2 and Nr. 11-3, 1981)

[7] A.D. Kovalenko, *Proceedings of "XV Advanced Research Workshop on High Energy Spin Physics. (DSPIN-13)", Dubna, October 8-12, 2013*, edited by A.V. Efremov, S.V. Goloskokov, JINR, E1,2-2014-12 (JINR, Dubna, 2014) 345-350

Toward centrality determination at NICA/MPD

A.S. Galoyan[1,⋆] and *V.V.* Uzhinsky[2]

[1] *VBLHEP, Joint Institute for Nuclear Research, Dubna, Russia.*
[2] *LIT, Joint Institute for Nuclear Research, Dubna, Russia.*

Abstract. Geometrical properties of nucleus-nucleus interactions at various centralities are calculated for the NICA energy range. A modified version of the Glauber Monte Carlo simulation code has been used for the calculations. It is shown that the geometrical properties of nucleus-nucleus interactions at the energies 5 – 10 GeV (NICA/MPD) and at energy 200 GeV (RHIC) are quite close to each other. A possible determination of centrality at NICA/MPD experiment using calculations of various Monte Carlo event generators are considered.

Introduction

As it is known, a new accelerator facility – the Nuclotron based Ion Collider facility (NICA) at JINR (Dubna) will start to operate in the nearest future. The main scientific goal of the NICA/MPD is to explore the mixed phase of quarks, gluons and hadrons in the nucleus-nucleus interactions at the CMS energy range from 4 to 11 GeV. To reach the main goal of MPD (Multi-Purpose Detector), the following studies are foreseen for different ions from H to Au: event-by-event fluctuation in hadron productions (multiplicity and $< P_T >$ fluctuations), femtoscopic correlation, directed and elliptic flows for various hadrons, multi-strange hyperon production, yield and spectra of hadrons, photon and electron, charge asymmetry and so on.

Nucleus-nucleus collisions with various centralities, with various impact parameters, will be studied at NICA/MPD experiment. For determination of the centrality, many experiments (ATLAS, CMS, FOBOS and others) use Glauber Monte Carlo model. It allows one to determine main parameters of interactions such as the impact parameter (b), a number of interacting nucleons (N_{part}), a number of binary collisions (N_{coll}), an eccentricity – anisotropy of interaction region, masses of residual nuclei and so on. A well-known Glauber Monte Carlo code [1] created by B. Alver, M. Baker et al. for FOBOS collaboration is used for these purposes. The code is used by many modern experiments at high energies: CMS, ALICE, STAR and others. A simplified Glauber approximation is implemented in the code. We will consider a more correct Glauber approximation which can be applied for beam energy scan program realized at NICA/MPD.

A centrality determination of collisions under the study - determination of impact parameter range, is a key problem in experimental investigations. Usually, a mapping of Glauber calculations and experimental observables (n_{obs}) is used for the problem solution. In a simple form, it looks like that: $< n_{obs} >= [\alpha < N_{part} > +(1-\alpha) < N_{coll} >]*\beta$, where α and β are tuned parameters. A question is what

⋆e-mail: galoyan@lxpub01.jinr.ru

observable is the most appropriate for this? It maybe a multiplicity of charged particles registered in a detector, or multiplicity of spectator neutrons, or energy carried by spectator nuclear fragments, and so on. We shortly consider them below.

1 Main ingredients of Glauber approach

Cross section of new particle production in nucleus-nucleus interactions is given in the Glauber approximation by the well-known expression:

$$\sigma_{AB}^{prod} = \sigma_{AB}^{tot} - \sigma_{AB}^{el} - \sigma_{AB}^{q.el} = \int d^2b \, \{1 - \prod_{i=1}^{A}\prod_{j=1}^{B}(1 - p_{ij}(\mathbf{b} - \mathbf{s}_i + \tau_j))\} \cdot$$

$$|\psi_A(\mathbf{r}_1, \ldots, \mathbf{r}_A)|^2 \; |\psi_B(\mathbf{t}_1, \ldots, \mathbf{t}_B)|^2 \prod_{i=1}^{A} d^3r_i \prod_{i=1}^{B} d^3t_i, \tag{1}$$

where A and B are mass numbers of the nuclei; $\mathbf{b}$ is the impact parameter;

$$p_{ij}(\mathbf{b}) = C_iC_j\gamma(\mathbf{b}) + C_iC_j\gamma^*(\mathbf{b}) - (C_iC_j)^2\gamma(\mathbf{b})\gamma^*(\mathbf{b});$$

$\gamma(\mathbf{b})$ is an amplitude of elastic nucleon-nucleon scattering in the impact parameter representation; C_i is shower enhancement coefficient in the vertex of pomeron-nucleon interaction; $\{\mathbf{s}_i\}$, $i = 1, 2, ..., A$ and $\{\tau_j\}$, $j = 1, 2, ..., B$ are sets of coordinates of nucleons of A and B nuclei in the impact parameter plane; $\mathbf{r}_i = (\mathbf{s}_i, z_i)$; $\mathbf{t}_i = (\tau_i, \zeta_i)$; ψ_A and ψ_B are wave functions of the nuclei in the ground states. At $C_i = 1$ one has the standard Glauber expression. At $C_i \neq 1$ the Gribov's inelastic screenings are taken into account. In this case each nucleon can be in an "active" state with a probability $1/C$ and $C_i = C$, or in a "passive" state with a probability $1 - 1/C$.

Equation (1) can be re-written in a form where each term of the expansion can be interpreted as a cross section of processes with fixed multiplicity of nucleon-nucleon collisions:

$$\sigma_{AB}^{prod} = \int d^2b \left\{ \sum_{i=1}^{A}\sum_{j=1}^{B} \frac{p_{ij}(\mathbf{b} - \mathbf{s}_i + \tau_j)}{1 - p_{ij}(\mathbf{b} - \mathbf{s}_i + \tau_j)} \prod_{k=1}^{A}\prod_{l=1}^{B}(1 - p_{kl}(\mathbf{b} - \mathbf{s}_k + \tau_l)) + \right.$$

$$+\frac{1}{2} \cdot \sum_{\substack{i=1,j=1\\ i\neq j}}^{A} \sum_{k=1}^{B} \frac{p_{ik}(\mathbf{b} - \mathbf{s}_i + \tau_k)}{1 - p_{ik}(\mathbf{b} - \mathbf{s}_i + \tau_k)} \frac{p_{jk}(\mathbf{b} - \mathbf{s}_j + \tau_k)}{1 - p_{jk}(\mathbf{b} - \mathbf{s}_j + \tau_k)} \cdot$$

$$\left. \prod_{l=1}^{A}\prod_{m=1}^{B}(1 - p_{lm}(\mathbf{b} - \mathbf{s}_l + \tau_m)) + \ldots \right\} \cdot$$

$$|\psi_A(\mathbf{r}_1, \ldots, \mathbf{r}_A)|^2 \; |\psi_B(\mathbf{t}_1, \ldots, \mathbf{t}_B)|^2 \prod_{i=1}^{A} d^3r_i \prod_{i=1}^{B} d^3t_i. \tag{2}$$

For example, the first term is a cross section of processes with one inelastic nucleon-nucleon collision. The second term is a cross sections of processes with two inelastic nucleon-nucleon collision, and so on.

A complexity of (1) and (2) prevent their analytical evaluation at a realistic choice of the scattering amplitude and nuclear densities with an exception A, $B \leq 4$. Thus, it was proposed in the papers [2] to use Monte Carlo averaging methods for the aim.

 DOI: 10.1051/epjconf/201713811003

In particular, in the paper [1] a program code was proposed which uses modern computational tools – the ROOT system. The code is intensively used in experiments at high energies, and is called "Glauber Monte Carlo". Though, the code is not free from some drawbacks: the profile of inelastic nucleon-nucleon collisions is chosen in a simplified form ($p_{ij}(\mathbf{b}) = \theta(\mathbf{r_{NN}} - |\mathbf{b}|)$, $r_{NN} = \sqrt{\sigma^{in}_{NN}/\pi}$), a set of allowed nuclei is strongly restricted. These mean that one needs to point out working with the code a cross section of inelastic nucleon-nucleon collisions at a given energy of the collisions, and nuclei from a defined list. Some drawbacks of the code were erased in the paper [3].

Recently, we have proposed [4, 5] the following parameterization of the nucleon-nucleon elastic scattering amplitude which can be used in the Glauber calculations:

$$\gamma(b) = A_1 \left\{ \frac{1}{1 + e^{(b-\tilde{R})/d}} + \frac{1}{1 + e^{-(b+\tilde{R})/d}} - 1 \right\} -$$
$$-i\,\rho A_1 \frac{\tilde{R}^2/2 + \pi^2 d^2/6}{\tilde{R}d} \left\{ \frac{e^{-(b-\tilde{R})/d}}{\left[1 + e^{-(b-\tilde{R})/d}\right]^2} + \frac{e^{-(b+\tilde{R})/d}}{\left[1 + e^{-(b+\tilde{R})/d}\right]^2} \right\} -$$
$$-i\,\frac{A_2}{2\pi B_2\, 25.64} e^{-b^2/(2\,B_2\,25.64)} \qquad (3)$$

$$\tilde{R} = R \;+\; (0.07 + d + 0.2d^2)\, e^{-1.2R/d}. \qquad (4)$$

The real part of the amplitude (3) is the symmetrized Fermi distribution. The imaginary part of the amplitude which is small at high energies has rather complicated form due to usage of dispersion relations in differential forms. The last term of the amplitude is needed for a description of nucleon-nucleon elastic scattering at large momentum transfers ($|t| > 1.75$ (GeV/c)^2). Energy dependencies of $\tilde{R}$, R, d, A_1, A_2, B_2, ρ are given in [6]. The parameterization allows one to describe rather well experimental data on differential cross sections of proton-proton elastic scattering at $\sqrt{s} \geq 3$ GeV [5].

Important structure elements of (1) and (2) are squared modules of the nuclear wave functions in the ground states. Very often they are represented as:

$$|\psi_A(\mathbf{r}_1, \ldots, \mathbf{r}_A)|^2 = \left[\prod_{i=1}^{A} \rho_A(\mathbf{r}_i)\right] \delta\left(\frac{1}{A}\sum_{i=1}^{A} \mathbf{r}_i\right). \qquad (5)$$

A systematical analysis of electron-nuclear data and a comparison with theoretical calculations (DHB approximation) have been done in the paper [7] for an extraction of one-particle densities of nuclei (ρ_A). As a result, the densities were parameterized in the standard form:

$$\rho_A(r) = \frac{\rho_0}{1 + e^{\frac{r-R_A}{d_A}}}, \quad R_A = 1.31\, A^{1/3} - 0.84 \quad [fm], \quad d_A = 0.5 \quad [fm]. \qquad (6)$$

The paper [7] is well-known in low and intermediate energy physics, especially in physics of exotic nuclei because parameterizations of one-particle densities of protons and neutrons in nuclei were also presented in the paper.

All specifications mentioned above were introduced in our Monte Carlo program of Glauber calculations. Results of calculations with the code are presented below.

Figure 1a shows dependencies of average multiplicities of the participating nucleons on a centrality of Au+Au collisions at $\sqrt{s_{NN}} = 9$ and 200 GeV (closed and open points, respectively). Dispersions of the quantities are presented as "experimental error bars". As it is seen, the multiplicities of the participating nucleons at low and high energies differ in 3 – 5 %. It is explained by the fact that high energy collisions differ from low energy one only due to larger radius of *NN* interactions from geometrical point of view. The radii are small compared with nuclear sizes. Though, the change of the radius at a transition from low to high energies has significant influence on the average multiplicities of the binary collisions (see figure 1b) and the anisotropy of nuclear interaction region (see figure 1c).

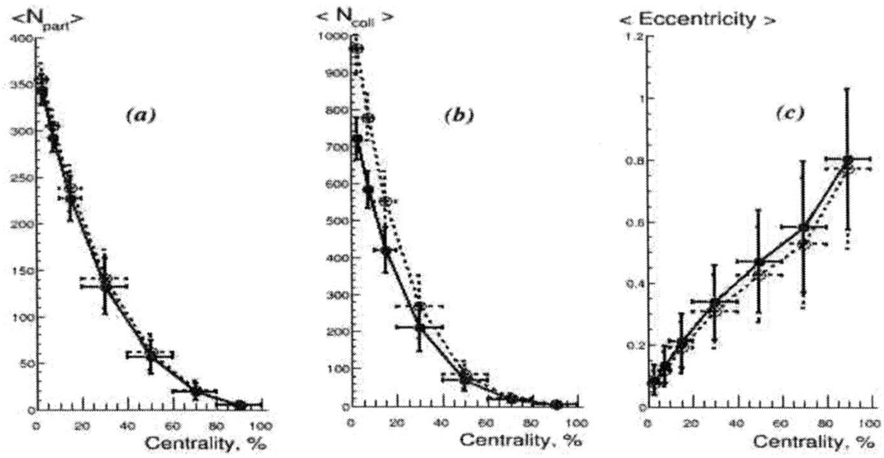

Figure 1. Mean multiplicities of the participating nucleons, binary collisions and the eccentricity as functions of the centrality (figures a, b and c, respectively). Dispersions of the quantities are presented as error bars.

Usually, events of nucleus-nucleus interactions are subdivided on the following centrality classes: 0–5, 5–10, 10–20, 20–40, 40–60, 60–80, 80–100 % at experimental studies of super high energies. As it is seen in Figs. 1a and 1b, distributions on multiplicities of the participating nucleons and the binary collisions are inessential overlapping at high and low energies in neighboring intervals of the centrality at such subdivisions. Thus, the corresponding quantities for neighboring intervals – $< N_{part} >$ and $< N_{bin} >$ can be used for revelations of scaling properties of the interactions and the jet quenching in the collisions. Such applications become problematic at a more narrow interval subdivision due to a possible strong overlap of the distributions. As it is also seen, distributions on anisotropy of the interaction regions are strongly overlapping at all centrality classes (see figure 1c).

The anisotropy of the interaction region is characterized by the eccentricity in the last decade – $\varepsilon_{part.} = (< y^2 > - < x^2 >)/(< y^2 > + < x^2 >)$, where x and y are coordinates of a participating nucleon in the impact parameter plane counted from a center of the interaction region. We have used the definition in our calculations. $\varepsilon_{part.}$ in super high energy interactions is proportional to a magnitude of an elliptic flow which is unambiguously connected with the properties of the quark-gluon plasma. One can expect a violation of the regularity and strong fluctuations of the collective flows at low energies and at an appearance of the mixed phase. Studies of such fluctuations are topics of modern experimental efforts of the last time.

2 Choose of observables for centrality determination

The geometrical property of interactions cannot be measured directly. They must be connected with experimental observables such as a multiplicity of charged particles in a detector, or an energy carried by spectator nuclear fragments measured by Zero Degree Calorimeter. Possible locations of the detectors are shown in figure 2 by rectangles. We present in figure 2 rapidity distributions of particles produced in Au+Au interactions at $\sqrt{s_{NN}}$ = 10 GeV predicted by UrQMD [8], QGSM [9], and FTF [10] models.

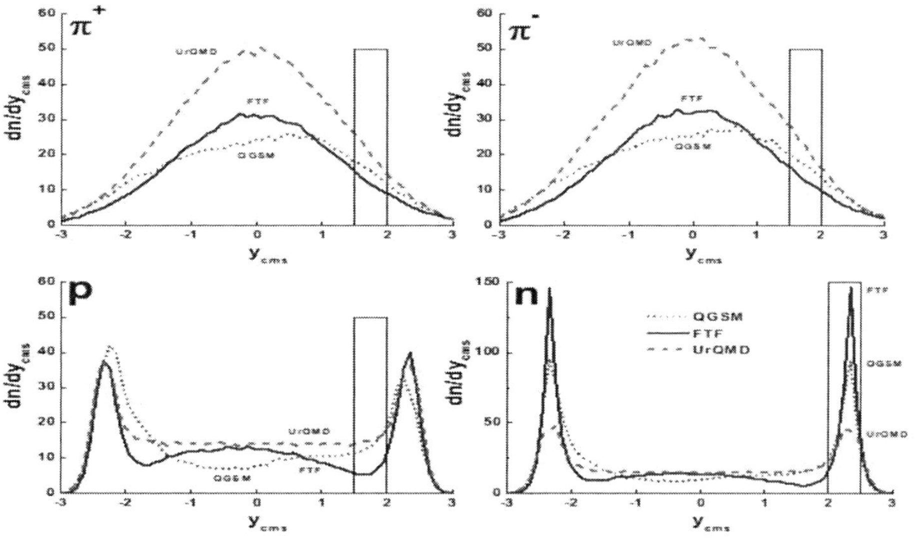

Figure 2. Rapidity distributions of hadrons in Au+Au interactions at $\sqrt{s_{NN}}$ = 10 GeV. Solid, dashed and dotted lines are FTF, UrQMD and QGSM model calculations, respectively.

As it is seen, the model predictions are in a disagreement with each other. UrQMD predicts the largest production of charged pi-mesons in central region. QGSM predicts the low multiplicity of produced mesons. FTF gives intermediate results. The FTF, UrQMD and QGSM predictions for proton production in Au+Au interactions at 10 GeV/per NN collision are close to each other.

However, UrQMD gives small number of evaporated neutrons as seen in the bottom right figure. To obtain the calculation results, we enlarged the UrQMD by the excitation-de-excitation model. We use for simulation of the de-excitation the well-known Statistical Multi-fragmentation Model (SMM).

Production of nuclear fragment is implemented in the QGSM model.

FTF gives the largest number of produced neutrons in fragmentation region. We checked the FTF predictions for neutron production in nucleus-nucleus interactions in [11].

Conclusion

- According to our calculations, the geometrical properties of nucleus-nucleus interactions at RHIC and NICA energies are quite close to each other.
- Predictions of various Monte Carlo models for meson production in AA interactions are varied in the range ±50 % for E_{cms} = 5 - 10 GeV. A large disagreement between the model predictions is observed for spectator neutron productions.
- According to the models, largest particle density must be for neutrons in very narrow interval of rapidity. This predictions can be used for design of centrality detector.

References

[1] B. Alver, M. Baker, C. Loizides and P. Steinberg, arXiv:0805.4411 [nucl-exp] (2005)
[2] A.M. Zadorozhnyi, V.V. Uzhinsky and S.Yu. Shmakov, Comp. Phys. Commun. **54**, 125 (1989)
[3] W. Broniowski, M. Rybczynski and P. Bozek, Comp. Phys. Commun. **180**, 69 (2009)
[4] V. Uzhinsky and A. Galoyan, JETP Lett. **94**, 499 (2011). Pisma v Zh.ETP **94**, 539 (2011)
[5] V. Uzhinsky and A. Galoyan, arXiv:1111.4984 [hep-ph] (2011). arXiv:1210.7338 [hep-ph] (2012)
[6] A.S. Galoyan and V.V. Uzhinsky, Phys. Part. Nucl. Lett. **12**, 166 (2015)
[7] L.C. Chamon et al., Phys. Rev. C **66**, 014610 (2002)
[8] S.A. Bass et al., Prog. Part. Nucl. Phys. **41**, 225 (1998). M. Bleicher et al., J. Phys. G **25**, 1859 (1999)
[9] V.D. Toneev, N.S. Amelin, K.K. Gudima, and S.Yu. Sivoklokov, Nucl. Phys. A **519**, 463 (1990)
[10] http://geant4.cern.ch/support/userdocuments.shtml/PhysicsReferenceManual.pdf
[11] A. Galoyan, A. Ribon and V. Uzhinsky, JETP Letters **102**, 324 (2015)

EPJ Web of Conferences **138**, 11004 (2017) DOI: 10.1051/epjconf/201713811004
Baldin ISHEPP XXIII

Simulation of the GEM detector for BM@N experiment

Dmitriy Baranov[1,*] and *Oleg* Rogachevsky[1,**]

[1] *Joint Institute for Nuclear Research, 6 Joliot Curie St., Dubna, Moscow region, 141980 Russia*

Abstract. The Gas Electron Multiplier (GEM) detector is one of the basic parts of the BM@N experiment included in the NICA project. The simulation model that takes into account features of signal generation process in an ionization GEM chamber is presented in this article. Proper parameters for the simulation were extracted from data retrieved with the help of Garfield++ (a toolkit for the detailed simulation of particle detectors). Due to this, we are able to generate clusters in layers of the micro-strip readout that correspond to clusters retrieved from a real physics experiment.

1 Introduction

The BM@N[1] [1] is one of the first heavy-ion experiments of the NICA[2] project [2, 3] at the Joint Institute for Nuclear Research (JINR) in Dubna. The main distinctive feature of this experiment is an available facility to accelerate heavy ions to an energy between 2 and 6A GeV. Such energy range covers the region of the maximum baryon density. Thus the BM@N experiment with a fixed target gives the great ability to study baryonic matter on relativistic heavy ion collisions at the Nuclotron.

The BM@N setup consists of a tracking system, a time of flight system for particle identification and some detectors to measure event characteristics.

2 BM@N GEM tracker

The BM@N central tracking system is comprised of GEM chambers with two-coordinate micro-strip readout [4]. Each GEM chamber registers the ionization trail left by charged particle in the sensitive gas volume. The overall structure of the detector is a set of separated GEM stations that are located along the beam-axis at a certain distance from each other. Full GEM tracker configuration (figure 1) includes 12 such stations, each of these has a certain type of the GEM chamber.

GEM plane is divided into areas. The areas called "hot zones" are generally placed at the center of the station. Thus we have parts with strip readout electrically independent from each other in one chamber. It allows us to process high multiplicity events.

*e-mail: dbaranov@jinr.ru
**e-mail: rogachevsky@jinr.ru

[1] Baryonic Matter at Nuclotron
[2] Nuclotron Based Ion Collider fAcility

DOI: 10.1051/epjconf/201713811004

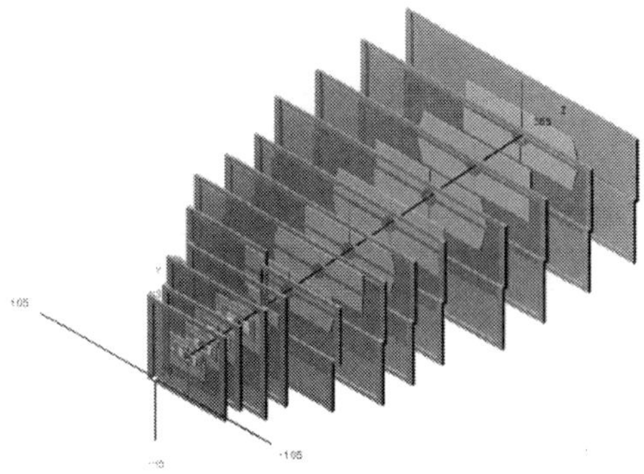

Figure 1. BM@N GEM tracker: full configuration.

2.1 Gas Electron Multiplier in BM@N

There are used triple-GEM chambers in the inner tracking detector for the BM@N experiment. Chamber capacity is filled with a gas mixture of Ar and CO_2 in the ratio 70:30. Each triple-GEM has multi-layer structure (figure 2): *a drift gap*, where multiple ionization processes happen, *two transfer gaps*, where drifted electrons passing through the GEM foils are multiplied by impact ionization in high electric inside foil holes, and *an induction gap*, where produced electron avalanches drift to the readout plane. The signal registered by readout strips from one charged particle has a cluster structure.

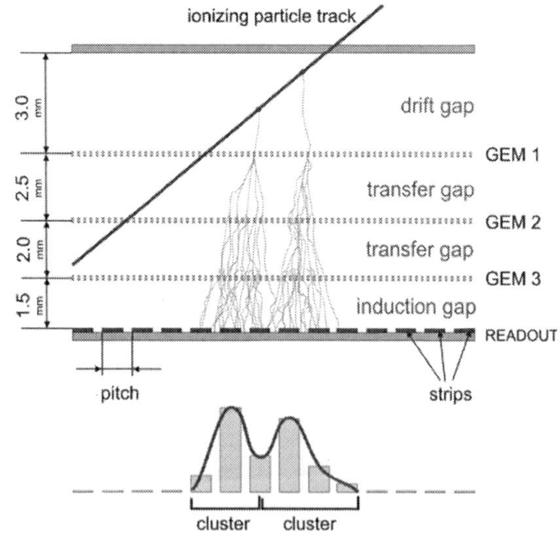

Figure 2. BM@N triple-GEM structure and signal formation scheme.

More detailed information about working principles and signal formation in GEM detectors are described in the works [5, 6].

DOI: 10.1051/epjconf/201713811004

3 Simulation model

In order to obtain proper parameters for the strip cluster generation (such as mean shift of electrons in magnetic field, electron diffusion and gas ionization characteristics) we used the Garfield++ toolkit [7].

Since the magnetic field inside the magnet is directed perpendicular to electric field inside the GEM, produced electron avalanches are shifted from its initial positions by Lorentz force. The dependence of the mean electron shift in the transverse drift direction on its traversed distance for different GEM gaps is shown below (figure 3, left). The dependence of the mean electron shift on the distance to the readout plane for various magnetic field strength is also shown (figure 3, right).

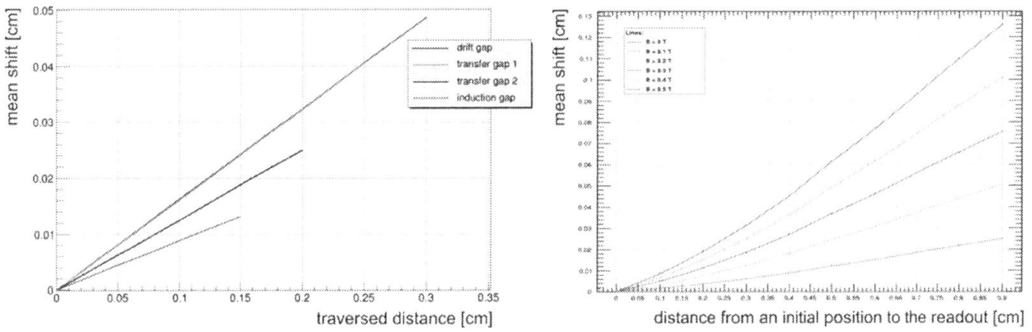

Figure 3. (Left) Dependence of the mean electron shift in the transverse drift direction on its traversed distance for all GEM gaps; (right) dependence of the mean electron shift on the distance to the readout plane for various magnetic field strength.

An electron cloud drifting towards the anode along field lines in electric field spreads due to random thermal motion and repulsion. Electron diffusion affects the size of complete electron avalanches that induce output signal on the readout plane. The graph below shows the dependence of the standard deviation (sigma) of electrons in the transverse drift direction on its traversed distance for all GEM gaps (figure 4, left). The cumulative distribution (on the readout plane) of electrons drifted through the thickness of the GEM chamber (0.9 cm) taking into account the electron shift and electron diffusion is also shown (figure 4, right).

We also obtained ionization characteristics of our gas media for different types of charged particles. The graph below presents an example of the distribution of primary electron-ion clusters along 1 cm track for particle pi+ with initial momentum 1 GeV/c (figure 5, left). In the other graph you can observe the distribution of collision distances between the electron-ion clusters for the same particle (figure 5, right).

Entire simulation model of BM@N GEM tracker includes the *geometric model* implemented as a ROOT [8] geometry file and the *program logic model* to generate and process output data. The geometric model describes the spatial dimensions and material properties of the basic detector elements. This model is used to simulate the passage of particles through the detecting planes with Geant4 software [9] based on Monte Carlo methods. The MC-points, as the result of Geant4 production, allow us to generate output data (strip clusters) that correspond to the real data.

The following data transformation has several steps:

- From MC-point (obtained by the Geant4 simulation engine) we have parameters, such as: an initial position of the entry point, momentum and type of a particle;

DOI: 10.1051/epjconf/201713811004

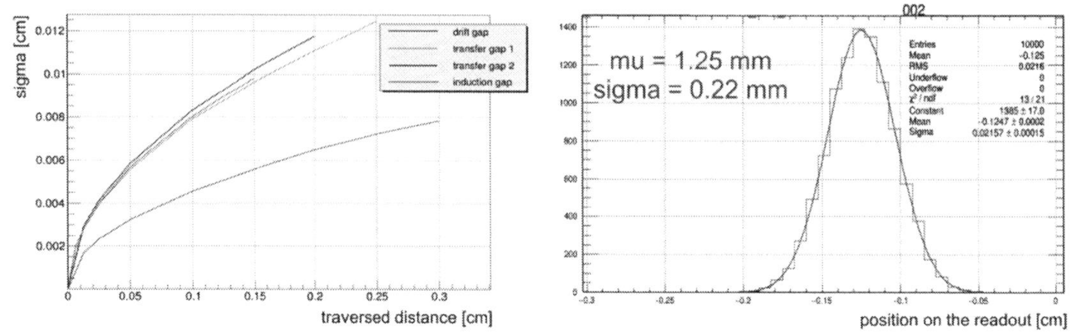

Figure 4. (Left) Dependence of the standard deviation (sigma) of electrons in the transverse drift direction on its traversed distance for all GEM gaps; (right) density distribution of electrons passed through the thickness of the GEM chamber (0.9 cm).

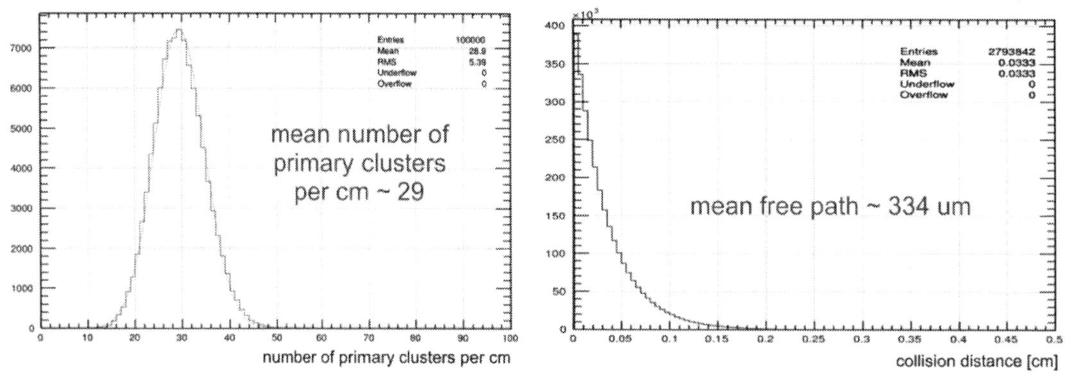

Figure 5. (Left) Distribution of primary electron-ion clusters along 1 cm of the track for the particle pi+ with initial momentum 1 GeV/c; (right) distribution of collision distances for the particle pi+ with initial momentum 1 GeV/c.

- Further, define a particle track through the GEM volume;
- Due to distributions of ionization parameters for the current particle we determine size and position of electron clusters along the track;
- According to dependencies of the mean electron displacement and standard deviation we generate electrons in each layer to calculate the strip signal.

The proposed algorithm allows to reduce considerably the time of data simulation in comparison to the full microsimulation by means of the Garfield++. Such performance is achieved by excluding non-essential details that have a little effect on the resulting signal, but take too much time to process.

Thus the output data received with our simulation model of the BM@N GEM tracker take into account features of the real signal formation in the GEM chamber. Relying on these data we can get preliminary evaluation of detector efficiency for various multiplicity events. The figure beneath (figure 6) demonstrates an example of a strip cluster registered by the readout in the model for the particle track with 15 degrees inclination angle from the beam axis.

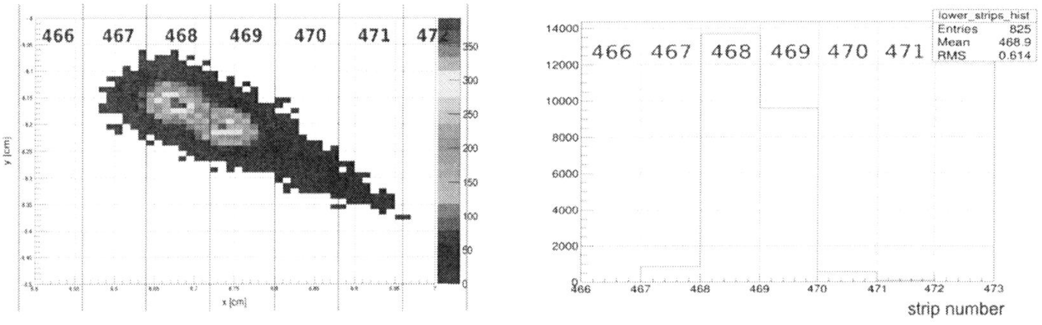

Figure 6. (Left) Cluster on the readout plane; (right) strip cluster signal.

The signal values on the strips are in abstract units, because subsequent processing steps do not depend on specific measurement units.

Simulation and subsequent reconstruction of the signal using the model was carried out to estimate model adequacy basing on sufficient statistics. The resulting accuracy of coordinate reconstruction can be evaluated on the distribution of residuals for both coordinates (figure 7).

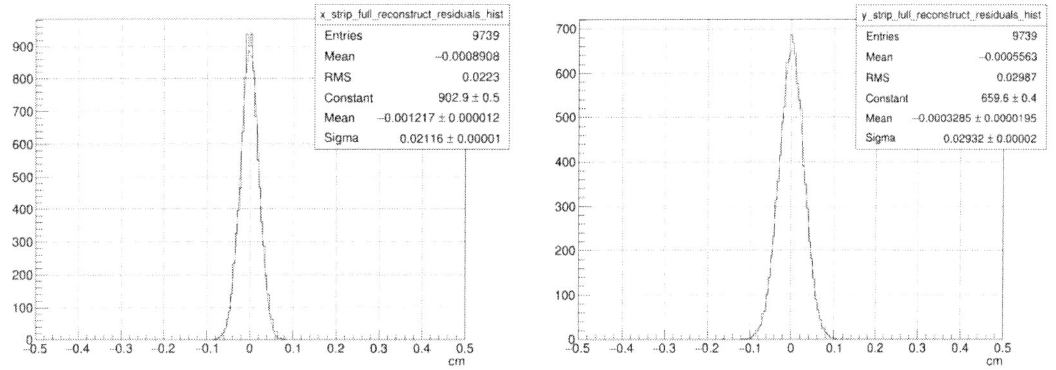

Figure 7. (Left) Residual error of X-reconstructed coordinate; (right) residual error of Y-reconstructed coordinate. The results are presented for particle tracks with inclination angle from 0 to 15 degrees from beam axis.

4 Conclusion

The algorithm described in the article allows to generate data equivalent to realistic. The detailed simulation of physics processes occurring in the GEM chamber to form the output signal can be carried out with Garfield++ toolkit. However this approach is unacceptable for high multiplicity events because of the huge computation time. Therefore, this toolkit was used for extracting parameters required to develop our model. Presented method takes into account factors such as particle track inclination and magnetic field influence.

References

[1] M. Kapishin (BM@N), Eur. Phys. J. A **52**, 213 (2016)
[2] V. Kekelidze et al., Nucl. Phys. A **956**, 846 (2016)
[3] *The NICA Complex*, http://nica.jinr.ru
[4] M. Ziegler et al., Nucl. Instrum. Meth. A **471**, 260 (2000)
[5] F. Sauli, Nuclear Instruments and Methods in Physics Research Section A: Accelerators, Spectrometers, Detectors and Associated Equipment **805**, 2 (2016)
[6] R. Bouclier et al., IEEE Transactions on Nuclear Science **44**, 646 (1997)
[7] *Garfield++*, http://garfieldpp.web.cern.ch
[8] *ROOT*, https://root.cern.ch
[9] *Geant4*, http://geant4.cern.ch

Drift chambers in BM@N Experiment

Ján Fedorišin[1,*]

[1] *Veksler and Baldin Laboratory of High Energy Physics, Joint Institute for Nuclear Research, 141980 Dubna, Moscow region, Russia*

Abstract. Drift chambers (DCH) constitute an important part of the tracking system of BM@N experiment designed to study the production of baryonic matter at the Nuclotron energies.
The method of particle hit and track reconstruction in the drift chambers is proposed and tested on the BM@N deuteron beam data.
In first step, the radius vs drift time calibration curve is estimated and applied to calculate DCH hit closest approach coordinates. These coordinates are used to construct hits in each DCH under the assumption of track linearity. Hits in both the DCHs are subsequently aligned and fitted to produce global linear track candidates. Eventually the hit and track reconstruction is optimized by the autocalibration method. The coordinate resolutions are estimated from Gaussian fits of the DCH hit residual spectra for different DCH planes. Furthermore, the deuteron beam momentum value is reconstructed in order to check reliability of the employed track reconstruction algorithm.

1 Introduction

BM@N (Baryonic Matter at Nuclotron) [1] is a fixed target experiment at the JINR NICA-Nuclotron complex proposed to study the hot compressed nuclear matter produced in collisions of relativistic heavy nuclei with emphasis on strange matter production. The detailed description comprising both the physical goals and the experimental apparatus of BM@N experiment can be found in [1, 2] or [3]. The outer tracker including two identical drift chambers DCH1 and DCH2 [3, 4] is placed outside the magnetic field of SP41 dipole magnet [5]. Each DCH consists of four segments measuring different track coordinates in a transverse plane relative to the beam axis. The measured coordinates are y,x,u,v where u and v are y coordinates rotated by angles $\pm45°$ around z axis. Actually, the directly measured experimental observables are the electron drift times that are subsequently by the proper calibration technique [3, 6–8] converted to the distances of closest approach (DCA) of tracks to the anode wires. The typical drift time spectrum for the BM@N deuteron beam data along with the corresponding radius-time (or r–t) calibration curve is shown in figure 1. Once the y,x,u,v coordinates are calculated using the DCAs provided by the calibration, the missing transverse coordinates in all the DCH planes are evaluated by solving the set of linear equations as described in [3]. Finally, the track candidates from both the drift chambers are combined and fitted by straight lines to produce the global track candidates.

*e-mail: fedorisin@jinr.ru

DOI: 10.1051/epjconf/201713811005

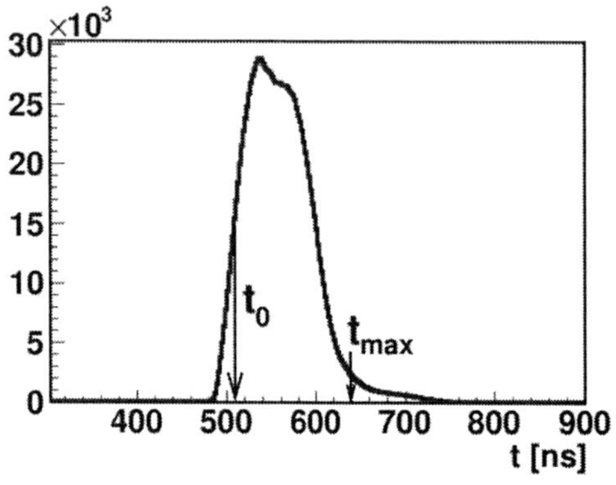

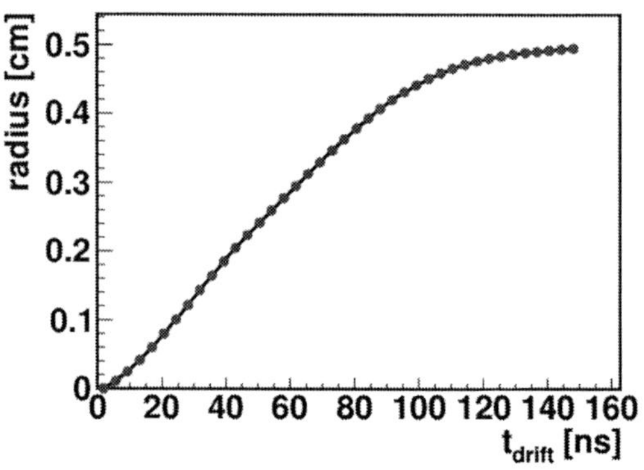

Figure 1. Experimental drift time spectrum for deuteron beam data (left) and the corresponding *r*–*t* calibration curve (right).

2 DCH autocalibration

At the end of the track reconstruction procedure, we construct the dependencies of hit residuals Δ on drift times for all the DCH planes where the residuals are defined as

$$\Delta = |r_{\text{fitted}} - r_{\text{anode}}| - |r_{\text{measured}} - r_{\text{anode}}|, \quad (1)$$

providing that the substitute symbol r represents y, x, u, v coordinates in all the four DCH planes respectively. The initial Δ vs drift time spectra are averaged over the residual axes and the resulting distributions are subtracted from the corresponding calibration curves in order to suppress their systematic uncertainties. The corrected calibration curves are applied again to the experimental data and the whole tuning process including

1. calibration;
2. DCH alignment;
3. hit estimation;
4. track candidates fitting;
5. residual estimation

is iteratively repeated until the averaged residual spectra are minimized, i.e. the mean values of averaged Δ distributions converge to 0.

Figure 2 presents the dependencies of hit residuals Δ on drift times for all the DCH1 planes after the last iteration. The analogous plots are constructed for DCH2 as well but they are not presented in this paper.

The averaged Δ distributions after the first and the last iterations for both the drift chambers are presented in figure 3 where the shown statistics parameters correspond to the last iterations. The spectra prove that the hit residuals are gradually suppressed during the iteration process.

3 Results

The hit residual spectra obtained by integration of figure 2 distributions along drift time axes are presented in figure 4. Standard deviations σ of Gaussian fits of the residual distributions suggest the

DOI: 10.1051/epjconf/201713811005

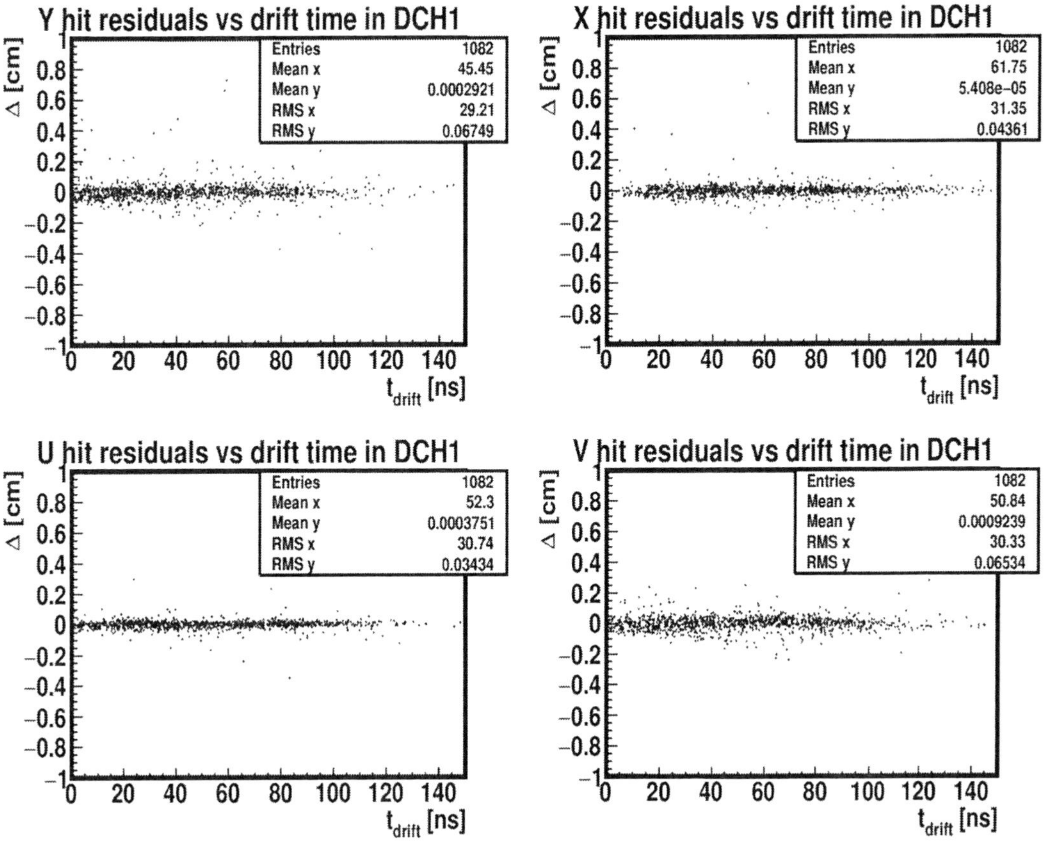

Figure 2. Hit residuals dependence on drift time for all the DCH1 coordinate planes after the last iteration.

coordinate resolutions in all the DCH1 planes. The coordinate resolutions for all the planes of DCH1 and DCH2 are listed in table 1. Taking into account that the arrangement of DCH planes along z axis

Table 1. Coordinate resolutions of all the DCH planes.

	V [μm]	U [μm]	X [μm]	Y [μm]
DCH1	350	160	330	330
DCH2	570	330	270	540

follows the order V,U,X,Y, one can see that the coordinate resolutions are better for inner DCH planes (U,X) than for the outer ones (V,Y). Furthermore, the tracks in DCH1 are reconstructed with higher precision than in DCH2, which can be explained by higher efficiencies of DCH1 planes as well as by the additional error brought to DCH2 track coordinates through the detector alignment procedure.

Evolution of the iterative process is demonstrated in figure 5 showing the dependence of Gaussian σ of residual spectra on iteration number for both the drift chambers: It is obvious that the saturation is achieved very fast, just after 2-3 iterations. However, the coordinate resolution does not get significantly improved by the autocalibration procedure since the observed coordinate error decreases only about 15%.

DOI: 10.1051/epjconf/201713811005

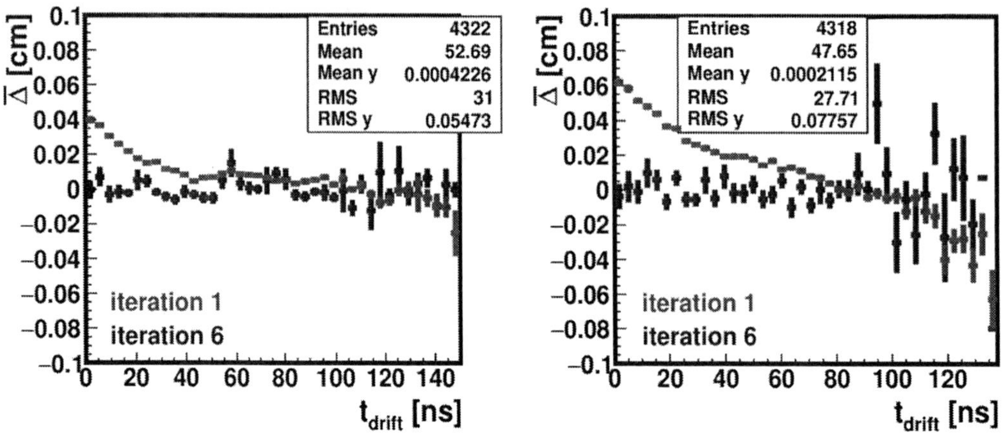

Figure 3. Averaged hit residuals $\bar{\Delta}$ for DCH1 (left) and DCH2 (right).

4 Consistency check of the track reconstruction algorithm

The DCH track reconstruction algorithm is verified through the momentum reconstruction of deuteron beam. In general if beam moves along z axis, its deflection in homogeneous magnetic field in xz plane is described by angle φ_{xz} which depends on magnetic field integral and beam momentum p:

$$\varphi_{xz}(\text{rad}) = q\frac{\int Bdl}{p} \quad (2) \qquad \Longrightarrow \qquad p = q\frac{\int Bdl}{\varphi_{xz}}, \quad (3)$$

where B is B_y component of magnetic field and q is beam particle charge [9]. Figure 6 displays the dependencies of angle φ_{xz} and beam momentum p on magnetic field integral. The different values of φ_{xz} and $\int Bdl$ come from the different investigated runs. The presented dependencies on the left and on the right agree with the functions 2 and 3 respectively. The figure 6 on the left shows also the extrapolated deflection angle φ_{xz} for $B = 0$ which is interpreted as an inclination angle of DCH transverse planes and therefore subtracted from all the φ_{xz} angles in 3 employed to estimate the beam momentum p. The final value of deuteron momentum is thus 8.4 ± 0.1 GeV/c which is somewhat lower than nominal value 8.68 GeV/c, i.e. the reconstructed momentum value seems a little underestimated. The difference can be attributed to non-homogeneities of BM@N magnetic field [5].

5 Summary

1. The complex method of particle hit and track reconstruction in the drift chambers is tested on BM@N deuteron beam data.
2. The r–t calibration curve is estimated and applied to calculate DCA coordinates of the DCH hits.
3. DCH hits are estimated under the assumption of track linearity. Hits in both the DCHs are aligned and fitted to produce linear track candidates.

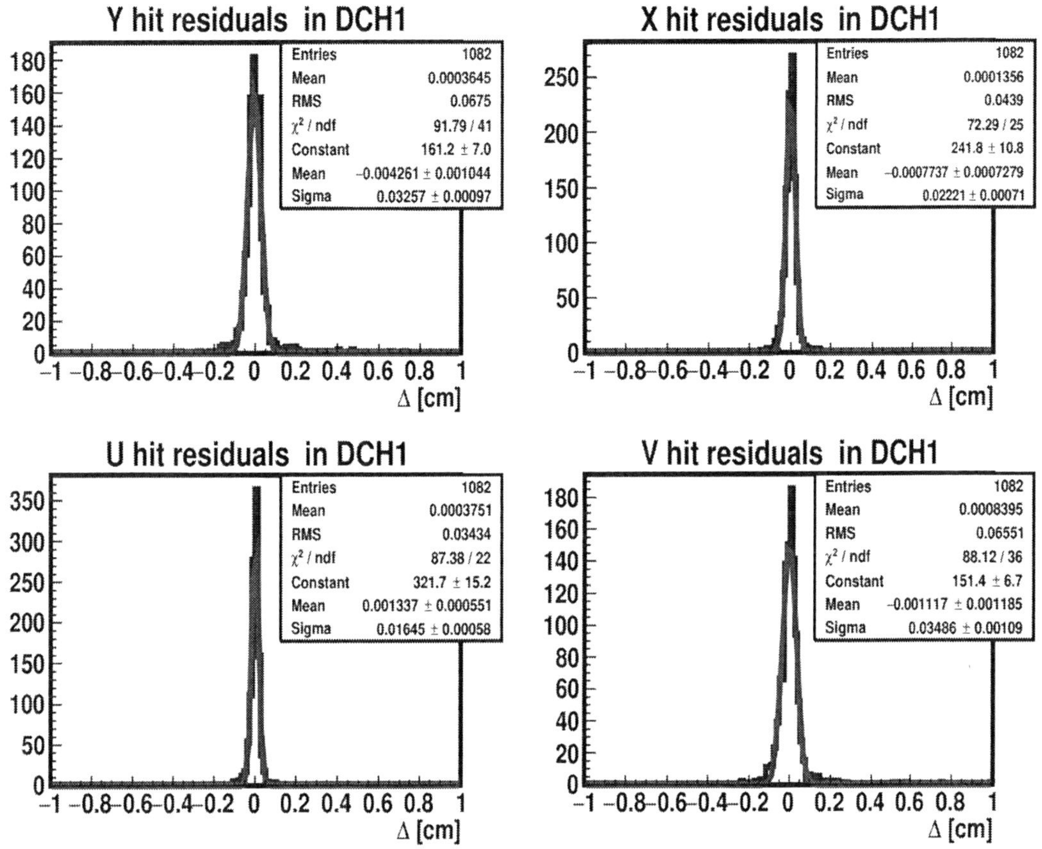

Figure 4. Hit residuals in all the DCH1 coordinate planes after the last iteration.

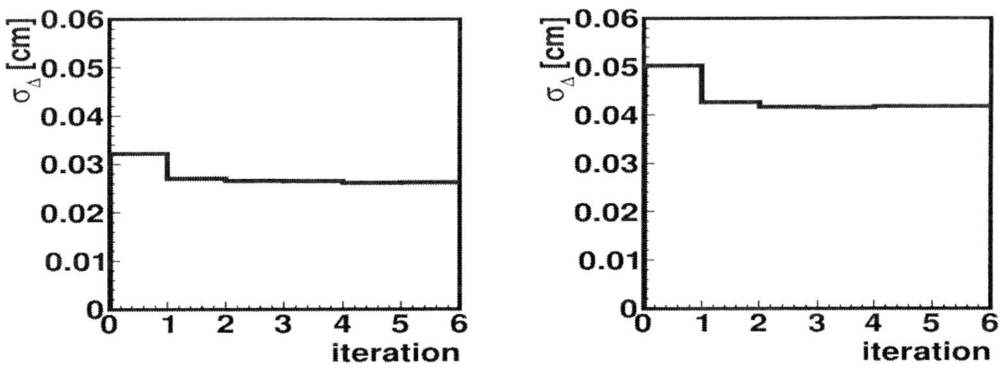

Figure 5. Gaussian σ of residual spectra for DCH1 (left) and DCH2 (right).

4. The hit and track reconstruction is optimized by the autocalibration method.

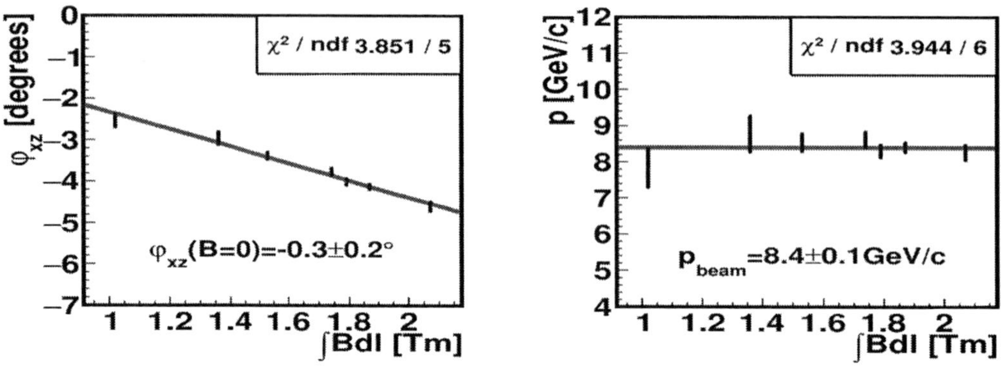

Figure 6. Deflection angle of deuteron beam in *xz* plane (left) and deuteron momentum (right) as functions of magnetic field integral.

5. The DCH hit residual spectra fitted by Gaussian functions imply the coordinate resolutions ranging from $\sigma \approx 160\,\mu$m up to $570\,\mu$m for different DCH planes.

6. The verification test based on the beam momentum estimation proves a reliability of the proposed method, however some refinement is still desirable.

References

[1] T.O. Ablyazimov et al., BM@N Collaboration, *BM@N – Baryonic Matter at Nuclotron* (Conceptual Design Report, 2012). http://nica.jinr.ru/files/BM@N/BMN_CDR.pdf
[2] M. Kapishin, Eur. Phys. J. A **52**, 213 (2016)
[3] J. Fedorišin, EPJ Web of Conferences **108**, 02021 (2016)
[4] D. Bédérede et al., Nucl. Instrum. Meth. A **367**, 88–91 (1995)
[5] P.G. Akishin et al., physics.ins-det arXiv:1407.7096v1 (2014)
[6] G. Avolio et al., Nucl. Instrum. Meth. A **523**, 309–322 (2004)
[7] W. Erni et al., Eur. Phys. J. **49**, 25 (2013)
[8] V. Glagolev et al., Instrum. and Exp. Tech. **56**, 387–397 (2013)
[9] *Tables of Physical & Chemical Constants (16th edition 1995)* (Kaye & Laby Online, version 1.0, 2005). http://www.kayelaby.npl.co.uk

EPJ Web of Conferences **138**, 11006 (2017) DOI: 10.1051/epjconf/201713811006
Baldin ISHEPP XXIII

Recent developments in track reconstruction and hadron identification at MPD

A. Mudrokh[1] and *A.* Zinchenko[1]

[1] *Joint Institute for Nuclear Research, Dubna, Russia*

Abstract. A Monte Carlo simulation of real detector effects with as many details as possible has been carried out instead of a simplified Geant point smearing approach during the study of the detector performance. Some results of realistic simulation of the MPD TPC (Time Projection Chamber) including digitization in central Au+Au collisions have been obtained. Particle identification (PID) has been tuned to account for modifications in the track reconstruction. Some results on hadron identification in the TPC and TOF (Time Of Flight) detectors with realistically simulated response have been also obtained.

1 Introduction

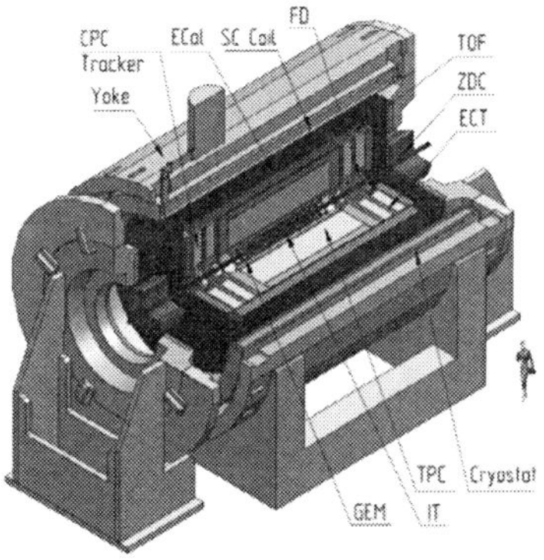

Figure 1. Schematic view of the MPD detector

The MPD detector is a large acceptance spectrometer for heavy-ion collisions study at NICA. The main scientific goal of the NICA/MPD project is to explore the phase diagram of strongly interacting matter in the region of highly compressed and hot baryonic matter [1]. For this purpose an excellent tracking and particle identification (PID) performance over a significant fraction of the final state phase-space are required.

The MPD detector (figure 1) consists of a barrel part and two endcaps located inside the magnetic field. The barrel part is a shell-like set of various detectors surrounding the interaction point and aimed to reconstruct and identify particles in the pseudorapidity region of $|\eta| < 1.2$. The endcaps are aimed for precise tracking over pseudorapidity range $1.2 < |\eta| < 2$.

The Time Projection Chamber (TPC) is the main MPD tracking detector. It covers the barrel and endcap parts of MPD: its geometrical acceptance limit $|\eta| = 2$. TPC is responsible for particle momentum measurements, vertex determination, two track separation, dE/dx measurements and particle identification.

Charged hadrons can be identified in MPD by the TPC and TOF detectors. The mass of the particle is used as an identification parameter, which can be obtained by combining the measurements of its time of flight (from TOF) and its momentum (from TPC).

A new approach based on the full simulation of the detector physics and response is required in order to get realistic estimates of the MPD performance instead of a simplified approach at the smeared hit production level. Most of the physics simulation results obtained so far have been produced for the midrapidity region $|\eta| < 1$. An extension of the MPD tracking (and PID) capability up to $|\eta| < 2$ would allow us, for example, to increase the proton yield in the event by 50% which is important in the search for the critical end point.

In this paper only the first results of the TPC tracking and TPC+TOF hadron identification performance studies are presented based on the realistic detector response simulation and respectively tuned cluster, hit, track finding and particle identification procedures.

2 Realistic simulation of MPD TPC

Data processing model during the MPD TPC simulation consists of the following steps: event generation, particle transport, TPC response simulation, cluster (hit) finding, tracking and dE/dx reconstruction and physics analysis.

The UrQMD event generator is used as an input for simulation in this study. The transport of particles through the detector is handled by the Geant3 package.

The following processes are considered in the TPC microsimulation procedure: primary ionization simulation, drift and diffusion of ionization electrons, gas gain fluctuations, pad response, electronics shaping and signal digitization.

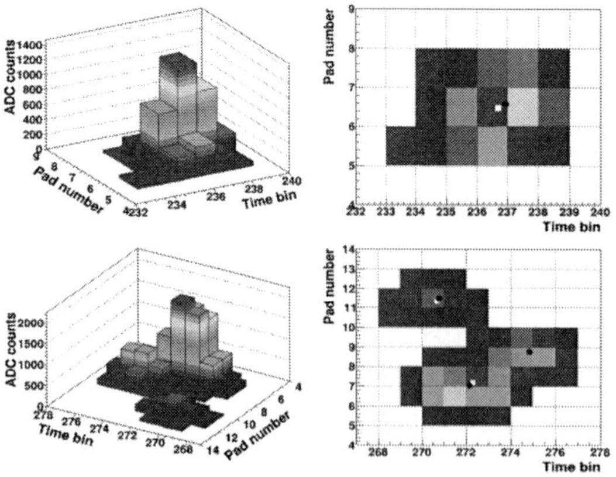

Figure 2. 3D and 2D views of single and triple track clusters

Ionization losses are simulated by Geant in the TPC volume along the trajectory of a charged track. They are converted into the number of electrons which are transported in the uniform electric field toward the anode wires. Both transverse and longitudinal diffusion are considered during the transport process. The charge collected on the sense wire plane is affected by the electron gas multiplication process described by the Polya distribution. The final charge is distributed over readout pads in accordance to the pad response function, which had been determined from a study of a TPC prototype.

The response of the readout electronics (*i.e.* shaping preamplifier with 180 ns FWHM shaping time) is folded with the electron drift time distribution. Finally, all the pad signals are digitized in ADCs with the 12-bit accuracy.

The cluster (hit) finding procedure includes the following steps: finding preclusters, finding separated local maxima using the "peak-and-valley" approach, reconstruction of the coordinates of the found local maxima via center-of-gravities and edge effects corrections.

The "peak-and-valley" approach consists in finding local maxima ("peaks") in the charge *vs* time bin *vs* pad number domain diagram (figure 2) separated by sufficiently deep "valleys". The obtained results are shown in figure 3 where the position resolution along the pad row (transverse direction) is plotted as a function of the pad row crossing angle. The spatial resolution along this direction degrades faster in the outer part of TPC where the pad height is larger.

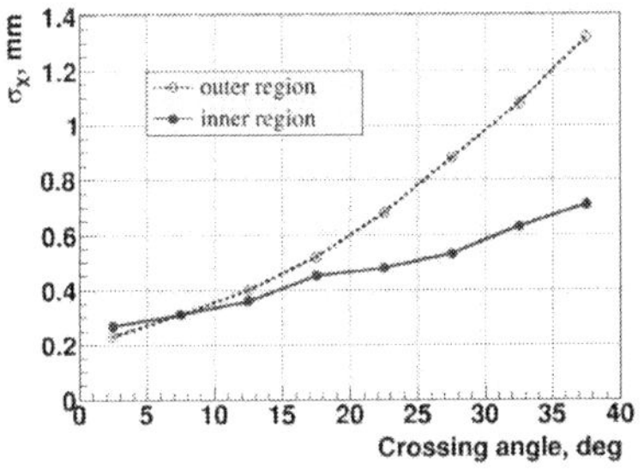

Figure 3. Position resolution along the pad rows for tracks with $p_t > 0.1$ GeV/c and dip angle below 45 deg

Figure 4 demonstrates 2-hit separation capability of the MPD TPC along the pad row direction. The efficiency weakly depends on whether hit is registered in the inner or outer part of the TPC.

The track reconstruction method is based on the Kalman filtering technique. In figure 4 the efficiency of track reconstruction is plotted as a function of transverse momentum for primary particles. The comparison of the realistic simulation with the one obtained with the simplified method, *i.e.* with Geant points smeared in the transverse direction, shows us the difference between two methods. The observed drop in the TPC tracking efficiency at transverse momenta $p_t < 200$ MeV/c is presumably due to a larger crossing angle of low-p_t tracks and a higher probability to cross inefficient regions at sector boundaries.

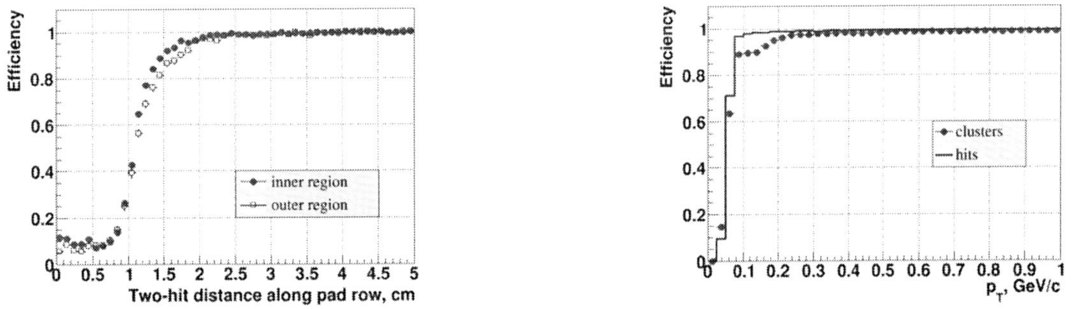

Figure 4. Two-hit resolution along the pad row direction (on the left) and track reconstruction efficiency as a function of p_t for primary particles (on the right)

dE/dx measurements are provided by the TPC up to its pseudorapidity acceptance limit (figure 5, left). A truncated mean value is computed for charges of the hits assigned to each track with the truncation level being chosen to obtain the mean energy loss equal to the most probable value (figure 5, right).

DOI: 10.1051/epjconf/201713811006

3 Particle identification (PID) in MPD detector

One of the main requirements to the MPD experiment to be successful is to provide efficient particle identification. For instance, hadrons ($p^{\pm}, \pi^{\pm}$ and $K^{\pm}$) should be identified over a momentum range from 100 MeV/c to 3 GeV/c. To ensure that, the energy deposit (from TPC) and mass of particle (from TPC and TOF) are used for reconstructed tracks to get partial (TPC) or combined (TPC+TOF) probabilities of being a particle of a certain specie as explained below. As it has been already mentioned, the detailed microsimulation of the TPC required additional PID tuning which was done on the simulated sample of central (impact parameter $b = 0 - 3$ fm) Au + Au collisions at $\sqrt{s} = 6.5$ GeV/c from the UrQMD (ver. 3.4) event generator.

The general idea of particle identification can be illustrated by the definition of the partial probability from the TPC. Charged particles can be identified by the value of their truncated mean energy deposit dE/dx plotted as a function of the full momentum (figure 5, left). For each small interval of momentum (50 MeV/c) we can produce a normalized energy deposit distribution which is a dE/dx-profile for all considered particles in units of the mean pion dE/dx expected for this momentum. As it can be seen from the right plot of figure 5, the normalized energy deposit distribution is a set of (almost) normal distributions, and the main part of the PID task is to parameterize the mutual arrangement of these normal distributions in the whole momentum interval, *i.e.* to determine the expected mean values, widths and amplitudes of the distributions. Given the PID parameterizations exist, for a reconstructed track with momentum p and energy deposit dE/dx we can obtain a vector of probabilities to belong to each of the particle species.

Currently, such parameterizations have been produced for TPC and TOF identification, but the set of used detectors can be expanded in future.

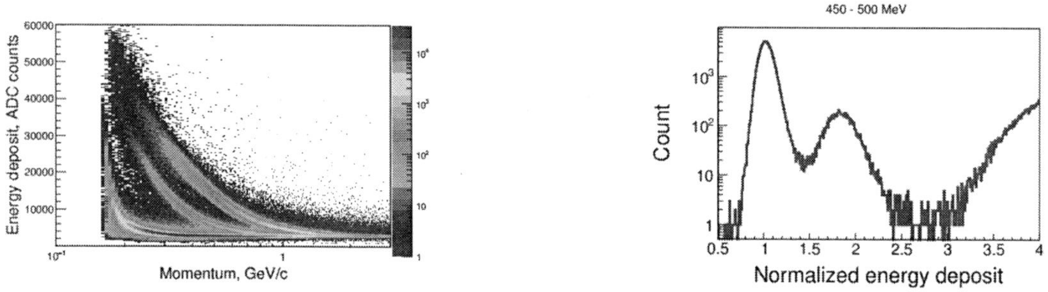

Figure 5. dE/dx reconstruction: energy deposit in ADC counts *vs* momentum (on the left); projection of energy deposit for particles with 0.45 – 0.50 GeV/c momentum (on the right)

According to the general idea described above, hadron identification in MPD can be achieved using the **ionization energy loss** (dE/dx) from the TPC. The mean dE/dx coincides with the most probable energy deposit after suppressing Landau tail using the truncation method. According to this algorithm, the projection of energy loss at a fixed momentum has a Gaussian form, which is completely defined by three parameters. The mean value of energy loss as a function of full momentum can be described by the Bethe-Bloch function. PID uses simplified Bethe-Bloch function with 5 parameters:

$$-\frac{dE}{dx} = a_0 \left\{ \frac{p}{\sqrt{p^2 + m^2}} \right\}^{-a_3} \left\{ a_1 - \frac{p}{\sqrt{p^2 + m^2}}^{a_3} - \ln\left(a_2 + \left(\frac{m}{p} \right)^{a_4} \right) \right\}, \quad (1)$$

where p - full momentum, a_i - fit parameters.

The parameterization was done for primary particles with the number of hits per track ≥ 10 and $|\eta| < 1$. The ratio of the reconstructed mean dE/dx to the mean value expected from the approximation (1) (figure 6) was determined in order to be sure that parameterization has been done correctly. The ratio has the largest deviation from 1 on the edges of momentum spectrum because of two reasons. First, the simplified Bethe-Bloch function does not well describe behavior of energy deposit for low momenta. And second, the data set had low statistics at high momenta. From the plotted ratio one can obtain also the relative dE/dx width, which is about 7% in the whole momentum interval.

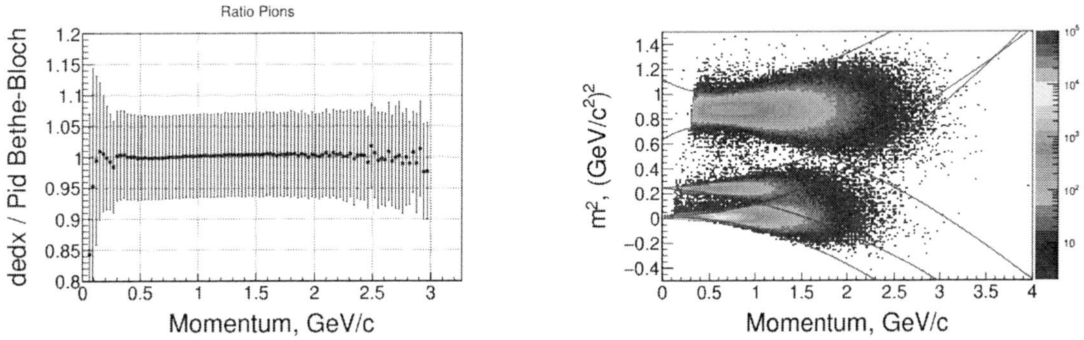

Figure 6. Ratio of the reconstructed mean dE/dx to the value from the approximation 1 (on the left) and dependence of mass-squared on particle momentum (on the right).

Particle identification works much better, if **TOF identification** is available. It significantly improves PID results for high momenta, where particles cannot be identified by the dE/dx value. TOF identification is based on separation of particles by their reconstructed mass-squared value, which can be calculated from particle momentum and measured time of flight (figure 6 right). Using the procedure, similar to the one described for dE/dx-identification, the mass-squared distributions have been produced for different momentum intervals and fitted to the gaussian functions. The obtained widths were fitted by polynomials for every particle specie as it can be seen in figure 6, where the red lines show 3σ mass intervals for different particles.

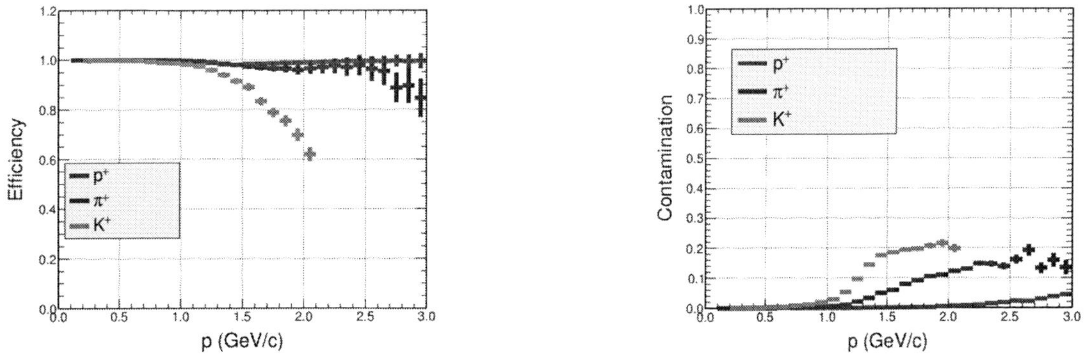

Figure 7. Combined efficiency (on the left) and contamination (on the right) of positive charged particles as a function of full momentum

The PID performance can be characterized by the efficiency and contamination dependencies on the full momentum (figure 7). The combined efficiency and contamination were obtained for positively charged particles which were identified by the PID probability above 70%. As it can be seen, this condition is too strong for kaons, causing their efficiency dropping at high momenta.

DOI: 10.1051/epjconf/201713811006

4 Conclusions

The MPD TPC cluster finder method based on the detailed microsimulation procedure has been developed. Simulation results have been compared with the ones obtained with the gaussian smearing approach. Further work on the reconstruction algorithm fine tuning is foreseen.

PID parameterizations have been also obtained. Efficient π/K separation up to 1.5 GeV/c and π/p separation up to 3 GeV/c has been demonstrated.

As a continuation of this work, the extension of the track reconstruction and particle identification up to $|\eta| = 2$ is planned.

References

[1] K.U. Abraamyan et al., Nucl. Instrum. Meth. A **628**, 99 (2011)

EPJ Web of Conferences **138**, 12001 (2017) DOI: 10.1051/epjconf/201713812001
Baldin ISHEPP XXIII

Superconducting dipole magnet for the CBM experiment at FAIR

P. Kurilkin[1,★], *P.* Akishin[1], *A.* Bychkov[1], *E.* Floch[1], *Yu.* Gusakov[1], *V.* Ladygin[1], *A.* Malakhov[1], *G.* Moritz[2], *H.* Ramakers[2], *P.* Senger[2], *A.* Shabunov[1], *P.* Szwangruber[2], and *F.* Toral[3]

[1]*JINR, Joliot-Curie 6, 141980, Dubna, Russia*
[2]*GSI, Planckstrae 1, 64291 Darmstadt, Germany*
[3]*CIEMAT, 28040, Madrid, Spain*

Abstract. The scientific goal of the CBM (Compressed Baryonic Matter) experiment at FAIR (Darmstadt) is to explore the phase diagram of strongly interacting matter at highest baryon densities. The physics program of the CBM experiment is complimentary to the programs to be realized at MPD and BMN facilities at NICA and will start with beam derived by the SIS100 synchrotron. The 5.15 MJ superconducting dipole magnet will be used in the silicon tracking system of the CBM detector. The magnet will provide a magnetic field integral of 1 Tm which is required to obtain a momentum resolution of 1% for the track reconstruction. The results of the development of dipole magnet of the CBM experiment are presented.

1 Introduction

The Compressed Baryonic Matter (CBM) will be one of the four major scientific experiments that are planned to be performed at the Facility for Anti-proton and Ion Research (FAIR) at Darmstadt, Germany [1]. The goal of the CBM research program is to perform the systematic investigation of the phase diagram of nuclear matter in the region of high baryon densities [2] and moderate temperature. These conditions will be created in the heavy-ion beam collisions with nuclear targets in the beam energy range from 2 AGeV to 45 AGeV. The proton-nucleus interaction will be also studied. The start of CBM data taking is expected for 2021.

Earlier, the study of the QCD phase diagram at large baryon-chemical potentials have been curried out at AGS in Brookhaven [3] and at low CERN-SPS beam energies [4]. The STAR collaboration at RHIC scanned the beam energies in order to search for QCD critical endpoint [5]. For the same reason, the NA61/SHINE experiment at CERN-SPS continues to search the first-order phase transition using light and medium size beams [6]. Neither NA61 at SPS nor STAR at RHIC will be able to run at beam energies of 10 AGeV and less. At the Joint Institute for Nuclear Research in Dubna, a fixed target experiment is under preparation at the Nuclotron in order to investigate the heavy-ion collisions at gold-beam energies up to 4 AGeV. Moreover, the collider facility NICA with multi-purpose detector(MPD) is planed at JINR [7]. However, due to luminosity or detector limitations these experiments are constrained to the investigation of abundantly produced particles.

In contrast, the Compressed Baryonic Matter (CBM) experiment [1] is designed to run at extremely high interaction rates (up to 10 MHz) [8].

★e-mail: pkurilkin@jinr.ru

2 CBM detector setup

The CBM experiment is a fixed target heavy-ion experiment. It will be set up in an underground cave to which the beam from the SIS100/300 synchrotrons will be extracted. The detector system of CBM is depicted schematically in figure 1. The geometrical acceptance of the CBM detector systems is full 2π for azimuth (ϕ) and from 2.5° to 25° for polar (Θ) angles. The target, Silicon Tracking System (STS) and Micro Vertex detector (MVD) are placed in the large gap of the superconducting dipole magnet [9] for charged particle tracking and the detection of short-lived decays. Downstream the beam follow the Ring Imaging Cherenkov Detector (RICH) and the Transition Radiation Detector (TRD) for electron identification and pion separation. The Time-of-Flight system identifies hadrons. The Projectile Spectator Detector (PSD) serves event plane characterization.
The CBM detector system will provide both electron and muon detection systems. Muon and electron pairs created during particle decays will be measured during separate runs in order to control the systematic approach of the measurements. In electron configuration following detectors will be used: MVD, STS, RICH, TRD, Resistive Plate Chambers for time-of-flight measurements, Electromagnetic Calorimeter (ECAL) and Projectile Spectator Detector (PSD). In muon configuration the RICH detector will be replaced by the Muon Detection System (MuCh) and ECAL will be removed.

The non-interacting beam particles, 99% of all with a typical target of 1% nuclear interaction length, are guided in a vacuum beam pipe through the experiment and are absorbed in the beam dump [8]. The HADES detector [10] will be placed in front of the CBM detector as shown in figure 1 to conduct a complementary physics program in separate data taking periods. The CBM setup may be operated in a minimal mode at FAIR startup with only key subsystems and the necessary part of the data acquisition and online computing installed, which are later on completed. Similarly, different optimizations of the experiment to perform measurements at SIS 100 and SIS 300 are possible.

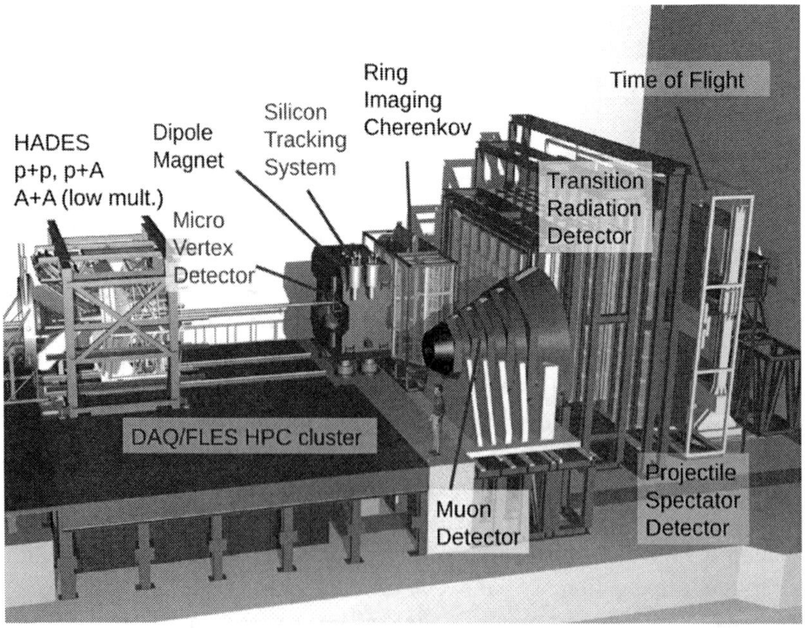

Figure 1: Schematic view of the CBM cave with the detector setup of CBM (right) and HADES (left).

 DOI: 10.1051/epjconf/201713812001

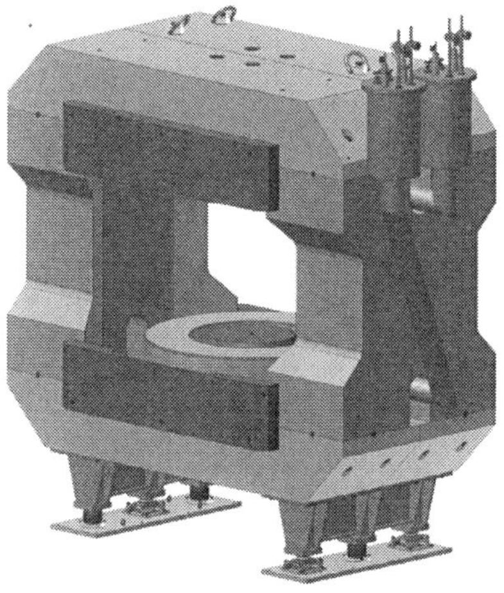

Figure 2: 3D view of CBM dipole magnet with the support.

Table 1: Parameters of the CBM dipole magnet.

type	H-type, SC magnet
Number of turns per coil	1749
Windings of coil	Impregnated close coiling
Maximum current	686 A
Magnetomotive force	1.2 MA turns/coil
Current density,	58 A/mm^2
Maximum field at coil	3.25 T
Central field	1.08 T
Field integral	1 Tm
Inductance	33-19 H
Stored energy	5.15 MJ
Coil cross section (at 4K)	158.8x131.1 mm^2
Yoke (widh/depth/height)	4.4/2.0/3.7 m
Pole type	Tapered
Pole sizes (Rout/Rin/H)	1200/800/500 mm

3 CBM dipole magnet

The H-type superconducting magnet of the CBM experiment is designed to provide a vertical magnetic field with bending power of 1 Tm over a length of 1 m from the target. The maximum energy storage is evaluated as 5.15 MJ when the operating current of the CBM magnet is rated at 686 A. Calculations have been performed to design the coil case, the coil vessel, the support links and the quench protection scheme for the CBM superconducting dipole magnet [11]- [17]. The code TOSCA [18] was used for calculating electromagnetic forces exerted on the coil, while the structural analysis was made using the code ANSYS [19]. The 3D quench calculations for the CBM magnet coil have been carried out by two numerical programs based on the finite-difference method. One of them is used currently at GSI and its detailed description can be found in [20].The other (SQUID) based on CIEMAT numerical code [21] was adapted to perform the quench simulation for the CBM magnet.

A 3D model of the magnet with the support is presented in figure 2. The parameters of the magnet are listed in table 1. The magnet yoke consists of two conical poles and two horizontal and two vertical parts to close a flux return. Two cylindrical superconducting coils will be installed in their respective cryostats [13] and will be cooled separately by helium bath based cooling system like the SAMURAI magnet at RIKEN [22]. Each of the sections includes an inner helium vessel, thermal shield and vacuum vessel [14]. The vacuum vessel (see figure 3) has flanges for connecting to the external cryogenic line, the turbo molecular pump, two 1000 A current leads, the multipin feed through with plug connectors, the safety valve of the vessel and the warm multipurpose helium line (quench gas collection). Also, the vacuum vessel and the heat shield have a large diameter access hole to connect pipes of the external cryogenic line with helium pipes of coils. All these connections are made by means flexible metal hoses. Internal helium vessel contains 10 liters of liquid helium, and has four flanges.

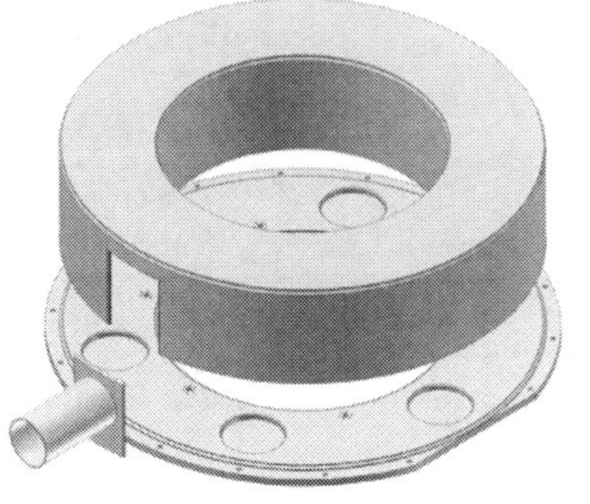

Figure 3: 3D view of the vacuum vessel.

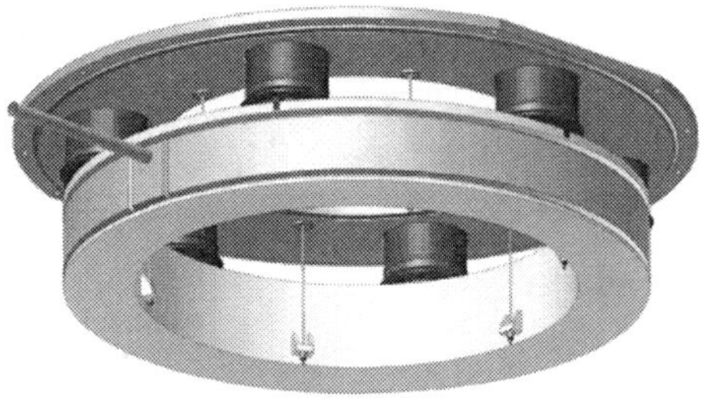

Figure 4: 3D view of the coil case support.

3.1 Coil case and winding

The design of coil for the CBM magnet is based on the design of the Super-FRS dipole [23]. The basic structure of the coil consists of the layer type helical winding on a stainless steel bobbin. Each coil has 1749 turns and 53 layers. The superconducting wire is a Nb/Ti wire and its specification is determined from the quench analysis. The conductor insulation consists of 0.1 mm polyimide type and 0.2 mm glass fiber (G10) material. Each layer is additionally insulated from the others by three layers of 0.1 mm fiber glass fabric with epoxy resin. The epoxy resin is carefully applied during the winding to the gap between coil turns to restrict movement of the conductor.

The coil case is designed considering two main functions: one is to protect the windings against magnetic forces during operation, and the other is to use the case as a container for liquid helium (LHe) to cool the winding. The coil case is supported by six main cylindrical support strats and six tie rods. To reduce the heat flux to the helium system, the outer surface of the casing will be wrapped with 20 layers of a multi-layer insulation. The suspension during the working cycle has two types of loading. When the magnet is switched off only the weight of the cold mass is applied to the suspension. In this case the vertical force is about 2000 kg. When the magnet is switched on, the vertical component of the Lorenz forces should be added to the weight of the cold mass. The maximum vertical force in this case is 254 tons. Figure 4 shows the suspension of the coil case on the flange of the vacuum vessel.

3.2 Quench protection

The 3D quench calculations [15]- [17] have been performed to optimize the behaviour of the winding of the magnet during the quench. The magnet coil is represented as a straight slab with a length equal to the average turn length. A regular mesh is applied to the slab so that the crosssection of the one mesh element equal to the cross-section of the insulated conductor. Longitudinally, the coil is divided into a number of slices which define a mesh element size. The material properties are homogeneous within each mesh element. Two possible quench protection systems for the CBM magnet were considered [17]. The final one is based on the extraction of the energy stored in the magnet via a dump resistor (*Rd*). The details can be found in [9]. The dump resistor is connected in parallel to the magnet and it is always on. In case of the quench the magnet will be disconnected from the power supply and the magnet current will be dumped via the dump resistor. The dump resistor with resistance of 1.9 – 2.1Ω will be used. Figure 5 presents the quench calculation results using the 2.1Ω dump resistor for the energy extraction. Simulation shows that the maximal "hot-spot" temperature in the coil in this case is about 70 K. The maximum voltage across the magnet (1441 V)

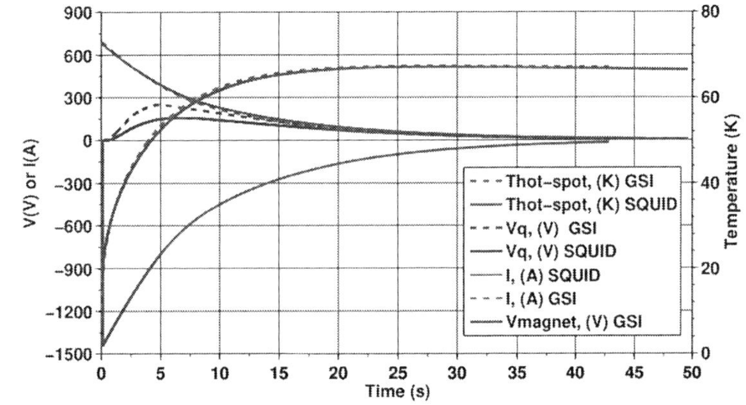

Figure 5: 3D quench calculations for the CBM magnet: magnet current, magnet voltage and the maximum ("hot-spot") temperature. A dump resistor of $Rd = 2.1\ \Omega$ is used for quench protection.

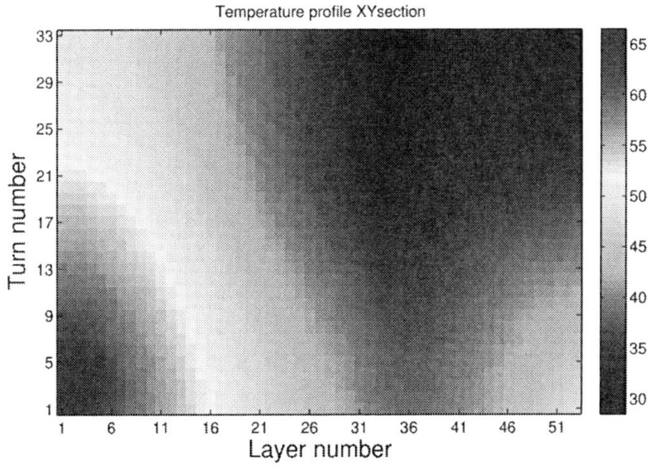

Figure 6: 3D quench calculations for the CBM magnet: the temperature distribution in the cross section of CBM magnet coil. The 85% of magnet storage energy is evacuated via $2.1\ \Omega$ dump resistor.

occurs when the dump resistor turns on. The temperature profile distribution in the coil cross section when the "hot-spot" temperature has a maximum value is shown in figure 6.

4 Conclusion

Compressed Baryonic Matter experiment is devoted to the third generation of the heavy-ion collision experiments. Its research program aims at the exploration of the structure of high density matter including. For these purposes the advanced experimental setup will be build for high counting rate conditions expected at FAIR.

The designed superconducting dipole magnet will be a unique component of CBM detector setup due to a large pole gap and wide horizontal opening. Technical design report for the Superconducting Dipole Magnet has already been approved by FAIR.

The design of the CBM magnet, its support, the cryogenic input lines, current leads and feed boxes has been performed to satisfy the working conditions. The quench protection system will be based on the energy evacuation from the 2.1 Ω dump resistor.

Acknowledgement

The work has been supported in part by the BMBF-JINR grant.

References

[1] Compressed Baryonic Matter (CBM) http://www.fair-center.eu/public/experiment-program/nuclear-matter-physics/cbm.html
[2] C. Hoehne et al., J. Phys. **420**, 012016 (2003)
[3] C. Pinkerburg et al., Phys. Rev. Lett. **83**, 1295 (1999)
[4] C. Alt et al., Phys. Rev. C **77**, 024903 (2008)
[5] A. Schmah, arXiv:1202.2389v1 (2012)
[6] A. Laszlo et al. (The NA61 Coll.), Pos(CPOD07)054 (2007)
[7] http://nica.jinr.ru
[8] E. Heuser, EPJ Web of Conferences **95**, 01006 (2015)
[9] P. Akishin et al. (The CBM collaboration), *Technical Design Report for the CBM Superconducting Dipole Magnet* (2014)
[10] G. Agakishiev *et al.* (HADES collaboration), Eur. Phys. J. A **41**, 243 (2009)
[11] A. Bychkov et al., *CBM Progress Report 2014* (Darmstadt, 2015) 12
[12] Yu. Gusakov et al., *CBM Progress Report 2015* (Darmstadt, 2016) 4
[13] Yu. Gusakov et al., *CBM Progress Report 2015* (Darmstadt, 2016) 5
[14] P. Akishin et al., *CBM Progress Report 2013* (Darmstadt, 2014) 5
[15] P. Kurilkin et al., *CBM Progress Report 2015* (Darmstadt, 2016) 7
[16] P. Kurilkin et al., *CBM Progress Report 2013* (Darmstadt, 2014) 8
[17] P. Kurilkin et al., J. Phys.: Conf. Ser. **742**, 012017 (2016)
[18] Vector Fields Ltd., *Tosca. 3d electromagnetics code* (Kidlington, 1984)
[19] http://www.ansys.com/
[20] P. Szwangruber et al., IEEE Trans. Appl. Supercond. **23**, 4701704 (2013)
[21] F. Toral, *Design and Calculation Procedure for Particle Accelerator Superconducting Magnets: Application to an LHC Superconducting Quadrupole*, Ph. D. Thesis. (Madrid, 2001)
[22] H. Sato et al., IEEE Trans. Appl. Supercond. **23**, 4500308 (2013)
[23] L. Z. Ma et al., IEEE Trans. Appl. Supercond. **22**, 9502104 (2012)

AUTHOR INDEX

AUTHOR INDEX

AUTHOR INDEX

AUTHOR INDEX